Chemical Science of π-Electron Systems

Takeshi Akasaka • Atsuhiro Osuka
Shunichi Fukuzumi • Hideki Kandori
Yoshio Aso
Editors

Chemical Science of π-Electron Systems

Springer

Editors
Takeshi Akasaka
Foundation for Advancement
of International Science
Tsukuba, Japan

Shunichi Fukuzumi
Department of Chemistry and Nano Science
Ewha Womans University
Seoul, Korea

Faculty of Science and Engineering
ALCA and SENTAN
Japan Science and Technology Agency (JST)
Meijo University
Nagoya, Japan

Atsuhiro Osuka
Kyoto University
Kyoto, Japan

Hideki Kandori
Nagoya Institute of Technology
Nagoya, Japan

Yoshio Aso
Osaka University
Suita, Japan

ISBN 978-4-431-55356-4 ISBN 978-4-431-55357-1 (eBook)
DOI 10.1007/978-4-431-55357-1

Library of Congress Control Number: 2015954085

Springer Tokyo Heidelberg New York Dordrecht London

Printed on acid-free paper

Springer Japan KK is part of Springer Science+Business Media (www.springer.com)

Preface

The versatility and wide potential of π-electron-based molecules, as typified by fullerenes and porphyrins, have continued to provide unique and sometimes unexpected properties and functions being continuously discovered and utilized in functional devices. In order to facilitate the investigation on highly elaborated π-spaces with more sophisticated and complex orders and functionality, a research project entitled "Emergence of Highly Elaborated π-Space and Its Function" has been launched as a Grant-in-Aid for Scientific Research on Innovative Areas (2008–2012) by The Ministry of Education, Culture, Sports, Science and Technology of Japan (MEXT). In this joint project, highly elaborated π-spaces are defined as highly complex orders/systems that stem from the π-electron-based interactions. Cutting-edge research has been performed based on clearly defined strategic objectives as well as an emergent approach that comes from innovation of unpredicted discovery. The research project consists of four groups (A01–04). The research group A01 led by Prof. Takeshi Akasaka has constructed novel π-electron systems of nonplanarity including metal-endofullerenes, whereas the research group A02 led by Prof. Atsuhiro Osuka has developed highly elaborated π-space functionality through the assembly of π-electron-based molecules. Based on new types of π-electron-based molecules and assemblies, the revolutionary functions have been explored by clarifying the electronic, optical, and magnetic interactions of the highly elaborated π-spaces in research group A03 led by Prof. Shunichi Fukuzumi. Research group A04 led by Prof. Hideki Kandori has studied the interactions between biological molecules with highly elaborated π-space to discover new phenomena and create new concepts in biology.

This book is composed of eight parts. Part I: Extension of Planar π-Electron Systems, Part II: Curved π-Electron Systems, Part III: Porphyrinoids, Part IV: Open-Shell π-Electron Systems, Part V: Heteroatom-Conjugated π-Electron Systems, Part VI: Supramolecular Chemistry of π-Electron systems, Part VII: Innovative Function of π-Electron Systems, and Part VIII: π-Electron Systems in Biosystems and Biomimetics. It summarizes cutting-edge research in the MEXT project of

"Emergence of Highly Elaborated π-Space and Its Function", covering the research area of A01–A04 and also interdisciplinary fields created by intensive collaborations between the different research groups to provide the current status and future perspective of highly elaborated π-spaces with high orders and functionality.

Suita, Japan Shunichi Fukuzumi

Contents

Contributors

Manabu Abe Department of Chemistry, Graduate School of Science, Hiroshima University, Higashi-Hiroshima, Hiroshima, Japan

Institute for Molecular Science, Okazaki, Aichi, Japan

CREST, JST, Chiyoda-ku, Tokyo, Japan

Takeshi Akasaka Department of Chemistry, Tokyo Gakugei University, Tokyo, Japan

Life Science Center of Tsukuba Advanced Research Alliance, University of Tsukuba, Ibaraki, Japan

School of Materials Science and Technology, Huazhong University of Science and Technology, Wuhan, P. R. China

Foundation for Advancement of International Science, Ibaraki, Japan

Midori Akiyama Department of Chemistry and Biotechnology, Graduate School of Engineering, The University of Tokyo, Tokyo, Japan

Koji Ando Department of Chemistry, Graduate School of Science, Kyoto University, Sakyo-ku, Kyoto, Japan

Lai Feng Life Science Center of Tsukuba Advanced Research Alliance, University of Tsukuba, Tsukuba, Japan

Jiangsu Key Laboratory of Thin Films and School of Energy, Soochow University, Suzhou, P. R. China

Hisashi Fujihara Department of Applied Chemistry, Kinki University, Higashi-Osaka, Japan

Kazuhiro J. Fujimoto Department of Computational Science, Graduate School of System Informatics, Kobe University, Kobe, Japan

Shunichi Fukuzumi Department of Chemistry and Nano Science, Ewha Womans University, Seoul, Korea

Faculty of Science and Engineering, ALCA and SENTAN, Japan Science and Technology Agency (JST), Meijo University, Nagoya, Japan

Mina Furukawa Department of Chemistry and Biology, Graduate School of Science and Engineering, Ehime University, Matsuyama, Japan

Hiroyuki Furuta Department of Chemistry and Biochemistry, Graduate School of Engineering and Education Center for Global Leaders in Molecular Systems for Devices, Kyushu University, Fukuoka, Japan

Taku Hasobe Department of Chemistry, Faculty of Science and Technology, Keio University, Yokohama, Japan

Sayaka Hatano Department of Chemistry, Graduate School of Science, Hiroshima University, Higashi-Hiroshima, Hiroshima, Japan

Shuhei Higashibayashi Institute for Molecular Science, Myodaiji Okazaki, Japan

Koichi Higashimine Center for Nano Materials and Technology, Japan Advanced Institute of Science and Technology, Nomi, Ishikawa, Japan

Shu Hotta Department of Macromolecular Science and Engineering, Graduate School of Science and Technology, Kyoto Institute of Technology, Kyoto, Japan

Hiroshi Ikeda Graduate School of Engineering, Osaka Prefecture University, Sakai, Osaka, Japan

Masatoshi Ishida Department of Chemistry and Biochemistry, Graduate School of Engineering and Education Center for Global Leaders in Molecular Systems for Devices, Kyushu University, Fukuoka, Japan

Shinobu Itoh Department of Material and Life Science, Division of Advanced Science and Biotechnology, Graduate School of Engineering, Osaka University, Suita, Japan

Takeaki Iwamoto Department of Chemistry, Graduate School of Science, Tohoku University, Sendai, Japan

Takuya Kamimura Institute for Materials Chemistry and Engineering, Kyushu University, Fukuoka, Japan

Hideki Kandori Department of Frontier Materials, Nagoya Institute of Technology, Nagoya, Japan

Masayuki Kanehara Research Core for Interdisciplinary Sciences, Okayama University, Okayama, Japan

Kiyohiko Kawai The Institute of Scientific and Industrial Research (SANKEN), Osaka University, Ibaraki, Osaka, Japan

Naoki Kawamoto Integrated Center for Sciences, Ehime University, Matsuyama, Japan

Takeshi Kawase Graduate School of Engineering, University of Hyogo, Himeji, Japan

Jun-ichi Kikuchi Graduate School of Materials Science, Nara Institute of Science and Technology, Ikoma, Nara, Japan

Chitoshi Kitamura Department of Materials Science, School of Engineering, The University of Shiga Prefecture, Hikone, Shiga, Japan

Hirotaka Kitoh-Nishioka Department of Chemistry, Graduate School of Science, Nagoya University, Nagoya, Japan

Junji Kobayashi College of Liberal Arts, International Christian University, Mitaka, Japan

Nagao Kobayashi Faculty of Textile Science and Technology, Shinshu University, Ueda, Japan

Akihito Konishi Department of Chemistry, Graduate School of Science, Osaka University, Toyonaka, Osaka, Japan

Takashi Kubo Department of Chemistry, Graduate School of Science, Osaka University, Toyonaka, Osaka, Japan

Yutaka Maeda Department of Chemistry, Tokyo Gakugei University, Tokyo, Japan

Tetsuro Majima The Institute of Scientific and Industrial Research (SANKEN), Osaka University, Ibaraki, Osaka, Japan

Yutaka Majima Materials and Structures Laboratory, Tokyo Institute of Technology, Yokohama, Japan

Yoshihiro Matano Department of Chemistry, Faculty of Science, Niigata University, Nishi-ku, Niigata, Japan

Yasunori Matsui Graduate School of Engineering, Osaka Prefecture University, Sakai, Osaka, Japan

Tsukasa Matsuo Department of Applied Chemistry, Faculty of Science and Engineering, Kinki University, Higashi-Osaka, Japan

Yutaka Matsuo Department of Chemistry, School of Science, The University of Tokyo, Tokyo, Japan

Yoshiyuki Mizuhata Institute for Chemical Research, Kyoto University, Uji, Kyoto, Japan

Hitoshi Mizuno Graduate School of Materials Science, Nara Institute of Science and Technology, Ikoma, Nara, Japan

Interdisciplinary Graduate School of Science and Engineering, Shimane University, Matsue, Shimane, Japan

Shigeki Mori Integrated Center for Sciences, Ehime University, Matsuyama, Japan

Yuma Morimoto Department of Material and Life Science, Division of Advanced Science and Biotechnology, Graduate School of Engineering, Osaka University, Suita, Japan

Hideyuki Murata School of Materials Science, Japan Advanced Institute of Science and Technology, Nomi, Ishikawa, Japan

Yasujiro Murata Institute for Chemical Research, Kyoto University, Uji, Japan

Shigeru Nagase Fukui Institute for Fundamental Chemistry, Kyoto University, Kyoto, Japan

Tsukasa Nakahodo Department of Applied Chemistry, Kinki University, Higashi-Osaka, Japan

Koji Nakano Department of Organic and Polymer Materials Chemistry, Tokyo University of Agriculture and Technology, Koganei, Japan

Tohru Nishinaga Department of Chemistry, Graduate School of Science and Engineering, Tokyo Metropolitan University, Hachioji, Tokyo, Japan

Hirofumi Nobukuni Institute for Materials Chemistry and Engineering, Kyushu University, Fukuoka, Japan

Takumi Noguchi Division of Material Science, Graduate School of Science, Nagoya University, Nagoya, Japan

Kyoko Nozaki Department of Chemistry and Biotechnology, Graduate School of Engineering, The University of Tokyo, Tokyo, Japan

Keisuke Ohdaira Green Devices Research Center, Japan Advanced Institute of Science and Technology, Nomi, Ishikawa, Japan

Eisuke Ohta Graduate School of Engineering, Osaka Prefecture University, Sakai, Osaka, Japan

Tetsuo Okujima Department of Chemistry and Biology, Graduate School of Science and Engineering, Ehime University, Matsuyama, Japan

Atsuhiro Osuka Department of Chemistry, Graduate School of Science, Kyoto University, Sakyo-ku, Kyoto, Japan

Akinori Saeki Department of Applied Chemistry, Graduate School of Engineering, Osaka University, Suita, Osaka, Japan

Ken-ichi Sakaguchi Institute for Materials Chemistry and Engineering, Kyushu University, Fukuoka, Japan

Hayato Sakai Department of Chemistry, Faculty of Science and Technology, Keio University, Yokohama, Japan

Hidehiro Sakurai Department of Applied Chemistry, Graduated School of Engineering, Osaka University, Suita, Osaka, Japan

Fumio Sasaki Electronics and Photonics Research Institute, National Institute of Advanced Industrial Science and Technology, Tsukuba, Ibaraki, Japan

Satoru Sato Life Science Center of Tsukuba Advanced Research Alliance, University of Tsukuba, Tsukuba, Japan

Shu Seki Department of Molecular Engineering, Graduate School of Engineering, Kyoto University, Nishikyo-ku, Kyoto, Japan

Tomokazu Shibata Department of Chemistry, University of Tsukuba, Tsukuba, Japan

Soji Shimizu Department of Chemistry and Biochemistry, Graduate School of Engineering, Kyushu University, Fukuoka, Japan

Hiroshi Shinokubo Department of Applied Chemistry, Graduate School of Engineering, Nagoya University, Nagoya, Japan

Shuichi Suzuki Department of Chemistry, Graduate School of Science, Osaka City University, Osaka, Japan

Kazunari Tagawa Department of Chemistry and Biology, Graduate School of Science and Engineering, Ehime University, Matsuyama, Japan

Keishiro Tahara Graduate School of Materials Science, Nara Institute of Science and Technology, Ikoma, Nara, Japan

Yuta Takano Life Science Center of Tsukuba Advanced Research Alliance, University of Tsukuba, Tsukuba, Japan

Institute for Integrated Cell-Material Sciences (WPI-iCeMS), Kyoto University, Kyoto, Japan

Jun Takeya Graduate School of Frontier Sciences, The University of Tokyo, Kashiwa, Chiba, Japan

Takayuki Tanaka Department of Chemistry, Graduate School of Science, Kyoto University, Sakyo-ku, Kyoto, Japan

Fumito Tani Institute for Materials Chemistry and Engineering, Kyushu University, Fukuoka, Japan

Kentaro Tashiro International Center for Materials Nanoarchitectonics (WPI-MANA), National Institute for Materials Science, Tsukuba, Ibaraki, Japan

Hayato Tsuji Department of Chemistry, The University of Tokyo, Tokyo, Japan

JST-PRESTO, Kawaguchi, Saitama, Japan

Shogo Tsuzaki School of Materials Science, Japan Advanced Institute of Science and Technology, Nomi, Ishikawa, Japan

Hidemitsu Uno Department of Chemistry and Biology, Graduate School of Science and Engineering, Ehime University, Matsuyama, Japan

Varun Vohra School of Materials Science, Japan Advanced Institute of Science and Technology, Nomi, Ishikawa, Japan

Department of Engineering Science, University of Electro-Communications, Tokyo, Japan

Hajime Watanabe Department of Chemistry and Biology, Graduate School of Science and Engineering, Ehime University, Matsuyama, Japan

Michio Yamada Department of Chemistry, Tokyo Gakugei University, Tokyo, Japan

Shigehiro Yamaguchi Institute of Transformative Bio-molecules and Graduate School of Science, Nagoya University, Nagoya, Japan

Yasuhiko Yamamoto Department of Chemistry, University of Tsukuba, Tsukuba, Japan

Hisao Yanagi Graduate School of Materials Science, Nara Institute of Science and Technology, Ikoma, Nara, Japan

Kazuhiro Yanagi Department of Physics, Tokyo Metropolitan University, Hachioji, Japan

Kazuma Yasuhara Graduate School of Materials Science, Nara Institute of Science and Technology, Ikoma, Nara, Japan

Part I
Extension of Planar π-Electron Systems

Chapter 1
Dibenzopentalenes and Related Compounds

Takeshi Kawase

Abstract Conjugated hydrocarbons with polycyclic frameworks have attracted much attention because of their utility in organic electronic devices. Although dibenzopentalenes possess a $4n\pi$-electron system, they are fairly stable compounds with a planar structure. Thus, appropriate modification would provide them with desirable electronic properties. Recently, we have developed a novel reaction yielding dibenzopentalene derivatives from readily available o-bromoethynylbenzenes using commercially available nickel complexes. The method is accessible to π-extended pentalene derivatives bearing various functional groups and aromatic ring systems. Among them, a dinaphthylpentalene derivative showed a very high hole mobility as amorphous materials. Furthermore, it was employed as a p-type material for organic heterojunction photovoltaic cells. The results indicate that π-extended pentalenes would serve as a good platform for materials applicable to organic electronics.

Keywords Fused ring systems • Cross-coupling • Aromaticity • Density functional calculation • Solid-state properties

1.1 Introduction

Recently, conjugated polycyclic hydrocarbons have been the subject of intense research for their potential use in a variety of materials applications such as organic light-emitting diodes, thin-film transistors, and solar cells. Higher oligoacenes such as rubrene **1** and pentacene **2** are essential as the active layer in high-performance OFETs (Scheme 1.1) [1]. However, higher oligoacenes are highly susceptible to light and oxygen due to their reduced aromaticity, which decreases device stability under ambient conditions. To overcome the difficulty of higher acenes, researchers have focused on more stable, but structurally related π-extended, polycyclic compounds. In this context, polycyclic compounds with heterocyclic

T. Kawase (✉)
Graduate School of Engineering, University of Hyogo, 2167 Shosha, Himeji, Hyogo 671-2280, Japan
e-mail: kawase@eng.u-hyogo.ac.jp

T. Akasaka et al. (eds.), *Chemical Science of π-Electron Systems*,
DOI 10.1007/978-4-431-55357-1_1

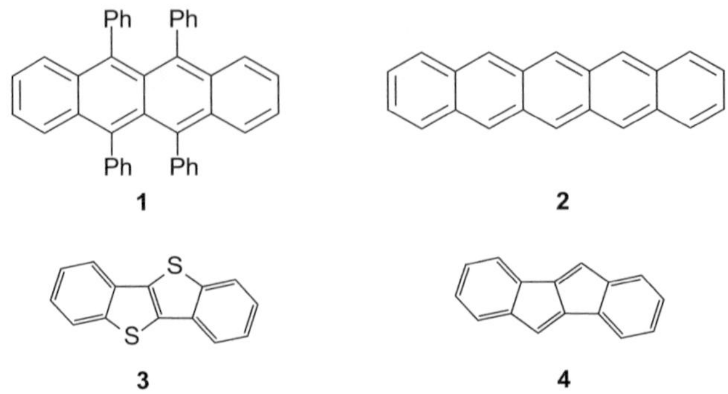

Scheme 1.1 Rubrene **1**, pentacene **2**, benzothienobenzothiophene **3,** and dibenzopentalene **4**

rings have extensively been explored as post-pentacene materials. Actually, considerably high hole mobility in a thin-film-based OFET has been reported for benzothienobenzothiophene derivative **3** (DNTT) [2, 3]. On the other hand, the application of dibenzopentalene **4** with the same 6–5–5–6 fused ring system to organic semiconductor devices has remained underdeveloped, probably due to the limited synthetic routes.

Pentalene possessing an 8π-electron periphery is highly elusive due to their high reactivities [4]. In contrast, dibenzo[*a,e*]pentalene **4** is a fairly stable compound despite possessing a 16π-electron periphery and a planar structure [5–7]. Recently, we found a novel reaction yielding dibenzopentalene derivatives from readily available *o*-bromoethynylbenzenes using commercially available nickel complexes [8]. Although the yields are not so high, the reaction can realize various π-extended pentalene derivatives. Especially, a dinaphthopentalene derivative is the first pentalene system applicable to an organic thin-film transistor [9]. The results suggested that π-extended pentalenes should provide a good platform for materials applicable to organic electronics. In this article, we discuss the syntheses, structures, and electronic properties of newly obtained π-extended pentalene derivatives.

1.2 Syntheses of π-Extended Pentalenes

Since the first synthesis of dibenzopentalenes by Brand in 1912 [10], various dibenzopentalenes with functional groups at positions 5 and 10 have been prepared [11]. The derivatives functionalized at the aromatic rings are rarely prepared probably because of the limitation of the known synthetic methods. In 1999, Youngs reported that an *o*-iodoethynylbenzene **5** was subjected to the Sonogashira coupling conditions to afford a dibenzopentalene derivative **6** (Scheme 1.2) [12]. Unfortunately, this reaction seems to be an exceptional case because the Sonogashira reaction of *o*-iodoethynylbenzenes generally affords phenylene–ethynylene macrocycles [13]. From a synthetic viewpoint, a dibenzopentalene synthesis starting from an *o*-substituted phenylacetylenes using a transition metal catalyst is considerably

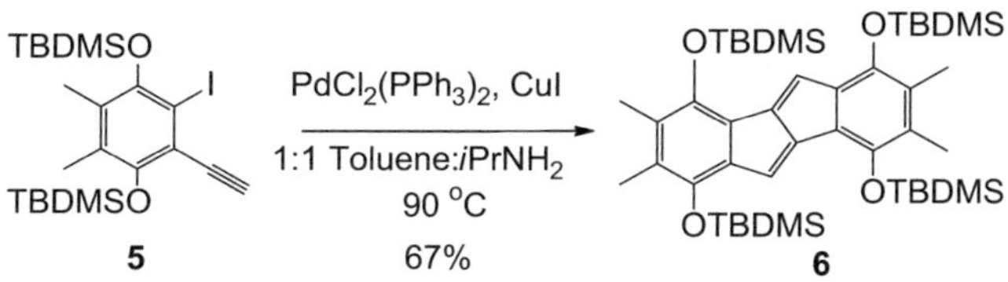

Scheme 1.2 Youngs' dibenzopentalene synthesis

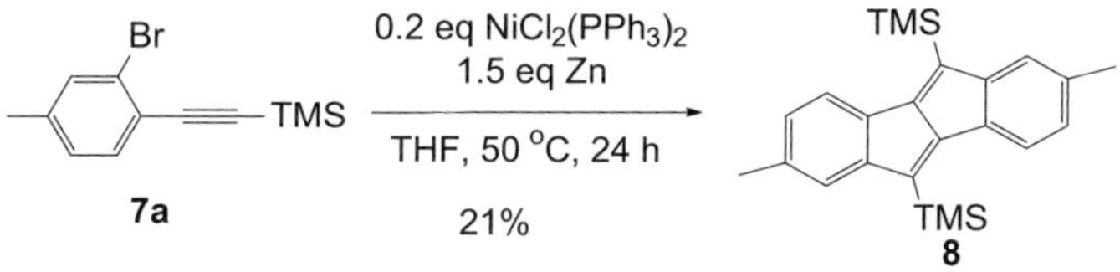

Scheme 1.3 Nickel-catalyzed reaction of **7a** to dibenzopentalene **8**

attractive due to the availability of the starting materials, the formation of three C–C bonds in one process, and potential applicability to π-extended pentalene derivatives.

The nickel(0) complexes have been known to effectively catalyze the Ullmann-type coupling of aryl halides [14, 15]. To prepare a biphenyl derivative, we examined the reaction of 2-bromo-4-methyl-1-(2-trimethylsilylethynyl)benzene **7a** with a Ni(0) complex as a catalyst as shown in Scheme 1.3. The reaction leads to dibenzopentalene **8** in 21 % yield. X-ray crystallographic analysis confirmed the structure of **8**. To elucidate the reaction mechanism and to optimize the yield of the reactions, we investigated the reaction conditions.

The reaction using $Ni(cod)_2$ was investigated, because the reaction of $Ni(cod)_2$ with ligands (L) cleanly generates active $Ni(0)L_2$ complexes under mild conditions [16]. Hence, the reactions of **7a** with $Ni(cod)_2$ in the presence of two equivalents of triphenylphosphine (PPh_3) afford arylnickel(II) complex **9** (21 %) as an air-stable pale-yellow crystals (Scheme 1.4). The structure established by X-ray crystallography confirms that the oxidative addition of $Ni(PPh_3)_2$ to **7a** produces **9**. This type of arylnickel(II) complex is postulated as an intermediate of homocoupling reactions and ordinarily is a thermally labile and air-sensitive compound. The molecular structure of **9** reveals that the C–Ni bond is effectively submerged by a bulky trimethylsilyethynyl group and two triphenylphosphine ligands [17]. Heating **9** at 80 °C in toluene gives **8** in 83 % yield (Scheme 1.4). The result indicates that thermal decomposition of **9** is the final stage in the formation of **8**. Heating a solution of **7a** in toluene (0.2 M) at 110 °C in the presence of $Ni(cod)_2$, PPh_3, and zinc dust with a 1:2:1 stoichiometry gives the best result (46 %). Although sufficient reaction mechanisms cannot be determined at this stage, the high thermal stability of **9** should play a key role in this reaction.

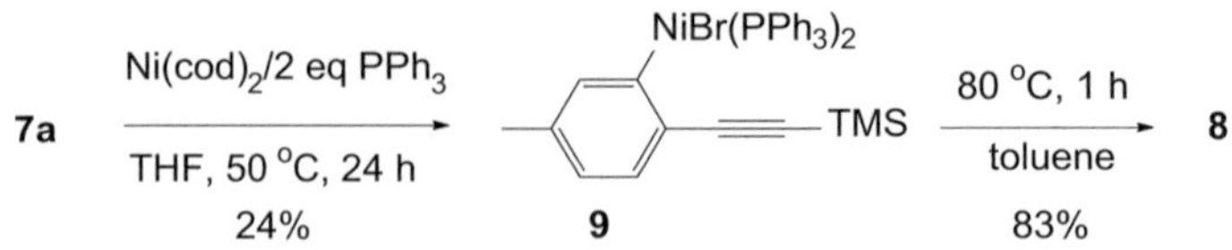

Scheme 1.4 Reactions of **7a** with $Ni(cod)_2$ complexes

Table 1.1 Synthesis of dibenzopentalenes **8** and **10–16**

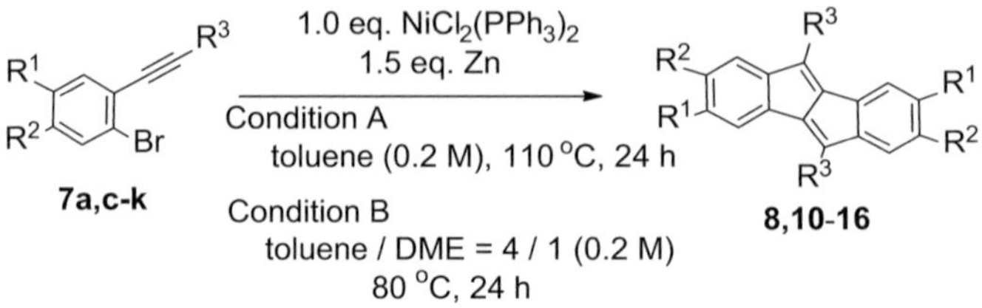

Entry	Compound	R^1	R^2	R^3	Products	Yield (%)
1[a]	**7a**	H	CH_3	$Si(CH_3)_3$	**8**	35
2[b]	**7a**	H	CH_3	$Si(CH_3)_3$	**8**	41
3[a]	**7c**	H	H	$Si(CH_3)_3$	**10**	35
4[a]	**7d**	OCH_3	OCH_3	$Si(CH_3)_3$	**11**	13
5[a]	**7e**	OCH_3	OCH_3	C_6H_5	**12**	24
6[b]	**7f**	H	$COOCH_3$	$Si(CH_3)_3$	**13**	24
7[a]	**7g**	H	CH_3	$C_6H_4CH_3$	**14**	16
8[a]	**7h**	H	CH_3	$C_6H_4OCH_3$	**15**	13
9[a]	**7i**	H	CH_3	$C_6H_4COOCH_3$	**16**	13
10[a]	**7j**	H	CH_3	H		—
11[a]	**7k**	H	CH_3	$COOCH_3$		—

[a]Condition A was employed
[b]Condition B was employed

For convenience, commercially available $NiCl_2(PPh_3)_2$ was employed, and the complex was reduced by zinc duct in situ to form the expected Ni(0) complex. When toluene is used as the solvent (condition A), **8** is obtained in 35 % yield (Table 1.1, entry 1). The presence of 1,2-dimethoxyethane (DME) as a cosolvent (toluene/DME = 4:1) (condition B) accelerates the reaction rate, and the yield of **8** increases to 41 % (entry 2). Then, *o*-bromoethynylarenes **7b–j** bearing various substituents were subjected to the reaction under condition B or A to afford dibenzopentalenes **10–16** in low but acceptable yields (entries 3–9). Unfortunately, the reaction of **7i** or **7j** forms a complex mixture (entries 10 and 11).

Prior to our report, Levi and Tilley independently reported another efficient dibenzopentalene synthesis using a Pd(0) complex (Scheme 1.5) [18]. Through these synthetic methods, various pentalene derivatives should be readily avail-

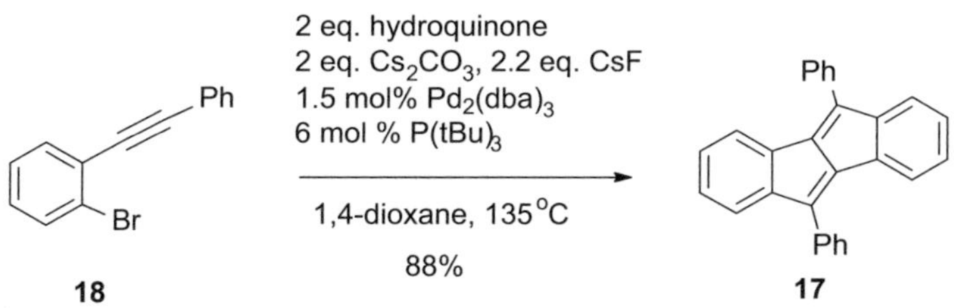

Scheme 1.5 Tilley's dibenzopentalene synthesis

able. Scheme 5 shows their typical conditions of the synthesis of **17** from 2-bromodiphenylacetylene **18**. Although the Ni-mediated reaction is inferior to the Pd-mediated reaction from the viewpoint of the product yields, wide applicability to bromoalkynes with various functionalities is advantageous for this reaction.

1.2.1 Synthesis of π-Extended Pentalenes with Thiophene and Naphthalene Rings

Although various dibenzopentalene derivatives are known, few π-extended pentalene derivatives with heteroaromatic or polycyclic rings in lieu of benzene rings have been prepared to date. Tilley's and Li and Zu's groups synthesized pentalene derivatives bearing thiophene and indole rings by their Pd-catalyzed reaction condition, respectively [18, 19]. Thus, we aimed to prepare di-(3,4)-thienopentalene **19** and dinaphthopentalene derivatives **20–21** using our method. The reactions of 3-bromo-4-ethynylthiophenes **22a–b** under the condition A afford **19b** in 21 % yield and **19c** in 18 % yield, respectively (Scheme 1.6A) [20].

The corresponding bromoethynylnaphthalenes **23a–b** and **24a–b** are subjected to the nickel(0)-assisted reaction under condition A and condition B, respectively, to furnish **20b–c** and **21b–c** in 11–20 % yields (Scheme 1.6B and 1.6C) [9]. The compounds are obtained as fairly stable crystalline substances.

1.3 Molecular Structures

Figure 1.1 shows the molecular structures of **8** [8], **19c** [20], **20b** [9], and **21b** [9], which were determined by X-ray crystal structure analysis. Figure 1.2 summarizes the bond lengths of the π-extended pentalene moieties. The π-extended pentalene skeletons are fairly planar. Analogous to known dibenzopentalenes [6], the molecular structure of **8** is characterized by large bond alternation in the pentalene moiety, but the bond alternation in the six-membered rings is relatively small. The large bond alternation greatly suppresses the anti-aromatic character in the pentalene

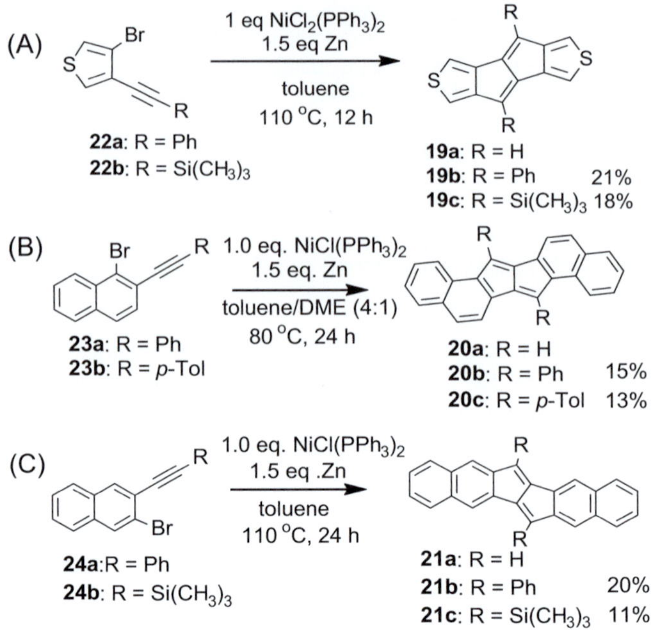

Scheme 1.6 Synthesis of π-extended pentalenes **19–21**

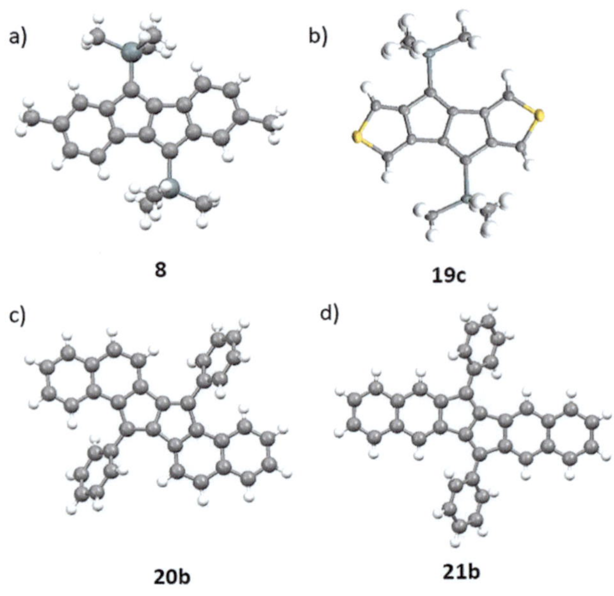

Fig. 1.1 Molecular structures of (**a**) **8**, (**b**) **19c**, (**c**) **20b**, and (**d**) **21b**

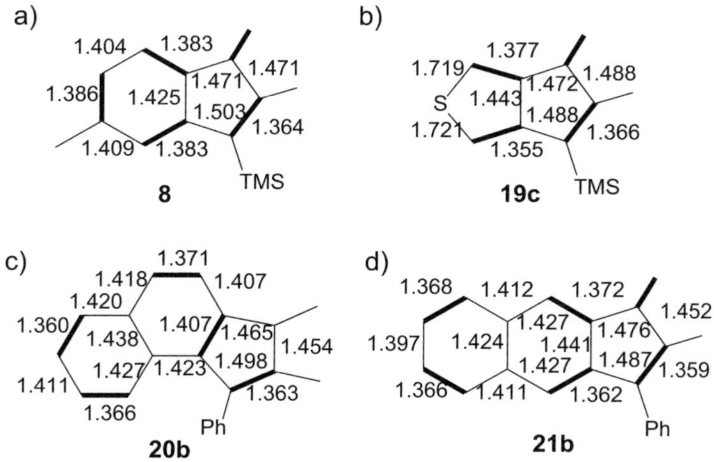

Fig. 1.2 Observed bond lengths (Å) of π-extended pentalenes (**a**) **8**, (**b**) **19c** (average values of two independent molecules), (**c**) **20b**, and (**d**) **21b**. Bold lines indicate relatively short bond lengths (<1.41 Å)

moieties. On the other hand, the 8π-electron system may affect the geometry of the benzene rings. The large bond length of the benzene–pentalene ring fusion (1.425 Å) suggests that the pentalene moiety is unfavorable for 4nπ-cyclic conjugation, but the smaller changes in the other bonds indicate that the aromatic character of the benzene rings is relatively unaffected.

Figure 1.2 summarizes observed bond lengths (Å) of π-extended pentalenes. Compared to **8**, the average bond length for the thiophene–pentalene ring fusion of **19c** (1.443 Å) is longer than that of the benzene–pentalene ring fusion of **8**, and the bond lengths of the thiophene ring of **19c** are typical for 3,4-substituted thiophene derivatives [21]. The result suggests that the contribution of pentalene π-system is almost negligible.

Additionally, the bonds in the pentalene moieties in dinaphthopentalenes **20b** and **21b** are significantly affected. Analogous to **8**, **21b** possesses exo-butadiene conjugations with regard to the pentalene skeleton, and the average bond length of the pentalene–naphthalene ring fusion of **21b** (1.441 Å) is longer than that in **8**. The longer bond length reflects the unfavorable effect from the 4nπ-cyclic conjugation of the pentalene π-system. In contrast, the bond length of the pentalene–naphthalene ring fusion of **20b** (1.407 Å) is much shorter than that of **8**, indicating that **20b** possesses a pentalene 8π-electron system. Considering the resonance contribution, *exo*-butadiene conjugations in regard to the pentalene skeleton destroy the aromatic sextet in the exterior six-membered rings. The counterbalance between aromatic stabilization of the exterior benzene and anti-aromatic destabilization of pentalene π-system should determine the bond alternation.

1.4 Electronic Properties

1.4.1 Electronic Properties of Dibenzopentalenes

Figure 1.3 shows the absorption spectra of functionalized 5,10-bis(trimethylsilyl) dibenzopentalenes **8**, **10**, **11**, and **13**, which consist of three characteristic absorption bands: the relatively weak longest absorption band (450–700 nm), the second longest absorption band (350–450 nm), and the third longest but most intense absorption band (250–350 nm) [8]. According to the density functional theory (DFT) calculations at the RB3LYP/6-31G** level of theory [22], the first absorption bands ($S_0 \rightarrow S_1$ bands) are due to HOMO → LUMO transitions, which are typically symmetry forbidden in 4nπ-electron systems. The second and third absorption bands are also assigned to HOMO–1 → LUMO and HOMO → LUMO + 1 transitions. 2,3,7,8-Tetramethoxy derivative **11** shows a considerably large bathochromic shift in the HOMO → LUMO transition. As supported by the TD-DFT calculation, the small HOMO–LUMO gap of **11** compared to **8**, **10**, and **13** stems from the large positive energy shift of the HOMO, which agrees with the electrochemical results (*vide infra*).

1.4.2 Electronic Properties of π-Extended Pentalenes

Figure 1.4 depicts the absorption spectra of **10**, **19c**, **20b**, and **21b** in CH_2Cl_2. Dibenzopentalene **10** and dithienopentalene **19c** have different aromatic rings and colors (dark red and orange), and **20b** and **21b**, which are both dinaphthopentalenes, have different fusion patterns and colors (reddish brown and orange, respectively). **20b** exhibits the characteristic weak long-wavelength absorption band; however, the band for **21b** disappears.

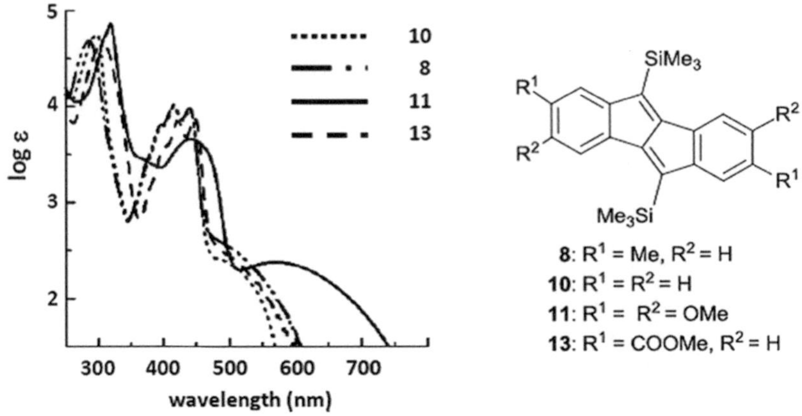

Fig. 1.3 Absorption spectra in CH_2Cl_2 and molecular structures of **8**, **10**, **11**, and **13**

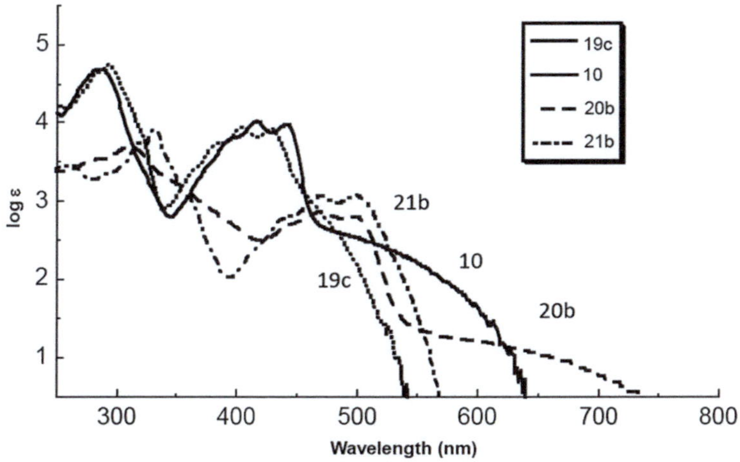

Fig. 1.4 Absorption spectra of **10** and **19c**, **20b**, and **21b** in CH_2Cl_2

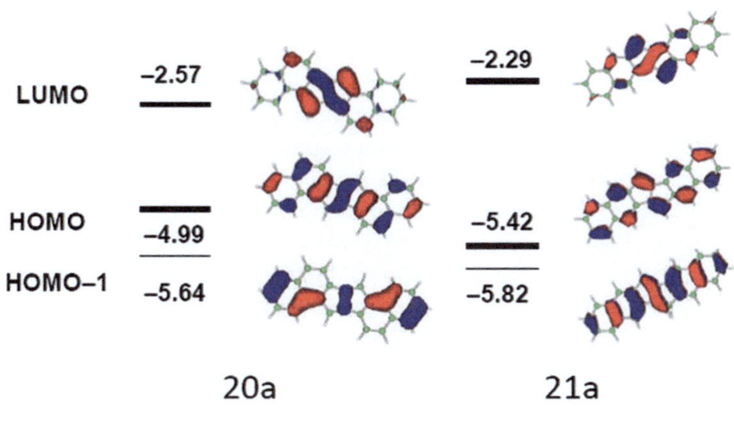

Fig. 1.5 Energy diagrams (eV), molecular orbitals of the LUMO, HOMO, and HOMO–1 of **20a** and **21a** calculated by the TD-DFT (RB3LYP/6-31G**) method

According to the DFT (RB3LYP/6-31G**) calculations, the HOMO–LUMO gap of **19a** (3.60 eV) is much larger than that of parent dibenzopentalene **2** (3.13 eV). The TD calculation shows that the longest absorption band of **19a** (S0 → S1 band) is due to a HOMO–1 → LUMO transition, which is typically symmetry forbidden as in 4nπ-electron systems. However, an allowed HOMO → LUMO transition occurs at a higher energy, and the proximity between HOMO–1 and HOMO should cause both transitions to overlap, which is consistent with the observed absorption spectrum of **19a**. The electronic properties of **20a** and **21a** are also determined using the DFT(RB3LYP/6-31G**) calculations (Fig. 1.5). For **20a**, the longest absorption bands ($S_0 \rightarrow S_1$ bands) are due to HOMO → LUMO transitions, which are typically symmetry forbidden in 4nπ-electron systems. On the other hand, HOMO–1 of **21a** has the same symmetry as HOMO of **20a**. Thus, the fusion pattern reverses the order of the energy levels. An allowed HOMO → LUMO transition occurs at higher energy than symmetry-forbidden HOMO–1 → LUMO transition, but the proximity of the HOMO–1 and HOMO energies cause both transitions to overlap. The value of **21a** (3.13 eV) is comparable to that of parent dibenzopentalene **4** (3.13 eV), while the HOMO–LUMO gap of **20a** (2.58 eV) is considerably smaller than those of **4** and **21a**.

1.5 Electrochemical Properties

The electrochemical properties of π-extended pentalenes were examined by cyclic voltammetry (CV) in DMF. Table 1.2 summarizes the redox potentials. In general, they exhibit four-stage redox waves in the measurable range, and their redox potentials sensitively vary with the electronic properties of the substituents, ring systems and ring fusions, respectively. The CV of tetramethoxy derivative **12** [23] indicates a highly electron-donating property; the ${}^{ox}E_1$ value (+0.10 V for **12**) is higher than those of dibenzopentalene **17** and benzothienobenzothiophene **3** [24], and is almost comparable to that of pentacene **2** ($E_{ox} = 0.22$ V) [25]. Combined with the relatively long-wavelength absorption, the energy of HOMO of **12** is considerably elevated. In contrast, **13**, which contains ester groups, exhibits higher electron affinities. Oxidation waves of **13** are not detected in the measurable range, indicating that the HOMO has a much lower energy. The redox properties of **13** are similar to those of perfluoropentacene ($E_{ox} = 0.79$ V, $E_{red} = -1.13$ V) [26].

The CV of thienopentalene **19c** exhibits a reversible reduction wave (–2.04 V) and an irreversible oxidation wave (+0.90 V) in CH_2Cl_2 solution [20]. The numerical sum of E_{ox} and E_{red} (2.94 V) is larger than that of **17** (2.50 V). These results demonstrate that **19c** has a larger HOMO–LUMO gap than **17**. On the other hand, the CVs of **20b** and **21b** display reversible first oxidation and reduction waves in CH_2Cl_2 solutions, indicating that they have amphoteric redox properties [9]. The electrochemical properties vary with the fusion pattern; **20b** possesses higher electron-donating and electron-accepting properties than **21b** and **17**.

1.6 Solid-State Properties

The solid-state properties of **12**, **17**, **20b**, and **21b** were investigated. These pentalenes readily form quality thin films on n-doped Si wafers. The films were examined by X-ray diffraction (XRD). No peaks in the spectra revealed that amorphous films form on the surfaces. OFETs of **17**, **20b**, and **21b** can be fabricated in a "top-contact" configuration, and that of **12** does in a "bottom-contact" configuration with solution method. The characteristics of the OFET devices were measured at

Table 1.2 Redox potentials (V) of π-extended pentalenes in DMF[a] and CH_2Cl_2[b]

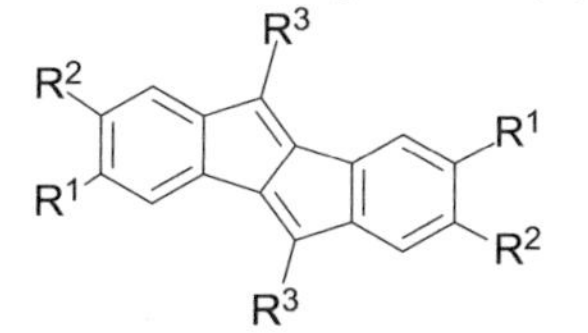

8: R^1 = Me, R^2 = H, R^3 = $SiMe_3$
10: R^1 = R^2 = H, , R^3 = $SiMe_3$
11: R^1 = R^2 = OMe, , R^3 = $SiMe_3$
12: R^1 = R^2 = OMe, , R^3 = Ph
13: R^1 = COOMe, R^2 = H, R^3 = $SiMe_3$
17: R^1 = R^2 = H, R^3 = Ph

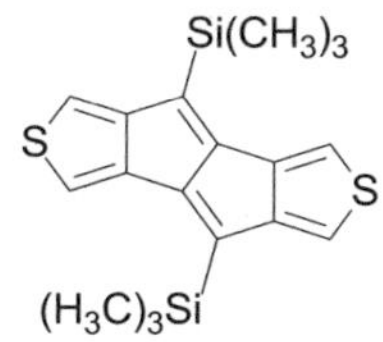

19c

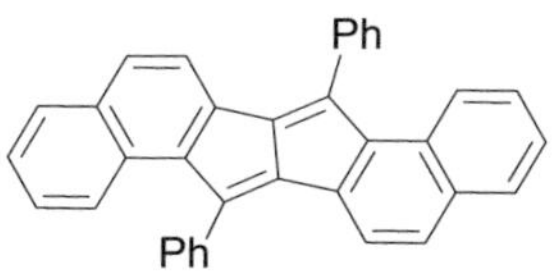

20b

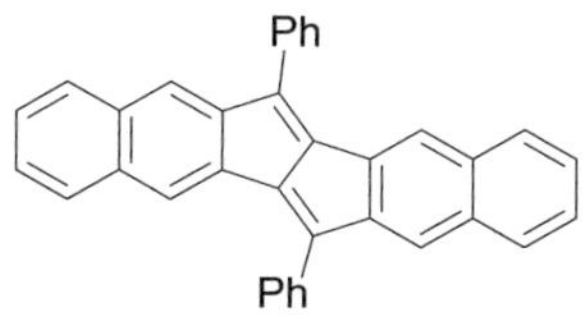

21b

Compounds	$^{ox}E_{2,1/2}$	$^{ox}E_{1,1/2}$	$^{red}E_{1,1/2}$	$^{red}E_{2,1/2}$
8[a]	1.01[c]	0.82[c]	−1.65	−2.22[c]
10[a]	1.11[c]	0.92[c]	−1.58	−2.22[c]
11[a]	0.67[c]	0.30	−1.68	−2.26[c]
12[b]	0.55[c]	0.10	−1.97	—[d]
13[a]	—[d]	—[d]	−1.28	−1.90[c]
17[b]	0.98[c]	0.87[c]	−1.62	−2.18[c]
19c[b]	—[d]	0.90[c]	−2.04	—[d]
20b[b]	1.14[c]	0.55	−1.52	−2.06[c]
21b[b]	1.11[c]	0.73	−1.79	−2.26[c]

Conditions: V vs. Ag/Ag^+, in 0.1 M $n$$Bu_4NClO_4$/solvent, scan rate 100 mV/s, +25 °C, ferrocene used as a standard
[a]Solvent = DMF
[b]Solvent = CH_2Cl_2
[c]Half-wave potentials
[d]Not detected

room temperature in air. Although **17** and **20b** show negligible mobilities under the measured conditions, **21b** has a hole mobility of 1.8×10^{-3} cm^2 V^{-1} s^{-1} with current on/off ratio (I_{on}/I_{off}) of 10^5, which is a relatively high value for amorphous materials [9]. Moreover, **12** exhibits a hole mobility of 4×10^{-5} cm^2 V^{-1} s^{-1} with current on/off ratio (I_{on}/I_{off}) of 600 [23]. The HOMO energy level of **21b** (–5.44 eV) is lower than that of **20b** (–5.26 eV), but is comparable to that of **17** (–5.52 eV). Despite possessing a fairly high electron-donating property, the hole mobility of **12** is relatively poor. These results indicate that the linearly extended fusion pattern in polycyclic conjugated systems plays an important role in the solid-state properties [27].

Moreover, **21b** has been applied to an electron-donor layer for heterojunction organic thin-film solar cells [28]. When fullerene and BCP (2,9-dimethyl-4,7-diphenyl-1,10-phenanthroline) are used as electron-acceptor and exciton-blocking layers, the device has a PCE value of 0.94 % and an open-circuit voltage (V_{OC}) of 0.96 V. Although the PCE value is lower than those of pentacene (2.7 %) and tetracene (2.3 %), the V_{OC} value is considerably higher than those of pentacene (0.58 V) and tetracene (0.36 V) [1, 9].

1.7 Conclusions

Nickel-mediated reactions of 1-bromo-2-ethynylarenes successfully afford several new π-extended pentalene derivatives. The dependences of the electronic and structural properties on the substituents as well as the fusion patterns are consistent with theoretical calculations. Among the prepared π-extended pentalenes, **21b** is the first pentalene derivative for organic thin-film transistors. Although the application of fully conjugated systems, the so-called non-benzenoid aromatic compounds, remains underdeveloped to date, the results herein demonstrate that π-extended pentalenes can serve as a good platform for materials applicable to organic electronics. Recently, the unique electronic properties of fulvenes [29], fulvalenes [30], and indenofluorenes [31–33] have received renewed interest, and these materials may serve as functional dyes from a materials science viewpoint.

Acknowledgments This study was partly supported by a Grant-in-Aid for Scientific Research on Innovative Areas (Nos. 21108521, 23108718, and 25109515) from the Ministry of Education, Culture, Sports, Science and Technology, Japan.

References

1. Anthony JE (2008) Angew Chem Int Ed 47:452–483
2. Yamamoto T, Takimiya K (2007) J Am Chem Soc 129:2224–2225
3. Takimiya K et al (2011) Adv Mater 23:4347–4370
4. Hafner K et al (1973) Angew Chem Int Ed Engl 12:337–339
5. Hopf H (2000) Classics in hydrocarbon chemistry. Wiley-VCH, Weinheim, New York, Chichester, Brisbane, Singapore, Toronto, pp 277–290
6. Saito M et al (2008) Chem Eur J 14:6062–6068
7. Saito M (2010) Symmetry 2:950–969
8. Kawase T et al (2009) Chem Eur J 15:2653–2661
9. Kawase T, Takimiya K et al (2010) Angew Chem Int Ed 49:7728–7732
10. Brand K (1912) Ber Dtsch Chem Ges 45:3071
11. Kawase T et al. (2011) Eur J Org Chem 4885–4891; Kawase T (2012) J Synth Org Chem Jan 70:354–365
12. Youngs WJ et al (1999) J Org Chem 64:2947–2949
13. Iyoda M et al (2011) Angew Chem Int Ed 50:10522–10553
14. Hassan J et al (2002) Chem Rev 102:1359–1470
15. Iyoda M, Oda M et al (1990) Bull Chem Soc Jpn 63:80–87

16. Ittel SD, Springer W et al (1990) Inorg Synth 28:98–104
17. Lee HB, Sharp PR (2005) Organometallics 24:4875–4877
18. Levi ZU, Don Tilley T (2009) J Am Chem Soc 131:2796–2797
19. Li Y, Zhu D et al (2011) Org Lett 13:1520–1523
20. Kawase T et al (2010) Chem Lett 39:300–301
21. Li ZY et al (2007) Synth Commun 37:71–78
22. The molecular geometries of the dibenzopentalenes were fully optimized at the RB3LYP/6-31G** level of theory embedded in the Gaussian 03 software package
23. Nishida J, Kawase T et al. unpublished results
24. Takimiya K et al (2006) J Am Chem Soc 128:12604–12605
25. Anthony JE et al (2005) Org Lett 7:3163–3166
26. Tokito S et al (2004) J Am Chem Soc 126:8138–8140
27. Takimiya K et al (2010) Bull Chem Soc Jpn 83:120–130
28. (a) Böuerle P et al (2007) Angew Chem Int Ed 46:1679–1683, (b) Tolbert SH, Schwartz BJ, Rubin Y et al (2008) J Am Chem Soc 130:17290–17292; (c) Matsuo Y, Nakamura E et al (2009) J Am Chem Soc 131:16048–16050
29. (a) Swager TM et al (2010) Org Lett 12:5302–5305; (b) Kato S, Diederich F (2010) Chem Commun 46:1994–2006; (c) Diederich F et al (2012) J Am Chem Soc 134:18139 − 18146
30. Heeger AJ, Wudl F et al (2010) Angew Chem Int Ed 49:532–536
31. (a) Shimizu A, Tobe Y (2011) Angew Chem Int Ed 50:6906–6910; (b) Tobe Y et al (2013) Angew Chem Int Ed 52:6076–6079; (c) Tobe Y et al (2012) Chem Sci 5:163–168
32. (a) Haley MM et al (2011) Angew Chem Int Ed 50:1127–1162; (b) Haley MM et al (2011) Angew Chem Int Ed 50:11103–11106; (c) Nuckolls C, Haley MM et al (2012) J Am Chem Soc 134:10349 − 10352; (d) Haley MM et al (2014) Chem Sci 5:1008–1014
33. Nishida J, Yamashita Y et al (2012) Chem Eur J 18:8964–8970

Chapter 2
Synthesis of a Porphyrin-Fused π-Electron System

Hidemitsu Uno, Kazunari Tagawa, Hajime Watanabe, Naoki Kawamoto, Mina Furukawa, Tetsuo Okujima, and Shigeki Mori

Abstract Oligoporphyrins with a large fused π-electron system were successfully synthesized by the fusion of porphyrin chromophores with polycyclic aromatic hydrocarbons (PAHs). Oligoporphyrins fused with benzene units and zinc diporphyrins fused with PAHs were prepared and their electronic properties were examined. Electronic spectra showed strong absorption in the red to near-infrared (NIR) region. Not only the absorption maxima but also the number of absorption bands was affected by the fusion mode of the porphyrin chromophores. In the UV-vis spectra of the oligoporphyrins, three major absorption bands corresponding to Soret and Q bands were observed. A linearly fused trizinc triporphyrin with two benzene moieties showed strong Q-band absorption at 680 nm, while split Q-band absorptions at 663 and 674 nm were observed in the L-shaped triporphyrin. Preparation based on the retro-Diels-Alder reaction of the precursors with a bicyclo[2.2.2]octadiene skeleton did not give naphthacene- and pentacene-fused diporphyrins but instead gave naphthalene-, anthracene-, and chrysene-fused diporphyrins. In the chrysene-fused diporphyrin, the absorption maximum in the Q band was 624 nm, which was similar to that of naphthalene-fused diporphyrin.

Keywords Porphyrinoid • Benzene-fused diporphyrin • Doubly benzene-fused triporphyrin • Triply benzene-fused tetraporphyrin • Quadruply benzene-fused pentaporphyrin • Retro-Diels-Alder reaction • Bicyclo[2.2.2]octadiene • Acene-fused diporphyrin • Pericyclic reaction • Near infrared • Two-photon absorption • [3 + 1] porphyrin synthesis • Cyclo-tetramerization

H. Uno (✉) • K. Tagawa • H. Watanabe • M. Furukawa • T. Okujima
Department of Chemistry and Biology, Graduate School of Science and Engineering, Ehime University, Matsuyama 790-8577, Japan
e-mail: uno@ehime-u.ac.jp

N. Kawamoto • S. Mori
Integrated Center for Sciences, Ehime University, Matsuyama 790-8577, Japan

T. Akasaka et al. (eds.), *Chemical Science of π-Electron Systems*,
DOI 10.1007/978-4-431-55357-1_2

2.1 Introduction

NIR dyes have garnered increasing attention due to their applicability as singlet oxygen sensitizers for photodynamic therapy [1], two-photon microscopic imaging agents [2], sensitizers for TiO_2-based organic photovoltaics [3], cutoff filters [4], and so on. For the former two applications, good permeability of light in living cells is essential. As water and hemoglobin are the major light-absorbing materials in mammalian cells, there is a light-permeable region with an optical window roughly from 600 to 1300 nm [5]. The use of NIR light for organic solar cells was keenly investigated in order to improve their performance [6], as more than 40 % of the solar energy reaching to the earth's surface is from NIR light.

These compounds are rather difficult to synthesize due to their instability under oxidizing conditions caused by their intrinsically high HOMO energy levels. They are also insoluble in common solvents due to the stacking nature of their large flat chromophores. The solubility is commonly improved by introducing solubilizing substituents such as long, branched alkyls [7] or bulky groups [8], which prevent the stacking of the flat chromophores. From the point of view of a synthetic strategy, fusion by construction of aromatic rings between the chromophore units at the very end of the synthesis is more advantageous than the successive expansion of chromophores. When the porphyrin system is employed, fusion method is thought to be ideal because the system is robust and reactive. There are two methods for the construction of fusing aromatic units (Fig. 2.1). In the first method, the targeted π-systems **2** are created by the construction of the fusing aromatic system such as benzene [9], thiophene [10], *etc.* [11] between the singly connected chromophore units **1** (Fig. 2.1a). In the second method, the targeted compounds **4** bearing the benzene unit are formed from the precursors **3** by pericyclic cycloreversion [12], oxidation [13], or rearrangement (Fig. 2.1b). In these methods, the purification of the target products is still a severe problem. Nanographenes are prepared by the

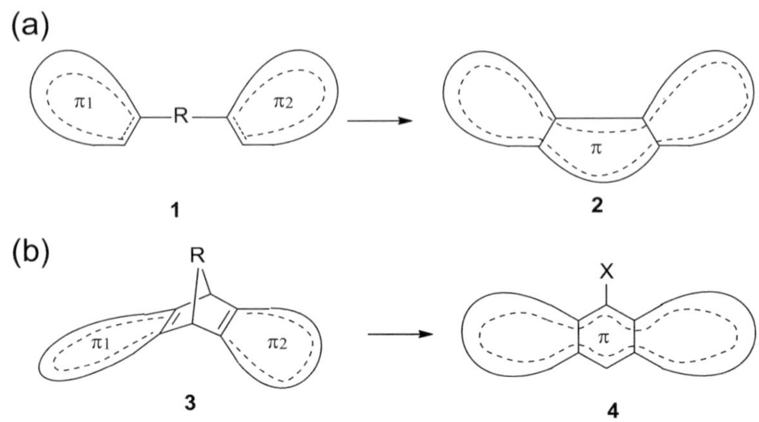

Fig. 2.1 Strategy for the construction of fusing aromatic units

former strategy from the well-designed polyphenyl derivatives based on the 6π-electrocyclic reaction, followed by thermal dehydrogenation [14]. The pericyclic cycloreversion in the latter method is one of the most advantageous reactions for the construction of such fusing π-systems, because this reaction can be conducted without reagents or solvents. Only the activation energy is required for the reaction, and no purification is needed, if the reaction occurs quantitatively. In this session, we have discussed our recent results on the preparation of large π-system based on the fusion of chromophores in the last stage of the synthesis.

2.2 Oligoporphyrins Fused with Benzene Units

Oligoporphyrins with fused π-systems are very attractive not only as NIR absorbing dyes but also as two-photon absorbing materials applicable to laser scanning fluorescence microscopy [15] and fine focused photodynamic therapy for inner tissues [16]. An important factor for large two-photon absorbing cross-section values is a coplanar geometry between neighboring porphyrin chromophores in order to maximize π-electron delocalization throughout the molecular network [17]. In order to know the effect of π-system shapes, various kinds of porphyrin oligomers fused with benzene moieties were prepared.

Bicyclo[2.2.2]octadiene (BCOD)-fused oligoporphyrins were synthesized and converted to the targeted fully fused oligoporphyrins by the retro-Diels-Alder reaction in the final step. First, the synthesis of porphyrins fused with pyrrole through the BCOD moiety (Scheme 2.1) was carried out. A pyrrole ring was constructed at the double bond of the BCOD-fused pyrrole-2-carboxylate ethyl

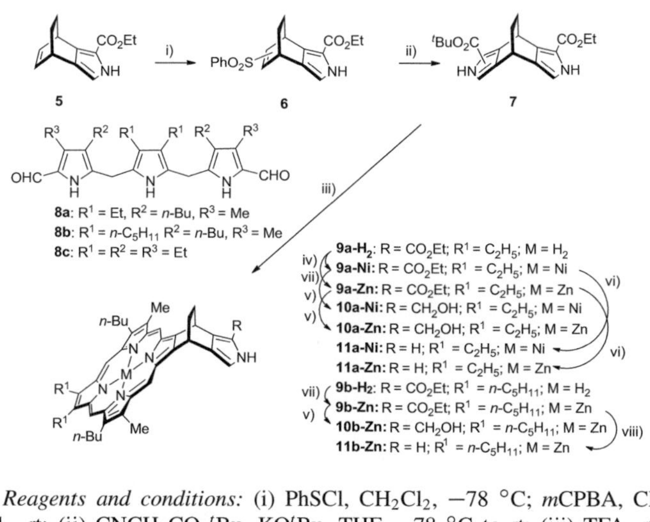

Scheme 2.1 *Reagents and conditions:* (i) PhSCl, CH_2Cl_2, −78 °C; *m*CPBA, CH_2Cl_2, 0 °C; DBU, CH_2Cl_2, rt; (ii) $CNCH_2CO_2{}^tBu$, KO^tBu, THF, −78 °C to rt; (iii) TFA, rt; **8**, CH_2Cl_2, rt; Et_3N, DDQ, rt; (iv) $Ni(OAc)_2 \cdot 4H_2O$, CH_2Cl_2, reflux; (v) $LiAlH_4$, THF, rt; (vi) NaOH, ethylene glycol; 170 °C; (vii) $Zn(OAc)_2 \cdot 2H_2O$, CH_2Cl_2, rt; (viii) NaOH, ethylene glycol; 170 °C; $Zn(OAc)_2 \cdot 2H_2O$, CH_2Cl_2, rt

ester **5** [18] by consecutive treatment with PhSCl, *m*-chloroperbenzoic acid (*m*CPBA), 1,4-diazabicyclo[5.3.0]undec-11-ene (DBU), and a potassium salt of *tert*-butyl isocyanoacetate. A regioisomeric mixture of *tert*-butyl ethyl dipyrroledicarboxylate **7** [19] was obtained with a good overall yield. The tripyrranes **8a** and **8b** were prepared according to procedures described in the literature [20] and then condensed with **7** by treatment with trifluoroacetic acid (TFA) followed by oxidation with DDQ. The pyrroloporphyrins **9a-H$_2$** and **9b-H$_2$** were each obtained with a yield of 27 %. The free-base pyrroloporphyrin **9a-H$_2$** was metalated with nickel acetate and zinc acetate to give the nickel and zinc porphyrins **9a-Ni** and **9a-Zn**, respectively, with good yields. Similarly, the zinc porphyrin **9b-Zn** was obtained with a yield of 93 %. The ester moiety of **9** was reduced with $LiAlH_4$ at 0 °C to give hydroxymethyl derivatives **10**, which were purified by short-column chromatography on silica-NH_2 gel. The ethyl ester group of **9** was removed by heating at 170 °C in an alkaline solution of ethylene glycol to give α-free pyrroloporphyrins **11** with a good yield.

2.2.1 Preparation of Benzene-Fused Diporphyrins

For the preparation of BCOD-connected diporphyrins, free-base pyrrole-connected porphyrins **11** were subject to inverse [3 + 1] porphyrin synthesis with tripyrranedicarbaldehydes **8**. Metalation with $Zn(OAc)_2 \cdot 2H_2O$ and $Cu(OAc)_2 \cdot H_2O$ gave dimetallodiporphyrins **12** with a moderate yield (12–33 %) (Scheme 2.2) [19, 21]. As the nickel porphyrin was stable under the acidic conditions employed in the [3 + 1] porphyrin synthesis [21], diporphyrins were prepared with two different metals. On the other hand, the zinc porphyrin was labile under the reaction conditions, such that only dizinc derivatives could be prepared.

The absorption and fluorescence maxima of the dimers are summarized in Table 2.1. The Soret and Q bands of the zinc-porphyrin monomer **9a-Zn** appeared at slightly longer wavelength [403 (log $\varepsilon = 5.49$), 532 (4.19), and

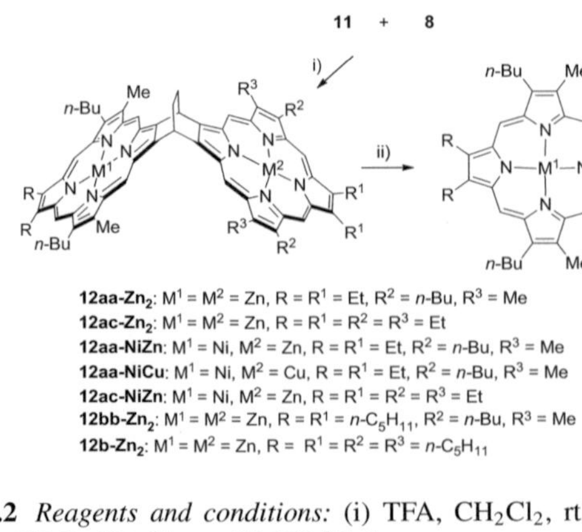

Scheme 2.2 *Reagents and conditions:* (i) TFA, CH_2Cl_2, rt; Et_3N; DDQ; $Zn(OAc)_2 \cdot 2H_2O$ or $Cu(OAc)_2 \cdot 2H_2O$, CH_2Cl_2, rt; (ii) 170 °C, in vacuo

Table 2.1 Absorption and emission maxima of benzene-fused oligoporphyrins

	λ_{max} /nm (log ε)						
Oligoporphyrin	Band I		Band II		Band III		Fluorescence/nm
12aa-Zn$_2$[a]	399	414			533	574	579
	(5.65)	(5.65)			(4.65)	(4.74)	624
12aa-NiZn[a]	394	410			555	572	578
	(5.42)	(5.49)			(4.53)	(5.72)	
13aa-Zn$_2$[b]	390	412	444	474	621	636	641
	(4.98)	(4.91)	(4.90)	(5.40)	(4.78)	(5.27)	
13aa-NiZn[b]	389		470	477	619	635	–
	(4.94)		(5.13)	(5.15)	(4.73)	(5.21)	
13aa-NiCu[a]	400		466		580	634	–
	1.00[c]		0.92[c]		0.22[c]	0.89[c]	
15b-Zn$_3$[b]	409	416	493		626	680	683
	(5,14)	(5,14)	(5.40)		(4.58)	(5.55)	
19b-Zn$_3$[b]	412	427	486		663	674	677
	(5.18)	(5.18)	(5.36)		(5.39)	(5.45)	
20b-Zn$_4$[b]	412	427	496		682	716	721
	(5.41)	(5.18)	(5.53)		(4.76)	(5.71)	
21b-Zn$_5$[b]	414	437	497	579	722	762	767
	(5.62)	(5.32)	(5.48)	(4.85)	(4.90)	(5.80)	

[a]In $CHCl_3$
[b]1% pyridine/ $CHCl_3$
[c]Intensity ratio

570 (4.24) nm] than those of **9a-Ni** [394 (5.28), 516 (4.04), and 554 (4.37) nm]. The bicyclo[2.2.2]octadiene-bridged dimers **12aa-NiZn** and **12 ac-NiZn** exhibit two Soret bands at 394 (5.42) and 410 (5.49) nm and three Q bands at 535 (4.40), 555 (4.53), and 572 (4.50) nm. The shorter-wavelength Soret-band absorption appeared at almost the same position as for the parent nickel porphyrin **9a-Ni** with a slightly larger intensity. In contrast, the longer-wavelength Soret band was absorbed at a lower energy field compared to that of zinc porphyrin **9a-Zn**, with almost the same intensity. Even in the case of the symmetric zinc-porphyrin dimers **12aa-Zn$_2$** and **12bb-Zn$_2$**, two separated Soret bands with the similar intensity (log $\varepsilon = 5.65$) were observed at 399 and 414 nm. This phenomenon is due to the exciton coupling between the porphyrin rings through the bicyclo[2.2.2]octadiene moiety [22].

In the UV-vis spectra of the benzene-fused diporphyrins **13**, there are three major absorption bands. The two bands (band I: 380–440 and band II: 440–580 nm) in the shorter-wavelength region correspond to Soret bands and the remaining major band (band III) is the Q band. The absorption and emission maxima in the UV-vis-NIR spectra of the dimers are summarized in Table 2.1. The longest-wavelength absorptions in the Q-band region (band III) become quite strong due to the fusion of porphyrin chromophores. In the fluorescence spectra, small Stokes' shifts (3–5 nm) are observed [23].

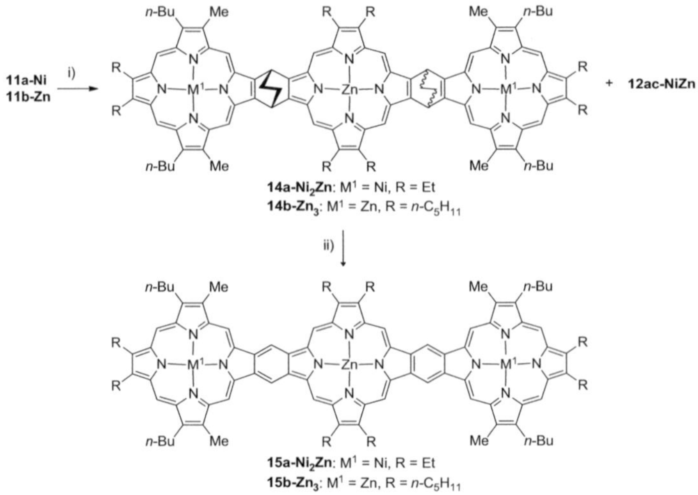

Scheme 2.3 Reagents and conditions: (i) 3,4-diethyl-2,5-bis(hydroxymethyl)pyrrole or 3,4-dipentyl-2,5-bis(hydroxymethyl)pyrrole, $BF_3 \cdot OEt_2$, CH_2Cl_2, rt; chloranil, $Zn(OAc)_2$,$\cdot 2H_2O$, 50 °C; (ii) 200 °C, in vacuo

2.2.2 *Preparation of Benzene-Fused Tri-, Tetra-, and Pentaporphyrins*

BCOD- and benzene-fused oligoporphyrins were prepared using the following method. The [2 + 2] porphyrin synthesis using α-free pyrroloporphyrins **11** and 2,5-bis(hydroxymethyl)pyrroles was first employed for the construction of linearly connected triporphyrins. An equal molar mixture of nickel porphyrin **11a-Ni** and 3,4-diethyl-1,5-bis(hydroxymethyl)pyrrole was successively treated with $BF_3 \cdot OEt_2$, chloranil, and zinc acetate to give a *syn*- and *anti*-mixture (*ca.* 1:1) of linear triporphyrin **14a-Ni$_2$Zn** at a yield of 45 % with a small amount (5 %) of diporphyrin **12 ac-NiZn** (Scheme 2.3). The *syn*- and *anti*-isomers (C_{2v} and C_{2h} point groups, respectively) of **14a-Ni$_2$Zn** were separated by fractional recrystallization after preparative GPC, and the structures were unambiguously determined by X-ray crystallography. Thermogravimetric (TG) analysis revealed that the triporphyrin **14a-Ni$_2$Zn** expelled two molecules of ethylene between 170 and 200 °C. No spectroscopic evidence of triporphyrin **15a-Ni$_2$Zn** was obtained due to its insoluble nature. In order to solubilize the linear triporphyrin, its ethyl and nickel moieties were changed to pentyl and zinc groups, respectively. Similar reactions of the nickel porphyrin **11b-Zn** and 3,4-dipentyl-1,5-bis(hydroxymethyl)pyrrole, however, did not produce the intended **14b-Zn$_3$** at all. Furthermore, the synthesis of **14a-Ni$_2$Zn** could not be reproduced. All trials, except for the first one mentioned above, produced only a trace amount of **14a-Ni$_2$Zn**. Instead, only zinc octaethylporphyrin was obtained in moderate yields.

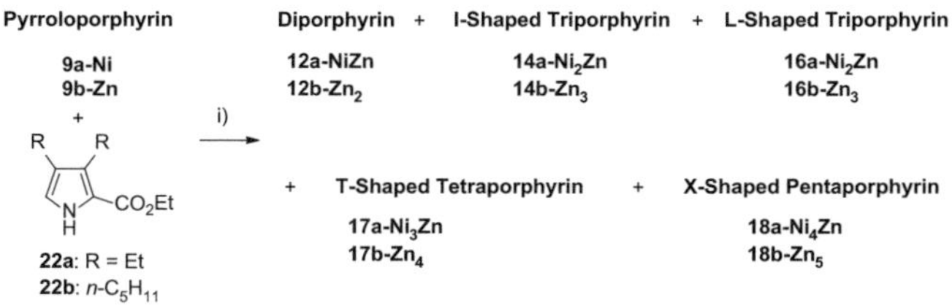

Scheme 2.4 Reagents and conditions: (i) $LiAlH_4$, THF, 0 °C; $BF_3 \cdot OEt_2$, CH_2Cl_2, rt; chloranil, $Zn(OAc)_2 \cdot 2H_2O$, 50 °C

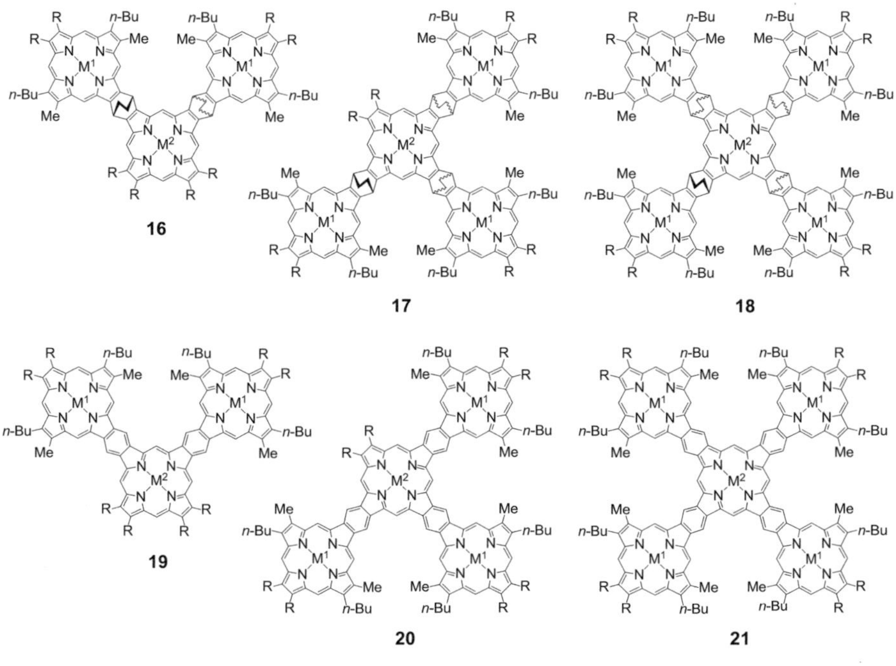

Fig. 2.2 Prepared oligoporphyrins; **a-Ni$_n$Zn**: $M^1 = Ni$, $M^2 = Zn$, R = Et; **a-Zn$_n$**: $M^1 = M^2 = Zn$, R = Et; **b-Zn$_n$**: $M^1 = M^2 = Zn$, R = n-C_5H_{11}

Oligoporphyrins were prepared by the random combination of (hydroxymethyl)pyrroloporphyrins **10** with 3,4-diethyl-2-(hydroxymethyl)pyrrole and 3,4-dipentyl-2-(hydroxymethyl)pyrrole (Scheme 2.4 and Fig. 2.2). An equal molar mixture of **9a-Ni** and ethyl 3,4-diethylpyrrole-2-carboxylate (**22a**) was reduced with $LiAlH_4$ to give a 1:1 mixture of **10a-Ni** and 3,4-diethyl-2-(hydroxymethyl)pyrrole. The mixture was treated with $BF_3 \cdot OEt_2$, oxidized with chloranil, and then reacted with zinc acetate to give a mixture of oligoporphyrins as well as octaethylporphyrin. Diporphyrin **12aa-NiZn** (14 %), I-shaped triporphyrin **14a-Ni$_2$Zn** (10 %), L-shaped triporphyrin **16a-Ni$_2$Zn** (7 %), T-shaped tetraporphyrin **17a-Ni$_3$Zn** (12 %), and X-shaped pentaporphyrin **18a-Ni$_4$Zn** (1 %) were separated by column chromatography followed by preparative GPC. Isomers (C_s and C_2 point groups) of the L-shaped triporphyrin **16a-Ni$_2$Zn** were separated by a combination of recrystallization and recycled GPC, and the structures were determined by X-ray crystallography. A similar reaction sequence starting from **9b-Zn** and ethyl 3,4-dipenylpyrrole-2-carboxylate (**22b**) provided diporphyrin **12b-Zn$_2$**, I-shaped triporphyrin **14b-Zn$_3$**, L-shaped triporphyrin **16b-Zn$_3$**, T-shaped tetraporphyrin **17b-Zn$_4$**, and X-shaped pentaporphyrin **18b-Zn$_5$**. The pentaporphyrins **18a-Ni$_4$Zn** and **18b-Zn$_5$** were more efficiently prepared by the cyclic tetramerization of **10a-Ni** and **10b-Zn** under conditions similar to random cyclization, with yields of 24 % and 16 %, respectively [24]. All of the zinc oligoporphyrins were quantitatively converted to the corresponding benzene-fused oligoporphyrins **13b-Zn$_2$**, **15b-Zn$_3$**, **19b-Zn$_3$**, **20b-Zn$_4$**, and **21b-Zn$_5$** by heating at 170–200 °C for 30 min in vacuo. TG analysis of **15b-Zn$_3$**, **19b-Zn$_3$**, **20b-Zn$_4$**, and **21b-Zn$_5$** showed single steep weight losses from 150 to 190 °C, corresponding to the expected weights of expelled ethylene molecules.

In the absorption spectra of the benzene-fused tri-, tetra-, and pentaporphyrins, three bands similar to those of the benzene-fused diporphyrins are observed (Table 2.1). As the number of porphyrin rings increased, the absorption maxima of the lowest energy transitions and the intensities also increased. In the X-shaped pentaporphyrin **21b-Zn$_5$**, very strong absorption was observed at 761 (log $\varepsilon = 5.80$) nm in the NIR region. In the case of the L-shaped triporphyrin **19b-Zn$_3$**, sprit Q-band absorptions were observed at 663 and 674 nm with similar intensities of log $\varepsilon = 5.39$ and 5.45, respectively. A single strong Q-band absorption was observed at 680 nm with log $\varepsilon = 5.55$ for the I-shaped triporphyrin **15b-Zn$_3$**. The most striking feature of the L- and I-shaped triporphyrins is the two-photon absorption ability. The two-photon absorbing cross-section value for the I-shaped triporphyrin **15b-Zn$_3$** (2200 GM) is about four times larger than that of the L-shaped triporphyrin **19b-Zn$_3$** (580 GM), which is similar to that of the diporphyrin **12aa-Zn$_2$** (520 GM) [23].

2.3 Diporphyrin Fused with Polycyclic Aromatic Hydrocarbons

Diporphyrins fused with polycyclic aromatic hydrocarbons were then targeted. Scheme 2.5 shows the preparation of naphthalene-fused diporphyrin **27**. The Diels-Alder reaction of triene **22** with tosylacetylene gave a tricyclic vinyl sulfone, which was then subjected to the modified Barton-Zard reaction. Dihydroethanonaphthalene-fused pyrrole **23** was obtained with a good yield [25]. Pyrrole ring construction was achieved at the double bond of **23** by addition of PhSCl, oxidation with mCPBA, dehydrochlorination with DBU, and the Barton-Zard reaction with ethyl isocyanoacetate. Dipyrrole **24** was obtained with a good overall yield. The ester groups of **24** were then removed by thermal treatment in alkaline ethylene glycol to produce α-free dipyrrole **25**. The double [3 + 1] porphyrin synthesis of **25** with tripyrranedicarbaldehyde **8a** gave BCOD-fused porphyrin-benzoporphyrin dyad **26**. Dyad **26** was finally converted to naphthalene-

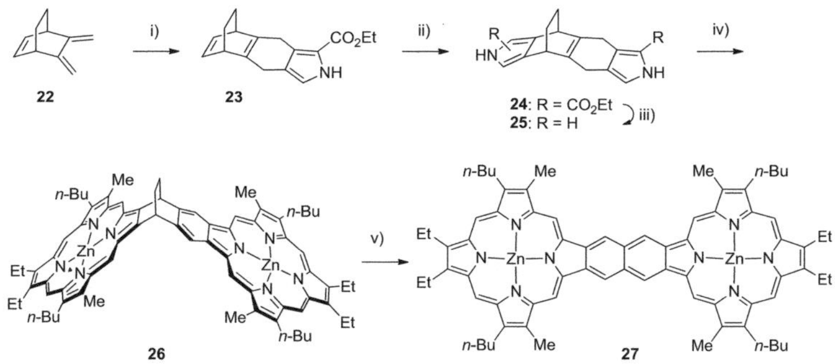

Scheme 2.5 Reagents and conditions: (i) tosylacetylene, $CHCl_3$; $CNCH_2CO_2Et$, KO^tBu, THF, 0 °C; (ii) PhSCl, CH_2Cl_2, 0 °C; *m*CPBA, CH_2Cl_2, rt; $CNCH_2CO_2Et$, KO^tBu, THF, 0 °C; (iii) KOH, ethylene glycol, 170–190 °C; (iv) **8a**, TFA, CH_2Cl_2; DDQ, CH_2Cl_2; Et_3N, $Zn(OAc)_2 \cdot 2H_2O$; (v) 270 °C, in vacuo

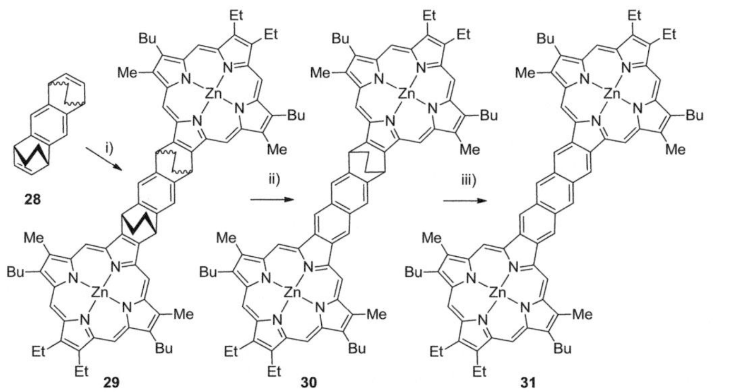

Scheme 2.6 Reagents and conditions: (i) PhSCl, CH_2Cl_2, 0 °C; *m*CPBA, CH_2Cl_2, rt; $CNCH_2CO_2Et$, KO^tBu, THF, 0 °C; KOH, ethylene glycol, 170–190 °C; **8a**, TFA, CH_2Cl_2; DDQ, CH_2Cl_2; Et_3N, $Zn(OAc)_2 \cdot 2H_2O$; (ii) 240 °C for *syn*-**29**, 210 °C for *anti*-**29**, in vacuo; (iii) 320 °C, in vacuo

fused diporphyrin **27** by the retro-Diels-Alder reaction (270 °C, 1 h, in vacuo). TG analysis of **26** revealed that extrusion of the ethylene molecule started from 240 °C and ceased around 270 °C. This conversion temperature was higher than that at which the benzene-fused oligoporphyrins were formed 170 °C.

Diporphyrin fused with anthracene was prepared starting from diethanoanthracene **28** (Scheme 2.6). There are two isomers (*syn*-(C_{2v}) and *anti*-(C_{2h})) in diethanoanthracene **28**, which were selectively prepared. Two porphyrin rings were similarly constructed at the edge of the double bonds of *syn*-**28** and *anti*-**28** to give *syn*-**29** and *anti*-**29**, respectively [26]. From the TG experiments, the first extrusion of ethylene from *syn*-**29** and *anti*-**29** to *anti*-**30** started at about 240 °C and 210 °C, respectively. This difference may be ascribed to the packing of their crystal structures [27]. The second extrusion of the ethylene molecule from **30**

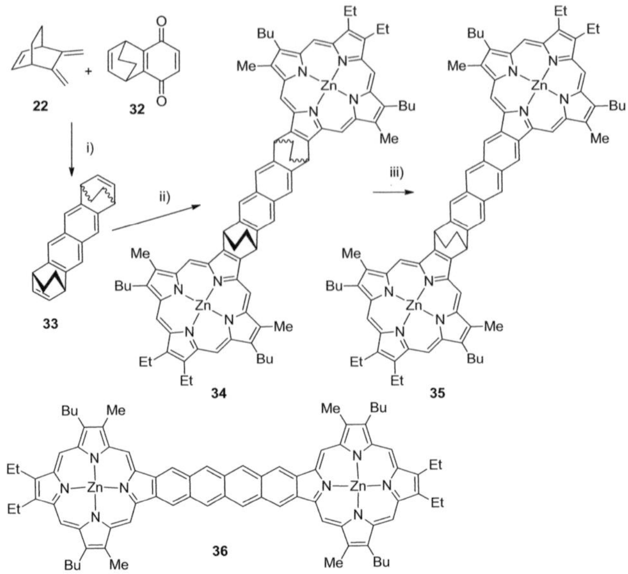

Scheme 2.7 Reagents and conditions: (i) $CHCl_3$, 40 °C, 3 days; $CeCl_3 \cdot 7H_2O$, $NaBH_4$, dioxane/MeOH; 4-M HCl, $CHCl_3$, reflux; DIBALH, CH_2Cl_2, −50 °C; aq-HCl; DDQ, dioxane; (ii) PhSCl, CH_2Cl_2, 0 °C; mCPBA, CH_2Cl_2, 0 °C; DBU, CH_2Cl_2, rt; $CNCH_2CO_2Et$, KO^tBu, THF, rt; KOH, ethylene glycol, 170–190 °C; **8a**, TFA, CH_2Cl_2; DDQ, CH_2Cl_2; Et_3N, $Zn(OAc)_2 \cdot 2H_2O$; (iii) 295 °C, in vacuo, 30 min

occurred at around 310 °C to produce the target anthracene-fused diporphyrin **31**. The intermediate **30** was purified by recrystallization, and the targeted **31** was prepared directly from **29** by heating at 320 °C.

Naphthacene-fused diporphyrins were prepared using Scheme 2.7. The Diels-Alder reaction of triene **22** with ethanonaphthoquinone **32** proceeded smoothly at 40 °C to produce a hexacyclo-endione compound, which was converted to a diastereomeric mixture (1:1) of diethanonaphthacene **33**, by sequential treatment with $NaBH_4$ in the presence of $CeCl_3$, hydrochloric acid, diisobutylaluminium hydride (DIBALH), hydrochloric acid, and 2,3-dichloro-5,6-dicyano-1,4-benzoquinone (DDQ). After the construction of two pyrrole rings at the distal double bonds of **33**, the double [3 + 1] porphyrin synthesis produce the target diethanonaphthacene-fused diporphyrin **34** in moderate yield [26]. The TG experiment of **34** revealed that the extrusion of one ethylene molecule occurred between 250 and 295 °C. The second retro-Diels-Alder reaction, however, occurred from *ca.* 320 °C with concomitant decomposition. No identifiable compounds were obtained, although peaks due to the molecular ions of fully conjugated naphthacene-fused diporphyrin **36** as well as naphthacenoporphyrin were observed in the MALDI-TOF experiment.

Scheme 2.8 shows the synthesis route of diethanopentacene-fused diporphyrin **40**. Triene **22** was reacted with benzadiyne generated by the reaction of

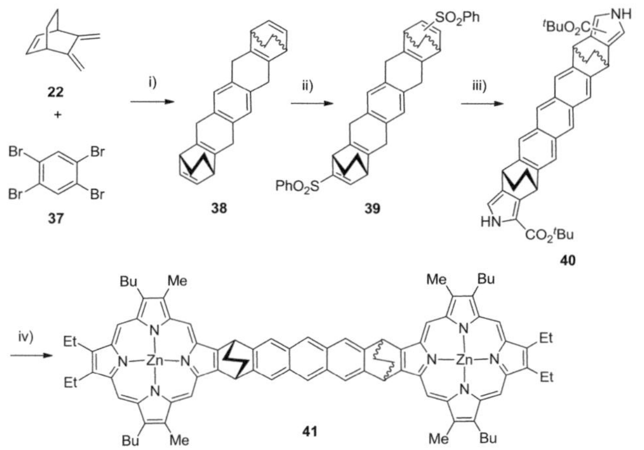

Scheme 2.8 Reagents and conditions: (i) ^{n}BuLi, toluene, 0 °C; (ii) PhSCl, CH_2Cl_2, 0 °C; mCPBA, CH_2Cl_2, 0 °C; DBU, CH_2Cl_2, rt; DDQ, dioxane, rt; $CNCH_2CO_2{}^{t}Bu$, $KO^{t}Bu$, THF, rt; (iii) TFA, rt, 10 min; **8a**, TFA, CH_2Cl_2, 50 °C; Et_3N, chloranil, CH_2Cl_2; $Zn(OAc)_2 \cdot 2H_2O$

tetrabromobenzene **37** with *n*-butyllithium to give a heptacyclic compound **38**. As the anthracene moiety was also reactive toward a phenylsulfenyl cation, compound **38** was successively treated with phenylsulfenyl chloride, mCPBA, and DBU to give disulfone **39**. Oxidation of **39** with DDQ followed by pyrrole ring construction using *tert*-butyl isocyanoacetate gave dipyrrole **40** as a mixture of four diastereomers. The [3 + 1] porphyrin synthesis of **40** with **8a** gave the target diethanopentacene-fused diporphyrin **41**. The TG experiment of **41** revealed that diporphyrin **41** did not undergo a clean retro-Diels-Alder reaction. Even the first extrusion of the ethylene molecule starting at *ca*. 350 °C was accompanied by decomposition of other parts of the molecule. No identifiable compound was obtained in the thermal treatment of **41**.

Considering that the anthracene fusing unit was the largest linear PAH of the fully conjugated diporphyrins in the preparation based on the retro-Diels-Alder reaction, synthesis of zigzag polycyclic aromatics was targeted next. We planned to prepare chrysene-fused diporphyrin (Scheme 2.9). Diethanochrysene **44** was thought to be the key compound for the preparation of the target chrysene-fused diporphyrin **48**. 1,5-Naphthalenediol (**42**) was reacted with bromine in the presence of iodine to give 2,6-dibromo-1,5-naphthalenediol. After the diol was protected by trimethylsilyl (TMS) groups, bromine-lithium exchange with *n*-butyllithium followed by quenching with TMSCl afforded 2,6-bisTMS-1,5-bisTMSoxynaphthalene, which was converted to bistrifluoromethanesulfonate derivative **43** by deprotection of the TMS ethers followed by quenching the resulting dilithium salt with trifluoromethanesulfonic anhydride. Naphthal-1,5-diyne generated by treatment of **43** with KF with the aid of 18-crown-6 reacted with 1,3-cyclohexadiene to give diethanochrysene **44** in a low yield. In spite of all of our efforts, the yield could not be improved. This low yield was attributed to the intrinsic low reactivity of 1,3-cyclohexadiene,

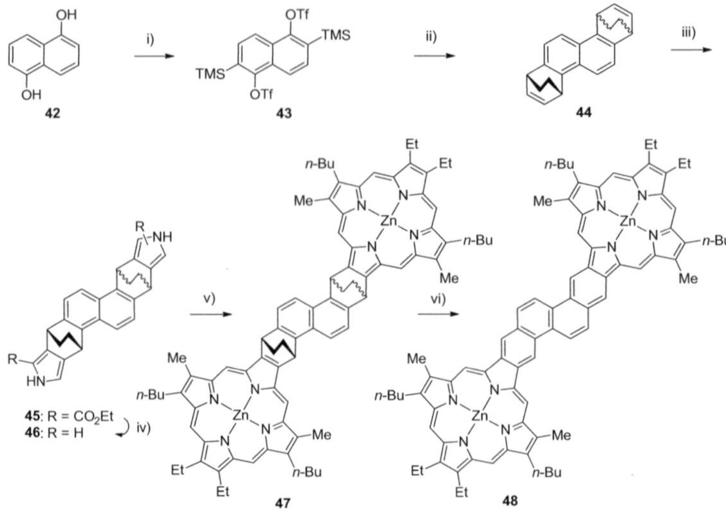

Scheme 2.9 Reagents and conditions: (i) Br_2, I_2 (cat.), AcOH, 80 °C; TMSCl, pyridine, toluene; ^{n}BuLi, TMSCl, THF; ^{n}BuLi, Tf_2O, THF, ether; (ii) 1,3-cyclohexadiene, KF, 18-crown-6; (iii) PhSCl, CH_2Cl_2, 0 °C; mCPBA, CH_2Cl_2, 0 °C; DBU, CH_2Cl_2, rt; DDQ, dioxane, rt; $CNCH_2CO_2Et$, KO^{t}Bu, THF, rt; (iv) KOH, ethylene glycol, 170–190 °C; (v) **8a**, TFA, CH_2Cl_2, 50 °C; Et_3N, DDQ, CH_2Cl_2; $Zn(OAc)_2 \cdot 2H_2O$; (vi) 200 °C, in vacuo

because the naphthaldiyne had smoothly reacted with 1,3-cyclopentadiene to give the corresponding dimethanochrysene in a good yield. The pyrrole ring construction at the distal double bonds of **44** followed by the [3 + 1] porphyrin synthesis afforded diethanochrysene-fused diporphyrin **47** [28]. In the TG experiment with **47**, only one steep weight loss was observed at 180 °C due to two ethylene molecules, and chrysene-fused diporphyrin **48** was obtained.

Electronic spectra of the PAH-fused dizinc diporphyrins are summarized in Table 2.2. In the case of the diethano-PAH-fused dizinc diporphyrins **29**, **34**, **41**, and **47**, no interaction between two zinc-porphyrin chromophores was observed and the absorption spectra were quite similar. This is quite different from the BCOD-fused dizinc porphyrin **12aa-Zn₂**, where *homo*-conjugation between the porphyrin chromophores was observed. The Soret- and Q-band absorptions of diethanopentacene-fused dizinc diporphyrin **41** showed a slight bathochromic shift (*ca.* 5 nm), which is probably due to the *homo*-conjugation between the porphyrin and anthracene moieties through the BCOD skeleton. *Homo*-conjugation between porphyrin and acene-fused porphyrin chromophores through the BCOD skeleton was clearly observed in **26**, **30**, and **35**. The absorption maxima of the longer-wavelength Soret absorptions in **26**, **30**, and **35** showed slight bathochromic shift compared to those of benzoporphyrin (416 nm in pyridine) [18], naphthoporphyrin (423 nm in pyridine) [29], and anthroporphyrin (440 nm in pyridine) [30], respectively.

Absorption and TD-DFT calculation (B3-LYP · 6-31 + G(d)/LANL2DZ) spectra of the acene-fused dizinc diporphyrins **13aa-Zn**$_2$, **27**, and **31** are shown in Fig. 2.3

Table 2.2 Absorption and emission maxima of PAH-fused diporphyrins

	λ_{max}/nm (log ε)							
Diporphyrin	Soret band			Q band				Fluorescence
26[a]	414	429			544	582		588
	(5.56)	(5,54)			(4.51)	(4.58)		638
27[a]	414	453	485		584	631		632
	(5.14)	(4.89)	(5.37)		(4.53)	(5.31)		681
29[b]	400				532	570		573
	(5.62)				(4.42)	(4.46)		625
30[c]	414	440		542	575	593	610	591
	(5.54)	(5.60)		(4.42)	(4.49)	(4.56)	(4.51)	638
31[d]	398	484			626	662		665
	1.00[e]	0.88[e]			0.77[e]	0.28[e]		
34[a]	402				532	569		576
	(5.61)				(4.42)	(4.45)		626
35[a]	420	441			607			608
	(5.03)	(4.96)			(4.44)			
41[b]	406				532	571		—[f]
	(5.73)				(4.49)	(4.57)		
47[a]	401				530	568		575
	(5.54)				(4.40)	(4.47)		626
48[c]	429	461			569	604		612
	(5.12)	(5.25)			(4.48)	(4.95)		667

[a] In $CHCl_3$
[b] 1% pyridine/$CHCl_3$
[c] In pyridine
[d] In THF
[e] Absorption ratio
[f] Not measured

[31]. In the calculated spectra of the acene-fused dizinc diporphyrins **13aa-Zn$_2$**, **27**, and **31**, the strongest Q-band transitions were at the wavelength close to 595 nm (594–598 nm), while the weaker Q-band transitions showed a bathochromic shift as the acene became higher. In these diporphyrins, the energy levels of the fully delocalized HOMO were higher than those of OMOs, while those of the delocalized LUMO and LUMO + 1 were close. Therefore, these three Frontier orbitals should be considered similar to the cases of phthalocyanines [32]. In the case of **13aa-Zn$_2$**, the LUMO and LUMO + 1 were degenerated so that the calculated Q-band absorptions were similar to those of phthalocyanines: the allowed strong transition from HOMO to LUMO + 1 and the forbidden weak transition from HOMO to LUMO had similar energies. As the fusing units became larger, the energy levels of HOMO and LUMO + 1 increased, while the energy levels of LUMO decreased. Therefore, the strong Q-band absorptions were at similar wavelengths, while the weak Q-band absorptions were at shorter wavelengths in **13aa-Zn$_2$**, at similar wavelengths in **27**, and at longer wavelengths in **31**.

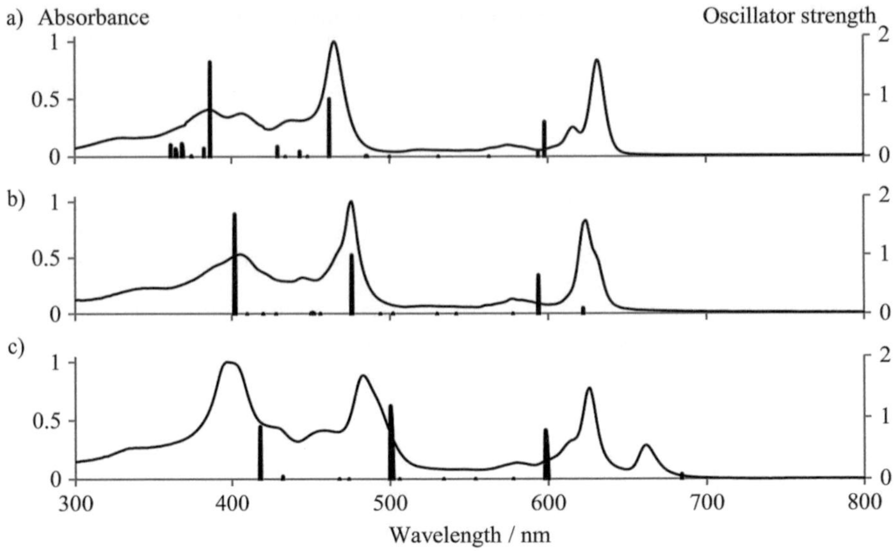

Fig. 2.3 Absorption (1 % pyridine/$CHCl_3$) and TD-DFT calculation (B3-LYP · 6-31 + G(d)/LANL2DZ) spectra of **13aa-Zn$_2$** (**a**), **27** (**b**), and **31** (**c**)

2.4 Conclusion

Oligoporphyrins fused with benzene units and dizinc diporphyrins fused with PAHs were prepared and their electronic properties were examined. The electronic spectra showed strong absorption in the red to NIR region. In the UV-vis spectra of the oligoporphyrins fused with benzene units, three major absorption bands were observed corresponding to two Soret bands and a Q band. A linearly fused trizinc triporphyrin with two benzene moieties showed a strong Q-band absorption, while split Q-band absorptions were observed in an L-shaped triporphyrin. The two-photon absorbing cross-section value for the I-shaped triporphyrin was about four times larger than that for the L-shaped triporphyrin, which was similar to that for diporphyrin. Naphthalene-, anthracene-, and chrysene-fused diporphyrins were successfully prepared, although preparation based on the retro-Diels-Alder reaction of the precursors failed to give naphthacene- and pentacene-fused diporphyrins. The absorption spectra of the benzene-, naphthalene-, and anthracene-fused diporphyrins were well rationalized by the TD-DFT calculations.

2.5 Experiments

Melting points are uncorrected. Unless otherwise specified, NMR spectra were obtained with a JEOL JNM AL-400 spectrometer at room temperature by using $CDCl_3$ as a solvent and tetramethylsilane as an internal standard for 1H and 13C.

FAB and MALDI-TOF mass spectra were measured with an MStation spectrometer (JEOL MS-700) and Voyager DE Pro (Applied Biosystems), respectively. UV-vis and fluorescence spectra were measured on JASCO V-570 and HITACHI F-4500, respectively. Absolute quantum yields were measured on a Hamamatsu Photonics C9920-02. TG analysis was done with SII Exstar 600 TG/DTA 6200. Elemental analysis was performed on a Yanaco MT-5 elemental analyzer. Preparative GPC was performed on JAI-8201 with combination of JAI-gel 1H and 2H columns.

2.5.1 I-Shaped BCOD-Fused Triporphyrin 14a-Ni_2Zn

Red crystals, mp >170 °C (decomp.); *syn*-isomer (C_{2v} isomer): ^{1}H NMR (400 MHz, $CDCl_3$): $\delta = 1.11$ (12H, m), 1.25 (8H, m), 1.72 (8H, m), 1.83 (12H, t, $J = 7.4$ Hz), 2.22 (12H, t, $J = 7.4$ Hz), 2.80 (8H, m), 3.77 (12H, s), 3.95 (16H, m), 4.46 (8H, m), 7.70 (4H, br-s), 9.82 (4H, s), 10.48 (4H, s), 10.82 (4H, s); MALDI-TOF: $m/z = 1657$ ($M^+ - 2C_2H_4$); UV-vis ($CHCl_3$): $\lambda_{max} = 391$, 414, 518, 554, 572 nm; CCDC No 102255 and 1022554.

2.5.2 L-Shaped BCOD-Fused Triporphyrin 16-Ni_2Zn

Red crystals, mp >170 °C (decomp.); *syn*-isomer (C_s isomer): ^{1}H NMR (400 MHz, $CDCl_3$): $\delta = 1.08$ (6H, t, $J = 7.4$ Hz), 1.14 (6H, t, $J = 7.4$ Hz), 1.72 (6H, m), 1.79 (18H, m), 1.99 (6H, m), 2.15 (12H, m), 2.29 (4H, m), 2.80–2.94 (8H, m), 3.76 (6H, s), 3.93 (12H, s), 4.01 (6H, s), 4.07 (4H, m), 4.20 (4H, m), 4.40 (4H, m), 7.74 (2H, br-s), 8.05 (2H, br-s), 9.80 (2H, s), 9.84 (2H, s), 10.27 (1H, s), 10.51 (2H, s), 10.70 (2H, s), 10.82 (2H, s), 11.47 (1H, s); MALDI-TOF: $m/z = 1657$ ($M^+ - 2C_2H_4$); UV-vis ($CHCl_3$): $\lambda_{max} = 394$, 411, 554 nm; CCDC No 1022529.

2.5.3 T-Shaped BCOD-Fused Tetraporphyrin 17-Ni_3Zn

Red crystals, mp >170 °C (decomp.); one isomer: ^{1}H NMR (400 MHz, $CDCl_3$): $\delta = 1.06$–2.97 (66H, m), 3.76 (6H, s), 3.95 (12H, s), 4.00 (16H, m), 4.46 (12H, m), 7.75 (2H, br-s), 9.89 (2H, br-s), 10.49 (2H, s), 10.68 (2H, s), 10.76 (2H, s), 10.87 (2H, s), 11.41 (2H, s); MALDI-TOF: $m/z = 2187$ ($M^+ - 3C_2H_4$); UV-vis ($CHCl_3$): $\lambda_{max} = 394$, 415, 555 nm. Anal. Calcd for $C_{104}H_{112}N_{12}Ni_3Zn \cdot C_6H_{14}$: C, 73.47; H, 6.89; N, 9.45. Found: C, 73.56; H, 7.39; N, 9.19 %.

2.5.4 Diporphyrin Fused with Ethanonaphthalene 26

To a solution of tripyrrane **8a** (109 mg, 0.22 mmol) and dipyrrole **25** (27 mg, 0.11 mmol) in $CHCl_3$ (30 mL) was added trifluoroacetic acid (0.8 mL) under an argon atmosphere. After being stirred for 3 h, the reaction mixture was quenched with triethylamine (1.4 mL) at 0 °C. DDQ (130 mg, 0.57 mmol) was added and the mixture was stirred at room temperature for 5 h. The reaction was quenched with aqueous sat. $NaHCO_3$ and the mixture was extracted with $CHCl_3$. The organic extract was washed with aqueous sat. $NaHCO_3$, water, and brine, dried over Na_2SO_4, and concentrated in vacuo. The residue and $Zn(OAc)_2 \cdot 2H_2O$ (500 mg, 2.27 mmol) were dissolved in a mixture of $CHCl_3$ (50 mL) and MeOH (5 mL), and the mixture was stirred at room temperature for 6 h. The reaction mixture was washed with aqueous sat. $NaHCO_3$, water, and brine, dried over Na_2SO_4, and concentrated in vacuo. The residue was chromatographed on silica gel ($CHCl_3$). The red fractions were collected to leave a red solid after removal of the solvent. Recrystallization of the residue from $CHCl_3$/MeOH gave 34 mg (0.027 mmol, 25 %) of the title compound as dark red crystals: mp > 250 °C (decomp.); ^{1}H NMR (400 MHz, pyridine-d_5): $\delta = 1.12$–1.19 (12H, m), 1.80–1.89 (8H, m), 1.98–2.06 (12H, m), 2.38–2.52 (8H, m), 2.57–2.64 (2H, m), 2.78–2.85 (2H, m), 3.92 (6H, s), 4.02 (6H, s), 4.15–4.40 (16H, m), 7.08 (2H, s), 10.08 (2H, s), 10.59 (2H, s), 10.99 (2H, s), 11.21 (2H, s); UV-vis ($CHCl_3$): λ_{max} (log ε) = 341 (4.69), 414 (5.56), 429 (5.54), 544 (4.51), 582 (4.58) nm; MALDI-TOF: $m/z = 1238$ ($M^+ + 1$); CCDC No 1022865.

2.5.5 Diporphyrin Fused with Naphthalene 27

Ethanonaphthalene-fused diporphryrin **26** was placed in a micro-sample tube, and the tube was placed in a cylinder, which was evacuated by an oil rotary pump. The cylinder was placed in a preheated glass-tube oven (270 °C) for 1 h. The cylinder was cooled to room temperature, and the title compound was taken out from the sample tube: green powder, mp >350 (decomp.); ^{1}H NMR (400 MHz, THF-d_8): $\delta = 1.17$ (12H, t, $J = 7.4$ Hz), 1.80–1.89 (8H, m), 1.94 (12H, t, $J = 7.7$ Hz), 2.38–2.52 (8H, m), 3.84 (12H, s), 4.01–4.22 (16H, m), 10.09 (4H, s), 10.74 (4H, s), 10.79 (4H, s); UV-vis ($CHCl_3$): λ_{max} (log ε) = 349 (4.79), 414 (5.14), 453 (4.89), 485 (5.37), 584 (4.53), 631 (5.31) nm; MALDI-TOF m/z 1209 ($M^+ + 1$).

2.5.6 Diporphyrin Fused with Ethanoanthracene 30

syn-Diethanoanthracene-fused diporphryrin **29** was placed in a micro-sample tube, and the tube was placed in a cylinder, which was evacuated by an oil rotary pump. The cylinder was placed in a preheated glass-tube oven (320 °C) for 1 h. The cylinder was cooled to room temperature, and the title compound was taken out from the sample tube: dark purple powder, mp > 350 °C (decomp.); ^{1}H NMR (400 MHz, pyridine-d_5): $\delta = 1.07$–1.14 (12H, m), 1.71–1.84 (8H, m), 1.91–1.99 (12H, m), 2.31–2.46 (10H, m), 2.60–2.67 (2H, m), 3.69 (6H, s), 3.85 (6H, s), 40.9–4.31 (16H, m), 6.75 (2H, s), 9.04 (2H, s), 10.43 (2H, s), 10.46 (2H, s), 10.52 (2H, s), 10.92 (2H, s), 10.98 (2H, s); UV-vis (pyridine): λ_{max} (log ε) = 341 (4.75), 414 (5.54), 440 (5.60), 542 (4.42), 575 (4.49), 593 (4.56), 610 nm (4.51); MALDI-TOF: *m/z* = 1259 (M^{+} +1); CCDC No 1024855.

2.5.7 Diporphyrin Fused with Anthracene 31

syn-Diethanoanthracene-fused diporphryrin **29** was placed in a micro-sample tube, and the tube was placed in a cylinder, which was evacuated by an oil rotary pump. The cylinder was placed in a preheated glass-tube oven (240 °C) for 1 h. The cylinder was cooled to room temperature, and the title compound was taken out from the sample tube: dark purple powder, mp >320 °C (decomp.); ^{1}H NMR (400 MHz, THF-d_8): $\delta = 1.03$ (12H, m), 1.82 (12H, m), 2.20–2.40 (8H, m), 2.80–3.40 (8H, m), 3.80–4.20 (16H, m), 4.59 (12H, s), 8.81 (2H, s), 9.94 (4H, s), 10.05 (4H, s), 10.42 (4H, s); UV-vis (pyridine): $\lambda_{max} = 398$, 484, 626, 662 nm; MALDI-TOF: *m/z* = 1288 (M^{+} +1). Anal. Calcd for $C_{80}H_{86}N_8Zn_2$: C, 74.46; H, 6.72; N, 8.68. Found: C, 74.15; H, 7.04; N, 8.82 %.

2.5.8 Diporphyrin Fused with Ethanonaphthacene 35

Diethanonaphthacene-fused diporphyrin **34** [28] (2.60 mg) was placed in a micro-sample tube, and the tube was placed in a cylinder, which was evacuated by an oil rotary pump. The cylinder was placed in a preheated glass-tube oven (295 °C) for 1 h. The cylinder was cooled to room temperature, and the title compound was taken out from the sample tube: green powder, mp >350 (decomp.); ^{1}H NMR (400 MHz, pyridine-d_5): $\delta = 1.02$–1.17 (12H, m), 1.25–1.32 (4H, m), 1.75–1.84 (8H, m), 1.92–1.98 (12H, m), 2.33–2.43 (8H, m), 3.64 (6H, s), 3.80 (6H, s), 4.13–4.20 (16H, m), 9.37 (2H, br-s), 9.86 (2H, s), 9.95 (2H, s), 10.38 (2H, s), 10.45 (2H, s), 10.52 (2H, s), 10.82 (2H, m), 10.99 (2H, s); MALDI-TOF: *m/z* = 1341 (M^{+} + 5), 1313 (M^{+} +5 − C_2H_4); UV-vis ($CHCl_3$): λ_{max} (log ε) = 420 (5.03), 441 (4.96), 6.7 (4.44).

2.5.9 Diporphyrin Fused with Diethanopentacene 41

Trifluoroacetic acid (0.51 mL, 6.48 mmol) was added to dipyrrole **40** (73.6 mg, 0.12 mmol) at 0 °C, and the solution was stirred at room temperature for 10 min. Tripyrrane **8a** (114.6 mg, 0.24 mmol) in dry CH_2Cl_2 (15 mL) was added, and the reaction mixture was stirred at 50 °C for 1 h under nitrogen in the dark. After being neutralized with triethylamine (0.80 mL, 5.76 mmol) at 0 °C, the reaction mixture was oxidized with chloranil (59.0 mg, 0.24 mmol) at room temperature for 1 h. The reaction mixture was quenched with an aqueous solution of $Na_2S_2O_3$ and extracted with CH_2Cl_2. The organic layer was washed successively with water, an aqueous solution of $NaHCO_3$, and brine, then dried over Na_2SO_4, and concentrated. The residue was dissolved in $CHCl_3$ (15 mL) and $Zn(OAc)_2.2H_2O$ (52.7 mg, 0.24 mmol) was added. The mixture was stirred at room temperature overnight under N_2. Following that, the reaction mixture was washed with water and an aqueous solution of $NaHCO_3$, dried over Na_2SO_4, and concentrated. The residue was chromatographed on silica gel ($CHCl_3$, Rf = 0.88) and recrystallized from THF/MeOH to afford **41** as a purple powder (22.7 mg, 13 %): mp 257 °C (decomp.); UV-vis ($CHCl_3$): λ_{max} (log ε) 406 (5.73), 532 (4.49), 571 (4.57); ^{1}H NMR (400 MHz, THF-d_8): $\delta = 1.08$ (12H, t $J = 7.8$ Hz), 1.74 (8H, m), 1.89 (12H, t $J = 8.1$ Hz), 2.18 (4H, m), 2.28 (8H, m), 2.34 (4H, m), 3.70 (12H, s), 4.12 (16H, m), 6.18 (4H, s), 8.19 (4H, s), 8.36 (2H, s), 10.09 (4H, s), 10.33 (4H, s); ^{13}C NMR (100 MHz, THF-d_8): $\delta = 11.09$, 11.15, 13.75, 13.83, 18.31, 18.39, 19.64, 19.72, 22.94, 23.02, 26.13, 26.19, 29.80, 35.66, 35.73, 40.31, 78.54, 96.29, 96.40, 97.48, 97.56, 107.00, 120.71, 125.00, 130.98, 135.90, 140.73, 140.83, 141.74, 141.84, 142.08, 142.18, 143.22, 143.24, 147.43, 147.47, 148.10, 148.24; IR (KBr): $\nu_{max} = 2952$, 2929, 2860, 771 cm^{-1}; MALDI-TOF: *m/z* =1418 (M^+ +4), 1392, 1362; HRMS (FAB^+) Calcd for $C_{90}H_{94}N_8Zn_2 + H^+$: 1415.6263. Found: 1415.6262. Anal. Calcd for $C_{90}H_{94}N_8Zn_2 + C_4H_8O$: C, 75.74; H, 6.90; N, 7.52. Found: C, 75.74; H, 6.89; N, 7.33 %.

2.5.10 *Diporphyrin Fused with Chrysene 48*

Diethanochrysene-fused diporphyrin **47** [28] (3.0 mg) was placed in a micro-sample tube, and the tube was placed in a cylinder, which was evacuated by an oil rotary pump. The cylinder was placed in a preheated glass-tube oven (250 °C) for 2 h. The cylinder was cooled to room temperature and the title compound was taken out from the sample tube: green powder, mp >350 (decomp.); ^{1}H NMR (400 MHz, pyridine-d_5): $\delta = 1.13$ (12H, t, $J = 7.4$ Hz), 1.96–2.00 (20H, m), 2.42(8H, q, $J = 7.1$ Hz), 3.84 (6H, s), 3.85 (6H, s), 4.07–4.32 (16H, m), 9.02 (2H, d, $J = 9.2$ Hz), 10.03 (2H, d, $J = 9.2$ Hz), 10.51 (4H, s), 10.54 (2H, s), 11.08 (2H, s), 11.40 (2H, s), 11.72 (2H, s); IR (KBr) $\nu_{max} = 2954$, 2925, 2859, 1538, 1010 cm^{-1}; UV-vis (pyridine): λ_{max} (log ε) = 429 (5.12), 461 (5.25), 569 (4.48), 604 (4.95) nm; MALDI-TOF: *m/z* = 1313.02 (M^+ +5); CCDC No 1023537.

References

1. Cui S, Yin D, Chen Y, Di Y, Chen H, Ma Y, Achilefu S, Gu Y (2013) ACS Nano 22:676–688; Xia L, Kong X, Liu X, Tu L, Zhang Y, Chang Y, Liu K, Shen D, Zhao H, Zhang H (2014) Biomaterials 35:4146–4156
2. Jechow A, Seefeldt M, Kurzke H, Heuer A, Menzel R (2013) Nat Photonics 7:973–976; Yang P, Shi L, Pratavieira S, Alfano RR (2013) J Appl Phys 114:153102
3. Hamamura T, Nakazaki J, Uchida S, Kubo T, Segawa H (2014) Chem Lett 43:655–657; Kinoshita T, Dy JT, Uchida S, Kubo T, Segawa H (2013) Nat Photonics 7:535–539
4. Fabian J, Nakazumi H, Matsuoka M (1992) Chem Rev 92:1197–1226
5. Anderson RR, Parrish JA (1981) J Investig Dermatol 77:13–19
6. Shalav A, Richards BS, Trupke T, Krämer KW, Güdel HU (2005) Appl Phys Lett 86:013505
7. Wicklein A, Lang A, Muth M, Thelakkat M (2009) J Am Chem Soc 131:14442–14453
8. Nakamura M, Tahara H, Takahashi K, Nagata T, Uoyama H, Kuzuhara D, Mori S, Okujima T, Yamada H, Uno H (2012) Org Biomol Chem 10:6840–6849
9. Yamamoto A, Ohta E, Kishigami N, Tsukahara N, Tomiyori Y, Sato H, Matsui Y, Kano Y, Mizuno K, Ikeda H (2013) Tetrahedron Lett 54:4049–4053; Wasserfallen D, Kastler M, Pisula W, Hofer WA, Fogel Y, Wang Z, Müllen K (2006) J Am Chem Soc 128:1334–1339
10. Oyaizu K, Iwasaki T, Tsukahara Y, Tsuchida E (2004) Macromolecules 37:1257–1270; Shinamura S, Osaka I, Miyazaki E, Takimiya K (2011) Heterocycles 83:1187–1204
11. Yamaguchi S, Xu C, Yamada H, Wakamiya A (2005) J Org Chem 690:5365–5377; Matano Y, Saito A, Fukushima T, Tokudome Y, Suzuki F, Sakamaki D, Kaji H, Ito A, Tanaka K, Imahori H (2011) Angew Chem Int Ed 50:8016–8020; London G, von Wantoch Rekowski M, Dumele O, Schweizer WB, Gisselbrecht J-PL, Boudon C, Diederich F (2014) Chem Sci 5:965–972; Wang X-Y, Zhuang F-D, Wang R-B, Wang X-C, Cao X-Y, Wang J-Y, Pei J (2014) J Am Chem Soc 136:3764–3767; Levine DR, Siegler MA, Tovar JD (2014) J Am Chem Soc 136:7132–7139; Kato S, Furuya T, Kobayashi A, Nitani M, Ie Y, Aso Y, Yoshihara T, Tobita S, Nakamura Y (2012) J Org Chem 77:7595–7606
12. Nakamura M, Kitatsuka M, Takahashi K, Nagata T, Mori S, Kuzuhara D, Okujima T, Yamada H, Nakae T, Uno H (2014) Org Biomol Chem 12:1309–1317; Yamada H, Okujima T, Ono N (2008) Chem Commun 2957–2974
13. Takase M, Enkelmann V, Sebastiani D, Baumgarten M, Müllen K (2007) Angew Chem Int Ed 46:5524–5527
14. Müllen K, Rabe JP (2008) Acc Chem Res 41:511–520
15. Ghoroghchian PP, Frail PR, Susumu K, Blessington D, Brannan AK, Bates FS, Chance B, Hammer DA, Therien MJ (2004) Proc Natl Acad Sci U S A 102:2922–2927
16. Balaz M, Collins HA, Dahlstedt E, Anderson HL (2009) Org Biomol Chem 7:874–888
17. Kim KS, Lim JM, Osuka A, Kim D (2008) J Photochem Photobiol C 9:13–28
18. Uoyama H, Takiue T, Tominaga K, Ono N, Uno H (2009) J Porphyrins Phthalocyanines 13:122–135
19. Uno H, Hashimoto M, Fujimoto A (2009) Heterocycles 77:887–898
20. Ito S, Nakamoto K, Uno H, Murashima T, Ono N (2001) Chem Commun 2696–2697; Ono N, Kuroki K, Watanabe E, Ochi N, Uno H (2004) Heterocycles 62:365–373
21. Uno H, Nakamoto K, Kuroki K, Fujimoto A, Ono N (2007) Chem Eur J 13:5773–5784
22. Bhattacharya S, Hashimoto M, Fujimoto A, Kimura T, Uno H, Komatsu N (2008) Spectrochim Acta A 70A:416–424
23. Kim P, Ham S, Oh J, Uoyama H, Watanabe H, Tagawa K, Uno H, Kim D (2013) Phys Chem Chem Phys 15:10612–10615
24. Uoyama H, Kim KS, Kuroki K, Shin J-Y, Nagata T, Okujima T, Yamada H, Ono N, Kim D, Uno H (2010) Chem Eur J 16:4063–4074
25. Okujima T, Jin G, Hashimoto Y, Yamada H, Uno H, Ono N (2006) Heterocycles 70:619–626
26. Uno H, Furukawa M, Fujimoto A, Uoyama H, Watanabe H, Okujima T, Yamada H, Mori S, Kuramoto M, Iwamura T, Hatae N, Tani F, Komatsu N (2011) J Porphyrins Phthalocyanines 15:951–963

27. Uno H, Masumoto A, Ono N (2003) J Am Chem Soc 125:12082–12083
28. Mori S, Kawamoto N, Okujima T, Uno H. manuscript in preparation
29. Ito S, Ochi N, Uno H, Murashima T, Ono N (2000) Chem Commun 2000:893–894
30. Cai C, Uoyama H, Nakamura M, Uno H (2012) Heterocycles 84:829–841; Uoyama H, Yamada H, Okujima T, Uno H (2012) Heterocycles 86:515–534
31. Frisch MJ, Trucks GW, Schlegel HB, Scuseria GE, Robb MA, Cheeseman JR, Scalmani G, Barone V, Mennucci B, Petersson GA, Nakatsuji H, Caricato M, Li X, Hratchian HP, Izmaylov AF, Bloino J, Zheng G, Sonnenberg JL, Hada M, Ehara M, Toyota K, Fukuda R, Hasegawa J, Ishida M, Nakajima T, Honda Y, Kitao O, Nakai H, Vreven T, Montgomery JA, Peralta JE Jr, Ogliaro F, Bearpark M, Heyd JJ, Brothers E, Kudin KN, Staroverov VN, Keith T, Kobayashi R, Normand J, Raghavachari K, Rendell A, Burant JC, Iyengar SS, Tomasi J, Cossi M, Rega N, Millam JM, Klene M, Knox JE, Cross JB, Bakken V, Adamo C, Jaramillo J, Gomperts R, Stratmann RE, Yazyev O, Austin AJ, Cammi R, Pomelli C, Ochterski JW, Martin RL, Morokuma K, Zakrzewski VG, Voth GA, Salvador P, Dannenberg JJ, Dapprich S, Daniels AD, Farkas O, Foresman JB, Ortiz JV, Cioslowski J, Fox DJ (2010) Gaussian 09, Revision C.01; Gaussian, Inc., Wallingford
32. Ishii K, Kobayashi N (2000) The photophysical properties of phthalocyanines and related compounds. In:Kadish KM, Smith KM, Guilard G (eds) The porphyrin handbook, vol 16. Academic Press, San Diego, pp 1–40

Chapter 3
Higher-Order π-Electron Systems Based on Helicene Molecules

Midori Akiyama, Koji Nakano, and Kyoko Nozaki

Abstract Owing to their helical topology, helicenes, polycyclic aromatic compounds with nonplanar screw-shaped skeletons, provide high optical rotations, high circular dichroism values, and several enhanced physical organic properties. Supramolecular chirality generated from regular arrangement of helicenes further amplifies their inherent chiroptical properties. Here in this chapter, examples of higher-order molecular architectures of helicenes and their related compounds in aggregates, in crystals, and in macromolecules connected by covalent bonds are summarized, and their unique chiroptical properties are discussed.

Keywords Helicenes • Helical structure • Supramolecular assembly • Chiroptical property

3.1 Introduction

[*n*]Helicenes are polycyclic aromatic compounds with nonplanar screw-shaped skeletons formed by n *ortho*-fused aromatic rings. Since its first example was synthesized in 1903 [1], continuous studies have been devoted on their synthesis and properties. In 1955, Newman et al. reported the synthesis of enantiomerically pure [6]helicene as the first example of helicene with stable helical chirality [2]. Since then, various synthetic methods have been developed including photo-cyclization, the Diels–Alder reaction, olefin metathesis, and so on. In addition to carbohelicenes whose aromatic rings consist of only carbon atoms, various examples were reported for heterohelicenes which include aromatic heterocycles, such as thiophene, pyrrole, furan, and pyridine, as member(s) of the helicene skeletons. The advances in helicene synthesis can be found in several excellent review articles [3].

M. Akiyama • K. Nozaki (✉)
Department of Chemistry and Biotechnology, Graduate School of Engineering, The University of Tokyo, 7-3-1 Hongo, Bunkyo-ku, Tokyo 113-8656, Japan
e-mail: nozaki@chembio.t.u-tokyo.ac.jp

K. Nakano
Department of Organic and Polymer Materials Chemistry, Tokyo University of Agriculture and Technology, 2-24-16 Naka-cho, Koganei, Tokyo 184-8588, Japan

T. Akasaka et al. (eds.), *Chemical Science of π-Electron Systems*,
DOI 10.1007/978-4-431-55357-1_3

The helical topology of helicenes provides high optical rotations, high circular dichroism values, and several enhanced physical organic properties. This is especially true if supramolecular chirality is generated from a self-assembly of helicenes. Here in this chapter, examples of higher-order molecular architectures of helicenes and their related compounds in aggregates, in crystals, and in macromolecules connected by covalent bonds will be summarized.

3.2 Aggregates of Helicene Molecules

Katz reported nonracemic helicene **1** form aggregates in dodecane (Fig. 3.1) [4]. The aggregation was confirmed by enhanced circular dichroism (CD) and g-values, red-shifted UV–Vis absorptions, increased specific rotations, increased light scattering, red-shifted fluorescence emissions, and upfield-shifted ^{1}H NMR resonances. In a solid state, nonracemic helicene **1** is organized in hexagonally packed column to form very long fibrous structure, which was confirmed by transmission electron microscopy and X-ray diffraction. The g-value of CD and specific rotation were further increased in a spin-coated film. Notably, this supramolecular organization in Langmuir–Blodgett films induced a second-order nonlinear optical (NLO) susceptibility about 30 times larger for nonracemic **1** than for racemic **1** [5]. Formation of hexagonal columnar liquid crystal phase was confirmed for nonracemic **2**, in which one of the quinone moieties of **1** is reduced and esterified [6]. The changes that occur in specific rotation and absorption, emission, CD, and ^{1}H NMR spectra are similar to those occur in **1**. This is the first columnar mesophase with a nonracemic helical core.

In 2011, Kühnle et al. succeeded in observation of self-assembly of racemic [7]helicene-2-carboxylic acid **3** into nanowire-like aggregates on a calcite (10–14) by noncontact atomic force microscopy (NC-AFM) (Fig. 3.2) [7]. The supramolecular nanostructure exceeded 100 nm length and was well aligned along the [010] crystallographic direction of the calcite (10–14) cleavage phase. The NC-AFM images and density functional theory calculations indicated that the nanowire structure formed through the columnar stacks of dimers of helicene **3**.

Takeuchi et al. reported that an enantiopure phthalhydrazide-functionalized helicene **4** aggregated into fiber structures (Fig. 3.3) [8]. Helicene **4** forms a trimeric disk with a multiple-hydrogen bonding site of its phthalhydrazide unit, and the

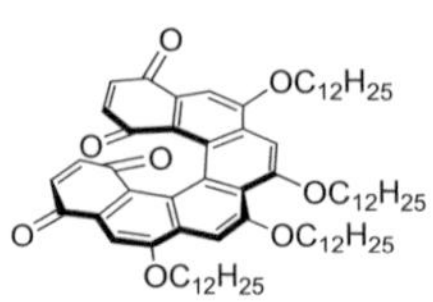

1

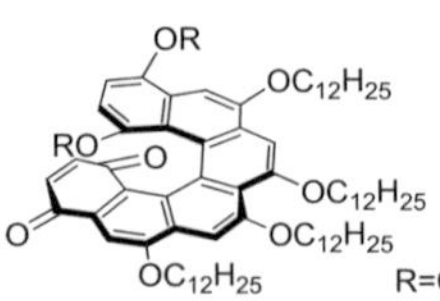

2

R=$COCH_2OCH_2Ph$

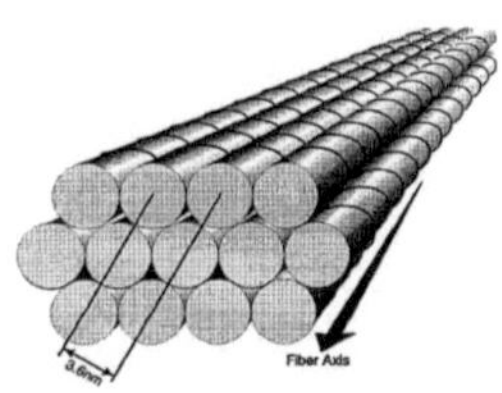

Fig. 3.1 Structure of helicenes **1** and **2** and schematic representation showing how helicene **1** is organized in the fibers (Reprinted with permission from Ref. [4b]. Copyright 1998, American Chemical Society)

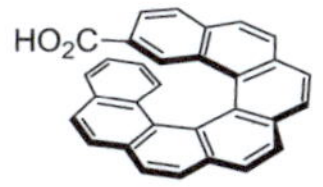

3

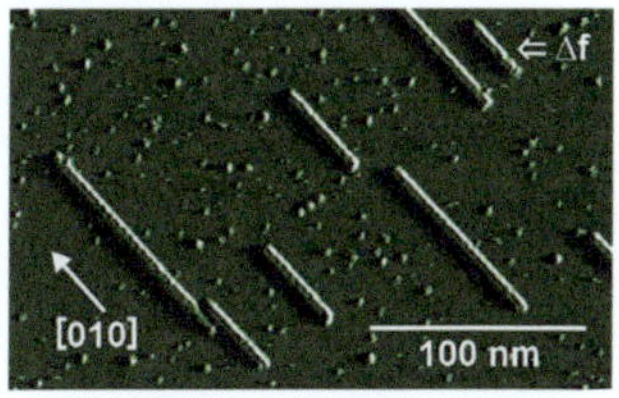

Fig. 3.2 Nanowire-like aggregate of racemic **3** on calcite (10–14) (Reprinted with permission from Ref. [7]. Copyright 2011, Wiley-VCH Verlag GmbH & Co. KGaA, Weinheim)

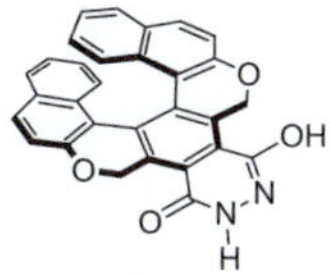

4

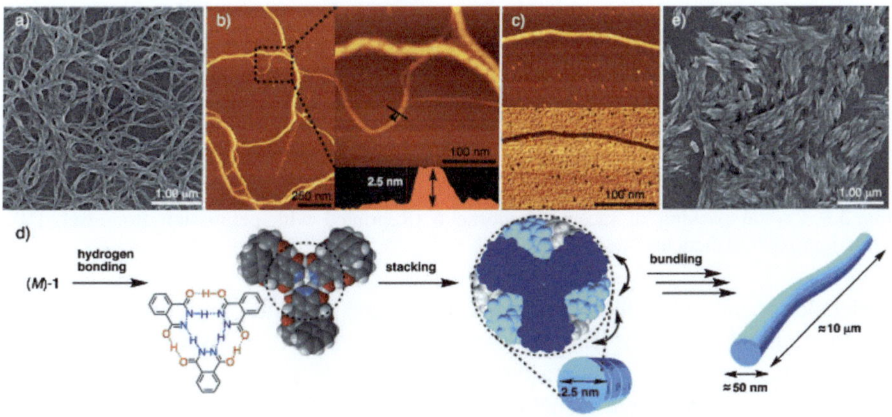

Fig. 3.3 Chiral fibers through one-dimensional stack of disk-like trimer of helicene **4**. [(**a**) SEM image of *M*-**4**; (**b**) and(**c**) AFM images of *M*-**4** prepared in toluene; (**d**) plausible mechanism for the formation of fibrous aggregates from the trimeric disk of *M*-**4**; and (**e**) SEM image of *rac*-**4**] (Reprinted with permission from Ref. [8]. Copyright 2011, Wiley-VCH Verlag GmbH & Co. KGaA, Weinheim)

trimer stacks in a one-dimensional manner. The trimers twist with respect to each because of the interaction between the helicene moieties, forming chiral fibers. The fibrous assembly was not prepared from polar solvents such as methanol, which interrupt hydrogen bonding. The assembled **4** exhibited superior circularly polarized luminescence (CPL) properties.

3.3 Helicene Molecules Aligned in Crystals

Single crystals would also be suitable for creating such architectures if the π-conjugated helical molecules are appropriately stacked. In 1993, Diederich reported the synthesis of diethyl [7]helicene-2,17-dicarboxylate **5** (Fig. 3.4) [9]. In a crystal,

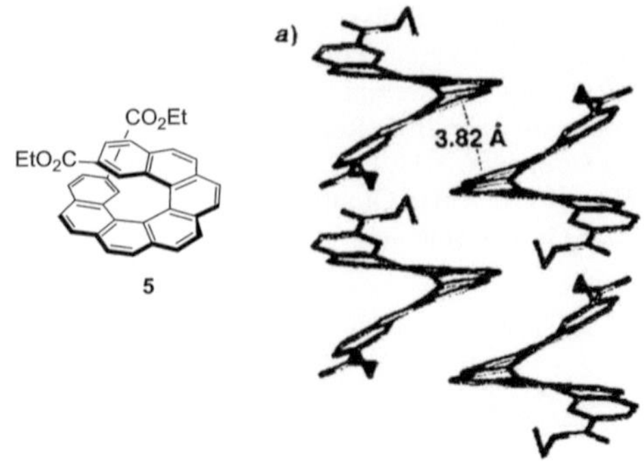

Fig. 3.4 Crystal packing of racemic helicene **5** (Reprinted with permission from Ref. [9]. Copyright 1993, Verlag GmbH & Co. KGaA, Weinheim)

Fig. 3.5 Crystal packing of racemic helicene **6** (Reprinted with permission from Ref. [10]. Copyright 2000, American Chemical Society)

molecules of the same chirality formed stacks, and two stacks of opposite chirality are interlocked in a pair. The average face-to-face contact distance between the benzene rings of different enantiomers is 3.82 Å.

In a crystal of racemic [7]helicene **6** bearing pyridinone hydrogen bond sites on the ends of the twisted backbone, columnar stacks are also observed (Fig. 3.5) [10]. Four stereochemically unique dimers are possible from the racemic mixture of helicene **6**, since it is substituted in a dissymmetric manner. However, only homochiral dimers with acyl group in a *cis* relationship appear in the crystal and arrange into offset racemic columns. The benzene rings of *PP* and *MM* diastereomers are oriented in a face-to-face-type orientation with average distance of 3.64 Å between aromatic rings.

Recently, our group reported the synthesis of λ^5-phospha[7]helicene **7** (Fig. 3.6) [11]. In the solid state, they form one-dimensional columnar stacks, which would be induced by their dipole moment parallel to the helical axes. It is notable that columns with one-direction dipole moment consist of single enantiomer (*P*- or *M*-isomer) and columns with the opposite dipole moment consist of the other enantiomer (*M*- or *P*-isomer). That is, each column is one-way street in terms of chirality.

Nakamura et al. synthesized azaboradibenzo[6]helicene **8** and showed that the racemate and single enantiomer of **8** were p- and n-type semiconductors, respectively (Fig. 3.7) [12]. *P*- and *M*-isomers are arranged in an alternating manner with a face-to-face π–π distance of 3.4–3.6 Å in the heterochiral crystal, while one-dimensional columnar arrangement is observed in the homochiral crystal. Electronic coupling calculations indicate that such difference in the packing structures of the hetero- and homochiral crystals causes the carrier inversion.

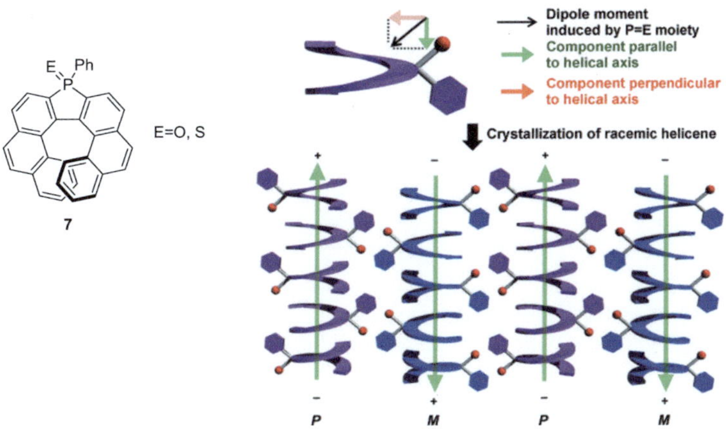

Fig. 3.6 Schematic representation of the columnar packing in a single crystal of racemic helicene **7** (E = S) (Reprinted with permission from Ref. [11]. Copyright 2012, Wiley-VCH Verlag GmbH & Co. KGaA, Weinheim)

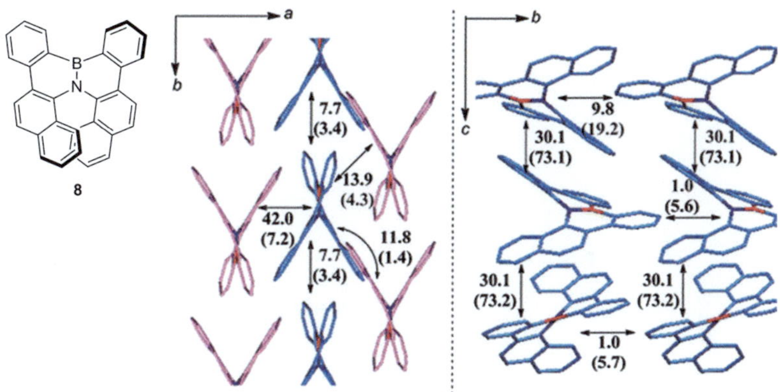

Fig. 3.7 Crystal packing of racemic and enantiopure helicene **8** (Reprinted with permission from Ref. [12]. Copyright 2012, American Chemical Society)

A heterohelicene **9a** containing both thiophene and pyridine rings was also reported to form helical columnar stacking structure in a crystal (Fig. 3.8) [13]. A single crystal from racemate contains both enantiomers, while each column consists of one enantiomer with an antiparallel stacking. Several intermolecular π–π interactions are observed, including benzene–thiophene and benzene–benzene stacking interaction. In a single crystal from racemate of **9b**, *P*- and *M*-isomers are arranged in an alternating antiparallel manner through a face-to-face π–π interaction between the central benzene rings, leading to the one-dimensional stacking structure.

Most recently, Hasobe et al. reported the synthesis of quinoxaline-fused [7]carbohelicene **10** (Fig. 3.9) [14]. In crystal state, *P*-enantiomer of **10** formed dimers, while

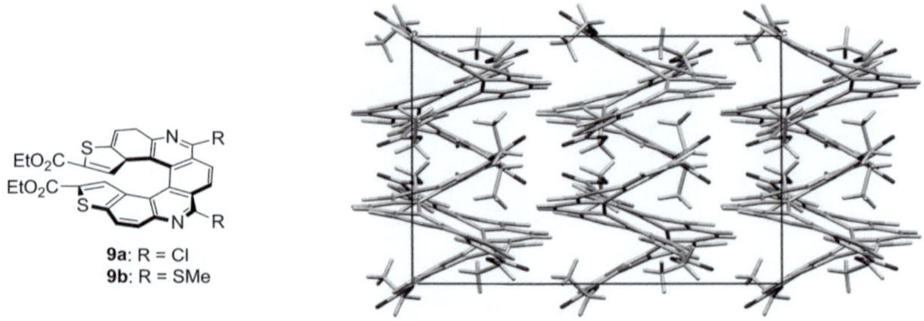

Fig. 3.8 Crystal packing of racemic and enantiopure helicene **9a** (Reprinted with permission from Ref. [13]. Copyright 2013, Wiley-VCH Verlag GmbH & Co. KGaA, Weinheim)

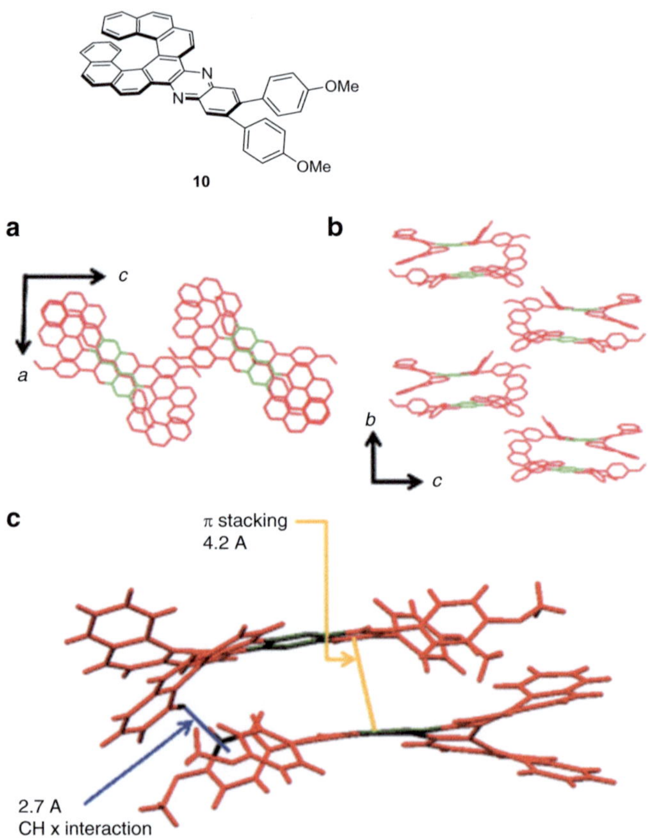

Fig. 3.9 Homochiral dimer of *P*-**10** and helical columnar stack from racemic **10** [(**a**) Packing mode along the *b* axis; (**b**) Packing mode along the *a* axis; and (**c**) π-stacking between two neighboring *P*-**10** with intermolecular distance. *Red*: *P* enantiomer, *green*: pyrazine ring] (Reprinted with permission from Ref. [14]. Copyright 2014, Wiley-VCH Verlag GmbH & Co. KGaA, Weinheim)

racemic **10** stacked into helical columnar structures in which *P*- and *M*-isomers are alternately stacked in the same column. The π-stacking distances were 4.2 Å and 3.6 Å, respectively. The smaller value of the heterochiral **10** might cause singlet–singlet annihilation and the subsequent excimer-like delocalized excited state, which were observed by time-resolved fluorescence spectra.

3.4 Molecules with Multiple Helicenes Connected by Covalent Bonds and Their Assemblies

In 1986, Katz reported metallocene oligomers **11** having cobaltocene moieties between two helicene units (Fig. 3.10) [15]. The oligomer length depends on the solvent used for preparation, and the longest oligomer was estimated to contain 6–7 helicene units on an average. Longer oligomer length resulted in larger optical rotation. Despite that the helicene moieties were connected by a cobalt ion, the interactions between both helicene metallic units were rather weak.

Helical conjugated ladder polymer **12** in which helicene units were connected by Ni/salophen units was also reported (Fig. 3.11) [16]. In this structure, Ni/salophen units constrain the π-orbital of the benzene ring they connect to be almost parallel,

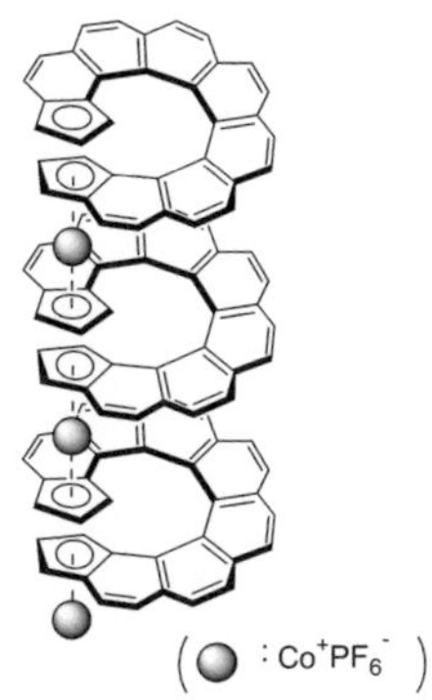

11

Fig. 3.10 Helical metallocene oligomer **11**

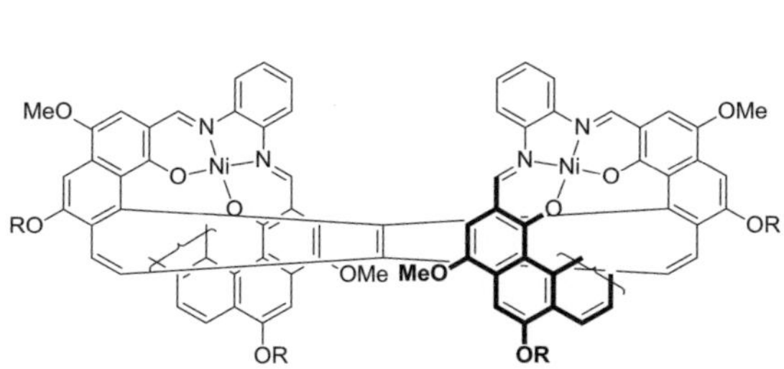

12

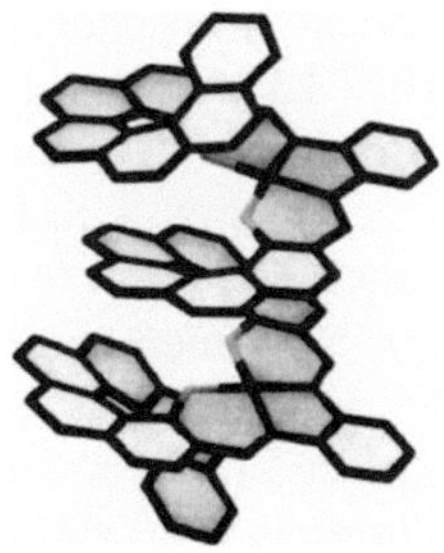

Fig. 3.11 Structure of helical conjugated ladder polymer **12** consisting of three helicene units and two metal/salophen units (Reprinted with permission from Ref. [16a]. Copyright 1996, VCH Verlagsgesellschaft mbH, D-69451 Weinheim)

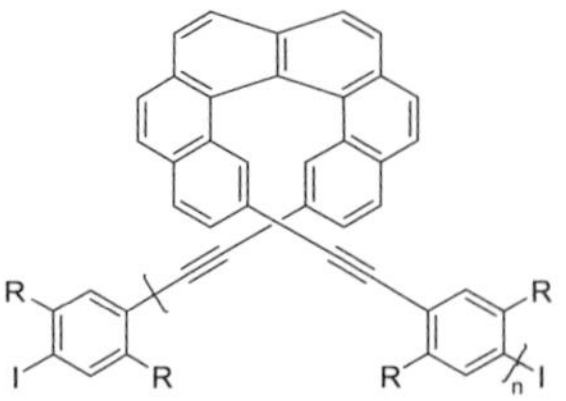

13

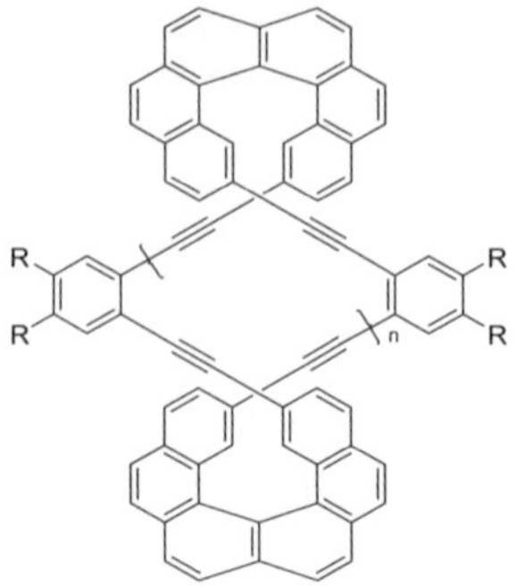

14

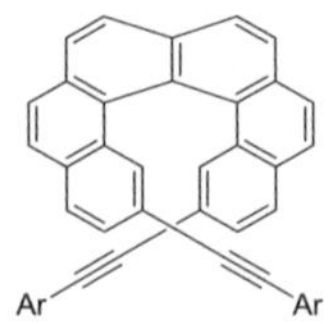

15

Fig. 3.12 Structures of polymer **13**, cyclophane **14**, and the corresponding monomeric compound **15**

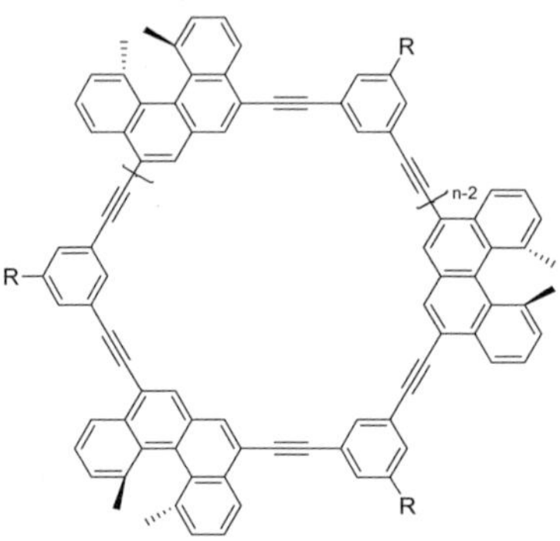

16

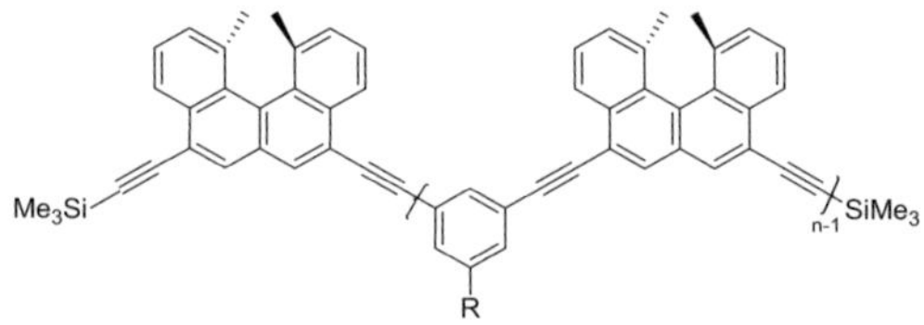

17

Fig. 3.13 Structures of macrocyclic helicene oligomer **16** and acyclic analog **17**

resulting in a conjugated link from one helicene unit to the next. Number average molecular weight of the resulting polymer was estimated to be *ca*. 7000.

In 1998, Fox et al. reported the synthesis of polymers **13** with helicenylene–ethynylene–para-phenylene units and cyclophanes **14** with helicenylene–ethynylene–ortho-phenylene units (Fig. 3.12). Based on FAB mass spectra, polymers **13** with n = 1–5 and cyclophanes **14** with n = 1–3 should be obtained [17]. Polymers **13** showed greater optical rotation and red-shifted UV and CD spectra compared with the corresponding monomeric compound **15**. On the other hand, the maxima in the UV and CD spectra of cyclophane **14** (n = 1) were like those of monomeric compound **15**. This is attributed to the rigid conformation of **14** (n = 1) in which two helicene moieties twisted with respect to one another and π-conjugation between them is inhibited. The $\Delta\varepsilon$ values in CD spectra of **14** (n = 1) were larger than **13** and **15**.

In 2006, Yamaguchi reported macrocyclic helicene oligomer **16** (n = 3–8) with meta-phenylene spacer (Fig. 3.13) [18]. Absorbance in UV–Vis spectra of

18

Fig. 3.14 Helical structure of polymer **18** based on molecular mechanics calculation (Reprinted with permission from Ref. [20]. Copyright 2014, The Royal Society of Chemistry)

16 increases in proportion to the number of helicene units. In contrast, larger macrocycles (n = 4–8) demonstrated smaller $\Delta\varepsilon$ values in CD spectra than the smallest one (n = 1). Based on the temperature effect on CD spectra, the authors concluded that, for R = $COOC_{10}H_{21}$, the macrocycles with n = 4, 5, 6, and 7 have rigid structure while the macrocycle with n = 8 has a flexible one. For acyclic analog **17**, they proposed formation of a stable double helical structure when at least seven helicene units are involved [19]. A double helical structure was indicated to be in equilibrium with two monomeric random coils. The stability of the double helical structure depends on solvents and concentration.

Recently, Yashima et al. reported a preparation of stereoregular helical polyacetylenes **18** with optically active helicene groups as a pendant (Fig. 3.14) [20]. These helicene pendant units are speculated to be overlapped via π–π interactions, which results in the formation of the one-handed helical array along the helical polymer backbone. These polymers showed a high chiral recognition ability toward racemic 1,1′-binaphthyl derivatives.

References

1. Meisenheimer J, Witte K (1903) Ber Dtsch Chem Ges 36:4153
2. Newman MS, Lutz WB, Lednicer D (1955) J Am Chem Soc 77:3420
3. (a) Martin RH, (1974) Angew Chem Int Ed 13:649; (b) Shen Y, Chen C-F (2012) Chem Rev 112:1463; (c) Gingras M (2013) Chem Soc Rev 42:968; (d) Gingras M, Felix G, Peresutti R (2013) Chem Soc Rev 42:1007; (e) Gingras M (2013) Chem Soc Rev 42:1051; (f) Kamikawa K (2014) J Synth Org Chem Jpn 72:58
4. (a) Nuckolls C, Katz TJ, Castellanos L (1996) J Am Chem Soc 118:3767; (b) Lovinger AJ, Nuckolls C, Katz TJ (1998) J Am Chem Soc 120:264; (c) Nuckolls C, Katz TJ, Verbiest T, Elshocht SV, Kuball H-G, Kiesewalter S, Lovinger AJ, Persoons A (1998) J Am Chem Soc 120:8656; (d) Nuckolls C, Katz TJ, Katz G, Collings PJ, Castellanos L (1999) J Am Chem Soc 121:79
5. Verbiest T, Elshocht SV, Kauranen M, Hellemans L, Snauwaert J, Nuckolls C, Katz TJ, Persoons A (1998) Science 282:913
6. Nuckolls C, Katz TJ (1998) J Am Chem Soc 120:9541
7. Rybáček J, Huerta-Angeles G, Kollárovič A, Stará IG, Starý I, Rahe P, Nimmrich M, Kühnle A (2011) Eur J Org Chem 2011:853
8. Kaseyama T, Furumi S, Zhang X, Tanaka K, Takeuchi M (2011) Angew Chem Int Ed 50:3684

9. Owens L, Thilgen C, Diederich F, Knobler CB (1993) Helv Chim Acta 76:2757
10. Murguly E, McDonald R, Branda NR (2000) Org Lett 2:3169
11. Nakano K, Oyama H, Nishimura Y, Nakasako S, Nozaki K (2012) Angew Chem Int Ed 51:695
12. Hatakeyama T, Hashimoto S, Oba T, Nakamura M (2012) J Am Chem Soc 134:19600
13. Waghray D, Cloet A, Van Hecke K, Mertens SFL, De Feyter S, Van Meervelt L, Van der Auweraer M, Dehaen W (2013) Chem Eur J 19:12077
14. Sakai H, Shinto S, Araki Y, Wada T, Sakanoue T, Takenobu T, Hasobe T (2014) Chem Eur J 20:10099
15. (a) Sudhakar A, Katz TJ, Yang BW (1986) J Am Chem Soc 108:2790; (b) Katz TJ, Sudhakar A, Teasley MF, Gilbert AM, Geiger WE, Robben MP, Wuensch M, Ward MD (1993) J Am Chem Soc 115:3182; (c) Gilbert AM, Katz TJ, Geiger WE, Robben MP, Rheingold AL (1993) J Am Chem Soc 115:3199
16. (a) Dai Y, Katz TJ, Nichols DA (1996) Angew Chem Int Ed 35:2109; (b) Dai Y, Katz TJ (1997) J Org Chem 62:1274
17. Fox JM, Lin D, Itagaki Y, Fujita T (1998) J Org Chem 63:2031
18. Takahira Y, Sugiura H, Yamaguchi M (2006) J Org Chem 71:763
19. (a) Sugiura H, Nigorikawa Y, Saiki Y, Nakamura K, Yamaguchi M (2004) J Am Chem Soc 126:14858; (b) Yamaguchi M (2011) J Synth Org Chem Jpn 69:17
20. Anger E, Iida H, Yamaguchi T, Hayashi K, Kumano D, Crassous J, Vanthuyne N, Roussel C, Yashima E (2014) Polym Chem 5:4909

Chapter 4
Planar Cyclooctatetraenes and Related Ring Systems: Antiaromaticity and Applications

Tohru Nishinaga

Abstract Antiaromaticity of planar cyclooctatetraene (COT) is high based on magnetic criteria of aromaticity in contrast to its small antiaromatic destabilization. The planar structures of typical COTs are the transition states of the dynamic processes and hence hardly accessible by spectroscopic means. However, some molecules have been designed to have a planar COT core. The planar structure of the COT core potentially causes a narrow HOMO–LUMO (H–L) gap, which is expected to be useful for various applications. In this chapter, the syntheses, antiaromaticity, and optical and electronic applications of planar COTs and related ring systems are reviewed.

Keywords Antiaromaticity • Cyclooctatetraene • Field-effect transistor • NICS • Vapochromism

Abbreviations

B3LYP	Becke, three-parameter, Lee–Yang–Parr hybrid functional
CASSCF	Complete active space self-consistent field
COT	Cyclooctatetraene
CTOCD-DZ	Continuous transformation of origin of current density-diamagnetic zero
FET	Field-effect transistor
GIAO	Gauge-including atomic orbitals
HF	Hartree–Fock
H–L	HOMO–LUMO
HMO	Hückel molecular orbital
HOMO	Highest occupied molecular orbital
LUMO	Lowest unoccupied molecular orbital
NICS	Nucleus-independent chemical shift

T. Nishinaga (✉)
Department of Chemistry, Graduate School of Science and Engineering, Tokyo Metropolitan University, Hachioji, Tokyo 192-0397, Japan
e-mail: nishinaga-tohru@tmu.ac.jp

T. Akasaka et al. (eds.), *Chemical Science of π-Electron Systems*,
DOI 10.1007/978-4-431-55357-1_4

NMR	Nuclear magnetic resonance
OLED	Organic light-emitting diode
SOMO	Singly occupied molecular orbital
TMS	Trimethylsilyl

4.1 Introduction

4.1.1 Brief History of Cyclooctatetraene (COT) Chemistry

Cyclooctatetraene (COT) has drawn attention since the dawn of research on aromaticity. In the late nineteenth century and the early twentieth century, the structure of benzene was gradually revealed, which caused a desire to prepare vinylogs of benzene. Under the background, the first synthesis of parent COT **1** was reported by Willstätter et al. in 1911 [1]. They synthesized **1** from pseudopelletierine with repetitive Hofmann eliminations (Scheme 4.1). However, the multistep route was not practical despite later improvements [2]. Thus, the studies on the synthesis and properties of COT and its derivatives had made little progress until Reppe's one-step synthesis of **1** reported in 1948 [3], i.e., cyclotetramerization of acetylene with nickel cyanide.

In the 1960s, valence isomerizations of C_8H_8 such as barrelene **2**, semibullvalene **3**, and tricyclo[4.2.0.0^{2,5}]octa-3,7-diene **4** (Fig. 4.1) to COTs were demonstrated [4]. Especially, bicyclo[4.2.0]octatriene **5** was found to be a good precursor for the construction of COT ring, because the equilibrium between **1** and **5** was shown to shift to **1**. Paquette et al. developed the synthetic route for various COT derivatives via ring contraction of five-membered cyclic sulfone to generate bicyclo[4.2.0]octatriene structures [5].

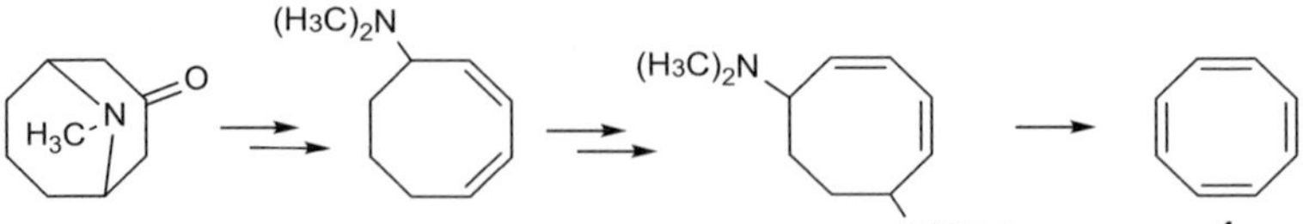

1

Scheme 4.1 The first synthesis of COT **1**

Fig. 4.1 Some C_8H_8 valence isomers

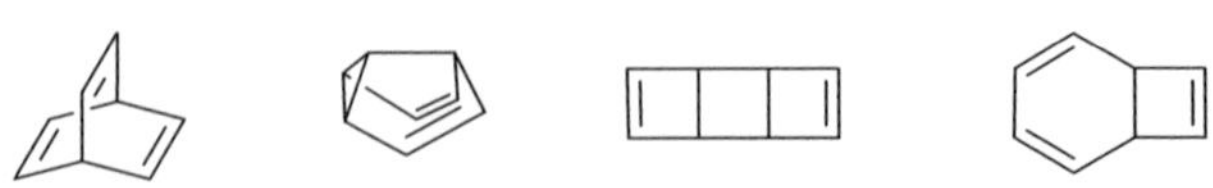

2 3 4 5

Scheme 4.2 Redox behavior of **1**

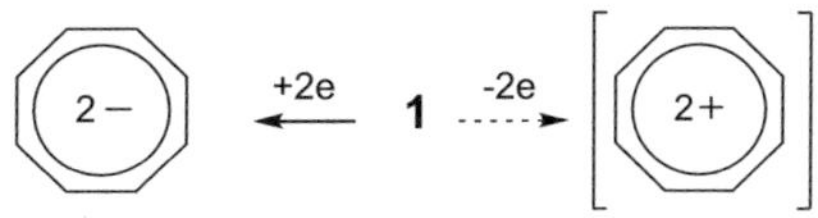

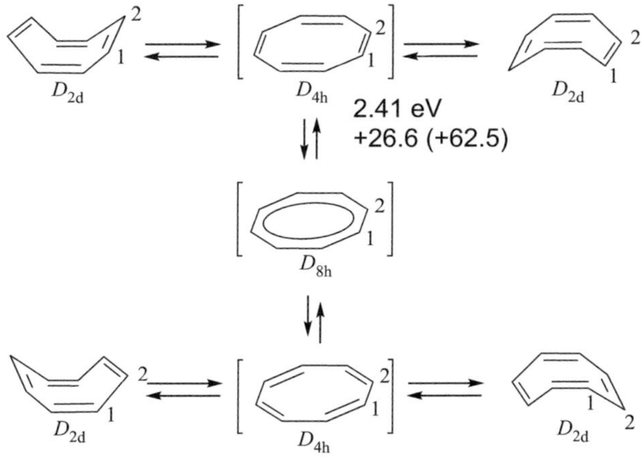

Scheme 4.3 Dynamic behavior of **1** with calculated H–L gap (B3LYP/6-31G(d,p)), NICS(0)$_{iso}$, and NICS(1)$_{zz}$ values in parenthesis (GIAO/HF/6-311 + G(d,p)) of **1**(D_{4h})

At the same period of the early study on the valence isomerizations, the generation of dianion of **1** was demonstrated by Katz (Scheme 4.2) [6] based on the prediction of Hückel rule that a high degree of aromatic stabilization is associated with $4n + 2$ π-electrons for $n = 2$. Later, COT dianions became one of the important sterically demanding ligands for lanthanide complexes [7]. On the other hand, in contrast to the rich anion chemistry, examples of stable cationic COTs in condensed phases have been scarce. The radical cation of **1** is highly reactive and the generation of persistent dication of **1** remains unsuccessful. Some alkyl-substituted COT derivatives can be converted into 6π aromatic COT dications [8, 9], and isolable radical cation salt has so far been obtained only by the full annelation of COT with bicyclo[2.2.2]octene framework [10].

The other issues on COT chemistry that researchers have addressed are the ground-state structure and dynamic behaviors, i.e., ring inversion and bond shift (Scheme 4.3) [11, 12]. For **1** and its typical derivatives, the eight-membered rings adopt a nonplanar tub-shaped structure. The structures of transition states of ring inversion and bond shift of **1** are considered to be planar with bond-alternated D_{4h} and bond-equalized D_{8h} symmetries, respectively. The barriers for the ring inversion (10–14 kcal mol^{-1}) and the bond shift (2–4 kcal mol^{-1} higher than the inversion barrier) of COTs are small. These energy barriers are related to energetic criteria of antiaromaticity of planar COT as described in next section.

4.1.2 Antiaromaticity of Planar COT

The Hückel molecular orbital (HMO) theory, which succeeded in theoretical explanation on the special stability of benzene, also predicted that cyclic conjugation of 4nπ-electrons is energetically unfavorable. In fact, some 4nπ-electron ring systems are destabilized in contrast to the stabilization characteristic of aromatic compounds. Breslow proposed the term of antiaromaticity to describe such systems [13, 14].

In the HMO theory, **1** with bond-equalized and planar D_{8h} structure has a pair of degenerate SOMOs (Fig. 4.2). Thus, **1**(D_{8h}) is predicted to have a triplet ground state according to Hund's rule. However, CASSCF calculations of **1** revealed that open-shell singlet state is much more stable than the triplet state [15]. The calculated energy differences between the ground-state D_{2d} structure (Scheme 4.3) and the transition state D_{4h} (10.6 kcal mol^{-1}) or D_{8h} structure (14.7 kcal mol^{-1}) were in good agreement with the observed energy barriers for the ring inversion and the bond shift of COTs. Taking a ring strain at the planar structures into consideration, the destabilization energy due to antiaromaticity in either **1**(D_{4h}) or **1**(D_{8h}) is much smaller than the aromatic stabilization of benzene (30 kcal mol^{-1}).

The other important criteria of aromaticity are related to the magnetic properties of compounds. In a magnetic field, diatropic and paratropic ring currents are induced in aromatic and antiaromatic rings, respectively. These effects cause downfield and upfield shifts in peripheral protons in ^{1}H NMR, respectively, while much greater shifts in the opposite direction are brought about in the ring center. However, the direct observation of the shifts in the ring center is impossible for most compounds. Thus, nucleus-independent chemical shift (NICS) [16] was proposed by Schleyer. NICS is obtained by quantum chemical calculations of the absolute magnetic shielding at the ring center. Negative and positive NICS values indicate the presence of aromatic and antiaromatic ring current, respectively, although care must be taken due to an inclusion of other effects such as from σ-bond framework of the ring. Both singlet **1**(D_{4h}) [17] and **1**(D_{8h}) [18] were calculated to show very large positive NICS

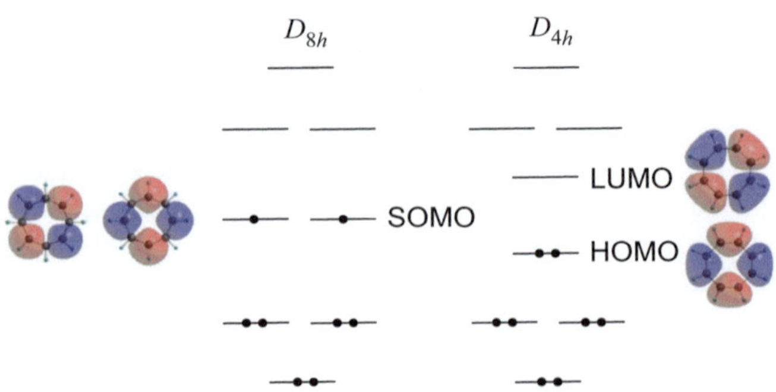

Fig. 4.2 Schematic MO diagrams of planar COT with D_{8h} and D_{4h} symmetries

values in sharp contrast to the small antiaromatic destabilization of planar COT. Rather, **1**(D_{4h}) has the largest NICS among bond-alternated [4n]annulenes [19], indicating planar COT has one of the highest antiaromaticity based on magnetic criteria.

In this chapter, design principles, synthesis, and properties of stable π-systems bearing planar COT structure [20, 21] and related eight-membered rings are reviewed. The designed π-systems bearing a planar COT ring have a bond-alternated structure. As shown in MO diagrams of bond-alternated planar **1**(D_{4h}) (Fig. 4.2), the degeneracy of SOMO observed in bond-equalized planar **1**(D_{8h}) is removed by a Jahn–Teller distortion to split into two nondegenerate components. Thus, π-systems with a bond-alternated COT core are expected to have a narrow HOMO–LUMO (H–L) gap. The narrow H–L gap significantly affects the NICS calculations of planar COTs, as the ring current of the COT ring is dominated by the H–L transition caused by the external magnetic field [22]. For most synthetic chemists studying (anti)aromaticity with NICS, the GIAO model through the HF or B3LYP methods was usually selected. However, in **1**(D_{4h}), GIAO-HF and GIAO-B3LYP gave quite different NICS(0)$_{iso}$ values (GIAO/HF/6-311 + G(d,p), +26.6, GIAO/B3LYP/6-311++G(d,p), +41.3) [23]. From the comparison of calculated and observed ^{1}H NMR shifts of other (4n)π (dehydro)annulenes that have inner ring protons, GIAO-HF method was found to give much more realistic values [23]. Accordingly, if necessary, the NICS calculations at the GIAO/HF/6-311 + G(d,p) level were performed here to compare the NICS values under the same level of theory. Furthermore, such a narrow H–L gap is expected to be useful for optical and electronic applications. Thus, recent advances in optical and electronic applications of COT and related ring systems are also briefly described.

4.2 Planar COTs and Related Ring Systems and Their Antiaromaticity

4.2.1 Dehydro[8]annulenes

In the initial synthetic study on the planarization of COT ring, the replacement of one or two double bonds by triple bonds was tested. The triple bond consisted of *sp*-hybridized carbons, which typically adopts a linear geometry, is somewhat flexible. Thus, the insertion of triple bond(s) causes widening of the inner angle, and, as a result, a planarization in the eight-membered ring is expected. However, unsubstituted dehydro[8]annulenes **6** [24] and **7** [25] and their monobenzo-derivatives **8** [26] and **9** [27] shown in Fig. 4.3 were found to be unstable at ambient conditions. Wong et al. achieved the synthesis of stable dehydro[8]annulenes as the dibenzo-derivatives **10** and **11** via dehydrobromination of bromides (Scheme 4.4) [28–30].

Fig. 4.3 Structures of **6–9**

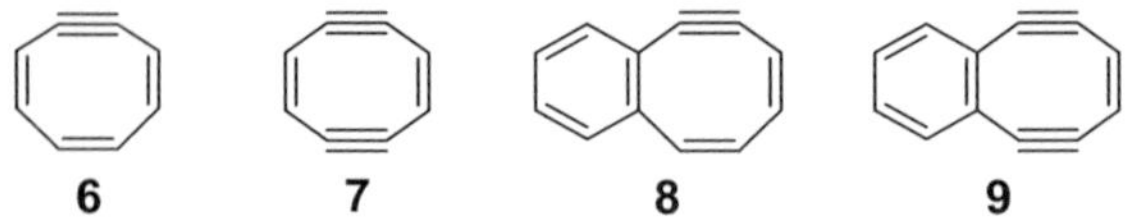

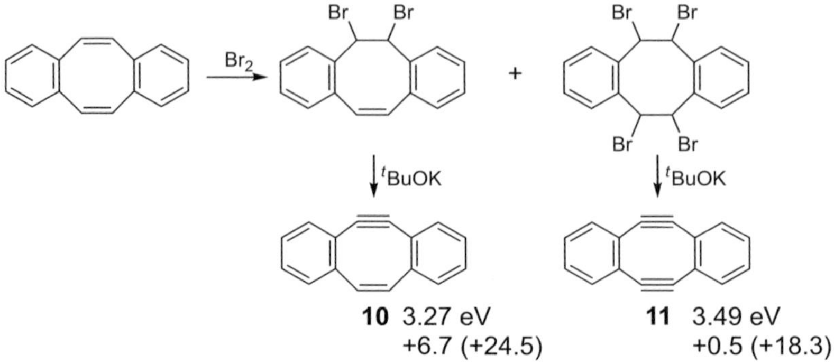

Scheme 4.4 Synthesis of **10** and **11** with their calculated H–L gap (B3LYP/6-31G(d,p)), NICS(0)$_{iso}$, and NICS(1)$_{zz}$ values in parenthesis (GIAO/HF/6-311 + G(d,p))

The NICS(0)$_{iso}$ of the eight-membered ring in **11** at the GIAO/HF/6-311 + G(d,p)//B3LYP/6-31G(d,p) level was +0.5 (+4.04 at the GIAO/B3LYP/6-31G(d) level [31]). Judging from this NICS value, the antiaromatic ring current in **11** seems to be negligibly small at first glance. However, NICS(1)$_{zz}$ (out-of-plane zz component of the NICS tensor at 1 Å above the ring center) [32] was +18.3, which is ca. 30 % of that (+62.5) of **1**(D_{4h}). Therefore, a weak antiaromatic ring current would be present in **11**. The small NICS(0)$_{iso}$ of **11** is considered to be caused by the triple bond closely located to the ring center. The in-plane xx and yy components of **1**(D_{4h}) (−2.6) were very small and hence brought about a smaller effect in its NICS(0)$_{iso}$ value. In contrast, the relatively large in-plane yy component (direction to the triple bond) of the NICS(0) tensor (−30.9) reduced the NICS(0)$_{iso}$ of **11**. As for **10**, the higher NICS(0)$_{iso}$ (+6.7) and NICS(1)$_{zz}$ (+24.5) than those of **11** were obtained. The reduction of the number of triple bond seems to be responsible for this result, and it can be concluded that the insertion of triple bond in the [8]annulene core reduces the NICS antiaromaticity.

In the tribenzo-derivatives of dehydro[8]annulene, steric repulsion between the protons of neighboring benzene rings brought about the instability of the compounds and hampered the isolation [30]. Wong et al. introduced carbon bridges to solve this problem and synthesized **12** as a stable compound (Scheme 4.5) [33]. The X-ray crystallography of **12** confirmed the planar structure of the [8]annulene core. The NICS(0)$_{iso}$ and NICS(1)$_{zz}$ were +6.2 and +22.0, respectively. Thus, in comparison with **10**, annelation of one additional benzene caused a slight decrease in the NICS antiaromaticity in the eight-membered ring.

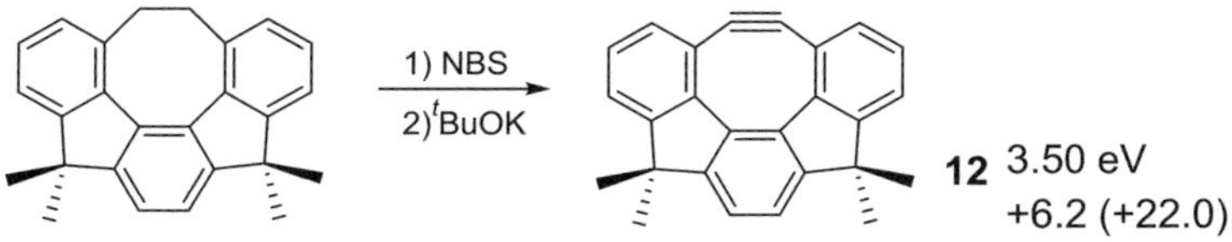

Scheme 4.5 Synthesis of **12** with its calculated H–L gap (B3LYP/6-31G(d,p)), NICS(0)$_{iso}$, and NICS(1)$_{zz}$ values in parenthesis (GIAO/HF/6-311 + G(d,p))

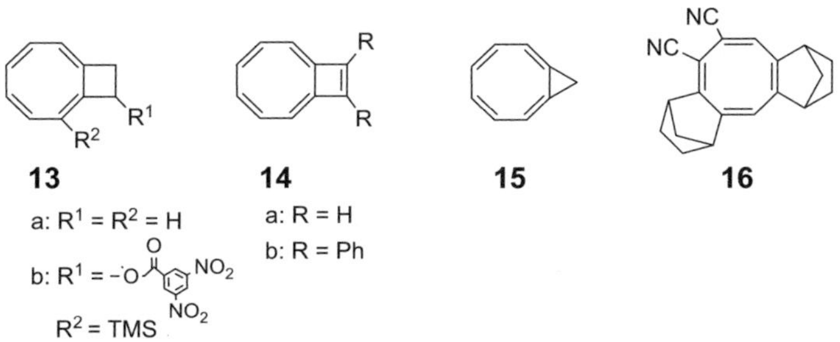

Fig. 4.4 Structures of **13–16**

As described in 4.1.2, the H–L gap and the magnetic antiaromaticity in the COT rings are closely related to each other. The H–L gap of **10–12** is 0.9–1.1 eV wider than that of **1**(D_{4h}), which is consistent with the lower NICS antiaromaticity of **10–12** in spite of the π-expansion. Thus, the annelation of aromatic benzene reduces the antiaromaticity of dehydro[8]annulenes accompanied by widening of the H–L gap.

4.2.2 COTs Planarized by Annelation with Small Rings

The ideal bond angle for sp^2-hybridized carbon is 120°, while the average bond angle of octagon is 135°. Therefore, to planarize the inherently tub-shaped COT ring, widening of the inner angles of the eight-membered ring is required. Small cyclic alkenes having a smaller inner angle than 120° should have a wider exocyclic bond angle, and its annelation to a COT ring would cause some flattening of the COT ring. Based on this strategy, syntheses of mono-annelated and bis-annelated COT derivatives **13–16** (Fig. 4.4) were attempted.

Cyclobutene-annelated COT **13a** [34, 35] is unstable in condensed phases. Thus, the X-ray structure was determined with **13b** bearing the bulky TMS and benzoate substituents [36], but the effect of the annelation of cyclobutene on the planarization of COT ring was found to be rather small. Consistent with the nonplanar structure of **13**, the NICS(0)$_{iso}$ (+5.7) [21] was rather small. In contrast, X-ray analysis of **14b** revealed that the annelation of cyclobutadiene makes the eight-membered ring more flattened [37, 38]. However, structural analysis [37] as well as the negative

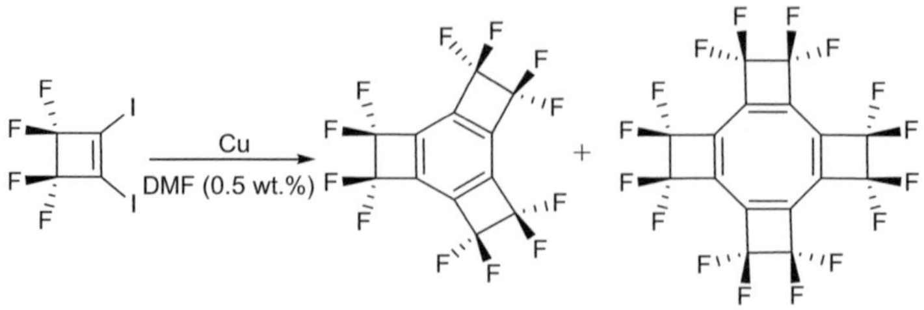

17 2.56 eV
+17.7 (+44.4)

Scheme 4.6 Synthesis of **17** with its calculated H–L gap (B3LYP/6-31G(d,p)), $NICS(0)_{iso}$, and $NICS(1)_{zz}$ values in parenthesis (GIAO/HF/6-311 + G(d,p))

$NICS(0)_{iso}$ (−0.4) [21] suggested that **14** has a 10π aromatic character in the peripheral conjugation that outweighs the 8π antiaromatic character in the COT ring. For cyclopropene-annelated derivative **15**, the paratropic ring current was predicted to survive in the more flattened but bent COT structure [39]. However, the isolation of **15** was not conducted, and only the generation and NMR observation were reported [40]. In the case of **16** bis-annelated with norbornene, the effect of this structural modification on the planarization of COT ring was found to be rather small [41].

The first synthesis of non-π-expanded planar COT was reported by Soulen et al. in1973 [42]. They conducted copper-mediated coupling reaction of 3,3,4,4-tetrafluoro-1,2-diiodo-1-cyclobutene in the presence of 0.5 wt.% of DMF to give COT **17** tetra-annelated with tetrafluorocyclobutene as a by-product of the corresponding cyclotrimerized benzene derivative (Scheme 4.6). The X-ray structure of **17** revealed the planar COT core with a bond-alternated structure [43]. Interestingly, the shorter "double bonds" in the COT ring locate in the endocyclic position of the cyclobutene ring, even though the larger strain may be released when the "double bonds" locate in the exocyclic position of the small cyclobutene. The unique structural feature was reproduced by theoretical calculations [44].

The color of **17** is deep red in spite of an 8π-electron system, suggesting that **17** has a narrow H–L gap. The calculated H–L gap is 2.56 eV which is only 0.15 eV wider than that of **1**(D_{4h}). The narrow gap would cause a strong paratropic ring current in the COT ring. CTOCD-DZ calculations predicted the presence of a strong paratropic ring current in **17** [45]. Similarly, the $NICS(0)_{iso}$ and $NICS(1)_{zz}$ were +17.7 and +44.0, respectively, which are 67 % and 71 % of those of **1**(D_{4h}).

The synthesis of dibenzo-COT **18** bis-annelated with cyclopropene ring was reported by Dürr et al. [46]. As shown in Scheme 4.7, stepwise photochemical elimination of N_2 from bis-3H-pyrazole gave the objective compound **18**. The X-ray analysis of **18** confirmed that the eight-membered ring is almost planar. However, dibenzo-annelation caused a significant reduction of magnetic antiaromaticity, and the $NICS(0)_{iso}$ and $NICS(1)_{zz}$ of **18** were +10.2 and +27.1, respectively, and the

Scheme 4.7 Synthesis of **18** with its calculated H–L gap (B3LYP/6-31G(d,p)), NICS(0)$_{iso}$, and NICS(1)$_{zz}$ values in parenthesis (GIAO/HF/6-311 + G(d,p))

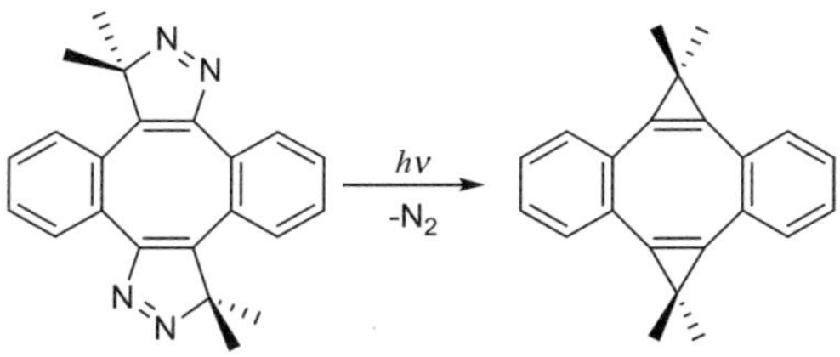

18 2.94 eV
+10.2 (+27.1)

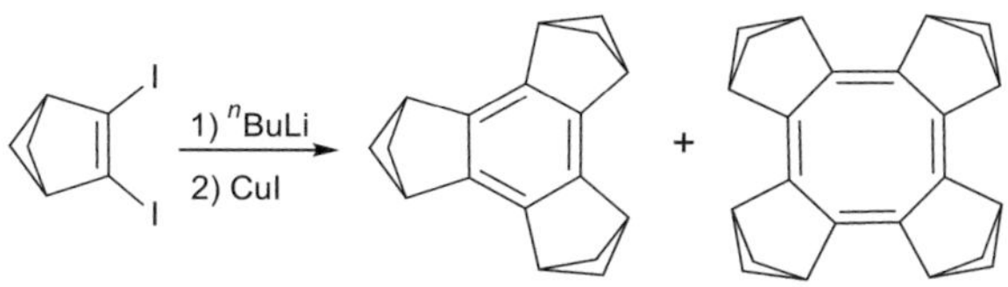

19 3.05 eV
+10.5 (+27.3)

Scheme 4.8 Synthesis of **19** with its calculated H–L gap (B3LYP/6-31G(d,p)), NICS(0)$_{iso}$, and NICS(1)$_{zz}$ values in parenthesis (GIAO/HF/6-311 + G(d,p))

H–L gap was 2.94 eV. Thus, **18** has the slightly higher NICS antiaromaticity with the narrower H–L gap than those of **10**, probably due to the lack of triple bond in the [8]annulene core of **18**.

The first and so far sole hydrocarbon derivative of non-π-expanded planar COT was realized by Komatsu et al. in 2001 [47]. The cyclotetramerization of diiodobicyclo[2.1.1]hexene was achieved with lithiation of the precursor followed by treatment with copper iodide to give COT **19** tetra-annelated with bicyclo[2.1.1]hexene as a by-product of the corresponding benzene derivative (Scheme 4.8). The X-ray structure of **19** showed the planar COT structure with a large bond-alternated structure. The shorter "double bonds" locate in the exocyclic position of bicyclo[2.1.1]hexene ring in contrast to the opposite bond-alternated structure for **17**. This structural feature of **19** can be simply interpreted by the release of strain in the small annelated rings, and the structure was also reproduced well by theoretical calculations [44].

The color of **19** is red as observed for **17**, suggesting that **19** also has a narrow H–L gap. However, the calculated H–L gap is 3.05 eV which is 0.64 eV wider than that of D_{4h} COT and even 0.49 eV wider than that of **17**. The wider gap would reduce the paratropic ring current in the COT ring of **19**. CTOCD-DZ calculations predicted the weaker paratropic ring current in **19** than that of **1**(D_{4h}) and **17** [45]. Similarly, the NICS(0)$_{iso}$ and NICS(1)$_{zz}$ were +10.5 and +27.3, respectively, which are only 39 % and 44 % of those of **1**(D_{4h}). We demonstrated that the relatively wider H–L gap was brought about principally by the raised LUMO level owing to a characteristic σ– π* orbital interaction between the LUMO of COT and an MO including highly strained C–C σ-bond in the annelated bicyclo[2.1.1]hexene unit nearly parallel to the p-orbitals [48].

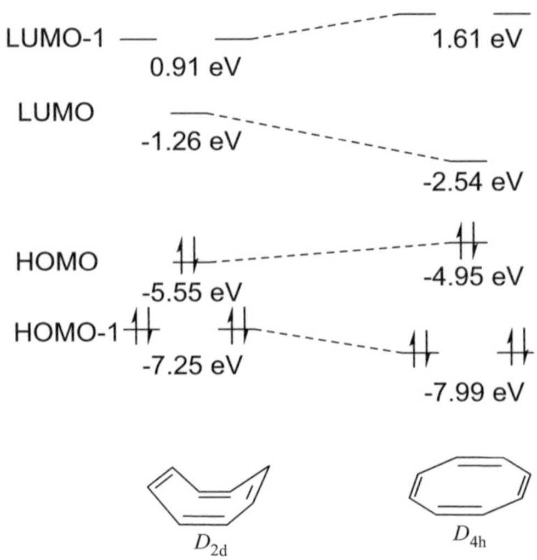

Fig. 4.5 MO diagrams of D_{2d} and D_{4h} COT at the B3LYP/6-31G(d,p) level

Planar COT **19** gave a fairly stable radical cation salt [49]. Interestingly, the longest wavelength absorption (630 nm) of radical cation of planar **19** was blueshifted from that (745 nm) of radical cation of tub-shaped COT fully annelated with bicyclo[2.2.2]octene units [10] contrary to the common sense that a planar structure maximizes the effective π-conjugations. The assignment of the absorption band is HOMO–SOMO transition, and the observed unique phenomenon occurred because of the widening of the HOMO–SOMO gap caused by the raised SOMO (HOMO in neutral COT) and lowered HOMO (HOMO-1 in neutral COT) accompanying the flattening of the COT ring. These results suggest that the stabilization of degenerate HOMO-1 levels for planar neutral COT is the principal contributor to the small inversion barrier which is directly related to the antiaromatic destabilization energy for planar COT. In fact, the similar change in the MO levels was observed in the calculated **1**(D_{2d}) and **1**(D_{4h}) as shown in Fig. 4.5. In these MO diagrams, two electrons in HOMO are destabilized by 0.6 eV upon the planarization, while four electrons in degenerate HOMO-1 are stabilized by 0.74 eV. The total 1.76 eV (40.6 kcal mol^{-1}) of stabilization energy is gained in the changes of these MOs upon the planarization.

4.2.3 COTs Planarized by Annelation with Rigid π-Systems

The other design strategy to construct a planar COT structure is annelation of rigid π-systems to the COT core. In 1970, Reiff et al. [50] showed that tetraphenylene (tetrabenzo-derivative of COT) can be planarized by carbonyl or methylene bridges. The planar tetraphenylene **21** was synthesized by condensation of tetraester of tetraphenylene, followed by reduction of the resultant cyclic tetraketone **20**

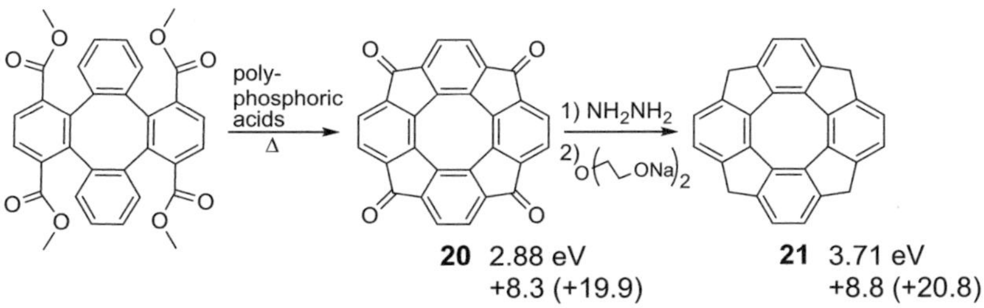

Scheme 4.9 Synthesis of **20** and **21** with their calculated H–L gap (B3LYP/6-31G(d,p)) and $NICS(0)_{iso}$ and $NICS(1)_{zz}$ values of the COT ring (GIAO/HF/6-311 + G(d,p))

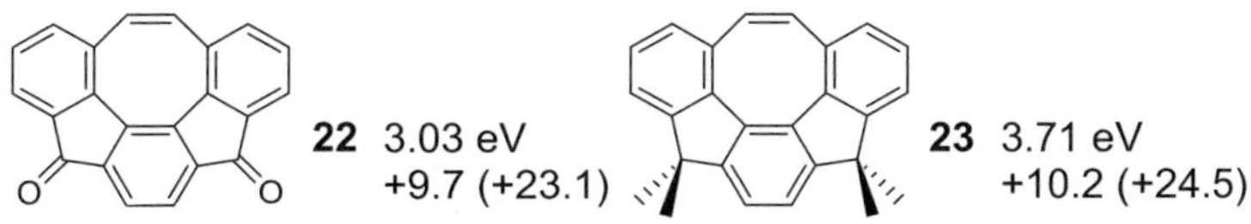

Fig. 4.6 Structures of **22** and **23** with their calculated H–L gap (B3LYP/6-31G(d,p)) and $NICS(0)_{iso}$ and $NICS(1)_{zz}$ values of the COT ring (GIAO/HF/6-311 + G(d,p))

(Scheme 4.9). The antiaromaticity of the octagon core in **20** and **21** is considerably reduced due to the tetra-annelation of aromatic benzene rings as indicated by NICS (**20**: $NICS(0)_{iso}$, +8.3; $NICS(1)_{zz}$, +19.9) (**21**: $NICS(0)_{iso}$, +8.8; $NICS(1)_{zz}$, +20.8). These NICS values are 31–33 % of those of **1**(D_{4h}).

As related carbonyl- and methylene-bridged compounds, tribenzo-derivatives **22** and **23** (Fig. 4.6) were synthesized by Wong et al. [51] in the course of their pursuit of dehydro[8]annulenes (see 4.2.1). Compounds **22** and **23** have olefinic protons directly connected to the [8]annulene ring. The ^{1}H NMR chemical shifts of the proton are δ 5.83 and δ 5.91 ppm. These values are ca. 0.7–1 ppm upfield in comparison with those of non-bridged tribenzo-derivatives of COT, indicating the presence of some paratropic ring current in the eight-membered ring. The NICS values of the octagon core in **22** and **23** (**22**: $NICS(0)_{iso}$, +9.7; $NICS(1)_{zz}$, +23.1) (**23**: $NICS(0)_{iso}$, +10.2; $NICS(1)_{zz}$, +24.5) are slightly larger than those of **20** and **21** owing to the reduction of the number of annelating aromatic benzene rings, but the NICS antiaromaticity is still considerably reduced in comparison with **1**(D_{4h}) (36–39 % of those of **1**(D_{4h})).

Wilcox et al. reported the synthesis of cycloocta[*def*]biphenylene **24** in 1972 [52] as the first example of planar COT with olefinic protons directly connected to the COT ring. The synthesis was attained by double Wittig reaction of biphenylene bisylide with glyoxal (Scheme 4.10), although the isolated yield was quite low (<1 %) [52, 53]. The ^{1}H NMR chemical shifts of the olefinic proton were δ 4.74 and δ 4.61 ppm, which are further more than 1 ppm upfield compared with that of **22** and **23**. The $NICS(0)_{iso}$ and $NICS(1)_{zz}$ of the COT ring were +13.8 and +32.5, respectively, [23] which are 52 % of those of **1**(D_{4h}).

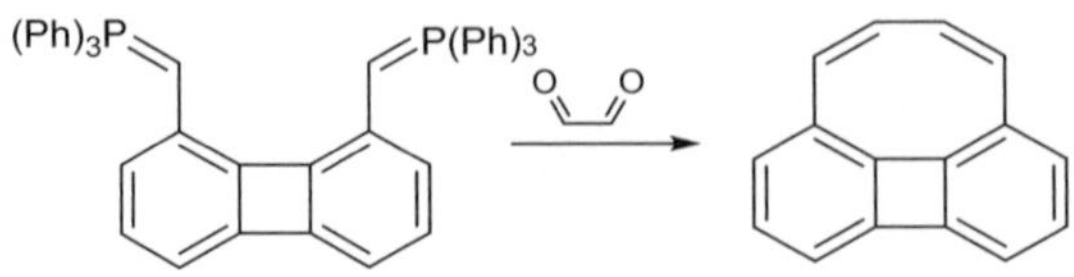

Scheme 4.10 Synthesis of **24** with its calculated H–L gap (B3LYP/6-31G(d,p)) and NICS(0)$_{iso}$ and NICS(1)$_{zz}$ values of the COT ring (GIAO/HF/6-311 + G(d,p))

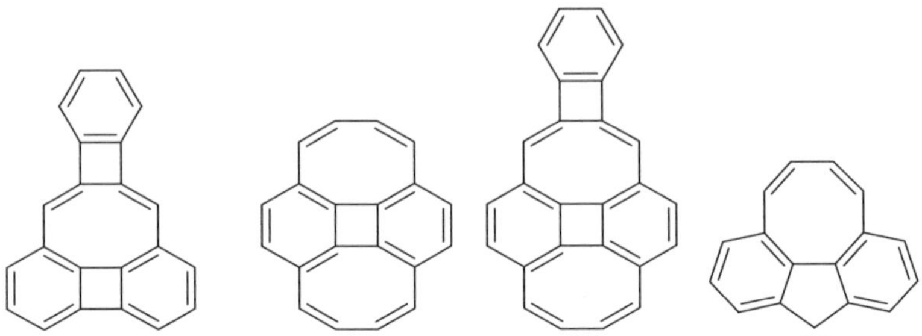

Fig. 4.7 Structures of **25–28** with calculated H–L gap (B3LYP/6-31G(d,p)) and NICS(0)$_{iso}$ and NICS(1)$_{zz}$ values of the COT ring (GIAO/HF/6-311 + G(d,p)) of **25**

As related biphenylene or fluorene-annelated COT derivatives, **25–28** were synthesized (Fig. 4.7). Among these compounds, **25** was isolated as stable species [54], and its X-ray structure was determined [55]. However, the paratropicity of the COT ring was considerably reduced in comparison with **24**, probably due to the similar 10 π aromatic contribution within the COT ring and the added four-membered ring as observed for **14** [37, 38]. Compound **26** showed a higher paratropicity than **24**, but **26** is quite unstable at ambient conditions [56]. The annelation of additional cyclobutene ring such as in **27** [57] caused some stabilization of the π-system due to the similar stabilization effect for **25**. Compound **27** sustained a higher paratropicity than **24**, yet the isolation of **27** was not achieved. On the other hand, Rabinovitz et al. synthesized **28** [58] as a stable compound. In the COT core, some paratropic contribution was confirmed by ^{1}H NMR spectroscopy. They claimed that this is due to a planar COT moiety. However, the later DFT calculations showed that the COT ring in **28** is not planar, though flattened to some extent, and hence has a rather reduced NICS antiaromaticity [21, 23].

Osuka et al. reported the synthesis of tetrameric porphyrin sheet bearing planar COT core. A stepwise Ag(I)-mediated homo-coupling of 5,10-diaryl Zn(II) porphyrin and oxidation with DDQ-Sc(OTf)$_3$ of the resultant cyclic tetramer afforded the porphyrin sheet **29** (Fig. 4.8) [59, 60]. Demetallation of **29** with TFA and H_2SO_4 gave free base porphyrin sheet **30b**. In the host–guest complexes between host **29** and guests 1,4-bis(1-methylimidazol-2-ylethynyl)benzene and 5,15-bis(1-methylimidazol-2-yl)-10,20-dihexylporphyrin, protons of guests located above the

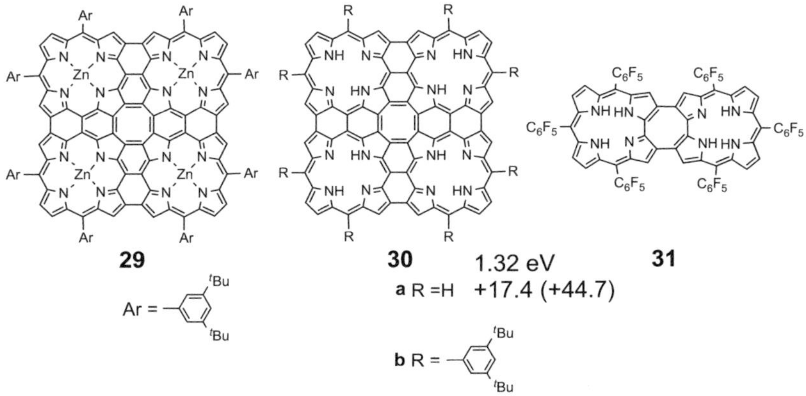

Fig. 4.8 Structures of **29**–**31** with calculated H–L gap (B3LYP/6-31G(d,p)) and NICS(0)$_{iso}$ and NICS(1)$_{zz}$ values of the COT ring (GIAO/HF/6-311 + G(d,p)) of **30a**

COT core of **29** showed a significant downfield shift in the ^{1}H NMR spectra. Thus, the presence of a paratropic ring current in the central COT core was experimentally proved. Interestingly, unlike other normal free porphyrin, the pyrrolic protons in **30b** are frozen at the position as the depicted structure. The NICS(0)$_{iso}$ and NICS(1)$_{zz}$ of the COT ring of **30a** were +17.4 and +44.7, respectively. The reported NICS(0)$_{iso}$ value for the COT of **30**a (+21.7 at the GIAO/B3LYP/6-31G(d)// B3LYP/6-31G(d)) [59, 60] may be somewhat overestimated with the GIAO-B3LYP method as described in 4.1.2. Nevertheless, the NICS(0)$_{iso}$ and NICS(1)$_{zz}$ values with the GIAO-HF method are large, which are 65 % and 72 % of those of **1**(D_{4h}). These results are in sharp contrast to the results that the annelation of aromatic benzene rings significantly reduced the NICS antiaromaticity of the COT ring as described above. On the contrary, the aromaticity of porphyrin ring was shown to be lost by the NICS calculations [59, 60]. As a related compound, doubly linked corrole dimer **31** has a similar octagon core [61, 62]. However, the NICS antiaromaticity (+8.95 at the GIAO/B3LYP/6-31G(d)// B3LYP/6-31G(d)) is small probably due both to the nonplanar structure and to the lack of the double-bond character within the octagon core.

We synthesized a series of cyclic tetrathiophenes **32** bridged by dimethylsilyl, sulfur, and sulfone units via homo-coupling of the corresponding bridged dithiophene [63]. As shown in Fig. 4.9, the X-ray structure of **32b** has a planar COT core, while the octagon in **32a** has a bent structure. Thus, the bent angle of the central COT core was found to be finely adjusted by the small differences in the bond lengths between the bridging units and thiophene rings. The planarity is enhanced in the order of **32b** > **32c** > **32a**. The NICS(0)$_{iso}$ and NICS(1)$_{zz}$ of the COT ring of **32b** were +17.4 and +44.8, respectively, which are 65 % and 72 % of those of **1**(D_{4h}). The NICS(0)$_{iso}$ was gradually reduced (**32b**, +17.4; **32c**, +15.4; **32a**, +12.7) with widening of the calculated H–L gaps (**32b**, 2.62 eV; **32c**, 2.72 eV; **32a**, 2.87 eV) as predicted by the structure–property relationship in parent COT **1**. In accord with

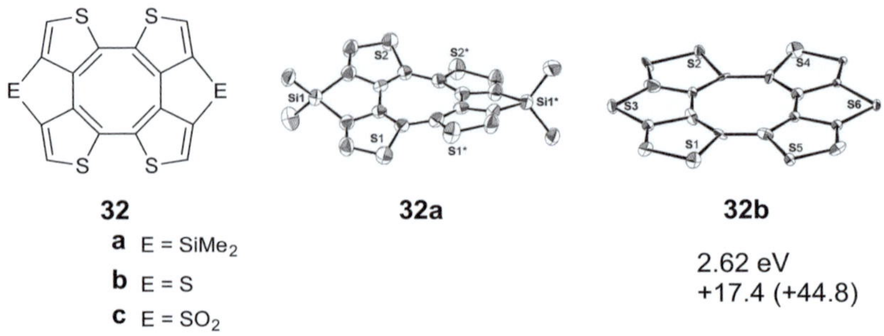

Fig. 4.9 Chemical structure of **32** and X-ray structures of **32a,b** with calculated H–L gap (B3LYP/6-31G(d,p)) and NICS(0)$_{iso}$ and NICS(1)$_{zz}$ values of the COT ring (GIAO/HF/6-311 + G(d,p)) of **32b**

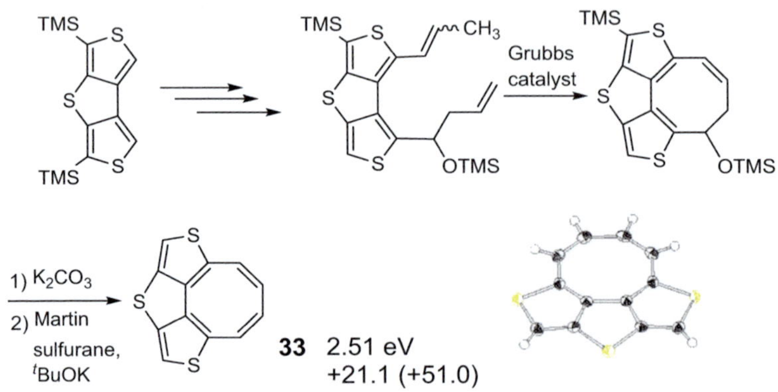

Scheme 4.11 Synthesis of **33** with its calculated H–L gap (B3LYP/6-31G(d,p)) and NICS(0)$_{iso}$ and NICS(1)$_{zz}$ values of the COT ring (GIAO/HF/6-311 + G(d,p)) and X-ray structure

these theoretical predictions, ^{1}H NMR chemical shifts of **32** showed upfield shifts in comparison with that of the corresponding bridged dithiophene precursors due to a paratropic ring current in the COT core. Also the longest absorption maxima of **32** showed bathochromic shifts with increasing planarity of the COT ring.

The reduction of the number of aromatic thiophene units annelated to COT ring is expected to enhance the antiaromaticity of the COT ring as described above. Thus, we designed and synthesized COT **33** mono-annelated with dithienothiophene unit via ring-closing metathesis as a key step as shown in Scheme 4.11 [23]. The X-ray analysis of **33** confirmed the planar COT structure. The NICS(0)$_{iso}$ and NICS(1)$_{zz}$ of the COT ring of **33** were +21.1 and +51.0, respectively, which are approximately 80 % of those of **1**(D_{4h}). In fact, ^{1}H NMR chemical shifts of the olefinic protons in **33** (δ 4.71 and δ 4.41 ppm) are in the more upfield region in comparison with **24** (δ 4.74 and δ 4.61 ppm) [52, 53], and so far, **33** has the most antiaromatic character among the stable planar COTs bearing the olefinic protons, although unstable compound **26** has a greater paratropicity judging from its NMR chemical shifts in the octagons (δ 3.59 and δ 3.63 ppm) [56]. In addition, it is interesting to note that the calculated H–L gap of **33** (2.51 eV) is narrower than that of **32b** (2.62 eV) in spite of its smaller π-system of **33**. This is also due to the higher antiaromaticity of **33** than **32b**.

4.2.4 Hetero[8]circulenes

[8]Circulene **34a** (Fig. 4.10) is a macrocyclic arene in which central octagon is completely surrounded and annelated by eight benzene rings. Recently, the syntheses of substituted derivatives **34b–d** [64] and tetrabenzo-derivative **35** [65, 66] of [8]circulene were reported, but the NICS antiaromaticity of the octagon core in **34a** (NICS(0)$_{iso}$: +9.8 at the GIAO/HF/6-31 + G(d)//ωB97X-D/6-31G(d,p) level) [64] is small due both to the nonplanar structure and to the annelation of aromatic benzene ring to the octagon core.

Hetero[8]circulenes are heterocyclic analogues of **34a** in which some of the surrounding benzene rings or all of them are replaced by heteroaromatics such as furan, pyrrole, and thiophene. Erdtman et al. reported a facile acid-mediated cyclotetramerization of α-naphthoquinone in 1968 [67] and *p*-benzoquinone in 1970 [68] to give **36** and **37a**, respectively (Fig. 4.11). Later, compound **37** was recognized as tetraoxa[8]circulene, and some derivatives were prepared [69, 70]. In addition, Pittelkow et al. reported the synthesis of azatrioxa- and diazadioxa[8]circulenes **38–40** recently [71, 72]. These compounds have a planar COT core. However, in view of the NICS index, the annelation of aromatic benzene rings considerably reduced the antiaromaticity of the octagon core as observed in the methylene-bridged analogue **21**.

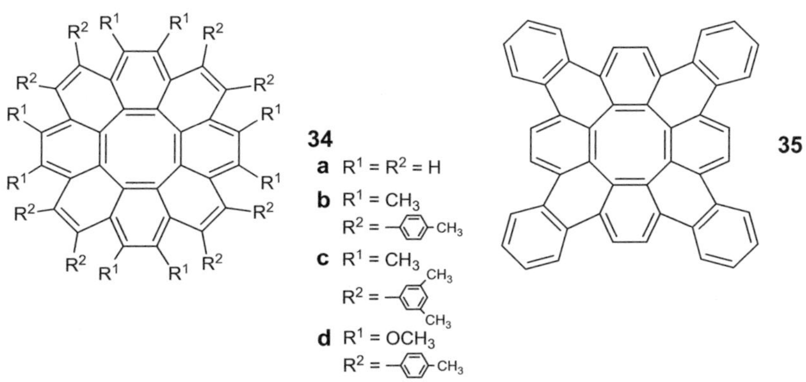

Fig. 4.10 Structure of **34** and **35**

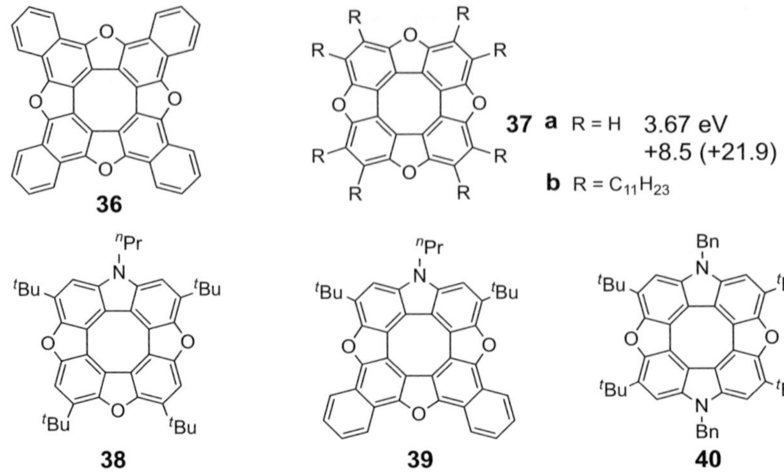

Fig. 4.11 Structures of **36–40** with calculated H–L gap (B3LYP/6-31G(d,p)) and NICS(0)$_{iso}$ and NICS(1)$_{zz}$ values of the COT ring (GIAO/HF/6-311 + G(d,p)) of **37a**

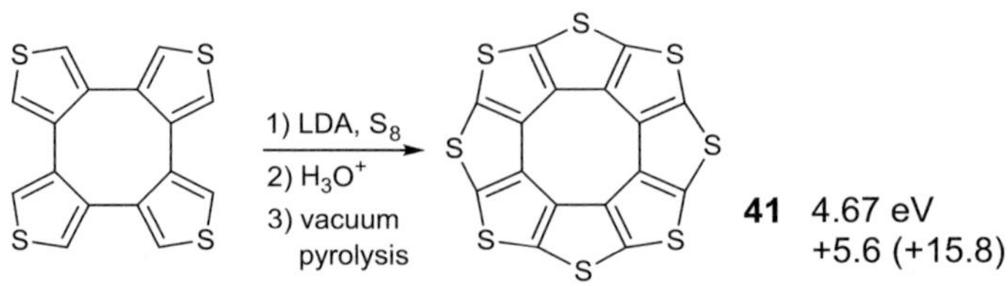

Scheme 4.12 Synthesis of **41** with its calculated H–L gap (B3LYP/6-31G(d,p)) and NICS(0)$_{iso}$ and NICS(1)$_{zz}$ values of the COT ring (GIAO/HF/6-311 + G(d,p))

The other interesting hetero[8]circulene is octathio[8]circulene **41**. In 2006, Nenajdenko et al. reported the efficient synthesis of **41** (Scheme 4.12) which was named "sulflower" [73]. This compound has also a planar octagon core [73, 74], yet the NICS-based antiaromaticity is small due to the radialene-type structure.

4.3 Optical and Electronic Applications

Since COT itself is a small π-system, an appropriate π-expansion is required to display attractive optical properties. In 2010, Pittelkow et al. reported organic light-emitting diode (OLED) based on tetraoxa[8]circulene **37b** and its benzo-derivatives **42–45** (Fig. 4.12) having long alkyl side chains [75]. They fabricated OLED devices in which the emitting layer of soluble **37b** and **42–45** with the (4,4′-N,N′-dicarbazole)biphenyl host was formed by spin coating. Among these materials, **37b** was found to show a blue light intensity of 276 cd m^{-2} at 10 V.

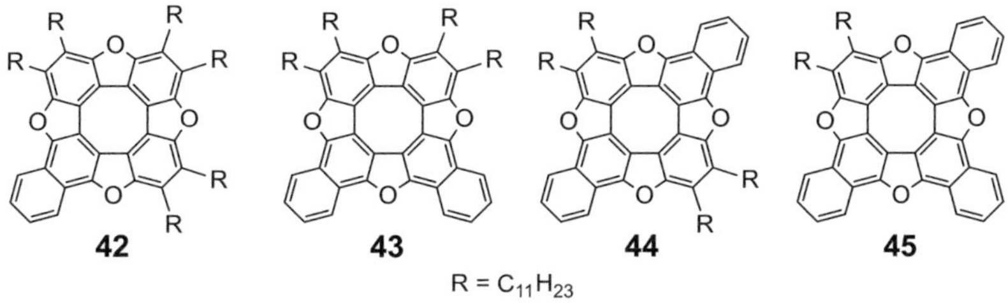

Fig. 4.12 Structures of **42**–**45**

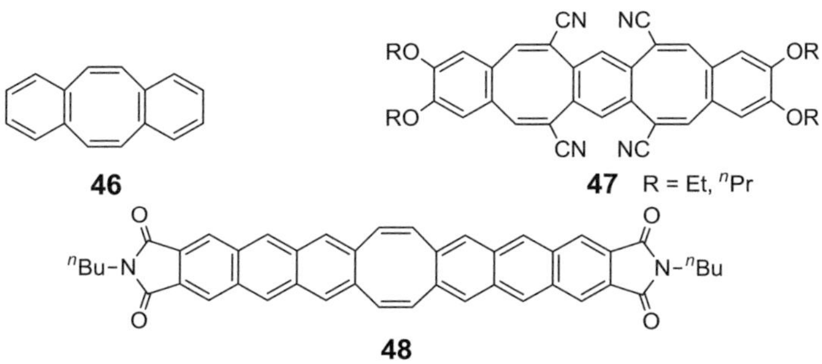

Fig. 4.13 Structures of **46**–**48**

Dibenzo[a,e]cyclooctene **46** (Fig. 4.13) is a dibenzo-derivative of COT, which has a ground-state structure (S_0) similar to parent COT **1**. Recently, we showed the results of DFT calculations that the excited S_1 state of **47** has COT cores with a shallow bent angle [76]. Furthermore, Saito and Yamaguchi et al. revealed that the excited S_1 state of π-extended systems **48** was optimized to be planar [77, 78]. The energy difference between S_0 and S_1 states was calculated to be quite sensitive to the bent angle of COT core. By taking advantage of the properties, we demonstrated that a unique vapochromism of **47** in which the emission colors were changed in the crystalline state reflecting the small difference in the bent angle of COT cores caused by the absorbed solvent molecules. On the other hand, Saito and Yamaguchi et al. demonstrated a unique red, green, and blue emission from a single component of **48** in crystals, solution, and polymer matrix, respectively.

Electrochemical study of octathio[8]circulene **41** and tetraselenotetrathio[8]circu lene **49** (Fig. 4.14) in solid states was conducted, and the energy levels of HOMO of these materials were shown to be −5.7 eV and −5.2 eV, respectively [79]. An interesting electrochromic behavior corresponding to the formation of radical cation and dication of **41** in the film was also demonstrated [80]. Field-effect transistor (FET) devices were fabricated based on vacuum-deposited thin films of **41** and **49**. The highest mobility of **41** (9×10^{-3} cm^2 V^{-1} s^{-1}) was shown to be better than that of **49** (1×10^{-3} cm^2 V^{-1} s^{-1}) [79]. FET devices based on films of **41** formed through electric double layers at the interface between organic semiconductor and the ionic

Fig. 4.14 Structures of **49**

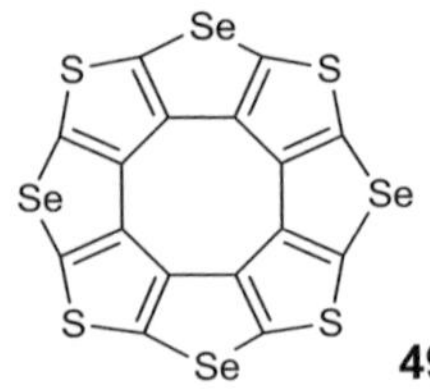

49

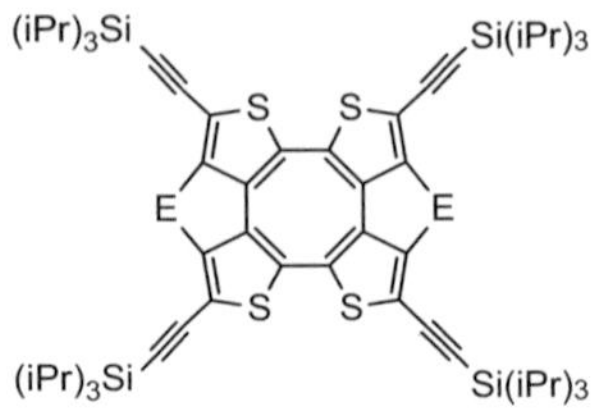

50

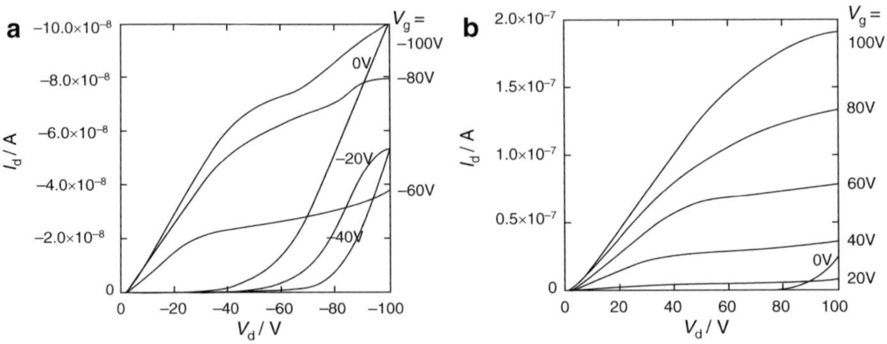

Fig. 4.15 Structures of **50** and output curves for a **50**-based single-crystal FET device with carbon source and drain electrodes under (**a**) negative- and (**b**) positive-biased conditions (Ref. [82] Reproduced by permission of The Royal Society of Chemistry)

liquid were also fabricated [81]. The mobilities were in the order of 2×10^{-2}–2×10^{-3} cm^2 V^{-1} s^{-1} depending on a kind of ion liquids.

The reduction of **41** was difficult due to its relatively large H–L gap caused by the low antiaromaticity of the octagon core. In contrast, cyclic tetrathiophene **32b** has a narrow H–L gap thanks to the substantially high antiaromaticity in the COT core [63]. Furthermore, various substituents can be introduced at the thiophene unit. Thus, we synthesized a derivative of **32b** radially π-extended with (triisopropylsilyl)ethynyl group, i.e., **50** (Fig. 4.15) [82]. Cyclic voltammetry of **50** showed not only oxidation wave ($E^{\text{ox1}} = 0.62$ V vs. Fc/Fc$^+$) but also reduction waves ($E^{\text{red1}} = -1.28$ V vs. Fc/Fc$^+$). The estimated HOMO (–5.4 eV) and LUMO (–3.5 eV) levels fulfilled a requirement for an ambipolar FET transport with conventional electrode such as gold and carbon. In practice, a single-crystal FET device showed ambipolarity with relatively high hole and electron mobilities of up to 0.40 and 0.18 cm^2 V^{-1} s^{-1}, respectively.

4.4 Conclusions

Magnetic criterion of antiaromaticity of planar COT is varied, which highly depends on the nature of the surrounding substituents. Some structural modifications can maintain a comparable magnetic antiaromaticity of parent D_{4h} COT. The high magnetic antiaromaticity of planar COT is accompanied by a narrow H–L gap, while the inherent small antiaromatic destabilization allows it to be handled as an isolable compound. The narrow H–L gap of planar COT is potentially useful for various applications, and in practice, it has started to apply for the development of organic electronics. Recently, similar developments have also been taking place in other π-systems bearing antiaromatic rings [83–85]. However, not only designed planar COTs but many other antiaromatic rings remain unexplored in detail. Thus, the π-systems bearing antiaromatic rings are potentially a gold mine of research exploration.

References

1. Willstätter R, Waser E (1911) Ber Dtsch Chem Ges 44:3423–3445
2. Cope AC, Overberger CG (1947) J Am Chem Soc 69:976
3. Reppe W, Schlichting O, Klager K, Toepel T (1948) Justus Liebigs Ann Chem 560:1–92
4. Nishinaga T (2009) Sci Synth 45a:383–406
5. Paquette LA (1975) Tetrahedron 31:2855–2883
6. Katz TJ (1960) J Am Chem Soc 82:3784–3785
7. Edelmann FT (1995) Angew Chem Int Ed Engl 34:2466–2488
8. Olah GA, Staral JS, Liang G, Paquette LA, Melega WP, Carmody MJ (1977) J Am Chem Soc 99:3349–3355
9. Nishinaga T, Komatsu K, Sugita N (1994) J Chem Soc Chem Commun 2319–2320
10. Nishinaga T, Komatsu K, Sugita N, Lindner HJ, Richter J (1993) J Am Chem Soc 115:11642–11643
11. Anet FAL, Bourn AJR, Lin YS (1964) J Am Chem Soc 86:3576–3577
12. Paquette LA (1993) Acc Chem Res 26:57–62
13. Breslow R (1968) Angew Chem Int Ed Engl 7:565–570
14. Breslow R (1973) Acc Chem Res 6:393–398
15. Hrovat DA, Borden WT (1992) J Am Chem Soc 114:5879–5881
16. Chen Z, Wannere CS, Corminboeuf C, Puchta R, Schleyer PR (2005) Chem Rev 105:3842–3888
17. Schleyer PR, Maerker C, Dransfeld A, Jiao H, Hommes NJRVE (1996) J Am Chem Soc 118:6317–6318
18. Karadakov PB (2008) J Phys Chem A 112:12707–12713
19. Wannere CS, Moran D, Allinger NL, Hess BA, Schaad LJ, Schleyer PR (2003) Org Lett 5:2983–2986
20. Klärner F (2001) Angew Chem Int Ed 40:3977–3981
21. Nishinaga T, Ohmae T, Iyoda M (2010) Symmetry 2:76–97
22. Steiner E, Fowler PW (2001) Chem Commun 2220–2221
23. Aita K, Ohmae T, Takase M, Nomura K, Nishinaga T (2013) Org Lett 15:3522–3525
24. Krebs A (1965) Angew Chem Int Ed Engl 4:953–954
25. Kiesewetter MK, Reiter RC, Stevenson CD (2005) Org Lett 7:2623–2626

26. Wong HNC, Sondheimer F (1976) Angew Chem Int Ed Engl 15:117–118
27. Wong HNC, Chan T-L, Sondheimer F (1978) Tetrahedron Lett 19:667–670
28. Wong HNC, Garratt PJ, Sondheimer F (1974) J Am Chem Soc 96:5604–5605
29. Huang NZ, Sondheimer F (1982) Acc Chem Res 15:96–102
30. Han J-W, Li X, Wong HNC (2015) Chem Rec 15:107–131
31. Matzger AJ, Vollhardt KPC (1998) Tetrahedron Lett 39:6791–6794
32. Fallah-Bagher-Shaidaei H, Wannere CS, Corminboeuf C, Puchta R, Schleyer PR (2006) Org Lett 8:863–866
33. Wang X, Wang R, Mak TCW, Wong HNC (1990) J Am Chem Soc 112:7790–7791
34. Elix JA, Sargent MV, Sondheimer F (1967) J Am Chem Soc 89:180
35. Elix JA, Sargent MV, Sondheimer F (1970) J Am Chem Soc 92:969–973
36. Pirrung MC, Krishnamurthy N, Nunn DS, McPhail AT (1991) J Am Chem Soc 113:4910–4917
37. Kabuto C, Oda M (1980) Tetrahedron Lett 21:103–106
38. Oda M, Oikawa H (1980) Tetrahedron Lett 21:107–110
39. Havenith RWA, Fowler PW, Jenneskens LW (2006) Org Lett 8:1255–1258
40. Kiesewetter MK, Reiter RC, Stevenson CD (2005) J Am Chem Soc 127:1118–1119
41. Klärner F-G, Ehrhardt R, Bandmann H, Boese R, Bläser D, Houk KN, Beno BR (1999) Chem Eur J 5:2119–2132
42. Soulen RL, Choi SK, Park JD (1973) J Fluor Chem 3:141–150
43. Einstein FWB, Willis AC, Cullen WR, Soulen RL (1981) J Chem Soc Chem Commun 526–528
44. Baldridge KK, Siegel JS (2001) J Am Chem Soc 123:1755–1759
45. Fowler PW, Havenith RWA, Jenneskens LW, Soncini A, Steiner E (2002) Angew Chem Int Ed 41:1558–1560
46. Dürr H, Klauck G, Peters K, von Schnering HG (1983) Angew Chem Int Ed Engl 22:332–333
47. Matsuura A, Komatsu K (2001) J Am Chem Soc 123:1768–1769
48. Nishinaga T, Uto T, Inoue R, Matsuura A, Treitel N, Rabinovitz M, Komatsu K (2008) Chem Eur J 14:2067–2074
49. Nishinaga T, Uto T, Komatsu K (2004) Org Lett 6:4611–4613
50. Hellwinkel D, Reiff G (1970) Angew Chem Int Ed Engl 9:527–528
51. Hou XL, Wong HNC (1987) J Am Chem Soc 109:1868–1869
52. Wilcox CF Jr, Uetrecht JP, Grohman KK (1972) J Am Chem Soc 94:2532–2533
53. Wilcox CF Jr, Uetrecht JP, Grantham GD, Grohmann KG (1975) J Am Chem Soc 97:1914–1920
54. Wilcox CF Jr, Grantham GD (1975) Tetrahedron 31:2889–2895
55. Obendorf SK, Wilcox CF Jr, Grantham GD, Hughes RE (1976) Tetrahedron 32:1327–1330
56. Wilcox CF Jr, Farley EN (1983) J Am Chem Soc 105:7191–7192
57. Wilcox CF Jr, Farley EN (1985) J Org Chem 50:351–356
58. Willner I, Rabinovitz M (1980) J Org Chem 45:1628–1633
59. Nakamura Y, Aratani N, Shinokubo H, Takagi A, Kawai T, Matsumoto T, Yoon ZS, Kim DY, Ahn TK, Kim D, Muranaka A, Kobayashi N, Osuka A (2006) J Am Chem Soc 128:4119–4127
60. Nakamura Y, Aratani N, Furukawa K, Osuka A (2008) Tetrahedron 64:11433–11439
61. Hiroto S, Furukawa K, Shinokubo H, Osuka A (2006) J Am Chem Soc 128:12380–12381
62. Cho S, Lim JM, Hiroto S, Kim P, Shinokubo H, Osuka A, Kim D (2009) J Am Chem Soc 131:6412–6420
63. Ohmae T, Nishinaga T, Wu M, Iyoda M (2010) J Am Chem Soc 132:1066–1074
64. Feng C-N, Kuo M-Y, Wu Y-T (2013) Angew Chem Int Ed 52:7791–7794
65. Sakamoto Y, Suzuki T (2013) J Am Chem Soc 135:14074–14077
66. Miller RW, Duncan AK, Schneebeli ST, Gray DL, Whalley AC (2014) Chem Eur J 20:3705–3711
67. Erdtman H, Högberg H-E (1968) Chem Commun 773–774
68. Erdtman H, Högberg H-E (1970) Tetrahedron Lett 11:3389–3392
69. Eskildsen J, Reenberg T, Christensen JB (2000) Eur J Org Chem 2000:1637–1640
70. Rathore R, Abdelwahed SH (2004) Tetrahedron Lett 45:5267–5270

71. Nielsen CB, Brock-Nannestad T, Hammershøj P, Reenberg TK, Schau-Magnussen M, Trpcevski D, Hensel T, Salcedo R, Baryshnikov GV, Minaev BF, Pittelkow M (2013) Chem Eur J 19:3898–3904
72. Hensel T, Trpcevski D, Lind C, Grosjean R, Hammershøj P, Nielsen CB, Brock-Nannestad T, Nielsen BE, Schau-Magnussen M, Minaev B, Baryshnikov GV, Pittelkow M (2013) Chem Eur J 19:17097–17102
73. Chernichenko KY, Sumerin VV, Shpanchenko RV, Balenkova ES, Nenajdenko VG (2006) Angew Chem Int Ed Engl 45:7367–7370
74. Fujimoto T, Suizu R, Yoshikawa H, Awaga K (2008) Chem Eur J 14:6053–6056
75. Nielsen CB, Brock-Nannestad T, Reenberg TK, Hammershøj P, Christensen JB, Stouwdam JW, Pittelkow M (2010) Chem Eur J 16:13030–13034
76. Nishiuchi T, Tanaka K, Kuwatani Y, Sung J, Nishinaga T, Kim D, Iyoda M (2013) Chem Eur J 19:4110–4116
77. Yuan C, Saito S, Camacho C, Irle S, Hisaki I, Yamaguchi S (2013) J Am Chem Soc 135:2884–2885
78. Yuan C, Saito S, Camacho C, Kowalczyk T, Irle S, Yamaguchi S (2014) Chem Eur J 20:2193–2200
79. Dadvand A, Cicoira F, Chernichenko KY, Balenkova ES, Osuna RM, Rosei F, Nenajdenko VG, Perepichka DF (2008) Chem Commun 5354–5356
80. Fujimoto T, Matsushita MM, Yoshikawa H, Awaga K (2008) J Am Chem Soc 130:15790–15791
81. Fujimoto T, Matsushita MM, Awaga K (2012) J Phys Chem C 116:5240–5245
82. Nishinaga T, Ohmae T, Aita K, Takase M, Iyoda M, Arai T, Kunugi Y (2013) Chem Commun 49:5354–5356
83. Kawase T, Fujiwara T, Kitamura C, Konishi A, Hirao Y, Matsumoto K, Kurata H, Kubo T, Shinamura S, Mori H, Miyazaki E, Takimiya K (2010) Angew Chem Int Ed Engl 49:7728–7732
84. Nishida J, Tsukaguchi S, Yamashita Y (2012) Chem Eur J 18:8964–8970
85. Chase DT, Fix AG, Kang SJ, Rose BD, Weber CD, Zhong Y, Zakharov LN, Lonergan MC, Nuckolls C, Haley MM (2012) J Am Chem Soc 134:10349–10352

Chapter 5
Synthesis, Crystal Structures, and Solid-State Optical Properties of Substituted Tetracenes

Chitoshi Kitamura

Abstract Various tetracenes possessing substituents, such as alkyl and ester groups, were prepared. In solution, their optical properties exhibited no marked differences. However, in the solid state, their photophysics differed with respect to the length, shape, number, substitution pattern, and kind of substituent groups, which is called crystallochromy. 1,4,7,10-Tetra(*n*-alkyl)tetracenes exhibited the solid-state colors of yellow, orange, and red. Both their molecular structures and molecular arrangements were classified into three groups. The molecular packing correlated well with the solid-state color. In 1,4,7,10-tetra(isoalkyl)tetracenes, the increase of fluorescence quantum yield in the solid state was remarkable. The isopropyl derivative showed the highest fluorescence quantum yield of 0.90 among any tetracene derivatives. In the case of *anti-/syn*-regioisomeric tetracene mixtures, the change in the solid-state color before and after recrystallization was observed. The X-ray analysis revealed that more crystalline molecules were *anti*-regioisomers. Molecular packing of 1,4-di(*n*-propyl)tetracene was a herringbone fashion which was different from that of 1,4,7,10-tetra(*n*-propyl)tetracene. Both 2,3-di(*n*-alkyl)- and 2,3,8,9-tetra(*n*-alkyl)tetracenes had low crystallinity and exhibited slight crystallochromy. In the crystal of 2,3-tetra(*n*-butyl)tetracene, there were not only a 1:1 disorder of two components of the molecule but also the repeating bilayer structure of the tetracene rings and alkyl groups. In tetracene octaester derivatives, the molecule with shorter alkyl chain length in the ester moiety was more redshifted. Relationship between the solid-state color and the interplanar distance between two neighboring tetracene rings in the crystal was observed. In conclusion, the alkyl side chains could act as not only spacer but also molecular orientation adjusters, to control the photophysical properties in the solid state.

Keywords Acenes • Substituent effects • Aggregate • Solid-state optical properties

---

C. Kitamura (✉)
Department of Materials Science, School of Engineering, The University of Shiga Prefecture, 2500 Hassaka-cho, Hikone, Shiga 522-8533, Japan
e-mail: kitamura.c@mat.usp.ac.jp

T. Akasaka et al. (eds.), *Chemical Science of π-Electron Systems*,
DOI 10.1007/978-4-431-55357-1_5

Acenes are linearly annelated polycyclic aromatic hydrocarbons (PAHs) and have received widespread attention because their electronic properties make them promising compounds as organic semiconducting materials [1]. Of acenes, anthracene plays an important role in emission [2] and photoconductivity [3], and tetracene and pentacene are the most promising as organic conductors because of their strong electronic interactions in the solid state [4]. However, owing to their low solubility in organic solvents, their practical use has been limited to the fabrication of low-cost devices over large areas. To address this problem, the functionalization of acenes has been performed by many groups to improve solubility, molecular ordering, and device performance. In recent years, we have prepared a variety of substituted acenes, and we found that the introduction of alkyl side chains into the acene framework facilitated their solubility in common organic solvents, even *n*-hexane. During the course of the work, we observed a most interesting and unexpected result: the color in the solid state differed for compounds with different alkyl side chains. Specifically, the materials ranged in color from yellow to red, whereas they were all orange in solution. Further, the solid-state fluorescence was also tuned by alkyl side chains. Part of our studies has been recently reviewed [5, 6]. This chapter will highlight the synthesis, solid-state optical properties, and crystal structures of substituted tetracenes.

5.1 Substituted Tetracenes

We synthesized various alkyl- and ester-substituted tetracenes and observed a very interesting result: in solution, there were no significant differences in their spectroscopic properties, absorption or fluorescence wavelengths, or fluorescence quantum yields. However, in the solid state, the substituted tetracenes showed significantly different optical properties that depended on the alkyl groups. Alkyl groups have little effect on the electronic structure at the molecular level. In MO calculations, researchers often try to reduce calculation costs by choosing the methyl group as representative alkyl group because they know that a long alkyl group takes long time to obtain the same result with the case of a short alkyl group. We speculated that not only the molecular electronic structure but also the electronic structure of the aggregate was an important factor that could explain our finding. When we searched the literature, we stumbled upon a nice term "crystallochromy," which defined as the dependence of solid-state color on molecular packing. This term was first brought by a BASF group [7], who studied a series of *N*,*N*′-substituted perylene bisimide pigments. Historically, the phenomenon of crystals of the same composition that exhibited different colors was first published in 1907 by Hantzsch, who used the term "chromoisomerism" [8]. In the dye and pigment industry, the term "polymorphism" is used [9] to describe variations in color that are strongly correlated with crystallochromy. In this chapter, we use the term crystallochromy to describe our results.

5.2 Synthesis of Substituted Tetracenes

5.2.1 *1,4,7,10-Tetra(*n*-alkyl)tetracenes*

We assumed that a strategy employing Diels–Alder reaction would be effective for the preparation of alkyl-substituted tetracenes because Gribble and co-workers had already reported the synthesis of 1,4,7,10-tetramethyltetracene [10] by a sequence involving treatment of a mixture of 3,6-dibromo-2,7-bis(tosyloxy)naphthalene and 2,5-dimethylfuran with PhLi, hydrogenation of the resulting bis(furan) adduct, and acid-catalyzed dehydration of the reduced product. However, the usage of Gribble's procedure led to the results of low yields of the Diels–Alder adducts (~10 %) with increasing alkyl side-chain length, and then we improved the method. We developed a new 2,6-naphthodiyne equivalent, that is, 3,6-bis(trimethylsilyl)naphthalene-2,7-bis(triflate) (**1**) [11, 12] (Scheme 5.1). Then, 1,4,7,10-tetra(*n*-alkyl)tetracenes **5a–e** ($R = C_2H_5$ to n-C_6H_{13}) were prepared as shown in Scheme 5.1 [11]. Reaction of **1** (0.5 equiv.) with 2,5-di(*n*-alkyl)furans **2** in the presence of KF and 18-crown-6 in THF at room temperature yielded bis(furan) adducts **3** by way of a double Diels–Alder reaction in 27–89 % yield. Hydrogenation of **3** over Pd/C and successive treatment of the reduced product **4** with acid furnished tetra(*n*-alkyl)tetracenes **5a–e** in 27–55 % two-step yield from **3**.

5.2.2 *1,4,7,10-Tetra(isoalkyl)tetracenes*

We thought to introduce isoalkyl groups as branched alkyl side chains instead of *n*-alkyl groups because we expected that the effects of the bulkiness and shape of the isoalkyl groups can modulate the molecular arrangement in the solid state. Similar to the preparation of **5**, 1,4,7,10-tetra(isoalkyl)tetracenes **9a–c** ($R = i$-C_3H_7 to i-C_5H_{11}) were synthesized starting from reaction of **1** with 2,5-di(isoalkyl)furans **6** in the presence of KF and 18-crown-6 (Scheme 5.2) [13].

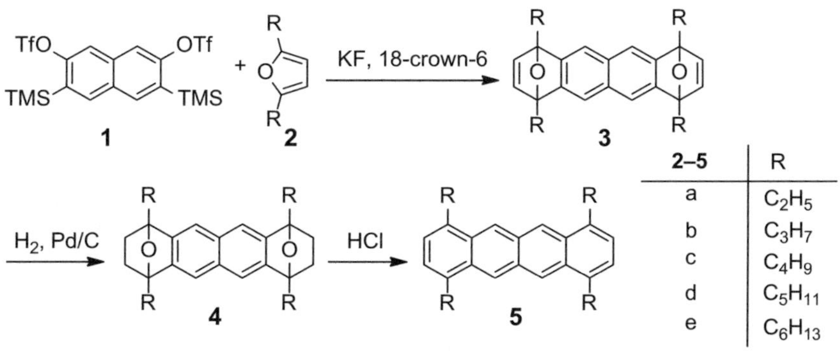

Scheme 5.1 Preparation of 1,4,7,10-tetra(*n*-alkyl)tetracenes (**5**)

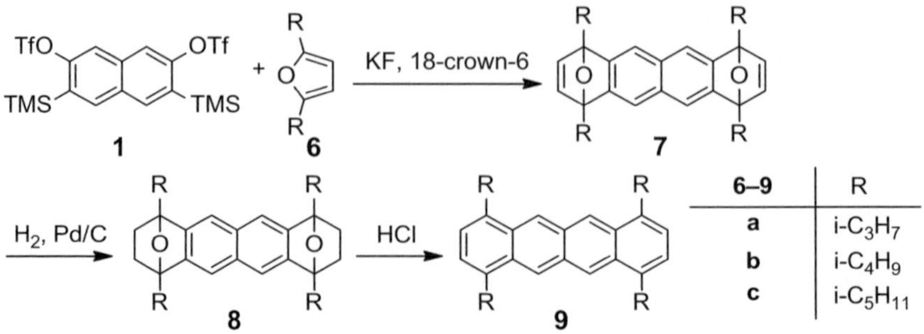

Scheme 5.2 Preparation of 1,4,7,10-tetra(isoalkyl)tetracenes (**9**)

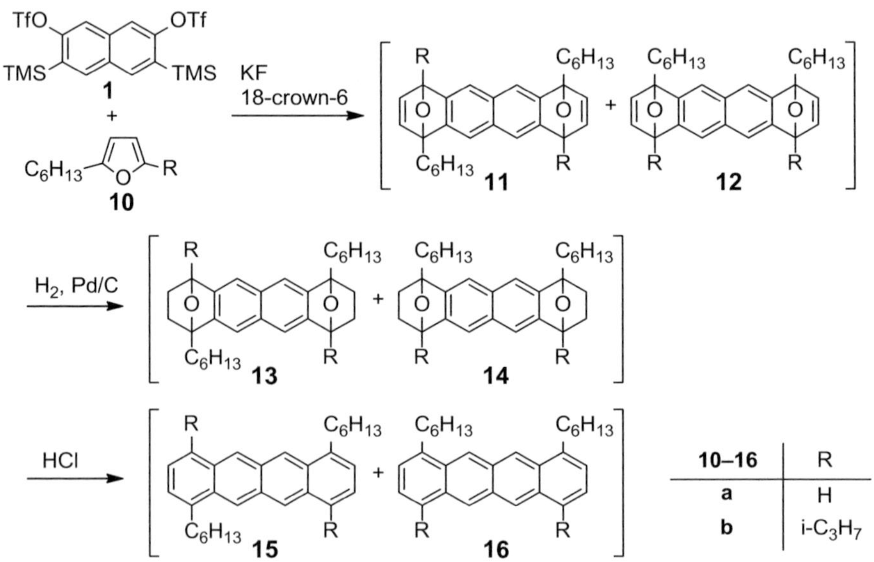

Scheme 5.3 Preparation of *anti-/syn*-regioisomeric tetracene mixtures **15/16**

5.2.3 **Anti – /syn-*regioisomeric Tetracene Mixtures***

When an unsymmetric furan was used in Scheme 5.1, two different types of products should be formed simultaneously. We tried the sequence starting from the reaction of **1** with unsymmetric furans, such as 2-*n*-hexylfuran (**10a**) and 2-*n*-hexyl-5-isopropylfuran (**10b**) in the presence of KF and 18-crown-6, and obtained bimolecular mixtures composed of **15** (*anti*-form) and **16** (*syn*-form), which were referred to as **15/16** (Scheme 5.3) [14]. Although the mixtures could not be separated due to the same properties, we found an interesting change in the solid-state color before and after recrystallization as stated later.

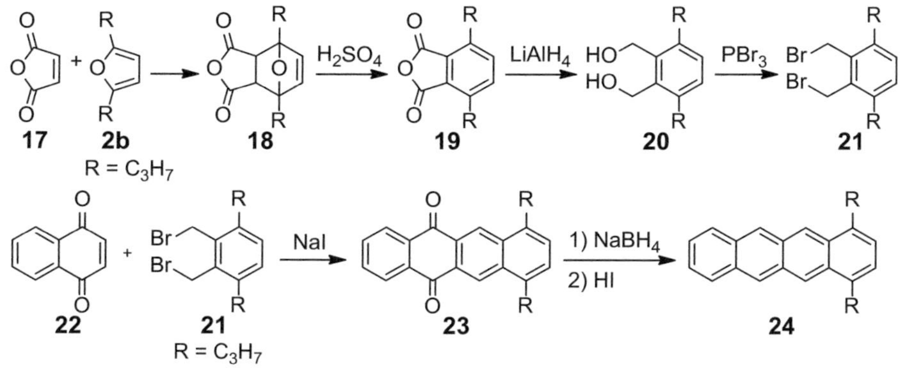

Scheme 5.4 Preparation of 1,4-di(*n*-propyl)tetracene (**24**)

5.2.4 1,4-Di(n-alkyl)tetracene

To examine the substituent effect of reducing the number of alkyl side chains, we next studied 1,4-di(*n*-alkyl)tetracene, which is a low-symmetry structure compared with 1,4,7,10-tetra(*n*-alkyl)tetracene. We performed the preparation of 1,4-di(*n*-propyl)tetracene (**24**) by way of McOmie's quinone synthesis [15]. 1,2-Bis(bromomethyl)-4,6-di(*n*-propyl)benzene (**21**) was synthesized using four-step reactions starting from a Diels–Alder reaction between maleic anhydride (**17**) and 2,5-di(*n*-propyl)furan (**2b**) in a nonpolar solvent such as Et_2O (Scheme 5.4) [16]. McOmie's tetracenequinone synthesis mediated by the reaction between 1,4-naphthoquinone (**22**) and 1,2-bis(bromomethyl)benzene **21** in the presence of NaI provided tetracenequinone **23**. A sequence involving hydride reduction of **60** with $NaBH_4$ and subsequent treatment of the resulting diol with HI gave 1,4-di(*n*-alkyl)tetracene **24** ($R = n\text{-}C_3H_7$).

5.2.5 2,3-Di(n-alkyl)tetracenes

To investigate the effects of the substitution position, we attempted to prepare new tetracenes having alkyl side chains at locations other than the 1-, 4-, 7-, and 10-positions. However, we noticed that the synthesis of the requisite 3,4-di(*n*-alkyl)furans was very difficult, and we developed another process for preparing alkyl-substituted tetracenes. A Diels–Alder reaction between 1,4-anthraquinone (**25**) and thiophene 1,1-dioxides **26** in refluxing acetic acid followed by loss of sulfur dioxide and oxidation afforded 2,3-di(*n*-alkyl)-5,12-tetracenequinones **27** (Scheme 5.5) [17]. Finally, 2,3-di(*n*-alkyl)tetracenes **28a–d** ($R = n\text{-}C_3H_7$ to $n\text{-}C_6H_{13}$) were obtained by a one-pot two-step sequence involving hydride reduction of quinones **27** with $NaBH_4$ and subsequent addition of $SnCl_2$ in conc. HCl in excellent yields [18].

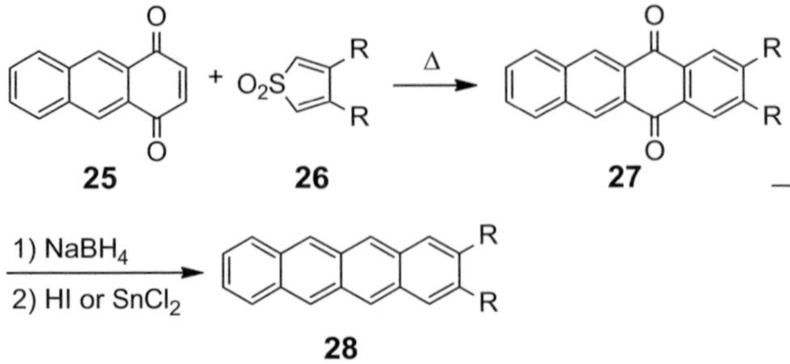

26–28	R
a	C_3H_7
b	C_4H_9
c	C_5H_{11}
d	C_6H_{13}

Scheme 5.5 Preparation of 2,3-di(*n*-alkyl)tetracenes (**28**)

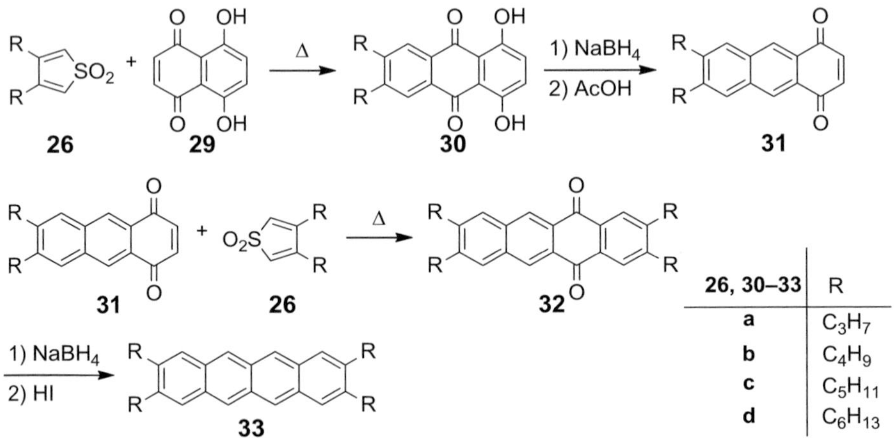

26, 30–33	R
a	C_3H_7
b	C_4H_9
c	C_5H_{11}
d	C_6H_{13}

Scheme 5.6 Preparation of 2,3,8,9-tetra(*n*-alkyl)tetracenes (**33**)

*5.2.6 2,3,8,9-Tetra(*n*-alkyl)tetracenes*

With the development of above 2,3-di(*n*-alkyl)tetracenes, the synthesis of 2,3,8,9-tetra(*n*-alkyl)tetracenes was carried out. A Diels–Alder reaction between thiophene 1,1-dioxides **26** and 5,8-dihydroxy-1,4-naphthoquinone (**29**) followed by loss of sulfur dioxide and oxidation furnished 6,7-di(*n*-alkyl)-1,4-dihydroxy-9,10-anthraquinones **30**, which were then transformed into 6,7-di(*n*-alkyl)-1,4-anthraquinones **31** by a one-pot sequence involving hydride reduction and acid-catalyzed dehydration (Scheme 5.6) [17]. Then, a second Diels–Alder reaction between **26** and anthraquinones **31** followed by loss of sulfur dioxide and oxidation produced 2,3,8,9-tetra(*n*-alkyl)-5,12-tetracenequinones **32**. These tetracenequinones were converted into 2,3,8,9-tetra(*n*-alkyl)tetracenes **33a–d** ($R = n\text{-}C_3H_7$ to $n\text{-}C_6H_{13}$) using a procedure similar to that described in Scheme 5.5.

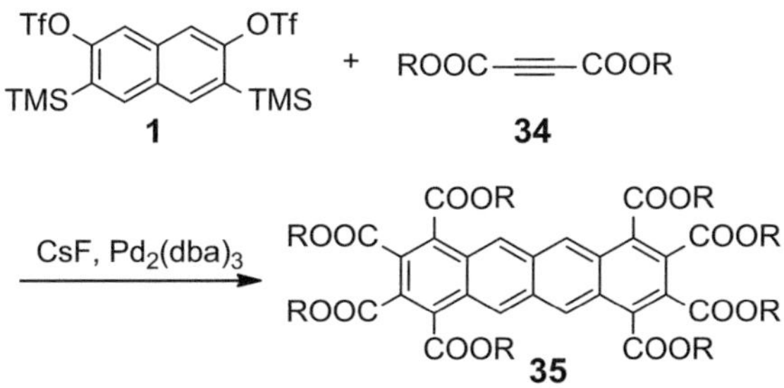

34,35	R
a	CH_3
b	C_2H_5
c	C_3H_7

Scheme 5.7 Preparation of octa(*n*-alkyl)tetracene-1,2,3,4,7,8,9,10-octacarboxylates (**35**)

*5.2.7 Octa(*n*-alkyl)tetracene-1,2,3,4,7,8,9,10-octacarboxylates*

To investigate the effects of substituents other than alkyl groups, we attempt to synthesize new tetracenes having ester groups. Treatment of **1** with di(*n*-alkyl) acetylenedicarboxylates **34** in the presence of CsF and catalytic amounts of $Pd_2(dba)_3$ afforded tetracene octaester derivatives **35a–c** ($R = CH_3$ to n-C_3H_7) in, at best, 16 % yield (Scheme 5.7) [12].

5.3 Optical Properties of Substituted Tetracenes

5.3.1 Solid-State Color

In solution, all the alkyl-substituted tetracenes, except for ester derivatives, showed no significant differences in their spectroscopic properties containing absorption or fluorescence wavelengths and fluorescence quantum yields. On the other hand, in the solid state, each substituted tetracene had an intrinsic color, depending upon the substituent groups, with exception that only 1,4,7,10-tetra(*n*-butyl)tetracene (**5c**) had two polymorphs, such as red and yellow solids [19]. The solid-state color of all the tetracenes ranged through yellow, orange, and red. For comparison, Kubelka–Munk spectra of tetracene derivatives were measured. Their wavelengths of absorption edge (λ_{edge}) and colors are listed in Table 5.1.

In 1,4,7,10-tetra(*n*-alkyl)tetracenes **5a–e**, the ethyl and *n*-butyl derivatives (**5a** and **5c**-yellow) were yellow, the *n*-propyl and *n*-pentyl derivatives (**5b** and **5d**) were orange, and the other *n*-butyl and the *n*-hexyl derivatives (**5d**-red and **5e**) were red [11]. The λ_{edge} values were in the following order: **5a** (537 nm) < **5c**-yellow (552 nm) < **5b** (580 nm) < **5d** (588 nm) < **5c**-red (619 nm) < **5e** (620 nm). The structural regularity of the alkyl side-chain length resembled the odd–even effect that has frequently been observed in liquid crystals [20].

In 1,4,7,10-tetra(isoalkyl)tetracenes **9a–c**, their color had little correlation to those of their corresponding *n*-alkyl derivatives **5b–d** [13]. Thus, the isopropyl,

Table 5.1 Solid-state absorption and fluorescence properties of substituted tetracenes

	Absorption[a]		Fluorescence	
Compd	λ_{edge} (nm)	Color	λ_{em}[b] (nm)	Φ_F[c]
5a	537	Yellow	564	0.40
5b	580	Orange	588	0.22
5c-red	619	Red	637	0.16
5c-yellow	552	Yellow	569	0.34
5d	588	Orange	581	0.13
5e	620	Red	623	0.20
9a	535	Yellow	560	0.90
9b	584	Red	590	0.42
9c	554	Orange-yellow	569	0.18
15a/16a before recryst	545	Orange-yellow	551	0.04
15a/16a after recryst	530	Yellow	563	0.08
15b/16b before recryst	546	Orange-yellow	569	0.31
15b/16b after recryst	565	Orange	560	0.27
24	560	Orange	557	0.23
28a	560	Orange-yellow	593	0.01
28b	565	Orange	624	0.01
28c	570	Orange	574	0.04
28d	564	Orange	589	0.05
33a	566	Orange	610	0.07
33b	546	Yellow	570	0.12
33c	551	Yellow	574	0.04
33d	534	Yellow	569	0.18
35a	605	Red	625	0.06
35b	582	Orange	580	0.05
35c	554	Orange-yellow	563	0.13

[a]Kubelka–Munk spectra in KBr pellets
[b]Excited at 325 nm
[c]Absolute quantum yield in powder form

isobutyl, and isopentyl derivatives (**9a**, **9b**, and **9c**) were yellow, red, and orange yellow, respectively. The order of increase of λ_{edge} was as follows: **9a** (535 nm) < **9c** (554 nm) < **9b** (584 nm).

In the case of bimolecular mixtures of **15a/16a** and **15b/16b**, the change in the solid-state color before and after recrystallization with Et_2O was observed [14]. For example, the solids of **15a/16a** before and after recrystallization were orange yellow (λ_{edge} = 545 nm) and yellow (λ_{edge} = 530 nm), respectively. On the other hand, those of **15b/16b** before and after recrystallization were orange yellow (λ_{edge} = 546 nm) and orange (λ_{edge} = 565 nm), respectively. These results probably showed an unbalanced distribution of *anti-/syn*-regioisomers in which the more crystalline component would dominate the solid-state optical properties.

The 1,4-di(*n*-alkyl)tetracene **24** (R = *n*-C_3H_7) was orange [16]. Although 1,4,7,10-tetra(*n*-propyl)tetracene (**5b**) was orange, there was a difference in absorption edge (λ_{edge}) between **24** ($\lambda_{edge} = 560$ nm) and **5b** ($\lambda_{edge} = 580$ nm), suggesting different molecular arrangements between both derivatives.

Both 2,3-di(*n*-alkyl)tetracenes (**28a–d**) and 2,3,8,9-tetra(*n*-alkyl)tetracenes (**33a–d**) showed slight changes in the solid-state color with alkyl chain length, in contrast with 1,4,7,10-tetra(*n*-alkyl)tetracenes **5a–e** [17]. In 2,3-di(*n*-alkyl)tetracenes **28a–d**, the *n*-butyl, *n*-pentyl, and *n*-hexyl derivatives (**28b**, **28c**, and **28d**) were orange (λ_{edge} for **28b**, 565 nm; for **28c**, 570 nm; for **28c**, 564 nm), while the *n*-propyl derivative (**28a**) was orange yellow ($\lambda_{edge} = 560$ nm). Then, in 2,3,8,9-tetra(*n*-alkyl)tetracenes **33a–d**, the *n*-butyl, *n*-pentyl, and *n*-hexyl derivatives (**33b**, **33c**, and **33d**) were yellow (λ_{edge} for **33b**, 546 nm; for **33c**, 551 nm; for **33d**, 534 nm), while the *n*-propyl derivative (**33a**) was orange ($\lambda_{edge} = 566$ nm).

In tetracene octaester derivatives **35a–c**, the methyl, ethyl, and *n*-propyl ester derivatives (**35a**, **35b**, and **35c**) were red ($\lambda_{edge} = 605$ nm), orange ($\lambda_{edge} = 582$ nm), and orange yellow ($\lambda_{edge} = 554$ nm), respectively, showing that the color was redshifted with shorter alkyl chain length in the ester moiety [12].

In the results above, we witnessed interesting variations in the solid-state color of tetracene derivatives, which can be explained by crystallochromy. Unfortunately, no general approach exists for understanding crystallochromy because it deals with complicated intermolecular interactions that are too difficult to formulate. Chemists can narrowly understand Kasha's simple point–dipole molecular exciton model [21, 22], which is employed to explain the electronic properties of *J* and *H* aggregates, by using the exciton-displacement energy ($\Delta E = |\mu|^2(1 - 3\cos^2\theta)/r^3$, where μ is the transition dipole and θ and r are the angle and distance between two transition dipoles, respectively). However, we require attention that the treatment is user-friendly but provides qualitative predictions in three-dimensional molecular arrangement [11]. At the present, the calculations using the transition–density–fragment interaction combined with transfer integral method [23] exhibited good agreement with our experimental data [24], suggesting that the treatment of not only exciton coupling but also π-overlap is important for understanding crystallochromy.

5.3.2 Solid-State Fluorescence

While the solid-state color depended upon the length, shape, number, and site of the alkyl side chains, the changes in the fluorescence properties in the solid state were also found. Recently, the solid-state fluorescence of organic molecules has received significant attention due to its importance in optoelectronic devices [25]. Accordingly, the development of intensely luminescent molecules is of great interest. The solid-state fluorescence maxima (λ_{em}) and fluorescence quantum yields (Φ_F) of tetracene derivatives are also listed in Table 5.1.

In accordance with their solid-state colors, the substituted tetracenes emitted yellow, orange, and red light. In solution, the tetracene derivatives did not exhibit marked differences with respect to fluorescence peaks and fluorescence quantum yield (0.03–0.13). On the other hand, the Φ_F values covered a wide range (0.01–0.90). Of particular note is the highest quantum yield ($\Phi_F = 0.90$) for 1,4,7,10-tetra(isopropyl)tetracene (**9a**) among any tetracene derivatives [13]. The value was far superior to the values for the second largest groups of 1,4,7,10-tetra(isobutyl)tetracene **9b** ($\Phi_F = 0.42$) and 1,4,7,10-tetra(ethyl)tetracene **5a** ($\Phi_F = 0.40$). Many other tetracene derivatives had low Φ_F values (<0.20).

In general, many organic chromophores are nonluminescent in the solid state, even if they are highly emissive in solution [25]. This phenomenon, in which the excited states of aggregated molecules decay through non-radiative pathways, is commonly referred to as "concentration quenching" [26]. As for **9b**, the structural features of the isopropyl group, which has bulkiness and rigidity, may cause the enhancement of fluorescence quantum yield. In addition, the molecular arrangement of the tetracene moieties would be another important factor. Our results showed that the fluorescent behavior in the solid state can be adjusted by the introduction of substituents.

With respect to the wavelengths of maximum fluorescence (λ_{em}), the order similar to the case of solid-state absorption was observed in many molecules.

In 1,4,7,10-tetra(*n*-alkyl)tetracenes **5a–e**, the order of increase of λ_{em} was as follows: **5a** (564 nm) < **5c**-yellow (569 nm) < **5d** (581 nm) < **5b** (588 nm) < **5e** (623 nm) < **5c**-red (637 nm) [11].

In 1,4,7,10-tetra(isoalkyl)tetracenes **9a–c**, that of λ_{em} was as follows: **9a** (560 nm) < **9c** (569 nm) < **9b** (590 nm) [13].

In the case of bimolecular mixtures of **15a/16a** and **15b/16b**, the change in the λ_{em} before and after recrystallization with Et_2O was observed [14]. In **15a/16a**, the λ_{em} values before and after recrystallization were 551 nm and 563 nm, respectively. On the other hand, in **15b/16b**, the λ_{em} values before and after recrystallization were 569 nm and 560 nm, respectively.

In 1,4-di(*n*-propyl)tetracene **24**, the λ_{em} value was 557 nm [16], which was smaller than that for 1,4,7,10-tetra(*n*-propyl)tetracene **5b** ($\lambda_{em} = 588$ nm). Similar to the solid-state absorption, the substance with more redshifted color exhibited a longer fluorescence maximum.

In 2,3-di(*n*-alkyl)tetracenes (**28a–d**) and 2,3,8,9-tetra(*n*-alkyl)tetracenes (**33a–d**), direct relations were not observed between the λ_{edge} and the λ_{em} values [17]. In the case of 2,3-di(*n*-alkyl)tetracenes **28a–d**, the *n*-butyl derivative **28b** had a more redshifted λ_{em} of 624 nm than those for **28a** ($\lambda_{em} = 593$ nm), **28c** ($\lambda_{em} = 574$ nm), and **28d** ($\lambda_{em} = 589$ nm). On the other hand, in 2,3,8,9-tetra(*n*-alkyl)tetracenes **33a–d**, the *n*-propyl derivative **33a** showed a more redshift ($\lambda_{em} = 610$ nm) compared with **33b** ($\lambda_{em} = 570$ nm), **33c** ($\lambda_{em} = 574$ nm), and **33d** ($\lambda_{em} = 569$ nm).

In tetracene octaester derivatives **35a–c**, the order of increase of λ_{em} was as follows: **35c** (563 nm) < **35b** (580 nm) < **35a** (625 nm), correlating well with the order of the λ_{edge} [12].

5.4 Crystal Structure of Substituted Tetracenes

5.4.1 Molecular Structure

It is well known that non-substituted acenes crystallize in the herringbone structure, which is caused by the C–H interactions [1]. However, the introduction of substituents into acene core would change the stacking structure because of the addition of new intermolecular interactions between substituents. We were interested in how the substituted tetracene exists in solid. Fortunately, recrystallization of many of the substituted tetracenes gave good single crystals, and X-ray crystallographic analysis was carried out to clarify the essence of crystallochromy. All the crystals measured had no solvents, and the disorder in the alkyl chains was scarcely observed. In addition, the tetracene rings were essentially planar.

In 1,4,7,10-tetra(*n*-alkyl)tetracenes **5a–e**, all of the molecules possessed a center of symmetry [11]. Further, all of the alkyl side chains adopted and all-*trans*-planar (zigzag) conformation. Interestingly, all of the zigzag planes were either coplanar with, or perpendicular to, the tetracene ring (Fig. 5.1). This configuration resulted in

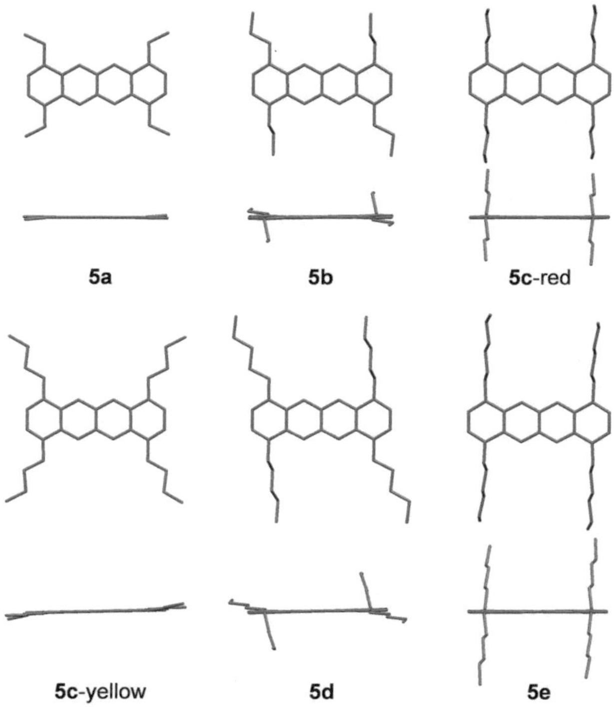

Fig. 5.1 Molecular structures of 1,4,7,10-tetra(*n*-alkyl)tetracenes (**5**), showing top view (*upper*) and side view (*lower*)

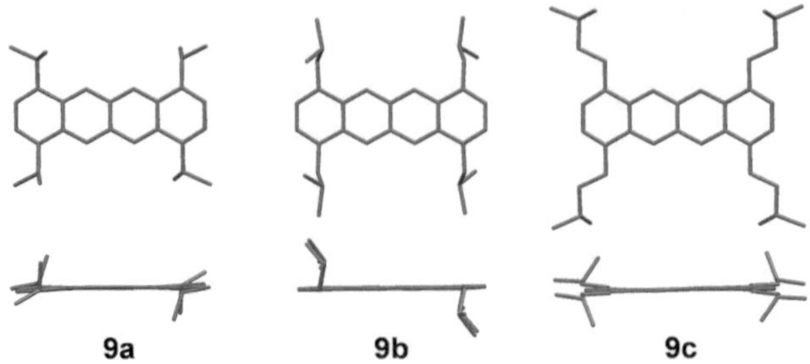

Fig. 5.2 Molecular structures of 1,4,7,10-tetra(isoalkyl)tetracenes (**9**), showing top view (*upper*) and side view (*lower*)

the formation of three different types of molecular structures. We categorized them into a planar form for **5a** and **5c**-yellow, a semi-chair form for **5b** and **5d**, and a chair form for **5c**-red and **5e**. In the planar form, the alkyl chains took a coplanar conformation with the tetracene ring. In the chair form, a pair of alkyl groups at the 1- and 10-positions extended upward and another pair of alkyl groups at the 4- and 7-positions extended downward out of the tetracene ring. The semi-chair form looked like an intermediate conformer between the planar and chair forms. Thus, a pair of alkyl groups at the 1- and 7-positions took a coplanar conformation with the tetracene ring and another pair of alkyl groups at the 4- and 10-positions extended upward and downward out of the tetracene plane.

In 1,4,7,10-tetra(isoalkyl)tetracenes **9a–c**, all of the molecules had a center of symmetry [13]. The isoalkyl groups were characterized by alkyl conformations that significantly varied in length (Fig. 5.2). In molecule **9b**, four isobutyl groups had a *gauche* conformation in which the two groups at the 1- and 4-positions extended upward and the others at the 7- and 10-positions extended downward. In molecules **9a** and **9c**, the conformations of the four isoalkyl groups looked similar. However, there were small differences in the directions of the two terminal methyl groups of the isoalkyl groups. Thus, in **9a**, two terminal methyl groups of the two isopropyl groups as the 1- and 4-positions pointed downward and two methyl groups of the other two isopropyl groups at the 7- and 10-positions pointed upward. On the other hand, in **9c**, two terminal methyl groups of the two isopropyl groups as the 1- and 10-positions pointed upward and two methyl groups of the other two isopropyl groups at the 7- and 10-positions pointed downward.

In the case of bimolecular mixtures of **15a/16a** and **15b/16b**, slow evaporation from Et_2O solutions gave single crystals, whose X-ray analysis revealed that the molecules in the crystals were *anti*-regioisomers **15a** and **15b** (Fig. 5.3) [14]. Since we observed that each **15a** and **15** had disordered sets of alkyl groups, minor sets were omitted in Fig. 5.3 for clarity. We could not obtain any single crystals of the *syn*-regioisomer, which might have low crystallinity. In the crystals, all of the molecules possessed a center of symmetry. In molecules **15a** and **15b**, their *n*-hexyl groups, except for the terminal methyl groups in **15a**, adopted zigzag conformations,

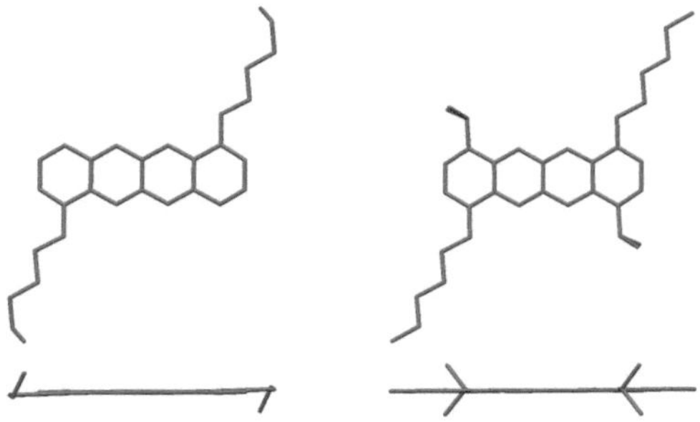

15a 15b

Fig. 5.3 Molecular structures of *anti*-regioisomeric tetracene derivatives (**15**), showing top view (*upper*) and side view (*lower*)

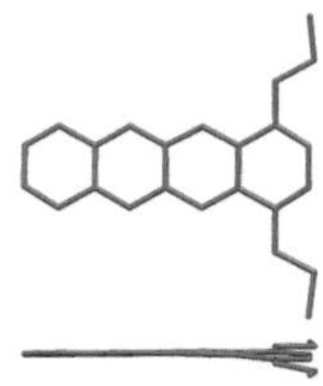

24

Fig. 5.4 Molecular structure of 1,4-di(*n*-propyl)tetracene (**24**), showing top view (*upper*) and side view (*lower*)

and the zigzag planes were almost coplanar with the tetracene ring. The *n*-hexyl conformations in **15a** and **15b** were completely different from those in 1,4,7,10-tetra(*n*-hexyl)tetracene (**5e**).

In 1,4-di(*n*-propyl)tetracene (**24**), the molecule had no symmetry in contrast to other substituted tetracenes. The two *n*-propyl groups took a coplanar conformation with the tetracene ring as shown in Fig. 5.4 [16], whose structural feature was different from that of 1,4,7,10-tetra(*n*-propyl)tetracene (**5b**), in which a pair of two *n*-propyl groups at the 1- and 7-positions took a coplanar conformation with the tetracene ring whereas another pair of two *n*-propyl groups at the 4- and 10-positions took a perpendicular conformation. The molecular structure of **24** was rather similar to that of 1,4-di(*n*-propyl)anthracene [27].

Both the 2,3-di(*n*-alkyl)tetracenes (**28a–d**) and 2,3,8,9-tetra(*n*-alkyl)tetracenes (**33a–d**) had lower crystallinity than other alkyl-substituted tetracenes. We attempted to prepare single crystals; however, almost all attempts were unsuccessful. In only the case of 2,3-di(*n*-butyl)tetracene (**28b**), we could isolate one single-crystal-like red prism by slow evaporation from a toluene solution and performed X-ray structural analysis [18]. The molecule had a center of symmetry, and half of the formula unit was crystallographically independent similar to many other alkyl-substituted tetracenes. However, the molecular structure seemed strange since the tetracene nucleus had four *n*-butyl side chains (Fig. 5.5). We considered the structure as a 1:1 disorder of two components of 2,3-di(*n*-butyl)tetracene, which were positioned to each other in the opposite direction: the *n*-butyl groups were disordered with 50 % occupancies for each set. Interestingly, the molecular structure was a new form. The two *n*-butyl groups took a zigzag conformation with

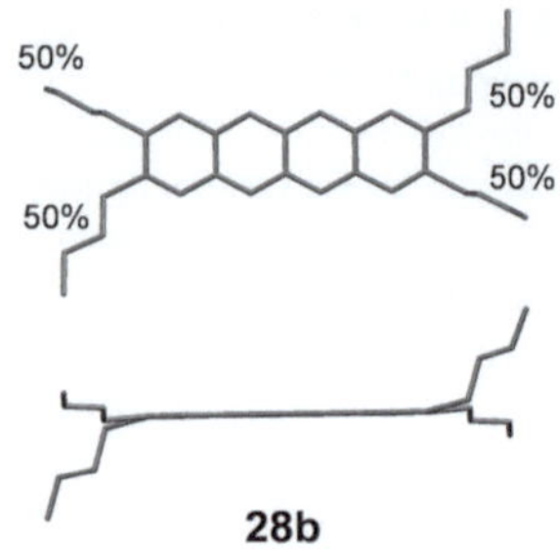

28b

Fig. 5.5 Molecular structure of 2,3-di(*n*-butyl)tetracene (**28b**), showing top view (*upper*) and side view (*lower*)

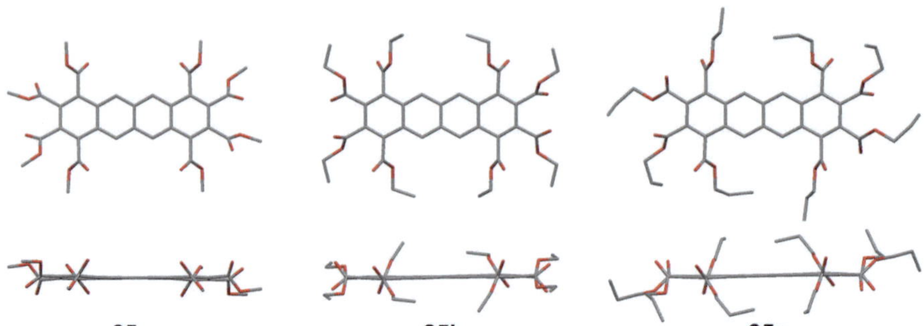

35a 35b 35c

Fig. 5.6 Molecular structures of octa(*n*-alkyl)tetracene-1,2,3,4,7,8,9,10-octacarboxylates (**35**), showing top view (*upper*) and side view (*lower*)

the tetracene ring. However, the one *n*-butyl group adopted a coplanar conformation with the tetracene ring, whereas the other neighboring *n*-butyl group was almost perpendicular to the tetracene ring.

In tetracene octaester derivatives **35a–c**, all of the molecules had a center of symmetry [12]. Different conformational changes in the ester moiety, caused by the rotation around the C–C bond between tetracene and the ester moieties, were observed (Fig. 5.6). In all the molecules, half of the carbonyl oxygen atoms were directed upward, and the other half were directed downward. The direction of the carbonyl group differed with respect to the position of the ester moiety in the molecule. Differences in the direction of the carbonyl oxygen atom are known to affect the electronic properties of the molecular surface [28].

5.4.2 *Molecular Packing*

We observed a variety of conformational changes in substituent groups within the molecule. However, the molecular structure alone could not explain the difference in the solid-state color because MO calculations based on one molecule gave

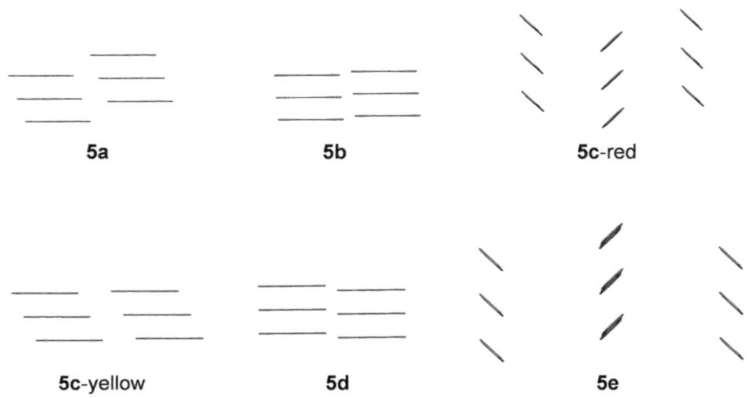

Fig. 5.7 Packing diagrams of the tetracene rings of 1,4,7,10-tetra(*n*-alkyl)tetracenes (**5**)

no fruitful information on crystallochromy. To understand crystallochromy, the analysis between molecular packing of tetracene units and the solid-state color is essential.

In 1,4,7,10-tetra(*n*-alkyl)tetracenes **5a–e**, we observed interesting information on molecular packing (Fig. 5.7) [11]. Molecules of **5a**, **5b**, **5c**-yellow, and **5d** adopted a slipped-parallel arrangement. In contrast, molecules of **5c**-red and **5e** were stacked in a herringbone-like fashion. The interplanar tilt angles between tetracene rings in two adjacent columns were 92.8° and 93.1° for **5c**-red and **5e**, respectively. In the stacking direction, the interplanar distance between two neighboring tetracene rings (*d*) was 3.43–3.56 Å (*d* for **5a**, 3.53 Å; for **5b**, 3.51 Å; for **5c**-red, 3.43 Å; for **5c**-yellow, 3.56 Å; for **5d**, 3.51 Å; for **5e**, 3.43 Å). There was a π-overlap in only **5a**, although **5a** and **5c**-yellow were similar in color, suggesting that not a small part of molecular packing but large part of molecular packing is important for crystallochromy. The nature of molecular packing probably depended on the balance of intermolecular interactions between tetracene moieties, such as face-to-face (π–π stacking) and edge-to-face (CH–π) interactions, and on the degree of self-assembly of the alkyl side chain. Similarities between the appearance of compounds **5a** and **5c**-yellow, **5b** and **5d**, and **5c**-red and **5e** were observed. Thus, it was certain that similar molecular packing correlated with similar color in the solid state. We thought that the color in the solid state closely related to alkyl side-chain length, alkyl conformation, and molecular arrangement. In other words, the length of the alkyl side chain could control alkyl conformation to give conformational polymorphs. Further, the individual polymorphs behaved to assemble unique molecular aggregates, which ultimately defined the solid-state color.

In 1,4,7,10-tetra(isoalkyl)tetracenes **9a–c**, the molecular packing was completely different to each other and also different from that for *n*-alkyl tetracene derivatives **5a–e** (Fig. 5.8) [13]. Molecules **9a** and **9c** appeared to be stacked in herringbone-like structures. The interplanar tilt angles between tetracene rings in two adjacent

9a 9b 9c

Fig. 5.8 Packing diagrams of the tetracene rings of 1,4,7,10-tetra(isoalkyl)tetracenes (**9**)

15a 15b

Fig. 5.9 Packing diagrams of the tetracene rings of *anti*-regioisomeric tetracene derivatives (**15**)

columns were 66.9° and 115.0° for **9a** and **9c**, respectively. In contrast, the tetracene rings in **9b** adopted a slipped-parallel arrangement of stacked molecular sheets. In molecules of **9a**, there was a slight π-overlap. Molecules of **9b** exhibited a large π-overlap, which would be a major factor contributing to the redshifts observed in the absorption and fluorescence spectra in the solid state. In contrast, there was no π-overlap in molecule **9c**. The interplanar distance between two neighboring tetracene rings (*d*) was close to 3.5 Å (*d* for **9a**, 3.46 Å; for **9b**, 3.46 Å; for **9c**, 3.55 Å). We already stated that the fluorescence quantum yield in the solid state of **9a** ($\Phi_F = 0.90$) was the highest among all tetracene derivatives. In the molecular packing of **2a**, molecules were slightly further away from neighboring molecules in one column along the stacking direction, although the center-to-center distances between adjacent columns were relatively short compared with those of **9b** and **9c**. A longer distance between neighboring molecules in one column along the stacking direction might interrupt concentration quenching leading to a higher quantum yield in the solid state. In addition, we believe that crystal rigidity is another possible factor [29]. Shorter isoalkyl groups should increase crystal rigidity. We concluded that both the relative positional relationship between neighboring tetracene units and the crystal rigidity were critical factors for controlling the solid-state fluorescence quantum yield.

In *anti*-regioisomers **15a** and **15b**, the molecules crystallized in layered structures, while **15a** and **15b** appeared to be stacked in a staircase-type fashion and a brickwork-type fashion, respectively (Fig. 5.9) [14]. Within the layer, one molecule of **15a** was surrounded by six neighboring molecules, and one molecule of **15b** was environed by eight neighboring molecules. There was no π-overlap in both molecules of **15a** and **15b**. The existence of isopropyl groups in **15b** broadened the interplanar distance between two neighboring tetracene rings (*d*) (*d* for **15a**, 3.56 Å; for **15b**, 3.79 Å). The differences in the molecular packing would probably cause the differences in the solid-state optical properties.

Fig. 5.10 Packing diagrams of the tetracene rings of 1,4-di(*n*-propyl)tetracene (**24**)

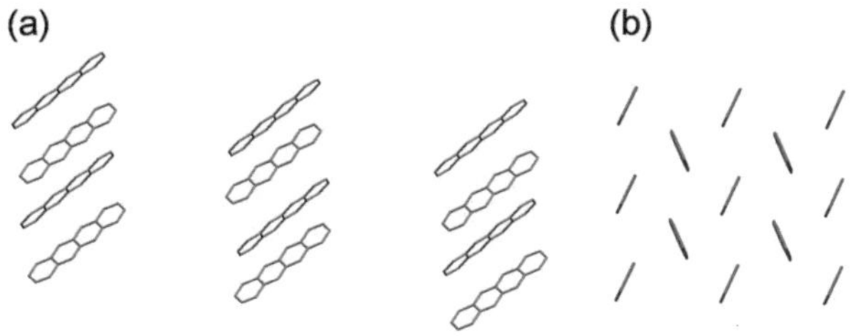

Fig. 5.11 Packing diagrams of the tetracene rings of 2,3-di(*n*-butyl)tetracene (**28b**), showing (**a**) layer-by-layer structure and (**b**) herringbone structure within a layer

In 1,4-di(*n*-propyl)tetracene (**24**), molecules adopted a herringbone arrangement (Fig. 5.10) which was completely different from that of 1,4,7,10-tetra(*n*-propyl)tetracene (**5b**) having a slipped-parallel arrangement [16]. The interplanar tilt angle between tetracene rings in two adjacent columns was 72.4°. There was no π-overlap between the two neighboring molecules. The tetracene rings along the column direction slipped relative to each other along the long molecular axis by 0.65 Å and along the short molecular axis by 4.48 Å. In addition, the interplanar distance between two neighboring tetracene rings (*d*) was 3.27 Å, which was shorter than those for other alkyl-substituted tetracenes (3.43–3.79 Å). The difference in molecular packing arises from the nonexistence of alkyl side chains at the 7- and 10-positions in **24**. Thus, the presence of aromatic hydrogen atoms at the 7- and 10-positions led to an edge-to-face arrangement in the crystal.

In molecular packing of 2,3-di(*n*-butyl)tetracene (**28b**), the repeating bilayer structure of the tetracene rings and alkyl groups was observed (Fig. 5.11a) [18]. The thickness of the layer of the tetracene rings was ca. 8.6 Å, and that of the alkyl groups was ca. 9.8 Å. Within each layer of tetracene rings, which looked like a monolayer of tetracene rings, the long molecular axes of the tetracene rings were aligned parallel to each other and inclined by ca. 56° with respect to the layer plane and were packed in an in-plane herringbone fashion (Fig. 5.11b). Since two *n*-butyl groups were located on one terminal side of the tetracene ring, the directionally distinguishable molecules should be arranged in an antiparallel orientation within each of the layers. Although no π-overlap was observed, the two-dimensional tight packing of the tetracene rings was reminiscent of the crystal structure of unsubstituted pentacene [30].

In tetracene octaester derivatives **35a–c**, molecular packing differed to each other considerably (Fig. 5.12) [12]. Molecules of **35a** and **35c** took a slipped-parallel

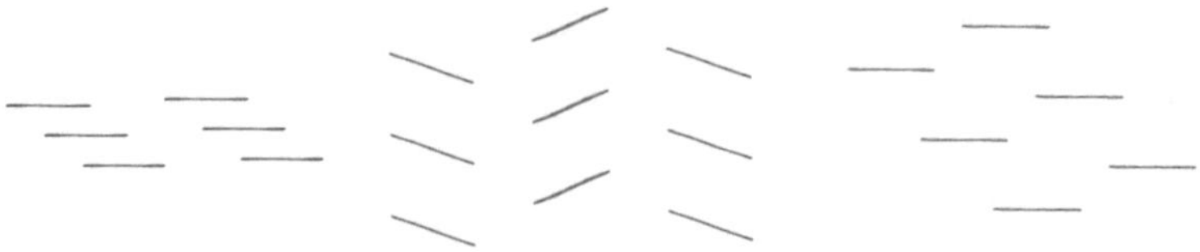

Fig. 5.12 Packing diagrams of the tetracene rings of octa(*n*-alkyl)tetracene-1,2,3,4,7,8,9,10-octacarboxylates (**35**)

arrangement. In contrast, molecules of **35b** adopted a herringbone-like structure with an interplanar tilt angle of 138.9° between two neighboring tetracene rings. In methyl ester derivative **35a**, the molecules formed face-to-face π-overlap with an interplanar distance between two neighboring tetracene rings (*d*) of 3.79 Å. In ethyl ester derivative **35b**, there was no π-overlap, and the *d* distance was 7.61 Å. In *n*-propyl ester derivative **35c**, the molecules seemed to form partial π-overlap. However, the long interplanar distance (*d*) of 8.15 Å indicated the presence of only weak intermolecular interactions along the stacking direction. These results demonstrated that bulkier ester substituents increased the distance between neighboring molecules.

5.5 Conclusions

Substituted tetracenes were synthesized using the Diels–Alder reaction as a key reaction. Unexpectedly, the molecules displayed unique optical properties in the solid state, namely, crystallochromy. We thought that the alkyl side chains could act as not only spacer but also molecular orientation adjusters, to control the photophysical properties in the solid state. On the other hand, alkyl side chains did not affect the optical properties in solution. The solid-state color varied through yellow, orange, and red, depending upon the length, shape, substitution pattern, and kind of substituent groups. Different substituent groups led to different molecular structures and intrinsic molecular arrangements, the latter of which is responsible for unique intermolecular interactions. To understand crystallochromy accurately, a higher level of theory dealing with intermolecular interactions would be necessary. The 1,4,7,10-tetraisopropyl derivative showed a quantum yield of $\Phi_F = 0.90$, which is the highest reported for any tetracene derivatives. We demonstrated that the introduction of substituent groups was effective for the changes in not only the absorption but also the fluorescence in the solid state.

Acknowledgment This work was supported by Grant-in-Aid for Scientific Research from JSPS and MEXT and by the fund of the Hyogo Science and Technology Association.

References

1. Bendikov M, Wudl F, Perepichka DF (2004) Chem Rev 104:4891–4945
2. Zhang Z, Zhang Y, Yao D, Bi H, Javed I, Fan Y, Zhang H, Wang Y (2009) Cryst Growth Des 9:5069–5076
3. Pope M, Swenberg CE (1999) Electronic processes in organic crystals and polymers, 2nd edn. Oxford University Press, New York
4. Anthony JE (2008) Angew Chem Int Ed 47:452–483
5. Kitamura C (2012) Chem Rec 12:506–514
6. Kitamura C (2014) In: Nishiwaki N (ed) Methods and applications of cycloaddition reactions in organic syntheses. Wiley, Hoboken, pp 407–428
7. Klebe G, Graser F, Hädicke E, Berndt J (1989) Acta Crystallogr Sect B 45:69–77
8. Hantzsch A (1907) Angew Chem 20:1889–1892
9. Bernstein J (2002) Polymorphism in molecular crystals. Oxford University Press, New York, pp 257–274
10. Gribble GW, Perni RB, Onan KD (1989) J Org Chem 50:2934–2939
11. Kitamura C, Abe Y, Ohara T, Yoneda A, Kawase T, Kobayashi T, Naito H, Komatsu T (2010) Chem Eur J 16:890–898
12. Kitamura C, Takenaka A, Kawase T, Kobayashi T, Naito H (2011) Chem Commun 47: 6653–6655
13. Kitamura C, Tsukuda H, Yoneda A, Kawase T, Kobayashi T, Naito H (2010) Eur J Org Chem 3033–3040
14. Kitamura C, Kano H, Tsukuda H, Kawase T, Kobayashi T, Naito H (2011) Heterocycles 83:1621–1629
15. McOmie JFW, Perry DH (1973) Synthesis 416–417
16. Kitamura C, Matsumoto C, Yoneda A, Kobayashi T, Naito H, Komatsu T (2010) Eur J Org Chem 2571–2575
17. Kitamura C, Ohara T, Yoneda A, Kawase T, Kobayashi T, Naito H (2011) Chem Lett 40:58–59
18. Kitamura C, Ohe G, Kawase T, Saeki A, Seki S (2014) Bull Chem Soc Jpn 87:915–921
19. Kitamura C, Ohara T, Kawatsuki N, Yoneda A, Kobayashi T, Naito H, Komatsu T, Kitamura T (2007) CrystEngComm 9:644–647
20. Kikkawa Y, Koyama E, Tsuzuki S, Fujiwara K, Miyake K, Tokuhisa H, Kanesato M (2007) Chem Commun 1343–1345
21. Kasha M, Rawls HR, El-Bayoumi MA (1965) Pure Appl Chem 11:371–792
22. Kasha M (1976) Spectroscopy of the excited state. Plenum Press, New York
23. Fujimoto KJ (2012) J Chem Phys 137:034101
24. Fujimoto KJ, Kitamura C (2013) J Chem Phys 139:084511
25. Shimizu M, Hiyama T (2010) Chem Asian J 5:1516–1531
26. Yoshida K, Ooyama Y, Miyazaki H, Watanabe S (2002) J Chem Soc Perkin Trans 2:700–707
27. Kitamura C, Matsumoto C, Kawatsuki N, Yoneda A, Asada K, Kobayashi T, Naito H (2008) Bull Chem Soc Jpn 81:754–756
28. Hisaki I, Sakamoto Y, Shigemitsu H, Tohnai N, Miyata M (2009) Cryst Growth Des 9:414–420
29. Mizobe Y, Ito H, Hisaki I, Miyata M, Hasegawa Y, Tohnai N (2006) Chem Commun 2126–2128
30. Holmes D, Kumaraswamy S, Matzger AJ, Vollhardt KPC (1999) Chem Eur J 5:3399–3412

Part II
Curved π-Electron Systems

Chapter 6
Chiral Sumanene, Triazasumanene, and Related Buckybowls

Shuhei Higashibayashi and Hidehiro Sakurai

Abstract Syntheses of many buckybowls, bowl-shaped aromatic compounds, have been achieved from 1990s, including corannulene and sumanene. After the establishment of these preparative methods, the syntheses of functionalized buckybowls possessing additional factors became important for the elucidation of the fundamental properties of curved π-electron systems as well as the exploration of science and application. The syntheses of chiral sumanene derivatives and chiral triazasumanene have been recently achieved, which possess "chirality" and "heteroatom" as the additional factors. These chiral buckybowls and azabuckybowl were synthesized by employing the strategy to construct the conjugated bowl from nonconjugated bowl through the conventional synthetic method under mild conditions, amenable enough to control the chirality, the reactivity of heteroatom, and high strain of bowl structure. As one of intrinsic properties of curved π-electron system, the first stereoelectronic effect of bowl-shaped π-electron system controlling the conformation of substituted sumanene is also described.

Keywords Buckybowl • Azabuckybowl • Bowl chirality • Heteroatom • Sumanene

6.1 Introduction

Bowl-shaped aromatic compounds, buckybowls, are oldest curved π-electronic molecules as the synthesis of C_{5v} symmetric corannulene (**1**) (Fig. 6.1) was first reported in 1966 [1]. Although the curved structure inspired Osawa to propose the existence of C_{60} already in 1970 [2], the idea had not been widely known until the discovery of C_{60} in 1985 [3], because of the writing only in Japanese. Discovery

S. Higashibayashi
Institute for Molecular Science, Myodaiji, Okazaki 444-8787, Japan

H. Sakurai (✉)
Department of Applied Chemistry, Graduated School of Engineering, Osaka University, 2-1 Yamada-oka, Suita, Osaka 565-0871, Japan
e-mail: hsakurai@chem.eng.osaka-u.ac.jp

T. Akasaka et al. (eds.), *Chemical Science of π-Electron Systems*,
DOI 10.1007/978-4-431-55357-1_6

Fig. 6.1 Corannulene and sumanene

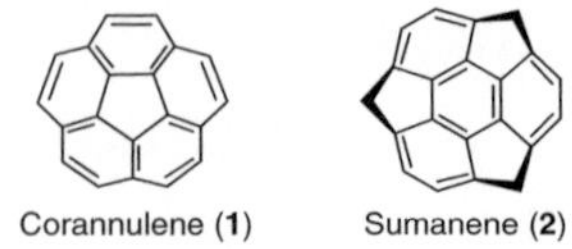

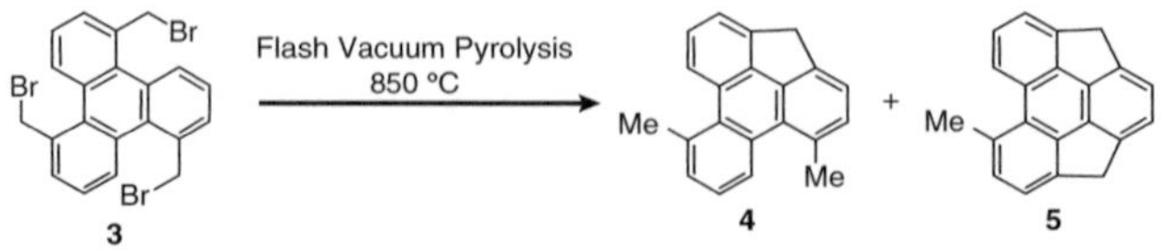

Fig. 6.2 Attempt to the synthesis of sumanene by flash vacuum pyrolysis

Fig. 6.3 Synthesis of sumanene

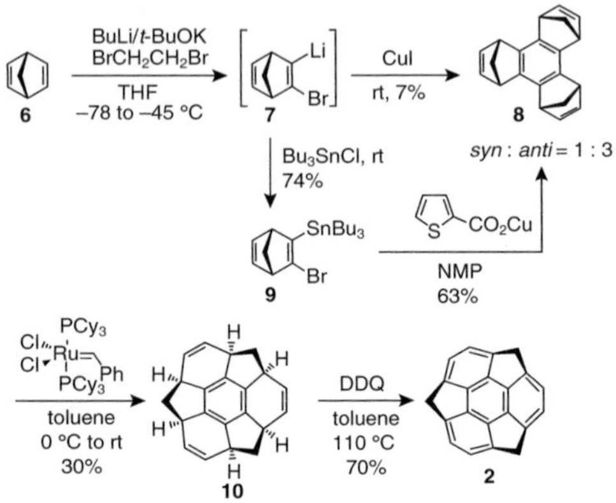

of C_{60} stimulated many organic chemists to study the synthesis and property of buckybowls. As a result, short efficient syntheses of buckybowls in gas or solution phase have been developed [4–22], which supplied enough amounts of compounds and enabled the investigation of interesting properties. The properties have been extensively studied to date due to the interest in the unique properties of bowl structure in addition to the comparison with C_{60}.

C_{3v} symmetric buckybowl, sumanene (**2**) (Fig. 6.1), first appeared in history in 1993 with the beautiful name which Mehta et al. named after "suman," the Hindi and Sanskrit word for a type of flower [23]. However, their attempt to the synthesis of **2** by flash vacuum pyrolysis (FVP), gas-phase synthesis, failed to construct the bowl structure (Fig. 6.2). Ten years later, we achieved the synthesis of **2** in 3–4 steps in solution-phase synthesis employing the strategy from nonconjugated to conjugated bowl (Fig. 6.3) [24]. This concise synthesis led to the elucidation of many unique properties of **2** and the derivatives such as bowl inversion [25–31], columnar crystal packing structure [28, 32–38], high electron conductivity [28, 39], metal complex [40–45], and so on [46–58].

In the early stage of buckybowl chemistry, synthetic organic chemists focused on the development of synthetic methods to construct the strained conjugated bowl structures themselves. Later, after the achievements of syntheses of pristine buckybowls such as **1** and **2**, the syntheses of functionalized buckybowls possessing additional factors also became one of the important topics for the elucidation of the fundamental properties of curved π-electron systems as well as the exploration of science and application. In this chapter, the syntheses and properties of chiral sumanene derivative [26–28] and chiral triazasumanene [29] are presented, which possess "chirality" and "heteroatom" as the additional factors. For the introduction of chirality or heteroatom, the scope of the synthetic strategy from nonconjugated to conjugated bowl through the conventional synthetic method under mild conditions was well demonstrated. As one example of intrinsic properties of substituted sumanene derivatives, the stereoelectronic effect of bowl-shaped π-electron system controlling the conformation of substituted sumanene is described [55].

6.2 Enantioselective Synthesis of Chiral Buckybowl

One of the structural features of three-dimensional curved π-electron systems is the existence of chirality. Fullerene and carbon nanotube are already well known to possess chirality because of the three-dimensional geometry of the structures [59, 60]. Buckybowls also possess inherent chirality due to the bowl structure [61]. We named the chirality to be "bowl chirality" for convenience. The chiral buckybowls possessing the inherent bowl chirality are classified into three types (Fig. 6.4): (A) the bowl structure itself possesses the chirality such as hemifullerene (**11**), (B) the introduction of substituent induces the bowl chirality into achiral bowl structure such as trimethylsumanene (**12**), and (C) the introduction of heteroatom in the bowl skeleton induces the bowl chirality into achiral bowl structure such as triazasumanene (**13**). In fact, many kinds of buckybowls are chiral, including

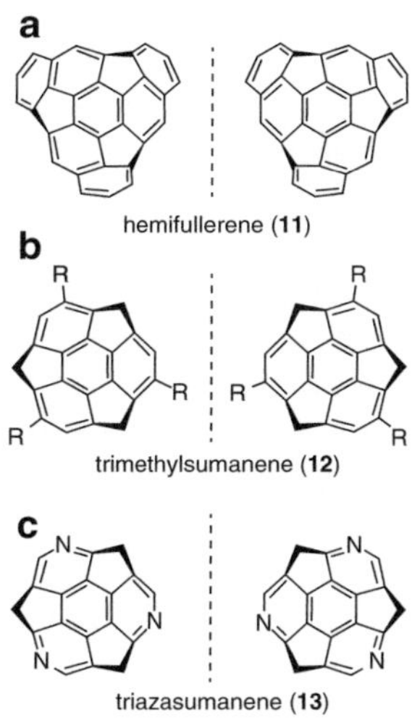

Fig. 6.4 Chiral buckybowls

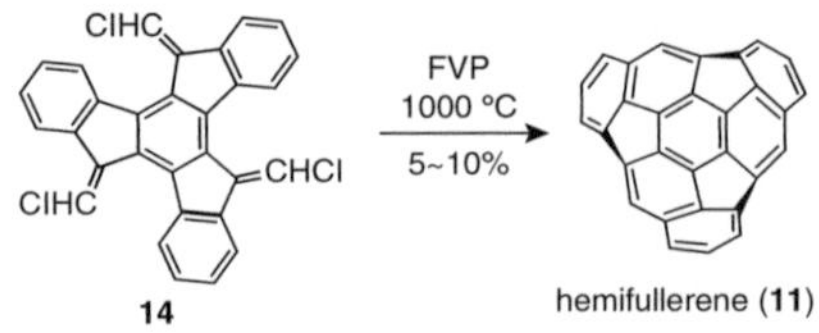

Fig. 6.5 FVP synthesis of hemifullerene (**11**)

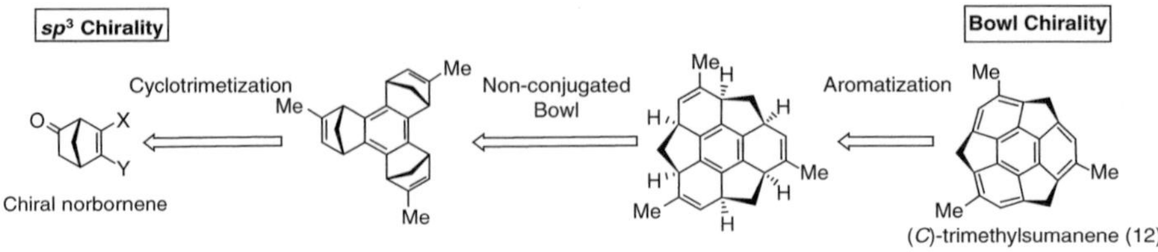

Fig. 6.6 Synthetic strategy for enantioselective synthesis of chiral trimethylsumanene (**12**)

simple buckybowls such as substituted corannulenes and sumanenes. Although the chirality is a common element in these three-dimensional curved π-electron systems (fullerene, carbon nanotube, buckybowl), there was no example of enantioselective synthesis controlling the chirality. As the first examples, we achieved the enantioselective syntheses of chiral trimethylsumanene (**12**) and triazasumanene (**13**) by stepwise synthesis employing the strategy from nonconjugated to conjugated bowl.

In the enantioselective synthesis of chiral buckybowls, there are two fundamental problems to overcome for the achievement, which are (1) the introduction of homochirality and (2) the control of racemization through bowl inversion. Regarding the first problem, the synthesis of racemic hemifullerene (**11**) was achieved by flash vacuum pyrolysis, where the planar π-conjugated precursor **14** was directly converted to conjugated bowl **11** under high temperature (1000 °C) (Fig. 6.5) [62]. However, the asymmetric induction during such harsh cyclization process would be hardly achieved. Our strategy for the induction of chirality depends on the stepwise synthesis under mild conditions (Fig. 6.6) employing the conversion from nonconjugated to conjugated bowl as demonstrated in the synthesis of achiral sumanene (**2**). In this strategy, chiral norbornene precursor possessing stereogenic center (sp^3 chirality) is used as the chiral source. After the conversion to nonconjugated bowl, the sp^3 chirality would be transmitted to the bowl chirality of **12**. However, in order to isolate the thus prepared chiral buckybowl, the second problem, the bowl inversion, must be considered. Buckybowls possess a dynamic behavior that the bowl structure is thermally flipped (Fig. 6.7). In chiral buckybowls, this bowl inversion corresponds to the racemization process. Thus, a chiral buckybowl with relatively small bowl inversion energy would be quickly racemized through bowl inversion even at room temperature. For example, the substituted corannulene derivatives generally have bowl inversion energy around 11 kcal/mol [63], in which the isolation of the enantiomer would be almost impossible. In contrast, the bowl inversion energy of sumanene is around 20 kcal/mol [25] and that of **12** was estimated to be ca. 21 kcal/mol by DFT calculation and the experimental value of

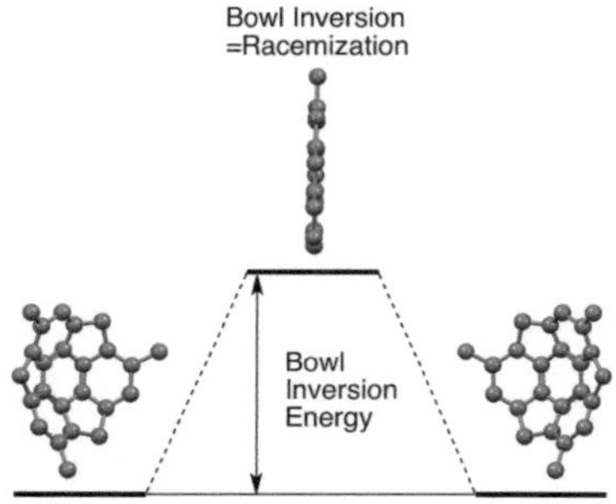

Fig. 6.7 Racemization of chiral buckybowls through bowl inversion

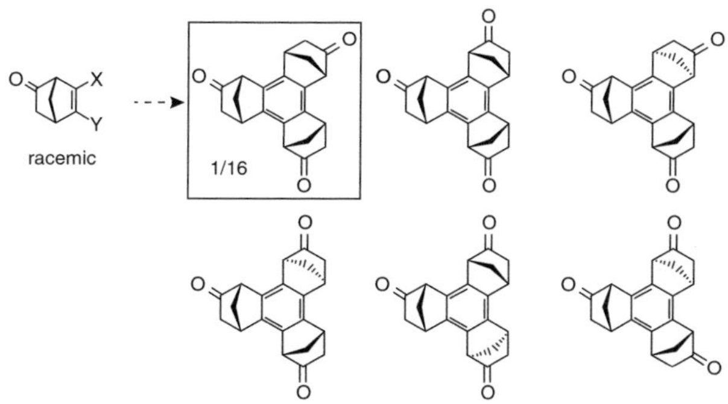

Fig. 6.8 Cyclotrimerization of racemic substituted norbornene derivative

sumanene. Since the energy barrier corresponds to 2 h half-life time at 0 °C, the isolation of an enantiomer would be possible if the last aromatization step is carried out at low temperature.

Although the strategy is reasonable, another synthetically severe problem in the first key step, cyclotrimerization of chiral norbornene derivatives, should be solved to realize the synthetic plan [64–67]. In general, the cyclotrimerization of norbornene derivatives affords the mixture of *syn* and *anti* isomers. For example, the Cu-mediated cyclotrimerization of norbornene derivative **9** afforded the *syn* and *anti* mixture of **8** with 1:3 ratio in sumanene synthesis. However, only *syn* isomer is utilized for the synthesis, while the major *anti* isomer is useless. In the case of substituted norbornene derivatives, possible isomers are increased up to six isomers, and the possibility of the formation of desired C_3 symmetric *syn* isomer becomes only 1/16 without selectivity (Fig. 6.8). Due to the lack of versatile *syn*-selective cyclotrimerization reaction, the new reaction for our purpose should be developed. In order to achieve *syn*-selective cyclotrimerization, two requirements have to be satisfied: (1) the use of enantiopure norbornene derivative and (2) the selective cross-coupling reactions (Fig. 6.9). By starting from the enantiopure norbornene derivatives, the possible isomers can be reduced to two isomers, *syn* and *anti*. The selective cross-coupling of enantiopure norbornene derivative only forms *syn* isomer, while the contamination of homo-coupling gives *anti* isomer. After screening of reaction conditions to satisfy these requirements, we finally succeeded to develop the Pd-catalyzed *syn*-selective cyclotrimerization from simple

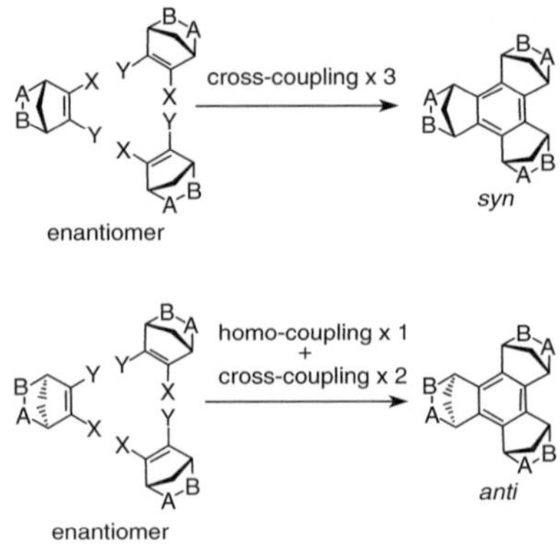

Fig. 6.9 Strategy for *syn*-selective cyclotrimerization

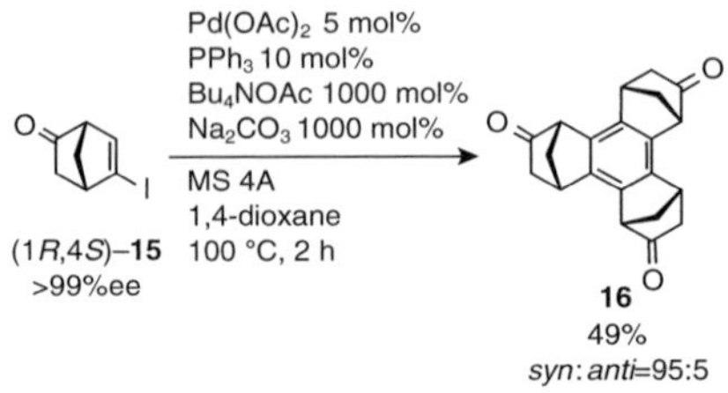

Fig. 6.10 Pd-catalyzed *syn*-selective cyclotrimerization

iodonorbornene (Fig. 6.10). The chiral *syn*-tris(norbornene)benzene **16** for chiral trimethylsumanene was synthesized from enantiopure iodonorbornene **15**. This Pd-catalyzed cyclotrimerization was versatile enough to synthesize other *syn*-tris(norbornene)benzenes **17–21** (Fig. 6.11).

The thus obtained **16** was converted to the nonconjugated bowl **24** through **22** and **23** by introduction of methyl groups and tandem ring-opening/closing metathesis reaction (Fig. 6.12). The final step was the aromatization from nonconjugated bowl to conjugated bowl. As mentioned above, the racemization by bowl inversion has to be prevented. Thus, the aromatization of **24** was conducted using excess DDQ in very short reaction time (1 min) at 0 °C (Fig. 6.13), and the resulting (*C*)-**12** [61] was quickly purified at <–20 °C. The obtained (*C*)-**12** was stored at –80 °C. The bowl chirality of (*C*)-**12** was confirmed by the CD spectra (Fig. 6.14). At –40 °C, the CD spectrum of (*C*)-**12** showed the Cotton effect deriving from the bowl chirality, which scarcely changed at the temperature in several hours. In contrast, the intensity of CD signals gradually decreased at 10 °C, indicating the racemization of (*C*)-**12** through bowl inversion. The enantiomeric excess of (*C*)-**12** was determined to be 89–90 % ee by the derivatization of (*C*)-**12** to **25** and **26** (Fig. 6.15). The introduction of substituents such as (*R*)-MTPA(=PhC(CF_3)(OMe)CO) or $SiMe_3$ induces additional stereogenic centers at the benzylic positions, which prevented the racemization of **25** and **26** through bowl inversion and enabled the determination of ee by NMR or HPLC. The bowl inversion energy of **12** was also determined to be 21.6 kcal/mol by the racemization rate.

While we succeeded in the first enantioselective synthesis of chiral buckybowl, the first optical resolution of racemic chiral buckybowls was also carried out.

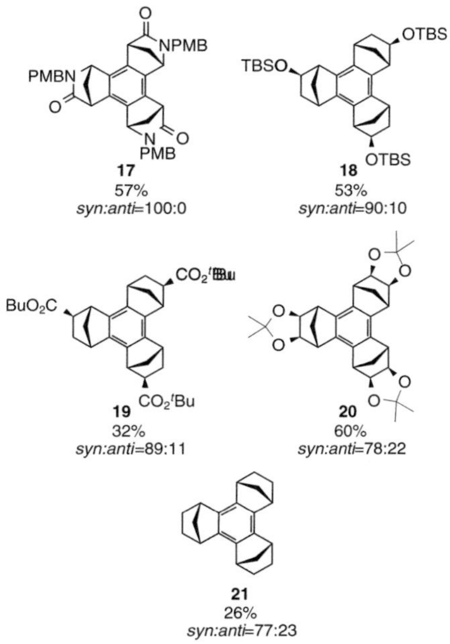

Fig. 6.11 Pd-catalyzed *syn*-selective cyclotrimerization

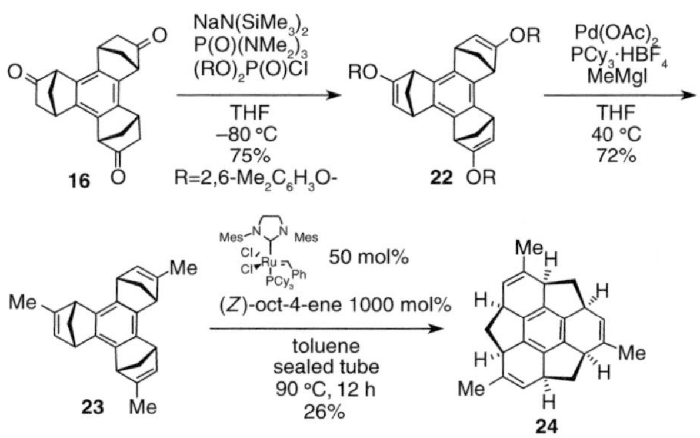

Fig. 6.12 Synthesis of chiral nonconjugated bowl

Fig. 6.13 Enantioselective synthesis of chiral (*C*)-trimethylsumanene (**12**)

The racemization of **12** was too fast to resolve the enantiomers. In contrast, trimethylsumanenetrione **27**, easily prepared from **12** by oxidation (Fig. 6.16), was predicted to possess higher bowl inversion energy as 23.5 kcal/mol, 2 kcal/mol higher than that of **12** by DFT calculation. The value seemed to be high enough

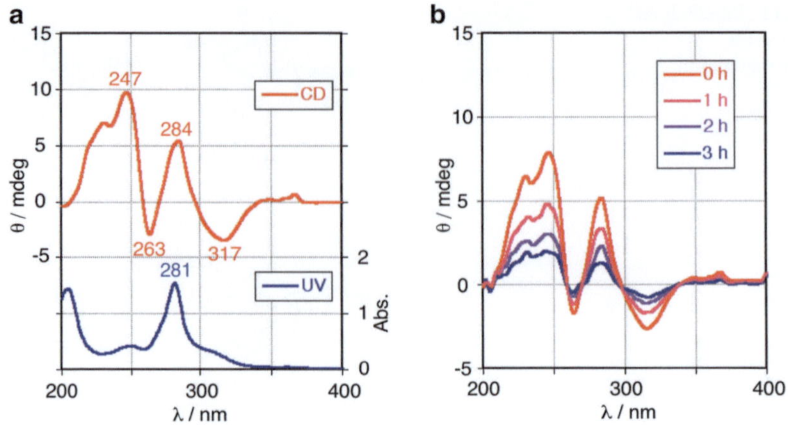

Fig. 6.14 CD spectra of (*C*)-**12** (**A**) at −40 °C and (**B**) at 10 °C

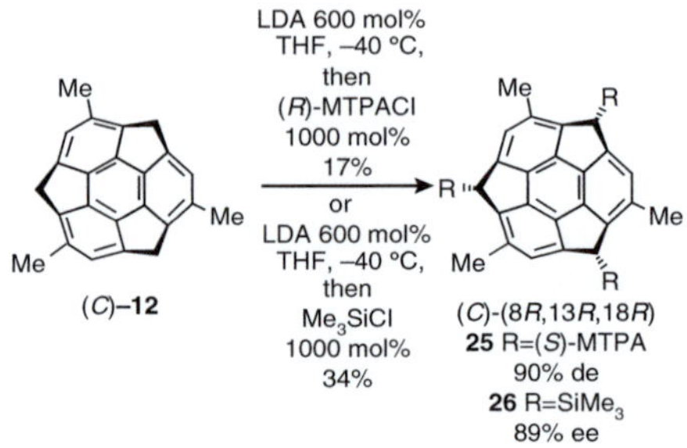

Fig. 6.15 Derivatization of **12** for the determination of ee

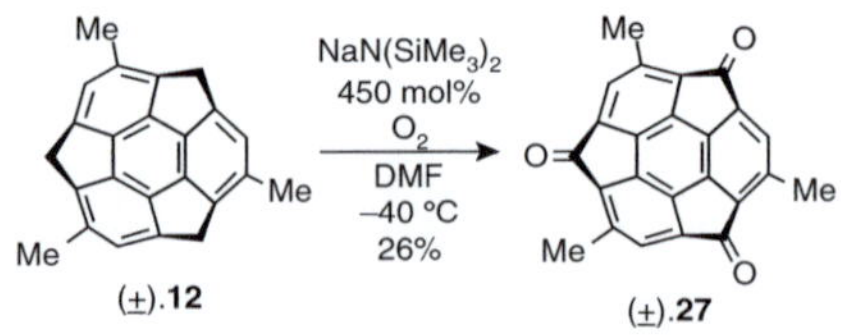

Fig. 6.16 Synthesis of trimethylsumanenetrione (**27**)

for separation by chiral HPLC. Indeed, the racemic **27** was finely resolved by DAICEL Chiralpak IA at 9 °C (Fig. 6.17). The bowl inversion energy of **27** was also determined to be 23.4 kcal/mol from the racemization rate. In both cases of **12** and **27**, the introduction of methyl and carbonyl substituent groups caused the higher bowl inversion energy than 20.3 kcal/mol of sumanene. These results led to the elucidation of the substituent effect and the relationship between bowl structure and bowl inversion energy [28].

As demonstrated in the enantioselective synthesis of chiral trimethylsumanene, the strategy to utilize nonconjugated curved structure for the construction of chiral curved conjugated π-electron system is useful and versatile. This strategy could be applicable to other curved π-electron systems in principle. The homochiral curved π-electron systems are still unexplored field to be addressed.

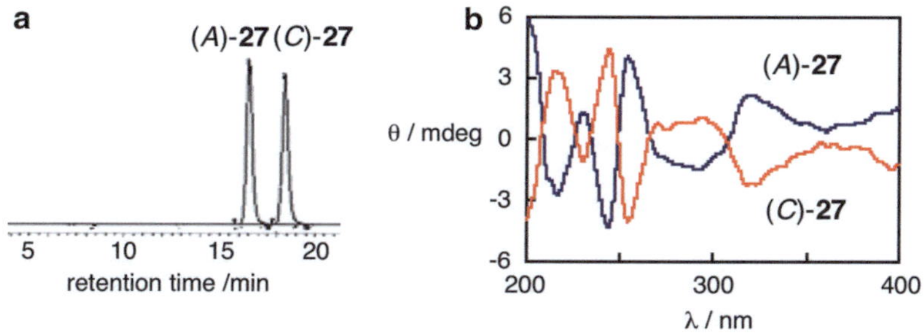

Fig. 6.17 Chiral HPLC analysis and CD spectra of **27**

6.3 Enantioselective Synthesis of Chiral Azabuckybowl

Compared with the buckybowls possessing only carbon bowl skeleton, the studies on heterobuckybowls with heteroatoms in bowl core structure are still very limited. The reported heterobuckybowls are shown in Fig. 6.18 [68–74]. One of the features of these heterobuckybowls is the shallower bowl structure than pristine sumanene except for triazasumanene, because of the longer bond lengths of between carbon and heteroatoms S, Si, Sn, Ge, and Se than the corresponding C-C bond length. For example, the bowl depths of trithiasumanene **28** and triselenasumanene **35** are 0.65 and 0.47 Å, and trisilasumanenes are almost planar structures. In contrast, only triazasumanene was expected to have a reverse property that the bowl depth of triazasumanene is deeper than sumanene due to the shorter bond length of C-N and C = N. This difference on bowl structure might cause significant difference on the synthetic method, because a deeper bowl has more strained structure, which would be more hardly constructed. Indeed, while the heterosumanenes **28**–**35** were accessible from planar precursors such as triphenylene, strained deeper azabuckybowl containing nitrogen, even consisting of corannulene framework, has never been synthesized. Based on the structural characteristics, our strategy was the conversion from nonconjugated to conjugated bowl (Fig. 6.19) as employed in the syntheses of sumanene and chiral trimethylsumanene. Since the triazasumanene also possesses bowl chirality, the chirality can be also controlled in the same manner as trimethylsumanene.

The *syn*-tri(azanorbornene)benzene derivative **17** was prepared from an enantiopure azanorbornene derivative by our developed Pd-catalyzed cyclotrimerization reaction (Figs. 6.11 and 6.20). Amide metathesis of **17** involving the hydrolysis and the following condensation reaction afforded the bowl structure **37. 37** was converted to the hexahydro-precursor **39** for aromatization through **38** in several steps. The final aromatization was achieved by oxidation with Ph_3CBF_4, giving the triazasumanene **36**. The deeper bowl structure of triazasumanene (1.30 Å) than sumanene (1.11 Å) was proved by the X-ray crystallographic analysis of (*C*)-**40**. Another notable feature of chiral triazasumanene is the high stability of bowl

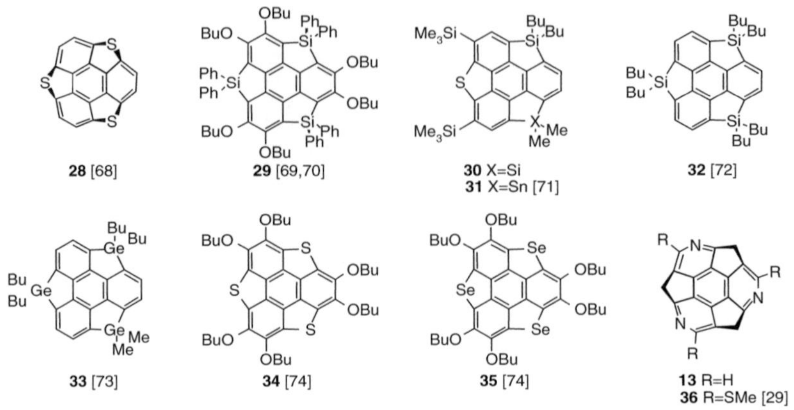

Fig. 6.18 Reported heterobuckybowls

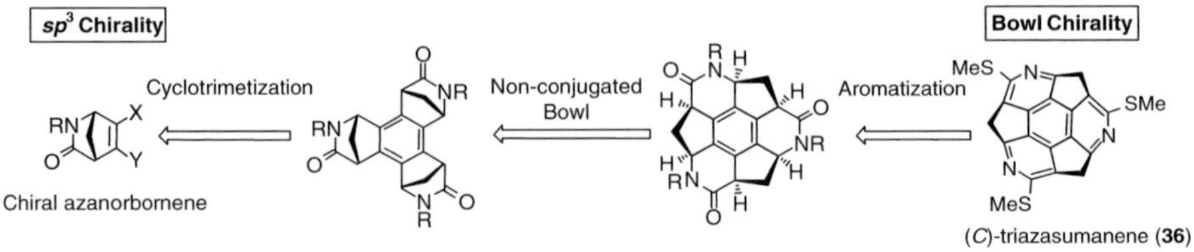

Fig. 6.19 Synthetic scheme of chiral triazasumanene (**36**)

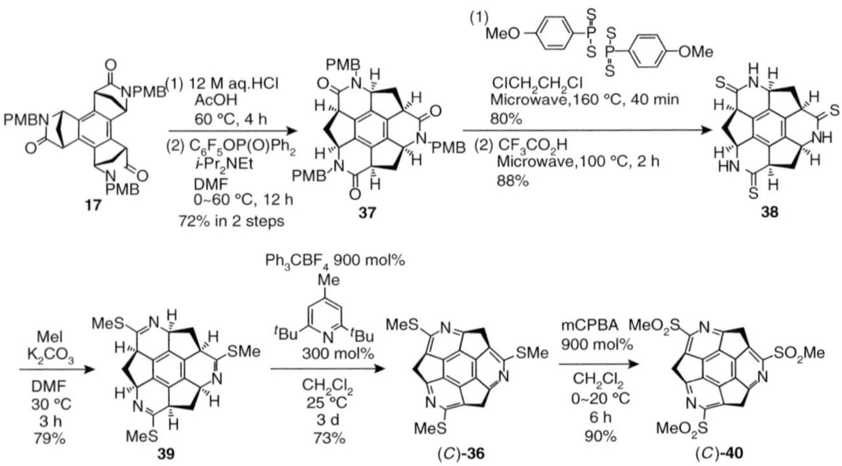

Fig. 6.20 Enantioselective synthesis of chiral (*C*)-triazasumanene (**36**)

chirality. The racemization of triazasumanenes **36** and **40** was not observed at all at room temperature, while the chiral trimethylsumanene **12** possessing all-carbon framework was quickly racemized. This high stability of chirality derives from the high bowl inversion energy of triazasumanene. The relationship between bowl depth (x) and bowl inversion energy (ΔE) is expressed by the $\Delta E = -ax^4$ [28], where a is a constant, indicating that the deeper bowl depth of triazasumanene causes high bowl inversion energy in 4th order. Indeed, the racemization of **36** was observed at 215 °C, and the bowl inversion energy was determined to be 42.2 kcal/mol, which is nearly twice the value of pristine sumanene (20.3 kcal/mol). This extremely high bowl inversion energy of **36** clearly demonstrates the significant influence of heteroatoms.

Our strategy from nonconjugated to conjugated bowl was also applicable to the synthesis of highly strained triazasumanene with reactive heteroatoms, which demonstrated the versatility of the strategy. The wide applicability of the strategy to introduce such as chirality, heteroatom, and functional substituents will produce more unexploited curved π-electron systems. The growing of the heterobucky-bowls will bring drastic effects on curved π-electron systems as demonstrated above.

6.4 Stereoelectronic Effect of Curved π-System

One principal question regarding the curved aromatic compounds (fullerene, carbon nanotubes, buckybowls, etc.) revolves around the differences between the concave face and the convex face. The curvature affects the nature of the two faces, thus resulting in differences of the through-space [75, 76] and through-bond effects. The latter would appear as a stereoelectronic effect between the curved aromatic structure and a connected functional group. In general, the stereoelectronic effect is reflected in the conformational stability and chemical reactivity of a molecule. Therefore, it is important to understand such conformation or chemical reactivity as a consequence of stereoelectronic effects. The interpretation of the stereoelectronic effect of curved aromatic compounds would lead to understanding the differences of the two faces. However, no example of stereoelectronic effects of curved aromatic compounds has been found and studied to date. We have found the first example of the stereoelectronic effect of a curved aromatic structure, an effect that dominates the *endo-/exo*-R conformational stability of benzylic-substituted sumanenes (**41**) (Fig. 6.21) [55].

Benzylic-substituted sumanene (**41**) possesses two conformations, *endo*-R and *exo*-R, where the substituent R is directed to the concave face or convex face of the bowl structure, respectively. During the course of study on the properties of substituted sumanenes, we found that the ratios of the *endo*-R and *exo*-R are changed, depending on the substituent. Thus, we synthesized benzylic-substituted sumanenes with representative substituents R = $SiMe_3$ (**41a**), Me (**41b**), OH (**41c**),

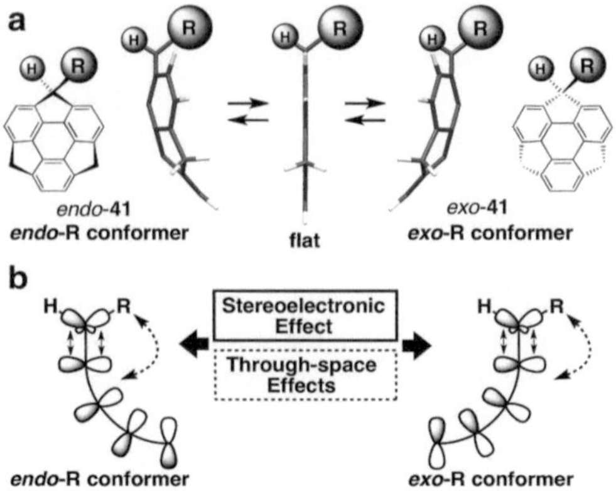

Fig. 6.21 Stereoelectronic effect and through-space effect on benzylic-substituted sumanenes (**41**)

Table 6.1 Experimental conformation of benzylic-substituted sumanenes **41**

	R	*endo*/*exo* ratio	Major conformer	ΔE
2	H	–	–	–
41a	$SiMe_3$	0:100	*exo*-R	–
41b	Me	81:19	*endo*-R	−0.9
41c	OH	90:10	*endo*-R	−1.3
41d	2,6-$Me_2C_6H_3$	82:18	*endo*-R	−0.9
41e	C_6H_5	49:51	*endo*-/*exo*-R	0.0

2.6-$Me_2C_6H_3$ (**41d**), and C_6H_5 (**41e**) and investigated the conformation ratio by ^{1}H NMR spectra (Table 6.1). In case of R = $SiMe_3$ (**41a**), only *endo*-R conformer was observed. This selectivity can be simply explained by the steric repulsion between $SiMe_3$ groups and neighboring aromatic CH groups in *endo*-R conformer (Fig. 6.22). In contrast, the *endo*-R conformer was observed with the minor *exo*-R in R = Me (**41b**), OH (**41c**), and 2.6-$Me_2C_6H_3$ (**41d**), and almost even ratio was observed in R = C_6H_5 (**41e**), which cannot be explained by the through-space steric factors. In order to elucidate the factors for this selectivity, DFT calculation (M06-2x/6-311++G(d,p)) was conducted, which nicely reproduced the relative conformational stability of **41** (Table 6.2). The optimized structure of **41d** implied the existence of stronger CH-π interaction between 2.6-$Me_2C_6H_3$ groups and bowl structure in the *endo*-R conformer (Fig. 6.22), which was further confirmed by atoms-in-molecules (AIM) analysis. On the other hand, the structures of **41b** and **41c** do not indicate any through-space interaction for the *endo*-R selectivity. This selectivity was clearly explained by stereoelectronic effect, which was indicated by natural bond orbital (NBO) analysis (Table 6.2). After the screening of the possible interactions between orbitals given by NBO analysis on the optimized structure of **41b** and **41c**, the hyperconjugation between the CH bond at benzylic position and π^*-orbitals was found to be supporting the *endo*-R selectivity of **41b** and **41c**. Originally, the convex and concave CH bonds at the benzylic position of pristine sumanene **2** have different strength of hyperconjugation. The CH bond at

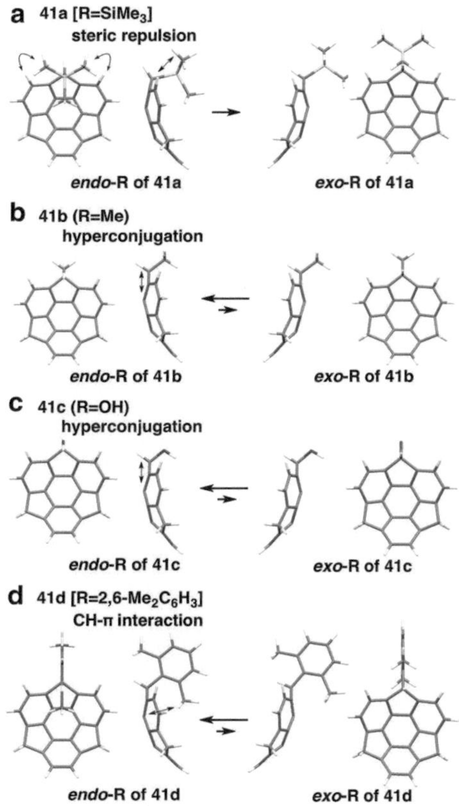

Fig. 6.22 The *endo*-R and *exo*-R conformers of **41** by DFT calculations [M06-2x/6-311++G(d,p)]

Table 6.2 Calculated[a] conformation of benzylic-substituted sumanenes **41**

	R	Major conformer	ΔE	NBO analysis of C-H bond	
				Major conformer	ΔE
2	H	–	–	–	−0.5
41a	$SiMe_3$	*exo*-R	2.0	*exo*-R	4.4
41b	Me	*endo*-R	−1.4	*endo*-R	−1.1
41c	OH	*endo*-R	−2.3	*endo*-R	−0.9
41d	2,6-$Me_2C_6H_3$	*endo*-R	−4.6	*exo*-R	4.2
41e	C_6H_5	*endo*-R	−2.1	*endo*-R	−0.3

[a]M06-2x-6-311++G(d,p)

the convex position of **2** has stronger hyperconjugation (−0.5 kcal/mol) than that at the concave position (Table 6.2). Similarly, the hyperconjugation of CH bonds of *endo*-R and *exo*-R conformers of **41b** and **41c** shows the different strength. NBO analysis showed that the convex CH bond has −1.1 and −0.9 kcal/mol stronger energy than the concave one, supporting the *endo*-R conformational stability of **41b** and **41c** (Table 6.2).

6.5 Perspectives

While only three recent examples on the synthesis and property of buckybowls were shown in this chapter, many interesting challenges on buckybowls are undergoing. For example, the application of buckybowl toward the synthesis of carbon nanotubes (CNTs) [77, 78] is one of the unsolved topics in this field because the cap structure, equal to half buckybowl, consisting of 6 pentagons (five-membered ring) and the rest of hexagons would lead to an identical CNT after the extension by adding hexagonal graphene units [79]. Since the regioselective allocation of 6 pentagons in the buckybowl framework is equivalent to the perfect control of chirality index of CNT, we believe that further development of buckybowl synthesis can also contribute to the CNT science and technology.

References

1. Barth WE, Lawton RG (1966) J Am Chem Soc 88:380–381
2. Osawa E (1970) Kagaku 25:854–863
3. Kroto HW, Heath JR, O'Brien SC, Curl RF, Smalley RE (1985) Nature 318:162–163
4. Scott LT (1996) Pure Appl Chem 68:291–300
5. Rabideau PW, Sygula A (1996) Acc Chem Res 29:235–242
6. Mehta G, Prakash Rao HS (1998) Tetrahedron 54:13325–13370
7. Scott LT, Bronstein HE, Preda DV, Ansems RBM, Bratcher MS, Hagen S (1999) Pure Appl Chem 71:209–219
8. Scott LT (2004) Angew Chem Int Ed 43:4994–5007
9. Petrukhina MA, Scott LT (2005) Dalton Trans 2969–2975
10. Wu Y-T, Siegel JS (2006) Chem Rev 106:4843–4867
11. Tsefrikas VM, Scott LT (2006) Chem Rev 106:4868–4884
12. Sygula A, Rabideau PW (2006) In: Haley MM, Tykwinski RR (eds) Carbon-rich compounds: from molecules to materials. Wiley-VCH, Weinheim, pp 529–565
13. Petrukhina MA (2007) Coord Chem Rev 251:1690–1698
14. Higashibayashi S (2009) J Synth Org Chem Jpn 67:38–50
15. Filatov AD, Petrukhina MA (2010) Coord Chem Rev 254:2234–2246
16. Higashibayashi S, Sakurai H (2011) Chem Lett 40:122–128
17. Amaya T, Hirao T (2011) Chem Commun 47:10524–10535
18. Sygula A (2011) Eur J Org Chem 1611–1625
19. Higashibayashi S (2014) J Synth Org Chem Jpn 72:26–38
20. Schmidt BM, Lentz D (2014) Chem Lett 43:171–177
21. Petrukhina MA, Scott LT (eds) (2012) Fragments and carbon nanotubes. Wiley, Hoboken
22. Wen B, Wang KK (2012) Pure Appl Chem 84:893–905
23. Mehta G, Shah SR, Ravikumar K (1993) J Chem Soc Chem Commun 1006–1008
24. Sakurai H, Daiko T, Hirao T (2003) Science 301:1878
25. Amaya T, Sakane H, Muneishi T, Hirao T (2008) Chem Commun 765–767
26. Higashibayashi S, Sakurai H (2008) J Am Chem Soc 130:8592–8593
27. Tsuruoka R, Higashibayashi S, Ishikawa T, Toyota S, Sakurai H (2010) Chem Lett 39:646–647
28. Higashibayashi S, Tsuruoka R, Soujanaya Y, Purushotham U, Sastry GN, Seki S, Ishikawa T, Toyota S, Sakurai H (2012) Bull Chem Soc Jpn 85:450–467
29. Tan Q, Higashibayashi S, Karanjit S, Sakurai H (2012) Nat Commun 3:891
30. Shrestha BB, Karanjit S, Higashibayashi S, Sakurai H (2014) Pure Appl Chem 86:747–753

31. Jaafar R, Pignedoli CA, Bussi G, Aït-Mansour K, Groening O, Amaya T, Hirao T, Fasel R, Ruffeux P (2014) J Am Chem Soc 136:13666–13671
32. Sakurai H, Daiko T, Sakane H, Amaya T, Hirao T (2005) J Am Chem Soc 127:11580–11581
33. Amaya T, Hifumi M, Okada M, Shimizu Y, Moriuchi T, Segawa K, Ando Y, Hirao T (2011) J Org Chem 76:8049–8052
34. Mebs S, Weber M, Luger P, Schmidt BM, Sakurai H, Higashibayashi S, Onogi S, Lentz D (2012) Org Biomol Chem 10:2218–2222
35. Shrestha BB, Karanjit S, Panda G, Higashibayashi S, Sakurai H (2013) Chem Lett 42:386–388
36. Schmidt BM, Topolinski B, Higashibayashi S, Kojima T, Kawano M, Lentz D, Sakurai H (2013) Chem Eur J 19:3282–3286
37. Shrestha BB, Higashibayashi S, Sakurai H (2014) Beilstein J Org Chem 10:841–847
38. Shreshta BB, Morita Y, Kojima T, Kawano M, Higashibayashi S, Sakurai H (2014) Chem Lett 43:1294–1296
39. Amaya T, Seki S, Moriuchi T, Nakamoto K, Nakata T, Sakane H, Saeki A, Tagawa S, Hirao T (2009) J Am Chem Soc 131:408–409
40. Amaya T, Sakane H, Hirao T (2007) Angew Chem Int Ed 46:8376–8379
41. Sakane H, Amaya T, Moriuchi T, Hirao T (2009) Angew Chem Int Ed 48:1640–1643
42. Amaya T, Wang W-Z, Sakane H, Moriuchi T, Hirao T (2010) Angew Chem Int Ed 49:403–406
43. Zanello P, Fedi S, de Biai FF, Giorgi G, Amaya T, Sakane H, Hirao T (2009) Dalton Trans 9192–9197
44. Hirao T, Amaya T, Moriuchi T, Hifumi M, Takahashi Y, Nowik I, Herber RH (2011) J Organomet Chem 696:3895–3899
45. Amaya T, Takahashi Y, Moriuchi T, Hirao T (2014) J Am Chem Soc 136:12794–12798
46. Amaya T, Mori K, Wu H-L, Ishida S, Nakamura J, Murata K, Hirao T (2007) Chem Commun 1902–1904
47. Amaya T, Nakata T, Hirao T (2009) J Am Chem Soc 13:10810–10811
48. Halling MD, Orendt AM, Strohmeter M, Solum MS, Tsefrikas VM, Hirao T, Scott LT, Pugmire RJ, Grant DM (2010) Phys Chem Chem Phys 12:7934–7941
49. Higashibayashi S, Baig RBN, Morita Y, Sakurai H (2012) Chem Lett 41:84–86
50. Chen J-J, Onogi S, Hsieh Y-C, Hsiao C-C, Higashibayashi S, Sakurai H, Wu Y-T (2012) Adv Synth Catal 354:1551–1558
51. Morita Y, Nakao S, Haesuwanakij S, Higashibayashi S, Sakurai H (2012) Chem Commun 48:9050–9052
52. Mack J, Morita Y, Higashibayashi S, Sakurai H, Kobayashi N (2013) Chem Phys Lett 556:188–194
53. Kunishige S, Kawabata M, Baba M, Yamanaka T, Morita Y, Higashibayashi S, Sakurai H (2013) J Chem Phys 139:044313
54. Topolinski B, Schmidt BM, Higashibayashi S, Sakurai H, Lentz D (2013) Dalton Trans 42:13809–13812
55. Higashibayashi S, Onogi S, Srivastava HK, Sastry GN, Wu Y-T, Sakurai H (2013) Angew Chem Int Ed 52:7314–7316
56. Fujita J, Tachi M, Murakami K, Sakurai H, Morita Y, Higashibayashi S, Takeguchi M (2014) Appl Phys Lett 104:043107
57. Higashibayashi S, Shrestha BB, Morita Y, Ehara M, Ohkubo K, Fukuzumi S, Sakurai H (2014) Chem Lett 43:1297–1299
58. Amaya T, Inada Y, Shimizu Y, Saeki A, Tsuji R, Seki S, Hirao T (2014) Chem Asian J 9:2568–2575
59. Thilgen C, Diederich F (2006) Chem Rev 106:5049–5135
60. Pen X, Komatsu N, Bhattacharya S, Shimawaki T, Aonuma S, Kimura T, Osuka A (2007) Nat Nanotechnol 2:361–365
61. For the stereodescriptor system of chiral buckybowls, see the references 26–28
62. Abdourazak AH, Marcinow Z, Sygula A, Sygula R, Rabideau PW (1995) J Am Chem Soc 117:6410–6411
63. Seiders TJ, Baldridge KK, Grube GH, Siegel JS (2001) J Am Chem Soc 123:517–525

64. Higashibayashi S, Sakurai H (2007) Chem Lett 36:18–19
65. Reza AFGM, Higashibayashi S, Sakurai H (2009) Chem Asian J 4:1329–1337
66. Higashibayashi S, Reza AFGM, Sakurai H (2010) J Org Chem 75:4626–4628
67. Yamanaka M, Morishima M, Shibata Y, Higashibayashi S, Sakurai H (2014) Organometallics 33:3060–3068
68. Imamura K, Takimiya K, Aso Y, Otsubo T (1999) Chem Commun 1859–1860
69. Furukawa S, Kobayashi J, Kawashima T (2009) J Am Chem Soc 131:14192–14193
70. Furukawa S, Kobayashi J, Kawashima T (2010) Dalton Trans 39:9329–9336
71. Saito M, Tanikawa T, Tajima T, Guo JD, Nagase S (2010) Tetrahedron Lett 51:672–675
72. Tanikawa T, Saito M, Guo JD, Nagase S (2011) Org Biomol Chem 9:1731–1735
73. Tanikawa T, Saito M, Guo JD, Nagase S, Minoura M (2012) Eur J Org Chem 7135–7142
74. Li X, Zhu Y, Shao J, Wang B, Zhang S, Shao Y, Jin Z, Yao X, Fang R, Shao X (2014) Angew Chem Int Ed 53:535–538
75. Vijay D, Sakurai H, Subramanian V, Sastry GN (2012) Phys Chem Chem Phys 14:3057–3065
76. Sygula A, Fronczek FR, Sygula R, Rabideau PW, Olmstead MM (2007) J Am Chem Soc 129:3842–3843
77. Scott LT, Jackson EA, Zhang Q, Steinberg BD, Bancu M, Li B (2012) J Am Chem Soc 134:107–110
78. Sanchez-Valencia JR, Dienel T, Gröning O, Shorubalko I, Mueller A, Jansen M, Amshavov K, Ruffienux P, Fasel R (2014) Nature 512:61–64
79. Sugai T, Sakurai H submitted

Chapter 7
Heteroatom-Containing Sumanene

Junji Kobayashi

Abstract Sumanene ($C_{21}H_{12}$) is a C_{3v} symmetric buckybowl, which corresponds to a partial fullerene structure. On the other hand, hetero-π-conjugated molecules, in which one or more carbon atoms of π-system are replaced by main group element, have attracted much attention due to their unique optoelectronic properties. In these molecules, an interaction between π-orbitals and atomic orbitals of main group elements occurs electronic perturbation leading fine-tuning of HOMO and/or LUMO energy levels. Recent advances in the syntheses of heteroatom-containing sumanene are summarized in this review.

Keywords Hetero-π(pi) conjugated molecules • Heterasumanenes • Fluorescent • sigma*-pi* conjugation

7.1 Introduction

Buckybowls are bowl-shaped aromatic hydrocarbons corresponding to a partial structure of fullerene. These buckybowls have attracted much attention due to their physical properties, not only as model compounds for fullerenes, but also as unique bowl-shaped aromatic compounds. Sumanene (**1**) is a C_{3v} symmetric buckybowl (Fig. 7.1), which Mehta et al. named after "suman", the Hindi and Sanskrit word for a type of flower. They first reported their attempts to synthesize **1** by a FVP technique in 1993 [1]. However, it failed to construct the strained bowl structure. The first example of synthesis of **1** has been reported by Sakurai et al. in 2003 [2, 3].

On the other hand, replacement of carbon atoms of π-conjugated compounds with other main group element leads to electronic and optical perturbation into the original hydrocarbons. Such kinds of change come from electronic interactions between π-orbitals of π-conjugations and atomic orbitals of main group elements. Therefore, introduced main group element is significantly affecting the properties of resulting derivatives. For example, Si or P atoms work as accepter atom because

J. Kobayashi (✉)
College of Liberal Arts, International Christian University, Mitaka, Japan
e-mail: junji@icu.ac.jp

T. Akasaka et al. (eds.), *Chemical Science of π-Electron Systems*,
DOI 10.1007/978-4-431-55357-1_7

Fig. 7.1 Sumanene (**1**)

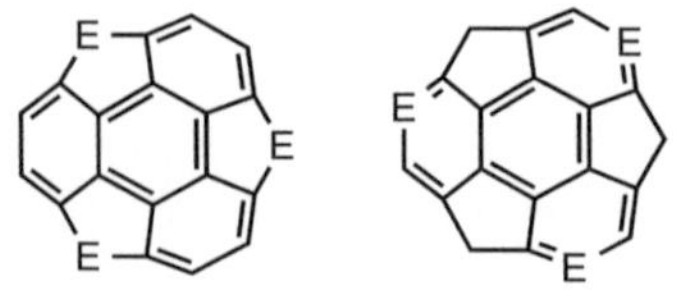

Fig. 7.2 Heterosumanenes. (1) Type 1: replacement of the bridging CH_2 groups (*left*). (2) Type 2: replacement of the aromatic carbons (*right*)

σ* orbital of these elements is lower lying and acts as acceptor orbital. On the other hand, lone pair orbital of group 15 or 16 elements works as donor orbital.

Looking into the sumanene skeleton, there are two types of place, where carbon atom can be replaced with main group element: (1) the bridging CH_2 group in the peripheral five-membered rings and (2) the aromatic carbon atoms in the peripheral six-membered benzene rings. In this review, the examples of heterosumanenes are reviewed classifying by type of structures and elements (Fig. 7.2).

7.2 Replacement of Bridging CH_2 Groups by Heteroatoms

7.2.1 Group 14 Elements

7.2.1.1 Silicon Derivatives (Trisilasumanene)

Replacement of bridging CH_2 groups by silicon atoms leads trisilasumanene containing silole skeletons as a partial structure. Silole-based π-electron systems are currently receiving much attention as new organic optical materials because of their low-lying LUMOs [4], and π-extended silole derivatives involving spiro-type and ladder-type molecules have desirable properties, such as high fluorescence quantum yields and high glass-transition temperatures. Kawashima and Kobayashi et al. reported the novel synthetic method for the construction of silole skeleton; Friedel-Crafts-type silylation of aromatic hydrocarbons. Using hydrosilanes **1** as starting materials, treatment of oxidation agent resulted the corresponding dibenzosilole in a good yield (Scheme 7.1) [5, 6].

Utilizing the intramolecular Friedel-Crafts silylation, the first examples of trisilasumanene **7** were reported (Scheme 7.2). X-ray structural analysis of trisilasumanene **7** indicated that the main framework was almost planar. The UV-vis absorption spectrum of trisilasumanene **7** was measured in dichloromethane. UV-vis of **7** showed that an intense absorption band ($\lambda max = 299$ nm, $\log \varepsilon = 4.67$) was slightly redshifted from that of sumanene ($\lambda max = 278$ nm). In addition, a

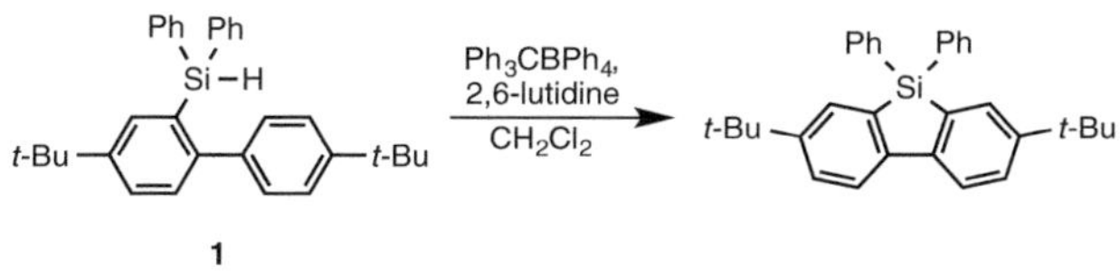

Scheme 7.1 Friedel-Crafts type silylation

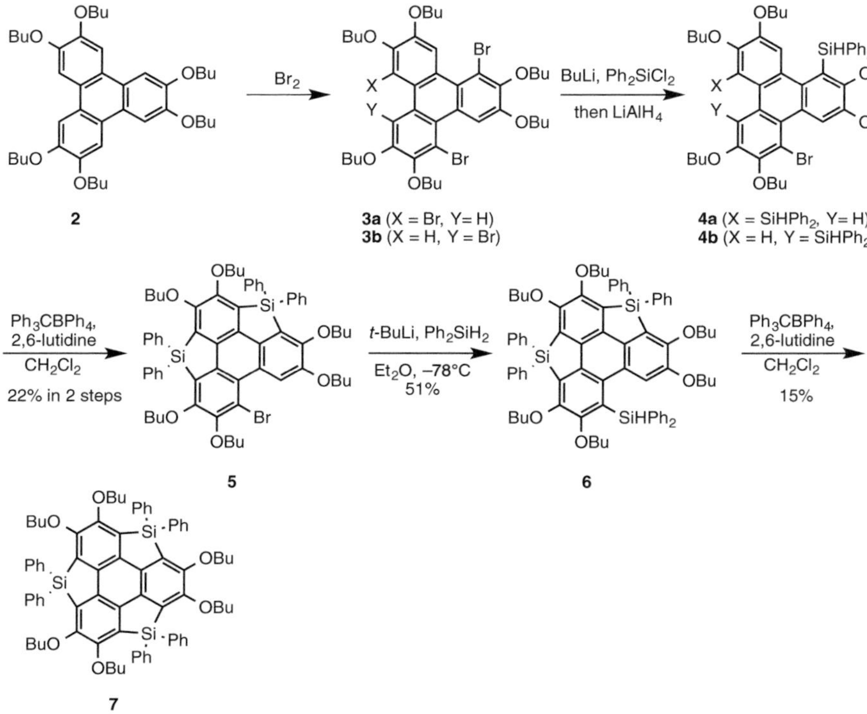

Scheme 7.2 Synthesis of trisilasumanene **7** using Friedel-Crafts type silylation

weak absorption band was observed in the longer-wavelength region (>350 nm) in **7**. This longer-wavelength absorption was mainly attributed to the HOMO-LUMO transition of **7** and suggested the existence of σ^*-π^* conjugation on the silicon atoms. In the emission spectrum, **7** showed a blue fluorescence in dichloromethane solution ($\lambda max = 427$ nm) and in the solid state ($\lambda max = 447$ nm).

Triphenylene is known to be deprotonated at bay positions giving 1,2-dilithiotriphenylene. These compounds can be regarded as a good precursor for the construction of dibenzoheterole moiety. Saito et al. reported the synthesis of trisilasumanene **14** using stepwise lithiation introduction of heteroatom procedure (Scheme 7.3) [7]. The π-framework of **14** is almost planar. **14** showed absorption

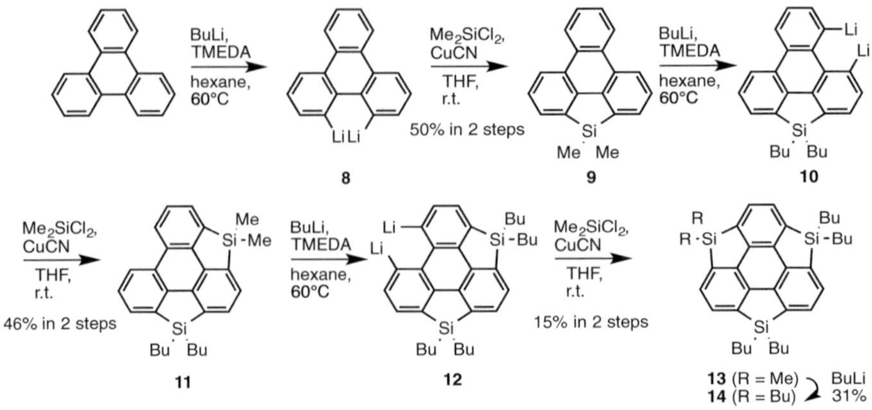

Scheme 7.3 Synthesis of trisilasumanene **14** using sequential lithiation procedure

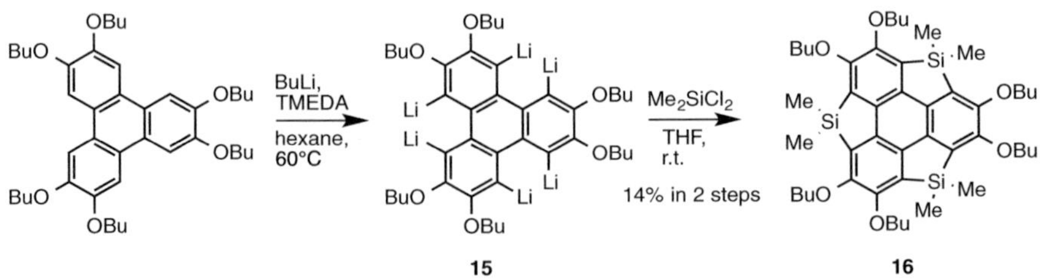

Scheme 7.4 Synthesis of trisilasumanene **16** using hexalithiation of hexaalkoxytriphenylene

maxima at 260, 270, and 294 nm, respectively. The intense absorption band of **14** at 270 nm is slightly blue shifted from that of sumanene (278 nm) and substantially blue shifted compared to that of Kawashima's hexabutoxytrisilasumanene **7** (299 nm). The difference in absorption maxima between **7** and **14** can be attributed to the existence of alkoxy groups on the periphery of the molecules.

Kawashima and Kobayashi et al. have also succeeded the alternative synthesis of trisilasumanene by using similar methodology via lithiation of hexaalkoxytriphenylene (Scheme 7.4) [8]. The presence of peripheral alkoxy groups allowed to proceed a sixfold lithiation upon single step generating hexalithiotriphenylene. Treatment of dichlorosilanes to this hexalithiotriphenylene afforded the corresponding trisilasumanene. By using this method, trisilasumanene can be obtained only by two steps from known hexaalkoxytriphenylene.

7.2.1.2 Germanium Derivative (Trigermasumanene)

Saito et al. succeeded in the synthesis of the first example of trigermasumanene **21** by using their stepwise methodology (Scheme 7.5) [9]. Absorption maxima of **21**

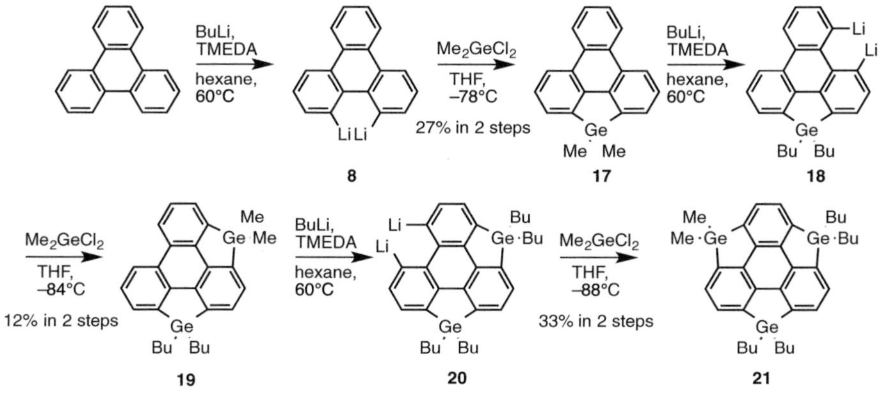

Scheme 7.5 Synthesis of trigermasumanene **21** using sequential lithiation procedure

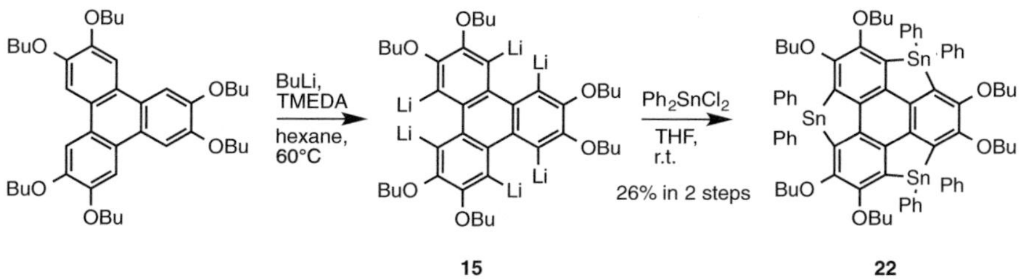

Scheme 7.6 Synthesis of tristannasumanene **22** using hexalithiation of hexaalkoxytriphenylene

(260, 270, and 294 nm) were almost the same as those of trisilasumanene **14**. It can be concluded that heteroatom functionalities have little effect on the electronic absorption of the triphenylene framework.

7.2.1.3 Tin Derivative (Tristannasumanene)

Kawashima and Kobayashi et al. also succeeded the synthesis of tristannasumanene **22** by using hexalithiotriphenylene as a key intermediate (Scheme 7.6) [8]. The absorption maxima of **22** were slightly blue shifted compared to those of trisilasumanene **16**, showing the difference of σ^*-π^* interaction in these compounds.

7.2.2 Group 15 Elements

For group 15 element analogues, only triphosphasumanene has been obtained as an example of triheterasumanene [10]. Kawashima and Kobayashi et al. succeeded in the synthesis of triphosphasumanene via hexalihiotriphenylene as a precursor.

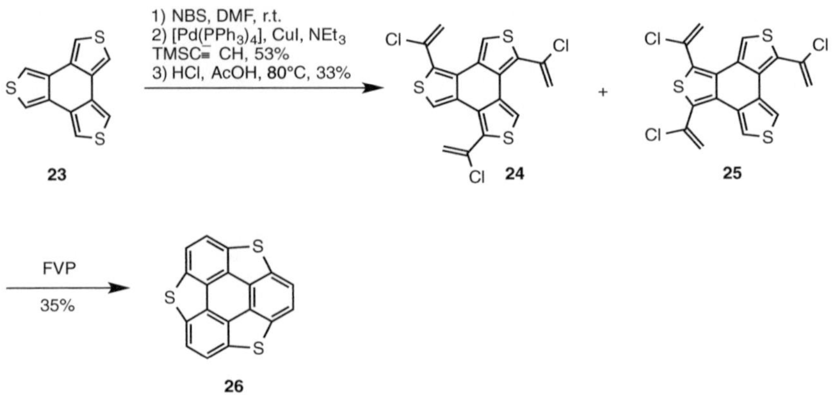

Scheme 7.7 Synthesis of trithiasumanene **26** *via* flash vacuum pyrolysis

7.2.3 Group 16 Elements

7.2.3.1 Sulfur Derivative (Trithiasumanene)

Imamura et al. reported the FVP synthesis of trithiasumanene **26**, the first synthetic heterabuckybowl, in 1999 (Scheme 7.7) [11]. The precursor in the FVP is a mixture of tris(chloroethenyl)benzotrithiophenes **24** and **25** prepared from benzotrithiophene **23**. Under FVP conditions at 1000 °C, **24** and **25** are converted to trithiasumanene **26**. In contrast to the unsuccessful synthesis of sumanene **1** by the FVP method, trithiasumanene is tolerant to the harsh conditions of FVP, possibly owing to the replacement of the reactive benzylic carbons by sulfur atoms and the shallower bowl structure. X-ray crystallographic analysis of **26** indicates a shallower bowl depth (0.79 Å) compared to that of **1** (1.11 Å). The molecules are stacked in a convex-to-concave fashion but alternatively slipped from side to side with intermolecular SS contacts.

7.2.3.2 Selenium Derivative (Triselenasumanene)

Triselenasumanene was synthesized via hexalithiotriphenylene as a key intermediate (Scheme 7.8) [12]. In the first step, hexaalkoxytriphenylene was lithiated with *n*-BuLi in tetramethylethylenediamine (TMEDA) at 60 °C to form the hexanionic intermediate; the resulting solution was then cooled down to −78 °C, followed by the addition of selenium powder in one portion and slowly warming up to room temperature. This one-pot reaction constructs two selenophene rings and one 1,2-diselenin ring (**27**) at the bay regions of triphenylene in a yield of 70 %. In the

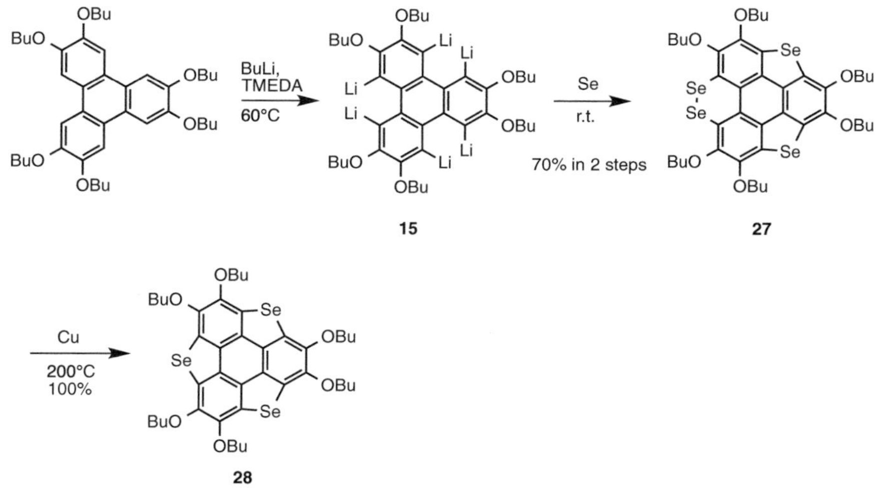

Scheme 7.8 Synthesis of triselenasumanene **28** using hexalithiation of hexaalkoxytriphenylene

second step, the deselenation of **27** was conducted by simply heating a mixture of powdered 8 and nanopowdered copper (80–100 nm grain size) in the solid state at 200 °C for 2 h, and the desired product **28** was quantitatively obtained. The same procedure was applicable to the alternative synthesis of trithiasumanene.

Although **28** possesses a bowl-shaped structure, its bowl depth (0.47 Å) is shallower than those of **26** (0.79 Å) and **1** (1.11 Å).

7.2.4 Others

Saito et al. succeeded in achieving the syntheses of heterasumanenes containing two or three kinds of heteroatoms by stepwise introduction of heteroatoms to triphenylene (Scheme 7.9) [13]. Triphenylenothiophene **29** is prepared from triphenylene. First, the more reactive α-positions of the dibenzothiophene moiety are capped by trimethylsilyl groups to give **30** in 71 % yield. Treatment of **30** with BuLi generates a dilithio species at the bay positions of **30**, which is trapped by dichlorodimethylsilane to afford **31** in 27 % yield. The next treatment of **31** with BuLi causes dilithiation at the bay positions as well as the replacement of two methyl groups of the dimethylsilole moiety by butyl groups. Addition of Me_2SiCl_2 or Me_2SnCl_2 furnishes heterasumanene **32** in 3 % yield or **33** in 0.4 % yield, respectively. X-ray crystallographic analysis of **32** revealed that it possesses a very shallow bowl structure (bowl depth = 0.23 Å) in the crystal packing.

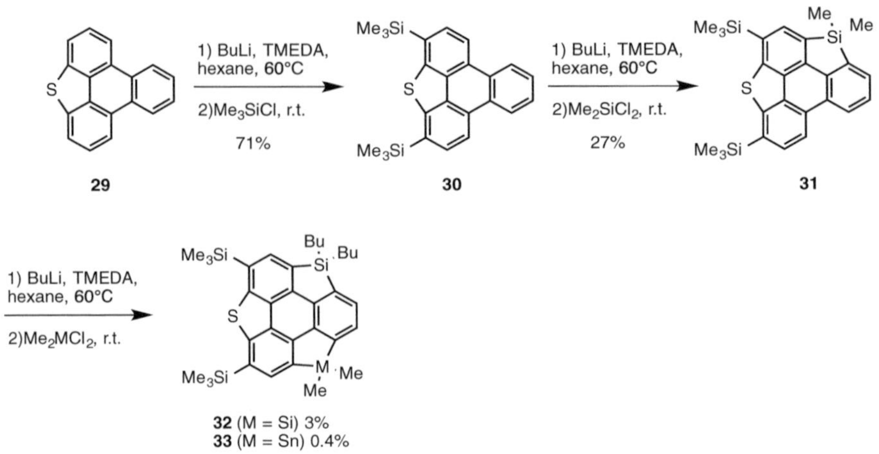

Scheme 7.9 Synthesis of triheterasumanene bearing different heteroatoms

7.3 Replacement of Carbon Atoms in the Peripheral Benzene Rings

Sakurai et al. succeeded in the synthesis of the first example of heterasumanene, in which aromatic carbons of sumanene were replaced by nitrogen atom (Scheme 7.10) [14]. Enantiopure (1*S*, 4*R*)-**34** was converted to syn-**35** exclusively under Pd-catalysed cyclotrimerization conditions. Compound **35** was hydrolyzed, followed by condensation, to afford the non-conjugated bowl-shaped lactam **36**. Many attempts to transform **36** directly to aromatic bowl-shaped triazasumanene were unsuccessful. Thus, it was converted to thioimidate **37** by a microwave-assisted thiation reaction utilizing Lawesson's reagent, deprotection of PMB, and treatment with MeI. Dehydrogenation of thioimidate **37** was successfully achieved by combination of Ph_3CBF_4 as an oxidant and a sterically hindered non-nucleophilic base DTBMP as an acid scavenger to afford (*C*)-(−)-**38** in 73 % yield after optimization. The opposite enantiomer (*A*)-(+)-**38** was similarly prepared from (1*R*, 4*S*)-**34**.

38 was oxidized by *m*CPBA to afford sulfone **39**, which gave single crystals suitable for X-ray crystallographic analysis. The bowl depth is measured as 1.30 Å based on the rim nitrogen atoms, which is deeper than that of sumanene (1.11 Å).

For a chiral buckybowl-possessing bowl chirality, the bowl inversion corresponds to the racemization process. From the racemization experiment, the rate constant and bowl inversion/racemization energy ($\Delta G^{\ddagger}$) of **38** were determined to be $1.24 \times 10^{-6}\ s^{-1}$ (488 K) and 42.2 kcal per mole according to the Eyring equation, respectively.

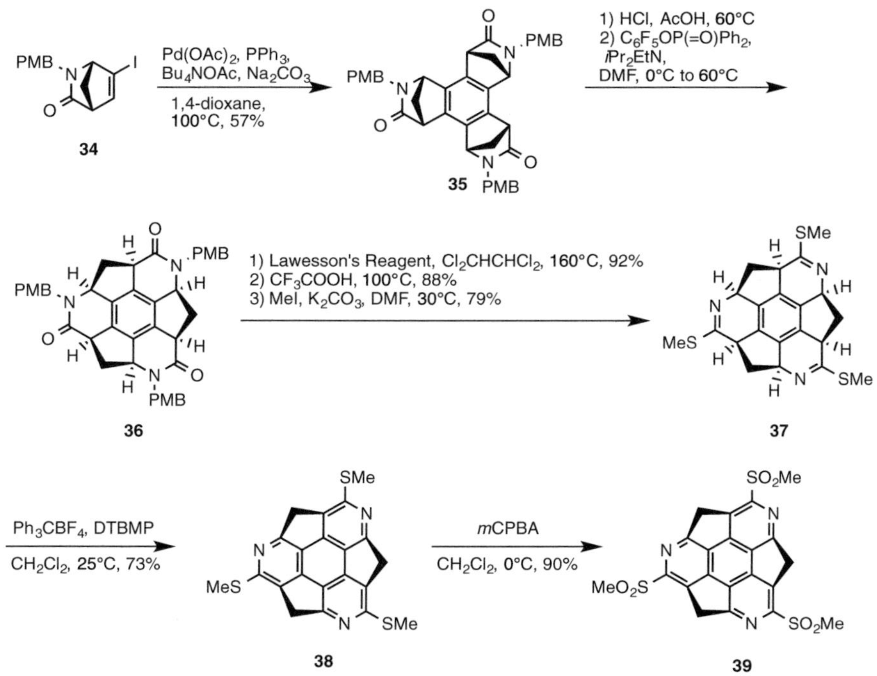

Scheme 7.10 Synthesis of triazasumanene (**38**)

7.4 Conclusion

Development of synthetic methodology for heteroatom-containing sumanene has opened up new attractive chemistry of bowl-shaped hetero-π-conjugated molecules. Various interesting unique physical properties according to the difference of main group elements have been reported. Further experimental and theoretical studies are expected to fully characterize them. Functionalization and derivatization modify the physical properties, leading to applications in material science. Recent synthetic advances of heterasumanenes will expand the range of properties of sumanene as well.

References

1. Mehta G, Shahk SR, Ravikumarc K (1993) J Chem Soc Chem Commun 1006
2. Sakurai H, Daiko T, Hirao T (2003) Science 301:1878
3. Higashibayashi S, Sakurai H (2011) Chem Lett 40:122
4. Tamao K, Uchida M, Izumizawa I, Furukawa K, Yamaguchi S (1996) J Am Chem Soc 118:11974
5. Furukawa S, Kobayashi J, Kawashima T (2009) J Am Chem Soc 131:14192

6. Furukawa S, Kobayashi J, Kawashima T (2010) Dalton Trans 39:9329
7. Tanikawa T, Saito M, Guo JD, Nagase S (2011) Org Biomol Chem 9:1731
8. Kobayashi J, Furukawa S, Shirokawa M, Kawashima T (2012) Abstract of 4th Asian Silicon Symposium in Tsukuba, p 140
9. Tanikawa T, Saito M, Guo JD, Nagase S, Minoura M (2012) Eur J Org Chem 7135
10. Furukawa S, Kobayashi J, Kawashima T. manuscripts in preparation
11. Imamura K, Takimiya K, Otsubo T, Aso Y (1999) Chem Commun 1859
12. Li X et al (2014) Angew Chem Int Ed 53:535
13. Saito M, Tanikawa T, Tajima T, Guo JD, Nagase S (2010) Tetrahedron Lett 51:672
14. Tan Q, Higashibayashi S, Karanjit S, Sakurai H (2012) Nat Commun 3:1

Chapter 8
Open-Cage Fullerene Derivatives: Synthesis, Reactions, and Encapsulation of a Small Molecule

Yasujiro Murata

Abstract The synthetic methods to create a large opening on the fullerene C_{60} cage are developed. A water molecule was encapsulated into the open-cage C_{60} derivative, and the opening was completely restored to afford $H_2O@C_{60}$ in a macroscopic scale under the mild reaction conditions. To realize an encapsulation of a large molecule, efficient way to create a larger opening is found by the reaction with nucleophilic oxidants, which was also applied to a precursor of azafullerenes, in which one of the carbon atoms in fullerene cages is replaced with a nitrogen atom. As another method to enlarge the opening, a sulfur atom was introduced on the rim of the openings. When this reaction was applied to an open-cage C_{60} derivative, a new open-cage $C_{59}S$ derivative was obtained in addition to the expected compound having a larger opening. Furthermore, when this reaction was applied to an open-cage C_{70} derivative, another new reaction was found to afford an open-cage $C_{69}S$ derivative. In this chapter, synthetic methods are described for these new carbon-cage molecular systems, including endohedral fullerenes, open-cage fullerenes encapsulating a molecule, and open-cage heterofullerene derivatives.

Keywords Open-cage fullerenes • Endohedral fullerenes • Small molecule • Azafullerenes • Thiafullerenes

8.1 Introduction

Open-cage fullerene derivatives [1] have been attracting considerable attention because they are interesting molecules for host–guest chemistry and can be used as precursors for endohedral fullerenes encapsulating atom(s) or small molecule(s) as well as heterofullerenes [2]. The inner space of fullerene C_{60}, which is spherical with a diameter of 3.7Å, is suitable to entrap a small molecule. When atoms or molecules are encapsulated in fullerenes, we have a chance to control the properties

Y. Murata (✉)
Institute for Chemical Research, Kyoto University, Uji, Japan
e-mail: yasujiro@scl.kyoto-u.ac.jp

T. Akasaka et al. (eds.), *Chemical Science of π-Electron Systems*,
DOI 10.1007/978-4-431-55357-1_8

of outer carbon cage as well as to study the isolated species. To generate endohedral fullerenes encapsulating metal ions, metal carbides, metal nitrides, metal oxides, rare gases, and N and P atoms, physical methods under harsh conditions are widely applied such as (i) arc discharge of carbon rods containing metal, (ii) ion implantation to empty C_{60} and C_{70}, and (iii) high-pressure/high-temperature treatments of empty fullerenes with rare gases [3]. However, these methods are not suitable to obtain endohedral fullerenes encapsulating small molecules. The fourth method to synthesize such endohedral fullerenes is molecular surgical approach [4], which includes construction of an opening on empty C_{60} or C_{70} cage, insertion of small molecules through it, and the following restoration of it with retention of the encapsulated species inside fullerenes [5, 6]. Using such an approach, hydrogen molecules were entrapped in C_{60} and C_{70} in macroscopic amounts [7, 8]. In order to apply this method to a larger molecule than H_2, it is important to develop open-cage fullerenes having an opening large enough for the small molecule to pass through [1]. Although it is reported that open-cage fullerenes having a large opening can encapsulate a molecule [9–18], restoration of the openings seems rather difficult. In this chapter, recent developments in our research group to give novel open-cage fullerene derivatives with a large opening are described.

8.2 C = C Bond Cleavage of Fullerene Cages

8.2.1 Diketo Derivative of C_{60} [19]

Applying our previous synthetic route [20, 21], diketone derivative of fullerene C_{60} **1a** and **1b** was synthesized from the reaction of C_{60} with pyridazine derivatives followed by photochemical cleavage of one of the C = C double bonds on the rim of the opening of the intermediates. DFT calculations show that the LUMO of open-cage C_{60} with a similar opening motif is mainly located on the conjugated butadiene moiety [20]; sulfur [20, 21] and hydrazine derivatives [9] were reported to react with that moiety. *N*-Methylmorpholine *N*-oxide (NMO) is known as a nucleophilic oxidant [22]. Thus, the reaction of **1** with 2.3 equivalents of NMO in wet THF at room temperature was conducted, and we found that **3a** and **3b** were obtained in good yields after purification with silica gel chromatography (Fig. 8.1).

The structure of **3a** was determined unambiguously by the single crystal X-ray analysis, showing that the opening is constructed by the 13-membered ring containing two hemiketal carbons in addition to two carbonyl carbons. This molecule is considered as the hydrate of tetraketone **2** having the 16-membered ring opening with four carbonyl carbons. Since wet THF was used in this reaction, a water molecule existing in the solution attacked one of the carbonyl carbons in **2** to give **3**. When hydrate **3b** was heated at reflux temperature in toluene for 30 min, complete transformation of **3b** to **2b** was observed. As the similar reaction is known in the literature [23], the conversion between **2** and **3** would be performed

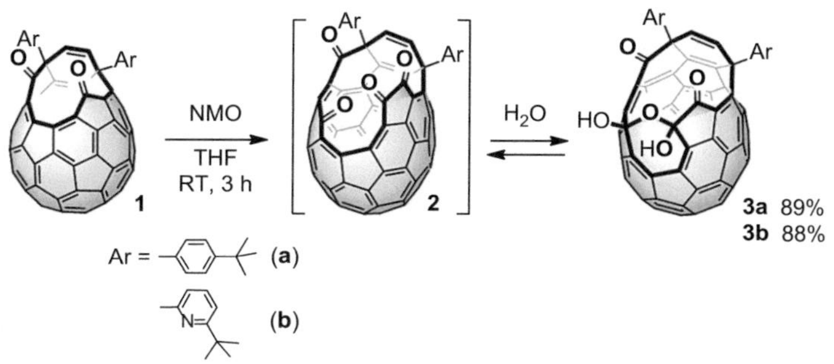

Fig. 8.1 Synthetic route of open-cage C_{60} **3**

Fig. 8.2 Encapsulation of a water molecule into open-cage C_{60} **3b**

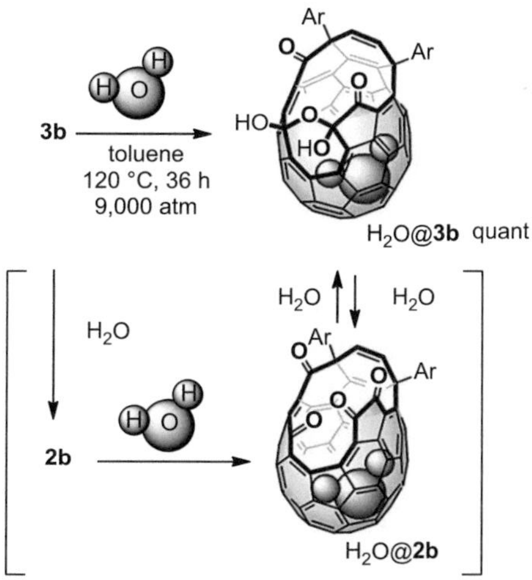

quantitatively. Actually, tetraketone **2b** was isolated and characterized [24]. The size of the opening on tetraketone **2** is larger than that of hydrate **3**, implying that insertion of an H_2O molecule into **2** could be possible.

When hydrate **3b** was refluxed in wet toluene, encapsulation of an H_2O molecule inside **3b** took place in 8 % yield. However, quantitative encapsulation of an H_2O molecule inside **3b** was achieved when this process was conducted under the high pressure of 9,000 atm at 120 °C for 36 h (Fig. 8.2). The high pressure is critical since 40 % encapsulation ratio was observed under 5,000 atm under the similar conditions. The formation of H_2O@**3b** is explained by the dynamic control of the opening size. Elimination of water from **3b** generates **2b** into which insertion of an H_2O molecule takes place through the 16-membered ring opening generated in situ, followed by addition of water to regenerate **3b** encapsulating an H_2O molecule.

After encapsulation of a water molecule inside **3b**, complete restoration of the opening was examined. As shown in Fig. 8.3, the reaction of H_2O@**3b** with excessive amounts of P(O*i*-Pr)$_3$ [25] in refluxing toluene gave H_2O@**2b** in 50 % isolated yield without the loss of the encapsulated H_2O molecule. The reaction

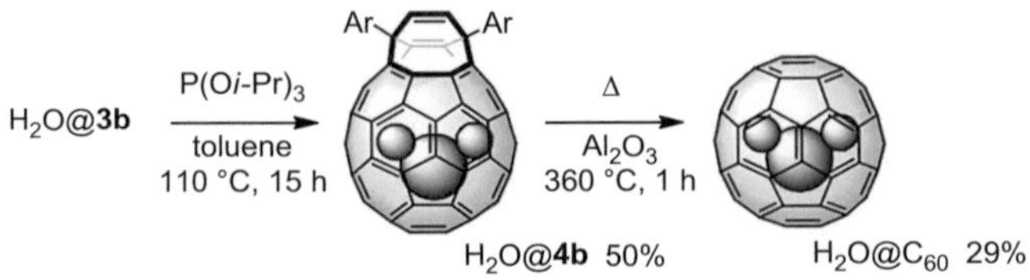

Fig. 8.3 Complete closure of the opening for the synthesis of $H_2O@C_{60}$

would include elimination of water from $H_2O@$**3b** to generate $H_2O@$**2b** followed by the two successive carbonyl couplings on $H_2O@$**2b** to give $H_2O@$**4b**. This one-pot process is very simple compared with our previous method for the synthesis of H_2-encapsulating endofullerenes, in which three reactions were needed to reduce the size of the openings from 13-membered rings to eight-membered rings with purification being necessary in each step [7, 8]. As the final step to realize the organic synthesis of $H_2O@C_{60}$, $H_2O@$**4b** was diluted with neutral Al_2O_3 powder, and the resulting mixture was heated at 360 °C for 1 h under vacuum to give $H_2O@C_{60}$ in 29 % yield after purification with silica gel chromatography.

8.2.2 *Ketolactam Derivative of C_{60} [26]*

Ketolactam derivative of C_{60} **5** is the first open-cage fullerene derivative reported by Wudl et al. in 1995 [27] and also known as a precursor for azafullerenes such as $(C_{59}N)_2$ [28] and $C_{59}NH$ [29], in which one of the carbon atoms of original C_{60} cage is replaced by a nitrogen atom. At the present time, these azafullerenes are the only example for heterofullerenes, for which rational synthetic pathways have been successfully accomplished [28–30]. We expected that magnetically interesting properties, which are different from those for endohedral C_{60}, would be observed due to the unpaired valence electron on the cage of $C_{59}N$ radical if a small molecule is entrapped inside the $C_{59}N$ cage. However, the size of the 11-membered ring opening of **5** is too small even for a helium atom to enter the cage. Thus, enlargement of the opening of **5** is necessary as the first step to obtain endohedral azafullerenes. Regarding the chemical modification of the opening of **5**, structural determination for its derivatives is critical because there are four conjugated reactive olefins **a**–**d** on the rim of the opening (Fig. 8.4). In 2003, the first example was reported by Iwamatsu et al. [31]. They synthesized an open-cage C_{60} derivative having a 15-membered ring opening by the reaction of **5** with aromatic hydrazines through the migration of two hydrogen atoms with the regioselective cleavage of the double bond **a**. They also performed the related reaction using *o*-phenylenediamine to give an open-cage C_{60} derivative having a 19-membered ring opening in 2004 [32]. On the other hand, Orfanopoulos et al. reported two ketolactam derivatives having 12- and 13-membered ring openings, of which contain sulfur atom(s) on the rim, in

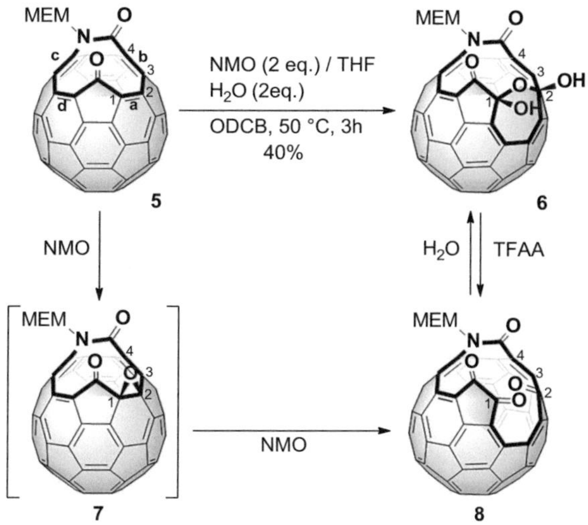

Fig. 8.4 Structure of ketolactam derivative of C_{60} **5** and its reactions

2004 and 2007, respectively [33, 34]. The development of the method to repair the enlarged opening on these compounds derived from the ketolactam remains as a challenging issue.

We conducted the reaction of **5** with NMO as a nucleophilic oxidant in *o*-dichlorobenzene (ODCB) in the presence of a small amount of water to give bis(hemiketal) derivative **6** in 40 % yield (Fig. 8.4). Under the same conditions, trimethylamine *N*-oxide (TMAO) instead of NMO resulted in a similar yield (37 %) of **6**. Since **6** was found to be reactive to methanol, ethanol, and 2-propanol (even trace amounts of ethanol as stabilizer in chloroform), these alcohols cannot be used as polar solvents for chromatographic purification of **6**.

DFT calculations explained the selective formation of bis(hemiketal) took place on bond **a**. The epoxidation of **5**, which occurs at one of the C = C double bonds **a**, **b**, **c**, and **d**, is considered to be the first step of this reaction. The theoretical calculations at the B3LYP/6-31G* level of theory demonstrated that the LUMO is relatively localized at the butadiene unit (C1 = C2–C3 = C4) including bonds **a** and **b**. Epoxide **7** formed by the reaction on bond **a** is 9.0 kcal mol^{-1} more stable than another epoxide formed by the reaction on bond **b**. To discuss more details of the epoxidation kinetically as well as thermodynamically, we investigated the reaction mechanism using model compound **5′**, in which the MEM group was replaced by a methyl group. The calculations were conducted at the M06-2X/6-31G* level of theory with the SMD solvation model [35] in THF at 25 °C (Fig. 8.5). Regarding the nucleophilic attack of TMAO to one of the double bonds on **5′**, formation of each **INT** on C1–C4 is calculated to be less selective. Then, the sequential elimination of trimethylamine (TMA) from each **INT** gives product **7′** or **7b′**, and the reaction pathway was found to be determined by the value of the barrier from **INT** to **TS2**. Among the possible reaction pathways, the reaction on C1 needs the least value of the barrier (16.2 kcal/mol), and formation of epoxide **7′** is more exothermic than **7b′**

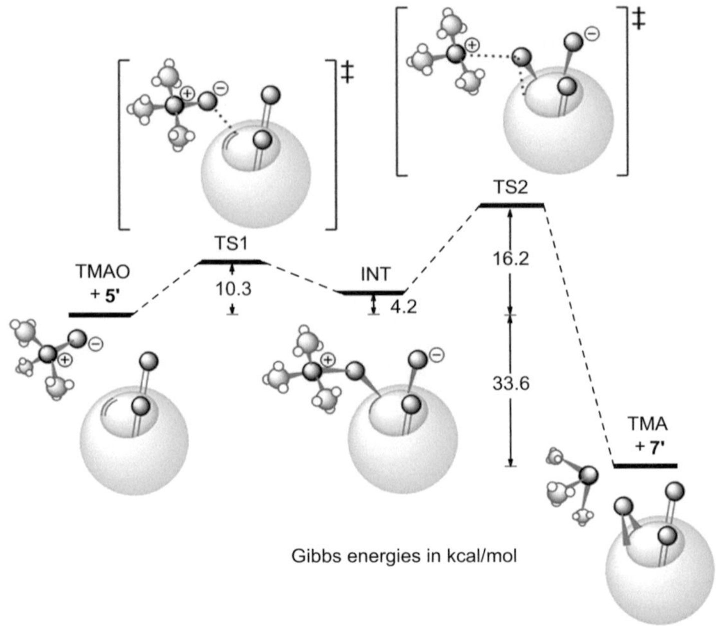

Fig. 8.5 Mechanism for the formation of epoxide **7′** (M06-2X/6-31G* with the SMD solvation model in THF at 25 °C)

(−33.6 kcal/mol for **7′** and −25.4 kcal/mol for **7b′**). These results can explain the regioselective formation of epoxide **7**, which reacts with the second oxidant to give **8**. The resulting **8** affords **6** by the addition of a water molecule to carbon C2 on the opening [19].

The enlargement of the opening of **6** was investigated through dehydration reaction from the bis(hemiketal) moiety. The reaction of **6** with an excessive amount of trifluoroacetic anhydride at room temperature gave tetraketo derivative **8**, quantitatively. In the presence of water or by column chromatography on silica gel, **8** readily turned back to **6**. The opening sizes of **6** and **8** were estimated by the theoretical calculations at the M06-2X/3-21G level of theory. By comparison of the energies required for insertion of a hydrogen molecule, it was shown that **8** has a larger opening than that of **6** (total electronic energies; 23.5 kcal mol^{-1} for **8** and 58.4 kcal mol^{-1} for **6**).

Thus, H_2 insertion experiments were conducted using both **6** and **8** as powders, as well as their ODCB solutions under high-pressure conditions of H_2. It was observed that **6** and **8** were not stable over 100 °C. However, we found that H_2 insertion into **8** was possible above 150 °C and that **6** can be also used as a substrate for H_2 insertion likely due to in situ dehydration to give **8** at this temperature. Under the typical conditions (165 atm, 150 °C, 24 h), formation of H_2@**8** and H_2@**6** was confirmed by ^{1}H NMR with the encapsulation ratio of 20–30 % (Fig. 8.6). After purification by column chromatography on silica gel, H_2@**6** was isolated in 48 % yield.

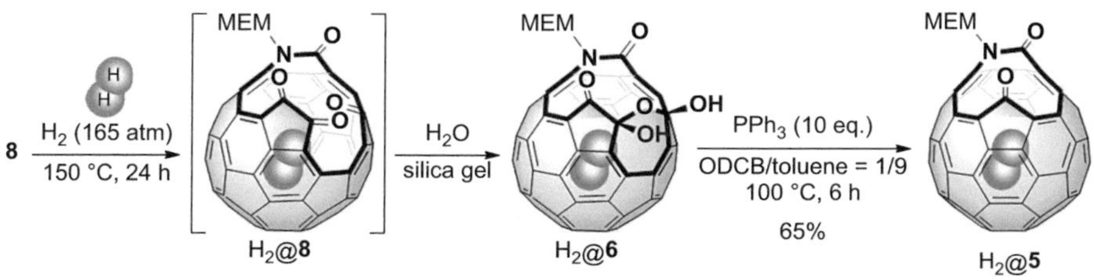

Fig. 8.6 Synthesis of ketolactam derivative of C_{60} encapsulating a hydrogen molecule

Finally, we attempted to reduce the opening size of **6**. In 2011, we reported a coupling reaction of two carbonyl groups on the rim of an open-cage C_{60} derivative by using triisopropyl phosphite [19]. Under the similar conditions, mono- and di-isopropylated **6** was formed as by-products together with desired **5**. In contrast, when triphenylphosphine was used in place of triisopropyl phosphite, **6** was converted to original **5** in 70 % yield without considerable amounts of by-products. We applied the same conditions to H_2@**6** (16 % encapsulation) in ODCB/toluene (Fig. 8.6). The isolation of H_2@**5** was successful in 65 % yield, maintaining the encapsulation ratio of the original 16 %. This is the first example for a possible precursor of endohedral azafullerenes such as $(H_2@C_{59}N)_2$ and $H_2@C_{59}NH$.

8.3 Insertion of a Sulfur Atom [24]

Sulfur insertion into the rim of the small opening on the C_{60} and C_{70} cage is a powerful method to construct a larger opening [20, 21, 36, 37]. The reaction of an open-cage C_{60} derivative having a 12-membered ring opening with sulfur in the presence of tetrakis(dimethylamino)ethylene (TDAE) as a π-donor was reported to give the sulfur-inserted products with a 13-membered ring opening in a good yield [20]. We applied this reaction condition to an open-cage C_{60} having a 16-membered ring opening [19], leading to construction of larger openings.

Open-cage C_{60} derivative **2b** with tetraketone moiety, which was used for the synthesis of $H_2O@C_{60}$ [19], was generated from its precursor **3b** with diketone and bis(hemiacetal) moieties by dehydration at 180 °C under vacuum (Fig. 8.7). Although generation of **2b** took place smoothly in solution [19], formation of **2b** in solid state needed several hours at higher temperature. It was found that **2b** is stable as power and in a dry solution and that **2b** gradually goes back to **3b** by hydration under ambient conditions in solution.

The reaction of **2b** with 1 equiv. of S_8 in *o*-dichlorobenzene (ODCB) at 180 °C in the presence of TDAE was conducted as shown in Fig. 8.7. It was found that the reaction using 0.2 equiv. of TDAE for 15 min afforded sulfide **9** having a 17-membered ring opening in 60 % isolated yield as an almost single product (entry 1) as expected. However, when the amount of TDAE was increased to 0.5 equiv.

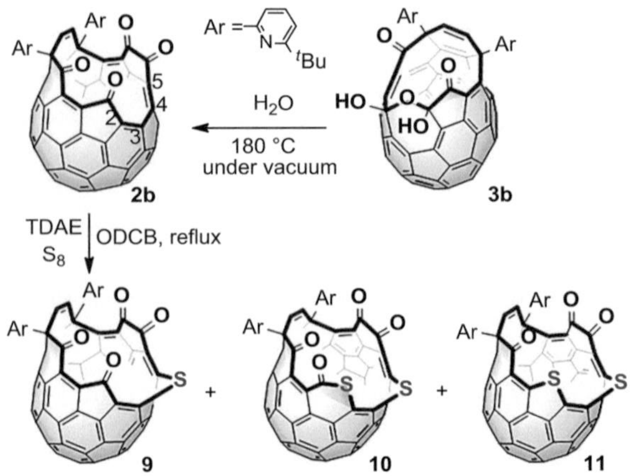

entry	TDAE (equiv)	time (h)	isolated yields 9	10	11
1	0.5	0.25	60	0	0
2	0.5	0.25	24	16	31
3	2.0	14	0	0	68

Fig. 8.7 Synthesis of new open-cage C_{60} derivatives **9–10** and open-cage $C_{59}S$ derivative **11**

(entry 2), unprecedented two products, sulfide thioester **10** (16 %) and open-cage $C_{59}S$ derivative **11** (31 %), were found to be obtained in addition to sulfide **9** (24 %). Furthermore, the reaction with 2.0 equiv. of TDAE for longer reaction time afforded **11** selectively in 68 % yield (entry 3). As a control experiment, we confirmed that the sulfur insertion reaction using **9** as the starting material in the presence of 1.0 equiv. of TDAE for 15 min afforded **10** and **11**, indicating the stepwise insertion of the second sulfur atom to the opening containing the first sulfur atom. The structures of **9–11** were determined unambiguously by the single crystal X-ray analyses.

In order to get insights into the size of the openings of these open-cage compounds, DFT calculations were conducted to estimate the required energies for the insertion of a nitrogen molecule into **9′**–**11′** at the M06-2X/3-21G level of theory, with the *t*-butyl groups replaced by methyl groups. The calculated energies are 27.3, 15.8, and 32.1 kcal/mol for **9′**, **10′**, and **11′**, respectively, indicating that the size of the openings is in the order of **11** < **9** < **10**, which is the same order of the length of the shorter axes of the openings observed in the single crystal X-ray structures. It is interesting to mention that the reason for the apparently smaller size of the opening of **11** with reference to that of **9** is the larger van der Waals radius of a sulfur atom (1.80 Å) [38] in **11** than that of a carbon (1.70 Å) [38] in **9**.

The encapsulation behavior of a water molecule inside **9–11** was reflected by the size of the openings. A sharp signal corresponding to the encapsulated water molecule inside **11** appeared at δ −10.96 ppm, whereas a broad signal was observed for that inside **9** at δ −11.48 ppm, both in $CDCl_3$ at room temperature. Although the signal for that inside **10** was not seen at room temperature, a broad signal was detected at δ −10.95 ppm at −25 °C, likely due to rapid exchange of water between the inside and the outside of the fullerene cage [39]. The encapsulation ratios in solution were ca. 70 % and 40 % for **11** and **9** at room temperature and ca. 25 % for **10** at −25 °C, respectively. The X-ray analyses of sulfide **9**, thioester **10**, and open-cage $C_{59}S$ **11** revealed that a water molecule is encapsulated in the solid state at the encapsulation ratios of 100 %, 52 %, and 75 % for **9**, **10**, and **11**, respectively. The larger encapsulation ratios of H_2O for **9–11** in solid state than those observed in solution can be ascribed to enrichment of the encapsulated H_2O during the recrystallization as reported [10]. It should be mentioned that the X-ray analysis of **10**, which has the largest opening, indicates the encapsulation of not only H_2O (52 %) but also N_2 molecule (48 %). Although the APCI-MS spectrum only exhibited molecular ion peak corresponding to empty **4**, N_2 was likely encapsulated during the course of the crystallization through slow evaporation of air-saturated benzene solution of **10**.

In these open-cage derivatives, all carbons originated from the pristine C_{60} cage retain sp^2-hybridization including carbonyl carbons and divinyl sulfides. Thus, compounds **9–11** still have spherical 60π systems with the functional groups directly introduced at the rim of the opening. To examine the electronic properties of these compounds, the redox behaviors were studied by cyclic voltammetry in *o*-dichlorobenzene. All derivatives **2b** and **9–11** exhibited pseudo-reversible three- or four-step reduction waves. The first reduction potential of tetraketone **2b** appeared at −0.80 V (vs. ferrocene/ferrocenium couple), which is anodically shifted by 0.29 V with reference to that of C_{60} (−1.09 V) [21]. On the other hand, those of sulfur-containing compounds **9–11** appeared at −1.10, −1.09, and −1.13 V, respectively, which are almost comparable to that of C_{60}. These electron-accepting characters were in good agreement with the LUMO levels calculated at B3LYP/6-31G* level of theory: −3.43 eV for **2b**, −3.30 eV for **9**, −3.35 eV for **10**, and −3.26 eV for **11**, respectively. The carbonyl groups directly connected to the π-system of the fullerene cage lower the LUMO levels, whereas the sulfide moieties raise them [20].

8.4 Reaction of Open-Cage C_{70} Derivative [40]

There are only a limited number of examples for open-cage C_{70} derivatives although some examples of open-cage C_{60} derivatives have been reported [1]. Until now, it has been acknowledged that C_{70} exhibits similar reactivity of C_{60}, that is, the typical reactivity of electron-deficient olefins [41]. The lower symmetry of C_{70} relative to that of C_{60} often causes difficulties in the separation and characterization (especially by ^{13}C NMR) of the usually isomeric reaction products [42]. Despite these difficulties, some examples of open-cage C_{70} derivatives have been isolated, which include an ester [43], ketolactams [44], and a diketosulfide [37]. Developing synthetic strategies toward open-cage C_{70} derivatives is desirable from an academic perspective as well as justified by their photophysical properties, which are very different from the C_{60} analogues. So is, for example, the absorption of visible light stronger [45], the longer lifetime of the excited triplet state by two orders of magnitude [46], and the better performance as an n-type semiconductor in polymer solar cells [47]. Moreover, open-cage C_{70} fullerene derivatives are also considered crucial for the generation of the corresponding hetero- and endofullerenes [8, 48]. In this section, we describe the synthesis of open-cage C_{70} diketone **12** and its reactions, which demonstrate that this C_{70} derivative exhibits a remarkably different reactivity compared to its C_{60} analogue. In addition, these reactions resulted in the formation of the first example of an open-cage $C_{69}S$ thiafullerene **16**.

Open-cage C_{70} diketone **12** was synthesized from the reaction of fullerene C_{70} with 3,6-bis(6-(*t*-butyl)pyridin-2-yl)pyridazine, followed by photochemical cleavage of one of the C = C double bond. After each reaction step, the resulting mixtures of isomers were carefully separated by SiO_2 chromatography [37]. The *t*-butyl groups on the pyridyl rings of the pyridazine reagent were introduced in order to gain enough solubility of **12** during the following reaction and separation steps [19]. The size of the cage opening in **12** was subsequently expanded by nucleophilic oxidation reactions (Fig. 8.8). The reaction of **12** with *N*-methylmorpholine *N*-oxide (NMO) at room temperature, i.e., under optimized reaction conditions described for the C_{60} analogue [19], resulted in the formation of complex product mixtures, implying the different reactivity patterns between the C_{70} and C_{60} derivatives. Likely due to the milder nucleophilicity of the oxidizing agent, reaction of **12** with 4-dimethylaminopyridine *N*-oxide (DMAPO), instead of NMO, yielded bis(hemiacetal) **15** in 86 % isolated yield [49]. We assumed that the reaction takes place via the initial formation of epoxide **13**, which would give tetraketone **14** after reaction with the second equivalent of the oxidant, followed by addition of a water molecule to one of the carbonyl groups to afford bis(hemiacetal) **15**. This mechanism was supported by the isolation of epoxide **13** in 60 % yield from the reaction of **12** with one equivalent of NMO at −78 °C. Consistent with the different reactivity observed for the C_{70} analogue **12**, the corresponding epoxide based on the

Fig. 8.8 Synthesis of open-cage C_{70} derivative **15**

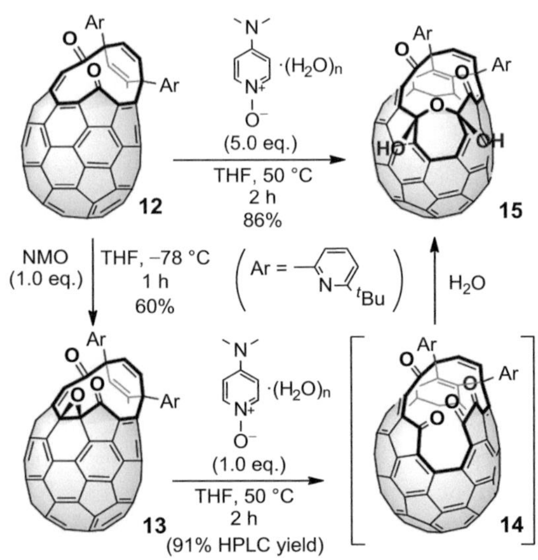

C_{60} analogue of the open-cage diketone was not detected under similar conditions [19]. A subsequent sequential reaction of epoxide **13** with the second equivalent of DMAPO afforded **15** in high yield (91 % estimated by HPLC analysis). The structures of diketone **12**, epoxide **13**, and bis(hemiketal) **15** were clearly determined by the single crystal X-ray analysis. The epoxide moiety in **13** is formed selectively next to the carbonyl group of the five-membered ring. The subsequent nucleophilic attack of the second equivalent of DMAPO should accordingly take place from the front because the back side is sterically hindered due to the cage structure [50]. The opening of **15** contains both a diketone and bis(hemiacetal) moieties, which adopt the same structural motif as in the C_{60} analogue [19]. This structural similarity prompted us to consider the possibility to get opening-expanded **14** by a dehydration reaction of **15** (Fig. 8.8), whereby a water molecule should be accommodated inside the cage of **14**, as reported for the C_{60} analogue [19].

In order to discuss the size of the opening in **14** by theoretical methods, DFT calculations at the M06-2X/6-31G* level of theory were conducted for the model compounds, C_{60}-diketone **2b′** and C_{70}-diketone **14′**, in which the *t*-butyl groups on the pyridyl rings were replaced with hydrogen atoms. To avoid unnecessary complexity potentially arising from hydrogen bond interactions in case of H_2O insertion, we decided to use the insertion process with an Ar atom instead. The required energies for an Ar atom to pass through the opening of **2b′** and **14′** were calculated to be 34.2 and 38.8 kcal/mol, respectively. These results show that the opening in the C_{70} derivative **14′** is smaller compared with that in the C_{60} analogue **2b′** in spite of the similarity of their structures. This result can be explained by the larger degree of strain release in the C_{60} cage [41] by the larger opening relative to the C_{70} derivative, in which the ten additional carbon atoms in the fullerene cage may already defuse the present strain to some extent. The smaller size of the opening in the C_{70} derivative is experimentally supported by the formation of only trace amounts of water encapsulated inside **15**, after treatment with water in toluene at 120 °C under 9,000 atm. Under the similar conditions, C_{60} analogue quantitatively encapsulated a molecule of water inside the cage [19].

For the open-cage C_{60} diketone [20, 21] and tetraketone [24] as well as for the open-cage C_{70} diketone [37], insertion of a sulfur atom on the rim of the opening has been known as an effective method to enlarge the cage opening. In these reactions, a sulfur atom is inserted into the C–C single bond of the conjugated butadiene units on the fullerene cage, where the coefficients of the LUMO are relatively localized in high degrees. Since open-cage C_{70} tetraketone **14** shares a similar opening motif with its C_{60} analogue **2b**, it was expected that a similar reaction product could be obtained from a simple insertion reaction with sulfur. However, when tetraketone **14** was subjected to similar reaction conditions as **2b**, an unprecedented reaction product was obtained. Thus, dehydration of bis(hemiacetal)-diketone **15** at 80 °C for 16 h resulted in the generation of **14**, which was dissolved in *o*-dichlorobenzene and heated to reflux for 15 min in the presence of one equiv. of elemental sulfur and 0.2 equiv. of tetrakis(dimethylamino)ethylene (TDAE) (Fig. 8.9) [20, 21, 24, 37]. After purification with column chromatography on silica gel, followed by recycling HPLC separation using Buckyprep columns, unprecedented open-cage $C_{69}S$ thiafullerene

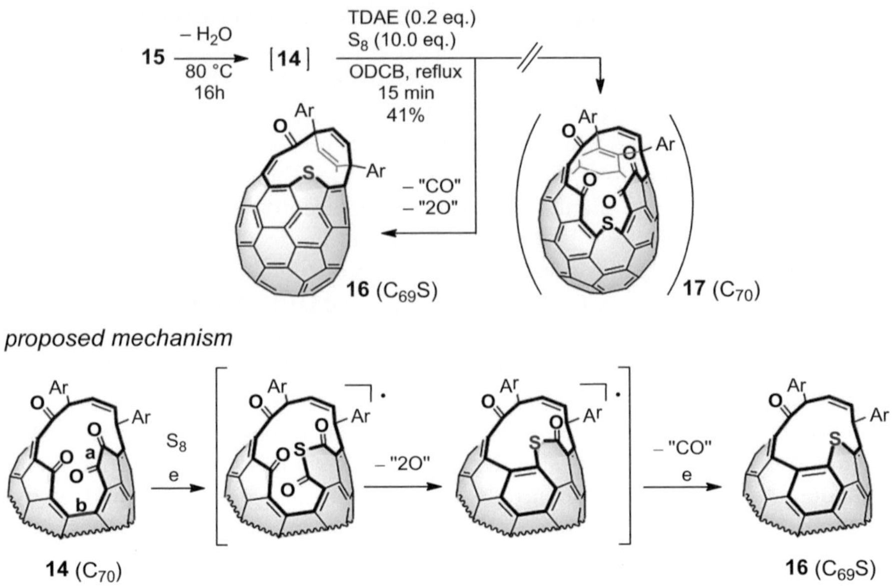

Fig. 8.9 Synthesis of open-cage $C_{69}S$ derivative **16**

derivative **16** was isolated in 41 % yield, whereas the expected product **17** with a 17-membered ring opening could not be observed. The structure of **16** thus obtained was clearly determined by the single crystal X-ray analysis. A possible explanation for the formation of **16** instead of **17** is also shown in Fig. 8.9. Although the LUMO of **2b**, the C_{60} analogue of **16**, is localized to a relatively high degree on the conjugated butadiene unit, the LUMO of the C_{70} derivative **16**, however, is not observed at the corresponding C–C bond "b" in **14**, thus explaining the absence of **17**. Instead, a sulfur atom should be inserted into the more reactive C(O)–C(O) bond "a" in **14**, probably during the anion radical state, which is generated by an electron transfer from TDAE to **14** under the reaction conditions. The insertion of sulfur should be followed by a carbonyl coupling reaction with concomitant loss of two oxygen atoms. A subsequent decarbonylation should then afford **16** after transfer of one electron to the starting material **14**, which is more electronegative than product **16**.

Because of the differences in size and electronegativity between carbon and sulfur, as well as the structural modification, i.e., the formation of an opening, sulfur-containing heterofullerenes are expected to show different electronic properties relative to closed carbon-based fullerenes. In order to study the effects of sulfur doping and decarbonylation on the redox behavior, **12** and **16** were examined by cyclic voltammetry and differential pulse voltammetry in *o*-dichlorobenzene with n-Bu_4NBF_4 as the supporting electrolyte. The cyclic voltammogram of diketone derivative of open-cage C_{70} **12** exhibited a pseudo-reversible first reduction wave and several other reduction waves at more negative potentials, suggesting electrochemical instability of the anionic species resulting from the electrochemical reduction of **12**. This electrochemical instability stems most probably from the two adjacent carbonyl groups on the rim of the opening, which should be susceptible to chemical reactions such as pinacol coupling under the conditions applied. However, for open-cage $C_{69}S$ thiafullerene **16**, four pseudo-reversible reduction waves were observed upon the cyclic voltammetry, indicating significantly higher electrochemical stability toward reduction process compared to **12**. The individual redox potentials of **12** were determined by differential pulse voltammetry as −1.03, −1.35, and −1.81 V (vs. ferrocene/ferrocenium couple). Under the same conditions, **16** showed more negative redox potentials at −1.16, −1.41, −1.89, and −2.21 V, respectively. The difference in redox potentials of **16** compared to **12** can be explained by the electropositive nature of the sulfur atom embedded onto the carbon cage and the loss of one of the electronegative carbonyl groups. DFT calculations at the B3LYP/6-31G* level of theory demonstrated close agreement with the observed trend: the calculated LUMO level of **16** (−3.05 eV) is significantly higher than that of **12** (−3.23 eV). Within the electrochemical window of *o*-dichlorobenzene, no oxidation peak was observed for **16**, even though the HOMO level of **16** (−5.54 eV) was calculated to be higher by 0.15 eV than that of **12** (−5.69 eV). In the UV–vis spectra, the similar absorption bands of **16** and **12** were observed, suggesting the absence of significant differences with respect to the HOMO–LUMO gap.

8.5 Summary

In this chapter, recent developments to synthesize novel open-cage fullerene derivatives, their encapsulation behavior, and their reactions are summarized. By the use of the diketo derivative of C_{60} **3b** as the host molecule for a water molecule, an isolated H_2O molecule was realized in a confined sub-nano space inside fullerene C_{60}. This molecule, considered as "wet fullerene" or "polar C_{60}," will allow us to study the intrinsic properties of a single molecule of H_2O, such as ortho and para conversion [51, 52]. Furthermore, dimerization reaction of $H_2O@C_{60}$ [53], permittivity measurements of single crystal of $H_2O@C_{60}$ [54], and application as a guest molecule toward carbon nanoring [55] were reported, and efficient synthetic methods were developed [56]. When this synthetic methodology is expanded to other small molecules into C_{60}, C_{70}, and higher fullerenes, research work on the isolated molecule or the control of physical properties of outer fullerene cages will be progressed in the near future. As one of the possible solutions of this issue, the methods for expanding the opening on C_{60} and C_{70} were developed, and during these studies, new reactions giving a precursor for endohedral azafullerenes were developed. Furthermore, a novel open-cage $C_{69}S$ derivative was obtained. These fullerene-based molecules would be nice host compounds for other small molecules as well as potential precursors for novel endohedral fullerenes in the near future. Research on the development of further molecular transformations as well as investigations into the solid-state properties of the compounds is currently in progress in our group.

References

1. Vougioukalakis GC, Roubelakis MM, Orfanopoulos M (2010) Chem Soc Rev 39:817
2. Vostrowsky O, Hirsch A (2006) Chem Rev 106:5191
3. Akasaka T, Nagase S (eds) (2002) Endofullerenes: a new family of carbon clusters. Kluwer, Dordrecht
4. Rubin Y (1997) Chem Eur J 3:1009
5. Murata M, Murata Y, Komatsu K (2008). Chem Commun 6083
6. Murata M, Murata Y, Komatsu K (2013) In: Torres T, Bottari G (eds) Organic nanomaterials: synthesis characterization, and device applications, vol 11. Wiley, West Sussex, pp 225–239
7. Komatsu K, Murata M, Murata Y (2005) Science 307:238
8. Murata M, Maeda S, Morinaka Y, Murata Y, Komatsu K (2008) J Am Chem Soc 130:15800
9. Iwamatsu S-I, Uozaki T, Kobayashi K, Re S, Nagase S, Murata S (2004) J Am Chem Soc 126:2668
10. Xiao Z, Yao J, Yang D, Wang F, Huang S, Gan L, Jia Z, Jiang Z, Yang X, Zheng B, Yuan G, Zhang S, Wang Z (2007) J Am Chem Soc 129:16149
11. Zhang Q, Jia Z, Liu S, Zhang G, Xiao Z, Yang D, Gan L, Wang Z, Li Y (2009) Org Lett 11:2772
12. Zhang Q, Pankewitz T, Liu S, Klopper W, Gan L (2010) Angew Chem Int Ed 49:9935
13. Liu S, Zhang Q, Yu Y, Gan L (2012) Org Lett 14:4002
14. Iwamatsu S-I, Stanisky CM, Cross RJ, Saunders M, Mizorogi N, Nagase S, Murata S (2006) Angew Chem Int Ed 45:5337
15. Stanisky CM, Cross RJ, Saunders M (2009) J Am Chem Soc 131:3392
16. Whitener KE Jr, Frunzi M, Iwamatsu S-I, Murata S, Cross RJ, Saunders M (2008) J Am Chem Soc 130:13996
17. Whitener KE Jr, Cross RJ, Saunders M, Iwamatsu S-I, Murata S, Mizorogi N, Nagase S (2009) J Am Chem Soc 131:6338
18. Shi LJ, Yang DZ, Colombo F, Yu YM, Zhang WX, Gan LB (2013) Chem Eur J 19:16545
19. Kurotobi K, Murata Y (2011) Science 333:613
20. Murata Y, Murata M, Komatsu K (2003) Chem Eur J 9:1600
21. Murata M, Morinaka Y, Kurotobi K, Komatsu K, Murata Y (2010) Chem Lett 39:298
22. Andres CJ, Spetseris N, Norton JR, Meyers AI (1995) Tetrahedron Lett 36:1613
23. Donohoe TJ, Raoof A, Linney ID, Helliwell M (2001) Org Lett 3:861
24. Futagoishi T, Murata M, Wakamiya A, Sasamori T, Murata Y (2013) Org Lett 15:2750
25. Borowitz IJ, Anschel M, Readio PD (1971) J Org Chem 36:553
26. Hashikawa Y, Murata M, Wakamiya A, Murata Y (2014) Org Lett 16:2970
27. Hummelen JC, Prato M, Wudl F (1995) J Am Chem Soc 117:7003
28. Hummelen JC, Knight B, Pavlovich J, González R, Wudl F (1995) Science 269:1554
29. Keshavarz-K M, González R, Hicks RG, Srdanov G, Srdanov VI, Collins TG, Hummelen JC, Bellavia-Lund C, Pavlovich J, Wudl F, Holczer K (1996) Nature 383:147
30. Nuber B, Hirsch A (1996). Chem Commun 1421
31. Iwamatsu S-I, Ono F, Murata S (2003) Chem Lett 32:614
32. Iwamatsu S-I, Murata S (2004) Tetrahedron Lett 45:6391
33. Vougioukalakis GC, Prassides K, Orfanopoulos M (2004) Org Lett 6:1245
34. Roubelakis MM, Vougioukalakis GC, Orfanopoulos M (2007) J Org Chem 72:6526
35. Marenich AV, Cramer CJ, Truhlar DG (2009) J Phys Chem B 113:6378
36. Chuang S-C, Murata Y, Murata M, Mori S, Maeda S, Tanabe F, Komatsu K (2007). Chem Commun 1278
37. Murata Y, Maeda S, Murata M, Komatsu K (2008) J Am Chem Soc 130:6702
38. Bondi A (1964) J Phys Chem 68:441
39. Frunzi M, Baldwin AM, Shibata N, Iwamatsu S-I, Lawler RG, Turro NJ (2011) J Phys Chem A 115:735
40. Zhang R, Futagoishi T, Murata M, Wakamiya A, Murata Y (2014) J Am Chem Soc 136:8193

41. Hirsch A, Brettreich M (2005) Fullerenes: chemistry and reactions. WILEY-VCH, Weinheim
42. Herrmann A, Diederich F, Thilgen C (1994) Helv Chim Acta 77:1689
43. Birkett PR, Avent AG, Darwish AD, Kroto HW, Taylor R, Walton DRM (1994). J Chem Soc Chem Commun 1869
44. Hasharoni K, Bellavia-Lund C, Keshavarz-K M, Srdanov G, Wudl F (1997) J Am Chem Soc 119:11128
45. Ajie H, Alvarez MM, Anz SJ, Beck RD, Diederich F, Fostiropoulos K, Huffman DR, Krätschmer W, Rubin Y, Schriver KE, Sensharma D, Whetten RL (1990) J Phys Chem 94:8630
46. Guldi DM, Kamat PV (2000) Fullerenes: chemistry, physics, and technology. Wiley-Interscience, New York, p 225
47. Li G, Zhu R, Yang Y (2012) Nat Photonics 6:153
48. Morinaka Y, Tanabe F, Murata M, Murata Y, Komatsu K (2010) Chem Commun 46:4532
49. Mukaiyama S, Inanaga J, Yamaguchi M (1981) Bull Chem Soc Jpn 54:2221
50. Tajima Y, Takeuchi K (2002) J Org Chem 67:1696
51. Beduz C, Carravetta M, Chen JY-C, Concistre M, Denning M, Frunzi M, Horsewill AJ, Johannessen OG, Lawler R, Lei X, Levitt MH, Li Y, Mamone S, Murata Y, Nagel U, Nishida T, Ollivier J, Rols S, Room T, Sarker R, Turro NJ, Yang Y (2012) Proc Natl Acad Sci U S A 109:12894
52. Goh K, Jimenez-Ruiz M, Johnson MR, Roles S, Ollivier J, Denning MS, Mamone S, Levitt ML, Lei X, Li Y, Turro NJ, Murata Y, Horsewill AJ (2014) Phys Chem Chem Phys 16:21330
53. Zhang R, Murata M, Wakamiya A, Murata Y (2013) Chem Lett 42:879
54. Aoyagi S, Hoshino N, Akutagawa T, Sado Y, Kitaura R, Shinohara H, Sugimoto K, Zhang R, Murata Y (2014) Chem Commun 50:524
55. Hitosugi S, Iizuka R, Yamasaki T, Zhang R, Murata Y, Isobe H (2013) Org Lett 15:3199
56. Krachmalnicoff A, Levitt MH, Whitby RJ (2014) Chem Commun 50:13037

Chapter 9
Endohedral Metallofullerenes: From Chemical Reactivity to Material Performance

Michio Yamada, Satoru Sato, Yuta Takano, Lai Feng, Shigeru Nagase, and Takeshi Akasaka

Abstract Endohedral metallofullerenes (EMFs) are hybrids of fullerenes and metal atoms or metal-containing clusters. The curved π-electron systems of the fullerene cages in EMFs are influenced strongly by the encaged metal species. The π-electron systems of EMFs, which stem from the electron transfer from the encaged metallic species, differ entirely from those of empty fullerenes. Accordingly, EMFs exhibit unique chemical reactivities and electronic properties that empty fullerenes never

M. Yamada
Department of Chemistry, Tokyo Gakugei University, Tokyo 184-8501, Japan

S. Sato
Life Science Center of Tsukuba Advanced Research Alliance, University of Tsukuba, Tsukuba, Ibaraki 305-8577, Japan

Y. Takano
Life Science Center of Tsukuba Advanced Research Alliance, University of Tsukuba, Tsukuba, Ibaraki 305-8577, Japan

Institute for Integrated Cell-Material Sciences (WPI-iCeMS), Kyoto University, Kyoto 606-8501, Japan

L. Feng
Life Science Center of Tsukuba Advanced Research Alliance, University of Tsukuba, Tsukuba, Ibaraki 305-8577, Japan

Jiangsu Key Laboratory of Thin Films and School of Energy, Soochow University, Suzhou 215006, P. R. China

S. Nagase
Fukui Institute for Fundamental Chemistry, Kyoto University, Kyoto 606-8103, Japan
e-mail: nagase@ims.ac.jp

T. Akasaka (✉)
Department of Chemistry, Tokyo Gakugei University, Tokyo 184-8501, Japan

Life Science Center of Tsukuba Advanced Research Alliance, University of Tsukuba, Ibaraki 305-8577, Japan

School of Materials Science and Technology, Huazhong University of Science and Technology, Wuhan 430074, P. R. China

Foundation for Advancement of International Science, Ibaraki 305-0821, Japan
e-mail: akasaka@tara.tsukuba.ac.jp

T. Akasaka et al. (eds.), *Chemical Science of π-Electron Systems*,
DOI 10.1007/978-4-431-55357-1_9

have. This chapter exclusively addresses the most recent achievements of our EMF research, ranging from basic chemical reactivity to material performance. This chapter includes four major parts. The first deals with the fascinating chemical reactivity of EMFs encountered very recently in our lab. The second part is devoted to EMF-based donor–acceptor conjugates, in which EMFs can display electron acceptor characters as well as electron donor characters, corresponding to the counterparts. In the third part, carrier transport properties of EMFs are summarized. The fourth part briefly describes the preparation and property of surface-grafted EMFs. Throughout, we hope that this chapter will stimulate readers to familiarize themselves with these fascinating molecules.

Keywords Endofullerene • Chemical functionalization • Donor–acceptor conjugate • Carrier transport property • Self-assembled monolayer

9.1 Introduction

Endohedral metallofullerenes (EMFs) are hybrids of fullerenes and metal atoms or metal-containing clusters in which the curved π-electron systems of the fullerene cage are influenced strongly by the encaged metal species. In EMFs, several electrons are transferred from the metal species to the fullerene cages in the ground state, engendering their unique electronic structures. Accordingly, the resulting anionic π-electron systems of EMFs differ entirely from those of empty fullerenes. Past research on EMFs has mainly concentrated on the search for new EMF species, their structural elucidation, and assessment of their basic chemical reactivity [1–6]. In fact, some EMF structures were characterized erroneously and were subsequently recognized by the fullerene society until recent corrective reports [7, 8]. Nevertheless, recent progress on the large-scale synthesis and separation of EMFs has elicited deeper insight not only into the fundamental aspects but also into applications for advanced materials [9, 10].

This chapter exclusively emphasizes the most recent achievements of our EMF research, ranging from basic chemical reactivity to material performance. It is earnestly hoped that this chapter will stimulate readers to familiarize themselves with these fascinating molecules.

9.2 Chemical Reaction of EMFs

Various chemical reactions developed for the functionalization of C_{60} have been extended to the chemistry of EMFs [11, 12]. However, as reflected by the different cage structures and the different electronic structures from those of C_{60}, some reactions have led to unprecedented products. Others have met with failure. In this section, their fascinating chemical reactivity of EMFs, encountered very recently in our lab, will be provided.

9.2.1 Diels–Alder Reaction

Actually, La@C_{2v}-C_{82} is a paramagnetic molecule. Recent DFT studies have predicted that the reaction between 1,3-butadiene radical cation and ethylene proceeds via a stepwise path involving an open chain intermediate [13–15]. Nevertheless, even qualitative concepts are lacking for the Diels–Alder (DA) reaction using paramagnetic molecules in the ground state. We found that the DA reaction of La@C_{2v}-C_{82} with cyclopentadiene proceeded smoothly to afford a single product [16]. Subsequently, single-crystal XRD analysis of the La@C_{2v}-C_{82}–1,2,3,4,5-pentamethylcyclopentadiene (Cp*) adduct was performed successfully, revealing that the addition position was the same as that in the Bingel–Hirsch adduct reported previously [17]. That was a surprising result because the carbon atoms of the addition site have neither large β-LUMO coefficients nor large spin densities. Instead, they have high positive charge densities. To explain the selectivity in the DA reaction of La@C_{2v}-C_{82}, DFT calculations were performed [18]. For a concerted path, the initial step is related to the formation of a complex of La@C_{2v}-C_{82} and Cp* (see Fig. 9.1a). The intermediate is more stable by 10.6 kcal mol^{-1} than the reactants. The energy barrier from the intermediate to the final product via a transition state is 12.2 kcal mol^{-1}. On the other hand, we found singly linked intermediate, which is 19.3 kcal mol^{-1} higher in energy than the reactants, for a stepwise path (Fig. 9.1b). Accordingly, results showed that the DA reaction of paramagnetic La@C_{2v}-C_{82} and Cp* proceeds via a concerted bond formation mechanism, through an intermediate formation of La@C_{2v}-C_{82} and Cp*, rather than via a stepwise addition mechanism. In view of the MO distribution, it was also suggested that the α-LUMO and β-LUMO + 1 of La@C_{2v}-C_{82} can play a main role in the interaction with the HOMO of Cp*.

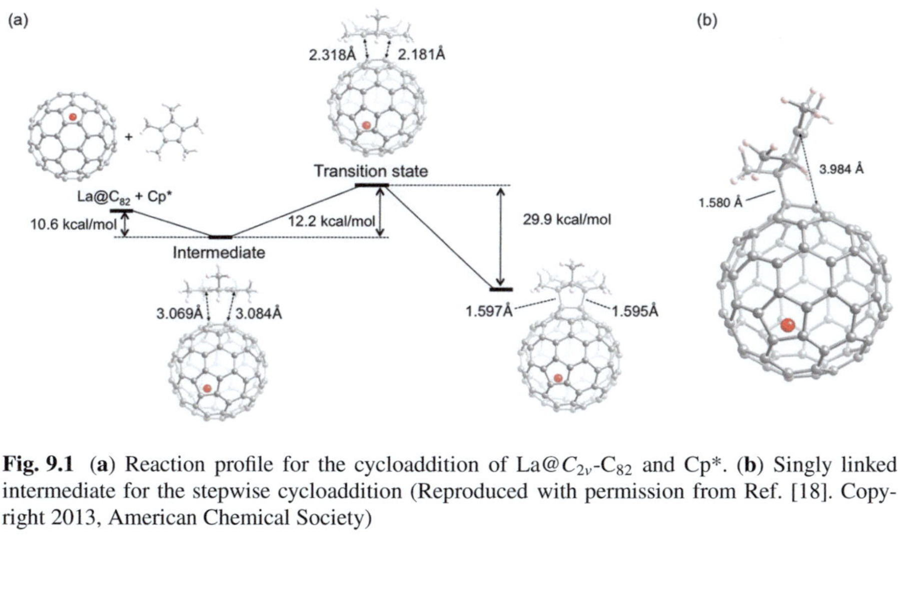

Fig. 9.1 (**a**) Reaction profile for the cycloaddition of La@C_{2v}-C_{82} and Cp*. (**b**) Singly linked intermediate for the stepwise cycloaddition (Reproduced with permission from Ref. [18]. Copyright 2013, American Chemical Society)

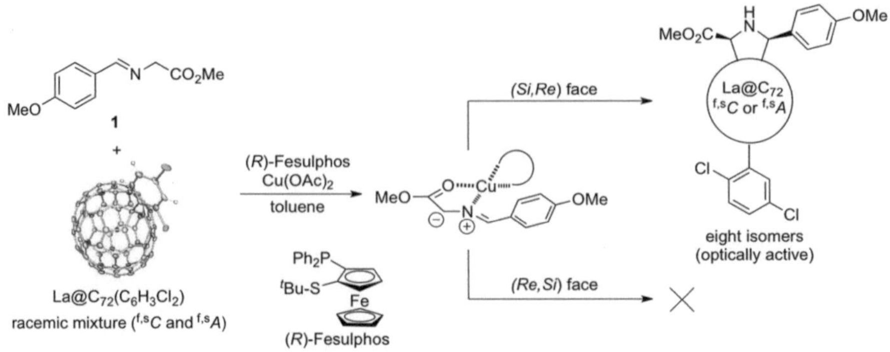

Fig. 9.2 Synthesis of optically active La@C_{72} derivatives

9.2.2 *Enantioselective 1,3-Dipolar Cycloaddition*

Preparation of enantiomerically pure fullerene derivatives has been an important issue because the stereochemical configuration is a key factor in molecular recognition. Nevertheless, racemic organofullerenes are often used even in medicinal studies. In this regard, Martín et al. established an enantioselective synthesis of pyrrolidino-[60] fullerenes using *N*-metallated azomethine ylide complexes [19]. Following that success, the enantioselective 1,3-dipolar cycloaddition was applied for the synthesis of optically active EMFs. Preliminary attempts to carry out the cycloaddition of *N*-metallated azomethine ylide on several EMFs including La@C_{2v}(9)-C_{82}, La_2@I_h(7)-C_{80}, and Sc_3N@I_h(7)-C_{80} were unsuccessful, probably because of the energy mismatch between the frontier orbitals of both reagents. The cycloaddition of racemic La@C_{72}($C_6H_3Cl_2$) [20] with methyl (*E*)-*N*-[(*p*-methoxyphenyl)methylene]glycinate (**1**) in the presence of a chiral copper complex, generated in situ from (*R*)-2-(*tert*-butylthio)-1-(diphenylphosphino)ferrocene and copper(II) acetate, proceeded at room temperature to yield eight optically active cycloadducts (Fig. 9.2) [21]. Results of ROESY NMR studies suggest that all the isomers have a *cis*-configuration at the methine hydrogens of the pyrrolidine ring. Therefore, it is readily apparent that the *cis*-configuration was fixed by virtue of the high enentioface selectivity on the *N*-copper azomethine ylide complex. In addition, the similarity in the absorption and CD spectra of the eight isomers provided positive evidence that the cycloaddition to La@C_{72}($C_6H_3Cl_2$) occurred with remarkable site selectivity. Only two site isomers were formed.

9.2.3 *Silylene Addition*

Generation of silylenes in the reaction with C_{60} requires intense photoirradiation using a low-pressure mercury lamp illumination to trisilane as the precursor.

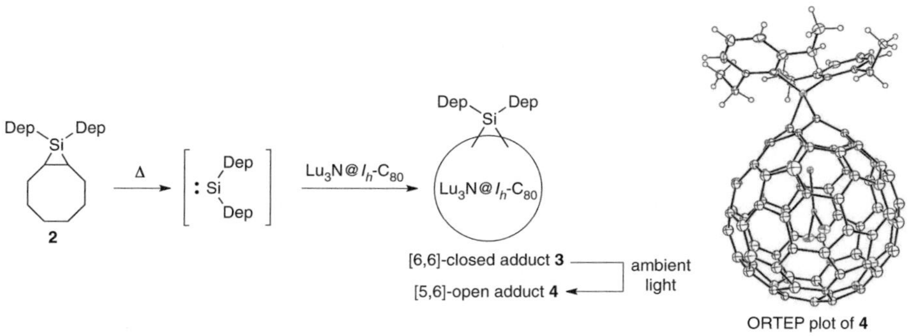

Fig. 9.3 Silylene addition to $Lu_3N@I_h$-C_{80}

However, it is noteworthy that EMFs tend to decompose under such harsh conditions. Therefore, the silylene addition methodology used for C_{60} could not be transferred directly into the chemistry of EMFs. Actually, Seyferth et al. were the first to report the discovery that hexamethylsilirane can generate silylene thermally [22]. Other cyclic silanes such as cyclohexene silacyclopropane have also been used as silylene precursors [23]. In fact, a new silylene precursor, 9,9-bis(2,6-diethylphenyl)-9-silabicyclo[6.1.0]nonane (**2**), can generate the corresponding silylene efficiently under both thermal and photochemical conditions [24]. The thermal reaction of $Lu_3N@I_h$-C_{80} with the silacyclopropane caused the dominant formation of the [6, 6]-closed silylene adduct (**3**). However, **3** was rather unstable and was transformed slowly into the [5, 6]-open silylene adduct (**4**) under ambient light (Fig. 9.3). Finally, **4** was isolated; its structure was resolved using single-crystal XRD. However, the silylene addition did not proceed under photochemical conditions. Apparently, the silylene adducts readily degrade to eliminate silylenes under photoirradiation. It is also noteworthy that the silylene addition with $Sc_3N@I_h$-C_{80} did not proceed efficiently; it gave the corresponding adducts in low yields.

9.2.4 Bisfulleroid Formation Reaction

Modification of the fullerene cage framework by replacing a cage carbon with another unit can alter the fullerene properties. Particularly, rationally designed cage-opening reactions enable the introduction of small guests into the cage as well as the ability to reclose the orifice to its original framework [25–27]. Such a strategy can sometimes be recognized as the "molecular surgery" concept. Accordingly, modification of the cage framework of EMFs persists as a challenging issue. As an initial trial, we adopted the reaction of $Sc_3C_2@I_h$-C_{80} with tetrazine to incorporate a four-membered ring on the cage framework, as Rubin accomplished on the C_{60} framework to prepare a stable C_{62} skeleton [28]. The reaction of $Sc_3C_2@I_h$-C_{80} and a large excess of 3,6-bis(4-*tert*-butylphenyl)-1,2,4,5-tetrazine (**5**) proceeded smoothly in *o*-DCB at 180 °C to afford a single product, although the reaction was much slower than that of C_{60} (Fig. 9.4a) [29]. The MALDI–TOF mass spectrum of the product displayed a single peak at *m*/*z* 1409, which is indicative of the successful addition of the $C_{22}H_{26}$ units. Subsequently, NMR spectroscopy was conducted for the anion form of the product because the neutral form of the product remained the paramagnetic character. Surprisingly, the NMR data showed that the product has C_1 symmetry despite the fact that the expected four-membered ring derivative (**6**) should have C_s symmetry if the addition took place at the expected [5, 6]-bond. In addition, the expected ^{13}C signals corresponding to the incorporated C_2 unit were not observed. Further NMR studies and theoretical calculations demonstrated that the product possesses a bisfulleroid structure (**7**). The plausible mechanism for the formation of **7** is described in Fig. 9.4b. The first two steps are [4 + 4] and retro-[2 + 2 + 2] cycloadditions. However, whereas the reaction with C_{60} proceeded via a double benzylic biradical intermediate that collapses to the corresponding C_{62} derivative, the larger C_{80} cage does not seem to favor the corresponding four-membered ring structure because larger carbon cages possess a lower degree of pyramidalization of the cage carbons. Instead, we assume that one of the benzylic

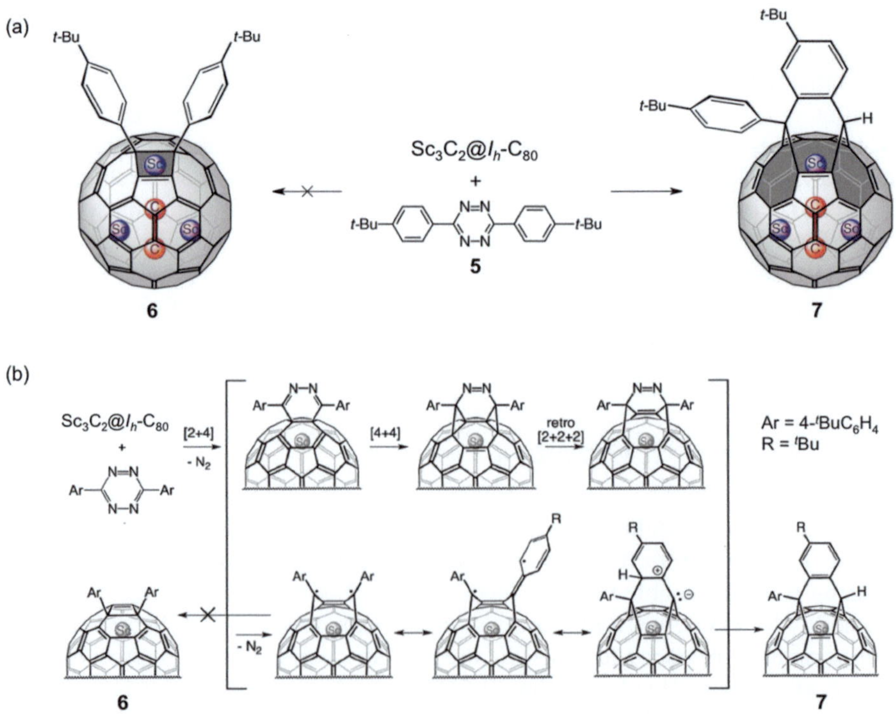

Fig. 9.4 (**a**) Reaction of tetrazine **5** with $Sc_3C_2@I_h$-C_{80} giving bisfulleroid **7** instead of the expected four-membered ring product **6**. (**b**) Formation mechanism for the formation of **7** (Reproduced with permission from Ref. [29]. Copyright 2012, American Chemical Society)

radicals of the intermediate recombines with the other radical through its hexadienyl radical resonance structure, forming a very stable six-membered ring within the bisfulleroid after 1,3-hydrogen migration (see Fig. 9.4b). The unexpected formation of the bisfulleroid is anticipated as a stepping-stone for additional ring opening.

9.3 EMF-Based Donor–Acceptor Conjugates

In the past two decades, the electron donor–acceptor assemblies designed for photosynthesis and photovoltaics have been of great interest in the field of solar energy conversion [30, 31]. Particularly, fullerenes such as C_{60} and C_{70} and their derivatives have been used widely as electron acceptors because of their unique three-dimensional structures, low reorganization energies, and rich reductive properties [32–34]. To date, hundreds of varieties of fullerene derivatives and fullerene-based assemblies have been designed and studied for mimicking the photosynthesis processes in nature or for developing better photovoltaic materials.

Recently, EMFs are emerging as new building blocks for photosynthesis and photovoltaics. Because of the metallic encapsulation, the electrochemical behaviors of EMFs differ greatly from those of empty fullerenes. For instance, the one-electron oxidation of $M_2@I_h$-C_{80} or $M_3N@I_h$-C_{80} is much more facile than most empty fullerenes, although its one-electron reduction potential is either lower or higher than that of C_{60} or C_{70}. Therefore, from an electrochemical perspective, EMFs can serve as an electron acceptor or electron donor. Accordingly, recent studies specifically examine not only the electron-accepting property but also the electron-donating property of EMFs [35]. In this section, we summarize the electron acceptor characters as well as electron donor characters of a series of EMFs.

9.3.1 Trimetallic Nitride Templated EMFs

Trimetallic nitride templated EMFs (TNT-EMFs) have been studied widely in recent years. The redox potentials of some typical species are presented in Table 9.1 [6]. The electrochemical behaviors of these TNT-EMFs are apparently different from one another, indicating their cluster-dependent redox properties. It appears that the electron acceptor characters of TNT-EMFs might be tunable by manipulating the composition of the endohedral cluster. Generally, the first reduction potentials of $M_3N@I_h$-C_{80} are 140–300 mV higher than that of C_{60}, suggesting their higher LUMO energy levels with respect to that of C_{60}. Additionally, it has been evidenced that using a $Lu_3N@I_h$-C_{80} derivative as an electron acceptor can minimize the LUMO energy offset between $Lu_3N@I_h$-C_{80} and the commonly used donor/polymer poly(3-hexyl)thiophene (P3HT), which enhanced organic photovoltaic performance through increase of the open circuit voltage [36, 37].

Table 9.1 Redox potentials[a] (V vs. Fc/Fc^{+}) of typical EMFs and TNT-EMFs

Fullerene	${}^{ox}E_2$	${}^{ox}E_1$	${}^{red}E_1$	${}^{red}E_2$	${}^{red}E_3$
C_{60}		$+1.21^{b,c}$	−1.12	−1.50	−1.95
La@C_{2v}-C_{82}	$+1.07^{b,c}$	+0.07	−0.42	−1.37	−1.53
La_2@I_h-C_{80}	+0.95	+0.56	−0.31	−1.71	$-2.13^{b,c}$
Sc_3N@I_h-C_{80}	$+1.09^{d}$	$+0.59^{d}$	$-1.26^{b,e}$	$-1.62^{b,e}$	$-2.37^{b,e}$
Lu_3N@I_h-C_{80}	$+1.11^{d}$	$+0.64^{d}$	$-1.42^{b,e}$	$-1.80^{b,e}$	$-2.26^{b,e}$
Pr_3N@I_h-C_{80}		+0.59	$-1.41^{b,e}$	$-1.84^{b,e}$	
$TiSc_2N$@I_h-C_{80}		+0.16	−0.94	−1.58	−2.21

[a]Half-wave potentials unless otherwise stated
[b]Irreversible process
[c]Values were obtained using differential pulse voltammetry
[d]Quasi-reversible process
[e]Peak potentials

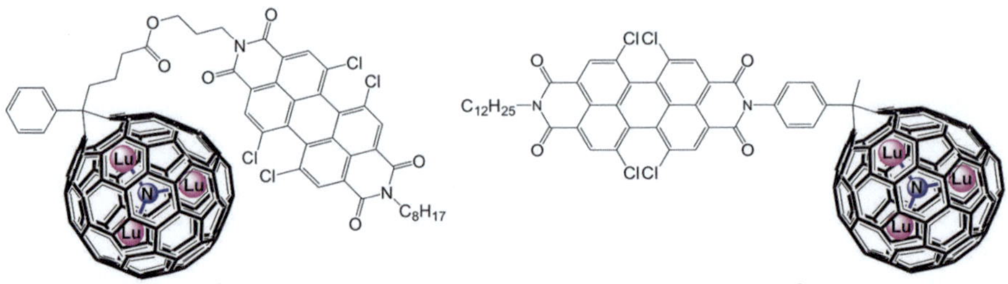

8

9

Fig. 9.5 Schematic structures of TNT-EMF-based conjugates **8** and **9**

The donor character of M_3N@I_h-C_{80} has been investigated and compared with that of C_{60}. In the case of C_{60}, its electron-donating property is extremely poor, which is related to the first oxidation of C_{60} at a high potential of 1.21 V vs. ferrocene/ferrocenium couple [34]. In fact, the electron transfer oxidation of C_{60} only occurs assisted by either strong oxidants/acceptors or scandium ions that can bind with the electron transfer product [38–41]. Such limitations might impede the application of C_{60} as a practicable electron donor. In contrast, the first oxidation potentials of M_3N@I_h-C_{80} are 500–600 mV lower than that of C_{60}, indicating the possible electron donation from M_3N@I_h-C_{80} under moderate condition. However, despite this simple proposal, experimental evidence was not available until recently. In 2012, we first reported a covalently linked Lu_3N@I_h-C_{80}–perylenebisimide (PDI) conjugate (see **8** in Fig. 9.5) [42], in which Lu_3N@I_h-C_{80} serves as the electron donor and PDI as the light harvester as well as the electron acceptor. Importantly, a series of photophysical studies, in terms of spectroscopy and kinetics, were performed, corroborating a photoinduced electron transfer evolving from the ground state of Lu_3N@I_h-C_{80} to the singlet excited state of PDI. This unusual electron transfer event caused the formation of a metastable radical ion pair (Lu_3N@I_h-C_{80})$^{+\bullet}$–$PDI^{-\bullet}$ with a lifetime of ca. 120 ps in toluene. In stark contrast, the photoreactivity of an analogous C_{60}–PDI conjugate is governed exclusively by a cascade of energy transfer process. Therefore, a better donor character of $Lu_3N@I_h$-C_{80} relative to that of C_{60} was revealed.

Moreover, another linear conjugate of $Lu_3N@I_h$-C_{80}–PDI (see **9** in Fig. 9.5) was designed and synthesized very recently [43]. The UV–Vis absorption and electrochemical studies revealed no evidence of electronic interaction between $Lu_3N@I_h$-C_{80} and PDI moieties in the ground state. However, photophysical studies demonstrated the initial formation of the singlet radical ion pair state, $^1(Lu_3N@I_h\text{-}C_{80})^{\bullet+}$–$PDI^{\bullet-}$, which undergoes radical ion pair intersystem crossing and produces the triplet radical ion pair state $^3(Lu_3N@I_h\text{-}C_{80})^{\bullet+}$–$PDI^{\bullet-}$ with a lifetime as long as 102 ns (toluene). This lifetime of the radical ion pair state increases about 1000-fold relative to that of the $^1(Lu_3N@I_h\text{-}C_{80})^{\bullet+}$–$PDI^{\bullet-}$ in the case of flexibly linked $Lu_3N@I_h$-C_{80}–PDI (**8**). It has been proposed that not only the morphology change but also the presence of the Lu_3N cluster might account for the unprecedented formation of the triplet radical ion pair state $^3(Lu_3N@I_h\text{-}C_{80})^{\bullet+}$–$PDI^{\bullet-}$ with long lifetimes.

Next, the $Lu_3N@I_h$-C_{80}-based donor was characterized further by a photovoltaic test, in which a bilayer heterojunction solar cell device was constructed with $Lu_3N@I_h$-C_{80}–PCBM (**10**) and a PDI derivative **11** as an electron donor and an electron acceptor, respectively (see Fig. 9.6) [42]. Particularly, a V_{oc} of 0.46 V and a J_{sc} of 0.38 mA cm^{-2} were measured for this PDI/$Lu_3N@I_h$-C_{80}–PCBM-based solar cell device (see *I–V* curve in Fig. 9.6). Although this untypical solar cell was not optimized, the sign of V_{oc} and the direction of photocurrent clearly

Donor

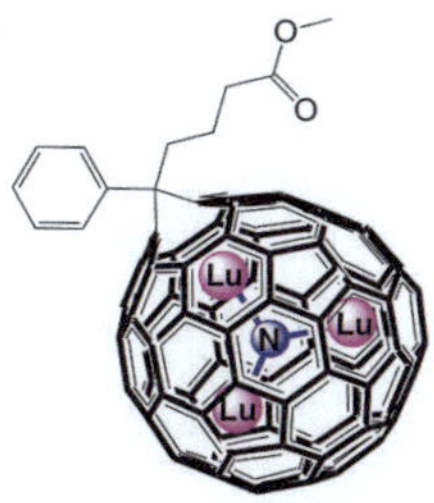

10

Acceptor

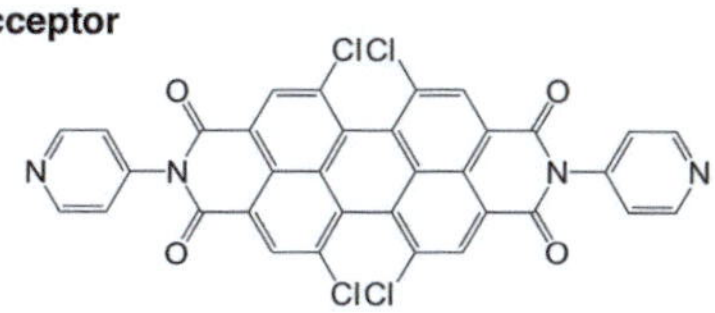

11

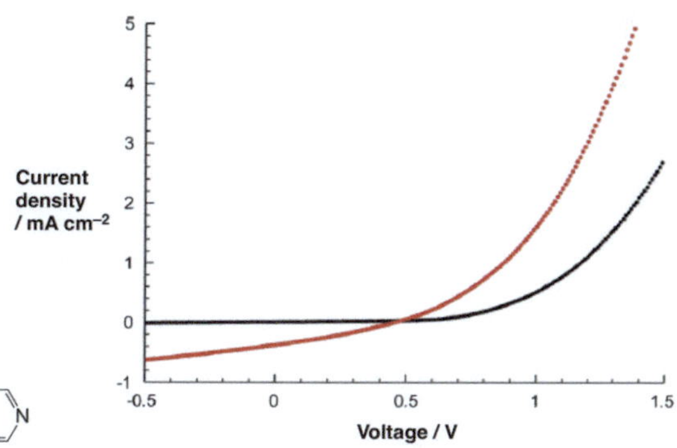

Fig. 9.6 Schematic structures of $Lu_3N@I_h$-C_{80}–PCBM (**10**) and the PDI acceptor **11** used for the solar cell device (*left panel*); current density–voltage (*J–V*) characteristics of PDI/$Lu_3N@I_h$-C_{80}–PCBM-based solar cell device under AM 1.5 radiation with the illuminated current density and dark current density are shown, respectively, as *red* and *black lines* (*right panel*) (Reproduced with permission from Ref. [42]. Copyright 2012, American Chemical Society)

underline the unique oxidative role of **10** when coupled with an electron acceptor. Therefore, in contrast to the acceptor character of $Lu_3N@I_h$-C_{80}–PCBH described above, where the P3HT was used as donor, the donor character of **10** was verified by either photophysical studies or a device testing experiment in which the PDI moiety was used as the counterpart: an electron acceptor.

9.3.2 Di-metallic EMFs

Furthermore, recent studies have demonstrated that di-metallofullerene $M_2@I_h$-C_{80} (M = La and Ce) might be another alternative electron acceptor or donor. In comparison with TNT-EMFs, di-metallic EMFs as well as monometallic EMFs possess remarkably high first reduction potentials [6]. In 2010, a report described the preparation of the conjugate $La_2@I_h$-C_{80}–ex-TTF, in which the ex-TTF moiety is combined with $La_2@I_h$-C_{80} via a 5,6-pyrrolidine ring (see **12** in Fig. 9.7) [44]. Upon photoexcitation, a fast charge separation was identified (6.0×10^{10} s^{-1} in THF and 5.0×10^{10} s^{-1} in toluene), followed by the formation of a radical ion pair $(La_2@I_h\text{-}C_{80})^{\bullet -}$–ex-$TTF^{\bullet +}$ with a lifetime as long as 3.3 ns in toluene, suggesting an acceptor character of $La_2@I_h$-C_{80}.

It is also worth mentioning that linking of TNT-EMFs with ex-TTF gives unstable products [45]. The difference in the electronic properties of EMFs can influence the stabilities of electron donor–acceptor moieties.

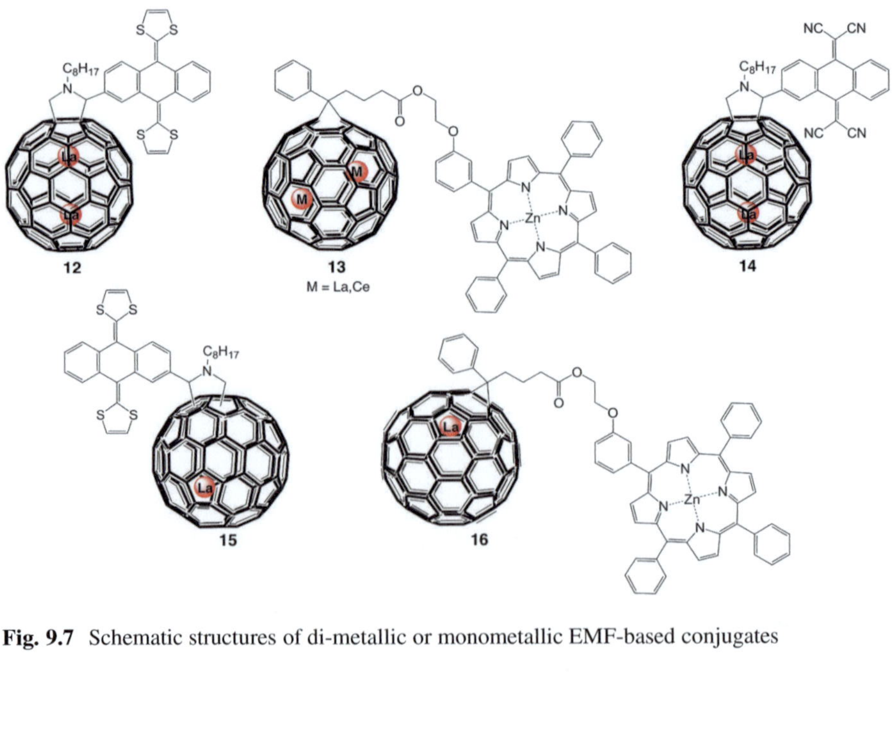

Fig. 9.7 Schematic structures of di-metallic or monometallic EMF-based conjugates

In another study, an electron donor ZnP that is weaker than Fc or ex-TTF has been attached to $M_2@I_h$-C_{80} (M = La and Ce) via a flexible linker, giving rise to conjugates **13**, as presented in Fig. 9.7 [46, 47]. It is particularly interesting that upon photoexcitation, these conjugates undergo switchable charge-transfer events. Specifically, as revealed by spectroscopic analyses, an unprecedented change occurs in photoinduced charge-transfer chemistry in toluene and THF vs. BzCN and DMF, producing radical ion pairs $(M_2)^{\bullet -}@I_h$-C_{80}–$(ZnP)^{\bullet +}$ in less-polar solvents and $M_2@(I_h$-$C_{80})^{\bullet +}$–$(ZnP)^{\bullet -}$ in polar solvents. The rationale is the following: (i) the stabilizing effect of polar solvents such as BzCN and DMF renders $(M_2@I_h$-$C_{80})$–^{1*}(ZnP) and $M_2@(I_h$-$C_{80})^{\bullet +}$–$(ZnP)^{\bullet -}$ comparable energy levels; (ii) strong electronic coupling between the C_{80} cage and ZnP, which is associated with the face-to-face arrangement of two moieties and their proximity, facilitates their mutual fast charge separation instead of that between M_2 and ZnP. In contrast, the analogous conjugate of $Sc_3N@I_h$-C_{80}–ZnP afforded only the $(Sc_3N@I_h$-$C_{80})^{\bullet -}$–$ZnP^{\bullet +}$ radical ion pair, irrespective of the solvent polarity. The $(Sc_3N@I_h$-$C_{80})^{\bullet +}$–$ZnP^{\bullet -}$ radical ion pair, however, is unlikely to be formed because of the thermodynamically uphill processes. Kinetically, the time absorption profiles corroborate that the $(Sc_3N@I_h$-$C_{80})^{\bullet -}$–$ZnP^{\bullet +}$ radical ion pair state decays at 2.3×10^{10} s^{-1} in toluene and 2.0×10^{10} s^{-1} in BzCN.

Moreover, we have demonstrated that $La_2@I_h$-C_{80} can be used as a pure electron donor when combined with a typical acceptor of 11,11,12,12-tetracyano-9,10-anthra-*p*-quinodimethane (TCNQ) (see **14** in Fig. 9.7) [48]. Because of the strong electron-accepting character as well as the weak electron-donating character of the TCNQ moiety, the EMF moiety no longer acts as an electron acceptor upon photoexcitation. According to spectroscopic analyses, the transient absorption features of one-electron-reduced TCAQ were observed clearly, suggesting that an electron transfer event occurred in **4**. From multiwavelength analyses, the lifetime of the photoinduced radical ion pair $(La_2@I_h$-$C_{80})^{\bullet +}$–$(TCAQ)^{\bullet -}$, derived from the singlet excited state, was ascertained as 230 ± 15 ps in THF and 110 ± 10 ps in toluene.

9.3.3 Monometallic EMFs

In contrast to di-metallic EMFs, monometallic EMFs such as $La@C_{2v}$-C_{82} are produced more abundantly but are seldom studied in the field of photosynthesis and photovoltaics. A likely rationale includes the presence of multiple isomeric products that are formed in most reactions. Nevertheless, it is noteworthy that the one-electron reduction potential of $La@C_{2v}$-C_{82} is almost comparable to that of $La_2@I_h$-C_{80}, which inspired our proposal of using $La@C_{2v}$-C_{82} as an electron acceptor (see Table 9.1). We conducted conjugation between $La@C_{2v}$-C_{82} and a TTF moiety via a pyrrolidine ring by 1,3-dipolar cycloaddition as analogous to the above-described $La_2@I_h$-C_{80}–ex-TTF. The reaction yielded two inseparable $La@C_{2v}$-C_{82}–ex-TTF conjugate diastereomers (see **15** in Fig. 9.7) [49]. Although structural identification was not achieved experimentally, we inferred that the two diastereomers have comparable redox properties based on the electrochemical measurements. In this regard, the excited-state interaction between La@C_{2v}-C_{82} and ex-TTF was investigated with the mixture. Accordingly, the (La@C_{2v}-C_{82})$^{\bullet -}$–(ex-TTF)$^{\bullet +}$ radical ion pair state was evolved from the ex-TTF-centered and La@C_{2v}-C_{82}-centered excited states with respective lifetimes (rate constants) of 4.3 ± 0.5 ps (2.3×10^{11} s^{-1}) and 6.3 ± 0.5 ps (1.5×10^{11} s^{-1}). The charge recombination was remarkably slow; the lifetime of the radical ion pair was determined as 2.4 ± 0.5 ns (4.1×10^{8} s^{-1}) in THF and 1.1 ± 0.3 ns (9.1×10^{8} s^{-1}) in cyclohexylisonitrile.

In another study, three isomeric conjugates of La@C_{2v}-C_{82}–H_2P were carefully prepared and characterized [50]. Although detailed spectroscopic analyses are lacking, very efficient fluorescence quenching in the major isomer of the porphyrin conjugate (see **16** in Fig. 9.7) suggests the occurrence of an electron transfer event upon photoexcitation.

9.4 Carrier Transport Properties of EMFs

9.4.1 Conductivity of EMF Crystals

Fullerene-based materials can be regarded as a promising candidate for *n*-type semiconductors in organic devices [51–59]. Regarding EMFs, early works [60–63] focused on the fabrication of thin films by vacuum deposition to prepare field emission transistors (FETs), which were subsequently subjected to conductivity measurements. However, the electron mobility of EMFs was estimated as being on the order of 10^{-4} cm^2 V^{-1} s^{-1}. The low mobilities in the EMF-based FETs might be caused by the low crystallinity of the thin films. Therefore, a well-ordered molecular arrangement is expected to enhance the carrier transport properties of EMFs. In this regard, recent advances in large-scale preparation and chemical functionalization of EMFs enable investigation of the carrier transport properties of EMF-containing crystals. We have found that adamantylidene (Ad) addition to EMFs strongly increases their crystallinity [64]. Accordingly, the electron mobility of single crystals of La@C_{2v}-C_{82}(Ad) was investigated using flash-photolysis time-resolved microwave conductivity (FP-TRMC) measurements [65]. At this point, it is noteworthy that the FP-TRMC method has served as an electrodeless measurement tool to evaluate intrinsic carrier transport properties [66, 67]. As a result, single crystals of La@C_{2v}-C_{82}(Ad) exhibited anisotropic and high electron mobility of μ >10 cm^2 V^{-1} s^{-1} along the *c* axis. As shown in Fig. 9.8a, the crystal packing of La@C_{2v}-C_{82}(Ad) shows that the shortest interatomic distance between the closest La@C_{2v}-C_{82}(Ad) molecules along the *c* axis reaches 2.60 Å. In addition, theoretical calculations suggest that the nearest carbon atom possesses a large LUMO coefficient. The mobility of a drop-cast film of La@C_{2v}-C_{82}(Ad) was obtained as 7×10^{-2} cm^2 V^{-1} s^{-1}. Consequently, the orderly alignment of the

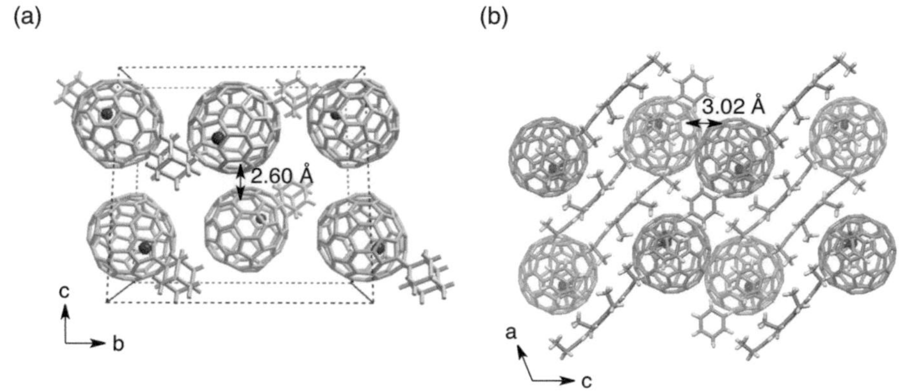

Fig. 9.8 Crystal packing structures of (**a**) the La@C_{2v}-C_{82}(Ad) single crystal and (**b**) the La@C_{2v}-C_{82} · Ni^{II}(OEP) · 1.5 benzene co-crystal

La@C_{2v}-C_{82}(Ad) molecules in crystals plays an important role in the excellent electron transporting. In addition, the La@C_{2v}-C_{82}(Ad) single crystal exhibited a considerable dark current in the *I*–*V* trace measurement, suggesting that the single crystal works as an organic conductor without light irradiation.

The drift mobilities of La@C_{2v}-C_{82}(Ad) and pristine La@C_{2v}-C_{82} were also compared to that of C_{60}(Ad) using time-of-flight (TOF) measurements to examine the effect of endohedral La atom doping. The electron drift mobilities of the drop-cast films of La@C_{2v}-C_{82}(Ad), La@C_{2v}-C_{82}, and C_{60}(Ad) were estimated, respectively, as $8.0 \pm 2.4 \times 10^{-2}$ cm^2 V^{-1} s^{-1}, $6.0 \pm 2.0 \times 10^{-2}$ cm^2 V^{-1} s^{-1}, and $5.0 \pm 1.5 \times 10^{-3}$ cm^2 V^{-1} s^{-1}. This observation indicated the importance of La atom doping in enhancing the charge-transporting property.

Another means of preparing EMF crystals is co-crystallization of EMFs with metal porphyrins such as nickel(II) octaethylporphyrin (Ni^{II}(OEP)). The co-crystal of La@C_{2v}-C_{82} · Ni^{II}(OEP) exhibited anisotropic and high electron mobility of 0.9 cm^2 V^{-1} s^{-1} along the *c* axis [68]. The crystal structure exhibits that the shortest intermolecular distance between the neighboring La@C_{2v}-C_{82} molecules is 3.02 Å, as presented in Fig. 9.8b. Such a short intermolecular distance between the La@C_{2v}-C_{82} units can engender strong orbital interaction and the formation of efficient charge career path. On the other hand, the electron mobility of the single crystal of Ni^{II}(OEP) was obtained as $\mu = 0.07$ cm^2 V^{-1} s^{-1} by the FP-TRMC measurements. Therefore, it can be inferred that the high electron mobility of the single crystal of La@C_{2v}-C_{82} · Ni^{II}(OEP) was derived from the presence of La@C_{2v}-C_{82} molecules.

To provide theoretical insight into the single-crystal properties of La@C_{2v}-C_{82}(Ad) and La@C_{2v}-C_{82} · Ni^{II}(OEP), local density functional calculations were conducted. The energy difference between the valence band top and the conduction band bottom (band gap) for the single crystal of La@C_{2v}-C_{82}(Ad) was estimated as 5 meV. At room temperature, the La@C_{2v}-C_{82}(Ad) single crystal has numerous valence electrons, which have sufficient thermal energy, excited across the very small band gap to the conduction band. Therefore, the La@C_{2v}-C_{82}(Ad) single crystal can behave as a semimetal. The band structure for the co-crystal of La@C_{2v}-$C_{82}\cdot Ni^{II}$(OEP) shows some bands cross the Fermi level along the $\Gamma > Y$ and $C > Z$ directions. Accordingly, these calculations supported that single crystals of La@C_{2v}-C_{82}(Ad) and La@C_{2v}-$C_{82}\cdot Ni^{II}$(OEP) have good electrical conduction.

For the La@C_{2v}-C_{82}(Ad) single crystal, the calculated effective masses of electrons of the conduction band bottom and the hole of the valence band top were 0.97 and 0.91 m_0 (where m_0 denotes the mass of free electrons), respectively, suggesting nearly free electron behavior. However, the effective masses of the electron and hole for the La@C_{2v}-$C_{82}\cdot Ni^{II}$(OEP) single crystal along the $\Gamma > Z$ direction, which corresponds to the c axis direction, were calculated, respectively, as 9.89 and 11.03 m_0. The higher electron mobility for the La@C_{2v}-C_{82}(Ad) single crystal than that for La@C_{2v}-$C_{82}\cdot Ni^{II}$(OEP) is associated with the difference in the effective masses.

To deepen our understanding about the carrier transport properties of EMFs, a series of I_h-C_{80}-based EMFs were examined using comparative assessment of FP-TRMC measurements and theoretical calculations. Drop-cast thin films of La_2@I_h-C_{80}, Sc_3N@I_h-C_{80}, and Sc_3C_2@I_h-C_{80} were conducted for the study (Fig. 9.9) [69]. These endohedrals comprise the same I_h-C_{80} cage. However, the difference in the encaged species engenders different electronic properties. First, La_2@I_h-C_{80} and Sc_3N@I_h-C_{80} are diamagnetic, whereas Sc_3C_2@I_h-C_{80} is paramagnetic. Secondly, the HOMO–LUMO gap of La_2@I_h-C_{80} is much smaller than that of Sc_3N@I_h-C_{80}.

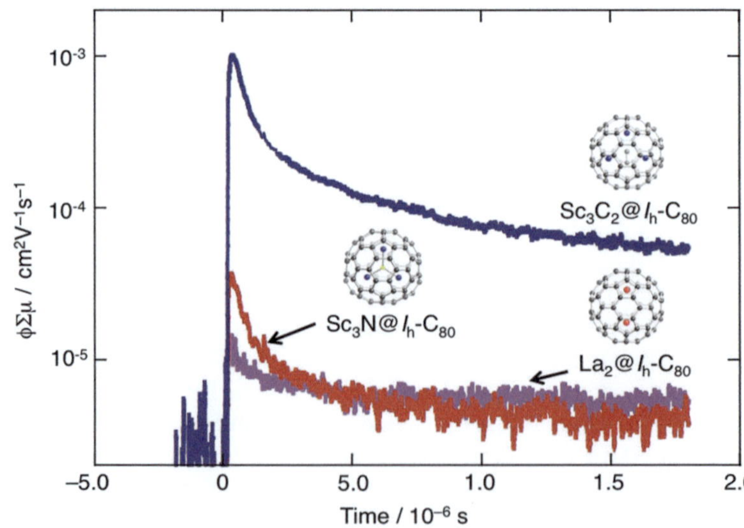

Fig. 9.9 Conductivity transients observed in air at room temperature for solvent-cast films of La_2@I_h-C_{80}, Sc_3N@I_h-C_{80}, and Sc_3C_2@I_h-C_{80} upon excitation at 355 nm and $(1.4–1.9) \times 10^{16}$ photons cm^{-2} (Reproduced with permission from Ref. [69]. Copyright 2012, American Chemical Society)

The observed electron mobility of the $Sc_3C_2@I_h$-C_{80} thin film reached 0.13 cm^2 V^{-1} s^{-1}, whereas those of the $La_2@I_h$-C_{80} thin film (5.0×10^{-3} cm^2 V^{-1} s^{-1}) and the $Sc_3N@I_h$-C_{80} thin film (5.7×10^{-3} cm^2 V^{-1} s^{-1}) displayed much less electron mobility. The I–V trace measurements with irradiation of light for the drop-cast thin films showed that the $Sc_3C_2@I_h$-C_{80} thin film had higher conductivity than either the $La_2@I_h$-C_{80} or the $Sc_3N@I_h$-C_{80} thin film. In addition, the $Sc_3C_2@I_h$-C_{80} thin film showed a considerable dark current, indicating that $Sc_3C_2@I_h$-C_{80} can conduct current without generation of charge carriers and that it can work as an organic conductor like the La@C_{2v}-C_{82}(Ad) single crystal. Results imply that the paramagnetic property of $Sc_3C_2@I_h$-C_{80} is responsible for the high electron mobility. Theoretical calculations revealed that the band gap of the $Sc_3C_2@I_h$-C_{80} is about 1/6 and 1/17 of those of $La_2@I_h$-C_{80} and $Sc_3N@I_h$-C_{80}, respectively. Also, the estimated intrinsic carrier concentration of $Sc_3C_2@I_h$-C_{80} is much higher than that of either of the two other I_h-C_{80} EMFs. Accordingly, the large conductivity of $Sc_3C_2@I_h$-C_{80} can be rationalized by the calculations.

9.4.2 Conductivity of a Single-Molecule EMF

Single-molecule electronics are a building block in ultrasmall electronic devices [70]. Recently, electron transport through a single-molecule EMF bridging two metal electrodes was studied [71]. The single Ce@C_{2v}-C_{82} molecule junction fabricated by direct binding of the metallofullerene to Ag electrodes shows high and fixed conductance. The obtained conductance of 0.28 ($\pm$0.05)G_0 ($G_0 = 2e^2/h$) was much higher than that of single-molecule junctions having anchoring groups (e.g., thiols, <0.01 G_0), but the obtained conductance is only half of the single C_{60} molecule junction of 0.5 G_0. Compared with the conductance of the single C_{60} molecule junction, the reduction of the conductance of the single Ce@C_{2v}-C_{82} molecule junction was explained by the first-principles quantum transport calculations. The electrons are markedly more delocalized in C_{60} than those in Ce@C_{2v}-C_{82}. The LUMO and LUMO + 1 of Ce@C_{2v}-C_{82} are more localized than those of C_{60} because the Ce atom is located at an off-cage center position inside the C_{82} cage. These are the main reasons why the C_{60} system conductance was greater than that of system involving Ce@C_{2v}-C_{82}. When using Au electrodes, the single Ce@C_{2v}-C_{82} molecule junction was not formed by the break junction technique because the Ce@C_{2v}-C_{82} molecule could not be trapped in the large Au nanogap formed immediately after breaking the Au contacts.

9.5 Surface-Grafted EMFs

Controlled deposition and immobilization of EMFs on certain substrates is necessary for their future implementation in molecular devices. The deposition by physisorption of EMFs on different surfaces has been prepared. Surface structures

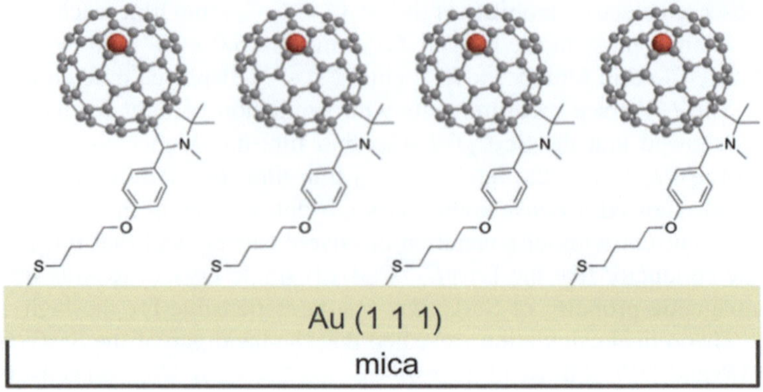

Fig. 9.10 Schematic structure of the La@C_{2v}-C_{82}-grafted SAM

have been investigated using scanning tunneling microscopy (STM) [72–74]. However, to fabricate robust surfaces, chemical bonding between the EMFs and the surfaces with a suitable grafting group might be prepared. We chose thiol moiety as an anchor for grafting La@C_{2v}-C_{82} onto the gold surface. The La@C_{2v}-C_{82} derivative, functionalized with a thioacetyl protecting group, was synthesized successfully and employed for preparing self-assembled monolayers (SAMs) (Fig. 9.10) [75]. The SAM-modified substrates revealed the same electrochemical and magnetic properties of the EMF in solution. Particularly, electrochemical switching between the neutral (paramagnetic) form and the anionic (diamagnetic) form of the grafted molecules was monitored by measuring the EPR spectrum after application of a fixed potential to the SAM. As a result, three consecutive reversible switching cycles were performed, giving rise to the expected appearance–disappearance of the EPR signal.

9.6 Concluding Remarks

Chemical reactions of EMFs offer numerous possibilities for transformation to fascinating derivatives. The DA reaction of paramagnetic La@C_{2v}-C_{82} with cyclopentadiene derivatives was investigated, in which the concerted mechanism was rationalized by DFT calculations. Enantioselective 1,3-dipolar cycloaddition to EMF was achieved to prepare optically active derivatives using La@C_{72}($C_6H_3Cl_2$) as the reactant in the aid of the chiral copper complex. Silylene addition to Lu_3N@I_h-C_{80} was demonstrated by thermal reaction with a 9-silabicyclo[6.1.0]nonane derivative. The reaction of Sc_3C_2@I_h-C_{80} with a tetrazine compound caused unprecedented formation of a bisfulleroid structure. These recently unveiled chemical reactivities of EMFs present high potential for additional chemical transformations.

In addition, recent photophysical studies display clearly that EMFs including TNT-EMFs as well as di-metallic and monometallic EMFs (i.e., $Lu_3N@I_h$-C_{80}, $M_2@I_h$-C_{80}, and $La@C_{2v}$-C_{82}) can serve unprecedentedly as electron donors or acceptors. However, the reduction roles of $M_2@I_h$-C_{80} and $La@C_{2v}$-C_{82} moieties have been investigated, respectively, in conjugates using ex-TTF and H_2P as electron donors. Despite the simplicity of these EMF-based conjugates, the occurrence of photoinduced electron transfer events can be identified clearly.

Carrier transport properties of EMFs were evaluated using FP-TRMC measurements and break junction techniques. Particularly, the single crystal of $La@C_{2v}$-C_{82}(Ad) exhibited anisotropic and quite high electron mobility of μ >10 cm^2 V^{-1} s^{-1} along the c axis. It was deduced from the results that a well-ordered molecular arrangement is expected to enhance the carrier transport properties of EMFs. Additionally, results show that the conductivity of $Sc_3C_2@I_h$-C_{80} reached the mobility of 0.13 cm^2 V^{-1} s^{-1}, even in the thin films. The unusually high mobility can be related to the paramagnetic character of $Sc_3C_2@I_h$-C_{80}.

Furthermore, immobilization of EMF derivatives on metal surfaces was developed, and the reversible electron uptake and release property was proven using EPR measurements. This technique is expected to be applicable for future implementation in molecular devices. However, to date, EMF-based devices remain limited in a conceptual framework and are very far from being practical. Future work must involve topological modification of the model conjugate as well as the optimization of related devices.

References

1. Akasaka T, Nagase S (eds) (2002) Endofullerenes: a new family of carbon clusters. Kluwer Academic Publishers, Dordrecht
2. Chaur MN, Melin F, Ortiz AL, Echegoyen L (2009) Chemical, electrochemical, and structural properties of endohedral metallofullerenes. Angew Chem Int Ed 48:7514–7538
3. Akasaka T, Wudl F, Nagase S (eds) (2010) Chemistry of nanocarbons. Wiley, Chichester
4. Yamada M, Akasaka T, Nagase S (2010) Endohedral metal atoms in pristine and functionalized fullerene cages. Acc Chem Res 43:92–102
5. Lu X, Feng L, Akasaka T, Nagase S (2012) Current status and future developments of endohedral metallofullerenes. Chem Soc Rev 41:7723–7760
6. Popov A, Yang S, Dunsch L (2013) Endohedral fullerenes. Chem Rev 113:5989–6113
7. Iiduka Y, Wakahara T, Nakahodo T, Tsuchiya T, Sakuraba A, Maeda Y, Akasaka T, Yoza K, Horn E, Kato T, Liu MTH, Mizorogi N, Kobayashi K, Nagase S (2005) Structural determination of metallofullerene Sc_3C_{82} revisited: a surprising finding. J Am Chem Soc 127:12500–12501
8. Yamada M, Kurihara H, Suzuki M, Guo JD, Waelchli M, Olmstead MM, Balch AL, Nagase S, Maeda Y, Hasegawa T, Lu X, Akasaka T (2014) $Sc_2@C_{66}$ revisited: an endohedral fullerene with scandium ions nestled within two unsaturated linear triquinanes. J Am Chem Soc 136:7611–7614
9. Yamada M, Akasaka T, Nagase S (2011) New vistas in endohedral metallofullerenes. In: D'Souza F, Kadish KM (eds) Handbook of carbon nano materials, vol 1. World Scientific, Singapore, pp 145–184

10. Yamada M, Lu X, Feng L, Sato S, Takano Y, Nagase S, Akasaka T (2013) Fundamental and applied aspects of endohedral metallofullerenes as promising carbon nanomaterials. In: Torres T, Bottari G (eds) Organic nanomaterials. Wiley, New Jersey, pp 241–258
11. Lu X, Akasaka T, Nagase S (2011) Chemistry of endohedral metallofullerenes: the role of metals. Chem Commun 47:5942–5957
12. Akasaka T, Lu X (2012) Structural and electronic properties of endohedral metallofullerenes. Chem Rec 12:256–269
13. Hofmann M, Schaefer HF III (1999) The $[C_6H_{10}]^{\bullet+}$ hypersurface: the parent radical cation Diels–Alder reaction. J Am Chem Soc 121:6719–6729
14. Haberl U, Wiest O, Steckhan E (1999) Ab initio studies of the radical cation Diels–Alder reaction. J Am Chem Soc 121:6730–6736
15. Hofmann M, Schaefer HF III (1999) Pathways for the reaction of the butadiene radical cation, $[C_4H_6]^{\bullet+}$, with ethylene. J Phys Chem A 103:8895–8905
16. Maeda Y, Miyashita J, Hasegawa T, Wakahara T, Tsuchiya T, Nakahodo T, Akasaka T, Mizorogi N, Kobayashi K, Nagase S, Kato T, Ban N, Nakajima H, Watanabe Y (2005) Reversible and regioselective reaction of La@C_{82} with cyclopentadiene. J Am Chem Soc 127:12190–12191
17. Maeda Y, Sato S, Inada K, Nikawa H, Yamada M, Mizorogi N, Hasegawa T, Tsuchiya T, Akasaka T, Kato T, Slanina Z, Nagase S (2010) Regioselective exohedral functionalization of La@C_{82} and its 1,2,3,4,5-pentamethylcyclopentadiene and adamantylidene adducts. Chem Eur J 16:2193–2197
18. Sato S, Maeda Y, Guo JD, Yamada M, Mizorogi N, Nagase S, Akasaka T (2013) Mechanistic study of the Diels–Alder reaction of paramagnetic endohedral metallofullerene: reaction of La@C_{82} with 1,2,3,4,5-pentamethylcyclopentadiene. J Am Chem Soc 135:5582–5587
19. Filippone S, Maroto EE, Martín-Domenech Á, Suarez M, Martín N (2009) An efficient approach to chiral fullerene derivatives by catalytic enantioselective 1,3-dipolar cycloadditions. Nat Chem 1:578–582
20. Wakahara T, Nikawa H, Kikuchi T, Nakahodo T, Rahman GMA, Tsuchiya T, Maeda Y, Akasaka T, Yoza K, Horn E, Yamamoto K, Mizorogi N, Slanina Z, Nagase S (2006) La@C_{72} having a non-IPR carbon cage. J Am Chem Soc 128:14228–14229
21. Sawai K, Takano Y, Izquierdo M, Filippone S, Martín N, Slanina Z, Mizorogi N, Waelchli M, Tsuchiya T, Akasaka T, Nagase S (2011) Enantioselective synthesis of endohedral metallofullerenes. J Am Chem Soc 133:17746–17752
22. Seyferth D, Annarelli DC (1975) Generation of dimethylsilylene under mild conditions by the thermolysis of hexamethylsilirane. J Am Chem Soc 97:7162–7163
23. Boudjouk P, Black E, Kumarathasan R (1991) Synthesis of 1,1-di-tert-butylsilirane, the first silirane with no substituents on the ring carbons. Organometallics 10:2095–2096
24. Sato K, Kako M, Suzuki M, Mizorogi N, Tsuchiya T, Olmstead MM, Balch AL, Akasaka T, Nagase S (2012) Synthesis of silylene-bridged endohedral metallofullerene $Lu_3N@I_h$-C_{80}. J Am Chem Soc 134:16033–16039
25. Komatsu K, Murata M, Murata T (2005) Encapsulation of molecular hydrogen in fullerene C_{60} by organic synthesis. Science 307:238–240
26. Kurotobi K, Murata T (2011) A single molecule of water encapsulated in fullerene C_{60}. Science 333:613–616
27. Morinaka Y, Sato S, Wakamiya A, Nikawa H, Mizorogi N, Tanabe F, Murata M, Komatsu K, Furukawa K, Kato T, Nagase S, Akasaka T, Murata Y (2013) X-ray observation of a helium atom and placing a nitrogen atom inside He@C_{60} and He@C_{70}. Nat Commun 4:1554
28. Qian W, Chuang SC, Amador RB, Jarrosson T, Sander M, Pieniazek S, Khan SI, Rubin Y (2003) Synthesis of stable derivatives of C_{62}: the first nonclassical fullerene incorporating a four-membered ring. J Am Chem Soc 125:2066–2067
29. Kurihara H, Iiduka Y, Rubin Y, Waelchli M, Mizorogi N, Slanina Z, Tsuchiya T, Nagase S, Akasaka T (2012) Unexpected formation of a $Sc_3C_2@C_{80}$ bisfulleroid derivative. J Am Chem Soc 134:4092–4095

30. Imahori H, Sakata Y (1997) Donor-linked fullerenes: photoinduced electron transfer and its potential application. Adv Mater 9:537–546
31. Gust D, Moore TA, Moore AL (2001) Mimicking photosynthetic solar energy transduction. Acc Chem Res 34:40–48
32. Imahori H, Sakata Y (1999) Fullerenes as novel acceptors in photosynthetic electron transfer. Eur J Org Chem 1999:2445–2457
33. Guldi DM (2000) Fullerenes: three dimensional electron acceptor materials. Chem Commun 321–327
34. Echegoyen L, Echegoyen LE (1998) Electrochemistry of fullerenes and their derivatives. Acc Chem Res 31:593–601
35. Rudolf M, Wolfrum S, Guldi DM, Feng L, Tsuchiya T, Akasaka T, Echegoyen L (2012) Endohedral metallofullerenes–filled fullerene derivatives towards multifunctional reaction center mimics. Chem Eur J 18:5136–5148
36. Ross RB, Cardona CM, Guldi DM, Sankaranarayanan SG, Reese MO, Kopidakis N, Peet J, Walker B, Bazan GC, Van Keuren E, Holloway BC, Drees M (2009) Endohedral fullerenes for organic photovoltaic devices. Nat Mater 8:208–212
37. Ross RB, Cardona CM, Swain FB, Guldi DM, Sankaranarayanan SG, Van Keuren E, Holloway BC, Drees M (2009) Tuning conversion efficiency in metallo endohedral fullerene-based organic photovoltaic devices. Adv Funct Mater 19:2332–2337
38. Fujitsuka M, Watanabe A, Ito O, Yamamoto K, Funasaka H (1997) Laser flash photolysis study on photochemical generation of radical cations of fullerenes C_{60}, C_{70}, and C_{76}. J Phys Chem A 101:7960–7964
39. Reed CA, Kim KC, Bolskar RD, Mueller LJ (2000) Taming superacids: stabilization of the fullerene cations $HC_{60}{}^{+}$ and $C_{60}{}^{\bullet +}$. Science 289:101–104
40. Fukuzumi S, Mori H, Imahori H, Suenobu T, Araki Y, Ito O, Kadish KM (2001) Scandium ion-promoted photoinduced electron-transfer oxidation of fullerenes and derivatives by *p*-chloranil and *p*-benzoquinone. J Am Chem Soc 123:12458–12465
41. Ohkubo K, Ortiz J, Martín-Gomis L, Fernández-Lázaro F, Sastre-Santos Á, Fukuzumi S (2007) Fullerene acting as an electron donor in a donor–acceptor dyad to attain the long-lived charge-separated state by complexation with scandium ion. Chem Commun 589–591
42. Feng L, Rudolf M, Wolfrum S, Troeger A, Slanina Z, Akasaka T, Nagase S, Martín N, Ameri T, Brabec CJ, Guldi DM (2012) A paradigmatic change: linking fullerenes to electron acceptors. J Am Chem Soc 134:12190–12197
43. Rudolf M, Feng L, Slanina Z, Akasaka T, Nagase S, Guldi DM (2013) A metallofullerene electron donor that powers an efficient spin flip in a linear electron donor–acceptor conjugate. J Am Chem Soc 135:11165–11174
44. Takano Y, Herranz MÁ, Martín N, Radhakrishnan SG, Guldi DM, Tsuchiya T, Nagase S, Akasaka T (2010) Donor–acceptor conjugates of lanthanum endohedral metallofullerene and π–extended TTF. J Am Chem Soc 132:8048–8055
45. Pinzón JR, Cardona CM, Herranz MÁ, Plonska-Brzezinska ME, Palkar A, Athans AJ, Martín N, Rodriguez-Fortea A, Poblet JM, Bottari G, Torres T, Gayathri SS, Guldi DM, Echegoyen L (2009) Metal nitride cluster fullerene $M_3N@C_{80}$ ($M = Y$, Sc) based dyads: synthesis, and electrochemical, theoretical and photophysical studies. Chem Eur J 15:864–877
46. Guldi DM, Feng L, Radhakrishnan SG, Nikawa H, Yamada M, Mizorogi N, Tsuchiya T, Akasaka T, Nagase S, Herranz MÁ, Martín N (2010) A molecular $Ce_2@I_h$-C_{80} switch – unprecedented oxidative pathway in photoinduced charge transfer reactivity. J Am Chem Soc 132:9078–9086
47. Feng L, Radhakrishnan SG, Mizorogi N, Slanina Z, Nikawa H, Tsuchiya T, Akasaka T, Nagase S, Martín N, Guldi DM (2011) Synthesis and charge-transfer chemistry of $La_2@I_h$-C_{80}/$Sc_3N@I_h$-C_{80}–zinc porphyrin conjugates: impact of endohedral cluster. J Am Chem Soc 133:7608–7618
48. Takano Y, Obuchi S, Mizorogi N, García R, Herranz MÁ, Rudolf M, Wolfrum S, Guldi DM, Martín N, Nagase S, Akasaka T (2012) An endohedral metallofullerene as a pure electron donor: intramolecular electron transfer in donor–acceptor conjugates of $La_2@C_{80}$ and 11,11,12,12-tetracyano-9,10-anthra-*p*-quinodimethane (TCAQ). J Am Chem Soc 134:19401–19408

49. Takano Y, Obuchi S, Mizorogi N, García R, Herranz MÁ, Rudolf M, Wolfrum S, Guldi DM, Martín N, Nagase S, Akasaka T (2012) Stabilizing ion and radical ion pair states in a paramagnetic endohedral metallofullerene/π-extended tetrathiafulvalene conjugate. J Am Chem Soc 134:16103–16106
50. Feng L, Slanina Z, Sato S, Yoza K, Tsuchiya T, Mizorogi N, Akasaka T, Nagase S, Martín N, Guldi DM (2011) Covalently linked porphyrin–La@C_{82} hybrids: structural elucidation and investigation of intramolecular interactions. Angew Chem Int Ed 50:5909–5912
51. Mirkin CA, Caldwell WB (1996) Thin film, fullerene-based materials. Tetrahedron 52: 5113–5130
52. Martín N, Sánchez L, Herranz MÁ, Illescas B, Guldi DM (2007) Electronic communication in tetrathiafulvalene (TTF)/C_{60} systems: toward molecular solar energy conversion materials? Acc Chem Res 40:1015–1024
53. Peet J, Heeger AJ, Bazan GC (2009) "Plastic" solar cells: self-assembly of bulk heterojunction nanomaterials by spontaneous phase separation. Acc Chem Res 42:1700–1708
54. Guldi DM, Illescas BM, Atienza CM, Wielopolski M, Martín N (2009) Fullerene for organic electronics. Chem Soc Rev 38:1587–1597
55. Liang Y, Yu L (2010) A new class of semiconducting polymers for bulk heterojunction solar cells with exceptionally high performance. Acc Chem Res 43:1227–1236
56. Anthony JE, Facchetti A, Heeney M, Marder SR, Zhan X (2010) n-Type organic semiconductors in organic electronics. Adv Mater 22:3876–3892
57. Dang MT, Hirsch L, Wantz G (2011) P3HT:PCBM, best seller in polymer photovoltaic research. Adv Mater 23:3597–3602
58. Pinzón JR, Villalta-Cerdas A, Echegoyen L (2012) Fullerenes, carbon nanotubes, and graphene for molecular electronics. Top Curr Chem 312:127–174
59. Wang C, Dong H, Hu W, Liu Y, Zhu D (2012) Semiconducting π-conjugated systems in field-effect transistors: a material odyssey of organic electronics. Chem Rev 112:2208–2267
60. Kanbara T, Shibata K, Fujiki S, Kubozono Y, Kashino S, Urisu T, Sakai M, Fujiwara A, Kumashiro R, Tanigaki K (2003) N-channel field effect transistors with fullerene thin films and their application to a logic gate circuit. Chem Phys Lett 379:223–229
61. Rikiishi Y, Kubozono Y, Hosokawa T, Shibata K, Haruyama Y, Takabayashi Y, Fujiwara A, Kobayashi S, Mori S, Iwasa Y (2004) Structural and electronic characterization of two isomers of Ce@C_{82}. J Phys Chem B 108:7580–7585
62. Nagano T, Kuwahara E, Takayanagi T, Kubozono Y, Fujiwara A (2005) Fabrication and characterization of field-effect transistor device with C_{2v} isomer of Pr@C_{82}. Chem Phys Lett 409:187–191
63. Kobayashi S, Mori S, Iida S, Ando H, Takenobu T, Taguchi Y, Fujiwara A, Taninaka A, Shinohara H, Iwasa Y (2003) Conductivity and field effect transistor of La_2@C_{80} metallofullerene. J Am Chem Soc 125:8116–8117
64. Maeda Y, Matsunaga Y, Wakahara T, Takahashi S, Tsuchiya T, Ishitsuka MO, Hasegawa T, Akasaka T, Liu MTH, Kokura K, Horn E, Yoza K, Kato T, Okubo S, Kobayashi K, Nagase S, Yamamoto K (2004) Isolation and characterization of a carbene derivative of La@C_{82}. J Am Chem Soc 126:6858–6859
65. Sato S, Seki S, Honsho Y, Wang L, Nikawa H, Luo G, Lu J, Haranaka M, Tsuchiya T, Nagase S, Akasaka T (2011) Semi-metallic single-component crystal of soluble La@C_{82} derivative with high electron mobility. J Am Chem Soc 133:2766–2771
66. Saeki A, Koizumi Y, Aida T, Seki S (2012) Comprehensive approach to intrinsic charge carrier mobility in conjugated organic molecules, macromolecules, and supramolecular architectures. Acc Chem Res 45:1193–1202
67. Seki S, Saeki A, Sakurai T, Sakamaki D (2014) Charge carrier mobility in organic molecular materials probed by electromagnetic waves. Phys Chem Chem Phys 16:11093–11113
68. Sato S, Nikawa H, Seki S, Wang L, Luo G, Lu J, Haranaka M, Tsuchiya T, Nagase S, Akasaka T (2012) A co-crystal composed of the paramagnetic endohedral metallofullerene La@C_{82} and a nickel porphyrin with high electron mobility. Angew Chem Int Ed 51:1589–1591

69. Sato S, Seki S, Luo G, Suzuki M, Lu J, Nagase S, Akasaka T (2012) Tunable charge-transport properties of I_h-C_{80} endohedral metallofullerenes: investigation of $La_2@C_{80}$, $Sc_3N@C_{80}$, and $Sc_3C_2@C_{80}$. J Am Chem Soc 134:11681–11686
70. Agraït N, Yeyati AL, van Ruitenbeek JM (2003) Quantum properties of atomic-sized conductors. Phys Rep 377:81–279
71. Kaneko S, Wang L, Luo G, Lu J, Nagase S, Sato S, Yamada M, Slanina Z, Akasaka T, Kiguchi M (2012) Electron transport through single endohedral $Ce@C_{82}$ metallofullerenes. Phys Rev B 86:155406
72. Deak DS, Silly F, Porfyrakis K, Castell MR (2006) Template ordered open-grid arrays of paired endohedral fullerenes. J Am Chem Soc 128:13976–13977
73. Zhao S, Zhang J, Dong J, Yuan B, Qiu X, Yang S, Hao J, Zhang H, Yuan H, Xing G, Zhao Y, Sun B (2011) Scanning tunneling microscopy investigation of substrate-dependent adsorption and assembly of metallofullerene $Gd@C_{82}$ on Cu(111) and Cu(100). J Phys Chem C 115: 6265–6268
74. Muthukumar K, Strózecha A, Myslivecek J, Dybek A, Dennis TJS, Voigtländer B, Larsson JA (2013) Endohedral fullerene $Ce@C_{82}$ on Cu(111): orientation, electronic structure, and electron-vibration coupling. J Phys Chem C 117:1656–1662
75. Crivillers N, Takano Y, Matsumoto Y, Casado-Montenegro J, Mas-Torrent M, Rovira C, Akasaka T, Veciana J (2013) Electrochemical and magnetic properties of a surface-grafted novel endohedral metallofullerene derivative. Chem Commun 49:8145–8147

Chapter 10
Tuning Physical Properties and Structures of π-Electron System Formed by Single-Wall Carbon Nanotubes with Selected Chiralities

Kazuhiro Yanagi

Abstract Single-wall carbon nanotubes (SWCNTs) are rolled graphene tubes with a diameter of approximately 1.0 nm and represent a model for one-dimensional π-electron systems. SWCNTs exhibit various remarkable physical properties depending on their chirality. In this chapter, the physical properties of bulk SWCNT assemblies with a selected chirality are discussed. First, several remarkable properties such as optical, conducting, and thermoelectric properties are described. Then, how to tune their properties by electric double-layer techniques (or electrochemical doping techniques) is described. Next, how to control the charge in molecules encapsulated inside the π-nanospace of SWCNTs is demonstrated. Finally, how to produce ordered π-electron system using SWCNTs, i.e., aligned assemblies of SWCNTs, with a selected chirality is discussed.

Keywords Carbon nanotube • Electric double layer • Electrochemical doping • Optical properties • Thermoelectric properties

Single-wall carbon nanotubes (SWCNTs) are rolled graphitic tubes with a diameter of approximately 1.0 nm. The typical length of a single rope of SWCNTs is approximately around 1.0 μm; thus, their electronic properties exhibit one-dimensional behaviors, such as physical properties reflecting van Hove singularities in density of states and Tomonaga-Luttinger liquid characteristics [1]. There are many structures in SWCNTs depending on how the graphene sheet is rolled. The way to roll the graphene sheet is classified using chirality (n, m). Remarkably, the electronic structures of SWCNTs exhibit various characteristics depending on their chirality. For example, when the n-m is the multiple of 3, the SWCNTs are metallic with closed band gap in Dirac cone structure. However, in other cases, the SWCNTs are semiconducting with finite band gap [2].

K. Yanagi (✉)
Department of Physics, Tokyo Metropolitan University, Hachioji, Japan
e-mail: yanagi-kazuhiro@tmu.ac.jp

T. Akasaka et al. (eds.), *Chemical Science of π-Electron Systems*,
DOI 10.1007/978-4-431-55357-1_10

Since the discovery of SWCNTs by Iijima et al. in 1993 [3], SWCNTs have been intensively investigated, and it has passed more than 20 years from their discovery; however, unveiled properties have still been remained. One of the reasons for that is that as-produced samples are in the mixed state of different chiralities. Usually, SWCNTs are produced in a mixture state with a ratio of semiconducting to metal = 2:1. Such mixture situation is a bottleneck impeding correct understanding of novel properties of SWCNTs in a π-electron bulk system composed by various SWCNTs.

To overcome the mixture situation, there are two approaches to prepare SWCNTs with selected chiralities: One is to differentiate SWCNTs with different chiralities from as-produced samples, and the other is to synthesize SWCNTs in a single-chiral state.

Regarding the differentiation of SWCNTs with different chiralities, recent advancement of purification techniques enables us to obtain high-purity metallic, semiconducting, and a single-chiral state of SWCNTs using density gradient centrifugation [4–6], gel chromatography methods [7, 8], and so on. In the synthesis approaches, there have been a lot of novel reports concerning production of basic ring structures of SWCNTs [9, 10]. Moreover, in 2014, two remarkable studies have been reported regarding single-chirality synthesis of SWCNTs [11, 12]. Therefore, now we have various techniques to prepare SWCNTs with selected electronic structures.

Thus, it is currently an attracting interest to clarify the physical properties of π-electron system formed by SWCNTs with selected chiralities. In this chapter, I describe several remarkable physical properties that will only be observed using SWCNTs with selected chiralities and describe how to control the properties (Fig. 10.1). First, I demonstrate how to control the colors, conductance, and thermoelectric properties of SWCNT networks with selected chirality. Next, I show how to control the charges in molecules inside the nano-hollow space of SWCNTs. Finally, I demonstrate how to produce the ordered π-electron system using SWCNTs.

10.1 Control of Colors of π-Electron Network System of SWCNTs [13]

After chirality purification of SWCNTs, the samples exhibit various colors depending on their chiralities. For example, metallic SWCNTs with diameters of 1.4, 1.0, and 0.8 nm exhibit cyan, magenta, and yellow colors (Fig. 10.2) [13, 14]. Here, we demonstrate how to tune the color of the colored SWCNT films.

One of the approaches to tune the optical absorption properties of SWCNTs is by electron or hole doping. Filling or depleting the electronic states known as van Hove singularities modifies the optical transitions between the states in SWCNTs.

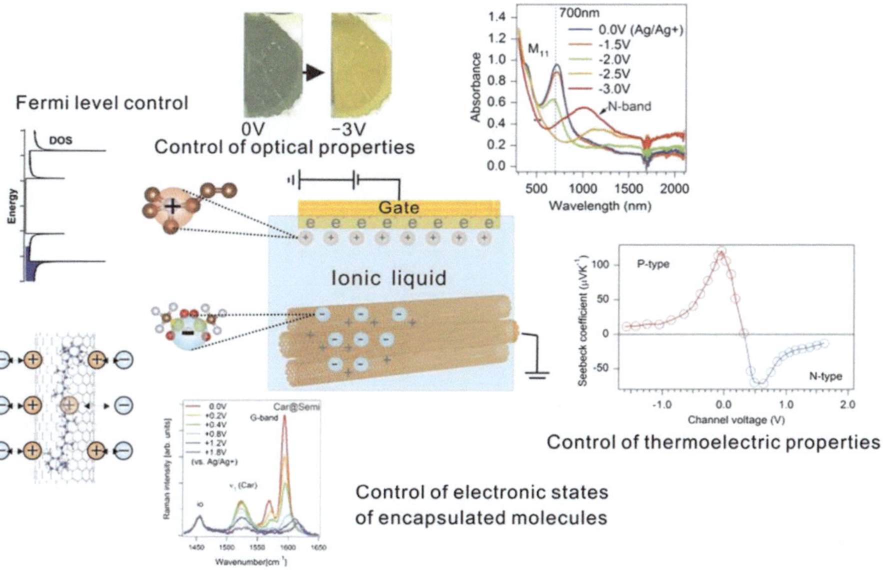

Fig. 10.1 Control of various physical properties of the π-electron system composed by SWCNTs with selected chiralities (These figures are re-depicted with permissions from Ref. [13] Copyright 2011 by John & Wiley & Sons, Inc., Ref. [19] Copyright 2014 American Chemical Society, Ref. [25] Copyright 2013 by the American Physical Society)

For efficient carrier injections, we utilized electric double-layer carrier injections using ionic liquids. The electric double layer is a layer formed on the liquid and solid interfaces. If there is a potential gradient on an interface between the solid and electrolyte solution, ions in the solution are assembled on the interface of the solution side, and charges in the solid side are assembled at the same time to produce counter electric field to compensate the potential gradient. As a result, a thin layer composed of ions and charges is formed on the interface. This layer is called as electric double layer (EDL). The thickness of the EDL is around few angstroms; thus, capacitance is huge, resulting in an efficient carrier injection.

A number of studies have reported changes in infrared optical absorption and emission spectra and in Raman spectra using electrochemical doping techniques [15–18]. However, control over color in the visible spectrum has proved elusive. Here, we demonstrate how to control the colors of SWCNT films with a selected chirality. The experimental setup is shown in Fig. 10.3. The three different colored SWCNT thin films, metallic SWCNTs with diameters of 1.4, 1.0, and 0.8 nm, were formed on a glass substrate, and the film edges were attached to Pt electrode. A Pt wire and a Ag/Ag + electrode were used for the counter and reference electrodes, respectively. Changes in the optical absorption spectra of the SWCNT films were detected as a function of the potential (vs. the Ag/Ag + electrode). Figure 10.3 shows the optical absorption spectra as a function of the film potential. The optical absorption peaks in the visible region (M_{11} bands) of metallic SWCNTs with

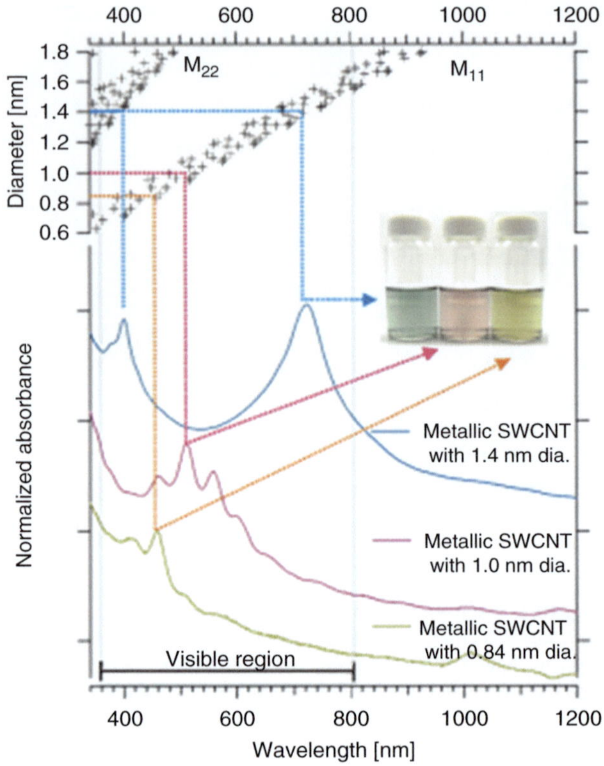

Fig. 10.2 The colors of metallic SWCNTs with diameters of 1.4, 1.0, and 0.84 nm. The relationships between diameters and optical absorption peaks calculated by tight-binding model are presented in the upper panel and corresponding optical absorption spectra (The figure is adapted with permission from Ref. [13] Copyright 2011 John Wiley & Sons, Inc)

diameters of 1.4, 1.0, and 0.84 nm are located at 700, 510, and 460 nm, respectively, corresponding to sample colors of blue green, magenta, and yellow. As shown in the figures, when the potential of the SWCNT film is around −2.0 V, the M_{11} band absorption peaks start to diminish. At a potential of around −3.0 V, the M_{11} bands in all three samples completely disappear. At the same time, the colors of the SWCNT films change as shown in Figure. The blue-green-, magenta-, and yellow-colored metallic SWCNT films change to yellow and light yellow. The electrochromic color changes are reversible and repetitive. When a pulsed-shape potential (on-state, −2.5 V for 60 s, and off-state, 0.0 V for 120 s) is applied, on/off switching was possible with almost no degradation, for more than 1000 repetitions.

The color change occurred uniformly across the SWCNT film when a constant potential was applied to the whole part of a film. However, we can also change the color in a selected area of the film by controlling the potential of each electrode in a similar manner to light-emitting ambipolar transistors. Figure 10.4 shows the color change in a film of metallic SWCNTs with a diameter of 1.4 nm by controlling the source and drain voltages applied to the contact electrodes on the film and the gate voltage on the counter electrode in the electrolyte solution. When VS = VD = VG = 0 V, the metallic SWCNT film was initially blue green. Then, VG was set to sufficiently dope electrons in the film from both VS and VD electrodes,

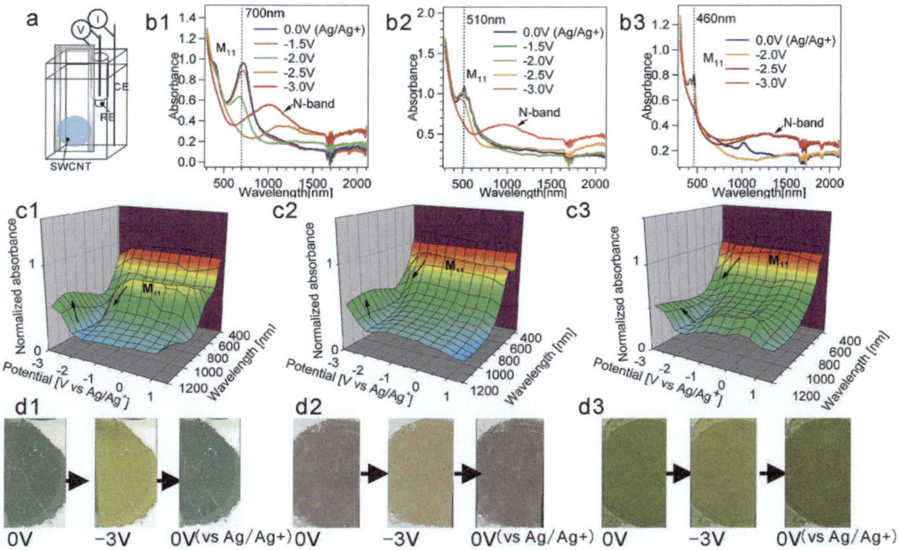

Fig. 10.3 (**a**) Schematic experimental setup. Optical absorption spectrum changes of metallic SWCNTs with diameters of 1.4 (**b1**, **c1**), 1.0 (**b2**, **c2**), and 0.8 nm (**b3**, **c3**). The pictures of color changes in metallic SWCNTs with diameters of 1.4 (**d1**), 1.0 (**d2**), and 0.8 (**d3**) by the shift of potential (The figure is adapted with permission from Ref. [13] Copyright 2011 John Wiley & Sons, Inc)

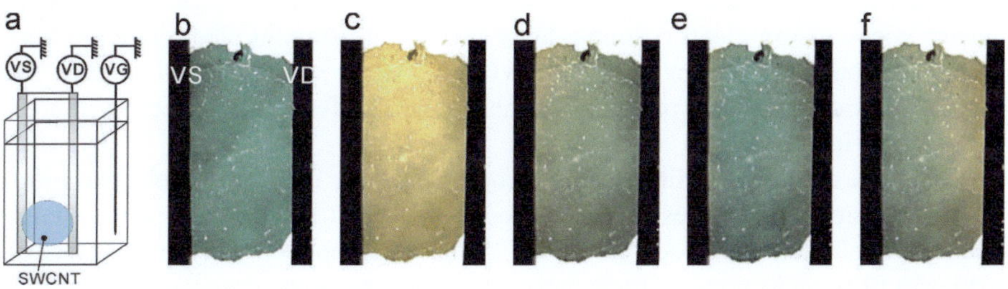

Fig. 10.4 (**a**) Schematic experimental setup for carrier injections of electron and holes at the same time. (**b**) Initial state, (**c**) all the electron doped, (**d**) VS (=0 V)<VD VG<5V, (**e**) VS (=0 V)<VG<VD<5 V, (**f**) VS (0 = V)<VG<VD 5 V (The figure is adapted with permission from Ref. [13] Copyright 2011 John Wiley & Sons, Inc)

VS = VD (=0 V)<VG<5 V. Then, we changed only the potential of the VD from 0 to 5 V. When VS (=0 V)<VD VG<5 V, the film reverted to the initial blue-green color in the area around the drain electrode. Because the voltage difference between VD and VG is negligible, that situation induces continuous dedoping of electrons around the drain electrodes. Figure 10.4e shows the case when VS (=0 V)<VG<VD<5 V. Both edges of the film remained yellow, and a blue-green stripe appeared in the center, indicating hole doping from the drain electrode. Importantly, the blue-green stripe indicates an area of equivalent electron and hole injection (non-doped area), and the yellow area indicates that either electrons or holes are dominant (doped area). Finally, when VS (0 = V) < VG < VD 5 V, the yellow area on the right-hand side increased, suggesting that we are able to intentionally control the color gradient.

10.2 Control of Conducting and Thermoelectric Properties of High-Purity Semiconducting SWCNTs [19]

In the previous section, we revealed that the colors of SWCNTs can be tuned by carrier injections using electric double layers. SWCNTs show clear ambipolar behavior, indicating that hole and electron conduction can be easily achieved by the shift of the potential. In this section, we demonstrate that conducting and thermoelectric properties are also tuned by the shift of the potential using electric double layers.

Thermoelectrics are a very important technology for efficiently converting waste heat into electric power. Hicks and Dresselhaus proposed two important approaches to innovate the performance of thermoelectric devices [20, 21]. One approach involves using low-dimensional materials, and the other approach involves properly tuning the Fermi level because the Seebeck coefficient, which is one of the essential parameters for characterizing the performance of thermoelectric materials, strongly depends on the Fermi level. As described in the previous section, carrier injections by electric double-layer formation using an electrolyte are very efficient to tune the Fermi level of SWCNTs. Here, we attempted to tune the thermoelectric properties, particularly the Seebeck effect, using electric double-layer carrier injections.

SWCNTs exhibit one-dimensional properties; thus, their thermoelectric properties have attracted interest and have been investigated. Recently, a very large Seebeck coefficient has been observed in the semiconducting type of SWCNT networks [22], and tuning of their thermoelectric properties, especially p-type and n-type control, is currently of great interest. Especially, determining the Seebeck coefficient as a function of the Femi level is very important for correctly tuning thermoelectric properties.

Figure 10.5a presents a schematic illustration of the experimental setup for controlling the Seebeck effect through electric double-layer techniques. A thin film of high-purity semiconducting SWCNTs with a diameter of 1.4 nm was used as the channel between the gold source and drain electrodes. A parylene layer was formed on the surface of the polyimide substrate for thermal isolation because of its extremely low thermal conductivity, and the device was formed upon the parylene layer. N,N,N-Trimethyl-N-propylammonium bis(trifluoromethane sulfonyl)imide (TMPA-TFSI) was used as the ionic liquid. Thermocouples were directly attached to the SWCNT film using silver paste, and these thermocouples were covered with insulating pastes to avoid electrochemical reactions. One end of the film was

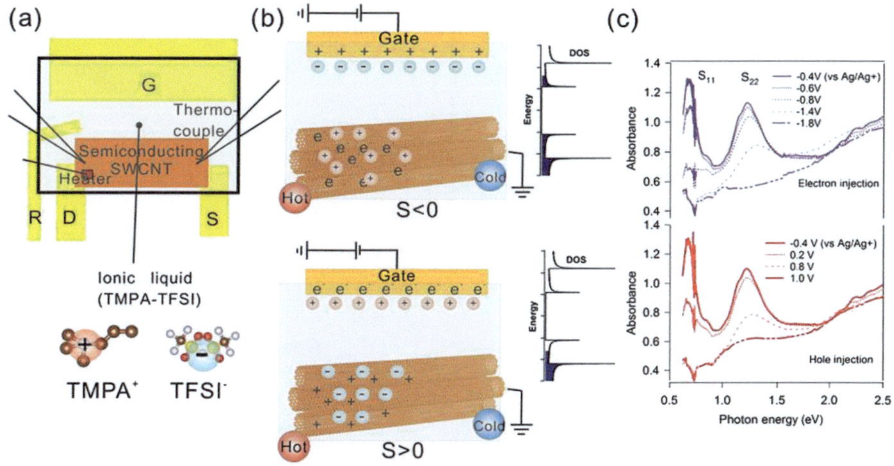

Fig. 10.5 (**a**) Schematic illustration of the experimental setup for controlling the Seebeck coefficient of the semiconducting SWCNTs through an electric double layer using an ionic liquid. S, D, R, and G represent the source, drain, and reference and gate electrodes, respectively. (**b**) Schematic illustrations of the electron injection state (*upper panel*) and hole injection state (*lower panel*) induced by shifting the gate voltage and of the corresponding Fermi level shift in the DOS of the semiconducting SWCNT. (**c**) Optical absorption spectra as a function of the potential shift in the semiconducting SWCNT film (The figure is adapted with permission from Ref. [19] Copyright 2014 American Chemical Society)

heated with a small heater attached to the film, and the actual voltage applied to the sample was measured as the channel voltage from the reference electrode. All of the measurements were performed under vacuum. As shown in a schematic illustration of Fig. 10.5b, electron or holes were injected by the shift in the potential of the gate voltages; consequently, the p-type and n-type behavior of the Seebeck coefficient was controlled. Indeed, the optical absorption spectra of the semiconducting SWCNT changed as a function of the potential shift, which was measured using a typical optical-electrochemical setup described in the previous section. The disappearance of the S_{11} and S_{22} bands due to both the positive and negative shifts in the potential of the semiconducting SWCNT film, corresponding to negative and positive potential shifts in gate voltage, respectively, was observed.

Figure 10.6 presents the experimental results of the Seebeck coefficient of the semiconducting SWCNT samples. The transport data obtained using the Seebeck coefficient measurement setup are presented in Fig. 10.6a. The data clearly indicated ambipolar behavior, suggesting that electrons and holes could be injected by a shift in the gate voltage. Figure 10.6b plots the measured Seebeck coefficient as a function of the shift of the gate voltage (the data are plotted by the channel voltage to evaluate the actual applied voltage). As observed in the figure, the p-type and n-type behavior of the Seebeck coefficient were clearly controlled by the shift in the potential. We performed measurements on four other samples, and the

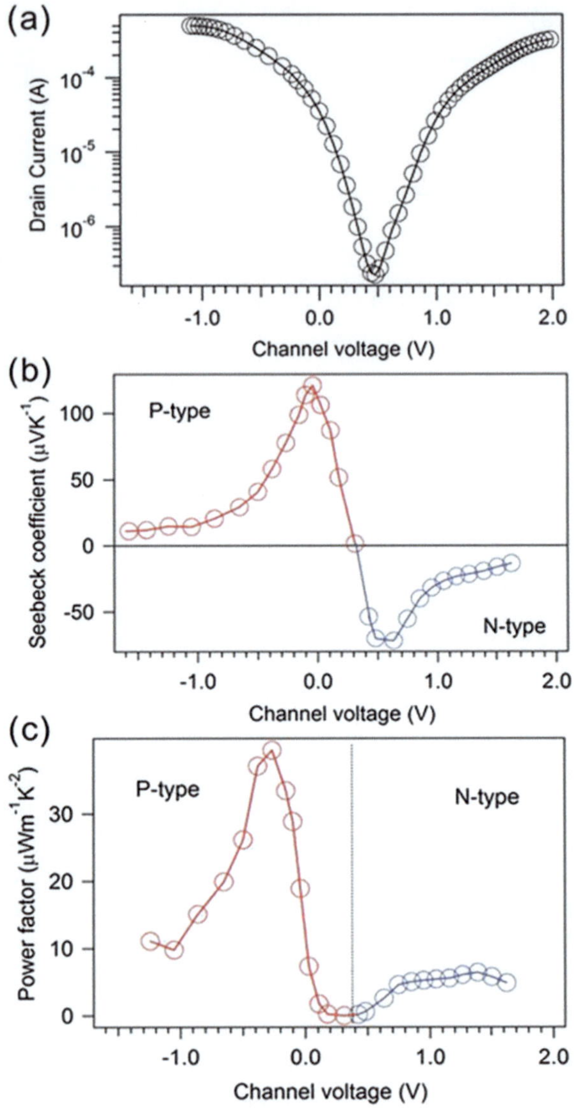

Fig. 10.6 (**a**) Transport properties of the device (at 0.4 V source-drain voltage), (**b**) Seebeck coefficient, and (**c**) power factor of the semiconducting SWCNTs as a function of channel voltage. The *dotted line* was inserted for the guide to identify the regions of p-type and n-type behavior (The figure is adapted with permission from Ref. [19] Copyright 2014 American Chemical Society)

averaged peak value of the p-type was $139 \pm 29\ \mu V\ K^{-1}$, and that of the n-type was $-92 \pm 27\ \mu VK^{-1}$. These values are consistent with those of bismuth antimony telluride alloys, typically ± 100–$200\ \mu V\ K^{-1}$ [23], indicating the good potential of semiconducting SWCNTs as thermoelectric materials.

The power factor σS^2 values are presented in Fig. 10.6c; these values were estimated from the observed Seebeck coefficient and the transport measurement data. In this calculation, the neutral points in carrier injection, which are the voltage point at the minimum current in the transport and the zero point at the Seebeck coefficient measurement, were adjusted to be the same. As observed in Fig. 10.6c, peak structures were present in the power factor, clearly indicating the importance of appropriately tuning the Fermi level of the semiconducting SWCNTs to achieve good performance of thermoelectric devices.

Next, we discuss the channel voltage dependence of the Seebeck coefficient using the phenomenological Mott's formula [24]:

$$S(E_F) = -\frac{\pi^2}{3}\frac{k_B^2 T}{\mathrm{e}}\frac{d\ln[\sigma(E)]}{dE}\bigg|_{E=E_F} \tag{10.1}$$

where $\sigma(E)$ is the electrical conductivity at carrier energy E and the Fermi level $E_F = \mathrm{e}V_{\text{channel}}$ was approximated as using the channel voltage V_{channel}. Figure 10.7 compared the observed Seebeck coefficient and that calculated using Mott's formula. The phenomenological expression in Eq. (10.1) is known to be suitable only for metallic states with a smooth DOS. In the present case, Eq. (10.1) is useful for capturing the behavior of the Seebeck coefficient, particularly in the high electron and hole injection regions. In these regions, the Seebeck coefficient decreases with increases in the channel voltage, and this behavior agrees well with that obtained from Mott's formula. In addition, the sign of the Seebeck coefficient that corresponds to the sign of the majority carrier (electron or hole) coincides with the result from Mott's formula. Thus, our experimental result is theoretically supported, and it is physically reasonable.

Here, we discuss the behavior of the Seebeck coefficient around the semiconducting gap region in more detail. Using the simple relationship $\sigma(E) \propto g(E)$, the Seebeck coefficient can be written as being proportional to the derivative of DOS, $S \propto \frac{\partial g(E)}{\partial E}\Big|_{E_F}$. The DOS of SWCNTs exhibits van Hove singularities because of their one-dimensional nature. Therefore, the Seebeck coefficient will possess a singularity value at the Fermi level near the van Hove singularities of SWCNTs. To understand the relationship between the DOS and the Seebeck coefficient, the Seebeck coefficients are compared with the capacitance of the semiconducting SWCNT film as a function of the reference voltage in Fig. 10.7. The capacitance of the semiconducting SWCNTs was measured in an electrochemical setup. The total capacitance of the film C_{tot} can be written as $C_{\text{tot}}^{-1} = C_{st}^{-1} + C_q^{-1}$. Here, C_{st} is the structural capacitance, and C_q is the quantum capacitance of the SWCNTs. C_q represents the DOS, $g(E_F)$, of the SWCNTs:

$$C_q = \mathrm{e}^2 g(E_F)$$

Therefore, when the Fermi level was located between the band gaps of the first van Hove singularities, the total capacitance was dominated by the quantum capacitance term, reflecting the smallness of the DOS of the semiconducting SWCNTs. Therefore, the position of the first van Hove singularities can be deduced from the capacitance measurements. As observed in the figure, a clear dip structure was observed in the capacitance, and the width of this capacitance dip is

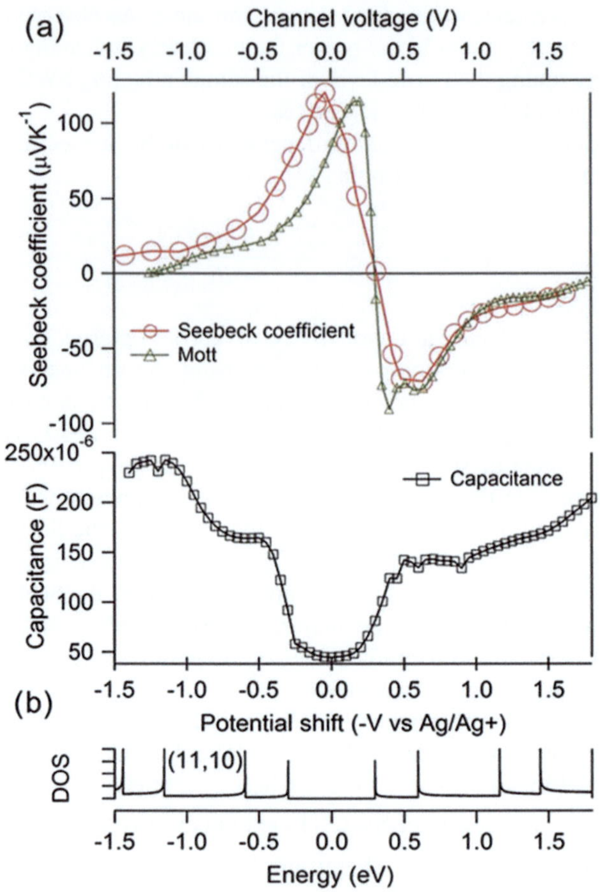

Fig. 10.7 (**a**) Comparison of Seebeck coefficient and the calculated line shape based on Mott's formula. Capacitance of the semiconducting SWCNT film as a function of reference voltage and (**b**) the density of states of (11,10) SWCNT calculated using the tight-binding model (The figure is adapted with permission from Ref. [19] Copyright 2014 American Chemical Society)

approximately 0.65 V, corresponding well to the expected energy gap, 0.6 eV. This value was calculated using tight-binding approximation with $\gamma = 3.0$ eV between the first van Hove singularities of (11,10) semiconducting SWCNTs, which was considered to be one of the main chiralities of our semiconducting sample [6]. The potential gap between the n-type and the p-type Seebeck peaks was estimated to be approximately 0.70 ± 0.08 V for all the samples we measured. The value was in good agreement with the width of the capacitance dip. Therefore, we concluded that the observed Seebeck peaks reflected the one-dimensional nature of the DOS of the semiconducting SWCNTs.

10.3 Control of Charges of Molecules Inside the π-Nanospace: Manipulation of Charges of Molecules Encapsulated Inside SWCNTs [25]

As described in the previous sections, electron or holes are easily injected to SWCNT using EDL techniques. The injected charges are distributed on the surface of the SWCNTs. Next, we investigated whether the surface charges influence the electronic states of molecules encapsulated inside SWCNTs. According to Gauss's law of electromagnetism, surface charges on conductive cylindrical tubes cannot induce electric fields inside hollow spaces (Fig. 10.8a, left). Therefore, charges on the materials located inside the cavity and electrically separated from the surface cannot be influenced by the amount of surface charges present. In general, surface charges are distributed to a depth of approximately 1 nm from the surface. SWCNTs are cylindrical tubes with a diameter of approximately 1.0 nm. Therefore, in this context, the surface charge might be formed using the charge of encapsulated molecules in the case of SWCNTs (Fig. 10.8a, right). A lot of studies about electrochemical doping upon peapods have been reported, but it is still unclear whether the charges of the molecules encapsulated inside the nanocavity can be controlled by the formation of the electric double layers on the nanotube surfaces [26–28]. Thus, we investigated whether the surface charges influence the molecules encapsulated inside SWCNTs or not.

In this study, β-carotene (Car) was employed as the encapsulated molecule (Fig. 10.8b) for the following three reasons: (1) Car has a simple linear polyene structure, which is a model for one-dimensional π-conjugated molecules. (2) The process of encapsulating Car is well established. The presence of Car inside SWCNTs has been well confirmed [29–31]. (3) The Raman signal of Car, such as C = C stretching mode, inside SWCNTs is relatively strong and comparable to the

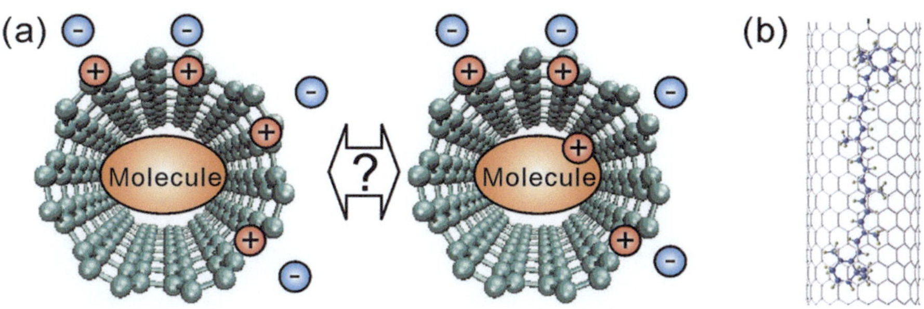

Fig. 10.8 (**a**) In a case, surface charges on SWCNT do not influence the encapsulated molecules (*left*), and surface charges do influence the encapsulated molecules (*right*). (**b**) A schematic illustration of β-carotene encapsulated in SWCNT (The figure is adapted with permission from Ref. [25] Copyright 2014 American Physical Society)

intensity of the G-band of the surrounding semiconducting SWCNTs. The strong Raman signals enable us to simultaneously evaluate doping properties of both Semi and encapsulated Car due to the strong relationship between the Raman spectra and the doping state of the SWCNTs and Car.

The Raman spectra of SWCNTs are very sensitive to electron or hole injection. The influence of electrochemical doping on the Raman spectra of the G-band of semiconducting SWCNTs has been thoroughly studied. Hole and electron injections to Semi induce a decrease in the intensity and a shift in the G-band. In the case of Car, Car can be oxidized when the potential is positive, and the dications and cation radicals of Car are formed by oxidation [32]. The optical absorption of neutral Car disappeared during the oxidation because the disappearance of the optical absorption of neutral Car resulted in an off-resonance state of the Raman excitation wavelength. Therefore, it is possible to evaluate the doping process from the changes in the Raman spectra.

Figure 10.9 shows the changes in the Raman spectra of Car@Semi as a function of the potential voltages. At the zero potential (versus Ag/Ag+), we can clearly identify the Raman peaks due to the C = C stretching mode (ν_1 mode) of Car and G-band of Semi. When the potential was shifted in the negative direction (i.e., electron injection), the intensity of the G-band decreased, and no significant change was observed in the ν_1 mode. These results indicate that electrons were introduced to the conduction band of Semi, but not to the Car. However, when the potential was shifted in the positive direction (i.e., electron extraction (hole injection)), the intensities of both the G-band and the ν_1 mode varied. When the potential was shifted to + 0.4 V, the intensity of the ν_1 mode was nearly the same as

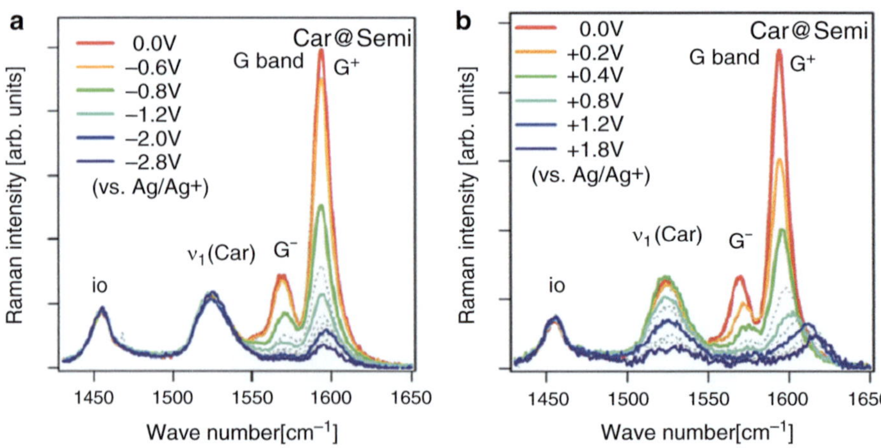

Fig. 10.9 Raman spectra of Car@Semi as a function of the potential voltages [(**a**) negative, (**b**) positive]. The spectra were obtained at a 488 nm excitation wavelength. "io" indicates the Raman peak of ionic liquid used in the measurements. The *dotted lines* are Raman spectra at potentials between those described with the colored solid lines (The figure is adapted with permission from Ref. [25] Copyright 2014 American Physical Society)

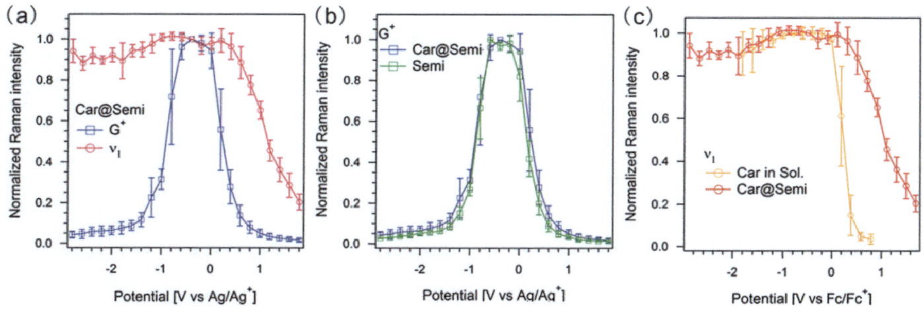

Fig. 10.10 Comparison of the potential dependence of the Raman intensities (**a**) between the G^+-band (*blue curve*) and the ν_1 mode (*red curve*) in Car@Semi, (**b**) between the G^+-bands of Car@Semi (*blue curve*) and Semi (*green curve*), and (**c**) between the ν_1 modes of Car in solution (*yellow curve*) and Car@Semi (*red curve*) (The figure is adapted with permission from Ref. [25] Copyright 2014 American Physical Society)

the initial intensity. However, the intensity of the G-band had decreased to half of the intensity at 0 V. At + 0.8 V, the intensity of the n1 mode began to decrease. Then, at 1.6 V, both the ν_1 mode and the G-band nearly disappeared. The decreased Raman signals of both the G-band and the ν_1 band were recovered when the potential was returned. The results indicate that the electron extraction occurred in both surrounding SWCNT and Car.

The intensities of the G^+-band and the ν_1 mode as a function of the potential voltages are shown in Fig. 10.10. The potential dependence of the intensities of the ν_1 mode of Car in Semi is significantly different from that of Car in solution. At 0.4 V (versus Fc/Fc+) in Fig. 10.10, the intensity of the ν_1 mode of Car in solution nearly disappeared. However, the intensity of the mode in Semi only decreased slightly. It is important to note that the decline in the mode of Car in Semi at a positive potential was gradual compared with the steep decline in the mode of Car in solution. This result suggests that the surrounding nanotube walls influenced the charge extraction processes of the encapsulated molecules. Moreover, the figure clearly indicates that the electron extraction occurred first in Semi and then next in Car.

To understand the physical backgrounds of the carrier injection processes of the Car@Semi system, electronic structures of Car@(17,0) were calculated using density functional theory (DFT) (Fig. 10.11). The highest occupied molecular orbital (HOMO) of Car is located in the gap of (17,0) SWCNTs when the system is un-doped (Fig. 10.11b). This electronic structure can explain the behavior of electron injection processes in Car@Semi. The experimental result indicated that electrons were injected into Semi, but not to Car. This reason can be explained by the electronic band structure as follows: The lowest unoccupied state of Car is located significantly above the conduction band. Therefore, electron can be injected to the conduction band of Semi, but not to Car.

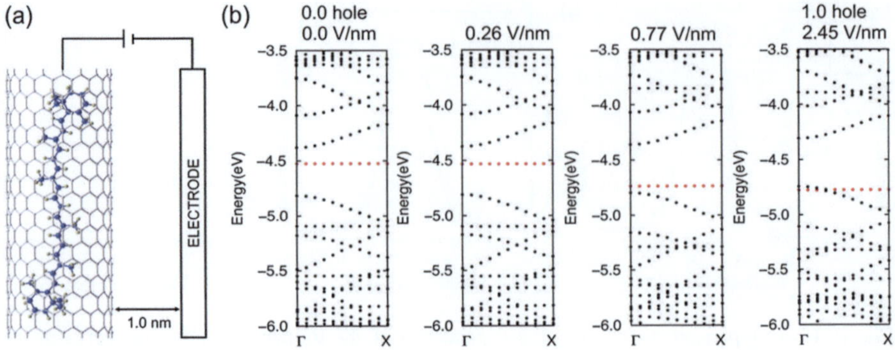

Fig. 10.11 (**a**) Schematic representation of the calculations. (**b**) Electronic structure of Car@(17,0) under various electric fields. The *red dots* correspond to the HOMO band of Car, and the *black dots* correspond to (17,0). At 2.45 V/nm, one electron extracted from Car (The figure is adapted with permission from Ref. [25] Copyright 2014 American Physical Society)

So what happens in the case of electron extraction processes? To understand the mechanisms, we additionally performed DFT calculations on Car@(17,0) which was subjected to an electric field (Fig. 10.11b). To simulate the Car@(17,0) under the EDL condition, we considered the structural model in which the Car@(17,0) is placed upon a two-dimensional electrode separated by 1 nm from the wall of a (17,0) SWCNT. The calculation showed that the HOMO shifts downward due to an increase in the electric field (or the number of holes). Then, the HOMO crosses the top of the valence band of the (17,0) under the electric field. From the results of the calculations, the following electron extraction mechanisms are deduced: The Fermi energy of Car@Semi was downshifted by the positive shift of the potential voltage. The electrons of the valence electronic bands of Semi located above the Fermi energy were totally extracted. However, the electrons of Car were just partially extracted due to the screening of the actual electric field by the hole carriers of Semi. Therefore, the electron first is extracted mainly from Semi and then Car.

One of the reasons for the substantial downward shift in the HOMO of Car was attributed to the presence of large on-site Coulomb repulsion energy arising from the size and dimensionality of the Car. The 1 nm size of the molecule leads to a large on-site energy for the occupied states. The calculated on-site Coulomb energy was a few hundred meV, which is substantial. Therefore, in principle, the electron extraction from the states results in a substantial downward shift with a decrease in energy. In sharp contrast, the π-electron states of the SWCNT extended throughout the nanotube axis of micrometer length. Therefore, the eigenvalue of the electronic states of the SWCNTs is insensitive to the electron extractions or injections and maintains its neutral energy. The results also suggest why the reduction of C_{60} encapsulated in SWCNTs was reported to be difficult [33]. The on-site Coulomb energy of C_{60} is several times larger than that of Car, resulting in the lowest unoccupied molecular orbital of C_{60} being higher than the conduction band of the

SWCNT under electron injection. Therefore, actual electric field upon the electron injection of C_{60} would be significantly screened due to the presence of electron carriers upon the surrounding SWCNTs.

In summary, we found that the carrier manipulation of charges in molecules encapsulated inside SWCNTs is possible using EDL techniques. In any cases, the electron or hole injections occur first on the SWCNT and next on the encapsulated molecules. One of the reasons for that is the shift of the HOMO or LUMO energy of encapsulated molecules due to the strong on-site Coulomb energy. The presence of the carriers upon the surrounding SWCNTs screens some part of the applied electric field; as a result, the actual field that encapsulated molecules feel is reduced from the applied field. However, to compensate the field, the carriers were injected to the encapsulated molecules.

10.4 Control of π-Electron Structure Formed by SWCNTs: Approaches to Make Ordered SWCNT Assemblies Through Self-Assembled Processes [34]

Formation of a well-ordered structure is an important requirement for anisotropic nanomaterials, such as liquid crystals and gold nano-rods, to ensure their optimal performance in devices. In previous sections, we discussed how to control the physical properties of π-space formed by random networks of SWCNTs with selected chirality. In the case of SWCNT, how to fabricate well-ordered structure of them is very important. For example, in the case of a random network film in which SWCNTs are randomly oriented and distributed, the tube-tube random contact functions as a scattering center and induces the localization of conducting carriers [35]. Therefore, even when we can use SWCNTs with a selected chirality for various kinds of devices, however, the presence of such scattering centers significantly degrades the device performance, implying that the intrinsic properties observed in a single rope of SWCNT cannot be observed in their network form. In addition, it is noteworthy that in the case of SWCNTs there is no report about the success of production of single crystals of SWCNTs. As one knows, only in the form of crystal the remarkable physical properties such as superconductivity can be observed in a bulk system. In the case of C_{60}, the first report about the crystallization of C_{60} enabled clear identification of the structure of C_{60} and also exploded highly intensive studies concerning superconducting state in the crystal of C_{60} [36]. Therefore, it is of great importance to develop a technique to produce ordered π-electron system using SWCNTs.

In this context, as a first step, we focused on how to produce aligned assembly of SWCNTs with a single chirality through self-assembled processes. Several studies have reported aligned arrays of purified SWCNTs utilizing liquid-air interfaces, surface modifications of the substrates, coffee ring phenomenon and temperature control, and so on [37–41]. Here, we report another simple technique for preparing one-dimensional assemblies of SWCNTs in which the SWCNTs are highly and densely aligned.

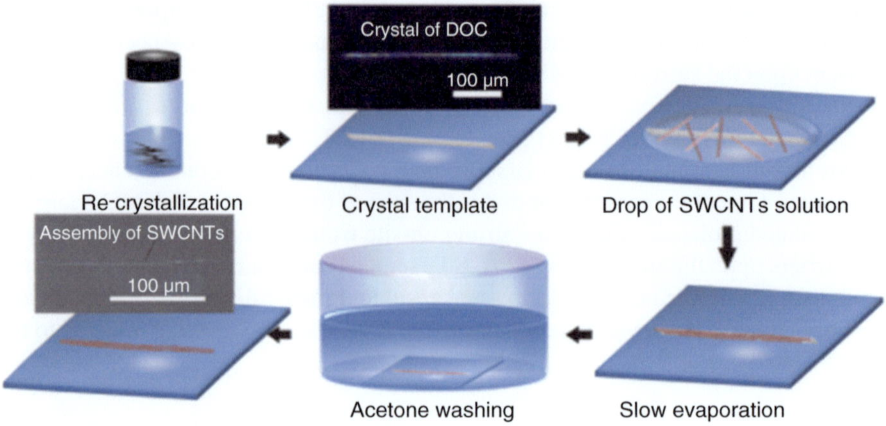

Fig. 10.12 Procedure to prepare aligned assembly of SWCNTs through crystal template method (This figure is reprinted with permission from Ref. [34]. Copyright 2014, AIP Publishing LLC)

Our method is to form aligned assembly on a crystal of surfactants. Figure 10.12 shows a schematic of the procedures used in our technique. First, we prepared high-purity semiconducting, metallic SWCNTs with a diameter of 1.4 nm and (6,5) SWCNTs by density gradient filtration methods and gel chromatography. The suspended surfactants were exchanged with 1 % sodium deoxycholate. Then, we prepared the needlelike crystals of DOC by recrystallization. The needlelike crystals of DOC, which were from 100 mm to 1 mm in length and from 1 mm to 2 mm in width, were placed on a substrate, and the liquid of the purified SWCNT solution was dispensed onto the crystal at 10 degree Celsius. Subsequently, the SWCNTs self-assembled upon the surface of the crystals during the evaporation of the solvents as shown in Fig. 10.11. After the evaporation step, the surfactant crystal was removed by soaking it in an acetone solution for 20 h, leaving only the needlelike assembly in which the SWCNTs were highly and densely aligned on the substrate. Scanning electron microscopy/energy-dispersive X-ray spectroscopy and scanning Auger mapping indicated that the presence of sodium was less than the detection limit, indicating that the amount was less than 4 at.%. These results suggest that the DOC was well removed by the washing procedures.

First, we discuss the characteristics of the needlelike assemblies of (6,5) SWCNTs. The alignment ratio of the assembly of SWCNTs obtained through the above procedure was determined by polarized Raman microscopy as shown in Fig. 10.13. Figure 10.12 shows the normalized G-band intensity of the assembly as a function of angle s between the polarization direction of the incident laser light and the main axis of the assembly. As a reference, the polarization dependence in random networks is also shown. The intensity of the G-band in the assembly reached the maximum intensity when the polarization of the incident radiation was parallel to the axis and reached the minimum intensity when the polarization was normal to the axis. This I_{max}/I_{min} was approximately 14, and the corresponding nematic order

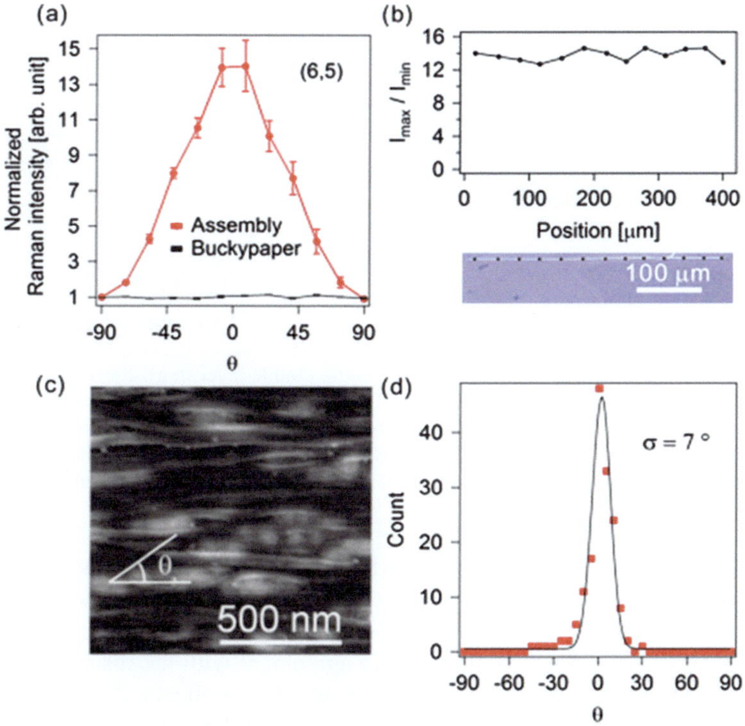

Fig. 10.13 (**a**) The normalized G-band intensities as a function of the angle θ between the axis of crystal template and the polarization vector of the incident laser light are shown. The *red line* represents the assembly made from (6,5) SWCNTs via the crystal template method; the *black line* is a buckypaper film composed of SWCNTs. (**b**) Distribution of the Imax/Imin value in the whole area of the assembly. (**c**) Scanning electron microscope image of the assembly of aligned (6,5) SWCNTs. (**d**) Distribution of the SWCNT bundle orientation obtained from the SEM image (This figure is reprinted with permission from Ref. [34]. Copyright 2014, AIP Publishing LLC)

parameter was 0.81, which are higher than those of previously reported aligned arrays of SWCNTs [37–41]. This result clearly indicates that the SWCNTs were highly aligned along the axis of the assembly.

For more detailed characterizations, we also investigated the distribution of alignment of SWCNTs within the assembly. Figure 10.13 shows a micrograph of the assembly with a length of approximately 400 μm and the I_{max}/I_{min} measured at intervals of approximately 30 mm on the assembly. The distribution of the I_{max}/I_{min} value ranged from 12 to 16, indicating that the SWCNTs were uniformly aligned. We estimated the angle distribution s of SWCNTs in the assembly from the Raman results, and the results indicate that approximately 70 % of the SWCNTs were aligned parallel to the axis of the assembly within 7–10 degrees. This value is consistent with the value evaluated from SEM image. From the SEM image, we estimated the angle distribution of SWCNT bundles as shown in Fig. 10.13. Here,

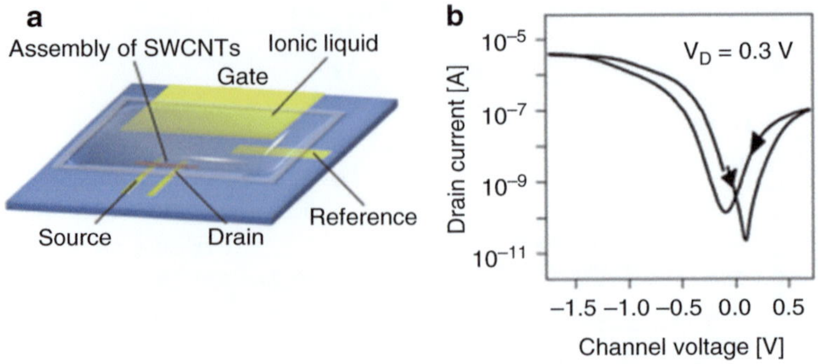

Fig. 10.14 (**a**) The experimental setup for the electric double-layer transistor using the assembly of semiconducting SWCNTs as the channel. The channel length and width were 80 μm and 2.16 μm, respectively. The height of the assembly was 201.0 nm. (**b**) Transport characteristics of the assembly of semiconducting SWCNT FET measured at source-drain voltage, $V_D = 0.3$ V (This figure is reprinted with permission from Ref. [34]. Copyright 2014, AIP Publishing LLC)

the angles between the directions of each bundle and the axis of the assembly, which were estimated from depicting several straight lines along the observed bundles, were evaluated and counted. The σ value was evaluated to be approximately 7 degree as shown in the figure, which is in agreement with the results of polarized Raman microscopy. This crystal template method is applicable for other SWCNT samples. For example, we prepared aligned assemblies of semiconducting and metallic SWCNTs with average diameters of 1.4 nm.

We measured the transport properties of the assembly of the semiconducting SWCNTs with average diameters of 1.4 nm. A schematic of the device structure is shown in Fig. 10.14. EDL technique was utilized for carrier injection using ionic liquid as a gate insulator. Figure 10.14 shows the transport characteristics of the assembly as a function of channel voltage. The on/off ratio was 1.3×10^5 at VD = 0.3 V. The on-state conductance and conductivity were 12.5μS and 2.31×10^3 Sm^{-1}, respectively. Cao et al. reported production of film in which SWCNTs are highly aligned and densely assembled using Langmuir-Schaefer method [37]. The on-state conductivity of their assemblies was evaluated to be 8.14×10^3. Thus, the on-state conductivity of our assembly is comparable to the best reported assemblies of aligned array of semiconducting SWCNTs [37]. This illustrates the high density and good alignment of SWCNTs in our assembly.

10.5 Summary

In this chapter, the author described several remarkable physical properties of π-electron systems formed by SWCNTs with selected chiralities and demonstrated how we can systematically control the physical properties by electric double-layer carrier injections. Recently, carrier injection using electric double layer is beginning to be noticed as a prominent tool to control various physical properties of two-dimensional materials, such as superconducting states [42], magnetizations [43], and metal-insulator transitions [44] (see a recent review by Ueno et al. [45]). In this chapter, the author clarified that this technique is also very useful for control of various physical properties of "bulk" π-electron systems formed by one-dimensional nanomaterials like SWCNTs. Electric double-layer carrier injection is, in principle, the same as electrochemical doping. Regarding SWCNTs, a plenty of electrochemical doping studies for SWCNTs in the mixture state of different chiralities have been reported since the discovery of SWCNTs [15–18, 26–28, 33]. However, studies upon SWCNTs with a selected chirality are still rare. As described in this chapter, the authors revealed that SWCNT with selected chirality can exhibit remarkable physical properties in their bulk form that cannot be observed in the mixed states. In addition, it is possible to tune the colors [13], conductivity, and thermoelectric properties [19] of SWCNTs with selected chiralities by carrier injection using electric double-layer techniques. Moreover, it is possible to manipulate the charges in the molecules encapsulated inside SWCNTs [25]. Such manipulation of physical properties in bulk network systems cannot be achieved by back-gating approaches, because electric field of back-gating just influences SWCNTs located on the first layer from the insulating substrates. Therefore, all-round gating approaches using electric double layer are a unique method for manipulation of one-dimensional π-electron bulk networks. The author wishes the revealed knowledge and techniques will stimulate further researches about control of other physical properties of not only one-dimensional π-electron system but also other π-electron bulk systems.

Acknowledgment KY would like to thank my students, H. Kawai, K. Hasegawa, Y. Oshima, Y. Kitamura, R. Moriya, and T. Igarashi, and to thank my collaborators, Profs. Y. Maniwa, Y. Nakai, T. Okada, T. Takenobu, T. Yamamoto, and R. Nakatsu and Drs. Y. Naitou and T. Cuong.

References

1. Bockrath M, Cobden D, Lu J, Rinzler A, Smalley R, Balents L, McEuen P (1999) Luttinger-liquid behaviour in carbon nanotubes. Nature 397:598–601
2. Saito R, Dresselhaus G, Dresselhaus MS (1998) Physical properties of carbon nanotubes. Imperial College Press, London
3. Iijima S, Ichihashi T (1993) Single-shell carbon nanotubes of 1-nm diameter. Nature 363(603)
4. Arnold M, Green A, Hulvat J, Stupp S, Hersam M (2006) Sorting carbon nanotubes by electronic structure using density differentiation. Nat Nanotechnol 1:60
5. Ghosh S, Bachilo S, Weisman R (2010) Advanced sorting of single-walled carbon nanotubes by nonlinear density-gradient ultracentrifugation. Nat Nanotechnol 5:443
6. Kawai M, Kyakuno H, Suzuki T, Igarashi T, Suzuki H, Okazaki T, Maniwa Y, Yanagi K (2012) Single chirality extraction of single-wall carbon nanotubes for the encapsulation of organic molecules. J Am Chem Soc 134:9545

7. Moshammer K, Hennrich F, Kappes M (2009) Selective suspension in aqueous sodium dodecyl sulfate according to electronic structure type allows simple separation of metallic from semiconducting single-wall carbon nanotubes. Nano Res 2:599
8. Liu H, Nishide D, Tanaka T, Kataura H (2011) Large-scale single-chirality separation of single-wall carbon nanotubes by simple gel chromatography. Nat Commun 2:1
9. Hitosugi S, Nakanishi W, Yamasaki T, Isobe H (2011) Bottom-up synthesis of finite models of helical (n, m)-single-wall carbon nanotubes. Nat Commun 2:492
10. Omachi H, Matsuura S, Segawa Y, Itami K (2010) A modular and size-selective synthesis of [n]cycloparaphenylenes: a step toward the selective synthesis of [n, n] single-walled carbon nanotubes. Angew Chem Int Ed 49:10202–10205
11. Yang F, Wang X, Zhang D, Yang J, Luo D, Xu Z, Wei J, Wang J, Xu Z, Peng F, Li X, Li R, Li Y, Li M, Bai X, Ding F, Li Y (2014) Chirality-specific growth of single-walled carbon nanotubes on solid alloy catalysts. Nature 510(522)
12. Sanchez-Valencia J, Dienel T, Groning O, Shorubalko I, Mueller A, Jansen M, Amsharov K, Ruffieux P, Fasel R (2014) Controlled synthesis of single-chirality carbon nanotubes. Nature 512(61)
13. Yanagi K, Moriya R, Yomogida Y, Takenobu T, Naito Y, Ishida T, Kataura H, Matsuda K, Maniwa Y (2011) Electrochromic carbon electrodes: controllable visible color changes in metallic single-wall carbon nanotubes. Adv Mater 23:2811–2814
14. Yanagi K, Miyata Y, Kataura H (2008) Optical and conductive characteristics of metallic single-wall carbon nanotubes with three basic colors; cyan, magenta, and yellow. Appl Phys Express 1:034003–034006
15. Kazaoui S, Minami N, Matsuda N, Kataura H, Achiba Y (2001) Electrochemical tuning of electronic states in single-wall carbon nanotubes studied by in situ absorption spectroscopy and ac resistance. Appl Phys Lett 78:3433
16. Kalbac M, Green A, Hersam M, Kavan L (2010) Tuning of sorted double walled carbon nanotubes by electrochemical charging. ACS Nano 4:459
17. Kavan L, Rapta P, Dunsch L, Bronikowski M, Willis P, Smally R (2001) Electrochemical tuning of electronic structure of single-walled carbon nanotubes: in-situ Raman and Vis-NIR study. J Phys Chem B 105:10764
18. Tanaka Y, Hirano Y, Niidome Y, Kato K, Saito S, Nakashima N (2009) Experimentally determined redox potentials of individual (n, m) single walled carbon nanotubes. Angew Chem Int Ed 48:7655
19. Yanagi K, Kanda S, Oshima Y, Kitamura Y, Kawai H, Yamamoto T, Takenobu T, Nakai Y, Maniwa Y (2014) Tuning of the thermoelectric properties of one-dimensional material networks by electric double layer techniques using ionic liquids. Nano Lett 14(6437)
20. Hicks L, Dresselhaus M (1993) Effect of quantum well structures on the thermoelectric figure of merit. Phys Rev B 47:12727
21. Hicks L, Dresselhaus M (1993) Thermoelectric figure of merit of a one-dimensional conductor. Phys Rev B 47:16631
22. Nakai Y, Honda K, Yanagi K, Kataura H, Kato T, Yamamoto T, Maniwa Y (2014) Giant seebeck coefficient in semiconducting single wall carbon nanotube film. Appl Phys Express 7:025103
23. Scherrer H, Scherrer S (2006) In: Rowe D (ed) Thermoelectric handbook: macro to nano. CRC Press, New York
24. Cutler M, Mott N (1969) Observation of Anderson localization in an electron gas. Phys Rev 181:1336
25. Yanagi K, Moriya R, Cuong T, Otani M, Okada S (2013) Charge manipulation in molecules encapsulated inside single-wall carbon nanotubes. Phys Rev Lett 110:86801
26. Kalbac M, Kavan L, Gorantla S, Gemming T, Dunsch L (2010) Sexithiophene encapsulated in a single-walled carbon nanotube: an in situ Raman spectroelectrochemical study of a peapod structure. Chem Eur J 16:11753
27. Kalbac M, Kavan L, Zukalova M, Dunsch L (2007) Influence of an extended fullerene cage: study of chemical and electrochemical doping of C_{70} peapods by in situ Raman spectroelectrochemistry. J Phys Chem C 111:1079

28. Kavan L, Dunshc L, Katura H (2002) In situ Vis-NIR and Raman spectroelectrochemistry at fullerene peapods. Chem Phys Lett 361:79
29. Yanagi K, Miyata Y, Kataura H (2006) Highly stabilized β-carotene in carbon nanotubes. Adv Mater 18:437
30. Yanagi K, Iakoubovskii K, Kazaoui N, Minami N, Maniwa Y, Miyata Y, Kataura H (2006) Light-harvesting function of β-carotene inside carbon nanotubes. Phys Rev B 74:155420
31. Liu Z, Yanagi K, Suenaga K, Kataura H, Iijima S (2007) Imaging the dynamic behaviour of individual retinal chromophores confined inside carbon nanotubes. Nat Nanotechnol 2(422)
32. Jeevarajan J, Wei C, Jeevarajan A, Kispert L (1996) Optical absorption spectra of dications of carotenoids. J Phys Chem 100:5637
33. Kavan L, Dunsch L, Kataura H, Oshiyama A, Otani M, Okada S (2003) Electrochemical tuning of electronic structure of C_{60} and C_{70} fullerene peapods: in situ visible near-infrared and Raman study. J Phys Chem B 107:7666
34. Kawai H, Hasegawa K, Oyane A, Naitoh Y, Yanagi K (2014) Self-formation of highly aligned metallic, semiconducting and single chiral single-walled carbon nanotubes assemblies via a crystal template method. Appl Phys Lett 105:093102
35. Yanagi K, Udoguchi H, Sagitani S, Oshima Y, Takenobu T, Katura H, Ishida T, Matsuda K, Maniwa Y (2010) Transport mechanisms in high-purity metallic and semiconducting single-wall carbon nanotube networks. ACS Nano 4:4027
36. Kratschmer W, Lamb L, Fostiropoulos K, Huffman D (1990) Solid C_{60}: a new form of carbon. Nature 347(354)
37. Li X, Zhang L, Wang X, Shimoyama I, Sun X, Seo W, Dai H (2007) Langmuir-Blodgett assembly of densely aligned single-walled carbon nanotubes from bulk materials. J Am Chem Soc 129:4890
38. Cao Q, Han S, Tulevski G, Zhu Y, Lu D, Haensch W (2013) Arrays of single-walled carbon nanotubes with full surface coverage for high-performance electronics. Nat Nanotechnol 8:180
39. Park H, Afzali A, Han S, Tulevski G, Franklin A, Tersoff J, Hannon J, Haensch W (2012) High-density integration of carbon nanotubes via chemical self-assembly. Nat Nanotechnol 7:787
40. Engel M, Small J, Steiner M, Freitag M, Green A, Hersam M, Avouris P (2008) Thin film nanotube transistors based on self-assembled, aligned, semiconducting carbon nanotube arrays. ACS Nano 2:2445
41. Kawai H, Hasegawa K, Nakatsu T, Naitoh Y, Takagi Y, Wada Y, Takenobu T, Yanagi K (2013) String like assembly of aligned single-wall carbon nanotubes in a single-chiral state. Appl Phys Express 6:065103
42. Ueno K, Nakamura S, Shimotani H, Ohtomo A, Kimura N, Nojima T, Aoki H, Iwasa Y, Kawasaki M (2008) Electric-field induced superconductivity in an insulator. Nat Mater 7:855
43. Shimamura K, Chiba D, Ono S, Fukami S, Ishiwata N, Kawaguchi M, Kobayashi K, Ono T (2012) Electrical control of Curie temperature in cobalt using an ionic liquid film. Appl Phys Lett 100:122402
44. Nakano M, Shibuya K, Okuyama D, Hatano T, Ono S, Kawasaki M, Iwasa Y, Tokura Y (2012) Collective bulk carrier delocalization driven by electrostatic surface charge accumulation. Nature 487(459)
45. Ueno K, Shimotani H, Yuan H, Ye J, Kawasaki M, Iwasa Y (2014) Field induced superconductivity in electric double layer transistors. J Phys Soc Jpn 83:032001

Chapter 11
Recent Progress on the Chemical Reactions of Single-Walled Carbon Nanotubes

Yutaka Maeda and Takeshi Akasaka

Abstract Single-walled carbon nanotubes (SWNTs) have received much attention because of their excellent mechanical and electronic properties. The structure of an SWNT, a tubular graphene sheet, is defined by the diameter and orientation of the carbon lattice. The electronic properties of SWNTs are strictly determined by their structure. The chemical functionalization of SWNTs has been widely studied to reveal the reactivity and property of their unique and new curved-π-electron system. Addition of new functionalities to SWNTs aids in enhancing their solubility in organic solvents.

In the first part of this chapter, the three main SWNT characterization methods—Vis-NIR absorption spectroscopy, Raman spectroscopy, and thermogravimetry—are presented. In the second subchapter, we focus on the reductive alkylation reactions of SWNTs. Methods for controlling the degree of functionalization by substituent effects in the two-step reductive alkylation are discussed in detail. Additionally, a method to estimate the ratio of two functional groups on the SWNT sidewall is described. The third subchapter is devoted to the photochemical reactions of SWNTs with organic compounds.

Keywords Single-walled carbon nanotubes • Chemical reaction • Reductive alkylation • Oxidation • Photoreaction

---

Y. Maeda (✉)
Department of Chemistry, Tokyo Gakugei University, Tokyo 184-8501, Japan
e-mail: ymaeda@u-gakugei.ac.jp

T. Akasaka
Department of Chemistry, Tokyo Gakugei University, Tokyo 184-8501, Japan

Life Science Center of Tsukuba Advanced Research Alliance, University of Tsukuba, Ibaraki 305-8577, Japan

School of Materials Science and Technology, Huazhong University of Science and Technology, Wuhan 430074, P. R. China

Foundation for Advancement of International Science, Ibaraki 305-0821, Japan

T. Akasaka et al. (eds.), *Chemical Science of π-Electron Systems*,
DOI 10.1007/978-4-431-55357-1_11

11.1 Introduction

Single-walled carbon nanotubes (SWNTs) have been studied extensively for numerous potential applications because of their excellent mechanical and electrical properties [1–6]. The structure of SWNTs can be visualized by rolling one graphene sheet seamlessly into a tube and is defined by its chiral vectors, the so-called chiral indices (n, m) [7–9]. According to the orientation of the hexagonal rings, SWNTs are classified as armchair, zigzag, and chiral tubes, as presented in Fig. 11.1.

SWNTs can possess metallic (n-m = 3 k, where k is an integer) or semiconducting (n-m $\neq$ 3 k) electrical properties (Fig. 11.1). However, SWNTs are typically grown as bundles of a mixture of metallic and semiconducting tubes. Thus, selective synthesis or development of a method to separate metallic and semiconducting SWNTs is an important subject [10, 11]. SWNTs show low dispersibility in organic solvents [12], which hinders their widespread application. To overcome this problem, several dispersion procedures have been developed using surfactants, polymers, π-conjugated compounds, and so on [13, 14].

In 1998, Haddon et al. reported that the dispersibility of SWNTs can be significantly increased by the addition of long alkyl functional groups (Scheme 11.1.) [15]. Since then, a variety of chemical reactions has been developed to modify the properties of SWNTs [13, 14]. Nowadays, it is well known that chemical functionalization can improve the field emission, field-effect transistor, and photoluminescence properties of these materials.

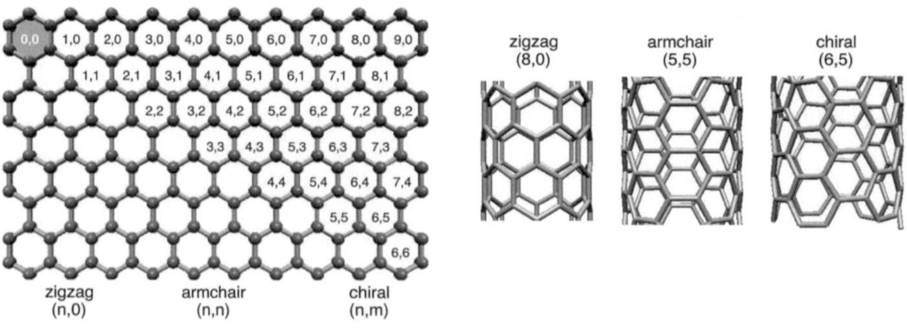

Fig. 11.1 Schematic plot of the chiral vector in graphene 2D lattice (*top*). Structures of zigzag, armchair, and chiral nanotubes (*bottom*)

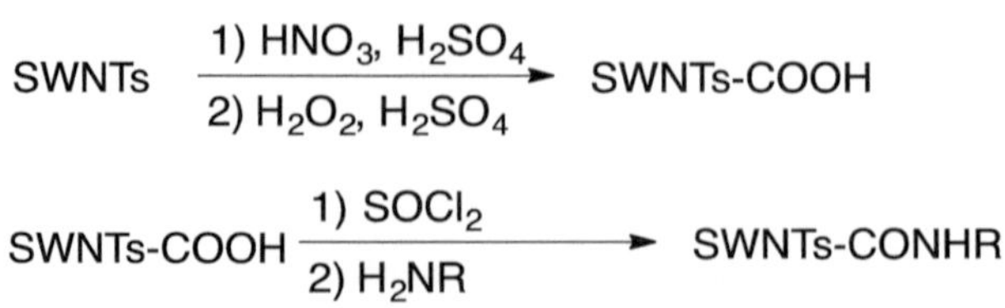

Scheme 11.1 Oxidation and following amidation reaction of SWNTs

In the first part of this chapter, we describe the three basic analysis methods for functionalized SWNTs and then discuss two types of reductive alkylation and two photochemical reactions of SWNTs.

11.2 Analysis of Sidewall-Functionalized SWNTs

Intensive and thorough structure analysis is essential for the precise characterization of functionalized SWNTs. Thus, in general, a combination of structural analysis methods is applied to characterize SWNTs. These methods mainly involve Vis-NIR absorption spectroscopy, Raman spectroscopy, and thermogravimetry (TGA). In this subchapter, we attempt to present these methods in brief.

11.2.1 Absorption Spectroscopy

SWNTs exhibit characteristic absorption peaks in the Vis-NIR region due to the interband electronic transitions between Van Hove singularities (VHSs) near the Fermi level (Fig. 11.2) [16–19]. (Hereafter, the transitions between the first and second pairs of VHSs of the semiconducting SWNTs are designated as S_{11} and S_{22}. The transition between the first pairs of VHSs of the metallic SWNTs is designated as M_{11} [20].) Absorption spectroscopy is one of the simplest and most versatile tool for the characterization of SWNTs. VHS singularities are characteristic to the

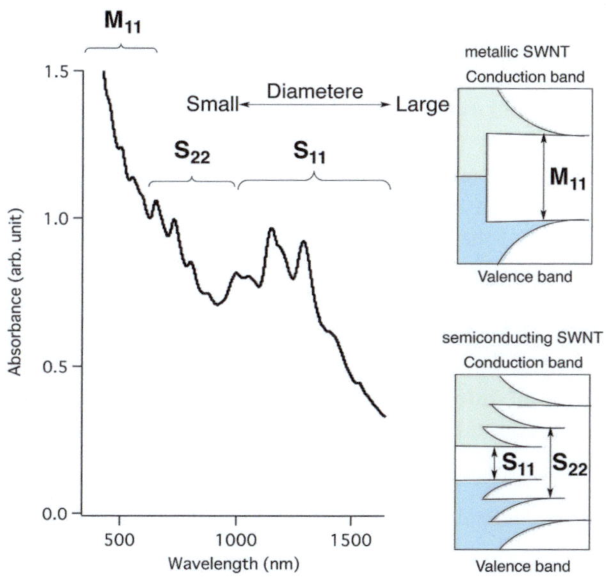

Fig. 11.2 Absorption spectrum of SWNTs (*left*). Schematic density of electronic states of metallic and semiconducting SWNTs (*right*)

SWNTs, and thus the absorption peaks can be used to determine their diameter and electronic state. Saito et al. reported that the electronic transitions in the absorption spectra are related to the mean diameter of the SWNT and proposed three equations to estimate the diameter of the tubes [21]. The correlation is limited to SWNTs with a diameter range of 1–2 nm (Eqs. 11.1, 11.2, and 11.3).

$$E\,(S_{11}) = 0.962/d_t \tag{11.1}$$

$$E\,(S_{22}) = 1.73/d_t \tag{11.2}$$

$$E\,(M_{11}) = 2.60/d_t \tag{11.3}$$

Haddon et al. reported absorption spectral changes upon dichlorocarbene functionalization on the SWNT sidewall. The absorption peak intensity assigned for S_{11}, S_{22}, and M_{11} decreased with increased dichlorocarbene addition [15, 22]. This spectral change was explained by the destruction of the extended π-network, which changes the electronic structure of the SWNTs. Other chemical reactions of SWNTs, such as protonation [23], oxidation, and reduction [24, 25], which cause reversible absorption spectral changes, have also been reported. Thus, it is difficult to distinguish whether or not the sidewall functionalization was caused by the absorption spectral changes.

11.2.2 Raman Spectroscopy

In general, three major peaks can be distinguished in the Raman spectra of SWNTs: the radial breathing mode (RBM, 100–350 cm^{-1}), the disorder-induced mode (D band), and the tangential mode (G band) (Fig. 11.3) [17]. The RBM peak frequency was found to be inversely proportional to the SWNT diameter (Eq. 11.4) [16, 26–28]. This correlation is used to estimate the diameter of SWNTs.

In addition, excitation energy allows for the assignment of chiral indices of SWNTs on the basis of the Kataura plot [19]. The relative intensity of the D band and G band (D/G ratio) indicates the degree of disorder in the sp^2 carbons. Thus, the D/G ratio is in fact an index of the SWNT purity. Haddon et al. reported that the intensity of the D band increased with decreased RBM intensity upon sidewall functionalization [22]. This result indicates that the D/G ratio can also be used to estimate the degree of sidewall functionalization. To use the latter method, high-purity SWNTs are required as starting materials so that the influence of impurities can be excluded.

$$\mathrm{wRBM} = \mathrm{A}d_t^{-1} + \mathrm{B} \tag{11.4}$$

($\mathrm{A} = 223.5\ \mathrm{cm}^{-1}$ and $\mathrm{B} = 12.5\ \mathrm{cm}^{-1}$) where ωRBM is the wavenumber of the RBM [26].

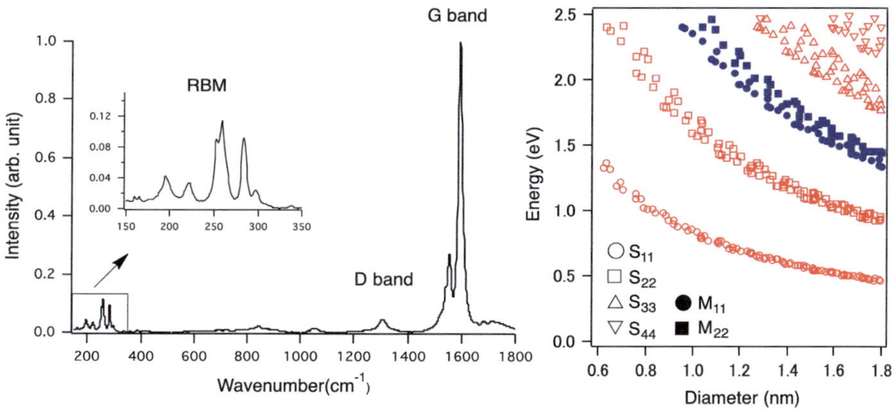

Fig. 11.3 Raman spectrum of SWNTs (*left*). Kataura plot (*right*)

11.2.3 Thermogravimetric Analysis

Since SWNTs are thermally stable under inert gas, thermogravimetry (TGA) is a key technique to determine some of their properties. Using TGA, the amount of metal catalyst and the degree of functionalization of SWNTs can be determined [29]. Although the absorption and Raman spectra peak intensities decrease upon functionalization, they can recover upon thermal treatment under N_2 atmosphere. Thus, the weight of a given substituent can be estimated from the thermal treatment of functionalized SWNTs under N_2 flow (Fig. 11.4). On the other hand, thermal treatment under oxygen or airflow helps in determining the weight of SWNTs [20]. The functional group coverage (FGC; number of carbon atoms of SWNTs per substituent) can be calculated from the weight of SWNTs and that of the substituent according to Eq. 11.1. For example, the FGC of bis-alkylated C_{60} ($C_{60}R_2$) is 30.

$$\mathrm{FGC_{exp}} = \frac{\mathrm{wt\%_{SWNTs}}/12.01}{\mathrm{wt\%_{substituent}} - Fw_{\mathrm{substituent}}} \tag{11.5}$$

11.3 Two-Step Reductive Alkylation of SWNTs

11.3.1 Reductive Alkylation and Nucleophilic Alkylation Using Organometallic Compounds

In 2004, reductive alkylation of SWNTs using lithium and alkyl halides was reported by Billups et al [29]. The reaction is highly effective for the sidewall functionalization of SWNTs. The bundled SWNTs are well exfoliated in the reductive step because of the repulsion of the negative charge on the SWNT anionic intermediates. It was also observed that the SWNT anions with alkyl halides could generate radical

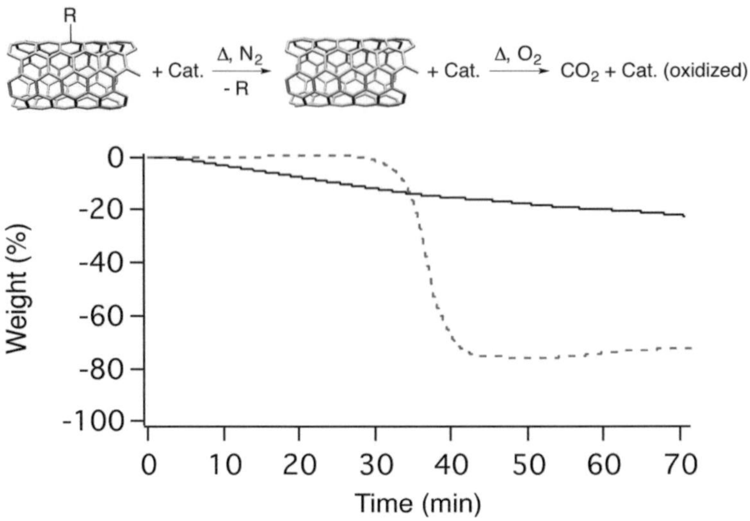

Fig. 11.4 TG curves of alkylated SWNTs from 100 to 800 °C at a heating rate of 10 °C/min in N_2 atmosphere (*solid line*). TG curves of the alkylated SWNTs after thermal treatment under N_2 atmosphere, from 100 to 800 °C at a heating rate of 10 °C/min under air atmosphere (*dotted line*)

intermediates. The formed highly reactive alkyl radical can attack the exfoliated SWNT surface efficiently. Besides alkylation reactions [30, 31], the addition of aryl [32, 33], imide [34], thioalkyl [32], peroxide [35], carbonyl groups [36], and carbon dioxide [37] was successively implemented through the reductive reaction route. Hirsch et al. reported that the small-diameter metallic SWNTs displayed higher reactivity in reductive alkylation than did the large-diameter semiconducting SWNTs [31]. Interestingly, highly selective addition of carbon dioxide [37] to the s-SWNT anion was also reported (Scheme 11.2) [11].

Nucleophilic alkylation of SWNTs using organometallic compounds was described by Hirsch et al. in 2006 [38]. The reaction of the nanotube with *tert*-butyllithium first affords the ionic ^{t}Bu-SWNT$^-$ Li$^+$ adduct. Then, the ^{t}Bu-SWNT$^-$ anions are oxidized by oxygen to give the neutral alkylated product, ^{t}Bu-SWNTs. The high reactivity of organometallic compounds and the high dispersibility of alkylated SWNT anions favor the addition reactions. This effective sidewall functionalization method was successfully applied for the incorporation of ethynyl [39], amino, [40], and alkyl groups [41] as well (Scheme 11.3).

11.3.2 Two-Step Reductive Alkylation of SWNTs

The SWNT anionic intermediates generated by organometallic compounds can be sufficiently applied for the functionalization of nanotubes, as presented in

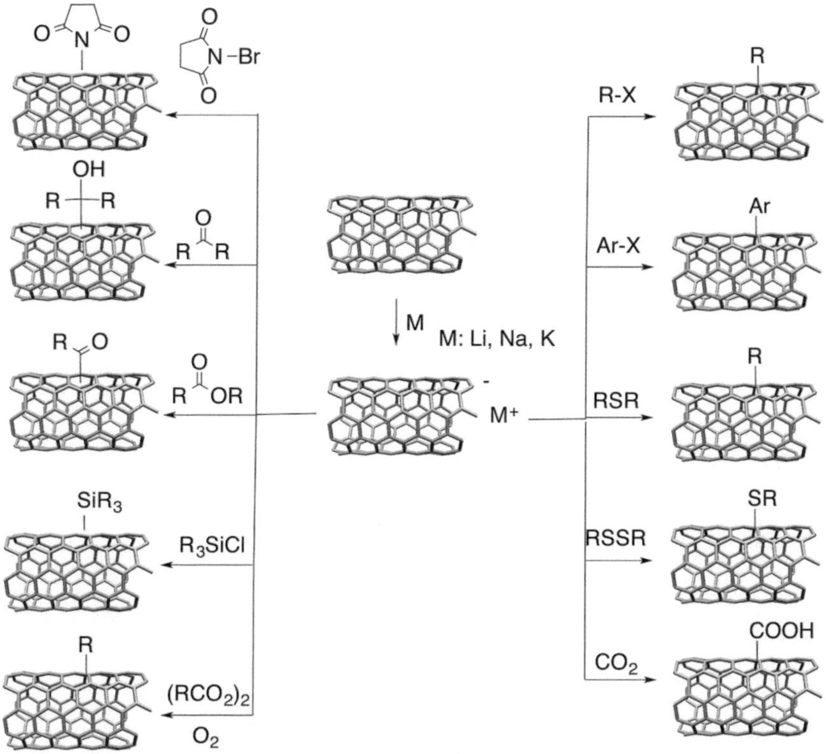

Scheme 11.2 Reaction of SWNTs anion with organic compounds

Scheme 11.4. The intermediate alkylated SWNT anion species can disperse easily in organic solvents and shows high reactivity because of the anionic surface. This two-step reductive alkylation method is an effective means of introducing different functionalities on the SWNT sidewall. The reaction of alkylated SWNT anion with alkyl bromide [42–44], aryl bromide [42], carbon dioxide [45], *N*-formylpiperidine [46], stylene [47], and carbonyl compounds [36] has been performed (Scheme 11.4).

Besides adding new functionality, substituents on the SWNT sidewall help in increasing the dispersibility of SWNTs and in controlling their electrical and optical properties. However, upon excessive functionalization, SWNTs lose their intrinsic electronic properties because of the partial disconnection of the π-conjugated systems [48, 49]. Hence, it is important to control the degree of functionalization on the SWNT sidewall. The substituent effect on functionalization has been extensively studied in two-step reductive alkylations.

Dialkylation of SWNTs (R^1-SWNTs-R^2) prepared by the consecutive addition of butyllithium (R^1Li) and butyl bromide (R^2Br) under argon atmosphere was analyzed by absorption and Raman spectroscopy. The significant spectral change in the spectra of ^{n}Bu-SWNT-R^2 ($R^2 = {}^n$Bu, ^{iso}Bu, ^{sec}Bu, ^{t}Bu) and ^{t}Bu-SWNTs-^{n}Bu

Scheme 11.3 Nucleophilic addition reaction of SWNTs with organometallic reagents

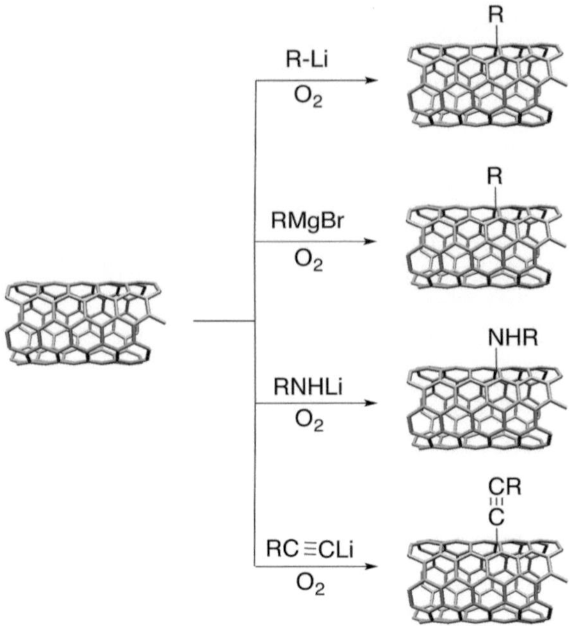

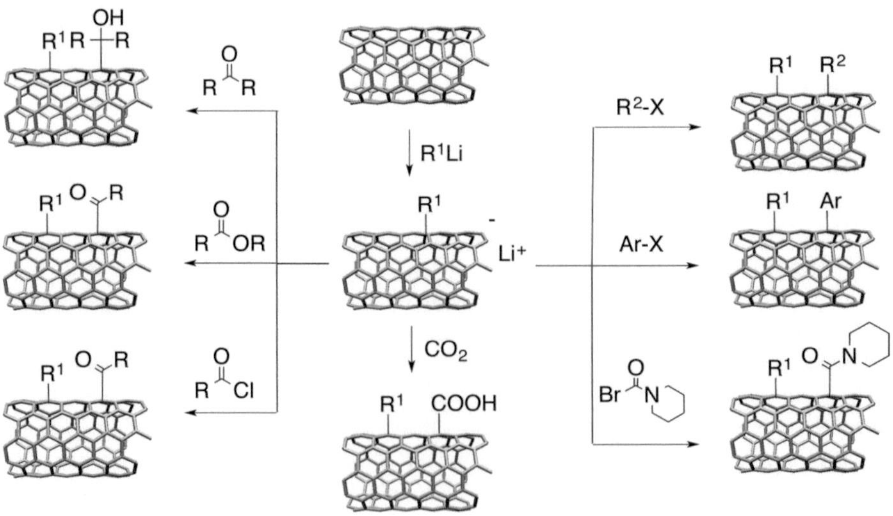

Scheme 11.4 Two-step reductive alkylation of SWNTs

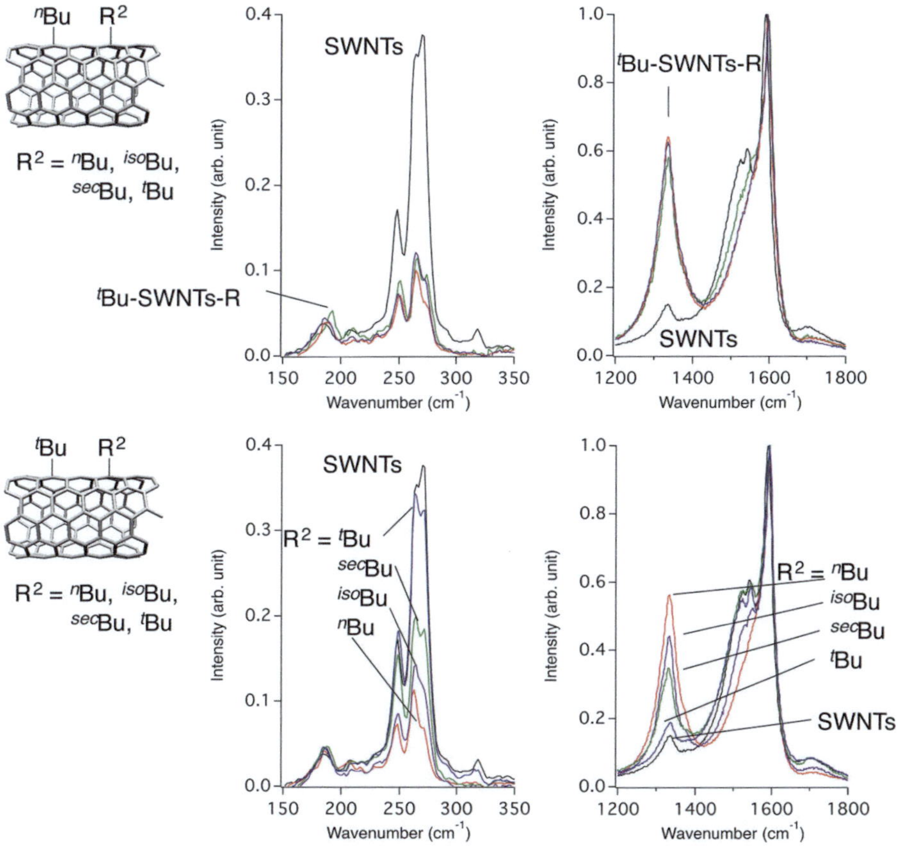

Fig. 11.5 Raman spectra of SWNTs and R-SWNTs-R (excitation wavelength 633 nm)

indicates high alkylation efficiency in the two-step reductive alkylation (Fig. 11.5). On the other hand, the spectral changes decreased in ^{t}Bu-SWNT-R^2 with the increased steric hindrance of R^2 (^{iso}Bu, ^{sec}Bu, ^{t}Bu). The result suggests that the degree of functionalization can be kinetically controlled by the steric effects of R^1 and R^2. In addition, the RBM peaks assigned to small-diameter SWNTs decreased compared to the peaks assigned to large-diameter SWNTs. The higher reactivity of small-diameter SWNTs is explained by their high strain energy and π-orbital misalignment angle, as proposed by Haddon et al [5]. These values increased inversely with the decrease in the nanotube diameter.

Similar steric effects in the reductive alkylation and silylation of SWNTs were observed by our research group. The degree of functionalization decreased with increased bulkiness of the alkyl bromide or silyl chloride. Previously, we revealed that the redox properties can be controlled by silylation [50]. In addition, theoretical studies predicted that the Fermi level of semiconducting SWNTs was tunable by the number of covalently introduced SiH_2 groups on the SWNT sidewall [51]. In fact, changes in the field-effect transistor and field emission properties of SWNTs upon silylation have been experimentally observed [51, 52]. This confirms that the SWNT electronic properties can be tuned by controlling the number of silyl groups on their sidewall.

Controlling the multiple functionalization is a challenging issue for the practical application of SWNTs, such as drug delivery [15, 53–55]. Determining the ratio of two different substituents in bis-functionalized SWNTs is difficult, when relying only on one analytical method. Nevertheless, it can be estimated when a combination of these methods is used, based on a correlation between the D/G ratio and FGC value of R^1-SWNTs-R^2 ($R^1 = {}^n$Bu, ^{sec}Bu, ^{t}Bu; $R^2 = {}^n$Bu, ^{iso}Bu, ^{sec}Bu, ^{t}Bu) [44]. Using the relation shown in Eq. 11.2, the ratio of the two functional groups ($R^1 \neq R^2$) can be estimated. The experimental results indicate that not only the degree of functionalization but also the bulkiness and length of R^1 and R^2 affect their ratio. It was found that the relative amount of R^2 decreased with increasing bulkiness and length of the substituents.

$$\begin{aligned} &\chi = \frac{\text{wt\%}_{\text{Sub}} \cdot \text{FGC}_{\text{cal}} \cdot 12.01}{\text{wt\%}_{\text{SWNTs}} \left(Fw_{R1} - Fw_{R2}\right)} - \frac{Fw_{R2}}{\left(Fw_{R1} - Fw_{R2}\right)} \\ &Fw_{\text{sub}} = \chi Fw_{R^1} + (1 - \chi)\, Fw_{R^2} \\ &(\text{R}_1 \neq R_2) \\ &\left(\frac{\text{R}_2}{\text{R}_1}\right) = \frac{1-\chi}{\chi} \end{aligned} \tag{11.6}$$

11.4 Photochemical Reaction of SWNTs

11.4.1 Photoinduced Electron Transfer Reactions with Amines

In recent years, nanocarbon materials have attracted special attention as photofunctional nanomaterials for the construction of artificial photosynthesis systems and photovoltaic cells. Electron transfer reactions of SWNTs with donor or acceptor molecules upon photoirradiation have been widely studied [56, 57]. The low dispersibility of SWNTs complicates the understanding of their photochemical properties since the SWNT bundles interact with each other. Therefore, in order to investigate the photochemical behavior of SWNTs, it is important to disperse them individually. Extensive studies have shown that the interaction of SWNTs with different amines in the ground state facilitates the dispersion of SWNTs in organic solvents [58, 59]. In addition, the ground-state interaction of SWNTs with amines can be applied as a purification method for SWNTs. Furthermore, the interaction can be used for the separation of metallic and semiconducting SWNTs [60–70].

The photochemical reactions of unsaturated hydrocarbons and fullerene are well known to afford the corresponding adducts via electron transfer [71–77]. Similarly, the photochemical behavior of SWNTs can be understood on the basis of their photochemical reactions with amines.

Maeda et al. found that the characteristic absorption of an SWNT tetrahydrofuran solution containing propylamine was decreased after photoirradiation [78]. The decrease of the absorption spectra was observed under degassed conditions but not under air. A significant decrease of the absorption peaks of metallic SWNTs and small-diameter semiconducting SWNTs was observed. Similar spectral changes were found in the solution-phase Raman spectra as well. These results indicated that the large-diameter semiconducting SWNTs possess lower reactivity under the applied reaction condition. After exposure to air, the characteristic absorption peaks of large-diameter SWNTs were recovered (Fig. 11.6). After the photoreaction, the Raman spectra of the small-diameter SWNTs showed an increase of the D band together with a decrease in the RBM peak. These results indicated that the sidewall functionalization of small-diameter SWNTs took place upon photoirradiation (Scheme 11.5). The higher reactivity of small-diameter SWNTs is explained by their strain energy and π-orbital misalignment angle, as mentioned before [79].

To confirm the reaction mechanism of the photoreaction, ESR spectroscopy was performed. The results indicated the presence of $^{+}NH_3CH_2CH_2CH_2C\bullet$, which was formed by the rearrangement of the propylamine radical cation. The detection of the radical intermediate strongly supports the possibility of a photoinduced electron transfer from the amine toward the SWNT. The reaction of small-diameter SWNT anion and the amine cation radical afforded the corresponding adducts, presented on Scheme 11.5. It is important to mention that the optical properties of large-diameter SWNTs were recovered after exposure to air. This observation indicates that the SWNT anion reacted with the oxygen to provide the neutral SWNTs [38, 80, 81]. It was revealed that metallic SWNTs have lower reduction potentials than do semiconducting SWNTs. This finding is consistent with the spectral changes during the photoreaction. The decreased reactivity of large-diameter semiconducting SWNTs (compared to the metallic SWNTs and small-diameter semiconducting SWNTs) during the photoirradiation might be explained by their high reduction potential and small strain energy. The large-diameter semiconducting SWNTs are

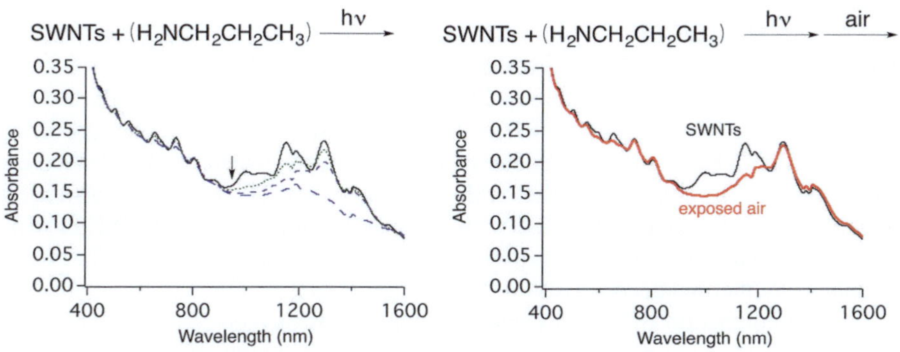

Fig. 11.6 Absorption spectra of SWNTs before and after photoreaction

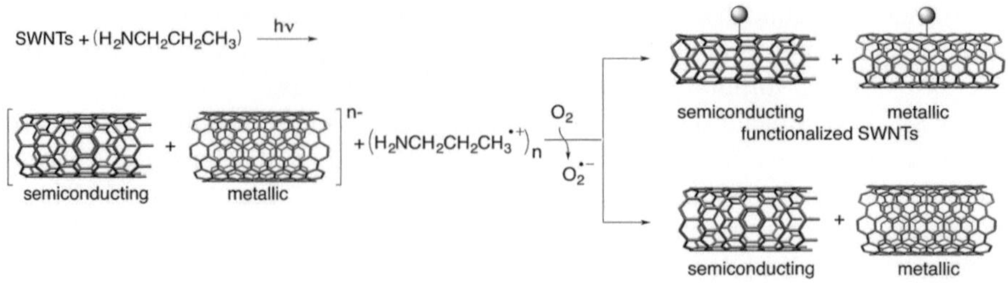

Scheme 11.5 Photoreaction of SWNTs with propylamine

Scheme 11.6 Reaction of SWNTs with disulfide

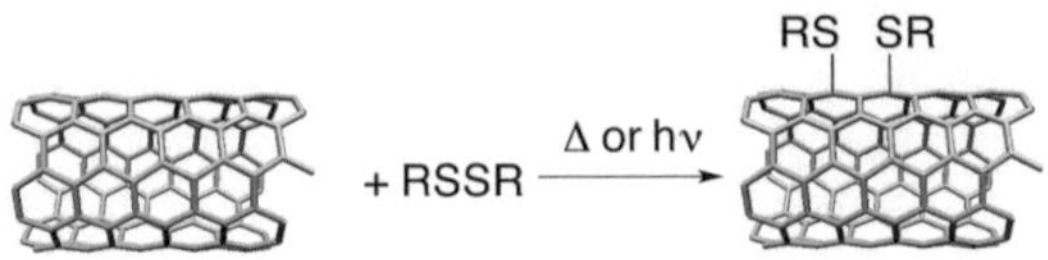

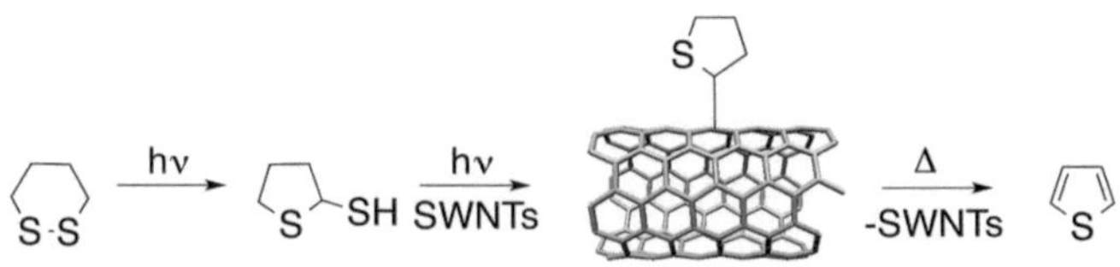

Scheme 11.7 Photoreaction of dithiane in the presence of SWNTs

probably good candidates for photosensitizers, since they are photoactive, but they generally show low reactivity in sidewall functionalization.

11.4.2 Helicity-Selective Photooxidations with Disulfides

Functionalization of SWNTs with organosulfur compounds has received a great deal of attention. The introduction of thiol groups to SWNTs was conducted for subsequent site-selective deposition of gold [82, 83]. Dai et al. reported a drug delivery system based on SWNTs having disulfide bonds which can be cleaved by enzymes [84]. The addition reaction of thiyl radical to SWNT sidewall has also been widely studied (Scheme 11.6) [32, 85, 86]. Recently, photoreaction of SWNTs with 1,2-dithiane was reported, as shown in Scheme 11.7. The sulfurized SWNTs afford thiophene by thermal decomposition (Scheme 11.7) [87].

The photoreaction of sulfides and disulfides with oxygen in the presence of photosensitizers gives persulfoxide and thiopersulfinate intermediates, respectively (Scheme 11.8) [88]. It is well known that these intermediates act as nucleophilic oxidizing agents. For example, the photoreaction of C_{60} in the presence of oxygen

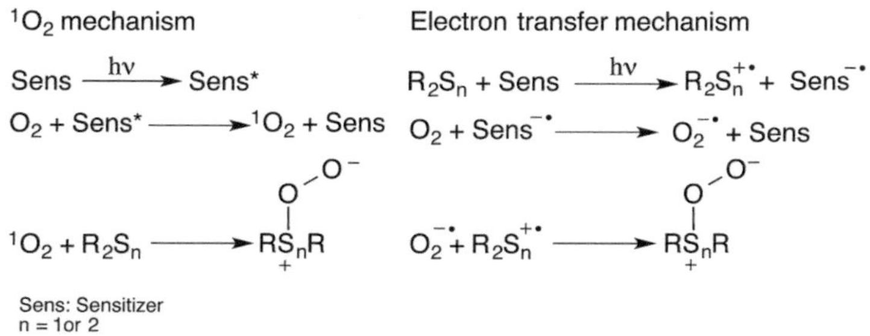

Scheme 11.8 Two reaction mechanisms of photooxidation of organosulfur compounds

Scheme 11.9 Photooxidation reaction of C_{60} in the presence of sulfide

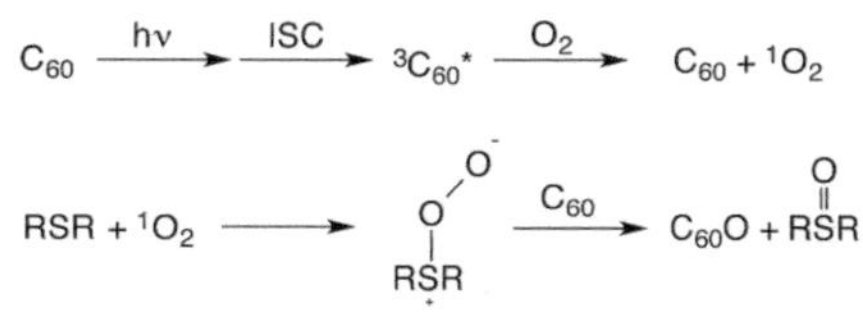

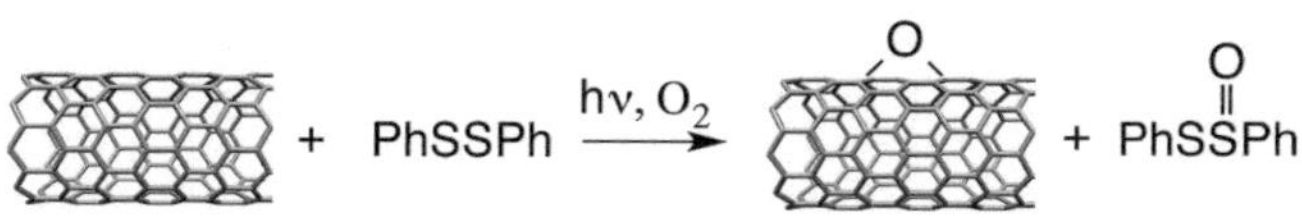

Scheme 11.10 Photooxidation reaction of SWNTs in the presence of disulfide

affords C_{60} epoxide ($C_{60}O$) via the formation of a persulfoxide intermediate (Scheme 11.9) [89]. In addition, $C_{60}O$ is also regarded as a promising intermediate for further functionalizations [90–96]. Thus, the photoreaction of SWNTs with disulfide in the presence of oxygen is conducted.

The photoreaction of SWNTs with diphenyl disulfide in the presence of oxygen was carried out by Maeda et al. (Scheme 11.10) [97]. Changes in the characteristic absorption and RBM peaks indicated that the reaction took place. Remarkably, the reaction was found to be helicity selective (metallic SWNTs/semiconducting SWNTs and diameter) (Fig. 11.7). The reactivity of the different SWNTs was estimated based on the spectral changes. It was revealed that the reactivity decreased in the order of metallic SWNTs > semiconducting SWNTs and small-diameter SWNTs > large-diameter SWNTs. Importantly, the photoreaction proceeded only if both disulfide and oxygen were present which demonstrates the key role of oxygen. The XPS measurements showed an increased O to C atomic ratio after the photoreaction. These results confirmed that oxygen atoms were introduced on the SWNT sidewall.

Noteworthy is that the photoreaction proceeded even when SWNTs were selectively excited. Two reaction mechanisms have been proposed for the generation of thiopersulfinate, as shown in Scheme 11.7. In laser photolysis and ESR experiments,

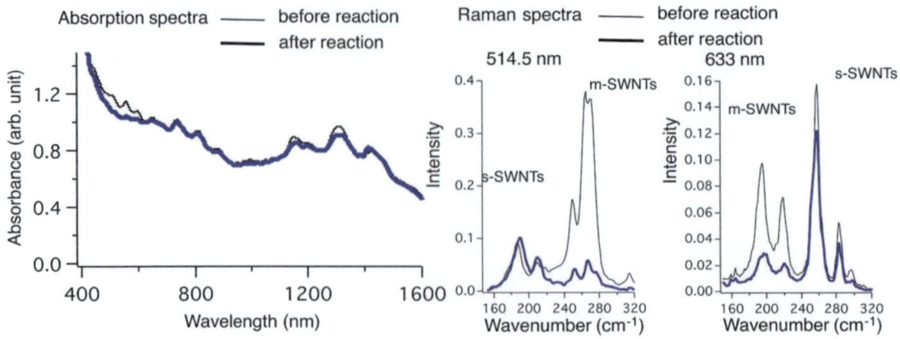

Fig. 11.7 Absorption spectra of SWNTs before and after the photoreaction. Raman spectra of SWNTs before and after the photoreaction

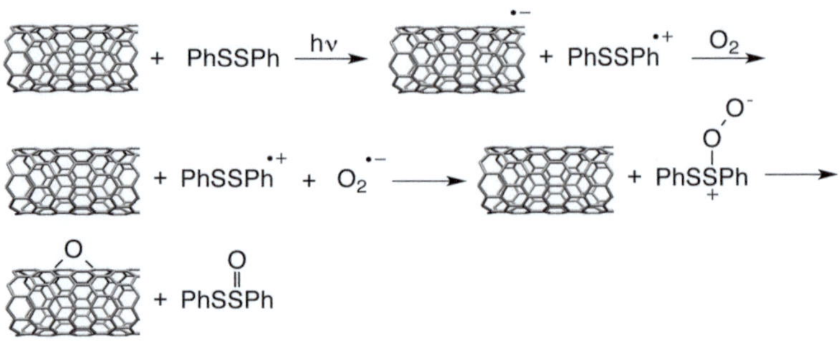

Scheme 11.11 Reaction mechanism of photooxidation reaction of SWNTs in the presence of disulfide

cationic radical species formed from the disulfide were detected. In addition, Murray et al. reported that the diphenyl disulfide showed low reactivity toward $^1O_2$ [98]. Thus, the plausible reaction mechanism involves a photoinduced electron transfer from the excited SWNTs to the disulfide (Scheme 11.11).

The highly selective reaction of diazonium compounds with metallic SWNTs was reported by Strano et al. [99]. The authors explained the observed high selectivity with the good electron-donating property of metallic SWNTs. Studies on the reaction of SWNTs with reductant in the ground state proved that metallic SWNTs have high electron-accepting ability compared to semiconducting SWNTs. Besides, it was also reported that thiopersulfinate intermediates act as nucleophile. Therefore, it is assumed that the high reactivity of metallic SWNTs in this photoreaction is caused by the high electrophilicity of metallic SWNTs toward the nucleophilic thiopersulfinate intermediates.

In 2002, the NIR region photoluminescence (PL) spectra of SWNTs were reported by Weisman et al. [26, 100]. In their work, the E_{11} emission was observed by the excitation of E_{22} wavelength. Photoluminescence is well known as a powerful

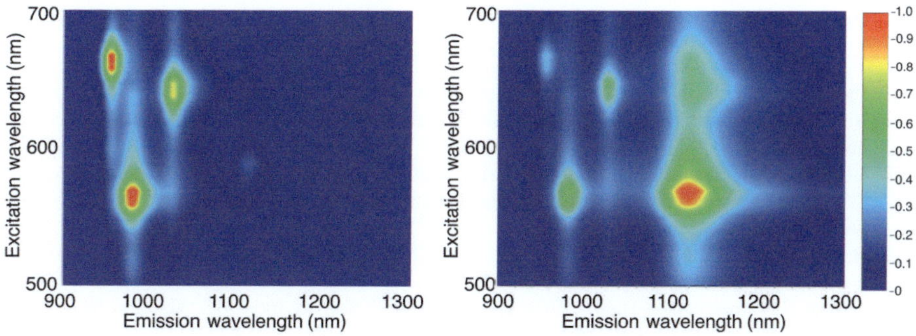

Fig. 11.8 Photoluminescence spectra of SWNTs (*left*) and oxidized SWNTs (*right*)

tool for the chiral assignment of semiconducting SWNTs. As a practical application of PL spectra of SWNTs, much attention has been focused on the biological probe, because the NIR light has high transmittance in biological tissues [101]. In 2010, it was reported that a new PL peak was detected after the photoreaction of SWNTs with O_3 [102]. The new PL peak intensity was significantly high compared to the E_{11} emission peak [103]. Interestingly, a new PL peak was also observed in the spectrum when disulfide was used in the oxidation reaction (Fig. 11.8). Thus, the photochemical reaction of SWNTs with disulfide in the presence of oxygen is useful not only for chirally selective functionalization but also for controlling the PL properties of semiconducting SWNTs.

11.5 Conclusion

Up to now, a wide range of chemical functionalizations on SWNT sidewalls have been developed. Chemical functionalization is an effective method to control the SWNT dispersibility in organic solvents. The degree of functionalization has to be controlled in order to achieve the desired electronic and optical properties. Excessive sidewall functionalization breaks the π-electron system and thus damages the beneficial properties of SWNTs. Organic chemistry offers a range of possibilities to prepare functionalized SWNT materials. In the two-step reductive alkylation method, the substituent properties can be exploited to control the degree of functionalization. Photochemistry also proved to be a powerful tool for SWNT functionalization. We believe that tuning the structure and properties of SWNTs via chemical functionalizations makes these materials particularly promising candidates for practical applications.

Acknowledgments This work was supported by Grants-in-Aid for Scientific Research (B) (26286012) and Grants-in-Aid for Challenging Exploratory Research (26600034) from the Ministry of Education, Culture, Sports, Science, and Technology of Japan.

References

1. Iijima S (1991) Helical microtubules of graphitic carbon. Nature 354:56–58
2. Iijima S, Ichihashi T (1993) Single-shell carbon nanotubes of 1-nm diameter. Nature 363:603–605
3. Bethune DS, Kiang CH, De Vries MS, Gorman G, Savoy R, Vazquez J, Beyers R (1993) Cobalt-catalysed growth of carbon nanotubes with single-atomic-layer walls. Nature 363:605–607
4. Baughman RH, Zakhidov AA, De Heer WA (2002) Carbon nanotubes – the route toward applications. Science 297:787–792
5. Niyogi S, Hamon MA, Hu H, Zhao B, Bhowmik P, Sen R, Itkis ME, Haddon RC (2002) Chemistry of single-walled carbon nanotubes. Acc Chem Res 35:1105–1113
6. Akasaka T, Wudl F, Nagase S (2010) Chemistry of nanocarbons. Wiley-Blackwell, Oxford
7. Saito R, Fujita M, Dresselhaus G, Dresselhaus MS (1992) Electronic structure of chiral graphene tubules. Appl Phys Lett 60:2204–2206
8. Hamada N, S-i S, Oshiyama A (1992) New one-dimensional conductors: graphitic microtubules. Phys Rev Lett 68:1579–1581
9. Mintmire JW, Dunlap BI, White CT (1992) Are fullerene tubules metallic? Phys Rev Lett 68:631–634
10. Yang F, Wang X, Zhang D, Yang J, Luo D, Xu Z, Wei J, Wang J-Q, Xu Z, Peng F, Li R, Li Y, Li M, Bai X, Ding F, Li Y (2014) Chirality-specific growth of single-walled carbon nanotubes on solid alloy catalysts. Nature 510:522–524
11. Sanchez-Valencia JR, Dienel T, Gröning O, Shorubalko I, Mueller A, Jansen M, Amsharov K, Ruffieux P, Rasel R (2014) Controlled synthesis of single-chirality carbon nanotubes. Nature 512:61–64
12. Bahr JL, Mickelson ET, Bronikowski MJ, Smalley RE, Tour JM (2001) Dissolution of small diameter single-wall carbon nanotubes in organic solvents? Chem Commun 193–194
13. Tasis D, Tagmatarchis N, Bianco A, Prato M (2006) Chemistry of carbon nanotubes. Chem Rev 106:1105–1136
14. Karousis N, Tagmatarchis N, Tasis D (2010) Current progress on the chemical modification of carbon nanotubes. Chem Rev 110:5366–5397
15. Chen J, Hamon MA, Hu H, Chen Y, Rao AM, Eklund PC, Haddon RC (1998) Solution properties of single-walled carbon nanotubes. Science 282:95–98
16. Hamon MA, Itkis ME, Niyogi S, Alvaraez T, Kuper C, Menon M, Haddon RC (2001) Effect of rehybridization on the electronic structure of single-walled carbon nanotubes. J Am Chem Soc 123:11292–11293
17. Rao AM, Richter E, Bandow S, Chase B, Eklund PC, Williams KA, Fang S, Subbaswamy KR, Menon M, Thess A, Smalley RE, Dresselhaus G, Dresselhaus MS (1997) Diameter-selective Raman scattering from vibrational modes in carbon nanotubes. Science 275:187–190
18. Reich S, Thomsen C (2002) Electronic band structure of isolated and bundled carbon nanotubes. Phys Rev B 65:155411
19. Kataura H, Kumazawa Y, Maniwa Y, Umezu I, Suzuki S, Ohtsuka Y, Achiba Y (1999) Optical properties of single-wall carbon nanotubes. Synth Met 103:2555–2558
20. Chiang IW, Brinson BE, Huang AY, Willis PA, Bronikowski MJ, Margrave JL, Smalley RE, Hauge RH (2001) Purification and characterization of single-wall carbon nanotubes (SWNTs) obtained from the gas-phase decomposition of CO (HiPco process). J Phys Chem B 105:8297–8301
21. Saito T, Ohmori S, Shukla B, Yumura M, Lijima S (2009) A novel method for characterizing the diameter of single-wall carbon nanotubes by optical absorption spectra. Appl Phys Express 2:095006
22. Hu H, Zhao B, Hamon MA, Kamaras K, Itkis ME, Haddon RC (2003) Sidewall functionalization of single-walled carbon nanotubes by addition of dichlorocarbene. J Am Chem Soc 125:14893–14900

23. Strano MS, Huffman CB, Moore VC, O'Connell MJ, Haroz EH, Hubbard J, Miller M, Rialon K, Kittrell C, Ramesh S, Hauge RH, Smalley RE (2003) Reversible, band-gap-selective protonation of single-walled carbon nanotubes in solution. J Phys Chem B 107:6979–6985
24. Kazaoui S, Minami N, Kataura H, Achiba Y (2001) Absorption spectroscopy of single-wall carbon nanotubes: effects of chemical and electrochemical doping. Synth Met 121:1201–1202
25. Jacquemin R, Kazaoui S, Yu D, Hasanien A, Minami N, Kataura H, Achiba Y (2000) Doping mechanism in single-wall carbon nanotubes studied by optical absorption. Synth Met 115:283–287
26. Bachilo SM, Strano MS, Kittrell C, Hauge RH, Smalley RE, Weisman RB (2002) Structure-assigned optical spectra of single-walled carbon nanotubes. Science 298:2361–2366
27. Kukovecz A, Kramberger C, Georgakilas V, Prato M, Kuzmany H (2002) A detailed Raman study on thin single-wall carbon nanotubes prepared by the HiPCO process. Eur Phys J B 28:223–230
28. Moore VC, Strano MS, Haroz EH, Hauge RH, Smalley RE, Schmidt J, Talmon Y (2003) Individually suspended single-walled carbon nanotubes in various surfactants. Nano Lett 3:1379–1382
29. Liang F, Sadana AK, Peera A, Chattopadhyay J, Gu Z, Hauge RH, Billups WE (2004) A convenient route to functionalized carbon nanotubes. Nano Lett 4:1257–1260
30. García-Gallastegui A, Obieta I, Bustero I, Imbuluzqueta G, Arbiol J, Miranda JI, Aizpurua JM (2008) Reductive functionalization of single-walled carbon nanotubes with lithium metal catalyzed by electron carrier additives. Chem Mater 20:4433–4438
31. Wunderlich D, Hauke F, Hirsch A (2008) Preferred functionalization of metallic and small-diameter single walled carbon nanotubes via reductive alkylation. J Mater Chem 18:1493–1497
32. Chattopadhyay J, Chakraborty S, Mukherjee A, Runtang W, Engel PS, Billups WE (2007) SET mechanism in the functionalization of single-walled carbon nanotubes. J Phys Chem C 111:17928–17932
33. Chattopadhyay J, Sadana AK, Liang F, Beach JM, Xiao Y, Hauge RH, Billups WE (2005) Carbon nanotube salts. Arylation of single-wall carbon nanotubes. Org Lett 7:4067–4069
34. Mukherjee A, Combs R, Chattopadhyay J, Abmayr DW, Engel PS, Billups WE (2008) Attachment of nitrogen and oxygen centered radicals to single-walled carbon nanotube salts. Chem Mater 20:7339–7343
35. Martínez-Rubí Y, Guan J, Lin S, Scriver C, Sturgeon RE, Simard B (2007) Rapid and controllable covalent functionalization of single-walled carbon nanotubes at room temperature. Chem Commun 48:5146–5148
36. Gebhardt B, Syrgiannis Z, Backes C, Graupner R, Hauke F, Hirsch A (2011) Carbon nanotube sidewall functionalization with carbonyl compounds—modified birch conditions vs the organometallic reduction approach. J Am Chem Soc 133:7985–7995
37. Gebhardt B, Hoffman F, Backes C, Müller M, Plocke T, Maultzsch J, Thomsen C, Hauke F, Hirsch A (2011) Selective polycarboxylation of semiconducting single-walled carbon nanotubes by reductive sidewall functionalization. J Am Chem Soc 133:19459–19473
38. Graupner R, Abraham J, Wunderlich D, Vencelová A, Lauffer P, Röhrl J, Hundhausen M, Ley L, Hirsch A (2006) Nucleophilic-alkylation-reoxidation: a functionalization sequence for single-wall carbon nanotubes. J Am Chem Soc 128:6683–6689
39. Gebhardt B, Graupner R, Hauke F, Hirsch A (2010) A novel diameter-selective functionalization of SWCNTs with lithium alkynylides. Eur J Org Chem 1494–1501
40. Syrgiannis Z, Hauke F, Röhrl J, Hundhausen M, Graupner R, Elemes Y, Hirsch A (2008) Covalent sidewall functionalization of SWNTs by nucleophilic addition of lithium amides. Eur J Org Chem 2544–2550
41. Wunderlich D, Hauke F, Hirsch A (2008) Preferred functionalization of metallic and small-diameter single-walled carbon nanotubes by nucleophilic addition of organolithium and -magnesium compounds followed by reoxidation. Chemistry 14:1607–1614

42. Roubeau O, Lucas A, Pénicaud A, Derré A (2007) Covalent functionalization of carbon nanotubes through organometallic reduction and electrophilic attack. J Nanosci Nanotechnol 7:3509–3513
43. Maeda Y, Kato T, Hasegawa T, Kako M, Akasaka T, Lu J, Nagase S (2010) Two-step alkylation of single-walled carbon nanotubes: substituent effect on sidewall functionalization. Org Lett 12:996–999
44. Maeda Y, Saito K, Akamatsu N, Chiba Y, Ohno S, Okui Y, Yamada M, Hasegawa T, Kako M, Akasaka T (2012) Analysis of functionalization degree of single-walled carbon nanotubes having various substituents. J Am Chem Soc 134:18101–18108
45. Chen S, Shen W, Wu G, Chen D, Jiang M (2005) A new approach to the functionalization of single-walled carbon nanotubes with both alkyl and carbonyl groups. Chem Phys Lett 402:312–317
46. Bayazit MK, Suri A, Coleman KS (2010) Formylation of single-walled carbon nanotubes. Carbon 48:3412–3419
47. Viswanathan G, Chakrapani N, Yang H, Wei B, Chung H, Cho K, Ryu CY, Ajayan PM (2003) Single-step in situ synthesis of polymer-grafted single-wall nanotube composites. J Am Chem Soc 125:9258–9259
48. Mickelson ET, Huffman CB, Rinzler AG, Smalley RE, Hauge RH, Margrave JL (1998) Fluorination of single-wall carbon nanotubes. Chem Phys Lett 296:188–194
49. Boul PJ, Liu J, Mickelson ET, Huffman CB, Ericson LM, Chiang IW, Smith KA, Colbert DT, Hauge RH, Margrave JL, Smalley RE (1999) Reversible sidewall functionalization of buckytubes. Chem Phys Lett 310:367–372
50. Wakahara T, Kako M, Maeda Y, Akasaka T, Kobayashi K, Nagase S (2003) Synthesis and characterization of cyclic silicon compounds of fullerenes. Curr Org Chem 7:927–943
51. Maeda Y, Sato Y, Kako M, Wakahara T, Akasaka T, Lu J, Nagase S, Kobori Y, Hasegawa T, Motomiya K, Tohji K, Kasuya A, Wang D, Yu D, Gao Z, Han R, Ye H (2006) Preparation of single-walled carbon nanotube – organosilicon hybrids and their enhanced field emission properties. Chem Mater 18:4205–4208
52. Kumashiro R, Hiroshiba N, Komatsu N, Akasaka T, Maeda Y, Suzuki S, Achiba Y, Hatakeyama R, Tanigaki K (2008) FET properties of chemically modified carbon nanotubes. J Phys Chem Solids 69:1206–1208
53. Dyke CA, Stewart MP, Tour JM (2005) Separation of single-walled carbon nanotubes on silica gel. Materials morphology and Raman excitation wavelength affect data interpretation. J Am Chem Soc 127:4497–4509
54. Wieckowski WW, Pastorin G, Benincasa M, LKlumpp C, Briand J-P, Gennaro R, Prato M, Bianco A (2005) Targeted delivery of amphotericin B to cells by using functionalized carbon nanotubes. Angew Chem Int Ed 44:6358–6362
55. Lee KM, Li L, Dai L (2005) Asymmetric end-functionalization of multi-walled carbon nanotubes. J Am Chem Soc 127:4122–4123
56. Umeyama T, Imahori H (2013) Photofunctional hybrid nanocarbon materials. J Phys Chem C 117:3195–3209
57. Guldi DM, Costa RD (2013) Nanocarbon hybrids: the paradigm of nanoscale self-ordering/self-assembling by means of charge transfer/doping interactions. J Phys Chem Lett 4:1489–1501
58. Choi N, Kimura M, Kataura H, Suzuki S, Achiba Y, Mizutani W, Tokumoto H (2002) Effect of amines on single-walled carbon nanotubes in organic solvents: control of bundle structures. Jpn J Appl Phys Part 1 41:6264–6266
59. Maeda Y, Kimura SI, Hiroshima Y, Kanda M, Lian Y, Wakahara T, Akasaka T, Hasegawa T, Tokumoto H, Shimizu T, Kataura H, Miyauchi Y, Maruyama S, Kobayashi K, Nagase S (2004) Dispersion of single-walled carbon nanotube bundles in nonaqueous solution. J Phys Chem B 108:18395–18397
60. Samsonidze GG, Chou SG, Santos AP, Brar VW, Dresselhaus G, Dresselhaus MS, Selbst A, Swan AK, Ünlü MS, Goldberg BB, Chattopadhyay D, Kim SN, Papadimitrakopoulos F (2004) Quantitative evaluation of the octadecylamine-assisted bulk separation of semicon-

ducting and metallic single-wall carbon nanotubes by resonance Raman spectroscopy. Appl Phys Lett 85:1006–1008
61. Chattopadhyay D, Galeska I, Papadimitrakopoulos F (2003) A route for bulk separation of semiconducting from metallic single-wall carbon nanotubes. J Am Chem Soc 125:3370–3375
62. Kim SN, Luo Z, Papadimitrakopoulos F (2005) Diameter and metallicity dependent redox influences on the separation of single-wall carbon nanotubes. Nano Lett 5:2500–2504
63. Maeda Y, Kimura SI, Kanda M, Hirashima Y, Hasegawa T, Wakahara T, Lian Y, Nakahodo T, Tsuchiya T, Akasaka T, Lu J, Zhang X, Gao Z, Yu Y, Nagase S, Kazaoui S, Minam N, Shimizu T, Tokumoto H, Saito R (2005) Large-scale separation of metallic and semiconducting single-walled carbon nanotubes. J Am Chem Soc 127:10287–10290
64. Lu J, Nagase S, Maeda Y, Wakahara T, Nakahodo T, Akasaka T, Yu D, Gao Z, Han R, Ye H (2005) Adsorption configuration of NH_3 on single-wall carbon nanotubes. Chem Phys Lett 405:90–92
65. Maeda Y, Kanda M, Hashimoto M, Hasegawa T, Kimura SI, Lian Y, Wakahara T, Akasaka T, Kazaoui S, Minami N, Okazaki T, Hayamizu Y, Hata K, Lu J, Nagase S (2006) Dispersion and separation of small-diameter single-walled carbon nanotubes. J Am Chem Soc 128:12239–12242
66. Maeda Y, Hashimoto M, Hasegawa T, Kanda M, Tsuchiya T, Wakahara T, Akasaka T, Miyauchi Y, Maruyama S, Lu J, Nagase S (2007) Extraction of metallic nanotubes of zeolite-supported single-walled carbon nanotubes synthesized from alcohol. NANO Brief Rep Rev 2:221–226
67. Maeda Y, Takano Y, Sagara A, Hashimoto M, Kanda M, Si K, Lian Y, Nakahodo T, Tsuchiya T, Wakahara T, Akasaka T, Hasegawa T, Kazaoui S, Minami N, Lu J, Nagase S (2008) Simple purification and selective enrichment of metallic SWCNTs produced using the arc-discharge method. Carbon 46:1563–1569
68. Lu J, Lai L, Luo G, Zhou J, Qin R, Wang D, Wang L, Mei WN, Li G, Gao Z, Nagase S, Maeda Y, Akasaka T, Yu D (2007) Why semiconducting single-walled carbon nanotubes are separated from their metallic counterparts. Small 3:1566–1576
69. Maeda Y, Yamada M, Hasegawa T, Akasaka T, Lu J, Nagase S (2012) Interaction of single-walled carbon nanotubes with amine. NANO Brief Rep Rev 7:1130001-1130001-1113001-1130010
70. Maeda Y, Komoriya K, Sode K, Kanda M, Yamada M, Hasegawa T, Akasaka T, Lu J, Nagase S (2010) Separation of metallic single-walled carbon nanotubes using various amines. Phys Status Solidi B 247:2641–2644
71. Cookson RC, De Costa SMB, Hudec J (1969) Photochemical addition of amines to styrenes. J Chem Soc Chem Commun 753–754
72. Bellas M, Bryce-Smith D, Gilbert A (1967) 1,4-Photoaddition of amines to benzene. Chem Commun 862–863
73. Cohen SG, Parola A, Parsons GHJ (1973) Photoreduction by amines. Chem Rev 73:141–161
74. Kawanisi M, Matsunaga K (1972) Novel photochemical reactions of diphenylacetylene and stilbenes in amines. J Chem Soc Chem Commun 313–314
75. Hirsch A, Li Q, Wudl F (1991) Globe-trotting hydrogens on the surface of the fullerene compound $C_{60}H_6(N(CH_2CH_2)_2O)_6$. Angew Chem Int Ed 30:1309–1310
76. Arbogast JW, Foote CS, Kao M (1992) Electron transfer to triplet C_{60}. J Am Chem Soc 114:2277–2279
77. Ma B, Lawson GE, Bunker CE, Kitaygorodskiy A, Sun YP (1995) Fullerene-based macromolecules from photochemical reactions of [60]fullerene and triethylamine. Chem Phys Lett 247:51–56
78. Maeda Y, Hasuike Y, Ohkubo K, Tashiro A, Kaneko S, Kikuta M, Yamada M, Hasegawa T, Akasaka T, Zhou J, Lu J, Nagase S, Fukuzumi S (2014) Photochemical behavior of single-walled carbon nanotubes in the presence of propylamine. ChemPhysChem 15:1821–1826
79. Chen Z, Thiel W, Hirsch A (2003) Reactivity of the convex and concave surfaces of single-walled carbon nanotubes (SWCNTs) towards addition reactions: dependence on the carbon-atom pyramidalization. ChemPhysChem 4:93–97

80. Jhi SH, Louie SG, Cohen ML (2000) Electronic properties of oxidized carbon nanotubes. Phys Rev Lett 85:1710–1713
81. Zhou J, Maeda Y, Lu J, Tashiro A, Hasegawa T, Luo G, Wang L, Lai L, Akasaka T, Nagase S, Gao Z, Qin R, Mei WN, Li G, Yu D (2009) Electronic-type- and diameter-dependent reduction of single-walled carbon nanotubes induced by adsorption of electron-donor molecules. Small 5:244–255
82. Liu J, Rinzler AG, Dai H, Hafner JH, Bradley RK, Boul PJ, Lu A, Iverson T, Shelimov K, Huffman CB, Rodriguez-Macias F, Shon Y-S, Lee TR, Colbert DT, Smalley RE (1998) Fullerene pipes. Science 280:1253–1257
83. Liu L, Wang T, Li J, Guo Z, Dai L, Zhang D, Zhu D (2003) Self-assembly of gold nanoparticles to carbon nanotubes using a thiol-terminated pyrene as interlinker. Chem Phys Lett 367:747–752
84. Kam NWS, Liu Z, Dai H (2005) Functionalization of carbon nanotubes via cleavable disulfide bonds for efficient intracellular delivery of siRNA and potent gene silencing. J Am Chem Soc 127:12492–12493
85. Nakamura T, Ohana T, Ishihara M, Tanaka A, Koga Y (2006) Sidewall modification of single-walled carbon nanotubes with sulfur-containing functionalities and gold nanoparticle attachment. Chem Lett 35:742–743
86. Syrgiannis Z, Parola VL, Hadad C, Lucío M, Vázquez E, Giacalone F, Prato M (2013) An atom-economical approach to functionalized single-walled carbon nanotubes: reaction with disulfides. Angew Chem Int Ed 52:6480–6483
87. Engel PS, Gudimetla VB, Gancheff JS, Denis PA (2012) Solution phase photolysis of 1,2-dithiane alone and with single-walled carbon nanotubes. J Phys Chem A 116:8345–8351
88. Lacombe S, Cardy H, Simon M, Khoukh A, Soumillion JP, Ayadim M (2002) Oxidation of sulfides and disulfides under electron transfer or singlet oxygen photosensitization using soluble or grafted sensitizers. Photochem Photobiol Sci 1:347–354
89. Maeda Y, Niino Y, Kondo T, Yamada M, Hasegawa T, Akasaka T (2011) Oxygen atom transfer from peroxide intermediates to fullerenes. Chem Lett 40:1431–1433
90. Smith AB III, Tokuyama H, Strongin RM, Furst GT, Romanow WJ, Chait BT, Mirza UA, Haller I (1995) Synthesis of oxo- and methylene-bridged C_{60} dimers, the first well-characterized species containing fullerene-fullerene bonds. J Am Chem Soc 117:9359–9360
91. Lebedkin SF, Ballenweg S, Gross J, Taylor R, Krätschmer W (1995) Synthesis of $C_{120}O$: a new dimeric [60]fullerene derivative. Tetrahedron Lett 1995:4971–4974
92. Balch AL, Costa DA, Fawcett WR, Winkler K (1996) Electronic communication in fullerene dimers. Electrochemical and electron paramagnetic resonance study of the reduction of $C_{120}O$. J Phys Chem 100:4823–4827
93. Tsyboulski D, Heymann D, Bachilo SM, Alemany LB, Weisman RB (2004) Reversible dimerization of [5,6]-$C_{60}O$. J Am Chem Soc 126:7350–7358
94. Shigemitsu Y, Kaneko M, Tajima Y, Takeuchi K (2004) Efficient acetalization of epoxy rings on a fullerene cage. Chem Lett 33:1604–1605
95. Tajima Y, Hara T, Honma T, Matsumoto S, Takeuchi K (2006) Lewis acid-assisted nucleophilic substitution of fullerene epoxide. Org Lett 8:3203–3205
96. Yang X, Huang S, Jia Z, Xiao Z, Jiang Z, Zhang Q, Gan L, Zheng B, Yuan G, Zhang S (2008) Reactivity of fullerene epoxide: preparation of fullerene-fused thiirane, tetrahydrothiazolidin-2-one, and 1,3-dioxolane. J Org Chem 73:2518–2526
97. Maeda Y, Higo J, Amagai Y, Matsui J, Ohkubo K, Yoshigoe Y, Hashimoto M, Eguchi K, Yamada M, Hasegawa T, Sato Y, Zhou J, Lu J, Miyashita T, Fukuzumi S, Murakami T, Tohji K, Nagase S, Akasaka T (2013) Helicity-selective photoreaction of single-walled carbon nanotubes with organosulfur compounds in the presence of oxygen. J Am Chem Soc 135:6356–6362
98. Murray RW, Jindal SL (1972) Photosensitized oxidation of dialkyl disulfides. J Org Chem 37:3516–3520

99. Strano MS, Dyke CA, Usrey ML, Barone PW, Allen MJ, Shan H, Kittrell C, Hauge RH, Tour JM, Smalley RE (2003) Electronic structure control of single-waited carbon nanotube functionalization. Science 301:1519–1522
100. O'Connell MJ, Bachilo SH, Huffman CB, Moore VC, Strano MS, Haroz EH, Rialon KL, Boul PJ, Noon WH, Kittrell C, Ma J, Hauge RH, Weisman RB, Smalley RE (2002) Band gap fluorescence from individual single-walled carbon nanotubes. Science 297:593–596
101. Weissleder R (2001) A clearer vision for in vivo imaging. Nat Biotechnol 19:316–317
102. Ghosh S, Bachilo SM, Simonette RA, Beckingham KM, Weisman RB (2010) Oxygen doping modifies near-infrared band gaps in fluorescent single-walled carbon nanotubes. Science 330:1656–1659
103. Miyauchi Y, Iwamura M, Mouri S, Kawazoe O, Ohtsu M, Matsuda K (2013) Brightening of excitons in carbon nanotubes on dimensionality modification. Nat Photonics 7:715–719

Part III
Porphyrinoids

Chapter 12
Novel π-Conjugated Systems Based on N-Confused Porphyrinoids

Masatoshi Ishida and Hiroyuki Furuta

Abstract N-confused porphyrin (termed as NCP) is an isomer of porphyrin in which a pyrrole is connected to the neighboring *meso*-carbon atoms at α,β′-positions. The specific π-conjugated scaffold present in NCPs provides unique photophysical and electrochemical properties as well as abilities of metal complexation and anion binding in many cases. That has inspired us to explore the further synthetic approaches for development of novel class of porphyrin analogues based on "N-confusion concept." This chapter introduces an overview of recent development in the chemistry of N-confused porphyrin and related macrocycles.

Keywords N-confusion • N-fusion • Neo-confusion • Porphyrinoids

12.1 Introduction

Porphyrin (**1**) is one of the most extensively studied macrocycle-based chromophores basically composing of four pyrrole rings linked with *meso*-methine carbon atoms (Chart 12.1) [1]. These naturally occurring tetrapyrrolic entities are well known because of the critical roles in the various biological processes and regulation of their functions. For instance, hemes (Fe porphyrins) and chlorophylls (Mg homologues) are included for transportation of gaseous molecules, chemical catalysis, light harvest, and electron transfer. By these fascinated abilities, porphyrins and the related macrocycles have been utilized as key components for functional dyes, catalysts, nonlinear optical materials, sensory receptors, and so on [2]. In pursuit of real applications in the fields that range from material chemistry to biomedicine, a large number of porphyrins and analogues have been synthesized almost exponentially in recent decades. The main features of porphyrins are simply macrocyclic, intensely colored, and aromatic. A demand, to understand and extend these characteristics, has inspired considerable interest in the preparation

M. Ishida • H. Furuta (✉)
Department of Chemistry and Biochemistry, Graduate School of Engineering and Education Center for Global Leaders in Molecular Systems for Devices, Kyushu University, Fukuoka 819-0395, Japan
e-mail: hfuruta@cstf.kyushu-u.ac.jp

T. Akasaka et al. (eds.), *Chemical Science of π-Electron Systems*,
DOI 10.1007/978-4-431-55357-1_12

Chart 12.1 Skeletal isomers of porphyrins (**1**–**6**)

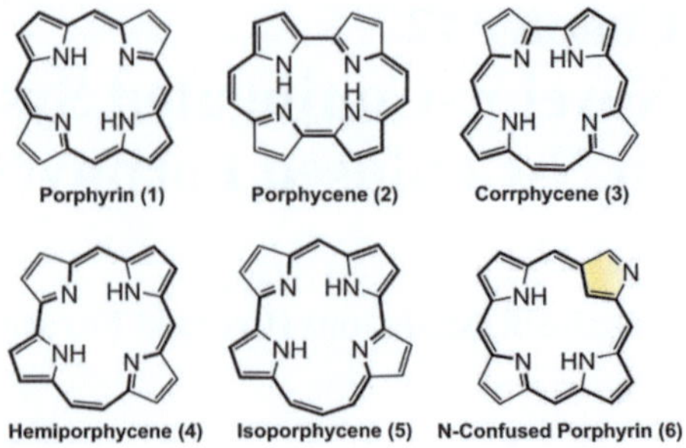

of the new type of porphyrin analogues by propagating the structural perturbations of π-conjugated cores. The synthesis of pure isomers of porphyrins constructed by altering the carbon atom bridging fashion between the pyrrole subunits, as porphycene (**2**) [3], corrphycene (**3**) [4], hemiporphycene (**4**) [5], and isoporphycene (**5**) [6], has been accomplished. We have in turn focused on an N-confused porphyrin (**6**) which is possessing inverted pyrrole subunits with different connection modes between *meso*-carbon atoms (Chart 12.1).

The structures of NCP in reality have been already speculated by Linus Pauling and Melvin Calvin in the 1940s [7]. After more than 50 years, the first synthesis of NCPs was achieved by the groups of author and Latos-Grażyński, independently [8]. The regular porphyrins possess four inward-orientated pyrrolic nitrogen atoms, whereas in the structural point of view, the position of one of the nitrogen atoms in NCPs is exchanged with a peripheral β-carbon atom of the macrocycle. Therefore, NCP incorporates a so-called N-confused pyrrole which is connected to *meso*-carbon atoms not in an α,α′-linkage fashion but in an α,β′-one in the constitutional tetrapyrrolic framework. Consequently, NCP **6** is formally classified as the families of carbaporphyrinoids [9].

The first synthesis procedure of NCPs was relied on a Rothemund-condensation methodology [10], namely, the acid-catalyzed condensation of pyrrole and arylaldehyde, despite affording an N-confused tetraarylporphyrin (**6′**) in 5–7 yields (Scheme 12.1). After the tremendous efforts were made, Lindsey et al. have reported convenient one-pot reactions for preferential preparation of NCPs by using methanesulfonic acid as an acid catalyst, and the yield was improved up to approximately 40% in the laboratory scale (Scheme 12.1) [11]. The specific target of *meso*-tetrakis(pentafluorophenyl)-substituted NCPs could be prepared by the McDonald [2 + 2]-type acid condensation of the appropriate pyrrolic precursors (i.e., dipyrromethanes) and followed by oxidation [12]. Concomitantly, this [2 + 2]-type stepwise synthetic manner can be applied for accessing "doubly N-confused porphyrins (termed as N_2CPs with *cis*-form **7** and *trans*-form **8**)" that contain two N-confused pyrrole rings in the macrocycle (Scheme 12.2) [13].

Various NCP families indeed share interesting features, such as NH tautomerism occurring in the inherent π-conjugated annulenic framework [14]. As a consequence of their lower symmetry relative to porphyrins, two independent NH tautomers exist in the NCPs depending on the choice of solvents. In other words, the inner 3H

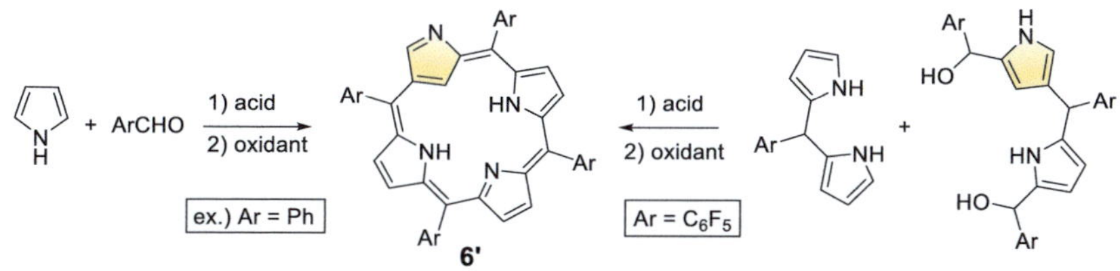

Scheme 12.1 Synthesis of *meso*-tetraaryl substituted N-confused porphyrins **6′**

Scheme 12.2 Synthesis of (**a**) *cis*-N_2CP **7** and (**b**) *trans*-N_2CP **8**

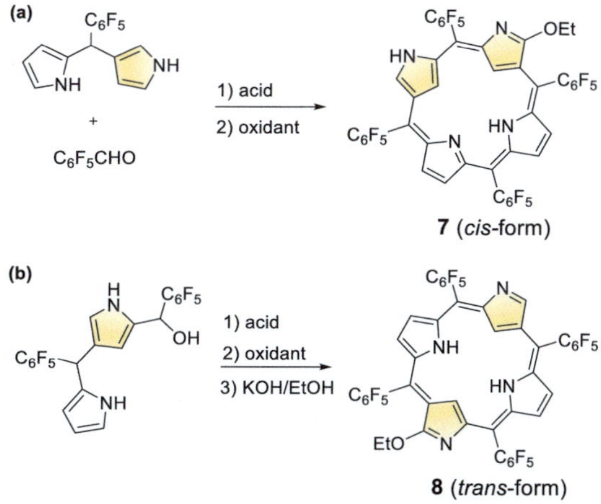

forms (**NCP3H**) and inner 2H form (**NCP2H**) are predominantly present in less proton-accepting solvents (e.g., toluene, CH_2Cl_2) and the strongly proton-accepting solvents (e.g., DMF, DMSO), respectively (Fig. 12.1). The interconversion of **NCP3H** and **NCP2H** being under chemical equilibrium in solution, distinct changes in the photophysical properties, metal coordination abilities, and aromatic nature were often observed. Each species exhibits different absorption spectral features; **NCP3H** has longer absorption bands in the Q-like region than **NCP2H**, while the emission efficiency is less ($\Phi_{FL} = 0.023$ in CH_2Cl_2 vs 0.046 in DMAc as **NCP2H** form). The shorter emission lifetime of **NCP3H** ($\tau_{FL} = 1.47$ ns) is thus elucidated in comparison with that of **NCP2H** ($\tau_{FL} = 1.84$ ns) [15]. These optical properties of NCPs could be reflected by their aromaticity. While the **NCP3H** has a complete aromatic 18π-electron-conjugated network, the corresponding π-circuit of **NCP2H** is incomplete. The stronger aromatic character of **NCP3H** was well rationalized by the NMR spectral analysis and theoretical assessments (*vide infra*).

From the viewpoint of coordination chemistry, the NCP can be served as unique coordination ligands in the inner coordination at a CNNN core [16]. Due to the presence of flexible tautomerism-dependent structures, multiplicity of the metal valence states was seen as the organometallic forms. As the representative example, the redox reactivity between a general copper(II) state and an unusual, higher

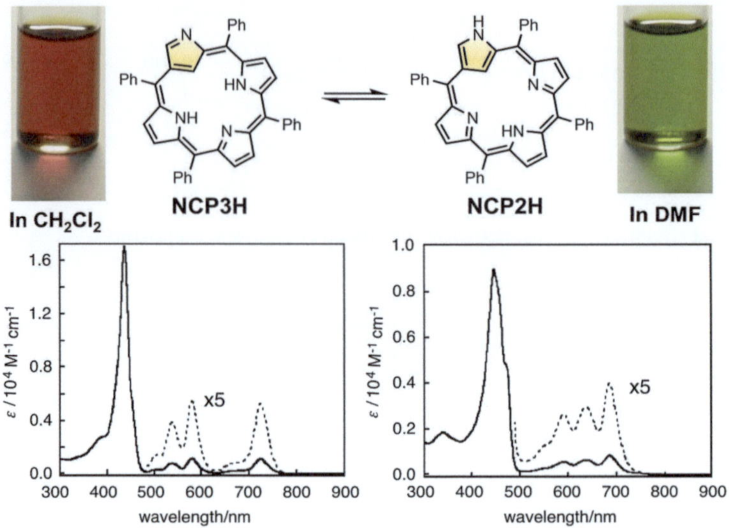

Fig. 12.1 Color of solutions and absorption spectra of NCP in CH_2Cl_2 (**NCP3H**, *left*) and in DMF (**NCP2H**, *right*) (Figures are reproduced from the Ref. [14a])

oxidation state of copper(III) center was well demonstrated in the NCP complex (Scheme 12.3) [17]. In this valence conversion, the protonation–deprotonation of an outward nitrogen site of the confused pyrrole occurred concomitantly, which is indicative of an important ability of NCP ligand for stabilization of the unusual valence metal states. Related to the tautomeric (or redox)-controlled acid–base properties of the peripheral nitrogen moiety of NCP complexes, the peripheral NH moiety can bind halide anions, e.g., fluoride, chloride, and so on [18, 17], whereas the imino N atom can accept the metal cation coordination to form, e.g., self-assembled dimers or trimers [19]. Furthermore, taking advantage of their acid–base properties of peripheral nitrogen sites, the copper(III) complexes of N_2CPs, **7** and **8**, form one-dimensional supramolecular rodlike polymeric chains and zigzag-type chained structures in solid states, respectively, via orientation-dependent intermolecular hydrogen bonding interactions between their peripheral amide-type NHs and imino N sites [13]. As shown briefly, these peculiar structures and properties of NCPs have not been shown in the studies of the regular porphyrins. The NCPs and their metal complexes can be regarded as 18π-electron embedded aromatic analogues, such as "expanded" pyridine, pyrrole, and imidazole, critically due to the presence of outward nitrogen atom(s) at the periphery.

This chapter focuses on recent advances in the development of novel N-confused porphyrinoids as artificial porphyrin analogues mostly contributed by us to keep the length of the review within reasonable limits [20]. As summarized in Fig. 12.2, the evolutionary advances have been made by further structural modifications of NCPs by incorporation of multiple N-confused rings, chemical fusions, skeletal expansions, ring contractions, and *neo*-confusion into parent NCP-

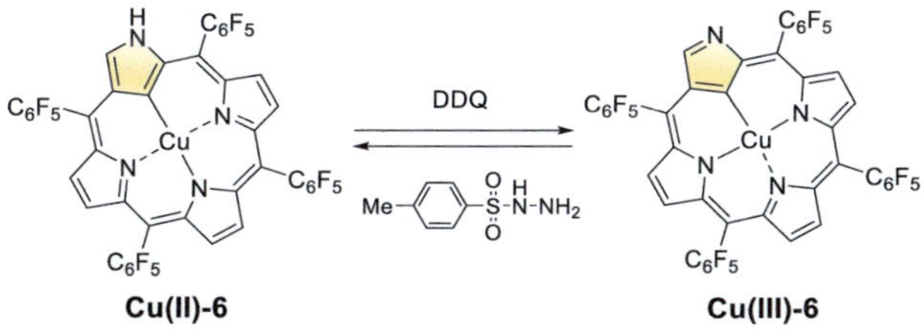

Scheme 12.3 Redox-coupled conversion between the complexes, **Cu(II)-6** and **Cu(III)-6**

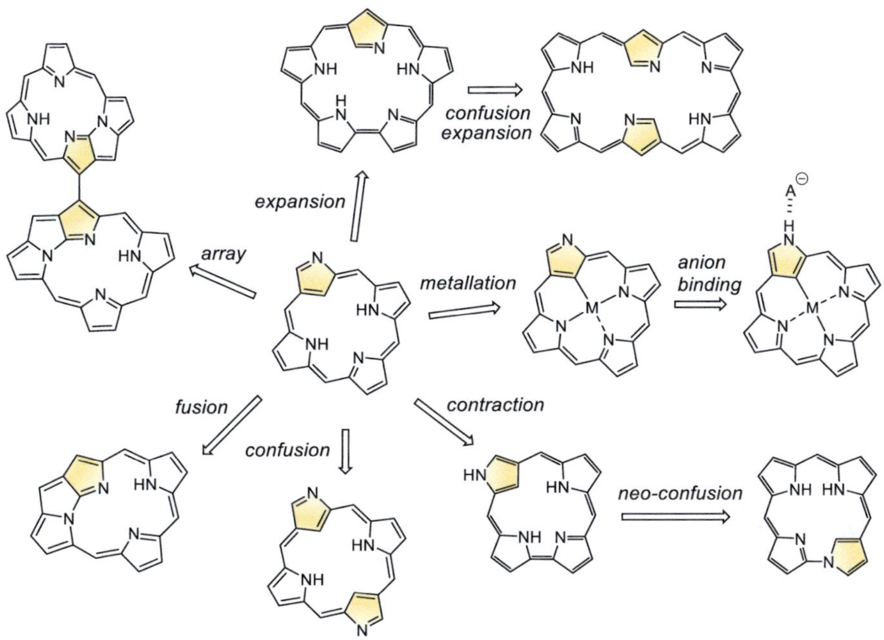

Fig. 12.2 Structural modifications of N-confused porphyrins

based frameworks. In particular, we herein present various synthetic approaches leading to specific π-conjugated systems created by N-confused porphyrinoids and highlight the electronic structures and properties of NCPs that are different from the original congeners. What if the core of NCPs is modified?

12.2 Reduction of NCPs to N-Confused Hydroporphyrins

The aromatic characteristics of porphyrins have intrigued chemists to generalize and understand the inherent properties of the macrocycles. General tetrapyrrolic macrocyclic system contains an 18π-electron substructure, which is responsible to be aromatic according to Hückel [4n + 2]π-electron rule [21]. In contrast, the evaluation of aromatic nature of NCPs is relatively complicated, since two bistable tautomers, **NCP3H** and **NCP2H** being under equilibrium, have distinct extents of aromaticity. The superior stability (by ca. 5 kcal/mol) of **NCP3H** to that of **NCP2H** reflects the dependent aromatic nature as evident from maximum difference (i.e., $\Delta\delta = 13.8$ ppm for **NCP3H** vs. 7.7 ppm for **NCP2H**) of the chemical shift values of peripheral CH and inner CH seen in the ^{1}H NMR spectra as well as the nucleus-independent chemical shift (NICS) values at the global center of the macrocycle to be −13.8 ppm contrast to −7.0 ppm for **NCP2H** [22]. On this basis, further interests on aromaticity–property relationship for NCP derivatives were shifted toward the targets of reduced porphyrinoids.

A phlorin is a class of dihydroporphyrin where a hydrogen atom is added to a pyrrolic nitrogen and *meso*-carbon positions of the porphyrin framework, thereby addressed as a nonaromatic compound due to the disruption of whole π-conjugated pathway [23]. The synthetic example of phlorin isomer containing an N-confused pyrrole unit, namely, N-confused phlorin (**9**), has been awaited to evaluate the structural effect of N-confusion (Fig. 12.3) [24]. Very recently, it was found that the simple chemical reduction of NCP **6** by sodium borohydride reagent afforded *bona fide* N-confused phlorins (**9a** or **10a**), which were characterized by NMR, mass spectroscopies, and X-ray crystallographic analysis, in sharp contrast to the porphyrin system generating a "chlorin (2,3-dihydroporphyrin)" as a dominant species [25]. Moreover, upon functionalization (e.g., methylation) at the inner

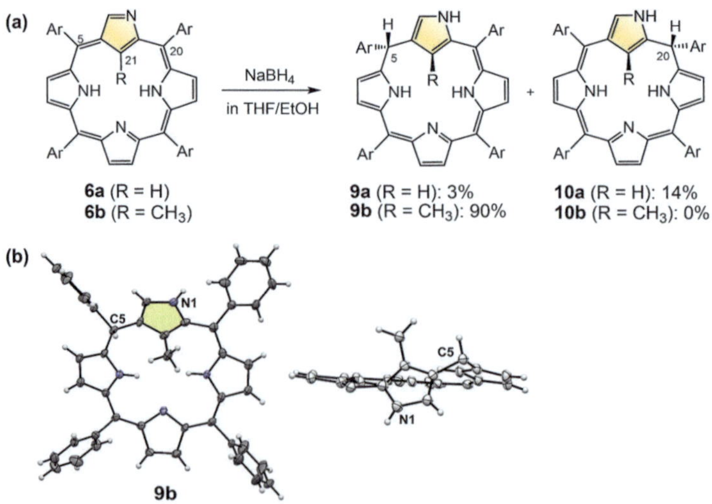

Fig. 12.3 (**a**) Reduction of NCP derivatives **6** to N-confused phlorin **9**. (**b**) *Top and side views* of X-ray crystal structure of **9b** showing with the thermal ellipsoids at a 50% probability level. *Meso*-phenyl groups in the side views are omitted for clarity (The structure was reproduced from Ref. [24] with permission from the World Scientific)

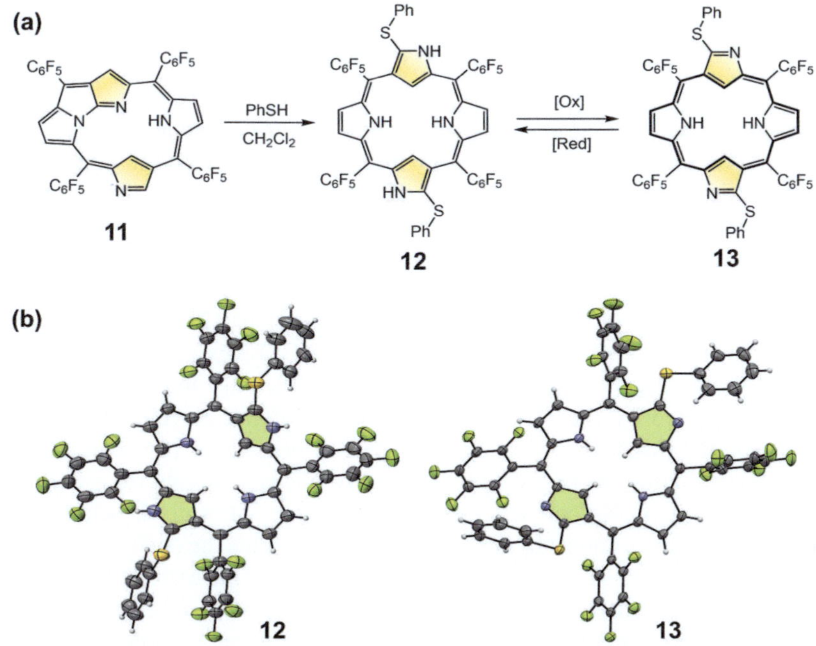

Fig. 12.4 (**a**) Synthesis of *trans*-doubly N-confused isophlorin **12** and subsequent oxidation to N_2CP derivative **13**. (**b**) X-ray crystal structures of **12** and **13** showing with the thermal ellipsoids at a 50% probability level (Reproduced from Ref. [26] with permission from RSC)

CH position, the N-confused phlorin derivative (**9b**) became thermodynamically stable, nevertheless retaining a nonaromatic character, which is responsible for the energetic benefit in the minimum loss upon the reduction reaction.

The above finding tempted us to prepare "isophlorin," another reduced porphyrin isomer; the reduction of doubly N-confused porphyrin (*trans*-N_2CP, **7**) being an implicitly destabilized analogue through additional N-confusion may be easier to access to the unstable 20π macrocycles, since the energy difference between the 18π- and 20π-electron macrocycles becomes smaller. In fact, the ring-opening reaction of N-confused, N-fused porphyrin (**11**) using sulfur nucleophile yielded a *trans*-doubly N-confused isophlorin derivative (**12**) wherein two arylthio groups were introduced to the α-positions of both N-confused pyrrole rings to protect the undesired side reactions (Fig. 12.4a) [26]. Although 20π-antiaromatic electronic structure could be considered according to the Hückel sense, the paratropicity of the isophlorin **12** is not prominent due to the disrupted cross-π-conjugation. The severely distorted core skeleton of **12** determined by X-ray crystallographic analysis should be reflected in the nonaromaticity, which is a marked contrast to the planar aromatic form of the substituted N_2CP (**13**) (Fig. 12.4b). The nearly zero NICS(0) value obtained by B3LYP method supports this conclusion.

12.3 Oxidation to N-Fused and C-oxo Analogues from N-Confused Porphyrinoids

The author's initial finding that an introduction of nitro group at the inner carbon of NCP **6** causes the confused pyrrole ring to be tilted drove to further studies on reactivity and electronic tunings of NCPs. Accordingly, the reaction for halogenation (e.g., bromination) in a specific solvent, pyridine, afforded another important mutant, N-fused porphyrin (termed as NFP, **14**), through rotation of the confused pyrrole ring and subsequent dehydrogenation for connection of the pyrrole to a neighboring inverted pyrrole (Scheme 12.4) [27]. This modification of π framework resulted in the narrower HOMO–LUMO gaps, indicating the NIR absorption band ($\lambda_{max} = 939$ nm vs 727 nm for NCP). This unusual bathochromic shift was expected by the electronic contribution of the bicyclic π-conjugation pathways on the fused tripentacyclic entity within the formal 18π annulenic core. This effect is more pronounced in the doubly N-fused porphyrin system [28]. The bromination of N-confused, N-fused porphyrin **11** with 1,3-dibromo-2,2-dimethylhydantoin, followed by treatment with base yields the desired product (**15**) (Scheme 12.4). The absorption spectrum of **15** exhibited the extremely low-energy band ($\lambda_{max} = 1240$ nm). The narrower theoretical HOMO–LUMO energy gap of 1.43 eV is attributed to the significant broken degeneration of HOMO and LUMO pairs, which is consistent with the observations of short excited lifetimes of 10 ps and 20 ps for **14** and **15**, respectively, by time-resolved spectroscopy [29].

While the incorporation of confused pyrrole moieties into the tetrapyrrolic frameworks made the derivatives unstable, the total energy difference between the N-confused π-expanded system and the regular one becomes comparable. For the pentapyrrolic N-confused analogues, a [5 + 0] condensation of N-confused

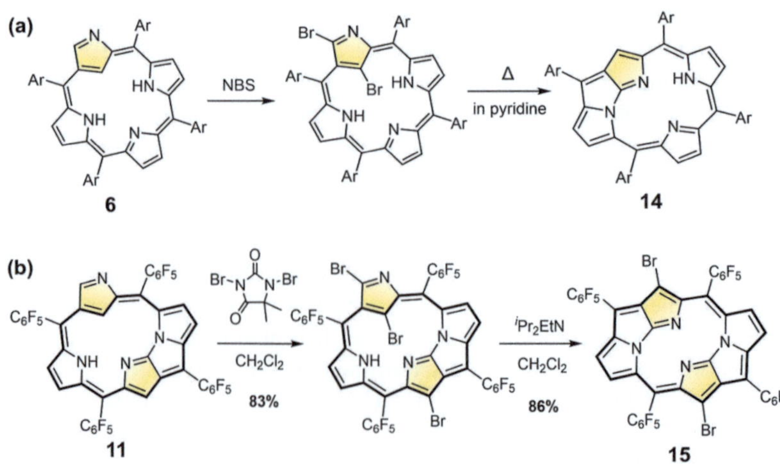

Scheme 12.4 Synthesis of (**a**) N-fused porphyrin **14** and (**b**) doubly N-fused porphyrin **15**

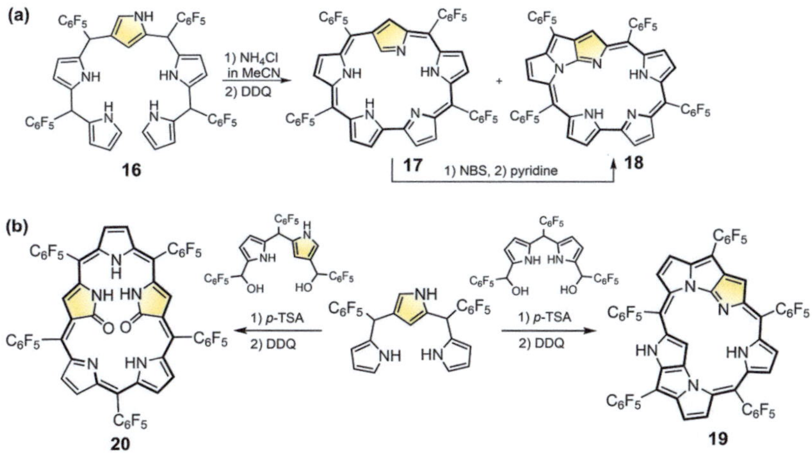

Scheme 12.5 Synthesis of (**a**) N-fused sapphyrin **18** via dehydrogenation of N-confused sapphyrin **17**. Synthesis of (**b**) N-fused pentaphyrin **19** and doubly N-confused dioxopentaphyrin **20** from an N-confused tripyrrane

pentapyrromethane (**16**) in the presence of ammonium chloride and followed by DDQ oxidation resulted in the formation of N-confused sapphyrin (**17**) as well as N-fused sapphyrin (**18**) in 6% and 2% yields, respectively (Scheme 12.5a) [30]. Again, the N-fused sapphyrin **18** revealed typical lower-energy absorption bands in the NIR region. In the same line of pentaphyrin families, doubly N-fused pentaphyrin(1.1.1.1.1) (**19**) has been also synthesized by a [3 + 2] acid condensation of N-confused tripyrrane and dipyrromethane dicarbinol and subsequent DDQ oxidation (Scheme 12.5b, right) [31].

Interestingly, upon further installing additional N-confused pyrrolic subunit in the pentaphyrin core, new structural analogue, namely, doubly N-confused dioxopentaphyrin (**20**), was isolated (Scheme 12.5b, left) [32]. The conventional oxidative condensation of N-confused tripyrrane and N-confused dipyrromethane dicarbinol resulted in the desired compound **20** along with the oxygenation of the pyrrolic α-positions instead of the N-fusion reaction as seen above. The comparative product analysis implies that the appropriate positions as well as directions of the pyrrolic NHs would lead to the dehydrogenative N-fusion with the neighboring C–Hs; otherwise the C–Hs will be oxygenated due to the intrinsic high reactivity. In accord with the pentaphyrin scaffold, a series of singly, doubly, and triply N-confused hexaphyrin (**21, 22**, and **23**, respectively) accompanied with the pyrrolic C(α)–H oxygenation have been synthesized from the appropriate combination of the poly(pyrromethane) precursors (Chart 12.2) [33]. All N-confused derivatives **21**, **22**, and **23** are recognized as aromatic species with 26π-electron conjugated cores as inferred from the ^{1}H NMR spectroscopy. The hexaphyrins **21** and **22** take rectangular conformations wherein the confused pyrrole rings are inverted, while the compound **23** has a C_3-symmetric triangular conformation dominantly. The

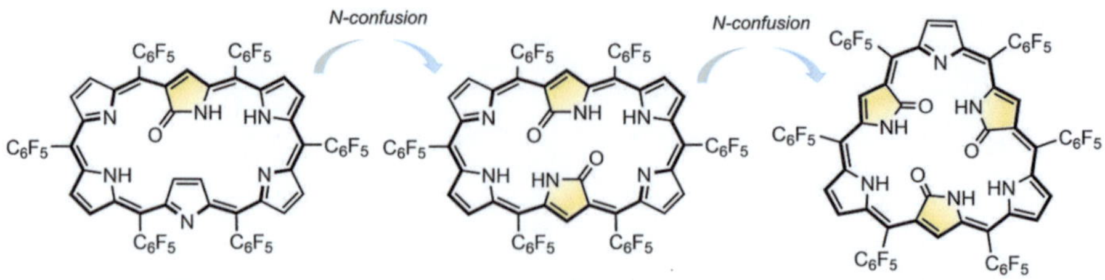

Chart 12.2 Chemical structures of a series of hexaphyrin derivatives **21**–**23** with a singly, doubly, and triply confused pyrroles

shape-dependent photophysical investigations of a series of N-confused hexaphyrins demonstrated the shorter excited lifetimes of **23** than that of **22** due to the reduced HOMO–LUMO energy gap [34].

12.4 Skeletal Rearrangement Toward N-Confused Porphyrinoids

As introduced in the above section, upon π-skeletal expansion of conjugated annulenic macrocycles, the electronic properties of the compounds would be perturbed significantly and can thus acquire the conformational flexibility as well as the ability to control the variable electronic states along with facile aromaticity switching [35]. The geometry alternation of large π-conjugated macrocycles governed by external stimuli (e.g., solvent polarity, temperature, (de)protonation, metallation, etc.) allows the distinct photophysical and electrochemical properties and reactivity. In particular systems, unique chemical transformations including the splitting reaction and skeletal rearrangement have been known in the expanded porphyrins (e.g., hexaphyrin, heptaphyrin, and octaphyrin) [36]. For example, Osuka et al. reported the thermal splitting reactions of bis-copper(II) octaphyrin complex into two copper(II) porphyrins in a metathesis-like reaction fashion [36a] and the boron(III)-induced rearrangement of a hexaphyrin(1.1.1.1.1.1) forming a hexaphyrin(2.1.1.0.1.1) [36d]. On this basis, it would be speculated that the metal complexation triggers an occurrence of these types of chemical reactions with NCP stuffs.

As an example of transformation to NCP derivatives, a treatment of palladium(II) salts for *meso*-hexakis aryl-substituted regular heptaphyrin (**24**) yielded a palladium(II) NCP derivative (**25**) bridging a tripyrromethene unit at the N-confused pyrrole ring in 9% yield as the side product (Fig. 12.5) [37]. Conducting the reaction in the polar reaction media such as acetone improved the yield up to 22%, and the coordination of palladium(II) ion causes this rearrangement in which the transannular interactions and energetic stabilization may participate in the formation of NCP derivative **25**. The UV-vis absorption of the NCP analogue **25**

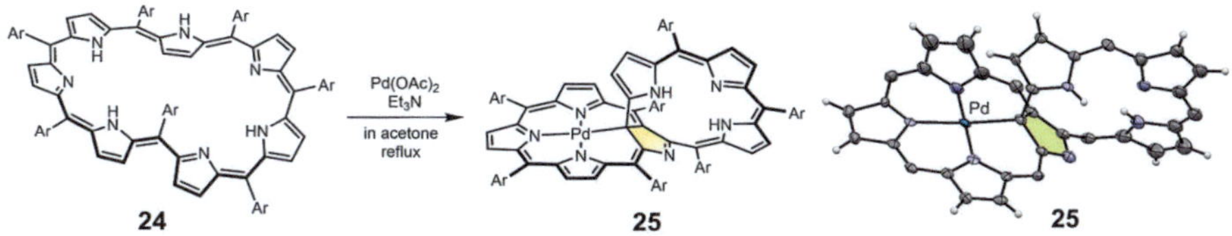

Fig. 12.5 Palladium-triggered rearrangement of *meso*-aryl-substituted heptaphyrin **24** to an N-confused porphyrin derivative **25**. The X-ray crystal structure of **25** is given showing with the thermal ellipsoids at a 50% probability level. The *meso*-aryl substituents are omitted for clarity (The structure was reproduced from the Ref. [37] with permission from Wiley)

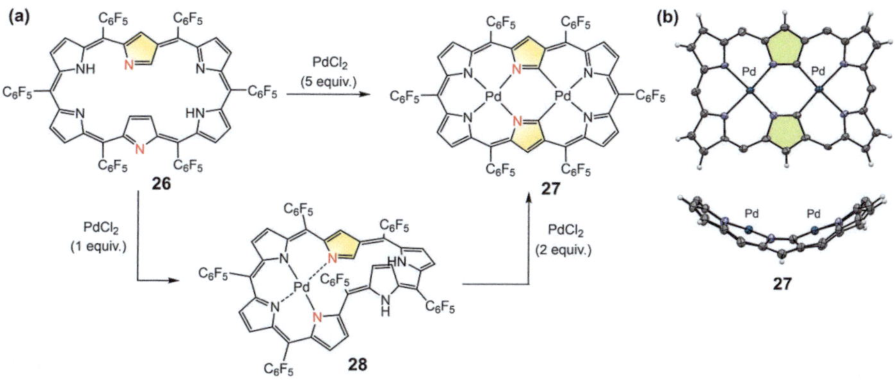

Fig. 12.6 (**a**) Formation of *cis*-doubly N-confused hexaphyrin bis-palladium(II) complex **27** by palladium-induced pyrrolic rearrangement of a singly N-confused hexaphyrin **26**. (**b**) *Front and side views* of X-ray crystal structure of **27** showing with the thermal ellipsoids at a 50% probability level (Reproduced from the Ref. [38] with permission from Wiley)

demonstrates a disjoint π-conjugation nature with significant hypsochromic shifts resembled to that of the Pd(II) complex of NCP derivative. Another representative example reported recently by the author is a palladium(II) complexation-induced skeletal rearrangement from a singly N-confused hexaphyrin (**26**) into a *cis*-doubly N-confused hexaphyrin (**27**) (Fig. 12.6) [38]. In the proposed reaction mechanism, the first palladium accommodation on 26π-system causes the reduction of the macrocycle to 28π with the deformation of the hexaphyrin core (**28**), and the second palladium metallation eventually triggered pyrrolic rearrangement through tautomeric intermediate including a *meso*-sp^3-hybridized carbon site. This is actually a similar reaction mode that the oxidative copper(I) metallation of a regular hexaphyrin in pyridine yields the bis-copper(II) doubly N-confused dioxohexaphyrin complex **22** through pyrrolic rearrangement concurrently, though the mechanism has not been cleared [39]. The generated organopalladium(II) hexaphyrin complex **27** is saddle shaped with a 26π-electron aromatic character. This reveals notable photophysical features; an NIR absorption band reaching to 1600 nm is fascinating despite the relatively compact hexapyrrolic scaffold (e.g., $\lambda_{max} < 1200$ nm of general hexaphyrin derivatives), which is consistent with the considerably narrower HOMO–LUMO energy gap of 1.3 eV based on the theoretical calculations.

12.5 Neo-confusion: N-Linked Porphyrinoids

The author's group has been actively involved in the synthesis of N-confused porphyrinoids to prove how the N-confusion modification alters their intrinsic properties, reactivities, and coordination abilities. The continuous synthetic efforts met a case to access to novel π-conjugated porphyrinoids that embedded a rare $N_{pyrrole}$–$C_{pyrrole}$ (or C_{meso}) linkage on the conjugated framework.

The oxidative cyclization of 1-aza-21-carbabilane (**29**) with 2,3-dichloro-5,6-dicyano-1,4-benzoquinone (DDQ) in acetonitrile provided an unprecedented N-linked isomer (viz., norrole **31**) of corrole [40] as a minor product accompanied with the formation of N-confused corrole (**30**) (Scheme 12.6a) [41]. The N-linked pyrrole unit present in **31** is found to be titled at 36.4° from the mean plane as determined by X-ray crystallographic analysis. This is reflected to the moderate aromaticity of **31** compared to the reference corrole as inferred from the ^{1}H NMR spectroscopy and NICS values ($\Delta\delta_{CH\text{-}NH} = 6.8$ and −7.28 ppm, respectively). This finding has started out a norrole-related chemistry in turn to open the synthetic way to the other type of N-linked analogues. As a matter of the fact, soon after the author's discovery, an N-linked isomer of benzo-fused porphyrin (**32**) was reported by Lash et al. (Scheme 12.6b) [42]. Their synthetic approach of neo-confused

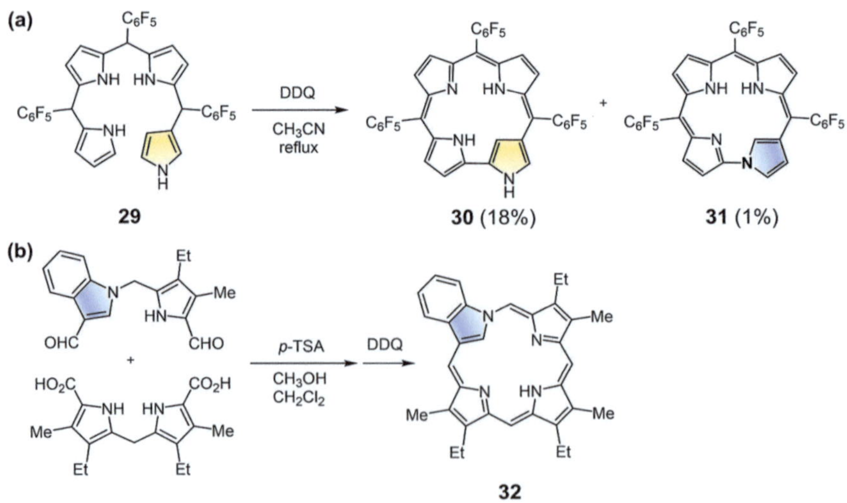

Scheme 12.6 Synthesis of (**a**) neo-confused corrole isomer **31** and (**b**) neo-confused porphyrin derivatives **32**

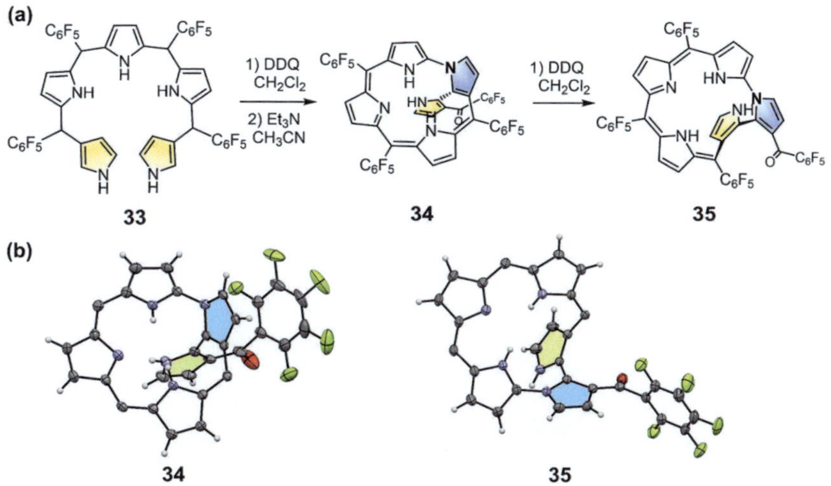

Fig. 12.7 (**a**) Synthesis of a pyrrolyl-appended norrole derivative **34** and oxidative transformation to neo-confused isosmaragdyrin **35**. (**b**) X-ray crystal structures of **34** and **35** showing with the thermal ellipsoids at a 50% probability level. The *meso*-pentafluorophenyl substituents were omitted for charity (Reproduced from the Ref. [43a] with permission from ACS)

porphyrin **32** was conducted by a [2 + 2]-type acid condensation of neo-confused dipyrromethane precursors wherein the diformyl–pyrrolomethylene N-linked indole derivatives incorporated and followed by DDQ oxidation. As similar to those of norrole **31**, the presence of diatropic ring current ($\Delta\delta_{CH\text{-}NH}$ value of 10.43 ppm) and typical porphyrin-like optical feature, namely, an intense Soret band and vibrated Q bands, was confirmed by NMR and UV-vis absorption spectroscopies, respectively.

Upon expansion of the skeletal N-linked core structure, unprecedented pyrrolic transformations of doubly terminal-confused pentapyrrane (**33**) and hexapyrrane (**36**) occur to form an N-linked isosmaragdyrin (**35**) and N-linked neo-fused hexaphyrin analogue (**38**), respectively, in the oxidative cyclization (Figs. 12.7 and 12.8) [43]. In the pentapyrrolic system, the treatment of a pentapyrrane **33** with 3.3 equiv of DDQ afforded a nonconjugated cyclized intermediate, and subsequent addition of triethylamine underwent macrocyclic contraction to give a pyrrole-appended norrole derivative (**34**). Interestingly, oxidation of **34** with DDQ again yielded an N-linked isosmaragdyrin **35** because the energetically unfavorable structure of **34** leads to an oxidative C–N bond recombination (Fig. 12.7). Although the [4n + 2]π-electronic pathways can be considered in the structures of **34** and **35**, these macrocycles are expected to be virtually nonaromatic (e.g., NICS values at their macrocyclic centers: +3.5 and −0.1 ppm, respectively) because of their large displacement of the confused pyrrole rings twisting out from the conjugating mean plane. In this regard, the conventional DDQ oxidation of larger hexapyrrane **36** containing two terminal-confused pyrrole units yielded the neo-fused hexaphyrin(1.1.1.1.1.0) **38** with double N-linkage modes (N–C(α) bonds) via formation of neo-confused hexaphyrin intermediate (**37**) (Fig. 12.8). These N-linked analogues, **37** and **38**,

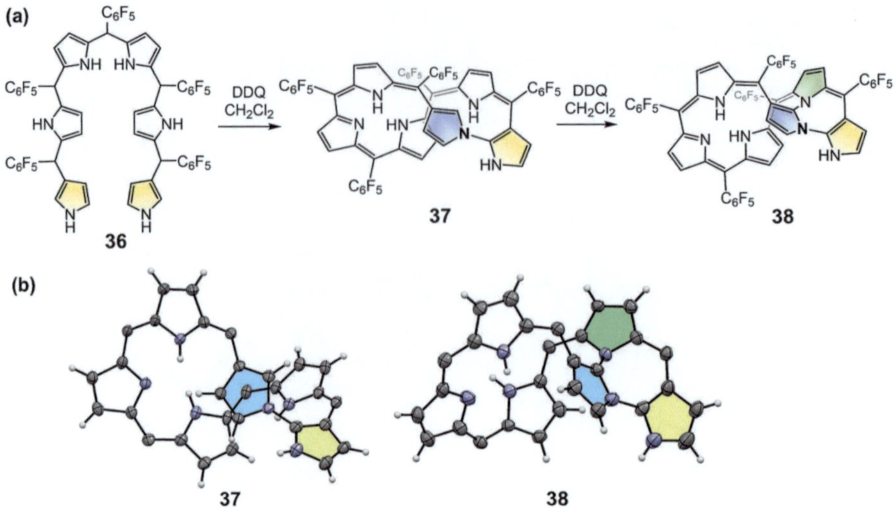

Fig. 12.8 (**a**) Synthesis of neo-confused hexaphyrin **37** and neo-fused hexaphyrin **38** analogues. (**b**) X-ray crystal structures of **37** and **38** showing with the thermal ellipsoids at a 50% probability level (Reproduced from the Ref. [43b] with permissions from Wiley)

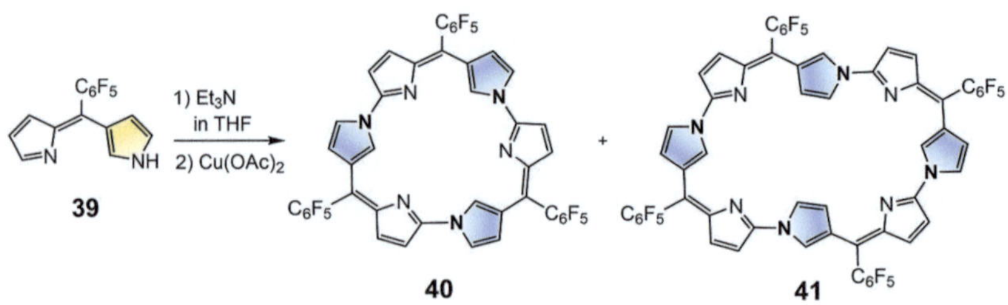

Scheme 12.7 Metal-assisted synthesis of multiply N-linked hexaphyrin **36** and octaphyrin **37** derivatives

are likewise nonaromatic due to the severely distorted core structures with figure-eight-like conformations determined by X-ray crystallographic analysis. The doubly N-linked neo-fused analogue **38** showed a relatively larger electrochemical HOMO–LUMO energy gap of $\Delta E_{\text{H-L}} = 1.62$ eV than that of **37** ($\Delta E_{\text{H-L}} = 1.51$ eV), which is reflected in the absorption spectrum of **38**, exhibiting a redshifted band in the NIR region. By theoretical aids, the stabilization of HOMO level attributed to the larger HOMO–LUMO energies of **38**.

Furthermore, the multiple N-linked hexaphyrin(1.0.1.0.1.0) (**40**) and octaphyrin(1.0.1.0.1.0.1.0) (**41**) have been reported by Anand et al. (Scheme 12.7) [44]. Their synthetic approach is actually unique as well as simpler; stirring a tetrahydrofuran solution of N-confused dipyrrins (32) in the presence of copper(II) acetate at room temperature afforded the corresponding cyclo-trimer **40** and tetramer **41** in 4% and 1% yields, respectively. With an aid of X-ray crystallographic analysis, both macrocycles are found to be planar, however, showing weak paratropic characters along with positive NICS(0) values of +4.5 and +1.5 ppm, respectively, presumably due to the weak π delocalization throughout the macrocycle via C–N linkages.

12.6 Longitudinally Stacked N-Confused Porphyrinoids

The molecular architectures with an interplane area sandwiched in a face-to-face manner by two or more π-building blocks have been attracting the chemist's long-standing interests because of the critical functions as π-spaces for biomimetic catalysis, recognition site for sensing, and so forth. To date, various cofacial porphyrin-based dimers have been utilized as building stuffs to explore unique interactions of the π-conjugated guest substrates [45]. In the theoretical sense, the larger π planes can be overlapped effectively, and strong intermolecular (interplane) interactions offer advantages in the diagnosis through the analysis of the π-molecular orbitals of the stacked dimer that will be perturbed with nondegenerated MOs. By an analogy to the typical porphyrin systems, the specific longitudinally stacked N-confused porphyrinoid systems were therefore gathered as an important family of porphyrin-based assemblies. The resulting three-dimensional extended π-system based on the N-confused porphyrinoids is presented in this section.

As shown above, NFP **14** possesses a characteristic [5.5.5]-type tricyclic heteroaromatic unit in the parent 18π annulenic scaffold [27]. This can serve as mono-negative tridentate nitrogen ligand upon deprotonation, which is isoelectronically similar to the tris(pyrazolyl)borate ligand. In particular, due to the smaller coordination sphere relative to the porphyrins, the metal cation is displaced out of the mean plane, allowing coordination of the second ligand accessing from the overhead with a close distance. On this basis, upon reaction of NFP **14** with iron(I)(cyclopentadienyl), bis-carbonyl reagent in benzonitrile afforded a double-decker-type iron(II) NFP dimer complex (**42**) in 14% yield as a single rotamer (Fig. 12.9) [46]. The X-ray crystal structure of iron complex **42** represents that the iron center is sandwiched between the two NFP ligands, and the distances between the iron and the NFP mean planes are 1.647 and 1.644 Å, respectively. These values are comparable to that of ferrocene (e.g., 1.643 Å), which is an indication that the complex **42** can be regarded as a model of π-extended ferrocene. In accord with the structural parameter, the electrochemical oxidation potential of **42** was determined to be $E_{ox} = -0.24$ V (vs. Fc/Fc^{+}), suggesting a facile oxidation due to the larger π-NFP ligands. It was found that there is a d–π conjugation character in HOMO and LUMO of **42** apparently as referred in ferrocene derivatives, giving the NIR absorption band in the longer wavelength [47]. Although it is not an N-confused porphyrin system, palladium(II) acetate complexation reaction of dicarbaporphyrin forms an unusual tripalladium sandwich

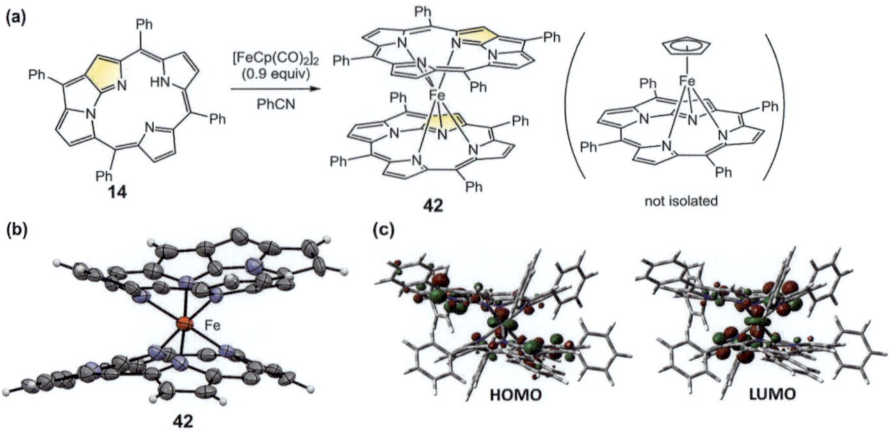

Fig. 12.9 (**a**) Synthesis of double-decker ferrocene-like iron(II) complex of NFPs **42**. (**b**) X-ray crystal structure of **42** showing with the thermal ellipsoids at a 50% probability level. The *meso*-phenyl substituents are omitted for clarity. (**c**) d–π-Conjugated orbitals of **42** present in HOMO and LUMO (Reproduced from the Ref. [46] with permission of Wiley)

complex consisting of two dianionic palladium(II) porphyrin units surrounding a palladium(IV) cation with η^5 coordination fashions [48]. The close interplane interaction of the dicarbaporphyrins (Pd–Pd bond length = 3.37 Å) leads us to learn the remarkable electronic perturbation in terms of the decreased aromaticity with NIR optical response.

Another example of a longitudinally perturbed π-system appeared in a covalently linked cofacial dimer of neo-confused corrole derivative, benzonorrole (**44** and **45**) (Fig. 12.10) [49]. As the more synthetically friendly derivative of norrole, the periphery-annulated norrole (**43**) prepared from an indole was developed for further studies. The chemical oxidation of **43** with DDQ gave unprecedented N-linked dimers with specific steric configurations in 75% yield as a *cis*-form **44** together with a *trans*-form (20%). Interestingly, the reduction of dimer **44** with tosylhydrazide ($TsNHNH_2$) afforded the reduced form of dimeric benzonorrole (**45**). From the comparative crystal structural analysis of two dimers **44** and **45**, the interplane distances are almost identical; however, the bonding mode of 2,2′-biindolyl moiety of the dimers exhibited the sharp difference; the C–C bond distances of the 2,2′-biindolyl in the dimers **44** and **45** are estimated to be 1.476 and 1.357 Å, respectively, and the angles between two individual indole rings are 55.1° and 10.0° as well. This critical difference is reflected in the longer-wavelength absorptions ($\lambda_{max} = 845$ nm for **44** vs. $\lambda_{max} = 633$ nm for **45**) and the narrower electrochemical energy gap (1.45 V vs. 1.66 V), which could be interpreted by the nondegenerated LUMO pairs found in the energy diagram of **44**. Weakened distinct aromaticity of the benzonorrole units is also figured out through this longitudinally perturbed π-stacking.

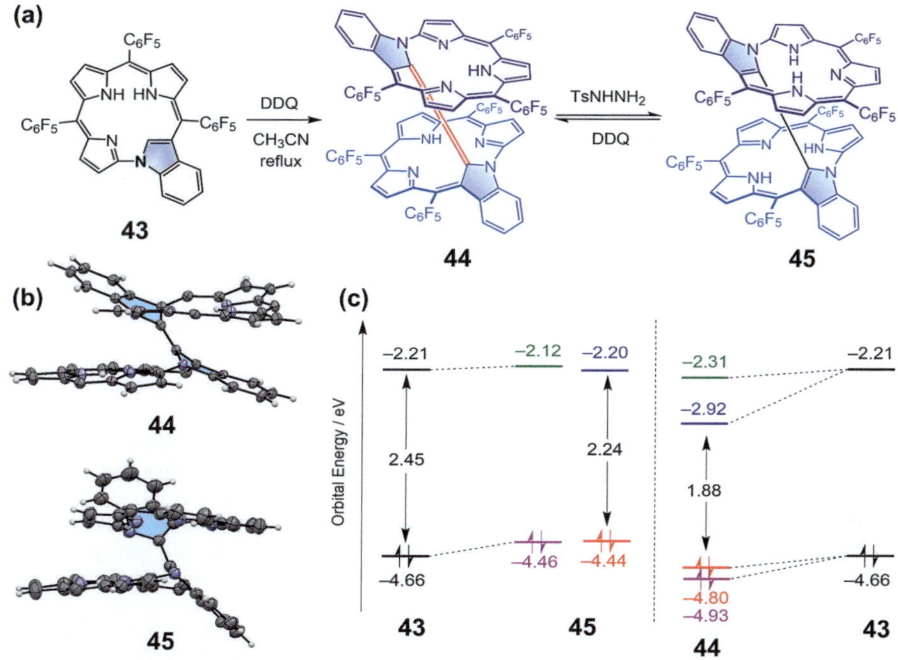

Fig. 12.10 (**a**) Oxidative dimerization of **43** and redox-induced interconversion of benzonorrole dimers **44** and **45**. (**b**) X-ray crystal structures of dimers **44** and **45** showing with the thermal ellipsoids at a 50% probability level and (**c**) comparative analysis of the MO energy interactions between the benzonorrole π-orbitals in **44** and **45** (Reproduced from the Ref. [49] with permission from Wiley)

12.7 Summary and Outlook

The landmark discovery of the synthetic route for NCPs by the author's group and Latos-Grażyński opened up the research fields of NCPs and related compounds. A diversity of the current NCP chemistry includes new synthesis, physical properties, metal coordination and anion-sensing abilities, and practical applications as catalysts or photosensitizers. In this chapter, the recent advances in synthetic N-confused porphyrin chemistry mostly contributed by the author's group has been described with a particular focus on the strategy for construction of novel π-conjugation systems derived from NCPs. The special series of N-confused porphyrinoids, such as (1) reduced dihydroporphyrin derivatives, (2) N-fused and C-oxo π-extended porphyrinoids, (3) π-conjugated organometallic species by pyrrolic rearrangements, (4) neo-confused scaffolds with N–C linkages, and (5) longitudinal π-expanded systems, are reviewed. From the synthetic viewpoints, a number of new structural isomers of NCPs, such as further multiplying confused homologues, can still be conceived, and it provides valuable π-conjugated platforms as the alternate porphyrin-based materials through the "N-confusion approach." The future efforts on the synthetic chemistry of novel N-confused porphyrinoids should thus be devoted to understand and generalize their intrinsic structure–function relationship for molecular-based material applications. It is the author's hope that the present review would contribute to widely enhance the richness and vitality of the field of NCPs.

Acknowledgment The authors are deeply indebted to all of our colleagues in the Furuta group for their indispensable contributions. We express our sincere thanks to Prof. Hidemitsu Uno and Prof. Shigeki Mori (Ehime University, Japan) for the help of X-ray crystallographic analysis. We acknowledge the financial support for the Grants-in-Aid (no. 21108518, 23108715 for "pi-space") from the Ministry of Education, Culture, Sports, Science, and Technology (MEXT), Japan.

References

1. (a) Kadish KM, Smith KM, Guillard R (eds) (2000) The porphyrin handbook. Academic Press, San Diego. (b) Kadish KM, Smith KM, Guillard R (eds) (2010) Handbook of porphyrin science. World Scientific, Singapore
2. (a) Fukuzumi S (2008) Development of bioinspired artificial photosynthetic systems. Phys Chem Chem Phys 10:2283–2297. (b) Che C-M, Huang J-S (2009) Metalloporphyrin-based oxidation systems: from biomimetic reactions to application in organic synthesis. Chem Commun 3996–4015. (c) D'Souza F, Ito O (2009) Supramolecular donor–acceptor hybrids of porphyrins/phthalocyanines with fullerenes/carbon nanotubes: electron transfer, sensing, switching, and catalytic applications. Chem Commun 4913–4928. (d) Medforth CJ, Wang Z, Martin KE, Song Y, Jacobsen JL, Shelnutt JA (2009) Self-assembled porphyrin nanostructures. Chem Commun 7261–7277. (e) Li L-L, Diau E-WG (2013) Porphyrin-sensitized solar cells. Chem Soc Rev 42:291–304. (f) Panda MK, Ladomenou K, Coutsolelos AG (2012) Porphyrins in bio-inspired transformations: light-harvesting to solar cell. Coord Chem Rev 256:2601–2627. (g) Ishihara S, Labuta J, Rossom WV, Ishikawa D, Minami K, Hill JP, Ariga K (2014) Porphyrin-based sensor nanoarchitectonics in diverse physical detection modes. Phys Chem Chem Phys 16:9713–9746
3. Vogel E, Köcker M, Schmickler H, Lex J (1986) Porphycene — a Novel Porphin Isomer. Angew Chem Int Ed Engl 25:257–259
4. (a) Sessler JL, Brucker EA, Weghorn SJ, Kisters M, Schäfer M, Lex J, Vogel E (1994) Corrphycene: a new porphyrin isomer. Angew Chem Int Ed Engl 33:2308–2312. (b) Aukauloo MA, Guilard R (1994) The "Etioporphycerin": synthesis and characterization of a new porphyrin isomer. New J Chem 18:1205–1207
5. (a) Callot HJ, Rohrer A, Tschamber T (1995) A novel porphyrin isomer: hemiporphycene. Formation and single-crystal x-ray diffraction structure determination of a hemiporphycene nickel complex. New J Chem 19:155–159. (b) Vogel E, Bröring M, Weghorn S, Scholz P, Deponte R, Lex J, Schmickler H, Schaffner K, Braslavsky SE, Müller M, Pörting S, Sessler JL, Fowler CJ (1997) Octaethylhemiporphycene: synthesis, molecular structure and photophysics. Angew Chem Int Ed Engl 36:1651–1654
6. Vogel E, Bröring M, Erben C, Demuth R, Lex J, Nendel M, Houk KN (1997) Palladium complexes of the new porphyrin isomers (Z)- and (E)-isoporphycene—PdII-induced cyclization of tetrapyrrole aldehydes. Angew Chem Int Ed Engl 36:353–357
7. Senge MO (2011) Extroverted confusion – Linus Pauling, Melvin Calvin, and Porphyrin Isomers. Angew Chem Int Ed 50:4272–4277
8. (a) Furuta H, Asano T, Ogawa T (1994) "N-confused porphyrin": a new isomer of tetraphenylporphyrin. J Am Chem Soc 116:767–768. (b) Chmielewski PJ, Latos-Grażyński L, Rachlewicz K, Głowiak T (1994) Tetra-p-tolylporphyrin with an inverted pyrrole ring: a novel isomer of porphyrin. Angew Chem Int Ed Engl 33: 779–781
9. (a) Stępień M, Lato-Grażyński L (2005) Benziporphyrins: exploring arene chemistry in a macrocyclic environment. Acc Chem Res 38:88–98. (b) Lash TD (2012) Carbaporphyrins,

porphyrin isomers and the legacy of Emanuel Vogel. J Porphyrins Phthalocyanines 16:423–433. (c) Lash TD (2014) Metal complexes of carbaporphyrinoid systems. Chem Asian J 9:682–705
10. Rothemund P (1939) Porphyrin studies. III. 1 The structure of the Porphine2 ring system. J Am Chem Soc 61:2912–2915
11. Geier GR III, Haynes DM, Lindsey JS (1999) An efficient one-flask synthesis of N-confused tetraphenylporphyrin. Org Lett 1:1455–1458
12. (a) Shin J-Y, Furuta H, Yoza K, Igarashi S, Osuka A (2001) meso-aryl-substituted expanded porphyrins. J Am Chem Soc 123:7190–7191. (b) Maeda H, Osuka A, Ishikawa Y, Aritome I, Hisaeda Y, Furuta H (2003) N-confused porphyrin-bearing meso-perfluorophenyl groups: a potential agent that forms stable square-planar complexes with Cu(II) and Ag(III). Org Lett 5:1293–1296
13. (a) Furuta H, Maeda H, Osuka A (2000) Doubly N-confused porphyrin: a new complexing agent capable of stabilizing higher oxidation states. J Am Chem Soc 122:803–807. (b) Maeda H, Osuka A, Furuta H (2003) Hydrogen bonding 1-D chain network of cis-doubly N-confused porphyrins. Supramol Chem 15:447–450. (c) Maeda H, Osuka A, Furuta H (2004) Synthesis of A2B2 type cis-doubly N-confused porphyrins from N-confused dipyrromethanes. Tetrahedron 60:2427–2432. (d) Maeda H, Osuka A, Furuta H (2003) Trans doubly N-confused porphyrins: Cu(III) complexation and formation of rodlike hydrogen-bonding networks. J Am Chem Soc 125:15690–15691
14. (a) Furuta, H, Ishizuka T, Osuka A, Dejima H, Nakagawa H, Ishikawa Y (2001) NH tautomerism of N-confused porphyrin. J Am Chem Soc 123:6207–6208. (b) Ghosh A (1995) N-confused porphyrins and singlet carbenes: is there a connection? Angew Chem Int Ed Engl 34:1028–1030; (c) Szterenberg L, Latos-Grażyński L (1997) Structure and stability of 2-Aza-21-carbaporphyrin tautomers prearranged for coordination. Inorg Chem 36:6287–6291
15. (a) Wolff SA, Alemán EA, Banerjee D, Rinaldi PL, Modarelli DA (2004) One-step synthesis and characterization of difunctionalized N-confused tetraphenylporphyrins. J Org Chem 69:4571–4576. (b) Shaw JL, Garrison SA, Alemán EA, Ziegler CJ, Modarelli DA (2004) Synthesis and spectroscopy of a series of substituted N-confused tetraphenylporphyrins. J Org Chem 69:7423–7427. (c) Alemán EA, Rajesh CS, Ziegler CJ, Moearelli DA (2006) Ultrafast spectroscopy of free-base N-confused tetraphenylporphyrins. J Phys Chem A 110:8605–8612
16. (a) Harvey JD, Ziegler CJ (2003) Developments in the metal chemistry of N-confused porphyrin. Coord Chem Rev 247:1–19. (b) Toganoh M, Furuta H (2010) In: Kadish KM, Smith KM, Guilard R (eds) Handbook of porphyrin science, vol 2. World Scientific, Singapore, pp 295–367. (c) Toganoh M, Furuta H (2012) Blooming of confused porphyrinoids – fusion, expansion, contraction, and more confusion. Chem Commun 48:937–954
17. (a) Maeda H, Ishikawa Y, Matsuda T, Osuka A, Furuta H (2003) Control of Cu(II) and Cu(III) states in N-confused porphyrin by protonation/deprotonation at the peripheral nitrogen. J Am Chem Soc 125:11822–11823. (b) Mitrikas G, Calle C, Schweiger A (2005) Asymmetric spin density distribution in the copper(II) complex of N-confused tetraphenylporphyrin: a multifrequency continuous-wave and pulse EPR study. Angew Chem Int Ed 44:3301–3303
18. (a) Liu J-C, Ishizuka T, Osuka A, Furuta H (2003) Modulation of axial coordination in N-confused porphyrin-antimony(V) dibromide complex by proton stimulus. Chem Commun 1908–1909. (b) Maeda H, Morimoto T, Osuka A, Furuta H (2006) Halide-anion binding by singly and doubly N-confused porphyrins. Chem Asian J 1:832–844. (c) Furuta H, Nanami H, Morimoto T, Ogawa T, Král V, Sessler JL, Lynch V (2008) Halide anion mediated dimerization of a meso-unsubstituted N-confused porphyrin. Chem Asian J 3:592–599
19. (a) Hung CH, Chen W-C, Lee GH, Peng S-M (2002) Dimeric iron n-confused porphyrin complexes. Chem Coomun 1516–1517. (b) Harvey JD, Ziegler CJ (2002) Dimeric and monomeric forms of manganese N-confused porphyrin. Chem Commun 1942–1943. (c) Furuta H, Morimoto T, Osuka A (2004) Structures and ligand exchange of N-confused porphyrin dimer complexes with group 12 metals. Inorg Chem 43:1618–1624. (d) Morimoto T, Uno H, Furuta H (2007) Benzene ring trimer interactions modulate supramolecular structures. Angew Chem Int Ed 46:3672–3675

20. (a) Furuta H, Maeda H, Osuka A (2002) Confusion, inversion, and creation – a new spring from porphyrin chemistry. Chem Commun 1795–1804. (b) Srinivasan A, Furuta H (2005) Confusion approach to porphyrinoid chemistry. Acc Chem Res 38:10–20. (c) Maeda H, Furuta H (2006) A dozen years of N-confusion: from synthesis to supramolecular chemistry. Pure Appl Chem 78:29–44.(d) Stepiń M, Sprutta N, Latos-Grażyński L (2011) Figure eights, Möbius bands, and more: conformation and aromaticity of porphyrinoids. Angew Chem Int Ed 50:4288–4340
21. Garratt PJ (1986) Aromaticity. Wiely, New York
22. Toganoh M, Furuta H (2009) Theoretical study on rotation of pyrrole rings in porphyrin and N-confused porphyrin. J Phys Chem A 113:13953–13963
23. (a) Mauzerall D (1962) The photoreduction of porphyrins: structure of the products. J Am Chem Soc 84:2437–2445. (b) Mauzerall D (1974) In: Smith KM (ed) Porphyrins and metalloporphyrins. Elsevier, Amsterdam, pp 704–705
24. Ishida M, Nakahara K, Sakashita R, Ishizuka T, Watanabe M, Uno H, Osuka A, Furuta H (2014) N-confused phlorin: a stable phlorin isomer containing a confused pyrrole ring. J Porphyrins Phthalocyanines 18:909–918
25. (a) Banfi S, Carusoa E, Caprioli S, Mazzagatti L, Canti G, Ravizza R, Gariboldi M, Monti E (2004) Photodynamic effects of porphyrin and chlorin photosensitizers in human colon adenocarcinoma cells. Bioorg Med Chem 12:4853–4860. (b) Balaban TS, Linke-Schaetzel M, Bhise AD, Vanthuyne N, Roussel C (2004) Green self-assembling porphyrins and chlorins: perfect mimics of the natural bacteriochlorophylls c, d, and e. Eur J Org Chem 3919–3930
26. Yan J, Takakusaki M, Yang Y, Mori S, Zhang B, Feng Y, Ishida M, Furuta H (2014) Doubly N-confused isophlorin: synthesis, structure and copper coordination. Chem Commun 50:14593–14596
27. (a) Ishikawa Y, Yoshida I, Akaiwa K, Koguchi E, Sasaki T, Furuta H (1997) Nitration of N-confused porphyrin. Chem Lett 26:453–454. (b) Furuta H, Ishizuka T, Osuka A, Ogawa T (1999) "N-fused porphyrin" from N-confused porphyrin. J Am Chem Soc 121:2945–2946. (c) Furuta H, Ishizuka T, Osuka A, Ogawa T (2000) "N-fused porphyrin": a new tetrapyrrolic porphyrinoid with a fused tri-pentacyclic ring. J Am Chem Soc 122:5748–5757
28. Toganoh M, Kimura T, Uno H, Furuta H (2008) Doubly N-fused porphyrin. Angew Chem Int Ed 47:8913–8916
29. Lee JS, Lim JM, Toganoh M, Furuta H, Kim D (2010) Comparative spectroscopic studies on porphyrin derivatives: electronic perturbation of N-confused and N-fused porphyrins. Chem Commun 46:285–287
30. Gupta I, Srinivasan A, Morimoto T, Toganoh M, Furuta H (2008) N-confused and N-fused meso-aryl sapphyrins. Angew Chem Int Ed 47:4563–4567
31. Srinivasan A, Ishizuka T, Furuta H (2004) Doubly N-fused pentaphyrin. Angew Chem Int Ed 43:876–879
32. Srinivasan A, Ishizuka T, Maeda H, Furuta H (2004) Doubly N-confused pentaphyrins. Angew Chem Int Ed 43:2951–2955
33. (a) Gokulnath S, Yamaguchi K, Toganoh M, Mori S, Uno H, Furuta H (2011) Singly N-confused [26]Hexaphyrin: a binucleating porphyrinoid ligand for mixed metals in different oxidation states. Angew Chem Int Ed 50:2302–2306. (b) Srinivasan A, Ishizuka T, Osuka A, Furuta H (2003) Doubly N-confused hexaphyrin: a novel aromatic expanded porphyrin that complexes bis-metals in the core. J Am Chem Soc 125:878–879. (c) Xie Y-S, Yamaguchi K, Toganoh M, Uno H, Suzuki M, Mori S, Saito S, Osuka A, Furuta H (2009) Triply N-confused hexaphyrins: near-infrared luminescent dyes with a triangular shape. Angew Chem Int Ed 48: 5496–5499
34. Lim JM, Lee JS, Chung HW, Bahng HW, Yamaguchi K, Toganoh M, Furuta H, Kim D (2010) Photophysical properties of N-confused hexaphyrins: effects of confusion of pyrrole rings and molecular shape on electronic structures. Chem Commun 46:4357–4359
35. (a) Sessler JL, Seidel D (2003) Synthetic expanded porphyrin chemistry. Angew Chem Int Ed 42:5134–5175. (b) Saito S, Osuka A (2011) Expanded porphyrins: intriguing structures, electronic properties, and reactivities. Angew Chem Int Ed 50:4342–4373. (c) Shin J-Y, Kim KS, Yoon M-C, Lim JM, Yoon ZS, Osuka A, Kim D (2010) Aromaticity and photophysical

properties of various topology-controlled expanded porphyrins. Chem Soc Rev 39:2751–2767. (d) Roznyatovskiy VV, Lee C-H, Sessler JL (2013) π-extended isomeric and expanded porphyrins. Chem Soc Rev 42:1921–1933

36. (a) Tanaka Y, Hoshino W, Shimizu S, Youfu K, Aratani N, Maruyama N, Fujita S, Osuka A (2004) Thermal splitting of Bis-Cu(II) octaphyrin(1.1.1.1.1.1.1.1) into two Cu(II) porphyrins. J Am Chem Soc 126:3046–3047. (b) Tanaka Y, Shinokubo H, Yoshimura Y, Osuka A (2009) Metathesis-like splitting reactions of metallated [36]octaphyrins(1.1.1.1.1.1.1.1): experimental and computational investigations. Chem Eur J 15:5674–5685. (c) Saito S, Kim KS, Yoon ZS, Kim D, Osuka A (2007) Extrusion of Boron(III) subporphyrin from meso-Heptakis(pentafluorophenyl)[32]heptaphyrin upon cooperative CuII and BIII metalation. Angew Chem Int Ed 46:5591–5593. (d) Moriya K, Saito S, Osuka A (2010) Boron(III) induced skeletal rearrangement of hexaphyrin(1.1.1.1.1.1) to hexaphyrin(2.1.1.0.1.1). Angew Chem Int Ed 49:4297–4300
37. Yoneda T, Saito S, Yorimitsu H, Osuka A (2011) Palladium(II)-triggered rearrangement of heptaphyrins to N-confused porphyrins. Angew Chem Int Ed 50:3475–3478
38. Gokulnath S, Nishimura K, Toganoh M, Mori S, Furuta H (2013) Palladium-induced pyrrolic rearrangement of a singly to a doubly N-confused [26]Hexaphyrin. Angew Chem Int Ed 52:6940–6943
39. Suzuki M, Yoon M-C, Kim DY, Kwon JH, Furuta H, Kim D, Osuka A (2006) A new entry to doubly N-confused [26]Hexaphyrins(1.1.1.1.1.1) from normal [26]Hexaphyrins(1.1.1.1.1.1) through an unprecedented double pyrrolic rearrangement. Chem Eur J 12:1754–1759
40. (a) Erben C, Will S (2000) In: The porphyrin handbook, vol 2. Academic Press, San Diego, p 233. (b) Gryko DT, Fox JP, Goldberg DP (2004) Recent advances in the chemistry of corroles and core-modified corroles. J Porphyrins Phthalocyanines 8:1091–1105. (c) Aviv I, Gross Z (2007) Corrole-based applications. Chem Commun 1987–1999
41. Fujino K, Hirata Y, Kawabe Y, Morimoto T, Srinivasan A, Toganoh M, Miseki Y, Kudo A, Furuta H (2011) Confusion and neo-confusion: corrole isomers with an NNNC core. Angew Chem Int Ed 50:6855–6859
42. Lash TD, Lammer AD, Ferrence GM (2011) Neo-confused porphyrins, a new class of porphyrin isomers. Angew Chem Int Ed 50:9718–9721
43. (a) Xie Y, Wei P, Li X, Hong T, Zhang K, Furuta H (2013) Macrocycle contraction and expansion of a dihydrosapphyrin isomer. J Am Chem Soc 135:19119–19122. (b) Wei P, Zhang K, Li X, Meng D, Ågren H, Ou Z, Ng S, Furuta H, Xie Y (2014) Neo-fused hexaphyrin: a molecular puzzle box with N-linked pentaphyrin inside. Angew Chem Int Ed 53:14069–14073
44. Gadekar SC, Reddy BK, Anand VG (2013) Metal-assisted cyclomerization of N-confused dipyrrins into expanded norroles. Angew Chem Int Ed 52:7164–7167
45. (a) Satake A, Kobuke Y (2007) Artificial photosynthetic systems: assemblies of slipped cofacial porphyrins and phthalocyanines showing strong electronic coupling. Org Biomol Chem 5:1679–1691. (b) Kim D (2012) In: Multporphyrin arrays: fundamentals and applications. Pan Stanford Publishing, Singapore. (c) Beletskaya I, Tyurin VS, Tsivadze AY, Guilard R, Stern C (2009) Supramolecular chemistry of metalloporphyrins. Chem Rev 109:1659–1713. (d) Valderrey V, Aragay G, Ballester P (2014) Porphyrin tweezer receptors: binding studies, conformational properties and applications. Coord Chem Rev 258–259:137–156
46. Toganoh M, Sato A, Furuta H (2011) Double-decker ferrocene-type complex of N-fused porphyrin: a model of π-extended ferrocene? Angew Chem Int Ed 50:2752–2755
47. Similar expanded model has been reported. See; Xue Z, Kuzuhara D, Ikeda S, Sakakibara Y, Ohkubo K, Aratani N, Okujima T, Uno H, Fukuzumi S, Yamada H (2013) η5-Cyclopentadienyliron(II)–[14]Triphyrin(2.1.1) sandwich compounds: synthesis, characterization, and stable redox interconversion. Angew Chem Int Ed 52:7306–7309
48. AbuSalim DI, Ferrence GM, Lash TD (2014) Synthesis of an adj-dicarbaporphyrin and the formation of an unprecedented tripalladium sandwich complex. J Am Chem Soc 136:6763–6772
49. Toganoh M, Kawabe Y, Uno H, Furuta H (2012) Unique interaction between directly linked laminated π planes in the benzonorrole dimer. Angew Chem Int Ed 51:8753–8756

Chapter 13
Heteroatom-Modified Porphyrinoids

Yoshihiro Matano

Abstract This chapter briefly summarizes the syntheses, structure–property relationships, and applications of some heteroatom-modified porphyrinoids, namely, (i) *meso*-phosphanyl-, *meso*-phosphinyl-, and *meso*-sulfinylporphyrins and (ii) β-unsubstituted 5,15-diazaporphyrins (DAPs), constructed by my research group. The *meso*-functionalized porphyrins were prepared by C–P and C–S cross-coupling reactions of *meso*-iodoporphyrins with Ph_2PH, $R_2P(O)H$, and RSH, and the DAPs were prepared from bis(dibromodipyrrin)–metal complexes and metal azides using metal-templating methods. Treatment of *meso*-phosphanylporphyrins with palladium(II) and platinum(II) salts produces two types of phosphametallacycle-fused diporphyrins via regioselective β-C–H activation. The palladium- and platinum-linked diporphyrins exhibit characteristic optical and electrochemical properties, resulting from the dπ–pπ interactions of the peripheral C–M–C covalent bonds (M = Pd, Pt), whereas the metalloporphyrin-appended phosphapalladacycles behave as precatalysts for high-temperature Heck reactions. *meso*-Phosphinyl- and *meso*-sulfinyl-zincporphyrins readily undergo self-assembly in solution to form cofacial dimers connected by complementary X=O–Zn coordination bonds (X = P, S). The *meso*-(phenylsulfinyl)porphyrin exists as a diastereomeric mixture of homo- and heterodimers, reflecting stable *S* chirality at the periphery. The β-unsubstituted DAPs are converted to several types of covalently linked DAP dimers using homo- and cross-coupling reactions. The optical and electrochemical properties of the π systems have been found to vary widely depending on the spacers connecting the two DAP rings.

Keywords Porphyrin • Phosphametallacycle • Diazaporphyrin • Dimerization • Heck reaction

---

Y. Matano (✉)
Department of Chemistry, Faculty of Science, Niigata University, Nishi-ku,
Niigata 950-2181, Japan
e-mail: matano@chem.sc.niigata-u.ac.jp

T. Akasaka et al. (eds.), *Chemical Science of π-Electron Systems*,
DOI 10.1007/978-4-431-55357-1_13

13.1 Introduction

Porphyrins are π-conjugated aromatic macrocycles consisting of four pyrrole subunits connected via methine (*meso*) carbons. There are many naturally occurring porphyrin derivatives that play vital roles as dyes and catalysts in processes such as photosynthesis, respiration, and cell metabolism. The unusual physical and chemical properties and beautiful colors of porphyrins have attracted much attention. In addition, the development and use of porphyrin-based artificial dyes and catalysts are important issues in the fields of materials chemistry, medical science, and pharmaceuticals. Various molecular design methods have been proposed for controlling the photophysical, electrochemical, and coordination properties of porphyrin π systems (Fig. 13.1). Among these, the peripheral substitution of a porphyrin ring with appropriate functional groups is widely used not only for fine-tuning of the π-electron energy but also for constructing structurally well-defined porphyrin assemblies; for example, the attachment of hydroxy, carboxy, amino, and imino groups at the periphery provides a promising way of constructing porphyrin dimers through hydrogen bonding, electrostatic interactions, and metal–ligand coordination. The internal functionalization of a porphyrin ring, i.e., the replacement of pyrrolic nitrogen atoms or *meso* carbon atoms with other elements, has also been frequently used. These two approaches, referred to as core modification and *meso* modification, respectively, significantly change the redox and coordination behaviors of the porphyrin π system.

To develop the interface between element chemistry and porphyrin chemistry, my research group has been investigating the syntheses, structure–property relationships, and coordination behaviors of heteroatom-containing porphyrin derivatives; these are divided into three classes: (i) porphyrins bearing phosphorus- or sulfur-derived functional groups on the *meso* carbons, (ii) porphyrins modified with a phosphorus atom at the core, and (iii) 5,15-diazaporphyrins (DAPs). In this chapter, some representative examples belonging to classes (i) and (iii) will be briefly

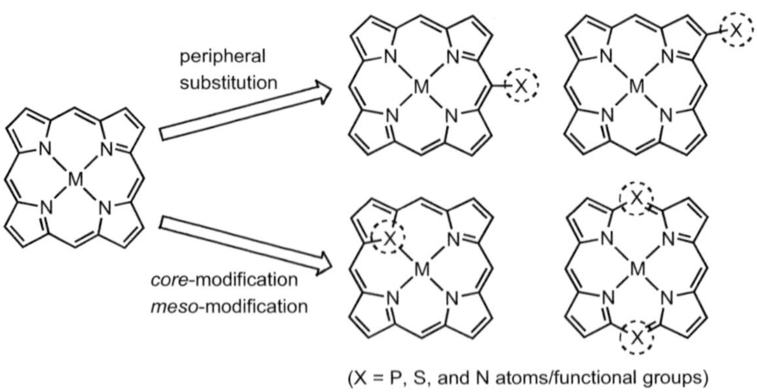

Fig. 13.1 Chemical functionalizations of porphyrin with heteroatoms

summarized with some experimental data. The main focus is to reveal the linkage effects on the optical and electrochemical properties of porphyrin dimers connected by the β-C–M covalent bonds (M = Pd, Pt) or *meso* X = O–Zn coordination bonds (X = P, S) and on DAP dimers linked by the β-C–C covalent bonds. The interactions between metalloporphyrin substituents and the palladium centers of the porphyrin-appended phosphapalladacycles in Heck reactions will also be described. The phosphorus-containing core-modified porphyrins belonging to class (ii) exhibit unusual redox properties, coordination behaviors, and aromaticities [1–5]. These topics have been covered in two accounts [6, 7] and are not included in this book. It is recommended that readers refer to the original references when the chemistry described here needs to be studied in detail.

13.2 Porphyrins Bearing Phosphorus or Sulfur Functional Groups on *meso* Carbons

13.2.1 meso-*Phosphanylporphyrins and Their Metal-Linked Dimers*

The construction of porphyrin arrays has long been important for understanding the mechanisms of electron and energy transports in photosynthetic reaction centers and for developing artificial photosynthesis. Various methods have been reported for integrating multiple porphyrin chromophores. The metal–ligand coordination bond is one promising tool, as it is easy to control the orientation and distance between pigments by choosing suitable metals and ligands; for example, peripherally attached pyridyl and imidazolyl groups have been widely used for the self-assembly of porphyrins through nitrogen-to-metal coordination [8–15]. To my knowledge, however, little attention had been paid to phosphorus- and sulfur-derived functional groups [16–18], partly because of the limited information available about C–P [19–24] and C–S [25–28] bond-forming reactions at the periphery of a porphyrin ring. Scheme 13.1 shows new protocols developed by my group for the syntheses of *meso*-phosphanylporphyrins **1** [29, 30], *meso*-phosphinylporphyrins **2** [31], *meso*-sulfanylporphyrins **3** [32], and *meso*-sulfinylporphyrins **4**. These protocols are based on palladium- or copper-catalyzed C–P and C–S cross-coupling reactions of *meso*-iodoporphyrins with Ph_2PH, $R_2P(O)H$, and RSH. *meso*-Phosphanylporphyrin **1** (R = Ph) and *meso*-sulfanylporphyrins **3** (R = Ph, 1-C_8H_{17}) are regarded as porphyrin-appended phosphines and sulfides, respectively. Phosphine **1** is gradually oxygenated under aerobic conditions to give phosphine oxide **2** (R = Ph). In contrast, sulfides **3** are air stable and converted to sulfoxides **4** by treatment with *m*-chloroperbenzoic acid (*m*CPBA).

The σ nucleophilicity of the tertiary phosphine **1** was evaluated from the ^{31}P NMR spectra of the Ph_2(Por)P = Se derivative **5** and the Ph_2(Por)P–W$(CO)_5$ complex **6** (Fig. 13.2). The observed ^{31}P–^{77}Se (for **5**) and ^{31}P–^{183} W (for **6**) coupling

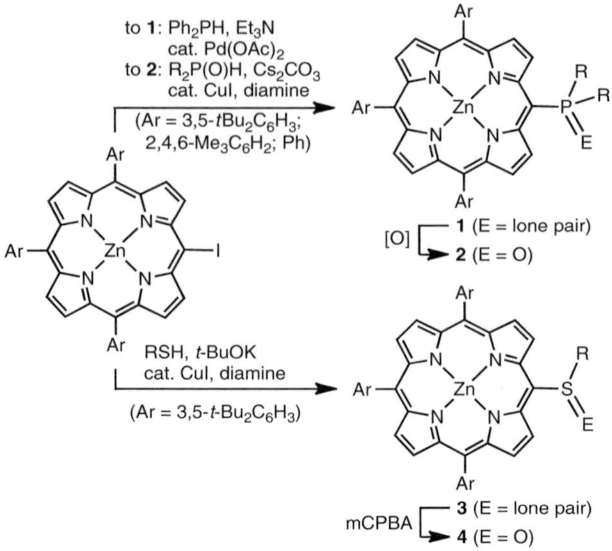

Scheme 13.1 Syntheses of *meso*-substituted porphyrins **1**–**4**

Fig. 13.2 ^{31}P NMR coupling constants of **5** and **6**

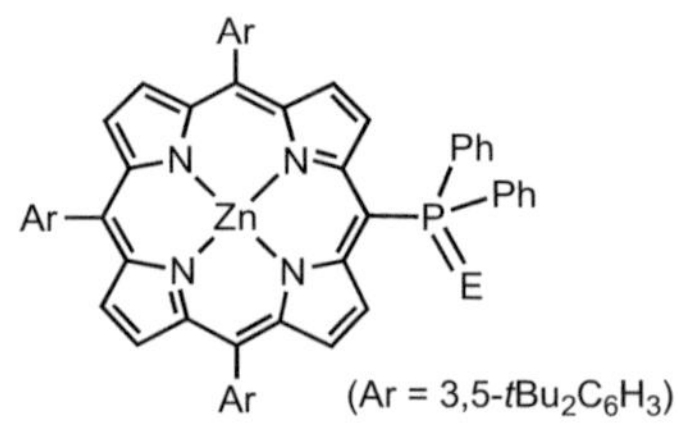

5 (E = Se; J_{P-Se} = 726 Hz)
6 (E = W(CO)$_5$; J_{P-W} = 250 Hz)
cf. Ph_3PSe (J_{P-Se} = 730 Hz)
$Ph_3PW(CO)_5$ (J_{P-W} = 280 Hz)

constants are smaller than the respective values for $Ph_3P{=}Se$ and $Ph_3PW(CO)_5$, indicating that the σ-donating ability of **1** is appreciably higher than that of triphenylphosphine [33, 34]. Indeed, **1** smoothly reacted with a half equivalent of a palladium(II) or platinum(II) salt to give a new class of porphyrin dimers (diporphyrins) **7 M**, in which the two porphyrin rings are linked by peripheral C–M–C σ bonds (Scheme 13.2). In both cases, regioselective β-C–H activation is considered to occur after coordination of the phosphine to the metal cation. As shown in Fig. 13.3, the palladium center in **7Pd** (Ar = 2,4,6-$Me_3C_6H_2$) adopts a distorted square-planar geometry, which forces the two porphyrin rings to be placed in almost the same plane, with a Zn•••Zn distance of 12.1 Å.

The literature contains only a few examples of metal-linked diporphyrins with peripheral carbon–metal σ bonds [35–37]. The phosphametallacycle-linked coplanar diporphyrins **7M** are suitable candidates for examining the mutual electronic

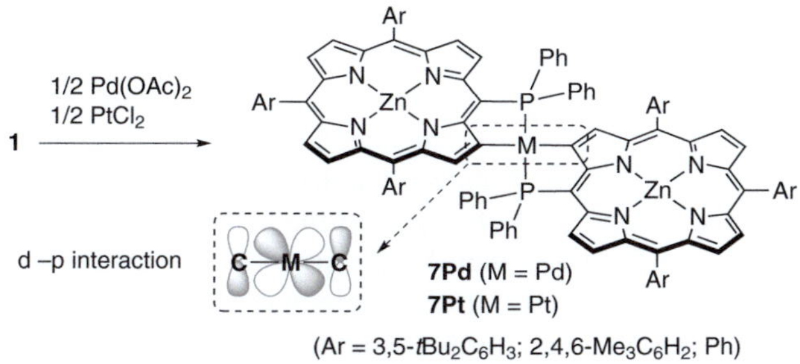

Scheme 13.2 Synthesis of metal-linked diporphyrins **7M**

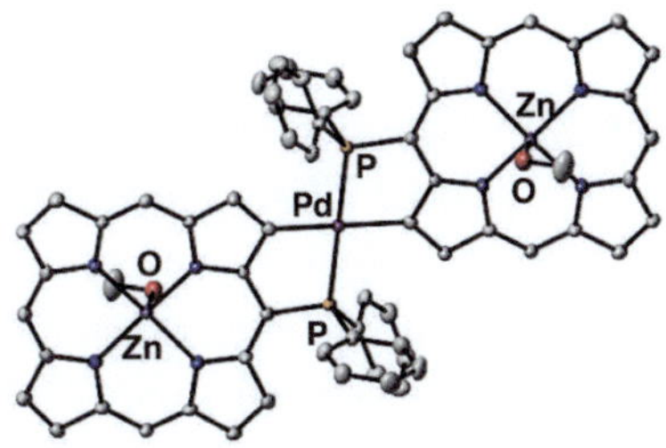

Fig. 13.3 Crystal structure of **7Pd** (Ar = mesityl). *meso*-Aryl groups and hydrogen atoms are omitted for clarity. MeOH coordinates to the zinc center

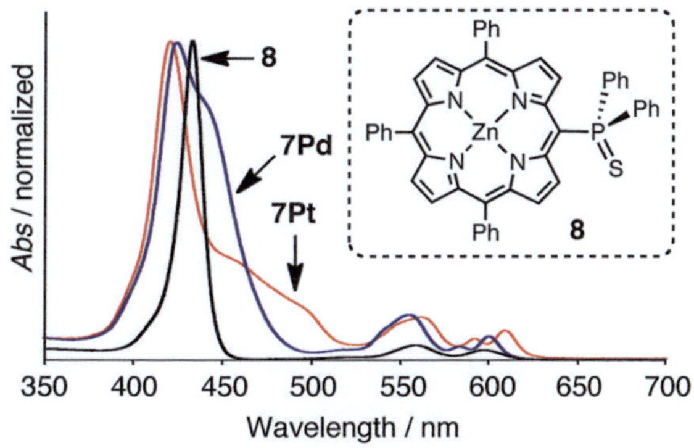

Fig. 13.4 UV/vis absorption spectra of **7Pd** (Ar = Ph), **7Pt** (Ar = Ph), and **8** in toluene

interactions between the two porphyrin π systems through dπ–pπ orbital interactions [38]. Figure 13.4 shows the UV/vis absorption spectra of the metal-linked dimers **7Pd** (Ar = Ph) and **7Pt** (Ar = Ph) and of the monomer reference **8**. As illustrated, the Soret bands of **7Pd** and **7Pt** are split and/or broadened compared with that of **8**. More importantly, the spectral features of **7Pd** and **7Pt** differ greatly from each other, suggesting that the metal dπ orbitals perturb the π–π* transitions of the porphyrin chromophores. The magnetic circular dichroism (CD) spectra and density functional theory (DFT) calculations of **7M** show that the observed electronic transitions originate from interstate mixing between the intrinsically weak, d-orbital-derived transitions and the symmetrically allowed porphyrin π–π* transitions through the C–M–C linkage (M = Pd, Pt). The electrochemical oxidation of **7Pt** progresses via two-step, one-electron oxidation, indicating that the **7Pt** radical cation is stabilized by delocalization of an unpaired electron through the peripheral Pt–C bonds. Furthermore, fluorescence lifetime and time-resolved transient pump–probe absorption measurements of **7M** and their monomer references show that intersystem crossing from the excited singlet (S_1) of **7M** to its triplet (T_1) reaches approximately 100 % because of the heavy-atom effects of palladium and platinum. It is worth noting that the dπ–pπ orbital interactions at the C–M–C linkages weakly but definitely affect the reciprocal electronic communication between the two porphyrin rings. The fundamental information obtained for **7M** will be helpful in the rational design of conceptually new multiporphyrin architectures based on peripheral metal–carbon σ bonds.

13.2.2 meso-*Phosphinyl- and* meso-*Sulfinylporphyrins and Their Cofacial Dimers*

Both *meso*-(phosphinyl)zincporphyrins **2** and *meso*-(sulfinyl)zincporphyrins **4** have polarized X=O groups (X = P, S) at the periphery, which are capable of coordinating intermolecularly to the zinc center of zincporphyrin. The ^{1}H NMR and electronspray ionization mass spectra of **2** show that P=O-tethered cofacial porphyrin dimers are the prevalent species in solution (Fig. 13.5a). X-ray crystallographic analysis of the dibutoxyphosphinyl derivative **2** (R = OBu) clearly supports complementary P=O–Zn coordination modes in the solid state (Fig. 13.5b). The UV/vis absorption spectra of the dimers in toluene show exciton coupling of

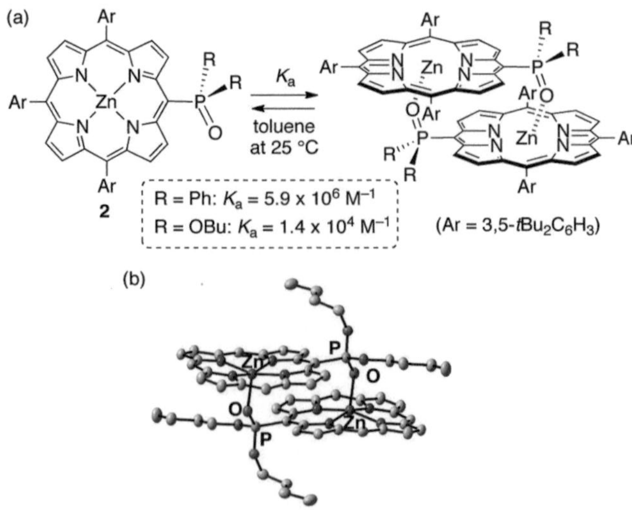

Fig. 13.5 (**a**) Formation of P=O-tethered cofacial porphyrin dimers. (**b**) Crystal structure of **2** (R = OBu). *meso*-Aryl groups and hydrogen atoms are omitted for clarity

the Soret bands. The self-association constants (K_a) of **2** in toluene at 25 °C were determined to be 5.9×10^6 M^{-1} (R = Ph) and 1.4×10^4 M^{-1} (R = OBu). As expected, the K_a values are strongly correlated with the coordinating ability of the *meso*-phosphinyl group, which is chemically tunable by changing the phosphorus substituents.

The sulfur center in *meso*-sulfinylporphyrin **4** has stable chirality; therefore optically active (*R*)-**4** and (*S*)-**4** (R = Ph) can be isolated by optical resolution of the racemate on a chiral high-performance liquid chromatography column. The aggregation modes of the S=O-tethered cofacial dimers were investigated using ^{1}H NMR and UV/vis absorption spectroscopies; the results show that the structures of the aggregates depend on the combination of the two *S* chiral centers (Fig. 13.6).The racemate of **4** [(*R*)/(*S*) = 50/50] readily undergoes self-aggregation in solution to form a diastereomeric mixture of hetero [(*R*)–(*S*)] and homo [(*R*)–(*R*) and (*S*)–(*S*)] dimers. In contrast, the optically pure (*S*)-**4** forms the (*S*)–(*S*) homodimer as the sole diastereomer/enantiomer. The UV/vis CD and absorption spectra of the homodimers (Fig. 13.7) and DFT calculations on their models indicate that they have helical structures, as shown in Fig. 13.6b. Self-aggregation greatly enhances the Cotton effects, enabling determination of the absolute configurations at the chiral sulfur centers using the exciton-coupled CD method. It should also be pointed out that the K_a value of (*S*)-**4** (R = Ph) in toluene is around ten times larger than that in CH_2Cl_2. This solvent effect can be interpreted in terms of the net dipole moments of the chromophores. The calculated dipole moments of the hetero- and homodimers are smaller than that of the monomer; therefore the relative stability of the dimer with respect to the monomer increases with decreasing solvent polarity.

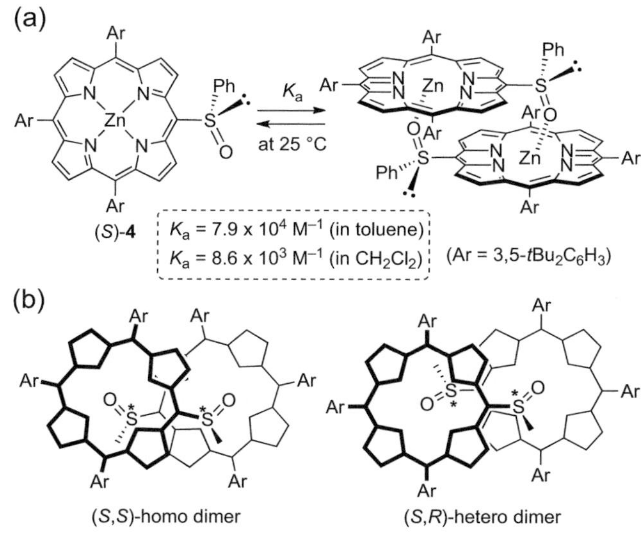

Fig. 13.6 (**a**) Formation of S=O-tethered cofacial porphyrin dimers. (**b**) Structures of homo- and heterodimers optimized by DFT calculations

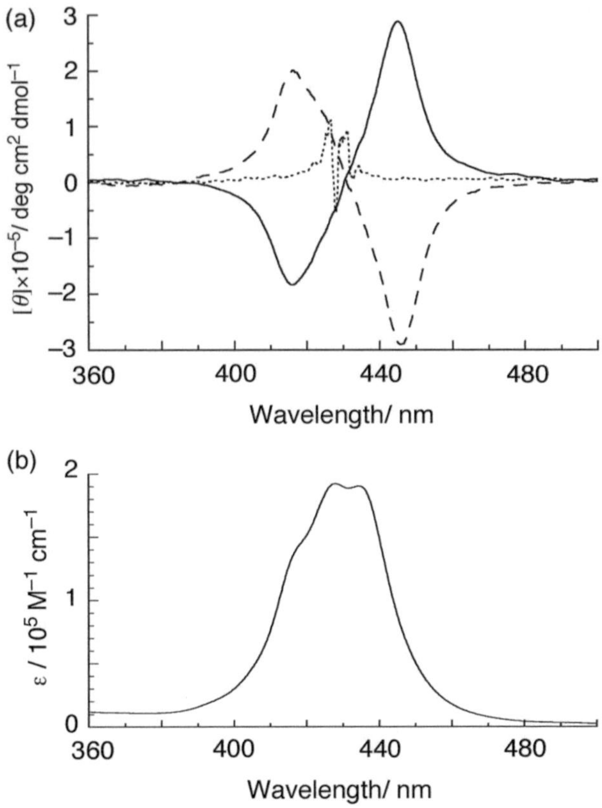

Fig. 13.7 (**a**) CD spectra of the optically pure (*S*)-**4** (*solid line*) and (*R*)-**4** (*dashed line*) in toluene at 1.0×10^{-4} M. *Dotted line*: CD spectrum of (*S*)-**4** in the presence of 10 % MeOH. (**b**) UV/vis absorption spectrum of (*S*)-**4** in toluene at 1.0×10^{-4} M

The cyclic and differential pulse voltammograms (CVs and DPVs) of **2** and **4** in CH_2Cl_2 (with Bu_4NPF_6 as the supporting electrolyte) show split, reversible one-electron-oxidation processes; this implies that the initially produced π-radical cations are stabilized by spatial interactions between the two porphyrin rings. The observed structure–property relationships of **2** and **4** show that the P=O- and S=O-tethered complementary porphyrin dimers could be new structural models for investigating the energy- and electron-transfer processes occurring at a photosynthetic reaction center with a weakly coupled special pair.

13.2.3 Porphyrin-Appended Phosphapalladacycles in Heck Reaction

The *meso*-(diphenylphosphanyl)porphyrin **1** (R = Ph) reacted with 1 equiv. of $PdCl_2(NCPh)_2$ in CH_2Cl_2 to afford the bis(μ-chloro)-bridged, zincporphyrin-appended phosphapalladacycle **9Zn** via regioselective β-C–H activation of the

Scheme 13.3 Synthesis of metalloporphyrin-appended phosphapalladacycles **9 M**

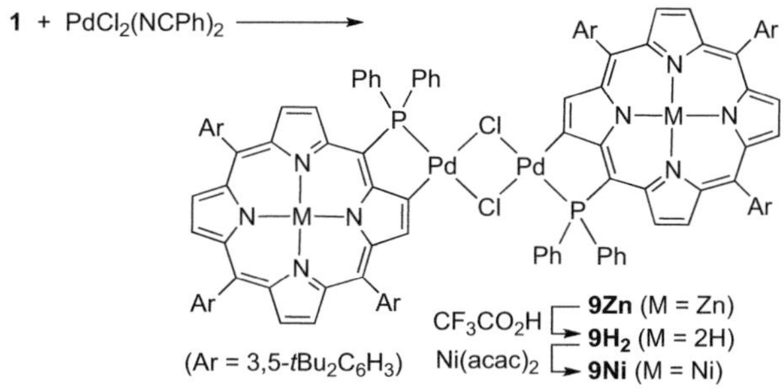

porphyrin ligand (Scheme 13.3). Demetalation of **9Zn** with trifluoroacetic acid gave the freebase-porphyrin-appended phosphapalladacycle **9H$_2$**, which was then converted to the nickel-porphyrin-appended phosphapalladacycle **9Ni** by treatment with Ni(acac)$_2$ (acac = acetylacetonato). All the phosphapalladacycles **9M** are air- and moisture-stable purple solids. The ^{31}P NMR spectra of **9M** in CDCl$_3$ displayed one singlet peak, indicating equivalent coordination of two phosphorus atoms to the palladium center. The ^{1}H NMR spectra of **9M** also verify that the two porphyrin rings are symmetrically equivalent. The UV/vis absorption spectra of **9M** show broadened or slightly split Soret bands, as a result of weak exciton coupling of the two chromophores.

The phosphapalladacycle structure of **9M** is similar to those of Herrmann–Beller palladacycles [39], which are well known as precatalysts for promoting various palladium-catalyzed cross-coupling reactions. Recent mechanistic studies of various palladacycles have clarified that nearly all palladium precatalysts, including the original Herrmann–Beller phosphapalladacycles, gradually decompose at high temperatures to liberate soluble "ligand-free" palladium colloids or nanoparticles [40–42]. This implies that the electronic nature of the phosphorus substituent correlates with the ability of the phosphapalladacycle to load active palladium(0) species with constant efficiency. With this in mind, the porphyrin-appended phosphapalladacycles **9M** were examined as precatalysts for the high-temperature Heck reaction between *p*-bromobenzaldehyde and butyl acrylate in the presence of sodium acetate in *N,N*-dimethylacetamide (DMA). It was envisioned that the σ-donating properties, i.e., the catalyst-loading abilities of **9 M**, would be tunable by changing the central metal of the porphyrin substituent. This concept for the design of porphyrin-appended palladium catalysts has also been substantiated by other groups [43, 44].

As shown in Fig. 13.8, C–C coupling accelerated significantly after induction periods of 10–35 min; this is typical of high-temperature Heck reactions using palladacycle precatalysts. At a **9M** loading of 0.05 mol%, the catalytic activity depended heavily on the central metal in the order **9H$_2$** < **9Zn** < **9Ni** (Fig. 13.8). The nickel-porphyrin-appended phosphapalladacycle **9Ni** showed the highest efficiency with respect to turnover frequencies. The effects of the central metals on the catalytic activities are more distinct at precatalyst loadings of 0.005 mol%; **9Ni** showed much

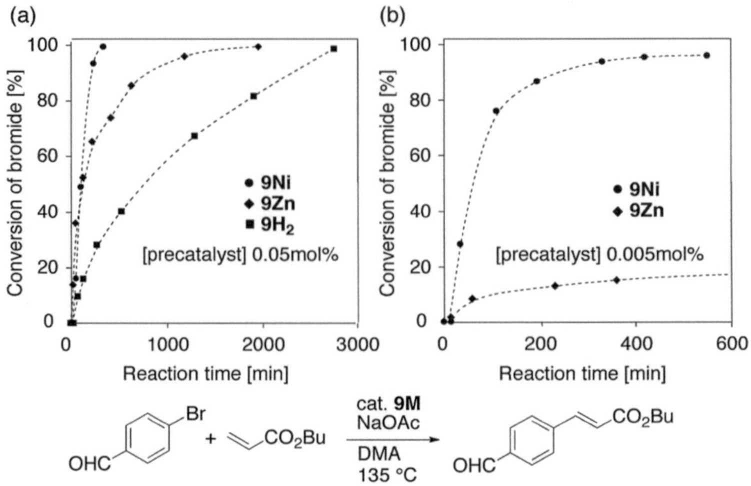

Fig. 13.8 Time profiles of the Heck reaction: (**a**) [**9M**] = 0.05 mol%; (**b**) [**9M**] = 0.005 mol%

higher catalytic activity than **9Zn**. It has been shown that the electron-donating ability of the porphyrin π system and the σ-donating ability of the phosphorus center increase in the order M = Ni < H_2 < Zn. This does not coincide with the order of the catalytic activities of **9M**, suggesting that an increase in the electron-donating ability of the porphyrin substituent does not simply enhance the catalyst-loading ability of **9 M**. The high catalytic activity observed for **9Ni** can be explained by considering its thermal stability under the conditions used for the Heck reaction. Prolonged heating of a DMA solution containing **9Ni** in the presence of an excess of sodium acetate at 130 °C showed negligible decomposition of **9Ni** after 8 h, implying that very small amounts of palladium were continuously released from **9Ni** into the solution. In contrast, the relatively electron-excessive **9Zn** decomposed much more rapidly under the same conditions, indicating that **9Zn** is more susceptible to protonolysis than **9Ni**. Presumably, the high catalytic activity observed for **9Ni** stems from its high thermal durability, which plays an important role in loading an adequate amount of active palladium(0) species at constant efficiency. These studies demonstrate the potential of phosphapalladacycles as visually detectable, thermally stable, and chemically tunable transition-metal catalysts.

Porphyrin-appended phosphonic acids and sulfonic acids can also be adsorbed onto glass substrates and organized at liquid–liquid interfaces. In this context, phosphorus- and sulfur-derived functional groups are promising for further development of element–porphyrin-based materials chemistry.

13.3 Porphyrin Derivatives Containing Nitrogen Atoms at *meso* Positions

13.3.1 β-Unsubstituted 5,15-Diazaporphyrins

Azaporphyrins are a class of peripherally modified porphyrins containing at least one nitrogen atom in place of the methine unit. The most popular azaporphyrin is phthalocyanine, which contains four *meso* nitrogen atoms and four fused benzo groups. Metal complexes of phthalocyanine have been widely used in the industry as sensitizers, semiconductors, and pigments, because of their high chemical and thermal stabilities [45–48]. However, except for phthalocyanine, the number of azaporphyrins is still limited. A DAP has two *meso* nitrogen atoms linking two dipyrromethene (dipyrrin) units and possesses an intrinsically D_{2h}-symmetric π system with nondegenerate highest occupied molecular orbitals (HOMOs) and lowest unoccupied molecular orbitals (LUMOs). DAPs therefore have characteristic optical and electrochemical properties that cannot be achieved with porphyrins and tetraazaporphyrins [49–53]; for example, 5,15-aza substitution greatly enhances the light-harvesting ability of the porphyrin chromophore in the long-wavelength region. Furthermore, DAPs have been used as redox-active and photoactive macrocyclic N_4-ligands [54–56]. However, applications of DAPs in medical science and materials chemistry require a systematic understanding of the effects of the peripheral substituents on the optical/electrochemical properties, aromaticities, and coordination behaviors of the DAP π skeletons. The synthesis and functionalization of β-unsubstituted DAPs were therefore chosen as the first topic for my research group.

The β-unsubstituted DAPs were successfully prepared using the metal-templating method [57], which was originally established for the high-yield synthesis of an octaalkyl-DAP–copper complex [58, 59]. As shown in Scheme 13.4, treatment of dibromodipyrrin complexes **10Ni** and **10Cu** with CuI–NaN_3 in *N*,*N*-dimethylformamide afforded NiDAP **11Ni** and CuDAP **11Cu**, respectively, in good yields. Shinokubo and coworkers also synthesized **11Ni** using a different protocol [60]. The free base **11H₂** was obtained from **10Pb** in two steps, but the yield was unsatisfactory (10–15 %) [61]. The complexation of **11H₂** with palladium(II) and platinum(II) salts afforded the respective metal complexes, **11Pd** and **11Pt**.

Table 13.1 summarizes selected optical, electrochemical, and structural data for **11M** and reference porphyrins **12M** (in Scheme 13.4). The characteristic properties of DAPs are as follows. (1) The Q bands of **11M** are observed as intense peaks at the longer wavelength than those of **12M**. (2) The π systems of **11M** are easier to reduce and harder to oxidize than those of the reference porphyrins. These properties stem from the intrinsically D_{2h}-symmetric DAP π system with nondegenerate HOMOs and LUMOs. The DAP ligand in **11M** provides a smaller coordination sphere compared with that in **12M**, as shown by the relatively short M–N bond lengths.

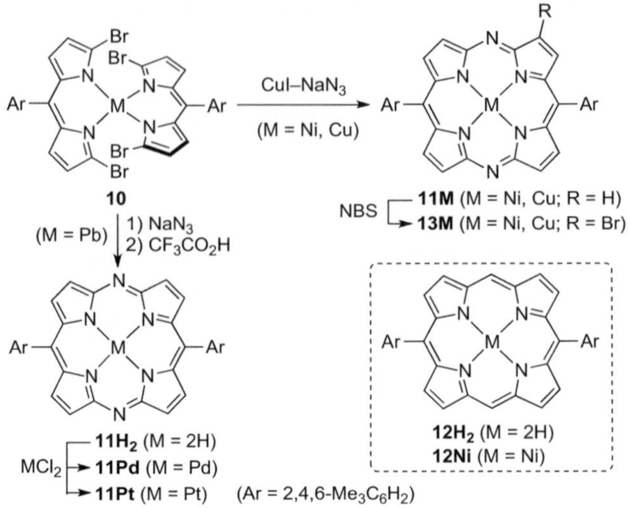

Scheme 13.4 Synthesis of DAPs **11M** and **13M** and structures of reference porphyrins **12 M**

Table 13.1 Optical, electrochemical, and structural data for **11M** and **12M**

	λ_Q [nm][a]	log ε[a]	E_{ox} [V][b]	E_{red} [V][b]	ΔE [V][c]	M–N [Å][d]
11H$_2$	627	4.6	+0.87	−1.26, −1.83	2.13	–
11Ni	571	4.8	+0.80	−1.40, −2.02	2.20	1.91
11Pd	564	5.0	+0.89	−1.32, −1.92	2.21	1.98
11Pt	557	4.7	+0.93	−1.34, −1.82	2.27	1.98
11Cu	577	4.9	+0.77	−1.37, −1.95	2.14	1.95
12H$_2$	627	3.2	+0.59, +1.00	−1.74, −2.14	2.33	–
12Ni	547	3.9	+0.59, +1.00	−1.77, −2.20	2.36	1.95

[a]Absorption maxima (λ_Q) and molar extinction coefficients (log ε) of the longest Q bands in CH_2Cl_2
[b]Redox potentials (vs. Fc/Fc$^+$) in CH_2Cl_2 in the presence of Bu_4NPF_6
[c]$\Delta E = E_{ox,1} - E_{red,1}$
[d]Average M–N bond lengths determined by X-ray crystallography

13.3.2 *Covalently Linked Bis(5,15-Diazaporphyrin)s*

The reaction of **11M** (M = Ni, Cu) with *N*-bromosuccinimide produced 3-bromo-DAPs **13M** (M = Ni, Cu) exclusively (Scheme 13.4). The high regioselectivity observed for the peripheral bromination probably results from the steric effects of the 2,4,6-trimethylphenyl (mesityl) groups at the 10 and 20 *meso* carbons. With **13M** as the starting materials, three kinds of β–β covalently linked DAP dimers, **14M**, **15**, and **16** (Fig. 13.9), were synthesized by palladium-catalyzed homo- and cross-coupling reactions [62]. The structures of these dimers have been fully characterized using ^{1}H NMR spectroscopy and X-ray crystallography. The ^{1}H NMR spectra

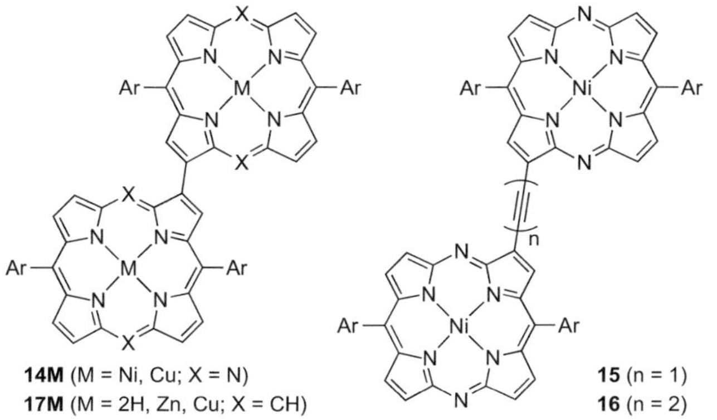

14M (M = Ni, Cu; X = N)
17M (M = 2H, Zn, Cu; X = CH)

15 (n = 1)
16 (n = 2)

Fig. 13.9 Covalently linked DAP dimers **14M**, **15**, and **16** and Osuka's diporphyrins **17M**

Fig. 13.10 Crystal structure of **14Ni**. Hydrogen atoms except for 2-C–H are omitted for clarity

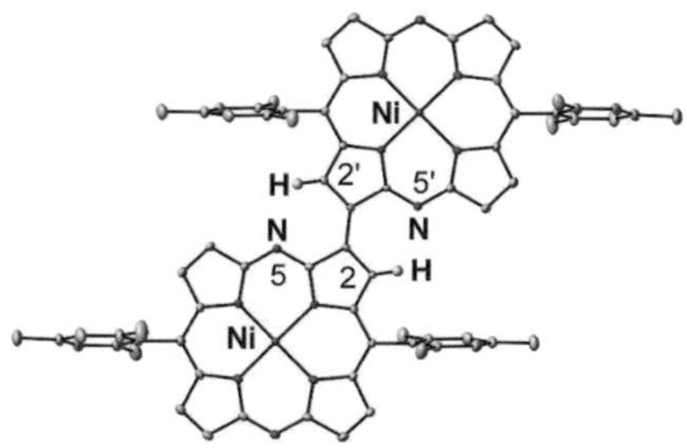

of **14Ni**, **15**, and **16** in CD_2Cl_2 show singlet peaks at $\delta = 11.00$ ppm for **14Ni**, 9.25 ppm for **15**, and 8.95 ppm for **16**, indicating pyrrolic β protons bound to the 2 (2′) carbons. It is worth noting that the downfield shift of the β-C_2–H of **14Ni** vs **11Ni** ($\delta = 8.78$ ppm) is considerably larger than those of **15** and **16** vs **11Ni**. It should also be noted that the β-C_2–H of **14Ni** is largely deshielded compared with the corresponding proton of Osuka's diporphyrin **17H₂** (M = 2H; Fig. 13.9) [63]. These results indicate that the two *meso* nitrogen atoms adjacent to the inter-ring bond play a crucial role in producing large deshielding effects on the β-C_2–H in **14Ni**. X-ray crystal structure analysis of **14Ni** shows that the two DAP rings are almost in the same plane with a *trans* conformation (Fig. 13.10). It should be noted that the analogous diporphyrin counterpart **17Zn** (M = Zn) adopts a more twisted conformation, as a result of steric repulsion between the *meso*-C–H and β-C–H groups [64]. The relatively short distance between the *meso*-N_5 and β-C_2–H atoms in **14M** (ca. 2.5 Å) implies that there is a weak hydrogen-bonding interaction between these two atoms. The steric and electronic effects of the *meso* nitrogen atoms contribute to the effective coplanarization of the two DAP π systems in **14M**.

The optical and electrochemical properties of **14Ni**, **15**, and **16** clearly show the effective π-conjugation through β–β covalent bonds. As shown in Fig. 13.11, the Soret and Q bands of **14Ni**, **15**, and **16** are broadened and redshifted compared with those of the monomer **11Ni**. The redshifts of the Q bands ($\Delta\lambda_Q$) relative

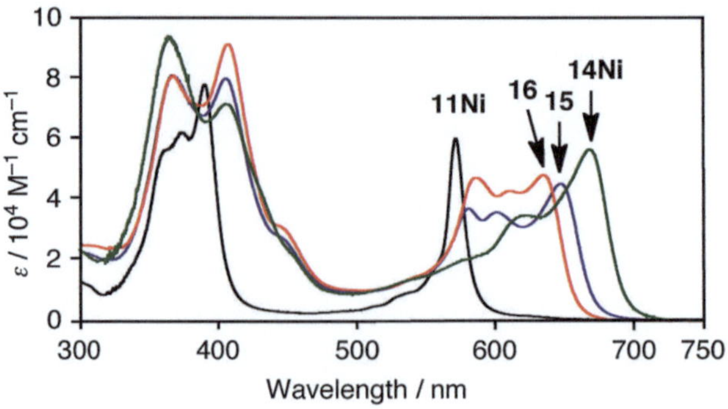

Fig. 13.11 UV/vis absorption spectra of **11Ni**, **14Ni**, **15**, and **16** in CH_2Cl_2

to the monomer reference increase in the order **16** < **15** < **14Ni**, which indicates that **14Ni** has the narrowest HOMO–LUMO gap. It is worth noting that the $\Delta\lambda_Q$ values observed for **14M** ($\Delta\lambda_Q = 89$–98 nm vs **11M**) are significantly larger than that observed for the β–β-linked diporphyrin **17Zn** ($\Delta\lambda_Q = 11$ nm vs monomer) [64]. The electrochemical oxidation/reduction processes of **14Ni**, **15**, and **16** were examined in CH_2Cl_2 using CV and DPV with Bu_4NPF_6 as the supporting electrolyte. The directly linked dimer **14Ni** displayed split and reversible one-electron redox processes, probably as a result of electronic coupling of the two DAP π systems through inter-ring covalent bonds. The coupling energy ($\Delta E_{ox} = E_{ox,2} - E_{ox,1}$) observed for the oxidation of **14Ni** ($\Delta E_{ox} = 0.19$ V) is larger than those for **15** ($\Delta E_{ox} = 0.14$ V) and **16** ($\Delta E_{ox} = 0.09$ V), indicating that the direct linkage is more effective than acetylene and butadiyne bridges in π-electron delocalization. The electrochemical HOMO–LUMO gaps ($E_{ox,1} - E_{red,1}$) increase in the order **14Ni** < **15** < **16**, which is in good accordance with the order of the optical HOMO–LUMO gaps.

The exchange interaction of electron spins between the two CuDAP π systems was investigated using electron spin resonance (ESR) spectroscopy for the copper(II) complexes **11Cu** and **14Cu** (Fig. 13.12). The DAP monomer **11Cu** gives an ESR signal ($g = 2.052$–2.168) with well-resolved hyperfine structures, which can be simulated based on the nuclear spins of one copper(II) and four core nitrogen atoms. The spectral features observed for **11Cu** resemble those observed for the copper(II)–porphyrin reference [65]. The spin densities on the two *meso* nitrogen atoms of **11Cu** are negligible compared with those on the core nitrogen atoms. In the ESR spectrum of the directly linked CuDAP dimer **14Cu**, the hyperfine splitting width for the copper nuclei is half that of **11Cu**; this can be reproduced by spectral simulation of two magnetically coupled CuDAPs. The degree of exchange interaction between the two copper(II) spins of **14Cu** was estimated from the temperature (T) dependence of the ESR signal intensity (I), which is proportional to the magnetic susceptibility. In each case, the IT values decreased sharply as the temperature dropped below 50 K, indicating that there is antiferromagnetic coupling between the two copper(II) centers. In other words, the ESR signals observed above

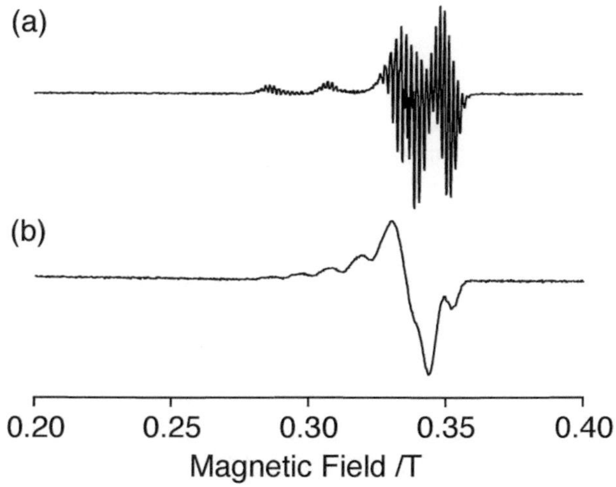

Fig. 13.12 ESR spectra of (**a**) **11Cu** and (**b**) **14Cu** in toluene at 40 K

50 K are probably the result of a thermally populated triplet state. The exchange coupling constant of **14Cu** ($J/k_B = -13.8$ K), which was determined by Bleaney–Bowers analysis of the observed *IT*–*T* plot, is ca. 15 times larger than that of the analogous copper–diporphyrin **17Cu** (M = Cu; $J/k_B = -0.94$ K) [66]. It is evident that direct connection through the β–β covalent bond is promising for effective antiferromagnetic coupling between the two coplanar CuDAP π systems.

13.4 Conclusion

The results presented in this chapter show that heteroatom-modified porphyrinoids have characteristic optical and electrochemical properties, derived from the incorporated heteroatoms, and verify that these porphyrin-based π systems are promising candidates as artificial sensitizers, redox mediators, and catalysts. Comprehensive studies of the applications of heteroatom-modified porphyrinoids in dye-sensitized solar cells, organic thin-film photovoltaics, and photodynamic therapy will continue to be investigated in porphyrin chemistry.

References

1. Matano Y, Nakabuchi T, Miyajima T, Imahori H, Nakao H (2006) Synthesis of a phosphorus-containing hybrid porphyrin. Org Lett 8:5713
2. Matano Y, Nakashima M, Nakabuchi T, Imahori H, Fujishige S, Nakao H (2008) Monophosohaporphyrins: oxidative p-extension at the peripherally-fused carbocycle of the phosphaporphyrin ring. Org Lett 10:553

3. Matano Y, Miyajima T, Ochi N, Nakabuchi T, Shiro M, Nakao Y, Sakaki S, Imahori H (2008) Syntheses, structures, and coordination chemistry of phosphole-containing hybrid calixphyrins: promising macrocyclic P,N_2,X-mixed donor ligands for designing reactive transition metal complexes. J Am Chem Soc 130:990
4. Matano Y, Nakabuchi T, Fujishige S, Nakano H, Imahori H (2008) Redox-coupled complexation of 23-phospha-21-thiaporphyrin with group 10 metals: a convenient access to stable core-modified isophlorin metal complexes. J Am Chem Soc 130:16446
5. Nakabuchi T, Nakashima M, Fujishige S, Nakano H, Matano Y, Imahori H (2010) Synthesis and reactions of phosphaporphyrins: dramatic alteration of p-electronic structures caused by chemical modification of core-phosphorus atom. J Org Chem 75:375
6. Matano Y, Imahori H (2009) Phosphole-containing calixpyrroles, calilxphyrins, and porphyrins: synthesis and coordination chemistry. Acc Chem Res 42:1193
7. Matano Y, Nakabuchi T, Imahori H (2010) Synthesis, structures, and aromaticity of phosphole-containing porphyrins and their metal complexes. Pure Appl Chem 82:583
8. Sanders JKM, Bampos N, Clyde-Watson Z, Darling SL, Hawley JC, Kim HJ, Mak CC, Webb SJ (2000) In: Kadish KM, Smith KM, Guilard R (eds) The porphyrin handbook, vol 3. Academic, San Diego, pp 3–48
9. Chambron JC, Heitz V, Sauvage JP (2000) In: Kadish KM, Smith KM, Guilard R (eds) The porphyrin handbook, vol 6. Academic, San Diego, pp 1–42
10. Sanders JKM (2000) In: Kadish KM, Smith KM, Guilard R (eds) The porphyrin handbook, vol 3. Academic, San Diego, pp 347–368
11. Wojaczyński J, Latos-Grażyński L (2000) Poly- and oligometalloporphyrins associated through coordination. Coord Chem Rev 204:113
12. Harvey PD (2003) In: Kadish KM, Smith KM, Guilard R (eds) The porphyrin handbook, vol 18. Academic, San Diego, pp 63–250
13. Drain CM, Varotto A, Radivojevic I (2009) Self-organized porphyrinic materials. Chem Rev 109:1630
14. Beletskaya I, Tyurin VS, Tsivadze AY, Guilard R, Stern C (2009) Supramolecular chemistry of metalloporphyrins. Chem Rev 109:1659
15. Aratani N, Kim D, Osuka A (2009) Discrete cyclic porphyrin arrays as artificial light-harvesting antenna. Acc Chem Res 42:1922
16. Darling SL, Stulz E, Feeder N, Bampos N, Sanders JKM (2000) Phosphine-substituted porphyrins as supramolecular building blocks. New J Chem 24:261
17. Stulz E, Scott SM, Ng YF, Bond AD, Teat SJ, Darling SL, Feeder N, Sanders JKM (2003) Construction of multiporphyrin arrays using ruthenium and rhodium coordination to phosphines. Inorg Chem 42:6564
18. Stulz E, Maue M, Scott SM, Mann BE, Sanders JKM (2004) Ru(II) and Rh(III) porphyrin complexes of primary phosphine-substituted porphyrins. New J Chem 28:1066
19. Evans B, Smith KM (1977) Novel *meso*-substitution reactions of zinc(II) octaethylporphyrin. Tetrahedron Lett 18:3079
20. Smith KM, Barnett GH, Evans B, Martynenko Z (1979) Novel *meso*-substitution reactions of metalloporphyrins. J Am Chem Soc 101:5953
21. Ruhlmann L, Gross M, Giraudeau A (2003) Bisporphyrins with bischlorin features obtained by direct anodic coupling of porphyrins. Chem Eur J 9:5085
22. Atefi F, Locos OB, Senge MO, Arnold DP (2006) *meso*-Iodo- and *meso*-iodovinylporphyrins via organopalladium porphyrins and the crystal structure of 5-iodo-10,20-diphenylporphyrin. J Porphyrins Phthalocyanines 10:176
23. Atefi F, McMurtrie JC, Turner P, Duriska M, Arnold DP (2006) *meso*-Porphyrinylphosphine oxides: Mono- and bidentate ligands for supramolecular chemistry and the crystal structures of monomeric {[10,20-diphenylporphyrinatonickel(II)-5,15-diyl]-bis-[P(O)Ph_2] and polymeric self-coordinated {[10,20-diphenylporphyrinatozinc(II)-5,15-diyl]-bis-[P(O)Ph_2]}. Inorg Chem 45:6479

24. Fujimoto K, Yoneda T, Yorimitsu H, Osuka A (2014) Synthesis and catalytic activities of porphyrin-based PCP pincer complexes. Angew Chem Int Ed 126:1145
25. Binstead RA, Crossley MJ, Hush NS (1991) Modulation of valence orbital levels of metalloporphyrins by β-substitution: evidence from spectroscopic and electrochemical studies of 2-substituted metallo-5,10,15,20-tetraphenylporphyrin. Inorg Chem 30:1259
26. Gao GY, Colvin AJ, Chen Y, Zhang XP (2004) Synthesis of *meso*-arylsulfanyl- and alkylsulfanyl-substituted porphyrins via palladium-mediated C-S bond formation. J Org Chem 69:8886
27. Shen DM, Liu C, Chen XG, Chen QY (2009) Facile and efficient hypervalent iodine(III)-mediated *meso*-functionalization of porphyrins. J Org Chem 74:206
28. Chen Q, Zhu YZ, Fan QJ, Zhang SC, Zheng JY (2014) Simple and catalyst-free synthesis of *meso*-O-, -S-, and -C-substituted porphyrins. Org Lett 16:1590
29. Matano Y, Matsumoto K, Nakao Y, Uno H, Sakaki S, Imahori H (2008) Regioselective β-metalation of *meso*-phosphanylporphyrins. Structure and optical properties of porphyrin dimers linked by peripherally fused phosphametallacycles. J Am Chem Soc 130:4588
30. Matano Y, Matsumoto K, Shibano T, Imahori H (2011) Porphyrin-appended phosphapalladacycle precatalysts: effects of central metals on the catalytic activity in a high-temperature Heck reaction. J Porphyrins Phthalocyanines 15:1172
31. Matano Y, Matsumoto K, Terasaka Y, Hotta H, Araki Y, Ito O, Shiro M, Sasamori T, Tokitoh N, Imahori H (2007) Synthesis, structures, and properties of *meso*-phosphorylporphyrins. Self-assembly through P -oxo-to-zinc coordination. Chem Eur J 13:891
32. Matano Y, Shinokura T, Matsumoto K, Imahori H, Nakano H (2007) Synthesis and aggregation behavior of *meso*-sulfinylporphyrins: Evaluation of the S-chirality effects on the self-organization to S -oxo-tethered cofacial porphyrin dimers. Chem Asian J 2:1417
33. Allen DW, Taylor BF (1982) The chemistry of heteroarylphosphorus compounds. Part 15. Phosphorus-31 nuclear magnetic resonance studies of the donor properties of heteroarylphosphines towards selenium and platinum(II). J Chem Soc Dalton Trans 51
34. Grim SO, Wheatland DA, McFarlane W (1967) Phosphorus-31 nuclear magnetic resonance study of tertiary phosphine derivatives of group VI metal carbonyls. J Am Chem Soc 89:5573
35. Hartnell RD, Arnold DP (2004) Peripherally h^1-platinated organometallic porphyrins as building blocks for multiporphyrin arrays. Organometallics 23:391
36. Yamaguchi S, Katoh T, Shinokubo H, Osuka A (2008) Pt(II)- and Pt(IV)-bridged cofacial diporphyrins via carbon-transition metal s-bonds. J Am Chem Soc 130:14440
37. Song J, Aratani N, Heo JH, Kim D, Shinokubo H, Osuka A (2010) Directly Pd(II)-bridged porphyrin belts with remarkable curvatures. J Am Chem Soc 132:11868
38. Matano Y, Matsumoto K, Hayashi H, Nakao Y, Kumpulainen T, Chukharev V, Tkachenko NV, Lemmetyinen H, Shimizu S, Kobayashi N, Sakamaki D, Ito A, Tanaka K, Imahori H (2012) Effects of carbon -metal -carbon linkages on the optical, photophysical, and electrochemical properties of phosphametallacycle-linked coplanar porphyrin dimers. J Am Chem Soc 134:1825
39. Herrmann WA, Brossmer C, Öfele K, Reisinger CP, Priermeier T, Beller M, Fischer H (1995) Palladacycles as structurally defined catalysts for the Heck olefination of chloro- and bromoarenes. Angew Chem Int Ed Engl 34:1844
40. Beletskaya IP, Cheprakov AV (2004) Palladacycles in catalysis – a critical survey. J Organomet Chem 689:4055
41. de Vries JG (2006) A unifying mechanism for all high-temperature Heck reactions. The role of palladium colloids and anionic species. Dalton Trans 421
42. Phan NTS, van der Sluys M, Jones CW (2006) On the nature of the active species in palladium catalyzed Mizoroki -Heck and Suzuki -Miyaura couplings – homogeneous or heterogeneous catalysis, a critical review. Adv Synth Catal 348:609
43. Suijkerbuijk BMJM, Herreras Martínez SD, van Koten G, Klein Gebbink RJM (2008) Hetero-multimetallic tetrakis(SCS-pincer palladium)-(metallo)porphyrin hybrids. Tunable precatalysts in a Heck reaction. Organometallics 27:534

44. Yamaguchi S, Katoh T, Shinokubo H, Osuka A (2007) Porphyrin pincer complexes: peripherally cyclometalated porphyrins and their catalytic activities controlled by central metals. J Am Chem Soc 129:6392
45. Kobayashi N (2000) In: Kadish KM, Smith KM, Guilard R (eds) The porphyrin handbook, vol 2. Academic, San Diego, pp 301–360
46. Rodríguez-Morgade MS, de la Torre G, Torres T (2003) In: Kadish KM, Smith KM, Guilard R (eds) The porphyrin handbook, vol 15. Academic, San Diego, pp 125–159
47. Ishii K, Kobayashi N (2003) In: Kadish KM, Smith KM, Guilard R (eds) The porphyrin handbook, vol 16. Academic, San Diego, pp 1–42
48. Mack J, Stillman MJ (2003) In: Kadish KM, Smith KM, Guilard R (eds) The porphyrin handbook, vol 16. Academic, San Diego, pp 43–116
49. Ogata H, Fukuda T, Nakai K, Fujimura Y, Neya S, Stuzhin PA, Kobayashi N (2004) Absorption, magnetic circular dichroism, IR spectra, electrochemistry, and molecular orbital calculations of monoaza- and opposite diazaporphyrins. Eur J Inorg Chem 1621
50. Pan N, Bian Y, Fukuda T, Yokoyama M, Li R, Neya S, Jiang J, Kobayashi N (2004) Homoleptic lanthanide triple-deckers of 5,15-diazaporphyrin with D_2h symmetry. Inorg Chem 43:8242
51. Pan N, Bian Y, Yokoyama M, Li R, Fukuda T, Neya S, Jiang J, Kobayashi N (2008) Sandwich-type heteroleptic *opposite*-(diazaporphyrinato)cerium complexes: synthesis, spectroscopy, structure, and electrochemistry. Eur J Inorg Chem 5519
52. Stuzhin PA, Nefedov SE, Kumeev RS, Ul-Haq A, Minin VV, Ivanova SS (2010) Effects of solvation on the spin state of iron(III) in 2,8,12,18-tetrabutyl-3,7,13,17-tetramethyl-5,10-diazaporphyrinatoiron(III) chloride. Inorg Chem 49:4802
53. Stuzhin PA, Ul-Haq A, Nefedov SE, Kumeev RS, Koifman OI (2011) Synthesis and study of the binuclear m-oxodiiron(III) complexes of 5-monoaza- and 5,15-diaza-substituted β-octaalkylporphyrins. Eur J Inorg Chem 2567
54. Shinmori H, Kodaira F, Matsugo S, Kawabata S, Osuka A (2005) Photosensitizing properties of diazaporphyrin derivatives for singlet oxygen generation. Chem Lett 34:322
55. Ohgo Y, Neya S, Uekusa H, Nakamura M (2006) An isocyanide probe for heme electronic structure: bis(*tert*-butylisocyanide) complex of diazaporphyrin showing a unique $(d_{xy})^2(d_{xz}, d_{yz})^3$ ground state. Chem Commun 4590
56. Okujima T, Jin G, Otsubo S, Aramaki S, Ono N, Yamada H, Uno H (2011) Tetrabenzodiazaporphyrin: the semiconducting hybrid of phthalocyanine and tetrabenzoporphyrin. J Porphyrins Phthalocyanines 15:697
57. Matano Y, Shibano T, Nakano H, Imahori H (2012) Nickel(II) and copper(II) complexes of β-unsubstituted 5,15-diazaporphyrins and pyridazine-fused diazacorrinoids: metal-template syntheses and peripheral functionalizations. Chem Eur J 18:6208
58. Harris RLN, Johnson AW, Kay IT (1966) A stepwise synthesis of unsymmetrical porphyrins. J Chem Soc C 22
59. Khelevina OG, Chizhova NV, Stuzhin PA, Semeikin AS, Berezin BD (1997) Acid forms of diazaporphyrines in non-aqueous media. Russ J Phys Chem 71:74
60. Horie M, Hayashi Y, Yamaguchi S, Shinokubo H (2012) Synthesis of nickel(II) azacorroles by Pd-catalyzed amination of a,a'-dichlorodipyrrin Ni^{II} complex and their properties. Chem Eur J 18:5919
61. Matano Y, Shibano T, Nakano H, Kimura Y, Imahori H (2012) Nickel(II) and copper(II) complexes of β-unsubstituted 5,15-diazaporphyrins and pyridazine-fused diazacorrinoids: metal-template syntheses and peripheral functionalizations. Inorg Chem 51:12879
62. Matano Y, Fujii D, Shibano T, Furukawa K, Nakano H, Higashino T, Imahori H (2014) Covalently linked 5,15-diazaporphyrin dimers: promising scaffolds for highly conjugated azaporphyrin π-system. Chem Eur J 20:3342
63. Hata H, Shinokubo H, Osuka A (2005) Highly regioselective Ir-catalyzed β-borylation of porphyrins via C-H bond activation and construction of β-β-linked diporphyrin. J Am Chem Soc 127:8264

64. Cho S, Yoon MC, Lim JM, Kim P, Aratani N, Nakamura Y, Ikeda T, Osuka A, Kim D (2009) Structural factors determining photophysical properties of directly linked zinc(II) porphyrin dimers: linking position, dihedral angle, and linkage length. J Phys Chem B 113:10619
65. Greiner SP, Rowlands DL, Kreilick RW (1992) EPR and ENDOR study of selected porphyrin- and phthalocyanine-copper complexes. J Phys Chem 96:9132
66. Ikeue T, Furukawa K, Hata H, Aratani N, Shinokubo H, Kato T, Osuka A (2005) The importance of a β-β bond for long-range antiferromagnetic coupling in directly linked copper(ii) and silver(ii) diporphyrins. Angew Chem Int Ed 44:6899

Chapter 14
Synthesis of Novel Porphyrinoids from Dipyrrins

Hiroshi Shinokubo

Abstract Most syntheses of *meso*-substituted porphyrinoid rely on acid-catalyzed condensation of pyrrole segments with aldehyde units followed by oxidation. This standard procedure is convenient but generally is low yielding. However, modern synthetic methodologies would offer an expeditious route to several intriguing porphyrinoids. In particular, the metal template strategy using dipyrrin metal complexes is effective to construct porphyrin-like cyclic π-conjugated systems, which are otherwise difficult to access by the conventional methodology.

Keywords Porphyrin • Dipyrrin • Catalysis • C–C bond formation • Cyclization

14.1 Introduction

Porphyrin is a representative functional molecule, which consists of four pyrrole units linked by four carbon atoms with an 18π aromatic cyclic conjugation. The material sciences related to porphyrin include functional dye, artificial photosynthesis, dye-sensitized solar cells, photodynamic therapy, near-infrared dyes, nonlinear optical materials, molecular wires, supramolecules, and so forth [1].

Aside from a number of such prominent properties and functions, most of porphyrin synthesis relies on classic reactions and often is not efficient even today. In contrast, modern organic synthesis has enabled highly elaborated synthesis of complex molecules such as naturally occurring compounds. Such new methodologies should be also beneficial in porphyrin chemistry to create exotic porphyrins with interesting structures and fascinating properties.

Dipyrrin is a monovalent and bidentate ligand, which can coordinate to a variety of metal ions to provide stable metal complexes [2]. Furthermore, many of these dipyrrin metal complexes have excellent photophysical and electrochemical properties. Recently, cyclic dipyrrins **2** bridged by some π-conjugated spacers such as aromatic, heteroaromatic, and alkynyl groups or some atoms have been

---

H. Shinokubo (✉)
Department of Applied Chemistry, Graduate School of Engineering, Nagoya University, Nagoya, Japan
e-mail: hshino@apchem.nagoya-u.ac.jp

T. Akasaka et al. (eds.), *Chemical Science of π-Electron Systems*,
DOI 10.1007/978-4-431-55357-1_14

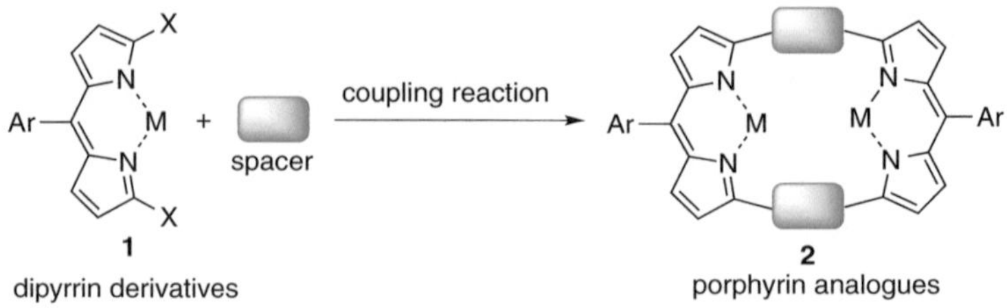

Scheme 14.1 Synthesis of porphyrinoids from dipyrrin precursors **1** through coupling reactions

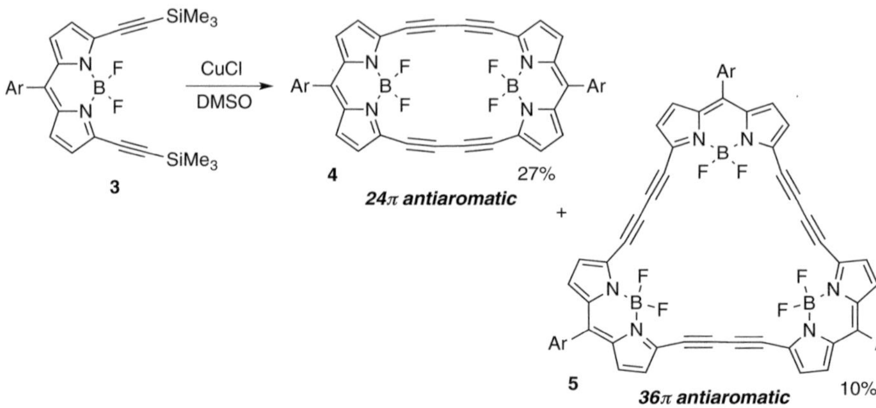

Scheme 14.2 Synthesis of butadiyne-bridged cyclic BODIPY oligomers

synthesized (Scheme 14.1). This kind of π-conjugated cyclic dipyrrins **2** can be regarded as porphyrinoids, which would exhibit intriguing photophysical and electronic properties.

14.2 Synthesis of Cyclic BODIPY Dimer and Trimer

Among a number of dipyrrin derivatives, boron dipyrrin (BODIPY) is most extensively investigated due to its high thermal and photochemical stability as well as strong absorption and emission in the visible region [3–6]. As an example of cyclic conjugated dipyrrin oligomers, butadiyne-linked BODIPY dimer **4** and trimer **5** have been created through the expeditious use of direct homo-coupling of alkynylsilanes (Scheme 14.2) [7]. The butadiyne-bridged cyclic BODIPYs should be considered as a hybrid of BODIPY and dehydroannulene. Dehydroannulene has received everlasting attention in terms of aromaticity of large cyclic conjugated systems [8–10]. The ethynylene moiety in the macrocycles enables effective extension of π-conjugation and offers rigidity to maintain planar molecular structures needed for

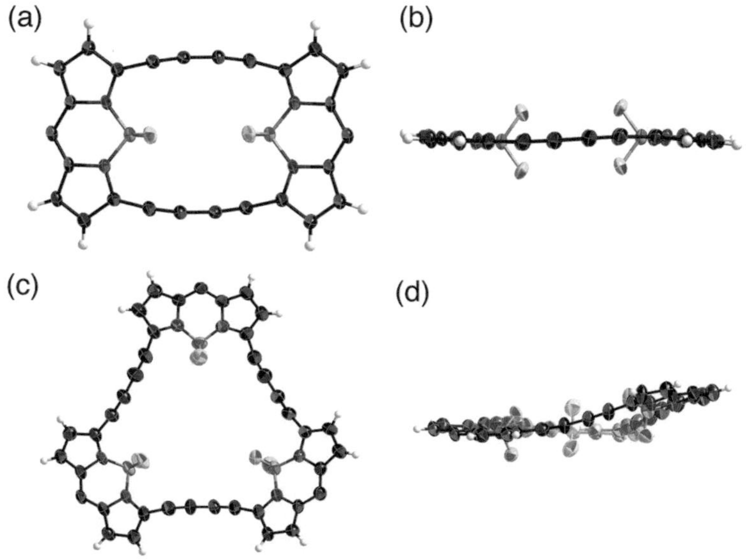

Fig. 14.1 X-ray structures of doubly butadiyne-bridged BODIPY dimer and trimers. (**a**) Top view and (**b**) side view of BODIPY dimer. (**c**) Top view and (**d**) side view of BODIPY trimer

Hückel's rule. Incorporation of BODIPY units into dehydroannulenes resulted in the formation of cyclic BODIPY dimer and trimer as stable antiaromatic porphyrinoids.

In these compounds, a flat structure of the whole macrocycle is enforced by the rigid butadiyne linker and the BF_2 unit. The mean plane deviation of the BODIPY dimer is only 0.058 Å, while the trimer is slightly distorted (Fig. 14.1). Antiaromatic character of these cyclic BODIPYs was elucidated by the ^{1}H and ^{19}F NMR spectra and the theoretical calculations of these compounds.

BODIPY is very famous for its bright fluorescence, and the emission quantum yield could be nearly unity. However, these cyclic π-conjugated BODIPY dimer and trimer have no emission due to their antiaromaticity.

14.3 Synthesis of Azaporphyrinoids from Bisdipyrrin

Porphyrin can be regarded as a type of dimers of dipyrrin. To obtain such dipyrrin dimers more efficiently, bisdipyrrin metal complexes **6** could be employed as a precursor (Scheme 14.3). The central metal of the complex should act as a template to enhance the reactivity of the cyclization reaction by taking advantage of the intramolecular reaction of the second step from **7** to **8**. In addition, this strategy is expected to regulate the formation of higher dipyrrin oligomers rather than the desired dimer.

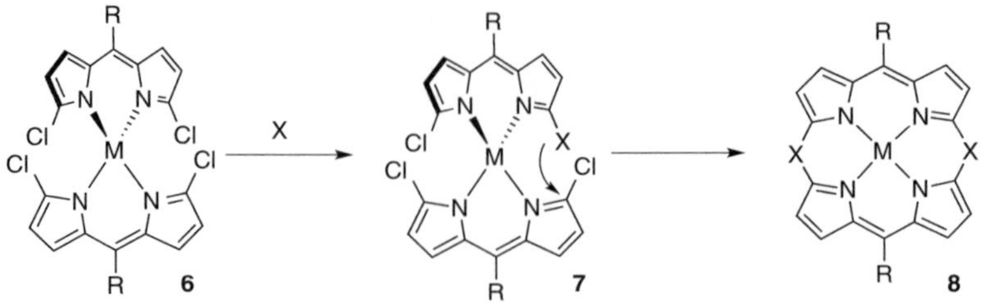

Scheme 14.3 Metal template strategy for efficient synthesis of dipyrrin dimers **8**

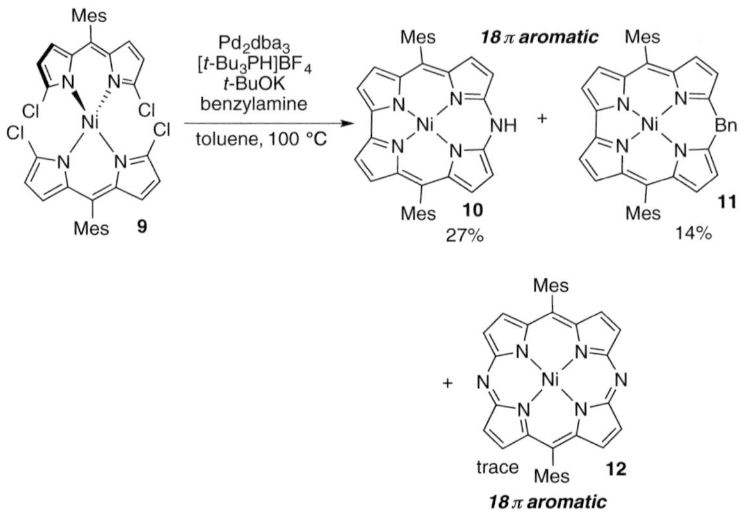

Scheme 14.4 Synthesis of azacorrole and diazaporphyrin through Pd-catalyzed amination

Palladium-catalyzed Buchwald–Hartwig amination [11–14] was employed to introduce benzylamine as a nitrogen source. However, the reaction of bisdipyrrin Ni(II) complex **9** unexpectedly afforded azacorroles [15] **10** and **11** as major products along with a trace amount of expected 5,15-diazaporphyrin **12** (Scheme 14.4) [16]. X-ray analysis of these azaporphyrinoids revealed their highly planar structures (Fig. 14.2). Furthermore, the ^{1}H NMR spectrum of azacorroles **10** and **11** displayed their β-protons in the aromatic region, thus confirming their strong aromaticity. This result indicates that azacorrole is considered to be [17]azaannulene with 18π aromatic π-conjugation.

Matano and co-workers also reported the metal template strategy to fabricate azaporphyrinoids by the use of bisdipyrrin precursor **13** [17, 18]. The use of sodium azide as a nucleophile in the presence of copper(I) iodide selectively afforded 5,15-diazaporphyrin **12** in excellent yields (Scheme 14.5).

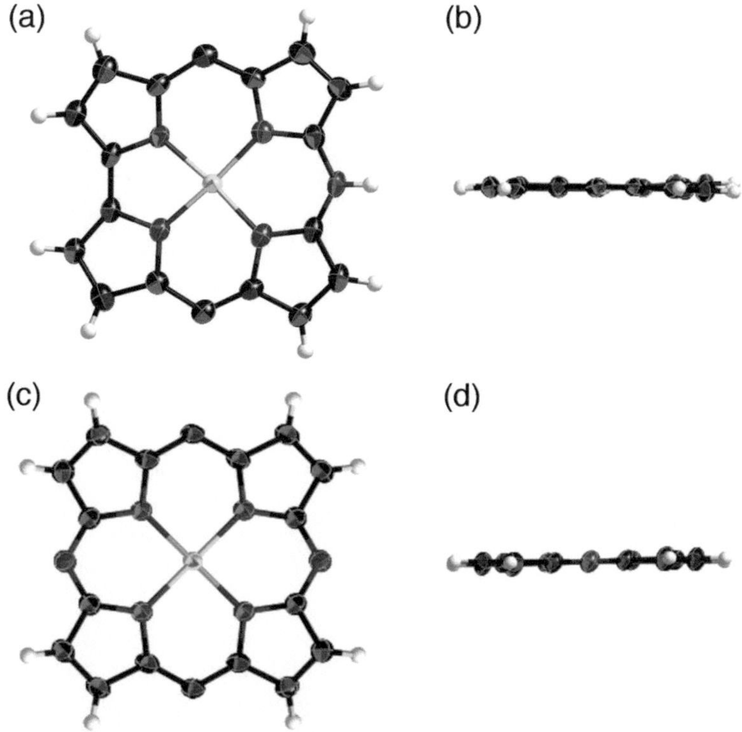

Fig. 14.2 X-ray structures of azacorrole Ni(II) and diazaporphyrin Ni(II)

Scheme 14.5 Efficient synthesis of 5,15-diazaporphyrin

14.4 Synthesis of Thiaporphyrinoids

Sulfur atoms were incorporated to the *meso*-positions of porphyrin and corrole by using a dipyrrin precursor (Scheme 14.6) [19]. Treatment of α,α'-dichlorodipyrrin Zn(II) complex **14** with sodium sulfide afforded dithiaporphyrin **15** in good yield. In contrast, the same reaction with α,α'-dichlorodipyrrin Ni(II) complex **9** yielded thiacorrole [20] Ni(II) complex **16**. While dithiaporphyrin **15** is nonaromatic, dithiaporphyrin Ni(II) **16** exhibits distinct aromatic character due to 18 π-electrons on its conjugation pathway.

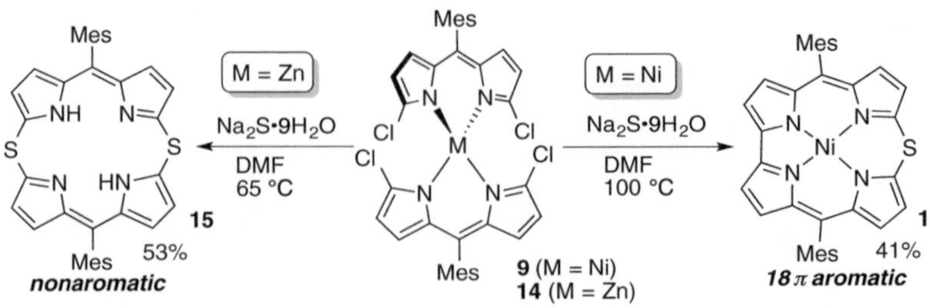

Scheme 14.6 Synthesis of thiacorrole and dithiaporphyrin

Scheme 14.7 Sulfur extrusion from dithiaporphyrin to thiacorrole

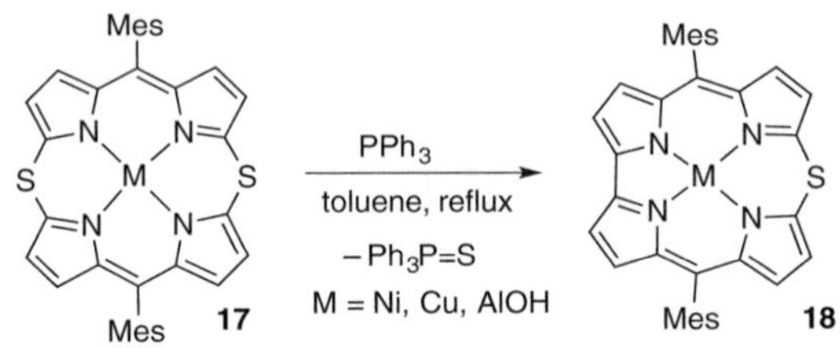

The reaction pathway to form thiacorrole involves sulfur extrusion of dithiaporphyrin (Scheme 14.7) [21, 22]. Treatment of dithiaporphyrin metal complex **17** with triphenylphosphine in refluxing toluene provided thiacorrole complexes **18** in good yield along with $Ph_3P{=}S$. Interestingly, however, the sulfur elimination reaction was not observed for freebase, Zn(II), Pd(II), and Pt(II) dithiaporphyrins. The small-size central metal ions such as Ni(II), Cu(II) [23], and Al(III) induce severe bending of nonplanar dithiaporphyrin metal complexes, and the resulting distortion facilitates the carbon–carbon bond formation via episulfide intermediates (Fig. 14.3).

14.5 Synthesis of Norcorrole Ni(II) Complex

Norcorrole is one of the attractive ring-contracted porphyrins, which lacks two *meso*-carbons from a regular porphyrin. In spite of its simple and beautiful structure, however, isolation of norcorrole as a stable molecule has been challenging. In 2008, Bröring and co-worker reported a pioneering work in the first detection of norcorrole **20** from oxidation of a tetrapyrrole iron complex **19** [24]. However, the norcorrole iron complex was unstable and easily dimerized to **21** (Scheme 14.8). Norcorrole should have 16π cyclic conjugation, thus inducing antiaromatic nature.

Recently, norcorrole Ni(II) complex **22** was synthesized in over 90 % yield through Ni(0)-mediated intramolecular homo-coupling of α,α'-dibromodipyrrin Ni(II) complex **13** (Scheme 14.9) [25]. Because of the facile synthetic procedure, the

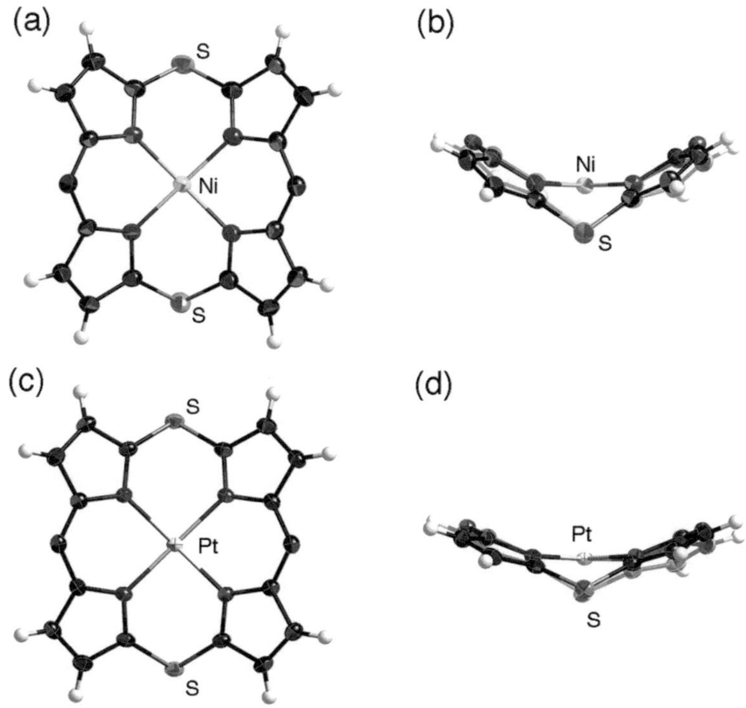

Fig. 14.3 X-ray structures of dithiaporphyrin Ni(II) and Pt(II)

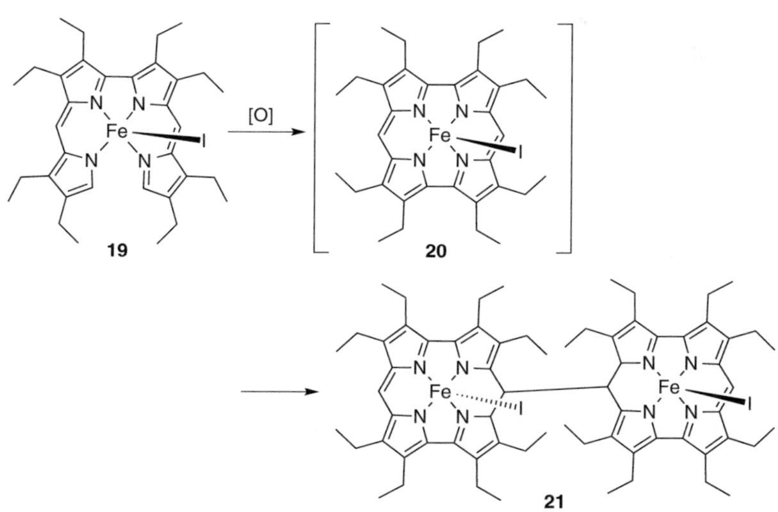

Scheme 14.8 The first detection of norcorrole complex

Scheme 14.9 Synthesis of norcorrole Ni(II) through homo-coupling of bisdipyrrin Ni(II) complex **23**

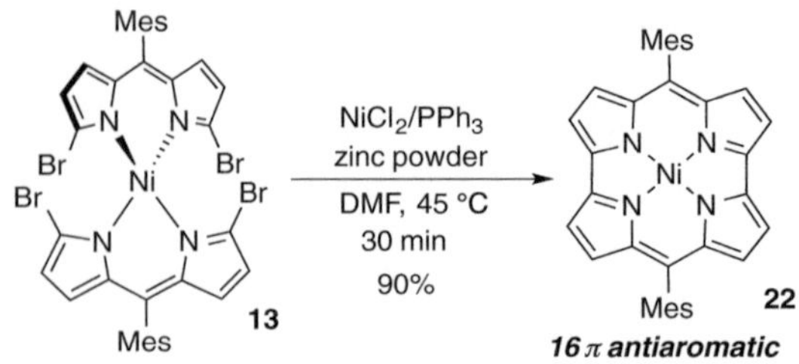

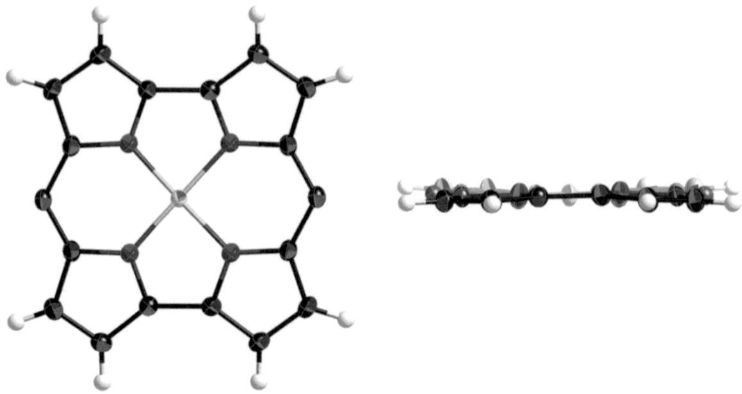

Fig. 14.4 X-ray structures of norcorrole Ni(II)

norcorrole **22** can be prepared in a gram quantity. X-ray crystallographic analysis elucidated that the main skeleton is perfectly planar (Fig. 14.4). The ^{1}H NMR spectrum displays two sets of double peaks for pyrrole β-protons in the far upfield region from $\delta = 1.45$ to 1.60 ppm, clearly indicating distinct antiaromaticity of Ni(II) norcorrole. Because of its antiaromaticity, norcorrole Ni(II) complex **22** has a small HOMO–LUMO gap, which was revealed by the electrochemical examination. Antiaromatic nature was supported by the UV/Vis absorption spectra and DFT calculations.

14.6 Reactivity of Antiaromatic Norcorrole Ni(II) Complex

The reactivity of antiaromatic porphyrinoids has not been fully understood. Norcorrole Ni(II) **22** is enough stable under air at room temperature to allow easy handling under the ambient conditions, thus being a good candidate in exploring the reactivity of antiaromatic porphyrinoids.

The reactivity of norcorrole Ni(II) complex **22** toward oxidation was investigated [25]. Interestingly, norcorrole Ni(II) was oxidized to oxacorrole [20, 26] Ni(II) **23** under oxygen at 100 °C or upon treatment with mCPBA at −78 °C (Scheme 14.10).

Scheme 14.10 Oxygen insertion to norcorrole affording oxacorrole

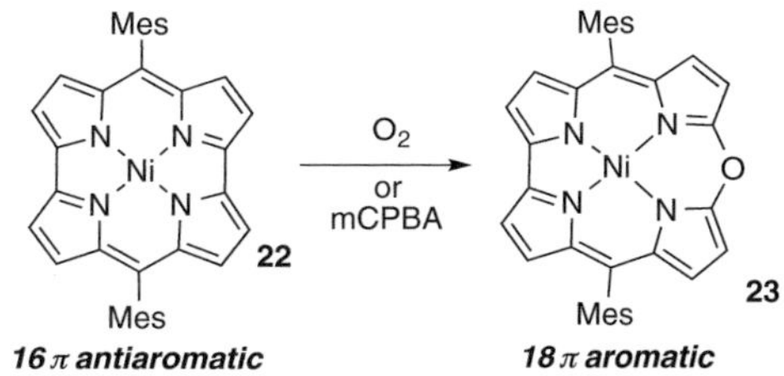

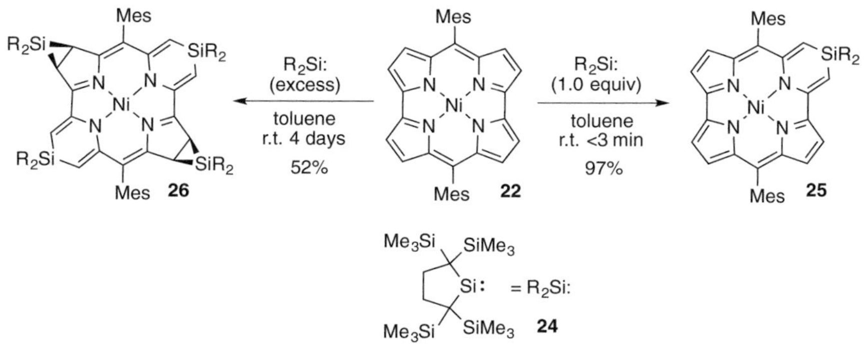

Scheme 14.11 Selective silicon insertion reaction to norcorrole Ni(II) complex

The 1H NMR spectrum of oxacorrole Ni(II) **23** revealed its aromatic character. The aromaticity of **23** can be explained by 18 π-electrons including the lone pair on the oxygen atom at the *meso*-position.

This interesting reactivity of oxygen insertion implies that norcorrole could be a potential precursor for various heteroatom-substituted porphyrinoids through an atom insertion reaction. Sterically stabilized silylene **24** [27–29] is reported to undergo regioselective insertion into the β–β pyrrole carbon–carbon bond of the norcorrole skeleton to furnish a novel silicon-containing porphyrinoid **25** excellently (Scheme 14.11) [30]. Interestingly, silicon-containing porphyrinoid **25** exhibits near-IR absorption bands reaching to 1300 nm. Because no reaction took place toward regular porphyrin Ni(II), antiaromaticity is likely the key for the high reactivity of norcorrole Ni(II) **22**. Furthermore, the reaction of **22** with an excess amount of silylene **24** provided fourfold addition product **26** as the major product. This kind of atom insertion reactivity into reactive antiaromatic porphyrinoids would enable efficient construction of novel porphyrinoids with fascinating structures and properties.

14.7 Synthesis of Octaphyrin from Bisdipyrrin

In the case of norcorrole formation, the C–C bond formation process selectively proceeds intramolecularly. However, the coupling reaction was reported to occur in an intermolecular fashion depending on the metal template or the reaction

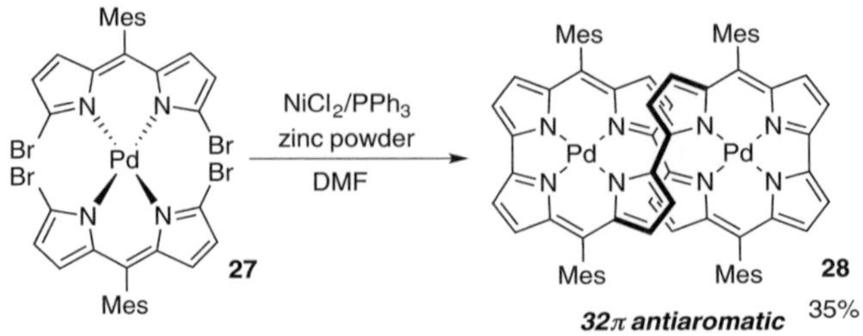

Scheme 14.12 Intermolecular homo-coupling of bisdipyrrin Pd(II) complex to octaphyrin Pd(II) complex

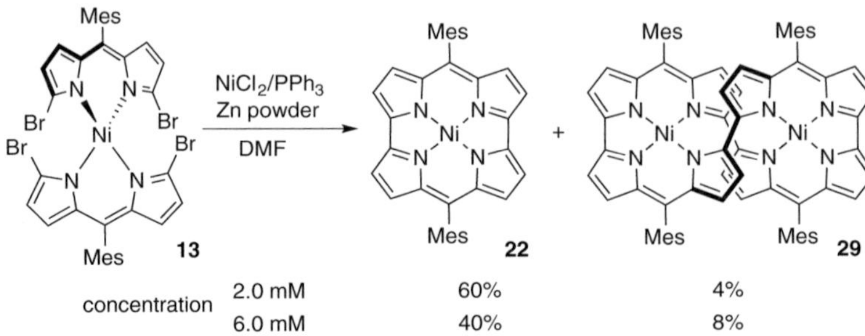

Scheme 14.13 Formation of norcorrole Ni(II) and octaphyrin bisNi(II) from bisdipyrrin Ni(II) complex

conditions. The use of Pd(II) ion as a central metal of the dipyrrin precursor **27** resulted in selective intermolecular homo-coupling reaction to provide [32]octaphyrin(1.0.1.0.1.0.1.0) bisPd(II) complex **28** exclusively (Scheme 14.12) [31]. In this case, the large ionic radius of Pd(II) seems not to fit to the small central cavity of norcorrole.

Under the concentrated reaction conditions, homo-coupling of Ni(II) bisdipyrrin **13** afforded [32]octaphyrin(1.0.1.0.1.0.1.0) bisNi(II) complex **29** as a by-product along with norcorrole Ni(II) complex **22** (Scheme 14.13) [31]. The yield of the octaphyrin complex **29** was dependent on the concentration. More concentrated conditions favored the formation of octaphyrin **29** through intermolecular dimerization reaction of the dipyrrin precursor **13**.

The different reactivity of Pd(II) and Ni(II) bisdipyrrin complex was accounted for by the structural difference in tetrapyrrole intermediates **30** resulting from the initial C–C bond formation (Scheme 14.14). DFT calculations suggested that the distance between two reacting carbon atoms in Ni(II) complex **31** is shorter than that in Pd(II) complex **32**. The intramolecular coupling to the norcorrole complex should be unfavorable in the case of **32**.

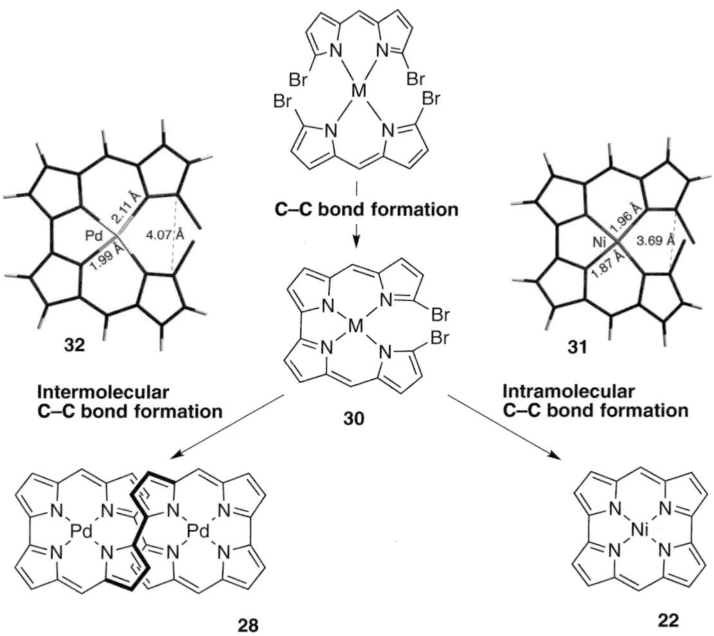

Scheme 14.14 Plausible reaction pathways for norcorrole Ni(II) and octaphyrin Pd(II)

14.8 Summary

Dipyrrin derivatives are nice precursors, which enable efficient synthesis of porphyrinoids such as heteroatom-incorporated porphyrins and antiaromatic porphyrinoids. Particularly in the case of norcorrole Ni(II) complex, the metal template strategy allows large-scale synthesis of a stable antiaromatic porphyrinoid. Scalability is one of the major issues to be solved in porphyrin synthesis when considering some practical applications [32]. In this regard, porphyrinoid synthesis on the basis of dipyrrin precursors is one of the candidates for the solution.

References

1. Kadish KM, Smith KM, Guilard R(eds) (2010) Handbook of porphyrin science, vol 1–35. World Scientific Publishing, Singapore
2. Wood TE, Thompson A (2007) Advances in the chemistry of dipyrrins and their complexes. Chem Rev 107(5):1831–1861. doi:10.1021/cr050052c
3. Loudet A, Burgess K (2007) BODIPY dyes and their derivatives: syntheses and spectroscopic properties. Chem Rev 107(11):4891–4932
4. Ulrich G, Ziessel R, Harriman A (2008) The chemistry of fluorescent bodipy dyes: versatility unsurpassed. Angew Chem Int Ed 47(7):1184–1201

5. Ni Y, Wu J (2014) Far-red and near infrared BODIPY dyes: synthesis and applications for fluorescent pH probes and bio-imaging. Org Biomol Chem 12(23):3774–3791. doi:10.1039/C3OB42554A
6. Lu H, Mack J, Yang Y, Shen Z (2014) Structural modification strategies for the rational design of red/NIR region BODIPYs. Chem Soc Rev 43(13):4778–4823. doi:10.1039/C4CS00030G
7. Sakida T, Yamaguchi S, Shinokubo H (2011) Metal-mediated synthesis of antiaromatic porphyrinoids from a BODIPY precursor. Angew Chem Int Ed 50(10):2280–2283. doi:10.1002/anie.201006314
8. Tahara K, Tobe Y (2006) Molecular loops and belts. Chem Rev 106(12):5274–5290. doi:10.1021/cr050556a
9. Zhang W, Moore JS (2006) Shape-persistent macrocycles: structures and synthetic approaches from arylene and ethynylene building blocks. Angew Chem Int Ed 45(27):4416–4439. doi:10.1002/anie.200503988
10. Spitler EL, Johnson CA, Haley MM (2006) Renaissance of annulene chemistry. Chem Rev 106(12):5344–5386. doi:10.1021/cr050541c
11. Wolfe JP, Wagaw S, Marcoux J-F, Buchwald SL (1998) Rational development of practical catalysts for aromatic carbon – nitrogen bond formation. Acc Chem Res 31(12):805–818. doi:10.1021/ar9600650
12. Hartwig JF (1998) Transition metal catalyzed synthesis of arylamines and aryl ethers from aryl halides and triflates: scope and mechanism. Angew Chem Int Ed 37(15):2046–2067
13. Hartwig JF (2008) Evolution of a fourth generation catalyst for the amination and thioetherification of aryl halides. Acc Chem Res 41(11):1534–1544. doi:10.1021/ar800098p
14. Surry DS, Buchwald SL (2008) Biaryl phosphane ligands in palladium-catalyzed amination. Angew Chem Int Ed 47(34):6338–6361. doi:10.1002/anie.200800497
15. Johnson AW, Kay IT, Rodrigo R (1963) 2,2′-bipyrrolic macrocyclic ring systems. J Chem Soc 2336–2342. doi:10.1039/JR9630002336
16. Horie M, Hayashi Y, Yamaguchi S, Shinokubo H (2012) Synthesis of nickel(II) azacorroles by Pd-catalyzed amination of α,α'-dichlorodipyrrin NiII complex and their properties. Chem Eur J 18(19):5919–5923. doi:10.1002/chem.201200485
17. Matano Y, Shibano T, Nakano H, Imahori H (2012) Nickel(II) and copper(II) complexes of beta-unsubstituted 5,15-diazaporphyrins and pyridazine-fused diazacorrinoids: metal-template syntheses and peripheral functionalizations. Chem Eur J 18(20):6208–6216. doi:10.1002/chem.201200463
18. Matano Y, Shibano T, Nakano H, Kimura Y, Imahori H (2012) Free base and metal complexes of 5,15-diaza-10,20-dimesitylporphyrins: synthesis, structures, optical and electrochemical properties, and aromaticities. Inorg Chem 51(23):12879–12890. doi:10.1021/ic301835c
19. Kamiya H, Kondo T, Sakida T, Yamaguchi S, Shinokubo H (2012) meso-thiaporphyrinoids revisited: missing of sulfur by small metals. Chem Eur J 18(50):16129–16135. doi:10.1002/chem.201203255
20. Sakow D, Böker B, Brandhorst K, Burghaus O, Bröring M (2013) 10-heterocorroles: ring-contracted porphyrinoids with fine-tuned aromatic and metal-binding properties. Angew Chem Int Ed 52(18):4912–4915. doi:10.1002/anie.201300757
21. Broadhurst MJ, Grigg R, Johnson AW (1970) Preparation of some sulphur-containing polypyrrolic macrocycles – sulphur extrusion from a meso-thiaphlorin. J Chem Soc D 13:807–809. doi:10.1039/c29700000807
22. Broadhurst MJ, Grigg R, Johnson AW (1972) Sulfur extrusion reactions applied to synthesis of corroles and related systems. J Chem Soc Perkin Trans 1(9–10):1124–1135. doi:10.1039/p19720001124
23. Wachi N, Kondo T, Ito S, Hiroto S, Shin J-Y, Shinokubo H (2014) Synthesis, reactivity, and property of 5,15-dithiaporphyrin copper(II) complex. J Porphyrins Phthalocyanines 18(8–9):675–678
24. Bröring M, Köhler S, Kleeberg C (2008) Norcorrole: observation of the smallest porphyrin variant with a N4 core. Angew Chem Int Ed 47(30):5658–5660. doi:10.1002/anie.200801196

25. Ito T, Hayashi Y, Shimizu S, Shin J-Y, Kobayashi N, Shinokubo H (2012) Gram-scale synthesis of Nickel(II) norcorrole: the smallest antiaromatic porphyrinoid. Angew Chem Int Ed 51(34):8542–8545. doi:10.1002/anie.201204395
26. Broring M, Brégier F, Cónsul Tejero E, Hell C, Holthausen MC (2007) Revisiting the electronic ground state of copper corroles. Angew Chem Int Ed 46(3):445–448. doi:10.1002/anie.200603676
27. Kira M, Ishida S, Iwamoto T, Kabuto C (1999) The first isolable dialkylsilylene. J Am Chem Soc 121(4–1):9722–9723. doi:10.1021/ja9925305
28. Kira M, Iwamoto T, Ishida S (2007) A helmeted dialkylsilylene. Bull Chem Soc Jpn 80(2):258–275. doi:10.1246/bcsj.80.258
29. Kira M (2010) An isolable dialkylsilylene and its derivatives. A step toward comprehension of heavy unsaturated bonds. Chem Commun 46(17):2893–2903. doi:10.1039/C002806A
30. Fukuoka T, Uchida K, Sung YM, Shin J-Y, Ishida S, Lim JM, Hiroto S, Furukawa K, Kim D, Iwamoto T, Shinokubo H (2014) Near-IR absorbing nickel(II) porphyrinoids prepared by regioselective insertion of silylenes into antiaromatic nickel(II) norcorrole. Angew Chem Int Ed 53(6):1506–1509. doi:10.1002/anie.201309921
31. Kido H, Shin J-Y, Shinokubo H (2013) Selective synthesis of a [32]octaphyrin(1.0.1.0.1.0.1.0) bis(palladium) complex by a metal-templated strategy. Angew Chem Int Ed 52(51):13727–13730. doi:10.1002/anie.201306905
32. Shin J-Y, Yamada T, Yoshikawa H, Awaga K, Shinokubo H (2014) An antiaromatic electrode-active material enabling high capacity and stable performance of rechargeable batteries. Angew Chem Int Ed 53(12):3096–3101. doi:10.1002/anie.201310374

Chapter 15
Möbius Aromatic and Antiaromatic Expanded Porphyrins

Takayuki Tanaka and Atsuhiro Osuka

Abstract Since our first report in 2008, *meso*-aryl-substituted expanded porphyrins have been recognized as a nice platform to realize Möbius aromatic and antiaromatic molecules. An increasing number of Möbius aromatic and antiaromatic expanded porphyrins have been produced from *meso*-aryl-substituted expanded porphyrins by metal complexation, fusion reaction, protonation, and deprotonation. [28]Hexaphyrin(1.1.1.1.1.1) and [32]heptaphyrin(1.1.1.1.1.1.1) have been shown to exist as a dynamic equilibrium between Hückel antiaromatic and Möbius aromatic conformers in solution, in which distribution of conformers is dependent upon temperature and solvent polarity. Regioselective peripheral functionalizations of Möbius aromatic expanded porphyrins have been also developed. Möbius aromatic and antiaromatic expanded porphyrins display optical and electronic properties that are analogous to those of usual aromatic and antiaromatic porphyrinoids, respectively.

Keywords Porphyrinoids • Expanded porphyrins • Möbius aromaticity • Hückel aromaticity

15.1 Möbius Aromaticity

The concept of Möbius aromaticity predicts aromatic nature for [4n]annulenes lying on a singly twisted Möbius strip, thus complementing the established Hückel aromaticity that predicts aromaticity for [4n +2]annulenes lying on a normal plain. This concept has been stimulating both theoretical and experimental researchers because of its simplicity [1–4], since this important concept was first proposed by Heilbronner in 1964 [5]. While there have been speculations on Möbius aromatic molecules as transition states, reactive intermediates, and conformational isomers [6], chemically stable and isolable Möbius aromatic molecules had not been synthesized for a long time. Difficulty in synthesis of Möbius aromatic molecules

T. Tanaka • A. Osuka (✉)
Department of Chemistry, Graduate School of Science, Kyoto University, Kitashirakawa Oiwake-cho, Sakyo-ku 606-8502, Kyoto, Japan
e-mail: osuka@kuchem.kyoto-u.ac.jp

T. Akasaka et al. (eds.), *Chemical Science of π-Electron Systems*,
DOI 10.1007/978-4-431-55357-1_15

has been stemming from two conflicting structural elements, a twisted Möbius topology and an overall cyclic conjugated [4n]π-electron system, which have to be implemented to realize Möbius aromaticity. The twisted nature of the Möbius topology necessitates the use of large annulenic systems to mitigate the distortion associated with molecular twists, in which substantial conformational flexibility makes fixation of a twisted conformation very difficult. Nearly 40 years have passed until we had a chance to meet the seminal work of Herges et al., who employed a sophisticated synthetic strategy by combining a usual planar conjugated system with a belt-like conjugated system to realize a twisted yet conjugated cyclic electronic system possessing moderate Möbius aromaticity [7].

Expanded porphyrins that consist of more than five pyrrolic subunits have emerged as a new class of heteroannulene variants that display a range of interesting electronic and conformational characteristics and versatile metal-coordinating abilities as multidentate ligands [8–11]. In addition, expanded porphyrins have been demonstrated to possess distinct advantages in the construction of Möbius-type aromatic systems such as (1) an overall conformational flexibility, (2) an ability to invert the constituent pyrrolic subunits, (3) the capacity to respond to two-electron oxidation and reduction through the facile release or capture of a pair of pyrrolic NH hydrogen atoms to balance the charge, and (4) the possibility of locking in twisted conformations through metalation or intramolecular fusion reaction. Using a similar but not fully conjugated macrocycle, di-*p*-benzi[28]hexaphyrin(1.1.1.1.1.1), Latos-Grażyński et al. reported temperature-dependent conformational switch between Hückel and Möbius topologies [12].

15.2 Metal Complexes

In 2008, we found spontaneous formations of Möbius aromatic molecules upon palladium metalation of expanded porphyrins (Fig. 15.1). Treatment of [26]hexaphyrin **1** with $Pd_2(dba)_3$ in the presence of sodium acetate gave [28]hexaphyrin palladium(II) complex **2**. Metalation of [32]heptaphyrin **3** and [36]octaphyrin **5** with $Pd(OAc)_2$ produced [32]heptaphyrin **4** and [36]octaphyrin **6**, respectively. The singly twisted Möbius structures of complexes **2**, **4**, and **6** were unambiguously confirmed by X-ray diffraction analyses. In complexes **2** and **4**, the palladium(II) ion is bound to the three pyrrolic nitrogen atoms and one carbon atom, and in complex **6**, the coordination of one palladium(II) ion is similar to those of **2** and **4**, and the other palladium(II) ions is bound to the two pyrrolic nitrogen atoms and the two carbon atoms. Importantly, the ^{1}H NMR spectra of **2**, **4**, and **6** exhibit the outer pyrrolic β-CH signals in the deshielded region and the inner pyrrolic signals in the shielded region. Chemical shift differences ($\Delta\delta$ values) between the most shielded proton and the most deshielded proton are calculated to be 7.05 ppm for **2**, 9.97 ppm for **4** and 11.12 ppm for **6**, respectively. These complexes display characteristic absorption features such as a sharp Soret-like band and distinct Q-band-like bands, which are characteristic of aromatic porphyrinoids. On the basis of these observations,

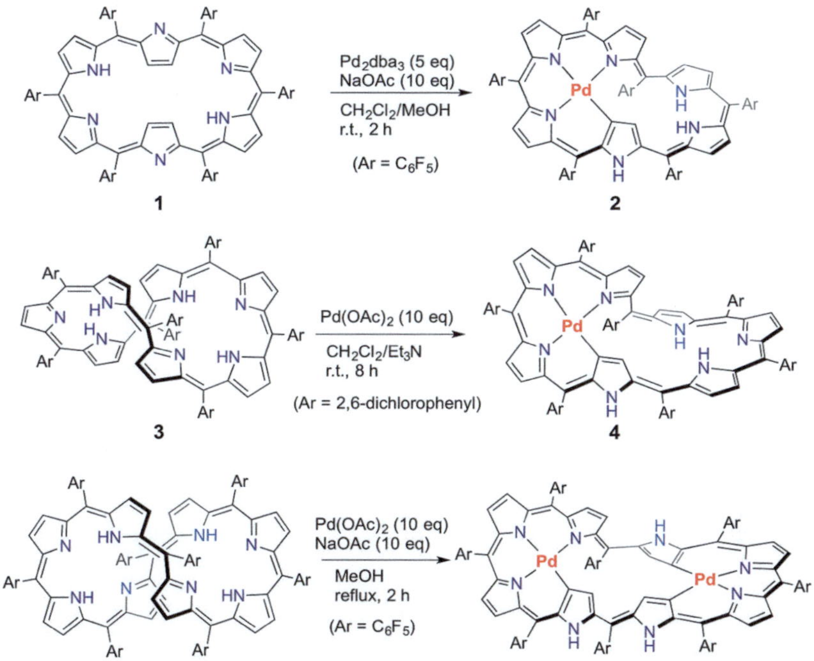

Fig. 15.1 Synthesis of Möbius aromatic expanded porphyrins by palladium(II) insertion

the complexes **2**, **4**, and **6** have been assigned as Möbius aromatic molecules [13]. These Möbius aromatic expanded porphyrins have been shown to display large two-photon absorption cross section as compared with nonaromatic and antiaromatic counterparts [14]. Palladium coordination enhances the molecular rigidity of the complexes, hence allowing the observation of the large $\Delta\delta$ values. Here it is to be noted that these complexes are the first Möbius aromatic molecules with distinct aromaticity.

Other group 10 metal ions such as nickel(II) and platinum(II) ions also served to form Möbius aromatic molecules such as [28]hexaphyrin nickel(II) complex **8** and platinum(II) complex **9** from [28]hexaphyrin **7** (Fig. 15.2). In complexes **8** and **9**, the nickel(II) and platinum(II) ions are bound to the three pyrrolic nitrogen atoms and one carbon atom, similar to complex **2** [15].

Formation of [26]hexaphyrin(2.1.1.0.1.1) **10**, namely, [26]isohexaphyrin, was found during boron(III) insertion into **7** and subsequent oxidative demetalation. This skeletal transposition rearrangement was unprecedented [16]. Isohexaphyrin **10** was reduced to its 28π congener and was then treated with $Pd(OAc)_2$ in the presence of triethylamine to give Möbius aromatic [28]hexaphyrin(2.1.1.0.1.1) palladium(II) complex **11** (Fig. 15.3). This complex was the first example of Möbius aromatic species formed from expanded porphyrins other than the regular *meso*-aryl expanded porphyrins possessing the regular alternate arrangement of pyrroles and methine carbons [17].

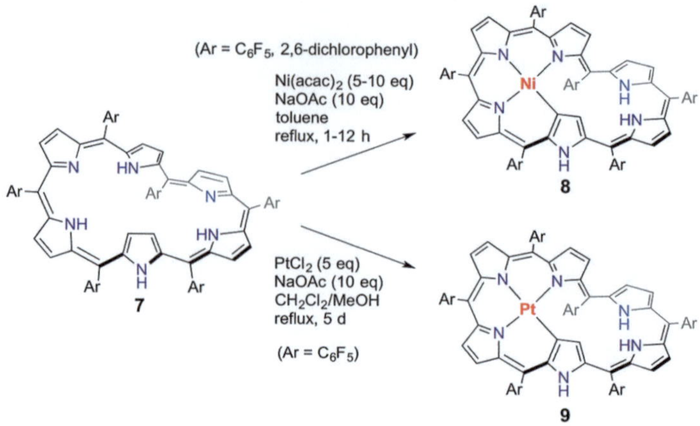

Fig. 15.2 Synthesis of [28]hexaphyrin nickel(II) and platinum(II) complexes

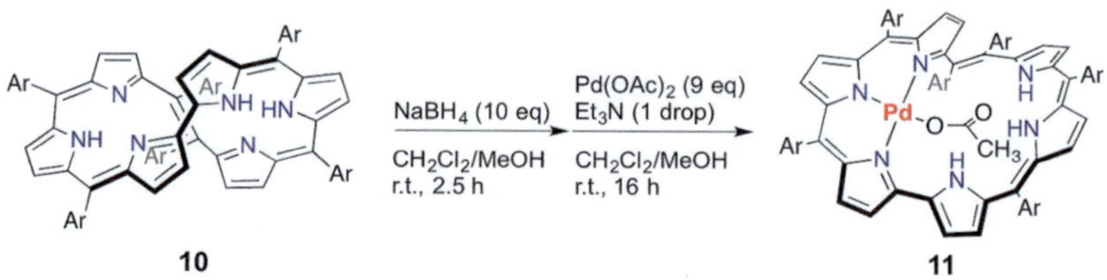

Fig. 15.3 Metalation of [26]hexaphyrin(2.1.1.0.1.1)

N-Fused [22]pentaphyrin **12** is a stable form of *meso*-aryl-substituted pentaphyrins that exhibits relatively weak Hückel aromaticity due to the overall structural distortion. Metalation of **12** with rhodium(I) ion produced two species that exhibit enhanced aromaticity as judged from their ^{1}H NMR chemical shifts and reduced bond length alternation in the solid-state structure (Fig. 15.4) [18]. These structures were unambiguously determined from X-ray diffraction analyses to be *N*-fused [22]pentaphyrin rhodium(I) complex **13** and *N*-fused [24]pentaphyrin rhodium(I) complex **14**, which are Hückel and Möbius aromatic molecule, respectively. The latter represents the smallest expanded porphyrin with a distinct Möbius aromaticity to date.

15.3 Fusion Reactions

Möbius aromatic molecules that are conformationally stable in solution without metal coordination have been highly desired to elucidate their intrinsic properties both in the ground state and the excited state without influence of a transition metal ion. In the meanwhile, we found the facile and spontaneous formation of

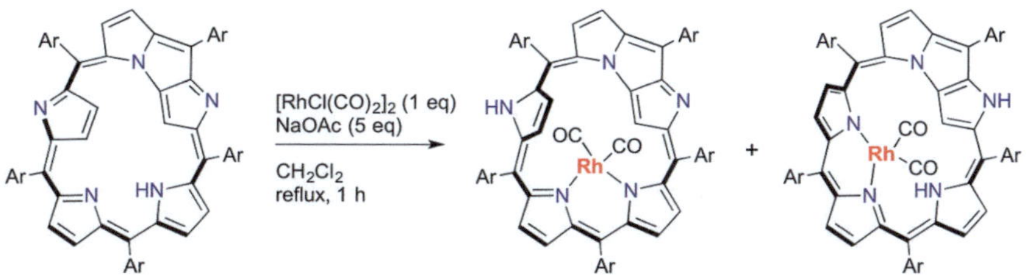

Fig. 15.4 Synthesis of Möbius aromatic *N*-fused [22]- and [24]pentaphyrin rhodium(I) complexes

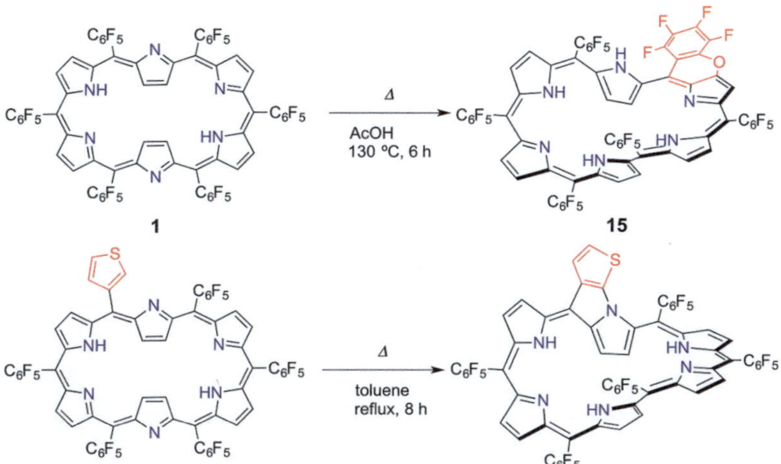

Fig. 15.5 Fusion reactions of hexaphyrins

benzopyrane-fused [28]hexaphyrin(1.1.1.1.1.1) **15** by simple heating of **1** in acetic acid (Fig. 15.5) [19]. Hexaphyrin **15** displays a distinct diamagnetic ring current effect with $\Delta\delta = 6.41$ ppm at room temperature, which was largely retained over a wide temperature range from −100 to 100 °C. Compared with the metal complexes, the conjugation is relatively smooth, with a largest dihedral angle of 36° in the solid state. In a different case, heating of 3-thienyl-substituted [26]hexaphyrin **16** in toluene resulted in quantitative formation of thienyl-fused [28]hexaphyrin **17** that is also a Möbius aromatic molecule [20]. The fused structures lead to a decreased conformational flexibility, thus allowing fixation of the conformationally stable Möbius aromatic molecules.

A singly *N*-fused [28]hexaphyrin **18** has been prepared with an expectation to possess similar Möbius aromatic character, but has been revealed to show rather complicated features reflecting conformational changes in the various Hückel and Möbius forms [21]. Palladium(II) metalation of **18** produced a single regioisomer of palladium(II) complex **19** with clear Möbius aromatic characters (Fig. 15.6).

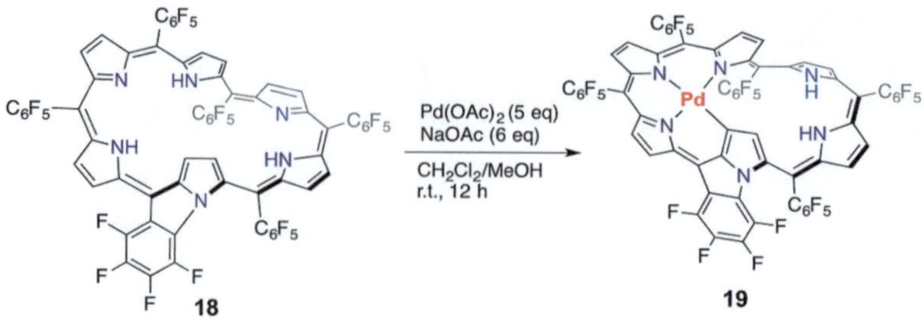

Fig. 15.6 Palladium(II) insertion to *N*-fused [28]hexaphyrin **18**

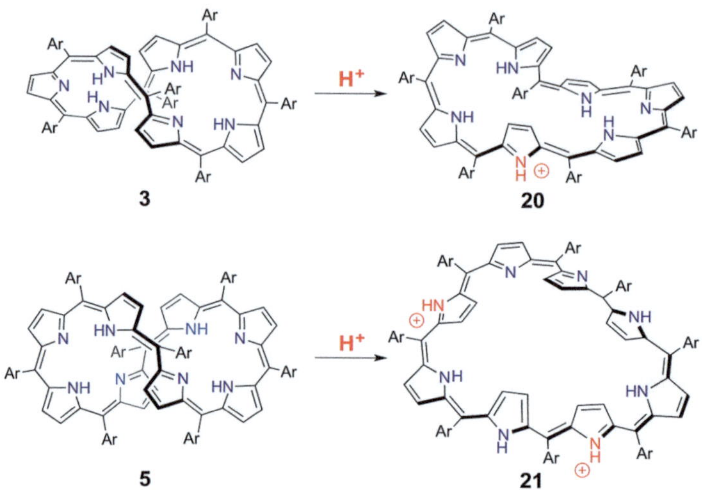

Fig. 15.7 Protonation of [32]heptaphyrin **3** and [36]octaphyrin **5**

15.4 Protonation and Deprotonation

As a means to switch [4n]π expanded porphyrins from antiaromatic Hückel species to Möbius aromatic species, protonation has been shown to be very effective. Protonation of [32]heptaphyrins induced a large conformational change, allowing for formation of twisted Möbius aromatic species **20** (Fig. 15.7) [22]. The ^{1}H NMR spectrum of **20** exhibits an evident diatropic ring current effect. The Möbius aromatic character was also supported by UV/Vis absorption spectroscopic studies (Fig. 15.8). The absorption spectrum of the neutral [32]heptaphyrin **3** is characterized by broad and ill-defined bands, reflecting its figure-of-eight conformation. On the other hand, the absorption spectrum of **20** is well resolved, featuring a Soret-like band at 653 nm with a shoulder at 707 nm as characteristic attributes of aromatic porphyrinoids [14]. A similar behavior was observed upon protonation of [36]octaphyrin **5** [23]. While diprotonated [38]octaphyrin displayed Hückel

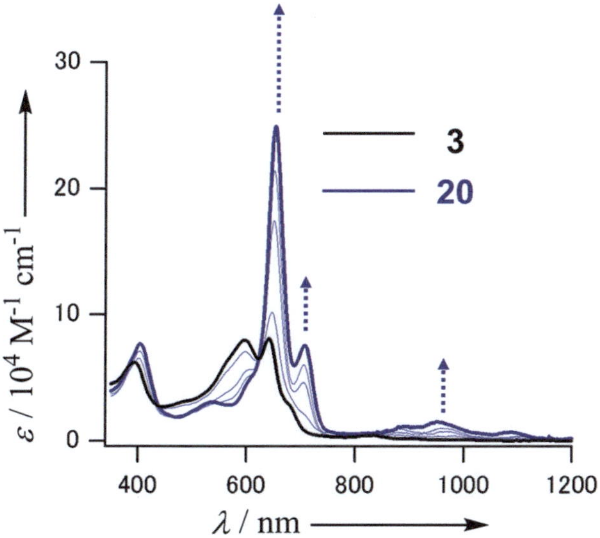

Fig. 15.8 UV/Vis absorption spectral change of [32]heptaphyrin **3** upon protonation

aromatic character, diprotonated [36]octaphyrin **21** exhibited distinct Möbius aromaticity only with a subtle difference in tilted pyrrole angles. So, this feature strongly suggests the possibility to control the macrocyclic topologies by pyrrole rotation to gain [4n]π Möbius or [4n + 2]π Hückel aromatic stabilization, depending on the number of π-electrons.

Protonation chemistry of [28]hexaphyrin **7** is more complicated, but has been finally elucidated. Addition of trifluoroacetic acid (TFA) to **7** caused a red shift of the Soret-like band from 591 nm to 621 nm with clear intensification and red shifts of the Q-like bands to 847 and 945 nm, indicating formation of Möbius aromatic [28]hexaphyrin mono-protonated species **22** predominantly (Fig. 15.9). On the other hand, treatment with methanesulfonic acid (MSA), a stronger acid than TFA, induced diprotonation, which caused a structural change to diprotonated triangular form **23**. Diprotonated [28]hexaphyrin **23** displayed strong Hückel antiaromatic character as indicated by a blue shift of the Soret-like band to 574 nm and a very broad absorption tail at up to 1200 nm (Fig. 15.10) [24]. Then, we designed 2,3,12,13,22,23-hexaphenyl [28]hexaphyrin **24**, which has an alternate arrangement of unsubstituted pyrroles and 3,4-diphenylpyrroles and may favor a triangular shape because of steric repulsion of the introduced phenyl groups. Protonation of **24** with MSA led to smooth formation of triangular antiaromatic hexaphyrin **25**. Interestingly, the amount of MSA needed for complete formation of **25** is much smaller (ca. 1000 eq.) as compared to that needed for complete formation of **25** (ca. 20000 eq.). In other case, 2,3,17,18-tetraphenyl-[28]hexaphyrin **26**, whose structure was identified as a figure-of-eight form in its freebase form, also showed a structural change to Möbius twisted form **27** upon protonation by TFA (Fig. 15.11) [25].

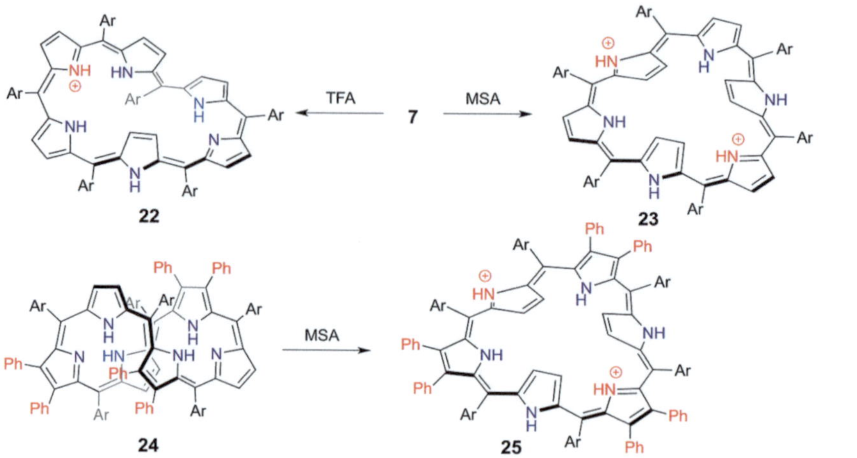

Fig. 15.9 Protonation of [28]hexaphyrin **7** and 2,3,12,13,22,23-hexaphenyl [28]hexaphyrin **24**

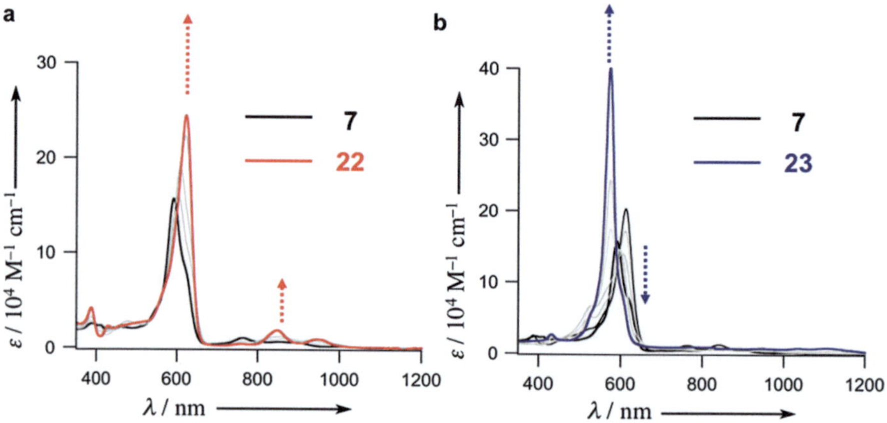

Fig. 15.10 UV/Vis absorption spectral change of [28]hexaphyrin **7** upon protonation with (**a**) TFA and (**b**) MSA

Although one of two isomers of **27** was confirmed by X-ray diffraction analysis, palladium(II) metalation furnished two structural isomers **28** and **29**, both of which were characterized by X-ray diffraction analysis [26].

Deprotonation has been also proven to be an effective means to induce twisted conformations with enhanced aromaticity. Upon treatment with tetrabutylammonium fluoride, [32]heptaphyrin **3** underwent a conformational twisting to form Möbius aromatic species as indicated by the appearance of a sharp Soret-like band and distinct Q-like bands and enhancement of two-photon absorption cross-section values [27]. It is conceivable that the removal of pyrrolic NH protons causes

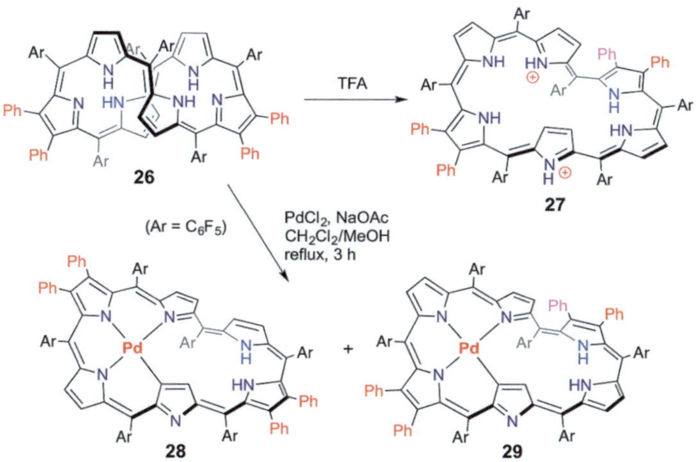

Fig. 15.11 Protonation and palladium(II) metalation of 2,3,17,18-tetraphenyl-substituted [28]hexaphyrin **26**

a disruption of intramolecular hydrogen bonding, hence driving a conformation change from a figure of eight to a twisted extended Möbius conformation.

15.5 Control of Equilibrium by Temperature and Solvent

^{1}H NMR spectroscopic behaviors of [28]hexaphyrin **7** have been long puzzling [28]. At room temperature, the ^{1}H NMR spectrum of **7** indicates a formal rectangular-shape symmetry (indicated as **7-Hückel**) with a moderate diatropic ring current in spite of its [4n]π electronic system. When [28]hexaphyrin takes a conformationally rigid rectangular shape, it should be strongly Hückel antiaromatic as demonstrated by corresponding gold(III) complexes [29]. To explain this contradictory situation, we considered the possibility of an equilibrium between Möbius aromatic conformer (**7-Möbius**) and planar Hückel antiaromatic conformer (**7-Hückel**) in freebase [28]hexaphyrins (Fig. 15.12). When the conformational dynamics are faster than ^{1}H NMR time scale, the observed ^{1}H NMR spectrum records averaged signals. Actually, the low-temperature NMR analyses revealed predominant population of Möbius conformation, which is slightly energetically lower (3.7 kcal/mol) than **7-Hückel**. In addition, several *meso*-aryl-substituted [28]hexaphyrins(1.1.1.1.1.1) were examined carefully by using NMR and UV/Vis absorption spectroscopy. Finally, the structures of both **7-Hückel** and **7-Möbius** have been indicated by X-ray diffraction analyses of different crystals obtained from *meso*-2,6-difluorophenyl-substituted [28]hexaphyrin. These crystals were obtained, just depending upon crystallization solvent.

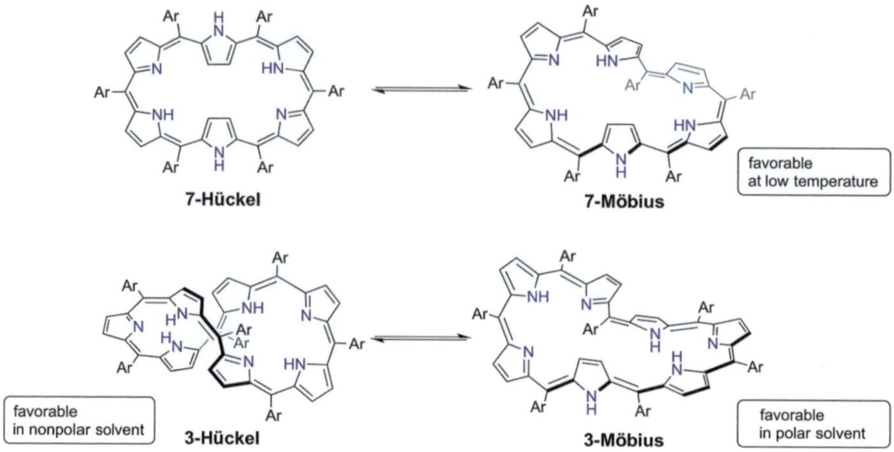

Fig. 15.12 Equilibrium between Hückel and Möbius conformers

In a different case, *meso*-pentafluorophenyl-substituted heptaphyrin **3** showed solvent-dependent conformational change between Hückel antiaromatic figure-of-eight conformer (**3-Hückel**) and Möbius aromatic conformer (**3-Möbius**) [30]. In nonpolar toluene, **3-Hückel** is dominant as judged by observed broad and weak absorption bands and moderate paratropic ring current in the ^{1}H NMR spectrum, while **3-Möbius** is dominant in polar solvent as revealed by a well-resolved sharp absorption spectrum, intense fluorescence spectrum, and a diatropic ring current.

15.6 Möbius Antiaromaticity

Möbius antiaromatic molecules should be energetically unfavorable both from electronic and structural viewpoints and thus are more difficult to prepare [31]. Oxidations of [4n]π Möbius aromatic molecules such as **2** and **15** were attempted with a hope to obtain corresponding Möbius antiaromatic molecules. However, these oxidations gave rise to formations of non-twisted [4n + 2]π Hückel aromatic species with concomitant topological changes.

In the course of phosphorus insertion into porphyrinoids, which is a useful means to realize isophlorin-type fully reduced annulenic π conjugation [32], we found that monophosphorus insertion into **7** afforded [28]hexaphyrin monophosphorus(V) complex **30** and successive phosphorus insertion furnished [30]hexaphyrin bisphosphorus(V) complex **31** (Fig. 15.13) [33]. Single-crystal X-ray diffraction analysis revealed that both **30** and **31** take similarly twisted structures in which P = O moieties were bound to the centers of NNN cavity or NNC cavity. ^{1}H NMR spectra of **30** and **31** exhibit a moderate diatropic ring current and a paratropic ring current, by showing the inner β-protons at 2.99 and 11.14 ppm, respectively. This

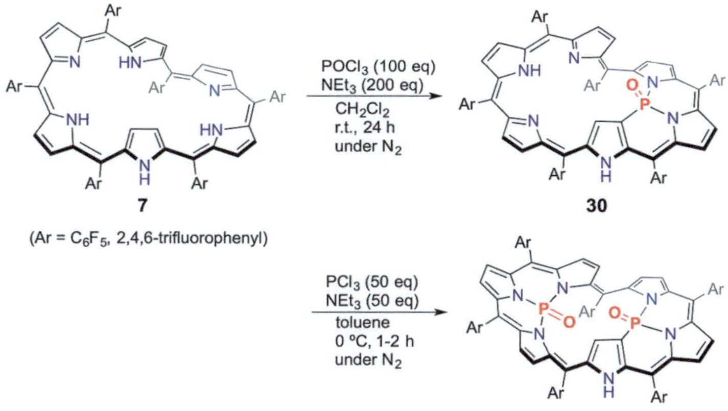

Fig. 15.13 Synthesis of [28]hexaphyrin monophosphorus(V) and [30]hexaphyrin bisphosphorus(V) complexes

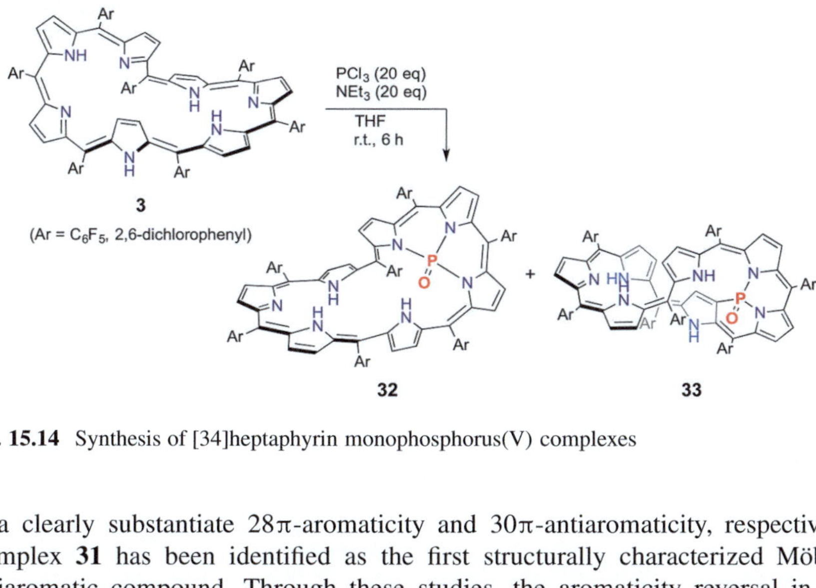

Fig. 15.14 Synthesis of [34]heptaphyrin monophosphorus(V) complexes

data clearly substantiate 28π-aromaticity and 30π-antiaromaticity, respectively. Complex **31** has been identified as the first structurally characterized Möbius antiaromatic compound. Through these studies, the aromaticity reversal in the Hückel rule upon changing the number of π-electrons between [4n + 2] and [4n] has been validated also for cyclic conjugated molecules that have a twisted Möbius topology. A highly reduced [30]hexaphyrin system is well stabilized owing to the presence of two electron-withdrawing phosphamide moieties, which also confer a certain structural rigidity to the molecule.

Phosphorus insertion to [32]heptaphyrin **3** led to formation of a Möbius antiaromatic phosphorus(V) complex of [34]heptaphyrin **32** as a kinetically controlled product (Fig. 15.14). This molecule underwent a thermal rearrangement to a more stable Hückel aromatic complex **33** quantitatively [34].

15.7 Peripheral Functionalizations

As mentioned above, oxidation of Möbius aromatic species to Möbius antiaromatic species has been unsuccessful. However, we found an interesting rearrangement reaction when [28]hexaphyrin palladium(II) complex **2** was oxidized with tris(bromophenyl)aminium hexachloroantimonate (TBAH) in acetonitrile (Fig. 15.15) [35]. This oxidation gave 26π Hückel aromatic rectangular palladium(II) complex **34** in good yield. Interestingly, NNNC palladium(II) coordination in **2** is changed to NNCC palladium(II) coordination in **34** with a molecular topology change from a twisted Möbius conformation to a planar Hückel conformation. Complex **34** was in turn reduced with $NaCNBH_3$ to regenerate **2**. This reduction can be considered to proceed with a regioselective hydride attack at the β-carbon of the central outer-pointing pyrrole. This consideration drove us to extend this reaction to regioselective halogenations. Actually, treatment of **34** with aqueous HCl in acetonitrile proceeded smoothly at room temperature to give chlorinated [28]hexaphyrin palladium(II) complex **35**, quantitatively [36]. The same reaction with KBr in the presence of *p*-toluenesulfonic acid gave β-brominated product **36**. With the halogenated Möbius aromatic molecules **35** and **36** in hand, we examined

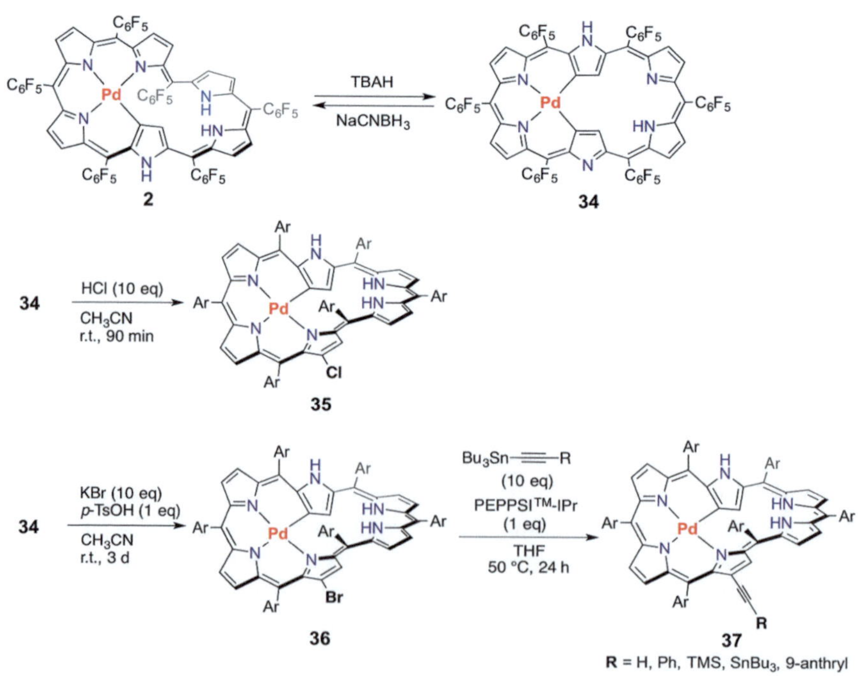

Fig. 15.15 Redox interconversion of hexaphyrin palladium(II) complexes and β-substitution reactions

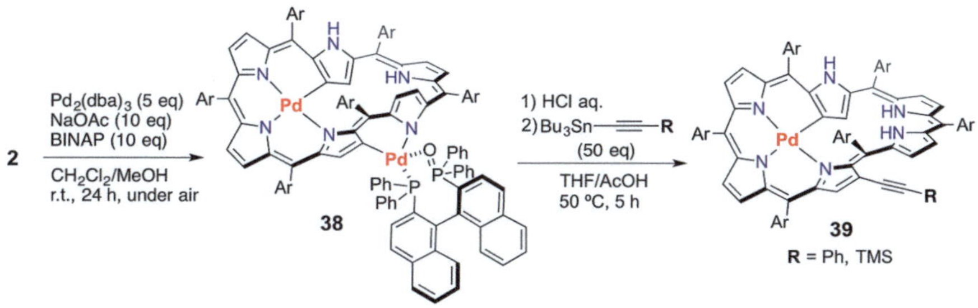

Fig. 15.16 Regioselective β-palladation of [28]hexaphyrin palladium(II) complex **2** and Stille coupling

further peripheral fabrications. While **35** was rather unreactive, **36** was found to be reactive under Stille coupling conditions to give various ethynylated products **37**.

A regioselective substitution reaction of **2** has been discovered by serendipity [37]. In the course of an enantioselective metalation study to form **2** in the presence of chiral BINAP ligand, we noticed formation of bis-palladium(II) complex **38** when the reaction was run under aerobic conditions (Fig. 15.16). The structure of **38** has been revealed by X-ray diffraction analysis, which indicated the preservation of a twisted structure of the [28]hexaphyrin fragment, to which the second palladium atom was bound at the pyrrolic β-position and nitrogen atom. The rigorous regioselectivity may be ascribed to least steric hindrance at this position and to the pyrrole N–palladium(II) coordination that serves to direct the following C–Pd bond formation. After a chloride substitution on palladium(II) in **38**, Stille coupling reaction was performed, which gave β-ethynylated products **39**. Thus, complementary substitution method at the 2- or 3-positions of Möbius aromatic [28]hexaphyrin has been developed. These regioselective fabrications are useful for further elaborations of Möbius aromatic expanded porphyrins.

Möbius aromatic [28]hexaphyrin phosphonium adducts **40** and **41** were obtained by additions of triphenylphosphine and tricyclohexylphosphine, respectively, to the peripheral β-position of [26]hexaphyrin **1** (Fig. 15.17) [38]. On the other hand, regioselective nucleophilic additions to [26]hexaphyrin palladium(II) complex **34** afforded other phosphonium adducts, [28]hexaphyrin palladium(II) complexes **42** and **43**. Possible resonance between the phosphonium ylide form and the phosphorane form as shown in Fig. 15.17 is an important feature in understanding their electronic network. The phosphonium ylide resonance structure allows one to delineate a cyclic 28π-conjugated Möbius aromatic electronic circuit, while the phosphorane structure is unfavorable for the overall macrocyclic conjugation due to its cross-conjugated electronic network. Interestingly, the phosphorane form contribution is more important in **40** and **41**, while the phosphonium ylide form contribution and hence the Möbius aromatic character is more important in **42** and **43**.

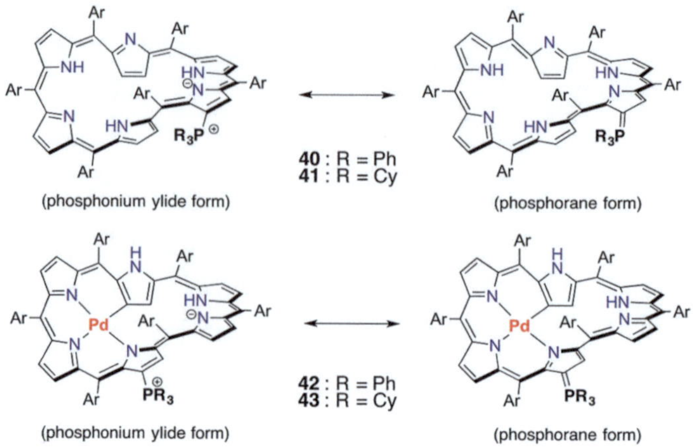

Fig. 15.17 [28]Hexaphyrin β-phosphorus adducts

15.8 Summary and Outlook

Expanded porphyrins have inherent advantages to realize Möbius aromaticity, such as (1) overall conformational flexibility that enables annulenic twisting without generating severe strain, (2) the capacity to respond to two-electron reduction and oxidation through the facile release or capture of a pair of pyrrolic NH hydrogen atoms, and (3) the possibility of locking in twisted conformations through metalation or intramolecular fusion reaction. By virtue of these advantages, many Möbius aromatic and antiaromatic expanded porphyrins have been explored during the past 5 years. Macrocycles with [4n]π electronic system can choose an appropriate conformation, Hückel or Möbius topology. The priority may be altered by peripheral fusion, protonation, and deprotonation or even by solvent or temperature control. Metal complexation should be the strongest method to fix Möbius twisted structure, so that it confers enough rigidity even to achieve Möbius antiaromatic species without causing conformational change to Hückel aromatic species. Despite such progress, Möbius aromatic expanded porphyrins have been limited to pentaphyrin, hexaphyrin, heptaphyrin, and octaphyrin, at the time of writing *this chapter*. Larger macrocyclic expanded porphyrins would be possible to take Möbius twisting more smoothly that may lead to the discovery of other stimuli to cause molecular twisting. Triply twisted [4n]π Möbius aromaticity is one of the most attractive targets to attack. Finally, effective peripheral fabrications of Möbius aromatic molecules would be beneficial to further elaboration and applications of these novel π-conjugated molecules in combination with their intrinsic chirality [39]. Such study is actively ongoing in our laboratory.

References

1. Rzepa HS (2005) Möbius aromaticity and delocalization. Chem Rev 105:3697–3715
2. Herges R (2006) Topology in chemistry: designing Möbius molecules. Chem Rev 106:4820–4842
3. Jux N (2008) The porphyrin twist: Hückel and Möbius aromaticity. Angew Chem Int Ed 47:2543–2546
4. Yoon ZS et al (2009) Möbius aromaticity and antiaromaticity in expanded porphyrins. Nat Chem 1:113–122
5. Heilbronner E (1964) Hückel molecular orbitals of Möbius-type conformations of annulenes. Tetrahedron Lett 5:1923–1928
6. Zimmerman HE (1966) On molecular orbital correlation diagrams, the occurrence of Möbius systems in cyclization reactions, and factors controlling ground- and excited-states reactions. I. J Am Chem Soc 88:1564–1565
7. Herges R (2007) Aromatics with a twist. Nature 450:36–37
8. Stępień M et al (2011) Figure eights, Möbius bands, and more: Conformation and aromaticity of porphyrinoids. Angew Chem Int Ed 50:4288–4340
9. Saito S, Osuka A (2011) Expanded porphyrins: Intriguing structures, electronic properties, and reactivities. Angew Chem Int Ed 50:4342–4373
10. Osuka A, Saito S (2011) Expanded porphyrins and aromaticity. Chem Commun 47:4330–4339
11. Stępień M et al (2007) Expanded porphyrin with a split personality: a Hückel–Möbius aromaticity switch. Angew Chem Int Ed 46:7869–7873
12. Shimzu S, Osuka A (2006) Metalation chemistry of *meso*-aryl-substituted expanded porphyrins. Eur J Inorg Chem 2006(7):1319–1335
13. Tanaka Y et al (2008) Metalation of expanded porphyrins: a chemical trigger used to produce molecular twisting and Möbius aromaticity. Angew Chem Int Ed 47:681–684
14. Shin J-Y et al (2010) Aromaticity and photophysical properties of various topology-controlled expanded porphyrins. Chem Soc Rev 39:2751–2767
15. Mori S et al (2005) Group 10 metal complexes of *meso*-aryl-substituted [26]hexaphyrins with a metal-carbon bond. Inorg Chem 44:4127–4129
16. Moriya K (2010) Boron(III) Induced skeletal rearrangement of hexaphyrin(1.1.1.1.1.1) to hexaphyrin(2.1.1.0.1.1). Angew Chem Int Ed 49:4297–4300
17. Moriya K et al (2011) A Möbius aromatic Pd(II) complex of [28]hexaphyrin(2.1.1.0.1.1). Chem Lett 40:455–457
18. Park JK et al (2008) Möbius aromaticity in *N*-fused [24]pentaphyrin upon Rh(I) metalation. J Am Chem Soc 130:1824–1825
19. Tokuji S et al (2009) Facile formation of a benzopyrane-fused [28]hexaphyrin that exhibits distinct Möbius aromaticity. J Am Chem Soc 131:7240–7241
20. Inoue M et al (2009) Thermal fusion reactions of *meso*-(3-thienyl) groups in [26]hexaphyrins to produce Möbius aromatic molecules. Angew Chem Int Ed 48:6687–6690
21. Higashino T et al (2010) Singly *N*-fused Möbius aromatic [28]hexaphyrins(1.1.1.1.1.1). J Org Chem 75:7958–7961
22. Saito S et al (2008) Protonation-triggered conformational changes to Möbius aromatic [32]heptaphyrins(1.1.1.1.1.1.1). Angew Chem Int Ed 47:9657–9660
23. Lim JM et al (2010) Protonated [4n] and [4n + 2] octaphyrins choose their Möbius/Hückel aromatic topology. J Am Chem Soc 132:3105–3114
24. Ishida S et al (2014) Diprotonated [28]hexaphyrins(1.1.1.1.1.1): triangular antiaromatic macrocycles. Angew Chem Int Ed 53:3427–3431
25. Koide T et al (2009) Multiple conformational changes of β-tetraphenyl *meso*-hexakis(pentafluorophenyl) substituted [26] and [28]hexaphyrins(1.1.1.1.1.1). Chem Commun (40):6047–6049
26. Koide T, Osuka A (2010) Möbius aromatic palladium(II) complexes of a β–tetraphenyl *meso*-hexakis(pentafluorophenyl) substituted hexaphyrin(1.1.1.1.1.1). Bull Chem Soc Jpn 83:877–879

27. Cha W-Y et al (2014) Deprotonation induced formation of Möbius aromatic [32]heptaphyrins. Chem Commun 50:548–550
28. Sankar J et al (2008) Unambiguous identification of Möbius aromaticity for *meso*-aryl-substituted [28]hexaphyrins(1.1.1.1.1.1). J Am Chem Soc 130:13568–13579
29. Mori S, Osuka A (2005) Aromatic and antiaromatic gold(III) hexaphyrins with multiple gold-carbon bonds. J Am Chem Soc 127:8030–8031
30. Yoon M-C (2011) Solvent-dependent aromatic versus antiaromatic conformational switching in *meso*-(heptakis)pentafluorophenyl [32]heptaphyrin. Chem Eur J 17:6707–6715
31. Pacholska-Dudziak E (2008) Palladium vacataporphyrin reveals conformational rearrangements involving Hückel and Möbius macrocyclic topologies. J Am Chem Soc 130:6182–6195
32. Miura T et al (2010) Phosphorus complexes of the first expanded isophlorins. Chem Eur J 16:55–59
33. Higashino T (2010) Möbius antiaromatic bisphosphorus complexes of [30]hexaphyrins. Angew Chem Int Ed 49:4950–4954
34. Higashino T (2012) A Möbius antiaromatic complex as a kinetically controlled product in phosphorus insertion to a [32]heptaphyrin. Angew Chem Int Ed 51:13105–13108
35. Inoue M, Osuka A (2010) Redox-induced palladium migrations that allow reversible topological changes between palladium(II) complexes of Möbius aromatic [28]hexaphyrin and Hückel aromatic [26]hexaphyrin. Angew Chem Int Ed 49:9488–9491
36. Yoneda T et al (2013) Regioselective fabrications of a Möbius aromatic [28]hexaphyrin palladium(II) complex. J Porphyrins Phthalocyanines 16:665–672
37. Tanaka T, Osuka A (2012) Regioselective palladation of a Möbius aromatic [28]hexaphyrin(1.1.1.1.1.1) Pd^{II} complex. Chem Eur J 18:7036–7040
38. Inoue M et al (2011) Möbius aromatic [28]hexaphyrin phosphonium adducts. Chem Eur J 17:9028–9031
39. Tanaka T et al (2010) Metal complexes of chiral Möbius aromatic [28]hexaphyrin(1.1.1.1.1.1): enantiomeric separation, absolute stereochemistry, and asymmetric synthesis. Angew Chem Int Ed 49:6619–6622

Chapter 16
Recent Advances in the Chemistry of Phthalocyanines as Functional Chromophores

Soji Shimizu and Nagao Kobayashi

Abstract Phthalocyanine (Pc), a conventional dyestuff exhibiting vivid blue or green color, has been utilized as a functional molecule for application in a variety of fields due to its excellent optical and electrochemical properties. Creation of novel π-conjugated systems based on the structure of Pc with the aim of further developing unique properties has, therefore, been of a prime importance. Despite the over a hundred-year history of phthalocyanine chemistry, progress in this regard has been fairly limited mainly due to the problems inherent in the synthesis of Pc. The authors have pioneered this research area by focusing on the control of the optical properties in the visible and near-infrared regions by (1) creation of the novel π-conjugated systems and (2) introduction of perturbation that is as simple as possible to the electronic structures and have reported a variety of unprecedented Pc analogues. Considerable attention has also been focused on the structure-optical property relationships of these novel Pc compounds. This chapter summarizes the contribution of the authors to the recent advances in the chemistry of phthalocyanine as functional chromophore systems.

Keywords Phthalocyanine • Near-infrared absorption • Magnetic circular dichroism (MCD) • Functional chromophore • vis/NIR spectroscopy

16.1 Introduction

Phthalocyanine (Pc, Fig. 16.1) is a classical dye and pigment exhibiting blue or green color, which has been utilized in various fields of industrial and medical applications, such as catalysts, deodorants, optical discs, photodynamic therapy, semiconductors, solar cells, nonlinear optics, and so forth [1]. The history of

S. Shimizu (✉)
Department of Chemistry and Biochemistry, Graduate School of Engineering, Kyushu University, Fukuoka 819-0395, Japan
e-mail: ssoji@cstf.kyushu-u.ac.jp

N. Kobayashi (✉)
Faculty of Textile Science and Technology, Shinshu University, Ueda 386-8567, Japan
e-mail: nagaok@shinshu-u.ac.jp

T. Akasaka et al. (eds.), *Chemical Science of π-Electron Systems*,
DOI 10.1007/978-4-431-55357-1_16

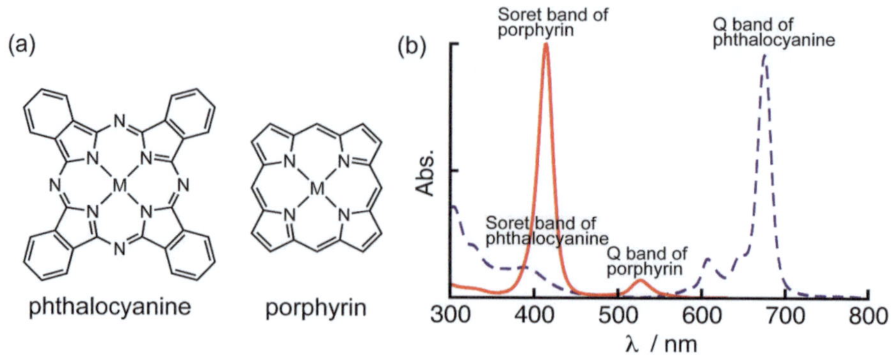

Fig. 16.1 (**a**) Porphyrin and phthalocyanine. (**b**) UV/vis absorption spectra of their metal complexes (*solid line*, nickel complex of *meso*-tetraphenyl-substituted porphyrin; *dashed line*, nickel complex of tetra-*t*-butyl-substituted phthalocyanine)

Pc chemistry can arguably be traced back to 1907 when Braun and Tcherniac observed a blue residue by chance when they heated *o*-phthalonitrile in an iron reactor [2]. The chemical structure of Pc was later determined in the 1930s during the pioneering research by Linstead [3]. Shortly thereafter, Robertson confirmed the structure of Pc by X-ray crystallography [4]. Pc adopts a macrocyclic 18π-electron conjugated system consisting of four isoindole units connected through aza-nitrogen bridges. This aza-annulene structure of Pc plays a crucial role in determining its unique optical and electrochemical properties. An intense absorption band at around 650 nm, which is referred to as the Q band, originating from the 18π-electron aromatic conjugated system and resulting in the characteristic blue or green color of Pc, is one of the main reasons why Pc has been central on functional chromophore molecules for over a century. Porphyrin, a *meso*-carbon counterpart of Pc (Fig. 16.1) [5], also exhibits Q bands, but their intensities are fairly weak due to the forbidden nature. In 1959, Gouterman developed a conceptual framework called the Gouterman's four orbital model using a symmetry-perturbed $C_{16}H_{16}^{2-}$ perimeter model for the 18π-electron aromatic conjugated systems of porphyrin and Pc and successfully rationalized the forbidden nature of the Q bands of porphyrins and the significant intensification of the Q bands of Pc [6]. In the case of porphyrin, the Q band absorption originates from transitions between the nearly degenerate highest occupied molecular orbitals (HOMOs) and the degenerate lowest unoccupied molecular orbitals (LUMOs) and configurational interactions of the HOMO-LUMO (or HOMO-LUMO + 1) and HOMO–1-LUMO (or HOMO–1-LUMO + 1) transitions. Since the intensity of the Q band absorption is generated by the difference of the configurational product of these transitions, the Q band absorption of porphyrin is forbidden due to the similar HOMO-LUMO (or HOMO-LUMO + 1) and HOMO–1-LUMO (or HOMO–1-LUMO + 1) energy gaps, which reflect the near orbital degeneracy of the ground and excited states of porphyrin. On the other hand, in the case of Pc, the accidental degeneracy of the ground states is lifted due to stabilization of the a_{2u} MO relative to the a_{1u} HOMO among the nearly degenerated ground states when electron-deficient nitrogen atoms occupy the *meso*-positions. This results in intensification of the Q band absorption of Pc due to the smaller configurational interaction compared to porphyrin. Since the Q band absorption of Pc, therefore, mainly consists of the HOMO-LUMO and HOMO-LUMO + 1 transitions with only minor or no contributions of n-π* and charge-transfer transitions in most cases, its energy and morphologies closely reflect the changes in the electronic structures of its ground and excited states [1, 7], and it can reasonably be anticipated that creation of π-conjugated systems that are formed based on the Pc structure through perturbations of its electronic structure will lead to novel chromophore systems. Synthetic investigations along this research direction have been fairly limited, however, compared to those in porphyrin chemistry mainly due to limitations in synthetic methodologies. With long years of experience and knowledge concerning the synthesis and properties of Pc and its related analogues, the authors have been at the cutting edge in this research area and have reported many unprecedented Pc-based chromophore systems. In this chapter, the contribution of the authors to the recent advances in (1) the synthesis of the novel Pc analogues and (2) the control of the optical properties of the Pc chromophore systems in the vis/NIR regions through simple perturbation is described.

16.2 Synthesis of Pc Analogues with Novel π-Conjugated Systems

The synthesis of Pc analogues with novel π-conjugated systems has not been as straightforward as it is in the case of porphyrins, for a number of reasons: (1) Pc is the most stable product in cyclization reactions of *o*-phthalonitrile or its derivatives due to the least-strained structure, and (2) a stepwise synthesis using building blocks cannot generally be applied to the synthesis of Pc [8]. The successful syntheses of Pc analogues have, therefore, been fairly limited. This is in marked contrast to porphyrin chemistry, in which a variety of porphyrin analogues, such as contracted, expanded, and isomeric porphyrin analogues, has been reported [9]. In order to solve this inherent problem in the synthesis of novel Pc analogues with unique π-conjugated systems, the authors reassessed the precursors and reaction conditions and identified novel synthetic precursors and templates, which had never been utilized in the Pc synthesis or had not previously been viewed as potential precursors in this regard. In this section, several unprecedented Pc analogues that were synthesized in this manner are described.

16.2.1 Phthalocyanine Analogues

Pc is in general formed as both a kinetically and thermodynamically favored product from reactions of *o*-phthalonitrile or its derivatives such as 1,3-diiminoisoindoline and phthalimide at high reaction temperatures in the presence of templating metal

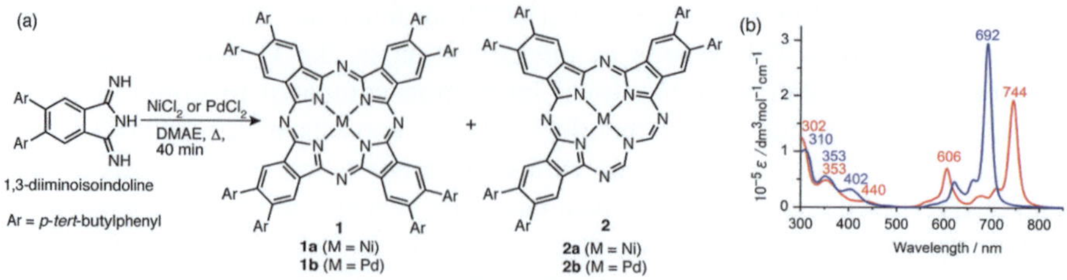

Fig. 16.2 (**a**) Synthesis of *seco*-tribenzoporphyrazine **2**. (**b**) UV/vis absorption spectra of **1a** (*blue line*) and **2a** (*red line*) in toluene (Reproduced from Ref. [10] with permission from Elsevier)

ions. Due to the modern progress in analytical techniques, the presence of Pc analogues in the reaction mixture, which adopt similar 18π-electron conjugation, but different structures, has recently been detected in some cases. The authors also unexpectedly encountered two examples of this type of Pc analogues.

In the general synthesis of nickel and palladium complexes of Pc **1** from 1,3-diiminoisoindoline in the presence of nickel chloride or palladium chloride, a blue-colored compound with a different absorption spectrum from that of **1** was obtained as a major by-product (Fig. 16.2) [10]. Based on the mass spectrometry and NMR spectroscopy, this compound was characterized as a *seco*-tribenzoporphyrazine **2**. Another group recently reported the same compound as a by-product during the synthesis of Pc analogues from 1,3-diiminoisoindoline and 2,5-diamino-3,4-dicyanothiophene [11]. Although the reaction mechanism of **2** still remains unclear, control experiments indicate that the elimination of a benzene ring from the Pc ring occurs during rather than the cyclization process. **2** exhibits an intense Q band originating from the 18π-electron conjugated systems that is similar to that of Pc, but the Q band absorption splits due to the lower molecular symmetry of **2**, which results in an electronic structure that is similar to that of a tetraazachlorin rather than that of Pc [12–14].

Another Pc analogue with a unique bridging unit between isoindole rings was obtained in the synthesis of a gold complex of Pc, which had been reported once previously in the literature but had not been fully characterized [15]. **4** was obtained along with a gold complex **3** from a reaction of 1,3-diiminoisoindoline in the presence of AuCl or [$KAuCl_4$] as a template, due to the templating effect of gold(I) or gold(III) ions (Fig. 16.3) [16]. X-ray single crystallographic analysis revealed that **4** has a planar structure with an 18π-electron macrocyclic conjugated system consisting of three isoindole units linked by a unique N-N-C-N bond. The absorption spectrum of **4** exhibits a Q band absorption that is somewhat similar to that of **3** with a broader bandwidth and the presence of two ill-defined bands at the higher energy. The absorption spectrum of **4** can again be explained in terms of a symmetry-perturbed 18π-electron conjugated system [6]. In addition to the unique structure and absorption spectrum, **4** also exhibits ring-expansion reactivity toward **3**. **4** can be converted into **3a** in the moderate yield (41 %), when **4** is heated at 100 °C in toluene in the presence of trifluoroacetic acid (Fig. 16.3). Although

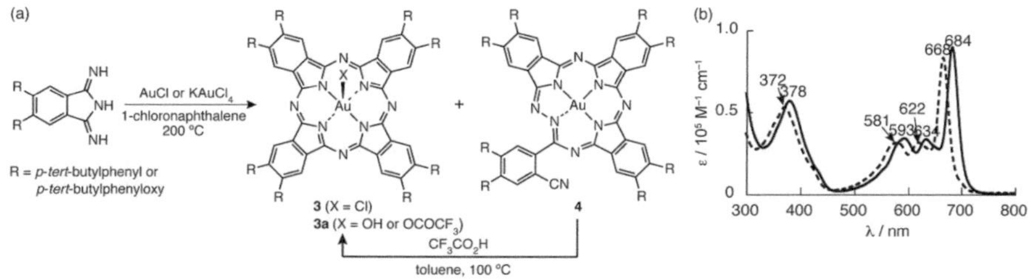

Fig. 16.3 (**a**) Synthesis of a Pc analogue **4**. (**b**) UV/vis absorption spectra of **4** (*solid line*, R = *p-tert*-butylphenyl; *dashed line*, R = *p-tert*-butylphenyloxy) in toluene (Reproduced from Ref. [16] with permission from Wiley)

the mechanisms of the formation of **4** during the synthesis of **3** and its thermal conversion have yet to be determined, it can be tentatively concluded that **4** is a kinetic product in the ring-formation reaction of Pc, which is plausibly stabilized by coordination of a gold(III) ion.

16.2.2 Core-Modified Phthalocyanines

Core modification including replacement of the inner five-membered ring units with other heteroaromatic ring units is one of the plausible strategies for generating Pc analogues with novel π-conjugated systems. This synthetic strategy has been intensively investigated ever since the early days of phthalocyanine chemistry. However, most of efforts have focused on a simple mixed condensation of 1,3-diiminoisoindoline and heteroaromatic diamines, which usually resulted in the formation of nonaromatic hemiporphyrazine molecules in most cases [17]. Since the inner five-membered ring of the isoindole unit is formed as a result of the 1,2-relationship of the positions of cyano groups in *o*-phthalonitrile, the authors utilized aromatic dicarbonitriles bearing cyano group in a 1,3- or 1,4-relationships in order to introduce core modification without any loss of the original conjugated system of Pc.

Mixed condensation reaction of *o*-phthalonitrile and 1,8-naphthalenedicarbonitrile provided a series of six-membered-ring-containing Pc analogues (Fig. 16.4, **6–8**) [18], whereas seven-membered-ring-containing Pc analogues were obtained from reactions of 2,2′-biphenyldicarbonitrile, 4,5-phenanthrenedicarbonitrile, and 2,2′-bipyridyl-3,3′-dicarbonitrile (Fig. 16.5, **10–12**) [19]. In the case of core modification with six-membered-ring units, the Q band absorption shifts to the red with respect to the number of six-membered-ring moieties that are incorporated into the structures, and its morphologies vary depending on the molecular symmetry, which have good correlation with what was observed for the benzo-annulated low-symmetry Pc systems [20]. These results indicate that the Pc analogues bearing six-membered rings retain Pc-like 18π-electron conjugated systems. In contrast, the Pc analogues with a seven-membered ring exhibit split Q band absorptions at both higher and lower

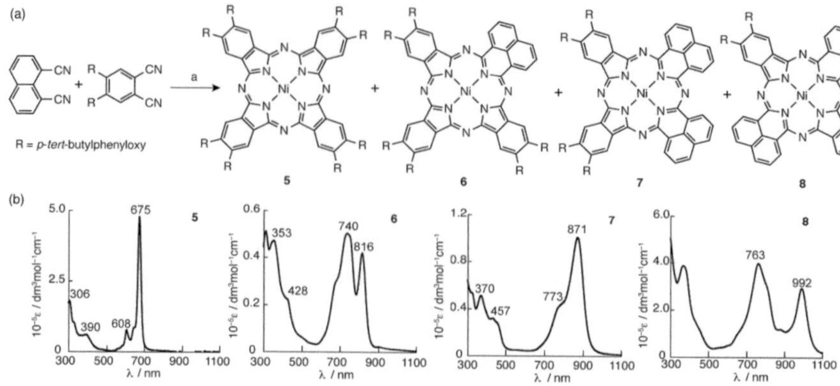

Fig. 16.4 (**a**) Synthesis of core-modified Pc analogues bearing six-membered rings. [a]Reagents and conditions: $Ni(OAc)_2$, $(NH_4)_2MoO_4$, quinoline, 330 °C. (**b**) Their UV/vis/NIR absorption spectra in chloroform (Reproduced from Ref. [18] with permission from Wiley)

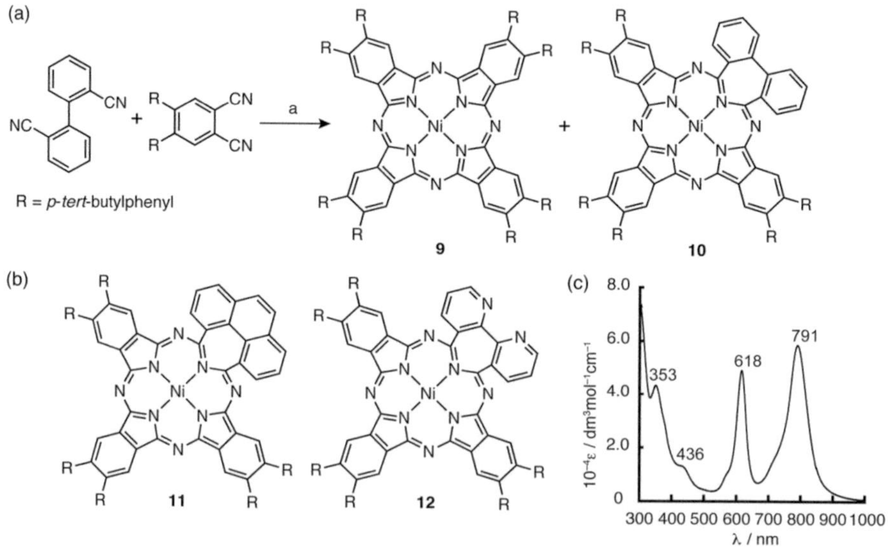

Fig. 16.5 (**a**) Synthesis of a core-modified Pc analogue bearing seven-membered rings. [a]Reagents and conditions: $Ni(OAc)_2$, $(NH_4)_2MoO_4$, quinoline, 330 °C. (**b**) Structures of other seven-membered-ring-containing Pc analogues. (**c**) UV/vis/NIR absorption spectrum of **10** in chloroform (Reproduced from Ref. [19] with permission from Wiley and Elsevier)

energy regions relative to that of Pc, which are similar to those of **2** and tetraazachlorin [10–14]. This results from the nearly perpendicular arrangement of the seven-membered-ring moiety, which eliminates any contribution of the peripheral double bonds to the macrocyclic conjugated system. These core-modification strategies can also be applied to the synthesis of subphthalocyanine, which is a contracted species of Pc, and the details have been presented elsewhere in literatures [21].

16.2.3 Expanded Pc Analogues

Porphyrin and Pc analogues with an extended conjugation larger than the 18π-electron system consisting of more than four pyrrole rings or heteroaromatic rings are referred to as expanded porphyrins and Pcs [9, 22]. For the same synthetic reasons as described above, expanded Pc analogues have not been reported with the exception of superphthalocyanine (SuperPc, Fig. 16.6) [23]. SuperPc consists of five isoindole units and a uranyl ion coordinated in the center of the molecule and adopts 22π-electron conjugated system. Due to the extended conjugation, this molecule exhibits redshifted Q band absorption in the NIR region. Despite the importance of this molecule as a NIR chromophore, only four papers on this molecule have been reported since the first report of its synthesis more than 40 years ago, and its derivatization remains unexplored [24]. The authors' group investigated the synthesis of superazaporphyrins (SuperAzP), benzo-abstracted analogues of SuperPc. Surprisingly, a series of SuperAzP **13** and **14** were more easily synthesized than expected from 3,4-di-*p*-tert-butylphenyl-2,5-pyrrolinediimine or from 3,4-di-*p*-tert-butylphenyl-2,5-pyrrolinediimine and *o*-phthalonitrile (Fig. 16.6) [25]. All of these novel SuperPc analogues exhibit the Q band absorption at ca. 880 nm, 40 nm to the blue of SuperPc, but still lie significantly to the red of Pc.

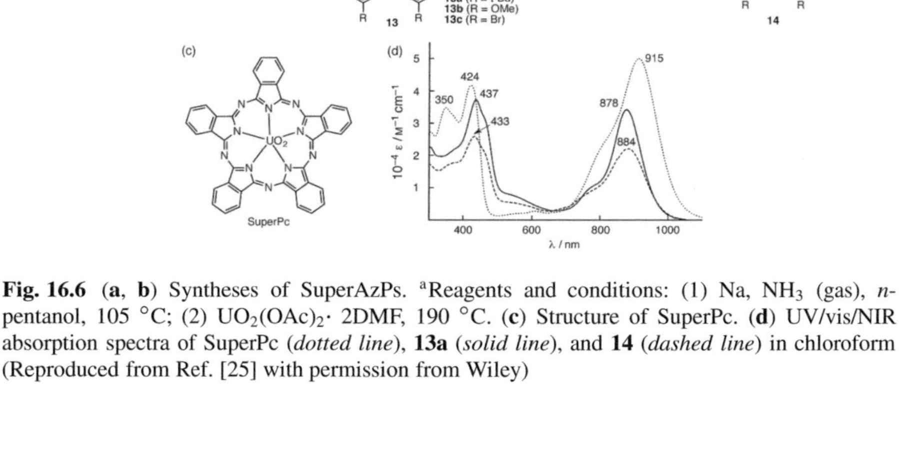

Fig. 16.6 (**a**, **b**) Syntheses of SuperAzPs. [a]Reagents and conditions: (1) Na, NH_3 (gas), *n*-pentanol, 105 °C; (2) $UO_2(OAc)_2 \cdot$ 2DMF, 190 °C. (**c**) Structure of SuperPc. (**d**) UV/vis/NIR absorption spectra of SuperPc (*dotted line*), **13a** (*solid line*), and **14** (*dashed line*) in chloroform (Reproduced from Ref. [25] with permission from Wiley)

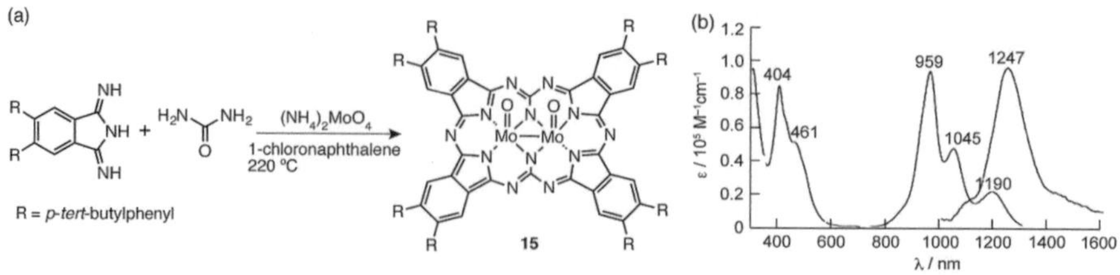

Fig. 16.7 (**a**) Synthesis of an expanded Pc analogue **15**. (**b**) UV/vis/NIR absorption and fluorescence spectra of **15** in CCl_4 (Reproduced from Ref. [26] with permission from ACS)

Recently, the authors' group also succeeded in the synthesis of an expanded Pc analogue bearing a novel structure [26]. Under the general synthetic conditions in the presence of molybdenum as a template, an expanded analogue **15** was obtained when excess amount of urea was utilized (Fig. 16.7). The bridging imino nitrogen atoms of Pc are replaced with guanidine units in the structure of **15**, and this results in an expansion of the π-conjugated systems from an 18π-electron to a 22π-electron system. Due to the expanded conjugation, the Q band absorption shifts markedly into the NIR region at 1190 and 959 nm. A similar reaction using ammonium tungstate in the place of ammonium molybdate also provides a similar tungsten complex, and very recently a similar compound was synthesized in higher yields by Muranaka and Uchiyama using guanidine hydrochloride in place of urea.

16.3 Control of Electronic Absorption in the Visible and Near-Infrared Regions: Perturbed Chromophore Systems of Pc

As described in the previous section, the Q band absorption of Pc is sensitive to any perturbations of the electronic structures. In the previous work of the authors' group, the changes in the positions and morphologies of the Q band absorption that are caused by benzo-annulation to and by benzo-abstraction from the periphery of Pc were demonstrated and clearly explained based on the perturbation of the energy levels of the frontier MOs when the molecular symmetry changes [20]. The next topic that was investigated is the introduction of further perturbations including push-pull effects to the conjugated system of Pc and its related analogues.

16.3.1 Push-Pull Tetraazaporphyrins

Push-pull effects, which are caused by the introduction of both electron-donating and electron-withdrawing substituents to the same molecules, have been widely utilized in order to tune the optical and electrochemical properties. Research based on this strategy has also been the focus of considerable attention within phthalocyanine chemistry. Initially, the authors thought that if push-pull substituents are introduced to the periphery of Pc, the effect would be marginal due to the presence of the benzene ring in between the substituents and the core 18π-electron conjugated system. Tetraazaporphyrin (TAP) [27] was, therefore, selected, since the direct perturbation of the push-pull substituents on the conjugated systems can be expected.

The synthesis of these kinds of push-pull TAP molecules from tricyanoethylene derivatives with an electron-donating substituent was pioneered by Luk'yanets et al. who reported the marked redshift of the Q band absorption of the push-pull TAP compared to that of regular TAP molecules at around 620 nm [28]. The authors also investigated this chromophore system to provide in-depth insight into how changes in the symmetries of molecular orbitals are induced by the perturbations of the push-pull substituents [29].

Four structural isomers with a C_{4h}, D_{2h}, C_{2v}, and C_s symmetry can be obtained from a cyclotetramerization reaction of a substituted tricyanoethylene (Fig. 16.8). The authors reported the first complete separation of all of these structural isomers. Single Q band absorption is observed for the C_{4h} and C_{2v} isomers, whereas the marked splitting of the Q band absorption is observed for the D_{2h} and C_s isomers. Magnetic circular dichroism (MCD) spectroscopy aided band assignment, since it is a powerful analytical tool for chromophore molecules that provides information such as the orbital degeneracies that cannot readily be deduced from other spectroscopies [30]. A Faraday *A* term, a derivative-shaped signal with an inflection point at the absorption maximum, is observed corresponding to the Q band absorption of the C_{4h} isomer, and this is indicative of degeneracy in the HOMO and HOMO–1, whereas a pair of Faraday *B* terms, a set of trough and peak with Gaussian-type shapes corresponding closely to the absorption maxima, is observed in the case of the D_{2h} and C_s isomers and is indicative of a nondegenerate HOMO and HOMO–1. The MCD spectrum of the C_{2v} isomer exhibits a Faraday *A* term-like MCD signal in the Q band region, which is often referred to as a pseudo Faraday *A* term [31]. This MCD signal is observed when molecules with lower symmetries than C_3 accidentally possess nearly degenerate excited states.

The redshift of these push-pull TAP molecules is ascribed to the synergetic effect of the push and pull substituents based on an analysis of the DFT calculations, since this causes decrease of the HOMO-LUMO gap due to the destabilization of the HOMO by the push substituents and stabilization of the LUMO by pull substituents. The theoretical calculations also reproduced the difference in the Q band morphologies observed for these structural isomers: the electronic perturbation of the push-pull substituents is significant enough to cause changes in the molecular orbital distribution pattern in the LUMO and LUMO + 1, resulting in the nondegenerate LUMO and LUMO + 1 in the case of the D_{2h} and C_s isomers. In contrast, the push-pull substituents of the C_{4h} and C_{2v} isomers almost equally perturb the LUMO and LUMO + 1, and this results in degeneracy and near degeneracy of these orbitals, respectively. The relationship between the molecular symmetries and the Q band morphologies is almost identical to that observed for Pc systems that are symmetry-perturbed by benzo-annulation and benzo-abstraction [20].

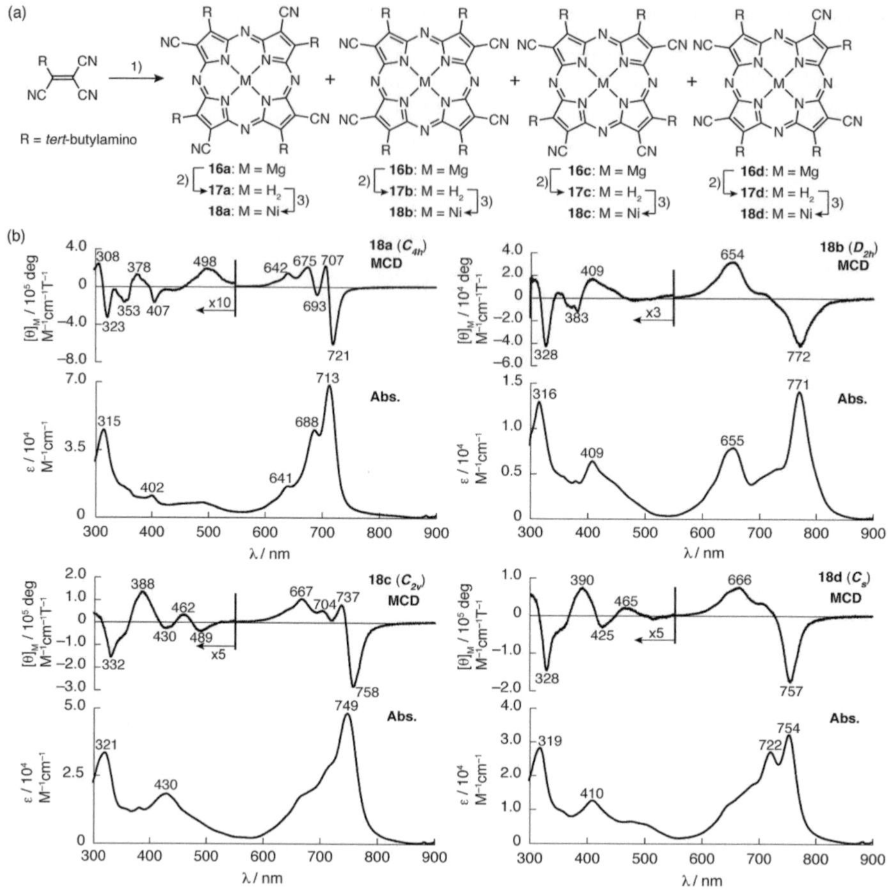

Fig. 16.8 (**a**) Synthesis of push-pull TAP compounds. Reagents and conditions: (1) $Mg(OC_6H_{13})_2$, 1-hexanol, 180 °C, 3 h; (2) CF_3CO_2H, $CHCl_3$, 30 min; (3) $NiCl_2 \cdot 6H_2O$, DMF, 180 °C, 1 h. (**b**) UV/vis absorption (*bottom*) and MCD (*top*) spectra of **18a–d** in chloroform, except for **18b** in pyridine (Reproduced from Ref. [29] with permission from Wiley)

16.3.2 Novel NIR-Absorbing Pc Using the Synergetic Effects of the Group 15 and 16 Elements

Recently, NIR-absorbing compounds have been of significant interest because of their potential applications, such as organic solar cells, photodynamic therapy, NIR imaging, and so forth. The chemistry of Pc is highly promising in this regard. Although Pc has greater potential than TAP as a platform for obtaining NIR absorption, the shift obtained in the absorption band wavelengths from simple perturbations, such as introduction of substituents, is less significant for Pcs than TAP as described above. The authors' group carefully calculated the frontier MOs of Pc and noticed that the MO coefficients of the HOMO are larger at the α-positions of the benzo moieties, whereas that of the LUMO is largely found on the pyrrole nitrogen atoms. Perturbation of these positions can, therefore, also be used to effectively control the HOMO-LUMO energy gap and the wavelength of the Q band absorption. The authors' group finally envisioned the rational design of a Pc analogue, in which an electron-donating group 16 element is introduced to the α-positions of the benzo moieties and lifts the HOMO energy level, while the coordination of a high valent group 15 element inside stabilizes the LUMO energy level [32]. Novel phosphorus(V) Pcs with a series of arylchalcogenyl substituents were synthesized, and the Q band absorption beyond 1000 nm was achieved (Fig. 16.9). Other metal complexes bearing the same substituents at the α-positions with the exception of the manganese(III) complex exhibit Q band at around 800 nm, which is similar to that of its free-base form, and the complexes of other group 15 elements also exhibit similar NIR Q band absorption; the synergetic effects of the group 15 and 16 elements have been clearly demonstrated. Recently, the authors' group has also applied this molecular design to TAP molecules [33]. Phosphorous TAP complexes with aryl substituent on the periphery were synthesized, which exhibit a characteristic intramolecular charge-transfer band in between the Soret and Q band absorption of TAP. The intramolecular CT bands shift depending on the nature of the substituents on the *para*-position of the aryl substituents, and there is a linear correlation of the shifts with the Hammett σ_p parameters (Fig. 16.10). It can be, therefore, concluded that utilization of push-pull effects of main group elements is a very effective strategy to finely tune the optical properties of Pc and its related compounds.

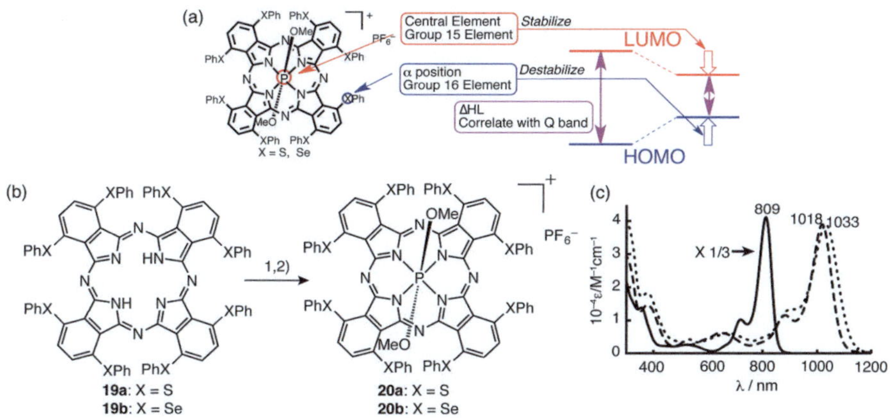

Fig. 16.9 (**a**) Molecular design toward NIR-absorbing Pc. (**b**) Synthesis of Pc bearing the group 15 and 16 elements. Reagents and conditions: (1) $POBr_3$ (excess), pyridine, reflux, 1.5 h (X = S), rt, 30 min (X = Se); (2) CH_2Cl_2/MeOH, rt, 30 min, then KPF_6 (4 equiv.), CH_2Cl_2, rt, 12 h. (**c**) UV/vis/NIR absorption spectra of **19a** (*solid line*), **20a** (*dashed line*), and **20b** (*dotted line*) in chloroform (Reproduced from Ref. [32] with permission from ACS)

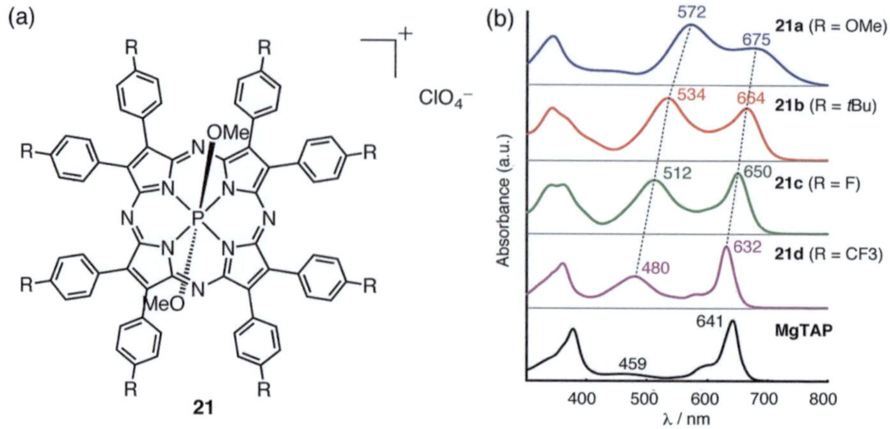

Fig. 16.10 (**a**) Structure of phosphorous(V) complex of TAP. (**b**) UV/vis absorption spectra of a series compounds (**MgTAP** is a magnesium complex of TAP bearing *tert*-butylphenyl substituents) in dichloromethane (Reproduced from Ref. [33] with permission from RSC)

16.3.3 Longitudinally Stacked Pc Systems

Longitudinally stacked Pc oligomers, such as μ-oxo-bridged silicon phthalocyanine oligomers and sandwich-type double- and triple-decker Pc complexes, have merged as an important class of molecular assemblies of Pc due to their potential properties for applications in semiconducting materials, electrochromic materials, and single-molecule magnets (SMM) [34]. From a viewpoint of optical properties, these longitudinally stacked oligomers can also be recognized as perturbed chromophore systems. In this subsection, two examples of this type of Pc systems that were developed by the authors' group are described.

μ-Oxo-bridged silicon Pc oligomers can be synthesized by simple dehydration reaction of axially hydroxy-substituted silicon Pc, and their potential utility as semiconducting materials has been investigated [35]. Since the interplanar interactions and orientation of the Pc rings of the oligomers are directly related to their conductivities, μ-oxo-bridged silicon Pc dimers have also attracted significant research interest. As a unique optical property observed for μ-oxo dimers, it is known that dimer molecules with peripheral alkoxy substituents are known to exhibit solvatochromic behavior with broad absorption in nonaromatic solvents, such as chloroform, and intense absorption in aromatic solvents, such as toluene [36, 37]. The solvatochromic behavior is not observed for the peripherally unsubstituted and alkyl-substituted dimers, which exhibit an intense, blue-shifted absorption relative to the Q band absorption of the corresponding monomer [38], and this can be simply explained using Kasha's exciton theory [39]. The solvatochromic behaviors are consistent with a certain degree of contribution of intramolecular interactions as has been demonstrated by Hanack et al. [37]. The charge resonance contribution has been introduced by Ishikawa and Kaizu as a theory for explaining the optical properties of longitudinally stacked Pc systems [40]. The authors recently observed similar solvatochromic behaviors in the case of a hetero μ-oxo dimer of Pc and naphthalocyanine (Nc) **25**. The effect of solvents on the morphologies of the absorption spectra was the opposite of that described above with intense absorption in chloroform and broader absorption bands in toluene (Fig. 16.11) [41]. The solvatochromic behaviors can be similarly explained in terms of the changes in the torsion angles between the Pc and Nc planes as was observed for the homo μ-oxo dimer of Pc, using variable-temperature absorption and ^{1}H NMR spectroscopic measurements and time-dependent DFT calculations. The conformation in a chloroform solution was plausibly assigned as a C_{4v} symmetric conformation with a torsion angle of $\theta = 0°$ or $45°$, whereas conformational isomers, a C_{4v} symmetric isomer, and less symmetric isomers with a torsion angle of $0° < \theta < 45°$ appeared to be in thermal equilibrium in toluene at room temperature, and a C_{4v} symmetric

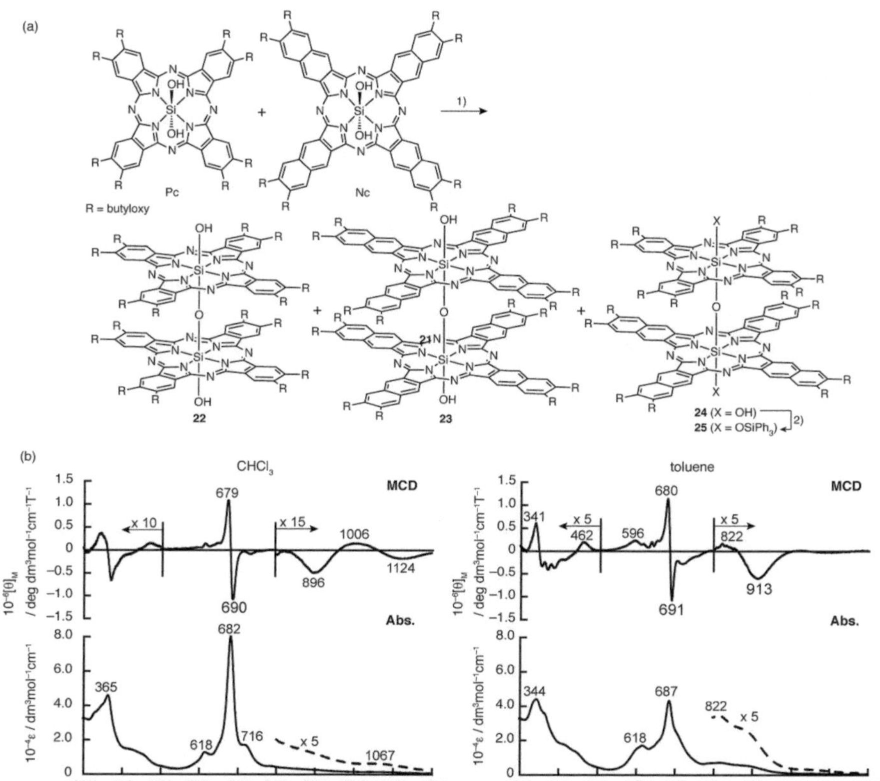

Fig. 16.11 (**a**) Synthesis of μ-oxo dimers. Regents and conditions: (1) quinoline, 180 °C, 3 h; (2) chlorotriphenylsilane, quinoline, 180 °C, 1 day. (**b**) UV/vis/NIR absorption (*bottom*) and MCD (*top*) spectra of **25** in chloroform (*left*) and toluene (*right*) at room temperature (Reproduced from Ref. [41] with permission from RSC)

isomer would become a predominant at low temperature. These results indicate that optical properties of the longitudinal Pc systems are significantly perturbed by the interactions with the neighboring Pc units, which varies with respect to the torsion angles.

Another unique example of a perturbed longitudinal Pc chromophore system is the first successful synthesis of a discrete quadruple-decker Pc by the authors' group [42]. Since Pc is a divalent ligand and lanthanide ions are generally present as trivalent cations, lanthanide complexes of double-decker Pc are known to adopt two forms, an anion $[LnPc_2]^-$ form and a neutral $[LnPc_2]^0$ radical form, which are also referred to as blue and green forms due to their colors in solution. Fukuda, one of the members of the authors' group, foresaw the potential utility of the blue form as a monovalent ligand to construct a dimer system of double-decker Pc by coordination with an appropriate dicationic metal ion. The first practical synthesis of quadruple-decker Pc was achieved in this manner from a reaction of $[LnPc_2]^-[NBu_4]^+$ and cadmium acetate at 400 °C under argon atmosphere (Fig. 16.12). The generality of this synthetic strategy toward multiple-decker Pcs opened access to further quintuple-decker and sextuple-decker Pcs thereafter, and the SMM behaviors of these molecules have been a recent hot research topic [43, 44]. The absorption spectra of these multiple-decker Pcs can be illustrated based on the theoretical approach developed by Ishikawa and Kaizu [40]. In the case of the quadruple-decker Pc, a lower-energy band with weaker intensity at 727 nm can be assigned as transitions with charge resonance character, whereas a greater energy band with more intensity at 623 nm arises primary from an exciton interaction (Fig. 16.12). Due to the closer proximity of the Pc chromophore relative to the abovementioned μ-oxo dimer systems, the interchromophore interaction is significantly larger. In addition to the unique investigation of the SMM behavior,

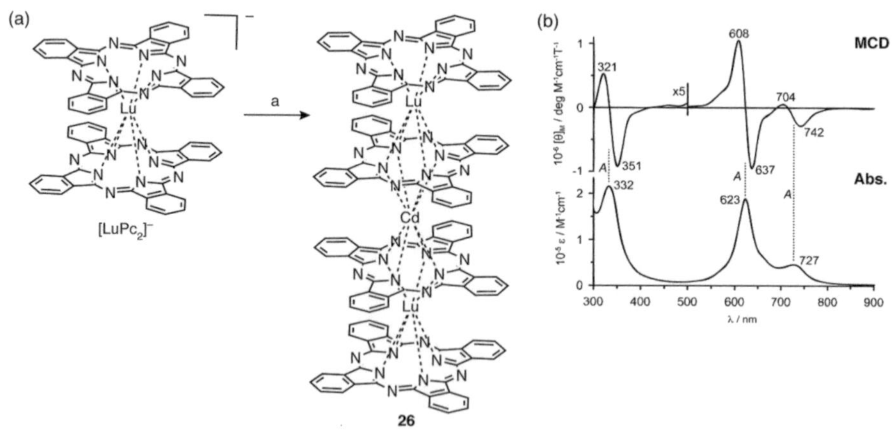

Fig. 16.12 (**a**) Synthesis of quadruple-decker Pc. Reagents and conditions: $Cd(OAc)_2 \cdot 2H_2O$, 400 °C, 30 min. (**b**) UV/vis/NIR absorption (*bottom*) and MCD (*top*) spectra of **26** in chloroform (Reproduced from Ref. [42] with permission from ACS)

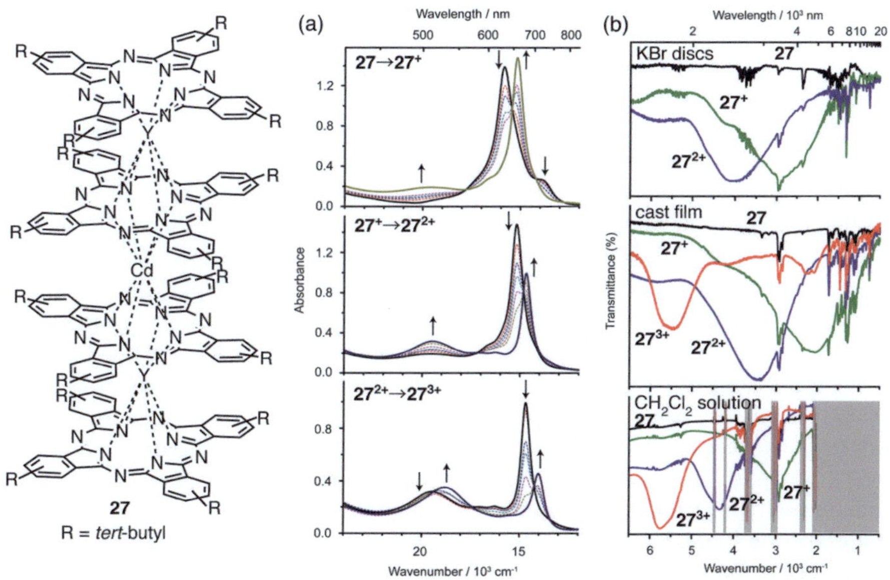

Fig. 16.13 (**a**) Developments of UV/vis absorption spectra of **27** in dichloromethane. (**b**) Electronic transitions of **27** observed in the NIR to IR energy region. The spectra were obtained by using KBr discs (*top*), cast films prepared on KBr crystals (*middle*), and dichloromethane solutions in a quartz cell (*bottom*) (Reproduced from Ref. [45] with permission from ACS)

Fukuda et al. reported the presence of exceptionally low-lying π-π* excited states in the oxidized forms of quadruple-decker Pc [45]. Figure 16.13 describes the absorption spectral developments observed upon oxidation. The stepwise oxidation processes from the neutral species to the radical monocation species and then from the radical monocation species to the singlet dication species and further to the radical trication species were confirmed by the spectral changes in the Q band region with isosbestic points and ESR measurements. The FT-IR spectroscopic measurements on the isolated, oxidized species revealed developments of new bands in the IR region. Based on the theoretical calculations, this was successfully assigned as bands arising from π-π* transitions between the HOMO and LUMO generated by oxidation. The exceptionally small gap between these MOs was ascribed to the fact that these MOs originate from the closely lying occupied frontier MOs of the quadruple-decker Pc. Longitudinally stacking of chromophore molecules and partial oxidation has, therefore, provided a new potential strategy for developing NIR chromophore systems.

16.4 Summary and Outlook

In this chapter, the recent advances in phthalocyanine chemistry contributed by the authors have been described with a special focus on novel structures of Pc analogues and strategies toward enhancement of NIR absorptions. Since the synthetic chemistry of Pc and its analogues have not been as thoroughly investigated as that in porphyrin chemistry, many opportunities to encounter Pc analogues with novel π-conjugated systems still remain, such as the real expanded homologues of Pc, which have never been realized with the exception of superphthalocyanine. On the other hand, the research topics related to the control of the optical properties in the visible and NIR regions through perturbations are now well established. Further investigations in this area are likely to focus on practical applications of a variety of perturbed vis/NIR chromophore systems of Pc in molecular electronics and materials science, such as organic photovoltaic cells. Despite the over a hundred-year history of Pc, there is still considerable research interests in this field due to the unusually high potential of the Pc chromophore, so further investigation on this "old but new" molecule will continue to contribute the advancement of materials chemistry in the year ahead.

References

1. (a) Kadish KM, Smith KM, Guilard R (eds) (2003) The porphyrin handbook, vol 15–20. Academic Press, San Diego; (b) Leznoff CC, Lever ABP (eds) (1989–1996) Phthalocyanines: properties and applications, vol 1–4. Wiley-VCH, New York; (c) McKeown NB (1998) In: Phthalocyanine materials: synthesis, structure and function. Cambridge University Press, Cambridge; (d) Kobayashi N, Fukuda T (2006) In: Kim S (ed) Functional dyes. Elsevier, Oxford
2. Braun A, Tcherniac J (1907) Über die Produkte der Einwirkung von Acetanhydrid auf Phthalamid. Ber Dtsch Chem Ges 40:2709–2714
3. (a) Linstead RP, Rydon HN (1934) Investigation of the olefinic acids. Part XV. The effect of peroxides on the orientation of the addition of hydrogen bromide to vinyl- and allyl-acefic acids. J Chem Soc 2001–2003; (b) Linstead RP (1934) Phthalocyanines. Part I. A new type of synthetic colouring matters. J Chem Soc 1016–1017; (c) Dent CE, Linstead RP, Lowe AR (1934) Phthalocyanines. Part VI. The structure of the phthalocyanines. J Chem Soc 1033–1039; (d) Dent CE, Linstead RP (1934) Phthalocyanines. Part IV. Copper phthalocyanines. J Chem Soc 1027–1031; (e) Byrne GT, Linstead RP, Lowe AR (1934) Phthalocyanines. Part II. The preparation of phthalocyanine and some metallic derivatives from o-cyanobenzamide and phthalimide. J Chem Soc 1017–1022
4. Robertson JM (1936) An X-ray study of the phthalocyanines. Part II. Quantitative structure determination of the metal-free compound. J Chem Soc 1195–1209
5. Kadish KM, Smith KM, Guilard R (eds) (2003) The porphyrin handbook, vol 1–10. Academic Press, San Diego
6. (a) Gouterman M (1959) Study of the effects of substitution on the absorption spectra of porphin. J Chem Phys 30:1139–1161; (b) Gouterman M (1961) Spectra of porphyrins. J Mol Spectrosc 6:138–163
7. Fukuda T, Kobayashi N (2010) In: Kadish KM, Smith KM, Guilard R (eds) Handbook of porphyrin science, vol 9. World Scientific, Singapore
8. Kadish KM, Smith KM, Guilard R (eds) (2003) The porphyrin handbook, vol 13. Academic Press, San Diego

9. (a) Sessler JL, Weghorn SJ (1997) Expanded, contracted & isomeric porphyrins. Pergamon, Oxford; (b) Sessler JL, Seidel D (2003) Synthetic expanded porphyrin chemistry. Angew Chem Int Ed 42:5134–5175
10. Sugita I, Shimizu S, Fukuda T, Kobayashi N (2013) Nickel and palladium complexes of seco-tribenzoporphyrazines derived from one-pot condensation of 1,3–diminoisoindoline. Tetrahedron Lett 54:1599–1601
11. Nemykin VN, Polshyna AE, Makarova EA, Kobayashi N, Lukyanets EA (2012) Unexpected formation of the nickel seco-tribenzoporphyrazine with a tribenzotetraazachlorin-type absorption spectrum. Chem Commun 48:3650–3652
12. Linstead RP, Whalley M (1952) Conjugated macrocycles. Part XXII. Tetrazaporphin and its metallic derivatives. J Chem Soc 4839–4846
13. Fukuda T, Kobayashi N (2008) Hydrogenated tetraazaporphyrins - old but new core-modified phthalocyanine analogues. Dalton Trans 4685–4704
14. (a) Mani NS, Beall LS, White AJP, Williams DJ, Barrett AGM, Hoffman BM (1994) Serendipitous desymmetrisation during porphyrazine synthesis - an X-ray crystallographic study of 2,3,7,8,12,13,17,18-octakis(Dimethylamino)-2-secoporphyrazine-2,3-dione. J Chem Soc Chem Commun 1943–1944; (b) Montalban AG, Lange SJ, Beall LS, Mani NS, Williams DJ, White AJP, Barrett AGM, Hoffman BM (1997) Seco-porphyrazines: synthetic, structural, and spectroscopic investigations. J Org Chem 62:9284–9289
15. MacCragh A, Koski WS (1965) The phthalocyanine of gold. J Am Chem Soc 87:2496–2497
16. Wong EWY, Miura A, Wright MD, He Q, Walsby CJ, Shimizu S, Kobayashi N, Leznoff DB (2012) Gold(II) phthalocyanine revisited: synthesis and spectroscopic properties of Gold(III) phthalocyanine and an unprecedented ring-contracted phthalocyanine analogue. Chem Eur J 18:12404–12410
17. Fernandez-Lazaro F, Torres T, Hauschel B, Hanack M (1998) Hemiporphyrazines as targets for the preparation of molecular materials: synthesis and physical properties. Chem Rev 98:563–575
18. Shimizu S, Zhu H, Kobayashi N (2010) Subazaphenalenephthalocyanine: a subphthalocyanine analogue bearing a six-membered ring unit. Chem Eur J 16:11151–11159
19. (a) Shimizu S, Zhu H, Kobayashi N (2011) Azepiphthalocyanine-an unprecedented large twist of a Π-conjugation system upon core-odification with a seven-membered ring unit. Chem Commun 47:3072–3074; (b) Shimizu S, Uemura K, Zhu H, Kobayashi N (2012) Core-modified phthalocyanine analogues with a seven-membered ring unit in place of a five-membered ring unit. Tetrahedron Lett 53:579–581
20. (a) Fukuda T, Makarova EA, Luk'yanets EA, Kobayashi N (2004) Synthesis and spectroscopic and electrochemical studies of novel benzo- or 2,3-Naphtho-fused tetraazachlorins, bacteriochlorins, and isobacteriochlorins. Chem Eur J 10:117–133; (b) Kobayashi N, Fukuda T (2002) Mono-aromatic ring-fused versus adjacently di-aromatic ring-fused tetraazaporphyrins: regioselective synthesis and their spectroscopic and electrochemical properties. J Am Chem Soc 124:8021–8034; (c) Kobayashi N, Miwa H, Nemykin VN (2002) *Adjacent* versus *opposite* type di-aromatic ring-fused phthalocyanine derivatives: synthesis, spectroscopy, electrochemistry, and molecular orbital calculations. J Am Chem Soc 124:8007–8020; (d) Kobayashi N, Mack J, Ishii K, Stillman MJ (2002) Electronic structure of reduced symmetry peripheral fused-ring-substituted phthalocyanines. Inorg Chem 41:5350–5363; (e) Miwa H, Ishii K, Kobayashi N (2004) Electronic structures of zinc and palladium tetraazaporphyrin derivatives controlled by fused benzo rings. Chem Eur J 10:4422–4435
21. (a) Shimizu S, Kobayashi N (2014) Structurally-modified subphthalocyanines: molecular design towards realization of expected properties from the electronic structure and structural features of subphthalocyanine. Chem Commun 50:6949–6966; (b) Zhu H, Shimizu S, Kobayashi N (2010) Subazaphenalenephthalocyanine: a subphthalocyanine analogue bearing a six-membered ring unit. Angew Chem Int Ed 49:8000–8003; (c) Shimizu S, Nakano S, Kojima A, Kobayashi N (2014) A core-expanded subphthalocyanine analogue with a significantly distorted conjugated surface and unprecedented properties. Angew Chem Int Ed 53:2408–2412

22. Saito S, Osuka A (2011) Expanded porphyrins: intriguing structures, electronic properties, and reactivities. Angew Chem Int Ed 50:4342–4373
23. Bloor JE, Schlabitz J, Walden CC, Demerdac A (1964) Organic complexes of uranium: Part I. The synthesis and spectrum of uranyl phthalocyanine. Can J Chem 42:2201–2208
24. (a) Marks TJ, Stojakovic DR (1975) Macrocycle contraction reactions of 5,35 : 14,19-Diimino-7,12 : 21,26 : 28,33-trinitrilopentabenzo [*c,h,m,r,w*][1,6,11,16,21]pentaazacyclopentacosinato-dioxouranium(VI). JCS Chem Commun 28–29; (b) Day VW, Marks TJ, Wachter WA (1975) Large metal ion-centered template reactions. Uranyl complex of cyclopentakis(2-iminoisoindoline). J Am Chem Soc 97:4519–4527; (c) Marks TJ, Stojakovic DR (1978) J Am Chem Soc 100:1695–1705; (d) Cuellar EA, Marks TJ (1981) Large metal ion-centered template reactions. Chemical and spectral studies of the "superphthalocyanine" dioxocyclopentakis(1-iminoisoindolinato)uranium(VI) and its derivatives. Inorg Chem 20:3766–3770
25. Furuyama T, Ogura Y, Yoza K, Kobayashi N (2012) Superazaporphyrins: meso-pentaazapentaphyrins and one of their low-symmetry derivatives. Angew Chem Int Ed 51:11110–11114
26. Matsushita O, Derkacheva VM, Muranaka A, Shimizu S, Uchiyama M, Luk'yanets EA, Kobayashi N (2012) Rectangular-shaped expanded phthalocyanines with two central metal atoms. J Am Chem Soc 134:3411–3418
27. Ficken GE, Linstead RP, Stephen E, Whalley M (1958) Conjugated macrocycles. Part XXXI. catalytic hydrogenation of tetrazaporphins, with a note on its stereochemical course. J Chem Soc 3879–3886
28. Kopranenkov VN, Goncharova LS, Lukyanets EA (1979) Cyano-substituted porphyrazines. Zh Obshch Khim 49:1408–1412
29. Shimizu S, Haseba Y, Yamazaki M, Kumazawa G, Kobayashi N (2014) Control of chromophore symmetry by positional isomerism of peripheral substituents. Chem Eur J 20:4822–4828
30. (a) Mack J, Stillman MJ, Kobayashi N (2007) Application of MCD spectroscopy to porphyrinoids. Coord Chem Rev 251:429–453; (b) Mack J, Kobayashi N, Shen Z (2012) In: Kadish KM, Smith KM, Guilard R (eds) Handbook of porphyrin science, vol 23, Chap. 109. World Scientific, Singapore; (c) Kobayashi N, Muranaka A, Mack J (2012) Circular dichroism and magnetic circular dichroism spectroscopy for organic chemists. Royal Society of Chemistry, Cambridge
31. (a) Kaito A, Nozawa T, Yamamoto T, Hatano M, Orii Y (1977) LCAO MO SCF Π-electron calculations on the magnetic circular dichroism of porphin, protoporphyrin, and porphyrin. Chem Phys Lett 52:154–160; (b) Tajiri A, Winkler J (1983) Magnetic circular-dichroism and molecular-orbital studies on condensed thiadiazoles. Z Naturforsch A 38:1263–1269
32. (a) Furuyama T, Satoh K, Kushiya T, Kobayashi N (2013) Design, synthesis, and properties of phthalocyanine complexes with main-group elements showing main absorption and fluorescence beyond 1000 nm. J Am Chem Soc 136:765–776; (b) Kobayashi N, Furuyama T, Satoh K (2011) Rationally designed phthalocyanines having their main absorption band beyond 1000 nm. J Am Chem Soc 133:19642–19645
33. Furuyama T, Yoshida T, Hashizume D, Kobayashi N (2014) Phosphorus(v) tetraazaporphyrins: porphyrinoids showing an exceptionally strong CT band between the soret and Q bands. Chem Sci 5:2466–2474
34. (a) Hanack M, Dini D (2003) In: Kadish KM, Smith KM, Guilard R (eds) The porphyrin handbook, vol 18, Chap. 114. Academic Press, San Diego; (b) Weiss R, Fischer J (2003) In: Kadish KM, Smith KM, Guilard R (eds) The porphyrin handbook, vol 16, Chap. 105. Academic Press, San Diego; (c) Buchler JW, Ng DKP (2003) In: Kadish KM, Smith KM, Guilard R (eds) The porphyrin handbook, vol 3, Chap. 20. Academic Press, San Diego
35. (a) Marks TJ (1985) Electrically conductive metallomacrocyclic assemblies. Science 227:881–889; (b) Dirk CW, Inabe T, Schoch KF, Marks TJ (1983) Metallomacrocycles as an approach to controlling lattice architecture in low-dimensional molecular solids. chemical, structural, oxidation state, transport, magnetic, and optical properties of halogen-doped [M(phthalocyaninato)O], macromolecules, where M = Si, Ge, and Sn. J Am Chem Soc 105:1539–1550; (c) Diel BN, Inabe T, Lyding JW, Schoch KF, Kannewurf CR, Marks TJ

(1983) Cofacial assembly of partially oxidized metallomacrocycles as an approach to controlling lattice architecture in low-dimensional molecular-solids - chemical and architectural properties of the face-to-face polymers [M(Phthalocyaninato)O]N, where M = Si, Ge, and Sn. J Am Chem Soc 105:1551–1567; (d) Dewulf DW, Leland JK, Wheeler BL, Bard AJ, Batzel DA, Dininny DR, Kenney ME (1987) Isolation, spectroscopic properties, and electrochemical properties of two oligomeric silicon phthalocyanines. Inorg Chem 26:266–270
36. (a) Ferencz A, Neher D, Schulze M, Wegner G, Viaene L, Deschryver FC (1995) Synthesis and spectroscopic properties of phthalocyanine dimers in solution. Chem Phys Lett 245:23–29; (b) OddosMarcel L, Madeore F, Bock A, Neher D, Ferencz A, Rengel H, Wagner G, Kryschi C, Trommsdorff HP (1996) Electronic states and relaxation dynamics of silicon phthalocyanine dimers. J Phys Chem 100:11850–11856
37. Kleinwachter J, Hanack M (1997) Rotational isomers in stacked macrocycles: synthesis and spectroscopic properties of peripherally substituted (μ-Oxo)bis(phthalocyaninatosilicon) compounds. J Am Chem Soc 119:10684–10695
38. (a) Ciliberto E, Doris KA, Pietro WJ, Reisner GM, Ellis DE, Fragala I, Herbstein FH, Ratner MA, Marks TJ (1984) Π-Π interactions and bandwidths in "molecular metals". A chemical, structural, photoelectron spectroscopic, and Hartree-Fock-Slater study of monomeric and cofacially joined dimeric silicon phthalocyanines. J Am Chem Soc 106:7748–7761; (b) Li ZY, Lieberman M (2001) Axial reactivity of soluble silicon(IV) phthalocyanines. Inorg Chem 40:932–939
39. Kasha M (1963) Energy transfer mechanisms and the molecular exciton model for molecular aggregates. Radiat Res 20:55–70
40. (a) Ishikawa N, Ohno O, Kaizu Y, Kobayashi H (1992) Localized orbital study on the electronic-structure of phthalocyanine dimers. J Phys Chem 96:8832–8839; (b) Ishikawa N (2001) Electronic structures and spectral properties of double- and triple-decker phthalocyanine complexes in a localized molecular orbital view. J Porphyrins Phthalocyanines 5:87–101
41. Oniwa K, Shimizu S, Shiina Y, Fukuda T, Kobayashi N (2013) A μ-oxo hetero dimer of silicon phthalocyanine and naphthalocyanine. Chem Commun 49:8341–8343
42. Fukuda T, Biyajima T, Kobayashi N (2010) A discrete quadruple-decker phthalocyanine. J Am Chem Soc 132:6278–6279
43. (a) Fukuda T, Kuroda W, Ishikawa N (2011) Observation of long-range f-f interactions between two f-electronic systems in quadruple-decker phthalocyanines. Chem Commun 47:11686–11688; (b) Fukuda T, Matsumura K, Ishikawa N (2013) Influence of intramolecular f-f interactions on nuclear spin driven quantum tunneling of magnetizations in quadruple-decker phthalocyanine complexes containing two terbium or dysprosium magnetic centers. J Phys Chem A 117:10447–10454
44. (a) Wang H, Qian K, Wang K, Bian Y, Jiang J, Gao S (2011) Sandwich-type Tetrakis(phthalocyaninato) Dysprosium-Cadmium Quadruple-Decker SMM. Chem Commun 47:9624–9626; (b) Wang H, Kobayashi N, Jiang J (2012) New sandwich-type phthalocyaninato-metal quintuple-decker complexes. Chem Eur J 18:1047–1049; (c) Wang H, Qi D, Xie Z, Cao W, Wang K, Shang H, Jiang J (2013) A sandwich-type phthalocyaninato metal sextuple-decker complex: synthesis and NLO properties. Chem Commun 49:889–891
45. Fukuda T, Hata K, Ishikawa N (2012) Observation of exceptionally low-lying Π–Π* excited states in oxidized forms of quadruple-decker phthalocyanine complexes. J Am Chem Soc 134:14698–14701

Part IV
Open-Shell π-Electron Systems

Chapter 17
Localized Singlet 1,3-Diradicals

Manabu Abe and Sayaka Hatano

Abstract Localized singlet diradicals are key intermediates in chemical reactions involving homolytic bond-cleavage and bond formation processes. In the past, the short-lived character of the reactive intermediate impeded the experimental investigations on the molecular and electronic structures of these intermediates. However, over the past 20 years, the chemistry of singlet diradicals has seen significant development following the pioneering studies of the generation of long-lived singlet diradicals. This chapter summarizes recent developments in the chemistry of localized singlet diradicals, which include (a) the effect of the substituent and the heteroatom on the ground-state spin multiplicity and the most stable electronic configuration of the singlet state (type 1 versus type 2), (b) the reactivity of localized singlet 1,3-diradicals, and (c) the generation and characterization of long-lived singlet diradicals with π-single-bonded character.

Keywords Singlet diradicals • π-Single bonds

17.1 Introduction

Bond-breaking and bond formation processes play a central role in chemistry. Anions [1] and cations [2, 3] are the key intermediates in heterolysis reactions, and they have been widely utilized for chemical transformations, for example, in materials synthesis. Fundamental research studies on reactive intermediates have enabled the design of appropriate pathways for synthesizing molecules. In contrast,

---

M. Abe (✉)
Department of Chemistry, Graduate School of Science, Hiroshima University (HIRODAI), 1-3-1 Kagamiyama, Higashi-Hiroshima, Hiroshima 739-8526, Japan

Institute for Molecular Science (IMS), Myoudaiji, Okazaki, Aichi 444-8787, Japan

CREST, JST, K's Gobancho, 7, Chiyoda-ku, Tokyo 102-0076, Japan
e-mail: mabe@hiroshima-u.ac.jp

S. Hatano
Department of Chemistry, Graduate School of Science, Hiroshima University (HIRODAI), 1-3-1 Kagamiyama, Higashi-Hiroshima, Hiroshima 739-8526, Japan

T. Akasaka et al. (eds.), *Chemical Science of π-Electron Systems*,
DOI 10.1007/978-4-431-55357-1_17

localized diradicals [4–7] are the key intermediates in processes involving homolytic bond-cleavage and formation reactions (Eq. 17.1). The singlet and triplet states are thermally equilibrated via the intersystem crossing process (ISC). The chemistry of triplet diradicals is well investigated, not only in their capacity as reactive intermediates but also in relation to their role in materials chemistry, because the triplet diradicals are relatively long-lived species [8–16]. The paramagnetic character of triplet diradicals enables them to function as organic magnets and other materials with novel properties. Thus, the chemistry of triplet diradicals is very rich, and their high-spin property has attracted continuous interest in many fields of chemistry [17, 18]. On the other hand, the singlet state of localized diradicals has been recognized as putative (undetectable) because the intramolecular radical–radical coupling reactions are supposed to be barrierless processes. In contrast to the rich chemistry of triplet diradicals, thus, the chemistry of localized singlet diradicals has received less attention, especially in experiments. However, a thorough investigation of these reactive intermediates would only be possible if detectable singlet diradicals could be generated. Fundamental research studies on the hitherto unknown character of localized singlet diradicals were expected not only to provide new insights into the mechanisms of homolytic reactions but also to reveal opportunities to discover new concepts and novel molecular functions (*vide infra* in Summary and Future Prospects).

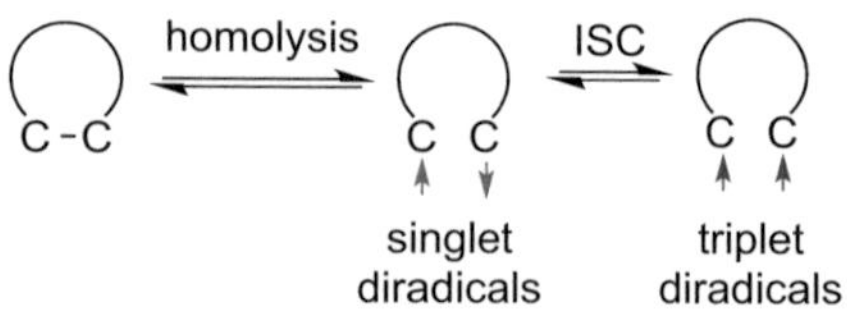

(17.1)

17.2 Generation of Cyclopentane-1,3-diyls and Cyclobutane-1,3-diyls

Singlet cyclopentane-1,3-diyls (n = 1) and cyclobutane-1,3-diyls (n = 0), which are promising intermediates in the thermal cycloreversion reaction of bicyclo[2.1.0]pentanes (n = 1) and bicyclo[1.1.0]butanes (n = 0), are good candidates for studying singlet diradical chemistry, because the singlet diradicals are kinetically stabilized as a result of the strain energies of the corresponding ring-closed bicyclic compounds (Scheme 17.1). The 1,3-diradicals of the cyclic systems are expected to be cleanly generated in the photochemical denitrogenation of the corresponding cyclic azoalkanes [19, 20]. This would enable their molecular and electronic structures to be thoroughly studied using UV–Vis, emission, and IR spectroscopic analyses.

The parent cyclopentane-1,3-diyl and cyclobutane-1,3-diyl diradicals (X = H in Scheme 17.1) were investigated by Closs, Schaefer, and Dougherty [21–24].

Scheme 17.1 Generation of cyclopentane-1,3-diyls and cyclobutane-1,3-diyls

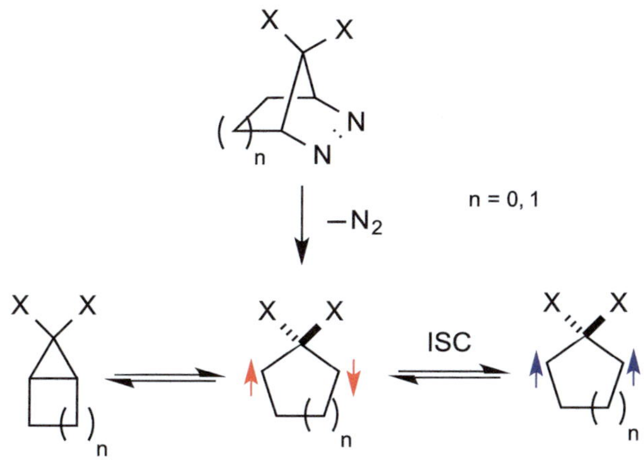

The reactivity of the triplet 2,2-dimethylcyclopentane-1,3-diyls (X = Me in Scheme 17.1) has also been well studied by Adam [25]. The triplet ground state of the 1,3-diradicals was confirmed on the basis of computational calculations and low-temperature spectroscopic analyses. The direct detection of the singlet state was not reported because (1) the triplet is the ground-state spin multiplicity and (2) the energy barrier for the ring-closing reaction is ~1 kcal mol^{-1}. The direct observation of singlet diradicals, and, hence, the investigation of their chemistry, has the indispensable minimum requirement that these diradicals be generated with a singlet ground state, because it is only the ground-state molecules that may be detected at the temperature of liquid nitrogen or helium. Thus, it is important for the investigation of singlet diradical chemistry to first understand the factors that determine the ground-state spin multiplicity in 1,3-diradicals. In the next section, the basic concept for the spin preference in diradicals, i.e., "through-space (TS)" and "through-bond (TB)" interactions, is explained. The scope of this chapter will not allow us to extend the discussion to delocalized diradicals [26–32].

17.3 Spin Preference in Diradicals: The Effects of "Through-Space (TS)" and "Through-Bond (TB)" Interaction

Diradicals are defined as molecules in which two electrons occupy two degenerate or nearly degenerate molecular orbitals [4, 16]. The ground-state spin multiplicity and the singlet–triplet energy spacing in diradicals are determined by two factors: (1) the energy gap between the two nonbonding molecular orbitals (NBMOs), which are together occupied by two electrons in total, and (2) the exchange repulsion energy. When the energy difference between the two NBMOs is sufficiently large, e.g., >

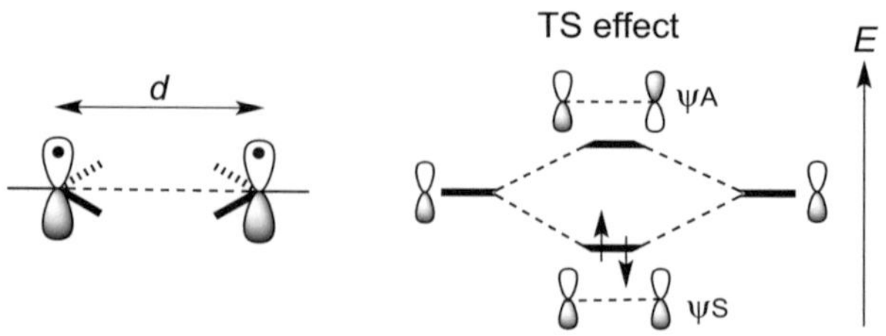

Fig. 17.1 TS interaction between the two methyl radicals in a π fashion, resulting in an energy gap between the two nonbonding molecular orbitals (NBMOs, ψ_S and ψ_A) and the most stable electronic configuration of the singlet state

~0.8 eV, the singlet is expected to be the ground-state spin multiplicity [15]. When the energy gap is small or zero, the ground-state spin multiplicity is determined by the magnitude of the exchange repulsion. Thus, as clarified by Dougherty, the singlet ground state of two adjacent methyl radicals in a π fashion is independent of the distance (d) between the two radicals, because the exchange integral is zero for two radicals that are far apart (Fig. 17.1) [24]. The TS interaction between the two radical p orbitals creates an energy gap between the two NBMOs (ψ_S and ψ_A), such that the paired electrons preferably occupy the bonding orbital ψ_S of which the energy has been lowered. Thus, the singlet state is always the ground-state spin multiplicity for the spatially orientated diradical.

In addition to TS interaction, TB interaction also plays a crucial role in adjusting the energy gap between the two NBMOs in 1,3-diradicals such as cyclopentane-1,3-diyls **DR1** (Fig. 17.2), as originally noted by Hoffmann for the stereo-mutation in the thermolysis of cyclopropanes [33]. This is because the symmetric nonbonding molecular orbital (ψ_S) is able to mix with both of the pseudo-π orbitals (π_{CX} and π^*_{CX}) of the CX_2 group in the 1,3-diradicals. The antisymmetric NBMO (ψ_A) is prevented by symmetry from mixing with the pseudo-π orbitals; hence, TB interaction leaves the energy level of ψ_A unchanged. The TB interaction causes the exchange integral between the two p orbitals to be substantial, such that the triplet is expected to be the ground-state spin multiplicity in 1,3-diradicals when the energy gap between the two NBMOs is small or zero in accordance with Hund's rule [34] (Fig. 17.3). Thus, the ground-state spin multiplicity and the singlet–triplet energy spacing in localized 1,3-diradicals are determined by the balance of magnitude between the TS interaction and the TB interaction, as shown in Fig. 17.2.

When the substituent X is an electron-withdrawing group (EWG), the TB interaction with π^*_{CX} becomes more important than that with π_{CX} to energetically stabilize ψ_S. The TB interaction increases the energy gap between ψ_S and ψ_A (type 1, red-colored TB interaction, Fig. 17.2). The ground state is supposed to be a singlet, because of the resulting large energy gap between the two orbitals. On the other hand, when X is an electron-donating group (EDG), the TB interaction

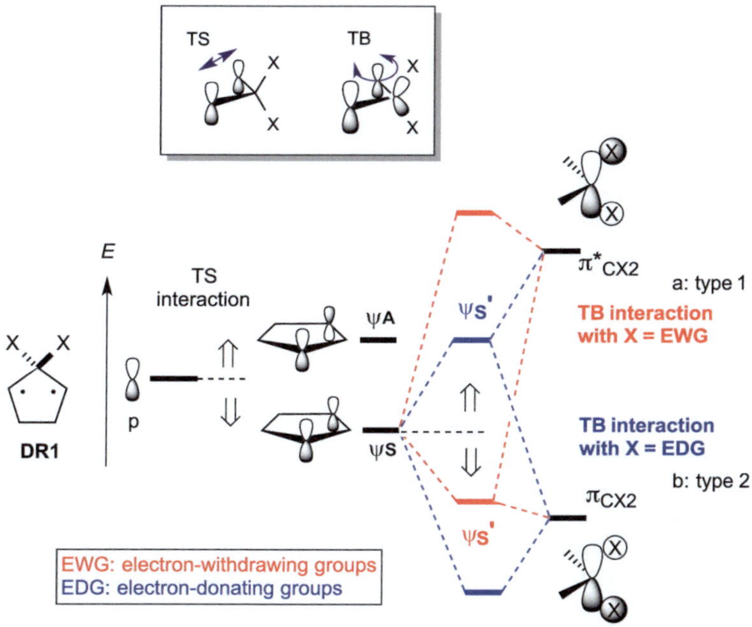

Fig. 17.2 TS and TB interaction between the two p orbitals in the 1,3-diradicals: orbital diagram showing the mixing between the symmetric NBMO ψ_S and the pseudo-π orbitals, π_{CX} and π^*_{CX}. (**a**) Type 1: the *red-colored* orbital interaction when X is an EWG, which stabilizes ψ_S. (**b**) Type 2: the *blue-colored* orbital interaction when X is an EDG, which destabilizes ψ_S

DR1

DR1	a	b	c	d	e	f
X	H	F		OH	-O(CH$_2$)$_2$O-	SiH$_3$
ΔE_{ST} (kcal mol^{-1})	+1.3	−11.2 (−9.8)*	−1.0	−6.7	−12.2	−5.2 (−5.8)*

$\Delta E_{ST} = E_S - E_T$ * at the CASPT2 level of theory

Fig. 17.3 Effect of the substituent (X) on the ground-state spin multiplicity of cyclopentane-1,3-diyls **DR1**, singlet versus triplet, $\Delta E_{ST} = E_S - E_T$, at the UB3LYP/6-31G(d) level of theory

with π_{CX} is larger than with π^*_{CX}, thereby increasing the energy level of ψ_S (type 2, blue-colored TB interaction). This is caused by the higher energy of the resulting π_{CX}. When the resulting energy gap between ψ_S and ψ_A is small or zero, the triplet state is the ground-state spin multiplicity. The existence of the triplet ground state of cyclopentane-1,3-diyl (**DR1a**, X = H) can be explained by the small energy gap between ψ_S and ψ_A. Thus, the ground state of the 1,3-diradicals **DR1** is expected to largely depend on the electronic properties of substituent X at the C(2) position.

17.4 Substituent Effect on Ground-State Spin Multiplicity and Singlet–Triplet Energy Gap in 1,3-Diradicals

The effect of the substituent X, located at the C(2) position on the ground-state spin multiplicity of cyclopentane-1,3-diyls **DR1**, is summarized in Fig. 17.3. In 1994, Borden and co-workers first reported a notable substituent effect when they observed the strong ability of electron-withdrawing fluorine atoms to energetically lower the singlet state to below the triplet state of 2,2-difluorocyclopentane-1,3-diyl (**DR1b:** X = F, $\Delta E_{ST} = E_S - E_T = -11.2$ kcal mol^{-1}, Fig. 17.3) [35, 36]. This pioneering study prompted us to investigate the effect of an alkoxy group (X = OR) on the ground-state spin multiplicity both computationally and experimentally [37]. From a synthetic point of view, alkoxy substituents present a more interesting opportunity to investigate the reactivity of 1,3 diradicals, because the oxygen atom is also highly electronegative, but has two covalent bonds. Thus, many derivatives can be designed and synthesized by modifying the substituents (R) on the oxygen atom. 2-Spiroepoxy- and 2,2-dihydroxycyclopentane-1,3-diyl (X = OH) were found to be singlet ground-state molecules with singlet–triplet energy gaps ($\Delta E_{ST} = E_S - E_T$) of −1.0 and −6.7 kcal mol^{-1}, respectively [38]. The singlet state was also calculated to be the ground-state spin multiplicity for 2,2-disilylcyclopentane-1,3-diyl **DR1f** (X = SiH_3), in which the electron-donating silyl groups are substituted at the C(2) position [39–41].

The notable substituent effect on the ground-state spin multiplicity has been explained in terms of TB interaction (Fig. 17.2) [42–44]. For the EWG-substituted diradicals (X = F or OR), as shown in Fig. 17.2a, the type 1 TB interaction of the symmetric NBMO ψ_S with the low-lying π^*_{CF} or π^*_{CO} orbital stabilizes ψ_S to increase the energy gap between ψ_S and ψ_A. Thus, the singlet was calculated to be the ground-state spin multiplicity. The smaller $|\Delta E_{ST}|$ value for 2,2-dihydroxycyclopentane-1,3-diyl **DR1d** compared to that of **DR1b** (X = F) is likely to be a consequence of the weaker electronegativity of oxygen compared to fluorine. Thus, the hyperconjugative electron delocalization of the p-π atomic orbitals (AOs) to π^*_{CX} plays an important role in stabilizing the singlet state. The complete active space multi-configuration self-consistent field (CASSCF) calculations for singlet 2,2-dihydroxycyclopentane-1,3-diyl **DR1d** revealed that 81 % of the two electrons selectively occupy ψ_S. Type 1 singlet diradicals containing electron-withdrawing substituents (i.e., where X is an EWG) possess the following electronic configuration: the highest occupied molecular orbital (HOMO) is the symmetric ψ_S, and the lowest occupied molecular orbital (LUMO) is the antisymmetric ψ_A. Thus, the π-single-bonding (−π−) character [15, 45] was proposed to be included in the singlet-state molecules with the hyperconjugative resonance structure of the zwitterions (C^+, X^-), as shown in Fig. 17.4a.

For the silyl-substituted diradicals (X = SiR_3), the type 2 TB interaction of ψ_S with the high-lying occupied π_{CSi} destabilizes ψ_S such that the energy of ψ_A is located below that of ψ_S (Figs. 17.2b and 17.4b). Thus, the hyperconjugative electron delocalization of π_{CSi} to the p-π AOs plays an important role in stabilizing

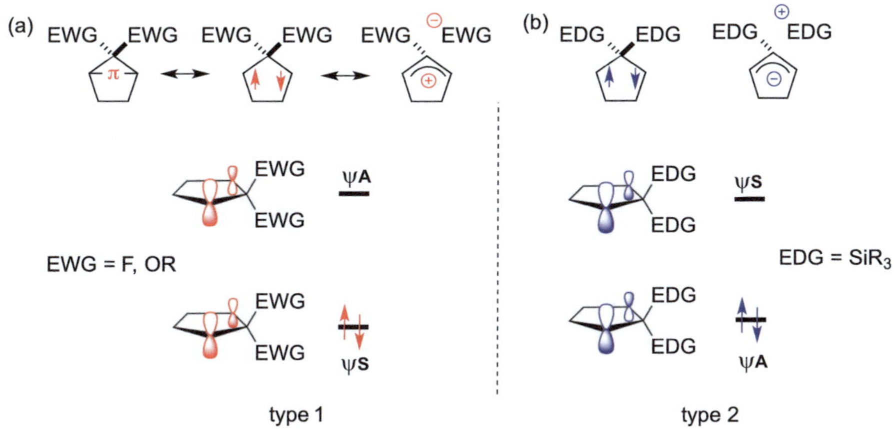

Fig. 17.4 Two types of singlet diradicals. The most stable electronic structure: (**a**) type 1, X = EWG, and (**b**) type 2, X = EDG

the singlet state. The CASSCF calculation clarified 72 % of the two electron occupied the antisymmetric NBMO ψ_A of the disilyl-substituted diradical **DR1e**. In contrast to the type 1 electronic configuration, the HOMO is ψ_A, and the LUMO is ψ_S for the type 2 singlet diradicals in Fig. 17.4b. The singlet-state molecule has the hyperconjugative resonance character of the zwitterions (C^-, X^+) and no π-single-bonding character between the two radical sites. Thus, the singlet diradicals can be categorized as being of type 1 and type 2 on the basis of the most stable electronic configuration of the singlet state. It is interesting and important not only to synthesize unusual molecules with π-single-bonded character but also to investigate the difference in reactivity of the two types of singlet 1,3-diradicals (*vide infra*).

17.5 Spiroconjugation Effect

As shown in Fig. 17.3, the singlet–triplet energy spacing ($\Delta E_{ST} = -12.2$ kcal mol^{-1}) of the ethylene ketal-substituted 1,3-diradical **DR1e** (X = –O(CH_2)$_2$O–) was found to be larger than those calculated for 2,2-difluorocyclopentane-1,3-diyl **DR1b** ($\Delta E_{ST} = -11.2$ kcal mol^{-1}) and 2,2-dihydroxycyclopentane-1,3-diyl **DR1d** ($\Delta E_{ST} = -6.7$ kcal mol^{-1}) [46]. The significant effect of the ethylene ketal substituent on the singlet–triplet energy spacing cannot be explained only by the TB interaction depicted in Fig. 17.2a. As shown in Fig. 17.5, the singlet–triplet energy gap in 2,2-dihydroxycyclopentane-1,3-diyl **DR1d** was found to be largely dependent upon the O–C–O–H dihedral angle (θ) in the 1,3-diradical. For example, at 0 ° the singlet–triplet energy gap was calculated to be −12.0 kcal mol^{-1} at the UB3LYP/6-31G(d) level of theory, which was nearly identical to the singlet–triplet energy spacing of the ethylene ketal-substituted diradical **DR1e** (Fig. 17.3).

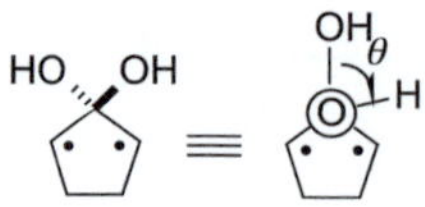

DR1d

θ /°	0	30	60	90	120	150	180
ΔE_{ST} /kcal mol^{-1}	–12	–9.1	–5.8	–4.2	–4.8	–7.8	–10
λ_{cal} /nm	383	408	435	452	458	435	418

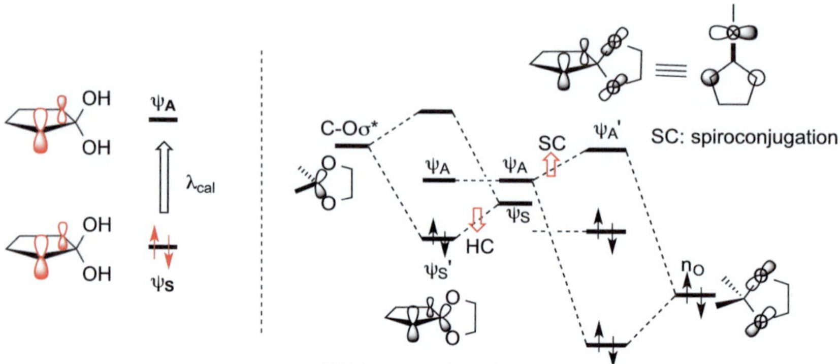

Fig. 17.5 Effect of the O–C–O–H (θ) dihedral angle on the singlet–triplet energy gap ΔE_{ST} (kcal mol^{-1}) and absorption wavelength (λ_{cal}) of 2,2-dihydroxycyclopentane-1,3-diyl **DR1d**. Spiroconjugation (SC) and hyperconjugation (HC). The spiroconjugation (SC) effect on increasing the energy gap between ψ_S and ψ_A in addition to hyperconjugative (HC) interaction

However, at 90 ° the energy gap drastically decreased and was calculated to be about -4.2 kcal mol^{-1}. Beyond 90 °, the singlet–triplet energy gap again increased to -10 kcal mol^{-1} at 180 °. The correlation suggests that the direction of the lone pair electrons (n orbital) of oxygen atoms plays a crucial role in controlling the singlet–triplet energy gap. Furthermore, the absorption wavelength (λ_{cal}, excitation energy) of the singlet diradical **DR1d** was also found to be largely dependent upon the dihedral angle (θ). The excitation energy, which corresponds to the HOMO (π) $\rightarrow$ LUMO (π^*) electronic excitation, increased from ~460 nm (~2.7 eV) at $\theta = 90$ ° to ~390 nm (~3.2 eV) at $\theta = 0$ °. The singlet–triplet energy gap of $|\Delta E_{ST}|$ also increased from 4 to 12 kcal mol^{-1}. The correlation is reasonable, because the widening of the energy gap between the HOMO and LUMO stabilizes the singlet state relative to the triplet state. The notable effect of the direction of the lone pair on the singlet–triplet energy spacing and the excitation energy is explained by spiroconjugation (SC) [47]. The SC of the n-orbital (n_O) of oxygen with the antisymmetric NBMO energetically destabilizes the antisymmetric NBMO (ψ_A), which leads to an increase in the energy gap between the two NBMOs. Thus, in addition to the TB interaction of ψ_S with π^*_{CO}, the SC plays a crucial role in increasing the energy gap (Fig. 17.5).

17.6 Heteroatom Effect

In 1995, Niecke and co-workers reported the isolation and the X-ray crystallographic analysis of C_2P_2 heterocyclic 1,3-diradicals (Fig. 17.6) [48]. In 2003, Yoshifuji and co-workers also reported the isolation of analogous diradicals of the C_2P_2 heterocyclic system [49]. The singlet was found to be the ground-state spin multiplicity of the diradicals. The zwitterionic resonance structure of C^-, P^+ was reported to be significant, which was experimentally proven by its reaction with methanol. Thus, the ground-state spin multiplicity of 1,3-diradicals was also influenced by the adjacent heteroatoms. The heteroatom effect of the phosphorus atoms can be explained by the interaction of the high-lying n-orbital of phosphorus atoms with the symmetrical NBMO ψ_S. The interaction destabilizes the energy level of ψ_S. Thus, the singlet diradicals of the C_2P_2 system are categorized as the type 2 diradicals (Fig. 17.6), i.e., HOMO = ψ_A and LUMO = ψ_S.

In 2002, Bertrand and co-workers reported the isolation of the type 1 singlet diradical B_2P_2 in the form of a four-membered ring system (Fig. 17.6) [50]. In 2004, Inagaki, Ma [51], and our group [52] found a silicon atom effect on the ground-state spin multiplicity in 1,3-diradicals, C_2Si_2. For example, 2-silacyclopentane-1,3-diyl was calculated to be the singlet ground state. Subsequent to the pioneering studies that were performed to isolate the four-membered singlet diradicals, several similar structures were reported, i.e., Sn_2N_2, Si_2N_2, Ge_2N_2, Ge_2NO [53].

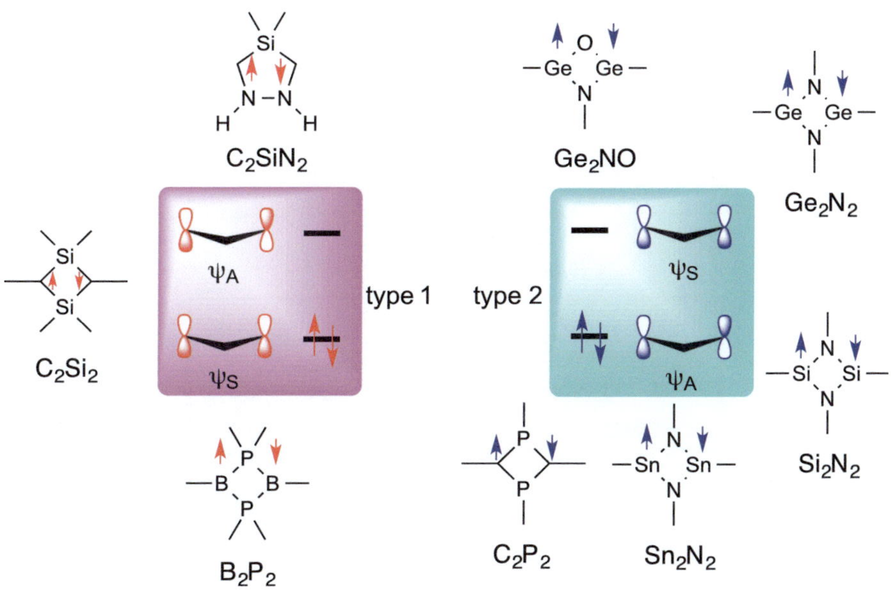

Fig. 17.6 Heteroatom effect on the ground-state spin multiplicity in the four-membered system

17.7 Reactivity of Type 1 and Type 2 Singlet Diradicals

As mentioned above, the computational calculations predict that the ground-state spin multiplicity of localized 1,3-diradicals and the most stable electronic configuration of the singlet state can be controlled by substituents on the C(2) position and the adjacent heteroatoms. Theoretical studies of the effect of the substituent and heteroatom revealed the existence of two types of singlet diradicals in cyclopentane-1,3-diyls, type 1 and type 2, which were categorized according to the most stable electronic configuration (Fig. 17.7). As the reactivity of molecules largely depends on their electronic configuration, an investigation of the difference in the reactivity of these two types of singlet diradicals is important for understanding the chemistry of singlet diradicals.

For dialkoxy-substituted diradicals with a type 1 configuration (EWG = OR), disrotatory ring closure was expected to be the energetically preferred mode to produce the *cis*-fused bicyclo[2.1.0]pentanes, since the two electrons preferably occupy the symmetric NBMO ψ_S (Fig. 17.4a). In contrast to the reactivity of type 1 diradicals, conrotatory ring closure is electronically required for type 2 diradicals, e.g., 2-silylcyclopentane-1,3-diyls (EDG = SIR_3), because the antisymmetric orbital ψ_A is preferably occupied by two electrons (Fig. 17.4b). The ring-closing mode "must" produce the transfused bicyclo[2.1.0]pentane. However, as can easily be imagined, the molecular strain of the transfused product is too high for it to be formed from the type 2 singlet diradicals. Thus, significant energy barriers are expected for the intramolecular cyclization reaction of type 2 molecules, because the ring-closing reaction must occur from the excited state, i.e., type 1 configuration (ψ_S) [41].

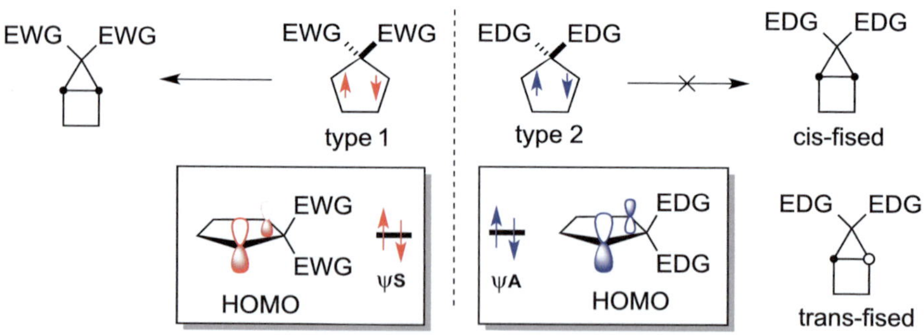

Fig. 17.7 Reactivity of the two different types of singlet diradicals: type 1 versus type 2

17.8 Experimental Studies on Localized Singlet Diradicals with Type 1 Electronic Configuration

The reactivity of 1,3-diradicals **DR2,3** with an alkoxy group at the C(2) position was investigated (Fig. 17.8) to understand (1) the substituent and heteroatom effects on the ground-state spin multiplicity of cyclopentane-1,3-diyls and (2) the reactivity of the type 1 singlet diradicals. The diradicals were cleanly generated by the denitrogenation of the corresponding azoalkanes **AZ2,3**. The generation and reactivity of the triplet dimethyl-substituted diradical **DR5** (X = Me) and its role in the photochemical denitrogenation of the corresponding azoalkane were thoroughly investigated by Adam and co-workers [54]. The triplet diradical was reported to show typical electron paramagnetic resonance (EPR) signals of triplet states and to absorb light at around ~330 nm. The absorption maximum in the ultraviolet region is known to be typical of those reported for other benzyl radicals [55]. As predicted for singlet diradicals, the absorption maximum of the model singlet 2,2-dihydroxycyclopentane-1,3-diyl **DR1d** (Fig. 17.5) was found to occur in the visible region. Moreover, the electronic excitation energy was found to correlate with the singlet–triplet energy spacing (Fig. 17.5). This valuable information, including the ground-state spin multiplicity and the impact of the substituents on the singlet–triplet energy spacing, was obtained by measuring the absorption spectra of the diradicals **DR2–5**. The results are shown in Figs. 17.8 and 17.9.

In 1998, Adam, Borden, and Wirz and their co-workers observed the localized singlet 1,3-diradical **DR2** (X = F) using a conventional nanosecond laser flash photolysis technique [56]. The diradical was generated by the denitrogenation of

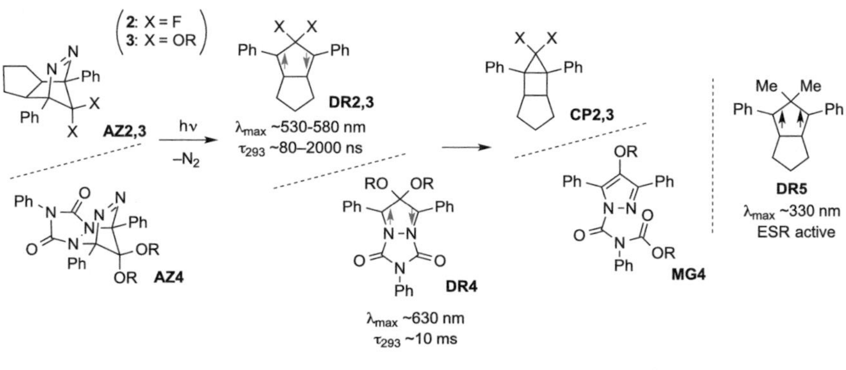

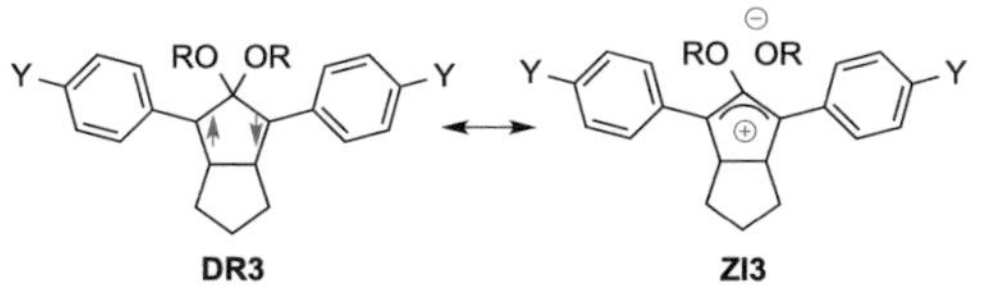

	OR	Y	τ_{293} in C_6H_6	τ_{293} in n-hex	τ_{293} in CH_3CN
a	OEt	H	880 ns	520 ns	1010 ns
c	OMe	H	300 ns		
d	OMe	OMe	1050 ns		
e	OMe	CN	625 ns		

Fig. 17.8 Generation of diradicals, **DR2–5**, from azoalkanes, **AZ2–5**

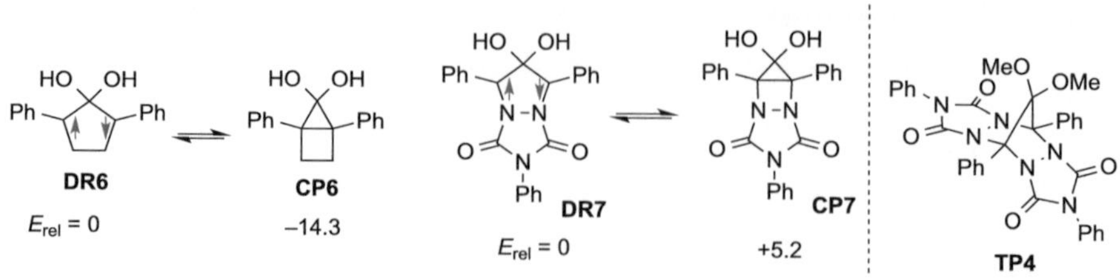

Fig. 17.9 Heteroatom effect on equilibration between singlet diradicals and the σ-bonded ring-closed product

the corresponding azoalkane **AZ2** (Fig. 17.8) and was found to persist at liquid nitrogen temperatures and absorb light in the visible region, λ_{max} ~530 nm. The lifetime of this diradical was determined to be ~80 ns at 293 K in *n*-pentane. The temperature dependence of the rate of disappearance of the diradical was measured. The activation parameters of $E_a = 7.8$ kcal mol^{-1} and log*A* 12.8 that were obtained were very different from those obtained for the disappearance of the triplet cyclopentane-1,3-diyl **DR5**, $E_a = 2.6$ kcal mol^{-1} and log*A* 6.7. At 77 K, the singlet diradical displayed no EPR signals, which clearly indicates that the energy of the singlet state is considerably lower than that of the triplet state. Indeed, a large energy spacing of 7.4 kcal mol^{-1} was calculated for the 2,2-difluoro-1,3-diphenylcyclopentane-1,3-diyl.

In 2000, a much longer-lived singlet diradical **DR3a** (OR = OEt, τ = 520 ns in *n*-hexane at 293 K, *E*a = 7.6 kcal mol^{-1}, log*A* = 12.0) was reported [38]. The singlet spin multiplicity was proven by the following observations: (1) the rate of the species to produce the ring-closed product **CP3a** (OR = OEt) quantitatively by the disrotatory ring-closing reaction was found to be first order; (2) a strong absorption maximum (λ_{max} 570 nm in 2-methyltetrahydrofuran (MTHF) at 77K, ~3500 M^{-1} cm^{-1}) was observed in the visible region; (3) the EPR displayed no signals at 77 K; (4) the lifetime of the diradical was insensitive to the presence of molecular oxygen, and the quenching rate constant, $k_{O2} = \sim 3 \times 10^6$ M^{-1} s^{-1}, was determined to be significantly lower than that of the triplet, ~3×10^9 M^{-1} s^{-1}; (5) the high log*A* value was indicative of a spin-allowed reaction, i.e., the ring closure to **CP3a**; (6) the calculated absorption spectra ($\lambda_{calcd} = 625$ nm at the TD-UB3LYP/6-31 + G* level of theory) of the singlet state of 2,2-dihydroxy-substituted diradical **DR3b** (OR = OH) were consistent with the singlet diradical **DR3a**; and (7) the absorption maximum of the corresponding triplet state was calculated to be 345 nm at the same level of theory.

This breakthrough for the detection of the localized singlet diradicals enabled us to design experiments to thoroughly investigate the reactivity of localized singlet diradicals. First, the solvent effect on the lifetime of the singlet diradical **DR3a** (OR = OEt) was examined to determine the character of the singlet 2,2-dialkoxy-1,3-diradicals. The lifetime was found to increase with increasing solvent polarity, i.e., 520 ns (*n*-hexane) < 880 ns (benzene) < 1010 ns (acetonitrile) [38]. The effect of the solvent on the lifetime indicates that the singlet state has the character of a zwitterion **ZI3a** (OR = OEt), i.e., it has the hyperconjugative resonance structure of singlet diradicals (Figs. 17.4 and 17.8). The zwitterionic character was further confirmed by examining the electronic substituent effect on the lifetime of the singlet diradicals **DR3c,d** (Fig. 17.8) [45]. The lifetime ($\tau = 1050$ ns in benzene at 293 K) of the singlet diradical **DR3d**, which has an electron-donating methoxy group (Y = OMe) at the *para*-position of the phenyl ring, was found to be longer than those of the parent diradical **DR3c** (Y = H) and the EWG-substituted diradical **DR3e** (Y = CN, $\tau = 625$ ns in benzene at 293 K). The substituent effect could be explained by the stabilization of the allylic cation in the hyperconjugative resonance structure. The radical character of the singlet diradical was also proposed to be significant, because the lifetime of **DR3e** (Y = CN) was found to be longer than that of the parent diradical **DR3c** (Y = H). The stabilization power of the *para*-cyano group on the benzyl-type radical was suggested to be responsible for the longer lifetime of **DR3e**, because the stabilizing effect of the substituent on the benzyl radical is in the following order: CN ($\sigma_C{}^\bullet = 0.47$) > OMe ($\sigma_C{}^\bullet = 0.27$) > H ($\sigma_C{}^\bullet = 0.00$) [57]. The strong absorption band around 570 nm was assigned to the HOMO(π)–LUMO(π_*) excitation in the π-single bond between the two radical sites.

17.9 Generation and Reactivity of DR7,8

In general, the ring-closed compounds (σ-bonded compounds), i.e., bicyclo[2.1.0]pentane derivatives **CP**, are energetically more stable than the singlet 2,2-dialkoxy-1,3-diphenylcyclopentane-1,3-diyls, which possess an open-shell character (Fig. 17.9). Indeed, the photochemical denitrogenation of the azoalkanes **AZ2,3** produced the ring-closed compounds **CP2,3** quantitatively by a disrotatory ring-closing mechanism from **DR2,3** [58]. In accordance with the experimental observation, the model ring-closed compound **CP6** was calculated to be more stable than the singlet diradical **DR6** by ca. 14 kcal mol^{-1} at the (U)B3LYP/6-31G(d) level of theory (Fig. 17.9). In contrast to the singlet cyclopentane-1,3-diyl diradicals, the singlet 1,2-diazacyclopentane-3,5-diyl diradicals and 1,2-diaza-4-silacyclopentane-3,5-diyl diradicals were calculated to be more stable than the corresponding ring-closed compounds by ~5 kcal mol^{-1} at the (U)B3LYP/6-31G(d) level of theory (Fig. 17.9). Although the UB3LYP method is known to overestimate the stability of the singlet diradicals, the computations clearly suggest that the diradicals **DR4** were expected to be longer lived than the singlet diradicals **DR2,3** (Fig. 17.8).

To examine the effect of the heteroatom on the ground-state spin multiplicity and the reactivity of the singlet diradicals, the azoalkane **AZ4** was prepared, after which photochemical denitrogenation was performed (Fig. 17.8). This produced the oxygen migration product **MG4** in 70% isolated yield. No trace of the intramolecular ring-closure compound **CP4** was detected. The generation of the singlet diradical **DR4** in the photodenitrogenation reaction was confirmed by performing a spectroscopic analysis of the photodenitrogenation of **AZ4** at 77 K in a MTHF glass matrix. A green-colored species, which displayed no EPR signals and was stable up to 140 K, was detected (λ_{max} 630 nm). The absorption maximum was similar to the calculated value of λ_{calcd} 710 nm for the model diradical **DR7**. The triplet state of this diradical was predicted to absorb light at λ_{calcd} 366 nm. The ground state of **DR7** was calculated to be a singlet, $\Delta E_{ST} = -3.9$ kcal mol^{-1}. Thus, the green-colored species absorbing at λ_{max} 630 nm was assigned to the singlet state of the diradical **DR4**. Upon addition of 4-phenyl-1,2,4-triazole-3,5-dione (PTAD) at 130K, the green color of **DR4** was observed to disappear and the PTAD-trapping product **TP4** was isolated in 42% yield [59]. The lifetime of the singlet diradical was found to be extremely long, ca. 10 ms at 293 K, which was determined by using laser flash photolysis in toluene solution.

17.10 Generation and Reactivity of Singlet Diradicals with a Type 2 Electronic Configuration

The generation and reactivity of the localized singlet diradical **DR8**, which has a type 2 electronic configuration, were investigated by studying the photochemical (hν ~340 nm) and thermal (100 °C) denitrogenation reactions of the silyl-substituted azoalkane **AZ8** (Fig. 17.10) [41]. The intramolecular cyclization product **CP8** was not observed to form during these reactions, and the silyl-migrated product 3-silylcyclopentene **MG8** was isolated in quantitative yields. The intramolecular cyclization product **CP8**, a bicyclo[2.1.0]pentane derivative, was reported to be stable below 150 °C. The thermal stability of the intramolecular cyclization product **CP8** suggests that 5-silylcyclopentene **MG8** was not derived from the intramolecular cyclization product, but was the primary product in the denitrogenation reactions of **AZ8**.

A quantitative understanding of the reactivity of 2-silylcyclopentane-1,3-diyl **DR8** was achieved by calculating the transition states for the ring-closing reactions and the silyl-migration step for the model compound ($SiR_3 = TMS$)

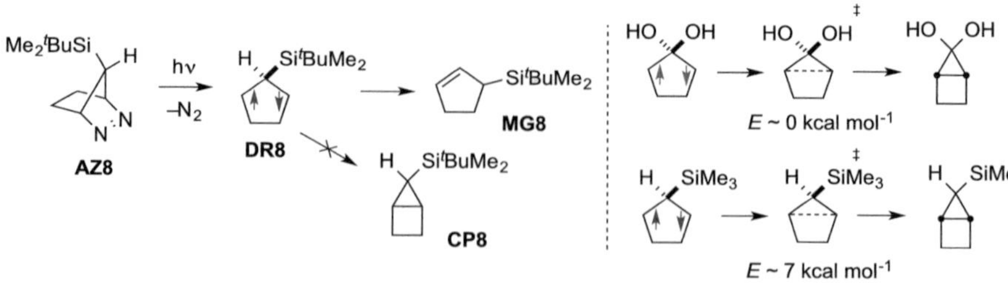

Fig. 17.10 Generation of a type 2 singlet diradical and its reactivity

at the UB3LYP/6-31G(d) level of theory. In contrast to the barrierless process followed by the disrotatory ring-closing mode of the type 1 diradicals, e.g., 2,2-dihydroxycyclopentane-1,3-diyl **DR2d**, significant energy barriers of ~7 kcal mol^{-1} were calculated for the formation of both the endo- and exo-configured bicyclo[2.1.0]pentane from the type 2 singlet diradical, i.e., 2-trimethylsilylcyclopentane-1,3-diyl. The singlet 2-silylcyclopentane-1,3-diyl was found to produce the silylcyclopentene derivative in a suprafacial manner via a transition state involving a concerted silyl-migration pathway. The energy barrier of the migration step was calculated to be ~3 kcal mol^{-1} at the UB3LYP/6-31G(d) level of theory, which was significantly lower than those of the ring-closing reactions.

17.11 Mechanism of the Denitrogenation Reaction of Cyclic Azoalkanes

As discussed above, the most stable electronic configuration of localized singlet diradicals plays a crucial role in determining their chemistry. In 2005, the most stable electronic configuration of the 1,3-diyls was also found to affect the thermal denitrogenation mechanism of the precursor azoalkanes, i.e., 2,3-diazabicyclo[2.2.1]hept-2-ene (DBH) derivatives (Fig. 17.11) [60]. Thus, when the HOMO of the lowest singlet state of the resulting diradicals is an antisymmetric NBMO (ψ_A), i.e., a type 2 configuration, the concerted denitrogenation is the symmetry-allowed process, as the phase of the HOMO matches the π^* of the nitrogen molecule (Fig. 17.11). However, when the HOMO is the symmetric NBMO (ψ_S), i.e., a type 1 configuration, the concerted denitrogenation is a symmetry-forbidden process, because the two phases do not match each other. Thus, the concerted denitrogenation would be the preferred denitrogenation mechanism for the electron-donating silyl group-substituted azoalkane (X = SiR_3). In contrast,

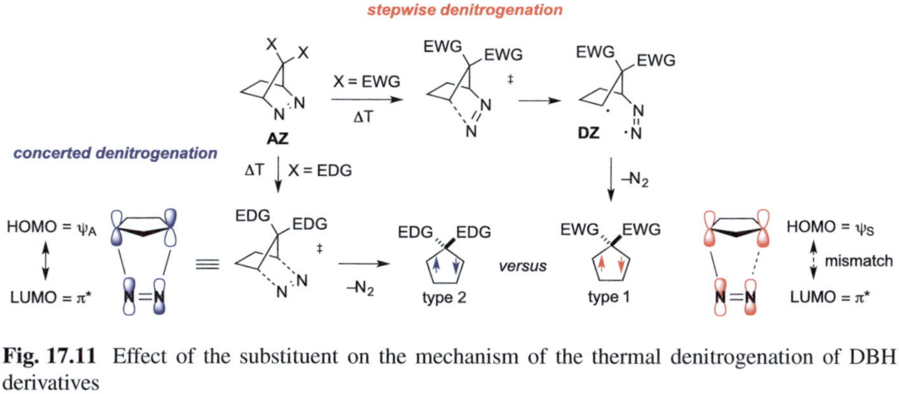

Fig. 17.11 Effect of the substituent on the mechanism of the thermal denitrogenation of DBH derivatives

the stepwise denitrogenation would be the preferred mechanism in the thermal denitrogenation of the EWG-substituted azoalkanes (X = F, OR). The substituent effect was proven by using a combination of quantum chemical calculations and experimental studies.

17.12 Summary and Future Prospects

Recent developments in the chemistry of localized 1,3-diradicals were summarized in this chapter, which include (1) the effect of the substituent and heteroatom on the ground-state spin multiplicity and the most stable electronic configuration of the singlet state, (2) the generation and characterization of long-lived singlet diradicals, and (3) the intra- and intermolecular reactivity of localized singlet diradicals.

The question is which aspects of chemistry were made accessible by the thorough computational and experimental studies of localized singlet diradicals? There are two types of singlet diradicals, which were categorized according to their most stable electronic configurations. For type 1 molecules, a new bonding concept "π-single bonding (−π−)" is proposed (Fig. 17.12), because the two electrons selectively occupy the π-bonding orbital ψ_S. The π-single-bonded species possess a small HOMO–LUMO gap, which was probed by examining the absorption band in the visible region. The small energy gap may play an important role in the redox property of the singlet diradicals. The π-single-bonded character of the type 1 singlet diradical **DR10** was suggested by measuring the transient IR spectra on a nanosecond time scale. The CN stretching mode of **DR10** occurs at a higher frequency than that of the triplet diradical **DR9**, indicating that the radical character of **DR10** is smaller than that of **DR9** [61].

The unusual molecular and electronic structure, i.e., the planar four-coordinate carbon atoms [62], is expected to stimulate future theoretical and experimental studies. For the type 2 diradicals, as mentioned for the C_2P_2 systems, the open-shell compounds were found to be surprisingly stable because of the high-energy barrier

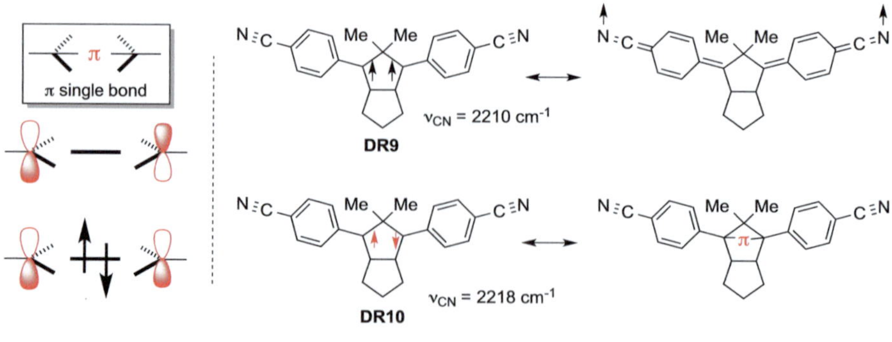

Fig. 17.12 π-Single-bonded compounds

for the intramolecular radical–radical coupling reaction. Thus, the unusual stability found in heteroatom-centered diradicals is expected to open new fields of materials chemistry, e.g., radical batteries, because the open-shell molecules are expected to undergo facile reduction and oxidation.

References

1. Amyes TL, Toteva MM, Richard JP (2004) In: Moss RA, Platz MS, Jones M (eds) Reactive intermediate chemistry. Wiley, Hoboken, pp 41–68
2. McClelland RA (2004) In: Moss RA, Platz MS, Jones M (eds) Reactive intermediate chemistry. Wiley, Hoboken, pp 3–40
3. Gronert S (2004) In: Moss RA, Platz MS, Jones M Jr (eds) Reactive intermediate chemistry. Wiley, Hoboken, pp 69–120
4. Borden WT (ed) (1982) Diradicals. Wiley, New York
5. Platz MS (ed) (1990) Kinetics and spectroscopy of carbenes and biradials. Plenum Press, New York
6. Salem L, Rowlaand C (1972) Electronic properties of diradicals. Angew Chem Int Ed 11:92–111
7. Namai H, Ikeda H, Hoshi Y, Mizuno K (2007) Thermoluminescence originating from the singlet excited state of 1,4-diarylcyclohexane-1,4-diyls: a potentially general strategy for the observation of short-lived biradicals. Angew Chem Int Ed 46:7396–7398
8. Scaiano JC (1982) Does intersystem crossing in triplet biradicals generate singlets with conformational memory. Tetrahedron 38:819–824
9. Doubleday C Jr, Turro NJ, Wang J-F (1989) Dynamics of flexible triplet biradicals. Acc Chem Res 22:199–205
10. Adam W, Grabowski S, Wilson RM (1990) Localized cyclic triplet diradicals – lifetime determination by trapping with oxygen. Acc Chem Res 23:165–172
11. Griesbeck AG, Mauder H, Stadtmueller S (1994) Intersystem crossing in triplet 1,4-biradicals – conformational memory effects on the stereoselectivity of photocycloaddition reactions. Acc Chem Res 27:70–75
12. Griesbeck AG, Abe M, Bondock S (2004) Selectivity control in electron spin inversion processes: regio- and stereochemistry of Paterno-Buchi photocycloadditions as a powerful tool for mapping intersystem crossing processes. Acc Chem Res 37:919–928
13. Johnston LJ, Scaiano JC (1989) Time-resolved studies of biradical reactions in solution. Chem Rev 89:521–547
14. Johnston LJ (1993) Photochemistry of radicals and biradicals. Chem Rev 93:251–266
15. Abe M, Ye J, Mishima M (2012) The chemistry of localized singlet 1,3-diradicals (biradicals): from putative intermediates to persistent species and unusual molecules with a pi-single bonded character. Chem Soc Rev 41:3808–3820
16. Abe M (2013) Diradicals. Chem Rev 113:7011–7088
17. Adam W, Harrer HM, Kita F, Nau WM (1997) Localized triplet diradicals as a probe for electronic substituent effects in benzyl-type radicals: the delta D scale. Pure Appl Chem 69:91–96
18. Lahti PM (ed) (1999) Magnetic properties of organic materials. Marcel Dekker, New York
19. Jain R, Sponsler MB, Coms FD, Dougherty DA (1988) Cyclobutanediyls: a new. Class of localized biradicals. Synthesis and EPR spectroscopy. J Am Chem Soc 110:1356–1366
20. Engel PS (1980) Mechanism of the thermal and photochemical decomposition of azoalkanes. Chem Rev 80:99–150
21. Buchwalter SL, Closs GL (1975) Electron-spin resonance study of matrix-isolated 1,3-cyclopentadiyl, a localized 1,3-carbon biradical. J Am Chem Soc 97:3857–3858

22. Buchwalter SL, Closs GL (1979) Electron-spin resonance and CIDNP studies on 1,3-cyclopentandiyls – localized 1,3 carbon biradical system with a triplet ground-state – tunneling in carbon-carbon bond formation. J Am Chem Soc 101:4688–4694
23. Conrad MP, Pitzer RM, Schaefer HF III (1979) Geometrical structure and energetics of Closs's diradical: 1,3-cyclopentadiyl. J Am Chem Soc 101:2245–2246
24. Dougherty DA (1991) Spin control in organic-molecules. Acc Chem Res 24:88–94
25. Adam W, Harrer HM, Kita F, Nau WM (1998) The EPR spectroscopic D parameter of localized triplet diradicals as probe for electronic effects in benzyl-type monoradicals. Adv Photochem 24:205–254
26. Boden WT, Iwamura H, Berson JA (1994) Violations of Hund's rule in non-Kekule hydrocarbons: theoretical prediction and experimental verification. Acc Chem Res 27:109–116
27. Berson JA (1997) A new class of non-Kekule molecules with tunable singlet-triplet energy spacings. Acc Chem Res 30:238–244
28. Matsuda K, Iwamura H (1998) Demonstration of the degeneracy of the singlet and triplet states in 2,3-dimethylenecyclohexane-1,4-diyl by measurement of its magnetic properties. J Chem Soc Perkin Trans 2: 1023–1026
29. (a) Nakano M, Minami T, Yoneda K, Muhammad S, Kishi R, Shigeta Y, Kubo T, Rougier L, Champagne B, Kamada K, Ohta K (2011) Giant enhancement of the second hyperpolarizabilities of open-shell singlet polyaromatic diphenalenyl diradicaloids by an external electric field and donor-acceptor substitution. J Phys Chem Lett 2:1094–1098. (b) Shimizu A, Nakano M, Hirao Y, Kubo T (2011) Experimental consideration of covalent bonding interactions in stacks of singlet biradicals. J Phys Org Chem 24:876–882
30. Hiroto S, Furukawa K, Shinokubo H, Osuka A (2006) Synthesis and biradicaloid character of doubly linked corrole dimers. J Am Chem Soc 128:12380–12381
31. Goldberg AH, Doughterty DA (1983) Effects of through-bond and through-space interactions on singlet-triplet energy gaps in localized biradicals. J Am Chem Soc 105:284–290
32. Borden WT, Davidson ER (1977) Some aspects of potential surface for singlet trimethylenemethane. J Am Chem Soc 99:4587–4594
33. (a) Hoffmann R (1968) Triethylene and addition of methylene to ethylene. J Am Chem Soc 90:1475–1485. (b) Hoffmann R (1971) Interaction of orbitals through bonds. Acc Chem Soc 4:1–9
34. Hund FZ (1925) Concerning the interpretation of complex spectra, especially the elements scandium to nickel. Physik 33:345–371
35. Xu JD, Hrovat DA, Borden WT (1994) Ab Initio calculations of the potential surfaces for the lowest singlet and triplet states of 2,2-difluorocyclopentane-1,3-diyl. The singlet diradical lies below the triplet. J Am Chem Soc 116:5425–5427
36. Getty SJ, Hrovat DA, Borden WT (1994) Ab-Initio calculations on the stereomutation of 1,1-difluorocyclopropane – prediction of a substantial preference for coupled disrotation of the methylene groups. J Am Chem Soc 116:1521–1527
37. Abe M, Adam W, Nau WM (1998) Photochemical generation and methanol trapping of localized 1,3 and 1,4 singlet diradicals derived from a spiroepoxy-substituted cyclopentane-1,3-diyl. J Am Chem Soc 120:11304–11310
38. Abe M, Adam W, Heidenfelder T, Nau WM, Zhang X (2000) Intramolecular and intermolecular reactivity of localized singlet diradicals: the exceedingly long-lived 2,2-diethoxy-1,3-diphenylcyclopentane-1,3-diyl. J Am Chem Soc 122:2019–2026
39. Skancke A, Hrovat DA, Borden WT (1998) Ab Initio calculations of the effects of geminal silyl substituents on the stereomutation of cyclopropane and on the singlet—triplet splitting in trimethylene. J Am Chem Soc 120:7079–7084
40. Johnson WTG, Hrovat DA, Skancke A, Borden WT (1999) Ab initio calculations find 2,2-disilylcyclopentane-1,3-diyl is a singlet diradical with a high barrier to ring closure. Theor Chem Acc 102:207–225
41. Abe M, Kawanami S, Ishihara C, Nojima M (2004) 2-Silyl group effect on the reactivity of cyclopentane-1,3-diyls. Intramolecular ring-closure versus silyl migration. J Org Chem 69:5622–5626

42. Borden WT (1998) Effects of electron donation into C–F σ* orbitals: explanations, predictions and experimental tests. Chem Commun 1919–1925
43. Abe M, Hattori M, Takegami A, Masuyama A, Hayashi T, Seki S, Tagawa S (2006) Experimental probe for hyperconjugative resonance contribution in stabilizing the singlet state of 2,2-dialkoxy-1,3-diyls: regioselective 1,2-oxygen migration. J Am Chem Soc 128:8008–8014
44. Abe M, Ishihara C, Nojima M (2004) 2-Silyl group effect on the reactivity of cyclopentane-1,3-diyls. Intramolecular ring-closure versus silyl migration. J Org Chem 68:1618–1621
45. Abe M, Adam W, Hara M, Hattori M, Majima T, Nojima M, Tachibana K, Tojo S (2002) On the electronic character of localized singlet 2,2-dimethoxycyclopentane-1,3-diyl diradicals: substituent effects on the lifetime. J Am Chem Soc 124:6540–6541
46. Abe M, Adam W, Borden WT, Hattori M, Hrovat DA, Nojima M, Nozaki K, Wirz J (2004) Effects of spiroconjugation on the calculated singlet-triplet energy gap in 2,2-dialkoxycyclopentane-1,3-diyls and on the experimental electronic absorption spectra of singlet 1,3-diphenyl derivatives. Assignment of the lowest-energy electronic transition of singlet cyclopentane-1,3-diyls. J Am Chem Soc 126:574–582
47. Simmons HE, Fukunaga T (1967) Spirocojugation. Am Chem Soc 89:5208–5215
48. (a) Niecke E, Fuchs A, Baumeister F, Nieger M, Schorller WW (1995) A P2C2 4-membered ring with unusual bonding – synthesis, structure, and ring-opening of A 1,3-diphosphacyclobutane-2,4-diyl. Angew Chem Int Ed 34:555–557. (b) Sebastian M, Nieger M, Szieberth D, Nyulászi L, Niecke E (2004) Synthesis and structure of a 1,3-diphosphacyclobutadienediide: an aniomesolytic fragmentation of a 1,3-diphosphetane-2,4-diyl in solution. Angew Chem Int Ed 43:637–641. (c) Niecke E, Fuchs A, Nieger M (1999) Valence isomerization of a 1,3-diphosphacyclobutane-2,4-diyl: photochemical ring closure to 2,4-diphosphabicyclo 1.1.0 butane and its thermal ring opening to gauche-1,4-diphosphabutadiene. Angew Chem Int Ed 38:3028–3031. (d) Schmidt O, Fuchs A, Gudat D, Nieger M, Hoffbauer W, Niecke E, Schoeller WW (1998) Valence isomerization in the solid state: from 1,3-diphosphacyclobutane-2,4-diyl to 1,2-dihydro-1,2-diphosphete. Angew Chem Int Ed 37:949–952
49. (a) Sugiyama CPH, Ito S, Yoshifuji M (2003) Synthesis of a 1,3-diphosphacyclobutane-2,4-diyl from Mes*. Angew Chem Int Ed 32:3802–3804. (b) Ito S, Miura J, Yoshifuji M, Arduengo AJ, III (2008) Poly(biradicals): oligomers of 1,3-diphosphacyclobutane-2,4-diyl units. Angew Chem Int Ed 47:6418–6421
50. (a)Scheschkewitz D, Amii H, Gornitzka H, Schoeller WW, Bourissou D, Bertrand G (2002) Singlet diradicals: from transition states to crystalline compounds. Science 295:1880–1881. (b) Amii H, Vranicar L, Gornitzka H, Bourissou D, Bertrand G (2004) Radical-type reactivity of the 1,3-dibora-2,4-diphosphoniocyclobutane-1,3-diyl. J Am Chem Soc 126:1344–1345. (c) Scheschkewitz D, Amii H, Gornitzka H, Schoeller WW, Bourissou D, Bertrand G (2004) Alpha-bond stretching: a static approach for dynamic process. Angew Chem Int Ed 43:585–587. (d) Gandon V, Bourg J.-B, Tham FS, Schoeller WW, Bertrand G, Angew Chem Int Ed 47:155–159. (e) Rodriguez A, Fuks G, Bourg JB, Bourissou D, Tham FS, Bertrand G (2008) 1,3-diborata-2,4-diphosphoniocyclobutane-1,3-diyls communicate through a para-phenylene linker. Dalton Trans 4482–4487. (f) Gandon V, Bourg J-B, Tham FS, Schoeller WW, Bertrand G (2008) The existence of two short-bond isomers for bicyclo[1.1.0]butane derivatives based on boron and phosphorus. Angew Chem Int Ed 47:155–159
51. Ma J, Ding Y, Hattori K, Inagaki S (2004) Theoretical designs of singlet localized 1,3-diradicals. J Org Chem 69:4245–4255
52. Abe M, Ishihara C, Takegami A (2004) Theoretical calculations of the effects of 2-heavier group 14 element and substituents on the singlet-triplet energy gap in cyclopentane-1,3-diyls and computational prediction of the reactivity of singlet 2-silacyclopentane-1,3-diyls. J Org Chem 69:7250–7255
53. (a) Takeuchi K, Ichinohe M, Sekiguchi A (2011) Access to a stable Si2N2 four-membered ring with non-kekule singlet biradical character from a disilyne. J Am Chem Soc 133:12478–12481. (b) Fuks G, Saffon N, Maron L, Bertrand G, Bourissou D (2009) Ionic-type reactivity

of 1,3-dibora-2,4-diphosphoniocyclobutane-1,3-diyls: regio- and stereoselective addition of hydracids. J Am Chem Soc 131:13681–13689. (c) Cui C, Brynda M, Olmstead MM, Power PP (2004) Synthesis and characterization of the non-kekule, singlet biradicaloid Ar'Ge(mu-NSiMe3)(2)GeAr ' (Ar '=2,6-dipp(2)C(6)H(3), dipp=2,6-i-Pr2C6H3). J Am Chem Soc 126:6510–6511. (d) Cox H, Hitchcock PB, Lappert MF, Peirssens LJ-M (2004) A 1,3-diaza-2,4-distannacyclobutanediide: synthesis, structure, and bonding. Angew Chem Int Ed 43:4500–4504
54. Adam W, Froehlich L, Nau WM, Korth H-G, Sustmann R (1993) Substituent effects on the zero-field splitting parameters of localized triplet 1,3-cyclopentanediyl biradicals. Angew Chem Int Ed 32:1339–1340
55. Tokumura K, Udagawa M, Ozaki T, Itoh M (1987) Doublet-doublet fluorescence of benzyl, *p*-methylbenzyl and *p*-chlorobenzyl radicals in solution. Chem Phys Lett 141:558–563
56. Adam W, Borden WT, Burda C, Foster H, Heidenfelder T, Jeubes M, Hrovat DA, Kita F, Lewis SB, Scheutzow D, Wirz J (1998) Transient spectroscopy of a derivative of 2,2-difluoro-1,3-diphenylcyclopentane-1,3-diyl - A persistent localized singlet 1,3-diradical. J Am Chem Soc 120:593–594
57. Creary X (2006) Super radical stabilizers. Acc Chem Res 39:761–771
58. (a) Abe M, Kubo E, Nozaki K, Matsuo T, Hayashi T (2006) An extremely long-lived singlet 4,4-dimethoxy-3,5-diphenylpyrazolidine-3,5-diyl derivative: a notable nitrogen-atom effect on intra- and intermolecular reactivity. Angew Chem Int Ed 45:7828–7831. (b) Abe M, Kubo E, Nozaki K, Matsuo T, Hayashi T (2012) Corrigendam, an extremely long-lived singlet 4,4-dimethoxy-3,5-diphenylpyrazolidine-3,5-diyl derivative: a notable nitrogen-atom effect on intra- and intermolecular reactivity. Angew Chem Int Ed 51:11924. (c) Nakamura T, Gagliardi L, Abe M (2010) Computational study of the cooperative effects of nitrogen and silicon atoms on the singlet-triplet energy spacing in 1,3-diradicals and the reactivity of their singlet states J Phys Org Chem 23:300–307
59. Nakamura T, Takegami A, Abe M (2010) Generation and intermolecular trapping of 1,2-Diaza-4-silacyclopentane-3,5-diyls in the denitrogenation of 2,3,5,6-tetraaza-7-silabicyclo 2.2.1 hept-2-ene: an experimental and computational study. J Org Chem 75:1956–1960
60. Abe M, Ishihara C, Kwanami S, Masuyama A (2005) Novel substituent effects on the mechanism of the thermal denitrogenation of 2,3-diazabicyclo[2.2.1]hept-2-ene derivatives, stepwise versus concerted. J Am Chem Soc 127:10–11
61. Maeda A, Oshita T, Abe M, Ishibashi T-A (2014) Time-resolved IR spectroscopy of 1,3-dicyanophenylcyclopentane-1,3-diyl diradicals: CN stretching wavenumber as a vibrational signature of radical character. J Phys Chem B 118:3991–3997
62. (a) Hoffmann R, Alder RW, Wilcox CF (1970) Planar tetracoordinate carbon. J Am Chem Soc 92:4992–4993. (b) Cooper OJ, Wooles AJ, McMaster J, Lewis W, Blake AJ, Liddle ST (2010) A monomeric dilithio methandiide with a distorted *trans*-planar four-coordinate carbon. Angew Chem Int Ed 49:5570–5573. (c) Collins JB, Dill JD, Jemmis ED, Schleyer PvR, Seeger R, Pople JA (1976) Stabilization of planar tetracoordinate carbon. J Am Chem Soc 98:5419–5427. (d) Priyakumar UD, Sastry GN (2004) A system with three contiguous planar tetracoordinate carbons is viable: a computational study on a C6H62þ isomer. Tetrahedron Lett 45:1515–1517. (e) Deva Priyakumar U, Reddy AS, Sastry GN (2004) The design of molecules containing planar tetracoordinate carbon. Tetrahedron Lett 45:2495–2498. (f) Sateesh B, Reddy AS, Sastry GN (2007) Towards design of the smallest planar tetracoordinate carbon and boron systems. J Comput Chem 28:335–343

Chapter 18
Unique Orbital Interactions in the Ground and Electronically Excited States of Biradicals Brought about by the Existence of "Twisted π-Space"

Hiroshi Ikeda, Yasunori Matsui, and Eisuke Ohta

Abstract The six-membered cyclic 1,4-radical cation $\mathbf{2}^{\bullet+}$ forms a "twisted π-space" due to noncoplanarity of the cumyl radical $\mathbf{3}^{\bullet}$ and cumyl cation $\mathbf{3}^{+}$ originating from a chair conformation of the central cyclohexane ring system. The results of substituent effects on absorption wavelengths of $\mathbf{2}^{\bullet+}$ indicate that both through-space (TS) and through-bond (TB) orbital interactions operate in a "twisted π-space" of $\mathbf{2}^{\bullet+}$, resulting in the theoretically maximal orbital interactions between $\mathbf{3}^{\bullet}$ and $\mathbf{3}^{+}$. An operation of similar TS and TB orbital interactions is also suggested in the excited-state biradical $^{1}\mathbf{2}^{\bullet\bullet*}$ by the results of substituent effects on thermoluminescence wavelengths.

Keywords Radical cation • Excited biradical • Substituent effect • Thermoluminescence

18.1 Introduction

Simple aromatic compounds such as benzene and naphthalene belong to the family of closed-shell molecular systems comprised of π-conjugated moieties in a single plane that forms a planar π-space. A question arises about what kind of π-space is formed in open-shell molecular systems comprised of multiple planar π-conjugated moieties. In recent studies of several photoinduced electron-transfer (PET) reactions, we observed that some open-shell molecular systems consisting of two planar π-conjugated moieties form a "twisted π-space," which promotes the formation of electronic states that exhibit a variety of unique spectroscopic and chemical properties.

H. Ikeda (✉) • Y. Matsui • E. Ohta
Graduate School of Engineering, Osaka Prefecture University, 1-1 Gakuen-cho, Sakai, Osaka 599–8531, Japan
e-mail: ikeda@chem.osakafu-u.ac.jp

T. Akasaka et al. (eds.), *Chemical Science of π-Electron Systems*,
DOI 10.1007/978-4-431-55357-1_18

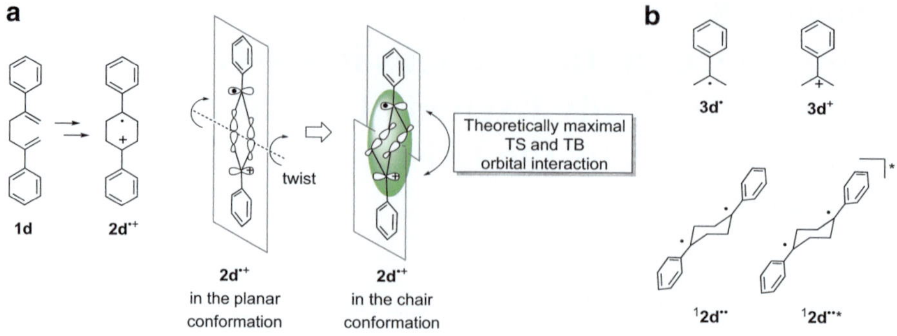

Fig. 18.1 (**a**) Existence of the "twisted π-space" in **2d**$^{\bullet+}$ generated by PET reaction of **1d** and (**b**) chemical structures of **3d**$^{\bullet}$, **3d**$^{+}$, ^{1}**2d**$^{\bullet\bullet}$, and ^{1}**2d**$^{\bullet\bullet}$*

The first open-shell system in which we observed this unusual phenomenon is the 1,4-diphenylcyclohexa-1,4-diyl radical cation (**2d**$^{\bullet+}$, Fig. 18.1a) that is produced by PET reaction of 2,5-diphenyl-1,5-hexadiene (**1d**) [1]. At first glance, **2d**$^{\bullet+}$ can be pictured as a planar species comprised of the cumyl radical (**3d**$^{\bullet}$, Fig. 18.1b) and cumyl cation (**3d**$^{+}$) residing in a common plane and connected by two σ-bonds. In fact, because the central cyclohexane ring system adopts a chair conformation, the two π-planes in **2d**$^{\bullet+}$ comprised of **3d**$^{\bullet}$ and **3d**$^{+}$ are parallel but uneven (i.e., noncoplanar). Thus, this conformational constraint causes the 1,4-diphenylcyclohexa-1,4-diyl radical cation to exists in a novel "twisted π-space." A similar "twisted π-space" is also expected for the ground (^{1}**2d**$^{\bullet\bullet}$) and electronically excited (^{1}**2d**$^{\bullet\bullet}$*) states of the six-membered cyclic singlet 1,4-biradical formed by one-electron reduction of **2d**$^{\bullet+}$ (Fig. 18.1b).

The presence of unique orbital interactions causes **2d**$^{\bullet+}$, ^{1}**2d**$^{\bullet\bullet}$, and ^{1}**2d**$^{\bullet\bullet}$* to exhibit a variety of interesting spectroscopic and chemical properties. In this chapter, we briefly describe some of these properties and their relationship to the orbital interactions present in **2d**$^{\bullet+}$ (Sect. 18.2) and ^{1}**2d**$^{\bullet\bullet}$* (Sect. 18.3).

18.2 Absorption Characteristics and Orbital Interactions in the Six-Membered Cyclic 1,4-Radical Cation 2$^{\bullet+}$ that Possesses "Twisted π-Space"

The transient radical cation **2**$^{\bullet+}$ is generated by laser flash photolysis (LFP) of diene **1** in methylene chloride (CH_2Cl_2) under PET conditions using *N*-methylquinolinium tetrafluoroborate ($NMQ^+BF_4^-$, Scheme 18.1) as a sensitizer and toluene as a co-sensitizer [1]. The absorption spectrum of **2d**$^{\bullet+}$ contains a maximum at $\lambda_{AB} = 476$ nm (Table 18.1), which is in a significantly longer wavelength region as compared to those of its component moieties, **3d**$^{+}$ ($\lambda_{AB} = 322$ nm in fluorosulfuric acid [2]) and **3d**$^{\bullet}$ ($\lambda_{AB} = 322$ nm in cyclohexane

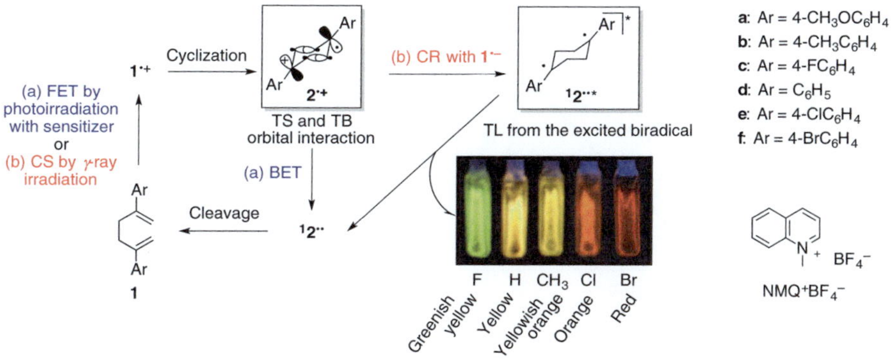

Scheme 18.1 Mechanisms for PET and γ-ray-induced electron-transfer reactions involving TL of **1** [4]

Table 18.1 Substituent constants for cation (σ^+) and radical ($\sigma^\bullet$); absorption wavelengths (λ_{AB}) of **2a–f**$^{\bullet+}$, **3a–f**$^+$, and **3a–f**$^\bullet$; TL wavelengths (λ_{TL}) of ^{1}**2a–f**$^{\bullet\bullet}$*; and the corresponding absorption and TL energies relative to those of the non-*para*-substituted derivatives (ΔE_{AB} and ΔE_{TL})

	Ar	σ^+	$\sigma^\bullet$	**2**$^{\bullet+}$		^{1}**2**$^{\bullet\bullet}$*		**3**$^+$		**3**$^\bullet$	
				λ_{AB} nm	ΔE_{AB} eV	λ_{TL} nm	ΔE_{TL} eV	λ_{AB} nm	ΔE_{AB} eV	λ_{AB} nm	ΔE_{AB} eV
a	4-$CH_3OC_6H_4$	−0.78	0.24	522	−0.230	–	–	362	−0.425	–	–
b	4-$CH_3C_6H_4$	−0.31	0.11	496	−0.105	586	−0.014	338	−0.182	322	0
c	4-FC_6H_4	−0.07	−0.08	473	0.017	572	−0.037	324	0.024	–	–
d	C_6H_5	0	0	476	0	582	0	322	0	322	0
e	4-ClC_6H_4	0.11	0.12	499	−0.120	590	−0.028	348	−0.288	318	−0.048
f	4-BrC_6H_4	0.15	0.14	509	−0.169	592	−0.036	364	−0.444	–	–

[3]). This observation strongly suggests that orbital interactions exist between the cation and radical moieties in **2d**$^{\bullet+}$ as a consequence of its existence in "twisted π-space."

In order to gain insights into the nature of these orbital interactions, we examined the effects of *para*-aryl ring substituents on λ_{AB} of **2d**$^{\bullet+}$ (Scheme 18.1 and Table 18.1). The results show that the λ_{AB} values of **2a–c**$^{\bullet+}$, containing *para*-electron-donating groups, and **2e–f**$^{\bullet+}$, possessing *para*-electron-withdrawing groups, are redshifted as compared to the absorption maximum of the parent **2d**$^{\bullet+}$. We have termed this redshift phenomenon Type I substituent effects.

The effects of *para*-aryl substituents on λ_{AB} of **2**$^{\bullet+}$ were subjected to linear free energy treatments using the Hammett equation. Inspection of the Hammett plot displayed in Fig. 18.2a shows that a two-phase, inverted V-shape, linear relationship exists between the cation substituent constants, σ^+, and the relative energies (ΔE_{AB}) of the absorption maximum of **2a–d**$^{\bullet+}$ and **2d–f**$^{\bullet+}$, which correspond to differences in electronic transition energies between the parent **2d**$^{\bullet+}$ and substituted derivatives. Moreover, the plot of ΔE_{AB} vs $\sigma^\bullet$ is purely linear. Finally, a two-phase, inverted

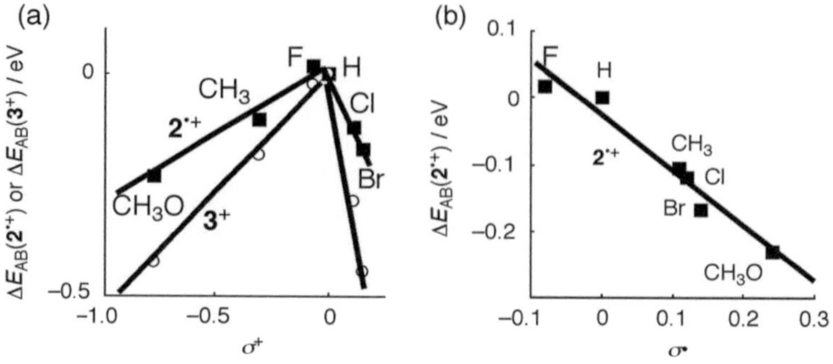

Fig. 18.2 Hammett plots of relative absorption energies (ΔE_{AB}) of $\mathbf{2}^{\bullet +}$ or $\mathbf{3}^{+}$ vs σ^{+} (**a**) and $\sigma^{\bullet}$ (**b**)

V-shape correlation is found for the Hammett plot of σ^{+} vs ΔE_{AB} for **3a–f**$^{+}$, which correspond to the component cation moieties in **2d–f**$^{\bullet +}$. Interestingly, the ρ values, obtained from the slopes of each of the phases of the ΔE_{AB} vs σ^{+} plots for **2d**$^{\bullet +}$ and **3a–f**$^{+}$, differ with each other. More importantly, the ratios of the ρ values of the transition energies of $\mathbf{2}^{\bullet +}$ and $\mathbf{3}^{+}$, $\rho(\mathbf{2}^{\bullet +})/\rho(\mathbf{3}^{+})$, which corresponds to the ratio of $\Delta E_{AB}(\mathbf{2}^{\bullet +})/\Delta E_{AB}(\mathbf{3}^{+})$ (see below), are 0.6 for the **2a–d**$^{\bullet +}$/**3a–d**$^{+}$ phases (left) and 0.4 for the **2d–f**$^{\bullet +}$/**3d–f**$^{+}$ phases (right) (Fig. 18.2a). These findings indicate that the magnitude of the effects of substituents on λ_{AB} for $\mathbf{2}^{\bullet +}$ is reduced to about one-half that for $\mathbf{3}^{+}$. We term this feature Type II substituent effects that correspond to a one-half reduction of the magnitude of the redshift.

Of equal significance is the fact that Hammett plots of ΔE_{AB} vs σ^{+} for both $\mathbf{2}^{\bullet +}$ and $\mathbf{3}^{+}$ are nonlinear. These features demonstrate that F, Cl, and Br substituents function as electron-donating rather than electron-withdrawing groups in governing the energies of the molecular orbitals of the radical cation and cation involved in the electronic transitions associated with light absorption. In this connection, it is interestingly that a purely one-phase linear relationship exists between $\Delta E_{AB}(\mathbf{2}^{\bullet +})$ and Arnold's radical substituent constants $\sigma^{\bullet}$ (Fig. 18.2b) [5]. Thus, the effects of substituents on the benzyl cation and benzyl radical are completely different, reflected in the fact that no proportional relationship exists between σ^{+} and $\sigma^{\bullet}$.

As described above, **2a–f**$^{\bullet +}$ have absorption maxima in longer wavelength regions than do **3a–f**$^{\bullet}$ and **3a–f**$^{+}$, the respective radical and cation moieties present in **2a–f**$^{\bullet +}$. Furthermore, the absorption maxima of **2a–f**$^{\bullet +}$ are dependent on *para*-aryl ring substituents in a manner, which we have termed Type I (redshift) and Type II (one-half reduction of redshift) effects. The origin of these unique substituent effects is thought to be associated with orbital interactions taking place between the cumyl radical and cation moieties in $\mathbf{2}^{\bullet +}$ caused by the existence of "twisted π-space." To probe this issue, we examined the molecular orbital consequence of orbital interactions that take place when the component moieties, **3d**$^{\bullet}$ and **3d**$^{+}$, approach one another to form **2d**$^{\bullet +}$ (Fig. 18.3). When only through-space (TS)

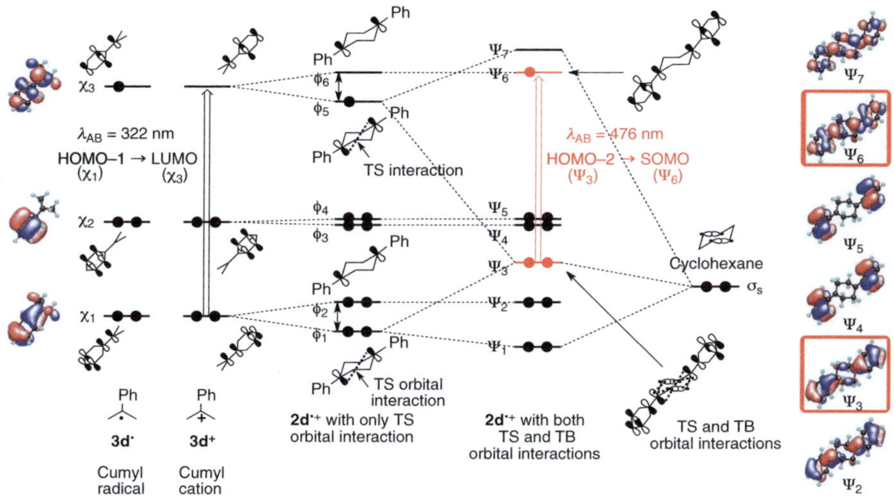

Fig. 18.3 TS and TB orbital interactions operating in formation of the orbitals of $\mathbf{2d}^{\bullet+}$ from those of $\mathbf{3d}^{\bullet}$ and $\mathbf{3d}^{+}$

orbital interactions between $\mathbf{3d}^{\bullet}$ and $\mathbf{3d}^{+}$ are taken into account, the transition from ϕ_2 to ϕ_5 in the parent phenyl derivative $\mathbf{2d}^{\bullet+}$ becomes lower in energy as compared with that from χ_1 to χ_3 in $\mathbf{3d}^{+}$. This change is a consequence of the fact that the energy differences between ϕ_1 and ϕ_2 and between ϕ_5 and ϕ_6 are large. This finding demonstrates why the observed absorption maximum of $\mathbf{2}^{\bullet+}$ is in a longer wavelength region than those of the component moieties. Furthermore, when a substituent is introduced, because the energies of χ_1 and ϕ_2, which have finite orbital coefficients at the benzylic positions, are increased, Type I substituent effects (i.e., the redshift) are observed.

Importantly, a consideration of only TS orbital interactions does not lead to an explanation of the Type II substituent effects (i.e., one-half reduction of the magnitude of the redshift). However, a reasonable rational for these effects becomes evident when through-bond (TB) orbital interactions, involving the two cyclohexane ring σ-bonds between the cumyl radical and cation moieties in $\mathbf{2d}^{\bullet+}$, are taken into account (Fig. 18.3). According to the orbital interaction theory suggested by Rauk [6], ratios of the relative absorption energies for $\mathbf{2}^{\bullet+}$ and $\mathbf{3}^{+}$, expressed as $\Delta E_{\mathrm{AB}}(\mathbf{2}^{\bullet+})/\Delta E_{\mathrm{AB}}(\mathbf{3}^{+})$, is 1 when TB orbital interactions do not exist and 0.5 when TB orbital interactions are maximal (Chart 18.1). As described above, the experimental value of $\Delta E_{\mathrm{AB}}(\mathbf{2}^{\bullet+})/\Delta E_{\mathrm{AB}}(\mathbf{3}^{+})$ is found to be very close to 0.5 (i.e., 0.6 for $\mathbf{2a{-}d}^{\bullet+}/\mathbf{3a{-}d}^{+}$, 0.4 for $\mathbf{2d{-}f}^{\bullet+}/\mathbf{3d{-}f}^{+}$), indicating that theoretically maximal orbital interactions take place in $\mathbf{2}^{\bullet+}$ as a consequence of the existence of "twisted π-space" [7].

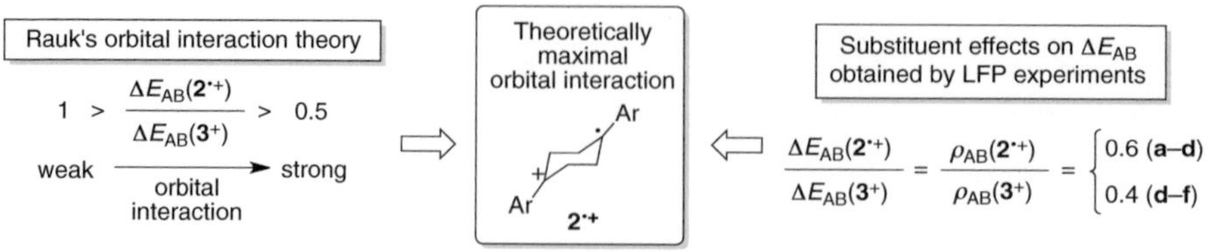

Chart 18.1 Evaluation of orbital interactions in $\mathbf{2}^{\bullet+}$ by using Rauk's theory

18.3 Luminescence Characteristics and Orbital Interactions in the Electronically Excited State of the Six-Membered Cyclic 1,4-Biradical $\mathbf{2}^{\bullet\bullet}$* that Possesses "Twisted π-Space"

TS and TB orbital interactions and, thus, an absorption maximum in the long wavelength region, similar to those observed for $\mathbf{2}^{\bullet+}$, are also expected for the ground-state biradical, $^1\mathbf{2}^{\bullet\bullet}$, which is generated by a sequence involving forward electron transfer (FET) of **1**, cyclization, and backward electron transfer (BET) (Scheme 18.1). Unfortunately, we were unable to observe the transient absorption spectrum of $^1\mathbf{2}^{\bullet\bullet}$ directly upon LFP with a time resolution of nanoseconds. This is likely a consequence of short-lived nature of $^1\mathbf{2}^{\bullet\bullet}$ which results in a lower concentration than sufficient concentration of the transient to give a detectable optical density for the $^1\mathbf{2}^{\bullet\bullet}$ to $^1\mathbf{2}^{\bullet\bullet}*$ transition (i.e., $^1\mathbf{2}^{\bullet\bullet} + h\nu_{\mathrm{AB}} \rightarrow {}^1\mathbf{2}^{\bullet\bullet}*$). To circumvent this problem, the γ-irradiation electron-transfer method was utilized to generate $^1\mathbf{2}^{\bullet\bullet}*$ and observe its thermoluminescence (TL) [8, 9] (i.e.,$^1\mathbf{2}^{\bullet\bullet}* \rightarrow {}^1\mathbf{2}^{\bullet\bullet} + h\nu_{\mathrm{TL}}$). Specifically, upon annealing a methylcyclohexane (MCH) matrix containing the parent diene **1d** after γ-irradiation at 77 K, an intense TL band at 582 nm, associated with $^1\mathbf{2d}^{\bullet\bullet}*$, is observed. This excited-state species is generated by a sequence involving charge separation (CS)–cyclization–charge recombination (CR) (Scheme 18.1 and Table 18.1). In this case, TL takes place in a longer wavelength region as do luminescence arising from the component moieties, i.e., the electronically excited state of the cumyl radical $\mathbf{3d}^{\bullet}*$ ($\lambda_{\mathrm{FL}} = 515$ nm in MCH [10]). This observation indicates that TS and TB orbital interactions operate in $^1\mathbf{2d}^{\bullet\bullet}*$. We next examined substituent effects on the wavelengths, λ_{TL}, of TL from substituted forms of $^1\mathbf{2d}^{\bullet\bullet}*$ by using the Hammett linear free energy treatment. A good linear plot of relative TL energies (ΔE_{TL}) vs $\sigma^{\bullet}$ was obtained (Fig. 18.4a). This finding strongly suggests that a radical (e.g., $^1\mathbf{2d}^{\bullet\bullet}*$) is the emitting species in this case. The assignment of singlet multiplicity to $^1\mathbf{2d}^{\bullet\bullet}*$ is supported by the results of time-dependent density functional theory calculations (UB3LYP/ccpVDZ level). Specifically, the absorption maximum for the electronic transition (HOMO → LUMO) of singlet $^1\mathbf{2d}^{\bullet\bullet}$ is calculated to be 623 nm, while that of triplet $^3\mathbf{2d}^{\bullet\bullet}$ does not occur in the 600-nm region (Fig. 18.4b). The calculated wavelength of the HOMO → LUMO transition in $^1\mathbf{2d}^{\bullet\bullet}$ is in good agreement with the observed λ_{TL} of 582 nm.

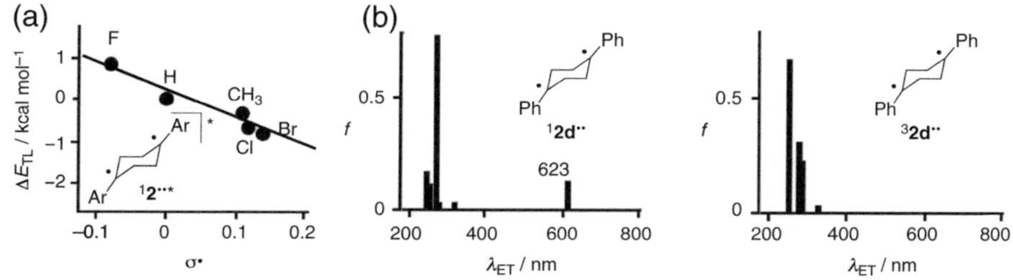

Fig. 18.4 (**a**) Hammett plot of relative TL energies (ΔE_{TL}) of ^{1}**2**$^{••}$* (Ar- = 4-XC_6H_4-) vs $\sigma^{•}$ and (**b**) electronic transitions (UB3LYP/ccpVDZ level) of singlet ^{1}**2d**$^{••}$ (*left*) and triplet ^{3}**2d**$^{••}$ (*right*)

As described above, because it possesses "twisted π-space," ^{1}**2**$^{••}$* has orbital interactions that are similar to those present in **2**$^{•+}$. We plan to evaluate this feature in the near future by using the latest theoretical calculations that are applicable to dealing with electronically excited states of open-shell, biradical species.

18.4 Conclusion

In this chapter, we described the absorption properties of the six-membered cyclic 1,4-radical cation **2**$^{•+}$ and luminescence properties of the corresponding electronically excited biradical ^{1}**2**$^{••}$*. Significantly, TS and TB orbital interactions between the cumyl radical and cation moieties in **2**$^{•+}$, which arise because of the existence of "twisted π-space," appear to operate to their theoretical maximum in governing the nature and energies of the molecular orbitals. These interaction causes light absorption (experiment) and the corresponding electronic transitions (theory) of **2**$^{•+}$ to take place at longer wavelengths than do those of the component cumyl radical (**3**$^{•}$) and cation (**3**$^{+}$) moieties. Because TL of ^{1}**2**$^{••}$* occurs in a similarly long wavelength region, it is expected that this transient also possesses "twisted π-space," which promotes effective TS and TB orbital interactions. Although the corresponding ground-state biradical, **2**$^{••}$, has not yet been detected in LFP studies, we anticipate that it will contain similar orbital interactions originating from "twisted π-space" and, consequently, that it will display a redshifted absorption maximum.

Observations made in this investigation provide useful insight into the absorption and luminescence spectra of transient, charged and neutral radical intermediates, and into how novel chromophores and fluorophores can be created. The results of experimental and theoretical studies that provide positive proof of the proposal coming from the current investigation are expected in the near future.

Acknowledgments H. I. gratefully acknowledges financial support from a Grant-in-Aid for Scientific Research on Priority Areas (Area No. 2007, Emergence of Highly Elaborated π-Space and Its Function) and others from the Ministry of Education, Culture, Sports, Science, and Technology (MEXT) of Japan.

References

1. Ikeda H, Minegishi T, Abe H, Konno A, Goodman JL, Miyashi T (1998) Photoinduced electron-transfer degenerate cope rearrangement of 2,5-diaryl-1,5-hexadienes: a cation-radical cyclization−diradical cleavage mechanism. J Am Chem Soc 120:87
2. (a) Boate DR, Scaiano JC (1989) Transient phenomena in the photochemistry of *trans*-azocumene. Tetrahedron Lett 30:4633. (b) Sumiyoshi T, Kamachi M, Kuwae Y, Schnabel W (1987) Laser flash photolysis of azocumenes. Direct observation of stepwise decomposition. Bull Chem Soc Jpn 60:77
3. (a) Sekuur TJ, Kranenburg P (1973) Electronic spectra of (di)arylalkyl- and (di)arylcyclopropylcarbonium ions. Spectrochim Acta A 29:803. (b) McClelland RA, Chan C, Cozens F, Modro A, Steenken S (1991) Laser flash photolysis generation, spectra, and lifetimes of phenylcarbenium ions in trifluoroethanol and hexafluoroisopropyl alcohol. On the UV spectrum of the benzyl cation. Angew Chem 103:1389. (1991) Angew Chem Int Ed Engl 30:1337. (c) Olah GA, Pittman CU Jr, Waack R, Doran M (1966) Laser flash photolysis generation, spectra, and lifetimes of phenylcarbenium ions in trifluoroethanol and hexafluoroisopropyl alcohol. On the UV spectrum of the benzyl cation. J Am Chem Soc 88:1488
4. (a) Ikeda H, Minegishi T, Takahashi Y, Miyashi T (1996) Phenyl substitution effects on rate acceleration of cation radical deazetation of 2,3-diazabicyclo[2.2.2]oct-2-enes. Tetrahedron Lett 37:4377. (b) Miyashi T, Ikeda H, Takahashi Y (1999) Photoinduced electron-transfer intramolecular cyclization of acyclic dienes with two styrene units. Acc Chem Res 32:815
5. Dust JM, Arnold DR (1983) Substituent effects on benzyl radical ESR hyperfine coupling constants. The $\sigma_\alpha{}^{\bullet}$ scale based upon spin delocalization. J Am Chem Soc 105:1221
6. Rauk A (1994) Chapter 3: Orbital interaction theory. In: Orbital interaction theory of organic chemistry. Wiley, New York. pp 57–93, ISBN13: 9780471593898
7. Ikeda H, Hoshi Y, Namai H, Tanaka F, Goodman JL, Mizuno K (2007) Evidence for significant through-space and through-bond electronic coupling in the 1,4-diphenylcyclohexane-1,4-diyl radical cation gained by absorption spectroscopy and DFT calculations. Chem Eur J 13:9207
8. (a) Namai H, Ikeda H, Hoshi Y, Kato N, Morishita Y, Mizuno K (2007) Thermoluminescence and a new organic light-emitting diode (OLED) based on triplet–triplet fluorescence of the trimethylenemethane (TMM) biradical. J Am Chem Soc 129:9032. (b) Ikeda H, Matsui Y, Akimoto I, Kan'no K-I, Mizuno K (2010) X-Ray-triggered thermoluminescence and density functional theory characterization of a *gem*-diphenyltrimethylenemethane biradical. Aust J Chem 63:1342. (c) Matsui Y, Namai H, Akimoto I, Kan'no K-I, Mizuno K, Ikeda H (2011) Twisted molecular geometry and localized electronic structure of the triplet excited *gem*-diphenyltrimethylenemethane biradical: substituent effects on thermoluminescence and related theoretical calculations. Tetrahedron 67:7431
9. Namai H, Ikeda H, Hoshi Y, Mizuno K (2007) Thermoluminescence originating from the singlet excited state of 1,4-diarylcyclohexane-1,4-diyls: a potentially general strategy for the observation of short-lived biradicals. Angew Chem Int Ed 46:7396
10. Irie M, Shimizu M, Yoshida H (1976) Charge transfer band of the benzyl radical-halide ion formed by dissociative electron attachment to benzyl halides in a rigid organic matrix. J Phys Chem 80:2008

Chapter 19
Recent Progress in Stable High-Spin Molecules Based on Nitroxide Radicals

Shuichi Suzuki

Abstract Organic-based open-shell π-electronic molecules are promising components for unique electronic devices because of the diversity in their design. Of these compounds, high-spin π-electronic molecules are attractive building blocks of molecule-based magnets and spintronic devices. However, there are only a few reports of high-spin π-electronic molecules with large intramolecular exchange interactions ($J_{\text{intra}}/k_B > +$ 300 K) and with isolable stability, although general guides of molecular designs to obtain high-spin π-electronic molecules have already been established. In this chapter, recent studies from our research group on high-spin π-electronic molecules are reported. First, stable trimethylenemethane analogues composed of nitroxide and nitronyl nitroxide or imino nitroxide moieties and their metal complexes are outlined. This is followed by the description of (nitronyl nitroxide)-substituted electron donor radical cations as new components for molecule-based magnets. Finally, a trinitroxide-substituted trioxytriphenylamine is discussed, which constitutes a novel spin-state conversion system.

Keywords High-spin π-electronic molecule • Nitroxide radical • Nitronyl nitroxide radical • Radical cation

19.1 Introduction

Open-shell π-electronic molecules are generally recognized as reaction intermediates or unstable chemical species. Recently, various stable open-shell π-electronic molecules have been designed and synthesized by kinetic protection with bulky substituents and by thermodynamic stabilization with extended π-conjugations [1–3]. Some of these molecules have attracted considerable interest for applications as electronic devices because of their physical properties derived from electron spins as well as their redox activities. These applications include molecule-based magnets [1, 4–7], electrical conductors [1, 8, 9], radical-based batteries [1, 10, 11], spintronic

S. Suzuki (✉)
Department of Chemistry, Graduate School of Science, Osaka City University, Osaka, Japan
e-mail: suzuki-s@sci.osaka-cu.ac.jp

T. Akasaka et al. (eds.), *Chemical Science of π-Electron Systems*,
DOI 10.1007/978-4-431-55357-1_19

devices [12, 13], molecular spin-based quantum computers [14, 15], and materials for dynamic nuclear polarization [16]. In general, control of the ferromagnetic or antiferromagnetic couplings between spin sources plays an important role in their magnetic properties [5, 17]. The first demonstration of a high-spin polycarbene by Ito's group and Wasserman's group [5, 18, 19] established a fundamental experimental guide to obtain high-spin π-electronic molecules [20]. Rajaca and coworkers succeeded in the generation of a high-spin poly(triphenylmethyl radical) having very large spin quantum numbers ($S \gg 5000$) [21]. However, very few stable high-spin π-electronic molecules with large intramolecular exchange interactions ($J_{intra}/k_B > + 300$ K) have been reported [22–25]. Thus, design and synthesis of new stable high-spin π-electronic molecules are fundamentally important for the development of new organic materials involving electronic spins. In this chapter, we report recent studies from our research group on stable high-spin π-electronic molecules with large intramolecular exchange interactions. Notably, the species described in this chapter can survive for a long time even under aerated conditions both in solution and in solid states.

19.2 Stable Trimethylenemethane Analogues

Trimethylenemethane (TMM, **1**), one of the simplest high-spin species, shows a large positive exchange interaction [17] (Fig. 19.1). However, it could not be isolated due to a facile ring closure reaction at temperatures above 123 K yielding methylenecyclopropane through a thermally accessible singlet diradical state [26]. Thus, only a few π-extended TMM analogues (for instance **2,** Yang's diradical) [4, 22, 27] have been investigated so far. Biradicals with a much shorter radical–radical distance are more attractive systems. We report here the synthesis of highly compact TMM-analogues, nitroxide-substituted nitronyl nitroxide **3,** and nitroxide-substituted imino nitroxide **4**, with large positive exchange interactions [25].

The TMM-analogue **3** was easily prepared from a parent nitronyl nitroxide **5** with lithium hexamethyldisilazide (LHMDS) and 2-methyl-2-nitrosopropane dimer [$(t$BuNO$)_2$], followed by oxidation using lead(IV) dioxide (Scheme 19.1). Diradical **3** was converted to **4** with sodium nitrate under acidic conditions. These diradicals

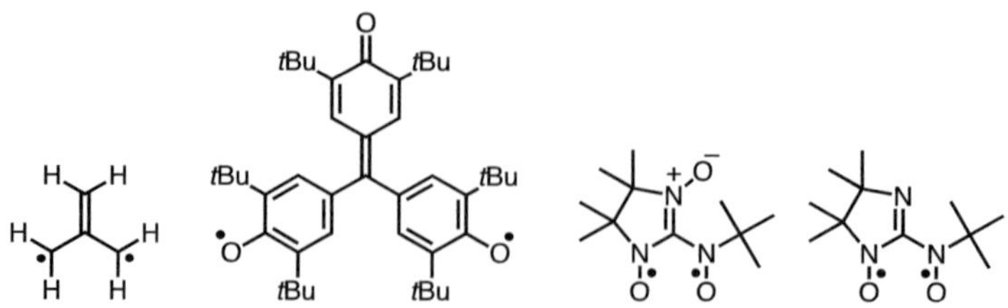

1 **2** **3** **4**

Fig. 19.1 Chemical structures of **1**–**4**

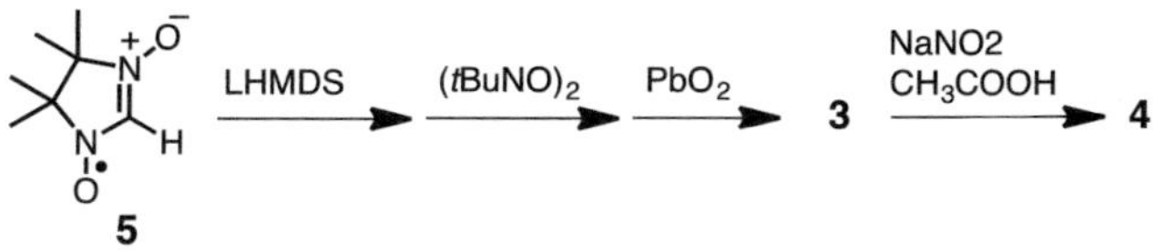

Scheme 19.1 Synthetic scheme of diradicals **3** and **4**

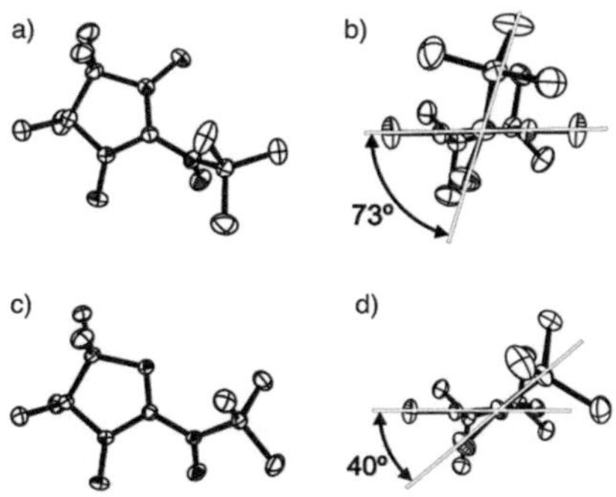

Fig. 19.2 ORTEP views of diradicals **3** and **4**: (**a**) top and (**b**) side views of diradical **3**. (**c**) Top and (**d**) side views of diradical **4**. Hydrogen atoms are omitted for clarity

are quite stable in spite of the close proximity of the two radical moieties. They can be readily purified by silica gel column chromatography, followed by recrystallization under aerated conditions. Furthermore, these diradicals are easily sublimed under reduced pressure without decomposition.

Figure 19.2 shows the molecular structures of **3** and **4** clarified by X-ray crystal structure analyses. The N–O bond lengths showed the typical values for nitroxide, nitronyl nitroxide, and imino nitroxide moieties. The dihedral angle between the nitronyl nitroxide and nitroxide moieties in **3** was estimated to be 73°. Steric repulsion between the *tert*-butyl group and nitroxide group on nitronyl nitroxide moiety induces a slightly twisted conformation. In the case of **4**, the dihedral angle between the imino nitroxide and nitroxide moieties was estimated to be 40°, which was quite smaller than that of **3**. In addition, the two nitroxide moieties of **4** can be found to be in the *cis* conformation.

To obtain insight into the spin–spin interactions through TMM-type cross-conjugation in **3** and **4**, the electron spin resonance (EPR) spectra of the diradicals were measured. The EPR spectrum of **3** in frozen diethyl phthalate showed a typical triplet pattern ($|D/hc| = 0.0250\ \text{cm}^{-1}$, $|E/hc| = 0.0016\ \text{cm}^{-1}$). A half-field signal assigned to the $\Delta m_s = \pm\ 2$ transition was clearly observed. The signal intensities of the $\Delta m_s = \pm\ 2$ transition increased at lower temperatures, indicating that the diradical is in its triplet ground state. An ESR spectrum with a triplet pattern was also observed for **4** ($|D/hc| = 0.0639\ \text{cm}^{-1}$, $|E/hc| = 0.0050\ \text{cm}^{-1}$). The larger $|D/hc|$ value in **4** than **3** can be explained by a closer distance between the nitroxide spin centers in the cisoid form.

The intramolecular exchange interactions of these diradicals were estimated by measuring their temperature-dependent paramagnetic susceptibility using a

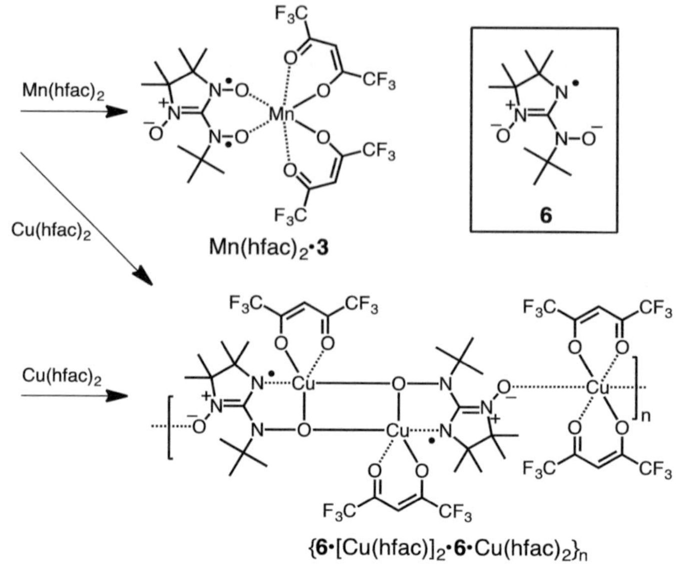

Scheme 19.2 Metal complexations of diradicals **3** and **4**

superconducting quantum interference device (SQUID). At room temperature, the $\chi_p T$ values of **3** and **4** were found to be nearly 1 (0.983 emu K mol^{-1} for **3** and 0.990 emu K mol^{-1} for **4**). In addition, the $\chi_p T$ values gradually increased to show broad maxima at 150 K for **3** and 230 K for **4**, respectively. These values decreased at lower temperatures. The increase in $\chi_p T$ values at higher temperatures indicates sizable intramolecular ferromagnetic interactions in both diradicals. The observed $\chi_p T$–T curves were simulated using a modified Bleaney–Bowers model, giving quite large ferromagnetic interactions of J_{intra}/k_B +390 K for **3** and J_{intra}/k_B +550 K for **4** with weak antiferromagnetic interactions. The larger J_{intra} value of **3** than **4** is attributed to the small dihedral angle between two radical moieties. The weak antiferromagnetic interactions can be assigned between the C(sp^3) and NO oxygen atoms.

We have also investigated the properties of metal complexes of **3** and **4** (Scheme 19.2) [28]. Diradical **3** underwent complexation with bis(hexafluoroacety lacetonato)manganese(II) to form a 1:1 complex, whose structure was determined to be [Mn(hfac)$_2$•**3**] by X-ray analysis (Fig. 19.3a). The $\chi_p T$ value at room temperature was approximately 2.05 emu K mol^{-1}. This value decreased at lower temperatures until 150 K, where it plateaued (1.94 emu K mol^{-1}) until 50 K. Finally, the $\chi_p T$ value rapidly decreased at temperatures lower than 50 K. Base on the simulation of $\chi_p T$–T curve, intramolecular magnetic interactions between Mn(II) and **3** were estimated to be $J/k_B = -217$ K. The rapidly decreasing $\chi_p T$ value at less than 50 K was derived from intermolecular interactions between the NO oxygen atom and the methyl carbon of *tert*-butyl group.

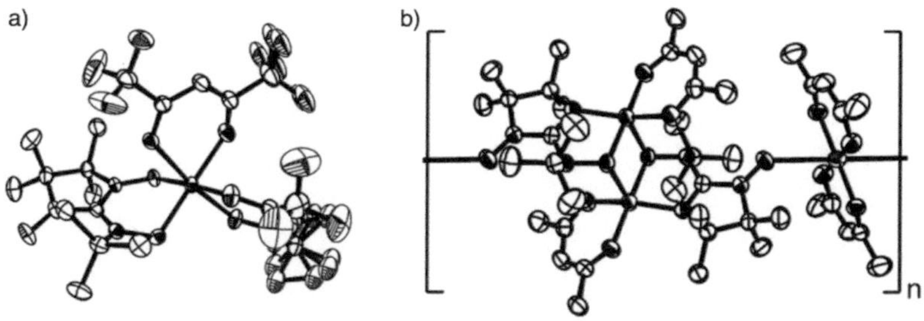

Fig. 19.3 ORTEP views of metal complexes (**a**) Mn(hfac)$_2$•**3** and (**b**) {**6**•[Cu(hfac)]$_2$•**6**• Cu(hfac)$_2$}. Hydrogen atoms are omitted for clarity

In contrast to the aforementioned Mn(hfac)$_2$ complex, the reaction of diradical **3** with bis(hexafluoroacetylacetonato)copper(II) produced a polymer chain complex [Cu$_2$(hfac)$_2$•**6**$_2$•Cu(hfac)$_2$]$_n$. In this polymer complex, radical anion **6** (one electron reductive species of **4**) is formed by the decomposition of diradical **3** (Scheme 19.2). The linear chain structure was determined by X-ray crystal structure analysis (Fig. 19.3b). The decomposition of diradical **3** proceeded in the presence of Cu(II) ions even under argon atmosphere and dry conditions. The conversion of diradical **3** to imino nitroxide **4** was confirmed by monitoring the reaction using TLC (although the detailed reaction pathway is uncertain). The χ_pT value at room temperature was estimated to be 2.448 emu K mol^{-1}. The χ_pT value increased to a broad maximum value (2.515 emu K mol^{-1}) at approximately 50 K. The χ_pT value then steeply decreased at temperatures below 50 K. The temperature dependence of χ_pT value indicated the presence of several magnetic interactions. As the results of detailed theoretical and experimental analyses, the magnetic interaction can be characterized by a strongly ferromagnetic intramolecular interaction ($J/k_B = +510$ K) between the imino nitroxide moiety **6** and the imino-coordinating Cu(hfac), a moderately large ferromagnetic intramolecular interaction ($J/k_B = +58$ K) between **6** and the imino nitroxide oxygen-coordinating Cu(hfac)$_2$, and a weak antiferromagnetic interaction between two Cu(hfac)$_2$•**6** moieties within the Cu$_2$O$_2$ framework [27].

19.3 Stable Radical-Substituted Radical Cation Species with Large Positive Exchange Interactions

Electron donor molecules substituted by stable neutral radicals are also good precursors for organic high-spin species [12, 13, 24, 29–41]. Selective oxidation of the electron donor moiety of these π-electronic systems gives a radical-substituted radical cation high-spin species. Yamaguchi and coworkers theoretically investigated magnetic interactions (especially, ferro- and ferromagnetic interactions) based on crystal-packing structures of charge transfer salts (or complexes) composed of radical-substituted radical cation species [42]. So far, various experimental

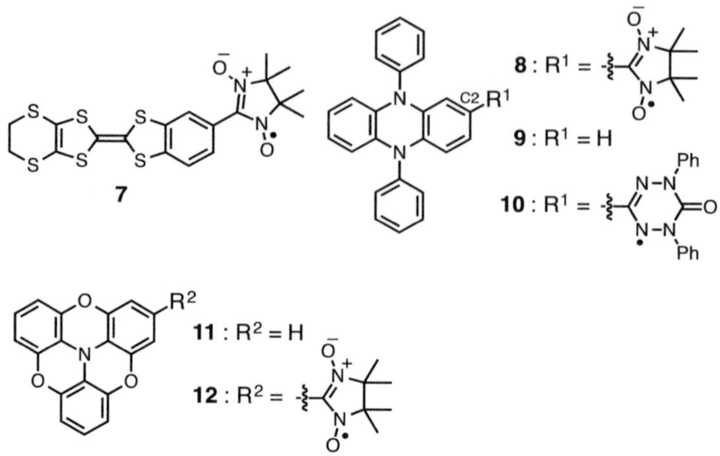

Fig. 19.4 Chemical structures of **7–12**

studies including aniline [30, 31], ferrocene [32], and tetrathiafluvalene derivatives [12, 13, 33–38] have been examined. However, the expected ferro- or ferrimagnetic behavior has not been observed. The difficulty in obtaining the expected properties for the radical-substituted radical cation systems may be due to small intramolecular exchange interactions (J_{intra}/k_B) in the examined systems compared to large intermolecular exchange interactions (J_{inter}/k_B). When $|J_{intra}/k_B|$ is quite smaller than $|J_{inter}/k_B|$, the system may behave like a radical-substituted diamagnetic ion pair, showing paramagnetic behavior [30, 38].

Recently, two types of radical cations substituted by a nitronyl nitroxide moiety have made breakthroughs in this research field [12, 13, 24, 29, 35, 36] (Fig. 19.4). First, a benzo-tetrathiafluvalene-based π-electronic system **7** [12, 13, 35, 36] reported by Sugawara and coworkers and second, a diphenyldihydrophenazine-based π-electronic system **8** [24, 29] reported by our group. The former exhibited giant negative magnetoresistance at low temperatures in spite of the absence of metal ions [12]. The latter has strong ferromagnetic interactions between a diphenyldihydrophenazine radical cation and a nitronyl nitroxide moiety ($J_{intra}/k_B > +300$ K) [24]. The diphenyldihydrophenazine radical cation $\mathbf{9}^{\bullet+}$ has large positive spin densities on the nitrogen atoms and C2 carbon atoms. The introduction of a nitronyl nitroxide moiety on the C2 carbon provides a high-spin system through a spin polarization mechanism [12]. In addition, the radical-substituted radical cations exhibit isolable stability under aerated conditions both in solution and in solid states. Similar behavior was also observed in an oxoverdazyl-substituted diphenyldihydrophenazine radical cation derivative $\mathbf{10}^{\bullet+}$ [39]. Because of their large intramolecular exchange interactions, these systems can be applied to molecule-based magnets by introducing a suitable spin as a counter anion.

Using a nitronyl nitroxide-substituted radical cation system $\mathbf{8}^{\bullet+}$ and introducing a third spin as a counter anion, we prepared tetrachloroferrate and tetrabromoferrate

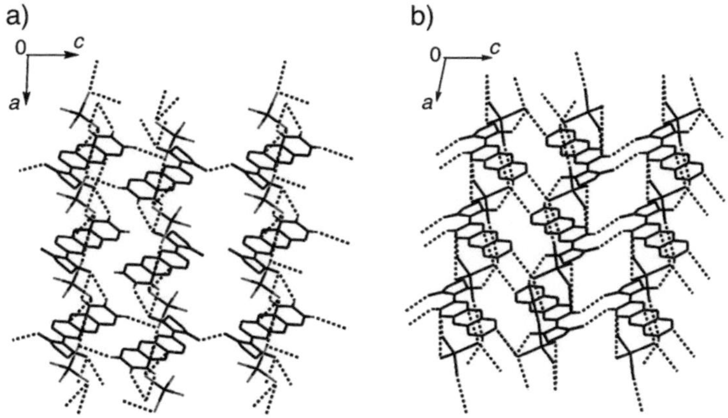

Fig. 19.5 Packing structures of (**a**) $\mathbf{8}^{\bullet+}\bullet FeCl_4^-$ and (**b**) $\mathbf{8}^{\bullet+}\bullet FeBr_4^-$ salts. Hydrogen atoms and phenyl groups are omitted for clarity

salts of $\mathbf{8}^{\bullet+}$ [29, 40]. Interestingly, $\mathbf{8}^{\bullet+}\bullet FeBr_4^-$ showed a phase transition at 6.7 K into a bulk ferrimagnet. Conversely, $\mathbf{8}^{\bullet+}\bullet FeCl_4^-$ showed no phase transition and its $\chi_p T$ values decreased at temperatures lower than 150 K. In the crystalline state, these two salts formed a similar columnar structure consisting of alternate stacking between $\mathbf{8}^{\bullet+}$ and $FeBr_4^-$ or $FeCl_4^-$ (Fig. 19.5). A striking difference in these salts was the arrangements of the nearest neighboring $\mathbf{8}^{\bullet+}$. Antiferromagnetic contacts between oxygen atoms of nitronyl nitroxide moieties were observed in $\mathbf{8}^{\bullet+}\bullet FeCl_4^-$, whereas ferromagnetic double CH–O contacts were observed in $\mathbf{8}^{\bullet+}\bullet FeBr_4^-$. Subtle differences in the sizes and positions of $FeCl_4^-$ and $FeBr_4^-$ induced the changes of relative orientation of $\mathbf{8}^{\bullet+}$.

Similar to diphenyldihydrophenazine, trioxytriphenylamine **11** (synthesized by our research group) [43, 44] is also a good electron donor. It provides a stable radical cation $\mathbf{11}^{\bullet+}$ in spite of a lack of substituents in the *para*-position [43, 44]. The stability and planarity of the molecule aided us in designing and synthesizing a (nitronyl nitroxide)-substituted trioxytriphenylamine **12** [41]. The radical-substituted radical cation salt $\mathbf{12}^{\bullet+}\bullet GaCl_4^-$ was obtained by chemical oxidation and is also stable under aerated conditions in solution and the solid state. In the crystalline state, $\mathbf{12}^{\bullet+}$ forms a tight linear chain structure by C–O and C–C intermolecular interactions (Fig. 19.6). The magnetic behavior of $\mathbf{12}^{\bullet+}\bullet GaCl_4^-$ was characterized by a strong intramolecular ferromagnetic interaction of $J_{intra}/k_B = +\ 400$ K and a weak antiferromagnetic intermolecular interaction of $J_{inter}/k_B = -1.85$ K. Surprisingly, an anomaly in the temperature dependence of the magnetic susceptibility was observed below 3 K. This was attributed to a magnetic phase transition from the antiferromagnetic state to the weak ferromagnetic state (spin-canted antiferromagnet) at 2.65 K. Magnetic heat capacity measurements exhibited a magnetic entropy value of $S_{mag} = 8.97$ J K^{-1} mol^{-1}, which is in good agreement with the theoretical value for an $S = 1$ spin system. There have been a few reports on spin-canting behavior

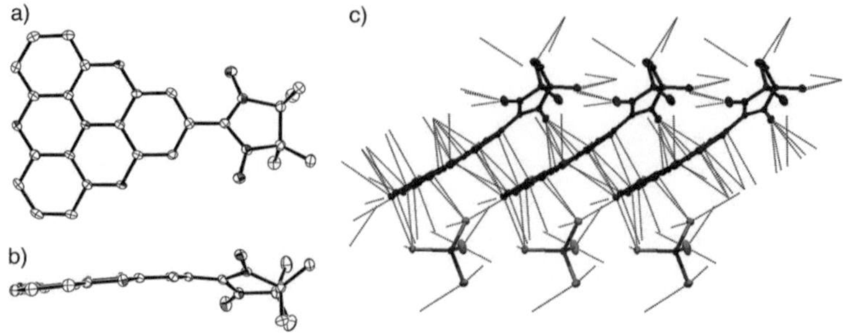

Fig. 19.6 ORTEP views of $\mathbf{12}^{\bullet+}\bullet GaCl_4^-$. (**a**) Top and (**b**) side views. Hydrogen atoms and counter anions are omitted for clarity. (**c**) Packing structure of $\mathbf{12}^{\bullet+}\bullet GaCl_4^-$. Hydrogen atoms are omitted for clarity

in organic compounds [45–48]. The present system with $\mathbf{12}^{\bullet+}$ has neither magnetic metal ions nor heavy-atom elements. This is classified as a 2p-electron-based weak ferromagnet, of which only very few examples have been reported [46–48].

19.4 Spin-State Conversion System by Oxidation: Trinitroxide-Substituted Trioxytriphenylamine

Oxidation of the electron donor in a radical-substituted donor can alter the π-conjugation network and affect intramolecular exchange interactions (Scheme 19.3). Tanaka and coworkers reported a tetraphenylethane derivative substituted by two nitroxide radicals and two diphenylamine moieties **13** [49]. In this system, the exchange interactions between the two nitroxide moieties switched from an antiferromagnetic interaction to a very weak interaction by two electron oxidation. Oxidation induces conversion of the π-conjugated system into a tetramethylenemethane system, which is known as a weak spin coupler [50]. Iwamura and coworkers reported that trinitroxide-substituted triphenylamine **14** is in a doublet ground state with $J_{\mathrm{intra}}/k_{\mathrm{B}} = -135$ K in the neutral state [51, 52]. The oxidized cationic species $\mathbf{14}^{\bullet+}$ is expected to have a TMM-type π-conjugated system in a triplet ground state with large intramolecular exchange interactions similar to the Yang's diradical π-electronic system **2** [22]. However, exchange interactions of the radical cation remain unknown, possibly due to its poor stability. In contrast, our triradical system **15** (consisting of trioxytriphenylamine and three nitroxide moieties) gives a stable radical cation species by electrochemical and chemical oxidation, which revealed ferromagnetic interactions in the cationic species [53].

The X-ray crystal structures show that the trioxytriphenylamine skeleton in neutral triradical species **15** has a shallow bowl structure similar to the parent trioxytriphenylamine **12** (Fig. 19.7a). Based on the magnetic susceptibility

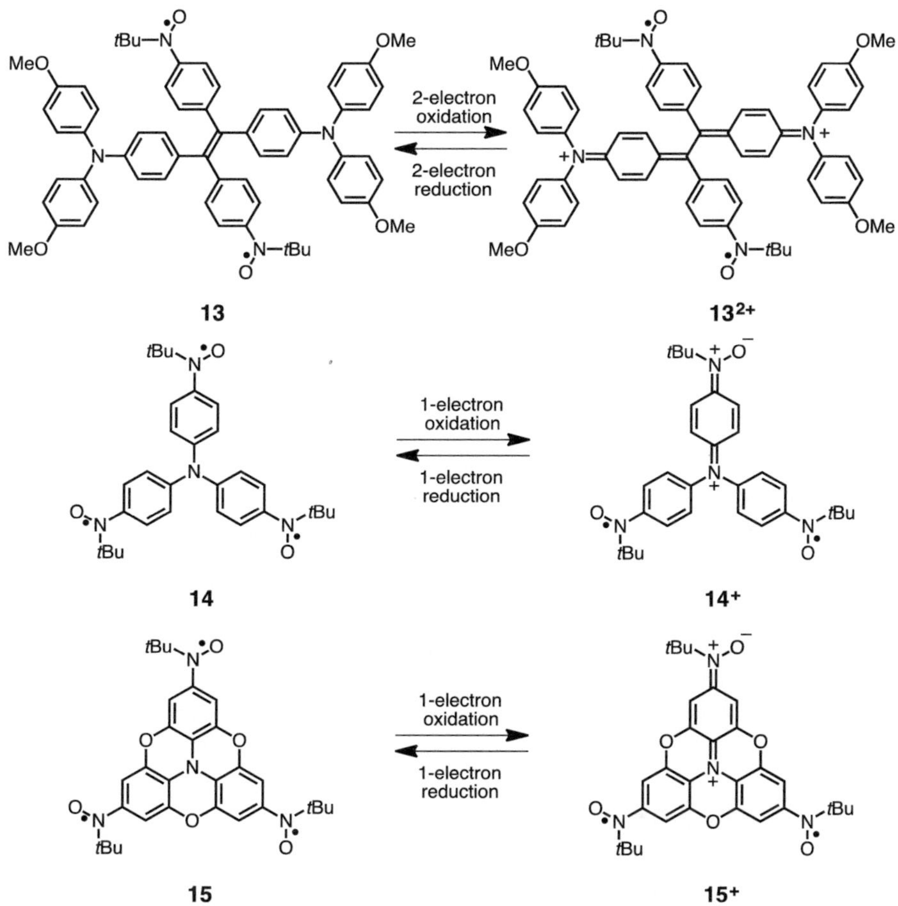

Scheme 19.3 Oxidation and reaction processes of compounds **13**, **14**, and **15**

measurements, the intramolecular interaction in the neutral triradical was obtained as $J_{\text{intra}}/k_{\text{B}} \sim -200$ K. Thus, the triradical was found to be in a doublet ground state similar to the triradical **14**. The diradical cationic state **15**$^{\bullet+}$•$SbCl_6^-$ was obtained as a blue solid by chemical oxidation using tris(4-bromophenyl)aminium hexachloroantimonate. X-ray crystal structure analysis of **15**$^{\bullet+}$ shows that the trioxytrophenylamine skeleton has a planar structure, in contrast to the neutral compound **15** (Fig. 19.7b). This is a typical structure change for the trioxytriphenylamine system by oxidation [43, 44]. Furthermore, the small dihedral angles between the nitroxides and the trioxytriphenylamine skeleton suggest that **15**$^{\bullet+}$ possesses a TMM-type π-conjugated network. The magnetic behavior of **15**$^{\bullet+}$ is best explained by assuming a triplet ground state with strong intramolecular exchange interactions of $J_{\text{intra}}/k_{\text{B}} >> +300$ K between the two nitroxide radicals. The large positive J value is consistent with results of theoretical calculations. Thus,

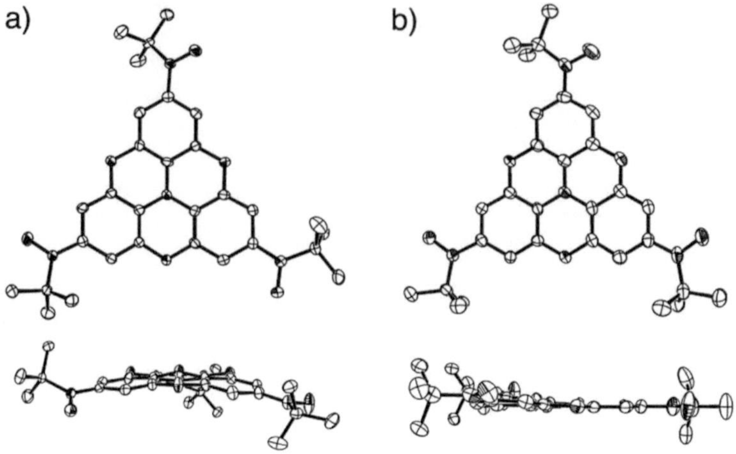

Fig. 19.7 ORTEP views of (**a**) top and side views of **15** and (**b**) top and side views of $\mathbf{15}^{\bullet+}\bullet SbCl_6^-$. Hydrogen atoms and counter anion are omitted for clarity

the triradical **15** showed a novel spin-state conversion from the neutral doublet state with the large negative exchange interaction ($J_{intra}/k_B \sim -200$ K) into the diradical cation triplet state with a large positive exchange interaction ($J_{intra}/k_B >> +300$ K) by the oxidative modulation of the π-conjugated network.

19.5 Conclusions

We have designed and synthesized new high-spin π-electronic molecules with large intramolecular exchange interactions and isolable stabilities. These high-spin π-electronic systems provide useful quantitative information for molecule-based magnets and spin-switching molecular systems. We believe that the wide diversity of these molecules could provide potential breakthroughs in related research fields involving molecular spins such as spintronics. These molecules exhibit large intramolecular ferromagnetic interactions with isolable stabilities in their oxidized states. Therefore, further applications to the related spintronics areas are also possible. In future studies we will continue to focus on multifunctional and controllable molecular spin systems using new synthetic methods [54, 55].

Acknowledgments These works were partially supported by a Grant-in-Aid for Scientific Research on Innovative Areas "Emergence of Highly Elaborated π-Space," "Stimuli-Responsive Chemical Species for the Creation of Functional Molecules," and "π-System Figuration: Control of Electron and Structural Dynamism for Innovative Functions" from MEXT.

The author gratefully acknowledges the invaluable discussions, many helpful suggestions, and encouragement of Professor Keiji Okada and Associate Professor Masatoshi Kozaki. The author would like to thank Professor Takeji Takui, Professor Kazunobu Sato, Associate Professor Daisuke Shiomi, Dr. Kazuo Toyota, and Dr. Kenji Sugisaki (Osaka City University) for their ESR and

SQUID measurements, theoretical calculations, and helpful discussions. The author would also like to thank Professor Yuko Hosokoshi (Osaka Prefecture University) for SQUID measurements and Associate Professor Yuji Miyazaki and Professor Akira Inaba (Osaka University) for their heat capacity measurements.

References

1. Hicks RG (ed) (2010) Stable radicals: fundamentals and applied aspects of odd-electron compounds. Wiley, Chichester
2. Hicks RG (2007) What's new in stable radical chemistry? Org Biomol Chem 5:1321–1338
3. Morita Y, Suzuki S et al (2011) Synthetic organic spin chemistry for structurally well-defined open-shell graphene fragments. Nat Chem 3:197–204
4. Lahti PM (ed) (1999) Magnetic properties of organic materials. Marcel Dekker, New York
5. Itoh K, Kinoshita M (eds) (2000) Molecular magnetism. Kodansha & Gordon and Breach, Tokyo
6. Miller JS, Drillon M (eds) (2001–2005) Magnetism: molecules to materials II–IV. Wiley-VCH, Weinheim
7. Veciana J (ed) (2001) π-electron magnetism from molecules to magnetic materials. Springer, Berlin
8. Pal SK, Itkis ME et al (2005) Resonating valence-bond ground state in a phenalenyl-based neutral radical conductor. Science 309:281–284
9. Iwasaki S, Hu R et al (2009) Interactive radical dimers in photoconductive organic thin films. Angew Chem Int Ed 48:4022–4024
10. Nishide H, Oyaizu K (2008) Toward flexible batteries. Science 319:737–738
11. Morita Y, Nishida S (2011) Organic tailored batteries materials using stable open-shell molecules with degenerate frontier orbitals. Nat Mater 10:947–951
12. Sugawara T, Komatsu H et al (2011) Interplay between magnetism and conductivity derived from spin-polarized donor radicals. Chem Soc Rev 40:3105–3118
13. Matsushita MM, Kawakami H et al (2008) Molecule-based system with coexisting conductivity and magnetism and without magnetic inorganic ions. Phys Rev B 77:195208
14. Coronad E, Epstein AJ (2009) Molecular spintronics and quantum computing. J Mater Chem 19:1670–1671
15. Nakazawa S, Nishida S (2012) A synthetic two-spin quantum bit: g-Engineered exchange-coupled biradical designed for controlled-NOT gate operations. Angew Chem Int Ed 51:9860–9864
16. Matsuki Y, Maly T et al (2009) Dynamic nuclear polarization with a rigid biradical. Angew Chem Int Ed 48:4996–5000
17. Boden WT (1982) Diradicals. Wiley, New York, pp 1–72
18. Itoh K (1967) Electron spin resonance of an aromatic hydrocarbon in its quintet ground state. Chem Phys Lett 1:235–238
19. Wasserman E, Murray RW et al (1967) Quintet ground states of *m*-dicarbene and *m*-dinitrene compounds. J Am Chem Soc 89:5076–5078
20. Lahti PM (1996) Design of organic-based materials with controlled magnetic properties. In: Turnbull MM, Sugimoto T, Thompson LK (eds) Molecule-based magnetic materials. American Chemical Society, Washigton, DC, pp 218–235
21. Rajca A, Wongsriratanakul J et al (2001) Magnetic ordering in an organic polymer. Science 294:1503–1505
22. Yang NC, Castro AJ (1960) Synthesis of a stable biradical. J Am Chem Soc 82:6208
23. Inoue K, Iwamura H (1995) 2-[*p*(*N*-*tert*-Butyl-*N*-oxyamino)phenyl]-4,4,5,5-tetramethyl-4,5-dihydroimidazol-3-oxide-1-oxyl, a stable diradical with a triplet ground state. Angew Chem Int Ed Engl 34:927–928

24. Hiraoka S, Okamoto T (2004) A stable radical-substituted radical cation with strongly ferromagnetic interaction: nitronyl nitroxide-substituted 5,10-diphenyl-5,10-dihydrophenazine radical cation. J Am Chem Soc 126:58–59
25. Suzuki S, Furui T (2010) Nitroxide-substituted nitronyl nitroxide and iminonitroxide. J Am Chem Soc 132:15908–15910
26. Dowd P, Chow M (1977) Generation and capture of common intermediates from proto-planar and proto-bisected trimethylenemethane precursors. Thermal rearrangement of a methylenepyrazoline. J Am Chem Soc 99:8507
27. Itoh T, Matuda K et al (2000) Tris[*p*-(*N*-oxyl-*N*-*tert*-butylamino)phenyl]amine, −methyl, and -borane have doublet, triplet, and doublet ground states, respectively. J Am Chem Soc 122:2567–2576
28. Furui T, Suzuki S et al (2014) Preparation and magnetic properties of metal-complexes from *N*-*t*-butyl-*N*-oxidanyl-2-amino-(nitronyl nitroxide). Inorg Chem 53:802–809
29. Masuda Y, Kuratsu M et al (2009) A new ferrimagnet based on a radical-substituted radical cation salt. J Am Chem Soc 131:4670–4673
30. Sugano T, Furusawa T et al (1991) Magnetic interactions among unpaired electrons in charge-transfer complexes of organic donors having a neutral radical. Synth Met 43:3281–3284
31. Sakurai H, Izuoka A (2000) Design, preparation, and electronic structure of high-spin cation diradicals derived from amine-based spin-polarized donors. J Am Chem Soc 122:9723–9734
32. Nakamura Y, Koga N et al (1991) Synthesis and characterization of 2-ferrocenyl-4,4,5,5-tetramethyl-2-imidazolin-l-oxyl 3-oxide and its CT-complex with DDQ. Chem Lett 20:69–72
33. Nakazaki J, Ishikawa Y et al (2000) Preparation of isolable ion-radical salt derived from TTF-based spin-polarized donor. Chem Phys Lett 319:385–390
34. Kumai R, Mastushita MM et al (1994) Intramolecular exchange interaction in a novel cross-conjugated spin system composed of π-ion radical and nitronyl nitroxide. J Am Chem Soc 116:4523–4524
35. Mastushita MM, Kawakami H et al (2007) Negative magneto-resistance observed on an ion-radical salt of a TTF-based spin-polarized donor. Chem Lett 36:110–111
36. Komatsu H, Matsushita MM et al (2010) Influence of magnetic field upon the conductance of a unicomponent crystal of a tetrathiafulvalene-based nitronyl nitroxide. J Am Chem Soc 132:4528–4529
37. Sugimoto T, Yamaga M et al (1993) Intramolecular spin-spin exchange in cation radicals of tetrathiafulvalene derivatives substituted with imino pyrolidine- and piperidine-1-oxyls. Chem Lett 22:1361–1364
38. Sugimoto T, Yamaga M et al (1993) Different magnetic properties of charge-transfer complexes and cation radical salts of tetrathiafulvalene derivatives substituted with imino pyrolidine- and piperidine-1-oxyls. Chem Lett 22:1817–1820
39. Masuda Y, Kuratsu M et al (2009) Preparation and magnetic properties of verdazyl-substituted dihydrophenazine radical cation tetrachloroferrate salts. Polyhedron 28:1950–1954
40. Masuda Y, Takeda H et al (2010) Radical-substituted dihydrophenazine radical cation salts: molecular packing structure and bulk magnetic property. Pure Appl Chem 82:1025–1032
41. Kuratsu M, Suzuki S et al (2012) (Nitronyl nitroxide)-substituted trioxytriphenylamine radical cation tetrachlorogallate salt: a 2p-electron-based weak ferromagnet composed of a triplet diradical cation. Chem Asian J 7:1604–1609
42. Yamaguchi K, Namimoto H et al (1990) Possibilities of organic ferromagnets and ferrimagnets by the use of charge-transfer (CT) complexes with radical substituents. Ab initio MO studies. Chem Phys Lett 166:408–414
43. Kuratsu M, Kozaki M et al (2005) 2,2′:6′,2′′,:6′′,6-Trioxytriphenylamine: synthesis and properties of the radical cation and neutral species. Angew Chem Int Ed 44:4056–4058
44. Kurastu M, Suzuki S et al (2007) Magnetic interaction of tri- and di-oxytriphenylamine radical cation $FeCl_4$ salts. Inorg Chem 46:10153–10157
45. Banister AJ, Bricklebank BK et al (1996) Spontaneous magnetization in a sulfur–nitrogen radical at 36 K. Angew Chem Int Ed 35:2533–2535

46. Tomiyoshi S, Yano T et al (1994) Weak ferromagnetism and antiferromagnetic ordering of 2p electrons in the organic radical compound 2,4,6-triphenylverdazyl. Phys Rev B 49:16031–16034
47. Kremer RK, Kanellakopulos B et al (1991) Weak ferromagnetism and magnetically modulated microwave absorption at low magnetic fields in 1,3,5-triphenyl-6-oxoverdazyl. Chem Phys Lett 230:255–259
48. Jamali JB, Achiwa N et al (1998) Single-crystal weak ferromagnetism of 1,3,5-triphenyl-6-oxoverdazyl free radical and ferromagnetic behavior of $(TOV)_{1-x}$ $(TOV\text{-}H)_x$ diluted system. J Magn Magn Mater 177–181:789–791
49. Ito A, Nakano Y et al (2005) Tetraarylethylene having two nitroxide groups: redox-switching of through-bond magnetic interaction by conformation change. Chem Commun 403–405
50. Mastuda K, Iwamura H (1997) Singlet and triplet states are degenerate in 2,3-dimethylenecyclohexane-1,4-diyl. J Am Chem Soc 119:7412–7413
51. Itoh T, Matsuda K et al (1999) A triphenylamine derivative with three *p*-(*N*-*tert*-butyl-*N*-oxylamino)phenyl radical units and yet a doublet ground state. Angew Chem Int Ed 38:1791–1793
52. Itoh T, Matsuda K et al (2001) The ground spin states of tris[*p*-(*N*-oxyl-*N*-*tert*-butylamino)phenyl] amine, —methyl, and -borane. Prospects of further studies. J Solid State Chem 159:428–439
53. Suzuki S, Nagata A et al (2012) Trinitroxide-trioxytriphenylamine: spin-state conversion from triradical doublet to diradical cation triplet by oxidative modulation of a π-conjugated system. Angew Chem Int Ed 51:3193–3197
54. Zhang X, Suzuki S et al (2012) NCN pincer–Pt complexes coordinated by (nitronyl nitroxide)-2-ide radical anion. J Am Chem Soc 134:17866–17868
55. Tanimoto R, Suzuki S et al (2014) Nitronyl nitroxide as a coupling partner: Pd-Mediated cross-coupling of (nitronyl nitroxide-2-ido)(triphenylphosphine)gold(I) with aryl halides. Chem Lett 43:678–680

Chapter 20
Organic Chemistry of Graphene Framework

Akihito Konishi and Takashi Kubo

Abstract Graphene, a sheet of carbon atoms arranged in a honeycomb lattice, has attracted much attention from chemists, physicists, and material scientists, because their peculiar electronic properties promise novel functions that radically change our lives in the coming years. Great contribution of physics has been disclosing electronic structures of graphene, whereas some of them can be understood by bottom-up approach, that is, by relating to chemistry of polycyclic aromatic hydrocarbons. This chapter mainly focuses on the origin of the magnetic properties of polycyclic aromatic hydrocarbons in terms of aromatic sextet formation and spin localization, in relation to "edge state" of graphene and graphene nanoribbons.

Keywords Polycyclic aromatic hydrocarbon (PAH) • Graphene • Edge state • Aromatic sextet

20.1 Introduction

Since its first isolation and characterization by Geim and Novoselov in 2004 [1], graphene, the newest member of the series of carbon allotropes, has developed the booming studies in a wide range of research areas, from physics, chemistry, and electronic devices [2]. Its two-dimensional form of pure sp^2-hybridized carbon, which provides us the fascinating electronic properties depending on the size and the peripheral shapes, has attracted the tremendous attention among both physicists and chemists [3].

Most of the interesting physics of graphene result from the linear energy dispersion in the vicinity of the Fermi level at the so-called Dirac point, where the conduction and valence bands cross in momentum space. In line with theoretical predictions, charge carriers in graphene behave like massless Dirac fermions at the Dirac point, which derives the ballistic electron transport [4, 5], the anomalous

---

A. Konishi • T. Kubo (✉)
Department of Chemistry, Graduate School of Science, Osaka University, 1-1 Machikaneyama, Toyonaka, Osaka 560-0043, Japan
e-mail: kubo@chem.sci.osaka-u.ac.jp

T. Akasaka et al. (eds.), *Chemical Science of π-Electron Systems*,
DOI 10.1007/978-4-431-55357-1_20

quantum hall effect [6–8], and the giant diamagnetism [9–14]. Furthermore, there has been widely growing interest in the chemical behavior of Dirac point of graphene [15–17].

Another interesting electronic feature is the localized nonbonding energy state for a nanoscaled graphene, which is referred to as "edge state" [18–22]. When graphene is cut down into a finite size, two types of edge are created, zigzag and armchair edges. The nonbonding π-electrons at the edge state localize only over the zigzag edges, giving a peculiar magnetic state. Investigating the electronic property of edges of graphene has been an important study in terms of understanding the relationship between electronic structure and carbon lattice topology.

From a perspective of its structure, graphene can be considered to be as indefinitely cumulative aromatic molecules, the limiting ones of flat polycyclic aromatic hydrocarbons (PAHs) [3, 23, 24]. PAHs have naturally distinct edges. Therefore, it is expected that PAHs become a good model for examining the edge state of graphene if a proper molecular design is given. Moreover, the precise constructions of the edges are controllable through bottom-up organic syntheses of PAHs, which means that the structure–property relationship of PAHs has an important role in the investigation of the edge state of graphene.

In this chapter, we describe the electronic properties and the chemical reactivity of nanoscale graphene frameworks, including the recent advances in organic chemistry of graphene frameworks. We also summarize our recent studies on the origin of the edge state by investigating anthenes in terms of the relationship between molecular structures and spin-localizing character.

20.2 Electronic Structures and Chemical Reactivity of Nanoscale Graphene Frameworks

20.2.1 Electronic Structure of Graphene and Nanographene

The honeycomb sp^2 carbon network of grapheme characterizes its electronic features. Graphene consists of chemically equivalent but crystallographically nonequivalent sp^2 carbon atoms depicted as either A or B in the unit cell (Fig. 20.1a) [3]. The sublattice built by either A- or B-carbon atom has the same threefold symmetry in an infinite graphene sheet, and therefore, the equivalency of the symmetry of these sublattices affords the linear energy dispersion, the Dirac corn, at around the Fermi level [2, 25]. The graphene valence band and conduction band cross linearly at the Dirac point, which defines the work function ($W = 4.6$ eV [26]). Consequently, the frontier orbitals of graphene form a degenerate pair at this point in momentum space with the same ionization potential (IP) and electron affinity (EA), and these states dominate the reactivity of graphene (see below).

A central concern of the study related to graphene is how graphene behaves when it is cut into nanoscale ribbons or flakes with distinct open edges. Nanographenes,

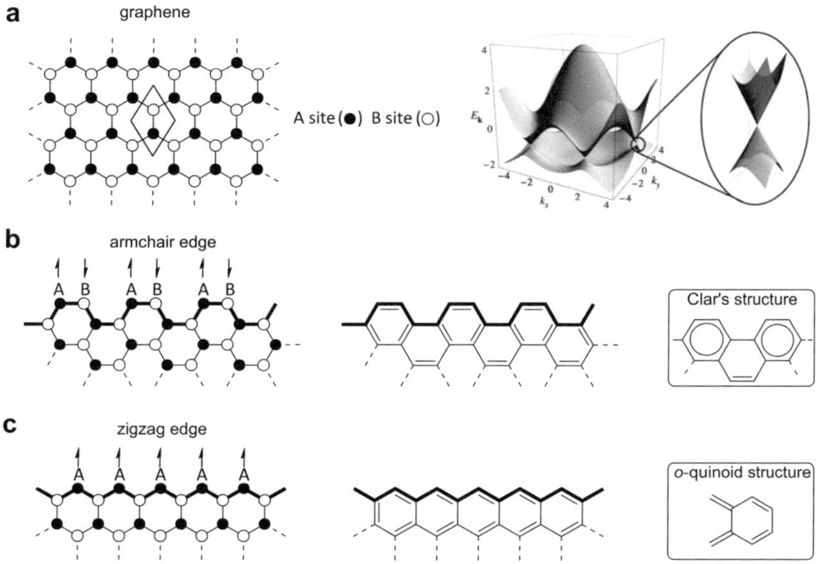

Fig. 20.1 (**a**) Sp^2 carbon atom of graphene network (carbon atoms depicted as A and B) and band structure of graphene. After ref [2]. (**b**) Armchair edge and (**c**) zigzag edge structures of GNRs modified from Ref. [3]

nanoscale graphenes, are an essential materials in terms of understanding the relationship between the electronic property and the carbon lattice topology [3], because they have a high ratio of edges to inner surface area compared to infinite graphene, which inherently leads to the significant effect of the edges on the electronic structure.

A nanographene results from breaking the equivalency of the grapheme sublattice. The boundary condition, which is to create the distinct edges, is introduced to the graphene sheet [3]. In the case of the armchair edges, the A- and B-carbon atoms alternately align, which means that a spin (e.g., α-spin) perfectly couples with an opposite signed one (e.g., β-spin) on the basis of the valence bond theory, giving nonmagnetic (closed-shell) ground state (Fig. 20.1b). In other words, the hexagonal rings in the armchair edge fuse in the full benzenoid π-conjugation manner, which leads to the stabilization of the system and affords the large HOMO–LUMO energy gap due to the Clar's aromatic sextet formation [27, 28]. On the other hand, only either A- or B-carbon atom localizes over the zigzag edge, which means that unpaired electrons reside on the zigzag edge and the nonbonding π-orbitals are generated (Fig. 20.1c).

Tanaka and co-workers demonstrated that one-dimensional zigzag-edged graphene nanoribbons (ZGNRs), not armchair-edged GNRs, possess the highly edge-localized nonbonding crystal orbitals by using a tight-binding theoretical calculation [18]. The wave function of the ZGNRs completely localizes over its edges at the Fermi level, with a quite similar distribution pattern seen at the edge of

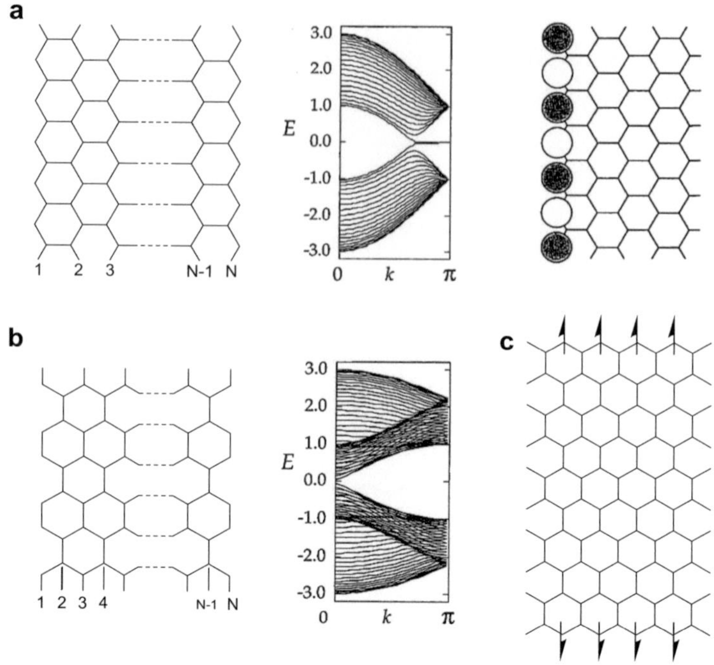

Fig. 20.2 (**a**) The band structure ($N = 20$) and the wave function ($k = \pi$) of GNR with zigzag edges. (**b**) The band structure ($N = 20$) of GNR with armchair edges. (**c**) The magnetic texture of ZGNR ($N = 10$). [19]

longer acenes. Fujita and co-workers also studied the electronic structure of GNRs with large width, by performing tight-binding band calculations, and they clarified a magnetic moment on the zigzag edges, by applying the Hubbard model to the edge-localized π-orbitals [19, 20]. It was estimated that unpaired electrons generated by the π-electrons localization on the zigzag edge ferromagnetically align along the ribbon and antiferromagnetically couple across the ribbon (Fig. 20.2). The localized nonbonding π-state, which is the spin-polarized state caused by multielectron correlation among unpaired electrons on the zigzag edges, is predicted to give GNRs magnetic activities. This peculiar electron localization in zigzag-edged GNRs is referred to as "edge state."

The electronic specificity of edge state has led scientists to observe the edge state. Coupled with the recent advances in the preparation of GNRs [29–31], including chemical [32–36] and lithographic methods [37, 38], as well as unzipping of carbon nanotubes [39–42], which enables the investigation of the physical and chemical properties of the edges of GNRs, various experimental measurements, such as scanning tunneling microscopy and spectroscopy (STM/S) [43–47], angle resolved photoemission spectroscopy [48], X-ray, and electron adsorption spectroscopy [49–51], have been attempted. Among them, STM/S is effective in investigating the correlation between the electronic state and the edge shape of the GNRs in atomic resolution [43, 52, 53]. Enoki and co-workers first achieved to propose the good evidence of the edge state around the zigzag edges in GNRs through STM/S measurements [43]. The STM images of the ZGNRs showed the higher electron density in the zigzag edge regions than the armchair edge, and the STS spectrum demonstrated the presence of the nonbonding π-state at the Fermi level, giving the evidence for the edge state. These broad theoretical and experimental studies of the edge state have clearly identified the precise fabrication of edge structures in GNRs as one key factor to understand the varied behaviors of the edge state.

20.2.2 Chemical Reactivity of Graphene

The chemical reactivity of graphene is a direct consequence of its electronic band structure. Haddon and co-workers summarized a unified treatment of the chemical reactivity of graphene based on its electronic band structure [16, 17, 54, 55]. As described above, the graphene valence band and conduction band linearly cross at the Dirac point, where these frontier orbitals perfectly degenerate as nonbonding molecular orbital (NBMO). The relationship of graphene frontier orbitals to the HOMOs and LUMOs of benzene is given in Fig. 20.3a [54, 55], which suggests that graphene can be written as the resonance hybrid of the Clar's representation. These structural and electronic features of graphene lead to the characteristic chemical reactivity.

One is the radical addition to the basal plane [56–62]. Because the frontier orbitals of graphene lie at a nonbonding state, the radical addition to the inner surface of graphene, as well as to the carbon atoms with high spin density, such as a dangling bond or zigzag edge, can take place, to afford the graphene with the rehybridized carbon atoms from sp^2 to sp^3. The thermodynamically (energetically) favorable processes involve the pairwise chemisorption of functional groups in different sublattices, rather than on the same sublattice [15, 63]. Theoretical calculations suggest that the pairwise chemisorption of a species in different sublattices is favored by 0.5 eV per addition [63, 64].

The other is the dual nature for the Diels–Alder reaction. The degenerate HOMO and LUMO orbitals of graphene, together with the symmetries of these frontier orbitals, enable graphene to behave as either diene or dienophile, when paired with the appropriate reaction partner in the Diels–Alder reaction [54, 65]. Graphene can function as either diene when paired with tetracyanoethylene (TCNE) and maleic anhydride (MA) or as dienophile when paired with 2,3-dimethoxybutadiene (DMBD) and 9-methylanthracene (9-MeA) (Fig. 20.3b). The unique electronic structure, the zero-band-gap electronic structure of graphene arising from the crossing of the valence and conduction bands at the Dirac point, allows graphene to participate in this surprisingly mild Diels–Alder reaction on the basal plane.

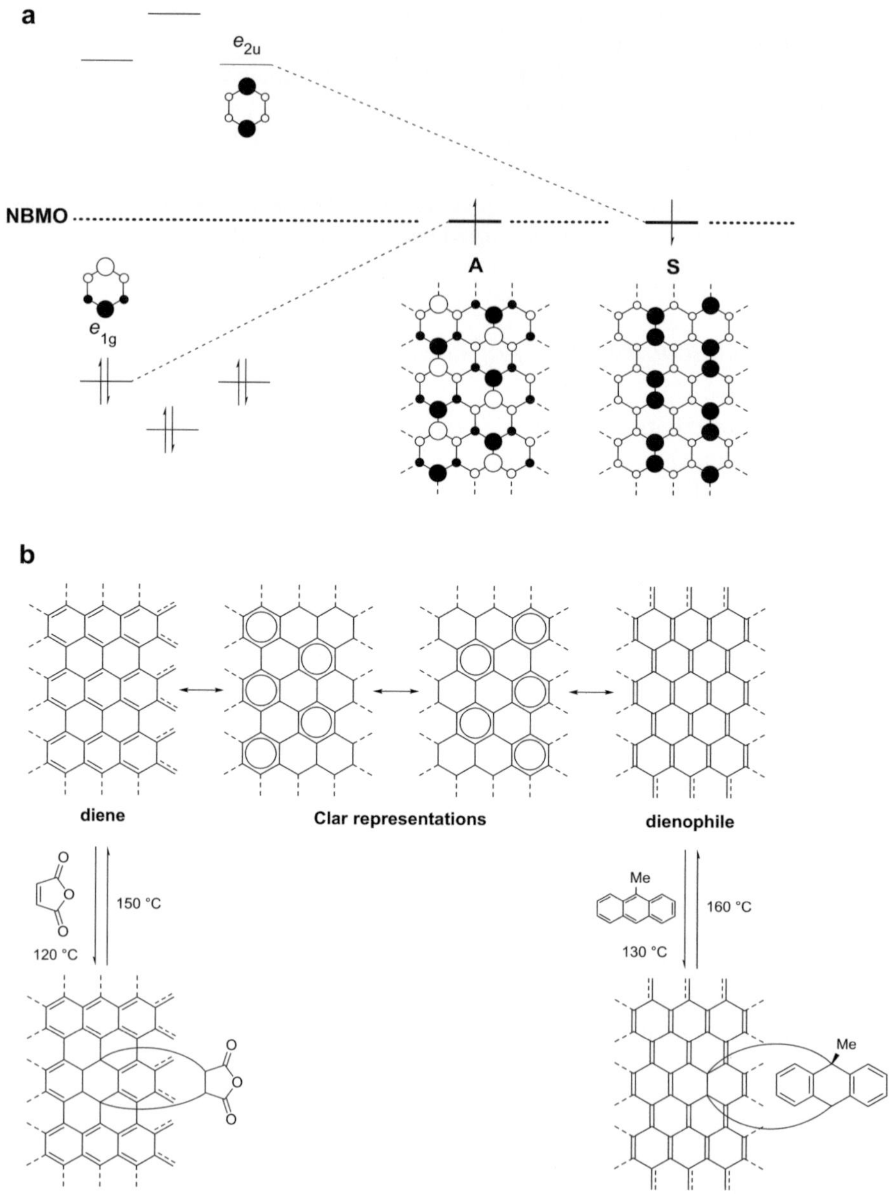

Fig. 20.3 (**a**) HMO energy levels of benzene and their symmetry, together with the orbital coefficients of HOMO and LUMO that map onto the degenerate conduction and valence bands of graphene at the Dirac point. (**b**) Resonance hybrids of graphene and schematic of the Diels–Alder reaction of graphene with maleic anhydride (MA, dienophile) and 9-methyantracene (MeA, diene). Modified from ref [55]

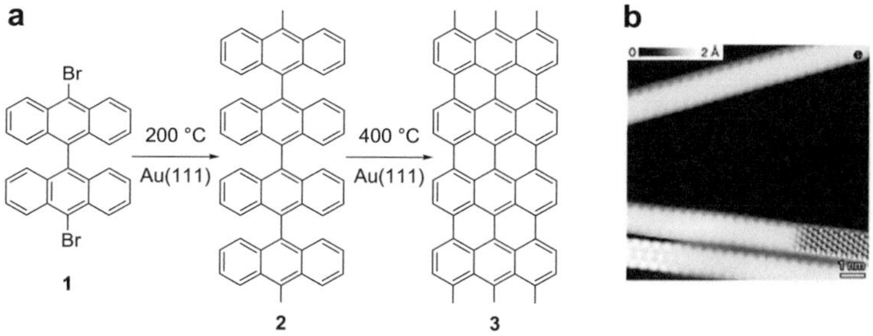

Fig. 20.4 (**a**) Reaction scheme from **1** to GNR **3** on an Au(111) surface. (**b**) High-resolution STM image of **3**. After ref. [35]

20.3 Recent Advances in Chemical Preparation of GNRs

Although there are several top-down methods to prepare thin GNRs, such as lithographic methods [37, 38] and unzipping of carbon nanotubes [39–42], the usefulness of the resulting GNRs is limited because of the atomically undefined width and uncontrolled edge structures. On the other hand, bottom-up approach, chemical preparation, is advantageous because it can give access to the atomically precise structure of GNRs without defects and to the chemical modification along the edges to tune the solubility and the electronic structure.

Müllen and co-workers have proposed the several methods to synthesize GNRs either in solution or metal surfaces [35, 36]. Atomically precise preparation of GNR was achieved on Au(111) surfaces (Fig. 20.4) [35]. Biradical species generated from **1** by heating at 200 °C on an Au(111) surface formed a polymeric compound **2**, and subsequent heating at higher temperature led to anthracene-ring fusion, giving a width-controlled GNR **3**. The width and the edge shape of **3** are defined by the structure of the molecular precursor **1**. The edge termination of the resulting GNR **3** by hydrogen atoms is confirmed by STM experiments, and the electron density distribution around the zigzag edges of the GNRs was estimated by the large-scale density functional theory calculations [66].

The bottom-up synthesis of GNRs in solution phase was also reported [34, 36]. The Diels–Alder polymerization of the cyclopentadienone derivative **4** as a key reaction afforded the oligophenylene precursor **5**, and subsequent intramolecular oxidative dehydrogenation led to the graphitization **6** (Fig. 20.5). Tailor-made synthesis of long (>200 nm) and chemically precise GNRs was achieved. The resulting GNRs **6** showed high dispersibility due to the installed long alkyl chains on the periphery, which enabled the liquid phase studies of **6**. The electronic and vibrational properties of **6** were characterized by UV–Vis, Raman, and infrared spectroscopies. The longest absorption band of **6** was observed around 550 nm, giving an optical band gap of 1.88 eV. The profile of the Raman spectrum of **6** is characterized by two strong bands (1600 cm^{-1}, 1350 cm^{-1}), which correspond to the G and D bands of graphite. Due to the high dispersibility, the AFM visualization

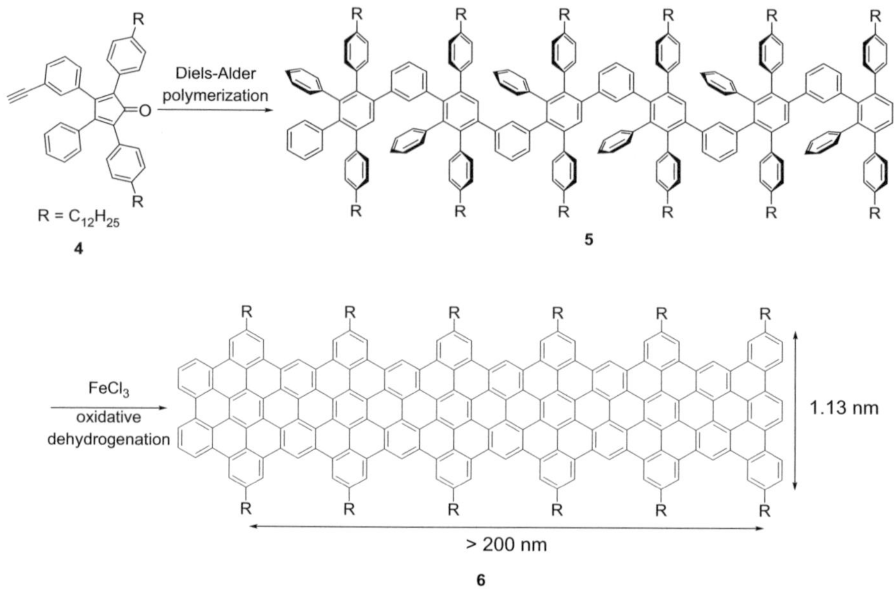

Fig. 20.5 Reaction scheme from **4** to GNR **6**. After ref. [36]

of **6** revealed the ordered self-assembled monolayers of uniform GNR on a HOPG surface. In these reports, the width and the edge shape of GNRs are defined by the structure of the molecular precursors **2** and **5**, respectively, and this means that GNRs might be fabricated and modified in response to the intended use.

20.4 Spin State of Polycyclic Aromatic Hydrocarbons (PAHs)

The synthesis of perfectly terminated and reliably size-controlled GNRs has been in high demand. Despite the several studies on the chemical preparations of GNRs [31, 34, 36], there is still the dearth of methods to synthesize the size-controlled and well-terminated zigzag-edged GNRs. Due to the accessibility of the defined structure and chemical modification, the elucidation of the edge state at the molecular level is growing in importance. Especially, polycyclic aromatic hydrocarbons (PAHs) are possible molecular candidates for the purpose since they are structural components of GNR and have the distinct edges [31]. Furthermore, they have been estimated to possess similar spin state to GNRs [66–70, 73–77]. In this section, the relationship between the edge topology and the electronic state of a given molecule is overviewed.

As described above, ZGNRs have unpaired electrons on the zigzag edges and generate the nonbonding π-orbitals. At a molecular level, the minimum unit of zigzag edges is phenalenyl, which is a neutral hydrocarbon radical with large

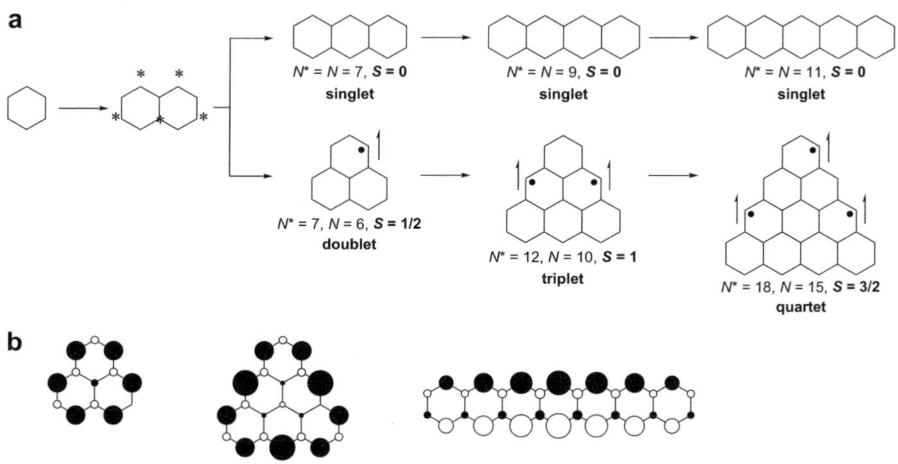

Fig. 20.6 (**a**) Prediction of spin quantum number *S* and spin multiplicity in the ground state of PAHs with zigzag edges. (**b**) Spin density distribution of linear and triangular PAHs. *Filled* and *open circles* denote α- and β-spin, respectively

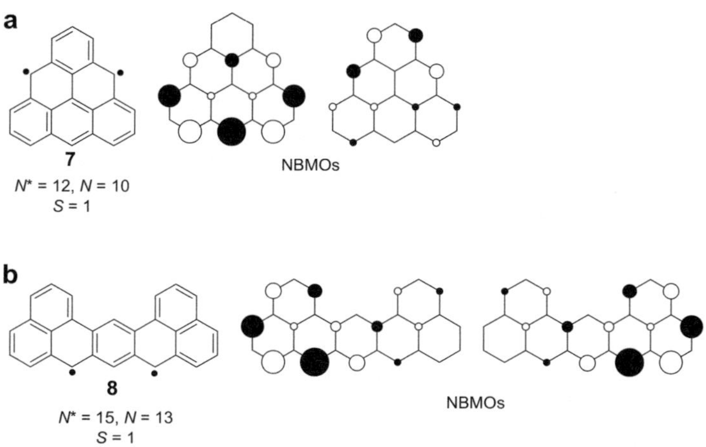

Fig. 20.7 Prediction of spin quantum number *S* and spin multiplicity in the ground state of **(a) 7** and **(b) 8**

spin density on the peripheral α-carbon atoms [78]. Larger triangularly ring-fused PAHs, which are solely decorated by zigzag edges, are also non-Kekulé PAHs and are predicted to possess multi-unpaired electrons localized over the periphery edges with a high-spin ground state according to Ovchinnikov's rule [79] and the non-disjoint nonbonding molecular orbital (Fig. 20.6) [3, 80, 81]. Experimental determination of the spin multiplicity has not been done so far, except for **7** and **8** (Fig. 20.7). Tri-*tert*-butyl derivative of **7** gives an ESR signal typical for triplet species with threefold symmetry, and a Curie plot of the signal intensity supports the

Fig. 20.8 Chemical structures of (*left*) acenes 9 and (*right*) [m.n]periacene 10

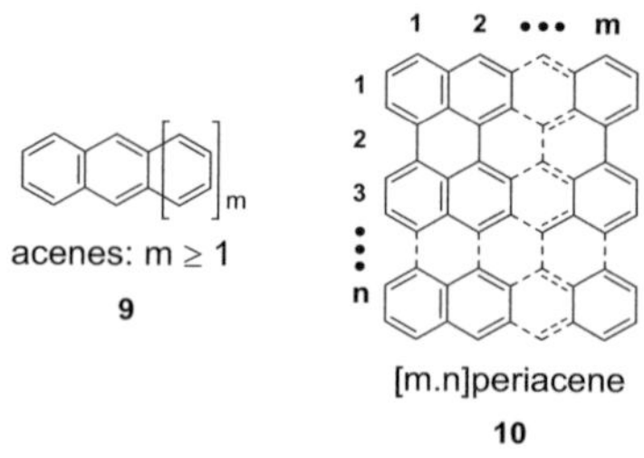

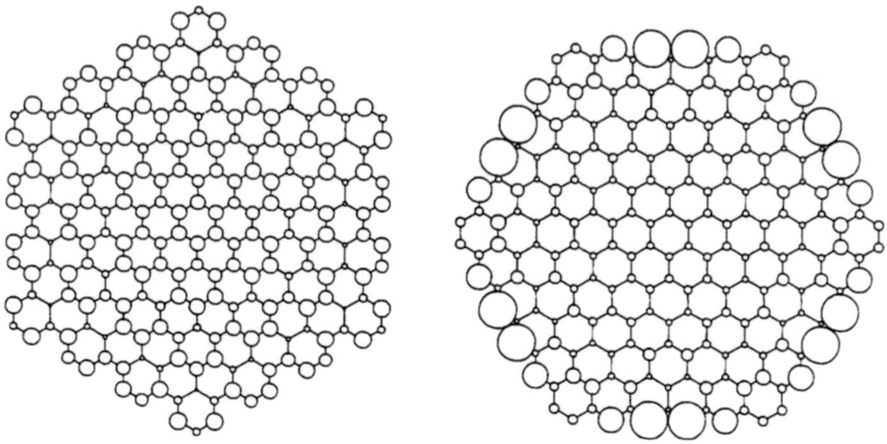

Fig. 20.9 Electronic density distributions of the HOMO for PAHs with (*right*) armchair and (*left*) zigzag edges. After ref. [67]

triplet ground state [82]. The mesityl derivative of **8**, which is an annulated *m*-xylene biradical, shows the triplet ground state confirmed by an electron spin transient nutation (ESTN) method [83]. These results of the triangular ring-fused non-Kekulé PAHs propose the adequate explanation of the ferromagnetic spin alignment along a zigzag edge in GNRs.

The other possible molecular structure for a zigzag edge is linear or two-dimensional ring annulations along with the formation of armchair edges, giving acenes **9** [68–72] and polyperiacenes ([m.n]periacene) **10** (Fig. 20.8) [73–77]. Although these molecules can be drawn as the Kekulé structures with no unpaired electron, recent theoretical calculations reveal that the same spin-polarized state as ZGNRs is generated in the ground state, which means that such a molecule has an antiferromagnetically coupled open-shell ground state [68–77].

Stein and co-workers demonstrated that the HOMO of zigzag-edged PAHs possesses the largest population of wave function in the zigzag edge regions (Fig. 20.9) [67]. The distribution pattern at the zigzag edge resembles that of longer acenes and the zigzag-edged GNR, which bears nonbonding character.

Recent calculations reveal that frontier orbitals of acenes bear nonbonding character and the HOMO–LUMO energy gap rapidly decreases, with increasing

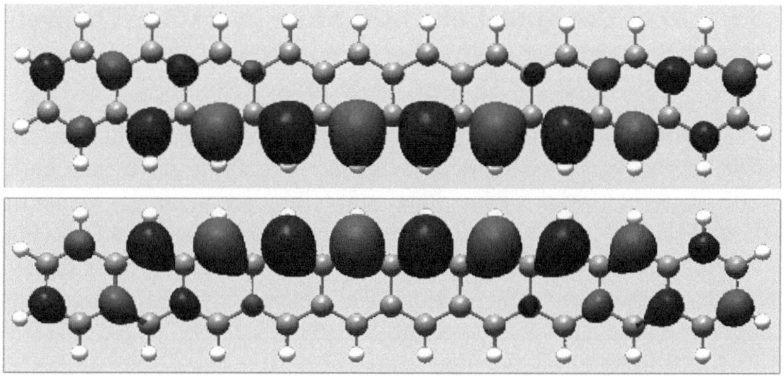

Fig. 20.10 SOMOs of decacene calculated by Bendikov and co-workers. After ref. [68]

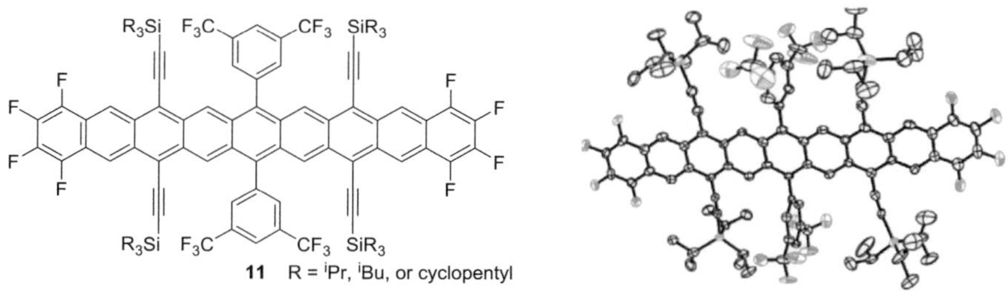

Fig. 20.11 Anthony's nonacenes **11** and crystal structure of the R =^{i}Pr derivative. After ref. [72]

ring annelation [68–70]. Bendikov, Houk, and co-workers estimated that the open-shell singlet is the lowest state in acenes larger than hexacene by using a broken symmetry UB3LYP/6-31G* method, and they behave as singlet biradicals, whose singly occupied molecular orbital (SOMO) is populated on their zigzag edges (Fig. 20.10) [68].

Anthony and co-workers determined the molecular structure of nonacene **11** by X-ray single-crystal crystallographic analysis (Fig. 20.11) [72]. Nonacene **11** showed a prominent HOMO–LUMO transition at 1014 nm, which was located at lower-energy region than that of non-substituted nonacene [71]. The optical HOMO–LUMO gap is determined to be 1.2 eV, which was almost the same as the electrochemical HOMO–LUMO gap (1.19 eV). These small HOMO–LUMO gaps suggest some extent of singlet biradical character of nonacene. Actually, ^{1}H-NMR spectrum of **11** gave no signal due to the existence of magnetic species, probably thermally excited triplet species.

The nonbonding frontier orbitals with a small gap are closely related to the biradical character, and a more sophisticated calculation reveals higher antiferromagnetic spin state (i.e., polyradical character) for longer acenes [69, 70].

20.5 Origin of the Spin-Polarized State in PAHs: Theoretical and Experimental Studies of Anthenes

Anthenes ([3.n]periacene) [84, 85] and polyperiacenes, in which two and more anthracenes or acenes are condensed in the *peri*-directions, have been good model systems for investigating the edge state, since they are small segments of GNRs and have well-defined two edge structures: zigzag and armchair edges (Fig. 20.8). Recent theoretical and experimental studies reveal that anthenes and polyperiacenes can inherently possess the same electronic structure as the edge state of ZGNRs, which depends on the edge shape and the molecular size [73–77]. Considering spin density, the terminology of the edge state can be translated into singlet biradical character for anthenes [86]. A great concern associated with the origin of the edge state at a molecular level is the effect of the molecular size on its open-shell character, including the effect of the inner π-conjugation on the peripheral electronic structure. Here, we summarize our recent studies on the origin of the edge state at a molecular level through the observation of magnetic, optical, and chemical behaviors of anthenes.

20.5.1 Theoretical Background

All anthenes can be drawn as a resonance hybrid of Kekulé and biradical forms (Scheme 20.1). In the resonance formula, anthenes are subjected to loss of one double bond when the structure is drawn with a maximum number of Clar sextets [27, 28]. In the biradical form, Clar sextets, the isolated double bonds, and the unpaired electrons have fixed positions, that is, the influence of sextet migration or spin delocalization is limited. Therefore, the discussion of the biradical character can be focused on the energy balance between the formal loss of the double bond and the aromatic sextet formation. The amplitudes of biradical character (y_0) of anthenes were theoretically estimated by using the index defined by Yamaguchi [87] coupled to the symmetry-broken UBHandHLYP/6-31G* calculation. $y_i = 1 - 2T_i/(1 + T_i^2)$,

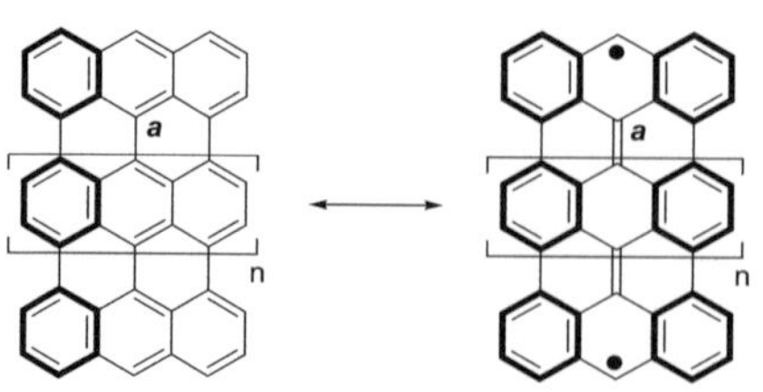

Bisanthene **12** (n=0), y_0 = 12%

Teranthene **13** (n=1), y_0 = 59%

Quateranthene **14** (n=2), y_0 = 84%

Scheme 20.1 Resonance formula of anthenes. The amplitudes of the biradical character (y_0) were calculated at the UBHandHLYP/6-31G* level. The six-membered rings depicted by bold lines represent the Clar sextets

Table 20.1 The degrees of open-shell character y_0 and y_1 of acenes and [m.n]periacenes

y_0		**m = 2**	**3**	**4**	**5**	**6**	**7**
	n = 1	0.00	0.00	0.01	0.07	0.20	0.40
	3	0.00	0.12	0.60	0.84	0.94	0.98
	5	0.01	0.59	0.91	0.98	0.99	1.00
	7	0.05	0.84	0.98	1.00	1.00	1.00
y_1		**m = 2**	**3**	**4**	**5**	**6**	**7**
	n = 1	0.00	0.00	0.00	0.00	0.01	0.02
	3	0.00	0.00	0.01	0.02	0.07	0.20
	5	0.00	0.00	0.01	0.05	0.23	0.54
	7	0.00	0.00	0.01	0.10	0.44	0.77

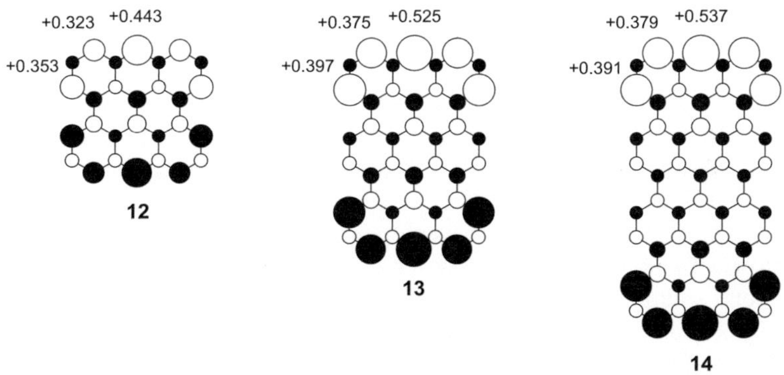

Fig. 20.12 Spin density map of **12**, **13**, and **14** (white circles, α-spin; black, β-spin). The numerical values are α-spin densities on the edge anthracene ring of each compound

$T_i = (n_{HOMO-i} - n_{LUMO+i})/2$, where n_{HOMO-i} and n_{LUMO+i} represent natural orbital occupation numbers of HOMO–i and LUMO + i, respectively. The y_0 and y_1, which are determined from the HOMO–LUMO pair and the HOMO–1–LUMO + 1 pair, are related to first and second π-bond cleavages, respectively, and vary continuously from zero to unity. A perfect biradical molecule has y_0 of unity and y_1 of zero, but for the molecule with large y_1, the biradical description is inaccurate, and tetraradical character should be taken into account. Table 20.1 shows the y_0 and y_1 values of acenes and polyperiacenes.

The calculation shows that the biradical character drastically increases with increasing inner anthracene moieties (n): small ($y_0 = 12$ %, **12**), intermediate (59 %, **13**), and large (84 %, **14**). This is because the difference in the number of the sextet between the Kekulé and biradical forms increases with increasing molecular size: two for **12**, three for **13**, and four for **14**. More sextets in the biradical form results in more dominant contribution of biradical electron configuration to the ground state. Spin density map shown in Fig. 20.12 indicates that unpaired electrons are more localized on the *meso*-carbons in larger anthenes.

With the increase in the molecular size, the y_0 of polyperiacene increases (Table 20.1). The y_0 of [3.7] (quateranthene), [4.n] (n: 5, 7), and [5.n]periacene (n:

3, 5, 7) is close to unity, whereas y_1 remains quite small. These open-shell degrees strongly suggest that the ground state of these polyperiacenes should be biradical. A multi-spin state becomes more likely in larger systems of polyperiacenes. The y_1 of [6.7]periacene is found to be large (0.44), suggesting an appreciable contribution of tetraradical structure to the ground state. The largest polyperiacene, [7.7]periacene, in this paper has very large y_0 (1.00) and y_1 (0.77), indicating the singlet tetraradical ground state.

20.5.2 *Syntheses of Anthenes*

20.5.2.1 Straightforward Stepwise Syntheses

Unsubstituted bisanthene **12** has been isolated as an air-sensitive crystalline powder [88, 89], and its crystal structure was recently determined by us [90]. Wu [91] and Scott [92] prepared *meso*-substituted derivatives and investigated their physical properties and chemical reactivity.

The *tert*-butyl derivative (**12a**) was prepared according to the reported procedure [93] and was found to be moderately stable in air under room light ($t_{1/2} = 19$ days in a toluene solution at room temperature). Teranthene **13** and quateranthene **14** have long been unknown compounds, but recently were successfully isolated. The key steps for the preparations are a partial cyclization with KOH/quinoline and a full cyclization with DDQ/$Sc(OTf)_3$. Scheme 20.2 shows the synthetic route to the derivatives of **13**. The compound **13a** was obtained as dark green plates by recrystallization from a CH_2Cl_2–hexane solution under argon flow [94]. A toluene solution of **13a** shows gradual decomposition with a half-life period of 3 days open to air under room light at room temperature. The isolation of **14a** was also achieved by careful recrystallization from an *o*-dichlorobenzene/mesitylene solution

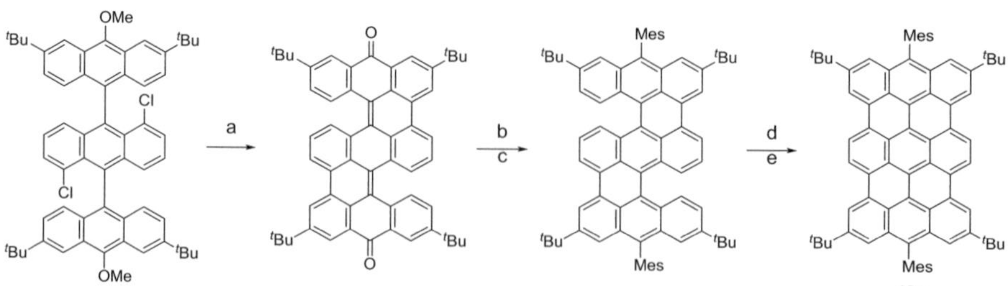

Scheme 20.2 Synthetic procedure for **13a**. Reagents and conditions: (**a**) KOH, quinoline, 190 °C, 78 %; (**b**) mestylceriumbromide, THF/ether (4:1), −30 °C; (**c**) NaI, $NaH_2PO_2 \cdot H_2O$, AcOH, 150 °C, 75 % (2 steps); (**d**) DDQ, $Sc(OTf)_3$, toluene, reflux; (**e**) $N_2H_4 \cdot H_2O$, toluene, rt, 72 %(2 steps)

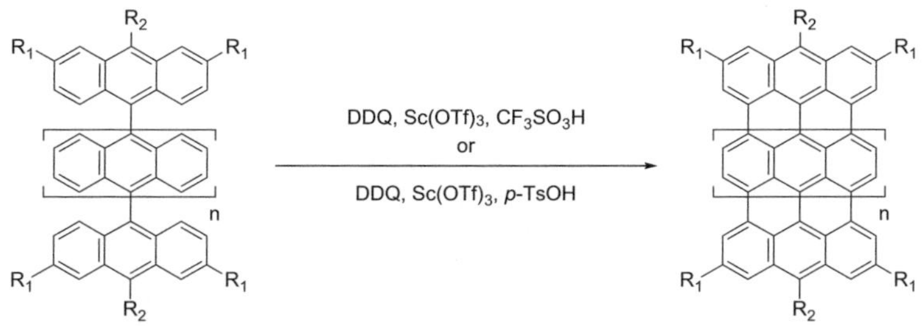

15a: m = 0, R_1 = H, R_2 = mesityl
15b: m = 0, R_1 = *tert*-butyl, R_2 = mesityl
16a: m = 1, R_1 = *tert*-butyl, R_2 = mesityl
17a: m = 2, R_1 = *tert*-butyl, R_2 = mesityl

12a: 57 %
12b: 74%
13a: 17%
14a: trace

Scheme 20.3 Direct synthesis of **12b**, **13a**, and **14a** from the corresponding anthracene oligomers

in a sealed degassed tube [95]. The half-life of **14a** at room temperature is only 15 h when exposed to air under room light.

20.5.2.2 Direct Cyclization of Anthracene Oligomer

Recently, a new method (Scheme 20.3) that yields anthenes directly from the corresponding anthracene oligomers by improving the Scholl reaction was developed [96]. As opposed to the conventional Scholl reaction, that new approach produced acceptable yields of **12b** and **13a** through the simultaneous combination of all three components (DDQ, Sc(OTf)$_3$, and Brønsted acids). Although the mechanistic details of the reaction have not been clarified yet, strong Brønsted acids may promote protonation of the anthracene ring and electron transfer to DDQ, and arenium-type cations may be generated as intermediate species [97].

20.5.3 Molecular Structure

The length of the C–C bond, indicated by ***a*** in Scheme 20.1, is well correlated to the singlet biradical character (y_0). The bond length should decrease with increasing the biradical character because the bond ***a*** has single and double bond character in the Kekulé and biradical forms, respectively. The bond lengths of anthenes were determined from the X-ray crystal structure of **12a**, **13a**, and **14a** (Fig. 20.13). The bond length of bisanthene **12a** is 1.451(2) Å, which is the same as the ordinary C(sp^2)–C(sp^2) single bond length (1.45 Å). Thus, **12a** has negligible biradical character in the ground state. On the other hand, teranthene **13a** and quateranthene **14a** show quite short bond lengths, indicating appreciable biradical character.

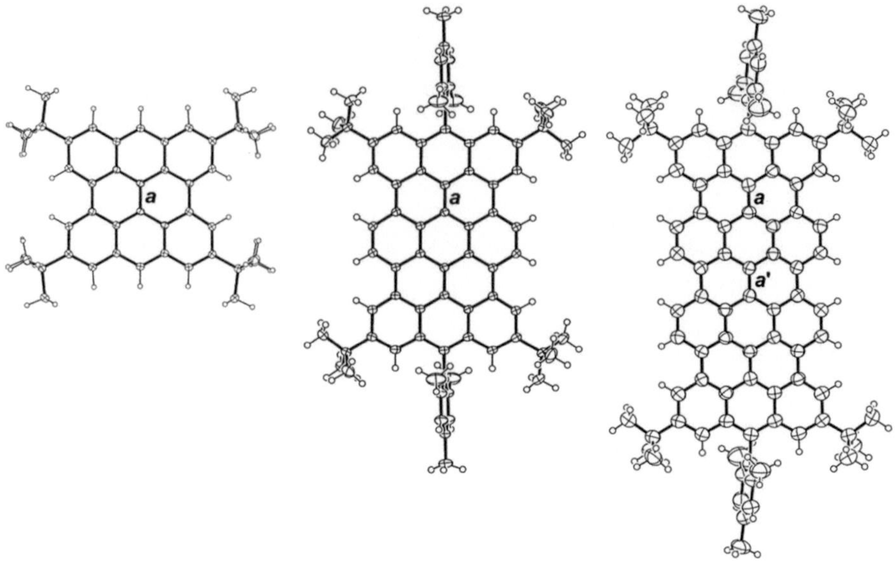

Fig. 20.13 ORTEP drawings of **12a** (*left*), **13a** (*middle*), and **14a** (*right*). The length of the bond: $\boldsymbol{a} = 1.451$ Å (**12a**), $\boldsymbol{a} = 1.424$ Å (**13a**), and $\boldsymbol{a} = 1.412$ Å and $\boldsymbol{a}' = 1.419$ Å (**14a**)

The resonance formula in Scheme 20.1 also suggests that the peripheral six-membered rings bear larger benzenoid character in larger anthenes. The harmonic oscillator model of aromaticity (HOMA) [98, 99] value is known to be a good aromaticity index based on a ring geometry. The HOMA values of the corner rings of **13a** (0.854) and **14a** (0.823) are quite large compared to **12a** (0.704), supporting the significant biradical character for **13a** and **14a**.

The aromatic stabilization energy of benzene based on the homodesmic stabilization energy is ca. 90 kJ/mol [100, 101], whereas the destabilization energy due to C–C π-bond cleavage is ca. 270 kJ/mol [102]. For bisanthene **12**, the destabilization energy of the π-bond cleavage is not fully compensated by the formation of aromatic sextets because the difference in number of sextets between the Kekulé form and biradical form is only two; thus, the Kekulé form has dominant contribution to the ground state. Teranthene **13** has three more sextets in the biradical form, and therefore, the destabilization energy (270 kJ/mol) is canceled with the aromatic stabilization energy ($90 \times 3 = 270$ kJ/mol). The aromatic stabilization energy of quateranthene **14** ($90 \times 4 = 360$ kJ/mol) exceeds the destabilization energy, indicating the dominant contribution of the biradical form to the ground state.

Considering these geometrical findings, the edge state would come from that the formation of aromatic sextets overwhelms the penalty for breaking one π-bond and that the resulting unpaired electrons are pushed out to the zigzag edges, especially to the *meso*-positions, as shown in Scheme 20.1.

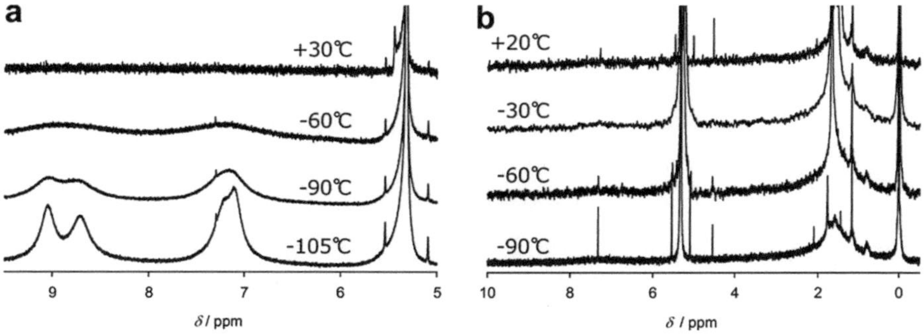

Fig. 20.14 (**a**) Variable temperature ^{1}H-NMR spectra of **13a** in CD_2Cl_2 in the aromatic region. (**b**) Variable temperature ^{1}H-NMR spectra of **14b** in CD_2Cl_2

20.5.4 Magnetic Properties

Intrinsically, magnetic properties of singlet biradical species cannot be measured due to strong antiferromagnetic coupling of unpaired electrons. Instead, the presence of thermally accessible triplet species would be a good criterion for the singlet biradical electronic structure. Thermally excited triplet species causes severe broadening of solution NMR signals due to electron–nuclear magnetic interaction.

Bisanthene **12a** showed sharp ^{1}H NMR signals even at +110 °C, indicating a negligible influence of the triplet species due to a large singlet–triplet energy gap ($\Delta E_{S\text{–}T}$). Indeed, the $\Delta E_{S\text{–}T}$ of bisanthene was estimated to be 6300 K at the B3LYP/6-31G* level of calculation. In contrast, the CD_2Cl_2 solution of teranthene **13a** showed no ^{1}H-NMR signals of the teranthene core at room temperature, whereas upon cooling, progressive line sharpening was observed (Fig. 20.14a). This behavior is associated with the presence of thermally excited triplet species at elevated temperatures. The influence of the thermally excited triplet species is more distinct in the ^{1}H-NMR measurements of quateranthene. Even at −92 °C, the CD_2Cl_2 solution of **14b** (4-*tert*-butyl-2,6-dimethylphenyl derivative instead of mesityl in **14a**) gave no ^{1}H-NMR signals of the quateranthene core (Fig. 20.14b), which suggests that **14b** possesses the large population of the thermally excited triplet species even at low temperature, due to the smaller $\Delta E_{S\text{–}T}$.

The small $\Delta E_{S\text{–}T}$ of **13** and **14** were confirmed by SQUID measurements. The measurements showed $\Delta E_{S\text{–}T}$ of 1920 K and 347 K for **13a** and **14a**, respectively. Generally, the $\Delta E_{S\text{–}T}$ of closed-shell compounds exceeds 10,000 K. These substantially small $\Delta E_{S\text{–}T}$, that is, very weak coupling of electrons, strongly support the edge localization of unpaired electrons in teranthene and quateranthene. Indeed, **14b** reacts with oxygen at less-protected carbon atoms with a large spin density around the zigzag edge, giving rise to an oxygen-adduct one-dimensional polymer **15** in a crystalline form (Fig. 20.15).

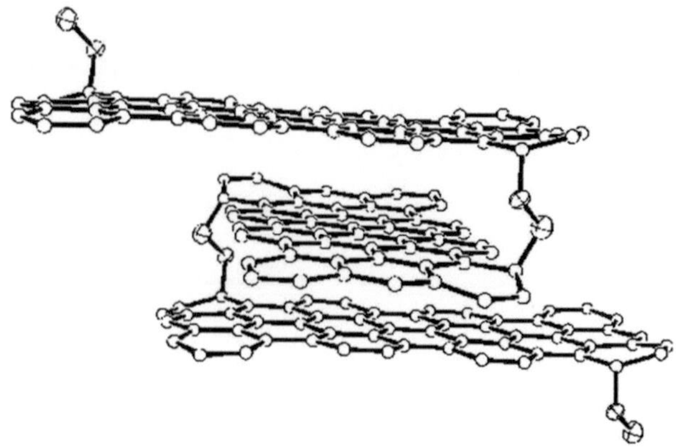

Fig. 20.15 ORTEP drawing of an oxygen-adduct one-dimensional polymer, **15**. 4-*tert*-Butyl-2,6-dimethylphenyl and *tert*-butyl groups and H atoms are omitted for clarity

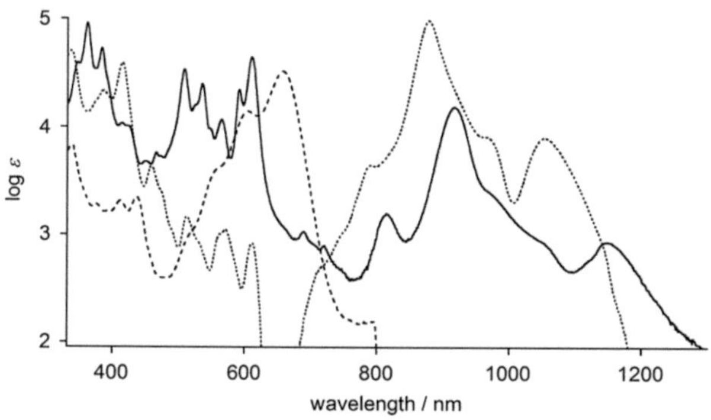

Fig. 20.16 One-photon electronic absorption spectra of **12a** (*hashed line*), **13a** (*dotted line*), and **14b** (*solid line*)

20.5.5 *Optical Properties*

Singlet biradicals feature the presence of a low-lying excited singlet state dominated by the doubly excited configuration [103]. As shown in Fig. 20.16, teranthene **13a** and quateranthene **14b** afford a weak low-energy band centered at 1054 and 1147 nm, respectively, whereas nonmagnetic **12a** gives an intense band at 686 nm whose profile is quite similar to that of nonmagnetic rylenes [104]. The weak bands observed in **13a** and **14b** are assignable to a HOMO → LUMO double excitation by a strongly contracted second-order *n*-electron valence state perturbation theory (NEVPT2) calculation that allows multielectron excitation [105]. The NEVPT2(8,8)/6-31G* calculation revealed that the first excited singlet state (S_1) of **13** and **14** is a $2A_g$ state dominated by the double excitation with a transition energy of 1.35 (916 nm) and 1.32 eV (940 nm), respectively. The transition to the $2A_g$ state is intrinsically a one-photon forbidden process due to the parity (g–g transition), whereas a two-photon allowed process. A two-photon absorption spectrum of **14b** measured in CS_2 showed a broad band at around 2300 nm, which is the same excitation energy as that of the one-photon band at 1147 nm. The appearance of the forbidden bands in the one-photon absorption spectra is due to the intensity borrowing from intense HOMO → LUMO single-electron transitions at 878 nm (**13a**, $\varepsilon = 97$ 800) and 917 nm (**14b**, $\varepsilon =$ 15 000). Thus, the weak bands observed in **13a** and **14b** are associated with the simultaneous excitation of the edge-localized electrons.

20.6 Concluding Remarks

Prominent electronic feature of graphene, which is given by its pure two-dimensional sp^2 carbon networks, has been explosively studied by many researchers. The role played by organic chemistry in graphene science is increasing importance in association with the global attention of graphene. Selective chemical modification and patterning of graphene wafers with atomic precision enable us to tune the electronic structure of graphene and provide the novel manufacturing technology necessary for carbon-based materials. Moreover, precise preparation of extraordinary large PAH molecules has brought about tailor-made nanographene materials in accordance with the intended use. Because of the accessibility of defined and terminated structure, investigation of the electronic structure and chemical behavior of large PAHs well designed as graphene mimics is growing in importance, which can promise the understanding of the electronic feature of graphene from molecular level.

In this regard, *peri*-condensations of anthracene discussed herein lead to the appearance of magnetic spins on zigzag edges, which is closely related to the edge state of ZGNRs. Studies on bisanthene **12** and teranthene **13** revealed that bisanthene **12** can be categorized as a nonmagnetic species, whereas teranthene **13** lies at the onset of the magnetic state. More distinct behaviors are observed in quateranthene **14**, whose physical and chemical properties are well explained by the edge localization of unpaired electrons. Aromatic sextet formation is a key mechanism of the magnetic state, and this classical and simple mechanism would indeed determine the electronic structure of nanographenes.

References

1. Novoselov KS, Geim AK, Morozov SV, Jiang D, Zhang Y, Dubonos SV, Grigorieva IV, Firsovet AA (2004) Electric field effect in atomically thin carbon films. Science 306:666–669
2. Neto AHC, Guinea F, Peres NMR, Novoselov KS, Geim AK (2009) The electronic properties of graphene. Rev Mod Phys 81:109–162
3. Enoki T, Takai K, Kiguchi M (2012) Magnetic edge state of nanographene and unconventional nanographene-based host-guest systems. Bull Chem Soc J 85:249–264

4. Baringhaus J, Ruan M, Edler F, Tejeda A, Sicot M, Taleb-Ibrahimi A, Li AP, Jiang Z, Conrad EH, Berger C, Tegenkamp C, De Heer WA (2014) Exceptional ballistic transport in epitaxial graphene nanoribbons. Nature 506:349–354
5. Berger C, Song Z, Li T, Li X, Ogbazghi AY, Feng R, Dai Z, Marchenkov AN, Conrad EH, First PN, de Heer WA (2004) Ultrathin epitaxial graphite: 2D electron gas properties and a route toward graphene-based nanoelectronics. J Phys Chem B 108:19912–19916
6. Novoselov KS, Geim AK, Morozov SV, Jiang D, Katsnelson MI, Grigorieva IV, Dubonos SV, Firsov AA (2005) Two-dimensional gas of massless Dirac fermions in graphene. Nature 438:197–200
7. Gusynin VP, Sharapov SG (2005) Unconventional integer quantum hall effect in graphene. Phys Rev Lett 95:146801
8. Zhang Y, Tan YW, Stormer HL, Kim P (2005) Experimental observation of the quantum Hall effect and Berry's phase in graphene. Nature 438:201–204
9. Sharapov SG, Gusynin VP, Beck H (2004) Magnetic oscillations in planar systems with the Dirac-like spectrum of quasiparticle excitations. Phys Rev B 69:075104
10. Fukuyama H (2007) Anomalous orbital magnetism and Hall effect of massless fermions in two dimension. J Phys Soc Jpn 76:043711
11. Koshino M, Ando T (2007) Diamagnetism in disordered graphene. Phys Rev B 75:235333
12. Koshino M, Ando T (2007) Orbital diamagnetism in multilayer graphenes: systematic study with the effective mass approximation. Phys Rev B 76:085425
13. Nakamura M, Hirasawa L (2008) Electric transport and magnetic properties in multilayer graphene. Phys Rev B 77:045429
14. Koshino M, Arimura Y, Ando T (2009) Magnetic field screening and mirroring in graphene. Phys Rev Lett 102:177203
15. Loh KP, Bao QL, Ang PK, Yang JX (2010) The chemistry of graphene. J Mater Chem 20:2277–2289
16. Bekyarova E, Sarkar S, Niyogi S, Itkis ME, Haddon RC (2012) Advances in the chemical modification of epitaxial graphene. J Phys D Appl Phys 45:154009
17. Sarkar S, Bekyarova E, Haddon RC (2012) Covalent chemistry in graphene electronics. Materialstoday 15:276–285
18. Tanaka K, Yamashita S, Yamabe H, Yamabe T (1987) Electronic properties of one-dimensional graphite family. Synth Met 17:143–148
19. Fujita M, Wakabayashi K, Nakada K, Kusakabe K (1996) Peculiar localized state at zigzag graphite edge. J Phys Soc Jpn 65:1920–1923
20. Nakada K, Fujita M, Dresslhaus G, Dresslhaus M (1996) Edge state in graphene ribbons: nanometer size effect and edge shape dependence. Phys Rev B 54:17954–17961
21. Okada S, Oshiyama A (2001) Magnetic ordering in hexagonally bonded sheets with first-row elements. Phys Rev Lett 87:146803
22. Yamashiro A, Shimoi Y, Harigaya K, Wakabayashi K (2003) Spin- and charge-polarized states in nanographene ribbons with zigzag edges. Phys Rev B 68:193410
23. Wu J, Pisula W, Müllen K (2007) Graphenes as potential material for electronics. Chem Rev 107:718–747
24. Fujii S, Enoki T (2013) Nanographene and graphene edges: electronic structure and nanofabrication. Acc Chem Res 46:2202–2210
25. Wallace PR (1947) The band theory of graphite. Phys Rev 71:622–634
26. Mathieu C, Barrett N, Rault J, Mi YY, Zhang B, de Heer WA, Berger C, Conrad EH, Renault O (2011) Microscopic correlation between chemical and electronic states in epitaxial graphene on SiC(0001). Phys Rev B 83:235436
27. Clar E (1964) Polycyclic hydrocarbons, vol I and II. Academic, London
28. Clar E (1972) The aromatic sextet. Wiley, London
29. Novoselov KS, Fal'ko VI, Colombo L, Gellert PR, Schwab MG, Kim K (2012) A roadmap for graphene. Nature 490:192–200
30. Allen MJ, Tung VC, Kaner RB (2010) Honeycomb carbon: a review of graphene. Chem Rev 110:132–145

31. Cheng L, Hernandez Y, Feng X, Müllen K (2012) From nanographene and graphene nanoribbons to graphene sheets: chemical synthesis. Angew Chem Int Ed 51:7640–7654
32. Datt SS, Strachan DR, Khamis SM, Johnson ATC (2008) Crystallographic etching of few-layer graphene. Nano Lett 8:1912–1915
33. Campos-Delgado J, Romo-Herrera JM, Jia X, Cullen DA, Muramatsu H, Kim YA, Hayashi T, Ren Z, Smith DJ, Okuno Y, Ohba T, Kanoh H, Kaneko K, Endo M, Terrones H, Dresselhaus MS, Terrones M (2008) Bulk production of a new form of sp2 carbon: crystalline graphene nanoribbons. Nano Lett 8:2773–2778
34. Yang XY, Dou X, Rouhanipour A, Zhi L, Räder HJ, Müllen K (2008) Two-dimensional graphene nanoribbons. J Am Chem Soc 130:4216–4217
35. Cai J, Ruffieux P, Jaafar R, Bieri M, Braun T, Blankenburg S, Muoth M, Seitsonen AP, Saleh M, Feng X, Müllen K, Fasel R (2010) Atomically precise bottom-up fabrication of graphene nanoribbons. Nature 466:470–473
36. Narita A, Feng X, Hernandez Y, Jensen SA, Bonn M, Yang H, Verzhbitskiy IA, Casiraghi C, Hansen MR, Koch AHR, Fytas G, Ivasenko O, Li B, Mali KS, Balandina T, Mahesh S, Feyter SD, Müllen K (2014) Synthesis of structurally well-defined and liquid-phase-processable graphene nanoribbons. Nat Chem 6:126–132
37. Chen ZH, Lin YM, Rooks MJ, Avouris P (2007) Graphene nano-ribbon electronics. Phys E 40:228–232
38. Han MY, Ozyilmaz B, Zhang YB, Kim P (2007) Energy band-gap engineering of graphene nanoribbons. Phys Rev Lett 98:206805
39. Jiao LY, Zhang L, Wang XR, Diankov G, Dai HJ (2009) Narrow graphene nanoribbons from carbon nanotubes. Nature 458:877–880
40. Kosynkin DV, Higginbotham AL, Sinitskii A, Lomeda JR, Dimiev A, Price BK, Tour JM (2009) Longitudinal unzipping of carbon nanotubes to form graphene nanoribbons. Nature 458:872–876
41. Elías AL, Méndez ARB, Rodríguez DM, González VJ, González DR, Ci L, Sandoval EM, Ajayan PM, Terrones H, Terrones M (2009) Longitudinal cutting of pure and doped carbon nanotubes to form graphitic nanoribbons using metal clusters as nanoscalpels. Nano Lett 10:366–372
42. Jiao L, Wang X, Diankov G, Wang H, Dai H (2010) Facile synthesis of high-quality graphene nanoribbons. Nat Nanotechnol 5:321–325
43. Kobayashi Y, Fukui K, Enoki T, Kusakabe K, Kaburagi Y (2005) Observation of zigzag and armchair edges of graphite using scanning tunneling microscopy and spectroscopy. Phys Rev B 71:193406
44. Niimi Y, Matsui T, Kambara H, Tagami K, Tsukada M, Fukuyama H (2005) Scanning tunneling microscopy and spectroscopy studies of graphite edges. Appl Surf Sci 241:43–48
45. Kobayashi Y, Fukui K, Enoki T, Kusakabe K (2006) Edge state on hydrogen-terminated graphite edges investigated by scanning tunneling microscopy. Phys Rev B 73:125415
46. Enoki T, Kobayashi Y, Fukui K (2007) Electronic structures of graphene edges and nanographene. Int Rev Phys Chem 26:609–645
47. Ritter KA, Lyding JW (2009) The influence of edge structure on the electronic properties of graphene quantum dots and nanoribbons. Nat Mater 8:235–242
48. Sugawara K, Sato T, Souma S, Takahashi T, Suematsu H (2006) Fermi surface and edge-localized states in graphite studied by high-resolution angle-resolved photoemission spectroscopy. Phys Rev B 73:045124
49. Suenaga K, Koshino M (2010) Atom-by-atom spectroscopy at graphene edge. Nature 468:1088–1090
50. Joly VLJ, Kiguchi M, Hao SJ, Takai K, Enoki T, Sumii R, Amemiya K, Muramatsu H, Hayashi T, Kim YA, Endo M, Campos-Delgado J, López-Urías F, Botello-Méndez A, Terrones H, Terrones M, Dresselhaus MS (2010) Observation of magnetic edge state in graphene nanoribbons. Phys Rev B 81:245428
51. Hou Z, Wang X, Ikeda T, Huang SF, Terakura K, Boero M, Oshima M, Kakimoto M, Miyata S (2011) Effect of hydrogen termination on carbon K-edge X-ray absorption spectra of nanographene. J Phys Chem C 115:5392–5403

52. Tao C, Jiao L, Yazyev OV, Chen YC, Feng J, Zhang X, Capaz RB, Tour JM, Zettl A, Louie SG, Dai H, Crommie MF (2011) Spatially resolving edge states of chiral graphene nanoribbons. Nat Phys 7:616–620
53. Pan M, Girão EC, Jia X, Bhaviripudi S, Li Q, Kong J, Meunier V, Dresselhaus MS (2012) Topographic and spectroscopic characterization of electronic edge states in CVD grown graphene nanoribbons. Nano Lett 12:1928–1933
54. Sarkar S, Bekyarova E, Haddon RC (2012) Chemistry at the Dirac point: Diels–Alder reactivity of graphene. Acc Chem Res 45:673–682
55. Bekyarova E, Sarkar S, Wang F, Itkis ME, Kalinina I, Tian X, Haddon RC (2013) Effect of covalent chemistry on the electronic structure and properties of carbon nanotubes and graphene. Acc Chem Res 46:65–76
56. Bon SB, Valentini L, Verdejo R, Fierro JLG, Peponi L, Lopez-Manchado MA, Kenny JM (2009) Plasma fluorination of chemically derived graphene sheets and subsequent modification with butylamine. Chem Mater 21:3433–3438
57. Ryu S, Han MY, Maultzsch J, Heinz TF, Kim P, Steigerwald M, Brus LE (2008) Reversible basal plane hydrogenation of graphene. Nano Lett 8:4597–4602
58. Ruffieux P, Groning O, Schwaller P, Schlapbach L, Groning P (2000) Hydrogen atoms cause long-range electronic effects on graphite. Phys Rev Lett 84:4910–4913
59. Bahr JL, Yang J, Kosynkin DV, Bronikowski MJ, Smalley RE, Tour JM (2001) Functionalization of carbon nanotubes by electrochemical reduction of aryl diazonium salts: a bucky paper electrode. J Am Chem Soc 123:6536–6542
60. Koehler FM, Jacobsen A, Ensslin K, Stampfer C, Stark WJ (2010) Selective chemical modification of graphene surfaces: distinction between single- and bilayer graphene. Small 6:1125–1130
61. Sharma R, Baik JH, Perera CJ, Strano MS (2010) Anomalously large reactivity of single graphene layers and edges toward electron transfer chemistries. Nano Lett 10:398–405
62. Bekyarova E, Itkis ME, Ramesh P, Berger C, Sprinkle M, de Heer WA, Haddon RC (2009) Chemical modification of epitaxial graphene: spontaneous grafting of aryl groups. J Am Chem Soc 131:1336–1337
63. Boukhvalov DW, Katsnelson MI (2009) Chemical functionalization of graphene. J Phys Condens Matter 21:344205
64. Boukhvalov DW (2011) Modeling of epitaxial graphene functionalization. Nanotechnology 22:055708
65. Sarkar S, Bekyarova E, Niyogi S, Haddon RC (2011) Diels–Alder chemistry of graphite and graphene: graphene as diene and dienophile. J Am Chem Soc 133:3324–3327
66. Talirz L, Sode H, Cai J, Ruffieux P, Blankenburg S, Jafaar R, Berger R, Feng X, Mullen K, Passerone D, Fasel R, Pignedoli CA (2013) Termini of bottom-up fabricated graphene nanoribbons. J Am Chem Soc 135:2060–2063
67. Stein SE, Brown RL (1987) π-Electron properties of large condensed polyaromatic hydrocarbons. J Am Chem Soc 109:3721–3729
68. Bendikov M, Duong HM, Starkey K, Houk KN, Carter EA, Wudl F (2004) Oligoacenes: theoretical prediction of open-shell singlet diradical ground state. J Am Chem Soc 126:7416–7417 and 10493.
69. Jiang D, Dai S (2008) Electronic ground state of higher acenes. J Phys Chem A 112:332–335
70. Hachmann J, Dorando JJ, Avilés M, Chan GKL (2007) The radical character of the acenes: a density matrix renormalization. J Chem Phys 127:134309
71. Tönshoff C, Bettinger HF (2010) Photogeneration of octacene and nonacene. Angew Chem Int Ed 49:4125–4128
72. Purushothaman B, Bruzek M, Parkin SR, Miller AF, Anthony JE (2011) Synthesis and structural characterization of crystalline nonacene. Angew Chem Int Ed 50:7013–7017
73. De J, Sumpter BG, Dai S (2007) First principles study of magnetism in nanographenes. J Chem Phys 127:124703
74. Hod O, Barone V, Scuseria GE (2008) Half-metallic graphene nanodots: a comprehensive first-principles theoretical study. Phys Rev B 77:035411

75. Moscardo F, San-Fabián E (2009) On the existence of a spin-polarized state in the n-periacene molecules. Chem Phys Lett 480:26–30
76. De J, Dai S (2008) Circumacenes versus periacenes: HOMO–LUMO gap and transition from nonmagnetic to magnetic ground state with size. Chem Phys Lett 466:72–75
77. Plasser F, Pašalić H, Gerzabek MH, Libisch F, Reiter R, Burgdçrfer J, Müller T, Shepard R, Lischka H (2013) The multiradical character of one- and two-dimensional graphene nanoribbons. Angew Chem Int Ed 52:2581–2584
78. Goto K, Kubo T, Yamamoto K, Nakasuji K, Sato K, Shiomi D, Takui T, Kubota M, Kobayashi T, Yakushi K, Ouyang J (1999) A stable neutral hydrocarbon radical: synthesis, crystal structure, and physical properties of 2,5,8-tri-tert-butyl-phenalenyl. J Am Chem Soc 121:1619–1620
79. Ovchinnikov A (1978) Multiplicity of the ground state of large alternant organic molecules with conjugated bonds. Theor Chim Acta 47:297–304
80. Borden WT, Davidson ER (1977) Effects of electron repulsion in conjugated hydrocarbon diradicals. J Am Chem Soc 99:4587–4594
81. Morita Y, Suzuki S, Sato K, Takui T (2011) Synthetic organic spin chemistry for structurally well-defined open-shell graphene fragments. Nat Chem 3:197–204
82. Inoue J, Fukui K, Kubo T, Nakazawa S, Sato K, Shiomi D, Morita Y, Yamamoto K, Takui T, Nakasuji K (2001) The first detection of a Clar's hydrocarbon, 2,6,10-tri-tert-butyltriangulene: a ground-state triplet of non-Kekulé polynuclear benzenoid hydrocarbon. J Am Chem Soc 123:12702–12703
83. Li Y, Huang KW, Sun Z, Webster RD, Zeng Z, Zeng W, Chi C, Furukawa K, Wu J (2014) A kinetically blocked 1,14:11,12-dibenzopentacene:a persistent triplet diradical of a non-Kekulé polycyclic benzenoid hydrocarbon. Chem Sci 5:1908–1914
84. For a recent summary of anthenes, see Konishi A, Hirao Y, Kurata H, Kubo T, M Nakano, Kamada K (2014) Anthenes: model systems for understanding the edge state of graphene nanoribbons. Pure Appl Chem 86:497–595. See also ref [85].
85. Konishi A, Hirao Y, Kurata H, Kubo T (2013) Investigating the edge state of graphene nanoribbons by a chemical approach: synthesis and magnetic properties of zigzag-edged nanographene molecules. Solid State Commun 175–176:62–70
86. For a recent review of open-shell polycyclic aromatic hydrocarbons, see: Sun Z, Wu J (2012) Open-shell polycyclic aromatic hydrocarbons. J Mater Chem 22:4151–4160
87. Yamaguchi K (1990) Chapter 7 Instability in Chemical Bonds. In: Carbo R, Klobukowski M (eds) Self-consistent field: theory and applications. Elsevier, Amsterdam, pp 727–823
88. Scholl R, Meyer K (1934) Der blaue aromatische Grundkohlenwasserstoff des meso-Naphtho-dianthrons und seine Überführung durch Maleinsäure-anhydrid in Anthro-dianthren. Ber Dtsch Chem Ges 67:1236–1238
89. Clar E (1948) Das Kondensationsprinzip, ein einfaches neues Prinzip im Aufbau der aromatischen Kohlenwasserstoffe (Aromatische Kohlenwassertoffe XLII. Mitteilung). Chem Ber 81:52–63
90. Hirao Y, Konishi A, Matsumoto K, Kurata H, Kubo T (2012) Synthesis and electronic structure of bisanthene: a small molecular-sized graphene with zigzag edges. AIP Conf Proc 1504:863–866
91. Li J, Zhang K, Zhang X, Huang KW, Chi C, Wu J (2010) meso-Substituted bisanthenes as soluble and stable near-infrared dyes. J Org Chem 75:856–863
92. Fort EH, Donovan PM, Scott LT (2009) Diels–Alder reactivity of polycyclic aromatic hydrocarbon bay regions: implications for metal-free growth of single-chirality carbon nanotubes. J Am Chem Soc 131:16006–16007
93. Arabei SM, Pavich TA (2000) Spectral-luminescent properties and photoinduced transformations of bisanthene and bisanthenequinone. J Appl Spectrosc 67:236–244
94. Konishi A, Hirao Y, Nakano M, Shimizu A, Botek E, Champagne B, Shiomi D, Sato K, Takui T, Matsumoto K, Kurata H, Kubo T (2010) Synthesis and characterization of teranthene: a singlet biradical polycyclic aromatic hydrocarbon having Kekulé structures. J Am Chem Soc 132:11021–11023

95. Konishi A, Hirao Y, Matsumoto K, Kurata H, Kishi R, Shigeta Y, Nakano M, Tokunaga K, Kamada K, Kubo T (2013) Synthesis and characterization of quarteranthene: elucidating the characteristics of the edge state of graphene nanoribbons at the molecular level. J Am Chem Soc 135:1430–1437
96. Konishi A, Hirao Y, Matsumoto K, Kurata H, Kubo T (2013) Facile synthesis and lateral π-expansion of bisanthenes. Chem Lett 42:592–594
97. Zhai L, Shukla R, Wadumethrige SH, Rathore R (2010) Probing the arenium-ion (proton transfer) versus the cation-radical (electron transfer) mechanism of scholl reaction using DDQ as oxidant. J Org Chem 75:4748–4760
98. Kruszewski J, Krygowski TM (1972) Definition of aromaticity basing on the harmonic oscillator model. Tetrahedron Lett 13:3839–3842
99. Krygowski TM (1993) Crystallographic studies of inter- and intramolecular interactions reflected in aromatic character of π-electron systems. J Chem Inf Comput Sci 33:70–78
100. Glukhovtesev MN, Bach RD, Laiter S (1997) Isodesmic and homodesmotic stabilization energies of [n]annulenes and their relevance to aromaticity and antiaromaticity: is absolute antiaromaticity possible? J Mol Struct 417:123–129
101. Slayden SW, Liebman JF (2001) The energetics of aromatic hydrocarbons: an experimental thermochemical perspective. Chem Rev 101:1541–1566
102. Douglas JE, Rabinovitch BS, Looney FS (1955) Kinetics of the thermal cis-trans isomerization of dideuteroethylene. J Chem Phys 23:315–323
103. Motta SD, Negri F, Fazzi D, Castiglioni C, Canesi EV (2010) Biradicaloid and polyenic character of quinoidal oligothiophenes revealed by the presence of a low-lying double-exciton state. J Phys Chem Lett 1:3334–3339
104. Koch KH, Müllen K (1991) Synthesis of tetraalkyl-substituted oligo(l,4-naphthylene)s and cyclization to soluble oligoberi-naphthy1ene)s. Chem Ber 124:2091–2100
105. Angeli C, Pastore M, Cimiraglia C (2007) New perspectives in multireference perturbation theory: the n-electron valence state approach. Theor Chem Acc 117:743–754

Part V
Heteroatom-Conjugated π-Electron Systems

Chapter 21
Boron-Containing π-Electron Systems

Shigehiro Yamaguchi

Abstract Developments of various types of boron-containing π-conjugated compounds are described in this chapter. Borole is one of the representative boracycles, which has a significant antiaromatic character. Several new borole compounds combined with heteroaryl rings have been synthesized and their unusual properties have been revealed. In the course of this chemistry, a new photoreactive boron compound, boryl-substituted dibenzoborepin, was obtained, which underwent a 4π-electrocyclic bora-Nazarov reaction. A series of planarized triphenylboranes has been also designed and synthesized as the stable organoboron compounds without steric protection of the boron atoms. Comprehensive studies on their chemical reactivity as well as photophysical and electronic properties have been conducted. A part of this compound class can be regarded as a model system of boron-doped graphenes and showed intriguing properties, such as the thermoresponsive complexation with Lewis basic species and photodissociation of the Lewis acid-base complex.

Keywords Borole • Borepin • Photochromism • Boron-doped graphene • Stabilization • Electron-transporting material • Sensor • Photodissociation

21.1 Introduction

Incorporation of main group elements into π-conjugated skeletons is a powerful strategy to develop new optoelectronic organic materials with unusual properties. Among various main group elements, group 13 boron is particularly useful to produce an unusual electronic structure [1]. There are several characteristic features of the boron atom from this point of view (Fig. 21.1a). First, boron is more electron positive than carbon, and thereby a boryl group acts as an inductively σ-donating group. On the other hand, secondly, as trivalent boron has an empty p orbital,

S. Yamaguchi (✉)
Institute of Transformative Bio-Molecules and Graduate School of Science, Nagoya University, Furo, Chikusa, Nagoya 464-8602, Japan
e-mail: yamaguchi@chem.nagoya-u.ac.jp

T. Akasaka et al. (eds.), *Chemical Science of π-Electron Systems*,
DOI 10.1007/978-4-431-55357-1_21

(a)

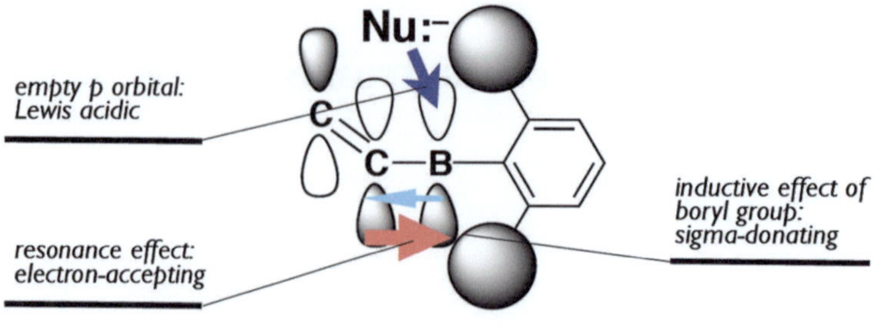

(b)

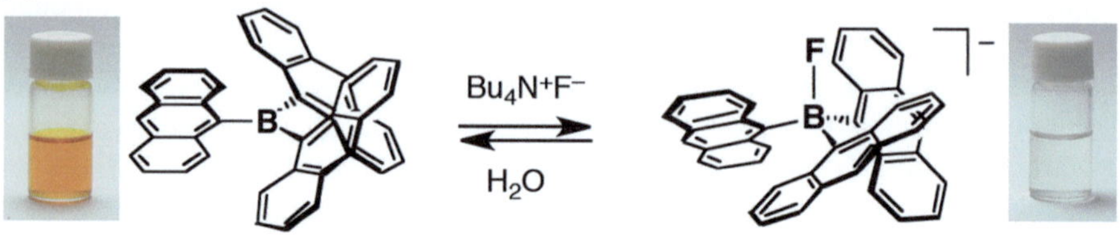

Fig. 21.1 (**a**) Key points for molecular designs of organoboron materials and (**b**) a fluoride ion sensing based on the coordination number change of the boron atom

a boryl group acts as a π-accepting group through p-π* conjugation with a π-conjugated carbon framework. Their electron-accepting ability is comparable to that of cyano group. Thirdly, a boryl group has a pronounced Lewis acidity due to the empty p orbital of the boron atom. In conjunction with this characteristic, triarylboranes readily react with certain nucleophiles to produce the corresponding tetracoordinate borates. The coordination number change from the trivalent borane to the tetracoordinate borates gives rise to a significant change in the electronic structure and thereby photophysical or electronic properties. Making use of these fundamental features of boron atoms, various types of functional organic molecules can be developed. As a representative one, we developed a fluoride anion sensor using trianthrylboranes **1** [2]. Upon capturing a fluoride ion to form tetracoordinate trianthylfluoroborates, the compounds show dramatic changes in the photophysical properties detectable by the naked eyes (Fig. 21.1b). Through this research, we demonstrated a new design principle of an anion sensor. In this section, we like to describe our recent results on the designs of various types of functional organoboron materials, including boracyclic π-conjugated systems, photoreactive organoboron materials, highly stabilized triarylboranes without steric protection, and boron-doped polycyclic aromatic hydrocarbons.

21.2 Boracycle-Based π-Conjugated Systems

By the incorporation of a boron atom into a cyclic π-conjugated skeleton, the orbital interaction between the empty p orbital of the boron atom and π or π* orbital of the π-conjugated moiety effectively occurs to produce unusual electronic structures. Therefore, boracyclic π-conjugated skeletons can serve as a unique building unit for π-electron materials. A representative example for such boracycles is borole, a 5-membered cyclic diene, which is isoelectronic to a cyclopentadienyl cation. Due to its 4π antiaromatic character, the borole derivatives are very unstable, but show intriguing photophysical properties. Although the first example of the borole derivatives, 1,2,3,4,5-pentaphenylborole, was reported by Eisch and coworkers in 1969 [3], its isolation without any coordination of base to the boron atom had not been achieved for a long time, and therefore, its chemistry had not been well matured. In 2008, two groups, including us, independently reported the crystal structures of a series of pentaarylborole derivatives [4, 5]. Since then, this old but new chemistry has attracted keen attention and in these few years several fascinating borole derivatives have been synthesized [6]. We have also devoted our efforts to this chemistry and succeeded in the synthesis of fascinating borole materials, such as heteroaryl-substituted boroles [7] and thiophene-fused ladder boroles [8].

We have synthesized 2,3,4,5-tetrathienyl-substituted borole **2** as the first example of a heteroaryl-substituted borole derivative. Its electrochemical property was investigated by cyclic voltammetry (Fig. 21.2), which showed both reversible reduction and oxidation processes with significantly low reduction and oxidation potentials of −1.57 V and −0.04 V (vs. Fc/Fc^+), respectively, indicative of a significantly narrow HOMO–LUMO gap. In agreement with this result, tetrathienylborole showed an absorption band at a rather long wavelength (λ_{max} 735 nm) in the visible and near-infrared region. Later, *B*-thienyl-substituted borole has been synthesized by the Braunschweig group [9].

A series of benzoheteroarene-fused boroles **3** and **4**, containing pyrrole, furan, and thiophene as the heteroarene rings (Fig. 21.2), has also been synthesized, and their structures were determined by X-ray crystallographic analysis [8]. Notably, the borole ring in the heteroarene-fused structures showed high antiaromaticity, in contrast to the fact that dibenzoborole has a reduced antiaromaticity. This enhanced antiaromaticity is due to the lower extent of bond alternation in the borole ring with the heteroarene-fused structures compared to that in the parent borole. Reflecting the antiaromatic character, the heteroarene-fused boroles exhibited characteristic properties, such as long-wavelength absorptions and low reduction potentials.

Among the fused borole derivatives, one of the intriguing target molecules was a boroloborole skeleton, as it should be a highly electron-accepting, antiaromatic scaffold. A dibenzo-fused boroloborole **5** was firstly synthesized by Piers and coworkers based on an elegant photoisomerization from a bi(benzoboracyclobutylidene) [10]. We also aimed at the synthesis of this compound independently. In our effort to synthesize this skeleton, however, we unexpectedly obtained a 10-borylated dibenzoborepin **6** via an anomalous skeletal rearrangement, as shown in Scheme 21.1.

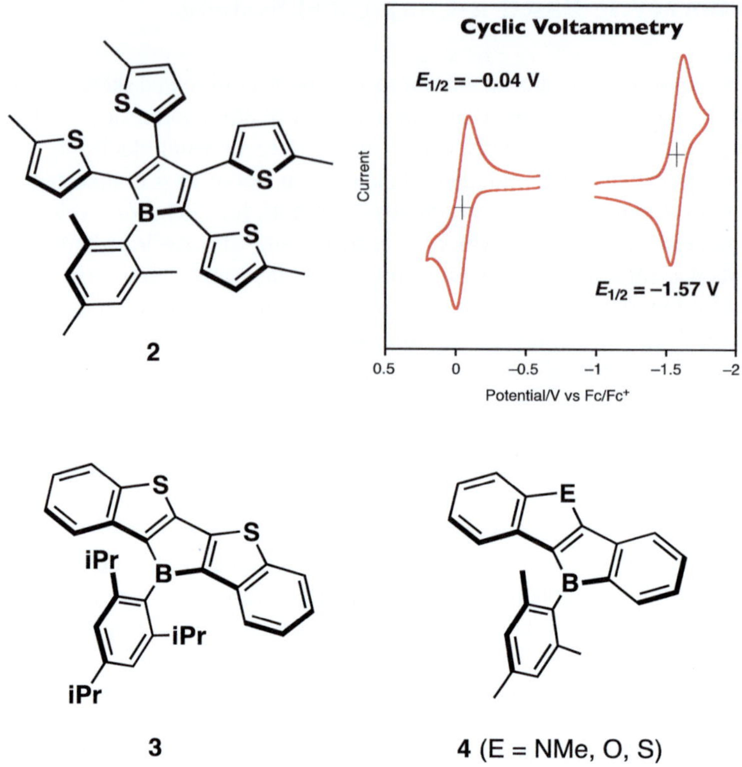

Fig. 21.2 Structures of compounds **2** and **3** and cyclic voltammetry of **2** in CH_2Cl_2 (vs. Fc/Fc^+)

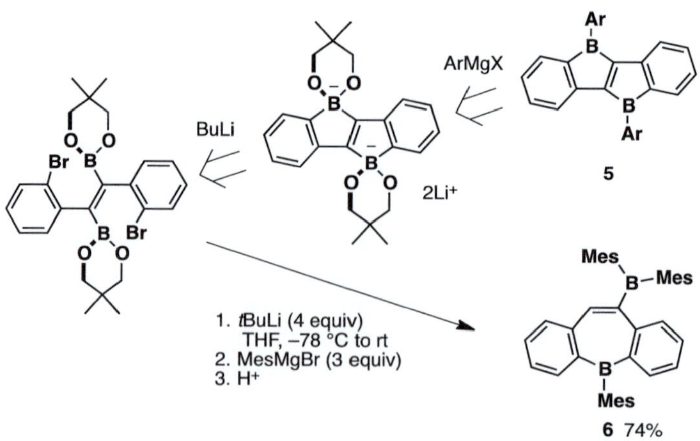

Scheme 21.1 Synthesis of boryl-substituted dibenzoborepin **6**

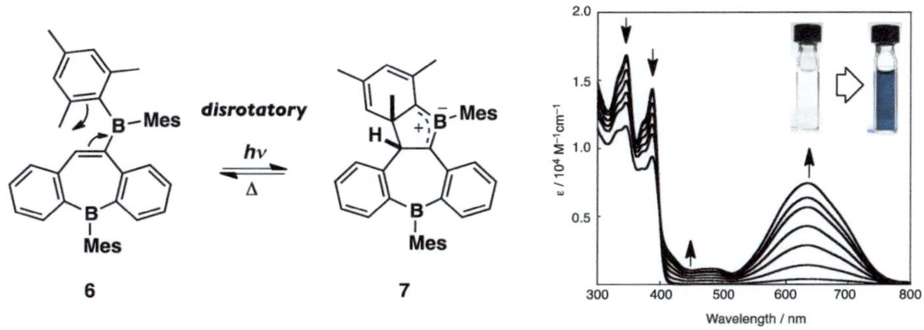

Fig. 21.3 Photoreaction of a boryl-substituted dibenzoborepin **6**

Borepin is the 7-membered boracycle with 6π electron aromaticity and thereby should be stable to some extent. In particular, dibenzoborepins with a bulky aryl group on the boron atom have sufficient stability enough to use as a building unit for emissive π-conjugated systems [11]. To our surprise, however, the obtained borepin **6** immediately underwent a photochemical reaction to produce a deep blue solution (Fig. 21.3). After careful isolation and characterization of the photoproduct, we realized that the compound underwent a boron version of Nazarov cyclization, namely, 4π electrocyclic reaction. While similar electrocyclic reactions of N-, O-, and S-containing substrates, namely, 6π electrocyclic reactions, have been well studied in the literature, the boron version of the Nazarov reaction is unprecedented. This reactivity is derived from the B/C^+ isosterism. The dimesitylborylborepin can be regarded to be isosteric with the intermediate of the Nazarov reaction. This reactivity may suggest the potential applicability as a new photochromic system. Moreover, the photoproduct **7** has an intriguing borata-allyl substructure, which is isoelectronic with ally cation and thereby of interest as a neutral but significantly electron-deficient building unit. Indeed, compound **7** has a deep blue color with the absorption maximum at 634 nm in benzene.

21.3 Planarized Triphenylboranes

The other issue to design organoboron materials is how to stabilize the inherently unstable Lewis acidic organoboron materials. A general strategy for the stabilization is to sterically protect the boron moiety. Most of the organoboron materials so far reported employ at least one or two bulky aryl groups on the boron atom. However, their steric bulkiness sometimes gives rise to a detrimental effect on the solid-state properties that rely on the intermolecular interaction, such as charge carrier transporting properties. We now introduce an idea for the stabilization of tricoordinate boron-containing π skeletons, which is "stabilization based on structural constraint." On the basis of this concept, we have designed and synthesized a

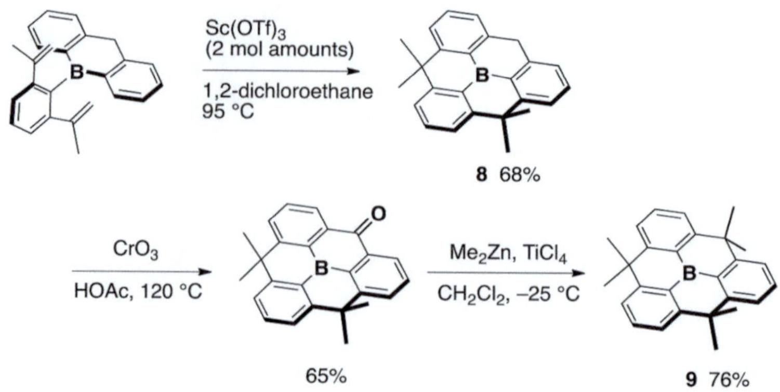

Scheme 21.2 Synthesis of planarized triphenylborane **9**

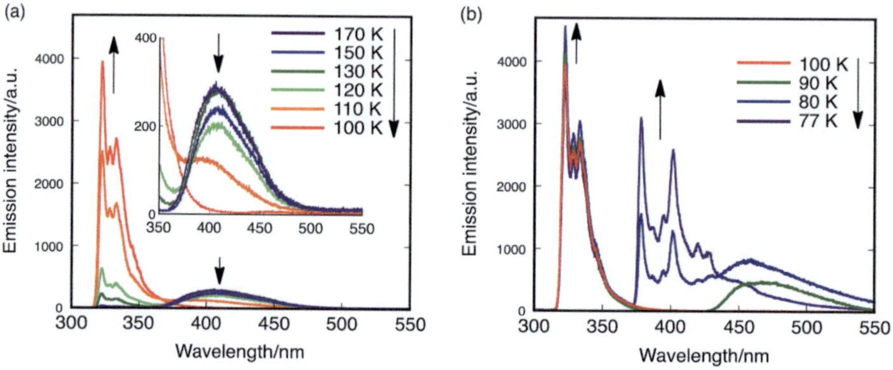

Fig. 21.4 Emission spectra of **9** in 3MP at (**a**) 170–100 K and (**b**) 100–77 K

planarized triphenylborane **9** [12]. This compound is isosteric to triangulenium ion, namely, planarized trityl cation. In light of the high stability of the triangulenium ion, compound **9** should be stable. Indeed, the compound, synthesized by double Friedel–Crafts cyclization from dipropenyl-substituted precursor with $Sc(OTf)_3$ (Scheme 21.2) showed high stability against moisture and oxygen and can be isolated by silica gel column chromatography. This is in sharp contrast to the fact that parent triphenylborane immediately decomposes on silica gel, demonstrating the efficacy of the stabilization by the structural constraint.

Despite the very simple structure, the planarized triphenylborane **9** showed dual fluorescence (Fig. 21.4) [13]. While the compound showed a broad emission band around 400 nm at room temperature, this band tends to decrease in intensity and a new band appeared at the shorter wavelength around 340 nm upon cooling to 100 K in 3-methylpentane. Moreover, further cooling the temperature to 77 K resulted in the emergence of a phosphorescence band at the longer wavelength. The total luminescence quantum yield at 77 K amounted to 0.77. According to the experimental and theoretical studies on the excited state behavior, the planarized

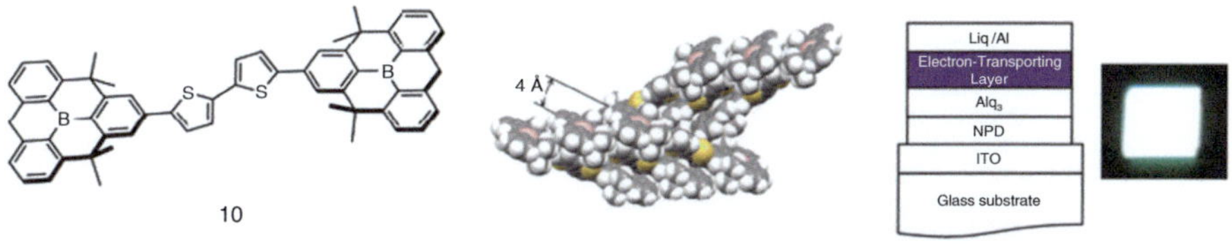

Fig. 21.5 Use of planarized triarylborane dimer **10** as an electron-transporting material

triphenylborane undergoes a structural change from a planar structure to a bowl-shaped structure in the excited singlet state, which is the origin of the intriguing dual fluorescence. This is in contrast to the widely accepted idea that the structural constraint is a way to render the structure rigid. In the planarized framework, the B–C bonds are significantly compressed in the ground state. Upon excitation, the charge transfer transition results in the elongation of the B–C bond and thereby induces the deformation into the bowl-shaped structure. These results provide an important implication that the structural constraint into a planar fashion is not only the strategy to construct a rigid skeleton but also a viable mechanism to impart the flexibility to the skeleton. Notably, chemical reduction of the structurally constrained triphenylborane **9** with K produced a radical anion that also has a bowl-shaped conformation [14].

The planarized triphenylborane derivatives have potential applicability to organic electronics. For example, we have synthesized bithiophene-connected dimer **10** (Fig. 21.5) [15]. The compound exhibited an intense fluorescence as well as characteristic redox properties with high reversibility. Moreover, the compound was stable enough to conduct vapor deposition despite the absence of the steric protection. The fabricated organic electroluminescent devices demonstrated that the planarized triphenylborane derivative has a high performance comparable to that of the widely used Alq_3.

The planarized triphenylboranes have been designed as a structurally constrained molecule. However, these compounds still preserve sufficient Lewis acidity to react with certain Lewis bases and form a nonplanar structure [12]. Upon reaction with a fluoride ion, the planar triarylboranes, such as diborane **11**, can be converted into bowl-shaped fluoroborates, which can be reverted to the original planar structure by treatment with a stronger Lewis acid than triarylborane (Fig. 21.6). It is noteworthy that this structural change is based on the coordination number change of the boron atom, and this is the advantage of the embedding of the boron atom. There are several examples of bowl-shaped scaffolds, such as corannulene and sumanene, which have attracted considerable attention as a fascinating core skeleton in materials chemistry. In our systems, the uniqueness is their capability to undergo the plane-to-bowl conversion in a controlled manner.

As mentioned previously, the planarized triphenylboranes have substantial chemical stability and react with Lewis bases to produce the corresponding borates without decomposition. Even when compound **8** was treated with bases such as KH, the triphenylborane moiety remains intact, and the deprotonation proceeded

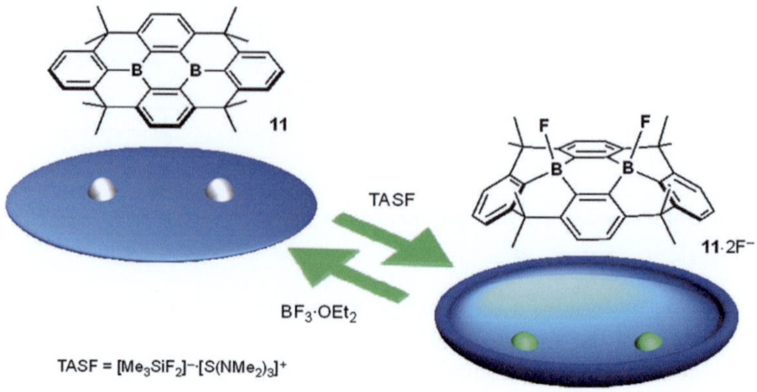

Fig. 21.6 Reversible plane-to-bowl conversion between **11** and **11**·2 F^{-}

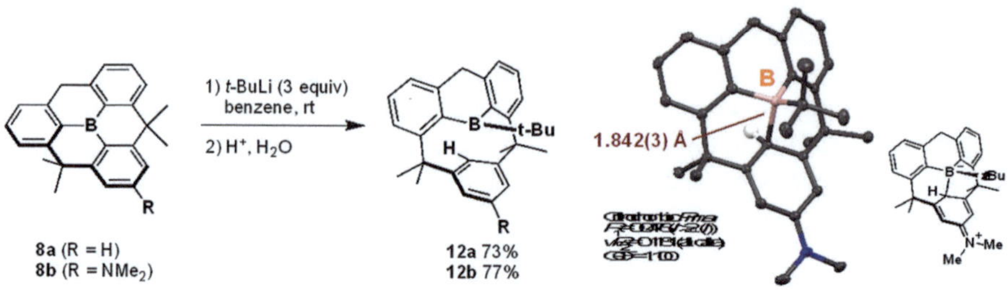

Fig. 21.7 Reaction of planarized triarylboranes **8** with *t*-BuLi and crystal structure of **12b**

instead at the benzylic position to produce an orange-emissive borataanthracene [16]. However, when the planarized triphenylboranes were treated with certain strong nucleophiles, such as alkyl or phenyllithiums, the compounds underwent a formal nucleophilic substitution via a tetracoordinate borate, producing unique boracyclophanes **12** (Fig. 21.7) [17]. X-ray crystallography demonstrated that the boracyclophanes have quite short contacts between the boron atom and the eliminated benzene ring. The calculated molecular orbitals, AIM analysis, and NBO analysis clearly demonstrated the bonding interaction between the boron atom and the benzene ring. This is an ideal skeleton to investigate the interaction between the boron atom and the benzene ring. The boron–benzene interaction in the boracyclophanes can be tuned from a π-complex character to a σ-complex character by increasing in the electron-donating character of the benzene ring (Fig. 21.7). In the Me_2N-substituted boracyclophane **12b**, the boron–benzene interaction can be modulated simply by the solvent polarity. As the solvent becomes more polar, the character of the boron–benzene interaction changes into a more σ-complex with a zwitterionic character.

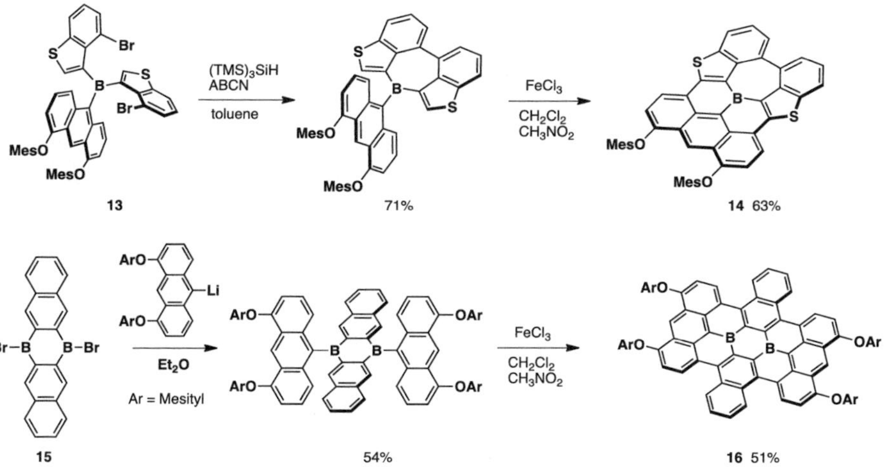

Scheme 21.3 Synthesis of expanded planarized triarylboranes **14** and **16**

21.4 Boron-Doped Nanographenes

As the triphenylborane skeleton in **9** is planarized with the sp^3-hybridized carbon atoms, the π conjugation is not fully expanded over the entire skeleton. In contrast, we have also succeeded in the preparation of all-sp^2 carbon-containing planarized boranes, which can be regarded as the models of boron-doped graphenes. Despite the remarkable electric and structural properties of graphene, the application to nanoscale transistors is impeded by its intrinsic zero-band-gap character. One of the solutions to overcome this issue is the chemical doping of B atoms into the graphene sheet. However, conventional physical preparation methods only produce a mixture of B-doped graphenes with various edge shapes, doping positions, and doping ratio. In this context, our synthesis produces the B-doped nanographene as a single kind of compound, which enabled us the precise elucidation of the molecular structure and electronic properties.

As the first example classified into the all-sp^2 carbon-containing planarized borane, compound **14** was synthesized from a precursor **13** by the radical annulation followed by the oxidative cyclization reaction (Scheme 21.3) [18]. Based on a similar synthetic route, we have also obtained compound **16** from a dibromodiborapentacene precursor **15** (Scheme 21.3) [19]. In this synthesis, the use of the aryloxy-substituted anthryl groups was crucial to facilitate the oxidative cyclization reaction [20]. It should be noted that both compounds show high stability towards oxygen, water, and silica gel, despite the absence of steric protection around the boron atom.

Compound **14** has a highly planar skeleton, which allows a π-stacked structure in the solid state. This is an unprecedented structure for triarylboranes. On the other hand, compound **16** consists of a 15-ring honeycomb structure composed of 48 sp^2-

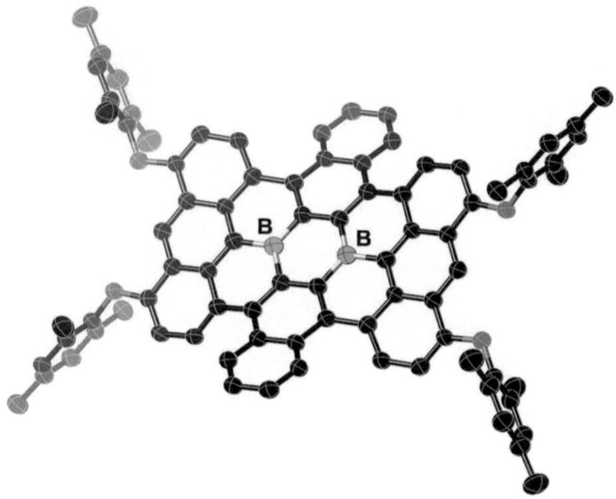

Fig. 21.8 X-ray crystal structure of **16** (50 % probability for thermal ellipsoids)

hybridized carbon atoms and 2 tricoordinate boron atoms (Fig. 21.8). The boron atoms are embedded at the center of the nanographene sheet with the remarkably short B–C bond lengths of 1.507(2), 1.531(2), and 1.535(2) Å, while maintaining their ideal tricoordinate characters.

The notable effect of the B-doping is the significant contribution of the p orbitals of the B atoms to both the relevant occupied and unoccupied orbitals, which leads to the broad absorption covering the entire visible region and the near-IR fluorescence spectra. Thus, while **14** showed a fluorescence band with the maximum wavelength of 729 nm, compound **16** exhibited a broad fluorescence band with the maximum wavelength of 679 nm, although the quantum yield was low ($\Phi = 0.04$). The fluorescence reaching the near-IR region is rare for the undoped nanographenes. The fluorescence comparable to that of **16** are only reported for supernaphthalene ($\lambda_{em} = 611$ nm) and quaterrylene ($\lambda_{em} = 680$ nm) [21]. The emission spectra of **14** and **16** as well as its absorption spectra showed negligible solvent effects.

In light of the intriguing electronic structure of boron-doped nanographenes, the important directions of their application should be the utilization as the electronic materials for high-performance transistors or lithium ion battery electrodes. Besides these possibilities, one potential application would be the use as a chemisorption material. The Lewis acidity and relevant coordination number change of the trivalent boron atoms should be beneficial in this aspect. This potential has been studied theoretically [22]. In this context, our planarized boranes should be good candidates to experimentally study this possibility. The Lewis basic species would readily form a complex with the boron center in the polycyclic skeletons and thereby perturb the electronic structures and properties. Indeed, compound **14** readily reacted with a fluoride ion to form a borate complex [18]. Its binding constant was determined to be 1.3×10^5 M^{-1} by the titration of its THF solution with n-Bu_4NF. This value is slightly lower than that of the triphenylborane derivative **9** (7.0×10^5 M^{-1}). This comparison demonstrated that **14** still preserves its Lewis acidity, but the degree is

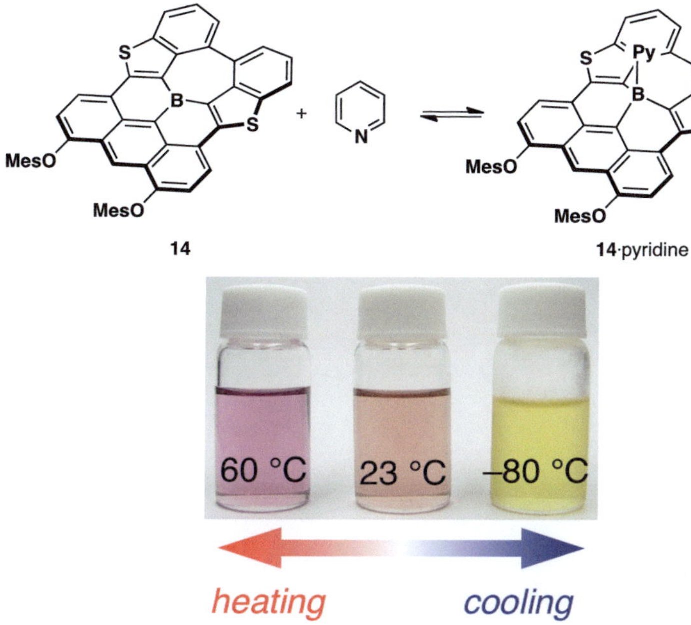

Fig. 21.9 Thermochromism of a 1:3 pyridine/THF solution of **14**

diminished by the structural constraint. Interestingly, **14** has a much lower-lying LUMO than **9**. Nevertheless, the Lewis acidity of **14** is weaker than that of **9**, indicating that the rigid structure of **14** gives a more dominant impact on the Lewis acidity than the LUMO energy level.

In relation to the reasonably high Lewis acidity, compound **14** showed an interesting thermochromic behavior in the presence of pyridine. The color of a 1/3 pyridine/THF solution of **14** was dramatically changed from purple at a higher temperature to yellow at a lower temperature (Fig. 21.9). According to the UV–vis titration, the binding constant of **14** with pyridine in THF was a very low value of $0.35\ M^{-1}$ at 23 °C. As the temperature decreased from 60 to −80 °C, the absorption bands at 563 and 668 nm decreased in intensity, and the intensity of the bands at 485 and 461 nm increased with the isosbestic points at 508 and 397 nm. Based on these results, the parameters for the equilibrium between **5** and the **5**·pyridine complex were estimated to be $\Delta H = -21.3\ \mathrm{kJ\ mol^{-1}}$ and $\Delta S = -76.2\ \mathrm{J\ K^{-1}\ mol^{-1}}$. As a result, the Gibbs energy change (ΔG) for this equilibrium becomes negative at low temperature, but positive at high temperature, and thereby this system shows the thermochromism. This result suggests the potential thermoresponsive chemisorption properties of the boron-doped nanographenes.

We also found unprecedented photoreactivity of a tetracoordinate boron-containing π-conjugated system, namely, the photodissociation of a Lewis acid-base complex between a fused polycyclic triarylborane and pyridine, which resulted in the emergence of dual fluorescence [23]. A key for this discovery was the successful

Scheme 21.4 Synthesis of partly fused planarized trinaphtylborane **18**

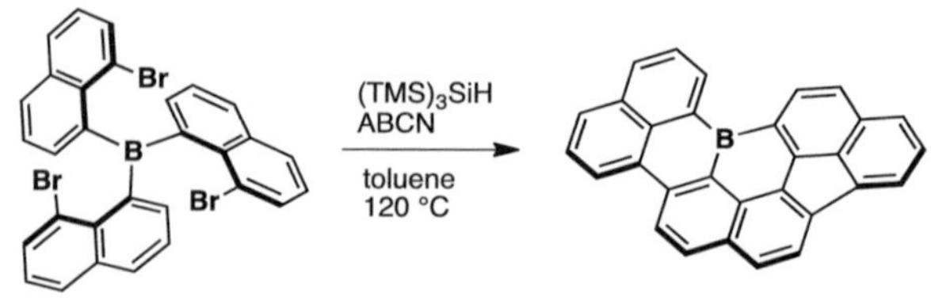

17 **18** 26%

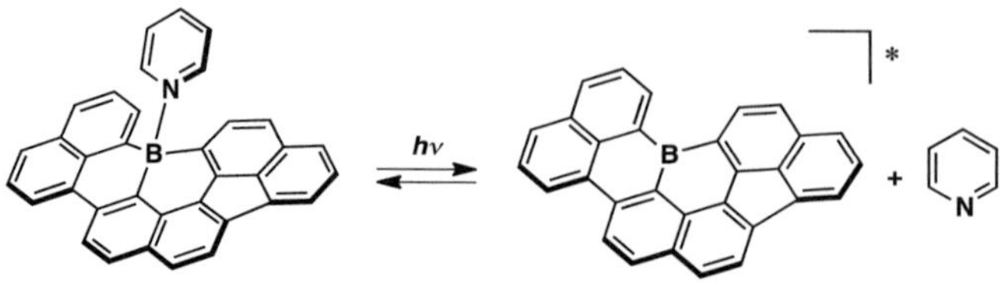

18·pyridine **18**

Scheme 21.5 Photodissociation of **18**·pyridine

synthesis of a highly stable, yet Lewis acidic planarized trinaphthylborane **18**. This compound was synthesized from tris(8-bromonaphthyl)borane **17** in one step by treatment with $(Me_3Si)_3SiH$ in the presence of 1,1′-azobis(cyclohexanecarbonitrile) (ABCN) as a radical initiator (Scheme 21.4). The reaction likely undergoes successive intramolecular radical cyclizations followed by a skeletal rearrangement. The obtained red compound forms a columnar π-stacked structure, where the average distance between planes is 3.45 Å. For air-stable triarylboranes, this is an unprecedented structural feature, and the ability to form columnar stacks would be beneficial for the construction of supramolecules or discotic liquid crystals. A top-contact OFET device was fabricated by vacuum vapor deposition of **18**, which showed ambipolar carrier-transporting abilities with a hole mobility of $\mu_h = 9.3 \times 10^{-6}$ cm^2 V^{-1} s^{-1} and an electron mobility of $\mu_e = 1.7 \times 10^{-5}$ cm^2 V^{-1} s^{-1}. Importantly, compound **18** underwent the complete formation of a Lewis adduct in solution even with weakly basic pyridine. The binding constant towards pyridine (toluene/25 °C) was determined to be 5.1×10^3 M^{-1}, which is much higher than that of **14** (0.35 M^{-1} in THF). When the association event was monitored by the fluorescence spectra, we noticed the photodissociation behavior of the B–N Lewis adducts (Scheme 21.5). Thus, even in the presence of an excess of pyridine, which resulted in a complete formation of the borate, the longer-wavelength emission band was dominantly observed. In particular, the B–N photodissociation took place more dominantly under the conditions of higher temperature, less polar solvent, and weaker Lewis base. The theoretical calculations suggested that a precise control of the frontier orbital distribution may be crucial for realization of this intriguing type of photoreactivity. The photoinduced B–N bond cleavage and the regeneration of the highly stable boron-embedded π systems is of importance not only because this behavior has a similarity to the classical photogeneration of a carbenium ion from triarylmethane leuco dyes but also because the simple combination of the planarized triarylborane and pyridines has a great potential for producing various functions, such as photochromism, organocatalysis, and photoresponsive supramolecular formation.

21.5 Conclusion and Outlook

In this chapter, we described our recent progress in the development of new boron-containing π-conjugated compounds. The most notable benefit to use group 13 boron atom is that the incorporation of a boron atom into a π-conjugated framework imparts the electron-deficient character, based on which we can produce highly electron-accepting π-electron systems with some intriguing properties and functions. One representative example is the borole derivatives, which have significant antiaromatic character due to the effective orbital interaction in the 5-membered ring. One key issue for the design of new boron-containing materials is how to stabilize the inherently Lewis acidic and thereby unstable molecules. As a complementary concept to the kinetic stabilization by bulky substituents, we now introduce an idea of the stabilization based on the structural constraint. Notably, this idea not only produces stable boron-based π-conjugated systems but also a new class of materials, namely, boron-doped nanographenes, which can be regarded as the models of boron-doped graphenes. The bottom-up synthesis of this compound class enabled us to study the fundamental properties and revealed several intriguing characteristics, such as thermoresponsive chemisorption and photodissociation of the Lewis acid-base complexes. These results suggest promising applications to various directions. The development of more precisely designed boron-containing frameworks would open a new avenue in this chemistry.

References

1. (a) Yamaguchi S, Tamao K (2005) A key role of orbital interaction in the main group element-containing π-electron systems. Chem Lett 34:2. (b) Yamaguchi S, Wakamiya A (2006) Boron as a key component for new π electron materials. Pure Appl Chem 78:1413. (c) Yamaguchi S, Fukazawa A (2011) Synthetic study of novel π-electron materials containing boron, nitrogen, and phosphorus as a key main group elements. J Syn Org Chem Jpn 69:661
2. (a) Akiyama S, Tamao K (2001) Colorimetric fluoride anion sensing by boron-containing π-electron systems. J Am Chem Soc 123:11372. (b) Yamaguchi S, Shirasaka T, Akiyama S, Tamao K (2002) Dibenzoborole-containing π-electron systems: remarkable fluorescence change based on the ON/OFF-control of the p-π* conjugation. J Am Chem Soc 124:8816
3. (a) Eisch JJ, Hotta KN, Kozima S (1969) Synthesis of pentaphenylborole, a potentially antiaromatic system. J Am Chem Soc 91:4575. (b) Eisch JJ, Galle JE, Kozima S (1986) Bora-aromatic systems. Part 8. The physical and chemical consequences of cyclic conjugation in boracyclopolyenes. The antiaromatic character of pentaarylboroles. J Am Chem Soc 108:379
4. Braunschweig H, Fernández I, Frenking G, Kupfer T (2008) Structural evidence for antiaromaticity in free borole. Angew Chem Int Ed 47:1951

5. So C-W, Watanabe D, Wakamiya A, Yamaguchi S (2008) Synthesis and structural characterization of pentaarylboroles and their dianions. Organometallics 27:3496
6. (a) Braunschweig H, Kupfer T (2008) Direct functionalization at the boron center of antiaromatic chloroborole. Chem Commun 4487. (b) Braunschweig H, Chiu C-W, Radacki K, Brenner P (2010) Platinum substituted boroles. Chem Commun 46:916. (c) Braunschweig H, Chiu C-W, Radacki K, Kupfer T (2010) Synthesis and structure of a carbene-stabilized π-boryl anion. Angew Chem Int Ed 49:2041. (d) Fan C, Piers WE, Parvez M (2009) Perfluoropentaphenylborole. Angew Chem Int Ed 48:2955. (e) Fan C, Mercier LG, Piers WE, Tuononen HM, Parvez M (2010) Dihydrogen activation by antiaromatic pentaarylboroles. J Am Chem Soc 132:9604. (f) Houghton AY, Karttunen VA, Fan C, Piers WE, Tuononen HM (2013) Mechanistic studies on the metal-free activation of dihydrogen by antiaromatic pentarylboroles. J Am Chem Soc 135:941
7. Araki T, Fukazawa A, Yamaguchi S (2012) Electron-donating tetrathienyl-substituted borole. Angew Chem Int Ed 51:5484
8. (a) Iida A, Yamaguchi S (2011) Thiophene-fused ladder boroles with high antiaromaticity. J Am Chem Soc 133:6952. (b) Iida A, Sekioka A, Yamaguchi S (2012) Heteroarene-fused boroles: what governs the antiaromaticity and lewis acidity of the borole skeleton? Chem Sci 3:1461
9. Braunschweig H, Dyakonov V, Engels B, Falk Z, Hçrl C, Klein JH, Kramer T, Kraus H, Krummenacher I, Lambert C, Walter C (2013) Multiple reduction of 2,5-bis(borolyl)thiophene: isolation of a negative bipolaron by comproportionation. Angew Chem Int Ed 52:12852
10. Araneda JF, Neue B, Piers WE, Parvez M (2012) Photochemical synthesis of a ladder diborole: a new boron-containing conjugate material. Angew Chem Int Ed 51:8546
11. (a) Mercier LG, Piers WE, Parvez M (2009) Benzo- and naphthoborepins: blue-emitting boron analogues of higher acenes. Angew Chem Int Ed 48:6108. (b) Caruso A Jr, Siegler M, Tovar JD (2010) Synthesis of functionalizable boron-containing π-electron materials that incorporate formally aromatic fused borepin rings. Angew Chem Int Ed 49:4213. (c) Levine DR, Caruso A Jr, Siegler MA, Tovar JD (2012) *Meta-B*-entacenes: new polycyclic aromatics incorporating two fused borepin rings. Chem Commun 48:6256
12. Zhou Z, Wakamiya A, Kushida T, Yamaguchi S (2012) Planarized triarylboranes: stabilization by structural constraint and their plane-to-bowl conversion. J Am Chem Soc 134:4529
13. Kushida T, Camacho C, Shuto A, Irle S, Muramatsu M, Katayama T, Ito S, Nagasawa Y, Miyasaka H, Sakuda E, Kitamura N, Zhou Z, Wakamiya A, Yamaguchi S (2014) Constraint-induced structural deformation of planarized triphenylboranes in the excited state. Chem Sci 5:1296
14. Kushida T, Yamaguchi S (2013) A radical anion of structurally constrained triphenylborane. Organometallics 32:6654
15. Shuto A, Kushida T, Fukushima T, Kaji H, Yamaguchi S (2013) π-Extended planarized triphenylboranes with thiophene spacers. Org Lett 15:6234
16. Kushida T, Zhou Z, Wakamiya A, Yamaguchi S (2012) Planarized B-phenylborataanthracene anion: structural and electronic impacts of coplanar-constraint. Chem Commun 48:10715
17. Kushida T, Yamaguchi S (2013) Boracyclophanes: modulation of the σ/π character in boron-benzene interactions. Angew Chem Int Ed 52:8054
18. Saito S, Matsuo K, Yamaguchi S (2012) Polycyclic π-electron system with boron at its center. J Am Chem Soc 134:9130
19. Dou C, Saito S, Matsuo K, Hisaki I, Yamaguchi S (2012) Boron-containing PAH as a substructure of boron-doped graphene. Angew Chem Int Ed 51:12206
20. (a) Davis NKS, Thompson AL, Anderson HL (2010) Bis-anthracene fused porphyrins: synthesis, crystal structure, and near-IR absorption. Org Lett 12:2124. (b) Davis NKS, Thompson AL, Anderson HL (2011) A porphyrin fused to four anthracenes. J Am Chem Soc 133:30
21. (a) Li J, Zhang K, Zhang X, Huang K-W, Chi C, Wu J (2010) Meso-substituted bisanthenes as soluble and stable near-infrared dyes. J Org Chem 75:856. (b) Wasserfallen D, Kastler M, Pisula W, Hofer WA, Fogel Y, Wang Z, Müllen K (2006) Suppressing aggregation in a large polycyclic aromatic hydrocarbon. J Am Chem Soc 128:1334

22. (a) Dai J, Yuan J, Giannozzi P (2009) Gas adsorption on graphene doped with B, N, Al, and S: a theoretical study. Appl Phys Lett 95:232105. (b) Kong X, Chen Q (2012) The positive influence of boron-doped graphene with pyridine as a probe molecule on SERS: a density functional theory study. J Mater Chem 22:15336
23. Matsuo K, Saito S, Yamaguchi S (2014) Photodissociation of B–N Lewis adducts: a partially fused trinaphthylborane with dual fluorescence. J Am Chem Soc 136:12580

Chapter 22
Doubly Bonded Silicon Compounds Showing Intramolecular Charge-Transfer Transitions

Takeaki Iwamoto

Abstract Recent study of doubly bonded silicon compounds showing intramolecular charge-transfer transition was reviewed. Trialkylaryldisilenes having 1-naphthyl, 9-phenanthyl, or 9-anthryl substituent were synthesized. X-ray diffraction analysis shows that π(Si=Si) and π(aryl) systems in the trialkylaryldisilenes are almost perpendicular to each other suggesting no significant conjugative interactions between two π systems, but they exhibit intramolecular charge-transfer (ICT) transitions from π(Si=Si) to π*(aryl). Especially, 9-anthryltrialkyldisilene shows a distinct ICT band at 525 nm, which redshifts to 535 nm in more polar 1,2-dichlorobenzene. Reaction of the anthryldisilene gives the corresponding [2 + 1] cycloadduct, disilacyclopropanimine, which undergoes further isomerization to give 3-silylene-2-silaaziridine with 9-anthryl substituent as a novel exocyclic silene. The anthrylsilene shows also a distinct ICT transition from π(Si=C) to π*(aryl). The high energy levels of π(Si=Si) and π(Si=C) and the low-lying π*(aryl) would be responsible for the distinct ICT band of the disilenes and the silene.

Keywords Disilene • Intramolecular charge transfer • π-electron system • Silene

Doubly bonded silicon compounds with unsaturated silicon atoms such as silenes ($R_2Si{=}CR_2$) and disilenes ($R_2Si{=}SiR_2$) (silicon π-electron compounds) have received much attention over the past three decades since the pioneering works of the isolable compounds $Mes_2Si{=}SiMes_2$ (Mes=2,4,6-trimethylsilyl) and $Me_2Si{=}C(OSiMe_3)R$ (R=*t*-Bu, 1-adamantyl) by West [1] and Brook [2, 3], respectively. These silicon π-electron compounds often show new structures and properties that have never been observed for the corresponding organic π-electron compounds comprising mainly second-low elements [4–14]. For instance,

T. Iwamoto (✉)
Department of Chemistry, Graduate School of Science, Tohoku University, Aoba-ku, Sendai 980-8578, Japan
e-mail: iwamoto@m.tohoku.ac.jp

T. Akasaka et al. (eds.), *Chemical Science of π-Electron Systems*,
DOI 10.1007/978-4-431-55357-1_22

Fig. 22.1 Planar and *trans*-bent geometry around double bonds of group-14 elements

while carbon–carbon double bond essentially adopts planar structure, silicon often forms *nonplanar* double bond (*trans*-bent double bond where substituents on each silicon atom forming the double bond are *trans* in respect to the double-bond plane; see Fig. 22.1) [15–19]. The π–π* gap of the silicon–silicon double bond is roughly half of that of carbon–carbon double bond (3 eV for $H_2Si{=}SiH_2$ (e.g., $\lambda_{max}(\pi \rightarrow \pi^*)$=344 nm for $Me_2Si{=}SiMe_2$ in argon matrix) [20] while 6 eV for $H_2C{=}CH_2$ (165 nm)). The rotational barrier of the Si=Si double bond is normally 60–100 kJ mol^{-1}, which is much smaller than that of C=C double bond (>200 kJ mol^{-1}). Because of these properties, doubly bonded silicon compounds as well as triply bonded compounds easily undergo polymerization and addition to various small molecules.

The reported isolable silicon π-electron compounds have normally bulky substituents around the double bond as protecting groups to keep from polymerization kinetically. Although the structures and properties of silicon π-electron systems should depend on the structure and electronic properties of the protecting groups on the double bond, there had been very few studies that focus on controlling the structure and properties of silicon π-electron systems (π_{Si}) by the protecting groups. Particularly, studies on understanding of the electronic communications between silicon π-electron systems (silenes and disilenes) and other electronic systems such as carbon π-electron systems (π_C) and transition-metal d-electron systems had been quite rare except for the extensive studies on stable silaaromatic compounds **Ia** and **Ib** (Fig. 22.2) by Tokitoh and coworkers [21]. In 2005, Scheschkewitz et al. synthesized disilenide $(Cp_2ClZn)(Tip)Si{=}SiTip_2$ **II** showing an absorption band at 715 nm assignable to the a ligand to metal charge-transfer transition [22]. In 2007, Tamao [23] and Scheschkewitz [24] independently reported synthesis and structures of the phenylene-bridged bis(disilene)s **III** which show significant interactions between Si=Si double bond and phenylene moieties. Tokitoh et al. have reported electronic communication between metalocenyl or ethynyl groups and disilenes in 1,2-metallocenyl disilenes **IV** and silicon versions of enediyne **V** [25–27]. Tamao also found that 1,2-(2-naphthyl)disilene **VI** shows electroluminescence in organic light-emitting devices [28]. Recently 1,2-disilabenzene derivatives **VII** [29, 30] and 1,4-diaza-2,3-disilabenzene derivatives **VIII** [31, 32] were synthesized by the reactions of stable disilyne with acetylenes and isocyanides, respectively.

Recently we have successfully synthesized new disilenes and silene with one polycyclic aromatic substituents (**1-4**) and found intramolecular charge-transfer (ICT) transitions from π(Si=Si) (or π(Si=C)) to π*(aryl) orbitals [33, 34]. In the present chapter, recent studies on disilenes and silenes showing ICT transitions will be discussed.

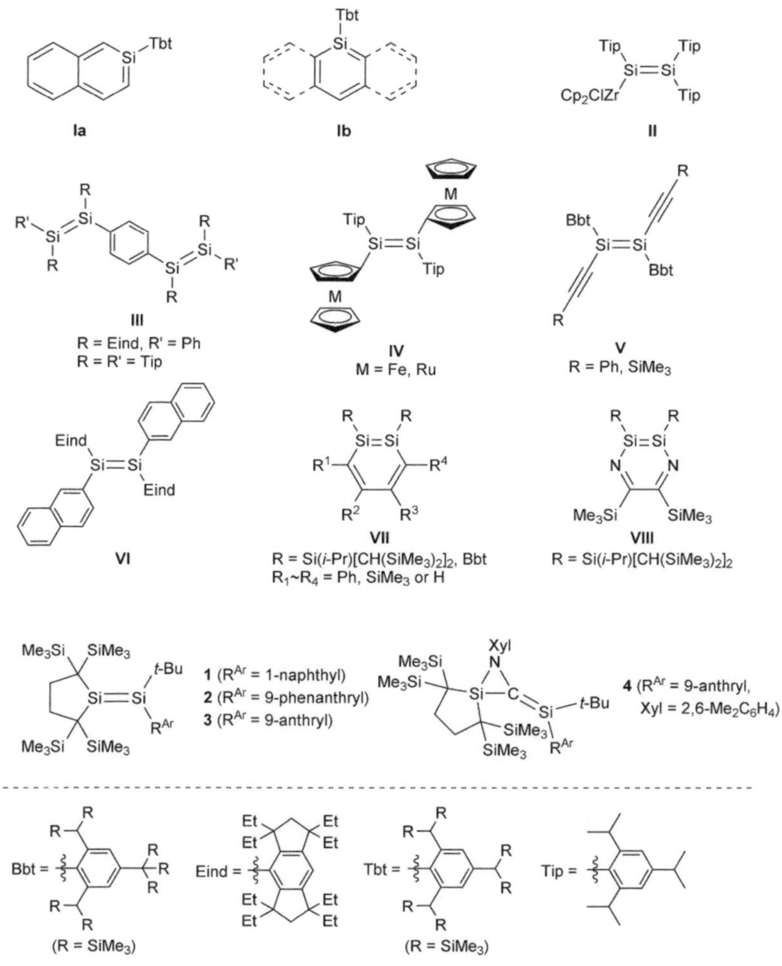

Fig. 22.2 Examples of isolable doubly bonded silicon compounds showing electronic communications with other electronic systems

22.1 Disilenes Showing Intramolecular Charge-Transfer (ICT) Transitions

Although various disilenes with bulky aromatic substituents (π_C systems) have been synthesized, all known disilenes bound to π_C systems have multiple π_{Si}–π_C interactions and, hence, are not suitable for exploring the nature of a single π_{Si}–π_C interaction. To investigate intrinsic interactions between π_{Si} and other electronic systems, we developed the new disilenide (a silicon analogue of vinyl anion) **7** without any π substituents. Potassium trialkyldisilenide **7** was prepared by reduction

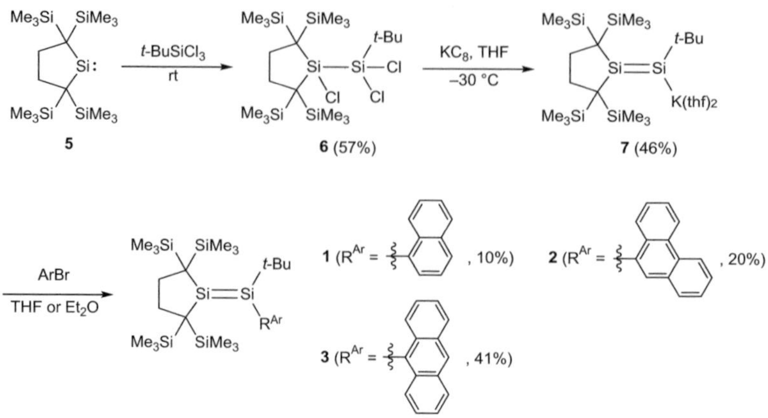

Scheme 22.1 Synthesis of disilenes **1–3**

of *t*-butyltrichlorosilane adduct of dialkylsilylene **5** [35] (**6**) with excess potassium graphite (KC_8) [33]. New disilenes with one aryl substituents, 1-naphthyl, 9-phenanthryl, and 9-anthryl-substituted trialkyldisilene **1-3** were synthesized by reactions of **7** with the corresponding aryl bromide as orange, orange, and blue-purple crystals in 10 %, 20 % and 41 % yields, respectively (Scheme 22.1).

X-ray diffraction (XRD) analysis disclosed that the silicon–silicon double bonds, and the aromatic ring planes in disilenes **1**, **2**, and **3** are almost perpendicular to each other with the dihedral angles of 83°, 80°, and 88°, respectively (Fig. 22.3 and Table 22.1). The perpendicular geometry would be due to severe steric repulsion between the *t*-Bu group introduced as a protecting group and the aromatic group and it is not suitable for effective conjugative interactions between silicon π systems (π_{Si}) and carbon π systems (π_C). Geometries around silicon–silicon double bond of **1-3** are *trans*-bent with the bent angles of 6.9°–26.7° around the Si1 and Si2 atoms of **1-3** (Table 22.1). The silicon–silicon double-bond distances of **1-3** (2.1754(12) Å-2.209(2) Å) are within the range of typical Si=Si double-bond distances of acyclic disilenes [4, 8, 14].

UV–vis spectra of disilenes **1** and **2** show three absorption bands I, II, and III assignable to ^1L_a transition of aromatic ring (I) and two transitions of the silicon–silicon double bond (II and III, Fig. 22.4). Band I having fine structures at 287 and 304 nm for **1** and **2**, respectively, are very similar to those of ^1L_a band maxima of 1-(pentamethyldisilanyl) naphthalene and 9-(pentamethyldisilanyl)phenanthrene (287 and 302 nm) [36]. An almost identical spectral feature of bands II and III (λ_{max}/nm 342 and 376 for **1** and 343 and 378 for **2**) indicates that these bands are due to transitions originating from the Si=Si double-bond moiety. These results suggest no remarkable interactions between π_{Si} and π_C systems in disilenes **1** and **2**.

9-Anthryldisilene **3** shows a spectral feature remarkably different from those of **1** and **2**. As shown in Fig. 22.4, the intense structured band between 320 and

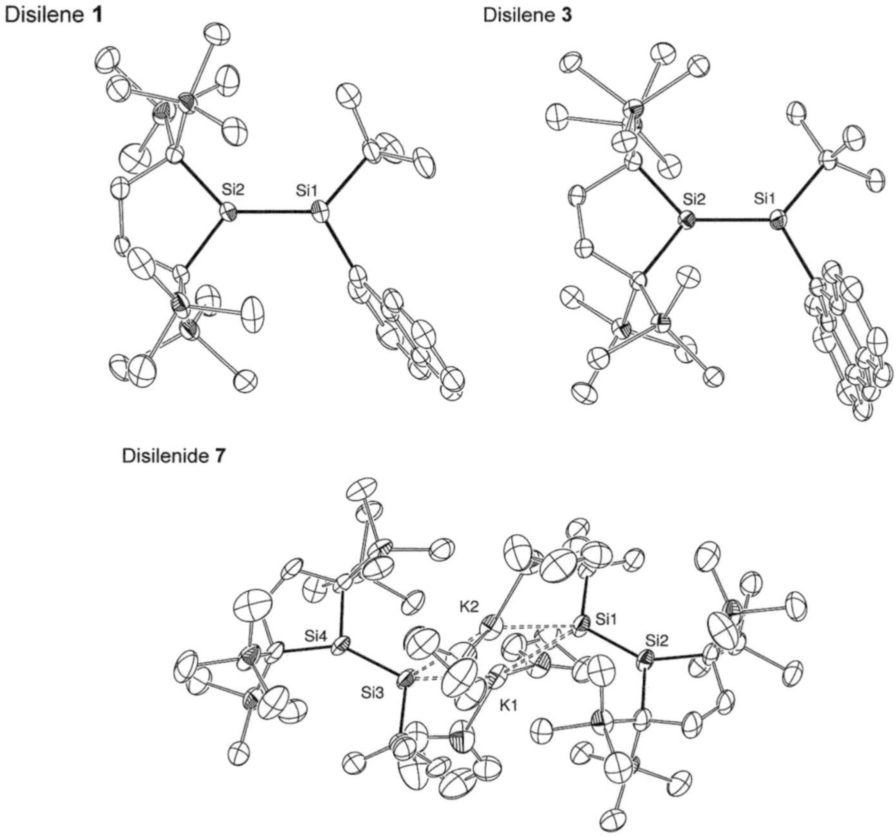

Fig. 22.3 ORTEP drawings of disilenes **1** and **3** and disilenide **7**. Thermal ellipsoids are shown at the 50 % probability level

Table 22.1 Selected structural parameters of disilenes **1–3**

Disilene (R^{Ar})	d(Si=Si)/Å	β^a/deg	δ^a/deg
1 (R^{Ar}=1-naphthyl)	2.1943(14)	23.8 (Si1), 12.2 (Si2)	83
2 (R^{Ar}=9-phenanthryl)	2.209(2)	26.7 (Si1), 12.9 (Si2)	80
3 (R^{Ar}=9-anthryl)	2.1754(12)	11.4 (Si1), 6.9 (Si2)	88

[a]Bent angle β: the angle between the axis through the Si=Si bond and R–Si(sp^2)–R plane. Dihedral angle δ: the angle between the two axes that bisect Si(sp^2)–Si(sp^2)–C(*t*-Bu) and C(sp^2)–C(ipso)–C(sp^2) angles as viewed along the Si(sp^2)–C(aryl) bond

450 nm and a weak but distinct absorption band at 525 nm were observed in 3-methylpentane. The former band can be regarded as band I overlapped with bands II and III, because the ^1L_a band of 9-(pentamethyldisilanyl)anthracene appears at 373 nm with an ε of 8.42×10^3 [36] and disilenes **1** and **2** show absorption bands

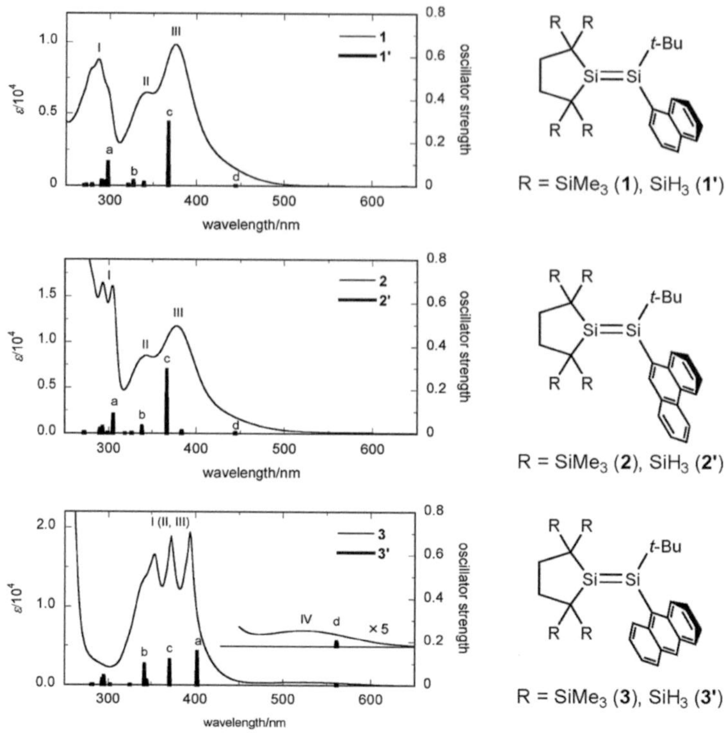

Fig. 22.4 UV–vis spectra of disilenes **1-3**

due to the Si=Si double-bond moiety at around 340–380 nm as shown above. Notably, the latter band (band IV) redshifts to 535 nm (ε 480) in more polar 1,2-dichlorobenzene suggesting the intramolecular charge-transfer (ICT) nature of this band.

DFT calculations support the assignment of the transitions. Calculated UV–vis spectra of model compounds for **1-3** (**1**′-**3**′) well reproduce the experimental spectra (calculated spectra are shown as vertical bars in Fig. 22.4). Bands I, II, and III of disilenes **1–3** can be assigned as transitions (a) [π(aryl)$\rightarrow\pi^*$(aryl)], (b) [π(Si=Si)$\rightarrow\sigma^*$(ring Si-C)], and (c) [π(Si=Si)$\rightarrow\pi^*$(Si=Si)]. The weak absorption band observed at 525 nm of anthryldisilene **3** (band IV) is due to an ICT transition π(Si=Si)$\rightarrow\pi^*$(aryl) (transition d). It should be noted that π(Si=Si)$\rightarrow\pi^*$(aryl) ICT transition bands are also predicted at 444 nm for both **1**′ and **2**′, suggesting the ICT absorption bands are overlapped with band III for **1** and **2**. These spectral and orbital feature indicate that π_C system with a lower-lying π^* LUMO such as anthracene is necessary for a longer-wavelength ICT band, and disilenes and πC systems work as electron donor and electron acceptor, respectively.

22.2 Silenes Showing Intramolecular Charge-Transfer (ICT) Transitions

Reaction of 1-naphthyldisilene **1** and 9-anthryldisilene **3** with xylyl isocyanide gave [2 + 1] cycloadducts, disilacyclopropanimines **8** and **9** in 76 % and 58 % yield, respectively (Scheme 22.2), similarly to the reaction of tetraxylyldisilene with xylylisocyanide [37]. Structures of both disilacyclopropanimines **8** and **9** were confirmed by XRD analysis (Fig. 22.5). While cycloadduct **8** is thermally stable, cycloadduct **9** bearing 9-anthryl substituent undergoes further isomerization to an exocyclic silene, 3-silylene-2-silaaziridine **4** quantitatively in 16 days at room temperature (or in 3 h at 60 °C) in benzene-d_6 (Scheme 22.3). Pure **4** was isolated as air-sensitive blue-purple crystals in 39 % yield. The structure of **4** was determined by NMR, MS, and XRD analysis. Recently Scheschkewitz reported reversible reactions of tetraaryldisilenes $Tip_2Si{=}SiTipPh$ and 1,4-$(Tip_2Si{=}SiTip)_2C_6H_4$ [Tip=2,4,6-$(i\text{-}Pr)_3C_6H_2$] with isocyanides to give 3-silylene-2-silaaziridines **10-13** (Fig. 22.6) via the corresponding disilacyclopropanimines as proposed intermediates [38]. In contrast, isomerization of cycloadduct **9** to silene **4** was clearly observed. Compound **4** is the first silene with polycyclic aromatic substituent. Similar isomerizations of silene, phosphasilene, and arsasilene upon addition of isocyanide were reported [39–41].

A noticeable structural feature of **4** disclosed by XRD analysis is that π(Si=C) and π(anthryl) moieties are almost perpendicular to each other with the dihedral angle of 81.7° similarly to anthryldisilene **3**. The perpendicular geometry would be owing to the severe steric hindrance around the anthryl moiety and suggest that no remarkable conjugative interactions between the two π systems exist. The other structural feature of **4** is the flat geometry of 3-silylene-2-silaaziridine moiety: the sum of the bond angles around the unsaturated silicon atom is 358.7(2)°. The Si=C distance in **4** of 1.730(5) Å is close to that of a reported 3-silylene-2-silaaziridine **10** (1.735(2) Å) [38] and longer than that in Wiberg's silene $Me_2Si{=}C(SiMe_3)\{SiMe(t\text{-}Bu)_2\}$ [1.702(5) Å] [42]. The C1–N1 distance of 1.447(6) Å in the C–Si–N three-membered ring elongates compared to those of the C=N double bonds in disilacyclopropanimine **8** [1.283(4) Å] but resembles to those of typical C–N single bonds (1.47–1.50 Å). These structural features indicate that neutral resonance structure **A** is the major contributor for the structure of **4** rather than the zwitterionic structure **B** that places negative charge on the silene Si atom (Fig. 22.6, right). The

Scheme 22.2 [2 + 1] cycloaddition of disilenes **1** and **3** with xylyl isocyanide

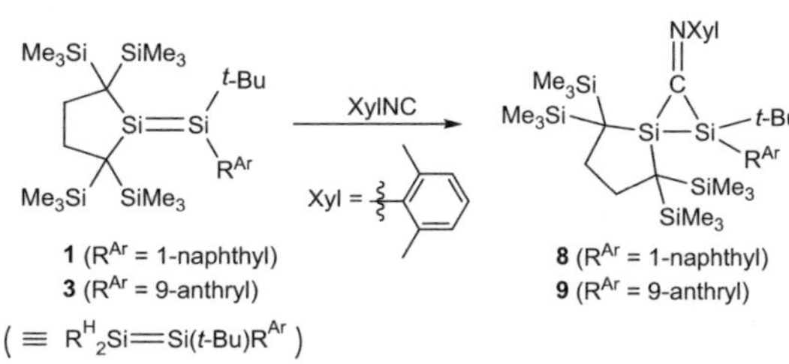

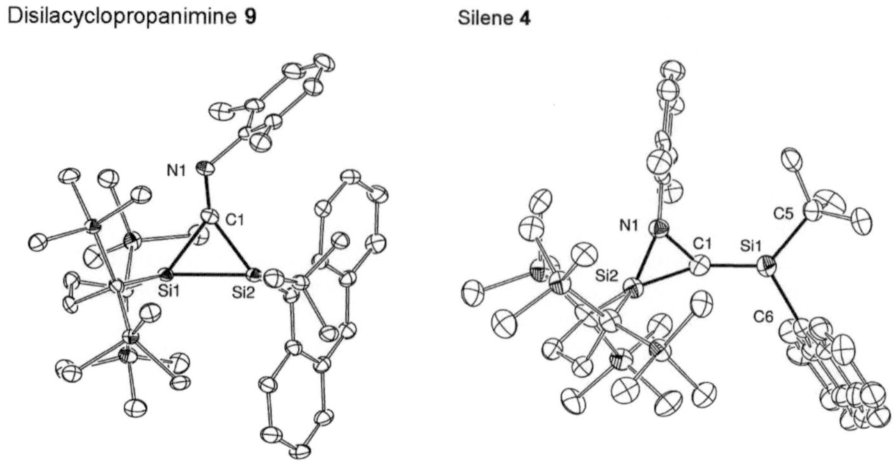

Fig. 22.5 ORTEP drawings of disilacyclopropanimine **9** and 9-anthrylsilene **4**

Scheme 22.3 Thermal isomerization of disilacyclopropanimine **9** with 9-anthryl substituent to 3-silylene-2-silaaziridine **4**

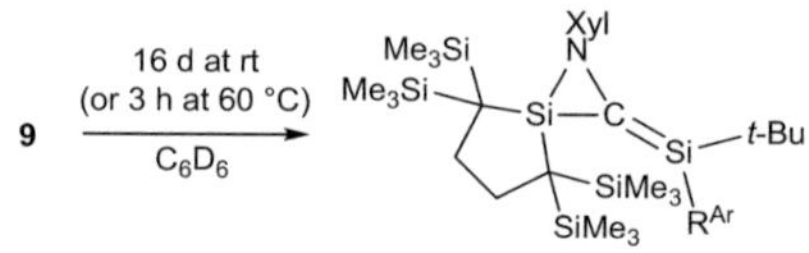

4 (R^{Ar} = 9-anthryl)

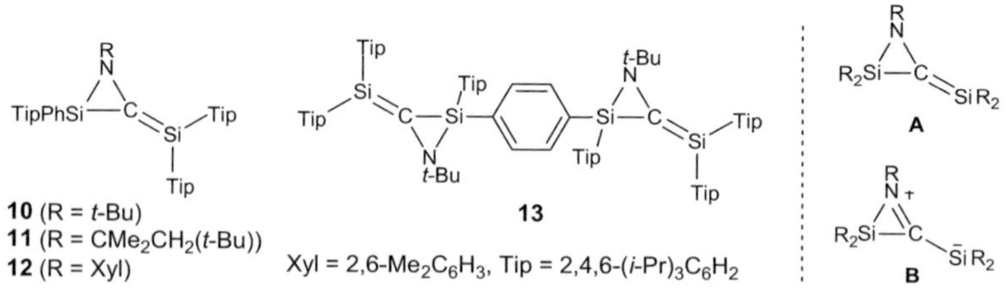

Fig. 22.6 Related 3-silylene-2-silaaziridines and possible resonance structures of 3-silylene-2-silaaziridine

^{29}Si resonance for doubly bonded Si nucleus of **4** was observed at +36.7 ppm, while the ^{13}C resonance for the unsaturated C nucleus appeared at +141.4 ppm. The slightly large $^{1}J_{Si-C}$ coupling constant of the Si=C double bond of **4** (98 Hz) compared to those for typical silenes (76–92 Hz) [5] would be due to the higher s character of the carbon orbital in the exocyclic Si=C bond of the three-membered ring.

Silene **4** shows two absorption bands; a structured band at 300-400 nm (band I) and a weak and apparent band at 550 nm (band II) (Fig. 22.7), which is in sharp contrast to those of typical silenes exhibiting an absorption band below 400 nm. Band I can be assigned to the $^{1}L_a$ band of the anthryl moiety overlapped with the $\pi \rightarrow \pi^*$ transition band of the Si=C bond (the $^{1}L_a$ band maximum of

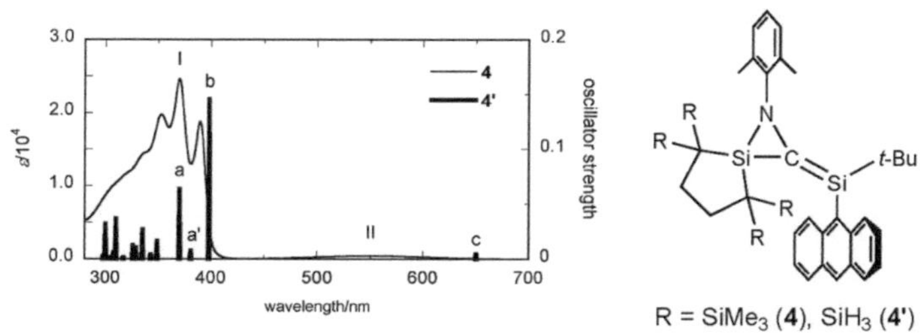

Fig. 22.7 UV–vis spectrum of 9-anthyrlsilene **4**

9-(trimethylsilyl)anthracene: 387.5 nm) [36]. Band II observed at 550 nm in hexane redshifts in polar but poorly coordinating solvents [556 nm in fluorobenzene, 560 nm in dichloromethane, and 561 nm in 1,2-dichlorobenzene] similarly to 9-anthryldisilene **6** and blueshifts in polar coordinating solvents [547 nm in tetrahydrofuran (THF) and 549 nm in 1,3-dimethylimidazolidin-2-one (DMI)] (dielectric constants, 1.88 (hexane); 5.42 (fluorobenzene); 7.58 (THF); 8.93 (dichloromethane); 9.93 (1,2-dichlorobenzene); 37.6 (DMI)) [43]. Although the detailed mechanism for the observed solvatochromism remains open, the redshift suggests the ICT nature of band II and the blueshift may indicate significant stabilization of the ground state by coordination of the polar solvent to the unsaturated Si atom of the Si=C bond as observed for other base-coordinate silenes.

DFT calculation of a model compound of **4** (**4**′) supports the assignment of the absorption bands I and II and disclosed the details of the electronic property of **4**. The calculated structure of **4**′ optimized at the B3LYP/6-311G(d) level reproduced well the structural parameters of **4**. The theoretical absorption spectrum of **4**′ calculated at the TD-B3LYP/6-311++G(d,p) level, which is shown in vertical bars in Fig. 22.7, qualitatively reproduces the experimental spectrum of **4**. The frontier Kohn–Sham (KS) orbitals of **4**′ are π orbital of Si=C double bond interacting with lone-pair orbital of the neighboring nitrogen ($\pi_{Si=C}$), π* orbital of Si=C double bond ($\pi^*_{Si=C}$) and π and π* orbitals of anthryl moiety (π_{aryl} and π^*_{aryl}). Judging from the comparison of experimental and theoretical spectra, bands I is assigned as a $\pi_{aryl} \rightarrow \pi^*_{aryl}$ transition (transition b) overlapped by a transition from $\pi_{Si=C}$ to $\pi^*_{Si=C}$ with small contribution of π^*_{aryl} (transitions a and a'), while band II is as a $\pi_{Si=C} \rightarrow \pi^*_{aryl}$ ICT transition (transition c). To the best of our knowledge, this is the first observation of a silene ICT transition. Although the π orbital level in silenes are usually predicted to be lower than those in disilenes (e.g., −8.41 for $H_2C{=}SiH_2$, while −7.64 eV for $H_2Si{=}SiH_2$ calculated at the MP2/6-31G(d)//B3LYP/6-31G(d) level] [44], the ICT band maxima of **4** (550 nm) is close to that of 9-anthryldisilene **3** (526 nm)). The distinct ICT band observed for **4** can be rationalized by the higher-lying $\pi_{Si=C}$ orbital level resulting from the significant interaction with the lone-pair orbital of nitrogen in the 3-silylene-2-silaaziridine moiety, which was mentioned by Scheschkewitz et al. [38]. The energy level of

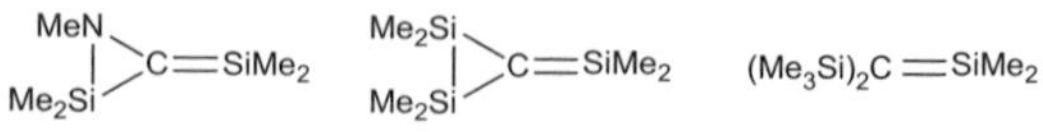

14 15 16

Fig. 22.8 Model compounds **14–16**

the π orbital of permethyl-substituted 3-silylene-2-azaaziridine **14** is −4.36 eV, while those of 2-silylene-1,3-disilacyclopropane (**15**) and $(Me_3Si)_2C{=}SiMe_2$ (**16**) bearing no nitrogen atoms are −5.81 and −5.91 eV, respectively, at the B3LYP/6-311++(d,p)//B3LYP/6-311G(d) level (Fig. 22.8).

22.3 Theoretical Study of Isomerization of Disilacyclopropanimine to 3-Silylene-2-Silaaziridine

The details of disilacyclopropanimine to 3-silylene-2-silaaziridine isomerization was investigated using the GRRM method [45–47] for model isomerization reactions of permethylated disilacyclopropanimine **17** to 3-silylene-2-silaaziridine **18** calculated at the B3LYP/6-311G(d) level. The GRRM method predicted three reaction routes from **17** (Scheme 22.4); a concerted 1,3-migration of the silyl group trans to the *N*-methyl group from silicon to nitrogen atoms with inversion of configuration of the silyl group to give **18** (route 1) similarly to acyclic disilanylimine to *N*-silylsilene [48], a stepwise route from **17** to **18** involving 1,2-migration via cyclic carbene intermediate **19** (route 2), a dissociation to give the corresponding isocyanide **21** and disilene **22** via the weak disilene–isocyanide complex **20**. The activation barrier for route 3 (+98.2 kJ/mol) is lower than those for route 1 (+114.0 kJ/mol) and the second step of route 2 (125.8 kJ mol^{-1}), which is in agreement with observation of free 9-anthryldisilene **3** and xylyl isocyanide during thermolysis of **9** at 55 °C. The activation barriers for routes 1 and 2 are comparable to each other. However, the observed geometry of **4**, where bulky silacyclopentane moiety is bonded to the nitrogen atom, indicates that the isomerization of **9** to **4** proceeds in a concerted mechanism close to route 1. While formation of **18** from **17** is calculated to be slightly endothermic (+21.5 kJ mol^{-1}), the more severe steric repulsion between bulky silacyclopentane and Si(*t*-Bu)(anthryl) moieties in **9** would make the reaction exothermic.

22.4 Conclusion

These results are examples that electronic properties of silicon π-electron systems are controllable by taking advantage of steric and electronic effects of the protecting groups on the silicon π-electron systems. Although synthetic routes to silicon π-electron systems are still limited compared to those of carbon π-electron systems, development of new synthetic routes for silicon π-electron systems under milder conditions may open new fields of silicon π-electron systems.

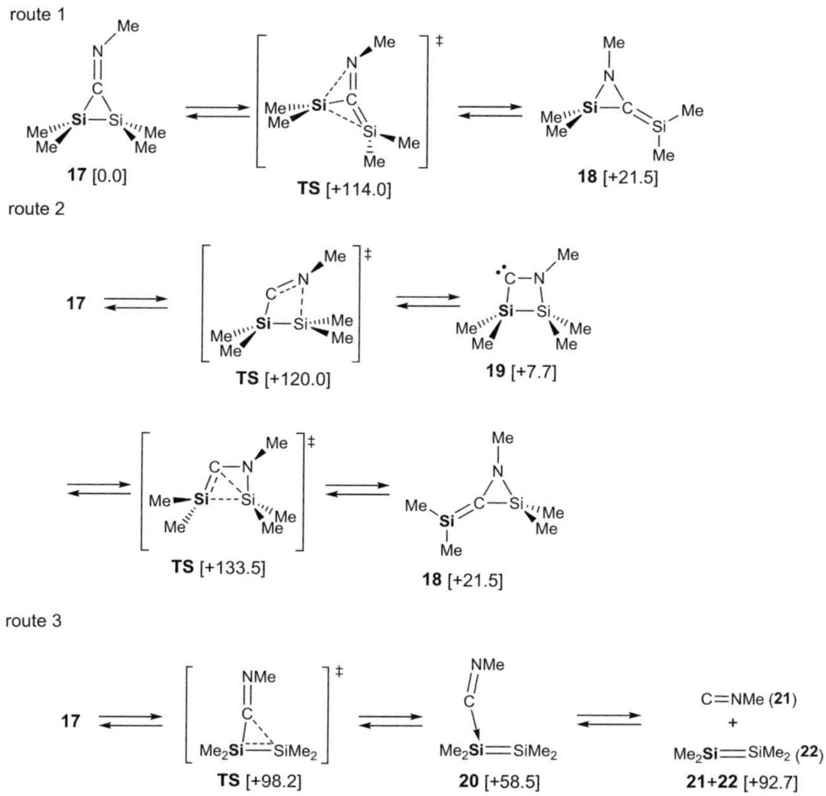

Scheme 22.4 Possible reaction routes from model disilacyclopropanimine **17** obtained by the GRRM (global reaction route mapping) method [45–47]. The values in the brackets are relative energies in kJ mol^{-1} at the B3LYP/6-311G(d) level

References

1. West R, Fink MJ, Michl J (1981) Tetramesityldisilene, a stable compound containing a silicon-silicon double bond. Science 214:1343–1344
2. Brook AG, Abdesaken F, Gutekunst B, Gutekunst G, Kallury RK (1981) A solid silaethene: isolation and characterization. J Chem Soc Chem Commun 17:191
3. Brook AG, Nyburg SC, Abdesaken F, Gutekunst B, Gutekunst G, Kallury RKMR, Poon YC, Chang Y-M, Wong-Ng W (1982) Stable solid sitalethylene. J Am Chem Soc 104:5667–5672
4. Okazaki R, West R (1996) Chemistry of stable disilenes. Adv Organomet Chem 39:231–273
5. Müller T, Ziche W, Auner N (1998) Silicon-carbon and silicon-nitrogen multiply bonded compounds. In: Rappoport Z, Apeloig Y (eds) The chemistry of organic silicon compounds. Wiley, Chichester, pp 857–1062
6. Weidenbruch M (2001) Recent advances in the chemistry of silicon-silicon multiple bonds. In: Rappoport Z, Apeloig Y (eds) Chemistry of Organic Silicon Compounds, vol 3. Wiley, Chichester, pp 391–428
7. Lee VY, Sekiguchi A (2004) Heteronuclear heavy alkenes E = E' (E, E'=Group 14 Elements): Germasilenes, silastannenes, germastannenes... Next stop? Organometallics 23:2822–2823

8. Kira M, Iwamoto T (2006) Progress in the chemistry of stable disilenes. Adv Organomet Chem 54:73–148
9. Scheschkewitz D (2011) The versatile chemistry of disilenides: Disila analogues of vinyl anions as synthons in low-valent silicon chemistry. Chem Lett 40:2–11
10. Ottosson H, Eklöf AM (2008) Silenes: Connectors between classical alkenes and nonclassical heavy alkenes. Coord Chem Rev 252:1287–1314
11. Scheschkewitz D (2009) Anionic reagents with silicon-containing double bonds. Chem Eur J 15:2476–2485
12. Lee VY, Sekiguchi A, Escudié J, Ranaivonjatovo H (2010) Heteronuclear double bonds E=E' (E=heavy group 14, E'=group 13-16 element). Chem Lett 39:312–318
13. Sasamori T, Tokitoh N (2013) A new family of multiple-bond compounds between heavier group 14 elements. Bull Chem Soc Jpn 86:1005–1021
14. Iwamoto T, Ishida S (2014) Multiple bonds to silicon: recent advances in synthesis, structure, and functions of stable disilenes. In: Scheschkewitz D (ed) Functional molecular silicon compounds II low oxidation state. Structure and Bonding, vol 156. Springer, New York, pp 125–202
15. Carter EA, Goddard WA III (1986) Relation between singlet-triplet gaps and bond energies. J Phys Chem 90:998–1001
16. Trinquier G, Malrieu J-P (1987) Nonclassical distortions at multiple bonds. J Am Chem Soc 109:5303–5315
17. Malrieu J-P, Trinquier G (1989) Trans bending at double bonds. Occurrence and extent. J Am Chem Soc 111:5916–5921
18. Karni M, Apeloig Y (1990) Substituent effects on the geometries and energies of the Si=Si double bond. J Am Chem Soc 112:8589–8590
19. Jacobsen H, Ziegler T (1994) Nonclassical double bonds in ethylene analogues: influence of Pauli repulsion on trans bending and π-bond strength. A density functional study. J Am Chem Soc 114:3667–3679
20. Sekiguchi A, Maruki I, Ebata K, Kabuto C, Sakurai H (1991) High-pressure synthesis, structure and novel photochemical reactions of 7,7,8,8-tetramethyl-7,8-disilabicyclo[2.2.2]octa-2,5-diene. J Chem Soc, Chem Commun 27:341–343
21. Tokitoh N (2004) New progress in the chemistry of stable metallaaromatic compounds of heavier group 14 elements. Acc Chem Res 37:86–94
22. Nguyen T-I, Scheschkewitz D (2005) Activation of a Si=Si bond by η^1-coordination to a transition metal. J Am Chem Soc 127:10174–10175
23. Fukawaza A, Li Y, Yamaguchi S, Tsuji H, Tamao K (2007) Coplanar oligo(*p*-phenylenedisilenylene)s based on the octaethyl-substituted s-hydrindacenyl group. J Am Chem Soc 129:14164–14165
24. Bejan I, Scheschkewitz D (2007) Two Si-Si double bonds connected by a phenylene bridge. Angew Chem Int Ed 46:5783–5786
25. Sasamori T, Yuasa A, Hosoi Y, Furukawa Y, Tokitoh N (2008) 1,2-Bis(ferrocenyl)disilene: a multistep redox system with an Si=Si double bond. Organometallics 27:3325–3327
26. Yuasa A, Sasamori T, Hosoi Y, Furukawa Y, Tokitoh N (2009) Synthesis and properties of stable 1,2-bis(metallocenyl)disilenes: novel d–π conjugated systems with a Si=Si double bond. Bull Chem Soc Jpn 82(7):793–805
27. Sato T, Mizuhata Y, Tokitoh N (2010) 1,2-Dialkynyldisilenes: silicon analogues of (*E*)-enediyne. Chem Commun 46:4402–4404
28. Tamao K, Kobayashi M, Matsuo T, Furukawa S, Tsuji H (2012) The first observation of electroluminescence from di(2-naphthyl)disilene, an Si=Si double bond-containing π-conjugated compound. Chem Commun 48:1030–1032
29. Kinjo R, Ichinohe M, Sekiguchi A, Takagi N, Sumimoto M, Nagase S (2007) Reactivity of a disilyne RSi≡SiR (R=Si*i*Pr[CH($SiMe_3$)$_2$]$_2$ toward π-bonds: Stereospecific addition and a new route to an isolable 1,2-disilabenzene. J Am Chem Soc 129:7766–7767
30. Han JS, Sasamori T, Mizuhata Y, Tokitoh N (2010) Reactivity of an aryl-substituted silicon-silicon triple bond: 1,2-disilabenzenes from the reactions of a 1,2-diaryldisilyne with alkynes. Dalton Trans 39(39):9238–9240

31. Takeuchi K, Ichinohe M, Sekiguchi A (2008) Reactivity of the disilyne RSi≡SiR (R=Si*i*Pr[CH(SiMe$_3$)$_2$]$_2$) toward silylcyanide: Two pathways to form the bis-adduct [RSiSiR(CNSiMe$_3$)$_2$] with some silaketenimine character and a 1,4-diaza-2,3-disilabenzene analogue. J Am Chem Soc 130:16848–16849
32. Takeuchi K, Ichinohe M, Sekiguchi A, Guo J-D, Nagase S (2010) Reactivity of the disilyne RSi≡SiR (R=Si*i*Pr[CH(SiMe$_3$)$_2$]$_2$) toward bis(silylcyanide) forming a 1,4-diaza-2,3-disilabenzene analog. J Phys Org Chem 23:390–394
33. Iwamoto T, Kobayashi M, Uchiyama K, Sasaki S, Nagendran S, Isobe H, Kira M (2009) Anthryl-substituted trialkyldisilene showing distinct intramolecular charge-transfer transition. J Am Chem Soc 2009:3156–3157
34. Iwamoto T, Ohnishi N, Akasaka N, Ohno K, Ishida S (2013) Anthryl-substituted 3-silylene-2-silaaziridine obtained by isomerization of disilacyclopropanimine: an exocyclic silene showing a distinct intramolecular charge transfer transition. J Am Chem Soc 135(29):10606–10609
35. Kira M, Ishida S, Iwamoto T, Kabuto C (1999) The first isolable dialkylsilylene. J Am Chem Soc 121:9722–9723
36. Shizuka H, Obuchi H, Ishikawa M, Kumada M (1984) Photochemical and photophysical behaviour of phenylsilanes and naphthylsilanes. J Chem Soc Faraday Trans 80:383–401
37. Yokelson HB, Millevolte AJ, Haller KJ, West R (1987) The synthesis and molecular structure of a disilacyclpropanimine. J Chem Soc Chem Commun 23:1605–1606
38. Majumdar M, Huch V, Bejan I, Meltzer A, Scheschkewitz D (2013) Reversible, complete cleavage of Si=Si double bonds by isocyanide insertion. Angew Chem Int Ed 52:3516–3520
39. Brook AG, Kong YK, Saxena AK, Sawyer JF (1988) Silaaziridines from the reaction of isonitrile with stable silenes. Organometallics 7:2245–2247
40. Brook AG, Saxena AK, Sawyer JF (1989) 1-Sila-3-azacyclobutanes: the insertion of isocyanides into silaaziridines. Organometallics 8:850–852
41. Driess M, Pritzkow H (1993) A novel type of oligofunctional macroheterocycle by reaction of silylidene-phosphane and -arsanes with 1,6-diisocyanohexane. J Chem Soc, Chem Commun 29:1585–1587
42. Wiberg N, Wagner G, Müller G (1985) Isolation and structure of a stable molecule containing a silicon-carbon double bond. Angew Chem Int Ed Engl 24:229–230
43. Reichardt C, Welton T (2011) Solvent and solvent effects in organic chemistry. Wiley-VCH, New York
44. Bendikov M, Kravchenko V, Apeloig Y, Becker JY (2004) The first electrochemical study of a silene. An unusually low oxidation potential, comparable to those of strong organic π-electron donors. Organometallics 23:921–923
45. Ohno K, Maeda S (2004) A scaled hypersphere search method for the topography of reaction pathways on the potential energy surface. Chem Phys Lett 384:277–282
46. Ohno K, Maeda S (2006) Global reaction route mapping on potential energy surfaces of formaldehyde, formic acid, and their metal substituted analogues. J Phys Chem A 110:8933–8941
47. Maeda S, Ohno K (2007) Structures of water octamers (H_2O)$_8$: Exploration on ab initio potential energy surfaces by the scaled hypersphere search method. J Phys Chem A 111:4527–4534
48. Eklöf AM, Guliashvili T, Ottosson H (2008) Relation between the π-contribution to reversed Si=C bond polarization and the reaction profile for the thermolytic formation of silenes. Organometallics 27:5203–5211

Chapter 23
π-Conjugated Disilenes and Tetrasilacyclobutadiene

Tsukasa Matsuo

Abstract The kinetic stabilization using sterically demanding bulky substituents or ligands provides us an exciting opportunity to explore a large variety of organoelement unsaturated compounds. We recently demonstrated that a new class of fused-ring bulky octa-R-substituted *s*-hydrindacenyl groups (Rind groups) can act as unique protecting groups in the main group chemistry as well as in the transition metal chemistry. This manuscript describes the recent advances in the chemistry of π-conjugated systems containing sp^2-hybridized silicon atoms stabilized by the bulky Rind groups. The synthesis and characterization of various π-conjugated disilene compounds as well as a planar rhombic charge-separated tetrasilacyclobutadiene incorporating the bulky Rind groups are covered.

Keywords Silicon • Disilene • Cyclobutadiene • π-Conjugated systems • Bulky protecting groups

23.1 Introduction

The stabilization of low-coordinated compounds has long been a key research theme in both the fundamental and applied chemistries. In 1981, West et al. introduced the concept of kinetic stabilization of the highly reactive Si=Si double bonds using bulky 2,4,6-trimethylphenyl (mesityl) groups, as shown in Fig. 23.1 [1]. Thereafter, a large variety of unsaturated compounds of the heavier main group elements have been successfully isolated using the sterically bulky protecting groups [2]. For example, in organosilicon chemistry, silaaromatics (silabenzene) [3], trisilaallenes ($R_2Si{=}Si{=}SiR_2$) [4], and disilynes ($RSi \equiv SiR$) [5] have been synthesized by taking advantage of the steric protection with the appropriate bulky substituents. In 2011, we reported the synthesis and characterization of a cyclobutadiene (CBD) silicon analog (tetrasilacyclobutadiene) with a planar rhombic charge-separated

T. Matsuo (✉)
Department of Applied Chemistry, Faculty of Science and Engineering, Kinki University, 3-4-1 Kowakae, Higashi-Osaka, 577-8502, Japan
e-mail: t-matsuo@apch.kindai.ac.jp

T. Akasaka et al. (eds.), *Chemical Science of π-Electron Systems*,
DOI 10.1007/978-4-431-55357-1_23

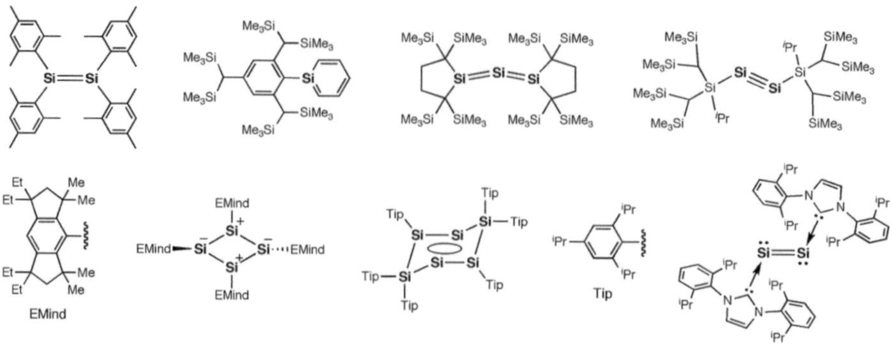

Fig. 23.1 Examples of stable unsaturated silicon compounds

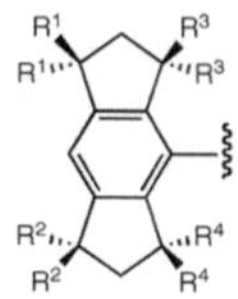

Chart 23.1 Rind groups

structure [6]. Also, a tricyclic aromatic valence isomer of hexasilabenzene [7] and a disilicon(0) fragment stabilized by the *N*-heterocyclic carbenes (NHCs) [8] were reported.

As shown in Chart 23.1, we recently developed a series of fused-ring bulky 1,1,3,3,5,5,7,7-octa-R-substituted *s*-hydrindacenyl groups, called the "Rind" groups [9]. The readily available Rind groups have several advantages over the existing bulky groups, as represented by the versatility and size controllability by the introduction of various R-substituents at the benzylic positions in the rigid *s*-hydrindacenyl skeleton. The abbreviation "Rind" in English denotes the thick outer skin of some types of fruit for protection of the inside fruity flesh, which is fully in accordance with our research concept. Actually, the bulky Rind groups enable us to isolate some unsaturated compounds of the main group elements as well as transition metals. In this manuscript, we describe the recent progress in the chemistry of π-conjugated silicon compounds featuring the bulky Rind groups.

23.2 π-Conjugated Disilenes

The incorporation of multiple bonds of the heavier main group elements into carbon π-conjugated architectures has attracted much attention in recent years, because of their potentially useful properties as functional materials [10]. However, this chemistry always suffers from a dilemma. While the steric protection by the bulky groups is crucial to stabilize the highly reactive heavier multiple bonds, it causes the π-conjugated framework to twist, which reduces the extension of the π-conjugation.

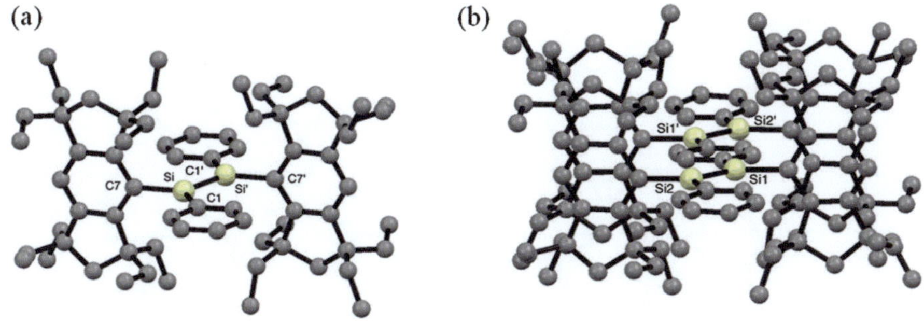

Fig. 23.2 Ball-and-stick model representations of (**a**) (Eind)PhSi=SiPh(Eind) and (**b**) [(Eind)PhSi=Si(Eind)]$_2$(C_6H_4)

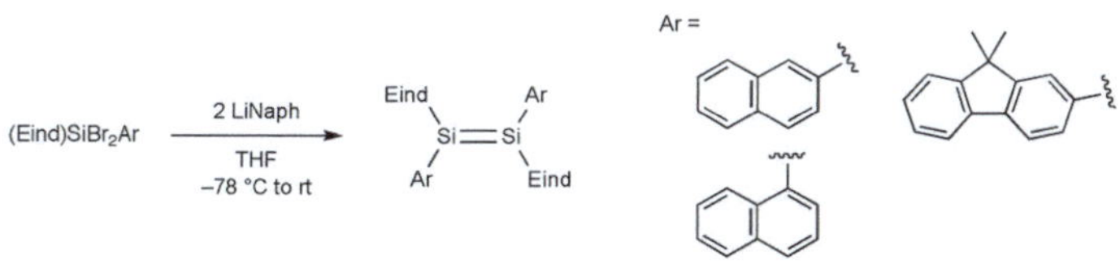

Scheme 23.1 Synthesis of (*E*)-1,2-diaryldisilenes

A key for the further evolution of this chemistry is to certainly attain a well-defined bulky substituent that can maintain the highly planar π-conjugated framework, in addition to providing sufficient steric protection.

In 2007, model systems of disilene analogs of the oligo(*p*-phenylenevinylene)s (Si–OPVs) have been synthesized using 2,4,6-triisopropylphenyl (Tip) and the Eind groups [11]. Figure 23.2 shows the molecular structures of the Eind-substituted 1,2-diphenyldisilene and 1,4-bis(disilenyl)benzene derivatives. The bulky Eind groups encapsulate the reactive Si=Si units and produce the highly coplanar π-conjugated frameworks by their perpendicular orientation to and the interlocking ethyl side chains above and below the π-frameworks. Accordingly, the 1,4-bis(disilenyl)benzene exhibits an orange emission at ambient temperature because of the efficient π-conjugation over the skeleton.

After the initial findings of the Si–OPVs, we have focused on the development of π-conjugated disilene compounds by the installation of two π-extended aromatic groups into the disilene core [12]. As shown in Scheme 23.1, the 1,2-diaryldisilenes protected by the bulky Eind groups, (Eind)ArSi=SiAr(Eind) (Ar=2-naphthyl, 2-fluorenyl, and 1-naphthyl), have been synthesized by the reductive coupling of the corresponding dibromosilanes, (Eind)ArSiBr$_2$. The resulting 1,2-diaryldisilenes are air stable in the solid state for more than several years with no detectable change, as confirmed by the ^{1}H NMR spectra, indicative of the outstanding protection ability of the Eind group. They decompose in a dilute hexane solution (ca. 10^{-5} mol L^{-1}) upon exposure to air, but slowly with a half-lifetime of ca. 2–4 h, which is much longer than that observed in Tip$_2$Si=SiTip$_2$ (17 min) [13], as estimated by the UV-vis absorption spectroscopy.

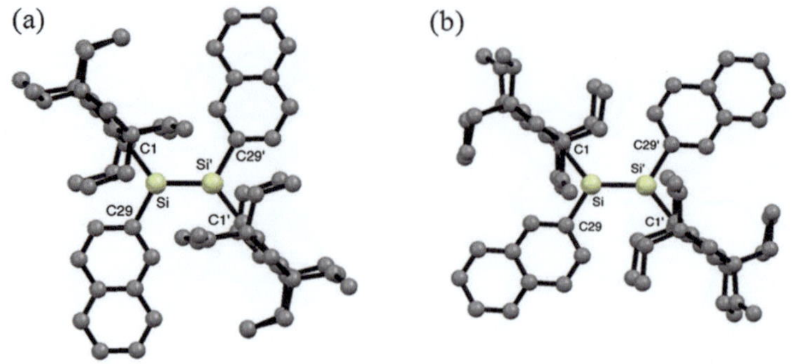

Fig. 23.3 Ball-and-stick representations of (**a**) *s-cis*, *s-cis* and (**b**) *s-trans*, *s-trans* conformers of (Eind)(2-Naph)Si=Si(2-Naph)(Eind)

Fig. 23.4 A ball-and-stick representation of (Eind)(1-Naph)Si=Si(1-Naph)(Eind)

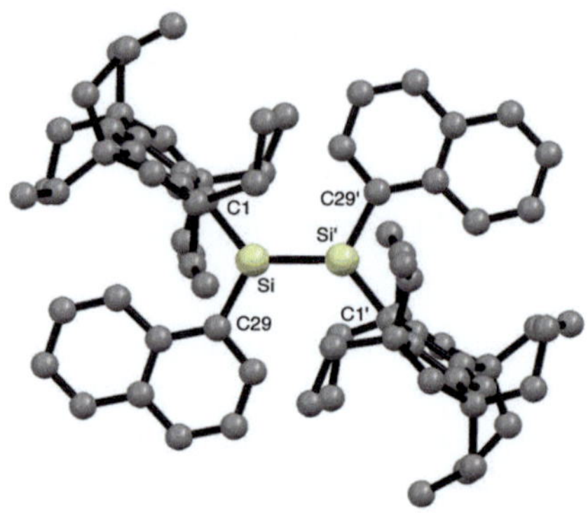

The structures of the two regioisomers of (*E*)-1,2-dinaphthyldisilenes have been determined by an X-ray crystallographic analysis. The 2-naphthyl counterpart exists as a mixture of two rotational isomers, *s-cis*, *s-cis* and *s-trans*, *s-trans* conformers, with the occupancy factors of ca. 0.4/0.6 in the crystals [12a], as shown in Fig. 23.3. Each rotamer exhibits the coplanar π-framework with the Si–Si–C–C torsion angle of 9.57(11)° for *s-cis*, *s-cis* and 2.25(14)° for *s-trans*, *s-trans*. For the structure of the 1-naphthyl counterpart shown in Fig. 23.4, the *peri*-H atoms do not cause any steric repulsion with the Eind-containing disilene moiety but participate in the CH/π interaction with the benzene ring of the perpendicularly oriented Eind groups [12b]. Thus, the 1-naphtyl groups are highly coplanar with the Si=Si bond having the Si–Si–C–C torsion angle of 4.93(12)°.

The (*E*)-1,2-dinaphthyldisilenes exhibit strong π–π* absorptions and distinct emissions at room temperature in THF due to the substantial contribution of the $3p_\pi$*(Si–Si)–$2p_\pi$*(carbon π-electron system) conjugation. This is in sharp contrast to the fact that the Eind-substituted 1,2-diphenyldisilene does not show any fluorescence in solution. The 1,2-dinaphthyldisilenes show stronger emissions in the solid state; the photographs in air are shown in Fig. 23.5. Thus, the fused-ring

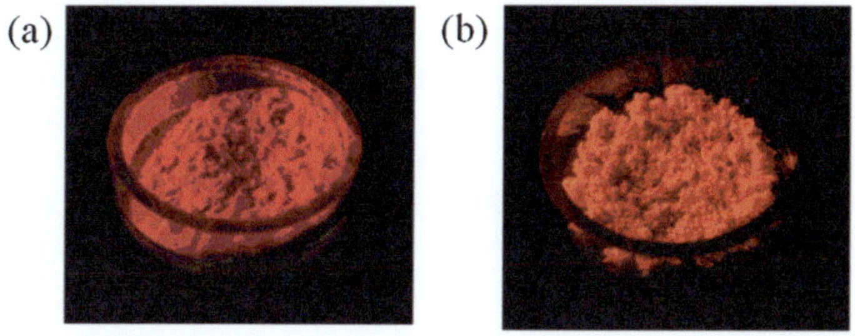

Fig. 23.5 Solid-state emissions of (**a**) (Eind)(1-Naph)Si=Si(1-Naph)(Eind) and (**b**) (Eind)(2-Naph)Si=Si(2-Naph)(Eind) at room temperature in the open air

bulky Eind groups effectively prevent the intermolecular interaction for attaining the intense solid-state emission.

The dinaphthyldisilenes also have a high thermal stability with a decomposition point of 282–285 °C for the 2-naphthyl isomer and 245–248 °C for the 1-naphthyl isomer under an argon atmosphere. The excellent stability would provide new opportunities for application to a range of organic electronic devices, since the Si=Si units have a narrower HOMO–LUMO gap compared to their carbon counterparts. We have found that the 1,2-di(2-naphthyl)disilene can emit light in an organic light-emitting diode (OLED) [14]. This is the first experimental observation of electroluminescence (EL) from the unsaturated compounds of the heavier group 14 elements, which has opened a new platform for the development of functional organoelement materials and devices.

Further investigations in synthesizing higher homologues of the Si–OPVs and a variety of π-conjugated disilenes with polycyclic aromatic groups are now in progress and will be reported.

23.3 Tetrasilacyclobutadiene

In 2011, we reported the isolation and characterization of a CBD silicon analog, tetrasilacyclobutadiene, $Si_4(EMind)_4$, stabilized by the bulky EMind groups [6]. The Si_4 ring has a planar rhombic charge-separated structure with alternating pyramidal and planar configurations at the four silicon atoms.

Figure 23.6 shows schematic representations of the molecular orbitals (MOs) of CBD with a cyclic 4π-electron system. Both experimental and theoretical investigations of the parent CBD, C_4H_4, demonstrated that a square-shaped triplet with two degenerate singly occupied molecular orbitals (SOMOs) is a less stable form, thus being subjected to a second-order Jahn–Teller (J–T) distortion for stabilization [15]. There are two major modes of the J–T distortion to compensate for the antiaromatic nature of CBD: One is a covalent J–T distortion, which makes a rectangular-shaped singlet bearing the two double bonds with the frontier MOs

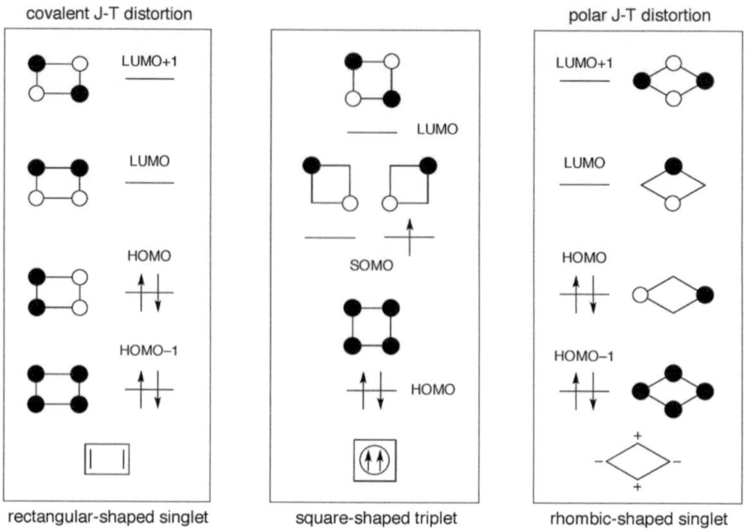

Fig. 23.6 Schematic representations of molecular orbital diagram of CBD

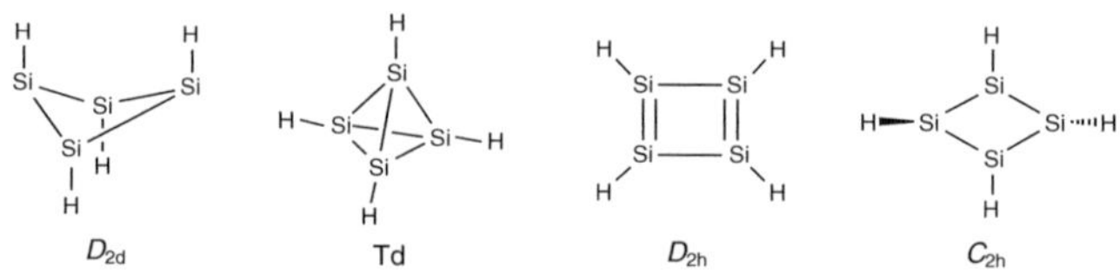

Fig. 23.7 Theoretically predicted molecules of $(SiH)_4$

delocalized on all four atoms; the other is a polar J–T distortion, producing a planar rhombic structure, which causes an alternating charge separation on four atoms with the out-of-phase frontier MOs localized on each set of diagonal atoms. For the experimental studies, some CBD derivatives having rectangular structures with two localized C=C double bonds have been isolated and structurally determined by taking advantage of the kinetic stabilization [16]. Also, the push–pull perturbed CBD derivatives provide the localized MOs as a result of the thermodynamic stabilization [17].

The parent molecule of tetrasilacyclobutadiene, Si_4H_4, has been the research theme of theoretical calculations [18]. Figure 23.7 shows some theoretically predicted isomers of $(SiH)_4$. While a puckered D_{2d} geometry and a cage molecule of tetrasilatetrahedrane are estimated to be local energy minima, a planar rectangular D_{2h} geometry with two Si=Si double bonds and a planar rhombic C_{2h} structure appear as saddle-point structures. Experimentally, the tetrasilatetrahedranes have been isolated by the introduction of the isotropically bulky silyl groups [19]. A square planar tetrasilacyclobutadienyl ligand is also observed in transition metal complexes [20].

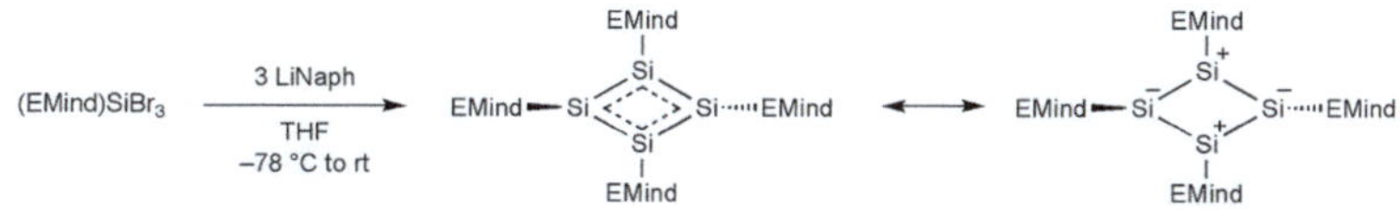

Scheme 23.2 Synthesis of tetrasilacyclobutadiene

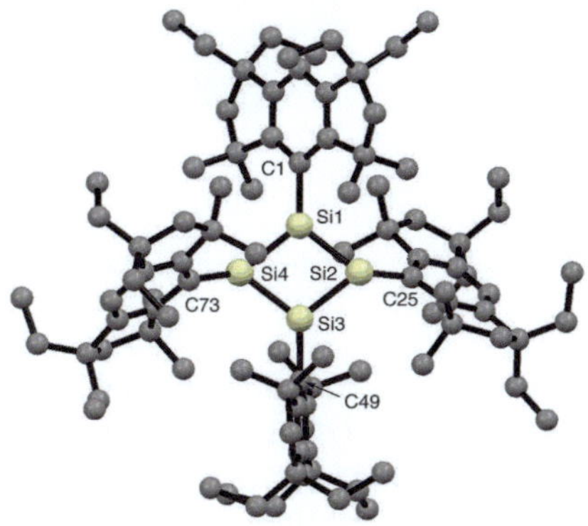

Fig. 23.8 A ball-and-stick representation of $Si_4(EMind)_4$

As shown in Scheme 23.2, the EMind-substituted tetrasilacyclobutadiene, $Si_4(EMind)_4$, has been synthesized as air- and moisture-sensitive orange crystals by the reductive coupling reaction of the corresponding tribromosilane, (EMind)$SiBr_3$, with 3 equiv. of LiNaph in THF. This compound gradually decomposes at room temperature even in the solid state under an inert atmosphere. The half-lifetime is ca. 3 days in benzene-d_6 at 298 K, monitored by ^{1}H NMR spectra. Nevertheless, single crystals suitable for an X-ray diffraction analysis have been obtained by recrystallization from a mixed solution of hexane and toluene at −30 °C.

As shown in Fig. 23.8, X-ray crystallography reveals a planar rhombic structure of the central Si_4 ring; the sum of the internal bond angles is 360.0°. The Si–Si bond lengths are 2.2671(8)–2.2924(8) Å (av. 2.283 Å), which are intermediate between the typical Si–Si single (2.34 Å) and double (2.14 Å) bond lengths. Whereas the Si1 and Si3 atoms are planar with sums of the surrounding bond angles of 360.0°, the Si2 and Si4 atoms are pyramidal with sums of the surrounding bond angles of 338.8° and 335.1°. Space-filling models show that the central Si_4 unit is efficiently covered by the four EMind groups, producing an EMind-meshed molecular gear.

In the solid-state ^{29}Si cross-polarization magic angle spinning (CP-MAS) NMR spectrum for $Si_4(EMind)_4$, two sets of two resonances appear in the higher-field region at −52 and −50 ppm and in the lower-field region at 300 and 308 ppm. The former signals are assignable to the pyramidal silicon atoms (Si2 and Si4) and the latter signals to the planar silicon atoms (Si1 and Si3). The large ^{29}Si chemical shift differences ($\Delta\delta > 350$ ppm) suggest a contribution of an alternately charge-separated structure associated with the configurational differences.

We have performed DFT computations for $Si_4(EMind)_4$ (B3LYP/6-31G(d,p) level) to elucidate the electronic nature of the CBD silicon analog. The optimized closed-shell structure well reproduces the X-ray structure; thus, the most stable

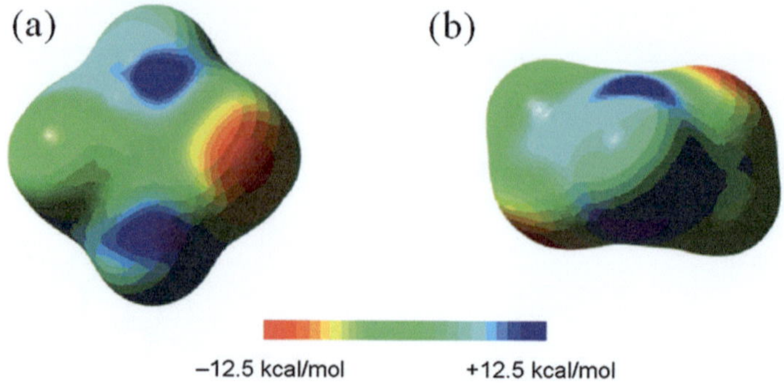

Fig. 23.9 Electrostatic potential maps of $(SiH)_4$ (C_{2h} geometry): (**a**) top view and (**b**) side view

isomer is a planar rhombic singlet. We have also found a planar rectangular singlet with two Si=Si double bonds and a planar square triplet at 6.41 and 2.09 kcal mol^{-1} higher energy levels, respectively. The Wiberg bond indices (WBI) of the four Si–Si bonds for Si_4(EMind)$_4$ are estimated to be nearly equal (1.070–1.179). The natural population analysis (NPA) charge distribution of Si_4(EMind)$_4$ shows a large charge difference on the four silicon atoms being +0.144 slightly positive for Si2 and +0.167 for Si4 and more highly positive at +0.647 for Si1 and +0.637 for Si3. The charge separation is much more evident in the parent Si_4H_4 molecule bearing the C_{2h} geometry with a positive +0.370 for Si1 and Si3 and a negative −0.077 for Si2 and Si4; the electrostatic potential maps are shown in Fig. 23.9. The NICS(1) value [21] is calculated to be −0.9 ppm at the center of the Si_4 ring of Si_4(EMind)$_4$, indicating the lack of a diatropic or paratropic ring current. Thus, the CBD silicon analog is neither aromatic nor antiaromatic, but nonaromatic, consistent with the charge-separated structure.

While the carbon-based CBD is largely stabilized by the bond-length alternation arising from the covalent J–T distortion, the silicon-based CBD is found to be stabilized as the planar rhombic charge-separated structure with the polar J–T distortion. This is probably due to the weaker Si=Si π-bond and the lower ability for s–p hybridization at the silicon[2] compared with carbon. We are now investigating the CBD germanium analog, which may provide further insights into the consequences of antiaromaticity in small rings. We are also very interested in the synthesis and properties of a benzene silicon analog, hexasilabenzene [7, 22, 23], which is the smallest fragment for silicene, a graphene-like allotrope of silicon with a two-dimensional honeycomb structure.

23.4 Summary

We have described our recent progress in the chemical science of π-conjugated systems containing sp^2-hybridized silicon atoms. The fused-ring bulky Rind groups successfully protect the reactive Si=Si double bonds, thus providing extraordinarily air-stable disilene molecules with a highly coplanar arrangement of the π-framework. For example, the two regioisomers of (*E*)-1,2-dinaphthyldisilenes demonstrate the strong π–π* absorptions and distinct emissions at ambient temperature arising from the π-conjugation between the Si=Si double bond and the carbon π-electron system. We have also observed an electroluminescence (EL) from the 1,2-di(2-naphthyl)disilene in an organic light-emitting diode (OLED).

The unique protection ability of the bulky Rind groups also enables us to isolate the hitherto unknown silicon-based CBD. The rhombic-shaped charge-separated singlet state with the polar J–T distortion stabilizes its cyclic 4π-electron antiaromaticity in a manner that sharply contrasts with the bond-length alternation with the covalent J–T distortion, characterizing the rectangular distortion of the carbon-based CBD.

The chemistry of the π-conjugated silicon systems has just begun, and further studies are needed for a deeper understanding of their properties and functionalities, which would open a new facet of functional organoelement materials and devices.

Acknowledgment The author thanks the Ministry of Education, Culture, Sports, Science, and Technology of Japan for the Grant-in-Aid for Specially Promoted Research (No. 19002008) and Scientific Research (B) (Nos. 24350031 and 15H03788). This work was also partially supported by MEXT-Supported Program for the Strategic Research Foundation at Private Universities 2014–2018, subsidy from MEXT and Kindai University. The calculations and experiments discussed in this article have been performed with many collaborators, whose names appear in the references. The author wishes to express his thanks to all of them for their kind collaboration, in particular, Prof. Kohei Tamao, for his continuous discussions.

References

1. West R, Fink MJ, Michl J (1981) Science 214:1343
2. For reviews: (a) Okazaki R, West R (1996) Adv Organomet Chem 39:231. (b) Power PP (1999) Chem Rev 99:3463. (c) Kira M, Iwamoto T (2006) Adv Organomet Chem 54:73. (d) Wang Y, Robinson GH (2009) Chem Commun 5201. (e) Fischer RC, Power PP (2010) Chem Rev 110:4877. (f) Lee YV, Sekiguchi A (2010) Organometallic compounds of low-coordinate Si, Ge, Sn and Pb. Wiley, Chichester (g) Scheschkewitz D (2011) Chem Lett 40:2 (h) Asay M, Sekiguchi A (2012) Bull Chem Soc Jpn 85:1245. (i) Sasamori T, Tokitoh N, Bull Chem Soc Jpn 86:1005
3. Wakita K, Tokitoh N, Okazaki R, Nagase S (2000) Angew Chem Int Ed 39:634
4. Ishida S, Iwamoto T, Kabuto C, Kira M (2003) Nature 421:725
5. Sekiguchi A, Kinjo R, Ichinohe M (2004) Science 305:1755
6. Suzuki K, Matsuo T, Hashizume D, Fueno H, Tanaka K, Tamao K (2011) Science 331:1306
7. Abersfelder K, White AJP, Rzepa HS, Scheschkewitz D (2010) Science 327:564
8. Wang Y, Xie Y, Wei P, King RB, Schaefer III HF, Schleyer PVR, Robinson GH (2008) Science 321:1069
9. Matsuo T, Suzuki K, Fukawa T, Li B, Ito M, Shoji Y, Otani T, Li L, Kobayashi M, Hachiya M, Tahara Y, Hashizume D, Fukunaga T, Fukazawa A, Li Y, Tsuji H, Tamao K (2011) Bull Chem Soc Jpn 84:1178
10. (a) Gates DP (2005) Top Curr Chem 250:107. (b) Baumgartner T, Réau R (2006) Chem Rev 106:4681. (c) Scheschkewitz D (2009) Chem Eur J 15:2476. (d) Matsuo T, Li B, Tamao K (2010) C R Chemie 13:1104. (e) Bates JI, Dugal-Tessier J, Gates DP, Dalton Trans 39:3151.

(f) Matsuo T, Kobayashi M, Tamao K (2010) Dalton Trans 39:9203. (g) Simpson MC, Protasiewicz JD (2013) Pure Appl Chem 85:801
11. Bejan I, Scheschkewitz D (2007) Angew Chem Int Edn 46:5783. (b) Fukazawa A, Li Y, Yamaguchi S, Tsuji H, Tamao K (2007) J Am Chem Soc 129:14164
12. (a) Kobayashi M, Matsuo T, Fukunaga T, Hashizume D, Fueno H, Tanaka K, Tamao K (2010) J Am Chem Soc 132:15162. (b) Kobayashi M, Hayakawa N, Nakabayashi K, Matsuo T, Hashizume D, FuenoH, Tanaka K, Tamao K (2014) Chem Lett 43:432
13. Watanabe H, Takeuchi K, Fukawa N, Kato M, Goto M, Nagai Y (1987) Chem Lett 16:1341
14. Tamao K, Kobayashi M, Matsuo T, Furukawa S, Tsuji H (2012) Chem Commun 48:1030
15. (a) Bally T, Masamune S (1980) Tetrahedron 36:343. (b) Maier G Angew Chem Int Edn Engl 27:309. (c) Bally T (2006) Angew Chem Int Edn 45:6616
16. (a) Kimling H, Krebs A Angew Chem Int Edn Engl 11:932. (b) Irngartinger H, Rodewald H (1974) Angew Chem Int Edn Engl 13:740. (c) Delbaere LTJ, James MNG, Nakamura N, Masamune S (1975) J Am Chem Soc 97:1973. (d) Maier G, Pfriem S, Schäfer U, Matusch R (1978) Angew Chem Int Edn Engl 17:520. (e) Petersson EJ, Fanuele JC, Nimlos MR, Lemal DM, Ellison GB, Radziszewski JG (1997) J Am Chem Soc 119:11122. (f) Wu JI, Evangelista FA, Schleyer PR (2010) Org Lett 12:768. (g) Legrand Y-M, van der Lee A, Barboiu M (2010) Science 329:299
17. Bonačić-Koutecký V, Koutecký J, Michl J (1987) Angew Chem Int Ed Engl 26:170
18. Yates BF, Clabo DA Jr, Schaefer HF III (1988) Chem Phys Lett 143:421
19. (a) Wiberg N, Finger CMM, Polborn K (1993) Angew Chem Int Edn Engl 32:1054. (b) Ichinohe M, Toyoshima M, Kinjo R, Sekiguchi A (2003) J Am Chem Soc 125:13328
20. (a) Takanashi K, Lee VY, Matsuno T, Ichinohe M, Sekiguchi A (2005) J Am Chem Soc 127:5768. (b) Lee VY, Takanashi K, Matsuno T, Ichinohe M, Sekiguchi A (2004) J Am Chem Soc 126:4758
21. Chen Z, Wannere CS, Corminboeuf C, Puchta R, Schleyer PVR (2005) Chem Rev 105:3842
22. Tsurusaki A, Iizuka C, Otsuka K, Kyushin S (2013) J Am Chem Soc 135:16340
23. Szilvási T, Veszprémi T (2012) Organometallics 31:3207

Chapter 24
Tin-Containing π-Electron Systems

Yoshiyuki Mizuhata

Abstract Introduction of tin atom into the carbon π-conjugated systems is expected to express the unique properties derived from their small HOMO–LUMO gaps and so on. However, the chemistry of tin-containing π-conjugated systems has been less studied in comparison with the silicon- and germanium-containing systems due to their instability. Recently, the development of the synthetic and stabilization methods enables us to access to the tin-containing systems. In this chapter, the chemistry of tin-containing π-electron systems, stannapentafulvenes, stannacyclopentadienides, and neutral stannaaromatic compounds will be discussed.

Keywords Tin • Kinetic stabilization • Steric protection • Aromaticity

24.1 Introduction

The multiple pπ–pπ bonds between second-row elements of the periodic table, such as boron, carbon, nitrogen, and oxygen, appear in nature and are investigated and utilized widely in chemistry. On the other hand, their heavier congeners had been considered as imaginary compounds for long years. This is because the long bond distances between heavier elements do not allow the sufficient overlapping of *n*p-orbitals, resulting in the formation of σ-bond by self-oligomerization or side reaction with other molecules [1].

The isolation of the stable distannene (Sn = Sn, **1**) in 1973 [2–5] was milestone in this chemistry, which was followed by those of phosphaalkene (P = C, **2**) in 1978 [6] and silene (Si = C, **3**) [7], diphosphene (P = P, **4**) [8], and disilene (Si = Si, **5**) [9] in 1981 (Chart 24.1). In all cases, introduction of bulky substituents on the central atoms prevented them from oligomerization to make such reactive species isolable as stable compounds. These researches revealed that the double-bond compounds of heavier main group elements could be isolated as stable species without oligomerization or other side reactions when they were well "kinetically stabilized" with bulky substituents. Subsequently, the chemistry of unsaturated

Y. Mizuhata (✉)
Institute for Chemical Research, Kyoto University, Gokasho, Uji, Kyoto 611-0011, Japan
e-mail: mizu@boc.kuicr.kyoto-u.ac.jp

T. Akasaka et al. (eds.), *Chemical Science of π-Electron Systems*,
DOI 10.1007/978-4-431-55357-1_24

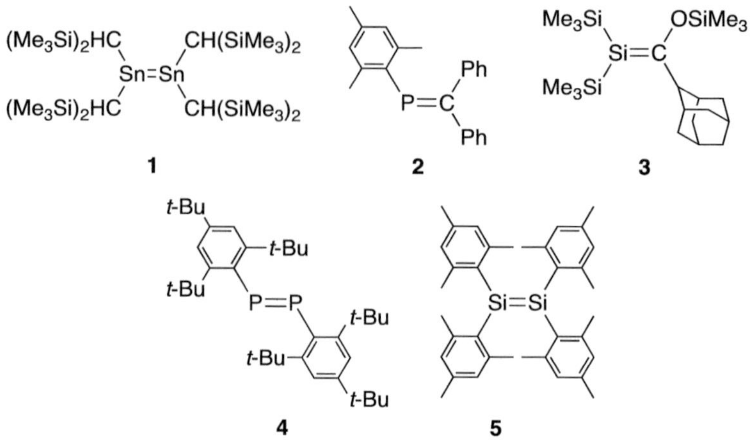

Chart 24.1 Pioneering works in the chemistry of doubly bonded systems containing (a) heavier main group element(s)

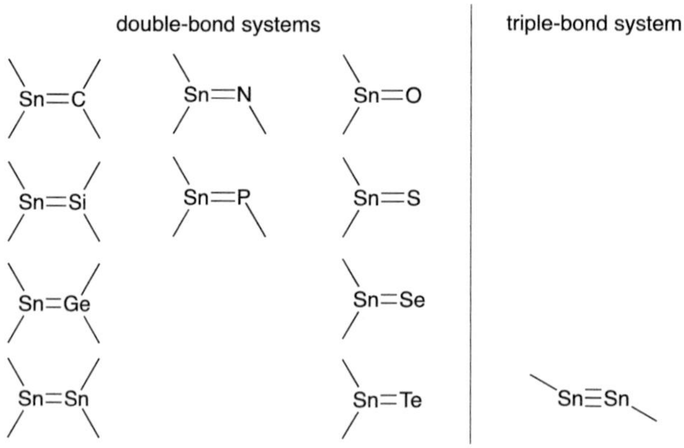

Chart 24.2 Tin-containing multiple-bond systems reported previously

compounds of heavier main group elements, especially in the field of group 14 elements, showed dramatic development by taking advantages of steric protection.

As for tin-containing double bonds, the bonds between tin and groups 14 (C, Si, Ge, Sn), 15 (N, P), and 16 (O, S, Se, Te) elements have been successfully synthesized and fully characterized (Chart 24.2). In addition, triply bonded systems between two tin atoms have been reported by Power and coworkers. These have been reviewed extensively elsewhere [10–12], and for the sake of brevity, I limit discussion to the compounds containing the tin atom cooperated in carbon π-conjugated systems.

24.2 Stannapentafulvenes

(Penta)fulvene is one of the valence isomers of benzene, C_6H_6, and regarded as the important model compound of cyclic cross-conjugated systems. Fulvene and its derivatives show characteristic colors and have high potential toward material science. While the replacement of skeletal carbon atom of a parent fulvene to tin atom has not been reported, several dibenzo derivatives have been synthesized and characterized by Escudié's and Tokitoh's groups.

The reported dibenzo derivatives, that is, fluorenylidene-stannenes **6–9**, were summarized in Chart 24.3. The generation method can be described as dehydrofluorination of the corresponding fluorostannane **10** by *t*-butyllithium (Scheme 24.1). At the low temperature, the reaction mixture showed red color derived from the corresponding anion **11**. Elimination of LiF occurred at around −10 °C, resulting in the color change to purple of stannene.

Stannene **6** bearing two Dis [bis(trimethylsilyl)methyl] substituents was transient, and its formation was suggested by only the trapping reactions as described below [13]. Stannene **7** bearing two Tip (2,4,6-triisopropylphenyl) groups on the tin atom was synthesized and characterized by its NMR spectroscopy [14, 15]. However, stannene **7** was unstable above 0 °C, resulting in the head-to-tail dimerization. These results showed remarkable difference to the germanium analogue: fluorenylidene-germenes could be isolated as thermally stable compounds by introduction of two Dis [16] or two smaller substituents than Tip, Mes (2,4,6-trimethylphenyl) [17, 18], on the germanium atom.

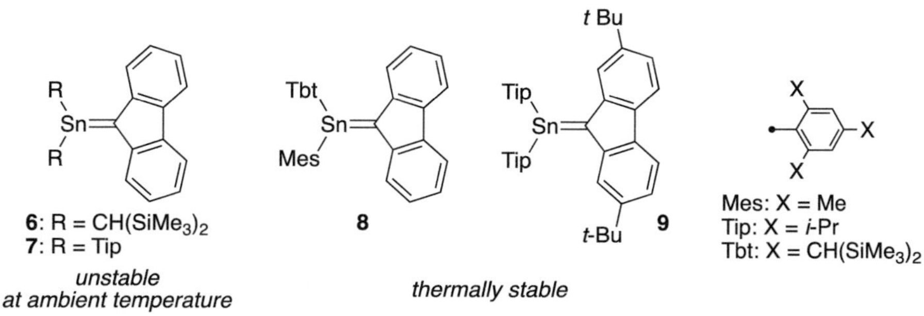

Chart 24.3 Fluorenylidene-stannenes

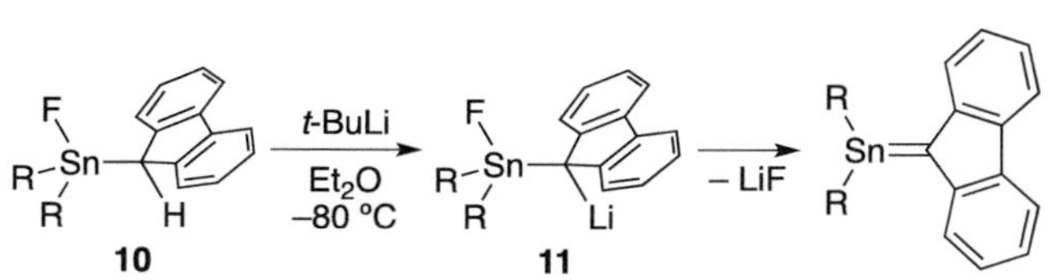

Schema 24.1 Generation of fluorenylidene-stannenes **6–9**

Introduction of bulkier substituent to the tin atom or additional substituents to the fluorenylidene skeleton was effective to stabilize stannene. Tokitoh and coworkers have succeeded in the synthesis and isolation of highly stable stannene **8** by the introduction of bulky substituent, Tbt (2,4,6-tris[bis(trimethylsilyl)methyl]phenyl) [19]. Stannene **8** showed no change at 80 °C in C_6D_6. Stannene **9** reported by Escudié and coworkers has 2,7-di-*tert*-butylfluorenylidene instead of fluorenylidene in **7** and was stable at 120 °C in diethyl ether [20].

In the ^{119}Sn NMR, the signals appeared at 288 (**7**), 270 (**8**), and 277 (**9**) ppm, respectively, suggesting they posses quite similar electronic structures to each other. The sp^2 carbon atoms resonated at 133.85 (**7**), 144.9 (**8**), and 146.9 (**9**) ppm, respectively. In the case of stannene **8**, $^1J_{SnC}$ could be observed, the value of which was 652 (^{119}Sn–^{13}C) and 624 (^{117}Sn–^{13}C) Hz [21].

These stannenes showed deep purple color, and the absorption maxima in the UV-vis spectra were 560 (**8**) and 552 (**9**) nm, respectively. The color was completely different from the germanium analogues (orange, no UV-vis data), suggesting the smaller HOMO–LUMO (π–π*) energy gaps by comparison with the germanium cases.

The structures of **8** and **9** were determined by X-ray crystallographic analysis (Fig. 24.1). The Sn=C bond lengths were 2.016(5) (**8**) and 2.003(5) (**9**) Å, which were shortened about 7 % to the Sn–C single bond (2.14–2.15 Å). These values were in good agreement with the computed values (1.95–2.06 Å) [22–24]. While they have completely trigonal planar geometry around the Sn and C atoms on the Sn=C moiety, torsion angles between two planes are quite different, 28.5° (**8**) and 10° (**9**).

The reactivity has been widely investigated (Schemes 24.2, 24.3, 24.4) [25]. Even thermally stable stannenes possess high reactivity toward small molecules. The reactions with proton or hydride sources afforded the corresponding adducts (Scheme 24.2).

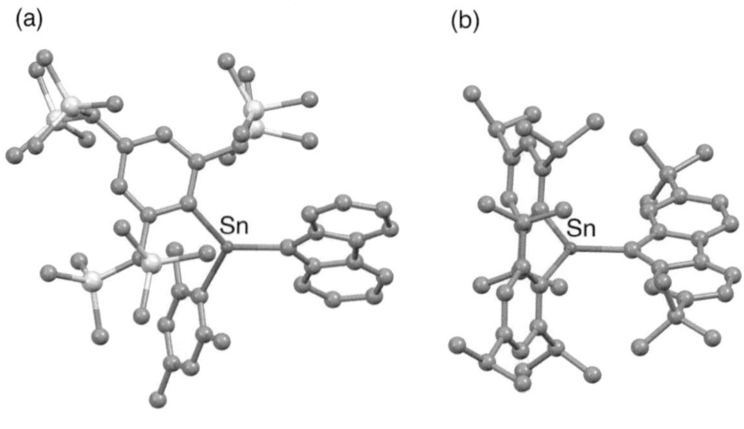

Fig. 24.1 Structures of stannenes **8 (a)** and **9 (b)**

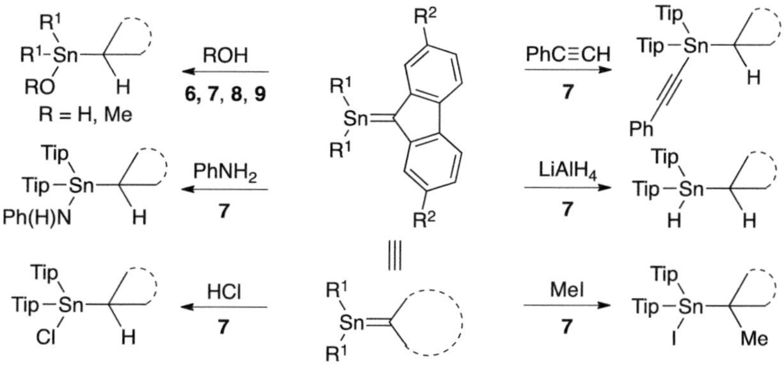

Scheme 24.2 Reactivity of stannenes (i)

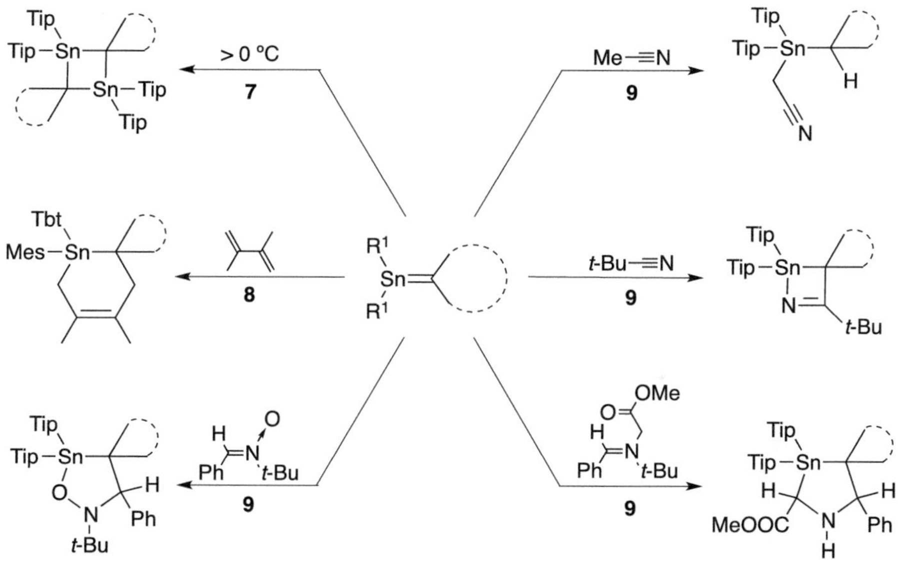

Scheme 24.3 Reactivity of stannenes (ii)

Other reactions mainly on cycloadditions were summarized in Schemes 24.3 and 24.4. As described above, less sterically hindered stannene **7** gave head-to-tail [2 + 2] dimer selectively. The reaction of **8** with 2,3-dimethyl-1,3-butadiene afforded the corresponding [2 + 4] cycloadduct, indicating that the tin–carbon bond in the stannene possesses a concrete π-bond (Scheme 24.3).

The reactions with carbonyl regents gave a variety of tin-containing heterocycles (Scheme 24.4). Especially, the reactions of **9** with *p*-benzoquinone derivatives gave unique bis-adducts regio- and stereo-selectively [26].

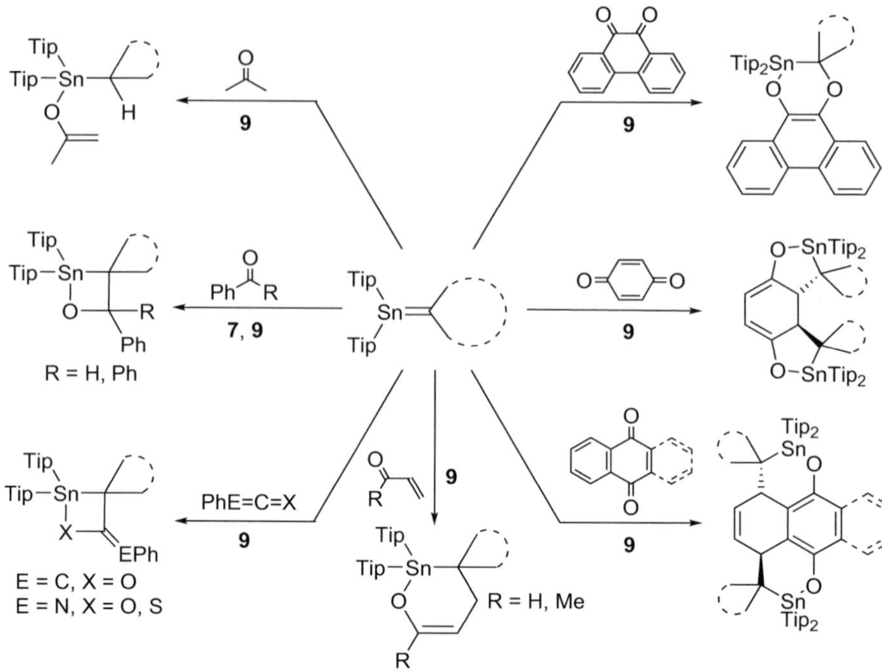

Scheme 24.4 Reactivity of stannenes (iii)

24.3 Stannacyclopentadienide

Cyclopentadienyl anions (Cp^-), needless to say, play a great role in organometallic chemistry as the ligand in transition-metal complexes. At the same time, its 6π-electron structure is important to understand the concept of "aromaticity"[27–29]. Therefore, the introduction of heavier group 14 elements to the Cp^- ring as (a) skeletal element(s) has been investigated for a long time. So far, all the families of group 14 metallole anions and dianions (EC_4^- and EC_4^{2-}, E = Si, Ge, Sn, Pb) have already been synthesized and structurally characterized [30–32]. It should be noted that group 14 metalloles can form more reduced species than Cp^-, metallole dianions ($EC_4R_4^{2-}$), which is in sharp contrast to the fact that such dianionic species have never been synthesized in the all-carbon systems. Importantly, while those dianions show aromatic characters, monoanions can be regarded as nonaromatic compounds.

Introduction of multiple heavier elements to Cp^- ring has been achieved in silicon and germanium cases, and compounds with $Si_3C_2^-$ [33] and $Si_2GeC_2^-$ [34] cores have been synthesized and characterized. Interestingly, these monoanions showed aromatic character on the contrary to the EC_4^- cases.

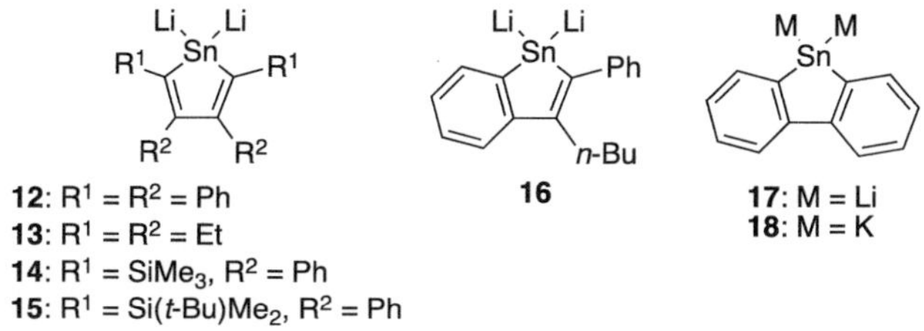

Chart 24.4 Stannole dianions

Scheme 24.5 Synthesis of stannole dianions

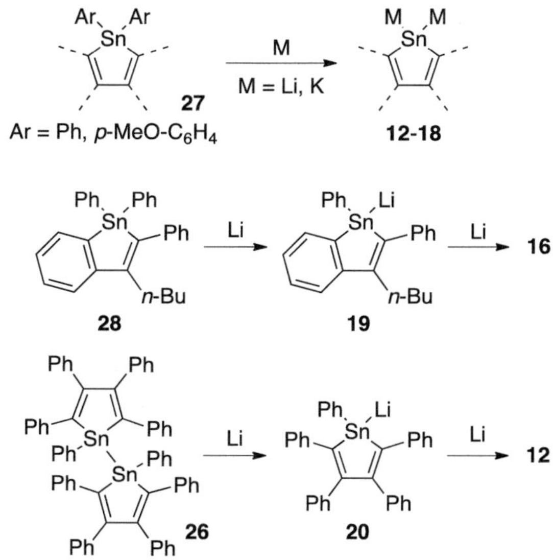

The application to the transition-metal complexes has been developed in the silicon and germanium systems, and η^5-sila-, germa-, trisila-, and disilagerma-cyclopentadienyl (SiC_4 [35], GeC_4 [36, 37], Si_3C_2 [38], and Si_2GeC_2 [39, 40], respectively) complexes have been synthesized by Tilley's and Sekiguchi's groups independently.

A series of stannole dianions **12–18** (Chart 24.4) and monoanions (including 1,2- or 1,3-dianions) **19–25** have been reported by Saito and coworkers. An excellent overview of this chemistry has been provided in a 2012 review article [41, 42].

Mainly, the synthesis of dianions was achieved by the reduction of bi(1,1-stannole)s **26** or of diarylstannoles **27** (Scheme 24.5). Monoanions were synthesized by the partial reduction of **26** or **28** (Scheme 24.5) or the reaction of the dianion with suitable electrophiles or aryl halides (Scheme 24.6). On the other hand, the reaction of tetraethyl-substituted dianion **13** with *t*-butylchloride afforded the 1,2-dianion **22** (Scheme 24.7). Interestingly, crystallization of **22** in the presence of 12-crown-6 gave sandwich-type compound, lithocene **23**, which is the first metallocene bearing the Cp-like ligands directly bonded without any bridging groups [43].

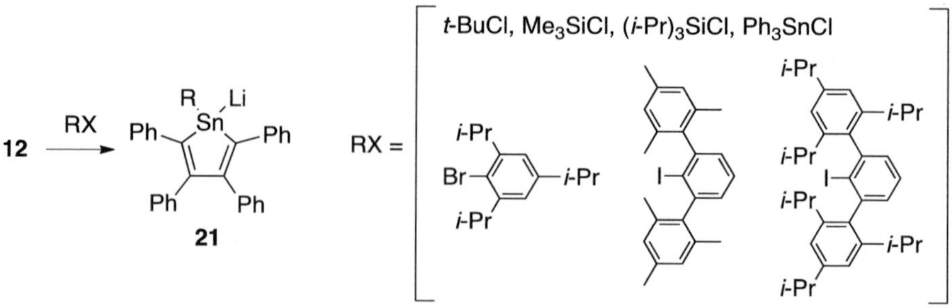

Scheme 24.6 Reactions of stannole dianion **12**

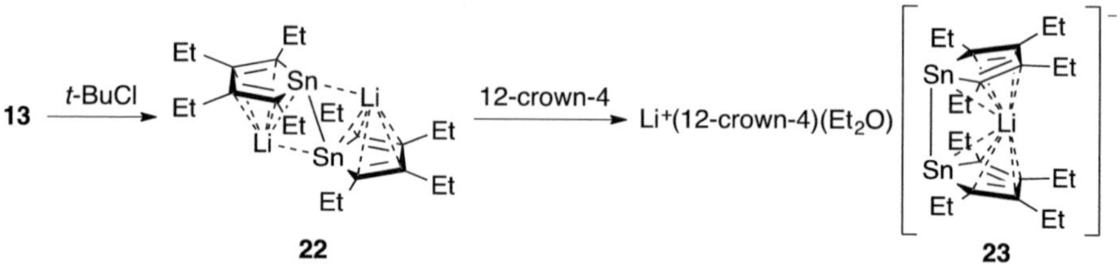

Scheme 24.7 Reaction of stannole dianion **13**

Scheme 24.8 Reversible redox behavior of **12**

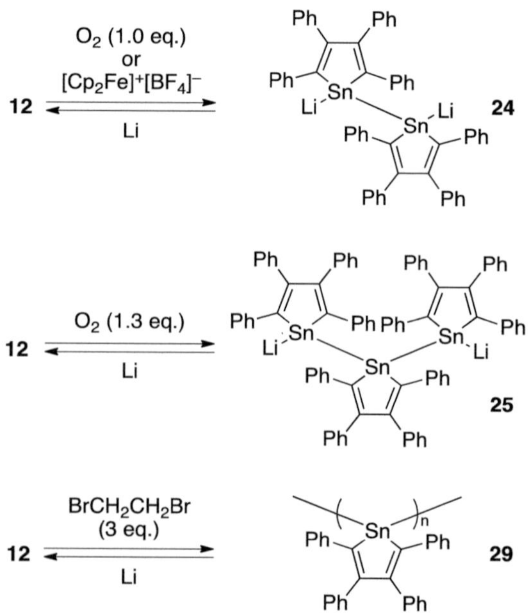

It is noteworthy that dianion **12** showed the reversible redox behavior (Scheme 24.8) [44, 45]. The oxidation of **12** by an equimolar amount of oxygen or $[Cp_2Fe]^+[BF_4]^-$ resulted in the formation of the 1,2-dianion **24**. The reaction of **12** with small excess of oxygen, that is, the oxidation with 1.3 equivalent of oxygen,

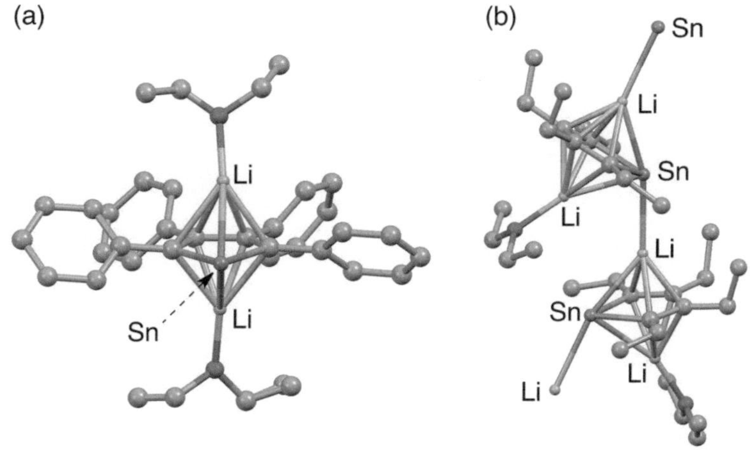

Fig. 24.2 Structures of dilithium stannoles **12** (**a**) and **13** (**b**)

afforded the terstannole-1,3-dianion **25**. On the other hand, the oxidation of **12** with 3 equivalent of 1,2-dibromoethane gave poly(1,1-stannole) **29**. The treatment of **24**, **25**, and **29** with lithium metal regenerates dianion **12**.

Crystallographic analysis of dianions **12–15** (**12** and **13** in Fig. 24.2) revealed the planarity of the stannole ring and the unique η^5,η^5-interaction mode. The lengths of the C–C bonds in the stannole rings were almost equal to each other, and the negative charges are considerably delocalized on the stannaaromatic ring. The lengths of the two Sn–C bonds in the rings were close to or longer than the typical Sn–C single-bond length (2.14–2.15 Å).

The ^{7}Li nuclei of **12–15** resonated at relatively upfield region, −4.4 (**12**), −5.2 (**13**), −5.8 (**14**), and −5.4 (**15**) ppm, respectively. This upfield resonance is caused by the strong shielding effect of the diatropic ring current resulting from the 6π-electron system, indicating their aromatic nature. On the other hand, their ^{119}Sn NMR chemical shifts were quite different, +163.3 (**12**), −27.8 (**13**), +446.4 (**14**), and +472.6 (**15**) ppm. The lower-field shifts in **14** and **15** can be explained in terms that silyl substitution enhanced the stannylene character of **14** and **15** due to greater interaction between 5p orbital on the Sn atom and LUMO on the butadiene moiety in comparison to those in alkyl and aryl derivatives. The Sn(II) character in **14** and **15** was supported by the ^{119}Sn Mössbauer spectra.

The structures of several monoanions including 1,2- or 1,3-dianions, that is, tin analogues of the cyclopentadienyl anion, were revealed (Figs. 24.3 and 24.4). Contrary to the stannole dianions, the bond alternation for the C–C bonds is observed in the five-membered ring, indicating that the monoanions have considerable diene character. The two lengths of Sn–C bonds are longer than not only those of dianions but also that of typical Sn–C single-bond length (2.14 Å). The geometry around the tin atoms is clearly pyramidalized between the SnC_4 plane and the Sn–X (X=Sn,

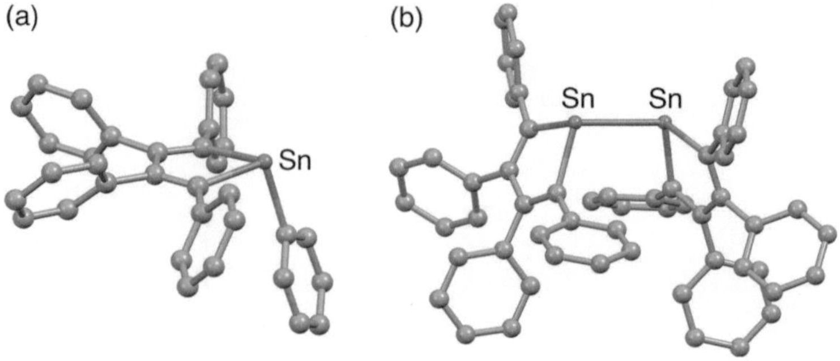

Fig. 24.3 Structures of **20** (**a**) and **24** (**b**). Counter cations and solvated molecules were omitted for clarity

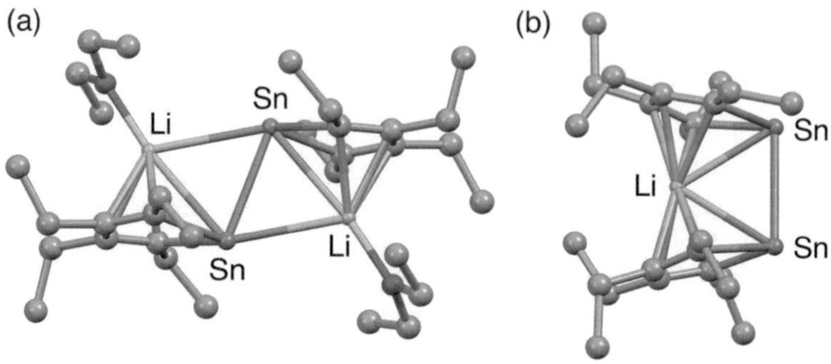

Fig. 24.4 Structures of **22** (**a**) and **23** (**b**). In (**b**), one of the counter cations and solvated molecules was omitted for clarity

Si, C) bond. These results indicated the lower aromaticity of monoanions than that of dianions.

While the ring structure of lithocene **23** showed bond alternation and diene character similarly to other monoanions, the signals in the ^{7}Li NMR appeared at −7.1 and −0.6 ppm, the former value of which was characteristic for the aromatic compounds. Theoretical calculations revealed that the lithium atom inside the sandwich structure is considerably shielded, as that on the aromatic ring.

Since 2013, the chemistry of transition-metal complexes of stannole anions has been developed dramatically.

The reaction with dilithium tetraethylstannole **13** with one equivalent of Cp_2TiCl_2 afforded the novel three-membered distannene complex **30** (Scheme 24.9, Fig. 24.5a) [46]. Complex **30** has unusual electronic structure with the contribution of a novel structure described as delocalized σ orbital over the $TiSn_2$ ring **30'**. The reaction of **13** with 0.5 equivalent of Cp_2TiCl_2 or that of **30** with **13** afforded the six-membered complex **31**. The latter reactivity indicated that compound **13** is a useful homogeneous one-electron reductant.

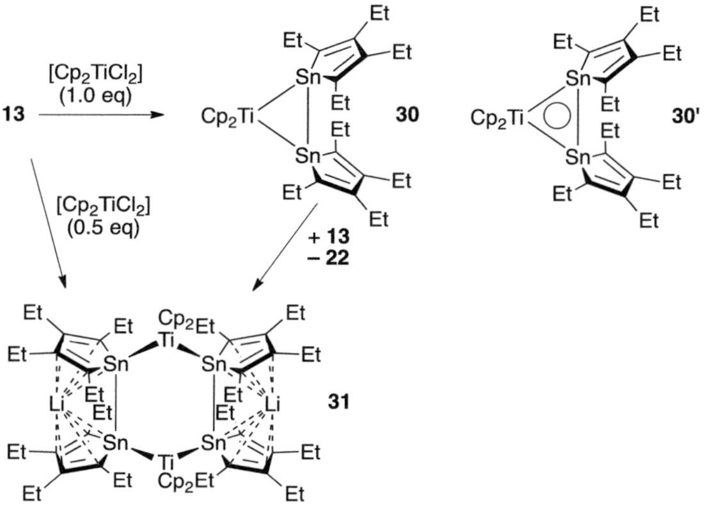

Scheme 24.9 Reactions of **13** with [Cp_2TiCl_2]

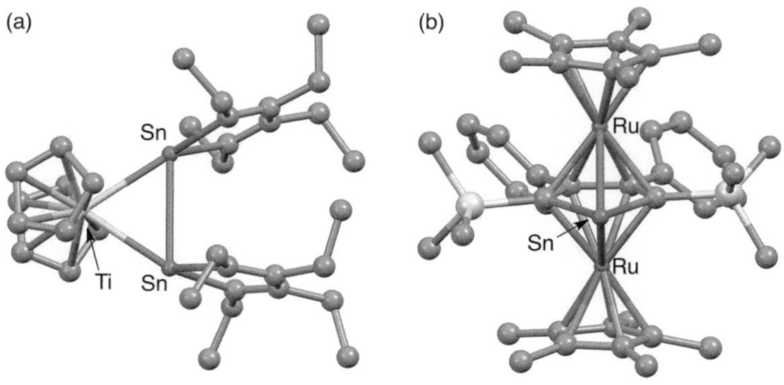

Fig. 24.5 Structures of **30** (**a**) and **35** (**b**)

The reactions of dilithium stannoles with [Cp*RuCl]$_4$ showed the unique substituent dependency. While those of tetraethyl-substituted derivative **13** afforded complexes **32** and **33** bearing a Ru_2Sn_4 core (Scheme 24.10) [47], η^5-type sandwich complexes **34**–**36** were obtained in the cases using silyl-substituted derivatives **14** and **15** (Scheme 24.11) [48]. The Ru–Ru bond length of the butterfly complex **32** [2.3428(6) Å] is the shortest among those of dinuclear ruthenium complexes having Cp or Cp*Ru units. The stannole rings **34**–**36** showed no C–C bond alternation, indicating their retainment of aromatic character (Fig. 24.5b for **35**).

Scheme 24.10 Reactions of 13 with $[Cp^*RuCl]_4$

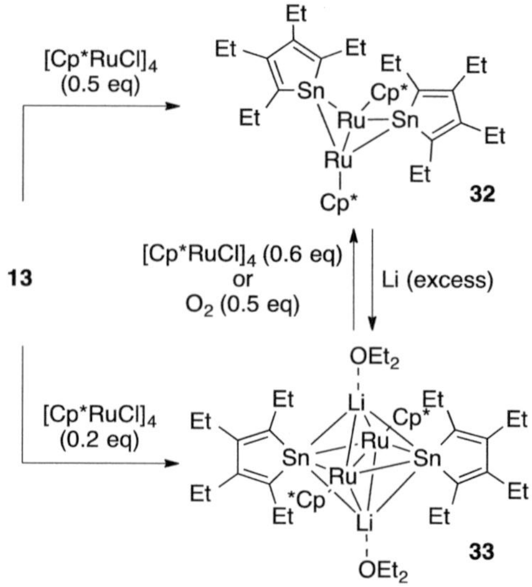

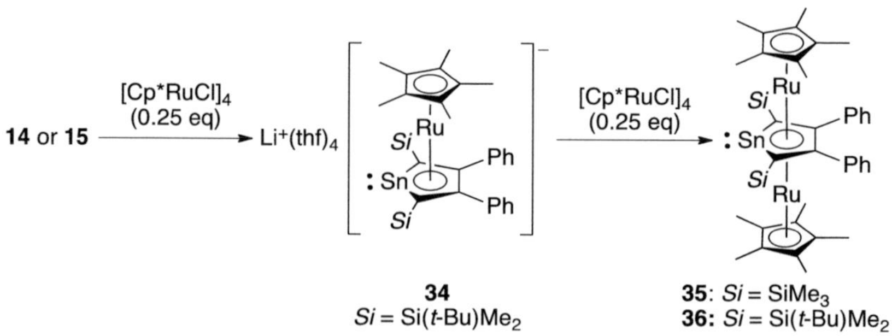

Scheme 24.11 Reactions of **14** and **15** with $[Cp^*RuCl]_4$

24.4 Neutral Stannaaromatic Compounds Based on a Six-Membered Ring

Benzenoid aromatic hydrocarbons as represented by benzene are one of the most important compounds in organic chemistry from the viewpoints of not only fundamental chemistry but also application to organic synthesis and material science. Their chemical behavior, structure, and energetic and magnetic features have been widely studied from both experimental and theoretical standpoints, and such findings have contributed to understand the concept, "aromaticity" [27, 28]. Although a variety of heteroaromatic compounds, in which a skeletal carbon atom of a benzenoid aromatic compound is replaced with a heteroatom, have been synthesized and focused for a long time to discuss "aromaticity," the analogues of heavier group 14 elements (Si, Ge, Sn, and Pb) have been recognized as highly reactive species [30]. For instance, the parent silabenzene ($HSiC_5H_5$) has been

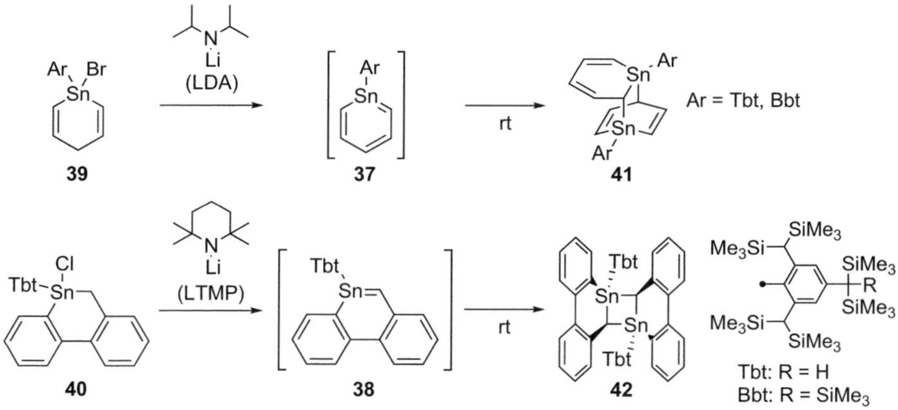

Scheme 24.12 Generation of stannabenzenes **37** and 9-stannaphenanthrene **38**

generated by flash thermolysis and studied in the gas phase and in low-temperature matrices, but it was found to oligomerize above 80 K [49–51]. This result marks a sharp contrast with the high thermal stability of phosphabenzene containing a phosphorus atom, which is the element in the same row as silicon [52].

Since late 1990s, the chemistry of heteroaromatic compounds of heavier group 14 elements showed remarkable development. Tokitoh and coworkers have succeeded in the synthesis and isolation of the first stable examples for neutral sila- and germaaromatic compounds [53–57] by taking advantage of kinetic stabilization afforded by Tbt group. Sekiguchi's and Tokitoh's groups have reported the synthesis and isolation of 1,2-disilabenzenes by the reaction of stable disilynes (–Si ≡ Si–) with acetylenes [58, 59]. In view of the recent progress in the chemistry of neutral sila- and germaaromatic compounds, the synthesis of neutral stannaaromatic compounds is of great interest from the standpoint of systematic elucidation of the properties of heteroaromatic systems of heavier group 14 elements.

Neutral stannaaromatic compounds, bearing a Tbt or 2,6-bis[bis(trimethylsilyl) methyl]-4-[tris(trimethylsilyl)methyl]phenyl (Bbt) group, have been reported by Tokitoh and coworkers. The generation of stannabenzenes **37** [60] and 9-stannaphenanthrene **38** [61] in the reaction of the corresponding halostannanes **39** and **40** with base was suggested by the generation of their dimer and/or the trapping experiments (Scheme 24.12). As mentioned above, compounds **37** and **38** were thermally unstable and readily undergo dimerization to give **41** and **42** at room temperature. The instability of **37** and **38** is in sharp contrast to the high stability of the corresponding sila and germa derivatives bearing a Tbt group, which are stable at 100 °C in C_6D_6.

By contrast, 2-stannanaphthalene **43** was thermally stable, showing no change under heat at 80 °C in C_6D_6 [62, 63]. The synthesis of **43** was achieved by the reaction of the corresponding bromostannane **44** with lithium diisopropylamide (Scheme 24.13). The X-ray crystallographic analysis of **43** (Fig. 24.6a) showed the

Scheme 24.13 Synthesis of 2-stannanaphthalene **43**

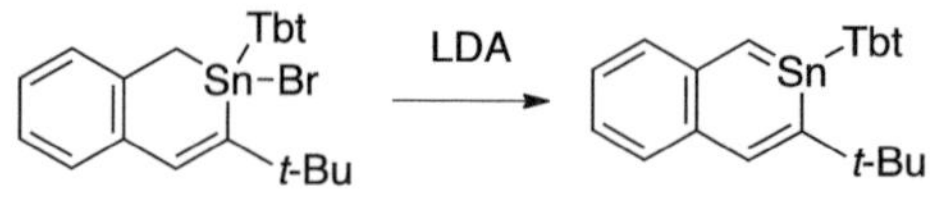

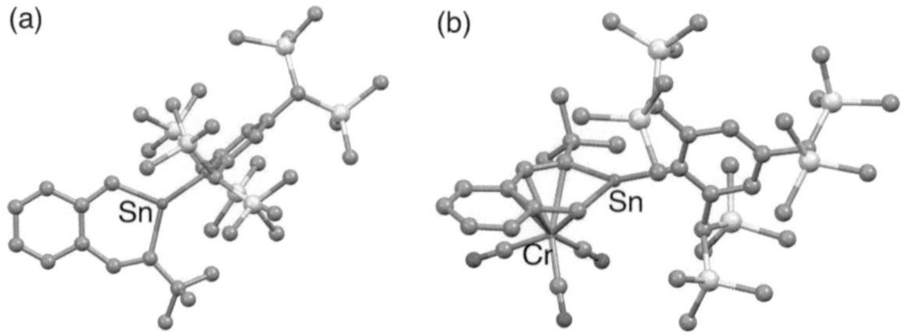

Fig. 24.6 Structures of **43** (**a**) and **45** (**b**)

perfect planarity of the 2-stannanaphthalene ring and the bond alternation, which are essentially the same features as those of a parent naphthalene. The lengths of two Sn–C bonds in the rings were 2.029(6) (Sn–1-C) and 2.081(6) (Sn–3-C) Å, respectively, the values of which were between those in stannenes **8** [2.016(5) Å] and **9** [2.003(6) Å] and typical Sn–C single-bond length (2.14–2.15 Å). In the ^{119}Sn NMR spectrum, the tin atom resonated at 264 ppm, which is very close to those of stannenes **7**–**9**. In the ^{1}H and ^{13}C NMR spectra, all the signals assigned for the 2-stannanaphthalene ring were observed in low-field region. The chemical shifts for 1-H and 4-H were 9.28 and 8.75 ppm, the former value of which was dramatically lower shifted in comparison with the corresponding values for silicon (7.40 ppm) and germanium [8.24 (for 3-H-derivative) and 8.34 (for 3-*t*-Bu-derivative) ppm] analogues. Similarly, that for 1-C (147.38 ppm) was lower shifted than those for silicon (128.19 ppm) and germanium (127.69 and 135.81 ppm, same as above) analogues.

The reactivity of compounds **38** and **43** showed their character as Sn=C double-bond compounds (Scheme 24.14) [64].

On the other hand, the ligand-exchange reaction of **43** with [$Cr(CH_3CN)_3(CO)_3$] resulted in the formation of the corresponding η^6-2-stannanaphthalene complex **45** (Scheme 24.15). The coordination occurred on the SnC_5 ring, and it was retained even in solution. The X-ray crystallographic analysis (Fig. 24.6b) revealed that the two endocyclic Sn–C bonds [2.035(5) (Sn–1-C) and 2.093(4) (Sn–3-C) Å] of the complex of **45** are somewhat elongated when compared to those of free 2-stannanaphthalene **43**. Complex **45** still keeps the planarity for the 2-stannanaphthalene moiety. In the ^{1}H, ^{13}C, and ^{119}Sn NMR spectra of **45**, signals corresponding to the atoms in the SnC_5 ring were shifted upfield relative to those for free 2-stannanaphthalene. The IR spectrum of the 2-stannanaphthalene complex recorded in the solid state showed the presence of three intense bands (1941, 1862, 1851 cm^{-1}) assigned to the carbonyl stretchings, which were observed in the region similar to those of the η^6-naphthalene(tricarbonyl)chromium complex (1941,

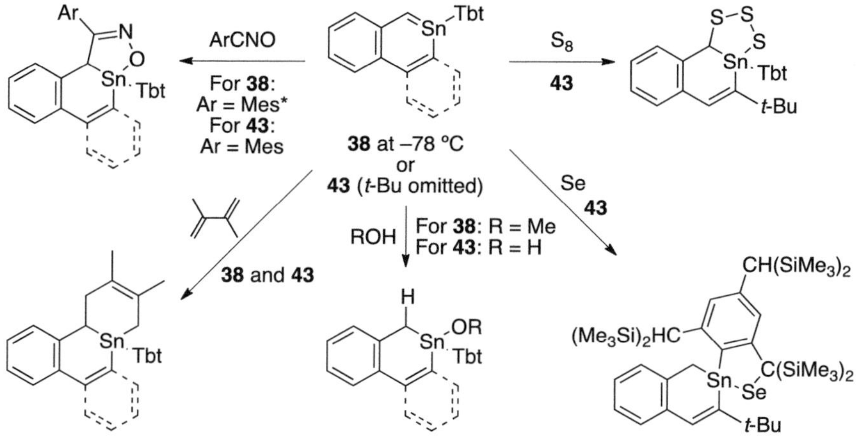

Scheme 24.14 Reactivity of **38** and **43**

Scheme 24.15
Ligand-exchange reaction of 2-stannanaphthalene **43**

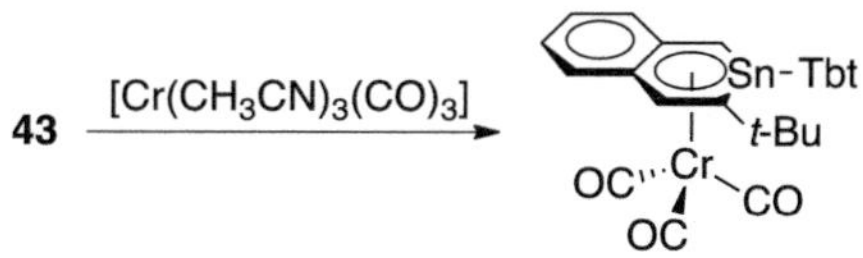

45

1864 cm^{-1}). This result suggested that 2-stannanaphthalene has a coordination ability as an arene ligand similar to that of naphthalene.

Judging from their structural features, NMR, UV-vis, and Raman spectra, and the chemical reactivities, 2-stannanaphthalene **43** was considered to have significant aromatic character.

Acknowledgments The author thanks "Emergence of highly elaborated π-space and its function" [Grant-in-Aid for Scientific Research on Innovative Areas (No. 23108707, "π-Space") from the Ministry of Education, Culture, Sports, Science, and Technology, of Japan] for giving him an overview of the recent development of tin-containing π-electron systems.

References

1. For a review, see: Guselnikov LE, Nametkin NS (1979) Formation and properties of unstable intermediates containing multiple pπ-pπ-bonded group-4b metals. Chem Rev 79:529–577
2. Davidson PJ, Lappert MF (1973): Stabilization of metals in a low coordinative environment using bis(trimethylsilyl)methyl ligand – colored Sn(II) and Pb(II) Alkyls, $M[Ch(SiMe_3)_2]_2$. J Chem Soc Chem Commun 9:317a–317a. http://pubs.rsc.org/en/content/articlelanding/1973/c3/c3973000317a#!divAbstract)
3. Goldberg DE, Harris DH, Lappert MF, Thomas KM (1976) New synthesis of divalent group 4b alkyls $M[Ch(SiMe_3)_2]_2$ (M=Ge or Sn), and crystal and molecular-structure of tin compound. J Chem Soc Chem Commun 12:261–262

4. Goldberg DE, Hitchcock PB, Lappert MF, Thomas KM, Thorne AJ, Fjeldberg T, Haaland A, Schilling BER (1986) Subvalent group-4b metal alkyls and amides.9. Germanium and tin alkene analogs, the dimetallenes M_2R_4 M = Ge or Sn, R = Ch$(SiMe_3)_2$ – X-Ray structures, molecular-orbital calculations for M_2H_4, and trends in the series $M_2R'_4$ M = C, Si, Ge, or Sn – R' = R, Ph, $C_6H_2Me_3$-2,4,6, or $C_6H_3Et_2$-2,6. J Chem Soc Dalton Trans 15:2387–2394
5. Zilm KW, Lawless GA, Merrill RM, Millar JM, Webb GG (1987) Nature of the tin–tin double-bond as studied by solid-state and solution nuclear-magnetic-resonance. J Am Chem Soc 109:7236–7238
6. Klebach TC, Lourens R, Bickelhaupt F (1978) Synthesis of P-mesityldiphenylmethylenephosphine: A stable compound with a localized P = C bond. J Am Chem Soc 100:4886–4888
7. Brook AG, Abdesaken F, Gutekunst B, Gutekunst G, Kallury RK (1981) A solid silaethene – Isolation and characterization. J Chem Soc Chem Commun 17:191–192
8. Yoshifuji M, Shima I, Inamoto N, Hirotsu K, Higuchi T (1981) Synthesis and structure of bis(2,4,6-tri-*tert*-butylphenyl)diphosphene – Isolation of a true phosphobenzene. J Am Chem Soc 103:4587–4589
9. West R, Fink MJ, Michl J (1981) Tetramesityldisilene, a stable compound containing a silicon–silicon double-bond. Science 214:1343–1344
10. Fischer RC, Power PP (2010) π-Bonding and the lone pair effect in multiple bonds involving heavier main group elements: Developments in the new millennium. Chem Rev 110:3877–3923
11. Power PP (1999) π-Bonding and the lone pair effect in multiple bonds between heavier main group elements. Chem Rev 99:3463–3503
12. Power PP (1998) Homonuclear multiple bonding in heavier main group elements. J Chem Soc Dalton Trans 27:2939–2951
13. Anselme G, Couret C, Escudié J, Richelme S, Satgé J (1991) Dibisyl(fluorenylidene)stannene – Evidence of its formation. J Organomet Chem 418:321–328
14. Anselme G, Ranaivonjatovo H, Escudié J, Couret C, Satgé J (1992) A stable compound with a formal tin carbon double-bond – the bis(2,4,6-triisopropylphenyl)(fluorenylidene)stannene. Organometallics 11:2748–2750
15. Anselme G, Declercq J-P, Dubourg A, Ranaivonjatovo H, Escudié J, Couret C (1993) A compound with a tin–carbon double bond, bis(2,4,6-triisopropylphenyl)(fluorenylidene)stannene: Aspects of its reactivity and X-ray structure of its dimer. J Organomet Chem 458:49–56
16. Lazraq M, Couret C, Escudié J, Satgé J, Soufiaoui M (1991) New stable germenes. Polyhedron 10:1153–1161
17. Lazraq M, Escudié J, Couret C, Satgé J, Drager M, Dammel R (1988) $(Mesityl)_2$Ge(fluorenylidene) – stabilization of a Ge–C double-bond by charge-transfer into an aromatic system. Angew Chem Int Ed Engl 27:828–829
18. Couret C, Escudié J, Satgé J, Lazraq M (1987) The 1st stable germene - A compound with a germanium carbon double-bond. J Am Chem Soc 109:4411–4412
19. Mizuhata Y, Takeda N, Sasamori T, Tokitoh N (2005) Synthesis and properties of a stable 6-stannapentafulvene. Chem Commun:5876–5878
20. Fatah A, El Ayoubi R, Gornitzka H, Ranaivonjatovo H, Escudié J (2008) A nearly planar stannene with a reactive tin–carbon double bond. Eur J Inorg Chem:2007–2013
21. Mizuhata Y, Tokitoh N (2010) The first observation of ^1J(Sn–C) coupling constants in tin–carbon double-bond compounds. Appl Organomet Chem 24:902–906
22. Windus TL, Gordon MS (1992) π-Bond strengths of $H_2X = YH_2$ – X = Ge or Sn, and Y = C, Si, Ge, or Sn. J Am Chem Soc 114:9559–9568
23. Jacobsen H, Ziegler T (1994) Nonclassical double-bonds in ethylene analogs – Influence of Pauli repulsion on trans bending and π-bond strength – A density-functional study. J Am Chem Soc 116:3667–3679
24. Dobbs KD, Hehre WJ (1986) π-Bond strengths in main-group olefin analogs. Organometallics 5:2057–2061

25. Fatah A, Ghereg D, Escudié J, Saffon N, Ladeira S, Ranaivonjatovo H (2012) Versatile reactivity of stannene $Tip_2Sn = CR_2$ with unsaturated compounds. Organometallics 31:6148–6153
26. Ghereg D, Ranaivonjatovo H, Saffon N, Gornitzka H, Escudié J (2009) From a stannene and quinones to various tin-containing heterocycles. Organometallics 28:2294–2299
27. Minkin VJ, Glukhovtsev MN, Simkin YB (1994) Aromaticity and antiaromaticity; electronic and structural aspects. Wiley, New York
28. Stanger A (2009) What is ... aromaticity: A critique of the concept of aromaticity-can it really be defined? Chem Commun 2009:1939–1947
29. Schleyer PV, Jiao HJ (1996) What is aromaticity? Pure Appl Chem 68:209–218
30. Lee VY, Sekiguchi A, Ichinohe M, Fukaya N (2000) Stable aromatic compounds containing heavier group 14 elements. J Organomet Chem 611:228–235
31. Lee VY, Sekiguchi A (2007) Aromaticity of group 14 organometallics: Experimental aspects. Angew Chem Int Ed 46:6596–6620
32. Saito M, Sakaguchi M, Tajima T, Ishimura K, Nagase S, Hada M (2010) Dilithioplumbole: A lead-bearing aromatic cyclopentadienyl analog. Science 328:339–342
33. Yasuda H, Lee VY, Sekiguchi A (2009) Si_3C_2-Rings: From a nonconjugated trisilacyclopentadiene to an aromatic trisilacyclopentadienide and cyclic disilenide. J Am Chem Soc 131:6352–6353
34. Lee VY, Kato R, Ichinohe M, Sekiguchi A (2005) The heavy analogue of CpLi: Lithium 1,2-disila-3-germacyclopentadienide, a 6π-electron aromatic system. J Am Chem Soc 127:13142–13143
35. Dysard JM, Tilley TD (1998) η^5-Silolyl and η^5-germolyl complexes of d(0) hafnium. Structural characterization of an η^5-silolyl complex. J Am Chem Soc 120:8245–8246
36. Freeman WP, Tilley TD, Rheingold AL, Ostrander RL (1993) A stable η^5-Germacyclopentadienyl complex – $[(\eta^5\text{-}C_5Me_5)Ru(\eta^5\text{-}C_4Me_4GeSi(SiMe_3)_3)]$. Angew Chem Int Ed Engl 32:1744–1745
37. Freeman WP, Dysard JM, Tilley TD, Rheingold AL (2002) Synthesis and reactivity of η^5-germacyclopentadienyl complexes of iron. Organometallics 21:1734–1738
38. Yasuda H, Lee VY, Sekiguchi A (2009) η^5-1,2,3-Trisilacyclopentadienyl – A ligand for transition metal complexes: Rhodium half-sandwich and ruthenium sandwich. J Am Chem Soc 131:9902–9903
39. Lee VY, Kato R, Sekiguchi A, Krapp A, Frenking G (2007) Heavy ferrocene: A sandwich complex containing Si and Ge atoms. J Am Chem Soc 129:10340–10341
40. Lee VY, Kato R, Sekiguchi A (2013) Heavy metallocenes of the group 8 metals: Ferrocene and ruthenocene derivatives. Bull Chem Soc Jpn 86:1466–1471
41. Saito M (2012) Challenge to expand the concept of aromaticity to tin- and lead-containing carbocyclic compounds: Synthesis, structures and reactions of dilithiostannoles and dilithioplumbole. Coord Chem Rev 256:627–636
42. Kuwabara T, Guo JD, Nagase S, Minoura M, Herber RH, Saito M (2014) Enhancement of stannylene character in stannole dianion equivalents evidenced by NMR and Mössbauer spectroscopy and theoretical studies of newly synthesized silyl-substituted dilithiostannoles. Organometallics 33:2910–2913
43. Saito M, Kuwabara T, Ishimura K, Nagase S (2011) Synthesis of a novel lithocene that has aromatic-like nature with nonaromatic rings. Chem Asian J 6:2907–2910
44. Haga R, Saito M, Yoshioka M (2006) Reversible redox behavior between stannole dianion and bistannole-1,2-dianion. J Am Chem Soc 128:4934–4935
45. Haga R, Saito M, Yoshioka M (2008) Stepwise oxidation of the stannole dianion. Chem Eur J 14:4068–4073
46. Kuwabara T, Guo JD, Nagase S, Saito M (2014) Diversity of the structures in a distannene complex and its reduction to generate a six-membered Ti_2Sn_4 ring complex. Angew Chem Int Ed 53:434–438
47. Kuwabara T, Saito M, Guo JD, Nagase S (2013) Unexpected formation of Ru_2Sn_2 bicyclic four-membered ring complexes with butterfly and inverse-sandwich structures. Inorg Chem 52:3585–3587

48. Kuwabara T, Guo JD, Nagase S, Sasamori T, Tokitoh N, Saito M (2014) Synthesis, structures, and electronic properties of triple- and double-decker ruthenocenes incorporated by a group 14 metallole dianion ligand. J Am Chem Soc 136:13059–13064
49. Maier G, Mihm G, Reisenauer HP (1980) Silabenzene. Angew Chem Int Ed Engl 19:52–53
50. Solouki B, Rosmus P, Bock H, Maier G (1980) Analysis of optimization of gas-phase reactions.15. Short-path pyrolysis – silabenzene. Angew Chem Int Ed Engl 19:51–52
51. Maier G, Mihm G, Reisenauer HP (1982) Thermal dehydrogenation of 1-sila-2,5-cyclohexadiene to silabenzene. Chem Ber Rec 115:801–803
52. Märkl G (1990) In multiple bonds and low coordination in phosphorus chemistry. In: Regitz M, Scherer OJ (eds) Multiple and low coordination in phosphorus chemistry. Thieme, Stuttgart, chapter D.5.
53. Tokitoh N (2004) New progress in the chemistry of stable metallaaromatic compounds of heavier group 14 elements. Acc Chem Res 37:86–94
54. Mizuhata Y, Tokitoh N (2011) Chemistry of aromatic compounds containing a heavier group 14 element. J Synth Org Chem Jpn 69:691–704
55. Sasamori T, Inamura K, Hoshino W, Nakata N, Mizuhata Y, Watanabe Y, Furukawa Y, Tokitoh N (2006) Synthesis and characterization of two isomers of 14π-electron germaaromatics: Kinetically stabilized 9-germaanthracene and 9-germaphenanthrene. Organometallics 25:3533–3536
56. Tokitoh N, Shinohara A, Matsumoto T, Sasamori T, Takeda N, Furukawa Y (2007) Synthesis and properties of a kinetically stabilized 9-silaphenanthrene. Organometallics 26:4048–4053
57. Tanabe Y, Mizuhata Y, Tokitoh N (2010) Synthesis and structure of a heavier congener of biphenyl: 1,1′-Disila-4,4′-biphenyl. Organometallics 29:721–723
58. Kinjo R, Ichinohe M, Sekiguchi A, Takagi N, Sumimoto M, Nagase S (2007) Reactivity of a disilyne $RSi \equiv SiR$ ($R = Si^{i}Pr[CH(SiMe_3)_2]_2$) toward π-bonds: Stereospecific addition and a new route to an isolable 1,2-disilabenzene. J Am Chem Soc 129:7766–7767
59. Han JS, Sasamori T, Mizuhata Y, Tokitoh N (2010) Reactivity of an aryl-substituted silicon-silicon triple bond: 1,2-Disilabenzenes from the reactions of a 1,2-diaryldisilyne with alkynes. Dalton Trans 39:9238–9240
60. Mizuhata Y, Noda N, Tokitoh N (2010) Generation of stannabenzenes and their properties. Organometallics 29:4781–4784
61. Mizuhata Y, Takeda N, Sasamori T, Tokitoh N (2005) Generation of 9-stannaphenanthrene and its reactivities. Chem Lett 34:1088–1089
62. Mizuhata Y, Sasamori T, Takeda N, Tokitoh N (2006) A stable neutral stannaaromatic compound: Synthesis, structure and complexation of a kinetically stabilized 2-stannanaphthalene. J Am Chem Soc 128:1050–1051
63. Mizuhata Y, Sasamori T, Nagahora N, Watanabe Y, Furukawa Y, Tokitoh N (2008) Synthesis and properties of stable 2-metallanaphthalenes of heavier group 14 elements. Dalton Trans 37:4409–4418
64. Mizuhata Y, Tokitoh N (2009) Reactions of a kinetically stabilized 2-stannanaphthalene with elemental sulfur and selenium: Synthesis of novel cyclic chalcogenides containing a tin atom. J Sulf Chem 30:385–392

Part VI
Supramolecular Chemistry of π-Electron Systems

Chapter 25
Supramolecular Structures and Photoelectronic Properties of π-Complexes Composed of Self-Assembling Cyclic Porphyrin Dimers and Fullerenes

Fumito Tani, Hirofumi Nobukuni, Ken-ichi Sakaguchi, and Takuya Kamimura

Abstract We have prepared nickel complexes and freebases of cyclic porphyrin dimers linked by butadiyne or phenothiazine groups. The porphyrin dimers have self-assembling pyridyl groups at the opposite *meso* positions and include fullerenes such as C_{60}, PCBM, C_{70}, and $Li^+@C_{60}$ to give π-complexes. In particular, the phenothiazine-bridged cyclic porphyrin dimers exhibit notable high affinities toward C_{60} ($K_{assoc} > 10^6$ M^{-1}) and C_{70} ($K_{assoc} > 10^7$ M^{-1}). In the crystal structures, the π-complexes of C_{60} and PCBM within the butadiyne-bridged nickel dimer and the phenothiazine-bridged freebase dimer afford self-assembled porphyrin nanotubes containing the linear arrays of the fullerenes. These self-assemblies are formed by the C-H···N hydrogen bonds and π-π interactions of the *meso* pyridyl groups. On the other hand, the π-complexes of C_{60} and C_{70} within the butadiyne-bridged freebase dimer gave the zigzag chains of the fullerenes through van der Waals contacts with each fullerene. These well-ordered C_{60} arrays yield high electron mobilities ($\Sigma\mu > 10^{-1}$ cm^2V^{-1} s^{-1}). The π-complexes of C_{60} and $Li^+@C_{60}$ within the butadiyne-bridged dimers perform photoinduced electron transfer from the porphyrins to the fullerenes to give the CS states. The π-complexes of $Li^+@C_{60}$ with the butadiyne-bridged dimers afford the CS states of remarkably long lifetimes reaching submillisecond order.

Keywords Porphyrin • Fullerene • Supramolecule • Nanotube • Self-assembly

---

F. Tani (✉) • H. Nobukuni • K.-i. Sakaguchi • T. Kamimura
Institute for Materials Chemistry and Engineering, Kyushu University, Hakozaki 6-10-1, Higashi-ku, Fukuoka 812-8581, Japan
e-mail: tanif@ms.ifoc.kyushu-u.ac.jp

T. Akasaka et al. (eds.), *Chemical Science of π-Electron Systems*,
DOI 10.1007/978-4-431-55357-1_25

25.1 Introduction

Supramolecular chemistry has been developed with a basic concept that well-organized self-assemblies of molecules realize unique structures and functions [1]. In particular, self-assemblies of molecules having large π-electron systems are expected to afford integrated useful photo- and electronic functions rather than simple sums of the properties of their components. From this viewpoint, we have tried to construct nanotube structures through the self-assembly of cyclic porphyrin dimers. Porphyrins are well-known compounds of large π-electron systems. For creating organic nanotubes, one of the most rational strategies is to modify rigid cyclic molecules with self-assembling substituents and align them via noncovalent interactions (Scheme 25.1) [2, 3]. The employed cyclic molecules (π-host molecules) have a cavity of nm order and include π-guest molecules by π-π interactions to form π-complexes in both solution and solid state. Further, the resulting π-complexes similarly are assembled by the self-assembling groups of the host molecules to produce highly ordered arrays of π-guest molecules in supramolecular nanotubes. This method possesses a merit that predictable assembled forms of functional π-complexes can be obtained straightforward based on rational molecular designs of π-host molecules having self-assembling groups.

In our study, nickel complexes and freebases of cyclic porphyrin dimers linked by rigid linkers have been prepared and 4-pyridyl groups have been adopted as self-assembling moieties (Fig. 25.1). In addition, fullerenes have been used as π-guest molecules (Fig. 25.2) because efficient complexation can be achieved by strong π-π interactions between curved π-surface of fullerenes and flat π-planes of porphyrins [4–6]. In the assembled structures of their π-complexes, fullerenes are linearly arranged within the inside channels of the self-assembled porphyrin nanotubes (Scheme 25.1). Fullerenes have been recognized as ideal electron acceptors due to their appropriate reduction potentials and small reorganization energy in electron

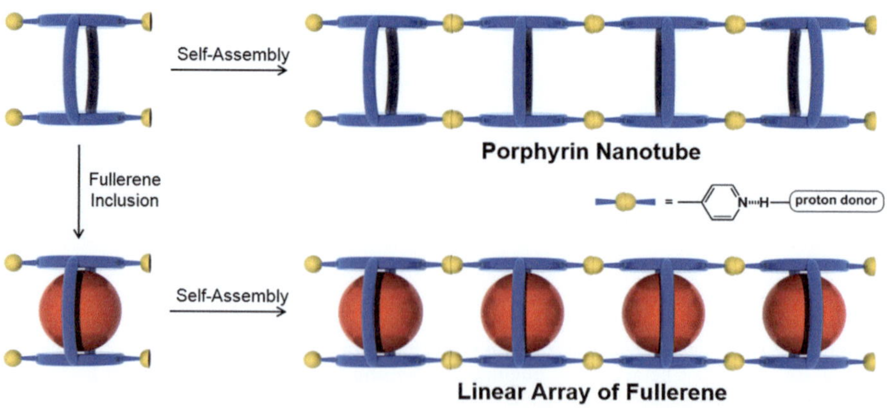

Scheme 25.1 Formation of self-assembled porphyrin nanotube and linear array of fullerene [29]

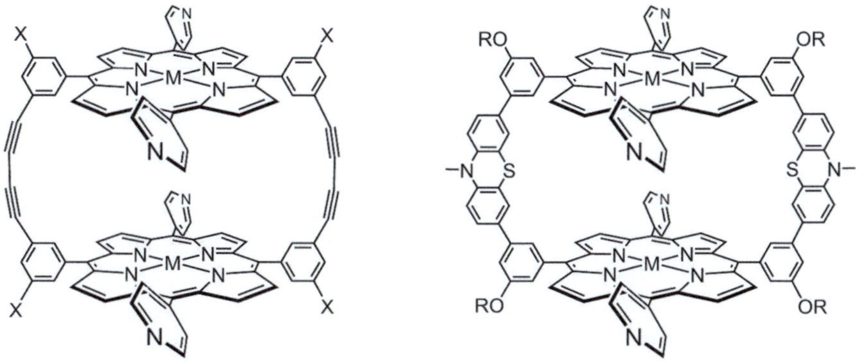

H_4-C_4-CPD_{Py}(H) : M = H_2, X = H
Ni_2-C_4-CPD_{Py}(H) : M = Ni, X = H
H_4-C_4-CPD_{Py}(OC_6) : M = H_2, X = OC_6H_{13}
Ni_2-C_4-CPD_{Py}(OC_6) : M = Ni, X = OC_6H_{13}

H_4-Ptz-CPD_{Py}(OC_6) : M = H_2, OR = OC_6H_{13}
H_4-Ptz-CPD_{Py}(OC_3) : M = H_2, OR = OC_3H_7
Ni_2-Ptz-CPD_{Py}(OC_6) : M = Ni, OR = OC_6H_{13}

Fig. 25.1 Molecular structures of cyclic porphyrin dimers [29]

Fig. 25.2 Fullerenes used in this study

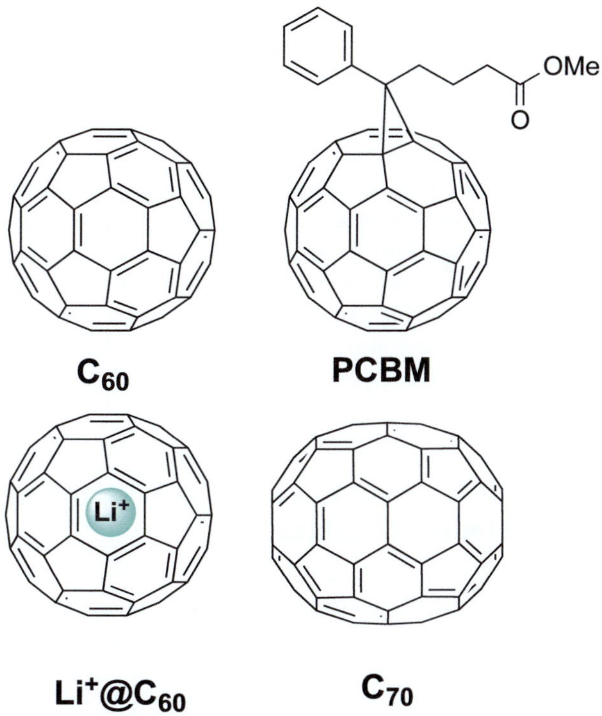

transfer reactions [7–11]. They have full of promise for applications such as acceptors of organic photovoltaics (OPV) and n-semiconductors of molecular electronics [7–11]. In these applications, it is required to produce well-ordered arrangements of fullerenes. Nevertheless, the control of fullerene arrangements has been a difficult problem especially for pristine fullerenes because they are composed of only sp^2-carbons and have no functional groups. Our supramolecular strategy for making arrays of fullerenes is anticipated to provide one solution for this problem. Moreover, porphyrins are good electron donors in their photoexcited states with very strong absorption bands in visible region [12, 13]. A lot of porphyrin-fullerene composites have been prepared as donor-acceptor models for photoinduced charge separation (CS) in the reaction center of natural photosynthesis and as materials for artificial solar energy conversion [12–18]. The conversion from photons to photocurrent of OPV includes three processes: light harvesting, charge separation, and charge transport. It has been suggested that a well-organized donor-acceptor arrangement in the active layer is a suitable structure to obtain the above three processes and improve the efficiency of OPV [19–22]. The π-complexes of our cyclic porphyrin dimers and fullerenes have potentiality for the materials of organic photovoltaics on the basis of the experimental results that they perform photoinduced charge separation and efficient charge transport along the linear arrangements of fullerenes in the self-assembled porphyrin nanotubes.

We have prepared nickel complexes and freebases of cyclic porphyrin dimers linked by butadiyne or phenothiazine groups (Fig. 25.1) [23–30]. The latter linker was subsequently chosen based on the study of the butadiyne-linked cyclic dimers (vide infra). The pyridyl groups were expected to induce self-assembly through complexation with hydrogen-bond donors or metal ions. Four alkoxyl chains were introduced onto the *meso*-phenyl groups to improve the low solubility of the pyridyl-substituted porphyrins. The freebases and nickel complexes of the hexyloxy-substituted cyclic dimers were used for the spectroscopic investigation of the fullerene inclusion in solution, while the non-substituted and propyloxy-substituted dimers were mainly employed for making single crystals for X-ray crystallographic analysis.

25.2 Inclusion of C_{60} and C_{70} Within Cyclic Porphyrin Dimers in Solution

The 1:1 association constants (K_{assoc}) of the cyclic porphyrin dimers with C_{60} and C_{70} were determined by titration on the basis of the change of UV-vis absorption and fluorescence (only for the freebases) spectra in toluene solution (Table 25.1) [29]. Upon addition of fullerenes to the solution of the dimers, the Soret bands (λ_{max} = ca. 420 nm) of the porphyrins were slightly redshifted with decrease of intensity. The fluorescence of the freebase dimers (λ_{max} = ca. 650 and 715 nm) was prominently reduced by addition of fullerenes. The K_{assoc} values of the butadiyne-linked freebase

Table 25.1 Association constants (M^{-1}) of cyclic porphyrin dimers for C_{60} and C_{70} in toluene at 25 °C [29]

Host compound	K_{assoc} for C_{60}	K_{assoc} for C_{70}
H_4-C_4-$CPD_{Py}(OC_6)$	$1.2 \pm 0.2 \times 10^5$	$2.2 \pm 0.5 \times 10^5$
Ni_2-C_4-$CPD_{Py}(OC_6)$	$2.9 \pm 0.4 \times 10^5$	$7.3 \pm 0.9 \times 10^5$
H_4-Ptz-$CPD_{Py}(OC_6)$	$3.9 \pm 0.7 \times 10^6$	[a] $7.4 \pm 0.8 \times 10^7$
Ni_2-Ptz-$CPD_{Py}(OC_6)$	$1.3 \pm 0.2 \times 10^6$	[b] $> 10^7$

[a]Determined by fluorescence method. Others by UV-vis absorption method

[b]Not determined due to high association constants over 10^7 M^{-1}

dimer (H_4-C_4-$CPD_{Py}(OC_6)$) for C_{60} and C_{70} are in the range of 10^5 M^{-1} order. The nickel counterpart (Ni_2-C_4-$CPD_{Py}(OC_6)$) has about two to three times higher affinities for C_{60} and C_{70} than the freebase dimer. C_{70} shows higher K_{assoc} than C_{60} within both butadiyne-linked dimers due to its larger π-plane. These C_{60} and C_{70} affinities are comparable to the average values of reported porphyrinic host molecules for fullerenes [4–6]. On the other hand, the affinity of H_4-Ptz-$CPD_{Py}(OC_6)$ for C_{60} is over 10^6 M^{-1} and much higher than that of H_4-C_4-$CPD_{Py}(OC_6)$. This is the highest C_{60} affinity among ever-reported host molecules which consist of freebase porphyrins. The nickel complex of the phenothiazine-linked dimer (Ni_2-Ptz-$CPD_{Py}(OC_6)$) also exhibits a high C_{60} affinity over 10^6 M^{-1}, although slightly lower than that of H_4-Ptz-$CPD_{Py}(OC_6)$. These remarkably high C_{60} affinities of H_4-Ptz-$CPD_{Py}(OC_6)$ and Ni_2-Ptz-$CPD_{Py}(OC_6)$ indicate that their phenothiazine-linked structures are very effective for C_{60} inclusion. Further, the K_{assoc} values of H_4-Ptz-$CPD_{Py}(OC_6)$ and Ni_2-Ptz-$CPD_{Py}(OC_6)$ for C_{70} are notably high (over 10^7 M^{-1}), revealing that their molecular structures are suitable also for C_{70} inclusion, probably due to the fact that the length of the shorter axis of C_{70} (7.1 Å) is very close to the diameter of C_{60} [31, 32]. The reason for the high affinities of the phenothiazine-linked dimers will be explained below based on the detailed structure analysis. The electrospray ionization mass (ESI-MS) and ^{13}C-NMR spectra of the 1:1 mixtures of fullerenes and the cyclic porphyrin dimers also supported the formation of the inclusion complexes (π-complexes) in solution. The ESI-MS spectra clearly exhibited the ion peak clusters of the π-complexes without those of the free porphyrins and fullerenes. The ^{13}C-NMR spectra showed high-field shifted signals of ^{13}C-enriched fullerenes on account of the ring-current effect of the porphyrins.

25.3 Supramolecular Structures and Photoelectronic Properties of π-Complexes of Ni_2-C_4-CPD_{Py}(H) with C_{60}

The X-ray crystallographic analysis of Ni_2-C_4-CPD_{Py}(H) revealed its molecular structure and self-assembled nanotube in accord with our expectation (Fig. 25.3) [23]. Ni_2-C_4-CPD_{Py}(H) has a molecular cavity of rectangular shape and the

Fig. 25.3 Crystal structure of tubular assembly of Ni_2-C_4-CPD_{Py}(H), (**a**) front view, (**b**) side view [30]

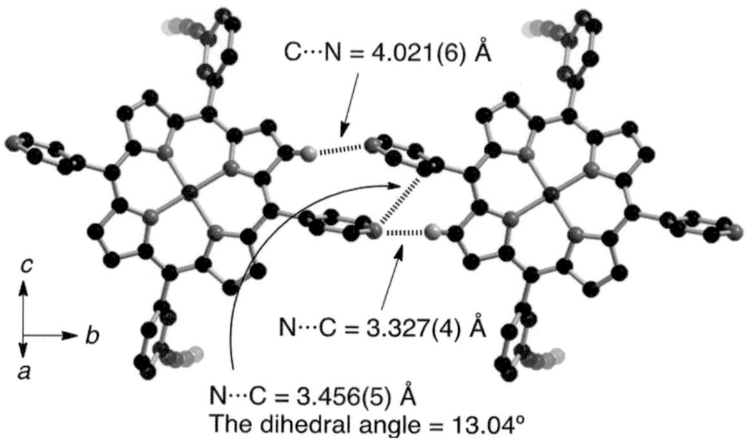

Fig. 25.4 Noncovalent interactions linking Ni_2-C_4-CPD_{Py}(H) for the formation of the self-assembled nanotube [30]

distance between the two nickel ions (center-to-center distance) is 11.635(1) Å. Two molecules of toluene of recrystallization are included in the cavity. The two porphyrin rings are crystallographically equivalent and show ruffled distortions. The *meso* carbon atoms with the phenyl groups are displaced inward and the *meso* carbon atoms with the pyridyl groups are displaced outward. In the crystal packing of Ni_2-C_4-CPD_{Py}(H), a self-assembled tubular structure is formed through two kinds of noncovalent interactions (Fig. 25.4): (1) nonclassical C-H···N hydrogen bonds between the pyrrole β-CH and the nitrogen atoms of the pyridyl groups and (2) π-π interactions of the pyridyl groups. Furthermore, the inclusion complex of Ni_2-C_4-CPD_{Py}(H) with C_{60} ($C_{60} \subset Ni_2$-C_4-CPD_{Py}(H)) gives a tubular assembly formed by the same C-H···N hydrogen bonds and π-π interactions of the pyridyl groups in crystalline state (Fig. 25.5) [23]. The included C_{60} molecules are aligned in the inner channel of this self-assembled porphyrin nanotube. The distance between the centers of the adjacent C_{60} molecules in the nanotube is 14.498 Å. This linear array of C_{60} in the porphyrin nanotube is the first crystal structure of a synthetic supramolecular peapod, which is a chemical analogue to fullerenes aligned

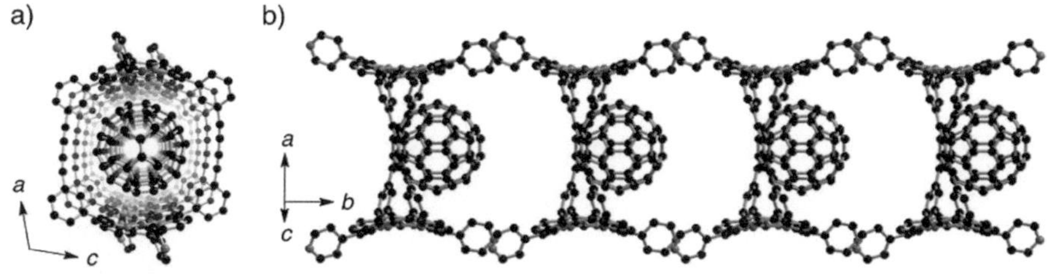

Fig. 25.5 Crystal structure of tubular assembly of $C_{60} \subset Ni_2\text{-}C_4\text{-}CPD_{Py}(H)$, (**a**) front view, (**b**) side view [30]

within carbon nanotubes. There are small changes in the framework of $Ni_2\text{-}C_4\text{-}CPD_{Py}(H)$ upon the inclusion. The two porphyrins have similar ruffled distortions. However, the center-to-center distance is increased to 12.596(2) Å, indicating that the inclusion occurs in an induced-fit fashion. In usual crystal structures of C_{60} cocrystallized with porphyrins, C_{60} molecules are placed just above the centers of porphyrins [4–6]. On the other hand, the C_{60} molecule of $C_{60} \subset Ni_2\text{-}C_4\text{-}CPD_{Py}(H)$ is not located above the centers of the porphyrins because the separation (8.318(1) Å) between the inside *ortho* hydrogen atoms of the opposite phenyl groups is too short for the center location of C_{60} in the cavity. The C_{60} molecule is captured by the two kinds of noncovalent interactions observed in the crystal structure of $C_{60} \subset Ni_2\text{-}C_4\text{-}CPD_{Py}(H)$: (i) $\pi-\pi$ interactions between the six-membered rings of C_{60} and the π-planes of the porphyrins and (ii) multiple C-H $\cdots \pi$ interactions by the C-H bonds of the phenyl and pyridyl groups.

Then, we have investigated the photoelectronic properties of this intriguing linear array of C_{60} in the self-assembled porphyrin nanotube [24]. It is worth noting that the single crystal of $C_{60} \subset Ni_2\text{-}C_4\text{-}CPD_{Py}(H)$ has a remarkably high anisotropic electron mobility ($\Sigma\mu = 0.72\ cm^2V^{-1}\ s^{-1}$) along the array of C_{60} determined by flash photolysis time-resolved microwave conductivity measurements [33]. The C_{60} molecule in the array plays a role of an electron carrier by hopping mechanism. This result clearly demonstrates that the control of fullerene arrangement by the use of host molecules is an effective methodology to realize excellent electronic properties of fullerenes.

Next, the photodynamics of $C_{60} \subset Ni_2\text{-}C_4\text{-}CPD_{Py}(H)$ was investigated by laser flash photolysis to confirm the photoinduced electron transfer from the porphyrins to C_{60} within the π-complexes [24]. Nonetheless, only the triplet exciplex of $C_{60} \subset Ni_2\text{-}C_4\text{-}CPD_{Py}(H)$ rather than the expected charge-separated (CS) state was observed in the time-resolved transient absorption spectra. The reasons for the absence of photoinduced electron transfer are as follows: (1) the rapid intersystem crossing of the nickel porphyrin to its triplet excited state and the following triplet energy transfer from the porphyrin to the ground state of C_{60} and (2) the energy (1.98 eV) of the estimated CS state much higher than the energy (1.57 eV) of the triplet excited state of C_{60} ($^3C_{60}$*). Therefore, we have prepared the corresponding freebase dimer which is considered to be more suitable for photoinduced electron transfer (vide infra).

25.4 Supramolecular Structures and Photoelectronic Properties of π-Complexes of H_4-C_4-CPD_{Py}(H) with C_{60}

In the crystal structure of H_4-C_4-CPD_{Py}(H) (Fig. 25.6) [25], it has a molecular cavity of a rectangular shape as Ni_2-C_4-CPD_{Py}(H). Nevertheless, H_4-C_4-CPD_{Py}(H) shows important differences in comparison with Ni_2-C_4-CPD_{Py}(H): (1) shorter center-to-center distance of the two porphyrins (10.785 Å) and (2) a higher planarity of the porphyrin planes. H_4-C_4-CPD_{Py}(H) is aligned by similar C-H···N hydrogen bonds of the pyridyl nitrogen atoms and π-π interactions of the pyridyl groups to afford a self-assembled porphyrin nanotube, again indicating the effectiveness of this pyridyl-substituted cyclic porphyrin dimer for the formation of tubular assemblies.

On the other hand, $C_{60} \subset H_4$-C_4-CPD_{Py}(H) in crystal shows a different assembled form (Fig. 25.7) [25]. The freebase dimer includes C_{60} in a clamshell conformation, wherein the two porphyrins are tilted with respect to each other. This conformation

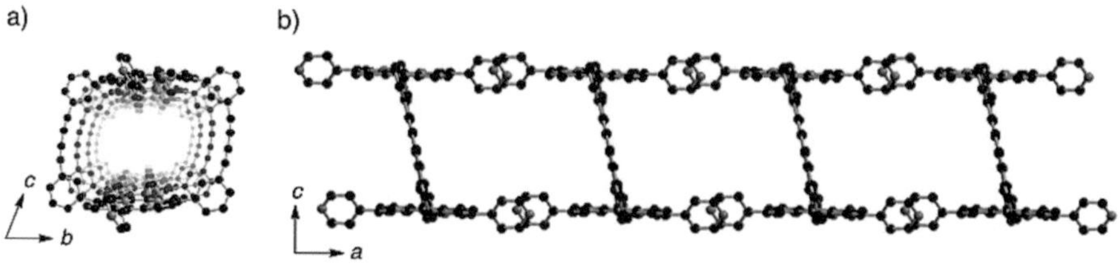

Fig. 25.6 Crystal structure of tubular assembly of H_4-C_4-CPD_{Py}(H), (**a**) front view, (**b**) side view [30]

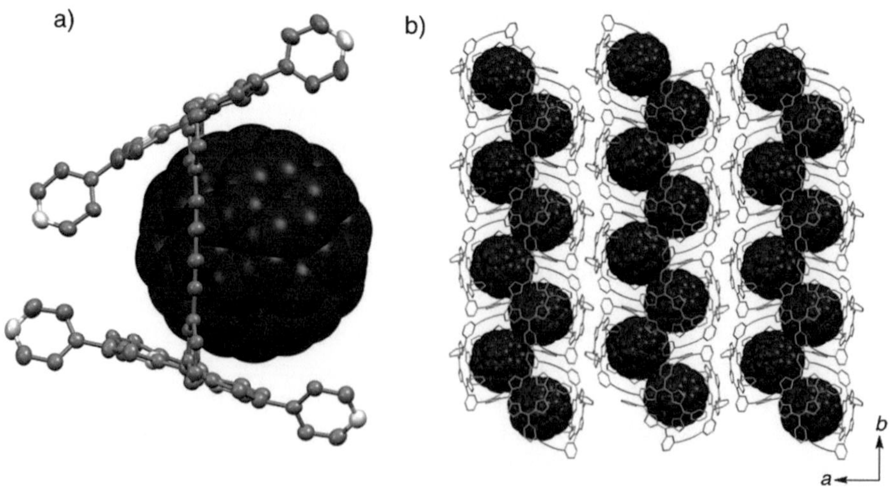

Fig. 25.7 Crystal structure of $C_{60} \subset H_4$-C_4-CPD_{Py}(H), (**a**) side view of the π-complex, (**b**) top view of C_{60} zigzag array [30]

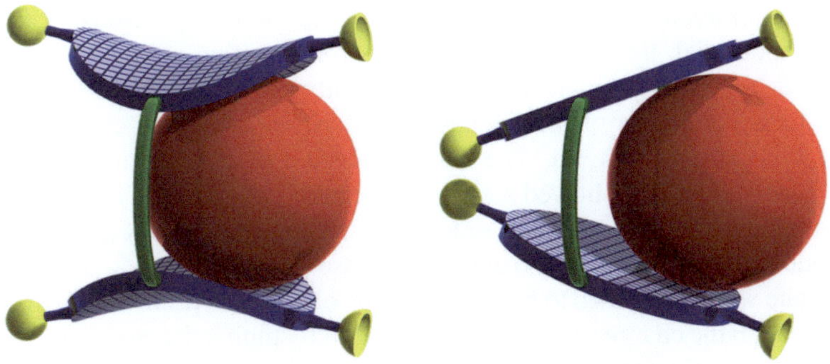

Fig. 25.8 Schematic illustrations of the crystal structures of π-complexes of C_{60} within the butadiyne-bridged cyclic porphyrin dimers [29]

is in clear contrast to the pseudoparallel conformations in the crystal structure of $C_{60} \subset Ni_2$-C_4-CPD_{Py}(H). The C_{60} inclusion induces a large structural change of the freebase dimer, because the center-to-center distance (11.126 Å) is not enough to accommodate C_{60} in a pseudoparallel conformation. Though Ni_2-C_4-CPD_{Py}(H) and H_4-C_4-CPD_{Py}(H) have the same molecular framework, the nickel porphyrin is distorted to afford the longer center-to-center distance (Fig. 25.8). The clamshell conformation of the inclusion complex is unable to give the self-assembled tubular structure through the C-H···N hydrogen bonds of the pyridyl nitrogen atoms and π-π interactions of the pyridyl groups. Instead, in the crystal packing of $C_{60} \subset H_4$-C_4-CPD_{Py}(H), the included C_{60} molecules constitute a zigzag array with the separation (10.018 Å) between the centers of adjacent C_{60}. This value is close to the outer diameter of C_{60} (ca. 10.3 Å), indicating that the C_{60} molecules are aligned with van der Waals contact with each other. The zigzag array in $C_{60} \subset H_4$-C_4-CPD_{Py}(H) results from the partially covered π-surface of C_{60} by H_4-C_4-CPD_{Py}(H). The uncovered π-surface of the spherical C_{60} molecule allows the mutual interaction of C_{60}.

The zigzag array of C_{60} of $C_{60} \subset H_4$-C_4-CPD_{Py}(H) inspired us to measure its charge mobility [25]. The single crystal of $C_{60} \subset H_4$-C_4-CPD_{Py}(H) was applied to flash photolysis time-resolved microwave conductivity measurements. Based on the experimental results, the main charge carrier is electron transported by the aligned C_{60} molecules and the charge mobility along the zigzag array of C_{60} is 0.13–0.16 $cm^2V^{-1}\ s^{-1}$. Although this value is lower than that of $C_{60} \subset Ni_2$-C_4-CPD_{Py}(H), it is a considerably high charge mobility for a crystal of an organic compound, demonstrating the importance of the fullerene alignment.

We investigated the photodynamics of $C_{60} \subset H_4$-C_4-CPD_{Py}(H) by laser flash photolysis to detect the photoinduced electron transfer from the porphyrins to C_{60} within the π-complexes [25]. The freebase dimer is considered to be more suitable for photoinduced charge separation for the following reasons: (1) the lower oxidation potential, that is, the stronger electron-donating ability of the freebase dimer than the nickel dimer, and (2) the greatly slower intersystem crossing of the singlet excited state of the freebase porphyrin to the triplet excited state than that of the nickel porphyrin. The time-resolved transient absorption spectra of $C_{60} \subset H_4\text{-}C_4\text{-}CPD_{Py}(H)$ showed a clear peak of the radical anion of C_{60}, indicating the formation of an expected CS state $\{H_4\text{-}C_4\text{-}CPD_{Py}(H)^{\bullet+} + C_{60}^{\bullet-}\}$ through the photoinduced electron transfer from the freebase porphyrin to C_{60}. The lifetime of the CS state was determined to be 0.47 ps based on the time profile of the C_{60} radical anion. Though the charge separation was realized by tuning the electronic properties of the porphyrin, the present CS lifetime is not satisfactorily long in comparison with the CS lifetime in natural photosynthesis. The improvement of CS lifetime will be described below.

In order to evaluate the efficiency of the photoelectric conversion by the π-complexes, $C_{60} \subset Ni_2\text{-}C_4\text{-}CPD_{Py}(H)$ and $C_{60} \subset H_4\text{-}C_4\text{-}CPD_{Py}(H)$ were employed as photosensitizers in photoelectrochemical cells composed of modified electrodes and I^-/I_3^-electrolyte (Fig. 25.9) [25]. The nanoclusters of the π-complexes were fixed on a transparent electrode by electrophoretic deposition. Upon the illumination of 350–800 nm UV-vis light, prompt photocurrent and steady voltage were observed. The incident photon to current efficiency was determined to be 4 % for $C_{60} \subset Ni_2\text{-}C_4\text{-}CPD_{Py}(H)$ and 17 % for $C_{60} \subset H_4\text{-}C_4\text{-}CPD_{Py}(H)$. The power conversion efficiency was also calculated to be 0.02 % for $C_{60} \subset Ni_2\text{-}C_4\text{-}CPD_{Py}(H)$ and 0.34 % for $C_{60} \subset H_4\text{-}C_4\text{-}CPD_{Py}(H)$. Thus, the photovoltaic activity of $C_{60} \subset H_4\text{-}C_4\text{-}CPD_{Py}(H)$ is much higher than that of $C_{60} \subset Ni_2\text{-}C_4\text{-}CPD_{Py}(H)$, as suggested by their photochemical behaviors described above. Though the photovoltaic efficiencies of these π-complexes are not enough for practical use at present, they are fundamentally candidate materials for photovoltaic applications.

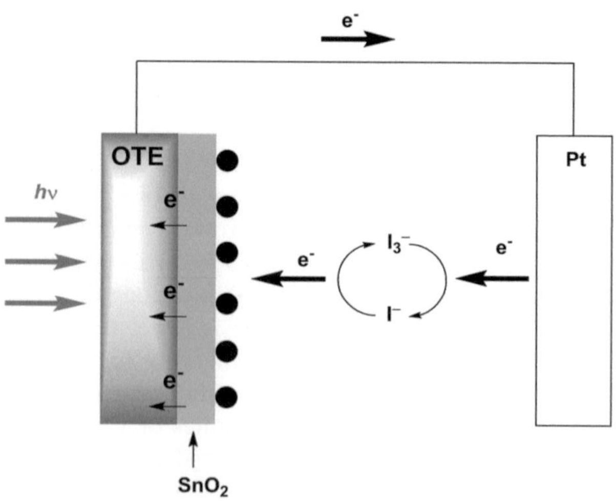

Fig. 25.9 Schematic illustration of the photoelectrochemical cell

25.5 Supramolecular Structures of π-Complexes of Ni_2-C_4-CPD_{Py}(H) and H_4-C_4-CPD_{Py}(H) with [6,6]-Phenyl-C_{61}-butyric Acid Methyl Ester (PCBM)

PCBM, the most representative C_{60} derivative, has been widely used as an electron acceptor in current studies of OPV on account of its suitable LUMO level and favorable solubility [34–36]. However, there have been very few examples for the analysis of supramolecular structures of PCBM at atomic level by X-ray crystallography [37, 38]. It is an intriguing question how the side chain of PCBM affects its supramolecular structures with donor molecules such as the cyclic porphyrin dimers. In the crystal structure of PCBM ⊂ Ni_2-C_4-CPD_{Py}(H) (Fig. 25.10) [27], an analogous nanotube assembly is formed to include the linear array of PCBM molecules as observed in the case of C_{60} ⊂ Ni_2-C_4-CPD_{Py}(H). The interactions for the formation of the self-assembled nanotube are the same as Ni_2-C_4-CPD_{Py}(H) and C_{60} ⊂ Ni_2-C_4-CPD_{Py}(H), C-H···N hydrogen bonds and π-π interactions of the pyridyl groups. The porphyrin moieties of PCBM ⊂ Ni_2-C_4-CPD_{Py}(H) have a high similarity of the molecular structure to the porphyrins in C_{60} ⊂ Ni_2-C_4-CPD_{Py}(H). The terminal part of the side chain is located in the outside space between the adjacent dimers without interfering with the formation of the nanotube. The formation of the similar porphyrin nanotubes in the crystal structures of Ni_2-C_4-CPD_{Py}(H), C_{60} ⊂ Ni_2-C_4-CPD_{Py}(H) and PCBM ⊂ Ni_2-C_4-CPD_{Py}(H) clearly implies that Ni_2-C_4-CPD_{Py}(H) is an effective framework for the construction of self-assembled nanotubular structures and linear arrangements of appropriate C_{60}-based fullerenes.

To investigate the effect of the PCBM side chain on the supramolecular structures, the π-complex of PCBM within H_4-C_4-CPD_{Py}(H) was subjected to X-ray crystallographic analysis [27]. The crystal structure of PCBM ⊂ H_4-C_4-CPD_{Py}(H) shows some features similar to those of C_{60} ⊂ H_4-C_4-CPD_{Py}(H) (Fig. 25.11): the high planarity of the porphyrin planes, the clamshell conformation of the dimer, and so on. However, an important difference is observed in the arrangement of the fullerenes between PCBM ⊂ H_4-C_4-CPD_{Py}(H) and C_{60} ⊂ H_4-C_4-CPD_{Py}(H). In the crystal packing of PCBM ⊂ H_4-C_4-CPD_{Py}(H), the included PCBM molecules make

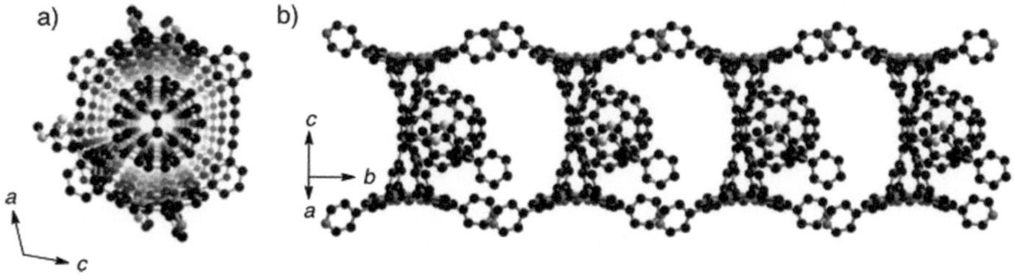

Fig. 25.10 Crystal structure of tubular assembly of PCBM ⊂ Ni_2-C_4-CPD_{Py}(H), (**a**) front view, (**b**) side view [30]

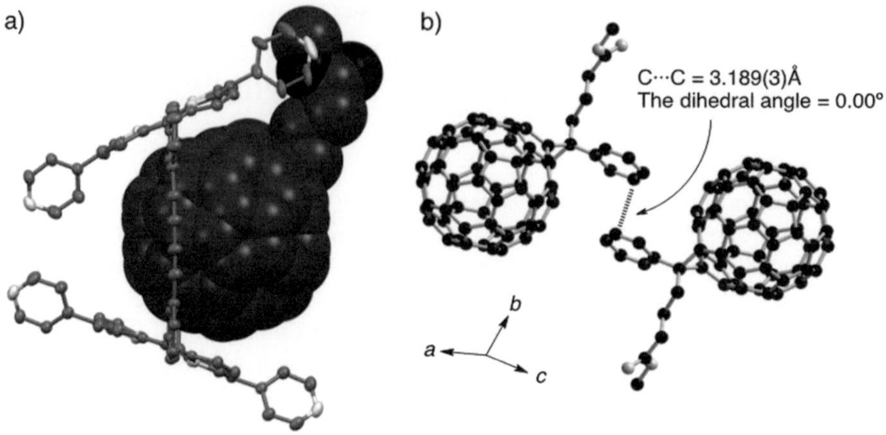

Fig. 25.11 Crystal structure of PCBM ⊂ H_4-C_4-CPD_{Py}(H), (**a**) side view of π-complex, (**b**) dimeric structure of PCBM [30]

a dimeric form through π-π stacking of the phenyl groups in the side chain. The PCBM dimer is separated from the adjacent dimer, leading to the absence of the continuous array of fullerenes such as the zigzag alignment of C_{60} in C_{60} ⊂ H_4-C_4-CPD_{Py}(H). For the construction of a zigzag array of fullerene, one molecule of fullerene needs mutual π-π interactions with two adjacent fullerene molecules. Although the PCBM molecule in PCBM ⊂ H_4-C_4-CPD_{Py}(H) also has the uncovered π-surface, the fullerene-fullerene interaction is inhibited by the side chain of PCBM in this case. This is a specific effect of the side chain of PCBM on the supramolecular structure of its π-complex with a host molecule.

25.6 Supramolecular Structures of π-Complexes of Ni_2-C_4-CPD_{Py}(H) and H_4-C_4-CPD_{Py}(H) with C_{70}

C_{70} is the most abundant fullerene next to C_{60}. However, unlike spherical C_{60}, C_{70} has an ellipsoidal shape with the longer (ca. 8.0 Å) and shorter (ca. 7.1 Å) axes [32]. Much attention has been denoted to the anisotropic structures and interactions of C_{70} [39–41]. There are two orientations of C_{70} in the π-π interactions with other π-systems: side-on and end-on orientations (Fig. 25.12). Side-on orientation of C_{70} is more energetically favorable than end-on orientation due to the larger π-π interaction of the former. A number of crystal structures of C_{70} cocrystallized with monomeric porphyrins have been reported [4, 42]. They all have side-on orientations of C_{70} toward porphyrin planes. Then, we investigated the supramolecular structures of C_{70} within the cyclic porphyrin dimers [26], which have the cavities of specific shapes and sizes to influence the orientation of included C_{70}. The ^{13}C-NMR

Fig. 25.12 Schematic illustrations of orientations of C_{70} included within porphyrin dimers [30]

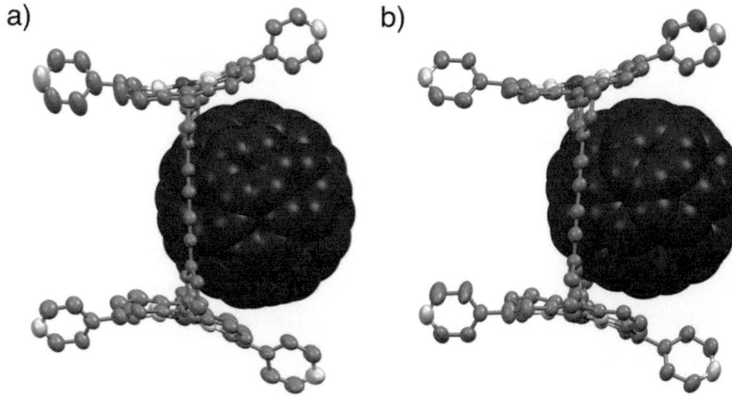

Fig. 25.13 Crystal structures of $C_{70} \subset Ni_2$-C_4-CPD_{Py}(H), (**a**) end-on-type structure, (**b**) side-on-type structure [30]

spectrum of $C_{70} \subset Ni_2$-C_4-CPD_{Py}(H) in solution indicated the existence of both side-on and end-on orientations of C_{70}. On the other hand, based on the ^{13}C-NMR spectrum of $C_{70} \subset H_4$-C_4-CPD_{Py}(H) in solution, the side-on orientation of C_{70} was highly dominant because of the smaller cavity of H_4-C_4-CPD_{Py}(H) than Ni_2-C_4-CPD_{Py}(H) (vide supra). The crystals of $C_{70} \subset Ni_2$-C_4-CPD_{Py}(H) showed polymorphism: two kinds of crystals, α and β [26]. Each crystal contained two crystallographically inequivalent inclusion complexes. Therefore, four kinds of crystal structures of $C_{70} \subset Ni_2$-C_4-CPD_{Py}(H) were obtained. It is worth noting that one of the complexes in crystal α showed an end-on orientation of C_{70} molecule toward the porphyrin planes (Fig. 25.13a). This is the first X-ray crystal structure having an end-on orientation of C_{70} cocrystallized with porphyrins. Other three crystal structures exhibited side-on orientations of C_{70} (Fig. 25.13b). The presence of both orientations in crystals was in good accord with the supramolecular structures in solution analyzed by ^{13}C-NMR.

In contrast, the crystal of $C_{70} \subset H_4$-C_4-CPD_{Py}(H) contained only a side-on orientation of C_{70} (Fig. 25.14), as observed by ^{13}C-NMR in solution [26]. The freebase

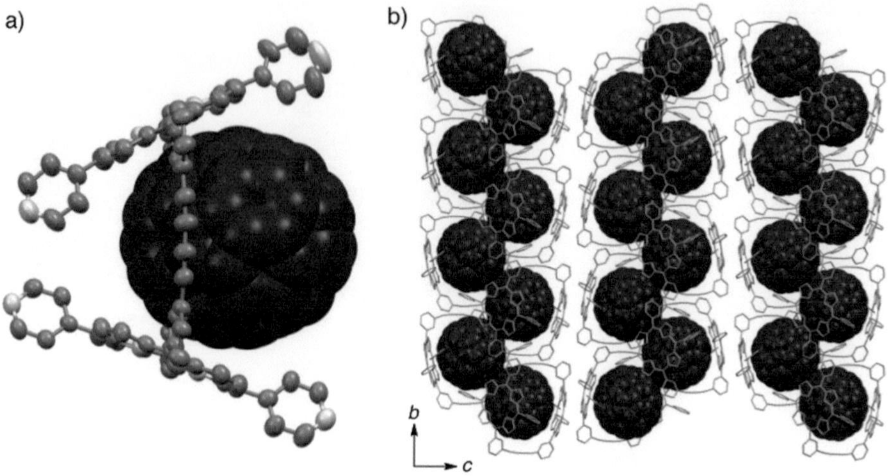

Fig. 25.14 Crystal structure of $C_{70} \subset H_4\text{-}C_4\text{-}CPD_{Py}(H)$, (**a**) side view of π-complex, (**b**) top view of C_{70} zigzag array [30]

dimer captures C_{70} in a clamshell conformation, wherein the two porphyrins are tilted with respect to each other. The included C_{70} molecules give a zigzag array in the crystal packing of $C_{70} \subset H_4\text{-}C_4\text{-}CPD_{Py}(H)$. The separation between the centers of the adjacent C_{70} molecules (10.6 Å), which is comparable to the shorter outer diameter of C_{70} (10.3 Å), indicates that the zigzag array is formed through van der Waals contacts of C_{70} molecules with each other. These structural features in the crystal of $C_{70} \subset H_4\text{-}C_4\text{-}CPD_{Py}(H)$ are similar to those of $C_{60} \subset H_4\text{-}C_4\text{-}CPD_{Py}(H)$.

25.7 Supramolecular Structures of π-Complexes of H_4-Ptz-$CPD_{Py}(OC_3)$ with C_{60}

As described above, $C_{60} \subset Ni_2\text{-}C_4\text{-}CPD_{Py}(H)$ affords the self-assembled porphyrin nanotube containing the linear arrangement of C_{60}, while $C_{60} \subset H_4\text{-}C_4\text{-}CPD_{Py}(H)$ lacks the formation of the tubular structure. Whether the porphyrin nanotube is constructed is dependent mainly upon the center-to-center distances of the porphyrin dimers. $C_{60} \subset Ni_2\text{-}C_4\text{-}CPD_{Py}(H)$ has the enough center-to-center distance (12.596 Å) to include C_{60} in the pseudoparallel conformation of the dimer (Fig. 25.8). This conformation allows the formation of the self-assembled nanotube through the C-H···N hydrogen bonds and the π-π interactions. On the other hand, these noncovalent interactions are not realized by the clamshell conformation, which is observed for $C_{60} \subset H_4\text{-}C_4\text{-}CPD_{Py}(H)$ having the shorter center-to-center distance (11.126 Å). Based on these previous results, we have designed a phenothiazine-bridged cyclic porphyrin dimer (Fig. 25.1) [29] as a new host molecule for fullerenes

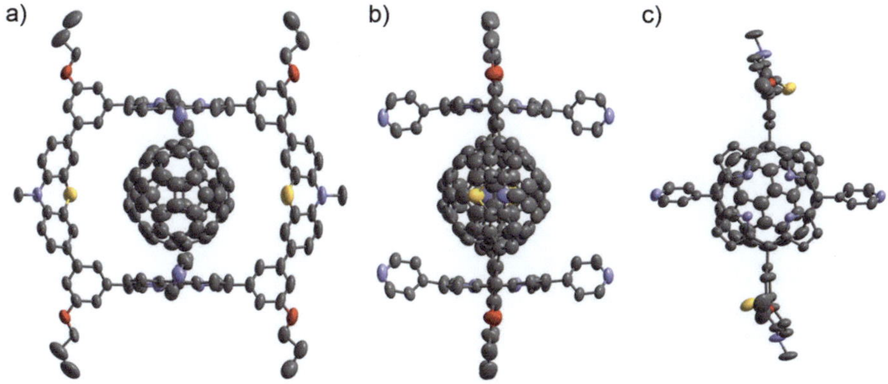

Fig. 25.15 Crystal structure of $C_{60} \subset H_4$-Ptz-CPD$_{Py}$(OC$_3$), (**a**) front view, (**b**) side view, (**c**) top view [29]

with the following prerequisites: (1) a longer rigid linkage than the butadiynyl group for the formation of a self-assembled porphyrin nanotube including a linear array of C_{60} molecules and (2) a freebase porphyrin rather than a nickel complex for photoinduced electron transfer from porphyrin to fullerene. On the basis of the DFT-calculated structure of the π-complex composed of H_4-Ptz-CPD$_{Py}$(H) and C_{60}, the phenothiazine linkages allow the dimer to take a parallel conformation with the center-to-center distance of 12.2 Å, whose value is acceptably close to the optimal value (ca. 12.5 Å) for C_{60} inclusion by cyclic porphyrin dimers as $C_{60} \subset Ni_2$-C_4-CPD$_{Py}$(H). We prepared also the phenothiazine-bridged nickel dimer (Ni_2-Ptz-CPD$_{Py}$(OC$_6$)) for the comparative analysis of the fullerene inclusion in solution. Both H_4-Ptz-CPD$_{Py}$(OC$_6$) and Ni_2-Ptz-CPD$_{Py}$(OC$_6$) show remarkably high affinities for C_{60} ($K_{assoc} > 10^6$ M^{-1}) and C_{70} ($K_{assoc} > 10^7$ M^{-1}) in solution (vide supra).

X-ray crystallographic analysis of $C_{60} \subset H_4$-Ptz-CPD$_{Py}$(OC$_3$) revealed its molecular structure and assembled form (Figs. 25.15, 25.16) [29], both of which fulfill our expectation. The planar porphyrin rings bite C_{60} in pseudoparallel conformation with the center-to-center distance of 12.454 Å, whose value is in the quite optimal range for C_{60} inclusion. The included C_{60} molecule is situated just above the centers of the porphyrins due to the enough separations (10.20(1) Å) between the opposite phenyl groups. This C_{60} location is largely different from the off-center position of C_{60} in $C_{60} \subset Ni_2$-C_4-CPD$_{Py}$(H). The 6:6 ring-juncture C-C bonds of both C_{60} poles are closest to the porphyrin planes as observed in most crystal structures of C_{60} cocrystallized with porphyrins, indicating strong π-π interactions of the two π-electron systems. These structural features account for the high affinities of H_4-Ptz-CPD$_{Py}$(OC$_6$) and Ni_2-Ptz-CPD$_{Py}$(OC$_6$) toward C_{60} and C_{70}. Moreover, in the crystal packing of $C_{60} \subset H_4$-Ptz-CPD$_{Py}$(OC$_3$), an expected self-assembled porphyrin nanotube is formed by analogous C-H···N hydrogen bonds and π-π interactions of the pyridyl groups. The C_{60} molecules are aligned in the inner

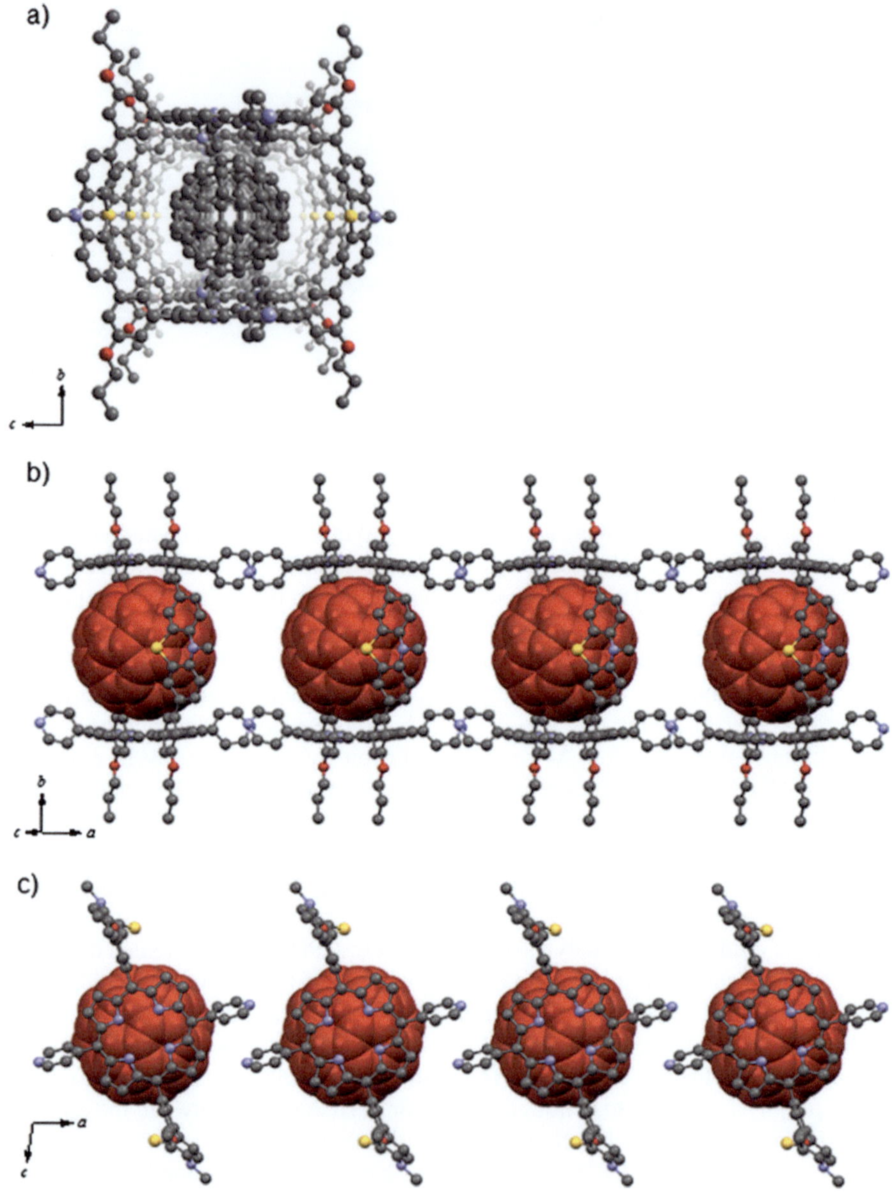

Fig. 25.16 Crystal structure of tubular assembly of $C_{60} \subset H_4$-Ptz-CPD$_{Py}$(OC_3), (**a**) front view, (**b**) side view, (**c**) top view [29]

channel of the nanotube to afford a supramolecular peapod with the distance between the C_{60} centers of 14.794 Å. It is confirmed that a linear array of fullerene is rationally produced by the use of a pyridyl-substituted cyclic porphyrin dimer having a suitable center-to-center distance (~12.5 Å) for fullerene inclusion.

25.8 Photoelectronic Properties of π-Complexes of H_4-C_4-$CPD_{Py}(OC_6)$ and Ni_2-C_4-$CPD_{Py}(OC_6)$ with $Li^+@C_{60}$

The π-complex of H_4-C_4-CPD_{Py}(H) with C_{60} ($C_{60} \subset H_4$-C_4-CPD_{Py}(H)) performed photoinduced electron transfer from the porphyrin to C_{60} to produce the CS state with the lifetime of 0.47 ps [25]. One of the reasons for the short lifetime is that the CS energy (1.83 eV) is higher than the triplet energy (ca. 1.50–1.60 eV) of the porphyrins and fullerenes. The resulting CS state decays to the triplet excited state of the chromophores. For achieving a long CS lifetime, CS energy is required to be lower than the triplet energy of the components.

Recently, lithium ion-encapsulated C_{60} ($Li^+@C_{60}$) has been reported to be a stronger electron acceptor than pristine C_{60} [43, 44]. It has a higher reduction potential by ca. 0.70 V than C_{60}. It is anticipated to make the energy level of the resulting CS state lower than the triplet energy of porphyrins and fullerene. Then, we investigated the photodynamics of the π-complexes of H_4, Ni_2-C_4-$CPD_{Py}(OC_6)$ with $Li^+@C_{60}$ by laser flash photolysis with the expectation that a long-lived photoinduced CS state would be obtained [28]. At first, the CS energy levels were determined by electrochemical method to be 1.07 eV for $Li^+@C_{60} \subset H_4$-C_4-$CPD_{Py}(OC_6)$ and 1.20 eV for $Li^+@C_{60} \subset Ni_2$-C_4-$CPD_{Py}(OC_6)$, being lower than the triplet energy of both $Li^+@C_{60}$ (1.53 eV) and the porphyrin dimers (1.51 eV). The time-resolved transient absorption spectra of the π-complexes clearly indicated the formation of the CS state in benzonitrile solution. The CS lifetimes were determined from the time profiles of the spectra to be 0.50 ms for $Li^+@C_{60} \subset H_4$-C_4-$CPD_{Py}(OC_6)$ and 0.67 ms for $Li^+@C_{60} \subset Ni_2$-C_4-$CPD_{Py}(OC_6)$. These CS lifetimes are the longest values among noncovalent porphyrin-fullerene supramolecules in solution and 10^6 times longer than that of $C_{60} \subset H_4$-C_4-CPD_{Py}(H). The quantum yields of the CS states were 0.32 for $Li^+@C_{60} \subset H_4$-C_4-$CPD_{Py}(OC_6)$ and 0.13 for $Li^+@C_{60} \subset Ni_2$-C_4-$CPD_{Py}(OC_6)$. These results indicate that the control of CS energy by the choice of chromophores in π-complexes is able to realize great extension of CS lifetime.

25.9 Summary

The freebases and nickel complexes of butadiyne- or phenothiazine-bridged cyclic porphyrin dimers having self-assembling pyridyl groups at the opposite *meso* positions efficiently include fullerenes such as C_{60}, PCBM, C_{70}, and $Li^+@C_{60}$ to give supramolecular π-complexes. In particular, the phenothiazine-bridged cyclic porphyrin dimers have remarkable high affinities toward C_{60} ($K_{assoc} > 10^6\ M^{-1}$) and C_{70} ($K_{assoc} > 10^7\ M^{-1}$). In crystalline state, $C_{60} \subset Ni_2$-C_4-CPD_{Py}(H), PCBM $\subset Ni_2$-C_4-CPD_{Py}(H), and $C_{60} \subset H_4$-Ptz-$CPD_{Py}(OC_3)$ form self-assembled porphyrin nanotubes containing the linear arrays of the fullerenes. The main factors for this self-assembly are the C-H···N hydrogen bonds and π-π interactions of the pyridyl groups. In contrast, $C_{60} \subset H_4$-C_4-CPD_{Py}(H) and $C_{70} \subset H_4$-C_4-CPD_{Py}(H) gave the zigzag arrays of the fullerenes. Based on these results, it is demonstrated that the self-assembly of a π-host molecule for fullerenes is an effective strategy for the control of fullerene arrangement. The above C_{60} arrays show high electron mobilities in the order of 10^{-1} cm^2V^{-1} s^{-1}. Further, $C_{60} \subset H_4$-C_4-CPD_{Py}(H), $Li^+@C_{60} \subset H_4$-C_4-$CPD_{Py}(OC_6)$, and $Li^+@C_{60} \subset Ni_2$-C_4-$CPD_{Py}(OC_6)$ undergo photoinduced electron transfer from the porphyrins to the fullerenes to produce the CS states, suggesting their possibilities for photovoltaic applications. It is noteworthy that $Li^+@C_{60} \subset H_4$-C_4-$CPD_{Py}(OC_6)$ and $Li^+@C_{60} \subset Ni_2$-C_4-$CPD_{Py}(OC_6)$ afford the CS states with the lifetimes of submillisecond order. Further study to apply these π-complexes to materials of photoelectric devices has now been under way.

Acknowledgments This work was supported by Grants-in-Aid (Scientific Research on Innovative Area "pi-Space" and the Global COE Program "Science for Future Molecular Systems") from the Ministry of Education, Culture, Sports, Science and Technology of Japan; by the Cooperative Research Program of Network Joint Research Center for Materials and Devices of Institute for Materials Chemistry and Engineering, Kyushu University; and by Research Grants to F.T. from Tokuyama Science and Technology Foundation and Iketani Science and Technology Foundation. H.N. acknowledges the Japan Society for the Promotion of Science (JSPS) for a Research Fellowship for Young Scientists. We are sincerely grateful to our coworkers for their kind cooperation whose names appear as coauthors in our papers.

References

1. Lehn JM (1995) Supramolecular chemistry: concepts and perspectives. Weinheim, Wiley-VCH
2. Shimizu T, Masuda M, Minamikawa H (2005) Supramolecular nanotube architectures based on amphiphilic molecules. Chem Rev 105:1401–1444
3. Pasini D, Ricci M (2007) Macrocycles as precursors for organic nanotubes. Curr Org Synth 4:59–80
4. Boyd PDW, Reed CA (2005) Fullerene—porphyrin constructs. Acc Chem Res 38:235–242
5. Tashiro K, Aida T (2007) Metalloporphyrin hosts for supramolecular chemistry of fullerenes. Chem Soc Rev 36:189–197
6. Canevet D, Pérez EM, Martín N (2011) Wraparound hosts for fullerenes: tailored macrocycles and cages. Angew Chem Int Ed 50:9248–9259
7. Kadish KM, Ruoff RS (eds) (2000) Fullerenes: chemistry physics and technology. Wiley Interscience, New York
8. Guldi DM, Martín N (eds) (2002) Fullerenes: from synthesis to optoelectronic properties. Kluwer, Dordrecht
9. Fukuzumi S, Guldi DM (2001) Electron-transfer chemistry of fullerenes. In: Balzani V (ed) Electron transfer in chemistry, vol 2. Wiley-VCH, Weinheim, pp 270–337
10. Hirsch A, Brettreich M (2005) Fullerenes: chemistry and reactions. Wiley-VCH, Weinheim
11. Langa F, Nierengarten J-F (eds) (2012) Fullerenes: principles and applications, 2nd edn. Royal Society of Chemistry Publishing, Cambridge
12. Kadish KM, Smith KM, Guilard R (eds) (2000) The porphyrin handbook. Academic, San Diego

13. Kadish KM, Smith KM, Guilard R (eds) (2010) Handbook of porphyrin science. World Scientific, Singapore
14. Fukuzumi S, Ohkubo K (2013) Long-lived photoinduced charge separation for solar cell applications in supramolecular complexes of multi-metalloporphyrins and fullerenes. Dalton Trans 42:15846–15858
15. Schuster DI (2013) Reflections on a fifty-year career in organic photochemistry: a personal perspective. J Org Chem 78:6811–6841
16. Gust D, Moore TA, Moore TL (2012) Realizing artificial photosynthesis. Faraday Discuss 155:9–26
17. Imahori H, Umeyama T, Kurotobi K, Takano Y (2012) Self-assembling porphyrins and phthalocyanines for photoinduced charge separation and charge transport. Chem Commun 48:4032–4045
18. D'Souza F, Ito O (2012) Photosensitized electron transfer processes of nanocarbons applicable to solar cells. Chem Soc Rev 41:86–96
19. Kira A, Umeyama T, Matano Y, Yoshida K, Isoda S, Park JK, Kim D, Imahori H (2009) Supramolecular donor-acceptor heterojunctions by vectorial stepwise assembly of porphyrins and coordination-bonded fullerene arrays for photocurrent generation. J Am Chem Soc 131:3198–3200
20. Matsuo Y, Sato Y, Niinomi T, Soga I, Tanaka H, Nakamura E (2009) Columnar structure in bulk heterojunction in solution-processable three-layered p-i-n organic photovoltaic devices using tetrabenzoporphyrin precursor and silylmethyl[60]fullerene. J Am Chem Soc 131:16048–16050
21. Moon JS, Lee JK, Cho S, Byun J, Heeger AJ (2009) "Columnlike" structure of the cross-sectional morphology of bulk heterojunction materials. Nano Lett 9:230–234
22. Hoppe H, Sariciftci NS (2006) Morphology of polymer/fullerene bulk heterojunction solar cells. J Mater Chem 16:45–61
23. Nobukuni H, Shimazaki Y, Tani F, Naruta Y (2007) A nanotube of cyclic porphyrin dimers connected by nonclassical hydrogen bonds and its inclusion of C_{60} in a linear arrangement. Angew Chem Int Ed 46:8975–8978
24. Nobukuni H, Tani F, Shimazaki Y, Naruta Y, Ohkubo K, Nakanishi T, Kojima T, Fukuzumi S, Seki S (2009) Anisotropic high electron mobility and photodynamics of a self-assembled porphyrin nanotube including C_{60} molecules. J Phys Chem C 113:19694–19699
25. Nobukuni H, Shimazaki Y, Uno H, Naruta Y, Ohkubo K, Kojima T, Fukuzumi S, Seki S, Sakai H, Hasobe T, Tani F (2010) Supramolecular structures and photoelectronic properties of the inclusion complex of a cyclic free-base porphyrin dimer and C_{60}. Chem Eur J 16:11611–11623
26. Nobukuni H, Kamimura T, Uno H, Shimazaki Y, Naruta Y, Tani F (2011) Supramolecular structures of inclusion complexes of C_{70} and cyclic porphyrin dimers. Bull Chem Soc Jpn 84:1321–1328
27. Nobukuni H, Kamimura T, Uno H, Shimazaki Y, Naruta Y, Tani F (2012) Supramolecular structures of inclusion complexes of cyclic porphyrin dimers and [6,6]-phenyl-C_{61}-butyric acid methyl ester (PCBM). Bull Chem Soc Jpn 85:862–868
28. Kamimura T, Ohkubo K, Kawashima Y, Nobukuni H, Naruta Y, Tani F, Fukuzumi S (2013) Submillisecond-lived photoinduced charge separation in inclusion complexes composed of Li + @C60 and cyclic porphyrin dimers. Chem Sci 4:1451–1461
29. Sakaguchi K, Kamimura T, Uno H, Mori S, Ozako S, Nobukuni H, Ishida M, Tani F (2014) Phenothiazine-bridged cyclic porphyrin dimers as high-affinity hosts for fullerenes and linear array of C_{60} in Self-assembled porphyrin nanotube. J Org Chem 79:2980–2992
30. Tani F, Nobukuni H, Kamimura T (2013) Finely designed arrangements and photoelectronic properties of supramolecular nanocarbons composed of fullerenes and cyclic porphyrin dimers. Tanso 260:284–291. http://dx.doi.org/10.7209/tanso.2013.284
31. David WIF, Ibberson RM, Matthewman JC, Prassides K, Dennis TJS, Hare JP, Kroto HW, Taylor R, Walton DRM (1991) Crystal structure and bonding of ordered C60. Nature 353:147–149

32. McKenzie DR, Davis CA, Cockayne DJH, Muller DA, Vassallo AM (1992) The structure of the C_{70} molecule. Nature 355:622–624
33. Saeki A, Koizumi Y, Aida T, Seki S (2012) Comprehensive approach to intrinsic charge carrier mobility in conjugated organic molecules, macromolecules, and supramolecular architectures. Acc Chem Res 45:1193–1202
34. Peet J, Heeger AJ, Bazan AC (2009) "Plastic" solar cells: self-assembly of bulk heterojunction nanomaterials by spontaneous phase separation. Acc Chem Res 42:1700–1708
35. Thompson BC, Fréchet JMJ (2008) Polymer–fullerene composite solar cells. Angew Chem Int Ed 47:58–77
36. Günes S, Neugebauer H, Sariciftci NS (2007) Conjugated polymer-based organic solar cells. Chem Rev 107:1324–1338
37. Rispens MT, Meetsma A, Rittberger R, Brabec CJ, Sariciftci NS, Hummelen JC (2003) Influence of the solvent on the crystal structure of PCBM and the efficiency of MDMO-PPV:PCBM 'plastic' solar cells. Chem Commun 2116–2118
38. Choi JH, Honda T, Seki S, Fukuzumi S (2011) Relationship between crystal packing and high electron mobility in the single crystal of thienyl-substituted methanofullerene. Chem Commun 47:11213–11215
39. Wang Y-B, Lin Z (2003) Supramolecular interactions between fullerenes and porphyrins. J Am Chem Soc 125:6072–6073
40. Goldoni A, Cepek C, Larciprete R, Sangaletti L, Pagliara S, Floreano L, Gotter R, Verdini A, Morgante A, Luo Y, Nyberg M (2002) C_{70} adsorbed on Cu(111): metallic character and molecular orientation. J Chem Phys 116:7685–7690
41. Okubo S, Okazaki T, Hirose-Takai K, Suenaga K, Okada S, Bandow S, Iijima S (2010) Electronic structures of single-walled carbon nanotubes encapsulating ellipsoidal C_{70}. J Am Chem Soc 132:15252–15258
42. Ishii T, Aizawa N, Kanehama R, Yamashita M, Sugiura K, Miyasaka H (2002) Cocrystallites consisting of metal macrocycles with fullerenes. Coord Chem Rev 226:113–124
43. Aoyagi S, Nishibori E, Sawa H, Sugimoto K, Takata M, Miyata Y, Kitaura R, Shinohara H, Okada H, Sakai T, Ono Y, Kawachi K, Yokoo K, Ono S, Omote K, Kasama Y, Ishikawa S, Komuro T, Tobita H (2010) A layered ionic crystal of polar Li@C_{60} superatoms. Nat Chem 2:678–683
44. Kawashima Y, Ohkubo K, Fukuzumi S (2012) Enhanced photoinduced electron-transfer reduction of Li^+@C_{60} in comparison with C_{60}. J Phys Chem A 116:8942–8948

Chapter 26
Sequence Control of π-Electron Systems

Kentaro Tashiro

Abstract In this chapter, examples of the sequence control of π-electron systems, recently conducted by the author and coworkers, are briefly reviewed. The concept can cover diverse research topics, e.g., design of organic semiconductors, switching of spin–spin interactions, construction of molecular switches, etc. Particularly highlighted are (1) the molecular design strategies to direct the spontaneous formation of desired supramolecular nanostructures such as columnar assemblies for liquid crystallization and heterojunctions for ambipolar photoconductions and (2) novel synthetic methodologies to allow the access to complex but well-defined structures which otherwise are inaccessible via conventional methodologies.

Keywords Sequence • Self-assembly • π-Electronic compounds • Amphiphilic molecular design • Solid-phase synthesis

26.1 Introduction

Precise control over the orientation of π-electronic moieties in molecular and supramolecular levels plays the key role to give excellent functions to their supramolecular assemblies. Trials on this research subject initially started with the systems of single component, which has still ample room to do something new. Meanwhile, double π-electronic component systems such as donor–acceptor hybrids have been also subjected to the studies in the past decade. More recently, control over the arrangement of more than two different π-electronic moieties in their assemblies has been attempted in spite of its difficulty due to the macroscopic phase separation and escalating numbers of possible structural outcomes. This review starts with works on two single component systems, moves on two examples with double component systems, and ends by showing trials on more complex systems.

K. Tashiro (✉)
International Center for Materials Nanoarchitectonics (WPI-MANA), National Institute for Materials Science, Tsukuba, Ibaraki, Japan
e-mail: TASHIRO.Kentaro@nims.go.jp

T. Akasaka et al. (eds.), *Chemical Science of π-Electron Systems*,
DOI 10.1007/978-4-431-55357-1_26

26.2 Directed Columnar Assembly of Fused Copper Porphyrin Dimers for the Design of Liquid Crystalline Organic Semiconductors

One of the well-known approaches to design liquid crystalline (LC) organic semiconductors is to induce columnar stacking of π-conjugated molecules in the LC state. The most common molecular design principle for this end is the introduction of long alkyl chains at the periphery of the π-conjugated core, which, however, does not always work sufficiently. This is actually the case when we tried to make an LC material from fused copper porphyrin dimers having an extra-large π-conjugated system. To address this issue, we introduced an amphiphilic molecular design strategy, where hydrophobic and hydrophilic side chains are attached to the core site specifically, to make the half side of the molecule hydrophobic, while the other half hydrophilic [1]. Chart 26.1 shows the molecular structures investigated, among which compound $\mathbf{1}_{\mathbf{C12/TEG}}$ bearing dodecyl and triethylene glycol chains formed a room-temperature columnar liquid crystal, while its two reference compounds having only either of these two types of side chains ($\mathbf{1}_{\mathbf{C12/C12}}$ and $\mathbf{1}_{\mathbf{TEG/TEG}}$) gave amorphous materials. X-ray diffraction (XRD) analysis at 30 °C of the LC assembly of amphiphilic $\mathbf{1}_{\mathbf{C12/TEG}}$ showed two distinct diffraction peaks at $2\Theta = 0.79$ and 1.60°, along with minor diffractions in a wider-angle region. With the help of 2D-XRD analysis on the sheared sample, these diffractions were successfully indexed to a rectangular columnar lattice with lattice parameters *a* and *b* of 65.3 and 37.2 Å, respectively. Based on these results, a 2D molecular packing pattern (Fig. 26.1) was proposed, which can help to understand why the amphiphilic molecular design of $\mathbf{1}_{\mathbf{C12/TEG}}$ works for the liquid crystallization of fused metalloporphyrin dimers. Namely, a nanoscale phase separation of two immiscible side chains plays the role to orient $\mathbf{1}_{\mathbf{C12/TEG}}$ in an ordered fashion. Coexistence of the hydrophobic and hydrophilic chains in the same molecule is essential, since mixing of hydrophobic $\mathbf{1}_{\mathbf{C12/C12}}$ and hydrophilic $\mathbf{1}_{\mathbf{TEG/TEG}}$ resulted in a macroscopic phase separation without any particular structural features.

Our incentive for the liquid crystallization of fused metalloporphyrin dimers is that their extra-large π-conjugated core could allow the access to solution-processable LC semiconductors with unusually large charge carrier mobilities. To evaluate the intrinsic charge carrier mobility with a minimum grain-boundary effect, we measured the flash-photolysis time-resolved microwave conductivity (FP-TRMC) of the LC state of $\mathbf{1}_{\mathbf{C12/TEG}}$. Photoexcitation of the LC film of $\mathbf{1}_{\mathbf{C12/TEG}}$ showed a transient absorption that corresponds to the formation of radical anion $\mathbf{1}_{\mathbf{C12/TEG}}^{\bullet-}$. From the conductivity value in TRMC along with the quantity of photochemically generated $\mathbf{1}_{\mathbf{C12/TEG}}^{\bullet-}$, the one-dimensional electron mobility of the LC film was evaluated to be 0.27 $cm^2/V \cdot s$ at 16 °C. Noteworthy, this was the largest electron mobility among those reported for room-temperature columnar LC materials, studied by TRMC at that time. This finding encouraged us to further investigate the behaviors of LC semiconductors based on fused metalloporphyrin dimers. In the following study analogous LC fused copper porphyrin dimers bearing semi-

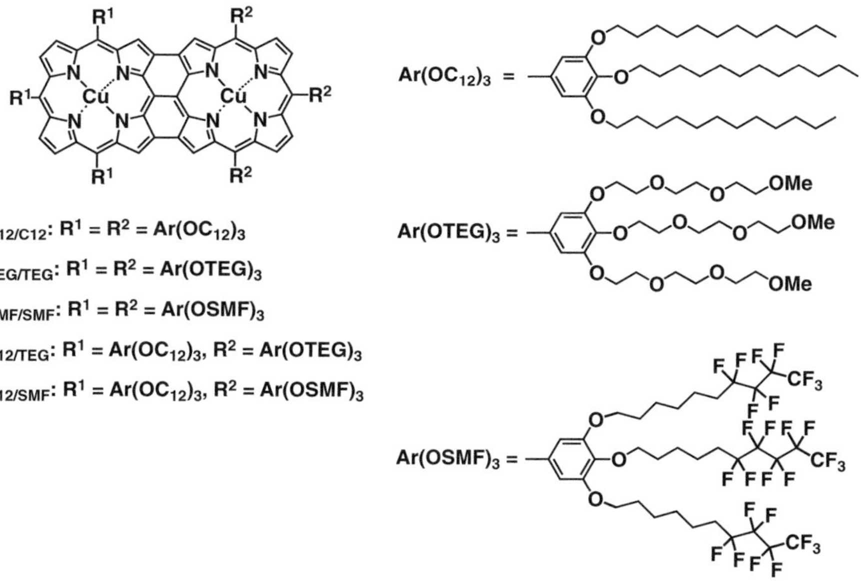

Chart 26.1 Molecular structures of fused copper porphyrin dimers **1**

Fig. 26.1 Schematic representation of a rectangular 2D molecular packing diagram for liquid crystalline $\mathbf{1}_{\mathbf{C12/TEG}}$

fluoroalkyl side chains fully or site specifically were newly designed (Chart 26.1), where, to our surprise, the former (**1**$_{\mathbf{SMF/SMF}}$) and the latter (**1**$_{\mathbf{C12/SMF}}$) exhibited p-type and n-type semiconductor properties, respectively [2]. As the π-electronic properties of their core units such as redox and absorption profiles in solution are substantially identical to one another, their contrasting charge carrier transport property should be originating from the difference in their π-stacking geometry at the LC state. Actually, **1**$_{\mathbf{SMF/SMF}}$ forms an orthorhombic mesophase, while **1**$_{\mathbf{C12/SMF}}$, similar to **1**$_{\mathbf{C12/TEG}}$, has a rectangular columnar mesophase. Moreover, their absorption spectral change profiles that accompanied by the isotropic-to-LC phase transition are clearly different, also demonstrating that the π-stacking geometry of the fused copper porphyrin dimer core unit is strongly dependent on the nature and substitution pattern of the side chains around the core. Although there has been a theoretical prediction that assembled π-conjugated motifs possibly switch their preference for carrier species by changing the π-stacking geometry, this would be, to the best of our knowledge, the first experimental demonstration for that idea.

26.3 The First Liquid Crystalline Corannulenes

Corannulene is a bowl-shaped π-conjugated molecule, whose carbon network skeleton is equivalent to a partial structure of fullerenes. Because of the nonplanarity of its π-conjugated system, columnarly π-stacked assemblies of corannulene have unique potentials for materials with ferroelectric properties or unidirectional charge carrier transport capability. On the other hand, due to the same structural reason, π-stacking interactions between corannulenes are usually less efficient than those of planar π-conjugated molecules of similar size. Therefore, the construction of columnar assemblies of corannulenes has been an interesting challenge for researchers, where its example in LC phase had been unprecedented before our work. In order to enhance the intermolecular interactions between corannulenes, we introduced amide functionality into each side chain of corannulene, which stabilizes the columnar assembly of corannulenes via hydrogen bonding interactions between them and allows the liquid crystallization of corannulene [3]. One of the newly designed corannulene derivatives **2**$_{\mathbf{10}}$ (Chart 26.2) having ten amidated long alkyl side chains exhibited an LC mesophase only in a very narrow and high temperature range, namely, two endothermic peaks at 169 and 163 °C at cooling from isotropic melt. In contrast, the other corannulene bearing thirty side chains (**2**$_{\mathbf{30}}$) became a room-temperature liquid crystal with two endothermic peaks at 148 and −14 °C at cooling. Introduction of the amide functionality into the corannulene periphery brought another fruit, as LC state of **2**$_{\mathbf{30}}$ is responsive to electric field. The fan texture, characteristic of hexagonal columnar LC assemblies, of **2**$_{\mathbf{30}}$ observed at its LC mesophase disappeared when a certain level (15 $V \cdot \mu m^{-1}$) of alternating electric field was applied to the sample at 125 °C. Since the same sample retrieved the birefringent texture after additional one heating–cooling cycle, the observed

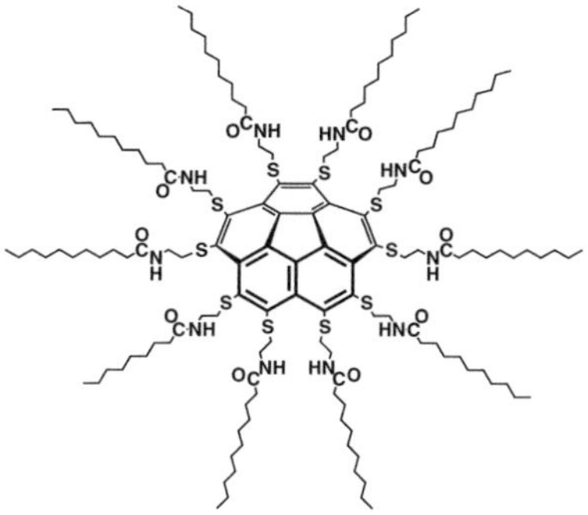

2_{10}

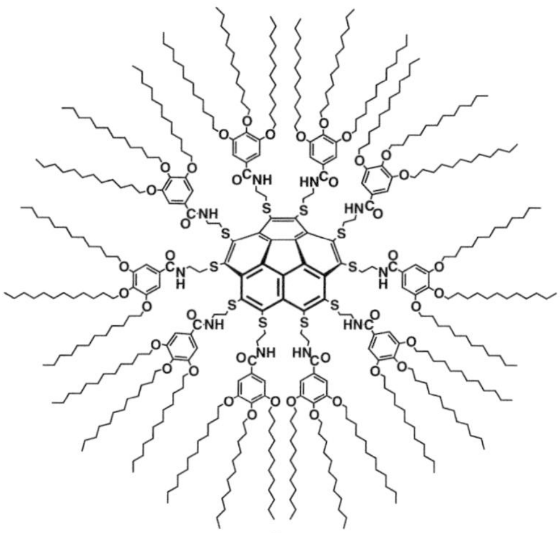

2_{30}

Chart 26.2 Molecular structures of liquid crystalline corannulenes **2**

phenomenon is not due to the decomposition of the sample but electronically induced homeotropic alignment of the columns of $\mathbf{2_{30}}$. Although LC materials responsive to the electric field have been well known, those of discotic LC materials are still rare. Together with the possible synchronized inversion of the corannulene bowls in the columns, electronically controlled homeotropic alignment of the LC columns of $\mathbf{2_{30}}$ is attractive for future application of it to ferroelectric materials.

26.4 A Supramolecular Spin Sequence That Exhibits Switchable Magnetic Interactions

Since ourfirst finding of the supramolecular complexation of fullerenes with cyclic dimers of metalloporphyrin in 1999, we tested various combinations of the host and guest molecules [4, 5]. Through these surveys, we found that a cyclic dimer of copper porphyrin forms a ferromagnetically coupled inclusion complex with La@C_{82}, a paramagnetic endohedral metallofullerene. This is a rare example of ferromagnetically coupled host–guest system, where the two spins on the two copper(II) centers of the host and one spin on the carbon cage of La@C_{82} interact each other to give a spin quartet ground state [$S = 3/2$], as determined by low-temperature ESR and electron spin transient nutation spectroscopies. In order to gain further insight into the spin–spin interactions between the host and guest system, we newly designed an analogous copper porphyrin cyclic dimer bearing alkyl side chains with olefinic termini ($\mathbf{3_{cyclo}}$ in Scheme 26.1) which is transformable into its caged analog to confine the included La@C_{82} by intramolecular ring-closing solefin metathesis (Scheme 26.1) [6]. While the host–guest complex before the metathesis

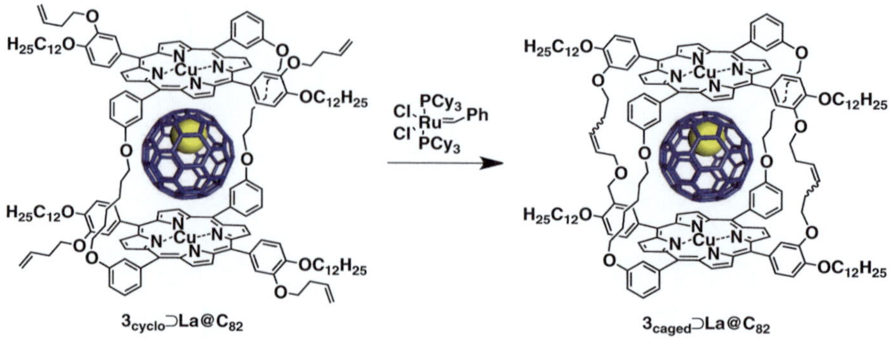

Scheme 26.1 Schematic illustrations of the structures of $\mathbf{3_{cyclo}} \supset La@C_{82}$ and $\mathbf{3_{caged}} \supset La@C_{82}$ and their transformation by ring-closing metathesis reaction

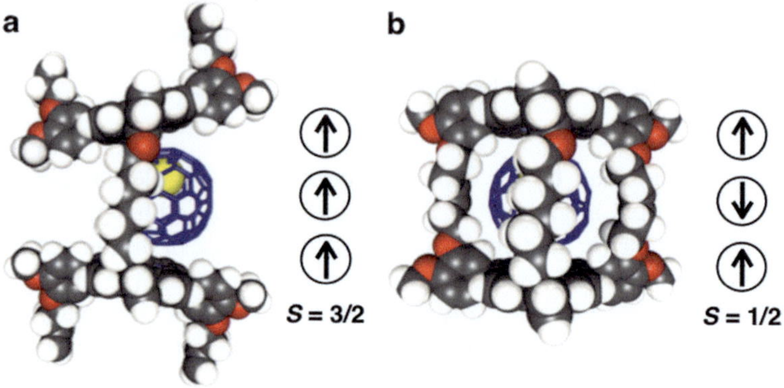

Fig. 26.2 Geometrically optimized molecular models of (**a**) $\mathbf{3_{cyclo}} \supset La@C_{82}$ and (**b**) $\mathbf{3_{caged}} \supset La@C_{82}$ with their spin ground states, calculated by means of DFT calculation at the BLYP level

reaction ($\mathbf{3_{cyclo}} \supset La@C_{82}$) underwent complete dissociation into the host and guest under the conditions for MALDI-TOF mass spectrometry and HPLC, the product ($\mathbf{3_{caged}} \supset La@C_{82}$) did not show any sign of the dissociation, demonstrating that trapped $La@C_{82}$ was successfully confined. As expected, the spin ground state of $\mathbf{3_{cyclo}} \supset La@C_{82}$ in frozen chlorobenzene was quartet (Fig. 26.2a). In contrast, the spin ground state of $\mathbf{3_{caged}} \supset La@C_{82}$ under the same conditions was determined to be a doublet [$S = 1/2$], suggesting that the host–guest spin–spin interaction switched to be ferrimagnetic upon confinement of $La@C_{82}$ (Fig. 26.2b). Because of the presence of guest $La@C_{82}$, both host–guest complexes showed a red-shifted NIR absorption band at $\sim$1000 nm. However, the red shift for $\mathbf{3_{caged}} \supset La@C_{82}$ was more explicit than that for $\mathbf{3_{cyclo}} \supset La@C_{82}$, suggesting that the host–guest geometry changes upon confinement of the guest. In accordance with these observations, a density functional theory (DFT) study of a simplified model of $\mathbf{3_{cyclo}} \supset La@C_{82}$ suggested that its two copper porphyrin units are not oriented parallel to one another and that the geometrical center of the included $La@C_{82}$ does not lie on the Cu–Cu axis of the host component due to the sizes of the host cavity and La@C_{82} (Fig. 26.2a). In contrast, in the optimized structure of $\mathbf{3}_{\mathbf{caged}} \supset$ La@C_{82}, La@C_{82} is in the geometrical center and the two copper porphyrin units are aligned parallel to one another (Fig. 26.2b). A host molecule with the ability to change its conformation in response to physical stimuli such as photo-irradiation might lead to the development of stimulus-responsive memory devices based on endohedral metallofullerenes.

26.5 Sequence Control of π-Electron Systems at Surfaces

Metalbisporphyrinate double-decker compounds have been regarded as interesting candidates for molecular rotors or gears for more than a decade [7, 8]. Recently we have achieved the single-molecular level control of the rotation of the upper porphyrin disk of a surface-attached cerium double-decker compound ($\mathbf{4}_{\mathbf{Ce}}$; Chart 26.3) by STM manipulation at very low temperatures [9]. To prepare the monolayer of $\mathbf{4}_{\mathbf{Ce}}$

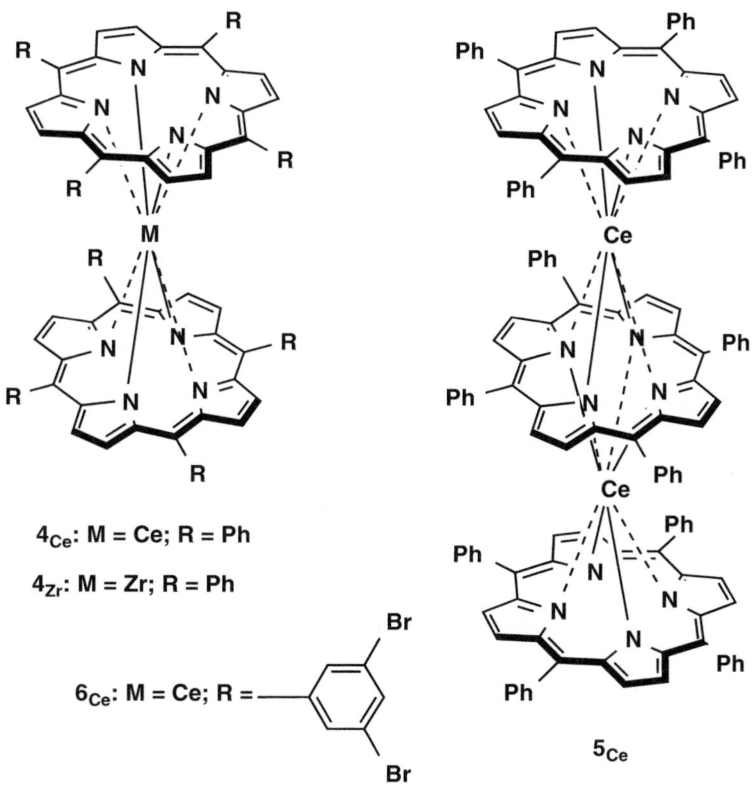

Chart 26.3 Molecular structures of metal bis- and trisporphyrin multi-decker compounds **4–6**

on a silver surface for the STM experiments, cerium atomic beam was irradiated to a precursor porphyrin multilayer on the substrate under vacuum, where not only the expected $\mathbf{4}_{Ce}$ but also another product higher than $\mathbf{4}_{Ce}$ can be detected. The height analysis as well as scanning tunneling spectroscopy of that unexpected species strongly suggested the production of biscerium triple-decker compound ($\mathbf{5}_{Ce}$; Chart 26.3), whose synthesis by the conventional solution phase approaches has kept failing so far.

Since porphyrin moieties in metal bisporphyrinate double-decker compounds are deformed from the planarity to give π-electron-rich concave surfaces, it is interesting to investigate their supramolecular interactions with fullerenes. In fact, we were able to obtain co-crystals of Ce or Zr double-decker compounds ($\mathbf{6}_{Ce}$ and $\mathbf{4}_{Zr}$; Chart 26.3) with C_{60}, where these double-decker compounds and C_{60} form alternatingly stacked infinite or discrete arrays of them to make concave–convex interactions between them effectively operative (Fig. 26.3) [10]. Encouraged by these observations, C_{60} was introduced onto the monolayer or isolated form of $\mathbf{4}_{Ce}$ on the silver substrate at 120 K by its sublimation under vacuum [11]. STM observation of the resultant hybrids at 6 K revealed that C_{60} is located on $\mathbf{4}_{Ce}$ with neither of any hopping migration to the neighboring $\mathbf{4}_{Ce}$ moieties nor three-dimensional rotations on the same $\mathbf{4}_{Ce}$. Thanks to such frozen dynamic motions of C_{60} on $\mathbf{4}_{Ce}$, two different orientations of C_{60}, with respect to $\mathbf{4}_{Ce}$, were visualized separately, among which more than half of C_{60} put one of its carbon atoms at the most proximal position to the underlying cerium metal center of $\mathbf{4}_{Ce}$ (Fig. 26.4a), while the remaining less than half made one of its 6:6 ring-juncture C–C bonds closest to the metal center (Fig. 26.4b). Application of a positive sample voltage over 1.8 V to the hybridized C_{60}–$\mathbf{4}_{Ce}$ dyad structure allowed to reorient the C_{60}

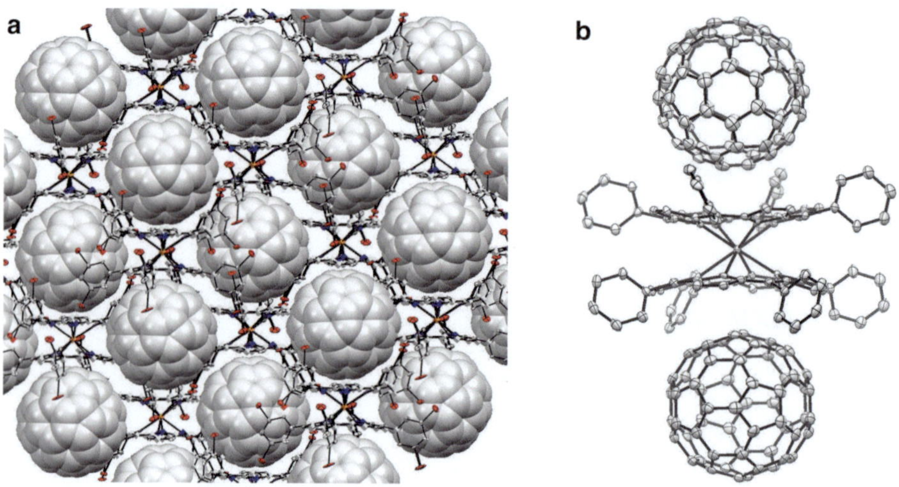

Fig. 26.3 Molecular packing diagrams of C_{60} and (**a**) $\mathbf{6}_{Ce}$ and (**b**) $\mathbf{4}_{Zr}$ in their co-crystals

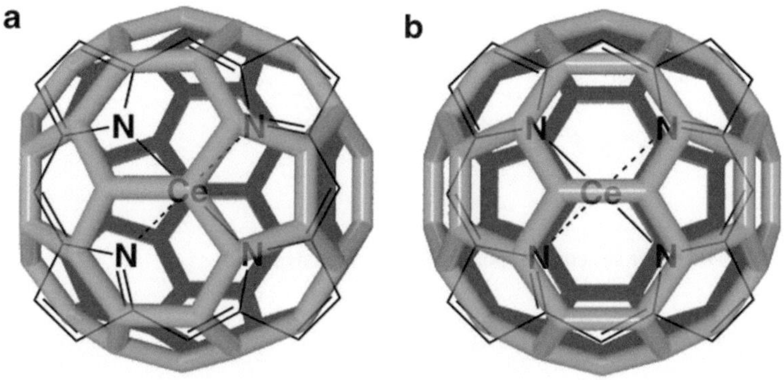

Fig. 26.4 (**a**) Major and (**b**) minor orientations of C_{60} with respect to the underlying $\mathbf{4}_{\mathbf{Ce}}$ on the silver surface

to the other configuration without any hopping migration of that C_{60}. Of interest, the conductivity through C_{60}–$\mathbf{4}_{\mathbf{Ce}}$ dyad, as evaluated by using the STM tip, was significantly dependent on the C_{60} orientation, where the major one expected to have stronger C_{60}–$\mathbf{4}_{\mathbf{Ce}}$ interactions exhibited distinctly larger tunnel current level than the minor one. Further detailed analysis revealed that the major orientation provided two distinguishable current levels, probably due to the symmetry lowering of the electronic configuration of $\mathbf{4}_{\mathbf{Ce}}$ on the silver surface. Therefore, the supramolecular C_{60}–$\mathbf{4}_{\mathbf{Ce}}$ dyad structure fabricated on a silver substrate can be regarded as a tristable molecular switch controlled by STM.

26.6 Controlled Self-Assembly of a Donor–Acceptor Dyad for the Construction of Supramolecular Heterojunction

Controlled self-assembly of donor–acceptor (D–A) dyads is an attracting approach for the construction of heterojunctions. We challenged it by designing a porphyrin–fullerene dyad (**6**; Chart 26.4) having a hydrophilic wedge at the porphyrin terminal with an expectation that this molecular design induced nanoscale segregation of porphyrin and fullerene moieties [12]. Unfortunately, however, this amphiphilic molecular design strategy turned out not to be enough to make an expected heterojunction, as we obtained only submicron-sized spheres (Fig. 26.5a) with intermolecularly stacked D–A moieties from the racemic mixture of **6** ($\mathbf{6}_{\mathbf{rac}}$). Because **6** has a stereogenic center in its fullerene unit, we further tried the synthesis and self-assembly of one of the enantiomers of **6** ($\mathbf{6}_{\mathbf{ena}}$). Scanning electron microscopy of the self-assembled $\mathbf{6}_{\mathbf{ena}}$, as obtained by slow diffusion of MeOH into a CH_2Cl_2 solution of $\mathbf{6}_{\mathbf{ena}}$, revealed that $\mathbf{6}_{\mathbf{ena}}$ behaves quite differently from $\mathbf{6}_{\mathbf{rac}}$, where the former formed bundles of very long nanofibers up to 10 μm (Fig. 26.5b). Absorption spectral change profiles of $\mathbf{6}_{\mathbf{ena}}$ upon assembly are typical of porphyrin homoaggregates, which are a good sign for the formation

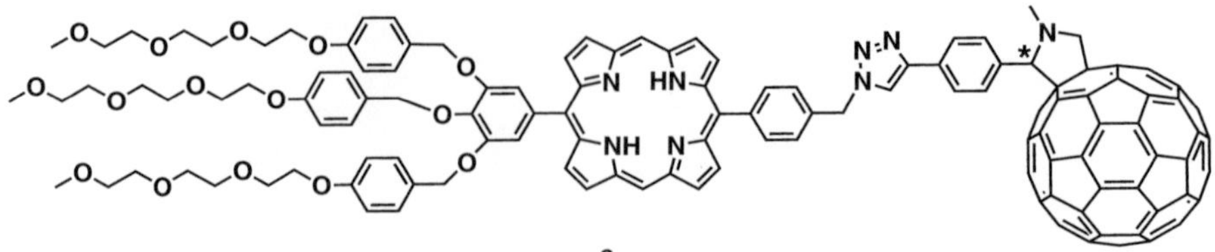

Chart 26.4 Molecular structure of porphyrin–fullerene dyad 6

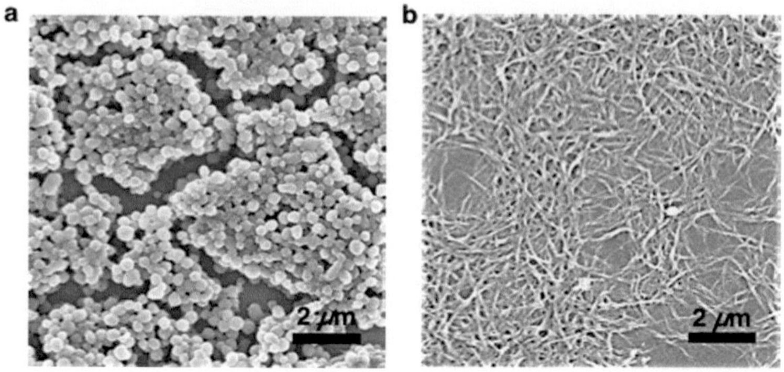

Fig. 26.5 SEM micrographs of cast films of assembled (**a**) $\mathbf{6}_{\mathbf{rac}}$ and (**b**) $\mathbf{6}_{\mathbf{ena}}$

of heterojunction structures in the fibers. Although many chiral molecules have been reported whose enantiomers and racemates self-assemble differently from one another, those giving rise to quite different morphologies such as spheres and fibers are very rare. Because of the morphological and molecular packing differences in the assemblies from the enantiomer and racemic mixture of **6**, the former exhibited an ambipolar charge-transporting character with electron and hole mobilities up to 0.14 and 0.10 $cm^2\ V^{-1}\ s^{-1}$, respectively, while the latter showed only small hole mobility ($\sim 10^{-4}\ cm^2\ V^{-1}\ s^{-1}$), as evaluated from their time-of-flight (TOF) profiles of transient photocurrents. It is noteworthy that its electron and hole mobilities in TOF were remarkably high and even better than those reported for top-class organic materials with a D–A heterojunction at that time.

26.7 Arrays of π-Electronic Metal Complexes with a Precisely Designed Sequence

Sequence-controlled organic arrays such as nucleic acids and proteins exhibit highly sophisticated functions. For their customized construction, solid-phase synthesis of these arrays at the surface of functionalized polymeric resin nowadays has become

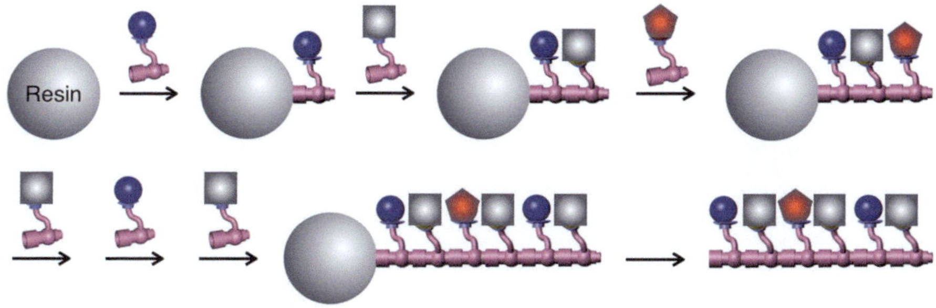

Scheme 26.2 Schematic illustrations of solid-phase synthesis of π-electronic metal complex arrays

an established and widely used methodology. In sharp contrast, its application to the metal complexes has been much less explored. In 2011 we reported our first success to control the sequence of π-electronic metal complex arrays via solid-phase synthetic technique (Scheme 26.2) [13, 14]. This approach could be a general method for producing multinuclear heterometallic complexes with controlled length, nuclearity, and sequence. To implement our idea, we functionalized tyrosine with a multidentate ligand, followed by metal binding, and then proceeded to link this unit through the tyrosine peptide bond in the usual manner. Consequently, a specific multimetallic unit of the desired length, nuclearity, and sequence can be obtained by simply controlling the number of the tyrosine units and the order with which they are added. Trace of the reactions at each step of the solid-phase synthesis was successfully done by MALDI-TOF mass spectrometry of the as-cleaved from the resin samples. It was critical to confirm that no metal exchange was taking place during the full reaction workup procedures, since any metal exchange would give unexpected sequences with disproportionate relative metal-to-metal content. Absence of metal exchange is supported by the fact that no other sequences with unexpected metal content were detected by mass spectroscopy. This approach allowed us to prove the construction of linear arrays containing up to six metal centers with a sequence such as Rh–Pt–Ru–Pt–Rh–Pt ($\mathbf{7}_{\mathbf{Rh\text{-}Pt\text{-}Ru\text{-}Pt\text{-}Rh\text{-}Pt}}$ in Chart 26.5). On the other hand, further increase of the number of metal centers in the structure looked impractical because of the nonquantitative coupling of the metal complex monomers on the resin. Moreover, the construction of branched forms is rather arduous due to the requirement of different protecting groups selectively cleavable. In order to address these issues, we newly introduce a modular approach, where the short linear arrays of metal complexes are prepared as modules by the solid-phase synthesis, and then they are coupled to make a longer array in solution (Scheme 26.3). By controlling the position of the functionality for the following coupling in the modules, not only linear but also T-branched and H-branched compounds can be also constructed. In our previous trials we experienced that the detection of the arrays by mass spectrometry became harder when they had large number of metal centers. To help the detection of the expected large

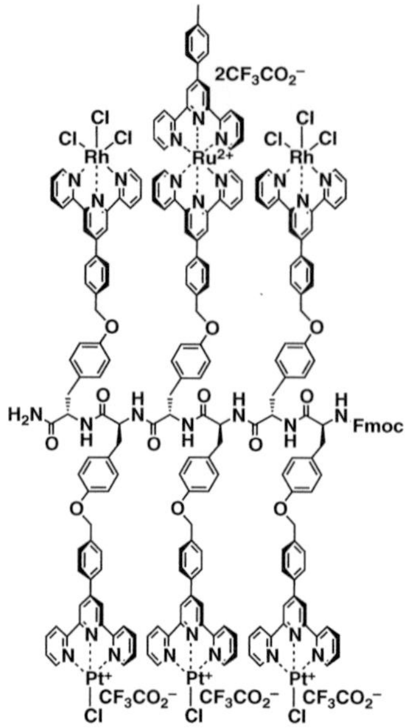

$\mathbf{7}_{\text{Rh-Pt-Ru-Pt-Rh-Pt}}$

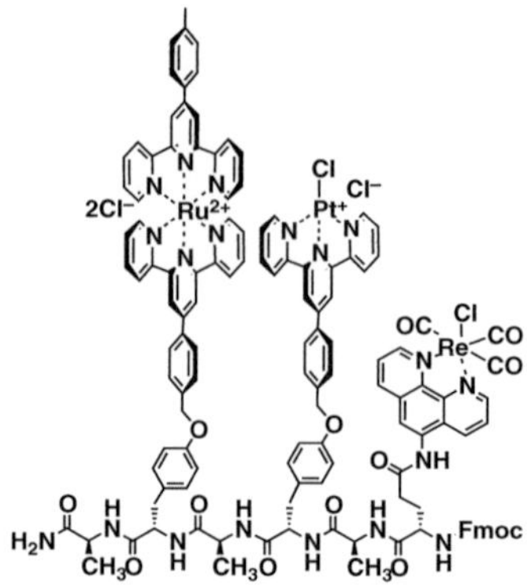

$\mathbf{8}_{\text{Ru-Pt-Re}}$

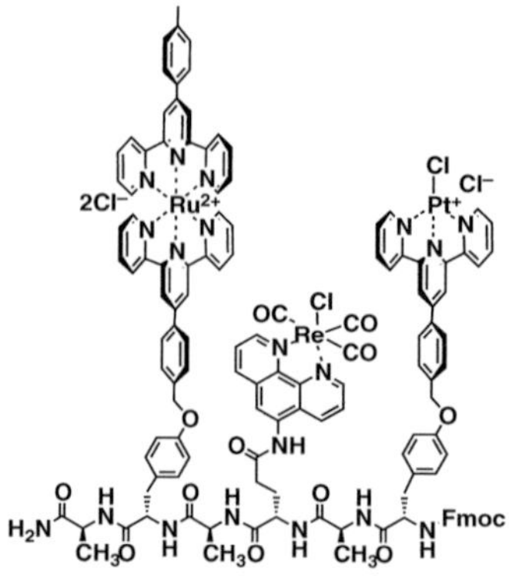

$\mathbf{8}_{\text{Ru-Re-Pt}}$

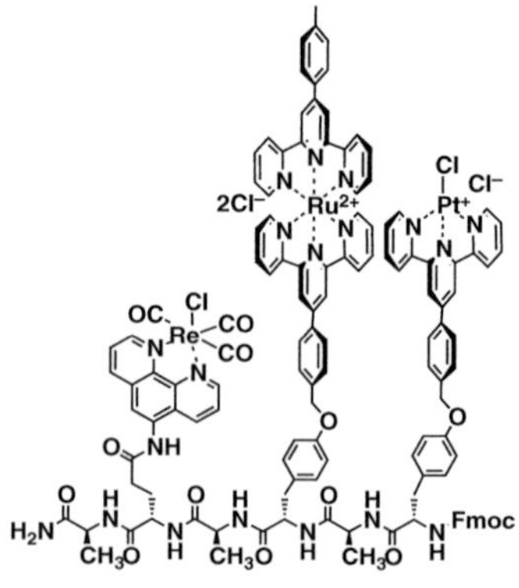

$\mathbf{8}_{\text{Re-Ru-Pt}}$

Chart 26.5 Molecular structures of π-electronic metal complex arrays **7** and **8**

arrays, we designed a porphyrin-appended glutamate derivative whose Cu(II) and Ni(II) complexes, easily detectable by MALDI-TOF mass spectrometry, were linked to Pt(II) or Ru(II) complex of terpyridine-tyrosine derivative for the construction of these modules. Based on such a design principle, we prepared six different modules bearing either of a carboxylic acid or amine functionality, from which four octanuclear arrays containing four different metals in linear, T-branched, and

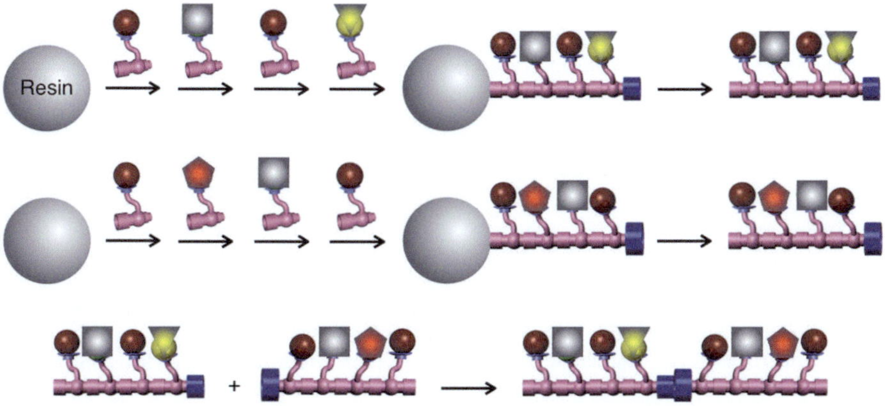

Scheme 26.3 Schematic illustrations of modular synthesis of π-electronic metal complex arrays

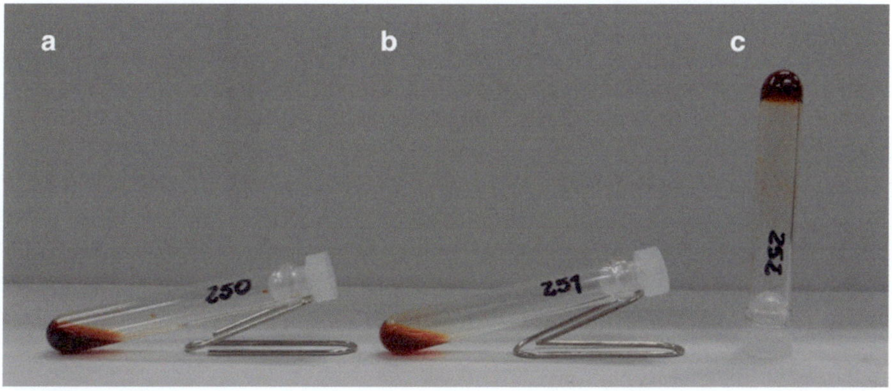

Fig. 26.6 Photograph that shows MeOH solutions prepared from (**a**) $\mathbf{8}_{\mathbf{Ru\text{–}Pt\text{–}Re}}$ and (**b**) $\mathbf{8}_{\mathbf{Ru\text{–}Re\text{–}Pt}}$ as well as (**c**) a gelified MeOH solution from $\mathbf{8}_{\mathbf{Re\text{–}Ru\text{–}Pt}}$. Concentrations of the triads in the samples are (**a**) 7.9 mM, (**b**) 7.6 mM, and (**c**) 7.9 mM

H-branched forms were obtained. It would be noteworthy that the molecular weights of the synthesized octads are close or even exceed those of small proteins such as ubiquitin (~8500).

By taking advantage of our synthetic methodology, we also synthesized three heterometallic triad species **8** with an isomeric sequence of metal centers (Chart 26.5) [15]. These are the first examples of isolated heteronuclear isomers with three or more kinds of metal centers. While the differences of the triads in their ^{1}H NMR and absorption spectra are only small, these isomers exhibit a contrasting self-assembling property (Fig. 26.6), where one of the three triads having Re–Ru–Pt sequence ($\mathbf{8}_{\mathbf{Re\text{–}Ru\text{–}Pt}}$) showed higher gelation capability than the other two isomers ($\mathbf{8}_{\mathbf{Ru\text{–}Pt\text{–}Re}}$ and $\mathbf{8}_{\mathbf{Ru\text{–}Re\text{–}Pt}}$). This observation is a good demonstration of how the sequence in a multimetallic complex can be finely controlled to impart a particular property on the complex.

Acknowledgment The author thanks all the colleagues appearing as coauthors in the referenced literatures for their important contributions to the works. These works were financially supported by the World Premier International Research Centre Initiative (WPI) on Materials Nanoarchitectonics as well as Grant-in-Aids for Scientific Research on Innovative Areas (No. 20108001, "pi-Space") and for Young Scientists (A) (No. 24685020) from MEXT, Japan.

References

1. Sakurai T, Shi K, Sato H, Tashiro K, Osuka A, Saeki A, Seki S, Tagawa S, Sasaki S, Masunaga H, Osaka K, Takata M, Aida T (2008) J Am Chem Soc 130:13812–13813
2. Sakurai T, Tashiro K, Honsho Y, Saeki A, Seki S, Osuka A, Muranaka A, Uchiyama M, Kim J, Ha S, Kato K, Takata M, Aida T (2011) J Am Chem Soc 133:6537–6540
3. Miyajima D, Tashiro K, Arakawa F, Takezoe H, Kim J, Kato K, Takata M, Aida T (2009) J Am Chem Soc 131:44–45
4. Tashiro K, Aida T, Zheng JY, Kinbara K, Saigo K, Sakamoto S, Yamaguchi K (1999) J Am Chem Soc 121:9477–9478
5. Tashiro K, Aida T (2007) Chem Soc Rev 36:189–197
6. Hajjaj F, Tashiro K, Nikawa H, Mizorogi N, Akasaka T, Nagase S, Furukawa K, Kato T, Aida T (2011) J Am Chem Soc 133:9290–9292
7. Tashiro K, Konishi K, Aida T (2000) J Am Chem Soc 122:7921–7926
8. Feringa BL (2000) Nature 408:151
9. Écija D, Auwärter W, Vijayaraghavan S, Seufert K, Bischoff F, Tashiro K, Barth JV (2011) Angew Chem 123:3958–3963; Angew Chem Int Edn 50:3872–3877
10. Tashiro K (2014) Fuller Nanotub Carbon Nanostruct 22:61–65
11. Vijayaraghavan S, Écija D, Auwärter W, Joshi S, Seufert K, Seitsonen AP, Tashiro K, Barth JV (2012) Nano Lett 12:4077–4083
12. Hizume Y, Tashiro K, Charvet R, Yamamoto Y, Saeki A, Seki S, Aida T (2010) J Am Chem Soc 132:6628–6629
13. Vairaprakash P, Ueki H, Tashiro K, Yaghi OM (2011) J Am Chem Soc 133:759–761
14. Jacoby M (2011) Chem Eng News 89:8
15. Fracaroli AM, Tashiro K, Yaghi OM (2012) Inorg Chem 51:6437–6439

Chapter 27
Integrated π-Electron Systems on Artificial Cell Membranes

Jun-ichi Kikuchi, Kazuma Yasuhara, and Keishiro Tahara

Abstract Highly elaborated π-spaces were constructed on lipid bilayer membranes by self-organization of functional π-conjugated molecules. The resulting artificial cell membranes showed dynamic performance as supramolecular devices mimicking information processing in biological systems. Membrane trafficking, including synthetic cell division and propagation of molecular capsule from sender to receiver vesicles, was achieved using π-conjugated molecules as a molecular signal or artificial receptor, respectively. Thermoresponsive and photoresponsive signal transduction behaviors were observed on supramolecular lipid bilayer membranes containing an artificial receptor with functional π-conjugated moiety, an enzyme as a signal amplifier, and a mediator species between the receptor and enzyme. The results indicated that lipid bilayer membranes are highly sophisticated platforms that can exhibit the potential of π-conjugated molecules.

Keywords Artificial cell membrane • Self-organization • Supramolecular device • Membrane trafficking • Signal transduction

27.1 Introduction

In biological systems, there are many fascinating examples of highly elaborated, molecular spaces, which perform systematic and hierarchical molecular-information processing. When we focus on cell membranes, we can notice that biomembranes, with basic lipid bilayer structures, serve as platforms for dynamic chemical processes and play important roles for the integration of various functional biomolecules and their efficient cooperation [31, 34]. Inspired by the wisdom of biological systems, it was possible here for us to use cell membrane models or lipid bilayer membranes, formed by self-assembly of lipid molecules in aqueous media,

J.-i. Kikuchi (✉) • K. Yasuhara • K. Tahara
Graduate School of Materials Science, Nara Institute of Science and Technology, 8916-5 Takayama, Ikoma, Nara 630-0192, Japan
e-mail: jkikuchi@ms.naist.jp

T. Akasaka et al. (eds.), *Chemical Science of π-Electron Systems*,
DOI 10.1007/978-4-431-55357-1_27

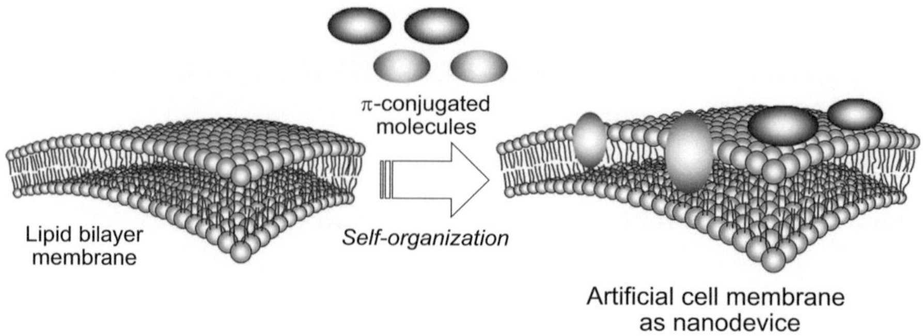

Fig. 27.1 Schematic illustration for the design of artificial cell membranes with integrated π-electron systems in lipid bilayer membranes

as a scaffold for the integration and organization of π-space. We have demonstrated some unique features of highly elaborated π-space that are induced by interactions of π-conjugated molecules with lipid bilayer membranes.

A general guidepost for designing artificial cell membranes with integrated π-space is schematically shown in Fig. 27.1. First, lipid bilayer membranes are formed by self-assembly of amphiphilic molecules in aqueous media. Second, π-conjugated molecules are integrated into the membranes through self-organization involving noncovalent intermolecular interactions. Depending on the potential of π-conjugated molecules, the resulting artificial cell membranes exhibit differing performance as intelligent molecular devices. In this chapter, we review the design, preparation, and characterization of lipid bilayer membranes, with organized π-spaces and the dynamic functions of artificial cell membranes, as supramolecular devices for membrane trafficking and signal transduction.

27.2 Artificial Cell Membranes as Platforms for Integrated π-Electron Systems

Typical examples of membrane-forming lipid molecules are shown in Chart 27.1. Lipid **1** is a well-known phospholipid, which forms bilayer vesicles called liposomes [39]. Lipids **2** and **3** are synthetic amphiphilic molecules with a hydrogen-belt-forming unit to increase the morphological stability of bilayer vesicles [7, 24]. Lipid **4** has a supplemental inorganic precursor-head moiety and forms cerasomes, liposomal membranes with ultrathin, ceramic surfaces [28, 29]. Cerasomes have unique characteristics as a new family of artificial cell membranes [14].

In general, multiwalled bilayer vesicles of submicrometer sizes are spontaneously formed upon vortex-mixing dispersion of membrane-forming lipids in aqueous media. Sonication of aqueous dispersions changes the vesicular structure from multiwalled bilayers to the corresponding small unilamellar vesicles (SUVs)

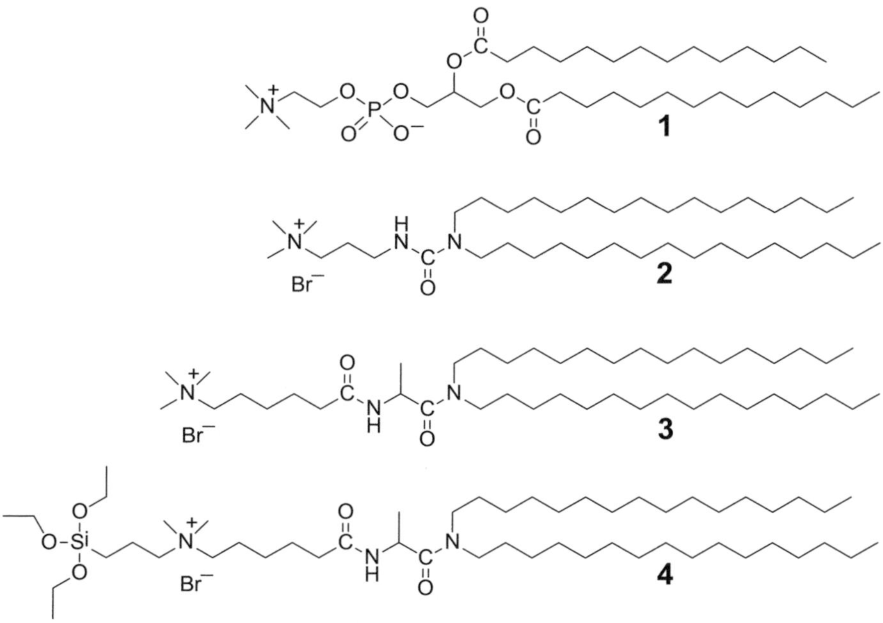

Chart 27.1 Membrane-forming lipid molecules

with diameters less than 100 nm. For sonication conditions, a few minutes use of a probe-type sonicator at 30 W power held above the lipid's phase transition temperature is generally recommended. On the other hand, large unilamellar vesicles (LUVs) with diameters of 100—200 nm are prepared from multiwalled bilayer vesicles by an extrusion technique. Cell-sized giant unilamellar vesicles (GUVs) with diameters larger than 1 μm are formed using preparation methods for giant liposomes [39]. Vesicular sizes can be evaluated by dynamic light scattering (DLS) measurement.

The aggregate structures of lipid bilayer membranes are evaluated by means of various microscopic measurements, such as transmission electron microscopy (TEM), scanning electron microscopy (SEM), atomic force microscopy (AFM), and optical microscopy. For TEM measurements of lipid membranes, we can use three kinds of imaging techniques [15]. First, negative staining with heavy metal ions is a simple and convenient technique, allowing an internal view of lipid vesicles with a bilayer thickness of 5 nm. However, this method might produce some ambiguity regarding the aggregate structure in aqueous media, as specimens must be examined after complete removal of the water phase in the presence of heavy metal-staining ions. A similar image is obtained by cryogenic TEM of frozen specimens without a staining agent. Thus, we have mainly adopted cryogenic TEM observation, which is a much more reliable imaging technique than negative staining TEM. However, freeze-fracture TEM imaging can provide information on vesicular surface structures. Typical images of SUVs and LUVs observed by

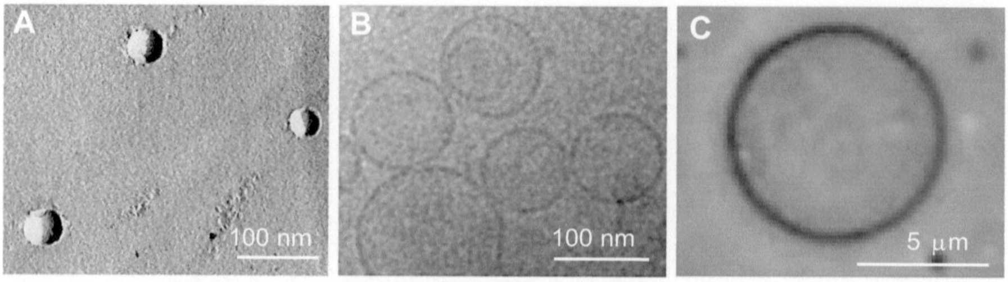

Fig. 27.2 Microscopic images of lipid bilayer vesicles: (**a**) SUVs, (**b**) LUVs, and (**c**) GUV taken by freeze-fracture and cryogenic TEM and phase-contrast optical microscopy, respectively

freeze-fracture and cryogenic TEM, respectively, are shown in Fig. 27.2a and b. In general, SEM is a challenging technique for imaging conventional liposomes formed with phospholipids because of poor sample structural integrity. In contrast, cerasomes can be visualized by SEM because of their high morphological stability. Also, the enhanced cerasome morphological stability allows clear AFM imaging. For micrometer-sized GUVs, optical microscopy is a powerful tool for dynamic imaging; Fig. 27.2c shows a typical phase-contrast image of GUV. Using an appropriate fluorescent probe, we can visualize the GUV lipid layer and inner aqueous phase, separately.

Functionalization of artificial cell membranes is achieved by integration of π-conjugated molecules on lipid membranes through noncovalent interactions. For example, water-insoluble fullerenes can be solubilized in liposomes through hydrophobic interactions in aqueous media. The resulting composites have potential as medical molecular devices for photodynamic therapy [10]. Cerasomes formed using lipid **4** show extremely low cytotoxicity and act as a cell-friendly drug or gene carrier [18, 28, 36]. Hybrid cerasomes formed from lipid **4** and hydrophobic vitamin B_{12}, the latter as a bioinspired, π-conjugated, molecular catalyst, have been successfully assembled on an electrode, maintaining vesicular structures, and the resulting interface exhibits catalytic and electrochemical activities [27]. A binary lipid system composed of a phospholipid with short hydrophobic double-chain segments and lipid **4** forms disk-type bilayer membranes, as silica bicelles, which can immobilize membrane proteins, such as rhodopsin, which has a functional π-conjugated element [43].

27.3 Dynamic Behavior of Artificial Cell Membranes as Molecular Devices

In this section, we describe characteristic functions of artificial cell membranes with integrated π-electron systems, focusing on dynamic membrane behaviors as supramolecular nanodevices for molecular-information processing.

27.3.1 Supramolecular Devices for Artificial Membrane Trafficking

In biological systems, membrane trafficking plays important roles in the inter- and intracellular transport of bioactive molecules and signals [3, 35]. Membrane trafficking consists of several different dynamic processes, such as membrane division, vesicle propagation, and membrane fusion. The detailed mechanism of each process has been clarified by biochemical and biophysical studies. However, the artificial reconstitution of membrane trafficking has not been achieved.

27.3.1.1 Synthetic Cell Division Induced by a Molecular Signal

We present here an example of the molecular signal-induced division of giant vesicles [40]. Currently, limited chemical triggers for membrane division have been reported [1, 6, 32, 33, 37, 42], which contrasts with the large number of fusion triggers reported [16, 25]. Consequently, a variety of division triggers are required to handle diverse biological information in artificial membrane traffic systems.

GUVs were prepared by gentle hydration of thin lipid films, which consisted of phospholipid **1** and synthetic cationic lipid **2** in a molar ratio of 7/3, in aqueous sucrose at 40 °C for 12 h. Initially we used a microscopy approach to observe morphological changes in GUVs induced by the addition of a water-soluble molecular signal, **5**. As the latter molecule displays green fluorescence at 510 nm with excitation at 400 nm in aqueous solution, fluorescence microscopy can directly visualize molecular signal localization on lipid membranes. The present procedure of GUV preparation quantitatively yielded GUVs with a diameter of 5 μm. The addition of **5** led to GUV membranes displaying a uniform green fluorescence (Fig. 27.3a), indicating that signal molecules were homogeneously adsorbed onto lipid membrane surfaces. However, over time, the signal molecules gradually localized to form domain structures (Fig. 27.3b). GUV morphology subsequently changed to an asymmetric pear-like shape that increased the local curvature of signal-rich domains on the lipid membranes. This GUV shape change continued until a small vesicle budded from the mother giant vesicle (Fig. 27.3c) and a small vesicle eventually released (Fig. 27.3d). Budding and fission of giant vesicles were observed with good reproducibility. Additionally, GUV morphological changes described above were not observed if GUVs were prepared with either **1** or **2** lipids, suggesting that the coexistence of at least two different lipid components was one of the requirements for the observed membrane division.

The effect of molecular signals on membrane properties has been investigated in detail using differential scanning calorimetry (DSC). DSC has been widely used to prove lateral phase separation of lipid membranes [8]. If an additive induces phase separation by accumulating a specific lipid, each lipid phase displays an independent transition temperature (T_m) closer to that of the pure lipid. In the absence of a molecular signal, we observed a broad endothermic peak at 25.4 °C for the phase

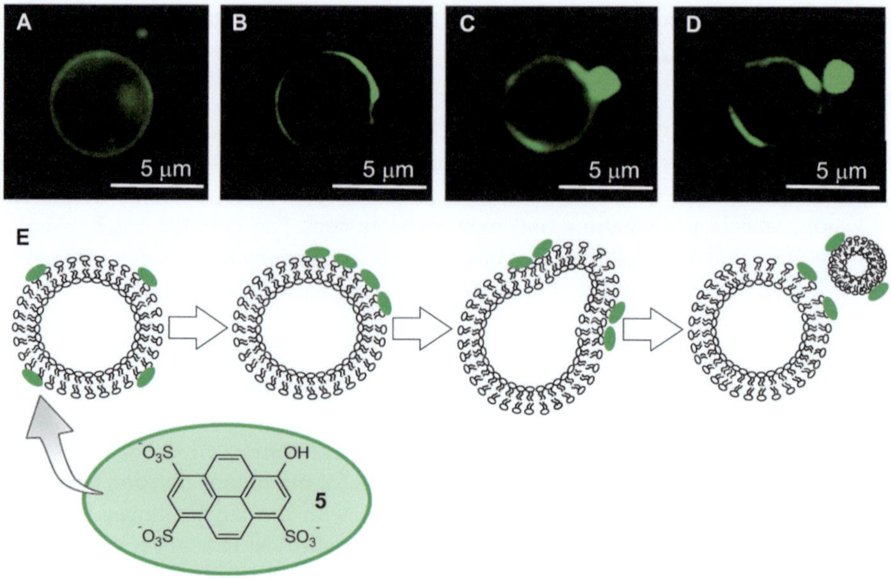

Fig. 27.3 Division of an artificial cell membrane induced by a molecular signal. Time-dependent fluorescence microscopic images (**a**, **b**, **c**, and **d**) and their schematic illustration (**e**)

transition of lipid bilayers from the gel to liquid-crystalline state, indicating that the two lipid components were miscible with each other. In contrast, addition of signal **5** led to a broadening of the endothermic peak with the concomitant appearance of a shoulder at higher temperatures. The temperature of the shoulder was close to the T_m of pure lipid **2** (32.9 °C), which showed no significant shift in the presence of signal **5**. Thus, at the observation temperature (25 °C), GUVs formed phase-separated structures comprised of lipid **1**-rich and lipid **2**-rich domains. Additionally, by increasing the signal **5** concentration, the phase transition enthalpy, with a higher T_m, increased. This response reflected an increase in the number of lipid molecules involved in the higher T_m phase transition. These results strongly suggested that the interaction of a molecular signal with mixed bilayer membranes promoted the formation of lipid **2**-rich domains in the membranes by the formation of complexes between lipid **2** and signal **5**.

Measurements of fluorescence spectra were used to evaluate the molecular signal's ionization state because **5**'s excitation spectrum reflected the 8-hydroxy group's ionization state. The excitation maximum displays a significant redshift when the 8-hydroxy group of **5** is ionized [12]. Upon the addition of mixed lipid vesicles, the molecular signal's excitation maximum shifted from 406 to 470 nm, indicating that **5**'s 8-hydroxy group was deprotonated by binding to the vesicle. This result is consistent with previous reports that showed that the surfaces of cationic liposomes display significantly higher pH values than the bulk phase [4, 45]. Thus, here, the molecular signal interacted with cationic lipid membranes as tetravalent anions, which were able to form a complex with up to four molecules of monovalent cationic lipid **2**.

Accordingly, we proposed a simple mechanism of membrane division induced by molecular signal **5** as follows. Initially, the signal homogeneously binds to the surfaces of positively charged lipid membranes. Subsequently, cationic lipids gradually accumulate to form domain structures through electrostatic interactions with the molecular signal. Formation of complexes between lipid **2** and signal **5** alters the molecular packing of lipid molecules in lipid **2**-rich domains with a concomitant decrease of local fluidity. Finally, lipid **2**-rich domains are excluded from the surrounding lipid membranes by the formation of small vesicles.

27.3.1.2 Vesicle Propagation Controlled by Artificial Receptors

We have been studying artificial, cell-based, membrane trafficking using lipid bilayer vesicles to establish bioinspired molecular communication systems. Figure 27.4 depicts propagation of molecular capsules from sender vesicles to receiver vesicles. We employed LUVs as molecular capsules because they encapsulate various information molecules in their aqueous compartments or in their hydrophobic lipid membrane domains, depending on their hydrophilicity. For sender and receiver vesicles, we chose cell-sized GUVs. Selective propagation of molecular capsules from sender to receiver was achieved by synthesizing artificial receptors and embedding them in lipid membranes to induce stimuli-responsive dynamic membrane processes.

For artifiial receptors that can control vesicle propagation, we designed gemini peptide lipids **6** and **7** (Chart 27.2). Molecular design of the gemini peptide lipids was inspired by a naturally occurring gemini lipid, cardiolipin, which possesses a unique dimeric lipid structure and interesting biological functions [9, 30]. Although

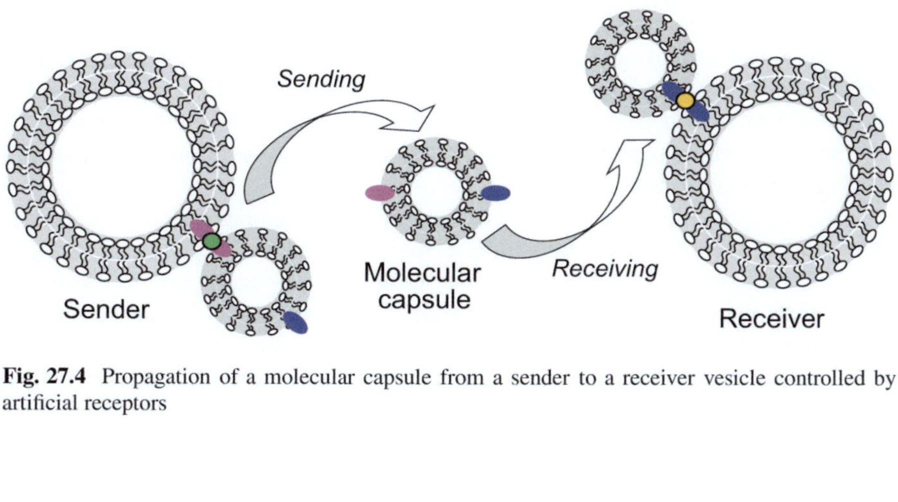

Fig. 27.4 Propagation of a molecular capsule from a sender to a receiver vesicle controlled by artificial receptors

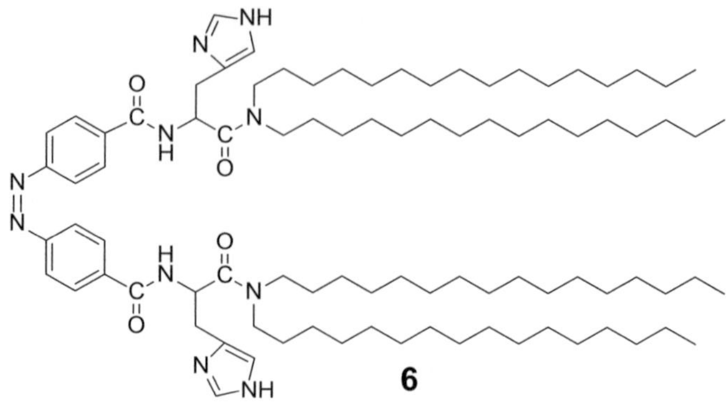

6

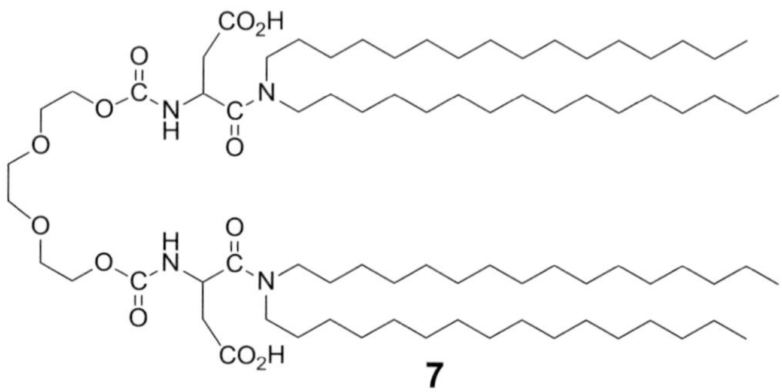

7

Chart 27.2 Gemini peptide lipids

much attention has been focused on physicochemical properties of synthetic gemini surfactants connecting two surfactant molecules via a spacer group [20, 44], there are few reports regarding synthetic gemini lipids with two double-chain segments, as in cardiolipin. These gemini peptide lipids have tripartite structures composed of two hydrophobic double-chain segments as hydrophobic anchors to the lipid membrane, two amino acid residues as ion recognition sites, and a spacer unit connecting the two amino acid moieties. Gemini peptide lipid **6** contained a photoresponsive azobenzene spacer connecting two chiral L-histidyl residues [11]. Alternatively, gemini peptide lipid **7** contained a tri(oxyethylene) spacer connecting two L-aspartate residues [26].

Gemini peptide lipid **6** embedded in lipid bilayer membranes shows ion recognition ability toward divalent transition metal ions. We have recently clarified that the Cu^{2+}-binding affinity of **6** in the *cis*-form of the azobenzene moiety was much higher than that of the corresponding *trans*-form, producing a photoresponsive assembling system of lipid vesicles [28]. Although vesicular assembly was observed by means of freeze-fracture TEM and dynamic light scattering measurements, fluorescence microscopy was unable to visualize such assembly because of fluorescence quenching by Cu^{2+} ions. When Cu^{2+} ions were replaced by Zn^{2+}

ions, the photoresponsive dynamic propagation behavior of molecular capsules to a molecular receiver was visualized by fluorescence microscopy.

Propagation of a molecular capsule to a target receiver vesicle was controlled by the photoresponsive artificial receptor **6** as follows [21]. Upon UV light irradiation, receptor molecules embedded in molecular capsules and the receiver radically change their conformation through photoisomerization of the azobenzene moieties from the *trans*-form to the corresponding *cis*-form. Judging from molecular modelling studies, metal-binding affinity of *cis*-**6** was greatly enhanced compared with the *trans*-form, as the former possesses an effective intramolecular coordination geometry of two imidazolyl ligands toward one metal ion. As Zn^{2+} ion provides four coordination sites, this ion provides stabilization by forming a 1/2 complex with *cis*-**6**. Thus, molecular capsules equipped with *cis*-**6** were capable of binding to a receiver possessing the same artificial receptor. Notably, visible light irradiation converts *cis*-**6** to the corresponding *trans*-form with less metal-binding affinity, thus prohibiting propagation of molecular capsules to the receiver.

The photoisomerization behavior of gemini peptide lipid **6** embedded in LUVs formed with lipid **1** was monitored by electronic absorption and circular dichroism (CD) spectroscopy. Upon UV light irradiation, absorbance at 333 nm by *trans*-**6** decreased with a concurrent increase in absorption at 258 and 440 nm, absorption assigned to the corresponding *cis*-isomer. A photostationary state was attained within 30 min at 30 °C, indicating efficient conversion from *trans*-**6** to *cis*-**6**, even in lipid bilayer membranes. In contrast, reverse isomerization took place upon visible light irradiation for 2 min. As this artificial receptor possessed chiral L-histidyl residues directly connected to the azobenzene moiety, the photoisomerization process was also monitored by CD spectroscopy. Induced CD spectral changes characteristic to *trans*-**6** isomerization to the *cis*-form were observed in bands of the azobenzene moiety absorption region. In lipid bilayer membranes, photoisomerization of **6** proceeded in the liquid-crystalline state but not in the gel state below the phase transition temperature. Additionally, thermal isomerization from *cis*-**6** to *trans*-**6** was extremely slow in the matrix membrane's liquid-crystalline state. For example, the half-life monitored by means of electronic absorption spectroscopy was 20 h in bilayer vesicles formed with **1** at 30 °C. Thus, all experimental data for *cis*-**6** were collected within 30 min after its photochemical formation to exclude the influence of thermal isomerization.

Interactions of Zn^{2+} ions with gemini peptide lipid **6** embedded in LUVs were evaluated by CD measurements at 30 °C. In general, metal-ligand interactions were greatly enhanced at lipid membrane-water interfaces rather than in aqueous solution. The CD spectra of receptor embedded in bilayer vesicles changed significantly upon addition of Zn^{2+} ions, reflecting Zn^{2+} ion binding to the L-histidyl residues of **6**. Titration isotherms obtained by monitoring CD intensity for *cis*-**6** and *trans*-**6** were used to determine the Zn^{2+} ion-receptor-binding constant. Job's plot analyses for the vesicular system revealed that Zn^{2+} ions bound to the molecular switch in a ratio of 1/2. Thus, the binding constant for the 1/2 complex of Zn^{2+} ions with *cis*-**1** in liposomes was determined to be 1.0×10^{10} dm^6 mol^{-2}. This was comparable to the binding constant of Cu^{2+} ions with *cis*-**6**, at 0.5×10^{10} dm^6 mol^{-2}. In contrast, the binding constant for the Zn^{2+} ion 1/2 complex with *trans*-**6** in vesicles was 9.0×10^7 dm^6 mol^{-2}, much weaker than for the corresponding *cis*-form. Accordingly, metal-ligand interactions were effectively turned off by photoisomerization of a molecular switch embedded in liposomal membranes.

The functions of the photoresponsive receptor for propagation of molecular capsules were clarified by evaluating the photoresponsive assembling behavior of LUVs equipped with **6** using DLS measurements. Hydrodynamic diameters of vesicles containing *cis*-**6** with high metal-binding affinity increased from 170 to 520 nm upon Zn^{2+} ion addition in aqueous media at 30 °C. However, vesicles containing *trans*-**6** with much lower metal-binding affinity retained their hydrodynamic diameters nearly unchanged in the presence and absence of Zn^{2+} ions under similar conditions. These results clearly indicated that differences in metal-binding ability between *cis*-**6** and *trans*-**6** in lipid bilayer membranes markedly reflected the photoresponsive assembly of liposomes.

On this basis, the photonic control of molecular capsule propagation from sender to receiver vesicles was evaluated by phase-contrast microscopic observations. LUVs and GUVs containing the receptor **6** were premixed and incubated in the presence of Zn^{2+} ions at 30 °C. Upon UV light irradiation of the mixture, LUVs, as molecular capsules, were observed as small black dots on GUV surfaces, whereas no molecular capsules were observed on GUVs with visible light irradiation. These results suggested that selective propagation of molecular capsules from sender to receiver vesicles could be attained if light irradiation to the sender and receiver vesicles was controlled separately. Photonic control of propagation behavior was confirmed by staining LUVs and GUVs with red and green fluorescent dyes, respectively. Fluorescence microscopic observations clearly revealed that red-labeled LUVs propagated onto green-labeled GUV upon UV light irradiation, while visible light irradiation yielded only images of green GUVs.

Propagation of molecular capsules from sender to receiver vesicles was also achieved using a combination of gemini peptide lipids **6** and **7** with Zn^{2+}- and Ca^{2+}-binding affinity, respectively. That is, we prepared two kinds of GUVs containing **6** or **7** as a sender or receiver, respectively, and LUVs equipped with both **6** and **7** as molecular capsules. Upon UV light irradiation of the mixture, LUVs were loaded on the sender vesicle in the presence of Zn^{2+} ions. When irradiation was changed from UV to visible light, molecular capsules were sent from vesicular surfaces into the bulk aqueous phase. Finally, molecular capsules collected on receiver vesicles upon addition of Ca^{2+} ions. Thus, molecular capsules were successfully delivered from sender to receiver vesicles by multiple photonic and ionic signals, which were recognized by gemini peptide lipids, as schematically shown in Fig. 27.4.

27.3.2 Supramolecular Devices for Artificial Signal Transduction

Recently, much effort has been devoted to develop nanoscale devices using molecules or molecular devices composed of molecular elements, such as switches, wires, and logic gates, and capable of extending current semiconductor technology to nanoscale information technology [2, 41]. However, integration of these functional elements to produce real molecular devices remains a challenge. A biologically inspired approach might present a unique solution for achieving integrated system architectures that will orchestrate large numbers of molecular devices inside future nanomachines. Thus, our recent attention was focused on functional simulation of biological signal transduction systems by employing self-organized molecular assemblies in aqueous media. A signal transduction system located in cell membranes is an example of naturally occurring nanodevices, in which signal transmission among functional biomolecules, such as receptors and enzymes, is efficiently achieved in membranes [17]. Recently, we have developed artificial cell membrane-type nanodevices, employing a concept inspired by biological signal transduction and entailing a system essentially comprised of three molecular components: an artificial receptor, enzyme, and lipid bilayer membrane (Fig. 27.5). The receptor and enzyme were self-assembled on membranes through noncovalent interactions and the enzyme's catalytic activity controlled by an external signal subsequent to receptor activation, with signal transduction mediated by metal ions. In this system, the receptor adjusted enzymatic activity, being dependent on molecular recognition of a specific signal.

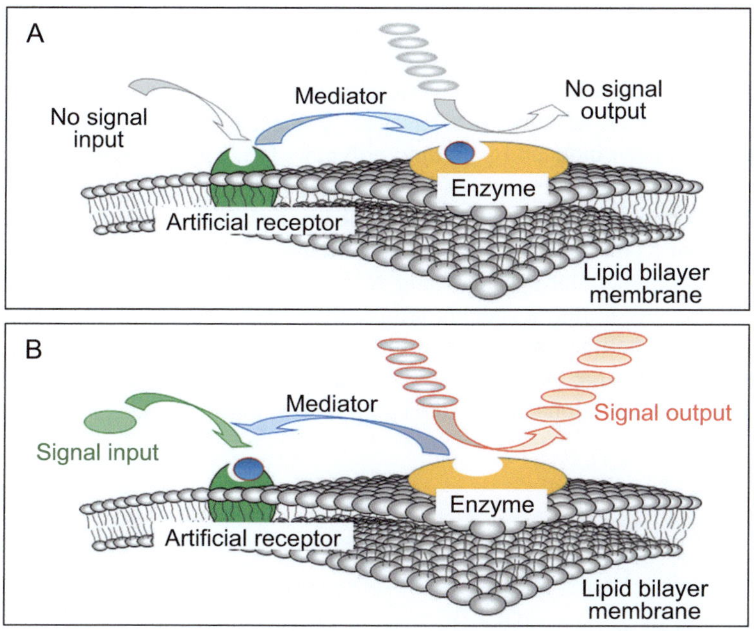

Fig. 27.5 Switching of enzymatic activity on an artificial cell membrane mediated by an artificial receptor. (**a**) and (**b**) represent the off and on enzyme state, respectively

27.3.2.1 Switching of Enzymatic Activity by a Thermoresponsive Artificial Receptor

In this section, we address the construction of a bioinspired molecular device that senses membrane fusion by changes in membrane-bound enzymatic activity [22]. Membrane fusion is one of the most fundamental processes in biological systems, being involved in cargo transport through secretory pathways, fertilization, organelle inheritance, and viral entry into host cells, but there have been few reports of molecular devices sensing membrane fusion. The present system functioned through cooperation of a thermoresponsive receptor and a natural enzyme with a signal mediator, as a means of converting lipid bilayer membrane state changes into a measurable enzyme response. A membrane platform was constructed using phospholipid **1**, cationic peptide lipid **3**, and three functional elements: a Schiff's base of pyridoxal 5'-phosphate with phosphatidylethanolamine as a thermoresponsive artificial receptor **8** (Chart 27.3), NADH-dependent L-lactate dehydrogenase (LDH) as an effector, and Cu^{2+} ions as the signal mediator.

The design strategy for this molecular device involved taking advantage of specific attributes of the three incorporated components. The receptor's ability to change its binding affinity toward metal ions, depending on the lipid bilayer membrane phase state, was one of the most important properties for the receptor molecule in this fusion-sensing system. When the membrane was in a gel state, the receptor possessed higher binding affinity for the signal mediator than did the enzyme, which resulted in an enzymatically active state, or "on state" (Fig. 27.5b). It is well known that lipid bilayer vesicle fusion can be induced by a membrane-interacting polymer, serving as a fusogen. For example, the fusion of cationic liposomes has been induced by anionic polymers [13, 19]. Here, when membrane fusion occurred in the presence of lipid bilayer vesicles functionalized with both receptor and enzyme, the matrix lipid composition shifted during fusion, changing the membrane's phase state from gel to liquid crystal. We previously clarified that a similar supramolecular system using receptor **8** and LDH markedly decreased enzymatic activity within a specific temperature range, reflecting the membrane's gel to liquid-crystalline phase transition [38]. This was due to increased mediator binding to the enzyme as a result of decreased mediator binding affinity toward the receptor. As Cu^{2+} ions are an LDH competitive inhibitor, this molecular device

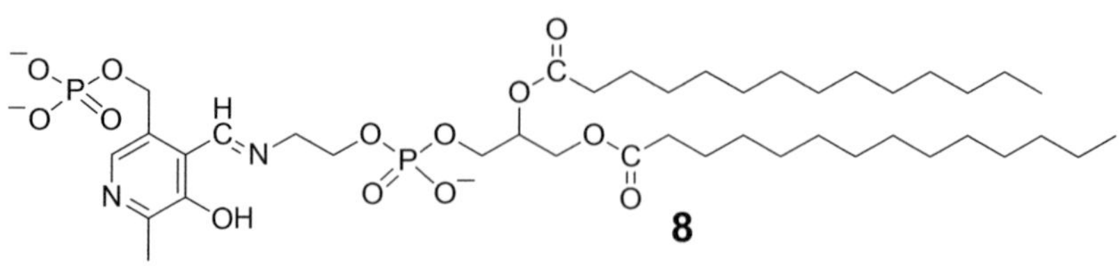

Chart 27.3 Thermoresponsive artificial receptor

turned the system to the enzymatically inactive state, or "off state" (Fig. 27.3a). Thus, the present membrane device could detect membrane fusion by translating the state change into an enzymatic response.

The effect of the membrane lipid matrix on enzymatic activity was evaluated using multilamellar liposomes formed from lipids **1** and **3** at various molar ratios. The transition temperature from the gel to liquid-crystalline phase was observed by DSC. The results indicated that membrane phase transitions were easily controlled by the lipid molar ratio. The activity of LDH immobilized on membranes, also functionalized with the receptor and mediator, was evaluated in aqueous media at 35 °C. The reduction rate of pyruvate to L-lactate was monitored spectrophotometrically by following the concurrent NADH consumption rate, and LDH activity represented as the magnitude of reactivity in the presence of Cu^{2+} ions relative to the copper-free system. The lipid matrix dependence of LDH activity showed good correlation with the phase transition temperature, and enzymatic activity was clearly capable of switching on and off below and above the phase transition temperature, respectively. Receptor **8** decreased the Cu^{2+} ion-binding constant in the liquid-crystalline state compared with the gel state. In the latter state, receptor **8** exhibited higher Cu^{2+} ion affinity than the enzyme, resulting in an enzymatically active state. However, the liquid-crystalline state allowed the enzyme to bind Cu^{2+} ions more strongly than the receptor and switched off activity, the Cu^{2+} ions competitively inhibiting LDH. These results indicated that if the present functionalized vesicles changed their phase state upon membrane fusion, fusion events could be detected by changes in enzymatic activity.

Fusion efficiency upon addition of water-soluble polymers to cationic bilayer vesicles formed with lipids **1** and **3** has been evaluated by fluorescence resonance energy transfer (FRET) using a pair of fluorescent phosphatidylethanolamine derivatives containing a rhodamine (Rh-PE) or *N*-(7-nitrobenz-2-oxa-1,3-diazol-4-yl) moiety (NBD-PE) [5]. The results indicated that an anionic polymer, poly(viny1 sulfate) potassium salt (PVSK), is bound on cationic vesicular surfaces through electrostatic interactions to induce membrane fusion, probably because of dehydration and disruption of membrane surfaces. However, a nonionic polymer, poly(vinyl alcohol) (PVA), exhibited little effect on cationic vesicle fusion, compared with anionic PVSK. Thus, in the present cationic vesicular system, anionic PVSK was found to be the more effective fusogen compared with nonionic PVA. The fusogenic behavior of PVSK was also confirmed by fluorescence microscopy and DSC measurements.

Based on the above results, the performance of functionalized vesicles as molecular devices for sensing membrane fusion was kinetically evaluated by means of electronic absorption spectroscopy in aqueous media at pH 7.0 and 35 °C. In homogeneous aqueous solutions, LDH activity was reduced to 4 % of the Cu^{2+}-free system. Upon addition of polymers, such as PVA or PVSK, enzymatic activity was comparable to polymer-free systems, indicating that these polymers did not influence enzymatic activity. The enzyme inhibition response to Cu^{2+} ions in the presence of polymers was also analogous to the polymer-free system. In a membrane system composed of lipids **1** and **3** at an 8/2 molar ratio containing receptor **8**, relative enzymatic activity in the presence of Cu^{2+} ions was 68 % of the Cu^{2+}-free system. The higher catalytic activity in membrane systems than that without membranes in the presence of Cu^{2+} ions strongly suggested that the pyridoxal moiety of receptor **8** acted as an effective metal-binding site. That is, as the vesicles were in a gel state under the present conditions, receptor **8** Cu^{2+} ion-binding affinity was much higher than the enzyme, and, thus, LDH activity drastically recovered in functionalized membranes compared with the homogeneous aqueous solution. In the mixed membrane system, containing vesicles composed of lipids **1** and **3** at both 8/2 and 5/5 molar ratios, LDH activity in the presence of Cu^{2+} ions and PVA was 62 % of the Cu^{2+}-free system. The Cu^{2+}-inhibition effect was comparable to the corresponding polymer-free system, as PVA scarcely induced membrane fusion. However, in the mixed membrane system containing PVSK instead of PVA, LDH activity in the presence of Cu^{2+} ions was 42 % of the Cu^{2+}-free system. The difference in LDH activities in the presence of Cu^{2+} ions upon addition of either polymer clearly reflected the difference in their fusion efficiencies. At 35 °C, membranes composed of lipids **1** and **3** at 8/2 and 5/5 molar ratios were in a gel and liquid-crystalline state, respectively. When PVSK induced fusion of these functionalized vesicles, fused vesicles transitioned from the gel to liquid-crystalline state. Thus, receptor-binding affinity toward the mediator decreased, enhancing mediator-caused enzyme inhibition. In summary, a novel, bioinspired molecular device was created in which enzymatic activity on lipid bilayer membranes was switched on or off by a thermoresponsive receptor sensitive to changes in the membrane phase state.

27.3.2.2 Switching of Enzymatic Activity by a Photoresponsive Artificial Receptor

We chose here the gemini peptide lipid **6** to serve as a receptor in an artificial signal transduction system [21, 23]. The basic platform was constructed with bilayer vesicles of cationic peptide lipid **3**. Receptor **6** and NADH-dependent LDH were immobilized on membranes mainly through hydrophobic and electrostatic interactions, respectively. As mentioned above, the receptor **6** embedded in such membranes drastically changed Cu^{2+}-binding affinity through photoisomerization of the azobenzene moieties. Thus, in this signalling system, photonic signal input to the receptor could be converted to an amplified chemical signal output at the enzyme through translocation of Cu^{2+} ions, as a mediator between them.

LDH activity was evaluated in terms of the magnitude of the initial velocity in the presence of Cu^{2+} ions relative to that of the corresponding metal-free system. The enzymatic activity in the signalling system containing *trans*-**6** was drastically decreased with an increase in Cu^{2+} concentration, mainly reflecting binding of metal ions to the enzyme active site, as the metal-binding ability of receptor in the *trans*-form was much lower than that of the enzyme. In contrast, the enzymatic activity in the signalling system containing *cis*-**6** showed higher activity compared with that containing *trans*-**6**. These results clearly indicated that signal amplification, as an enzymatic reaction, in the present membrane devices was switchable through photoisomerization of the gemini peptide lipid.

The present signalling system was highly specific to metal ions as the mediator species between the receptor and enzyme. LDH activities on membranes formed with peptide lipid **3** containing molecular switch **6** in the *cis*- or *trans*-form were evaluated in the presence and absence of various metal ions. While enzymatic activity on membranes was markedly altered from 18 to 74 % through photoisomerization of *trans*-**6** to *cis*-**6** in the presence of Cu^{2+} ions, other metal ions, such as Zn^{2+}, Ni^{2+}, and Co^{2+}, did not perform as effective mediators between the receptor and enzyme. The receptor possessed analogous binding affinity toward Cu^{2+} and Zn^{2+} ions, whereas enzymatic activity was specifically inhibited by Cu^{2+} but not by Zn^{2+} ions. Thus, Cu^{2+} ions, exhibiting strong affinity toward both the receptor and the enzyme, acted as a specific mediator in this signal transduction system.

The structural importance of the gemini peptide lipid, having an azobenzene moiety and two L-histidyl residues as the photoresponsive receptor, was clarified by replacing receptor **6** with three reference lipids: a non-gemini peptide lipid with an L-histidyl residue, a gemini peptide lipid with an azobenzene spacer unit but lacking two imidazolyl groups, and a gemini peptide lipid with two L-histidyl residues but lacking an azobenzene moiety. LDH activities on the membrane formed with **3** containing these molecular switch analogs were evaluated upon UV or visible light irradiation. The results clearly indicated that the gemini structure with both an azobenzene spacer unit and two L-histidyl residues was essential for performing as a photoresponsive receptor to control enzymatic activity on the lipid membrane.

27.4 Conclusions

Molecular information processing in cells includes membrane trafficking and signal transduction. Such biological processes were reconstituted on artificial cell membranes. For example, generation of carrier vesicles from a sender vesicle formed with bilayer-forming lipids was induced by the input of a polyanionic π-conjugated molecule. Selective propagation of carrier vesicles from sender to receiver vesicles was achieved by employing appropriate artificial receptors with π-conjugated functional units. Gemini peptide lipid having a photoresponsive azobenzene moiety and two imidazolyl groups served as an effective receptor, and membrane trafficking was controlled by dual input signals, as photonic and ionic signals. When an artificial receptor with a functional π-conjugated moiety, showing molecular recognition toward an input signal, was integrated in lipid bilayer membranes with a mediator species and an enzyme as an effector, signal transduction behavior from a receptor to an effector was successfully observed. Accordingly, lipid bilayer membranes can provide highly elaborated reaction fields to bring out the potential of π-conjugated molecules.

References

1. Andes-Koback M, Keating CD (2011) Complete budding and asymmetric division of primitive model cells to produce daughter vesicles with different interior and membrane compositions. J Am Chem Soc 133:9545–9555
2. Balzani V, Credi A, Venturi M (2008) Molecular devices and machines: concepts and perspectives for the nanoworld, 2nd edn. Wiley-VCH, Weinheim
3. Behnia R, Munro S (2005) Organelle identity and the signposts for membrane traffic. Nature 438:597–604
4. Fendler J, Hinze WL (1981) Reactivity control in micelles and surfactant vesicles – kinetics and mechanism of base-catalyzed-hydrolysis of 5,5'-dithiobis(2-nitrobenzoic acid) in water. J Am Chem Soc 103:5439–5447
5. Glaser PE, Gross RW (1994) Plasmenylrthanolamine facilitates rapid membrane fusion: A stopped flow kinetic investigation correlating the propensity of a major plasma membrane constituent to adopt an H_{II} phase with its ability to promote membrane fusion. Biochemistry 33:5805–5812
6. Hamada T, Miura Y, Ishii K, Atraki S, Yoshikawa K, Vestergaard M, Takagi M (2007) Dynamic processes in endocytic transformation of a raft-exhibiting giant liposome. J Phys Chem B111:10853–10857
7. Hashizume M, Kawanami S, Iwamoto S, Isomoto T, Kikuchi J (2003) Stable vesicular nanoparticle 'cerasome' as an organic-inorganic hybrid formed with organoalkoxysilane lipids having a hydrogen-bonding unit. Thin Solid Films 438–439:20–26
8. Heimburg T (2007) Thermal biophysics of membranes. Wiley-VCH, Weinheim
9. Hoch FL (1992) Cardiolipins and biomembrane function. Biochim Biophys Acta 1113:71–133
10. Ikeda A, Matsumoto M, Akiyama M, Kikuchi J, Ogawa T, Takeya T (2009) Direct and short-time uptake of [70]fullerene into the cell membrane using an exchange reaction from a [70]fullerene-γ-cyclodextrin complex and the resulting photodynamic activity. Chem Commun 12:1547–1549
11. Iwamoto S, Otsuki M, Sasaki Y, Ikeda A, Kikuchi J (2004) Gemini peptide lipids with ditopic ion-recognition site. Preparation and functions as an inducer for assembling of liposomal membranes. Tetrahedron 60:9841–9847
12. Kano K, Fendler J (1978) Pyranine as a sensitive pH probe for liposome interiors and surfaces. pH gradients across phospholipid vesicles. Biochim Biophys Acta 509:289–299
13. Keren-Zur M, Beigel M, Loyter A (1989) Induction of fusion in aggregated and nonaggregated liposomes bearing cationic detergents. Biochim Biophys Acta 983:253–258
14. Kikuchi J, Yasuhara K (2011) Cerasomes: a new family of artificial cell membranes with ceramic surface. In: George A (ed) Advances in biomimetics. Intech, Rijeka
15. Kikuchi J, Yasumhara K (2012) Transmission electron microscopy (TEM). In: Gale PA, Steed JW (eds) Supramolecular chemistry from molecules to nanomaterials, vol 2. Wiley, Chichester, pp 633–645
16. Kikuchi J, Ariga K, Sasaki Y (2002) Molecular recognition and functional connection in lipid membranes. In: Gokel GW (ed) Advances in supramolecular chemistry, vol 8. Cerberus Press, Miami, pp 131–173
17. Krauss G (2003) Biochemistry of signal transduction and regulation. Wiley-VCH, Weinheim
18. Matsui K, Sando S, Sera T, Aoyama Y, Sasaki Y, Komatsu T, Terashima T, Kikuchi J (2006) Cerasome as an infusible, cell-friendly, and Serum-compatible transfection agent in a viral size. J Am Chem Soc 128:3114–3115
19. Matsui K, Sasaki Y, Komatsu T, Mukai M, Kikuchi J, Aoyama Y (2007) RNAi gene silencing using cerasome as a viral-size siRNA-carrier free from fusion and cross-linking. Bioorg Med Chem Lett 17:3935–3938
20. Menger FM, Keiper JS (2000) Gemini surfactants. Angew Chem Int Ed 39:1907–1920

21. Mukai M, Marui K, Kikuchi J, Sasaki Y, Hiyama S, Moritani Y, Suda T (2009) Propagation and amplification of molecular information using a photo-responsive molecular switch. Supramol Chem 21:284–291
22. Mukai M, Sasaki Y, Kikuchi J (2012) Fusion-triggered switching of enzymatic activity on artificial cell membrane. Sensors 12:5966–5977
23. Mukai M, Maruo K, Sasaki Y, Kikuchi J (2012) Intermolecular communication on a liposomal membrane. Enzymatic amplification of a photonic signal with gemini peptide lipid as a membrane-bound artificial receptor. Chem Eur J 18:3253–3263
24. Murakami Y, Kikuchi J (1991) Supramolecular assemblies formed with synthetic peptide lipids. Functional models of biomembranes and enzymes. In: Dugas H (ed) Bioorganic chemistry frontiers, vol 2. Springer, Berlin, pp 73–113
25. Noble GT, Mart RJ, Liem KP, Webb SJ (2012) Supramolecular chemistry of membranes. In: Gale PA, Steed JW (eds) Supramolecular chemistry from molecules to nanomaterials, vol 4. Wiley, Chichester, pp 1731–1749
26. Otsuki M, Sasaki Y, Iwamoto S, Kikuchi J (2006) Liposomal sorting onto substrate through ion recognition by gemini peptide lipids. Chem Lett 35:206–207
27. Qiao Y, Tahara K, Zhang Q, Song XM, Hisaeda Y, Kikuchi J (2014) Construction of molecular communication interface formed from cerasome and hydrophobic vitamin B_{12} on glassy carbon electrode. Chem Lett 43:684–686
28. Sasaki Y, Matsui K, Aoyama Y, Kikuchi J (2006) Cerasome as an infusible and cell-friendly gene carrier: synthesis of cerasome-forming lipids and transfection using cerasome. Nat Protoc 1:1227–1234
29. Sasaki Y, Iwamoto S, Mukai M, Kikuchi J (2006) Photo- and thermo-responsive assembly of liposomal membranes triggered by a gemini peptide lipid as a molecular switch. J Photochem Photobiol A Chem 183:309–314
30. Schlame M, Rua G, Greenberg ML (2000) The biosynthesis and functional role of cardiolipin. Prog Lipid Res 39:257–288
31. Simons K (ed) (2011) The biology of lipids: trafficking, regulation, and function. Cold Spring Harbor Lab Press, New York
32. Sommerdijk NAJM, Hoeks THL, Synak M, Feiters MC, Nolte RJM, Zwaneburg M (1997) Stereodependent fusion and fission of vesicles: calcium binding of synthetic gemini phospholipids containing two phosphate groups. J Am Chem Soc 119:4338–4344
33. Staneva G, Seigneuret M, Koumanov K, Trugnan G, Angelova MI (2005) Detergents induce raft-like domains budding and fission from giant unilamellar heterogeneous vesicles – a direct microscopy observation. Chem Phys Lipids 136:55–66
34. Stillwell W (2013) An introduction to biological membranes from bilayers to rafts. Academic, London
35. Szostak JW, Bartel DP, Luisi PL (2001) Synthesizing life. Nature 409:387–390
36. Tahara K, Moriuchi T, Tsukui M, Hirota A, Maeno T, Toriyama M, Inagaki N, Kikuchi J (2013) Ceramic coating of liposomal gene carrier for minimizing toxicity to primary hippocampal neurons. Chem Lett 42:1265–1267
37. Takakura K, Sugawara T (2004) Membrane dynamics of a myelin-like giant multilamellar vesicle applicable to a self-reproducing system. Langmuir 20:3832–3834
38. Tian WJ, Sasaki Y, Fan SD, Kikuchi J (2005) Intermolecular communication on lipid bilayer membrane. Tuning of enzymatic activity with phase transition of the matrix membranes. Bull Chem Soc Jpn 78:715–717
39. Torchilin VP, Weissig V (2003) Liposomes, 2nd edn. Oxford University Press, Oxford
40. Wang Z, Yasuhara K, Ito H, Mukai M, Kikuchi J (2010) Budding and fission of cationic binary lipid vesicles induced by the incorporation of pyranine. Chem Lett 39:54–55
41. Waser R (ed) (2003) Nanoelectronics and information technology: advanced electronic materials and novel devices. Wiley-VCH, Weinheim

42. Yamashita Y, Masum SM, Tanaka T, Yamazaki M (2002) Shape changes of giant unilamellar vesicles of phosphatidylcholine induced by a de novo designed peptide interacting with their membrane interface. Langmuir 18:9638–9641
43. Yasuhara K, Miki S, Nakazono H, Ohta A, Kikuchi J (2011) Synthesis of organic-inorganic hybrid bicelles. Lipid bilayer nanodiscs encompassed by siloxane surfaces. Chem Commun 47(16):4691–4692
44. Zana R, Xia J (eds) (2004) Gemini surfactants. Marcel Dekker, New York
45. Zuidam NJ, Barenholz Y (1997) Electrostatic parameters of cationic liposomes commonly used for gene delivery as determined by 4-heptadecyl-7-hydroxycoumarin. Biochim Biophys Acta 1329:211–222

Chapter 28
Supramolecular Porphyrin Nanorods for Light Energy Conversion

Taku Hasobe and Hayato Sakai

Abstract Recent developments of supramolecular bar-shaped molecular architectures of porphyrins are widely described. The utilization of synthetic and supramolecular techniques enables one to effectively organize molecular aggregates, which possess light energy conversion functionalities such as light harvesting, charge separation, and carrier transport. In the photovoltaic and photocalytic (hydrogen evolution) measurements, these highly organized architectures quantitatively exhibit drastic enhancements of light energy conversion properties as compared to the reference non-organized system.

Keywords Porphyrin • Light energy conversion • Nanorods

28.1 Introduction

There is a growing need to transition from fossil fuels to renewable energy sources with the significant rise in global energy demand to maintain a clean atmosphere and reduce global warming. To meet the increasing energy demands in the future and solve the environmental problems, sustainable and clean energy sources are certainly required. Among renewable energy sources, sunlight is the most attractive and largest exploitable resource because the Sun provides a huge amount of light energy to the earth (annual basis at a rate of $\sim 10^{17}$ W). Therefore, recent attentions have been drawn toward the development of solar energy conversion such as photovoltaic [1–6] and photocalytic [7–10] systems using solar energy. Solar cells convert light energy directly into electricity by the photovoltaic effect. The mechanism of energy conversion is generally as follows: (1) light harvesting and exciton diffusion, (2) charge separation, and (3) resulting carrier transport [11, 12]. Photocatalytic systems such as hydrogen evolution are also composed of three-component units such as a photosensitizer, an electron relay, and a catalyst [13–15].

T. Hasobe (✉) • H. Sakai
Department of Chemistry, Faculty of Science and Technology, Keio University, Yokohama 223-8522, Japan
e-mail: hasobe@chem.keio.ac.jp

T. Akasaka et al. (eds.), *Chemical Science of π-Electron Systems*,
DOI 10.1007/978-4-431-55357-1_28

In both cases, light absorption and the subsequent electron transfer via the excited states are consequently essential for the final energy conversion processes [11].

Molecular architecture based on self-association is a simple and convenient way to construct organized assemblies [16]. The possibility of controlling macroscopic aggregate formations through a proper choice of the molecular components opens a way to design and synthesize materials capable of exhibiting specific properties. These organizations from molecular aggregates to controllable macroscopic architectures and materials with advanced functions may meet the requirement of many objectives such as photocatalytic systems and electronic and optical devices. Therefore, such a supramolecular technique is also considered to be one of the useful strategies to control the functionalities for light energy conversion such as light harvesting and charge separation properties in aggregate states [17, 12].

Porphyrins used as an electron donor as well as a sensitizer are also suitable for efficient electron transfer with small reorganization energies [18, 19]. In addition, rich and extensive absorption features of porphyrinoid systems guarantee increased absorption cross sections and an efficient use of the solar spectrum [20]. Moreover, porphyrins are also promising building blocks for supramolecular systems [17]. So far, various nanoassemblies composed of these derivatives with different sizes and shapes are widely reported [21–45]. In this section, the focus is on the recent advances in the photofunctional bar-shaped molecular architectures of porphyrin derivatives. The details on preparation and structural and photophysical properties of molecular assemblies are discussed here.

28.2 Results and Discussion

28.2.1 Sonication-Assisted Supramolecular Nanorods of Porphyrins

Ultrasound irradiation is useful for preparation of crystal nucleus and molecular crystals, but seldom favors the formation of ordered aggregates. Novel supramolecular nanorods and fibers controlled by intermolecular interactions such as π–π stacking interaction without additional bonds/interactions or aiding surfactants were prepared in a sonicated solution medium [46].

We employed 5,15-diaryl-substituted porphyrins (H_2DBuPP and H_2DPP) and the fully *meso*-substituted reference porphyrins, 5,10,15,20-tetraaryl-substituted porphyrins (H_2TBuPP and H_2TPP), to examine a substituent effect on the assembled structures (Fig. 28.1). Porphyrin nanorods were prepared by the following procedure. First, 3.5 mM porphyrin solution was prepared in toluene. Then, the toluene solution was simply mixed with 9 times volume of acetonitrile. The final concentration is 0.35 mM in a poor/good solvent = 9/1, v/v. The solution was sonicated (45 kHz) for 30 min at 15 °C to organize macroscopic assemblies of porphyrins by using the original setup [46].

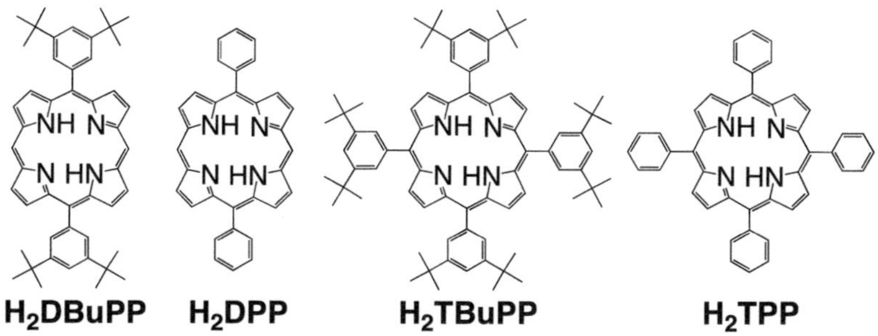

Fig. 28.1 Porphyrin compounds bearing different substituents at *meso*-positions in this study

Figure 28.2a–f shows electron and optical microscope images of H_2DBuPP and H_2DPP nanorods, reference systems, respectively. In contrast with large square crystal structures (~2–3 μm) of the H_2PTBuPP assembly (Fig. 28.2c), H_2PDBuPP nanorods (Fig. 28.2a) form a lot of long bar-shaped supramolecular structures. The structures of H_2PDBuPP nanorods are analyzed as 360 ± 130 nm in diameter and 5.02 ± 1.94 μm in length. The difference between H_2PDPP and H_2PTPP (Fig. 28.2b and d) is also similar to that between H_2PDBuPP and H_2PTBuPP. To construct much longer fiber structures, the suspended solution of nanorod is further allowed to stand for 6 days. Then, the crystallized porphyrins were stirred for 5 min at room temperature. The nanorod assemblies finally grow to become much longer fiber structures. In Fig. 28.2e and f, one can clearly see much longer fiber structures, which were analyzed as 890 ± 270 nm in diameter and 27.2 ± 6.9 μm in length. In the formation of fiber structures, a large increase of length from 5.02 to 27.2 μm (~6 times) relative to that of diameter (~2 times) indicates that anisotropic crystal growth largely occurs toward the length direction. Thus, we have successfully controlled the structure of porphyrin fibrous assemblies within a certain definite range [46]. Additionally, the internal structures of porphyrin aggregates were investigated by the XRD method. The strong diffraction peaks of *a* and *c* axes such as (100) and (002) in the unit cell were observed, whereas the diffraction intensity based on *b* axis is very weak. Considering the unit cell structure (monoclinic structure: dihedral angles 90° in axes *a*–*b* and *b*–*c*), growth direction of the rod crystals may be in the *b* axis direction [46].

Figure 28.3a shows the absorption spectrum of H_2DBuPP nanorods (spectrum *a*), which exhibits much broader and more intense absorption in the visible and near-infrared regions than those of the corresponding simple drop-cast film prepared by H_2DBuPP in toluene (spectrum *b*) and the monomeric form in toluene (spectrum *c*). This drastic change of absorption property is likely ascribed to the strong supramolecular $\pi - \pi$ interaction between a porphyrin and its nearest neighbor in nanorod assemblies.

To further investigate the photoresponse behavior, a photoelectrochemical solar cell composed of H_2DBuPP nanorod-modified optically transparent electrode [denoted as OTE/H_2P rod] was prepared by drop-casting method. Photoelectrochemical performance of OTE/H_2DBuPP nanorods was measured

Fig. 28.2 TEM images of (**a**) H_2DBuPP nanorod, (**b**) H_2DPP nanorod, (**c**) H_2TBuPP assembly, and (**d**) H_2TPP assembly. SEM and optical microscope images of H_2DBuPP fiber are (**e**) and (**f**), respectively (Reproduced from Ref. [46] with permission of the Royal Society of Chemistry)

in a standard two-compartment cell [47]. The incident photon to current conversion efficiency (IPCE) spectrum (spectrum *a* in Fig. 28.3b) shows a broad photoresponse in the visible region (~6 % at 460 nm) [46], which parallels the corresponding absorption (spectrum *a* in Fig. 28.3a) and excitation spectra (spectrum *b* in Fig. 28.3b). These experiments confirmed the role of porphyrin nanorods toward

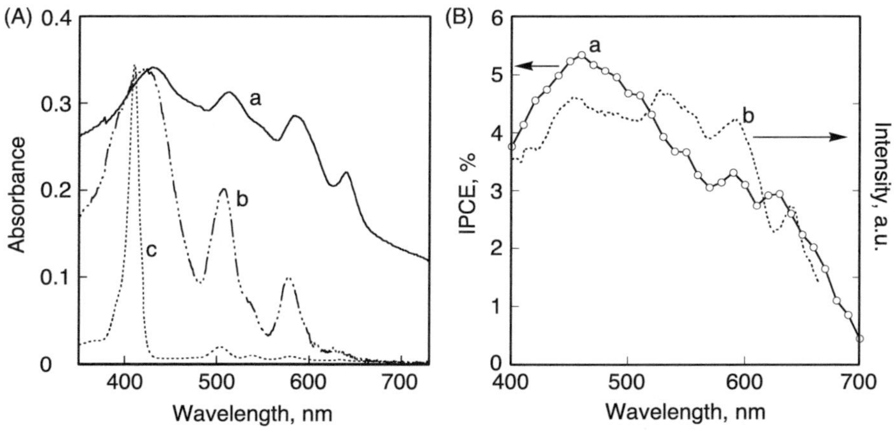

Fig. 28.3 (**a**) Absorption spectra of (a) H_2PDPP nanorod film, (b) H_2PDPP drop-cast film on quartz plates, and (c) H_2PDPP (4 μM) in toluene. Absorbance is normalized for comparison. (**b**) (a) Photocurrent (IPCE) action spectrum of H_2PDPP nanorod on an OTE. Electrolyte: 0.5 M LiI and 0.01 M I_2 in acetonitrile. (b) Excitation spectrum of H_2P nanorod film, observed at 710 nm (Reproduced from Ref. [46] with permission of the Royal Society of Chemistry)

harvesting solar energy and generating photocurrent during the operation of a photoelectrochemical cell.

28.2.2 Metal Coordination-Assisted Supramolecular Porphyrin Assemblies

In the above section, the driving force for supramolecular formation is mainly a single interaction such as $\pi-\pi$ interaction. In addition to utilization of $\pi-\pi$ interaction, one of the useful methods for controlling the organized structures is to employ coordination bonding since one can design a topology network of internal structures with metal ions and organic ligands as widely reported in the research fields of metal organic frameworks (MOF) [48, 49]. In this topic, we demonstrate synthesis and structural and photophysical properties of nano-/micro-sized architectures of porphyrins with controlled aspect ratios, which are effectively organized by metal coordination and intermolecular interaction.

The porphyrin assemblies are synthesized by the following method. First, the solutions of 5,10,15,20-tetrakis-(4-carboxyphenyl)-21H,23H-porphyrin ($H_2P(CO_2H)_4$, Fig. 28.4a) were prepared at the appropriate three different concentrations (52.5, 105, and 420 μM) in mixed solvents (DMF/dioxane/H_2O = 4/1/1, v/v/v). Then, $Zn(NO_3)_2$ 6 H_2O (2.0 equivalents) was added to the porphyrin solution. The solution was stirred for 3 h at 100 °C to form the aggregate structures. Finally, after centrifugation and washing, $H_2P(CO_2H)_4$ assemblies were very stable

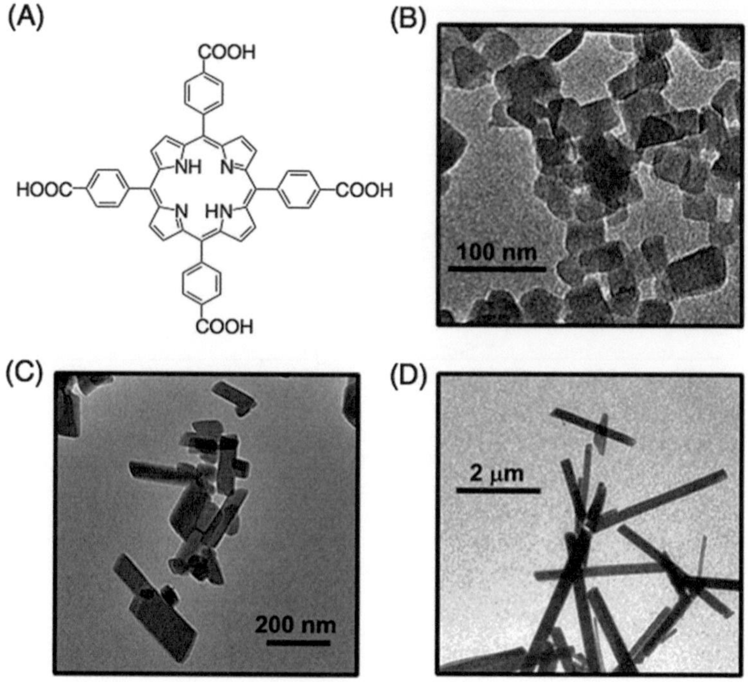

Fig. 28.4 (**a**) A chemical structure of $H_2P(CO_2H)_4$ and TEM images of (**b**) nanocubes (52.5 μM), (**c**) nanorods (105 μM), and (**d**) microrods (420 μM) (Reproduced from Ref. [50] with permission of the Royal Society of Chemistry)

in DMF. Transmission electron microscopic measurement (TEM) was performed to evaluate the sizes and shapes of these porphyrin aggregates (Fig. 28.4b–d). In the case of a low concentrated condition (52.5 μM), the $H_2P(CO_2H)_4$ assembly shows cubic structures (Fig. 28.4b), whereas longer bar-shaped structures with high aspect ratios (i.e., rods) are observed upon increasing the concentrations of $H_2P(CO_2H)_4$ (Fig. 28.4c and d). To discuss the quantitative information about structural sizes of these $H_2P(CO_2H)_4$ assemblies, we calculated several hundred samples, respectively, and the analyzed data are summarized in Table 28.1. The low concentration condition provides the nanoscale cubic formations with *ca.* 40–50 nm size (Fig. 28.4B). In contrast, a large increase of length direction from 50 nm to 2.5 μm (maximum 50 times) relative to that of width from 37 to 190 nm (5 times) is observed (Fig. 28.4c and d). This indicates that the anisotropic structural growth largely occurs in the length direction. Thus, the aspect ratios have been successfully controlled in the range of *ca.* 1 to 14 (Table 28.1) [50].

The possible organization process of $H_2P(CO_2H)_4$ assemblies is summarized in Fig. 28.5. The initial process is a planar checkerboard pattern formation of $H_2P(CO_2H)_4$ organized by paddle-wheel complexes between carboxylate and zinc ions, which is followed by the subsequent stacking formation of the checkerboard

Table 28.1 The summarized structural data of nanocubes, nanorods, and microrods

System	Width	Length	Aspect ratio[a]
52.5 μM	37 ± 11 nm	49 ± 17 nm	1[b]
105 μM	110 ± 15 nm	210 ± 63 nm	2.8 ± 0.2
420 μM	190 ± 66 nm	2.50 ± 0.99 μm	14 ± 5.1

Taken from Ref.[50] with permission of the Royal Society of Chemistry

[a]Aspect ratio = length (long side)/width (short side)

[b]The aspect ratio of nanocubes is assigned to be *ca.* 1

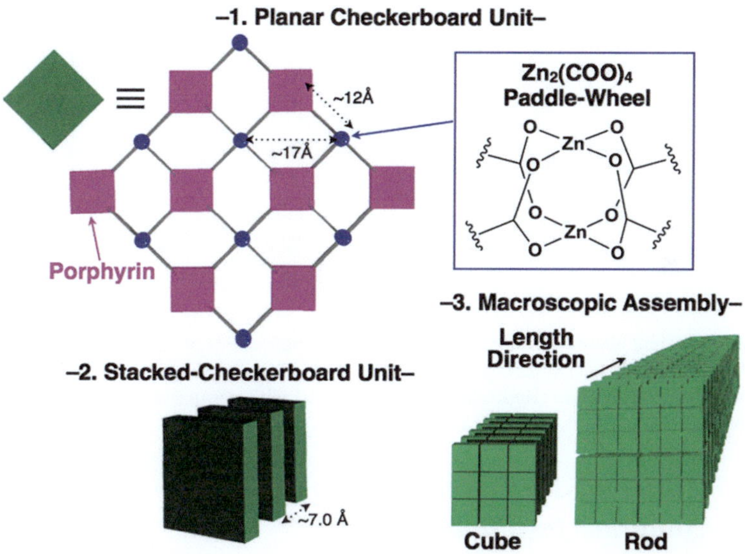

Fig. 28.5 A schematic illustration of the proposed structures of $H_2P(CO_2H)_4$ architectures via paddle-wheel complexes in this study (Reproduced from Ref.[50] with permission of the Royal Society of Chemistry)

units. In this case, with increasing concentrations of $H_2P(CO_2H)_4$ in the parent solution, the stacking formation of planar units becomes accelerated and grows into $H_2P(CO_2H)_4$ rodlike assemblies with high aspect ratios. The results of XRD and infrared absorption (IR) spectra also support a stacking formation of $H_2P(CO_2H)_4$ [50]. Additionally, the lifetimes of triplet excited states in $H_2P(CO_2H)_4$ assemblies with paddle-wheel complexes [0.59 μs (66 %), 4.2 μs (34 %)] are more than 20 times larger than that reported for 100 nm-sized $H_2P(CO_2H)_4$ self-assembled nanoparticles organized by relatively weak π–π interaction and hydrogen bond (26 ns) as shown in Fig. 28.6 [51]. Therefore, such organic nanoassemblies combined with a MOF strategy could pave a new way for developing novel organic–inorganic hybrid materials for solar energy conversion and electronics.

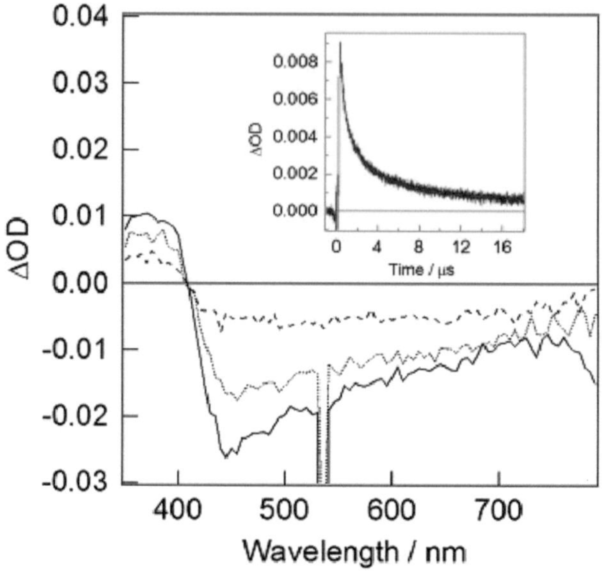

Fig. 28.6 Transient absorption spectra of $H_2P(CO_2H)_4$ nanocubes observed by 532 nm laser light (ca. 2 mJ/pulse) irradiation at 0.30 μs (*solid line*), 0.66 μs (*dotted line*), and 3.3 μs (*dashed line*). Inset: time profile at 370 nm (Reproduced from Ref.[50] with permission of the Royal Society of Chemistry)

28.2.3 Fullerene-Encapsulated Supramolecular Porphyrin Nanorods for Photovoltaics

In the recent years, attention and interest have been drawn to the development of inexpensive renewable energy sources. New approaches for the production of efficient and low-cost organic solar cells are necessary for future development of next-generation devices. As stated in the introduction section, the fundamental photovoltaic mechanism is (1) light harvest and exciton diffusion, (2) charge separation for carrier generation and (3) carrier transport [12, 52–54] A bar-shaped structure composed of two different molecules with separated inside and outside layers is a good candidate for photovoltaic applications since these processes are simultaneously and continuously performed within the molecular assembly. Therefore, construction of new molecular composites has been performed. The fullerene-encapsulated porphyrin hexagonal nanorods are composed of zinc *meso*-tetra(4-pyridyl) porphyrin [$ZnP(Py)_4$] and fullerene derivatives (denoted as fullerene–$ZnP(Py)_4$ nanorod), which were prepared by aiding a surfactant, cetyltrimethylammonium bromide (CTAB), in a DMF/acetonitrile mixed solvent (Fig. 28.7) [55, 56].

The synthetic procedure is as follows. A mixed proper ratio of $ZnP(Py)_4$ and fullerene derivatives in DMF solution was injected into 7.5 times volume of continuously stirred 0.20 mM acetonitrile solution containing CTAB surfactant at

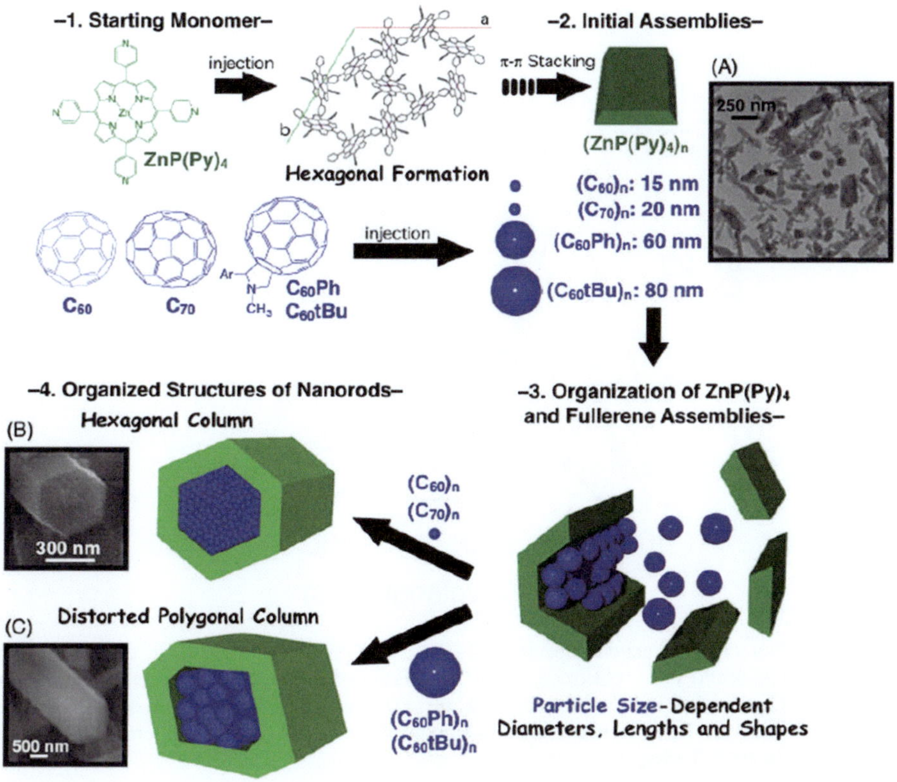

Fig. 28.7 An illustration of fullerene nanoparticle-promoted organization of $ZnP(Py)_4$ nanorods. CTAB surfactant is omitted for clarity. The electron micrographs show (**a**) immediate formations of $C_{60}t$Bu and $ZnP(Py)_4$ composites (1 min) after injection, (**b**) hexagonal C_{60}–$ZnP(Py)_4$ rod, and (**c**) distorted polygonal $C_{60}t$Bu–$ZnP(Py)_4$ rod (Reproduced from Ref. [55] with permission of the American Chemical Society)

room temperature. The final concentrations of $ZnP(Py)_4$ and fullerene derivatives are 0.03 and 0.02 mM in DMF/acetonitrile (2/15, v/v), respectively. On injecting, they largely form $ZnP(Py)_4$ flake assemblies and fullerene-based nanoparticles, separately, in the initial process (Fig. 28.7). With increasing the bulkiness of substituent groups of fullerene derivatives, the average diameters increased from 15 to 80 nm, which was confirmed by dynamic light scattering (DLS) measurement. In $ZnP(Py)_4$ assemblies, with the diffusion of DMF into acetonitrile, the Zn–N axial coordination of pyridyl N atoms to zinc atoms of $ZnP(Py)_4$ promotes the growth of aggregates, which continue to grow into a flake structure. In this case, the organization process of $ZnP(Py)_4$ moieties is initially derived from coordination bonds in contrast with C_{60} assemblies, which are based on relatively weak $\pi-\pi$ interactions. Then, fullerene–$ZnP(Py)_4$ nanorods are finally formed as shown in Fig. 28.7b and c. In scanning electron microscopy (SEM) measurement, $ZnP(Py)_4$ pristine hexagonal nanotubes with a large hollow structure (denoted as $ZnP(Py)_4$ tube) are observed, whereas the hollow is completely closed in the case of nanorods composed of

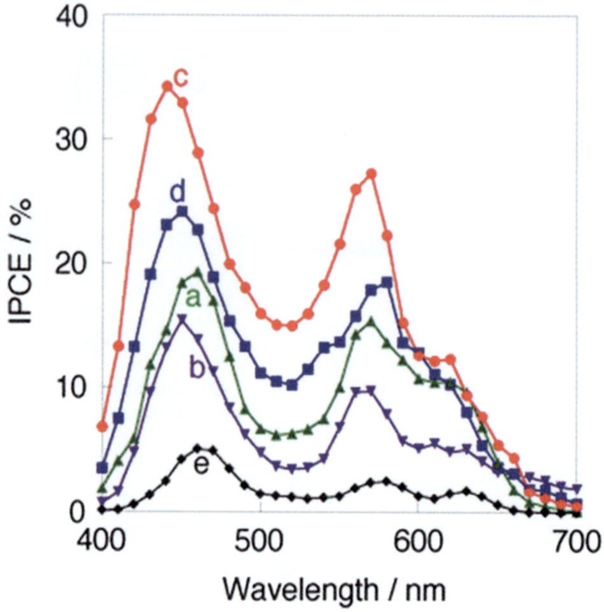

Fig. 28.8 Photocurrent action spectra of (**a**) OTE/SnO_2/C_{60}–ZnP(Py)$_4$ electrode, (**b**) OTE/SnO_2/C_{70}–ZnP(Py)$_4$ electrode, (**c**) OTE/SnO_2/C_{60}Ph–ZnP(Py)$_4$ electrode, (**d**) OTE/SnO_2/C_{60}tBu–ZnP(Py)$_4$ electrode, and (**e**) OTE/SnO_2/ZnP(Py)$_4$ electrode. Electrolyte 0.5 M LiI and 0.01 M I_2 in acetonitrile (Reproduced from Ref. [55] with permission of the American Chemical Society)

fullerenes (C_{60} and C_{70}) and ZnP(Py)$_4$ [fullerene–ZnP(Py)$_4$ rod] [55, 56]. In C_{60} derivative ZnP(Py)$_4$ rods, distorted polygonal columnar structures with large diameter and length, are formed, as compared to the hexagonal structures of C_{60}–ZnP(Py)$_4$ and C_{70}–ZnP(Py)$_4$ rods. The reference ZnP(Py)$_4$ hexagonal nanotube without C_{60} was also prepared in the same manner for comparison (denoted as ZnP(Py)$_4$ tube).

The absorption spectrum of the fullerene–ZnP(Py)$_4$ rods demonstrates much broader and more intense absorptions in the visible and near-infrared regions than that is seen for the corresponding monomers, ZnP(Py)$_4$ or C_{60} in DMF. This is probably due to the aggregate formation. Such a broad absorption property in the visible region is useful for solar energy conversion. The photocurrent action spectrum of OTE/fullerene–ZnP(Py)$_4$ nanorod shows a broad photoresponse in the visible region, and the maximum IPCE ($\sim$35 %) is much larger than the sum of two individual IPCE values ($\sim$6.5 %) of OTE/ZnP(Py)$_4$ tube and OTE/C_{60} assembly under the same conditions as the abovementioned electrolyte system (Fig. 28.8) [55, 56]. These results clearly indicate that an organized structure between C_{60} and ZnP(Py)$_4$ plays an important role for the improvement of light energy conversion properties.

28.2.4 Pt-/TiO$_2$-Encapsulated Supramolecular Porphyrin Nanorods for Photocatalytic Hydrogen Evolution

An efficient visible light-induced hydrogen evolution system has been developed by using an internal cavity of supramolecular porphyrin hexagonal nanotubes. The bar-shaped structure composed of Pt/TiO$_2$ (inside layer) and zinc *meso*-tetra(4-pyridyl)porphyrin [ZnP(Py)$_4$] (outside layer) was formed with the aid of a surfactant, CTAB, in a DMF/H$_2$O solution mixture [denoted as Pt/TiO$_2$–ZnP(Py)$_4$ rod] by the abovementioned method as shown in Fig. 28.9a.

The XRD and SEM results clearly indicate that Pt colloid-deposited TiO$_2$ nanoparticles (Pt/TiO$_2$) were successfully encapsulated within a ZnP(Py)$_4$ hexagonal nanotube. Pt/TiO$_2$–ZnP(Py)$_4$ rods also show a broadened absorption in the visible region because of aggregation of ZnP(Py)$_4$.

Figure 28.10 shows the time course of hydrogen evolution in the reaction system composed of ascorbic acid (AsA) or dihydronicotinamide adenine dinucleotide

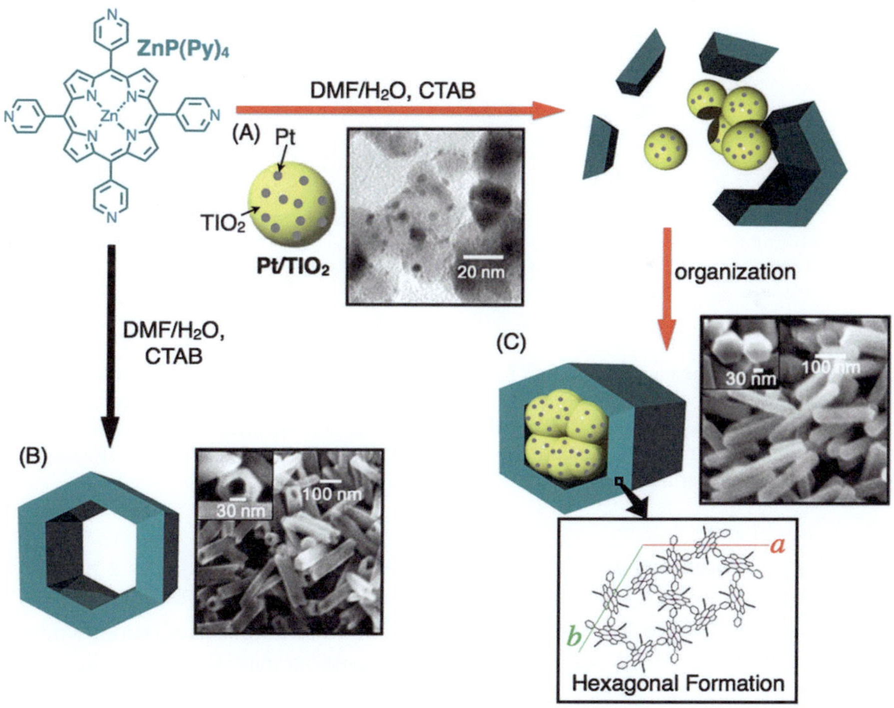

Fig. 28.9 A schematic illustration of organization procedure of Pt/TiO$_2$–ZnP(Py)$_4$ nanorods in this study. The electron micrograph images show (**a**) Pt/TiO$_2$ (TEM), (**b**) ZnP(Py)$_4$ nanotubes (SEM), and (**c**) Pt/TiO$_2$–ZnP(Py)$_4$ nanorods (SEM). The crystal structure of ZnP(Py)$_4$*Hexagonal Formation* was analyzed by the reported data (Reproduced from Ref. [57] with permission of the American Chemical Society)

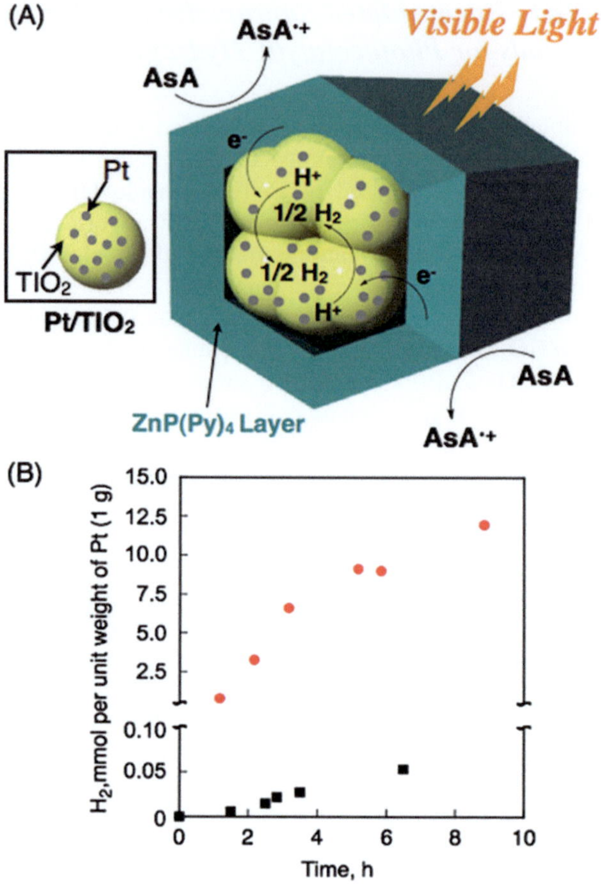

Fig. 28.10 (**a**) A schematic illustration for photocatalytic hydrogen evolution in this study. (**b**) Time dependence of hydrogen evolution per unit weight of Pt (1 g) under photoirradiation ($\lambda > 420$ nm) to the aqueous solution at pH 4.5 containing ascorbic acid [AsA] (3.8×10^{-2} M). ● Pt/TiO_2–ZnP$(Py)_4$ rods and ■ non-encapsulated Pt/TiO_2 + ZnP$(Py)_4$ composites (Reproduced from Ref. [57] with permission of the American Chemical Society)

(NADH) as a sacrificial electron donor and Pt/TiO_2–ZnP$(Py)_4$ rods as a photocatalyst, respectively, under visible light irradiation ($\lambda > 420$ nm). Pt/TiO_2–ZnP$(Py)_4$ rods exhibited efficient hydrogen evolution under visible light irradiation, whereas no hydrogen was evolved in the case of Pt/TiO_2 without ZnP$(Py)_4$. In addition, the hydrogen evolution efficiency of Pt/TiO_2–ZnP$(Py)_4$ rods per unit weight of Pt was two orders of magnitude greater than that of the non-encapsulated system: Pt/TiO_2 and ZnP$(Py)_4$ nanotube composites [Pt/TiO_2 + ZnP$(Py)_4$ composites] [57].

The photodynamics of the excited state of Pt/TiO_2–ZnP$(Py)_4$ rods were examined by femtosecond time-resolved transient absorption spectroscopy to clarify the photocatalytic mechanism. Judging from the ultrafast quenching process, photoinduced electron transfer from ^{1}ZnP$(Py)_4^*$ to TiO_2 is likely to be the initial event for

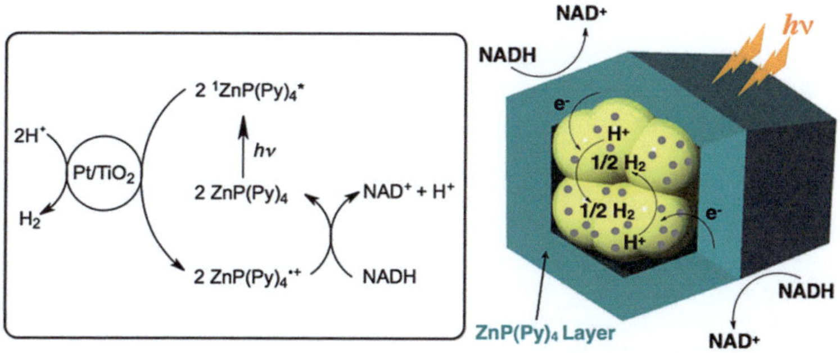

Fig. 28.11 A schematic illustration for mechanism of hydrogen evolution (Reproduced from Ref. [57] with permission of the American Chemical Society)

photocatalytic hydrogen evolution. Then, the injected electrons migrate toward Pt nanoparticles on the TiO_2 surface to reduce H^+ to H_2. The oxidized porphyrins [$ZnP(Py)_4^{\bullet+}$] undergo the electron transfer reduction with AsA or NADH. Based on the above results and electrochemical data of $ZnP(Py)_4$, hydrogen evolution may be initiated by photoinduced electron injection from the porphyrin singlet excited state [$ZnP(Py)_4^{\bullet+}/{}^1ZnP(Py)_4^* = -1.09$ V *vs.* SCE] to the conduction band of TiO_2 (~ -0.7 V *vs.* SCE) [58] in Pt/TiO_2–$ZnP(Py)_4$ nanorods. The injected electrons migrate toward Pt nanoparticles on the TiO_2 surface to reduce H^+ to H_2. The oxidized porphyrins [$ZnP(Py)_4^{\bullet+}$] undergo the electron transfer reduction with NADH, because the one-electron oxidation potential of NADH ($E^0{}_{ox} = 0.76$ V *vs.* SCE) [59, 60] is less positive than that of the one-electron oxidation potential of $ZnP(Py)_4$ [$ZnP(Py)_4^{\bullet+}/ZnP(Py)_4 = 0.98$ V *vs.* SCE]. The one-electron oxidation potential of ascorbic acid is reported to be $E^0{}_{ox} = 0.29$ V vs. SCE [61]. Alternatively photoinduced electron transfer from NADH to singlet excited state of porphyrins [${}^1ZnP(Py)_4^*/ZnP(Py)_4^{\bullet-} = +0.94$ V *vs.* SCE] is energetically feasible. In this case, however, the observed decay rate constant of the porphyrin singlet excited state with NADH is estimated as $\sim 2.5 \times 10^8$ s^{-1} at the concentration of NADH (2.5×10^{-2} M) assuming that the rate constant of the bimolecular photoinduced electron transfer is a diffusion-limited value ($\sim 10^{10}$ M^{-1} s^{-1}). Judging from the ultrafast quenching process, photoinduced electron transfer from ${}^1ZnP(Py)_4^*$ to TiO_2 is likely to be the initial event for photocatalytic hydrogen evolution (Fig. 28.11) [57].

28.3 Conclusion

In this chapter, recent developments of supramolecular porphyrin nanorods are widely described. The utilization of supramolecular techniques enables one to effectively organize molecular aggregates, which possess light energy conversion functionalities such as light harvesting, charge separation, and carrier transport. In the photovoltaic measurement, these highly organized architecture-based electrodes quantitatively exhibit a drastic enhancement of light energy conversion properties as compared to the reference non-organized system. The photocatalytic hydrogen evolution property of organized porphyrin nanorods is also two orders of magnitude greater than that of the non-organized system. This simple method of designing supramolecular assembly could pave the way for developing light energy conversion materials.

Acknowledgment The author thanks and expresses gratitude to his collaborators and coworkers whose names appear in the references. This work was partially supported by Grants-in-Aid for Scientific Research (23108721 & 23681025), PRESTO project, Japan Science and Technology Agency (JST), and MEXT-Supported Program for the Strategic Research Foundation at Private Universities, 2009–2013. All figures, schemes, and tables are reproduced from the related references with permission of the American Chemical Society and the Royal Society of Chemistry.

References

1. Dennler G, Scharber MC, Brabec CJ (2009) Polymer-fullerene bulk-heterojunction solar cells. Adv Mater 21(13):1323–1338
2. Grätzel M (2009) Recent advances in sensitized mesoscopic solar cells. Acc Chem Res 42(11):1788–1798
3. Kamat PV, Tvrdy K, Baker DR, Radich JG (2010) Beyond photovoltaics: semiconductor nanoarchitectures for liquid-junction solar cells. Chem Rev 110(11):6664–6688
4. Gunes S, Neugebauer H, Sariciftci NS (2007) Conjugated polymer-based organic solar cells. Chem Rev 107(4):1324–1338
5. Imahori H, Umeyama T, Ito S (2009) Large π-aromatic molecules as potential sensitizers for highly efficient dye-sensitized solar cells. Acc Chem Res 42(11):1809–1818
6. Martinez-Diaz MV, de la Torre G, Torres T (2010) Lighting porphyrins and phthalocyanines for molecular photovoltaics. Chem Commun 46(38):7090–7108
7. McEvoy JP, Brudvig GW (2006) Water-splitting chemistry of photosystem ii. Chem Rev 106(11):4455–4483
8. Yamada Y, Miyahigashi T, Kotani H, Ohkubo K, Fukuzumi S (2012) Photocatalytic hydrogen evolution with ni nanoparticles by using 2-phenyl-4-(1-naphthyl)quinolinium ion as a photocatalyst. Energy Environ Sci 5(3):6111–6118
9. Maeda K, Domen K (2010) Photocatalytic water splitting: recent progress and future challenges. J Phys Chem Lett 1(18):2655–2661
10. Kubacka A, Fernández-García M, Colón G (2011) Advanced nanoarchitectures for solar photocatalytic applications. Chem Rev 112(3):1555–1614
11. Hasobe T (2013) Porphyrin-based supramolecular nanoarchitectures for solar energy conversion. J Phys Chem Lett 4:1771–1780
12. Hasobe T (2010) Supramolecular nanoarchitectures for light energy conversion. Phys Chem Chem Phys 12(1):44–57
13. Kalyanasundaram K, Kiwi J, Grätzel M (1978) Hydrogen evolution from water by visible light, a homogeneous three component test system for redox catalysis. Helv Chim Acta 61(7):2720–2730
14. Barber J (2009) Photosynthetic energy conversion: natural and artificial. Chem Soc Rev 38(1):185–196
15. Ozawa H, Sakai K (2011) Photo-hydrogen-evolving molecular devices driving visible-light-induced water reduction into molecular hydrogen: structure-activity relationship and reaction mechanism. Chem Commun 47(8):2227–2242
16. Hasobe T, Sakai H (2013) Preparation and photoelectrochemical properties of supramolecular assemblies of nanoscale carbon material composites. ECS J Solid State Sci Technol 2(10):M3015–M3022

17. Hasobe T (2012) Photo- and electro-functional self-assembled architectures of porphyrins. Phys Chem Chem Phys 14(46):15975–15987
18. Imahori H, Yamada H, Guldi DM, Endo Y, Shimomura A, Kundu S, Yamada K, Okada T, Sakata Y, Fukuzumi S (2002) Comparison of reorganization energies for intra- and intermolecular electron transfer. Angew Chem Int Ed 41(13):2344–2347
19. Araki Y, Ito O (2008) Factors controlling lifetimes of photoinduced charge-separated states of fullerene-donor molecular systems. J Photochem Photobiol C Photochem Rev 9(3):93–110
20. Gust D, Moore TA, Moore AL (2001) Mimicking photosynthetic solar energy transduction. Acc Chem Res 34(1):40–48
21. Drain CM, Smeureanu G, Patel S, Gong X, Garnod J, Arijeloyea J (2006) Porphyrin nanoparticles as supramolecular systems. New J Chem 30:1834–1843
22. Gong X, Milic T, Xu C, Batteas JD, Drain CM (2002) Preparation and characterization of porphyrin nanoparticles. J Am Chem Soc 124(48):14290–14291
23. Drain CM, Varotto A, Radivojevic I (2009) Self-organized porphyrinic materials. Chem Rev 109(5):1630–1658
24. Radivojevic I, Likhtina I, Shi X, Singh S, Drain CM (2010) Self-organized nanofibers and nanorods of porphyrins bearing hydrogen bonding motifs. Chem Commun 46(10):1643–1645
25. Medforth CJ, Wang Z, Martin KE, Song Y, Jacobsen JL, Shelnutt JA (2009) Self-assembled porphyrin nanostructures. Chem Commun 47:7261–7277
26. Wang Z, Ho KJ, Medforth CJ, Shelnutt JA (2006) Porphyrin nanofiber bundles from phase-transfer ionic self-assembly and their photocatalytic self-metallization. Adv Mater 18(19):2557–2560
27. Wang H, Song Y, Medforth CJ, Shelnutt JA (2006) Interfacial synthesis of dendritic platinum nanoshells templated on benzene nanodroplets stabilized in water by a photocatalytic lipoporphyrin. J Am Chem Soc 128(29):9284–9285
28. Wang Z, Li Z, Medforth CJ, Shelnutt JA (2007) Self-assembly and self-metallization of porphyrin nanosheets. J Am Chem Soc 129(9):2440–2441
29. Martin KE, Wang Z, Busani T, Garcia RM, Chen Z, Jiang Y, Song Y, Jacobsen JL, Vu TT, Schore NE, Swartzentruber BS, Medforth CJ, Shelnutt JA (2010) Donor–acceptor biomorphs from the ionic self-assembly of porphyrins. J Am Chem Soc 132(23):8194–8201
30. Wang Z, Medforth CJ, Shelnutt JA (2004) Self-metallization of photocatalytic porphyrin nanotubes. J Am Chem Soc 126(51):16720–16721
31. Wang Z, Medforth CJ, Shelnutt JA (2004) Porphyrin nanotubes by ionic self-assembly. J Am Chem Soc 126(49):15954–15955
32. Lee SJ, Hupp JT, Nguyen ST (2008) Growth of narrowly dispersed porphyrin nanowires and their hierarchical assembly into macroscopic columns. J Am Chem Soc 130(30):9632–9633
33. Lee SJ, Mulfort KL, Zuo X, Goshe AJ, Wesson PJ, Nguyen ST, Hupp JT, Tiede DM (2008) Coordinative self-assembly and solution-phase x-ray structural characterization of cavity-tailored porphyrin boxes. J Am Chem Soc 130(3):836–838
34. Lee SJ, Jensen RA, Malliakas CD, Kanatzidis MG, Hupp JT, Nguyen ST (2008) Effect of secondary substituent on the physical properties, crystal structures, and nanoparticle morphologies of (porphyrin)sn(oh)2: diversity enabled via synthetic manipulations. J Mater Chem 18(31):3640–3642
35. Harada R, Matsuda Y, Okawa H, Kojima T (2004) A porphyrin nanotube: size-selective inclusion of tetranuclear molybdenum–oxo clusters. Angew Chem Int Ed 43(14):1825–1828
36. Fukuzumi S, Kojima T (2008) Photofunctional nanomaterials composed of multiporphyrins and carbon-based p-electron acceptors. J Mater Chem 18(30):1427–1439
37. Hu JS, Guo YG, Liang HP, Wan LJ, Jiang L (2005) Three-dimensional self-organization of supramolecular self-assembled porphyrin hollow hexagonal nanoprisms. J Am Chem Soc 127(48):17090–17095
38. Wang Z, Lybarger LE, Wang W, Medforth CJ, Miller JE, Shelnutt JA (2008) Monodisperse porphyrin nanospheres synthesized by coordination polymerization. Nanotechnology 19(39):395604

39. Floris H, Cameron CL, Marko MLN, Jeroen CG, Peter CMC, Antje L, George F, Philippe ELGL, Albertus PHJS, Meijer EW (2010) Dilution-induced self-assembly of porphyrin aggregates: a consequence of coupled equilibria. Angew Chem Int Ed 49(23):3939–3942
40. Ozawa H, Tanaka H, Kawao M, Uno S, Nakazato K (2009) Preparation of organic nanoscrews from simple porphyrin derivatives. Chem Commun 47:7411–7413
41. Elemans JAAW, Rv H, Nolte RJM, Rowan AE (2006) Molecular materials by self-assembly of porphyrins, phthalocyanines, and perylenes. Adv Mater 18(10):1251–1266
42. van Hameren R, Schon P, van Buul AM, Hoogboom J, Lazarenko SV, Gerritsen JW, Engelkamp H, Christianen PCM, Heus HA, Maan JC, Rasing T, Speller S, Rowan AE, Elemans JAAW, Nolte RJM (2006) Macroscopic hierarchical surface patterning of porphyrin trimers via self-assembly and dewetting. Science 314(5804):1433–1436
43. Lensen MC, Elemans JAAW, Dingenen SJTV, Gerritsen JW, Speller S, Rowan AE, Nolte RJM (2007) Giant porphyrin disks: control of their self-assembly at liquid-solid interfaces through metal-ligand interactions. Chem Eur J 13(28):7948–7956
44. de Pieter AJ, Witte CM, Cornelissen JJLM, Scolaro LM, Nolte RJM, Rowan AE (2003) Helical polymer-anchored porphyrin nanorods. Chem Eur J 9(8):1775–1781
45. Lensen MC, Takazawa K, Elemans JAAW, Jeukens CRLPN, Christianen PCM, Maan JC, Rowan AE, Nolte RJM (2004) Aided self-assembly of porphyrin nanoaggregates into ring-shaped architectures. Chem Eur J 10(4):831–839
46. Hasobe T, Oki H, Sandanayaka ASD, Murata H (2008) Sonication-assisted supramolecular nanorods of meso-diaryl-substituted porphyrins. Chem Commun 6:724–726
47. Hasobe T, Hattori S, Kamat PV, Urano Y, Umezawa N, Nagano T, Fukuzumi S (2005) Organization of supramolecular assemblies of fullerene, porphyrin and fluorescein dye derivatives on tio2 nanoparticles for light energy conversion. Chem Phys 319(1–3):243–252
48. Makiura R, Motoyama S, Umemura Y, Yamanaka H, Sakata O, Kitagawa H (2010) Surface nano-architecture of a metal-organic framework. Nat Mater 9(7):565–571
49. Motoyama S, Makiura R, Sakata O, Kitagawa H (2011) Highly crystalline nanofilm by layering of porphyrin metal-organic framework sheets. J Am Chem Soc 133(15):5640–5643
50. Sakuma T, Sakai H, Hasobe T (2012) Preparation and structural control of metal coordination-assisted supramolecular architectures of porphyrins. Nanocubes to microrods. Chem Commun 48(37):4441–4443
51. Sandanayaka ASD, Araki Y, Wada T, Hasobe T (2008) Structural and photophysical properties of self-assembled porphyrin nanoassemblies organized by ethylene glycol derivatives. J Phys Chem C 112(49):19209–19216
52. Schlenker CW, Thompson ME (2011) The molecular nature of photovoltage losses in organic solar cells. Chem Commun 47(13):3702–3716
53. Peet J, Heeger AJ, Bazan GC (2009) ‘äúplastic‘äù solar cells: self-assembly of bulk heterojunction nanomaterials by spontaneous phase separation. Acc Chem Res 42(11):1700–1708
54. Sun S-S, Sariciftci NS (2005) Organic photovoltaics. Taylor & Francis, Boca Raton
55. Sandanayaka ASD, Murakami T, Hasobe T (2009) Preparation and photophysical and photoelectrochemical properties of supramolecular porphyrin nanorods structurally controlled by encapsulated fullerene derivatives. J Phys Chem C 113(42):18369–18378
56. Hasobe T, Sandanayaka ASD, Wada T, Araki Y (2008) Fullerene-encapsulated porphyrin hexagonal nanorods. An anisotropic donor-acceptor composite for efficient photoinduced electron transfer and light energy conversion. Chem Commun (29):3372–3374
57. Hasobe T, Sakai H, Mase K, Ohkubo K, Fukuzumi S (2013) Remarkable enhancement of photocatalytic hydrogen evolution efficiency utilizing an internal cavity of supramolecular porphyrin hexagonal nanocylinders under visible-light irradiation. J Phys Chem C 117(9):4441–4449
58. Hasobe T, Imahori H, Fukuzumi S, Kamat PV (2003) Nanostructured assembly of porphyrin clusters for light energy conversion. J Mater Chem 13:2515–2520
59. Zhu X-Q, Yang Y, Zhang M, Cheng J-P (2003) First estimation of c4-bond dissociation energies of nadh and its radical cation in aqueous solution. J Am Chem Soc 125(50):15298–15299

60. Kotani H, Hanazaki R, Ohkubo K, Yamada Y, Fukuzumi S (2011) Size- and shape-dependent activity of metal nanoparticles as hydrogen-evolution catalysts: Mechanistic insights into photocatalytic hydrogen evolution. Chem Eur J 17(9):2777–2785
61. Imahori H, Kimura M, Hosomizu K, Sato T, Ahn TK, Kim SK, Kim D, Nishimura Y, Yamazaki I, Araki Y, Ito O, Fukuzumi S (2004) Vectorial electron relay at ito electrodes modified with self-assembled monolayers of ferrocene–porphyrin–fullerene triads and porphyrin–fullerene dyads for molecular photovoltaic devices. Chem Eur J 10(20):5111–5122

Chapter 29
Metal Nanoparticle/Porphyrinoid Hybrids

Masayuki Kanehara

Abstract In this chapter, electronic properties of Au/porphyrin or phthalocyanine (porphin) hybrid nanostructure are mentioned. The coordination ability of porphin onto metal nanostructure is relatively strong exploiting orbital hybridization between π-orbitals of porphin. The orbital hybridization gives some useful properties which cannot be obtained from isolated metal nanostructure or porphin.

Keywords Porphyrin • Phthalocyanine • Metal nanoparticle • Printed device

29.1 Au/Porphin Hybrid Nanostructures

29.1.1 Au/Porphyrin Hybrid Nanoparticles

Gold (Au) nanoparticles play important roles in different branches of science, such as in nanoelectronics, nonlinear optics, biological labeling, oxidation catalysis, etc. [1–4]. Many ligand-functionalized metal nanoparticles have been reported based on ligation using the chemical affinity of organic functional groups toward the nanoparticle surface in order to stabilize the nanoparticles. Polymers [5], linear molecules with long alkyl chains [6], and dendrimers [7] have all been effectively used for this purpose, relying on the σ-electrons of the functional groups. For the application of nanoparticles in nanoelectronic devices, exploiting the organoelectronic π-orbital interactions, which are generally used in electron-conductive polymers, organic charge–transfer complexes, organic transistors, etc., is quite important in light of the reduction of the tunneling resistance of the surrounding ligands. Though the conjugated π-orbital that provides various important electronic properties, the direct interaction of π-orbitals with metal nanoparticles is still rare. In this communication, we report the tuning of the optical property of a macrocyclic porphyrin by changing the distance between the porphyrin rings and the Au nanoparticle surface.

M. Kanehara (✉)
Research Core for Interdisciplinary Sciences, Okayama University, Okayama, Japan
e-mail: kaneha-m@cc.okayama-u.ac.jp

T. Akasaka et al. (eds.), *Chemical Science of π-Electron Systems*,
DOI 10.1007/978-4-431-55357-1_29

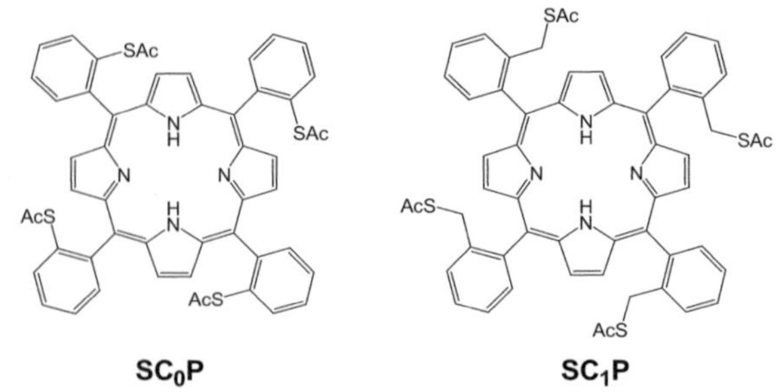

Fig. 29.1 Chemical structure of tetradentate porphyrin ligands, SC_0P and SC_1P

To obtain stable Au nanoparticles surrounded by π-orbitals perpendicular to the Au surface, we focused on Au nanoparticles formed through strong multidentate ligation using thiol derivatives. As shown in Fig. 29.1, multidentate macrocyclic porphyrin–thioester derivatives, tetrakis-5,10,15,20-(2-acetylthiophenyl)porphyrin (SC_0P), and tetrakis-5,10,15,20-(2-acetylthio-methylphenyl)porphyrin (SC_1P) were synthesized. The SC_1P ligand was designed with methylene groups between the benzene rings and the acetylthio groups, in order to increase the distance between the porphyrin ring and the Au surface. The SC_nP ($n = 0, 1$) ligands were synthesized from the corresponding aldehyde and pyrrole using Lindsey's method [8] in 15 % and 40 % yields, respectively. Since the acetylthio groups easily dissociate to form sulfur–Au bonds on bare Au surfaces in a slightly alkaline condition [9], these groups are considered an excellent thiol source to protect the Au surface. The SC_nP-protected Au (SC_nP–Au) nanoparticles were prepared by ligand-exchange reactions from citrate-protected Au (CA–Au) nanoparticles. After ligand exchange, the nanoparticles became insoluble in water but soluble in DMAc, indicating that ligand exchange was accomplished. In order to confirm the stability of SC_nP, these ligands were annealed with citrate and tannic acid in the absence of Au nanoparticles at 120 °C in DMAc/water mixed solvent. No UV-vis spectral change was observed, demonstrating that the SC_nP ligands are stable in the ligand-exchange condition.

Figure 29.2 shows the TEM images of CA–Au and SC_nP–Au nanoparticles. The sizes of the SC_nP–Au nanoparticles remained unchanged after the ligand exchange, and the DMAc solutions of SC_nP–Au nanoparticles are quite stable under ambient conditions, and no size change has been observed at least within 1 year due to the tetradentate nature of the SC_nP ligands. In order to confirm the existence of SC_nP on the Au surface and reveal the coordination geometry, laser Raman measurements and thermogravimetric analysis (TGA) were conducted. The laser Raman spectra of the SC_nP–Au nanoparticles were similar to those of the SC_nP ligands, indicating the existence of porphyrin rings on the Au nanoparticles. Further evidence was obtained by cyanide decomposition of SC_nP–Au nanoparticles [10]. When the SC_nP–Au nanoparticles were treated with an excess amount of sodium cyanide, the intensities of both the Soret band and Q band were recovered regardless of the decomposition of the SC_nP ligands. TGA analysis revealed that the weight losses of 3.1 % and 3.0 % for SC_0P–Au and SC_1P–Au, corresponding to the number of ligands on a single Au nanoparticle, were 246 and 226 for SC_0P–Au and SC_1P–Au, respectively.

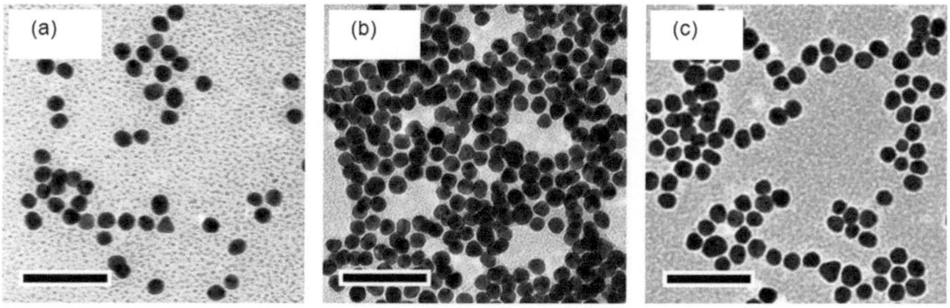

Fig. 29.2 TEM images of (**a**) citrate-, (**b**) SC_0P-, and (**c**) SC_1P-protected Au nanoparticles. Scale bar indicates 50 nm. Particle sizes are (**a**) 10.5 ± 1.0 nm, (**b**) 10.5 ± 1.0 nm, and (**c**) 10.0 ± 0.9 nm, respectively

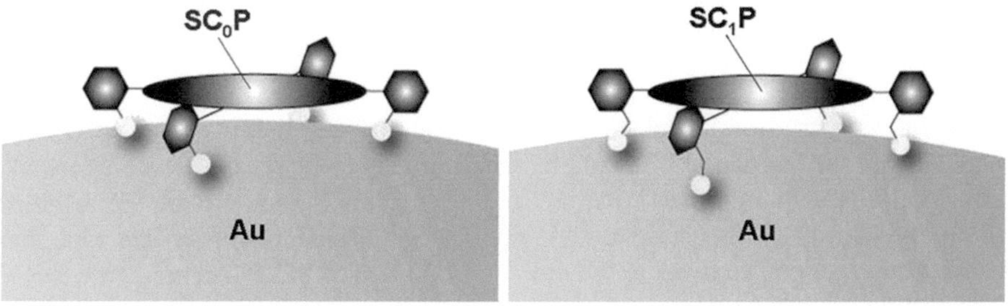

Fig. 29.3 Schematic illustrations of the coordination of SC_nP ligands on the Au surface

Considering the length of the neighboring meso-substituted phenyl groups that are about 1.25 nm, 222 SC_nP ligands can cover a single Au nanoparticle surface, suggesting that these porphyrin ligands densely protect the Au nanoparticles in a face-coordination fashion, as shown in Fig. 29.3.

Interestingly, these SC_nP–Au nanoparticles exhibited unique optical properties. Figure 29.4 shows the UV-vis spectra of the DMAc solutions containing SC_nP ligands or SC_nP–Au nanoparticles. The DMAc solutions of SC_0P and SC_1P gave Soret bands at 421 and 423 nm, respectively, and four distinct Q bands in the range from 515 to 646 nm. On coordinating SC_nP ligands on Au nanoparticles, the Q bands for SC_nP–Au nanoparticles completely disappeared, and the Soret bands for the SC_0P–Au and SC_1P–Au nanoparticles were broadened and redshifted to 432 and 427 nm, respectively. The molar absorption coefficients of the Soret bands for the SC_0P–Au and SC_1P–Au nanoparticles were calculated to be 2.5×10^4 and $7.0 \times 10^4\ M^{-1}\ cm^{-1}$, respectively, using TGA results and subtraction of the Au nanoparticle background. These values are smaller by one order of magnitude than those for SC_nP ligands themselves, $3.9 \times 10^5\ M^{-1}\ cm^{-1}$.

These results are completely different from those of the Au nanoparticles protected by meso-3-acetylthiophenyl-substituted porphyrin, where the porphyrin rings are tilting at the Au surface and the extensive diminishing of both the Soret

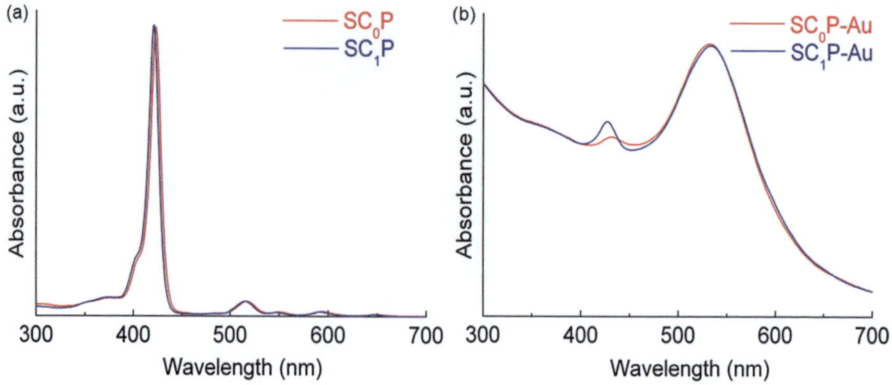

Fig. 29.4 UV-vis spectra of (**a**) SC_nP ligands and (**b**) SC_nP–Au nanoparticles. Each set of spectra was normalized by (**a**) ligand and (**b**) nanoparticle concentrations

band and Q band was not observed [11]. Generally, a redshift of the Soret band results from the formation of a side-by-side partially π–π stacked J-aggregate-like structure [12], although the formation of J-aggregates cannot explain the decrease in intensity of the Soret band and Q band of the SC_nP–Au nanoparticles. One may explain such unique changes in the optical properties of SC_nP–Au nanoparticles as a distortion of the porphyrin rings on the Au surface. To confirm this hypothesis, high-resolution XPS measurements were carried out. The SC_nP ligands exhibit two distinct chemically inequivalent N1s core-level spectra corresponding to free iminic (–C = N-) and pyrrolic (–NH-) nitrogens, respectively [13]. On the other hand, SC_nP–Au exhibits a single N1s peak with a binding energy similar to that of the iminic nitrogens coordinated to the Au(111) surface, as reported by Feringa and coworkers [13], who observed three distinct N1s peaks (free iminic, coordinating iminic, and free pyrrolic), indicating that the four nitrogen atoms of the SC_nP ligand dehydrate to coordinate to the Au surface. The possibility of the formation of Au(III)porphyrin was denied because no Au $4f_{7/2}$ peak assigned to oxidize gold was observed in SC_nP–Au. Since, to the best of our knowledge, this coordination mode has never been reported for porphyrins on bulk Au(111) or Ag(111) surfaces, we conclude this unique optical phenomenon as a nanoparticle effect. The porphyrin derivatives are expected to coordinate mainly to the step edges and plane boundaries that the large nanoparticles possess on their surfaces, centering at the porphyrin nitrogens. As a result, a certain degree of porphyrin ring distortion was generated to decrease the Soret-band intensity. A similar porphyrin ring distortion was observed on Au(111) by using STM [14]. Considering the longer porphyrin–Au distance for SC_1P–Au, as illustrated in Fig. 29.3, the Au–porphyrin interaction for SC_0P–Au is expected to be much larger than that for SC_1P–Au, resulting in the larger redshift and the broadening of the Soret band for SC_0P–Au. However, the drastic decrease in the intensity of the Soret band cannot be fully explained by the distortion model, because a drastic porphyrin ring distortion must lead to the considerable shift of the Soret band [13]. Though the details are not understood yet, the unique optical change exhibited by the porphyrin rings may be derived from the direct interaction between the porphyrin π and Au orbitals, in other words, the partial charge transfer between the porphyrin and Au [15], which is often observed between the porphyrin rings and fullerenes [16].

29.1.2 Au/Phthalocyanine Hybrid Nanoparticles

Electrodes are necessary in all types of electric devices such as transistors, photovoltaic cells, and light-emitting devices. These devices are commonly fabricated by employing expensive vacuum deposition processes. A promising alternative that is cheaper and faster for fabricating devices is wet printing using material from specially designed solutions. Many examples of solution processes for transistors, photovoltaic cells, and light-emitting devices have been reported. However, in order to achieve high efficiency and reliable devices, vacuum-processed metals and conductive metal oxide electrodes must be used, thereby producing only partially solution-processed devices. To realize the fabrication of all-solution-processed devices, electrodes must be fabricated at relatively low temperatures from a solution, and the resultant electrode must fulfill three requirements: they must have (1) good contacts, i.e., low resistance at the semiconductor–electrode interface, (2) high electroconductivity, and (3) high oxidation stability. In this study, we report that the Au nanoparticles (NPs) possessing orbital hybridization between the Au core and relatively large aromatic molecules meet these three requirements and, consequently, produce reliable electrodes that are comparable to vacuum-deposited Au electrodes.

For the fabrication of NP-based electrodes under ambient conditions, surface ligands of inorganic NPs must meet two requirements: they must (1) adhere firmly to the NP surface for colloidal stabilization and (2) provide stable and facile carrier transport between NPs. We found that orbital hybridization between π-orbitals of aromatic molecules and inorganic orbitals of the Au NPs improves the interparticle-carrier transport. The molecular orbital structure of aromatic compounds was altered when the π-conjugated plane of aromatic compounds closely bonded while parallel to the metal surface, and this interaction is referred to as a π-junction interaction. We propose a generalized approach for the efficient interparticle-carrier transport between inorganic NPs using large aromatic phthalocyanine (Pc) derivatives as ligands. A planar π-conjugated organic Pc ligand is capable of achieving a strong π-junction interaction with minimal interparticle-steric repulsion. In addition, Pc derivatives (including porphyrins) are known to coordinate with various types of metal atoms, including Au(0) atoms on Au NPs. Therefore, Pc derivatives are promising candidates as π-junction ligands.

We synthesized two H_2–Pc derivatives, 2,3,9,10,16,17,23,24,-octakis(2-*N,N*-dimethylaminoethylthio)phthalocyanine (OTAP) and 2,3,11,12,20,21,29,30-octakis (2-*N,N*-dimethylaminoethylthio)naphthalocyanine (OTAN). Eight substituted thioether groups elevate the highest occupied molecular orbital (HOMO) level of the ligands to provide the Pc ring with the ability to coordinate with Au NPs. Terminal dimethylamino groups cause Pc-coordinated Au NPs to dissolve in aqueous solution under acidic conditions. First, we synthesized OTAP and OTAN in two steps from their corresponding halogenated phthalonitrile derivatives. Then, we easily synthesized OTAP- and OTAN-protected Au (OTAP–Au and OTAN–Au) NPs by a ligand-exchange reaction using conventional citrate-capped Au NPs having a diameter of 14.3 ± 2.2 nm [17]. The synthesized NPs showed good solubility in water under acidic conditions and no solubility under basic conditions, which indicate a successful ligand exchange.

Figure 29.5 shows images of OTAP–Au and OTAN–Au NPs, with sizes of 14.3 ± 2.3 and 14.4 ± 2.3 nm, respectively. The sizes of Au NPs remained unchanged during ligand exchange. The completion of ligand exchange was confirmed by solubility behavior, laser Raman spectroscopy, and wavelength-dispersive X-ray spectroscopy (WDS). In particular, while the seed citrate-capped Au NPs show no obvious Raman peak from 600 to 1600 cm^{-1}, OTAN–Au and OTAP–Au NPs show distinct peaks that are assigned to the Pc ring. These peaks are in good agreement with those observed in Raman spectra of OTAP and OTAN. The Au/sulfur atomic ratios for OTAP and OTAN are 37.8 and 49.8, respectively, as determined by WDS. Assuming that the occupied areas of the molecules of OTAP

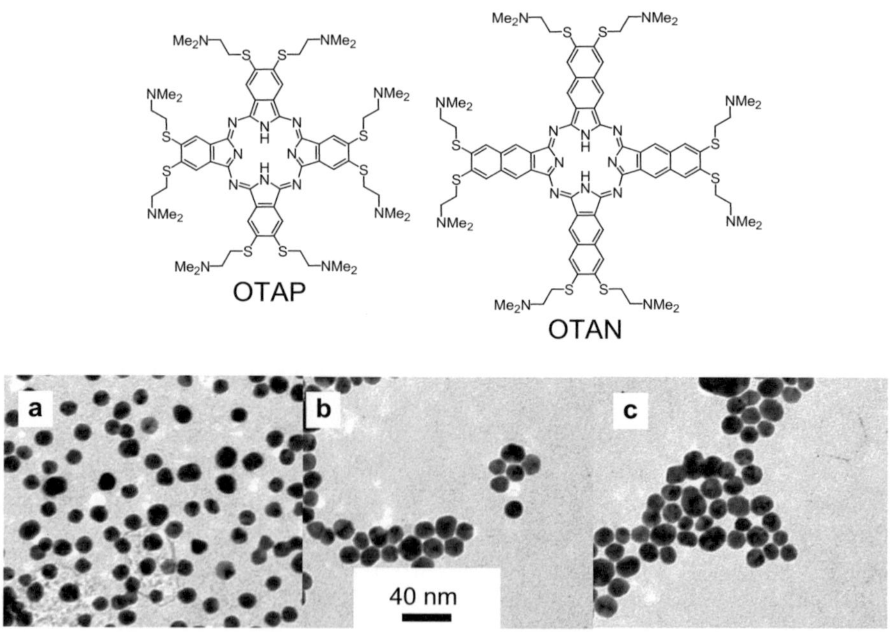

Fig. 29.5 Chemical structures of OTAP and OTAN ligands. TEM images of (**a**) citrate-protected (14.3 ± 2.2 nm), (**b**) OTAP-protected (14.3 ± 2.3 nm), and (**c**) OTAN-protected (14.4 ± 2.3 nm) Au NPs (scale bar = 40 nm)

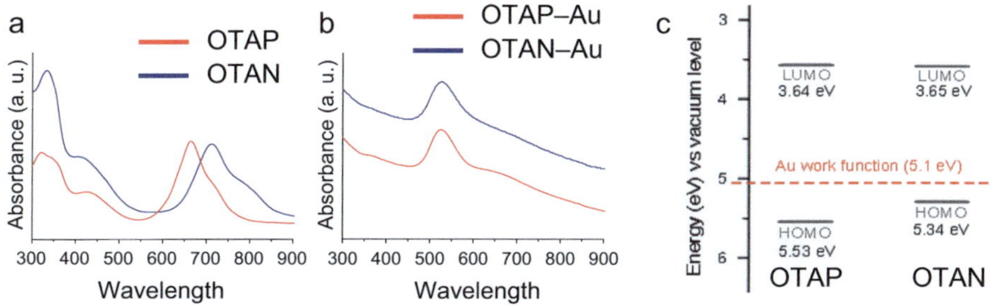

Fig. 29.6 OTAP and OTAN spectra and energy levels: (**a**) UV-vis spectra of OTAP and OTAN ligands; (**b**) UV-vis spectra of OTAP–Au and OTAN–Au NPs; and (**c**) energy levels of OTAP and OTAN ligands, showing the Fermi level of Au

and OTAN are 2.25 and 2.89 nm^2, respectively, 290 and 230 molecules of OTAP and OTAN, respectively, can densely protect a 14.4-nm Au NP in a face-coordination fashion. Thus, the Au NP is covered by 260 and 220 molecules of OTAP and OTAN, respectively, which is in good agreement with the calculated number of Pc molecules on an Au NP.

Figures 29.6a and b show UV-vis spectra of aqueous solution of OTAP and OTAN ligands and those of OTAP–Au and OTAN–Au NPs, respectively. The peak positions of OTAP and OTAN Q bands are 664 and 734 nm, respectively. The relatively narrower HOMO–lowest unoccupied molecular orbital (LUMO) energy gap for OTAN results in a redshifted Q band. Upon coordination with Au NPs, the OTAP Q band decreases significantly, and the OTAN Q band essentially disappears. These drastic spectral changes clearly show the strong π-junction interaction between Pcs and Au NPs [15]. This interaction leads to hybridization between the π-orbital of the Pc and the orbital of Au NPs. As a result, the π-orbital of the Pc becomes metallic in nature, and its continuous band-like nature broadens the UV-vis spectra of Au NPs. The disappearance of the Q band reflects the metallic nature of the π-orbital of Pc on Au NPs and is similar to the spectra of metallic NPs that have broad absorption from the UV to infrared regions [18]. Figure 29.6c shows energy level diagrams for OTAP, OTAN, and Au. The HOMO levels for OTAP and OTAN are 5.53 and 5.34 eV, respectively. The relatively large conjugation area for OTAN can increase the HOMO level to be close to the Fermi level of Au NPs, giving rise to relatively strong π-junction interaction and stabilizing the Au NPs.

Figure 29.7 shows various properties of OTAN–Au NPs and its thin film. We found that an acidic aqueous formic acid solution of OTAP–Au and OTAN–Au NPs could be used as a printable electroconductive material without posttreatment. The purple aqueous solutions of OTAP–Au and OTAN–Au NPs became metallic gold in color after deposition and drying on the substrate under ambient conditions (Fig. 29.7a). OTAP–Au and OTAN–Au NP thin films, dried at 55 °C for 10 min to remove the remaining solvent, exhibited electric conductivities of 1560 and 6600 S/cm, respectively. OTAP–Au and OTAN–Au NP thin films are extremely stable, and the electroconductivity of the films remains unchanged for at least two

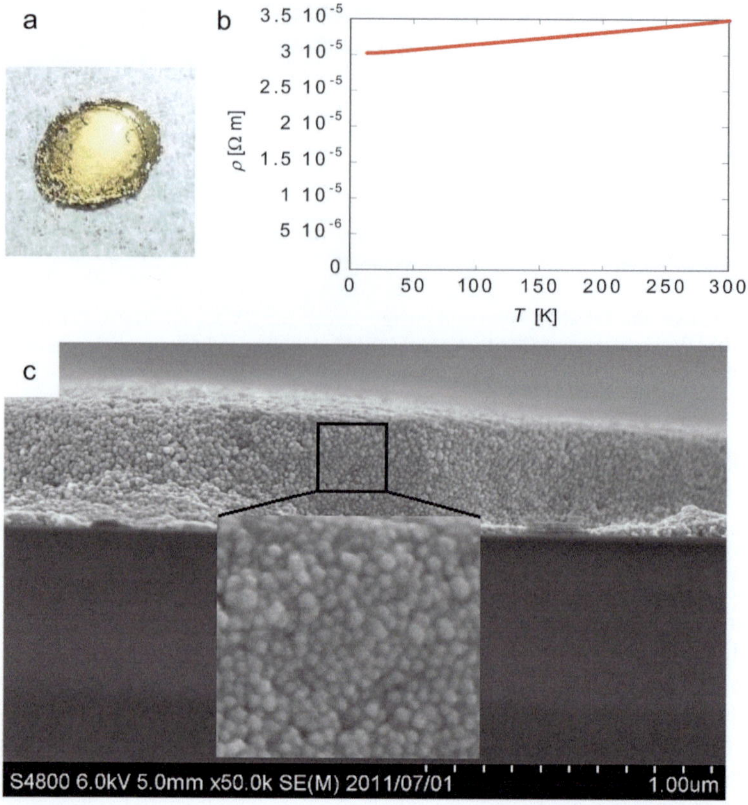

Fig. 29.7 Properties of OTAN–Au NPs and NP thin film: (**a**) image of NP thin film dried under ambient conditions. (**b**) Plot of resistivity against temperature for OTAN–Au NP thin film. (**c**) Cross-sectional SEM image of NPs on a glass substrate

years under ambient conditions. As shown in Fig. 29.6c, the energy gap between HOMO of OTAN and Fermi level of Au is smaller than that of OTAP and Au. This relatively small energy gap leads the strong orbital hybridization between OTAN and Au, which gave greater electric conductivity. The temperature-dependent conductivity of the OTAN–Au NP thin film exhibits a metallic nature, and the resistance decreases with temperature (Fig. 29.7b). The metallic electric junction of the interparticle interface of the OTAN–Au NP thin film was obtained only after drying the OTAN–Au NP solution under ambient condition. And the conductivity of the OTAN–Au NP thin film was restricted by the interface between the NPs. A cross-sectional scanning-electron-microscopy (SEM) image of an OTAN–Au NP thin film fabricated under ambient condition shows the absence of cracks (Fig. 29.7c). Usually, in order to obtain a carrier-transportable NP thin film, annealing at high temperatures (over 150 °C) or washing processes involving the use of chemical reagents such as hydrazine are required to remove insulating ligands, thereby inducing direct interparticle contact [19–21]. During such posttreatments, the reduction in the surface-ligand volume always causes cracks on the resultant NP thin film with highly concentrated carrier trap sites [22]. Because OTAP–Au and OTAN–Au NPs do not need such posttreatments, their dried thin films are densely packed structures that are essential for the fabrication of reliable devices.

29.2 OFET Device Fabrications

The properties required for electrodes in various types of devices are more stringent than those required for inter-device wiring because an electrode at a semiconductor interface must make good contact [23]. To demonstrate the effectiveness of OTAN–Au NPs, we fabricated all-solution-processed OFETs fabricated under ambient conditions, which are good candidate switching devices for both flexible and printable electronic circuits owing to the development of many solution-processed organic semiconductors [5]. The fabrication of both polymer and small-molecule semiconductors into stable OFETs by near-room-temperature processes have been reported [6–8, 24–26]. However, metal electrodes have not yet been fabricated using simple solution techniques. On the other hand, fabrication attempts have resulted in only partially solution-processed devices. We prepared organic semiconductor layers of 2,7-dioctyl[1]benzothieno[3,2-b]benzothiophene (C8–BTBT) using a crystallizing technique developed by one of the authors to enable an easy fabrication of large-domain crystalline films on an Si–SiO_2 substrate [26]. We then painted a solution of OTAN–Au NPs on the semiconductor film at room temperature to construct a top-contact, bottom-gate OFET. Figure 29.8 shows different views and characteristics of the constructed OFET. The channel length and width are 400 and 120 μm, respectively (Figs. 29.8a and b). The typical transistor characteristics demonstrate clear FET performance with mobility values as high as 2 cm^2/Vs (Figs. 29.8c and d). This value is comparable to the previously reported value for the mobility of devices having vacuum-deposited Au electrodes, which ranged from 3.5 to 5 cm^2/Vs [26]. We note that the structure and the channel dimensions are essentially the same for the previously reported solution-crystallized C8–BTBT OFETs as compared with the present devices incorporating the π-junction Au NPs. Therefore, the results indicate general usefulness of painted OTAN–Au NPs that were fabricated using a very simple room-temperature method described here. The mechanism that causes slightly smaller value of mobility in the present devices is under investigation in order to further increase the transistor performances. It can be suspected that the present paining process results in slightly more resistive metal–semiconductor contacts or that the vapors from the water-based solvent induced shallow hole-trapping levels attached to the surface of the C8–BTBT channels [27].

Printed electronics is an emergent subject for the low-cost and large-area fabrication of flexible electronic devices. Direct printing of organic thin-film transistors (OTFTs) using soluble organic semiconductors is a particularly promising fabrication method that offers lower production costs, reduced energy consumption, and a

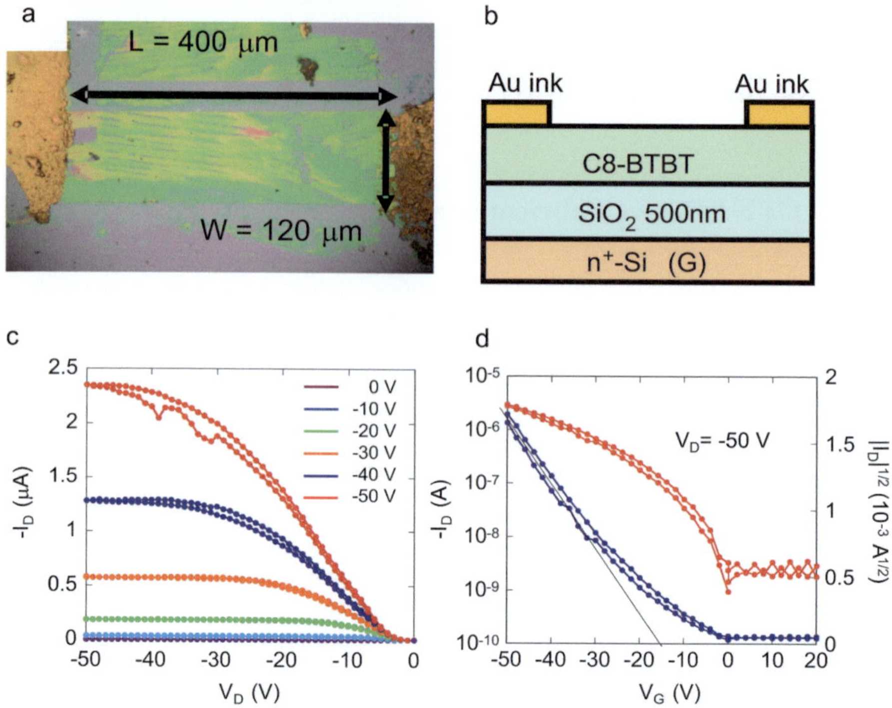

Fig. 29.8 Full solution-fabricated OFET made using a solution-deposited C8–BTBT semiconductor layer coated with OTAN–Au NPs and used for electrodes: (**a**) top view, (**b**) schematic illustration, (**c**) output characteristics, (**d**) transfer characteristics

smaller environmental burden. Fully printed OTFTs have been realized by several groups, but the proposed processes still involve some fundamental problems such as poor field-effect mobility (μ_{FET}) and low reproducibility [28–31]. We consider that the low reproducibility mainly belongs to the process of electrode formation, where Ag nanoparticle (NP) ink is commonly used as the electrode material. This process is actually not compatible to a common flexible substrate because the high processing temperature of 150 °C causes significant distortion to the substrate. For instance, a heat shrinkage rate of a plastic film such as polyethylene terephthalate (PET) or polyethylene naphthalate (PEN) is 2 % at 150 °C, resulting in 20-mm shrink for a 1-m-long substrate after annealing at 150 °C. The thermal deformation prevents high-resolution patterning and accurate alignment among the different layers of devices on a flexible substrate. This problem becomes more significant as the production size increases.

The critical difficulty in achieving a process for RT-printed electronics is the high-temperature annealing process required for metal electrode deposition using metal NP ink. A conventional metal NP ink contains a nonconductive material as the ligand. High-temperature annealing is required to remove it and sinter the metal

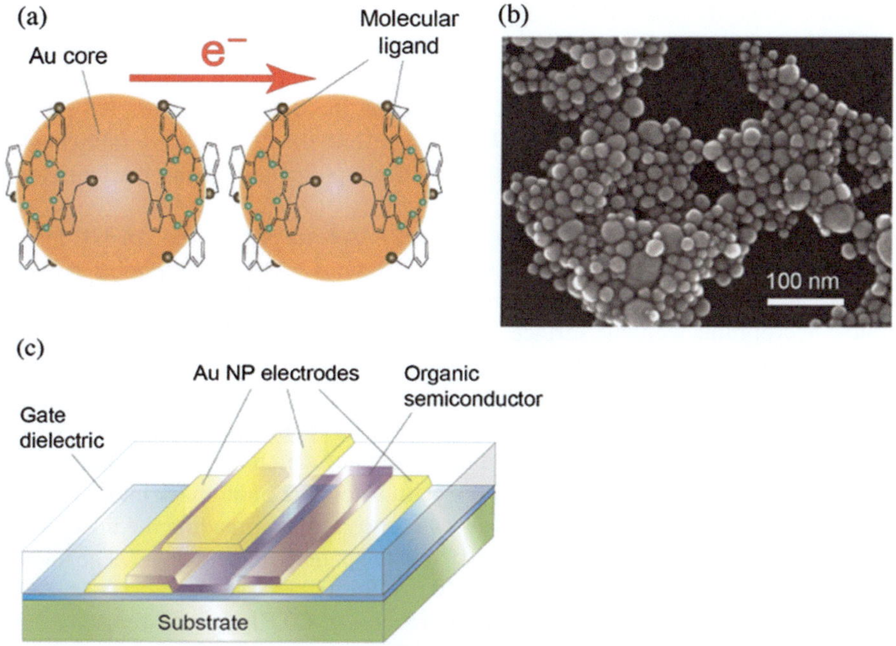

Fig. 29.9 (**a**) Schematic illustration of π-junction Au NPs. The metal core is surrounded by phthalocyanine molecular ligands. (**b**) Scanning electron microscope image of Au NPs deposited on a substrate. (**c**) Structure of OTFT devices fabricated by the RT printing process

cores to obtain a conductive metal film. Therefore, we developed π-junction Au NPs for the RT printing of the metal electrodes [27]. The concept for this process and the principle behind the Au NP ink are shown in Fig. 29.9. The Au NPs possess a metal core surrounded by phthalocyanine molecules as the conductive ligand (Fig. 29.9a), which is the biggest difference between the new inks and conventional metal NP inks. We employed a derivative of metal-free phthalocyanine as the conductive ligand, which has the highly planer molecular structure and large π-conjugated system. The planer structure enables a close contact between the NPs without acting as the steric barrier. The large π-conjugated system is also essential for improving the electrical pathway among the NPs. In the π-junction NPs, orbital hybridization between the π-orbitals of the aromatic ligand and orbitals of the metal core contributes charge transport among the NPs. Thus, conductive film can be obtained by RT deposition of the ink without removal of the ligand by annealing. A scanning-electron-microscopy image of the Au NPs deposited on a substrate is shown in Fig. 29.9b.

The Au NPs clearly have a spherical shape that is maintained in the film. However, it exhibits low resistivity of $\sim 9 \times 10\text{–}6\ \Omega$ cm, which is of the same order of magnitude as that of pure Au of $2.2 \times 10\text{–}6\ \Omega$ cm, without the sintering process because of the smooth charge transport among NPs through the conductive ligands. Thus, the Au NP film can be used as the electrode material for the RT printing method. For fabrication of the printed OTFTs using the Au NP ink, a stack of four layers, including source/drain, organic semiconductor, gate dielectric, and gate electrode layers, was prepared to complete the devices, and the RT process was employed for all layers (Fig. 29.9c). Therefore, the OTFTs were fabricated at RT by solution-based methods without any heat application during the fabrication processes.

A surface-selective deposition method was used to pattern the Au NP ink on the substrate [29]. For this method, selective regions on the hydrophobic polymer surface were exposed to vacuum ultraviolet (VUV) light to transform the surface into a hydrophilic, wettable surface. The VUV exposure under air generates oxygen radicals and ozone molecules, which cause chemical bond dissociation at the polymer surface and render the region hydrophilic, resulting in patterned wettability. The VUV irradiation system was composed of a Xe excimer lamp as the VUV light source (wavelength of 172 nm) and a mask aligner, which allows selected regions on the surface to be exposed to the VUV light through a photomask under ambient atmospheric conditions. The Au NP ink was deposited by spreading the ink onto the patterned surface using an applicator to obtain the patterned metal layer. The highest resolution achieved by this method was 10 μm for a line or space.

Figure 29.10 illustrates the RT printing process of OTFT arrays. We used a PEN film with a surface-planarizing layer of parylene as the flexible plastic substrate. The source/drain electrodes, an organic semiconductor layer, and gate electrodes were fully patterned using a solution-based printing method at RT. First, the source/drain electrode patterning was performed on the hydrophobic substrate surface. The source/drain regions on the surface were exposed to VUV irradiation through a photomask under ambient atmospheric conditions to render the exposed surface area hydrophilic. After surface wettability patterning, the Au NP ink was applied to the surface, resulting in a pattern of Au ink only on the hydrophilic regions. The Au ink dried in several seconds at RT, and the source/drain electrode layer was obtained using a RT process (Fig. 29.10a).

Then the organic semiconductor layer was added onto the channel region of the devices using a two-step printing processes, screen printing, and solution casting (Fig. 29.10b). On the source/drain electrode layer, first a fluorinated polymer layer of Cytop (Asahi Glass Co. Ltd.) was formed around the channel region of OTFTs by screen printing. Then, the Cytop was dried at RT without annealing, resulting in a uniform polymer layer 120-nm thick. The lyophobic nature of the fluorinated polymer surface inhibits deposition of organic semiconductor. Therefore, it acts as a guide for the organic semiconductor solution. Semiconductor layer deposition was performed next, through simple drop casting of 0.5 wt% chlorobenzene solution of dioctylbenzothienobenzothiophene (C8–BTBT) into the lyophilic channel region. Because of the highly repellent nature of the Cytop surface, the semiconductor solution was deposited only on the channel regions, and a patterned organic layer was formed after solvent evaporation. The high-boiling-point solvent allows formation of a uniform semiconductor layer with relatively large grain size of −100 μm.

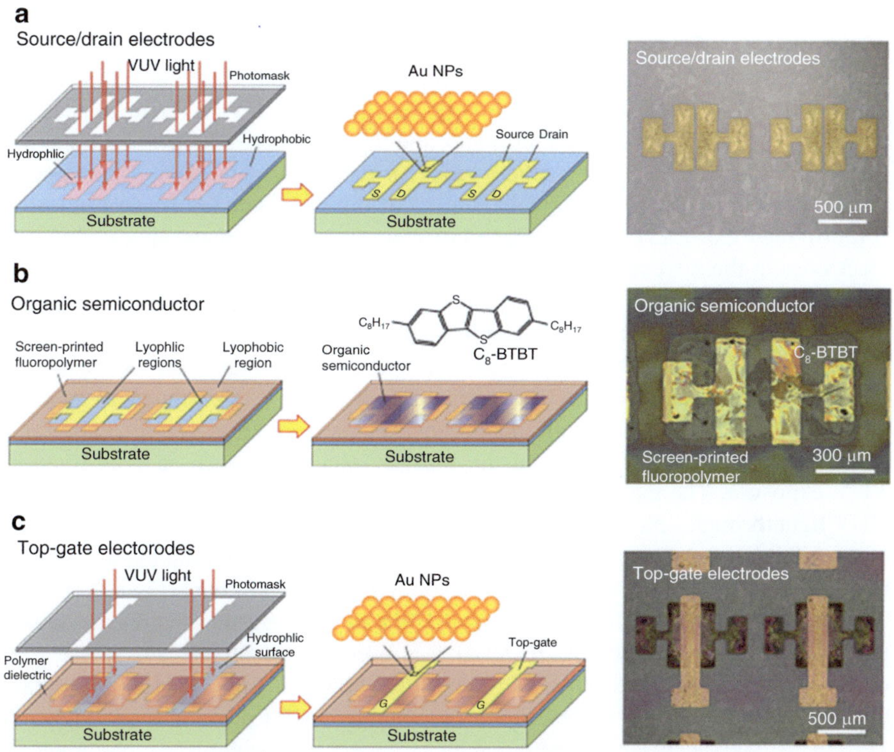

Fig. 29.10 Schematic illustration of OTFT array fabrication using the RT printing method. (**a**) Formation of source/drain electrodes on a hydrophobic surface using: (i) selective wettability patterning by VUV irradiation of the source/drain electrode regions to render them hydrophilic (*left*) and (ii) selective deposition of Au NP ink on the hydrophilic regions to form the source/drain electrodes (*center*). Picture on the right is an optical micrograph of the source/drain electrodes formed by the selective deposition method. (**b**) Formation of organic semiconductor layers using screen printing and solution casting: (iii) the fluorinated polymer layer was screen printed out of the channel region as a guide for the organic semiconductor deposition (*left*). (iv) Drop casting of a C8–BTBT solution onto the surface to selectively crystallize the molecules only in the channel region (*center*). Picture on the right is an optical image of the organic semiconductor film patterned by this method. (**c**) Patterning of the top-gate electrodes by (v) VUV irradiation of the gate electrode regions on the gate insulator to render them hydrophilic (left). (vi) Formation of top-gate electrodes by application of Au NP ink to the hydrophilic regions (*center*). Picture on the right is an optical image of the TFT devices after formation of the top-gate electrodes

The bilayer structure was used for the gate dielectric. Here, Cytop is used as the interfacial layer at the organic semiconductor/gate dielectric interface. Charges accumulate in the vicinity of the interface in OTFTs, and a hydrophobic polymer or self-assembled monolayer is useful for modifying the interface and reducing charge trapping in the channel region. Introduction of the Cytop layer greatly reduces the trap density at the interface and suppresses electrical hysteresis behavior of the devices. The role of the second layer is suppression of VUV penetration into the organic layer during top-gate electrode formation, which degrades device characteristics due to disruption of the organic layer by VUV, in the next step. The upper layer also offers better patterning of the top-gate electrode layer compared to that on the Cytop layer. Two choices exist for the upper gate dielectric: chemical vapor deposition of parylene or spin coating of the fluorinated polymer Fluorosurf (Fluoro Tech Co. Ltd.). The parylene layer provides better patterning of top-gate electrodes in the next step, but a chemical vapor deposition process under vacuum is necessary to form it. The fluorinated polymer can be formed by spin coating and dries naturally at room temperature without applying heat but requires a longer VUV irradiation time and results in lower resolution of the patterning of the top-gate layer, probably due to greater surface roughness and lower VUV absorbance. In contrast, both of the gate dielectrics result in similar OTFT performance.

The last step was patterning of the top-gate electrode layer, accomplished by VUV exposure. The gate electrode regions on the gate dielectric were exposed to VUV light through a photomask to render them hydrophilic, followed by application of Au NP ink using the applicator to selectively deposit it only onto the hydrophilic regions. The Au NPs were dried at RT without application of heat (Fig. 29.10c).

Results of RT printing of OTFT arrays on the flexible plastic substrate are shown in Fig. 29.11. Since a flexible material was used for all of the layers, the array obtained was a fully flexible OTFT array, as shown in Fig. 29.11a. The optical micrograph of the fully printed OTFT array obtained by RT printing is also shown in Fig. 29.3a. The electrode and semiconductor layers were completely patterned for complete isolation among the devices, which is essential for reducing gate leakage current and inter-device cross talk (Fig. 29.11b).

The results of electrical characterization for RT-printed OTFTs are shown in Fig. 29.11c and d. The nonlinear increase in drain current in the low-drain-voltage region in the output characteristics (Fig. 29.3c) was attributed to the relatively high contact resistance at the metal/semiconductor interface. Since C8–BTBT is wide bandgap material and the film has a deep valence band (VB) level (EV = 5.7 eV) [32], a high charge injection barrier can exist at the metal/organic interface due to the energy mismatch between the Fermi level of the Au NP electrodes and the VB of the C8–BTBT film. In contrast, the drain current in the high-drain-voltage regions shows saturation characteristics, which confirms an ideal MOSFET operation. Figure 29.11d shows that hysteresis-free transfer characteristics were achieved with a steep increase in drain current in the subthreshold region. Gate leak current was substantially reduced by complete patterning of the electrode and semiconductor layers (not shown), which allowed an on–off ratio of 10^6. The average μ_{FET} and threshold voltage (V_T) of the fully printed OTFTs calculated from 20 devices were 7.9 ± 1.1 cm^2/Vs and 1.1 ± 0.4 V, respectively. The variation in these two values may be related to the polycrystalline structure of the C8–BTBT film and the high contact resistance of the OTFTs as reflected in the nonlinear rise in drain current shown in Fig. 29.11c [33].

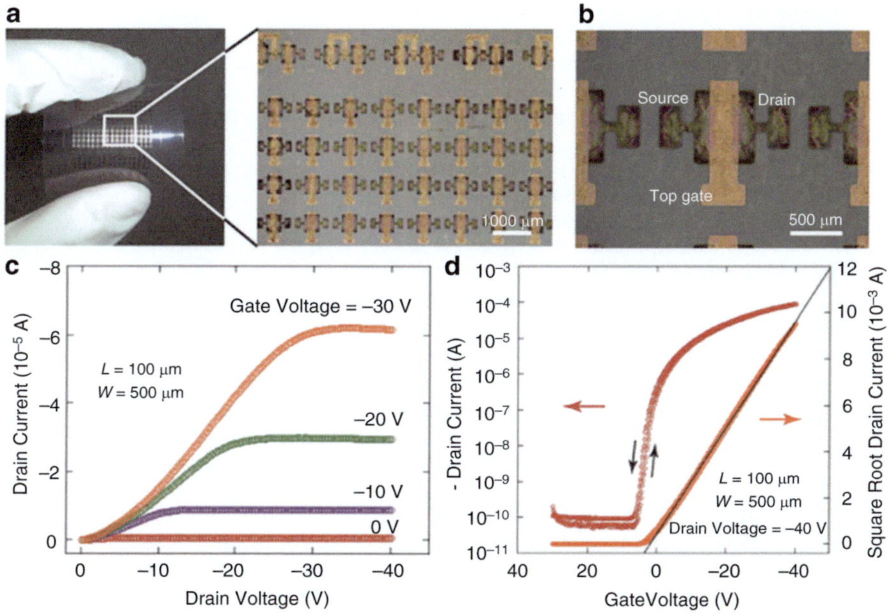

Fig. 29.11 OTFT devices fabricated using the RT printing process on a plastic substrate. (**a**) Flexing the printed OTFT arrays on the plastic substrate (*left*). Substrate size was 40×40 mm^2. An optical microscope image of the OTFT array (*right*). (**b**) Optical microscope image of the individual device. The electrode and organic semiconductor layers are fully patterned to reduce the off current and cross talk among the devices. (**c**) Typical output characteristics of the RT-printed OTFT. (**d**) Typical transfer characteristics of the OTFT device

In conclusion, flexible electronic devices were printed at RT using π-junction Au NP ink as the electrodes, which demonstrated the usefulness of this method as a core technology for flexible printed electronics. All printed devices will be commercialized soon.

References

1. Daniel M-C, Astruc D (2004) Chem Rev 104:293–346
2. Kanehara M, Oumi Y, Sano T, Teranishi TJ (2003) Am Chem Soc 125:8708–8709
3. Azuma Y, Kanehara M, Teranishi T, Majima Y (2006) Phys Rev Lett 98(016108):1–4
4. Kanehara M, Kodzuka E, Teranishi TJ (2006) Am Chem Soc 128:13084–13094
5. Teranishi T, Kiyokawa I, Miyake M (1998) Adv Mater 10:596–599
6. Brust M, Walker M, Bethell D, Schiffrin DJ, Whyman R (1994) Chem Commun 801–802
7. Zheng J, Petty JT, Dickson RMJ (2003) Am Chem Soc 125:7780–7781
8. Gryko DT, Clausen C, Roth KM, Dontha N, Bocian DF, Kuhr WG, Lindsey JSJ (2000) Org Chem 65:7345–7355
9. Yasseri AA, Syomin D, Malinovskii VL, Loewe RS, Lindsey JS, Zaera F, Bocian DFJ (2004) Am Chem Soc 126:11944–11953

10. Gopidas KR, Whitesell JK, Fox MAJ (2003) Am Chem Soc 125:6491–6502
11. Kanehara M, Teranishi T (2005) EJ Surf Sci Nanotechnol 3:30–32
12. Imahori H, Norieda H, Nishimura Y, Yamazaki I, Higuchi K, Kato N, Motohiro T, Yamda H, Tamaki K, Arimura M, Sakata YJ (2000) Phys Chem B 104:1253–1260
13. Katsonis N, Vicario J, Kudernac T, Visser J, Pollard MM, Feringa BLJ (2006) Am Chem Soc 128:15537–15541
14. Hill JP, Hewitt IJ, Anson CE, Powell AK, McCarty AL, Karr PA, Zandler ME, D'Souza FJ (2004) Org Chem 69:5861–5869
15. Kanehara M, Takahashi H, Teranishi T (2008) Angew Chem Int Ed 47:307–310
16. Hosseini A, Taylor S, Accorsi G, Armaroli N, Reed CA, Boyd PDW (2006) J Am Chem Soc 128:15903–15913
17. Ji X, Song X, Li J, Bai Y, Yang W, Peng X (2007) J Am Chem Soc 129:13939
18. Kanehara M, Kodzuka E, Teranishi T (2006) J Am Chem Soc 128:13084
19. Zabet-Khosousi A, Dhirani A-A (2008) Chem Rev 108:4072
20. Wakuda D, Hatamura M, Suganuma K (2007) Chem Phys Lett 441:305
21. Yamamoto M, Kakiuchi H, Kashiwagi Y, Yoshida Y, Ohno T, Nakamoto M (2010) Bull Chem Soc Jpn 83:1386
22. Law M, Luther JM, Song Q, Hughes BK, Perkins CL, Nozik AJ (2008) J Am Chem Soc 130:5974
23. Miyadera T, Minari T, Tsukagoshi K, Ito H, Aoyagi Y (2007) Appl Phys Lett 91:013512
24. Anthony JE, Brooks JS, Eaton DL, Parkin SR (2001) J Am Chem Soc 123:9482
25. Ebata H, Izawa T, Miyazaki E, Takimiya K, Ikeda M, Kuwabara H, Yui T (2007) J Am Chem Soc 129:15732
26. Uemura T, Hirose Y, Uno M, Takimiya K, Takeya J (2009) Appl Phys Express 2:111501
27. Kanehara M, Takeya J, Uemura T, Murata H, Takimiya K, Sekine H, Teranishi T (2012) Bull Chem Soc Jpn 85:957
28. Daami A, Bory C, Benwadih M, Jacob S, Gwoziecki R, Chartier I, Coppard R, Serbutoviez C, Maddiona L, Fontana E, Scuderi A (2011) IEEE Int Solid-State Circuits Conf 328–330
29. Minari T, Liu C, Kano M, Tsukagoshi K (2012) Adv Mater 24:299
30. Kang H, Kitsomboonloha R, Jang J, Subramanian V (2012) Adv Mater 24:3065
31. Fukuda K, Sekine T, Kobayashi Y, Kumaki D, Itoh M, Nagaoka M, Toda T, Saito S, Kurihara M, Sakamoto M, Tokito S (2012) Org Electron 13:1660
32. Kano M, Minari T, Tsukagoshi K (2009) Appl Phys Lett 94:143304
33. Minari T, Kanehara Y, Liu C, Sakamoto K, Yasuda T, Yaguchi A, Tsukada S, Kashizaki K, Kanehara M (2014) Adv Funct Mater 24:4886–4892

Part VII
Innovative Function of π-Electron Systems

Chapter 30
Control of Chiral π-Space in Highly Organized π-Conjugated Polymer Nanotube Composites and Their Functions

Hisashi Fujihara and Tsukasa Nakahodo

Abstract This chapter describes the following topics. (1) We have for the first time succeeded in the preparation of plasmonically active hybrid nanotubes (NTs) consisting of fullerene C_{60}, polythiophene and Ag or Au nanoparticles (NPs). In these hybrid NTs, the Ag NPs or Au NPs are confined as components of the wall thickness of the NTs. The multifunctional hybrid NTs consisting of fullerene C_{60}, polythiophene (PT) and Ag NPs or Au NPs will become a new type of SERS substrate using NTs. (2) We showed that the sequential template-based electropolymerization technique provides a novel approach to grow additional materials on the top of initially formed hybrid NT-deposited nanopores of an alumina membrane to form fascinating NT heterojunctions. The hybrid nanotube morphology can be controlled by the sequential electropolymerization of the different metal NPs in a nanoporous template. (3) The multifunctional chiral binaphthylcarboxylate-modified Fe_3O_4 NPs with a polymerizable group produce the polymer nanocomposite films on an electrode and new chiral hybrid nanotubes of Fe_3O_4 NPs and polythiophene by a template-based electropolymerization. (4) We demonstrated the encapsulation and release of an anionic fluorescent dye in the redox-active cationic polymer NTs containing viologen via an electrostatic interaction. We expect this research to yield a broad range of potential applications in catalysis, nanoelectrochemistry, molecular separation, drug delivery, controlled release devices and sensor developments.

Keywords Polythiophene • Fullerene • Hybrid • π-Nanotube

One-dimensional (1D) nanomaterials, i.e., tubes, wires and rods, and metal nanoparticles (NPs) have received considerable recent attention due to their fundamental significance and potential applications in nanoscale devices and sensors. In particular, the nanotubular structure has attracted significant interest because its wall thickness

H. Fujihara (✉) • T. Nakahodo
Department of Applied Chemistry, Kinki University, Kowakae, Higashi-Osaka 577-8502, Japan
e-mail: h-fuji@apch.kindai.ac.jp

T. Akasaka et al. (eds.), *Chemical Science of π-Electron Systems*,
DOI 10.1007/978-4-431-55357-1_30

and length can be used to control the physical and chemical properties. Generally, conductive polymer nanotubes of thiophene, pyrrole and aniline are fabricated by chemically or electrochemically using various templates. A large number of nanowires and nanorods have been synthesized, but the nanotube (NT) synthesis is more difficult and requires more delicate control of the experimental parameters. The detailed studies on the growth mechanism of the conductive polymer NTs of thiophene and pyrrole by template-based electropolymerization are quite limited, despite the importance and high applicability of the conductive polymer NTs in displays, energy storage devices, actuators and sensors. The functionalization of the conductive polymer NTs with metal NPs or organic molecules has received less attention. Similarly, only a few examples of the synthesis of metal NTs have been reported, which deal with the reduction of metallic complexes and immobilization of metal NPs within the pores of a nanoporous alumina template. We recently reported a new metal–polymer NT composite from the electropolymerization of thiophene-capped metal NPs. It is worth pointing out that our approach to preparing a metal–polymer NT composite completely differs from all previous studies on deposition of metal ions onto the surface of polymer NTs [1], although metal–polymer NT composites are only slightly known. The development of a generalized method for the fabrication of multifunctional metal–polymer NT composites remains an important and challenging issue.

30.1 Plasmonic Hybrid Nanotubes

Recently, there has been great interest in the synthesis and property of fullerene-based nanostructured materials. The nanostructured materials of fullerenes with various morphologies are yet to be explored for various applications. The development of a generalized method for the fabrication of fullerene–polythiophene nanotubes (NTs) and their metal NP hybrid NTs remains an important and challenging issue. Although only a few examples of the synthesis of fullerene NTs are known, the nanotube composite of fullerene and polythiophene (PT) and its metal hybrid NTs, i.e., SERS (surface-enhanced Raman scattering)-active hybrid NTs, have never been reported.

The hybridization of the fullerene C_{60}–polythiophene NTs (**2**-C_{60}-PT-NT, Fig. 30.1b) with metal NPs is envisaged to provide a valuable and multifaceted platform for the development of new plasmonic NTs [2]. However, if C_{60} and noble metal NPs were to be dispersed into two solvent phases, they would more likely concentrate at the interface due to their strong tendency to form aggregates. In order to overcome this limitation, we have developed a new method for the fabrication of a C_{60}/Ag or Au NP/PT nanotube composite (Ag-C_{60}-PT-NT and Au-C_{60}-PT-NT, Fig. 30.1b) by the template-based electrocopolymerization of terthiophene-linked C_{60} (**2**-C_{60}, Fig. 30.1a) and terthiophene-modified silver or gold NPs (**1**-AgNP, **1**-AuNP: Fig. 30.1a) without aggregation of the silver or gold NPs. To our knowledge, this is the first example for the template-based electrocopolymerization

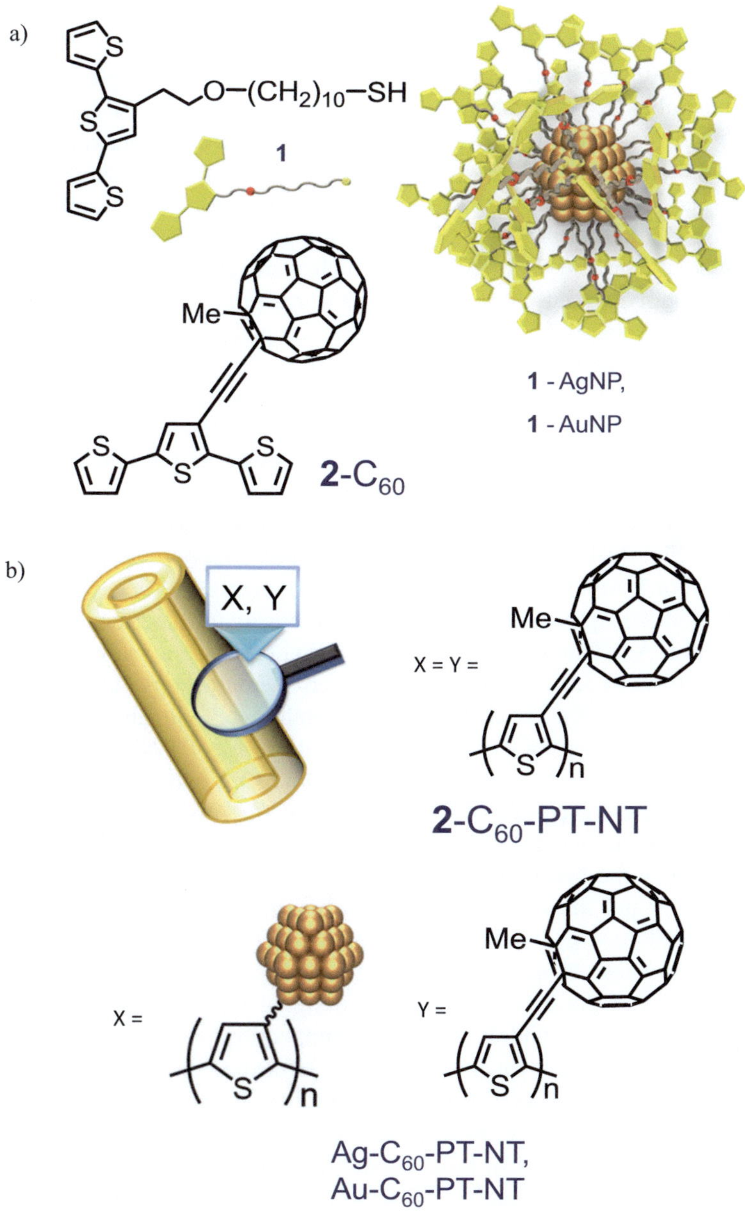

Fig. 30.1 (**a**) Structures of **1**, **1**-AgNP, **1**-AuNP and **2**-C_{60}. (**b**) Structures of **2**-C_{60}-PT-NT, Ag-C_{60}-PT-NT and Au-C_{60}-PT-NT

of two different monomers such as the C_{60}-thiophene and the thiophene-modified metal NPs. It has been known that polythiophene NTs are fabricated by template-based electropolymerization using one monomer such as thiophene. The novel hybrid nanotubes (**2**-C_{60}-PT-NT, Ag-C_{60}-PT-NT and Au-C_{60}-PT-NT) involve the following new aspects: (1) the template-based electropolymerization of **2**-C_{60} produced the fullerene C_{60}–polythiophene NTs (**2**-C_{60}-PT-NT) which showed the reversible redox behaviour of the fullerene C_{60}, and (2) the plasmonic hybrid NTs (Ag-C_{60}-PT-NT and Au-C_{60}-PT-NT) show a significant SERS compared to that of the NTs (**2**-C_{60}-PT-NT, Fig. 30.1b) which were prepared from the template-based electropolymerization of **2**-C_{60}. It is worth pointing out that our approach to preparing a metal–polymer NT composite completely differs from all previous studies on the deposition of metal NPs onto the surface of polymer NTs. This part presents the novel plasmonic hybrid NTs of fullerene C_{60}–PT–silver or gold NPs and their SERS effects. The nanotubular structure of fullerene and polythiophene is one of the ideal structures that can enhance the device performance by improving the charge transport rate as well as increasing the surface area.

We have succeeded in the fabrication of hybrid NTs (Ag-C_{60}-PT-NT and Au-C_{60}-PT-NT, Fig. 30.1b) from the template-based *electrocopolymerization* of **2**-C_{60} and **1**-AgNP or **1**-AuNP in a nanoporous alumina membrane (*ca.* 200–250 nm). No aggregation of the Ag NPs or Au NPs occurred during the electrocopolymerization of **2**-C_{60} and **1**-AgNP or **1**-AuNP. The hybrid nanotubes were analysed by scanning electron microscopy (SEM), transmission electron microscopy (TEM) and energy-dispersive X-ray (EDX). The SEM (Fig. 30.2a and Fig. 30.3a) and TEM (Fig. 30.2b and Fig. 30.3b) images of Ag-C_{60}-PT-NT and Au-C_{60}-PT-NT show the tubular structure. The EDX spectra show the presence of C, O, S and Ag for Ag-C_{60}-PT-NT (Fig. 30.2c) and the presence of C, O, S and Au for Au-C_{60}-PT-NT (Fig. 30.3c). The UV–Vis spectra of Ag-C_{60}-PT-NT (Fig. 30.2d) and Au-C_{60}-PT-NT (Fig. 30.3d) in CH_2Cl_2 show the broad absorption due to the surface plasmon of the Ag NPs or the Au NPs and the PT. We studied the potential of plasmonic NTs for Ag-C_{60}-PT-NT and Au-C_{60}-PT-NT using Raman spectroscopy. The Raman signal is typically very weak. When molecules are adsorbed onto the roughened surfaces or nanoparticles of Ag and Au, the Raman signal will be enhanced (*i.e.*, SERS effect); thus, it has been widely studied for applications regarding single-molecule detection. Much less is known about the SERS effect based on the nanotube itself of polymer NTs containing metal NPs. Accordingly, the Raman spectra were measured for **2**-C_{60}-PT-NT, Ag-C_{60}-PT-NT and Au-C_{60}-PT-NT using a 532 nm laser source. The Raman spectrum of **2**-C_{60}-PT-NT exhibited a peak at 1460 cm^{-1} (Fig. 30.3e, blue line). In contrast to **2**-C_{60}-PT-NT, the SERS effect (Fig. 30.3e, purple line and red line) was significantly found in the silver NP or gold NP composite (Ag-C_{60}-PT-NT, Au-C_{60}-PT-NT). As shown in Fig. 30.3e, purple and red lines, the Raman scattering peak at 1460 cm^{-1} of Ag-C_{60}-PT-NT and Au-C_{60}-PT-NT is enhanced due to the electromagnetic field enhancement. This band can be mainly assigned to the $A_g(2)$ mode for the Raman-active vibration mode of C_{60}, because **1**-Ag-PT-NT showed no enhancement for the Raman scattering of the PT as shown in Fig. 30.3e, green line. This is the first observation for the SERS effect based on the

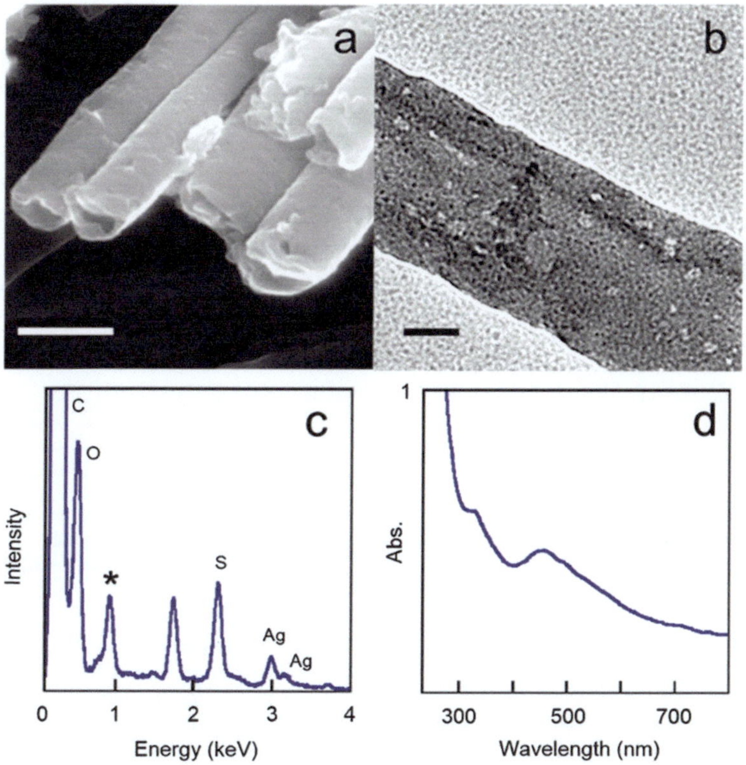

Fig. 30.2 (**a**) SEM image [scale bar = 250 nm], (**b**) TEM image [scale bar =50 nm] of Ag-C_{60}-PT-NT, (**c**) EDX spectrum of Ag-C_{60}-PT-NT (the peak indicated by *asterisk* is due to the supporting copper grid), (**d**) UV–Vis spectrum of Ag-C_{60}-PT-NT in CH_2Cl_2

nanotube itself of hybrid NTs consisting C_{60}, Ag NPs or Au NPs and PT. We will compare the power conversion efficiency of the hybrid nanotubes (Ag-C_{60}-PT-NT and Au-C_{60}-PT-NT) with that of the polymer bulk heterojunction solar cell, poly(3-hexylthiophene)/phenyl-C_{61}-butyric acid methyl ester.

30.2 Heterojunctioned Hybrid Polymer Nanotubes

There is growing interest in the synthesis of nanowires and nanorods containing heterojunctions; however, much less attention has been paid to the fabrication of NTs with a heterojunction due to the difficulty in preparing NTs.

We have prepared ligand **3**, because we found that the chiral binaphthylbis-phosphine acts as an excellent protective ligand for the stabilization of small metal NPs which are superior asymmetric nanocatalysts [3]. Ligand **3** provides

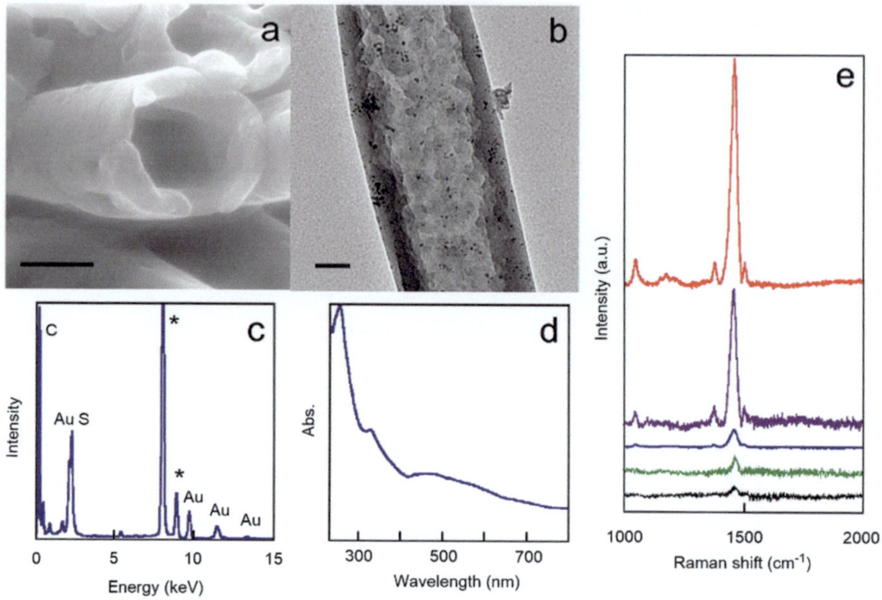

Fig. 30.3 (**a**) SEM image [scale bar = 150 nm], (**b**) TEM image [scale bar = 50 nm] of Au-C_{60}-PT-NT, (**c**) EDX spectrum of Au-C_{60}-PT-NT (the peak indicated by *asterisk* is due to the supporting copper grid), (**d**) UV–Vis spectrum of Au-C_{60}-PT-NT in CH_2Cl_2 and (**e**) Raman spectra of **1**-PT-NT (*black line*), **1**-Ag-PT-NT (*green line*), **2**-C_{60}-PT-NT (*blue line*), Ag-C_{60}-PT-NT (*purple line*) and Au-C_{60}-PT-NT (*red line*): they are listed at the same scale

the same particle size for the **3**-Au NPs and **3**-Pd NPs. The heterojunctioned metal NP–PT hybrid NTs were synthesized as illustrated in Fig. 30.4. A typical procedure for the fabrication of the **3**-Au/**3**-Pd–PT NTs is as follows (Fig. 30.4a–c). The electropolymerization was performed by attaching the alumina membrane (Whatman Anodisc; pore diameter of about 200–250 nm, thickness of 60 μm) to a Pt electrode, another Pt electrode was used as the counter electrode and Ag/0.1 M $AgNO_3$ was used as the reference electrode. The electrolysis solution contains **3**-Au NPs in 0.1 M Bu_4NPF_6-CH_2Cl_2 (1 mL). A voltage of 1.0 V (vs. Ag/0.1 M $AgNO_3$) was applied for 20 min. After the electropolymerization of **3**-Au NPs, the resulting **3**-Au–PT NT-deposited nanopores of the alumina membrane were rinsed in acetone and then CH_2Cl_2. Subsequently, using the resulting **3**-Au–PT NT-deposited nanopores of the alumina membrane as a second template, the electropolymerization (1.0 V) of **3**-Pd NPs in 0.1 M Bu_4NPF_6-CH_2Cl_2 (1 mL) was performed for 30 min. After the electropolymerization, removal of the alumina membranes with 1 M NaOH led to the release of the **3**-Au/**3**-Pd–PT NTs. The UV–Vis spectrum of the **3**-Au/**3**-Pd–PT NTs in $CHCl_3$ exhibits an absorption peak due to the polythiophene at 470 nm. Similar methods have provided the **3**-Au–PT NTs, **3**-Pd–PT NTs and **3**-Au/**3**-Pd/Au–PT NTs.

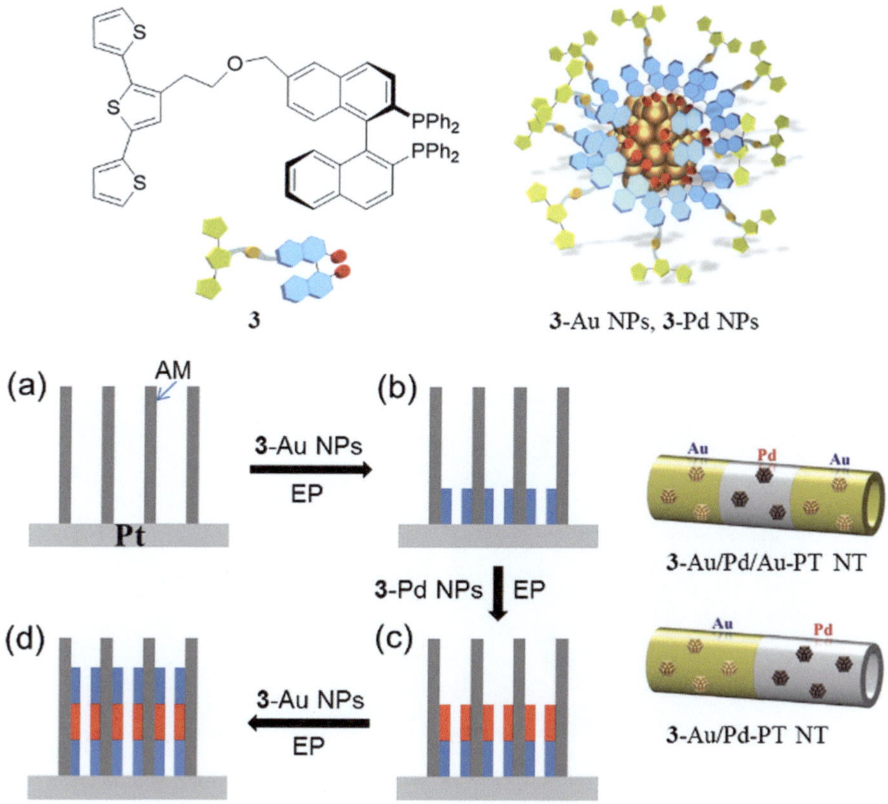

Fig. 30.4 Ligand **3** and metal NPs. Nanotubes. Fabrication steps: (**a**) bare nanoporous alumina membrane (AM) attached to a Pt electrode; (**b**) **3**-Au–PT NT-deposited nanopores of alumina membrane formed after electropolymerization (EP) of **3**-Au NPs; (**c**) **3**-Au/**3**-Pd–PT NT-deposited nanopores of alumina membrane formed after EP of **1**-Pd NPs; (**d**) **3**-Au/**3**-Pd/Au–PT NT-deposited nanopores of alumina membrane formed after EP of **3**-Au NPs

We can create heterojunctioned hybrid NTs consisting of heterometallic NPs and PT by the *sequential* electropolymerization of **3**-Au NPs and **3**-Pd NPs, as shown in Fig. 30.4a–c. The **3**-Au–PT NTs were initially formed by the template electropolymerization of the **3**-Au NPs. Subsequently, using the **3**-Au–PT NT-deposited nanopores of the alumina membrane as a second template, the electropolymerization of the **3**-Pd NPs led to the formation of the heterojunctioned hybrid NTs, **3**-Au/**3**-Pd–PT NTs, with a length of tens of micrometres and ca. 250 nm outer diameter (Fig. 30.5a, c, d). The SEM elemental profiles (Fig. 30.6) of the **3**-Au/**3**-Pd–PT NTs showed that P and S are uniformly distributed along the NT axes, while Au and Pd appear localized in the NT portions of the **3**-Au/**3**-Pd–PT NTs. Analogously, the TEM image of the **3**-Au/**3**-Pd–PT NTs indicated that the EDX analyses (Fig. 30.5b) performed at the two locations marked on Fig. 30.5a confirm that the blue-circled

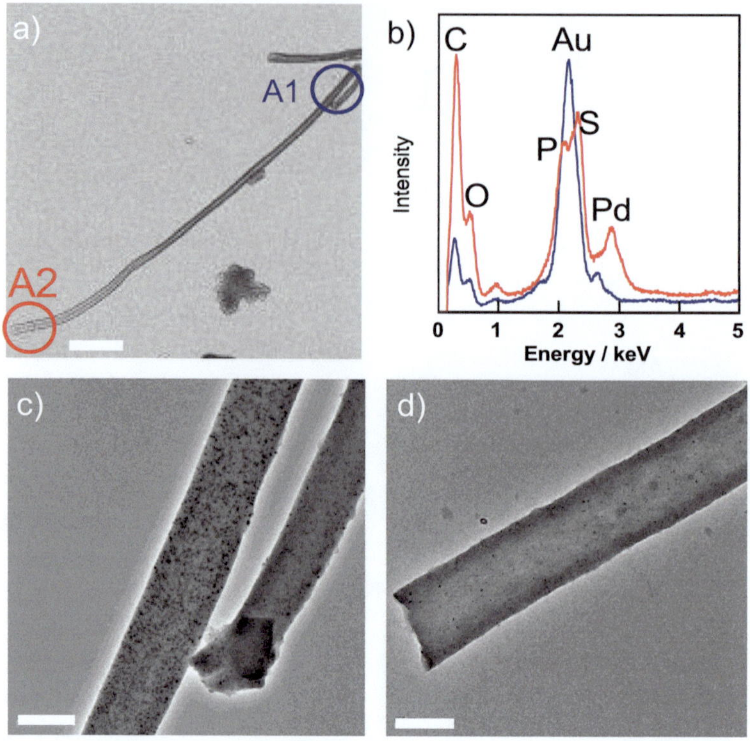

Fig. 30.5 (**a**) TEM image of **3**-Au/**3**-Pd–PT NTs (scale bar = 2 μm). (**b**) EDX spectra taken from selected areas marked in A1 (shown in *blue*) and A2 (*red*) of Fig. 30.5a. (**c**, **d**) TEM images from the areas marked A1 (*blue circle*) and A2 (*red circle*) indicate the Au–PT region (Fig. 30.5c) and Pd–PT region (Fig. 30.5d) of 3-Au/3-Pd–PT NTs, respectively [scale bar, (**c**, **d**) 200 nm]

area (A1) of the NTs consisted of the Au NPs and PT segment and the red-circled area (A2) consisted of the Pd NPs and PT segment where the wall thickness can be seen in Fig. 30.5d. The Au and Pd NPs did not aggregate in the part of the segments of the **3**-Au–PT NTs and **3**-Pd–PT NTs. It is noteworthy that none of the nanowires or nanorods were formed from the electropolymerization of the metal NPs, despite the use of the hybrid nanotube-deposited nanopores of the alumina template. This is an important result for the formation of polymer NTs by the template-based electropolymerization, because it has been reported that the template-based electropolymerization of thiophene in an alumina template leads to either nanowires (nanorods) or nanotubes, i.e., the NT synthesis is difficult and requires more delicate control of the experimental parameters. We have for the first time succeeded in the synthesis of a hybrid nanotube heterojunction by electropolymerization using the hybrid nanotube-deposited nanoporous alumina template.

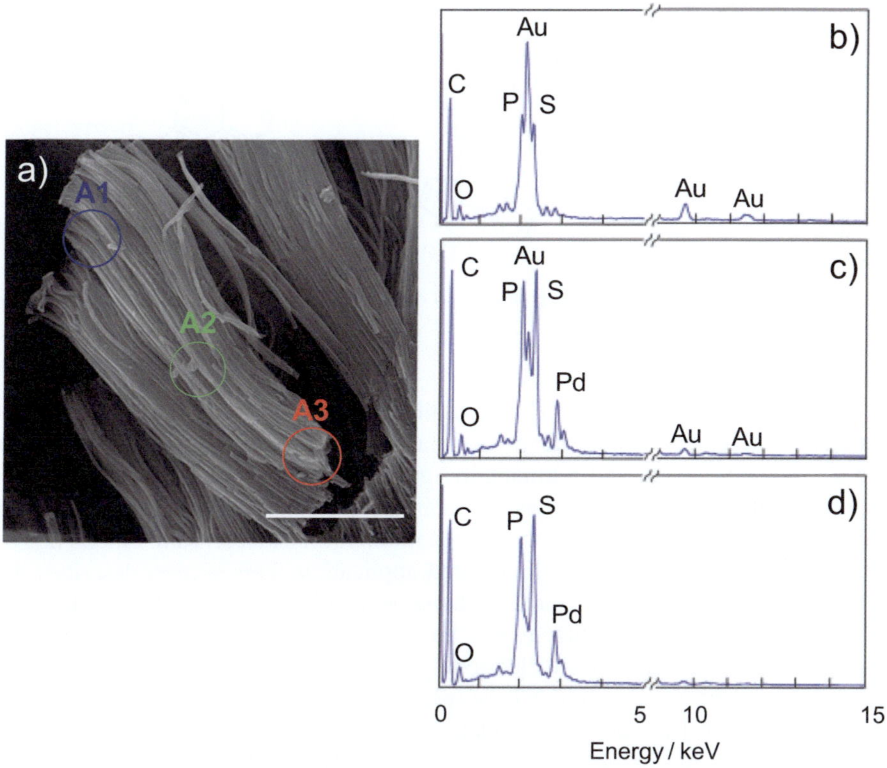

Fig. 30.6 **3**-Au/**3**-Pd–PT NTs: (**a**) SEM image [scale bar =10 μm]. (**b**) EDX spectrum taken from the area marked A1 (*blue circle* in Fig. 30.6a). (**c**) EDX spectrum taken from the area marked A2 (*green circle* in Fig. 30.6a). (**d**) EDX spectrum taken from the area marked A3 (*red circle* in Fig. 30.6a)

30.3 Chiral Magnetite–Polythiophene Nanotubes

The nanotubular structure of polythiophene is one of the ideal structures that can enhance the device performance by improving the charge transport rate as well as increasing the surface area. Thus, the nanocomposites based on conducting polymers and metal NPs, such as inorganic nanomaterials, are of interest due to their synergistic and hybrid properties as components. In contrast, so far there have been no reports of the synthesis of nanotube composites consisting of magnetite NPs and a conducting polymer as a constituent of hybrid NTs by the electropolymerization of thiophene-modified magnetite NPs and their related chiral nanocomposites. The combination of conducting polymers with magnetite NPs is the focus of research because materials having both a high conductivity and a high magnetic susceptibility can be used in different applications, such as electromagnetic interference shielding, magneto-optical storage, magnetic electrocatalysis and drug delivery.

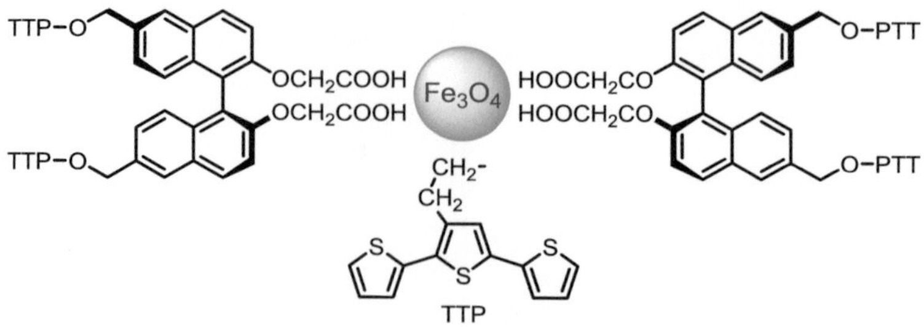

Fig. 30.7 Structure of chiral stabilizer **4** used to derive the (*R*)- and (*S*)-**4**-modified Fe_3O_4 NPs [(*R*)-, (*S*)-**4**-Fe_3O_4 NPs]

We have developed a new chiral stabilizer (**4**) which consists of a carboxyl group for the *coordination* site of the magnetite (Fe_3O_4) NPs, terthiophene for the *polymerization* site and a binaphthyl group as a *chiral* site (Fig. 30.7) [4]. It is noteworthy that in many cases the stabilizers can not only stabilize the NPs but also be used for further functionalization, for instance, with other NPs or various ligands, depending on the desired application. This section describes the synthesis of chiral **4**-modified small-size magnetite (Fe_3O_4) NPs [(*R*)-, (*S*)-**4**-Fe_3O_4 NPs], the chiral surface modification of an electrode with the chiral magnetite NP–polymer composite films by the electropolymerization of (*R*)-**4**-Fe_3O_4 NPs and the preparation of the chiral hybrid NTs of magnetite NPs and polythiophene, (*R*)- and (*S*)-**4**-Fe_3O_4-PT-NTs, from the template-based electropolymerization of (*R*)- and (*S*)-**4**-Fe_3O_4 NPs in a nanoporous alumina membrane. Much less attention has been paid to the preparation of magnetite NPs modified by an optically active stabilizer and to related chiral nanotube composites. In contrast to magnetite NPs, the chirality of extended metal NPs has become an emerging field of research in recent years triggered by potential applications in asymmetric catalysis. Despite a number of theoretical and experimental studies on optically active inorganic nanomaterials, such as metal NPs, the origin of the chirality is still under discussion.

A typical procedure for the template synthesis of the chiral hybrid nanotubes, (*R*)-**4**-Fe_3O_4-PT-NTs, is as follows. The electropolymerization was performed by attaching the alumina membrane (Whatman Anodisc; pore diameter of about 200–250 nm, thickness of 60 μm) to a Pt electrode. Another Pt electrode was used as the counter electrode, and Ag/0.1 M $AgNO_3$ was used as the reference electrode. The electrolysis solution contained (*R*)-**4**-Fe_3O_4 NPs (15 mg, 4.4 nm) in 0.1 M Bu_4NPF_6-CH_2Cl_2 (1 mL). A voltage of 0.55 V (*vs*. Ag/0.1 M $AgNO_3$) was applied for 20 min. After the electropolymerization, removal of the alumina membranes with 1 M NaOH led to the release of the hybrid nanotubes, (*R*)-**4**-Fe_3O_4-PT-NTs. The UV–Vis spectrum of the (*R*)-**4**-Fe_3O_4-PT-NTs in $CHCl_3$ exhibited a broad absorption peak at 440 nm which is ascribed to the polythiophene (Fig. 30.8). The TEM image of the (*R*)-**4**-Fe_3O_4-PT-NTs revealed an outer diameter of about 230 nm which corresponds to the pore diameter of the alumina membrane and a wall thickness of around 40 nm (Fig. 30.9a). The elemental composition of

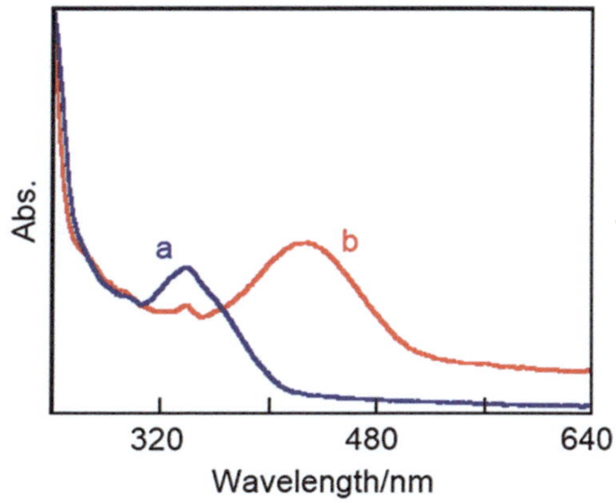

Fig. 30.8 UV–Vis spectra of (*R*)-**4**-Fe_3O_4 NPs (**a**, *blue*) and (*R*)-4-Fe_3O_4-PT-NTs (**b**, *red*) in $CHCl_3$

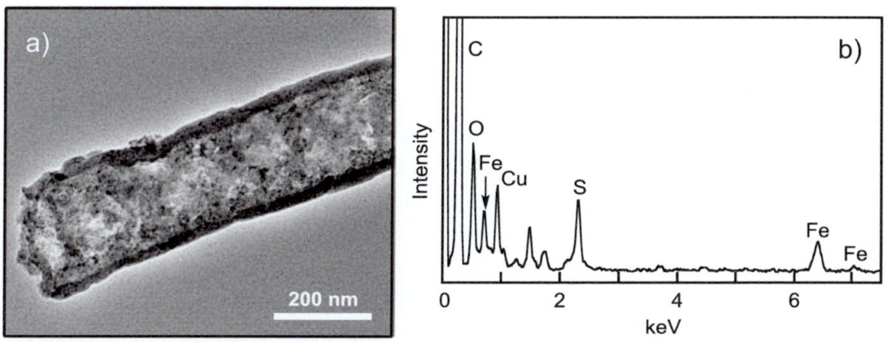

Fig. 30.9 (**a**) TEM image of (*R*)-**4**-Fe_3O_4-PT-NTs. (**b**) EDX spectrum of (*R*)-**4**-Fe_3O_4-PT-NTs. Cu peak is from the supporting copper grid

the (*R*)-**4**-Fe_3O_4-PT-NTs was confirmed by the EDX analysis of single nanotubes (Fig. 30.9a); its EDX spectrum (Fig. 30.9b) shows the presence of Fe, O, S and C elements in the hybrid NTs, which confirms that the Fe_3O_4 NPs have been successfully incorporated in the hybrid NTs during the template-based electropolymerization.

The CD spectra of the (*R*)-**4**-Fe_3O_4-PT-NTs/(*S*)-**4**-Fe_3O_4-PT-NTs show intense Cotton effects and a mirror image relationship in the measured range (Fig. 30.10a). Figure 30.10b shows that the (*R*)-**4**-Fe_3O_4-PT-NTs were attracted towards the external magnetic field located on the backside of the sample vial, demonstrating the high magnetic sensitivity of our hybrid nanotubes.

We have successfully developed a new multifunctional chiral stabilizer **4** for the synthesis of the chiral **4**-modified Fe_3O_4 NPs with a polymerizable group. The multifunctional chiral **4**-modified Fe_3O_4 NPs with a polymerizable group produce the polymer nanocomposite films on an electrode and new chiral hybrid nanotubes of Fe_3O_4 NPs and polythiophene by a template-based electropolymerization. The composition of our hybrid NTs is two components of Fe_3O_4 NPs and polythiophene

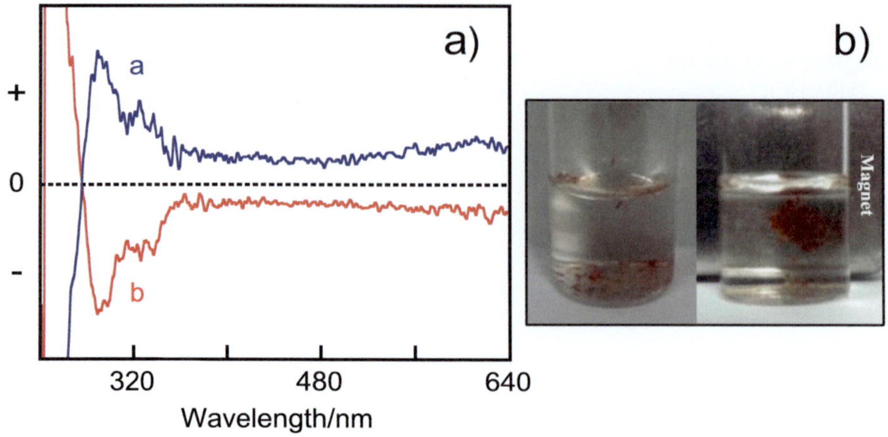

Fig. 30.10 (**a**) CD spectra of (*R*)-**4**-Fe_3O_4-PT-NTs (**a**, *blue*) and (*S*)-**4**-Fe_3O_4-PT-NTs (**b**, *red*) in $CHCl_3$. (**b**) Photographs of (*R*)-**4**-Fe_3O_4-PT-NTs in water (*left*) and their response to a magnet (*right*)

unlike decorating NTs with Fe_3O_4 NPs. To the best of our knowledge, this is the first example of chiral hybrid NTs consisting of conducting polythiophene and Fe_3O_4 NPs. Our developed synthetic method can be extended to the incorporation of other metal oxides in the chiral hybrid nanotubes. The combined functionalities will provide many exciting opportunities in developing applications, such as drug delivery, electro-magnetorheological polymer films, electromagnetic shielding and magnetically recoverable chiral catalysts. We believe that the chirality of the magnetite NPs and their hybrid NTs is producing more interest because of its novel properties and potential applications.

30.4 Encapsulation and Release in the Nanospace of Redox-Active Cationic Polymer Nanotubes

The template-synthesized conducting polymer NTs can have relatively large cavities and tunable interior chemical functionality. Especially, impregnation of the hollow NTs with guest materials provides a potentially fascinating route to create functional nanomaterials for a variety of applications. The development of a generalized method for the controlled encapsulation/release of a molecule in the conducting polymer NTs remains an important and challenging issue.

We have performed the synthesis of new polythiophene (PT) NTs with redox-active viologen and the nanotube membranes and the encapsulation or release of metal NPs or a guest molecule by the NTs (Figs. 30.11, 30.12). We have chosen a cationic viologen (*N,N*'-disubstituted-4,4'-bipyridinium salt, VL^{2+}) group to achieve the assembly in nanopores via *electrostatic attraction* as a noncovalent approach [5].

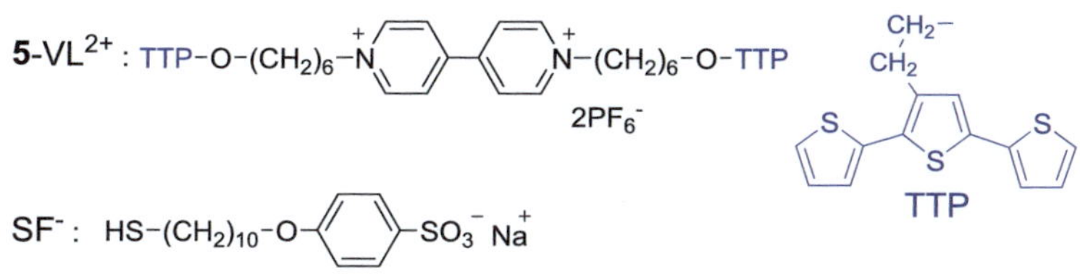

Fig. 30.11 Structures of 5-VL^{2+} and SF^{-}

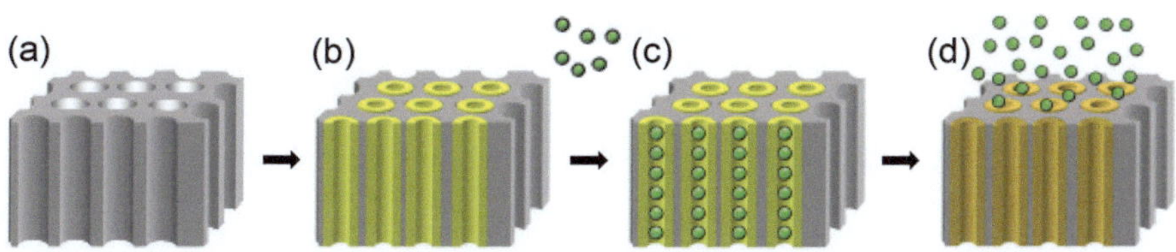

Fig. 30.12 Fabrication of PT (polythiophene) NTs and their composites: (**a**) bare alumina membrane (AM). (**b**) Modification of the nanopore of AM with PT NTs generated by template-based electropolymerization of **5**-VL^{2+} leads to **5**-VL^{2+}-PT-NT-AM as a polymer nanotube membrane. PT- T-modified nanopore of AM: **5**-VL^{2+}-PT-NT-AM. (**c**) SF^{-}-AuNPs or Pyra encapsulated in PT-NT-modified AM: SF^{-}-AuNP@**5**-VL^{2+}-PT-NT-AM and Pyra@**5**-VL^{2+}-PT-NT-AM. (**d**) Release of Pyra from Pyra@**5**-VL^{2+}-PT-NT-AM by reduction

The template-based electropolymerization of cationic viologen-linked terthiophene (**5**-VL^{2+}) in a nanoporous alumina membrane (AM) as a template gave the corresponding polythiophene NT-modified alumina membranes, which were converted into the redox-active cationic polythiophene NTs (**5**-VL^{2+}-PT-NTs) upon dissolution of the alumina membranes by a sodium hydroxide solution (Figs. 30.11, 30.12). The redox-active cationic polythiophene NT-modified alumina membrane (denoted as **5**-VL^{2+}-PT-NT-AM) as a *polymer nanotube membrane* was utilized to incorporate the anionic fluorescent dye, pyranine (8-hydroxyl-1,3,6-pyrene-trisulfonate, Pyra), in which the Pyra was expelled from the Pyra confined within **5**-VL^{2+}-PT-NT-AM (denoted as Pyra@**5**-VL^{2+}-PT-NT-AM) by electrochemical and chemical reductions of the viologen units in the cationic NTs (Fig. 30.12d). There is currently considerable interest in using functional nanotube membranes in molecular storage and release and sensing applications. This part describes the encapsulation of ionic metal NPs or Pyra into the cavity of the redox-active π-nanotubes via electrostatic interaction, and the preliminary realization of the electron supply induced the release of the Pyra from Pyra@**5**-VL^{2+}-PT-NT-AM.

The electropolymerization was performed by attaching the alumina membrane (AM) (Whatman Anodisc; pore diameter of about 200–250 nm, thickness of 60 μm) to a Pt electrode, another Pt electrode was used as the counter electrode and Ag/0.1 M $AgNO_3$ was used as the reference electrode. The electrolysis solution contains **5**-VL^{2+} (5 mM) in 0.1 M Bu_4NClO_4-MeCN (1 mL). A voltage of +0.75 V (vs. Ag/0.1 M $AgNO_3$) was applied for 20 min. **5**-VL^{2+}-PT-NT-modified nanopore

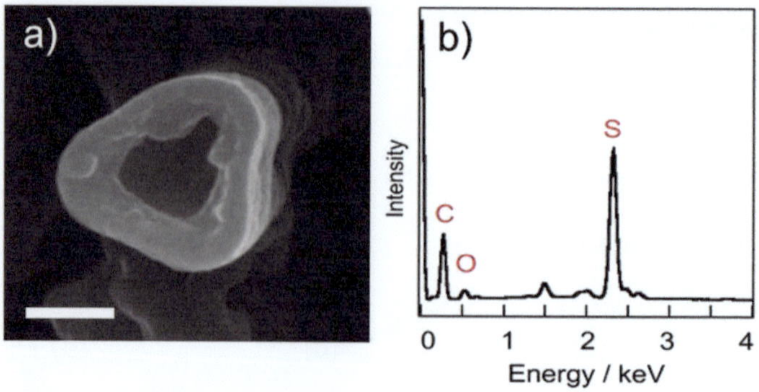

Fig. 30.13 (**a**) SEM image and (**b**) EDX spectrum of **5**-VL^{2+}-PT-NTs. Scale: 100 nm

of AM (**5**-VL^{2+}-PT-NT-AM) was formed. After the electropolymerization, removal of the alumina membrane of **5**-VL^{2+}-PT-NT-AM with 1 M NaOH led to the release of **5**-VL^{2+}-PT-NTs. Scanning and transmission electron microscopies (SEM and TEM) and energy-dispersive X-ray (EDX) spectroscopy were used for the analysis of the NTs. Figure 30.13a shows an SEM image of the **5**-VL^{2+}-PT-NTs, which were grown using the nanoporous alumina template; most of them have outer diameters of about 200–230 nm. The elemental composition of the **5**-VL^{2+}-PT-NTs was confirmed by an EDX analysis, which reveals the presence of C, O and S (Fig. 30.13b). In order to confirm the electrostatic interaction between an anionic guest material and the inner surfaces of the cationic NTs, 4-(10-mercapto-decyloxy)-benzenesulfonic acid sodium salt (SF^-, Fig. 30.11) and the SF^--stabilized gold NPs (SF^--AuNPs) were prepared. We used the NT-modified alumina membrane for the encapsulation of the Au NPs or Pyra into the inner surfaces of the NTs because the outer surfaces of the NTs are in contact with the pore wall of the alumina membrane and thus masked. The encapsulation of SF^--AuNPs (4.6 ± 0.5 nm) into **5**-VL^{2+}-PT-NTs was conducted as follows (Fig. 30.14). The UV–Vis spectra of SF^--AuNPs in H_2O showed a plasmon resonance at 533 nm in the absence and the presence of the bare alumina membrane (Fig. 30.14d-i). Interestingly, in contrast, in the presence of **5**-VL^{2+}-PT-NT-AM (Fig. 30.14d-ii), the absorption intensity decreased with the storage time, suggesting the continuing encapsulation of the Au NPs into the inner cavity of the **5**-VL^{2+}-PT-NT-AM. The colour of the solution turned from pink to colourless after overnight storage (Fig. 30.14c-0 and c-1).

The immersion of the **5**-VL^{2+}-PT-NT-AM into a solution of SF^--AuNPs was done, and any Au NPs remaining in solution were removed by filtration and washed several times with water. After removal of the alumina membrane template by the 1 M NaOH solution, the Au NPs confined inside the NTs (denoted as SF^--AuNP@**5**-VL^{2+}-PT-NTs) were obtained by filtration. The composite, SF^--AuNP@**5**-VL^{2+}-PT-NTs, contains Au and S, as evidenced by the EDX analysis. Thus, the electrostatic interaction should be effective for fabricating a metal

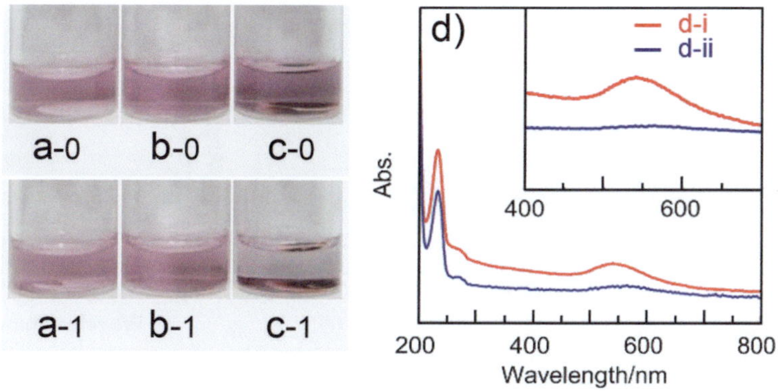

Fig. 30.14 Photographs of a solution of SF^{-}-AuNPs in (**a**) the presence of bare AM, (**b**) the absence of bare AM and (**c**) the presence of **5**-VL^{2+}-PT-NT-AM. (a-0, b-0, c-0) before and (a-1, b-1, c-1) after overnight storage. (**d**) Each UV–Vis spectrum (d-i, *red line*) corresponds to the photograph of c-0 and (d-ii, *blue line*) corresponds to the photograph of c-1

NP–PT NT composite in the nanopore. Importantly, this approach based on the electrostatic interaction provides an efficient method to attach other nanostructures to functionalized polythiophene NTs and can be used as an illustrative detection of functional groups on the inner walls of polythiophene NTs.

We have succeeded in the encapsulation and release of an organic molecule, such as Pyra, using the redox-active **5**-VL^{2+}-PT-NTs via electron transfer process. We postulate that the positively charged inner surfaces of the **5**-VL^{2+}-PT-NTs strongly accelerate the encapsulation of the negatively charged Pyra via an electrostatic interaction. In contrast, such an interaction between Pyra and the viologen neutral form generated by the two-electron reduction of the viologen dication is unfavourable; accordingly, the release of Pyra from the Pyra confined within the **5**-VL^{2+}-PT-NTs will be confirmed by the reduction. To prove this concept, the encapsulation of Pyra into the **5**-VL^{2+}-PT-NTs was examined. The UV–Vis spectra of a Pyra solution in the absence or the presence of a bare alumina membrane in water showed the characteristic absorption due to the Pyra (Fig. 30.15a), while no absorption in the UV–Vis region was observed in the presence of **5**-VL^{2+}-PT-NT-AM (Fig. 30.15b), indicating the formation of Pyra@**5**-VL^{2+}-PT-NT-AM by the encapsulation of Pyra into the NTs. On the other hand, the cyclic voltammogram of the **5**-VL^{2+}-PT-NTs in 0.1 M Bu_4NClO_4-MeCN at a glassy carbon (GC) electrode exhibited two reversible redox peaks corresponding to the VL^{2+} dication–$VL^{\cdot +}$ radical cation–VL neutral system at $E_{1/2} = -0.69$ and -1.13 V (vs. Ag/0.1 M $AgNO_3$). Thus, the electronic state of the **5**-VL^{2+}-PT-NTs can be electrochemically controlled. Consequently, at 1200 s, a potential of -1.25 V (Ag/0.1 M $AgNO_3$) was applied to the Pyra@**5**-VL^{2+}-PT-NT-AM. At this potential, the viologen is present in the neutral form by the reduction; consequently, the Pyra would be released from Pyra@**5**-VL^{2+}-PT-NT-AM. Actually, after the electrochemical reduction of Pyra@**5**-VL^{2+}-PT- NT-AM, the Pyra was released from the NT–alumina membrane

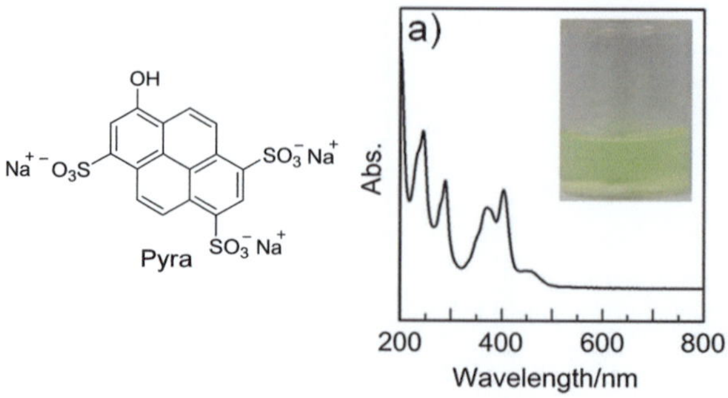

Fig. 30.15 Structure of Pyra. UV–Vis spectra and photographs of a solution of Pyra (**a**) after overnight storage in the presence of bare AM and (**b**) after overnight storage in the presence of **5**-VL^{2+}-PT-NT-AM in water. The photograph of (**a**) shows a typical colour of Pyra in water

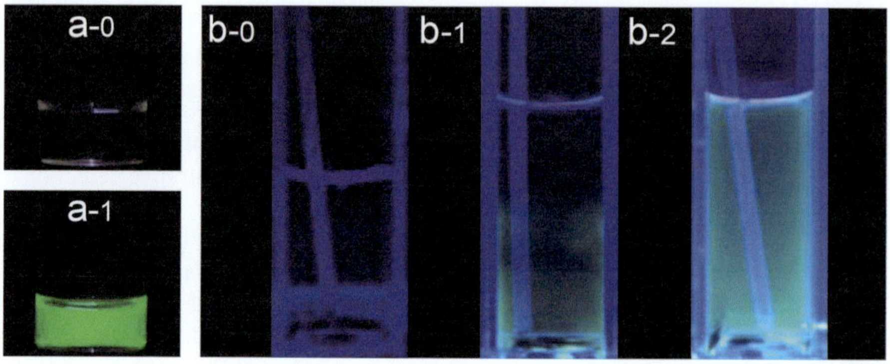

Fig. 30.16 Photographs of a solution of (a-0) before electrochemical reduction of Pyra@**5**-VL^{2+}-PT-NT-AM in water and (a-1) after electrochemical reduction of Pyra@**5**-VL^{2+}-PT-NT-AM in water. The fluorescence of Pyra was monitored by irradiation at 365 nm UV light. (**b**) Photographs for monitoring reaction of Pyra@**5**-VL^{2+}-PT-NT-AM with dithionite: (b-0) before addition, (b-1) after 120 s and (b-2) after 180 s. The fluorescence of Pyra was monitored by irradiation at 365 nm UV light. The photographs of b-1 and b-2 showed the fluorescence of the released Pyra

as evidenced from the UV–Vis spectra and the observed fluorescence due to the released Pyra (Fig. 30.16a-1). Moreover, the injection of sodium dithionite ($Na_2S_2O_4$) as a reductant into the Pyra@**5**-VL^{2+}-PT-NT-AM induced the release of the Pyra, i.e., the fluorescence of the released Pyra from Pyra@**5**-VL^{2+}-PT-NT-AM was observed when a solution of sodium dithionite in water was injected into the Pyra@**5**-VL^{2+}-PT-NT-AM using a syringe pump (Fig. 30.16b-1, b-2). It is noteworthy that the redox-active viologen-functionalized polythiophene NT membrane is very unique from the standpoint of electrochemical control of the reversible encapsulation/release system.

References

1. Umeda R, Awaji H, Nakahodo T, Fujihara H (2008) J Am Chem Soc 130(11):3240–3241
2. Yoshida R, Matsumura T, Nakahodo T, Fujihara H (2014) Chem Commun (Camb) 50(96):15183–15186
3. Awaji H, Nakahodo T, Fujihara H (2011) Chem Commun 47:3547–3549
4. Mitsumori M, Nakahodo T, Fujihara H (2012) Nanoscale 4(1):117
5. Sato K, Nakahodo T, Fujihara H (2011) Chem Commun 47(36):10067–10069

Chapter 31
Photoinduced Electron-Transfer Functions of π-Electron Donor–Acceptor Supramolecular Complexes

Shunichi Fukuzumi

Abstract This chapter describes the recent development of supramolecular π-electron donor–acceptor complexes linked by noncovalent bonds for light harvesting and charge separation. Carbon nanomaterials such as carbon nanohorns and graphene oxides act as π-electron donors to construct π-electron donor–acceptor supramolecular nanohybrids linked with π-electron acceptors such as fullerenes and perylenediimide, which undergo efficient charge separation. Multiple reaction center models have been constructed using multiporphyrin assemblies (porphyrin dendrimers and porphyrin polypeptides), which are noncovalently linked with π-electron acceptors, affording efficient energy migration, followed by long-lived charge separation. Bimolecular charge recombination of charge-separated states has been prohibited in supramolecular complexes between cationic and anionic porphyrins, affording the long-lived charge-separated states in solution.

Keywords Photoinduced Electron Transfer • Carbon nanomaterials • Supramolecular Nanohybrids • Charge Separation

31.1 Introduction

The rapid consumption of fossil fuel has caused unacceptable environmental problems such as the greenhouse effect by CO_2 emission, which could lead to disastrous climatic consequences [1, 2]. Thus, renewable and clean energy resources are definitely required in order to solve global energy and environmental issues particularly after the nuclear disaster in Fukushima on March 11 in 2011. Among renewable energy resources, solar energy is by far the largest exploitable resource [3, 4]. The ultimate goal to utilize solar energy is to develop a large-scale and

S. Fukuzumi (✉)
Department of Chemistry and Nano Science, Ewha Womans University, Seoul 120–750, Korea

Faculty of Science and Engineering, ALCA and SENTAN, Japan Science and Technology Agency (JST), Meijo University, Nagoya, Aichi 468-0073, Japan
e-mail: fukuzumi@chem.eng.osaka-u.ac.jp

T. Akasaka et al. (eds.), *Chemical Science of π-Electron Systems*,
DOI 10.1007/978-4-431-55357-1_31

cost-effective artificial photosynthetic system for production of solar fuels production, which can replace fossil fuels [5–10]. The development of such an artificial photosynthetic system requires combining five units: (1) the light-harvesting unit, (2) the charge-separation unit, (3) the catalytic unit for water reduction, (4) the catalytic unit for water oxidation, and (5) the catalytic CO_2 fixation unit [10].

Photosynthesis consists of two photosystems I and II (PSI and PSII), involving a number of steps [11–15]. Solar photons are captured in the light-harvesting antenna complexes composed of chlorophyll and other accessory pigments [16]. Energy transfer in antenna chlorophyll occurs in the femtosecond to picosecond time scale to produce the excited state of a chlorophyll dimer that is called a special pair in the photosynthetic reaction center [16]. Then, multistep electron-transfer reactions occur in the picosecond to nanosecond time scale to produce the final charge-separated state in the photosynthetic reaction center [11, 12]. The hole in the charge-separated state regains the lost electron from a water molecule through a process called photolysis, which results in release of O_2 in photosystem II [16]. On the other hand, electrons are shuttled through an electron transport chain to photosystem I, generating a chemiosmotic potential across the membrane [16].

Extensive efforts have so far been devoted to mimic the functions of the photosynthetic reaction center by synthesizing a variety of electron donor–acceptor linked molecules [17–30]. In this chapter, we focus on recent development of the combination of the light-harvesting and charge-separation units by noncovalent bonding for long-lived charge separation, which is essential for artificial photosynthesis.

31.2 Supramolecular Complexes of Carbon Nanohorns with Fullerenes

Among the wide range of carbon nanomaterials, carbon nanohorns (CNHs) consist of pseudo-cylindrical single-wall tubules with a conical tip [31–34]. The diameter of the tubular part of CNHs is 2–4 nm with a length of 30–50 nm, and the CNHs particles form a spherical assembly 80 nm in diameter. CNHs are generally synthesized by laser ablation of pure graphite without using metal catalyst with high production rate and high yield (around 75 %) and typically form radial aggregates (*dahlia-like*). In contrast to single-wall carbon nanotubes (SWCNTs) which contain metal particles, CNHs are essentially metal-free and pure samples of CNHs are available more easily than pure samples of SWCNTs, which favors their applications [35–37]. For example, a supramolecular complex was formed between a CNH with an NH_3^+ group (CNH-sp-$NH_3^+F^-$) and C_{60} with a crown ether moiety (crown-C_{60}) by ammonium–crown ether interaction (Fig. 31.1) [38]. The CNH-sp-NH^{3+}–C_{60} complex was characterized by Raman and XPS spectroscopy, thermogravimetry analysis (TGA), and high-resolution transmission electron microscopy (HR-TEM) [38]. The incorporation of the crown-C_{60} moiety

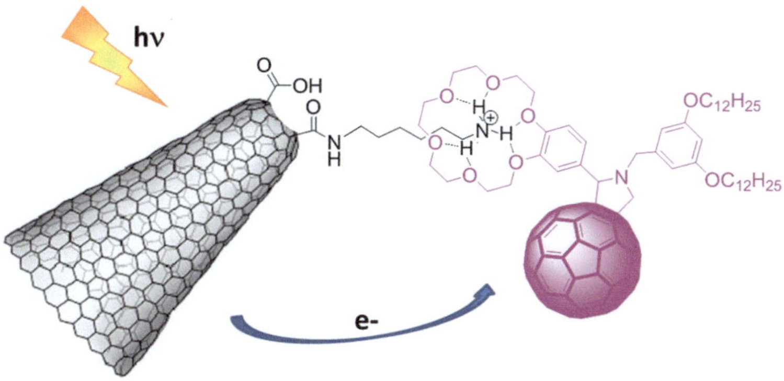

Fig. 31.1 A supramolecular complex of CNH-sp-NH^{3+} and crown-C_{60} and the photoinduced electron transfer

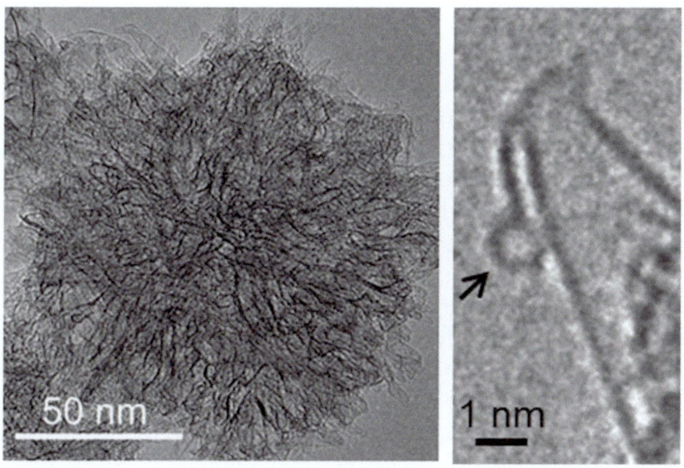

Fig. 31.2 HR-TEM images of CNH–C_{60} complexes. Aggregates kept the spherical shapes (*left panel*) and the C_{60} moieties showed spherical shapes (*arrow* in *right panel*)

onto CNH-sp-NH^{3+} was confirmed by the Raman spectrum [38]. The number of crown-C_{60} attached to a CNH was estimated as 1 crown-C_{60} per 110 carbon atoms of CNH based on the TGA analysis [38]. The CNH-sp-NH^{3+}–C_{60} complex was observed with high-resolution transmission electron microscopy (HR-TEM) as shown in Fig. 31.2, where the CNHs aggregate kept the spherical forms, and the C_{60} moieties were found in the hybrids [38].

In the supramolecular complex (CNH-sp-NH^{3+}–C_{60}), CNH-sp-NH_3^+ acts as an electron donor, whereas crown-C_{60} acts as an electron acceptor. Femtosecond laser flash photolysis measurements of CNH–C_{60} together with those of reference compounds in PhCN revealed that efficient electron transfer from CNH to the singlet excited state of the C_{60} moiety occurred with a rate constant of 6.5×10^{10} s^{-1} to produce the radical ion pair, which decayed by charge recombination with a lifetime of 1.0 ns to afford the triplet excited state of CNH-sp-$NH_3{}^+F^-$ and crown-C_{60} [38].

31.3 Supramolecular Complexes of Graphene Oxides with Perylenediimide

Graphene and graphene oxides are a new class of 2-D carbon nanomaterials made up entirely of conjugated sp^2 carbons arranged in a honeycomb structure, offering immense potential for practical applications [39–41]. Dispersibility of graphene-based composites in solvents is required for practical applications [39–43]. Chemical oxidation of graphite is an appealing approach to obtain exfoliated and dispersible graphene oxide (GO) sheets in aqueous media [44]. By this method, hydrophilic, oxygen-containing functional groups are introduced to π-surface as defects [45]. Some of these defects, such as –COOH groups, are also operative as anchoring groups for the noncovalent functionalization of GO with redox-active dye molecules for photoinduced applications. Dye functionalization of GO has been achieved in aqueous environments by the electrostatic attraction between the cationic substituents of the corresponding dye and the anchoring groups on GO sheets and hydrophobic π–π interactions between the π-systems of dyes and GO [46, 47]. Many hybrids of GO have been formed by employing various dyes with hydrophilic substituents (pyrenes, perylenes, porphyrins, coronenes, and semiconducting oligomers and polymers) [48–50].

Photoinduced electron-transfer dynamics of GO was reported for supramolecular complexes of GO with *N,N'*-di(2-(trimethylammoniumiodide)-ethylene) perylenediimide (PDI–GO) as shown in Fig. 31.3, where the cationic group of the PDI derivative enhanced the electrostatic interactions with the functional groups of GO [51]. In addition, the large aromatic core of PDI establishes strong π–π interactions with the π-surface of the GO sheet [51]. As a result of these interactions, PDI–GO forms large aggregates in water [51]. The fluorescence of PDI was quenched by formation of the supramolecular complex with GO due to

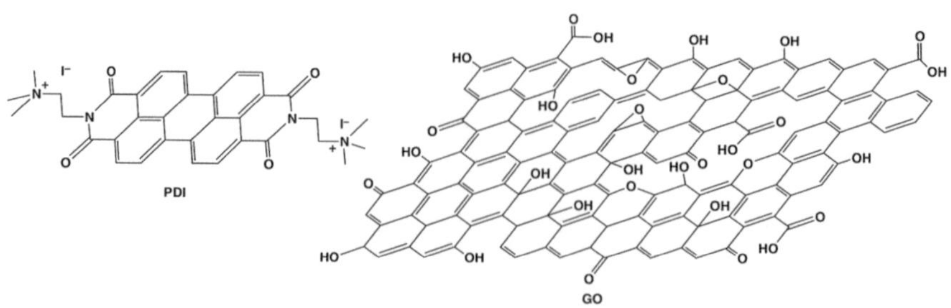

Fig. 31.3 Molecular structure of PDI and proposed structural model of a single-layer GO

electron transfer from GO to the singlet excited state of PDI ($^1PDI^*$). The driving force of the photoinduced electron transfer was determined to be 1.15 eV from the excitation energy of $^1PDI^*$ (2.28 eV), the one-electron oxidation potential of GO (0.88 V vs. SCE), and the one-electron reduction potential of PDI (−0.25 V vs. SCE) in water [51]. The occurrence of the photoinduced electron transfer was confirmed by the femtosecond transient absorption of aggregates of PDI–GO in water, which exhibits fast formation of broad transient absorption around 720 and 960 nm due to $PDI^{•-}$ after the selective excitation of PDI at 510 nm [51]. From the rise of the transient absorption at 720 nm, the rate constant of charge separation (k_{CS}) was determined to be 3.6×10^{11} s^{-1} [51]. The decay at the same wavelength comprises two components with lifetimes of 31 and 417 ps [51]. Such a two-component decay was also observed in the π-stacks PDI hosting porphyrins via π–π and electrostatic interactions in water, where the slower decay results from electron migration among the π-stacks of PDI [52, 53].

31.4 Supramolecular Complexes of Porphyrin Dendrimers with π-Electron Acceptors

Dendrimers with attached porphyrins have been synthesized mainly to study their light-harvesting and energy-transfer properties [54–64]. Porphyrin dendrimers [D(MP)$_n$ (M = 2H or Zn, n = 1, 4, 8, 16)] acting as light-harvesting units have been combined with a series of π-electron acceptors: pyridylnaphthalenediimide (PyNIm), 1,4-bis(*N*-hexyl-4-pyridinium)butadiene diperchlorate (BHV^{2+}), and fulleropyrrolidine bearing a pyridine ligand (PyC_{60}) [65] as shown in Fig. 31.4 [66, 67]. D(MP)$_n$ forms supramolecular complexes with the π-electron acceptors with 1:1 stoichiometry between the porphyrin moiety and π-electron acceptors as indicated by the UV–vis absorption spectral titrations [66–68]. The fluorescence of D(MP)$_n$ was quenched by the supramolecular formation with the π-acceptors.

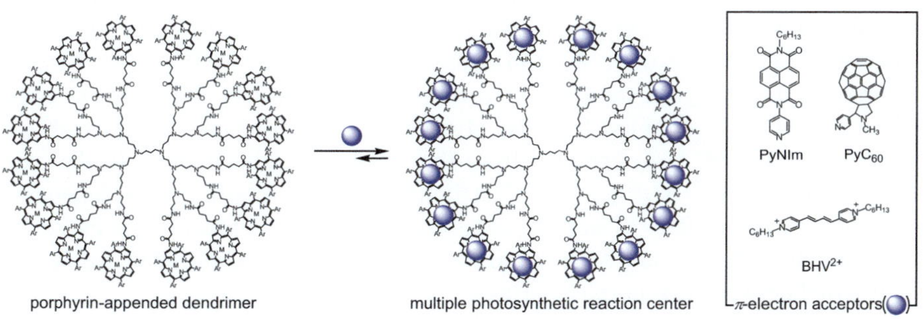

Fig. 31.4 Porphyrin dendrimer containing 16 porphyrin units [D(MP)$_{16}$, M = H_2 or Zn, Ar = 3,5-(t-Bu)$_2C_6H_3$] and π-electron acceptors

Table 31.1 Formation constants (K) and CS lifetimes (τ) of the supramolecular complexes of porphyrin dendrimers with π-electron acceptors in PhCN at 298 K

Porphyrin	Acceptor	UV visible,[a] K (M^{-1})	Fluorescence,[b] K (M^{-1})	τ, ms
$D(ZnP)_{16}$	PyC_{60}	1.6×10^4	5.0×10^4	0.25 ± 0.02
$D(ZnP)_8$	PyC_{60}	2.2×10^4	3.8×10^4	0.36 ± 0.02
$D(ZnP)_4$	PyC_{60}	8.3×10^3	8.4×10^4	0.34 ± 0.02
$D(ZnP)_{16}$	PyNIm	3.3×10^3	6.9×10^3	0.56 ± 0.02
ZnTPP	PyNIm	5.2×10^2	5.2×10^2	0.33 ± 0.02
$D(H_2P)_{16}$	BHV^{2+}	c	c	0.79 ± 0.06
$D(H_2P)_8$	BHV^{2+}	c	c	0.65 ± 0.04
$D(H_2P)_4$	BHV^{2+}	c	c	0.44 ± 0.06
$D(H_2P)_1$	BHV^{2+}	5.8×10^2	5.7×10^2	0.34 ± 0.04

[a]Determined from the absorption spectral change
[b]Determined from the fluorescence spectral change
[c]The K value varies depending on concentration of BHV^{2+}

The formation constants (K) of the supramolecular complexes were determined from the absorption and fluorescence spectral changes as listed in Table 31.1 [66, 67]. The K value ($5.2 \times 10^2\ M^{-1}$) determined from the absorption spectral change of the monomer (ZnTPP) with PyNIm agrees with the value ($5.2 \times 10^2\ M^{-1}$) determined from the fluorescence quenching. This indicates that the fluorescence is quenched by electron transfer from the singlet excited state of ZnP ($^1ZnP^*$) to bound PyNIm in the complex (static quenching) rather than by an intermolecular electron transfer from $^1ZnP^*$ to unbound PyNIm (dynamic quenching). In contrast to the case of the monomer, the K values of a zinc porphyrin dendrimer, [$D(ZnP)_{16}$], determined from the fluorescence quenching by PyNIm are significantly larger than the corresponding values determined from the absorption spectral changes (Table 31.1). This indicates that the excited energy migration between the porphyrin units occurs efficiently, when the fluorescence of unbound ZnP is quenched by PyNIm that is bound to a different ZnP moiety [67].

Free-base porphyrin dendrimers $D(H_2P)_n$ ($n = 4$, 8, 16) also form supramolecular complexes with BHV^{2+} via π–π interaction as indicated by the significant absorption and fluorescence spectral changes by addition of BHV^{2+} to a PhCN solution of $D(H_2P)_n$. In the case of monomer porphyrin (H_2TPP), the K value ($5.8 \times 10^2\ M^{-1}$) determined from the absorption spectral change agrees with the value ($5.7 \times 10^2\ M^{-1}$) determined from the fluorescence quenching. In the case of the dendrimers [$D(H_2P)_n$: $n = 4$, 8, 16], however, the K values vary depending on concentration of BHV^{2+} and the dendrimer generation number (n), because the structural deformation may be required for the π–π interaction of BHV^{2+} that may be inserted between two porphyrin rings.

The occurrence of photoinduced electron transfer in the supramolecular complex in PhCN was confirmed by the transient absorption spectra of the supramolecular complex of $D(ZnP)_{16}$ with PyNIm using nanosecond laser flash photolysis as shown

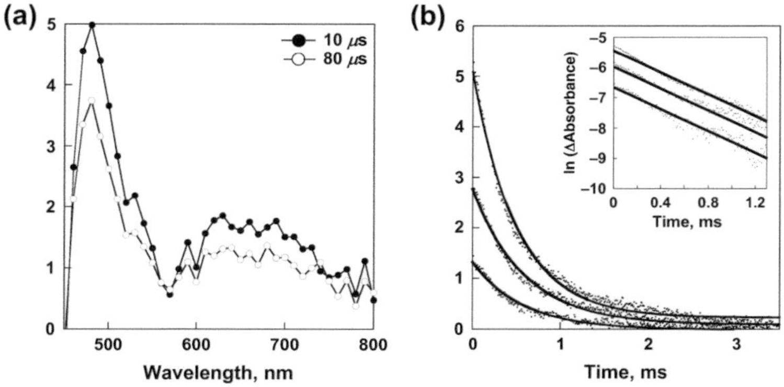

Fig. 31.5 (**a**) Transient absorption spectra of $D(ZnP)_{16}$ (2.9×10^{-6} M) in the presence of PyNIm (4.9×10^{-3} M) in deaerated PhCN at 298 K taken at 10 μs (*solid line* with *black circles*) and 80 μs (*solid line* with *white circles*) after laser excitation at 431 nm, respectively. (**b**) Time profiles of the absorption at 480 nm of $ZnP^{•+}$ with different laser intensities (22, 4, and 1 mJ/pulse) at 298 K. Inset: first-order plots

in Fig. 31.5 [67]. The transient absorption bands observed in Fig. 31.5a agree with those of superposition of the absorption bands due to $ZnP^{•+}$ produced by the electrochemical oxidation of monomer porphyrin with Zn and those of $PyNIm^{•-}$ produced by the one-electron reduction of PyNIm with tetramethylsemiquinone radical anion in PhCN [67]. Thus, the transient absorption spectrum in Fig. 31.5a clearly indicates formation of the CS state of the supramolecular complex by photoinduced electron transfer from the singlet excited state of $D(ZnP)_{16}$ to bound PyNIm in the supramolecular dendrimer complex. The free energy change of photoinduced electron transfer (ΔG_{ET}) in PhCN was determined to be −0.70 eV from the one-electron oxidation potential of $D(ZnP)_{16}$, the excitation energy, and the one-electron reduction potential of PyNIm in PhCN [67]. The CS state detected in Fig. 31.5a decayed obeying clean first-order kinetics as shown in Fig. 31.5b, where the first-order plots at different initial CS concentrations afforded linear correlations with the same slope (inset of Fig. 31.5b). If there is any contribution of intermolecular back electron transfer from unbound $PyNIm^{•-}$ to $ZnP^{•+}$, the second-order kinetics would be involved for the decay time profile. Thus, the first-order decay process is ascribed to back electron transfer in the supramolecular dendrimer complex rather than intermolecular back electron transfer between $ZnP^{•+}$ and $PyNIm^{•-}$ produced by intermolecular photoinduced electron transfer. The CS lifetime of the supramolecular complex of $D(ZnP)_{16}$ with PyNIm was determined to be 0.56 μs at 298 K [67].

The occurrence of the photoinduced electron transfer in the supramolecular π-complex of $D(H_2P)_{16}$ with BHV^{2+} was also confirmed by the transient absorption spectra composed of superposition of the absorption bands due to $H_2P^{•+}$ and those of $BHV^{•+}$ in PhCN. The decay of the CS state also obeyed clean first-order kinetics. The CS lifetimes of the supramolecular π-complexes of other dendrimers

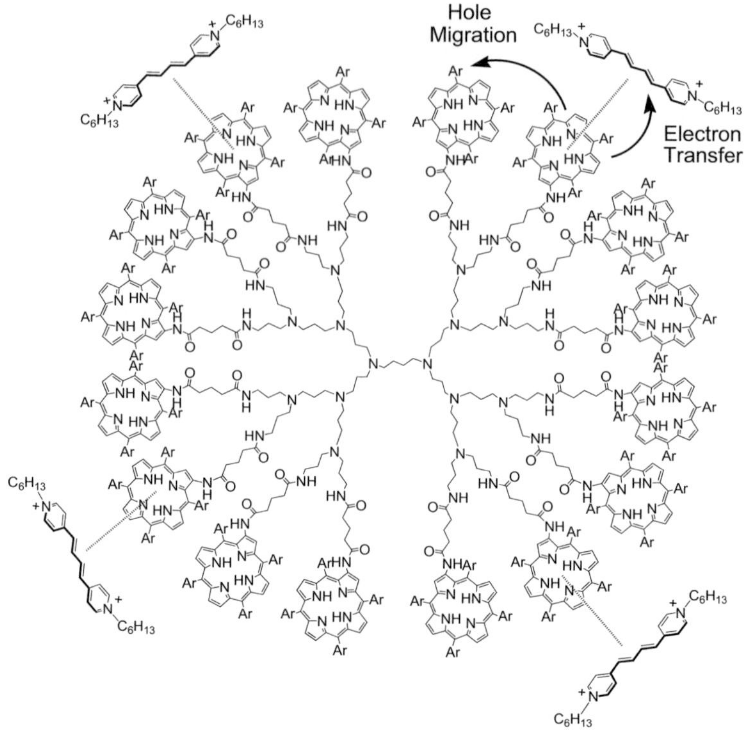

Fig. 31.6 Energy migration and charge separation in the supramolecular π-complex of $D(H_2P)_{16}$ with BHV^{2+}

and monomer [$D(H_2P^+)_n$: $n = 1, 4, 8$] became longer with increasing generation of porphyrin dendrimer. This results from efficient hole migration between the porphyrin units following photoinduced electron transfer in the supramolecular complex as shown in Fig. 31.6. The longest CS lifetime (0.75 ms) is attained for the supramolecular π-complex of $D(H_2P)_{16}$ with BHV^{2+}. Thus, the increase in the generation of porphyrin dendrimers resulted not only in improving the light-harvesting efficiency but also in attaining the longer CS lifetime. The laser photoexcitation of the supramolecular complex of $D(ZnP)_{16}$ with PyC_{60} also results in formation of the CS state as indicated by the transient absorption spectra [66]. The CS lifetime was determined to be 0.25 ms at 298 K [66]. Thus, multiple photosynthetic reaction centers combined with antenna complexes have been constructed using supramolecular complexes formed between porphyrin dendrimers and π-electron acceptors via coordination bond or/and π–π interaction [66, 67]. The excitation energy migration occurs between porphyrin units followed by long-lived charge separation [66, 67].

31.5 Supramolecular Complexes of Porphyrin Polypeptides with Fullerenes

Light-harvesting multiporphyrin systems have also been developed using zinc porphyrinic polypeptides with various number of porphyrin units [P(ZnP)$_n$; n = 2, 4, 8, 16] [69–72], which were combined with the reaction center units bearing molecular recognition sites with electron acceptors of fulleropyrrolidine bearing a pyridine (PyC$_{60}$) or imidazole coordinating ligand (ImC$_{60}$) as shown in Fig. 31.7 [73]. The ZnP moiety of P(ZnP)$_8$ forms a 1:1 supramolecular complex with PyC$_{60}$ as indicated by the UV–vis spectral titration [73]. The fluorescence of P(ZnP)$_8$ was strongly quenched by the electron transfer from the singlet excited state of the ZnP moiety (^{1}ZnP*) in P(ZnP)$_8$ to bound PyC$_{60}$ (i.e., static quenching) [73]. The values of the formation constants (*K*) of porphyrinic polypeptides and C$_{60}$ derivatives (PyC$_{60}$ and ImC$_{60}$) increased with increasing number of porphyrins in a polypeptide unit, indicating that the interaction between porphyrin and C$_{60}$ becomes stronger with increasing the number of the porphyrin units in the polypeptides [73]. This shows sharp contrast to monomer ZnP which forms no supramolecular complex with PyC$_{60}$ in PhCN [74]. The strong complex formation of P(ZnP)$_8$ with

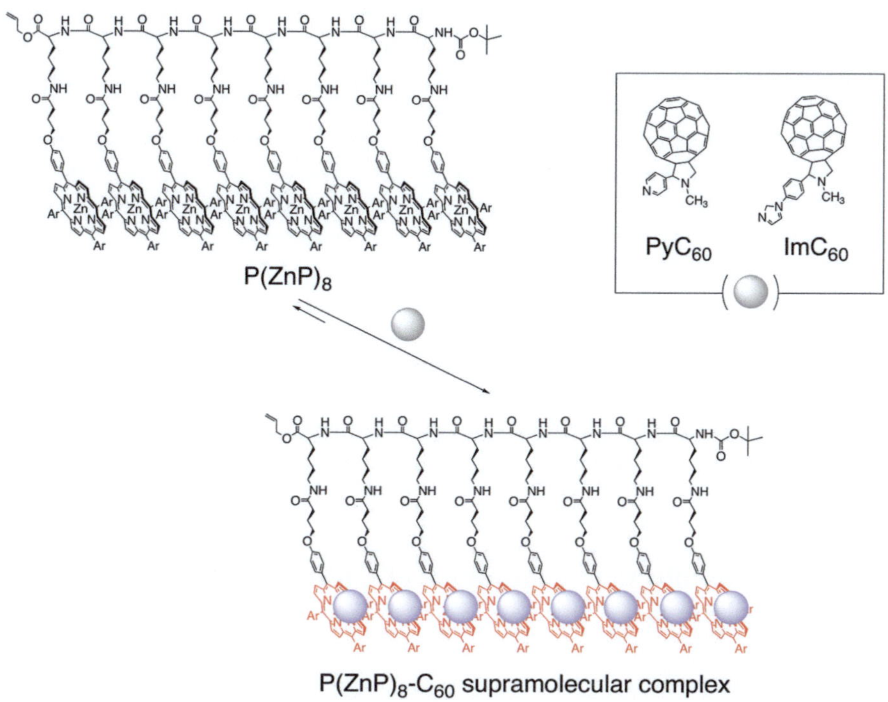

Fig. 31.7 Illustration of a supramolecular complex composed of porphyrin-peptide octamer [P(ZnP)$_8$, Ar = 3,5-(t-Bu)$_2$C$_6$H$_3$] and fulleropyrrolidine bearing pyridine (PyC$_{60}$) or imidazole (ImC$_{60}$) coordinating ligand

PyC_{60} and ImC_{60} results from the inclusion of C_{60} in the polypeptide backbone in addition to the coordination bond, leading to the stronger π–π interaction between porphyrins and fullerene as compared to the corresponding monomer system [73]. The apparent formation constant determined from the fluorescence quenching is significantly larger than that determined from the UV–vis spectral change and the difference of such association constants increases with increasing the number of the porphyrin units in the polypeptides [73]. This indicates that the efficient excited energy migration between the porphyrin units occurs efficiently prior to the electron transfer to the bound C_{60} moiety.

The structural change of polypeptidic backbone associated with formation of the supramolecular complex with C_{60} derivatives was observed in the CD spectrum of a PhCN solution of $P(ZnP)_8$, which shows cotton effect originating from the porphyrin Soret band at 431 nm [73]. The intensity of the CD signal corresponding to the Soret band of $P(ZnP)_8$ decreased with increasing the concentration of PyC_{60} [73]. This indicates that structural deformation occurs by forming the supramolecular complex between porphyrins and C_{60} and which may result from the structural change of backbone bearing porphyrin units toward the more suitable conformation to include the C_{60} guest with two porphyrin rings.

The occurrence of photoinduced electron transfer in the supramolecular complex in PhCN was confirmed by the nanosecond laser-induced transient absorption spectra of the supramolecular complex of $P(ZnP)_8$ with PyC_{60} [73]. The laser photoexcitation of the supramolecular complex of $P(ZnP)_8$ with PyC_{60} results in formation of the CS state as indicated by the transient absorption spectra [73]. The absorption band due to $PyC_{60}^{\bullet-}$ was clearly observed at 1000 nm together with that due to $ZnP^{\bullet+}$ at 630 nm after laser excitation at 561 nm where only the ZnP moiety was excited [73]. The CS state decayed obeying clean first-order kinetics, and the first-order plots at different initial CS concentrations afforded linear correlations with the same slope as observed in Fig. 31.5 for the supramolecular complex of $D(ZnP)_{16}$ with PyNIm [73]. Thus, the decay process is ascribed to back electron transfer in the supramolecular complex rather than intermolecular back electron transfer between $ZnP^{\bullet+}$ and $PyC_{60}^{\bullet-}$. The CS lifetimes became longer with increasing the number of the porphyrin units in the polypeptides [73]. The longest CS lifetime (1.11 ± 0.06 ms) was attained for the supramolecular complex of $P(ZnP)_{16}$ with PyC_{60} [73]. This may result from efficient hole migration between the porphyrin units following photoinduced electron transfer in the supramolecular complex.

31.6 Supramolecular Complexes of Cationic and Anionic Porphyrins

In solution, fast bimolecular charge recombination processes of the charge-separated states of electron donor–acceptor linked molecules usually predominate over much slower intramolecular charge recombination pathways [75–82]. Such a bimolecular charge recombination process has been retarded in a supramolecular

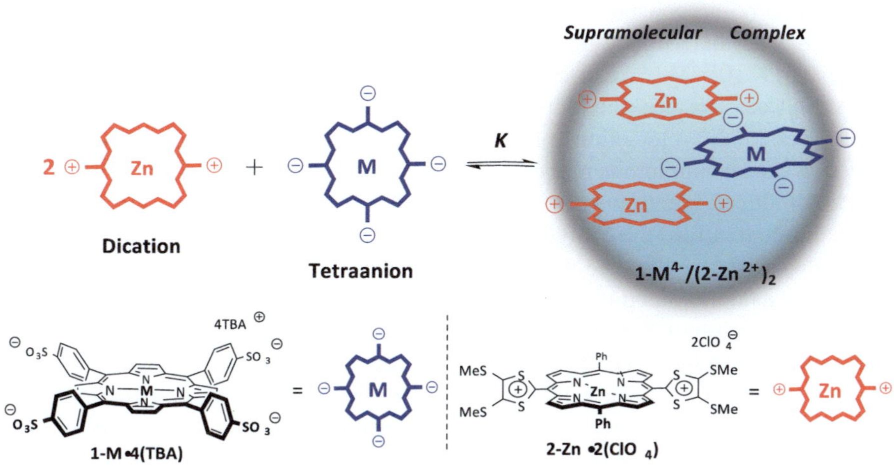

Fig. 31.8 Formation of supramolecular porphyrin complexes, **1**-M^{4-}/(**2**-Zn^{2+})$_2$; M = H_2 or Zn

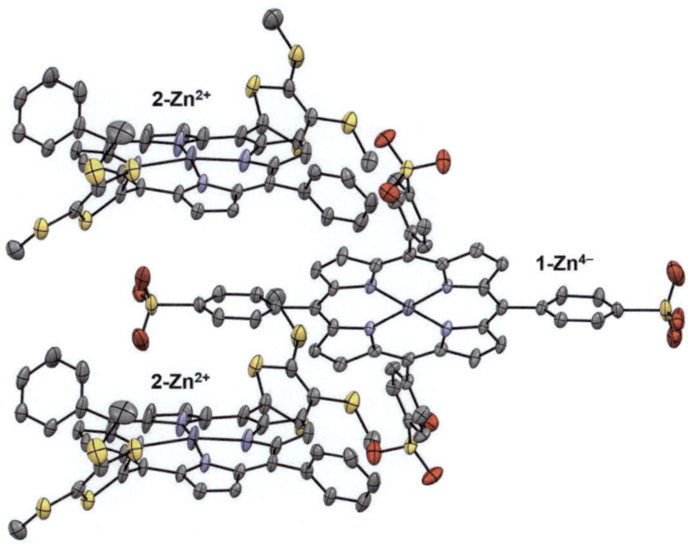

Fig. 31.9 X-ray crystal structure of the supramolecular complex **1**-Zn^{4-}/(**2**-Zn^{2+})$_2$. Thermal ellipsoids are scaled to the 50 % level. Hydrogen atoms and solvent molecules are omitted for clarity

electron donor–acceptor complex composed of one tetra-anionic porphyrin (**1**-M^{4-}, M = H_2 and Zn) and two dicationic porphyrins (**2**-Zn^{2+}) produced by the two-electron oxidation of a π-extended 1,3-dithiol-2-ylidene quinoidal porphyrin (porphyrin-bridged TTF) in Fig. 31.8 [82]. **2**-Zn^{2+} forms a 2:1 supramolecular complex with**1**-M^{4-} in PhCN, wherein both the top and bottom faces are covered by large charged molecules as shown in the X-ray crystal structure in Fig. 31.9

[82]. The binding constants of **1**-Zn^{4-}/(**2**-Zn^{2+})$_2$ and **1**-H_2^{4-}/(**2**-Zn^{2+})$_2$ in PhCN at 298 K were determined form the spectral titrations to be $(5.9 \pm 0.5) \times 10^{11}$ and $(6.6 \pm 0.5) \times 10^{12}$ M^{-2}, respectively [82]. The large binding constants result from the strong electrostatic interaction between the cationic and anionic porphyrins. Such electrostatic interactions have been utilized in formation of various electron donor–acceptor supramolecular complexes [83–89].

Photoinduced electron transfer from **1**-M^{4-} to **2**-Zn^{2+} occurs via the triplet excited state to afford the charge-separated (CS) state [82]. The transient absorption associated with the CS state was surprisingly persistent [82]. The spin state of the CS state was determined to be triplet as detected by EPR measurements [82]. The decay of the triplet CS state obeyed first-order kinetics and the slope of the first-order plots proved invariant to laser intensity and therefore the concentration of the photoinduced CS state. This indicates that the charge recombination occurs within the supramolecular complex and there is no contribution from intermolecular charge recombination between supramolecular complexes [82]. The intermolecular charge recombination processes are prohibited as the result of electrostatic repulsion. The two positively charged moieties, **2**-Zn^{2+} and **2**-$Zn^{\bullet+}$, between which the radical trianionic porphyrin **1**-$M^{\bullet 3-}$ is sandwiched, provide a positively charged buffer layer that prevents close contacts with either another supramolecular ensemble or a free electron donor moiety (i.e., **2**-$Zn^{\bullet+}$). The lifetimes of the CS states of**1**-$H_2^{\bullet 3-}$/(**2**-$Zn^{\bullet+}$) (**2**-Zn^{2+}) and **1**-$Zn^{\bullet 3-}$/(**2**-$Zn^{\bullet+}$) (**2**-Zn^{2+}) in PhCN at 298 K were determined to be 83 and 43 ms, respectively, which are the longest CS lifetimes ever reported for supramolecular π-electron donor–acceptor complexes in solution [82].

According to the Marcus theory of nonadiabatic electron transfer, the temperature dependence of the intramolecular electron transfer (k_{ET}) is given by Eq. (31.1),

$$\ln\left(k_{ET}T^{1/2}\right) = \ln\left(2\pi^{3/2}\,V^2/h(\lambda k_B)^{1/2}\right) - (\Delta G_{ET} + \lambda)^2/\,(4\lambda k_B T) \qquad (31.1)$$

where T is the absolute temperature, V is the electronic coupling constant, λ is the reorganization energy, ΔG_{ET} is the Gibbs free energy change of electron transfer, k_B is the Boltzmann constant, and h is the Planck constant [90]. The temperature dependence of intramolecular back electron transfer in the CS state of the supramolecular complex [**1**-$H_2^{\bullet 3-}$/(**2**-$Zn^{\bullet+}$) (**2**-Zn^{2+})] was analyzed based on Eq. (31.1) to afford the λ and V values as 0.24 eV and 0.16 cm^{-1}, respectively [82]. The small V value results from the spin-forbidden back electron transfer of the triplet CS state and small orbital interactions between **1**-$H_2^{\bullet 3-}$ and **2**-$Zn^{\bullet+}$ arising from the slipped-sandwich arrangement (Fig. 31.9) [82].

31.7 Conclusions

Supramolecular complexes between π-electron donors and acceptors linked by noncovalent bonding such as electrostatic and π–π interactions together with coordination bonding provide excellent combination of light-harvesting and charge-separation units for efficient energy migration followed by long-lived charge separation as observed in the photosynthetic reaction center. Such photofunctions of supramolecular complexes between π-electron donors and acceptors have been applied to construct efficient supramolecular solar cells [59, 71, 91]. Further development of supramolecular complexes between π-electron donors and acceptors is hoped to improve the energy conversion efficiency.

Acknowledgments The author gratefully acknowledges the contributions of their collaborators and coworkers mentioned in the cited references and support by a Grant-in-Aid (No. 20108010 to SF) from MEXT, Japan.

References

1. IPCC fifth assessment report: climate change 2013 (AR5) from IPCC website: www.ipcc.ch/report/ar5/wg1/
2. Leung DYC, Caramanna G, Maroto-Valer MM (2014) Renew Sust Energ Rev 39:426–443
3. Reddy KG, Deepak TG, Anjusree GS, Thomas S, Vadukumpully S, Subramanian KRV, Nairb SV (2014) A. Nair, A. S. Phys Chem Chem Phys 16:6838–6858
4. McFarland EW (2014) Energy Environ Sci 7:846–854
5. Kärkäs M, Johnston EV, Verho O, Åkermark B (2014) Acc Chem Res 47:100–111
6. Sherman BD, Vaughn MD, Bergkamp JJ, Gust D, Moore AL, Moore TA (2014) Photosynth Res 120:59–70
7. Wu L-Z, Chen B, Li Z-J, Tung C-H (2014) Acc Chem Res 47:2177–2185
8. Frischmann PD, Mahata K, Würthner F (2013) Chem Soc Rev 42:1847–1870
9. Gust D, Moore TA, Moore AL (2009) Acc Chem Res 42:1890–1898
10. Fukuzumi S, Ohkubo K, Suenobu T (2014) Acc Chem Res 47:1455–1464
11. Hoff AJ, Deisenhofer J (1997) Phys Rep 287:1–247
12. Nelson N, Yocum CF (2006) Annu Rev Plant Biol 57:521–565
13. Umena Y, Kawakami K, Shen J-R, Kamiya N (2011) Nature 473:55–61
14. Yano J, Yachandra V (2014) Chem Rev 114:4175–4205
15. Nelson N (2011) Biochim. Biophys Acta 1807:856–863
16. Scholes GD, Fleming GR, Olaya-Castro A, van Grondelle R (2011) Nat Chem 3:763–774
17. El-Khouly ME, Fukuzumi S, D'Souza F (2014) ChemPhysChem 15:30–47
18. Wasielewski MR (2009) Acc Chem Res 42:1910–1921
19. Fukuzumi S, Ohkubo KJ (2012) Mater Chem 22:4575–4587
20. Ito O, D'Souza F (2012) Molecules 17:5816–5835
21. Fukuzumi S, Ohkubo K, D'Souza F, Sessler JL (2012) Chem Commun 48:9801–9815
22. Bottari G, Trukhina O, Ince M, Torres T (2012) Coord Chem Rev 256:2453–2477
23. D'Souza F, Ito O (2012) Chem Soc Rev 41:86–96
24. Guldi DM, Sgobba V (2011) Chem Commun 47:606–610
25. Wrobel D, Graja A (2011) Coord Chem Rev 255:2555–2577
26. Lemmetyinen H, Tkachenko NV, Efimov A, Niemi M (2011) Phys Chem Chem Phys 13:397–412
27. Ohkubo K, Fukuzumi S (2009) Bull Chem Soc Jpn 82:303–315
28. Fukuzumi S, Kojima TJ (2008) Mater Chem 18:1427–1439
29. Fukuzumi S (2008) Phys Chem Chem Phys 10:2283–2297
30. Fukuzumi S, Ohkubo K (2013) Dalton Trans 42:15846–15858
31. Iijima S, Yudasaka M, Yamada R, Bandow S, Suenaga K, Kokai F, Takahashi K (1999) Chem Phys Lett 309:165–170

32. Vizuete M, Gómez-Escalonilla MJ, Fierro JLG, Yudasaka M, Iijima S, Vartanian M, Iehl J, Nierengarten J-F, Langa F (2011) Chem Commun 47:12771–12773
33. Karousis N, Sato Y, Suenaga K, Tagmatarchis N (2012) Carbon 50:3909–3914
34. Karousis N, Ichihashi T, Yudasaka M, Iijima S, Tagmatarchis N (2011) Chem Commun 47:1604–1606
35. Dai L (2013) Acc Chem Res 46:31–42
36. Sandanayaka ASD, Ito O, Tanaka T, Isobe H, Nakamura E, Yudasaka M, Iijima S (2009) New J Chem 33:2261–2266
37. Zhu S, Xu G (2010) Nanoscale 2:2538–2549
38. Vizuete M, Gómez-Escalonilla MJ, Fierro JLG, Ohkubo K, Fukuzumi S, Irie M, Yudasaka M, Iijima S, Nierengarten J-F, Langa F (2014) Chem Sci 5:2072–2080
39. Novoselov KS, Geim AK, Morozov SV, Jiang D, Zhang Y, Dubonos SV, Grigorieva IV, Firsov AA (2004) Science 306:666–669
40. Chabot V, Higgins D, Yu A, Xiao X, Chen Z, Zhang J (2014) Energy Environ Sci 7:1564–1596
41. Chua CK, Pumera M (2014) Chem Soc Rev 43:291–312
42. Li D, Müller MB, Gilje S, Kaner RB, Wallace GG (2008) Nat Nanotechnol 3:101–105
43. Su Q, Pang S, Alijani V, Li C, Feng X, Müllen K (2009) Adv Mater 21:3191–3195
44. Stankovich S, Dikin DA, Piner RD, Kohlhaas KA, Kleinhammes A, Jia Y, Wu Y, Nguyen ST, Ruoff RS (2007) Carbon 45:1558–1565
45. Banhart F, Kotakoski J, Krasheninnikov AV (2011) ACS Nano 5:26–41
46. Wojcik A, Kamat PV (2010) ACS Nano 4:6697–6706
47. Geng J, Jung H-TJ (2010) Phys Chem C 114:8227–8234
48. Lightcap IV, Kamat PV (2013) Acc Chem Res 46:2235–2243
49. Chang H, Wu H (2013) Energy Environ Sci 6:3483–3507
50. Tang Q, Zhou Z, Chen Z (2013) Nanoscale 5:4541–4583
51. Supur M, Ohkubo K, Fukuzumi S (2014) Chem Commun 50:13359–13361
52. Supur M, Fukuzumi S (2012) J Phys Chem C 116:23274
53. Supur M, Fukuzumi S (2013) Phys Chem Chem Phys 15:2539
54. Yeow EKL, Ghiggino KP, Reek JNH, Crossley MJ, Bosman AW, Schenning APH, Meijer EW (2000) J Phys Chem B 104:2596
55. Ballester P, Gomila RM, Hunter CA, King ASH, Twyman LJ (2003) Chem Commun 38–39
56. Zhang G-D, Nishiyama N, Harada A, Jiang D-L, Aida T, Kataoka K (2003) Macromolecules 36:1304
57. Frampton MJ, Magennis SW, Pillow JNG, Burn PL, Samuel IDW (2003) J Mater Chem 13:235
58. Kim Y, Mayer MF, Zimmerman SC (2003) Angew Chem Int Ed 42:1121
59. Hasobe T, Kashiwagi Y, Absalom MA, Sly J, Hosomizu K, Crossley MJ, Imahori H, Kamat PV, Fukuzumi S (2004) Adv Mater 16:975
60. Hasobe T, Kamat PV, Absalom MA, Kashiwagi Y, Sly J, Crossley MJ, Hosomizu K, Imahori H, Fukuzumi S (2004) J Phys Chem B 108:12865
61. Larsen J, Brüggemann B, Polívka T, Sundström V, Åkesson E, Sly J, Crossley MJ (2005) J Phys Chem A 109:10654
62. Cho S, Li W-S, Yoon M-C, Ahn TK, Jiang D-L, Kim J, Aida T, Kim D (2006) Chem Eur J 12:7576
63. Larsen J, Brüggemann B, Khoury T, Sly J, Crossley MJ, Sundström V, Åkesson E (2007) J Phys Chem A 111:10589
64. Zhang X, Zeng Y, Yu T, Chen J, Yang G, Li YJ (2014) Phys Chem Lett 5:2340–2350
65. D'Souza F, Deviprasad GR, Zandler ME, Hoang VT, Klykov A, VanStipdonk M, Perera A, El-Khouly ME, Fujitsuka M, Ito O (2002) J Phys Chem A 106:3243
66. Fukuzumi S, Saito K, Kashiwagi Y, Crossley MJ, Gadde S, D'Souza F, Araki Y, Ito O (2011) Chem Commun 47:7980–7982
67. Fukuzumi S, Ohkubo K, Saito K, Kashiwagi Y, Crossley MJJ (2011) J Porphyr Phthalocyanines 15:1292–1298

68. Sanders JKM, Bampos N, Clyde-Watson Z, Darling SL, Hawley JC, Kim H-J, Mak CC, Webb SJ (2000) In: Kadish KM, Smith KM, Guilard R (eds) The porphyrin handbook, vol 3. Academic, New York, pp 1–48
69. Solladié N, Hamel A, Gross M (2000) Tetrahedron Lett 41:6075–6078
70. Saito K, Qiu H, Troiani V, Solladié N, Sakata T, Mori H, Ohama M, Fukuzumi SJ (2007) Phys Chem C 111:1194–1199
71. Hasobe T, Saito K, Kamat PV, Troiani V, Qiu H, Solladié N, Ahn TK, Kim KS, Kim SK, Kim D, D'Souza F, Fukuzumi SJ (2007) Mater Chem 17:4160–4170
72. Fujitsuka M, Cho DW, Solladié N, Troiani V, Qiu H, Majima TJ (2007) J Photochem Photobiol A Chem 188:346–350
73. Fukuzumi S, Saito K, Ohkubo K, Troiani V, Qiu H, Gadde S, D'Souza F, Solladié N (2011) Phys Chem Chem Phys 13:17019–17022
74. D'Souza F, Smith PM, Gadde S, McCarty AL, Kullman MJ, Zandler ME, Itou M, Araki Y, Ito OJ (2004) Phys Chem B 108:11333–11343
75. Imahori H, Guldi DM, Tamaki K, Yoshida Y, Luo C, Sakata Y, Fukuzumi SJ (2001) Am Chem Soc 123:6617–6628
76. Imahori H, Sekiguchi Y, Kashiwagi Y, Sato T, Araki Y, Ito O, Yamada H, Fukuzumi S (2004) Chem Eur J 10:3184–3196
77. Guldi DM, Imahori H, Tamaki K, Kashiwagi Y, Yamada H, Sakata Y, Fukuzumi SJ (2004) Phys Chem A 108:541–548
78. Fukuzumi S, Kotani H, Ohkubo K, Ogo S, Tkachenko NV, Lemmetyinen HJ (2004) Am Chem Soc 126:1600–1601
79. Ohkubo K, Kotani H, Fukuzumi S (2005) Chem Commun 4520–4522
80. Murakami M, Ohkubo K, Fukuzumi S (2010) Chem Eur J 16:7820–7832
81. Murakami M, Ohkubo K, Nanjo T, Souma K, Suzuki N, Fukuzumi S (2010) ChemPhysChem 11:2594–2605
82. Fukuzumi S, Doi K, Suenobu T, Ohkubo K, Yamada Y, Karlin KD (2012) Proc Natl Acad Sci U S A 109:15572–15577
83. Bill NL, Ishida M, Kawashima Y, Ohkubo K, Sung YM, Lynch VM, Lim JM, Kim D, Sessler JL, Fukuzumi S (2014) Chem Sci 5:3888–3896
84. Honda T, Nakanishi T, Ohkubo K, Kojima T, Fukuzumi SJ (2010) Am Chem Soc 132:10155–10163
85. Fukuzumi S, Ohkubo K, Kawashima Y, Kim DS, Park JS, Jana A, Lynch VM, Kim D, Sessler JLJ (2011) Am Chem Soc 133:15938–15941
86. Ohkubo K, Kawashima Y, Fukuzumi S (2012) Chem Commun 48:4314–4316
87. Kawashima Y, Ohkubo K, Mase K, Fukuzumi SJ (2013) Phys Chem C 117:21166–21177
88. Sessler JL, Karnas E, Kim SK, Ou Z, Zhang M, Kadish KM, Ohkubo K, Fukuzumi SJ (2008) Am Chem Soc 130:15256–15257
89. Fukuzumi S, Mase K, Ohkubo K, Karnas E, Kim SK, Ou Z, Zhang M, Kadish KM, Sessler JLJ (2014) Phys Chem C 118:18436–18444
90. Marcus RA, Sutin N (1985) Biochim Biophys Acta 811:265–322
91. Ohkubo K, Kawashima Y, Sakai H, Hasobe T, Fukuzumi S (2013) Chem Commun 49:4474–4476

Chapter 32
Fused π-Electron Systems Containing Group 15 Elements and Their Application to Organic Electronics

Hayato Tsuji

Abstract This chapter discusses our recent progress on the synthesis, properties, and applications of fused π-electron systems containing group 15 heteroatoms, such as benzodipyrroles, bis-cinnolines, and benzophospholes. These compounds were obtained based on a modular approach using intramolecular cyclization reactions. Some of the thus obtained compounds were found to improve the performance of organic electronic devices by taking advantage of their high carrier mobility and orbital energy levels.

Keywords Benzodipyrrole • Bis-cinnoline • Benzophosphole

32.1 Introduction

The discovery of conductivity of doped perylene by Akamatu et al. [1] opened the possibility of electronic applications of organic molecules, which had been considered to be insulators. To date, an enormous number of small organic molecules and polymers have been developed as represented by aromatic amines, thiophenes, and phenylenevinylenes. Although organic electronics has become close to practical use, improvements in device performance and lifetime have remained an important issue to solve. To this end, the development of novel classes of materials, that is, the contribution of synthetic organic chemistry, has become increasingly significant. In this chapter, we focus on our recent progress on the synthesis, properties, and device applications of benzodipyrroles, bis-cinnolines, and benzophospholes, that is, heteroaromatic compounds containing group 15 elements.

H. Tsuji (✉)
Department of Chemistry, The University of Tokyo, 7-3-1 Hongo, Bunkyo-ku, Tokyo 113-0033, Japan

JST-PRESTO, 4-1-8 Honcho, Kawaguchi, Saitama 332-0012, Japan
e-mail: tsuji@chem.s.u-tokyo.ac.jp

T. Akasaka et al. (eds.), *Chemical Science of π-Electron Systems*,
DOI 10.1007/978-4-431-55357-1_32

32.2 Effect of Group 15 Heteroatoms (1): Nitrogen

Substitution of one or two carbon atoms of benzene by heteroatom(s) significantly affects the electronic state, as has been discussed theoretically and experimentally [2–4]. Major factors include electronegativity, lone pair, and orbital interaction. In the case of group 15 elements, such as nitrogen and phosphorus, the predominant factor drastically changes depending on the element and hybridization state. For example, in the case of pyrrole, the effect of the lone pair plays a decisive role: six electrons including the lone pair on the nitrogen atom are shared with five elements (Fig. 32.1a). This feature renders pyrroles electron rich (p-type character). Polypyrroles are used as electron-injection material, and porphyrins and phthalocyanines are used as donor material for organic solar cells. In contrast, pyridine and related compounds incorporating sp^2-nitrogen atom(s) show π-electron-deficient character because of the electronegativity of nitrogen (Fig. 32.1b). These compounds therefore exhibit n-type character and are used as electron-transport or hole-blocking materials in organic electronic devices. This section describes our recent studies on the synthesis of functionalized benzo[1,2-*b*:4,5-*b*′]

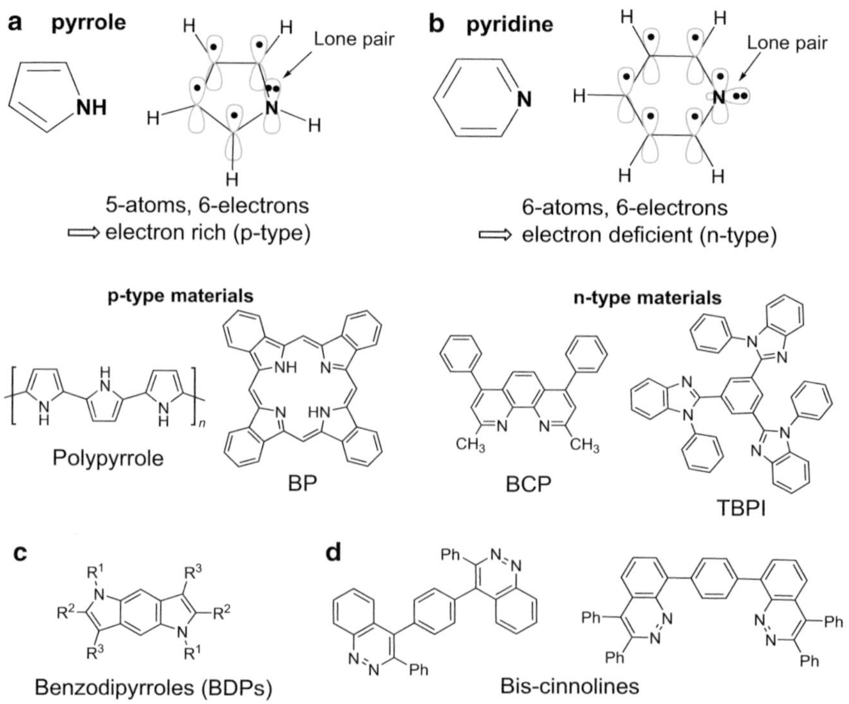

Fig. 32.1 Effect of nitrogen atoms. (**a**) Electronic structures of pyrrole and structures of representative p-type materials, (**b**) pyridine, and n-type materials. Structures of (**c**) benzodipyrroles and (**d**) bis-cinnolines

dipyrroles (simply called benzodipyrroles hereinafter and abbreviated as BDPs) as well as bis-cinnolines and their utility as a novel class of nitrogen-containing semiconducting materials (Figs. 32.1c, d).

32.2.1 Fused π-Electron Systems Containing Nitrogen Atoms

During the course of our studies on developing new reactions to build fused π-electron conjugated molecules based on a modular approach [5], we found a versatile synthesis of functionalized BDPs [6, 7] and their utility as p-type semiconducting materials (Fig. 32.1c). We recently developed a convenient synthetic route to benzotripyrroles (BTPs) as well [8]. Because their application is now under study, their synthesis and applications will be reported elsewhere. We also found that cinnoline derivatives can be obtained from a similar starting material (see Sect. 32.2.3 for details) and can work as new types of n-type semiconducting materials (Fig. 32.1d).

32.2.2 N-HexBn-BDPs and Their Application to OLEDs

The synthesis of BDPs is based on a zinc-mediated intramolecular cyclization reaction (Scheme 32.1). *N,N′*-Di-*p*-hexylbenzyl (HexBn) group-substituted 2,5-dialkynylphenylenediamine **1** was deprotonated with BuLi, followed by transmetalation using $ZnCl_2$. Upon heating at 120 °C, the double cyclization reaction proceeded quantitatively to afford the desired 3,7-dizincio BDP **2** as a thermally stable intermediate. Formation of the dizincio intermediate **2** was confirmed by quenching the reaction mixture with D_2O to obtain 98 % yield (93 % D-incorporation). This dizincio species is a versatile synthetic intermediate ("synthetic module") that can be functionalized to afford a variety of derivatives. Indeed, based on this modular synthetic approach, we obtained **3a–e** in which aromatic groups were introduced at the β,β′-positions using palladium-catalyzed Negishi cross-coupling reactions.

The thus obtained BDPs were found to possess suitable ionization potentials (IPs) and electron affinities (EAs) for p-type application in organic light-emitting diodes (OLEDs). Cyclic voltammetry (CV) of the BDPs showed reversible multistep oxidation, indicating their stability under oxidative conditions. However, these *N,N′*-HexBn-BDPs **3a–e** show rather low thermal stability, which has been overcome by hexaaryl-BDPs (see below).

Taking advantage of these properties, we fabricated OLEDs using tris(pentafluorophenyl)borane (PPB)-doped BDPs as the hole-injection layer formed by spin-coating (Fig. 32.2a) and compared the performance with a PEDOT:PSS-based OLED (Table 32.1). The OLED performance showed that the PPB-doped devices exhibited improved driving voltage and luminous efficiency compared with the PEDOT:PSS reference, except for **3a**, which showed an exceptionally low efficiency, most likely because of the aggregation of the material that produces a dark spot in the lighting.

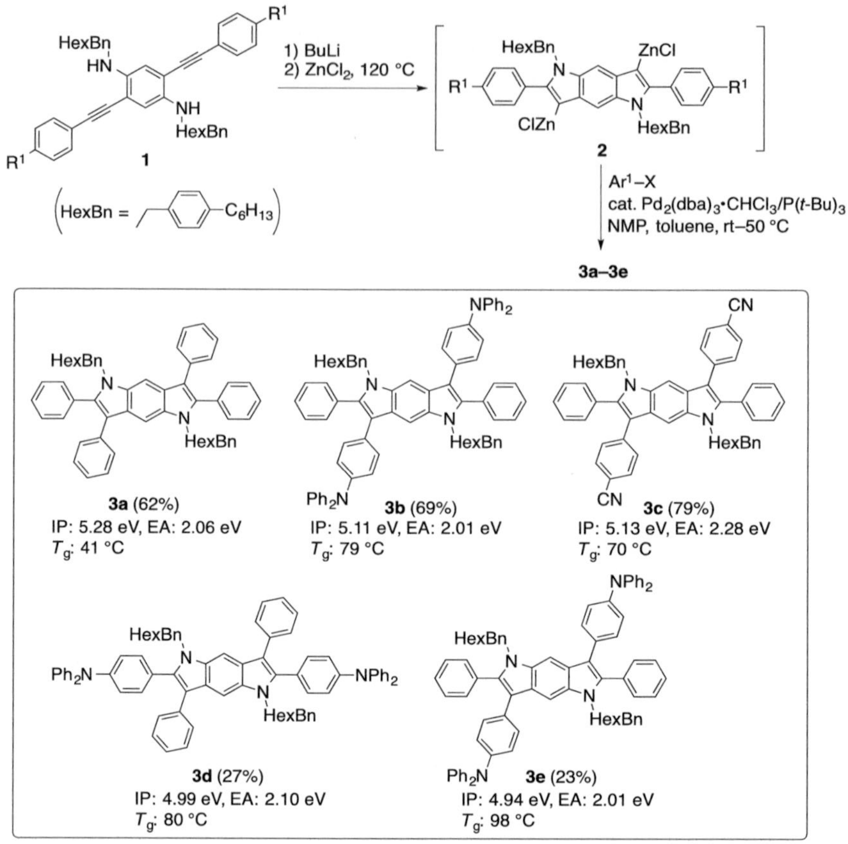

Scheme 32.1 Synthesis and properties of *N,N′*-HexBn-substituted BDPs. IPs were measured by photoemission yield spectroscopy (PYS) using BDP films, and electron affinities (EAs) were calculated from the IP and optical gap

Interestingly, the performance of the doped-BDP-based OLED devices was correlated with the IP of the BDPs (Fig. 32.2b). The V_{1000} values exponentially decrease as the IP values of the BDPs became closer to that of α-NPD, a hole-transport material in the devices. The data suggest that the hole-injection barrier from the BDP layer to the α-NPD layer is an important factor that determines the V_{1000} value rather than that from the ITO electrode to the hole-injection layer (HIL), probably because of good contact between the BDP layer and ITO.

32.2.3 Hexaaryl-BDPs and Their Application to OLEDs

To improve the thermal stability, we next introduced aryl groups on the nitrogen atom to obtain hexaaryl-BDPs (HABDPs, Scheme 32.2). Deprotection of *N,N′*-

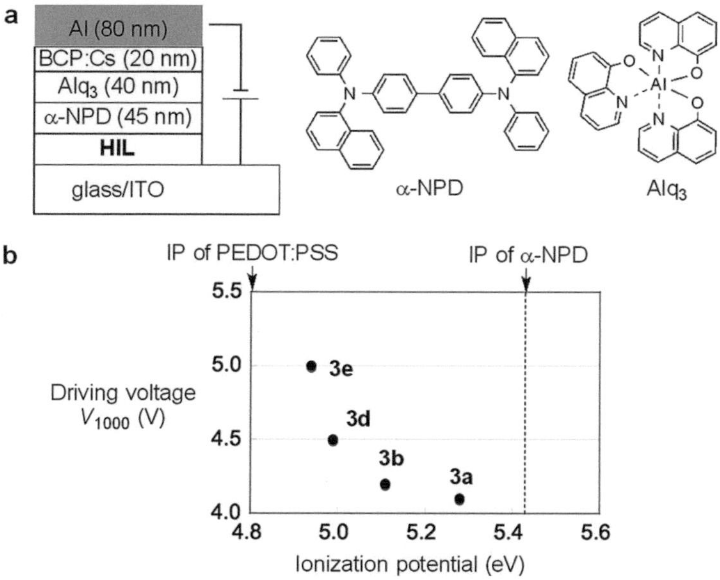

Fig. 32.2 Application of BDPs to OLEDs: (**a**) Device configuration. (**b**) Driving voltage vs IP plot

Table 32.1 Summary of OLED performance

HIL	V_{1000}[a] (V)	η_{1000}[b] (lm/W)	L/J_{1000}[c] (cd/A)
3a:PPB	4.1	2.4	3.0
3b:PPB	4.2	2.4	3.7
3d:PPB	4.5	2.3	3.6
3e:PPB	5.0	2.3	3.7
PEDOT:PSS	5.4	2.3	4.0

[a]Driving voltage
[b]Luminous efficiency
[c]Current efficiency

HexBn-protected BDPs **3** was performed using $AlCl_3$ in anisole, followed by a palladium-catalyzed *N*-arylation reaction to obtain the HABDPs. These compounds were also electrochemically stabile toward oxidation, having low IP and EA levels, which are suitable for p-type materials. Thermal stabilities were also high. Time-of-flight (TOF) carrier mobility measurements were performed using a vacuum-deposited amorphous film. The HPBDPs **5a** and **5b** showed nondispersive transient currents for holes, indicating trap-free carrier transportation, with hole mobilities of 2×10^{-3} cm^2/(V s) for both compounds at an electric field of 2.5×10^5 V/cm. The carbazole-substituted BDP **5c** showed ambipolar character, with hole and electron mobilities of 6×10^{-4} and 2×10^{-3} cm^2/(V s), respectively, at an electric field of 2.5×10^5 V/cm. It is noteworthy that carrier drift mobilities of the order of 10^{-3} cm^2/(V s) are considered as the upper limit for amorphous materials.

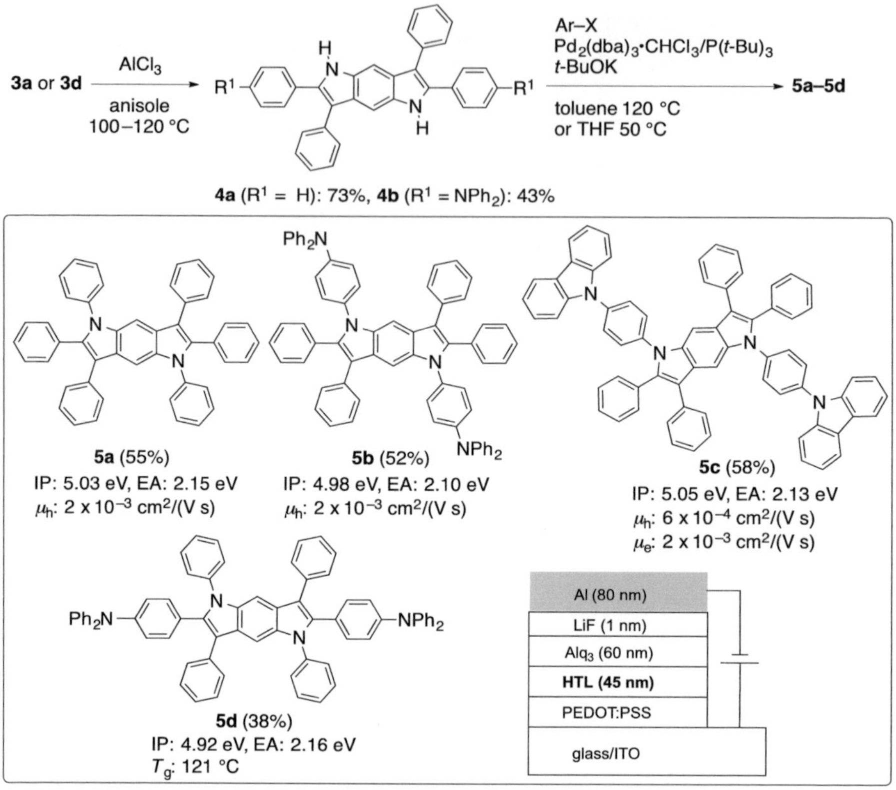

Scheme 32.2 Synthesis of hexaaryl-BDPs. Note that IP values were estimated using the oxidation potential in CV measurements and the EA values are from the IP and optical gap. OLED device configuration is also shown at the *bottom right*

Table 32.2 OLED performance using BDPs as an HTL

HTL	V_{1000}[a] [V]	η_{1000}[b] [lm/W]	L/J_{1000}[c] [cd/A]
5a	7.0	0.81	1.80
5b	8.5	0.84	2.29
α-NPD	8.5	0.59	1.60

[a]Driving voltage
[b]Luminous efficiency
[c]Current efficiency

The p-type BDPs **5a** and **5b** were used as a hole-transport layer (HTL) in OLEDs (Table 32.2, Scheme 32.2 bottom right). The driving voltage of the device using **5a** was significantly lower than that of the reference device using α-NPD, while that of the **5b**-based device was the same as for the reference. The luminance and the current efficiencies were also significantly improved compared with the reference device. The improved performance may be due to their suitable HOMO energy levels for carrier injection both from the anode and to Alq_3.

32.2.4 Bis(cinnoline)s and Their Application as a Buffer Layer in Solar Cells

Cinnoline is a molecule possessing two nitrogen atoms in the naphthalene framework. While significant attention has been paid to their bioactivities [9], their optoelectronic properties have remained unexplored. During our studies on BDP, it was noticed that 4-bromocinnoline **7** [10] can be prepared from *o*-alkynylaniline **6**, the common starting material to fused pyrrole (indole) compounds (Scheme 32.3 in parentheses). The 4-bromocinnoline **7** was transformed into bis(4-cinnoline) **8** via a cross-coupling reaction. Unfortunately, this molecule was found to give low efficiency in device applications despite its high electron mobility: an amorphous film showed electron drift mobility of 2×10^{-3} $cm^2/(V\ s)$ at an electric field of 2.5×10^5 V/cm using the TOF method.

In contrast, bis(8-cinnoline)s **10a–c** showed significant performance (Fig. 32.3, Table 32.3). Synthesis is shown in Scheme 32.4. Starting from 8-bromocinnoline **9**, prepared from a hydrazine derivative and benzil according to the known method, bis(8-cinnoline)s **10a–c** were prepared using cross- or homocoupling reactions [11]. The estimated EAs are 2.69–2.77 eV. Carrier mobility measurements using the

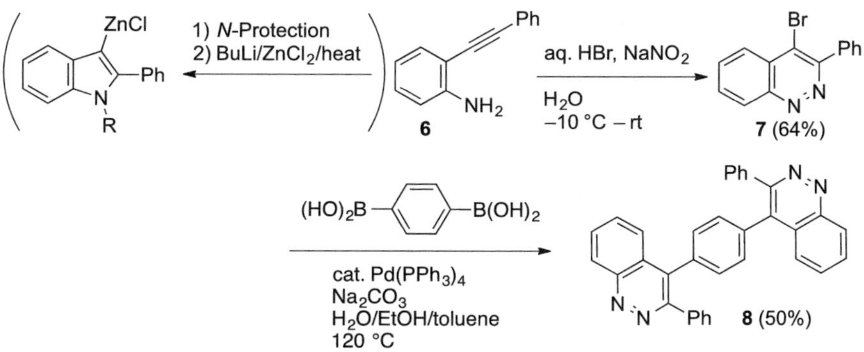

Scheme 32.3 Synthesis of a bis(4-cinnoline) **8**

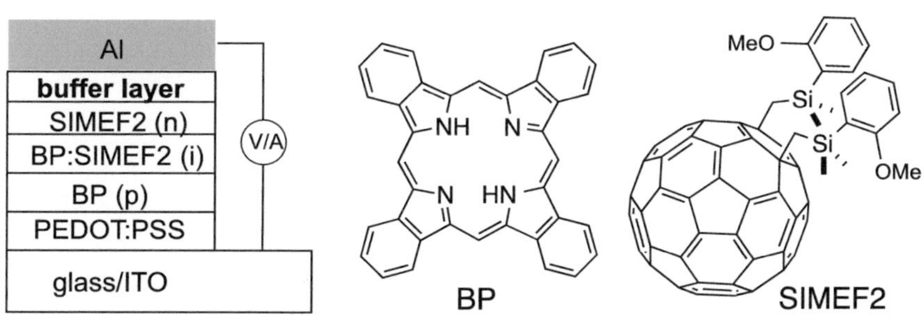

Fig. 32.3 Configuration of the organic solar cells using bis-cinnoline as a cathode buffer layer, together with the material structures for the active layer

Table 32.3 Performance of organic solar cells involving the bis-cinnolines as a cathode buffer layer, together with those of a BCP-based cell for a reference, under irradiation with 100 mW/cm^2 incident light, using an AM 1.5G filter

	V_{OC}[a] [V]	J_{SC}[b] [mA/cm^2]	FF[c]	PCE[d] [%]
10a	0.77 ± 0.01	8.64 ± 0.12	0.59 ± 0.04	4.0 ± 0.21
10b	0.77 ± 0.01	9.31 ± 0.16	0.56 ± 0.04	4.0 ± 0.37
10c	0.77 ± 0.01	8.30 ± 0.06	0.56 ± 0.05	3.6 ± 0.34
BCP	0.80 ± 0.00	7.82 ± 0.38	0.56 ± 0.03	3.5 ± 0.36

[a]Open-circuit voltage
[b]Short-circuit current density
[c]Fill factor
[d]Power conversion efficiency

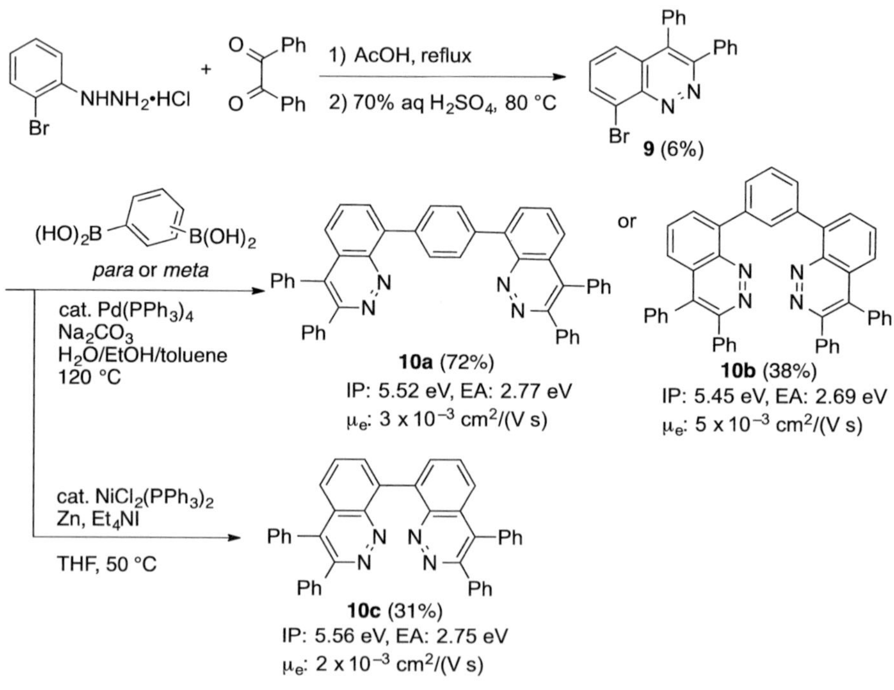

Scheme 32.4 Synthesis of bis(8-cinnoline)s **10**. EA values were estimated using the reduction potential in DPV measurements; the IP values are calculated from the EA and optical gap

TOF method showed electron mobilities of these bis-cinnolines of 2×10^{-3} to 5×10^{-3} cm^2/(V s), which are also very high for amorphous materials. Note that transient current signals for the electron were dispersive, and no hole transient current signal was observed. Thermal stabilities were also high. CV measurements in THF revealed that **10a** showed a pseudo-reversible reduction wave, while that of **10b** showed a reversible first reduction wave.

The bis(8-cinnoline)s **10** were used as a cathode buffer layer in a p–i–n heterojunction device using benzoporphyrin (BP) and a fullerene derivative (SIMEF2) as a donor and an acceptor [12], respectively. The power conversion efficiency (PCE) of the cells reached 4.0 %, which is higher than that using BCP (3.5 ± 0.36 %), a commonly used cathode buffer material. The cells composed of the BCs showed higher J_{SC} values than the cell composed of BCP, whereas V_{OC} and FF were comparable to those of the BCP-based cell.

32.3 Effect of Group 15 Heteroatoms (2): Phosphorus

Phospholes are analogues of pyrroles and have attracted increasing interest as functional materials in the past decade [13–19] because of their intriguing optoelectronic properties. Their striking feature is the low-lying LUMO character because of the orbital interaction between the P–C σ* orbital and C = C π* orbitals (σ*–π* interaction), similar to siloles [5]. The diversity of the valence (i.e., trivalent and pentavalent) and coordination of a metal on phosphorus are also characteristic features and significantly affect the orbital energy levels (Fig. 32.4). In this light, phosphole derivatives can serve as a new class of versatile n-type semiconducting materials. This chapter describes versatile syntheses of fused phosphole derivatives developed by the author, together with their properties as semiconducting materials and applications to organic solar cells.

32.3.1 Fused π-Electron Systems Including Phosphorus Atoms

We developed a synthesis of functionalized benzo[*b*]phospholes (simply called benzophospholes hereinafter) based on a modular approach (Scheme 32.5) [20, 21].

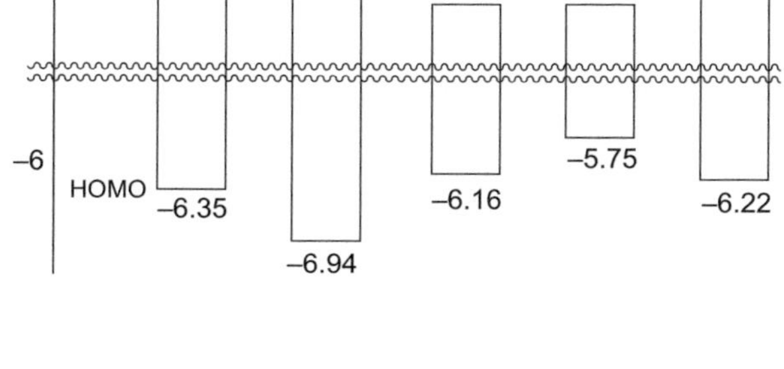

Fig. 32.4 Calculated orbital energy levels of phosphole derivatives (at the B3LYP/6-311G* level); those of siloles are shown as a reference

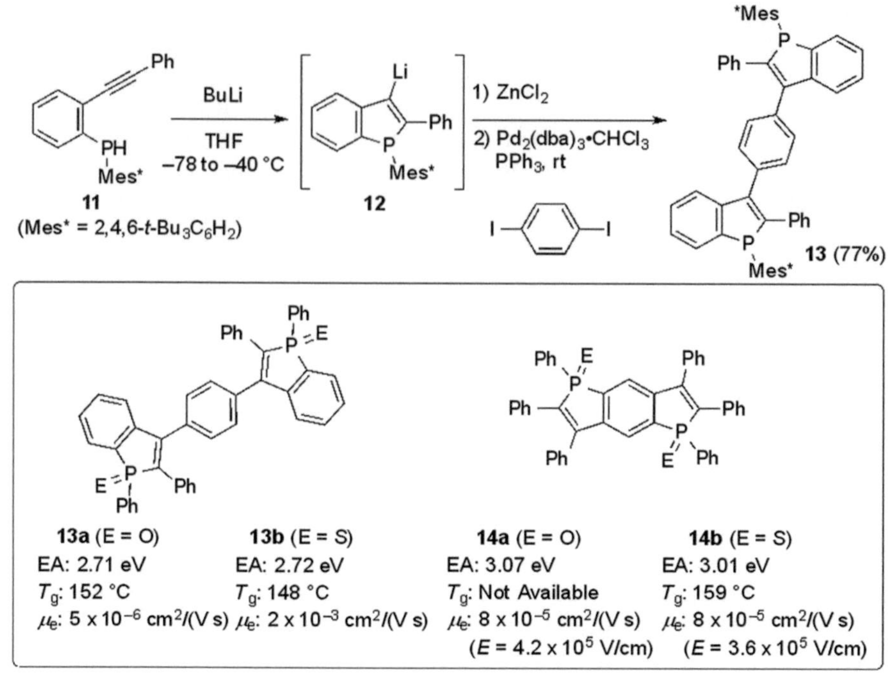

Scheme 32.5 Synthesis of benzophosphole derivatives. Electron affinities (EAs) were estimated from the reduction potential of the CV measurements

It should be mentioned that Fukazawa and Yamaguchi's group [22] and Sanji and Tanaka's group [23] concomitantly reported cyclization reactions to obtain similar benzophosphole derivatives.

In contrast to the pyrrole synthesis (see Sect. 32.2.2), deprotonation of (*o*-alkynylphenyl)hydrophosphine **11** induced intramolecular cyclization at low temperature to afford the 3-lithiobenzophosphole intermediate **12**, which was further functionalized using well-established methods. The bis-benzophospholes **13** and benzodiphosphole **14a** were obtained as 1:1 to 1:4 diastereomeric mixtures, which were inseparable by chromatography, because of the presence of chirality on the phosphorus atoms, while diastereomers of **14b** were separated by chromatography. These four compounds showed a reversible first reduction wave in CV measurements in THF, indicating high stability against reduction. The estimated EA values are 2.71–3.07 eV, which are suitable for a n-type material. No oxidation wave was observed within the solvent window. Their thin films showed stable amorphous character with high glass transition temperatures. Electron mobility of the bis-benzophosphole *P*-sulfide **13b** was as high as 2×10^{-3} cm^2/(V s) (measured using the TOF method at an applied electric field $E = 2.5 \times 10^5$ V/cm), whereas those of the other three remained low, indicating that phosphole itself has high potential as a semiconducting material providing the appropriate molecular design.

Table 32.4 Performance of organic solar cells involving the phosphole derivatives **13** and **14** as a cathode buffer layer,[a] under irradiation with 100 mW/cm^2 incident light, using an AM 1.5G filter

	V_{OC} [V]	J_{SC} [mA/cm^2]	FF	PCE [%]
13a	0.70	10.1	0.61	4.3
13b	0.71	10.8	0.61	4.6
14a	0.71	10.6	0.60	4.5
14b	0.72	10.4	0.61	4.6

[a]The whole device was annealed at 120 °C for 10 min

These benzophosphole derivatives were found to serve as a cathode buffer layer for organic solar cells possessing a similar configuration to that described in the previous section for bis-cinnoline: a p–i–n heterojunction structure composed of crystalline BP and SIMEF as an active layer with a device configuration: ITO/PEDOT:PSS/BP/BP:SIMEF/SIMEF/cathode buffer (5 nm)/Al (Table 32.4). We found that the efficiencies significantly improved by annealing the whole device at 120 °C for 10 min to improve the PCE. For example, a cell fabricated with **13b** showed PCE values of 2.9 % before annealing, while PCE increased to 4.6 % after annealing by a significant increase in the short-circuit current and the fill factor. The dark current also decreased significantly, suggesting improved contact between the phosphole and the neighboring layers. The other three benzophospholes also afforded good PCE values of 4.3–4.6 %. These data indicate that carrier mobility probably has little effect because of the thinness of the buffer layer (5 nm), and we tentatively attribute the improved contact to the high affinity of the P = O or P = S moieties to aluminum.

32.4 Summary

This chapter described our series of studies on the synthesis, properties, and device applications of heteroaromatic compounds containing group 15 elements. Effects of the incorporation of heteroatoms into the π-electron system reside in the control of the electronic properties as well as morphological, thermal, and electrochemical stabilities, which are decisive factors for the performance of electronic devices. Based on the series of studies, we note that the role of organic synthesis is becoming increasingly important in obtaining new classes of high-performance materials for organic electronic applications. Note that we also developed novel synthetic reactions of fused aromatic compounds including group 14 (carbon and silicon) and 16 (oxygen) elements and found that these compounds serve as innovative materials in optoelectronic applications, which are described in other manuscripts and accounts [24].

Acknowledgments This work was partly supported by a Grant-in-Aid for Scientific Research on Innovative Areas (No. 21108509 and 23108704, "pi-Space") and Grant-in-Aid for Young Scientists (A) (No. 20685005). The author is very grateful to his coworkers, Mr. Yuki Yokoi and Mr. Kosuke Sato, for their great effort in developing nitrogen- and phosphorus-containing compounds; to Dr. Yoshiharu Sato for device applications; and to Prof. Eiichi Nakamura for fruitful discussions.

References

1. Akamatu H, Inokuchi H, Matsunaga Y (1954) Electrical conductivity of the perylene–bromine complex. Nature 173:168
2. Salzner U, Lagowski JB, Pickup PG, Poirier RA (1998) Comparison of geometries and electronic structures of polyacetylene, polyborole, polycyclopentadiene, polypyrrole, polyfuran, polysilole, polyphosphole, polythiophene, polyselenophene, and polytellurophene. Synth Met 96:177–189
3. Yamaguchi S, Tamao K (2005) A key role of orbital interaction in the main group element-containing π-electron systems. Chem Lett 34:1
4. Yamaguchi S, Fukazawa AJ (2011) Synthetic study of novel pi-electron materials containing boron, nitrogen, and phosphorus as key main group elements. Syn Org Chem Jpn 69:661–670
5. Tsuji HJ (2010) Modular synthesis of organic electronic materials. Syn Org Chem Jpn 68:1057–1066
6. Tsuji H, Yokoi Y, Mitsui C, Ilies L, Sato Y, Nakamura E (2009) Tetraaryl-substituted Benzo[1,2-*b*:4,5-*b*′] dipyrroles: synthesis, properties, and applications to hole-injection materials in OLED devices. Chem Asian J 4:655–657
7. Tsuji H, Yokoi Y, Furukawa S, Nakamura E (2015) Hexaaryl-benzodipyrroles: properties and application as amorphous carrier-transporting materials. Heterocycles 90:261–270
8. Tsuji H, Cantagrel G, Ueda Y, Chen T, Wan L-J, Nakamura E (2013) Synthesis of benzotrifuran and benzotripyrrole derivatives and molecular orientations on the surface and in the solid. Chem Asian J 8:2377–2382
9. Lewgowd W, Stanczak A (2007) Cinnoline derivatives with biological activity. Arch Pharm Chem Life Sci 340:65–80
10. (a) Schlecker W, Huth A, Ottow E, Mulzer J (1995) Cinnolines and pyrazolopyridazines. – novel synthetic and mechanistic aspects of the Richter reaction. J Liebigs Ann Chem 775–779. (b) Allen CFH, VanAllen JA (1951) Some 3,4-diphenylcinnolines and related compounds. J Am Chem Soc 73:5850–5856
11. Tsuji H, Yokoi Y, Sato Y, Tanaka H, Nakamura E (2011) Bis-cinnolines as n-Type semiconducting material with high electron-mobility and thermal stability and their application to organic photovoltaic cells. Chem Asian J 6:2005–2008
12. Matsuo Y, Sato Y, Niinomi T, Soga I, Tanaka H, Nakamura E (2009) Columnar structure in bulk heterojunction in solution-processable three-layered p-i-n organic photovoltaic devices using tetrabenzoporphyrin precursor and silylmethyl[60]fullerene. J Am Chem Soc 131:16048
13. Crassous J, Réau R (2008) π-conjugated phosphole derivatives: synthesis, optoelectronic functions and coordination chemistry. Dalton Trans (48):6865–6876
14. Matano Y, Imahori H (2009) Design and synthesis of phosphole-based π systems for novel organic materials. Org Biomol Chem 7:1258–1271
15. Ren Y, Baumgartner T (2012) Combining form with function – the dawn of phosphole-based functional materials. Dalton Trans 41:7792–7800
16. Stolar M, Baumgartner T (2013) Organic n-type materials for charge transport and charge storage applications. Phys Chem Chem Phys 15:9007–9024
17. Baumgartner T (2014) Insights on the design and electron-acceptor properties of conjugated organophosphorus materials. Acc Chem Res 47:1613–1622
18. Stolar M, Baumgartner T (2014) Phosphorus-containing materials for organic electronics. Chem Asian J 9:1212–1225
19. Simpson MC, Protasiewicz JD (2013) Phosphorus as a carbon copy and as a photocopy: new conjugated materials featuring multiply bonded phosphorus. Pure Appl Chem 85:801–815
20. (a) Tsuji H, Sato K, Ilies L, Itoh Y, Sato Y, Nakamura E (2008) Modular synthesis of benzo[*b*]phosphole derivatives via BuLi-mediated cyclization of (*o*-alkynylphenyl)phosphine. Org Lett 10:2263–2265 (b) Tsuji H, Sato K, Sato Y, Nakamura E (2009) Benzo[*b*]phosphole sulfides. Highly electron-transporting and thermally stable molecular materials for organic semiconductor devices. J Mater Chem 19:3364–3366 (c) Tsuji H, Sato K, Sato Y, Nakamura E

(2010) Benzophosphole oxide and sulfide for thermally stable cathode buffer layers in organic thin-film photovoltaic devices. Chem Asian J 5:1294–1297
21. Kojima T, Furukawa S, Tsuji H, Nakamura E (2014) Synthesis of triphosphatruxene via sextuple aromatic nucleophilic substitution and simple isolation of stereoisomers. Chem Lett 43:676–677
22. Fukazawa A, Ichihashi Y, Kosaka Y, Yamaguchi S (2009) Benzo[*b*]phosphole-containing π-electron systems: synthesis based on an intramolecular trans-halophosphanylation and some insights into their properties. Chem Asian J 4:1729–1740
23. Sanji T, Shiraishi K, Kashiwabara T, Tanaka M (2008) Base-mediated cyclization reaction of 2-alkynylphenylphosphine oxides: synthesis and photophysical properties of benzo[*b*]phosphole oxides. Org Lett 10:2689–2692
24. (a) Tsuji H, Ilies L, Nakamura E (2014) Synthetic strategy for multi-substituted fused furan compounds using main group metal reagents. Synlett 25:2099–2110 (b) Mitsui C, Soeda J, Miwa K, Shoyama K, Ota Y, Tsuji H, Takeya J, Nakamura E (2015) Single-crystal organic field-effect transistors of naphthodifurans. Bull Chem Soc Jpn 88:776–783

Chapter 33
Fullerene Derivatives for Organic Solar Cells

Yutaka Matsuo

Abstract Functionalization of fullerenes can produce various fullerene-based electron acceptors for solution-processed organic thin-film solar cells. In this chapter, silylmethyl[60]fullerenes (SIMEFs), methano indene fullerenes (MIFs), fullerenyl esters, and lithium ion-encapsulated fullerene derivatives are introduced to discuss design concepts, synthetic methods, and photovoltaic properties of these fullerene electron acceptors.

Keywords Organic solar cells • Fullerene • Electron acceptors • Photovoltaics • Organic semiconductors

33.1 Introduction

Organic π-electron conjugated systems are essential for photoelectric functions of organic compounds. Fullerenes have curved and spherical π-conjugation systems with rich functions unavailable for ordinary organic π-conjugated compounds. Fullerene derivatives are commonly used electron-accepting materials in solution-processed organic solar cells because of their low-lying lowest unoccupied molecular orbital (LUMO) levels, small reorganization energy, and electron delocalization ability. Addition of organic addends to fullerenes enables tuning electronic properties of fullerene derivatives. We are focusing on the development of new fullerene derivatives that have high electron mobility, suitable LUMO levels for high open-circuit voltage (V_{OC}), and self-assembling ability to achieve high-performance organic photovoltaic cells. In this chapter, several photo-electronically functional fullerene derivatives are introduced to offer insight into the design, synthesis, and properties of fullerene-based electron-accepting materials used for organic thin-film solar cells.

Y. Matsuo (✉)
Department of Chemistry, School of Science, The University of Tokyo, Tokyo, Japan
e-mail: matsuo@chem.s.u-tokyo.ac.jp

T. Akasaka et al. (eds.), *Chemical Science of π-Electron Systems*,
DOI 10.1007/978-4-431-55357-1_33

33.2 1,4-Bis(Silylmethyl)[60]Fullerenes and Related Compounds

Phenyl-C_{61}-butyric acid methyl ester (PCBM) is a benchmark fullerene electron acceptor used in the research of solution-processed organic solar cells. This fullerene derivative has a 58π-electron conjugated system with a 1,2-addition pattern. We designed 1,4-bis(silylmethyl)[60]fullerenes with 1,4-58π-electron conjugated systems to realize lower electron affinity to have higher-lying LUMO levels for higher V_{OC} in organic thin-film solar cells [1–3]. The key reaction to have 1,4-bis(silylmethyl)[60]fullerenes is a dimethylformamide (DMF)-assisted efficient monoaddition reaction of Grignard reagents. Monoaddition of tri(organo)silylmethylmagnesium chlorides to C_{60} smoothly takes place in the presence of DMF to produce (organo)(hydro)[60]fullerenes, $C_{60}R^1H$ in good yield (Fig. 33.1; compounds **1**–**6**). Methyl and aryl Grignard reagents are also used in this reaction to give corresponding monoadducts (compounds 7 and 8). The monoadduct having an isopropoxy group on the silicon atom is used for further modification into chloro, amino, and fluoro silane compounds (Fig. 33.2). The obtained hydrofullerenes are then deprotonated to have the corresponding fullerene monoadduct anions, $C_{60}R^-$, which are subjected into nucleophilic substitution reaction with alkyl halides to afford 1,4-$C_{60}R^1R^2$ (compounds **12**–**31**) in good

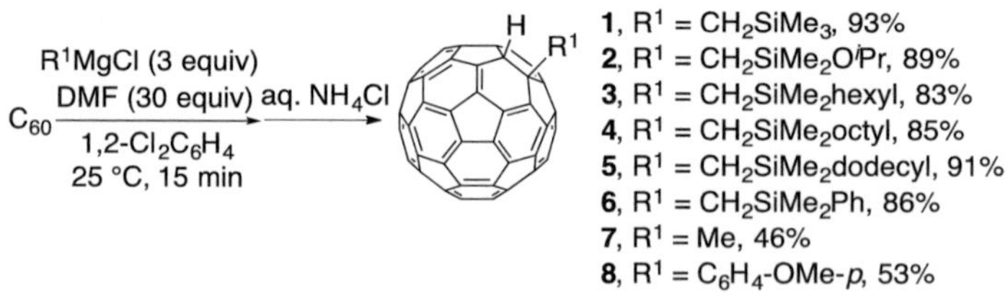

Fig. 33.1 DMF-assisted efficient monoaddition reactions to C_{60}

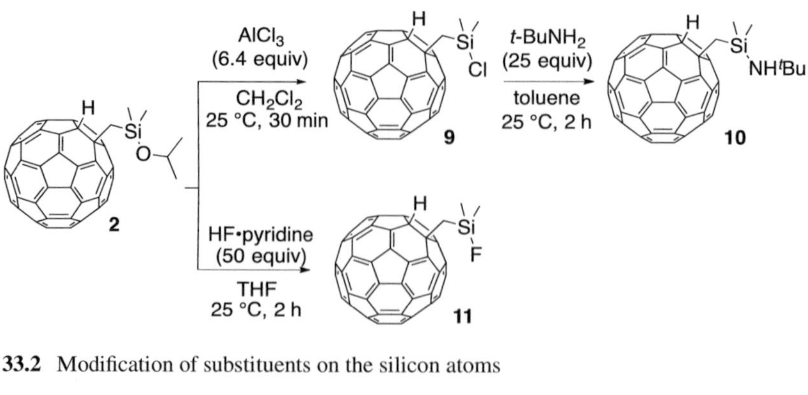

Fig. 33.2 Modification of substituents on the silicon atoms

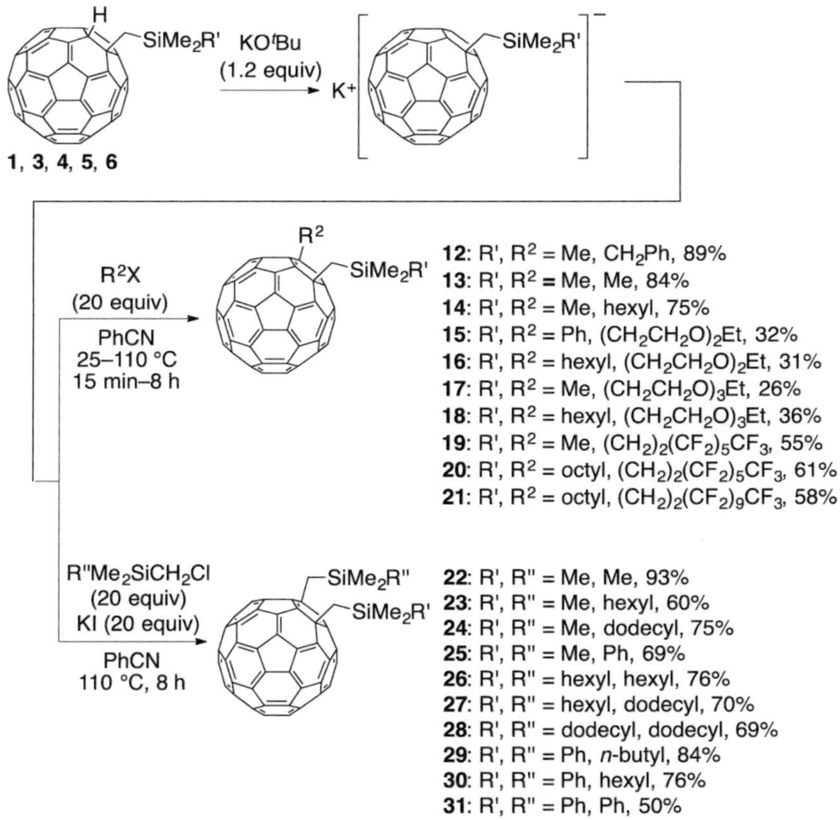

Fig. 33.3 Synthesis of various 1,4-bis-adducts by nucleophilic substitutions with alkyl halides

to high yield (Fig. 33.3). The two-step synthesis provides a wide variety of 1,4-di(organo)[60]fullerenes bearing the same or different organic addends on the [60]fullerene core. Chemical and physical properties of the compounds are tuned by changing the addends. We named the compound **31** as SIMEF as in "**si**lyl**me**thyl**f**ullerenes."

Photovoltaic performance of {dimethyl(*o*-anisyl)silylmethyl}(dimethylphenylsilylmethyl)[60]fullerene (SIMEF2) has been investigated on the inverted structured bulk heterojunction organic thin-film solar cells [4, 5]. A device with poly(3-hexylthiophene) (P3HT) and SIMEF2 exhibited 2.9 % power conversion efficiency (PCE; $J_{SC} = 7.9$ mA/cm^2, $V_{OC} = 0.66$ V, and FF = 0.56) in the device configuration ITO/ZnO/P3HT:SIMEF2/PEDOT:PSS/Au (Fig. 33.4) under AM1.5G illumination at 100 mW/cm^2. In comparison with a PCBM-based device (PCE = 2.4 %, $J_{SC} = 7.8$ mA/cm^2, $V_{OC} = 0.56$ V, and FF = 0.55), the SIMEF2-based device gave higher V_{OC} without a decrease in J_{SC}. Improvement

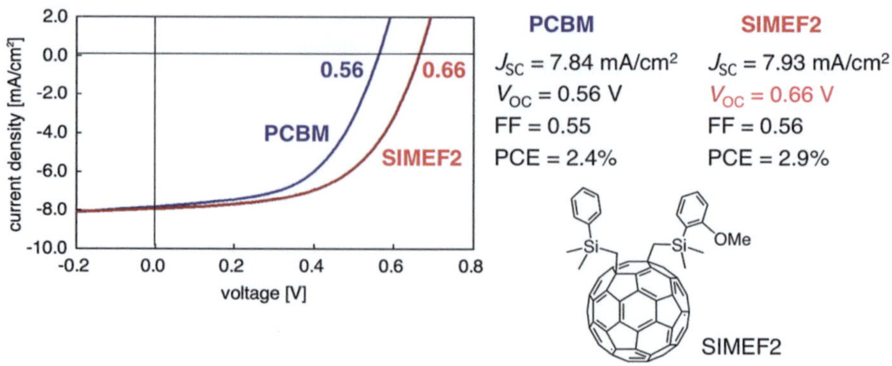

Fig. 33.4 Photocurrent-voltage curves of ITO/ZnO/P3HT:SIMEF2/PEDOT:PSS/Au and ITO/ZnO/P3HT:PCBM/PEDOT:PSS/Au cells

of V_{OC} is ascribed to higher-lying LUMO level of SIMEF2 (−3.72 eV) compared with PCBM (−3.80 eV). Similar J_{SC} in the SIMEF2 and PCBM devices is attributable to similar steric bulkiness of the organic addends in SIMEF2 and PCBM. This result marks progress toward overcoming the trade-off relationship between V_{OC} and J_{SC} in the development of organic solar cell devices.

33.3 1-Aryl-4-Silylmethyl[60]Fullerenes

Nucleophilic addition of aryl Grignard reagents (aryl = 4-$MeOC_6H_4$, 4-$Me_2NC_6H_4$, Ph, 4-$CF_3C_6H_4$, and thienyl) to C_{60} in the presence of dimethyl sulfoxide efficiently produced 1,2-arylhydro[60]fullerenes after acid treatment. The reaction of anions of the monoaryl[60]fullerene with dimethylphenylsilylmethyl iodide or dimethyl(2-isopropoxyphenyl)silylmethyl iodide yielded the target compounds, 1-aryl-4-silylmethyl[60]fullerenes (Fig. 33.5) [6]. Properties and structures of 1-aryl-4-silylmethyl[60]fullerenes (aryl = 4-$MeOC_6H_4$ and thienyl) were characterized by electrochemical studies, X-ray crystallography, flash-photolysis time-resolved microwave conductivity (FP-TRMC) measurements, and electron mobility measurement under space charge limited current (SCLC) model (**32b**, 1×10^{-5} cm^2/Vs; **36b**, 5×10^{-5} cm^2/Vs). Organic photovoltaic devices were fabricated using 1-aryl-4-silylmethyl[60]fullerenes with polymer-based bulk heterojunction, small molecule-based p-n heterojunction, and p-i-n junction structures, and the most efficient device exhibited power conversion efficiency (PCE) of up to 3.4 % (short circuit current density (J_{SC}) = 8.1 mA/cm^2 and open-circuit voltage (V_{OC}) = 0.69, fill factor (FF) = 0.59) (Fig. 33.6).

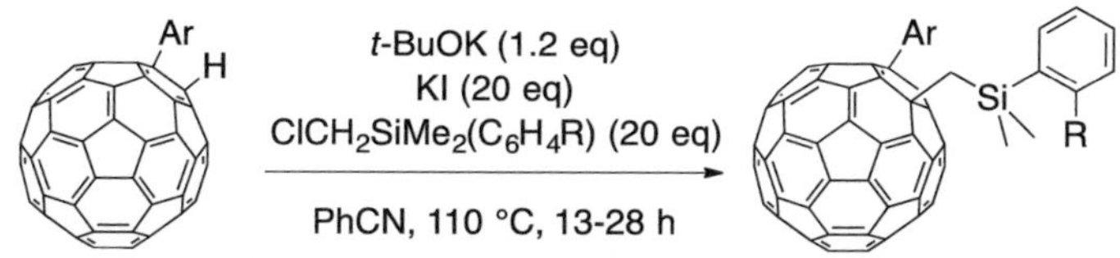

Ar	product yield (%)	solubility in toluene (wt%)	LUMO level (eV)
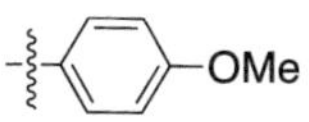	(**32a**: R = H) 21 (**32b**: R = O*i*Pr) 36	0.4 0.8	–3.72 –3.72
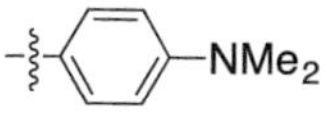	(**33**) 34	0.6	–3.72
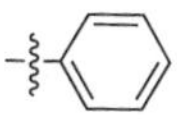	(**34**) 18	0.3	–3.81
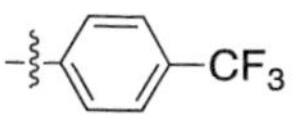	(**35**) 19	0.4	–3.86
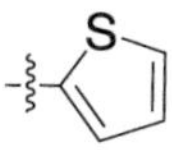	(**36a**: R = H) 24 (**36b**: R = O*i*Pr) 28	0.5 0.9	–3.71 –3.71

Fig. 33.5 Synthesis and properties of 1-aryl-4-silylmethyl[60]fullerenes

33.4 Methano Indene Fullerenes

In order to engineer fullerenes that possess LUMO levels that are higher than 58π fullerenes (PCBM and SIMEFs), construction of 56π-electron conjugated systems on the fullerene is considered. Bis-PCBM [7], indene-C_{60} bis-adduct (ICBA, $C_{60}(Ind)_2$) [8], and various other examples [9] have been known to possess high-LUMO levels and give high V_{OC}. However, these 56π fullerenes often cause a lowering of J_{SC} and FF, due to the large volume of organic addends that lowers electron mobility. In view of this, the authors have reported the use of the smallest organic addend, a methylene group (CH_2) to obtain 56π-electron dihydromethano[60]fullerenes, which minimizes steric demand for the purpose of obtaining good electron mobility [10–12]. Furthermore, 56π-electron dihydromethano-indene adduct of [60]fullerene, $C_{60}(CH_2)(Ind)$ (Ind = indene, Fig. 33.7; methano indene fullerene, MIF) has given high V_{OC} without an accompanying decrease in J_{SC} and FF to afford high PCE (5.9 %). In addition, 66π-electron [70]fullerene derivative that possesses both a dihydromethano group and the indene addend ($C_{70}(CH_2)(Ind)$) has shown 6.4 % PCE when used in a poly(3-hexylthiophene) (P3HT)-based bulk heterojunction OPV devices [12].

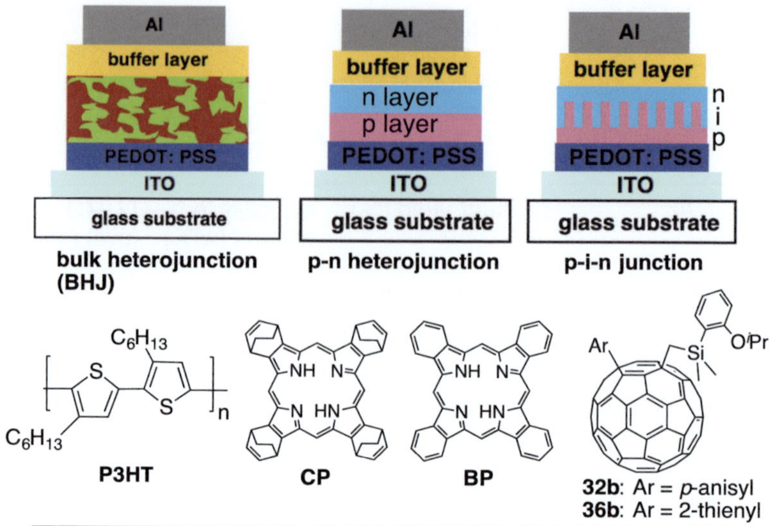

acceptor	donor	structure	J_{SC} [mA/cm^2]	V_{OC} [V]	*FF*	PCE [%]
32b	P3HT	BHJ	1.7	0.62	0.36	0.38
36b	P3HT	BHJ	8.7	0.61	0.53	2.8
SIMEF	P3HT	BHJ	9.0	0.65	0.62	3.6
PCBM	P3HT	BHJ	9.2	0.55	0.63	3.2
32b	BP	p-n	6.3	0.66	0.63	2.3
36b	BP	p-n	6.8	0.66	0.61	2.7
32b	BP	p-i-n	7.2	0.71	0.56	2.8
36b	BP	p-i-n	8.1	0.69	0.59	3.4

Fig. 33.6 Photovoltaic performance of 1-aryl-4-silylmethyl[60]fullerenes with various device structures

Starting from $C_{60}(CH_2)$ and $C_{70}(CH_2)$, these compounds were heated with excess indene (25–50 equiv.) in 1,2-dichlorobenzene to obtain 56π- and 66π-electron conjugated fullerenes and $C_{60}(CH_2)$(Ind) and $C_{70}(CH_2)$(Ind) in 47 % and 36 % yields, respectively. These products were obtained as mixtures of isomers. Small steric bulkiness of the dihydromethano group realizes efficient overlap of the π-electron conjugated systems of the fullerene cores in the solid state. From the mixture of isomers, C_S and C_1 isomers were separated by the preparative HPLC technique and characterized by NMR measurements. The X-ray crystallographic structure of the C_S isomer reveals small volume of the dihydromethano group (Fig. 33.8).

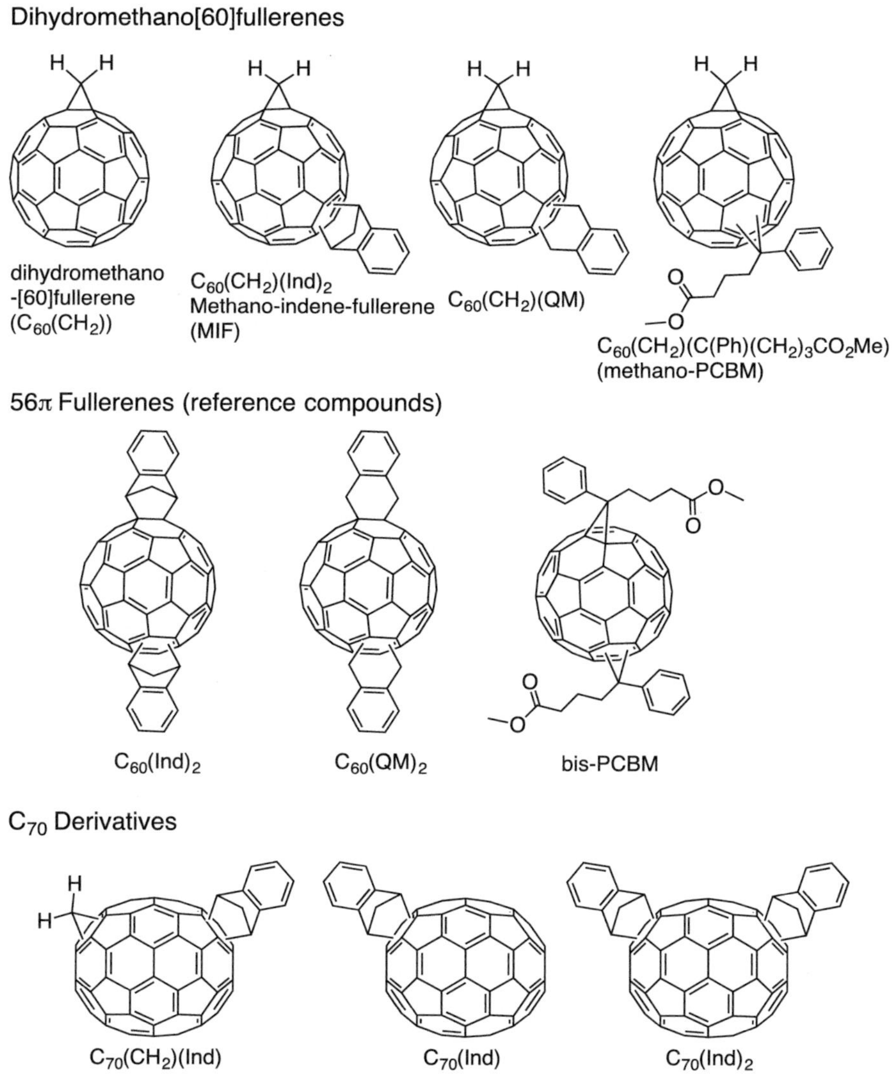

Fig. 33.7 Molecular structure of $C_{60}(CH_2)(Ind)$, $C_{70}(CH_2)(Ind)$, and related compounds

Estimated LUMO levels determined from the first reduction potentials of fullerene derivatives in the cyclic voltammetric analysis showed a 100-mV higher LUMO level of a 56π-electron fullerene $C_{60}(CH_2)(Ind)$ compared with a 58π-electron fullerene, $C_{60}(Ind)$, because of the smaller size of the π-electron conjugated system in $C_{60}(CH_2)(Ind)$. On the other hand, $C_{60}(CH_2)(Ind)$ showed a 30-mV lower LUMO level as compared to $C_{60}(Ind)_2$. This is likely because of the less electron-donating nature of the three-membered dihydromethano moiety than the five-membered bicyclic indene moiety. Solubility (4.0 wt%) of $C_{60}(CH_2)(Ind)$ in

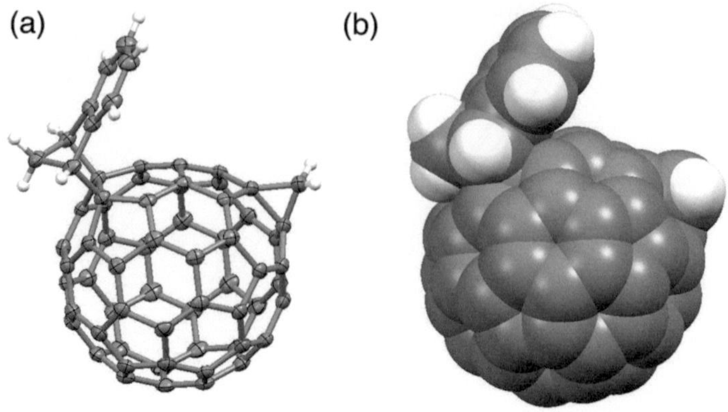

Fig. 33.8 X-ray crystallographic structure of the C_S isomer of $C_{60}(CH_2)(Ind)$. (**a**) ORTEP drawing. (**b**) Space-filling model

Table 33.1 LUMO levels, solubility, and FET electron mobility of fullerene derivatives

Fullerene	LUMO levels (eV)	Solubility in toluene (wt%)	FET electron mobility (cm^2/Vs)
$C_{60}(CH_2)(Ind)$	−3.66	4.0	3×10^{-3}
$C_{60}(CH_2)(QM)$	−3.66	4.0	–
methano-PCBM	−3.70	7.0	9.1×10^{-6}
$C_{70}(CH_2)(Ind)$	−3.65	1.5	–
$C_{60}(Ind)$	−3.76	0.5	–
$C_{60}(Ind)_2$	−3.63	7.5	7×10^{-4}
$C_{60}(QM)$	−3.78	0.3	–
$C_{60}(QM)_2$	−3.63	5.0	8×10^{-4}
PCBM	−3.80	1.5	–
bis-PCBM	−3.71	>10	–
$C_{70}(Ind)$	−3.76	0.1	–
$C_{70}(Ind)_2$	−3.62	3.0	–

toluene was fairly high, because of the presence of regioisomers. This value was eight times higher than the solubility of singly isomeric $C_{60}(Ind)$. In addition, solubility of $C_{70}(CH_2)(Ind)$ in toluene was respectable, at 1.5 wt% (Table 33.1).

Electron mobility of methano indene fullerene was evaluated by means of a field-effect transistor (FET) method (Table 33.1). Electron mobility of $C_{60}(CH_2)(Ind)$ was found to be 3×10^{-3} cm^2/Vs. This value is higher than that of $C_{60}(Ind)_2$ (7×10^{-4} cm^2/Vs) and $C_{60}(QM)_2$ (8×10^{-4} cm^2/Vs). The good electron mobility is attributed to the small dihydromethano moiety, which enables efficient electron hopping in the bulk fullerene film. This characteristic is favorable to obtain high performance in the organic thin-film devices, especially high J_{SC} and FF, because high electron mobility of materials raises electron transport efficiency and hence increases the charge collected at electrodes.

Table 33.2 Photovoltaic properties of P3HT-based bulk heterojunction devices

Entry	Fullerene	V_{OC} (V)	J_{SC} (mA/cm^2)	FF (−)	PCE (%)
1	$C_{60}(CH_2)$(Ind)	0.78	10.3	0.73	5.9
2	$C_{60}(CH_2)$(QM)	0.77	9.6	0.67	5.0
3	methano-PCBM	0.73	9.0	0.71	4.6
4	$C_{70}(CH_2)$(Ind)	0.79	11.1	0.73	6.4
5	C_{60}(Ind)	0.66	7.9	0.69	3.6
6	$C_{60}(Ind)_2$	0.84	8.3	0.69	4.8
7	C_{60}(QM)	0.65	9.5	0.68	4.2
8	$C_{60}(QM)_2$	0.85	8.3	0.67	4.7
9	PCBM	0.61	9.1	0.72	4.0
10	bis-PCBM	0.76	8.2	0.71	4.4
11	C_{70}(Ind)	0.70	8.6	0.70	4.2
12	$C_{70}(Ind)_2$	0.85	9.3	0.68	5.4

Photovoltaic performance of methano indene fullerenes and related reference compounds was evaluated on the standard bulk heterojunction devices with a configuration of ITO/PEDOT:PSS/P3HT:fullerenes/Ca/Al. The device using $C_{60}(CH_2)$(Ind) recorded PCE of 5.9 % (Table 33.2, entry 1). The V_{OC} (0.78 V) of this device was obviously higher than those of devices using 58π fullerenes such as C_{60}(Ind) ($V_{OC} = 0.66$ V, entry 5) and PCBM ($V_{OC} = 0.61$ V, entry 9). Undoubtedly, this is because of the higher LUMO level of 56π $C_{60}(CH_2)$(Ind) as compared to those of C_{60}(Ind) and PCBM. This V_{OC} value was slightly lower than that of a device using 56π $C_{60}(Ind)_2$ ($V_{OC} = 0.84$ V, entry 6), due to the slightly lower-lying LUMO level of $C_{60}(CH_2)$(Ind) as compared to that of $C_{60}(Ind)_2$. However, more importantly, J_{SC} (10.3 mA/cm^2) and FF (0.73) of the $C_{60}(CH_2)$(Ind) device were higher than those of the $C_{60}(Ind)_2$ device ($J_{SC} = 8.3$ mA/cm^2, FF = 0.69). This is attributed to the high electron mobility (3×10^{-3} cm^2/Vs vs. 7×10^{-4} cm^2/Vs for $C_{60}(Ind)_2$, Table 33.1) of $C_{60}(CH_2)$(Ind), which is derived from the smaller steric demand of the dihydromethano group. Consequently, the $C_{60}(CH_2)$(Ind) device showed higher PCE, as compared to the $C_{60}(Ind)_2$ device (4.8 %). Other 56π dihydromethano[60]fullerenes, $C_{60}(CH_2)$(QM) (entry 2) and methano-PCBM (entry 3), also showed higher V_{OC} (0.77 and 0.73 V, respectively) in comparison with their corresponding 58π fullerenes, C_{60}(QM) (0.65 V, entry 7) and PCBM (0.61 V, entry 9). Both exhibited higher J_{SC} with similar FF ($J_{SC} = 9.6$ mA/cm^2, FF = 0.67 for $C_{60}(CH_2)$(QM); $J_{SC} = 9.0$ mA/cm^2, FF = 0.71 for methano-PCBM) are compared with their 56π bis-adduct counterparts, $C_{60}(QM)_2$ ($J_{SC} = 8.3$ mA/cm^2, FF = 0.67, entry 8) and bis-PCBM ($J_{SC} = 8.2$ mA/cm^2, FF = 0.71, entry 10). More importantly, among these three 56π dihydromethano[60]fullerene derivatives (entries 1–3), $C_{60}(CH_2)$(Ind) showed the best performance. Although it is difficult to explain why $C_{60}(CH_2)$(Ind) is superior to $C_{60}(CH_2)$(QM), we can suggest that the better performance of $C_{60}(CH_2)$(Ind) as compared to methano-PCBM is due to the good electron mobility of $C_{60}(CH_2)$(Ind), arising from a smaller indene group as compared to the phenylbutyric acid ester group of PCBM.

The device using a 66π dihydromethano[70]fullerene, $C_{70}(CH_2)$(Ind) showed PCE of 6.4 % ($V_{OC} = 0.79$ V, $J_{SC} = 11.1$ mA/cm^2, FF = 0.73, entry 4). The V_{OC} and FF of the $C_{70}(CH_2)$(Ind) device were similar to that of its C_{60}-based counterpart (entry 1). However, comparatively higher J_{SC} was obtained because of higher light absorption coefficient of C_{70} derivatives as compared to corresponding C_{60} derivatives [13]. Similar to the trend in C_{60}-based devices, $C_{70}(CH_2)$(Ind) showed higher V_{OC} (0.79 V) than C_{70}(Ind) (0.70 V; entry 11) and higher J_{SC} (11.1 mA/cm^2) and FF (0.73) than $C_{70}(Ind)_2$ ($J_{SC} = 9.3$ mA/cm^2 and FF = 0.68; entry 12). From the analysis of the abovementioned data, the author suggests that the use of the dihydromethano group is an effective methodology to obtain high V_{OC} without an accompanying decrease in J_{SC} and FF. In addition, $C_{70}(CH_2)$(Ind) is shown to be one of the best fullerene-based electron acceptors for P3HT-based bulk heterojunction organic photovoltaic devices, which are commonly used, practical, and considered to be the standard organic thin-film solar cells.

33.5 Fullerene Derivatives Having Low-Lying LUMO Levels

In general, design guidelines on creating fullerene acceptors are based on raising LUMO levels to obtain a higher open-circuit voltage (V_{OC}) for OPV devices. These have been realized by introducing electron-donating groups [14] and decreasing the size or changing the shape of the fullerene π-electron conjugated systems [3] to decrease the electron affinity of the fullerene core. On the other hand, the lowering of LUMO levels of fullerenes is also a respected strategy. Low bandgap polymers [15], which were recently introduced for OPV application, have low-LUMO levels that tend to require low-LUMO fullerenes for efficient charge separation. Although obtaining high-LUMO fullerenes is relatively straightforward, since the addition of carbon functional groups to fullerene generally raises LUMO levels, the approach toward low-LUMO fullerenes is rather limited. Only a few low-LUMO fullerenes such as azafulleroid and fullerene ketolactams have been investigated for OPV devices [16, 17]. Further investigations on low-LUMO fullerenes are awaited to expand upon this field of research.

To lower the LUMO levels of fullerenes, electron-withdrawing groups, such as the cyano group, are installed on fullerene [18]. Another approach involves attaching electronegative atoms such as oxygen on fullerene to increase the electron affinity of the fullerene core. Various oxygen-attached fullerene derivatives such as hydroxylfullerenes [19], alchoxylfullerenes [20], peroxidized fullerenes [21], 1,3-dioxolane fullerenes [22], fullerooxazoles [23], fullerene isoxazolines [24], and fused furanylfullerenes [25] have been reported. Among these, fullerenyl esters [26] are considered to be the most suitable for OPV application: these compounds bear electron-withdrawing ester moieties and exhibit high solubility in organic solvents.

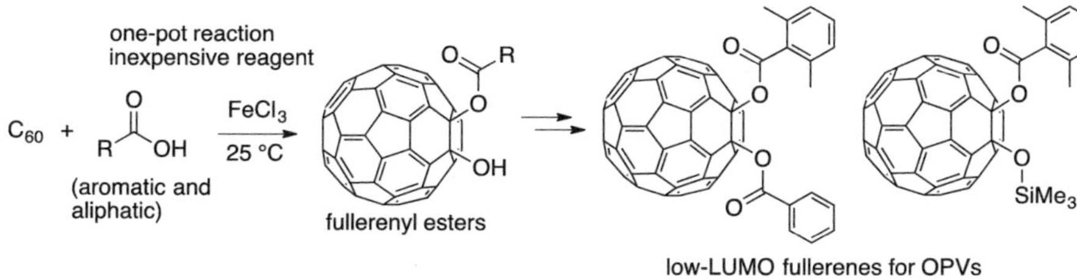

Fig. 33.9 $FeCl_3$-mediated synthesis of fullerenyl esters

The authors have reported facile one-pot syntheses of hydroxyfullerenyl esters C_{60}(OCOR)(OH) by using versatile carboxylic acids and inexpensive iron trichloride ($FeCl_3$) [27]. This reaction proceeds under mild conditions and possesses substrate diversity catering to both aromatic and aliphatic carboxylic acids. The hydroxyl functionality can be utilized to obtain a diester derivative C_{60}(OCOAr)(OCOPh) (Ar = 2,6-xylyl) and a siloxyl derivative C_{60}(OCOAr)($OSiMe_3$) (Fig. 33.9).

Electrochemical studies of fullerenyl esters C_{60}(OCOAr)(OCOPh) and C_{60}(OCOAr)($OSiMe_3$) (Ar = 2,6-xylyl) in 1,2-dichlorobenzene showed that the compounds have LUMO levels of −3.69 and −3.67 eV, respectively. These values are the same as that for the LUMO level of C_{60} (−3.69 eV, estimated from electrochemical measurement in 1,2-dichlorobenzene; note that estimated LUMO levels are dependent on the choice of the solvent used in the electrochemical measurements. Data can be used for comparison when the same solvent is used on the measurements.) and deeper than that of PCBM (−3.61 eV, determined on the same measurement condition). These compounds were examined in bulk heterojunction solar cells with poly(4,8-bis-octoxybenzo(1,2-b:4,5-b)dithiophene-2,6-diyl-alt-(dodecyl-thieno(3,4-b)thiophene-2-carboxylate)-2,6-diyl) (PBDTTT) [28] as the donor material in the active layer and showed PCE of 1.3 % (J_{SC} = 5.4 mA/cm^2, V_{OC} = 0.74 V, FF = 0.34).

33.6 Lithium Ion-Containing Fullerene Derivatives

Synthesis and isolation of lithium ion-encapsulated [60]fullerene ($Li^+@C_{60}$) was first reported in 2010 [29]. Recently, host-guest complexes of calix [4] pyrroles and [$Li^+@C_{60}$]PF_6^- had been investigated to study electron transfer processes involved [30]. However, covalent chemical modification of [$Li^+@C_{60}$]PF_6^- has been eagerly awaited, since such chemical functionalizations are expected to produce exotic functional materials for applications not only in chemistry, but also in applied physics and biology. Chemical modification of fullerenes plays a significant role in fullerene-based material research, including the expanding field of organic photovoltaics. In particular, PCBM could be considered as one of the most defining fullerene derivative, and it is widely used as an electron acceptor in organic thin-film solar cells. The authors demonstrated the synthesis of the Li^+-encapsulated

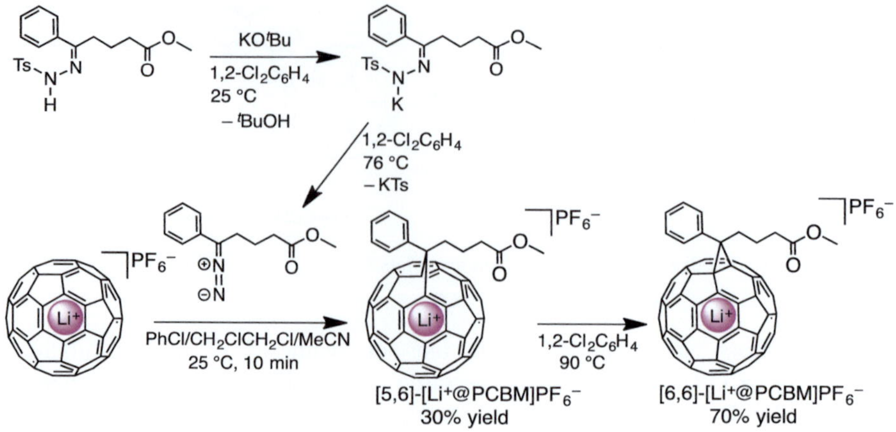

Fig. 33.10 Synthesis of lithium ion-encapsulated PCBM

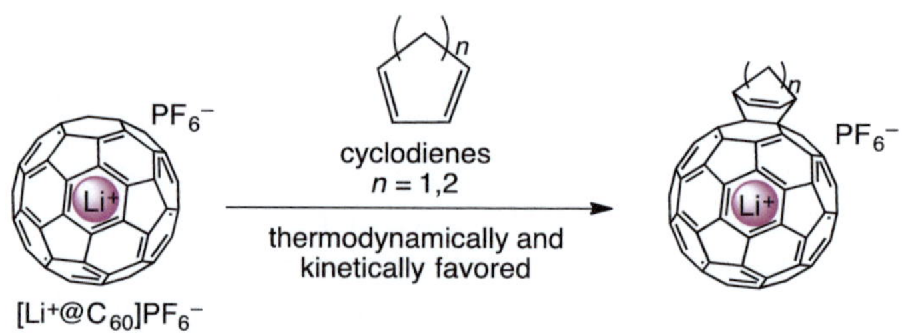

Fig. 33.11 DA reaction of lithium ion-encapsulated [60]fullerene with dienes

version of PCBM, since encapsulation of a lithium ion into the fullerene cage drastically alters the electronic property of PCBM [31]. The [5,6]- and [6,6]-isomers of [Li^+@PCBM]PF_6^- were successfully prepared, and the structure of [6,6]-[Li^+@PCBM]PF_6^- was elucidated by X-ray crystallography (Fig. 33.10).

The Diels-Alder (DA) reaction of fullerene is one of the most important reactions even in the research area of fullerene chemistry. However, the inherent equilibrium between the DA and retro-DA reactions lowers the yield and limits the scope of dienes. Considering the Lewis acid activation of dienophiles, we have focused on the DA reaction of Li^+@C_{60}. Comparing C_{60} with Li^+@C_{60}, it is emphasized that the steric effects can be ignored on the kinetics and thermodynamics. The authors studied the reaction of Li^+@C_{60} with cyclopentadiene (Fig. 33.11), which was chosen as a diene, and found that this reaction was highly efficient, with an equilibrium constant that was more than 1000-fold that for the reaction with empty C_{60} [32].

The DA reaction of [Li^+@C_{60}](PF_6^-) with 1,3-cyclohexadiene was also kinetically and computationally investigated to precisely determine the kinetic parameters such as E_a $\Delta H^{\ddagger}$, $\Delta S^{\ddagger}$, and $\Delta G^{\ddagger}$ through the reaction rate measurement [33]. The product was fully characterized by spectroscopic and single-crystal X-ray analyses. Compared with empty C_{60}, Li^+@C_{60} reacted 2400-fold faster at 303 K. The result was well explained by DFT calculation at the M06-2X/6-31G(d) level of theory considering the reactant complex with dispersion corrections as well as by the frontier molecular orbital theory. This is the first detailed report on the kinetics of a DA reaction catalyzed by "an intramolecularly encapsulated catalyst" without coordination to a heteroatom in the dienophile.

33.7 Conclusions

Several fullerene-based electron-accepting materials as well as electronically functionalized fullerene derivatives are discussed in this chapter. Modification of electronic structure such as LUMO levels of fullerene derivatives is of importance to achieve high power conversion efficiency in organic thin-film solar cell application. By reducing the size of π-electron conjugated systems of fullerene derivatives, electron affinity can be lowered, and hence larger energy difference between the LUMO levels of electron acceptors and the HOMO levels of electron donors is achieved for high V_{OC}. In this design, the volume of organic addends should be considered to have high electron mobility to obtain high J_{SC} and FF. In this chapter, fullerenyl esters having electron-withdrawing groups and lithium ion-encapsulated fullerene derivatives are also discussed for the design of low-LUMO level fullerene derivatives.

Acknowledgments The author sincerely thanks Professor Eiichi Nakamura (The University of Tokyo), Mitsubishi Chemical Corporation, and all collaborators listed in the literature cited. The author gratefully acknowledges the Funding Program for Next-Generation World-Leading Researchers (GR30).

References

1. Matsuo Y, Iwashita A, Abe Y, Li CZ, Matsuo K, Hashiguchi M, Nakamura E (2008) Regioselective synthesis of 1,4-Di(organo)[60]fullerenes through DMF-assisted mono-addition of silylmethyl Grignard reagents and subsequent alkylation reaction. J Am Chem Soc 130: 15429–15436. doi:10.1021/ja8041299
2. Matsuo Y, Sato Y, Niinomi T, Soga I, Tanaka H, Nakamura E (2009) Columnar structure in bulk heterojunction in solution-processable three-layered p-i-n organic photovoltaic devices using tetrabenzoporphyrin precursor and silylmethyl[60]fullerene. J Am Chem Soc 131:16048–16050. doi:10.1021/ja9048702
3. Matsuo Y (2012) Design concept for high-LUMO-level fullerene electron-acceptors for organic solar cells. Chem Lett 41:754–759. doi:10.1246/cl.2012.754
4. Matsuo Y, Hatano J, Kuwabara T, Takahashi K (2012) Fullerene acceptor for improving open-circuit voltage in inverted organic photovoltaic devices without accompanying decrease in short-circuit current density. Appl Phys Lett 100:063303. doi:10.1063/1.3683469
5. Tanaka H, Abe Y, Matsuo Y, Kawai J, Soga I, Sato Y, Nakamura E (2012) An amorphous mesophase generated by thermal annealing for high-performance organic photovoltaic devices. Adv Mater 24:3521–3525. doi:10.1002/adma.201200490

6. Matsuo Y, Oyama H, Soga I, Okamoto T, Tanaka H, Saeki A, Seki S, Nakamura E (2013) 1-Aryl-4-Silylmethyl[60]fullerenes: synthesis, properties, and photovoltaic performance. Chem Asian J 8:121–128. doi:10.1002/asia.201200726
7. Lenes M, Wetzelaer GJAH, Kooistra FB, Veenstra SC, Hummelen JC, Blom PWM (2008) Fullerene bisadducts for enhanced open-circuit voltages and efficiencies in polymer solar cells. Adv Mater 20:2116–2119. doi:10.1002/adma.200702438
8. He Y, Chen HY, Hou J, Li Y (2010) Indene $-C_{60}$ bisadduct: a new acceptor for high-performance polymer solar cells. J Am Chem Soc 132:1377–1382. doi:10.1021/ja908602j
9. Zhen Y, Obata N, Matsuo Y, Nakamura E (2012) Benzo[c]thiophene-C60 diadduct: an electron acceptor for p–n junction organic solar cells harvesting visible to near-IR light. Chem Asian J 7:2644–2649. doi:10.1002/asia.201200698
10. Zhang Y, Matsuo Y, Li CZ, Tanaka H, Nakamura E (2011) A scalable synthesis of methano[60]fullerene and congeners by the oxidative cyclopropanation reaction of silylmethylfullerene. J Am Chem Soc 133:8086–8089. doi:10.1021/ja201267t
11. Li CZ, Chien SC, Yip HL, Chueh CC, Chen FC, Matsuo Y, Nakamura E, Jen AKY (2011) Facile synthesis of a 56π-electron 1,2-dihydromethano-[60]PCBM and its application for thermally stable polymer solar cells. Chem Commun 47:10082–10084. doi:10.1039/C1CC14446D
12. Li CZ, Matsuo Y, Nakamura E (2011) Regioselective synthesis of tetra(aryl)-mono(silylmethyl)[60]fullerenes and derivatization to methanofullerene compound. Tetrahedron 67:9944–9949. doi:10.1016/j.tet.2011.09.125
13. Wienk MM, Kroon JM, Verhees WJH, Knol J, Hummelen JC, van Hal PA, Janssen RAJ (2003) Efficient methano[70]fullerene/MDMO-PPV bulk heterojunction photovoltaic cells. Angew Chem Int Ed 42:3371–3375. doi:10.1002/anie.200351647
14. Kooistra FB, Knol J, Kastenberg F, Popescu LM, Verhees WJH, Kroon JM, Hummelen JC (2007) Increasing the open circuit voltage of bulk-heterojunction solar cells by raising the LUMO level of the acceptor. Org Lett 9:551–554. doi:10.1021/ol062666p
15. Zhou H, Yang L, Price SC, Knight KJ, You W (2010) Enhanced photovoltaic performance of low-bandgap polymers with deep LUMO levels. Angew Chem Int Ed 49:7992–7995. doi:10.1002/anie.201003357
16. Yang C, Cho S, Heeger AJ, Wudl F (2009) Heteroanalogues of PCBM: N-bridged imino-PCBMs for organic field-effect transistors. Angew Chem Int Ed 48:1592–1595. doi:10.1002/anie.200805228
17. Nagamachi T, Takeda Y, Nakayama K, Minakata S (2012) Selective functionalization of fullerenes with N, N-dihalosulfonamides as an N1 unit: versatile syntheses of Aza[60]fulleroids and Aziridino[60]fullerenes and their application to photovoltaic cells. Chem Euro J 18:12035–12045. doi:10.1002/chem.201201680
18. Creegan KM, Robbins JL, Robbins WK, Millar JM, Sherwood RD, Tindall PJ, Cox DM, Smith AB III, McCauley JP Jr, Jones DR, Gallagher RT (1992) Synthesis and characterization of $C_{60}O$, the first fullerene epoxide. J Am Chem Soc 114:1103–1105. doi:10.1021/ja00029a058
19. Chiang LY, Swirczewski JW, Hsu CS, Chowdhury SK, Cameron S (1791–1793) Creegan K (1992) multi-hydroxy additions onto C_{60} fullerene molecules. J Chem Soc Chem Commun. doi:10.1039/C39920001791
20. Baranov V, Hopkinson AC, Bohme DK (1997) Isomer-specific trends with charge state in gas-phase reactions of fullerene cations, C60x + (x = 1–3), with nitromethane and methyl nitrite: polymethoxylation of C60 dications. J Am Chem Soc 119:7055–7060. doi:10.1021/ja962530u
21. Gan L, Huang S, Zhang X, Zhang A, Cheng B, Cheng H, Li X, Shang G (2002) Fullerenes as a tert-butylperoxy radical trap, metal catalyzed reaction of tert-butyl hydroperoxide with fullerenes, and formation of the first fullerene mixed peroxides $C_{60}(O)(OO^tBu)_4$ and $C_{70}(OO^tBu)_{10}$. J Am Chem Soc 124:13384–13385. doi:10.1021/ja027714p
22. Elemes Y, Silverman SK, Sheu C, Kao M, Foote CS, Alvarez MM, Whetten RL (1992) Reaction of C_{60} with dimethyldioxirane–formation of an epoxide and a 1,3-dioxolane derivative. Angew Chem Int Ed 31:351–353. doi:10.1002/anie.199203511

23. Averdung J, Mattay J, Jacobi D, Abraham W (1995) Addition of photochemically generated acylnitrenes to C_{60}. Tetrahedron 51:2543–2552. doi:10.1016/0040-4020(95)00013-X
24. Meier MS, Poplawska M (1996) The addition of nitrile oxides to C_{60}. Tetrahedron 52:5043–5052. doi:10.1016/0040-4020(96)00112-3
25. Jensen AW, Khong A, Saunders M, Wilson SR, Schuster DI (1997) Photocycloaddition of cyclic 1,3-diones to C_{60}. J Am Chem Soc 119:7303–7307. doi:10.1021/ja962509t
26. Bernstein R, Foote CS (1998) Activated carbon as the reagent for the oxidative cyclization of fullerene adducts. Tetrahedron Lett 39:7051–7054. doi:10.1016/S0040-4039(98)01527-5
27. Hashiguchi M, Obata N, Maruyama M, Yeo YS, Ueno T, Ikebe T, Takahashi I, Matsuo Y (2012) $FeCl_3$-mediated synthesis of fullerenyl esters as low-LUMO acceptors for organic photovoltaic devices. Org Lett 14:3276–3279. doi:10.1021/ol301186u
28. Hou J, Chen HY, Zhang S, Chen RI, Yang Y, Wu Y, Li G (2009) Synthesis of a low band gap polymer and its application in highly efficient polymer solar cells. J Am Chem Soc 131:15586–15587. doi:10.1021/ja9064975
29. Aoyagi S, Nishibori E, Sawa H, Sugimoto K, Takata M, Miyata Y, Kitaura R, Shinohara H, Okada H, Sakai T, Ono Y, Kawachi K, Yokoo K, Ono S, Omote K, Kasama Y, Ishikawa S, Komuro T, Tobita H (2010) A layered ionic crystal of polar Li@C60 superatoms. Nat Chem 2:678–683. doi:10.1038/nchem.698
30. Fukuzumi S, Ohkubo K, Kawashima Y, Kim DS, Park JS, Jana A, Lynch VM, Kim D, Sessler JL (2011) J Am Chem Soc 133:15938–15941. doi:10.1021/ja207588c
31. Matsuo Y, Okada H, Maruyama M, Sato H, Tobita H, Ono Y, Omote K, Kawachi K, Kasama Y (2012) Covalently chemical modification of lithium ion-encapsulated fullerene: synthesis and characterization of $[Li^+@PCBM]PF_6^-$. Org Lett 14:3784–3787. doi:10.1021/ol301671n
32. Kawakami H, Okada H, Matsuo Y (2013) Efficient Diels–Alder addition of cyclopentadiene to lithium ion encapsulated [60]fullerene. Org Lett 2013(15):4466–4469. doi:10.1021/ol4020046
33. Ueno H, Kawakami H, Nakagawa K, Okada H, Ikuma N, Aoyagi S, Kokubo K, Matsuo Y, Oshima T (2014) Kinetic study of the Diels–Alder reaction of $Li^+@C_{60}$ with cyclohexadiene: greatly increased reaction rate by encapsulated Li^+. J Am Chem Soc 136:11162–11167. doi:10.1021/ja505952y

Chapter 34
Efficient Organic Devices Based on π-Electron Systems: Comparative Study of Fullerene Derivatives Blended with a High Efficiency Naphthobisthiadiazole-Based Polymer for Organic Photovoltaic Applications

Varun Vohra, Koichi Higashimine, Keisuke Ohdaira, Shogo Tsuzaki, and Hideyuki Murata

Abstract This study focuses on a polymer based on quaterthiophene and naphthobisthiadiazole units (PNTz4T). Three fullerene derivatives, namely, $PC_{61}BM$, $PC_{71}BM$, and $IC_{60}BA$, were selected as potential electron acceptors in regular device architectures. The resulting average PCE are 7.52, 8.52, and 2.58 % for $PC_{61}BM$, $PC_{71}BM$, and $IC_{60}BA$, respectively. Through a careful and systematic study, we investigated the origins of the differences observed in devices'

The work was supported by a Grant-in-Aid for Scientific Research on Innovative Areas (No. 20108012 "pi-Space") from the Ministry of Education, Culture, Sports, Science, and Technology, Japan and by JSPS KAKENHI Grant number 26889029. The authors would like to acknowledge Prof. I. Osaka and Prof. K. Takimiya (CEMS RIKEN) for very constructive discussions on this project and for providing us with the polymer material.

V. Vohra
School of Materials Science, Japan Advanced Institute of Science and Technology, 1-1 Asahidai, Nomi, Ishikawa 923-1292, Japan

Department of Engineering Science, University of Electro-Communications, 1-5-1 Chofugaoka, Chofu, Tokyo 182-8585, Japan
e-mail: varun.vohra@uec.ac.jp

K. Higashimine
Center for Nano Materials and Technology, Japan Advanced Institute of Science and Technology, 1-1 Asahidai, Nomi, Ishikawa 923-1292, Japan

K. Ohdaira
Green Devices Research Center, Japan Advanced Institute of Science and Technology, 1-1 Asahidai, Nomi, Ishikawa 923-1292, Japan

S. Tsuzaki • H. Murata (✉)
School of Materials Science, Japan Advanced Institute of Science and Technology, 1-1 Asahidai, Nomi, Ishikawa 923-1292, Japan
e-mail: murata-h@jaist.ac.jp

T. Akasaka et al. (eds.), *Chemical Science of π-Electron Systems*,
DOI 10.1007/978-4-431-55357-1_34

performances. In particular, the higher Jsc and consequently higher PCE of the $PC_{71}BM$ devices (as compared to $PC_{61}BM$ devices) can be easily explained by the better light-harvesting properties in the visible range of the larger fullerene derivative. Furthermore, we demonstrate that the limiting factor in these devices is the electron collection which is closely related to the crystallinity of the fullerene derivative. The low crystallinity and resulting low electron-transporting properties of $IC_{60}BA$ is at the origins of the low performances of the $IC_{60}BA$-based devices. Through this comparative study, we confirm that developing new materials is the key to remarkably increase the PCE of polymer solar cells. However, in order to obtain PCE over 10 %, a particular attention should be given to material combination, process, and charge balance in the devices.

Keywords Polymer solar cells • Donor-acceptor polymers • Bulk heterojunctions • Fullerene derivatives

34.1 Introduction

Organic photovoltaic (OPV) devices represent one of the most innovative yet challenging technologies studied by various groups in the past decade [1, 2]. In fact, they have the potential to overcome their inorganic analogue both in terms of cost of production and light absorption efficiency [2–4]. However, the low charge-carrier mobilities in organic semiconductors lead to increased recombination and low fill factors (FF) which results in low power conversion efficiencies (PCE) [5]. Therefore, one of the key parameters to obtain highly efficient devices is to simultaneously increase the charge-carrier mobilities of both the electron-donor and acceptor materials. Optimizing the morphology of the active layer in bulk heterojunction OPV can provide such an increase of both charge-transport properties and FF in the devices. The intrinsic properties of the active materials as well as the device process are therefore both of significant importance.

To improve the intrinsic performances of both p- and n-type organic semiconductors, novel molecules have been synthesized [6]. More specifically, the development of wide absorption range polymers with high hole mobility to act as electron-donor materials has proved to be a good strategy to improve OPV PCE [7]. On the other hand, fullerene derivatives are a well-established electron-donor material in polymer-based OPV [6]. Tuning of the optoelectronic properties of fullerene derivatives can however result in original properties such as absorption in the visible and/or higher lowest unoccupied molecular orbital (LUMO) levels which increase the short-circuit current density (Jsc) and open-circuit voltage (Voc) in OPV devices, respectively [8, 9].

Furthermore, the effect of both solvent and thermal annealing as well as the active-layer deposition conditions in OPV has been widely studied [10, 11]. In organic semiconductors, charge-carrier mobility is closely related to the crystallinity of the material [12]. Consequently, annealing and deposition from high-boiling

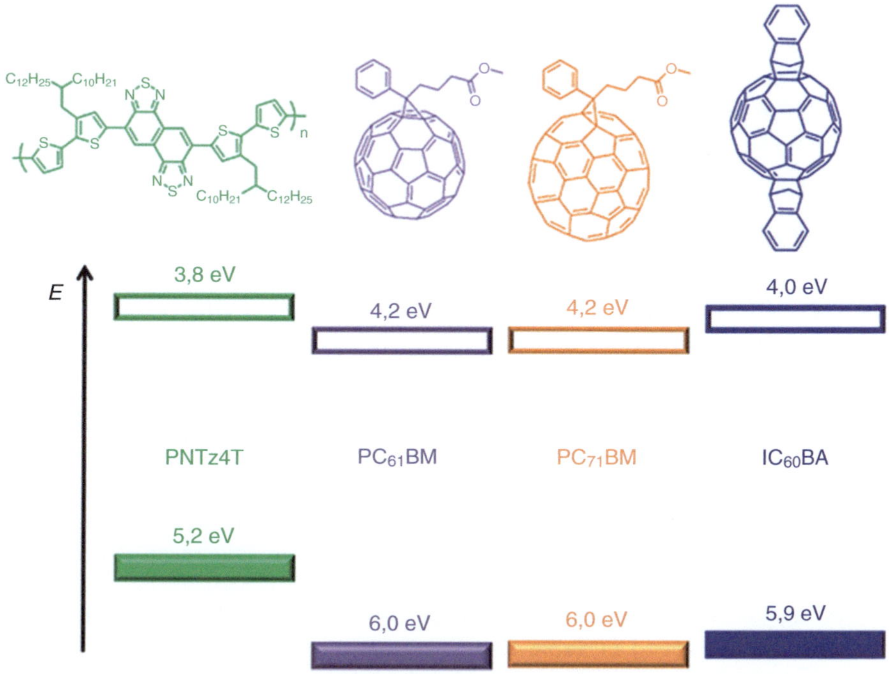

Fig. 34.1 Chemical structure and energetic diagram of the various active molecules

point solvents provide means to increase the charge-carrier mobilities as both donor and acceptor materials form larger crystalline domains. In the case of conjugated polymers, the orientation of the crystallites also plays an important role as charges are more efficiently transported along the $\pi-\pi$ stacking direction [13]. The annealing and film deposition conditions should nonetheless be carefully controlled in order to avoid large phase separation which would decrease the interface between donor and acceptor and lead to low charge-generation efficiencies.

Here, we present a comparative study of various fullerene derivatives blended with a naphthobisthiadiazole-based polymer (PNTz4T, Fig. 34.1) [13]. While the thin films of PNTz4T blended with phenyl-C_{61}-butyric acid methyl ester ($PC_{61}BM$) and phenyl-C_{71}-butyric acid methyl ester ($PC_{71}BM$) exhibit high photovoltaic performances (PCE of 7.5 and 8.5 %, respectively), the combination of the highly efficient p-type polymer with indene-C_{60} bis-adduct ($IC_{60}BA$) barely reaches PCE of 2.6 %. Through a systematic study of the three systems in terms of morphological and electrical characteristics, we demonstrate that the limiting factor in these devices is the electron transport. Furthermore, through the analysis of the data collected from hole- and electron-only devices, we demonstrate that high FF in PNTz4T-based OPV can be achieved when the charge-carrier conductivities are not only high but also well balanced.

This complete study confirms that intrinsic properties of donor and acceptor materials for OPV, even though important, do not necessarily lead to high-performing devices. In fact, the active layers of organic solar cells should be optimized both in terms of morphology and electrical parameters to be able to approach PCE of 10 %.

34.2 Optical Characteristics and Device Performances

34.2.1 Photovoltaic Performances of PNTz4T:Fullerene Derivatives

The photovoltaic devices we fabricate are based on simple regular architectures with the active layers sandwiched between a poly(3,4-ethylenedioxythiophene) poly(styrenesulfonate) (PEDOT:PSS)-covered indium tin oxide (ITO) anode and lithium fluoride (LiF) aluminum (Al) cathode (Fig. 34.2). The active layers consist of a 1:2 ratio of PNTz4T:fullerene derivative blends (chemical structures in Fig. 34.1).

Average device performances for each type of PNTz4T:fullerene derivative blends are summarized in Table 34.1. Their current density-voltage (J–V) characteristics are displayed in Fig. 34.2. Very high FF are obtained for $PC_{61}BM$ and $PC_{71}BM$ devices with values around 65 %. In association with high Jsc and Voc in the order of 0.75 V, we obtain average PCE of 7.52 and 8.52 for PNTz4T:$PC_{61}BM$ and PNTz4T:$PC_{71}BM$, respectively. On the other hand, $IC_{60}BA$ devices have a Jsc which is approximately 2.5 times lower than the two other fullerene derivatives, and

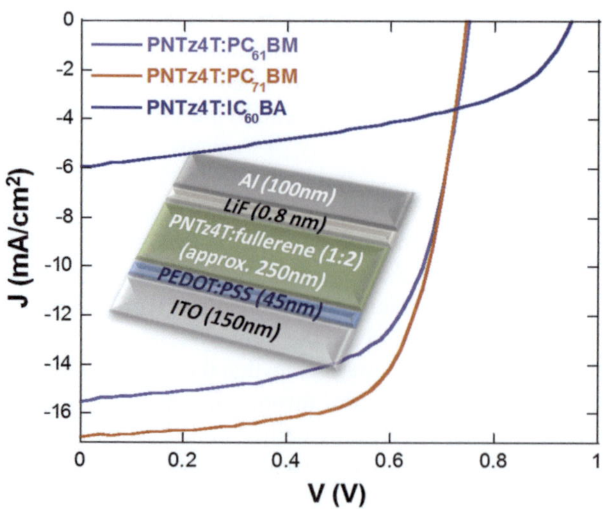

Fig. 34.2 Schematic representation of the device architecture and average J–V curves of PNTz4T-based devices

Table 34.1 Average device performances (8 devices per device type) of PNTz4T-based devices with active layers spin coated from chlorobenzene

Acceptor	Jsc (mA/cm^2)	Voc (V)	FF (%)	PCE (%)
$PC_{61}BM$	15.48	757	64.2	7.52
$PC_{71}BM$	16.98	751	66.8	8.52
$IC_{60}BA$	5.97	956	45.2	2.58

while the Voc increases to 0.96 V, the FF drops to 45 % and consequently, a PCE of only 2.58 % can be achieved.

Active layers in OPV represent a complicated system where various conditions apply to achieve high performances [14]. We will therefore verify which of these conditions are lacking in the PNTz4T:$IC_{60}BA$ devices step by step starting with the initial requirement in OPV light absorption.

34.2.2 Relationship Between Light Absorption and Jsc

Figure 34.3 displays the absorption spectra of the three types of active layers. While $PC_{61}BM$ and $IC_{60}BA$ mainly absorb light in the UV region, $PC_{71}BM$ has an absorption band between 400 and 600 nm (extends up to 700 nm). $PC_{71}BM$ is therefore not only the perfect complimentary of PNTz4T in terms of visible light absorption but its absorption also matches quite well the maximum intensity of the solar spectrum. The higher Jsc (in particular, compared to $PC_{61}BM$) observed in $PC_{71}BM$ devices can therefore partially be explained by this more adequate absorption. We further compare these absorption spectra with incident-photon-to-current efficiency (IPCE) spectra measured from the three types of device.

In Fig. 34.3, we can clearly observe a difference in IPCE between $PC_{61}BM$ and $PC_{71}BM$ devices. More specifically, the wavelength range extending from 350 up to 700 nm confirms that the absorption of $PC_{71}BM$ in the visible is beneficial to increase the charge generation in PNTz4T devices. However, even though in the absorption spectra (not normalized), the peak around 715 nm, corresponding to the maximum of absorption of the polymer, has the same intensity for all three fullerene derivatives, in the IPCE spectra, some differences can be observed. In fact, the IPCE of $PC_{61}BM$, $PC_{71}BM$, and $IC_{60}BA$ devices at 715 nm are 61, 63, and 38 %, respectively, which reflects very well the trend observed in FF.

By performing photoluminescence quenching measurements of the three systems, we have confirmed that when a photon is absorbed by PNTz4T in the three types of active layer, electron/hole pairs can be efficiently generated (99 % of the photoluminescence is quenched). However, we cannot obtain similar information about the charge-generation efficiency when the fullerene derivatives absorb light.

While the absorption and IPCE spectra give us a hint about the differences between $PC_{61}BM$ and $PC_{71}BM$, they do not explain the low Jsc observed in $IC_{60}BA$

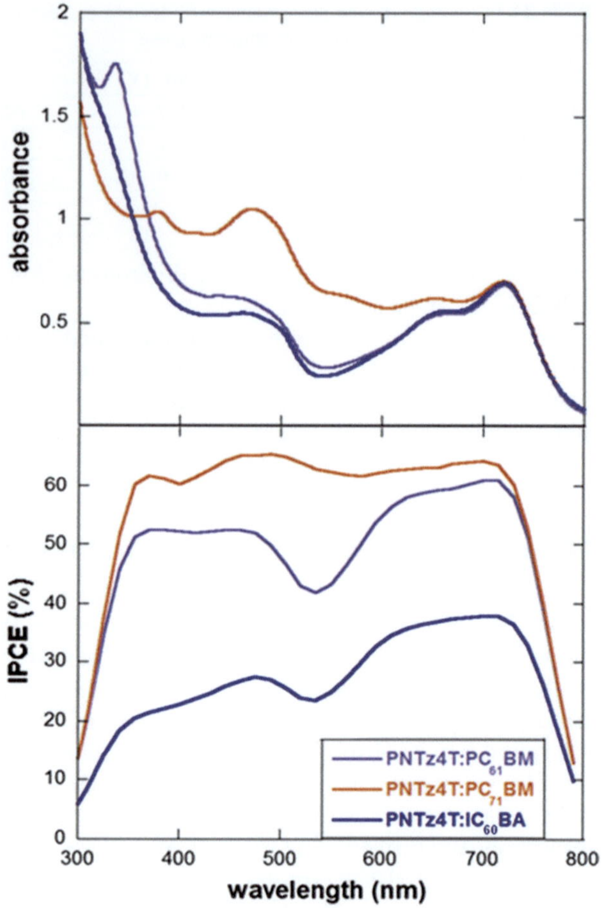

Fig. 34.3 Absorption spectra and IPCE of PNTz4T:fullerene active layers

devices. The low FF suggests that, once charges are generated in PNTz4T:IC_{60}BA films, they cannot be easily collected at the respective electrodes which implies either low or unbalanced charge-carrier collection at the electrodes or that there is no percolation due to morphological issues. To understand the reasons for the low Jsc and FF in PNTz4T:IC_{60}BA devices, we next investigate the thin films' morphologies of the three types of active layers.

34.3 Morphology of PNTz4T:Fullerene Derivatives' Thin Films

34.3.1 PNTz4T Crystallinity and Crystallite Orientation

In a previous study on PNTz4T using two-dimensional X-ray diffraction (2D-XRD), the fact that PNTz4T crystallites undergo a reorientation from edge on to face on upon addition of PC_{61}BM into the films has been demonstrated [13].

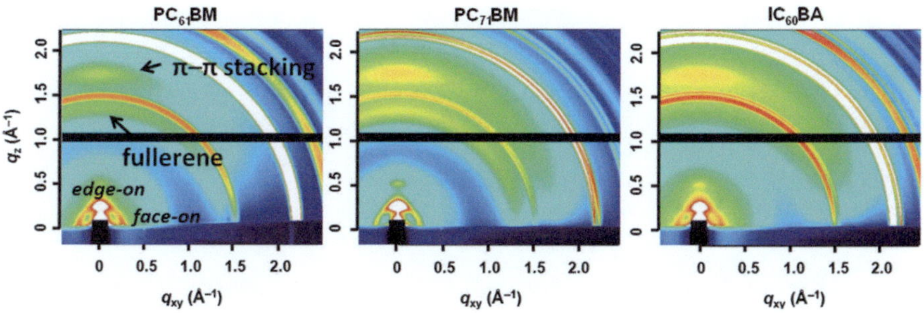

Fig. 34.4 2D-XRD patterns of PNTz4T:fullerene active layers

Figure 34.4 displays the 2D-XRD patterns of the PNTz4T:fullerene derivatives films. It is worth noticing that the diffraction patterns clearly show the presence of crystalline fullerene domains in PNTz4T:$PC_{61}BM$ and PNTz4T:$PC_{71}BM$ films with the crystallinity peak of $PC_{71}BM$ being more intense than that of $PC_{61}BM$. On the other hand, no evidence of crystalline $IC_{60}BA$ can be observed.

Furthermore, a mixture of face-on and edge-on crystallites is detected in the blends for all three fullerene derivatives. However, the face-on-to-edge-on ratio seems to vary between the films depending on the electron acceptor material used. More specifically, in the PNTz4T:$PC_{61}BM$ film, the ratio of face-on crystallites (with respect to the edge-on crystallites) seems to be slightly higher than in the two other active layers. Previous results demonstrated that PNTz4T has a high hole mobility up to 0.56 $cm^2/(V\ s)$ in organic field-effect transistors (OFET) configuration which may be related to the edge-on orientation in the pure polymer films. Indeed, edge-on crystallite thin films favor the horizontal charge transport, while face on promotes the vertical charge transport. We should therefore expect a better hole transport in the PNTz4T:$PC_{61}BM$ films as compared to PNTz4T:$PC_{71}BM$ and PNTz4T:$IC_{60}BA$. Phase separation also constitutes an important aspect to understand whether charges can easily percolate to the respective electrodes and qualitatively estimate the amount of charge recombination which could be expected in these thin films.

34.3.2 Formation of Vertical Donor-Acceptor Gradients for Efficient Charge Percolation

Energy-dispersive X-ray spectroscopy (EDS) elemental mapping provides such information by probing the sulfur concentration through the cross section of the various devices (Fig. 34.5) [15–17]. In Fig. 34.5, we compare the EDS profiles obtained with the three different PNTz4T:fullerene derivative active layers.

On the one hand, even though the vertical concentration gradient seems to be present in all types of active layers, it is more discernible in the case of

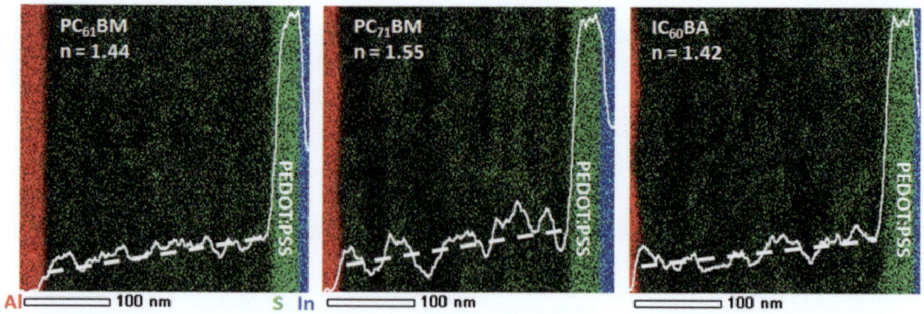

Fig. 34.5 EDS elemental mapping of the cross sections of the various devices. The sulfur counts are represented by the *white line*. *n* corresponds to the ideality factor extracted from the dark J–V curve fitting

PNTz4T:PC_{71}BM films. On the other hand, the PNTz4T:PC_{71}BM films are the only ones displaying an alternating donor-rich/acceptor-rich layered structure. Due to the high crystallinity of PNTz4T, annealing the active layers induces large-scale phase separation which results in a drop in device efficiencies. At room temperature, PC_{71}BM is known to crystallize faster than PC_{61}BM and IC_{60}BA which may be the reason for such alternating structure observed only in the PNTz4T:PC_{71}BM films [18]. In fact, the slow crystallization of IC_{60}BA is one of the main reasons why high performances in IC_{60}BA-based devices can only be achieved after annealing the active layers at 140 °C.

While the concentration gradient in these thin films is beneficial for regular-structure devices, the alternating structure observed in PNTz4T:PC_{71}BM films may lead to more charge recombination within the active layer. Indeed, by fitting the J–V curves obtained in the dark using a two-diode model following a procedure previously described [17], we extrapolated the ideality factor (n) in the photovoltaic cells. The values obtained are 1.44; 1.55 and 1.42 for PNTz4T:PC_{61}BM, PNTz4T:PC_{71}BM and PNTz4T:IC_{60}BA devices, respectively. While the variation between PC_{61}BM and IC_{60}BA is negligible, n_{PC71BM} is slightly higher which implies more recombination in the PNTz4T:PC_{71}BM devices. As expected, the layered structure therefore increases the probability for charge recombination. Once again, this is detrimental for device performances and should therefore lower the FF of the PNTz4T:PC_{71}BM devices.

In the following section, we will demonstrate that even though crystallite orientation and charge recombination may play a role, the limiting factor in these devices are the electron transport properties of the fullerenes and the balance in charge transport to the electrodes.

34.4 Charge Collection in PNTz4T:Fullerene Derivatives' Devices

34.4.1 Crystallite Orientation and Hole Conductivity

To evaluate the hole conductivities in our solar cells, we prepare hole-only devices with the following architecture: ITO/PEDOT:PSS/PNTz4T:fullerene derivatives/gold (Au). Using a simple approach, we calculate the hole conductivity (σ_+) from the dark J–V of the various hole-only devices at a voltage of 0.58 V. This corresponds to the voltage where maximum conversion efficiency can be obtained for PNTz4T:PC_{61}BM/PC_{71}BM active layers. Similar trends can be observed at a voltage of 0.72 V which corresponds to the maximum conversion efficiency voltage for PNTz4T:IC_{60}BA devices.

Table 34.2 summarizes the hole conductivities obtained in the various devices. We can clearly observe that $\sigma_{+,\text{PNTz4T:PC61BM}}$ is higher than the $\sigma_{+,\text{PNTz4T:PC71BM}}$ and $\sigma_{+,\text{PNTz4T:IC60BA}}$. This may result from the previously mentioned crystal orientation and how it influences the charge-transport properties with face-on oriented crystallites prone to increase the vertical hole-transport properties in the thin films. As we will now see, this effect of crystallinity over the charge-transport properties becomes even more obvious in electron-only devices.

34.4.2 Electron Conductivities of the Various Fullerene Derivatives in the Blends

Electron-only devices (ITO/titanium oxide (TiOx)/PNTz4T:fullerene derivatives/LiF/Al) were fabricated to obtain complimentary information about the electron conductivity in our devices. Similarly to the hole-only devices, we calculate the electron conductivity at 0.58 V for all active layers. In Table 34.2, the resulting electron conductivities (σ_-) display remarkable differences when changing the fullerene acceptors. Indeed, $\sigma_{-,\text{PNTz4T:PC71BM}}$ is one order of magnitude higher than $\sigma_{-,\text{PNTz4T:PC61BM}}$ which is almost one order of magnitude higher than $\sigma_{-,\text{PNTz4T:IC60BA}}$. Furthermore, as we previously intuited, electron conductivity in these thin films may be directly related to the crystallinity of the fullerene

Table 34.2 Hole and electron conductivities extracted from the various hole-/electron-only devices based on PNTz4T:fullerene active layers

Active layer	σ_+ (10^{-8} S/m)	σ_- (10^{-8} S/m)	σ_+/σ_-
PNTz4T:PC_{61}BM	513.6	1.44	0.003
PNTz4T:PC_{71}BM	124.3	14.4	0.116
PNTz4T:IC_{60}BA	102.7	0.25	0.002

derivative. The 2D-XRD profiles in Fig. 34.4 suggest that $PC_{71}BM$ is more crystalline than $PC_{61}BM$ which is well correlated with the trends observed in electron conductivities.

$IC_{60}BA$, on the other hand, is known to require higher annealing temperatures in order to crystallize and show high conduction properties. Annealing our PNTz4T:fullerene derivative films leads to large phase separation and is detrimental to the device performances. However, evidence of the increase in electron mobilities of $PC_{61}BM$- and $IC_{60}BA$-based blends was obtained by annealing the three types of active layers at relatively low temperatures (data not provided).

We now have all the elements in terms of charge-transport properties to fully understand both the high performances obtained with $PC_{61}BM/PC_{71}BM$ and the low performances from $IC_{60}BA$.

34.4.3 Limiting Factor in the PNTz4T:Fullerene Derivative Blends

In Fig. 34.6, a schematic representation of the various systems summarizes the charge-collection efficiency and morphology in the active layers. The results of charge conductivity calculations clearly show that the limiting factor in all devices is the electron transport. Indeed, the hole-transport properties vary slightly due to crystallite orientation but the main differences can be observed in electron conductivities which reflect the crystallinity of the fullerene acceptors.

While a higher hole conductivity can be observed in PNTz4T:$PC_{61}BM$ films compared to the one in PNTz4T:$PC_{71}BM$ active layers, the one order of magnitude lower electron conductivity and the resulting unbalanced charge collection leads to a slightly lower FF in the $PC_{61}BM$-based devices. However, the electron transport properties of $PC_{61}BM$ are still high enough to give rise to FF reaching over 60 %. On the other hand, similarly unbalanced charge collection is found in $IC_{60}BA$ devices with, in addition, a one order of magnitude lower electron conductivity which results in a large decrease of the FF (45 % in average).

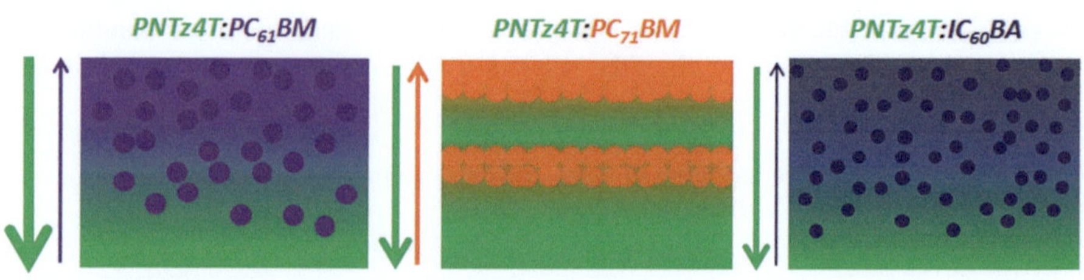

Fig. 34.6 Schematic representation of the three PNTz4T:fullerene derivatives. The width of the *arrow* corresponds to the efficiency of the charge transport in the displayed direction

This complete study of the morphology and electrical and charge-transport characteristics of PNTz4T-based photovoltaic devices demonstrates that to obtain highly functional devices, much attention should be given to the material selection and to the process. In fact, in the PNTz4T:$PC_{71}BM$ system, the intrinsic properties of the materials, together with a suitable morphology, provide means to have efficient light absorption and charge separation, as well as efficient and balanced charge transport to the respective electrodes.

34.5 Conclusion

Using a p-type semiconducting polymer (PNTz4T) with wide absorption range and high hole-transport properties, we fabricate devices based on a variety of fullerene derivatives. Through a systematic study of the PNTz4T:fullerene blends, we observed the influence of crystallite orientation and vertical phase separation over the performances of the various devices. In particular, in PNTz4T:$PC_{71}BM$, which displays a higher crystallinity of the fullerene derivative at room temperature, we obtain high electron and hole conductivities which results in devices with high FF. Furthermore, the alternating electron-donor-rich and electron acceptor-rich layered structure increases the probability for charge recombination. The PNTz4T:$PC_{71}BM$ system might therefore lead to even higher FF and performances if such layered structures can be avoided while maintaining its other electrical and morphological properties.

On the other hand, PNTz4T:$PC_{61}BM$ films display a higher hole conductivity due to the higher concentration of face-on polymer crystallites in the thin films. However, the lower electron conductivity becomes the limiting factor in these devices. Such unbalanced charge transport results in a slight decrease of the FF when compared to the PNTz4T:$PC_{71}BM$ system, even though lower charge recombination occurs in PNTz4T:$PC_{61}BM$-based thin films.

A similar unbalanced charge transport can be observed in PNTz4T:$IC_{60}BA$ films with lower values of both electron and hole conductivities. The lower absorption from $IC_{60}BA$ in the visible is therefore not the only explanation for the large decrease in Jsc observed in PNTz4T:$IC_{60}BA$ devices. $IC_{60}BA$ cannot crystallize at room temperature which remarkably decreases its electron transport properties and results in very low charge-collection efficiencies. The decrease in FF and Jsc of PNTz4T:$IC_{60}BA$ devices are a direct consequence of this low charge-collection efficiency.

In conclusion, the complete study to understand the origins of the high efficiencies in PNTz4T:fullerene derivatives' devices presented here highly contributes to OPV research by providing essential information for researchers aiming to develop new materials for efficient OPV devices.

34.6 Experimental Section

34.6.1 Device Preparation

For regular structures as well as hole-only devices, PEDOT:PSS (Clevios P Al4083) was spin coated at a speed of 4000 rpm for 30 s on pre-cleaned and UV-ozone-treated ITO substrates. The measured thickness of PEDOT:PSS was approximately 40 nm after a baking process (5 min at 200 °C). On the other hand, the TiOx layer for electron-only devices was prepared following previously reported fabrication methods [19].

PNTz4T (synthesized as described previously) [13] and the fullerene derivatives were blended with a 1:2 ratio in chlorobenzene. PC61BM and PC71BM were purchased from Luminescence Technology Corp., while IC60BA was obtained from 1-material. The polymer concentration was kept at 6 mg/ml and the solutions were stirred at 100 °C for 1 h prior to spin coating. The blend solution was spin coated on the substrates at 400 rpm for 20 s followed by 1500 rpm for 5 s. The resulting films were left to dry in vacuum overnight to ensure that solvent traces were completely removed. The active-layer thickness varies between 280 and 360 nm. The regular architecture and electron-only devices were finalized by evaporating the cathode consisting of 0.8 nm of LiF and 100 nm of Al, while a 100 nm thick Au layer was deposited on the active layers of hole-only devices.

34.6.2 Morphological Characterizations

Film thicknesses were measured by scratching the film and measuring the height difference across the scratch using an atomic force microscope from Keyence (Nanoscale Hybrid Microscope VN-8000). Some variations in active-layer thickness can be observed for each type of blends. Absorption spectra were measured using a Jasco V670 UV/vis spectrometer. PL spectra were collected using a photonic analyzer (Hamamatsu C747). The samples were excited with a xenon lamp (Asahi Spectra MAX-303). The displayed PL spectra were normalized by the absorption intensity. EDS for elemental analysis was carried out with a scanning transmission electron microscope, JEM-ARM200F, from JEOL. The acceleration voltage and probe current were 200 kV and 59 pA, respectively, to obtain 256×256 pixel images with a dwell time of 1 ms/pixel. Each image was accumulated for 120 s to improve the signal to noise ratio.

2D-XRD experiments were conducted at the SPring-8 on beamline BL46XU. The sample was irradiated with an X-ray energy of 12.39 keV ($\lambda = 1$ Å) at a fixed incident angle on the order of 0.12° through a Huber diffractometer. The 2D-XRD patterns were recorded with a two-dimensional image detector (Pilatus 300 K).

34.6.3 Device Characterization

The J–V characteristics of the encapsulated devices were measured in air, using a Keithley 2400 SourceMeter and a solar simulator at one sun light irradiance (AM 1.5, 100 mW/cm^2) at room temperature. Regular structures and hole-only devices are characterized as prepared. IPCE measurements were obtained using encapsulated devices and measuring Isc and input power for each monochromated incoming excitation.

References

1. Deibel C, Dyakonov V (2010) Polymer–fullerene bulk heterojunction solar cells. Rep Prog Phys 73:096401. doi:10.1088/0034-4885/73/9/096401
2. Scharber MC, Sariciftci NS (2013) Efficiency of bulk-heterojunction organic solar cells. Prog Polym Sci 38:1929–1940. doi:10.1016/j.progpolymsci.2013.05.001
3. Bagher AM (2014) Comparison of organic solar cells and inorganic solar cells. Int J Renew Sustain Energy 3:53–58. doi:10.11648/j.ijrse.20140303.12
4. Kalowekamo J, Baker E (2009) Estimating the manufacturing cost of purely organic solar cells. Sol Energy 83:1224–1231. doi:10.1016/j.solener.2009.02.003
5. Savoie BM, Tan S, Jerome JW, Shu C-W, Ratner MA, Marks TJ (2012) Ascertaining the limitations of low mobility on organic solar cell performance. In: 2012 15th international workshop on computational electronics (IWCE). doi: 10.1109/IWCE.2012.6242859
6. Wang ZX, Zhang FJ, Wang J, Xu XW, Wang J, Liu Y, Xu Z (2012) Organic photovoltaic cells: novel organic semiconducting materials and molecular arrangement engineering. Chin Sci Bull 57:4143–4152. doi:10.1007/s11434-012-5202-3
7. Bundgaard E, Krebs FC (2007) Low band gap polymers for organic photovoltaics. Sol Energy Mater Sol Cells 91:954–985. doi:10.1016/j.solmat.2007.01.015
8. Bijleveld JC, Zoombelt AP, Mathijssen SGJ, Wienk MM, Turbiez M, de Leeuw DM, Janssen RAJ (2009) Poly(diketopyrrolopyrrole-terthiophene) for ambipolar logic and photovoltaics. J Am Chem Soc 131:16616–16617. doi:10.1021/ja907506r
9. Li H, Wang J (2012) Layer-by-layer processed high-performance polymer solar cells. Appl Phys Lett 101:263901. doi:10.1063/1.4773515
10. Yang X, Uddin A (2014) Effect of thermal annealing on P3HT:PCBM bulk-heterojunction organic solar cells: A critical review. Renew Sust Energ Rev 30:324–336. doi:10.1016/j.rser.2013.10.025
11. Li G, Yao Y, Yang H, Shrotriya V, Yang G, Yang Y (2007) "Solvent annealing" effect in polymer solar cells based on poly(3-hexylthiophene) and methanofullerenes. Adv Funct Mater 17:1636–1644. doi:10.1002/adfm.200600624
12. Yang H, Shin TJ, Yang L, Cho K, Ryu CY, Bao Z (2005) Effect of mesoscale crystalline structure on the field-effect mobility of regioregular poly(3-hexyl thiophene) in thin-film transistors. Adv Funct Mater 15:671–676. doi:10.1002/adfm.200400297
13. Osaka I, Shimawaki M, Mori H, Doi I, Miyazaki E, Koganezawa T, Takimiya K (2012) Synthesis, characterization, and transistor and solar cell applications of a naphthobisthiadiazole-based semiconducting polymer. J Am Chem Soc 134:3498–3507. doi:10.1021/ja210687r
14. Bredas J-L, Norton JE, Cornil J, Coropceanu V (2009) Molecular understanding of organic solar cells: the challenges. Acc Chem Res 42:1691–1699. doi:10.1021/ar900099h
15. Vohra V, Higashimine K, Murakami T, Murata H (2012) Addition of regiorandom poly(3-hexylthiophene) to solution processed poly(3 hexylthiophene):[6,6]-phenyl-C61-butyric acid methyl ester graded bilayers to tune the vertical concentration gradient. Appl Phys Lett 101:173301. doi:10.1063/1.4761998

16. Vohra V, Arrighetti G, Barba L, Higashimine K, Porzio W, Murata H (2012) Enhanced vertical concentration gradient in rubbed P3HT:PCBM graded bilayer solar cells. J Phys Chem Lett 3:1820–1823. doi:10.1021/jz300710a
17. Vohra V, Higashimine K, Tsuzaki S, Ohdaira K, Murata H (2014) Formation of vertical concentration gradients in poly(3-hexylthiophene-2,5-diyl): Phenyl-C61-butyric acid methyl ester-graded bilayer solar cells. Thin Solid Films 554:41–45. doi:10.1016/j.tsf.2013.05.171
18. Mayer AC, Toney MF, Scully SR, Rivnay J, Brabec CJ, Scharber M, Koppe M, Heeney M, McCulloch I, McGehee MD (2009) Bimolecular crystals of fullerenes in conjugated polymers and the implications of molecular mixing for solar cells. Adv Funct Mater 19:1173–1179. doi:10.1002/adfm.200801684
19. Kuwabara T, Nakayama T, Uozumi K, Yamaguchi T, Takahashi K (2008) Highly durable inverted-type organic solar cell using amorphous titanium oxide as electron collection electrode inserted between ITO and organic layer. Sol Energy Mater Sol Cells 92:1476–1482. doi:10.1016/j.solmat.2008.06.012

Chapter 35
Solution-Processed Organic Thin-Film Transistors

Jun Takeya

Abstract Organic transistors are MOS-type field-effect transistors in which an organic semiconductor is used as the active layer. Some organic semiconducting compounds are highly soluble in certain solvents, so that it becomes possible to form a semiconductor thin film by solution coating. Thus, there are considerable expectations for innovative industrial applications, in which low-cost, large-area devices can be produced on plastic films using printing technology. In this report, I introduce the physics of carrier transport in an intermolecularly delocalized electronic state, which is a key factor for realizing high-performance organic transistors. I also describe prospects for the industrial application of printable single-crystal organic semiconductors with mobilities exceeding 10 cm^2/Vs.

Keywords Organic semiconductor • Organic transistor • Organic thin-film transistor • Organic single-crystal transistor • Solution crystallization

35.1 Introduction

Organic field-effect transistors (OFETs), in which an organic semiconductor is used as the active layer, are considerably expected for practical industrial applications. Their advantage is that a wide variety of low-cost, large-area devices can be manufactured in small quantities using printing technology, as some organic semiconducting compounds are highly soluble in common solvents. With inorganic semiconductors, high-temperature processes in the range of 300–1000 °C are necessary. In contrast, with organic semiconductors, low-temperatures processes in the range of room temperature to a few hundred degrees Celsius can be used. Another benefit of organic semiconductors is that they offer excellent mechanical flexibility.

J. Takeya (✉)
Graduate School of Frontier Sciences, The University of Tokyo, 5-1-5, Kashiwanoha, Kashiwa, Chiba 277-8561, Japan
e-mail: takeya@k.u-tokyo.ac.jp

T. Akasaka et al. (eds.), *Chemical Science of π-Electron Systems*,
DOI 10.1007/978-4-431-55357-1_35

Thus, research and development are being carried out toward the realization of new devices such as flexible displays and plastic electronic tags, by fabricating organic semiconductor circuits on plastic substrates.

Generally, the binding energy between atoms in inorganic semiconductors is several electron volts, whereas in low-molecular-weight organic semiconductors, the energy associated with molecular aggregation is more than one order of magnitude smaller. In the absence of a chemical change, they sublime at a few hundred degrees even under ordinary pressures. This makes it easy to deposit organic semiconductors on flexible substrates such as plastics. In particular, coating with π-conjugated molecules dissolved in an organic solvent at around room temperature offers excellent productivity, so this is directly linked to the development of low-cost semiconductor devices. The term "printed electronics" is also used for this process, and it is attracting a lot of attention from a wide range of industry sectors, including the electronics and printing industries.

Organic semiconductors composed of molecules with low molecular weights form soft solids due to weak intermolecular forces between generally a single type of π-conjugated molecules. This allows them to be prepared by simple methods at temperatures close to room temperature and provides them with mechanical flexibility. This is in contrast with inorganic semiconductors as typified by silicon, in which strong covalent bonds hold the atoms together, forming a hard solid. Presently, these unique semiconductors made of organic materials are drawing a lot of attention even from industry. However, the carrier transport in a "soft solid" is not well understood at present. Recently, progress has been made in the development of organic semiconductor transistors with high mobilities of the order of 10 cm^2/Vs. Since this is one order of magnitude higher than the value of about 1 cm^2/Vs for conventional organic semiconductors, it raises expectations for next-generation semiconductor devices. In this report, we focus on high-mobility organic semiconductors and investigate the mechanisms for achieving high-mobility carrier transport, utilizing the Hall effect, which is produced by the action of a magnetic field on a current.

In the case of silicon, which is strongly covalent, electrons are delocalized among the orbitals of covalent bonds as shown in Fig. 35.1a, and band conduction takes place. The energy gain through delocalization is several electron volts, which is the same as the binding energy. The effective mass of electrons is a few tenths of the mass of free electrons, depending upon the bandwidth. In an organic semiconductor, however, where molecules are held together by weak intermolecular forces, the degree of orbital overlap is much smaller and is more strongly affected by molecular vibrations even at room temperature. Such vibrations have a detrimental effect on carrier transport (Fig. 35.1b). In other words, although the softness of the material makes it flexible and allows easy low-temperature fabrication, it also makes it difficult to produce devices that exhibit both high mobility and high performance. In order to achieve this, it is first necessary to understand microscopic carrier transport.

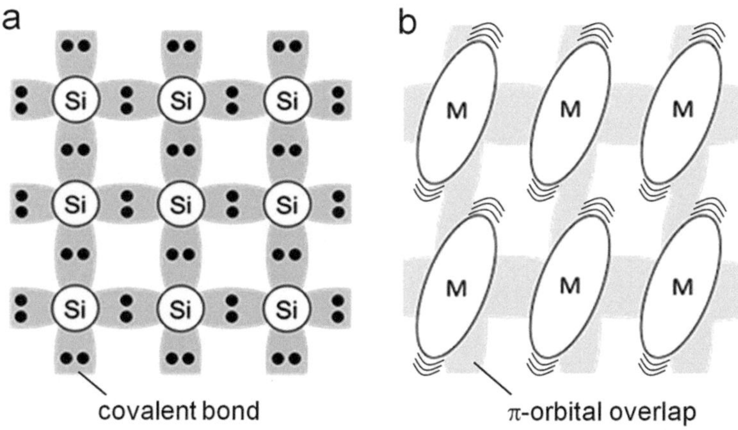

Fig. 35.1 Carrier transport paths in (**a**) silicon and (**b**) an organic semiconductor. In the organic semiconductor, strong vibrations of the organic molecules (M) occur even at room temperature

35.2 Hall Measurements on High-Mobility Organic Semiconductor Transistors

35.2.1 High-Mobility Organic Transistors

An organic transistor has a structure similar to that shown in Fig. 35.2. Carriers are injected by applying an electric field to the gate-insulator film. When a gate voltage V_G is applied between the gate electrode and the organic semiconductor active layer, an electric field appears in the gate-insulating layer, which gives rise to a buildup of charges on the surface of the organic semiconductor facing the gate electrode. Since the carriers injected into the organic semiconductor can move, if a drain voltage V_D is applied between the source and drain electrodes, a drain current I_D is produced when V_G is applied. The charge per unit area Q that is introduced into the organic semiconductor is given by $Q = C_i\,(V_G - V_{th})$, where C_i is the capacitance of the gate-insulator film and V_{th} represents the threshold voltage. The conductivity σ can then be expressed as $\sigma = Q\mu = ne\mu = \mu C_i\,(V_G - V_{th})$, where μ is the mobility and e is the elementary charge. Accordingly, the mobility μ can be determined from the change in σ with V_G.

The rate of change of I_D with respect to V_G is referred to as the transconductance and represents the degree of amplification to the output current. It is directly related to μ because $I_D/V_D \propto \sigma$.

Before the year 2000, the reported mobility values for organic transistors (organic FETs) were not very high, the largest being about 1 cm^2/Vs for a pentacene (Fig. 35.3a) polycrystalline thin film that was prepared by vacuum evaporation [1]. In many polycrystalline thin films, the mobility showed a tendency to decrease

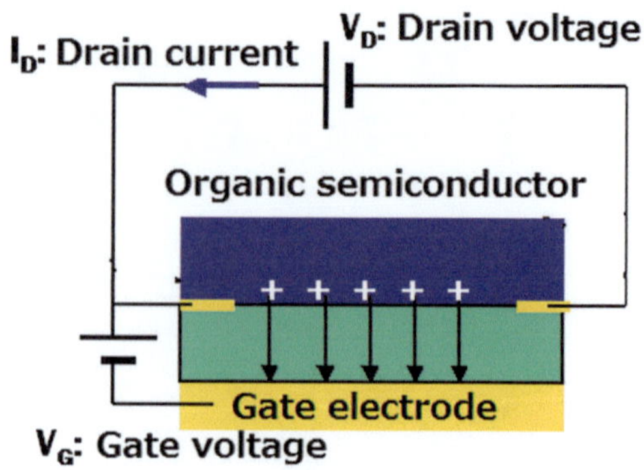

Fig. 35.2 Structure of an organic transistor

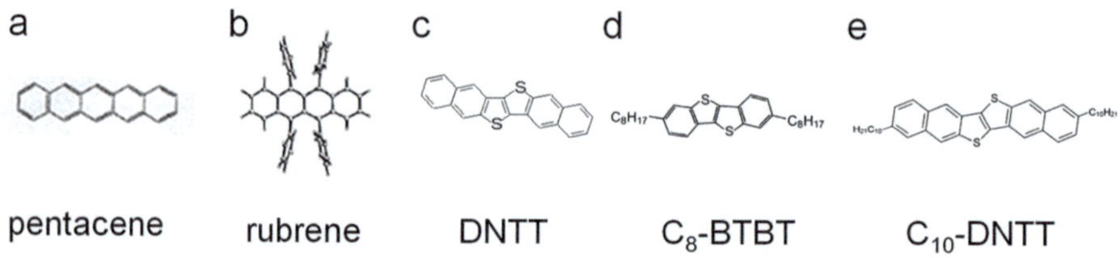

Fig. 35.3 High-mobility organic semiconductor compounds used in this study

with temperature. For this reason, the mobility was often considered in terms of a modified version of the Marcus equation [2] represented by

$$\mu = \frac{2\pi}{\hbar}\frac{ea^2}{kT}\frac{t^2}{\sqrt{4\pi\lambda kT}}\exp\left(-\frac{(\lambda/4 - t)}{kT}\right), \tag{35.1}$$

where $\hbar$ is the Planck constant, k is the Boltzmann constant, e is the elementary charge, a is the intermolecular distance, T is the temperature, t is the intermolecular charge-transfer integral, and λ is the reorganization energy. For example, when the carriers are holes, conduction is thought to occur by hopping transport. In this model, the holes become localized at π-conjugated molecules, causing them to become cations, and the holes sequentially jump to neighboring molecules. Thus, a neighboring molecule must be oxidized in each hopping step. The amount of energy required is obtained by subtracting the energy associated with the charge-transfer integral t due to overlap of neighboring molecular orbitals, from the reorganization energy λ due to the molecular deformation. If this energy is positive, the temperature change is thermal-activation type.

In the present century, numerous organic transistors with mobilities exceeding 1 cm^2/Vs have been reported, starting with single-crystal organic semiconductor transistors [3] and including transistors based on newly synthesized organic semiconductors [4–7]. For a rubrene (Fig. 35.3b) single-crystal transistor, the mobility was found to increase at temperatures lower than room temperature, which led to the suggestion that carrier transport was occurring through band conduction, in which carriers are delocalized between molecules [8]. It has been argued for some time that in bulk ultrahigh-purity organic semiconductor crystals, photoexcited carriers undergo band conduction, based on the temperature dependence of the mobility determined by time-of-flight measurements [9]. There has been increased interest in clarifying the transport mechanism for carriers in actual high-mobility transistors, which accumulate at the interface between the organic semiconductor and the gate insulator.

35.2.2 Hall Measurements on OFETs

Despite the fact that the Hall coefficient is one of the most basic carrier transport properties, a considerable amount of time passed before it was measured for organic transistors. In inorganic transistors such as those made from silicon, the carrier concentration and mobility have been accurately determined using Hall measurements, and important information concerning the conduction mechanism has been obtained for amorphous materials. Therefore, the development of effective measurement methods for organic FETs has been strongly desired. In past organic thin-film transistors, the conductivity was several orders of magnitude smaller than that for inorganic semiconductors and the impedance was at least 1 MΩ. As a result, satisfactory detection sensitivity could not be achieved, making measurements difficult.

In 2005, for the first time, Hall measurements using a high-mobility rubrene single-crystal FET were achieved [10, 11]. The measurement precision has subsequently been improved, and the technique has since been applied to a variety of high-mobility organic semiconductors [12–15]. In the present report, results will be presented for a range of high-mobility organic semiconductors used in organic FETs, for which Hall measurements have already been carried out. The carrier transport mechanism will then be discussed. In addition to pentacene and rubrene, these semiconductors are dinaphtho-[2,3-b:2′,3′-f]thieno[3,2-b]thiophene (DNTT), 2,7-dioctyl[1]benzothieno[3,2-b][1]benzothiophene (C_8-BTBT), and 2,9-didecyldinaphtho[2,3-b:2′,3′-f]thieno[3,2-b]thiophene (C_{10}-DNTT), which were recently synthesized by Takimiya et al. [5–7].

The Hall effect occurs due to the Lorentz force exerted by a magnetic field on a continuous flow of carriers (current); therefore, the presence of delocalized carriers is assumed. Thus, when the dominant carrier transport mechanism is hopping conduction, it is extremely difficult to measure the electromotive force due to the Hall effect. On a microscopic level, the electromotive force occurs along the direction perpendicular to both the current I_D and magnetic field B, based on the primary term in the Hamiltonian, in which the wavenumber operator $\widehat{\mathbf{k}}$ for an electron with an effective mass of m^* and the vector potential $\mathbf{A}$ are coupled, $\frac{1}{2m^*}\left(\hbar\widehat{\mathbf{k}} - e\mathbf{A}\right)^2$, which represents the motion of a charged particle [16]. The Hall coefficient R_H, which is determined from the transverse voltage (Hall voltage) V_H, using the expression $R_H = V_H/I_D B$, is linked to the charge per unit area Q, i.e., ne by the relation $R_H = 1/ne$ under conditions of low inelastic scattering, when the

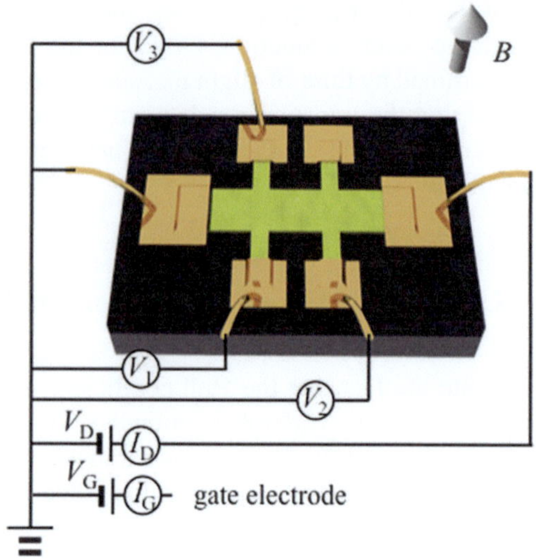

Fig. 35.4 Experimental setup for Hall measurements

carriers are free-electron-like and the state defined by the wavenumber **k** is realized. On the other hand, in the case of hopping conduction in which electrons propagate by a discontinuous tunneling process, the only influence of the magnetic field is to interfere with the hopping pathways. Therefore, V_H is extremely small, if it can be measured at all. In fact, for low-mobility amorphous silicon, V_H is only 1/10 of that for free-electron-like transport [17]. Thus, it is possible to investigate the carrier conduction mechanism using Hall measurements, and such measurements were carried out for various high-mobility organic semiconductors.

In Fig. 35.4, an experimental setup for high-sensitivity Hall measurements is shown [10, 13–16]. An organic FET is first fabricated by patterning a single-crystal or polycrystalline thin-film organic semiconductor on a gate-insulating layer such as silicon dioxide. In addition to the source and drain electrodes for current flow, four electrodes are formed for accessing the central section of the channel so that the Hall voltage and four-terminal conductivity can be measured simultaneously. A current is passed between the source and drain, while a magnetic field is applied perpendicular to the conduction channel, and the Hall voltage and the potential drop in the channel are measured.

35.3 Hall Measurements on Organic Single-Crystal Rubrene FET

Since no grain boundaries exist in a single-crystal FET, the intrinsic carrier transport coefficient for the organic semiconductor layer can be measured. Since a rubrene single-crystal FET was the first organic FET for which Hall measurements were carried out, we will begin by describing the results for this device [10, 13].

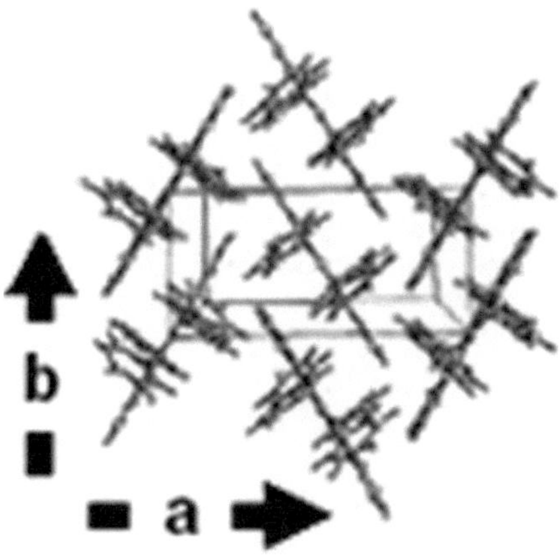

Fig. 35.5 Crystal structure of rubrene single crystal

A single crystal of rubrene was prepared by the physical vapor transport (PVT) method. This is a simple and commonly used method in which a temperature gradient is set up in a tubular furnace, the raw material is sublimed in the high-temperature section and transported to the low-temperature section by a flow of argon gas, and the organic molecules then become crystallized. This method produces a crystal with the structure shown in Fig. 35.5, where the molecules are arranged two dimensionally on the surface. The rubrene molecules constitute a conjugated electron system, in which the π electrons in the four central benzene rings become delocalized, and the molecular plane is perpendicular to the crystal surface. Accordingly, when the plate is viewed from the top, the molecular planes appear as straight lines running in directions inclined approximately 30° to the b-axis. Intermolecular electron transfer occurs due to the π orbitals that extend parallel to the surface. This in-plane conductivity offers advantages because the interface with the gate insulator can be formed along this plane. In this study, a single-crystal transistor was fabricated by selecting a thin rubrene single crystal (thickness of 1 μm or less) and laminating it onto the SiO_2 gate insulator by electrostatic attraction.

Figure 35.6 shows the results of the Hall measurements. The reciprocal of the Hall coefficient $1/R_H$ is plotted with respect to the gate voltage V_G, and it is compared with the charge per unit area Q estimated from the capacitance of the gate-insulating layer and V_G. It can be seen that the two sets of values are in good agreement and increase with negative gate voltage. This implies that holes injected by the field effect do not become localized in the molecules but are widely distributed among the molecules. From the dependence of the four-terminal conductivity on the gate voltage, also shown in Fig. 35.6, the mobility in this device was about 8 cm^2/Vs. The high degree of carrier delocalization implies that the transistor undergoes band-type conduction at room temperature. Based on the results of Hall measurements, Podzorov et al. also reported band-type conductivity for a rubrene single-crystal transistor at room temperature [11].

Carrier delocalization in rubrene single-crystal FETs was later also identified using other experimental methods. Li et al. reported low-energy Drude-type

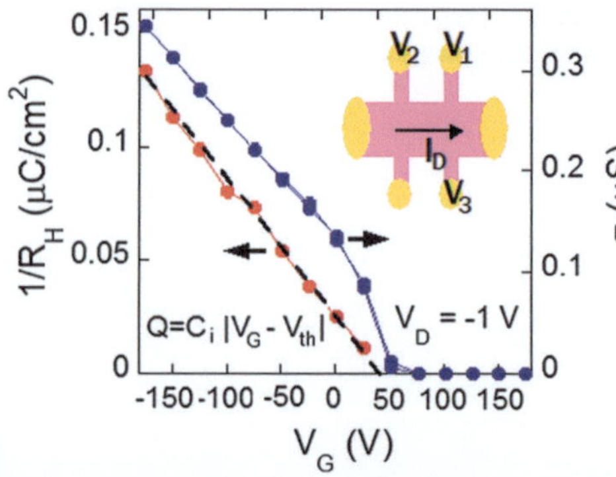

Fig. 35.6 Results of Hall measurements for a rubrene single-crystal transistor. The reciprocal of the Hall coefficient (*red circles*) is in good agreement with the charge per unit area *Q*" (*dotted line*) determined from the capacitance of the gate-insulating layer and the gate voltage. Also shown is the four-terminal conductivity

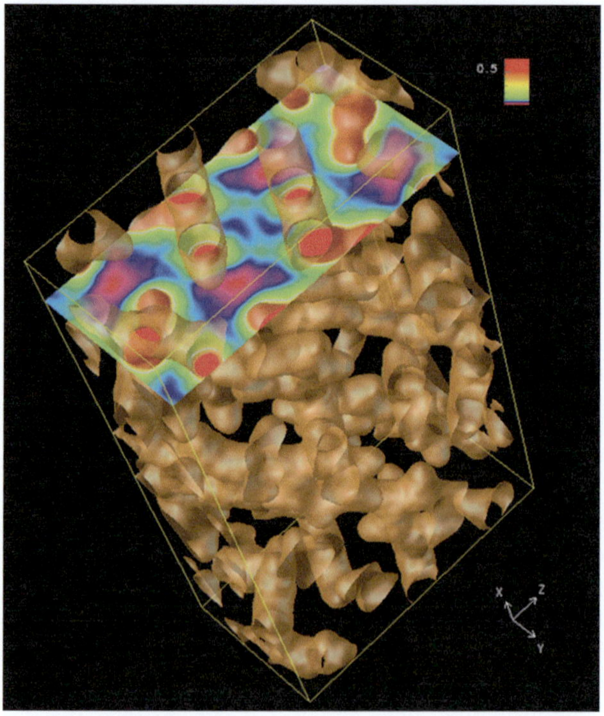

Fig. 35.7 Electron density distribution between molecules determined by high-intensity X-ray diffraction

conductivity based on the results of light absorption measurements [18]. Recently, Machida et al. carried out angle-resolved photoelectron spectroscopy of rubrene single crystals and observed a clear band dispersion [19]. In addition, the intermolecular electron density in rubrene was directly determined by high-intensity X-ray diffraction. As shown in Fig. 35.7, the presence of electrons "shared" among rubrene molecules was actually confirmed. Thus, it can be concluded that an intermolecularly delocalized electronic state is realized in the rubrene crystal, which gives rise to the high carrier mobility.

35.4 Transistors Based on New Organic Semiconductors DNTT and C_8-BTBT

Since a free-electron-like Hall effect was identified in the rubrene single-crystal FET, it is necessary to next investigate whether this is also the case for other high-mobility organic semiconductors. Takimiya et al. developed high-mobility organic semiconductor materials such as DNTT, C_8-BTBT, and C_{10}-DNTT, whose structures are shown in Fig. 35.3. They found that high mobilities of about 3, 4, and 8 cm^2/Vs, respectively, could be obtained in polycrystalline thin-film transistors [5–7]. Since these compounds are less easily oxidized than pentacene and are stable in air, there is considerable interest in their practical application. For C_8-BTBT and C_{10}-DNTT, single-crystal transistors were also fabricated by solution coating and the maximum mobility values obtained were 6 and 11 cm^2/Vs, respectively [16, 17]. Furthermore, although preparation of polycrystalline rubrene thin films is difficult, both single-crystal and polycrystalline films of DNTT, C_8-BTBT, and C_{10}-DNTT can be prepared, allowing a comparison of the conduction mechanism in each type of film.

35.4.1 DNTT Single-Crystal FET Prepared by PVT Method

A single-crystal DNTT FET fabricated using the same PVT method used for rubrene was found to exhibit a typical mobility of about 3 cm^2/Vs [21]. When prepared by other methods, a maximum mobility of 8 cm^2/Vs was obtained [22]. Figure 35.8 shows the results of Hall measurements for the FET fabricated using the PVT method. The measurements were carried out using the method described in Section 2. As was the case for rubrene, $1/R_H$ and Q are in extremely good agreement, so it can be concluded that the carriers are delocalized. This result suggests that rubrene is not the only organic semiconductor that exhibits band-type conduction and that an intermolecularly delocalized electronic state is more commonly obtained in high-mobility organic semiconductors [14].

35.4.2 C_8-BTBT and C_{10}-DNTT High-Mobility Single-Crystal FETs Prepared by Solution Crystallization Method

Single-crystal FETs based on C_8-BTBT and C_{10}-DNTT were prepared by a method in which a crystal is precipitated on a substrate from a solution [16, 21]. In the case of C_8-BTBT, a room-temperature process was possible because its solubility is sufficiently high [20, 23]. As shown in Fig. 35.9a, the crystal is precipitated by tilting the substrate, retaining liquid droplets at the end of the liquid-holding structure and drying for several minutes [20]. Unlike a normal spin-coating method, this method

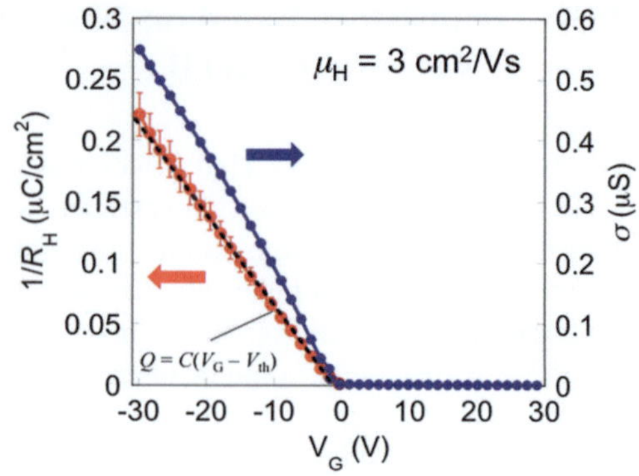

Fig. 35.8 Free-electron-like Hall effect in DNTT single-crystal transistor

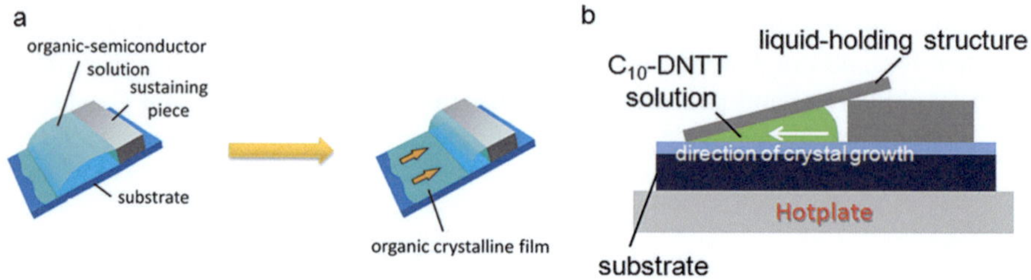

Fig. 35.9 Fabrication of high-mobility organic single-crystal FET by solution crystallization

has the advantage that it can produce thin films containing a large domain, in which the direction of crystal growth is controlled. On the other hand, since sufficient room-temperature solubility cannot be obtained for C_{10}-DNTT, the process was carried out at about 100 °C. As shown in Fig. 35.9b, a large domain can also be formed by precipitating a crystal with the use of a hot plate while controlling the direction. Transistors fabricated by this method have good properties as shown in Figs. 35.10 and 35.11. In addition, the maximum mobility obtained was more than 10 cm^2/Vs, and mobility comparable to that of amorphous oxide can be achieved by the solution process at about 100 °C. Thus, it is expected that this method will have practical industrial applications [15].

35.4.3 DNTT and C_8-BTBT Polycrystalline TFTs Prepared by Solution Crystallization Method

Hall measurements were carried out for a polycrystalline DNTT TFT using the same method, and the results were compared with those for the single-crystal FET. In a polycrystalline material, the presence of high-resistance grain boundaries is expected to have a detrimental effect on the FET characteristics. The measured Hall electromotive force is the sum of electromotive forces associated with each

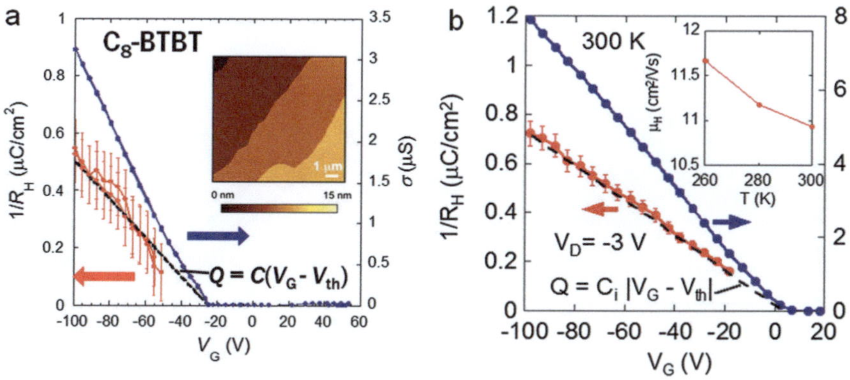

Fig. 35.10 Free-electron-like Hall effect in high-mobility solution-processed single-crystal transistors of (**a**) C_8-BTBT and (**b**) C_{10}-DNTT. The inset in (**a**) shows an atomic force micrograph illustrating the molecular-level flatness of the crystal surface. The inset in (**b**) shows the changes in mobility with temperature

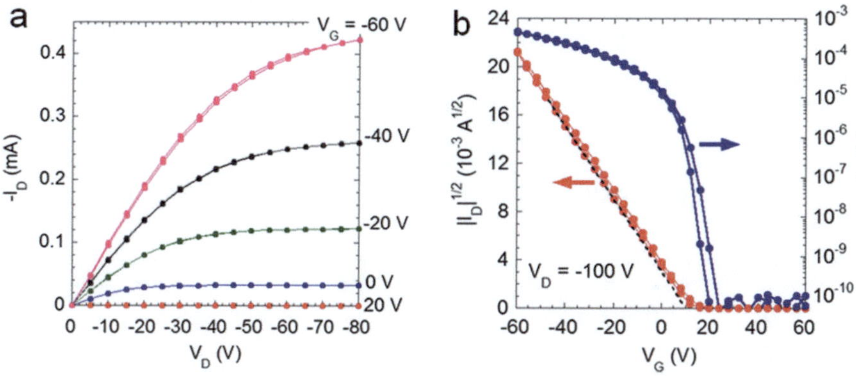

Fig. 35.11 (**a**) Output and (**b**) transfer characteristics for C_{10}-DNTT organic single-crystal FET produced by solution crystallization

crystal grain, even in the polycrystalline case, if the crystal grains are electrically connected. Therefore, it is expected that a Hall coefficient similar to that for a single-crystal organic semiconductor would be obtained if the carriers behave in a free-electron-like manner. Accordingly, the transport mechanism of carriers can be investigated in more detail than before by clarifying the electronic state inside the crystal grains [14].

Figure 35.12 shows the results of Hall measurements for a DNTT polycrystalline transistor. Again, there is extremely good agreement between $1/R_H$ and Q. Thus, even for a DNTT polycrystalline thin film prepared by vacuum evaporation, the microscopic arrangement of DNTT molecules was sufficiently regular and the carriers were delocalized inside the grains. The measured sample mobility was about 1 cm^2/Vs, which is lower than that for single-crystal DNTT.

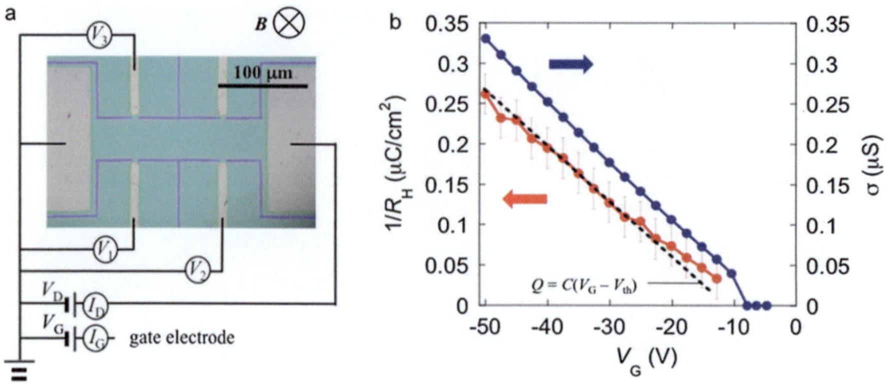

Fig. 35.12 (**a**) Image of sample used for Hall measurements on DNTT polycrystalline thin-film transistor and (**b**) measurement results indicating a free-electron-like Hall effect

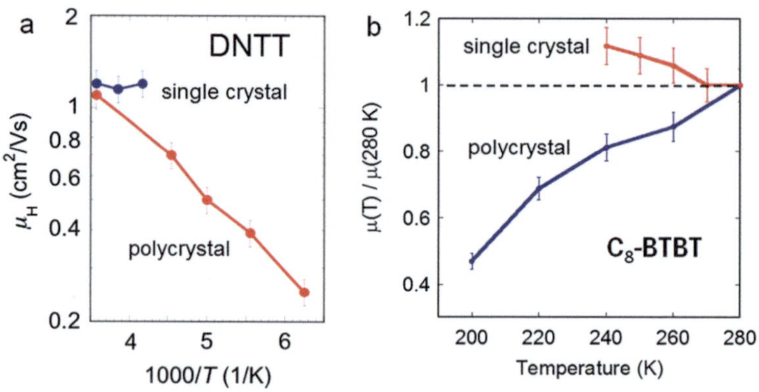

Fig. 35.13 Changes in mobility with temperature for polycrystalline and single-crystal transistors of (**a**) DNTT and (**b**) C_8-BTBT

In Fig. 35.13a, the temperature dependence of the mobility is compared with that for the single-crystal FET. For the polycrystalline sample, it was found that the mobility decreases with temperature. In the past, this would have been attributed to hopping conduction. However, the results of the Hall measurements indicate that band-type conduction occurs inside the crystal grains. Therefore, the mechanism is different from simple hopping transport, with carrier localization inside molecules, as expressed in Eq. (35.1). In fact, in the DNTT single-crystal case, very little decrease in the mobility is observed, at least above 200 K. The same results were obtained, as shown in Fig. 35.13b, for C_8-BTBT prepared by vapor deposition. This suggests that this is the general case for high-mobility polycrystalline organic thin films.

Regardless of whether band-type conduction occurs inside the crystal grains, the decrease in mobility at low temperature for the polycrystalline transistor may be the effect of shallow traps at grain boundaries. These are associated with disorders and defects at the grain boundaries, and their effect becomes more prominent as the temperature decreases. Thus, by combining Hall and four-terminal conductivity measurements, it has become possible to qualitatively separate the contributions to electron transport from inside crystal grains and grain boundaries. Thus, carrier transport in the high-mobility polycrystalline thin films can be understood in more detail.

35.5 Pentacene Single-Crystal and Polycrystalline FETs

Finally, Hall measurements were preformed for pentacene single-crystal and polycrystalline FETs. The single-crystal device was fabricated by laminating vapor-deposited thin crystals, and the polycrystalline device was produced by vacuum evaporation. Typical values were obtained for the mobility, namely, about 2 cm^2/Vs for the single-crystal device and about 1 cm^2/Vs for the polycrystalline device.

Although all of the high-mobility organic semiconductors described thus far exhibited a free-electron-like Hall effect, Fig. 35.14 shows that for the pentacene FETs, no agreement was found between $1/R_H$ and Q. Similar results were also obtained by Sekitani et al. for a polycrystalline pentacene thin film [12]. Since the measurement precision in Fig. 35.14 is the same as that for the other samples, this difference must reflect the intrinsic properties of pentacene. From Fig. 35.14, $\frac{1}{R_H} > C_i\,|V_G - V_{th}|$ for both single-crystal and polycrystal transistors, whereas the relationship $\frac{1}{R_H} < C_i\,|V_G - V_{th}|$ was reported by Podzorov et al. for a rubrene single-crystal FET at temperatures lower than room temperature [11].

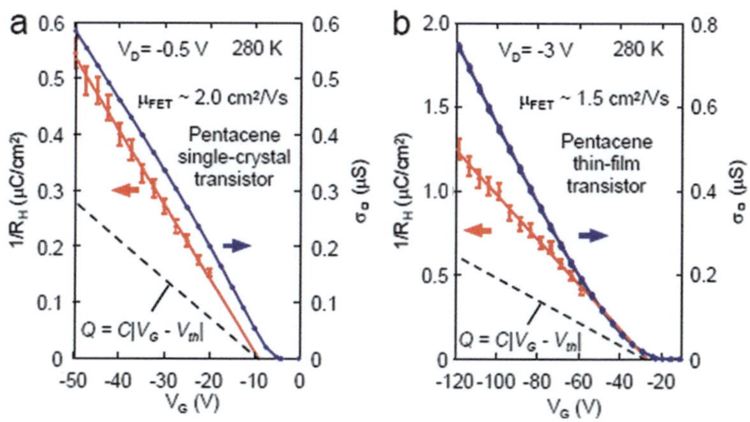

Fig. 35.14 Hall measurement results for pentacene (**a**) single-crystal FET and (**b**) polycrystalline FET

The number of high-conductivity carriers that contribute to the Hall effect (i.e., not localized in traps) is not larger than the number of carriers estimated from the capacitance and gate voltage. Therefore, the results indicate that the microscopic state of the carriers accumulated on pentacene deviates from the free-electron-like picture. Interestingly, the room-temperature values for the carrier coherence factor α, which is the ratio of $\frac{1}{R_H}$ to the slope of $C_i\,|V_G - V_{th}|$, namely, $\alpha = C_i/\left(\frac{\partial(1/R_H)}{\partial V_G}\right)$, were approximately the same for five pentacene samples with different mobilities, regardless of whether they were single crystals or polycrystalline. This suggests that α is determined by microscopic factors inside the molecules or crystal grains and is not affected by trap levels at grain boundaries or elsewhere.

If we recall that the Hall effect is due to the coherence of electron wave functions, the results for pentacene FETs imply that carrier coherence is not maintained over sufficient distances. Since there is no sample dependence, it is likely that this is due to molecule-specific factors rather than extrinsic factors such as disorders. Figure 35.15 shows that α increases with decreasing temperature. This suggests that molecular vibrations, which increase with temperature, strongly influence the carrier coherence. As described in the Introduction, molecular vibrations are large in organic semiconductors, even at room temperature, so this effect may become too large to ignore. In fact, it was theoretically shown by Troisi that when the effect of molecular vibrations is large, carriers may become localized [24].

It would be of interest to examine the reasons why, among all the semiconductor systems for which Hall measurements have been carried out to date, the effect of molecular vibrations is large only in the case of pentacene. The results may serve as guidelines for the design of organic semiconductors that exhibit a high mobility at room temperature. Comparing the molecules shown in Figure 35.3, it can be seen that the shape of the pentacene molecule has the highest symmetry. It is therefore plausible that it could easily vibrate along its long-axis direction even under the influence of intermolecular forces from neighboring molecules. An

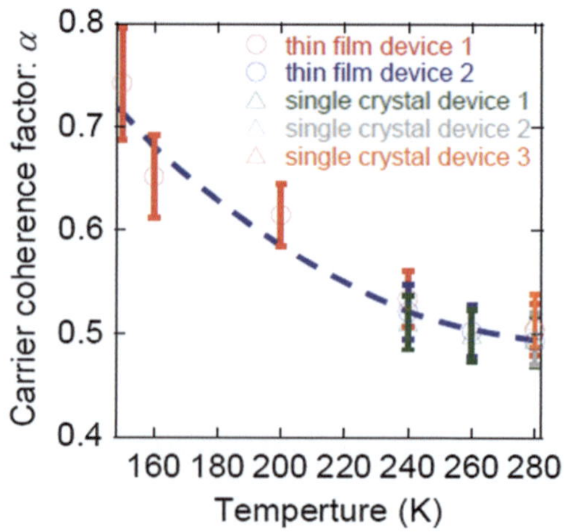

Fig. 35.15 Carrier coherence factor α for five kinds of single-crystal and polycrystalline pentacene FETs as function of temperature. The loss of coherence at higher temperatures is due to an increased amount of molecular vibrations

attempt was made to carry out Hall measurements for C_{60}, which is a molecule with even higher symmetry than pentacene, but the Hall coefficient was so small that the measurements were not possible. Based on the measurement sensitivity limit, this would imply $\alpha < 0.1$. Therefore, the effect of molecular vibrations for C_{60} is considered to be larger than for pentacene. In the future, experiments should be carried out to verify the above hypothesis for different molecules [15].

35.6 Summary and Prospects

A high-precision method for carrying out Hall measurements even for FETs based on organic semiconductors was developed, and measurements were carried out on different high-mobility organic FETs. The results highlighted the fact that Hall measurements are a powerful tool for investigating the essential mechanisms of carrier transport in single-crystal and polycrystalline thin films of organic semiconductors, as has long been the case for inorganic semiconductors. That is, the number of carriers accumulated by the field effect can be accurately determined as a quantity proportional to the capacitance. Therefore, essential information concerning carrier coherence can be obtained by comparing the reciprocals of the Hall coefficients.

In many organic semiconductors such as rubrene, DNTT, C_8-BTBT, and C_{10}-DNTT, which show mobilities of several cm^2/Vs or more, the Hall coefficient was similar to that for expected using the typical free-electron-like model. Thus, the high-mobility carrier transport in these systems was clarified to be based on a band mechanism due to an intermolecularly delocalized electronic state. On the other hand, this was different to the case for pentacene FETs, for which the electronic state was more strongly affected by molecular vibrations at room temperature. This leads to a reduction in electron coherence, and the Hall electromotive force that is produced by coupling with the magnetic field is weaker than that for a fully coherent free-electron-like state. Thus, suitable design of organic molecules may be an effective approach for developing high-mobility organic semiconductors. In the future, a more quantitative investigation should be carried out in combination with theoretical considerations, and further verification is also necessary in other systems.

Lately, new organic semiconductors are being developed at an impressive rate. Significant progress has also been made with regard to solution processes, which take maximum advantage of the fact that organic semiconductors can be fabricated by coating at temperatures near room temperature. Organic FETs exhibiting a mobility exceeding 10 cm^2/Vs can now be produced by the coating method. In the future, the potential industrial applications of high-mobility organic semiconductors will increase. We believe that understanding carrier transport in such semiconductors will promote further material development and the expansion of research into the electronic properties of interfacial two-dimensional electron systems.

References

1. Lin Y-Y et al (1997) IEEE Electron Device Lett 18:606
2. Bao Z, Locklin J (2007) Organic field-effect transistors: optical science and engineering series. CRC Press, London/New York
3. Podzorov V et al. (2003) Appl Phys Lett 82:1739; Takeya J et al. (2003) J Appl Phys 94:5800
4. Anthony JE et al. (2001) J Am Chem Soc 123:9482; Klauk H et al. (2002) J Appl Phys 92:5259
5. Yamamoto T, Takimiya K (2007) J Am Chem Soc 129:2224
6. Ebata H et al (2007) J Am Chem Soc 129:15732
7. Kang MJ et al. (2011) Adv Mater 23:1222
8. Podzorov V et al (2004) Phys Rev Lett 93:086602
9. Warta W et al (1985) Appl Phys A: Solids Surf 36:163
10. Takeya J et al (2005) Jpn J Appl Phys 44:L1393
11. Podzorov V et al (2005) Phys Rev Lett 95:226601
12. Sekitani T et al (2006) Appl Phys Lett 88:253508
13. Takeya J et al (2007) Phys Rev Lett 98:196804
14. Yamagishi M et al. (2010) Phys Rev B 81:161306(R)
15. Nakayama K et al. (2011) Adv Mater 23:1626–1629
16. Fukuyama H et al (1969) Prog Theor Phys 42:494
17. Friedman L (1971) J Non-Cryst Solids 6:329
18. Li ZQ et al (2007) Phys Rev Lett 99:016403
19. Machida SI et al (2010) Phys Rev Lett 104:156401
20. Uemura T et al (2009) Appl Phys Exp 2:111501
21. Uno M et al (2009) Appl Phys Lett 94:223308
22. Haas S et al (2009) Appl Phys Lett 95:022111
23. Liu C et al (2011) Adv Mater 23:523
24. Troisi A, Orlandi G (2006) J Phys Chem A 110:4065

Chapter 36
Unveiling Charge Carrier Transport in π-Conjugated Molecular Wire on Micro- and Macroscopic Scales

Akinori Saeki and Shu Seki

Abstract Molecular wires of π-conjugated polymers and molecular assemblies are cutting-edge motifs for evolving flexible and cost-effective organic electronics. They consist of various morphologies in length scale from a single-molecule, multi-molecular aggregate to a bulk film; however, the conventional evaluation of charge carrier motility cannot clarify the complicated interplay of length-dependent charge-transport processes. Thus, bridging the gap between microscopic and macroscopic electronic natures has been the subject of intense debate within the scientific community. In this regard, time-resolved microwave conductivity (TRMC) offers a versatile approach to unveiling the local charge carrier mobility with minimized effects of extrinsic impurities and barriers at the interfaces. This dissertation seeks to review the recent evolution of TRMC such as extension of excitation light sources, frequency modulation together with comprehensive analysis of complex conductivity, and combination with field-inductive metal/insulator/semiconductor device. We believe that this account is exploratory and interpretative for understanding the charge carrier transport in π-conjugated molecular wires.

Keywords Charge carrier mobility • Conjugated polymer • Time-resolved microwave conductivity

---

A. Saeki (✉)
Department of Applied Chemistry, Graduate School of Engineering, Osaka University, 2-1 Yamadaoka, Suita, Osaka 565-0871, Japan
e-mail: saeki@chem.eng.osaka-u.ac.jp

S. Seki
Department of Molecular Engineering, Graduate School of Engineering, Kyoto University, Nishikyo-ku, Kyoto 615-8510, Japan
e-mail: seki@moleng.kyoto-u.ac.jp

T. Akasaka et al. (eds.), *Chemical Science of π-Electron Systems*,
DOI 10.1007/978-4-431-55357-1_36

36.1 Introduction

A semiconducting π-conjugated material is increasingly recognized as a benefitting motif, which are potentially compatible with cost-effective printing on flexible substrate. The past few decades have seen the rapid development of organic electronics in a myriad of applications including organic field effect transistor (OFET) [1, 2], organic light emitting diode (OLED) [3, 4], and organic photovoltaics (OPV) [5, 6]. In spite of the flourishing inorganic electronics made of silicon and rare metals, the growing organic counterparts have drawn much attention to provide bio-friendly devices and sustainable circulation of abundant resources [7, 8]. As schematized in Fig. 36.1, a unique prospect of organic electronics is that they consist of a hierarchical structure from a single-molecule, multi-molecular assembly formed via intermolecular interaction, large-scale domains, to a bulk film. Therefore, the charge carrier transport is likely to differ over a range of these length scales. However, very little is known about the connection of electrical properties between microscopic and macroscopic spatial domains.

A principal entity of organic electronic materials is a π-conjugated molecule, where the π-electrons on sp and sp_2 carbon are mostly responsible for the optical

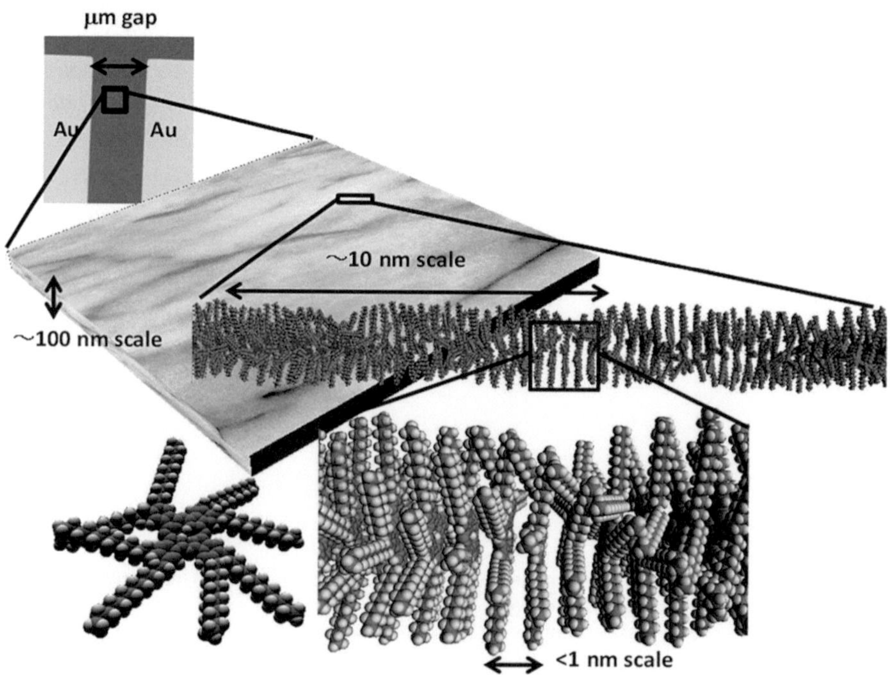

Fig. 36.1 Schematic drawing of organic semiconductors composed of molecules, assemblies, and their aggregates, alongside their corresponding length scales

and electronic properties. Upon photoexcitation, electronic transition takes place within a single molecule or a few neighboring molecules, promptly leading to the rearrangement of atomic coordinate and molecular orbitals. Recent computational techniques represented by density functional theory (DFT) are able to predict the electronic and structural features of a neutral and charged molecule at affordable cost. The generation of excited state is followed by energy migration and charge separation among the highest occupied molecular orbital (HOMO) and/or the lowest unoccupied molecular orbital (LUMO) at homogeneous or heterogeneous molecular junctions. In the case of charge separation, singly occupied molecular orbitals (SOMO) having opposite charge are formed at two molecules, referred to as radical cation (positive charge) and radical anion (negative charge). They are also termed by hole and electron, respectively, which are excess charges in valence and conduction bands in inorganic electronics. The similar rearrangement of atomic coordinate is involved in these charge-separated states, where the energy loss during the relaxation (reorganization energy) is essential for organic electronics based on the assembly of individual molecular components. Intra- and intermolecular charge carrier transport, quantified by charge carrier mobility, μ (cm^2/Vs), is of particular importance and impacted by the molecular conformation and spatial overlap of HOMO or LUMO orbitals, as well as reorganization energy of the charged molecules.

A general technique to evaluate charge carrier mobility is a direct-current (dc) method using metal electrodes on organic semiconductor and applying voltage to inject and transport charge carriers. The examples include time of flight (TOF) [9], FET [10], space-charge-limited current (SCLC) [11], Hall effect (HE) [12], and the photoinduced charge carrier extraction by linearly increasing voltage (photo-CELIV) [13]. Out of them, FET is the most commonly used method to determine the charge carrier mobility in the parallel direction to a substrate at a very thin (one or a few molecular) layer of organic semiconductor on an insulator. FET mobility is directly linked to the performance of practical organic thin-film transistor (OTFT) to drive a logical circuit in electronic devices such as a display. The pilot scaffold in OFET is a pentacene thin film fabricated by thermal evaporation process in a vacuum chamber. Pentacene OFETs have demonstrated the hole mobilities as much as 3–6 cm^2/Vs optimized by film processes and gate insulator modifications to ensure the ordered herringbone-packing and fine-grain features [14–16]. Surprisingly, the charge carrier mobility as high as 30–40 cm^2/Vs has been achieved in a single crystal rubrene [17, 18] and solution-processed single-crystalline C_8-BTBT films [19], where the mobility exceeds that of amorphous silicon and is rationalized by band transport.

As shown in Fig. 36.2, the spatial scale of source-drain electrode gap of FET is typically a few to hundreds of μm. Therefore, FET mobility is influenced by complicated interplay between molecular packing, morphology, disorder, impurities, and interfacial characteristics. Mobility on the order of 10^{-4}–10^{-8} cm^2/Vs has been often obtained for conjugated polymer films containing a lot of structural and chemical defects, which are analyzed by disorder formalism considering that energetic and spatial disorders limit the long-range translational motion with a

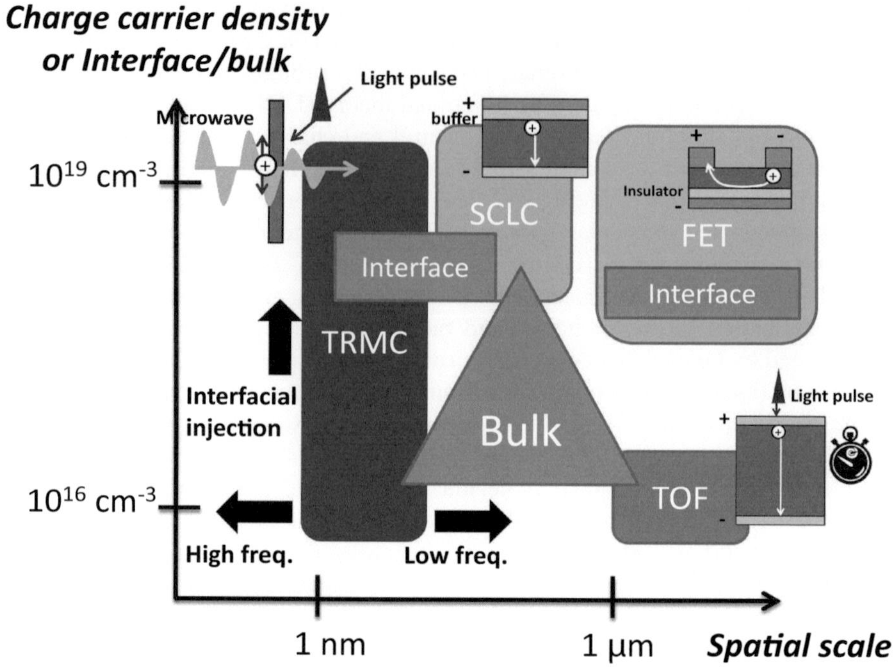

Fig. 36.2 Mapping of charge carrier density and spatial scale of various mobility measurement techniques

certain mobility. However, the trap associated with these disorders can be fulfilled if the trap density is less than the density of injected charge carrier typically of the order of 10^{18}–10^{19} cm^{-3} [20]. Furthermore, the trap density can be reduced by the use of structurally and chemically pure single crystal or an appropriate surface treatment of insulator underneath the organic semiconducting layer [21].

SCLC technique characterizes the mobility in the perpendicular (vertical) direction to the substrate, which is relevant to OPV and OLED operation (Fig. 36.2). Researches have often attempted to correlate SCLC hole and electron mobilities with short-circuit current density under light (J_{sc}) and fill factor (FF) of OPV device. In some cases, a good correlation is seen [22], but not always found [23], reiterating the complex factors convoluted in the long-range charge transport. A typical thickness of semiconducting layer of SCLC device is 100–200 nm. The resultant SCLC mobility is generally smaller than FET, on the order of 10^{-6}–10^{-3} cm^2/Vs, regardless of the equivalent charge carrier density to those of FET. This is likely due to the difference of charge carrier transporting region, i.e., at an interface (FET) and in a bulk (SCLC). In addition, the latter is largely affected by the charge trapping site which hampers the further injection of charge carrier by space-charge effect. TOF technique, another dc measurement, characterizes the vertical charge transport similar to SCLC; however, the charge carrier density is a few orders of magnitude smaller than those of FET and SCLC, because the charge carriers are produced from inefficient exciton splitting. The generated charge carrier density is calculated less than 10^{16} cm^{-3} from a typical photogeneration yield of charge carriers, incident photon density of light, and absorption coefficient. Excessive increase in charge carrier density is impossible due to the threshold of sample damage upon laser pulse irradiation. The charge carrier mobility is evaluated from the average flight time of photocurrent, which corresponds to the time required for the charge carrier on the surface to travel to the bottom counter electrodes. Therefore, TOF mobility reflects from the electronic property of bulk state, and accordingly the density and height of trapping sites as well as their spatial/energetic distribution, temperature, and electric field strength are considerably dominating factors. Overall, mobility of organic semiconductors suffers from several drawbacks partly inherent to organic molecules: chemical impurity, grain boundary, structural defects, and resistance at semiconductor/metal interface. While numerous researches have been carried out on the dc mobility measurements, the length-scale-dependent mechanism by which the charge carrier transport obeys has not been established explicitly.

36.2 Evolution of Microwave Conductivity Technique

Turning now to the alternating current (ac) measurement of charge carrier mobility, we can list up impedance spectroscopy (IS) [24], terahertz (THz) conductivity [25], and time-resolved microwave conductivity (TRMC) [26–30]. Among them, the frequency of IS is limited less than MHz and still requires an electrode contact to bias ac electric field. Thus, the situation is similar to the dc measurements. In contrast, TRMC using gigaHertz (GHz) microwave utilizes an ac electromagnetic (EM) wave. EM can transport even in a vacuum, air, and other gas media, allowing us to perform electrode-less measurement. The charge carriers can be injected into neutral or partly doped organic semiconductors by exposing pulsed laser like TOF or irradiating pulsed radiation such as electron beam and X-rays. The latter combination with TRMC is called as pulse-radiolysis (PR)-TRMC [28, 31]. PR-TRMC merits the quantitative and homogeneous charge generation in a sample, arisen from the determinable radiation chemical yield of ionization. Therefore, the change of conductivity ($\Delta\sigma$) evaluated by TRMC is readily separated into the sum of positive and negative charge carrier mobility, $\Sigma\mu$ ($= \mu_h + \mu_e$), and induced charge carrier concentration (Δn) through the definition of conductivity ($\Delta\sigma = e \Delta n \Sigma\mu$, e is an elementary of charge). On the other hand, the charge carrier concentration generated by light exposure is dependent on various factors originated from materials (single or mixture components, absorption coefficient, etc) and experimental setup (wavelength, light intensity, and sample geometry). Accordingly, the charge carrier generation yield, ϕ, has to be independently estimated to discern a $\Sigma\mu$ value from light-induced THz and GHz conductivity.

The continuing advance in a femtosecond laser technology has brought about a stable, intense, and high-repetition rate femtosecond THz pulse. Photoinduced time-domain THz spectroscopy using one pumping laser pulse and two probing THz pulse allows fast acquisition of THz spectrum at desired time delay between pump and probe [32]. Therefore, the transient complex conductivity composed of real and imaginary parts is obtained on picosecond time scale. Meanwhile, flash-photolysis (FP)-TRMC utilizes a resonant cavity to secure a high degree of sensitivity, which limits the instrumental response to a nanosecond time scale, typically from several to 100 ns. The power of continuous microwave is usually set at a few mW, so that the electric field of the microwave is sufficiently small to not disturb the motion of charge carriers. The third harmonic generation (THG; 355 nm) of a Nd:YAG laser (typically 5–8 ns pulse duration) or visible light pulses from an optical parametric oscillator seeded the laser is used as an excitation source. The photoconductivity transient ($\Delta\sigma$) is converted to the product of ϕ and $\Sigma\mu$, by $\phi\Sigma\mu = \Delta\sigma\ (eI_0F_{\text{light}})^{-1}$, where e and F_{Light} are the unit charge of a single electron and a correction (or filling) factor [33]. The noteworthy aspect of electrode-less laser-flash THz and GHz conductivity is that the charge carrier motion under such a high-frequency EM field is confined in a local area, which is assumed on a few to tens of nanometer scales (Fig. 36.2). This length scale corresponds to the polymer chain or intermolecular distance and is varied by charge carrier mobility itself as well as the frequency of probing EM wave. The charge carrier density of laser-flash TRMC is less than 10^{16} cm^{-3}, similar to TOF experiments; however, the density can be increased by combining TRMC with FET or metal/insulator/semiconductor device structures (Fig. 36.2, vide infra).

Using FP-TRMC, we have investigated charge carrier dynamics in organic molecular crystals, liquid crystals, conjugated polymers, self-assembled opto-electronic architecture, and inorganic semiconductors [26, 27, 34]. Evaluation of anisotropic photoconductivity in high angle-resolved two-dimensional polar coordinate [35] or three-dimensional axes (parallel, cross, and perpendicular to the long axis of a quartz substrate) [36, 37] is noteworthy in TRMC. In particular, three-dimensional evaluation using the identical single crystals is not accessible by conventional dc techniques. By virtue of direct access to the local charge carrier transport, FP-TRMC has revealed a rational structure–property relationship of abovementioned organic electronics, in which well-organized or phase-segregated samples demonstrated superior photoconductivity to those of random (amorphous) or phase-mixed ones. In some cases, simultaneous evaluation by transient absorption spectroscopy (TAS) of charged species of a sample [38] or incorporated electron acceptor (e.g., perylene bisimide, PDI) [39] enables to independently determine ϕ and therefore extract $\Sigma\mu$ value from $\Delta\sigma$. Typically 0.1–1 cm^2/Vs ordered mobilities have been revealed as $\Sigma\mu$ of conjugated polymers and self-organized electronics [26].

Figure 36.3 shows the several versions of TRMC setup we have developed so far. In addition to the normal FP-TRMC system using a nanosecond laser, a 10 μs-width white light pulse (pseudo-solar spectrum) from a Xe-flash lamp is used to evaluate photoelectric conversion efficiency of OPV device [40]. Comparative investigation on low bandgap polymers and p/n blend ratio revealed a good correlation between the transient photoconductivity maxima ($\Delta\sigma_{\text{max}}$) and OPV performance (PCE/V_{oc};

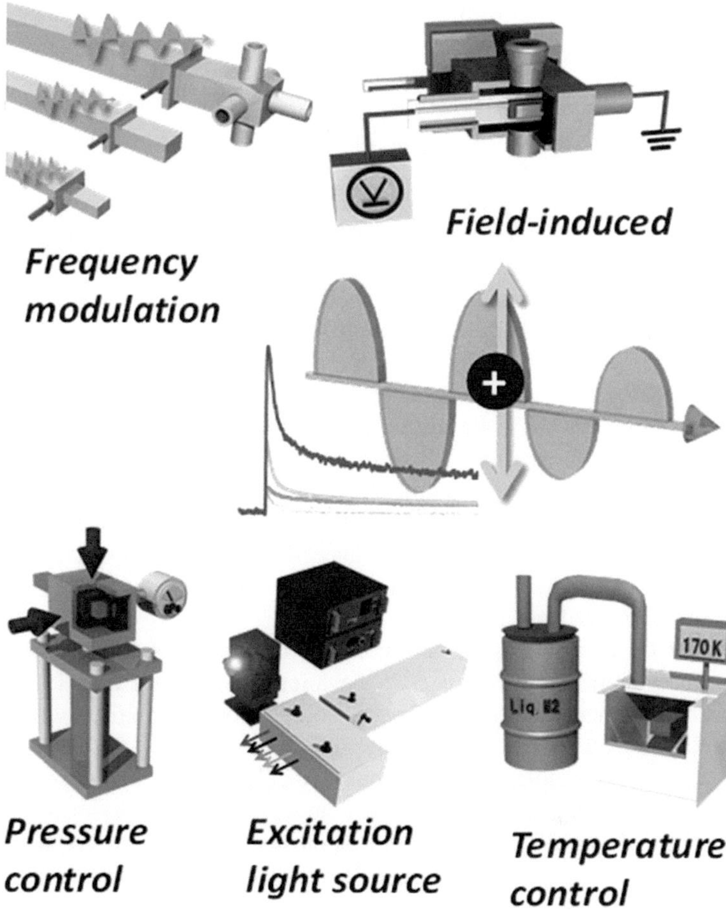

Fig. 36.3 Illustration of evolved TRMC systems

PCE, power conversion efficiency, and V_{oc}, open-circuit voltage). This is because $\Delta\sigma_{max}$ of Xe-flash TRMC includes information about the charge separation yield, local charge carrier mobility, their lifetimes, and sunlight absorption property of the films. Therefore, Xe-flash TRMC allows facile and versatile screening of process and material without fabricating tedious device characterization, and it has been practically utilized for the development of novel p-type polymers [41–43] and n-type molecules [44].

Another notable advance in TRMC family is its combination with field-induced (FI) metal/insulator/semiconductor (MIS) device [45]. The charge carrier is induced by biasing a voltage between the sandwiched gold electrodes, in which the polarity of charge carrier is easily controlled by the voltage polarity. To achieve a high sensitivity and gain time-resolved kinetics on millisecond scale, a rectangular-shaped voltage is applied for the MIS device. More importantly, the number of injected charge carriers is precisely measured by monitoring the current at the injection and extraction timing, leading to direct evaluation of charge carrier mobility (μ_h or μ_e) accumulated near the semiconductor/insulator interface. For instance, the hole mobility in pentacene film was up to 6.5 cm^2/Vs, in good agreement with the reported highest FET mobility [46].

Temperature dependence of $\Delta\sigma_{max}$ and decay speed is informative to gain mechanistic insight into the charge transport. The low-temperature experiments (TRMC@LT) are performed by flow of cold nitrogen gas or heat exchange via cooled metal plate in a vacuum chamber [47, 48]. With decreasing the temperature, $\Delta\sigma_{max}$ of organic semiconductors in hopping transport framework is decreased, due to the decreased thermal energy necessary to leap from the localization site [47]. In contrast, the band-transport semiconductors indicate a more intense $\Delta\sigma_{max}$ at low temperature, resulted from the decreased phonon scattering. Organic cation/metal/halogen perovskite materials, which have increasingly occupied a prestigious position in OPV, are an example of band-transport behavior on temperature [48]. High local mobility as much as 60–75 cm^2/Vs and low charge recombination rates have been identified as the important feature of the excellent photovoltaic performance of perovskite solar cell. Its PCE has been approaching 20 % comparable to those of inorganic solar cells [49]. Rapid progress in material and process is ongoing worldwide along with the fundamental studies by TRMC and other time-resolved spectroscopies.

X-band (9 GHz) is the most common and convenient frequency for TRMC setup, because of the moderate size of the microwave circuit (ca. $4 \times 2 \times 1$ cm^3 cavity) and rich lineup of the microwave components compatible with electron magnetic resonance (ESR). Continuous scanning of microwave frequency like IS measurement is not possible, while distribution of THz complex conductivity allows for information to be gained on the transport model in semiconductors. Accordingly, we have recently developed new TRMC circuits, of which resonant frequencies were discretely changed from X-band (9 GHz), K_u-band (15 GHz), and K-band (23 GHz), to Q-band (33 GHz) [50]. By increasing the frequency, the circuit becomes compact and the time resolution is improved from 40 ns for X-band to 6 ns for Q-band. The frequency-modulated (FM) TRMC assisted by short frequency scanning method to evaluate complex GHz conductivity is able to access the depth and population of shallowly trapped charge carriers. We have clarified that the shallowly trapped electrons have a large impact on the negative imaginary conductivity, while the free electron significantly contributes to the positive real part in spite of its small population. The kinetic mismatch of real and imaginary conductivities suggests the involvement of different decay processes, which is a sharp contrast to a blend film of poly(3-hexylthiophene) (P3HT) and methanofullerene (PCBM), where both real and imaginary transients were identical over the microsecond time range. Accordingly we suggest that GHz complex photoconductivity contains rich fingerprints regarding interplay between free and trapped electrons, which can facilitate the rational design of high-performance optoelectronic devices.

Our latest interest in "soft" conjugated materials is on the modulation of conformation and intermolecular distance controlled by pressure. Synthetic approach via side-chain modification can tailor the backbone feature, leading to the modulations of π-conjugation length in isolated phase [51] and intermolecular arrangement in condensed matters [52]. However, synthetic modifications require tedious procedures, where the steric hindrance and solubility of substituent also impact the variation of molecular weight and film quality. In contrast, applying pressure to a single polymer in a hydrostatic chamber or diamond anvil cell is a straightforward and facile way to *continuously* tune its intrachain conformation and interchain distance without chemical engineering. Therefore, high-pressure (HP)-TRMC is expected to be highly useful for investigating the impact of polymer conformation on the intramolecular charge carrier transport [53]. Thus, the HP-TRMC provides an exciting opportunity to advance our knowledge of the hidden nature of soft organic electronics.

In this section, we introduced the five new options of TRMC evaluation: Xe-flash TRMC, FI-TRMC, FM-TRMC, TRMC@LT, and HP-TRMC. The evolution of TRMC would make a major contribution to research on electronic nature of molecular wire by demonstrating the local charge carrier mobility and connection with the long-range mobility and device performance. The next section will review the recent studies on TRMC, FET, and SCLC mobilities of conjugated materials.

36.3 Unveiling Charge Carrier Transport of Molecular Wires

Multicomponent conjugated polymers consisting of donor–acceptor- or push–pull-type moieties are highly promising materials as color-tunable emitting layers in OLEDs and bulk heterojunction frameworks with low band gap. Therefore, considerable efforts have been devoted to the development of novel copolymers possessing unique electronic and optical properties. Polyfluorene and its copolymers form an integral part of OLEDs, as they have high absorption coefficient, photoluminescence yield, and chemical stability and also due to highly pure blue emission [54]. We have investigated the intramolecular charge-transport property of fluorene–thiophene copolymers by FP-TRMC technique and independent ϕ evaluation by transient photocurrent technique (Fig. 36.3a) [55]. The important observation was that the charge carrier mobilities of copolymers with the even-numbered oligothiophene units with composition ratios of 50 and 25 % are higher than those of the odd-numbered polymers (Fig. 36.3b). This is correlated with the tilted bond direction of thiophene and fluorene rings and the resultant alternating symmetry of the alkyl chains at the fluorene unit. Interestingly, this odd–even effect disappeared, and the one-dimensional mobility ($\Sigma\mu_{1D}$) converged to ca. 0.7 cm^2/Vs once the oligothiophene composition ratio was decreased to less than 10 %. The rotation of the fluorene units in an odd-numbered copolymers to make the alkyl chains in opposite directions to maximize the interaction with solvents distorts the polymer backbone as shown in Fig. 36.4c and hence reduces the charge carrier motion parallel to the direction of electric field of probing microwave. These findings are extremely important from the point of view of designing copolymers for various applications like OPVc and OLEDs.

The similar type of donor–acceptor conjugated polymer has been investigated for fluorene–triphenylamine copolymers, F_nTPA_m (Fig. 36.5a) [56]. In Fig. 36.5b, we

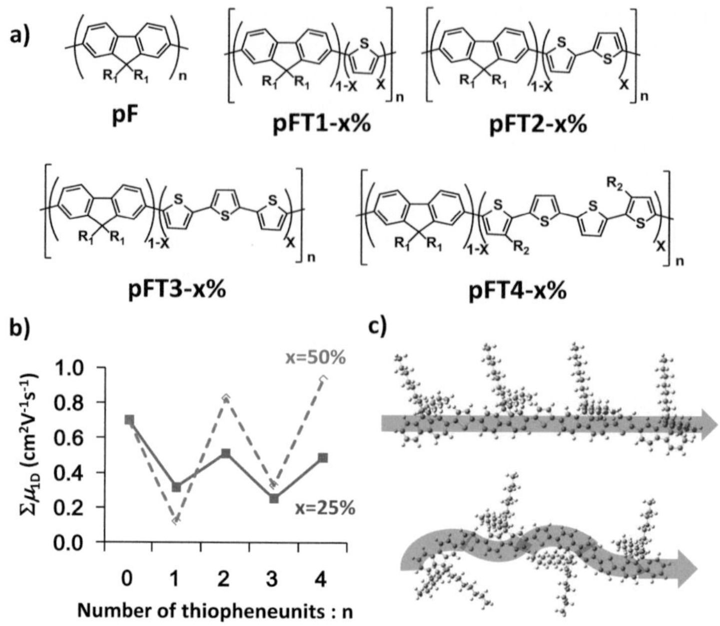

Fig. 36.4 (**a**) Chemical structures of fluorene–thiophene copolymers (pFT$_n$–x%). (**b**) One-dimensional TRMC mobility ($\Sigma\mu_{1D}$) of pFT$_n$–x% ($n = 0$–4, $x = 25$, and 50 %). (**c**) Illustration of backbone conformation of pFT$_1$-50 % for (*upper*) one side and (*lower*) alternating alkyl chains of fluorene unit. The opaque *arrows* represent the intramolecular charge transport (Reproduced with permission from Ref. [55]. Copyright 2011 American Chemical Society)

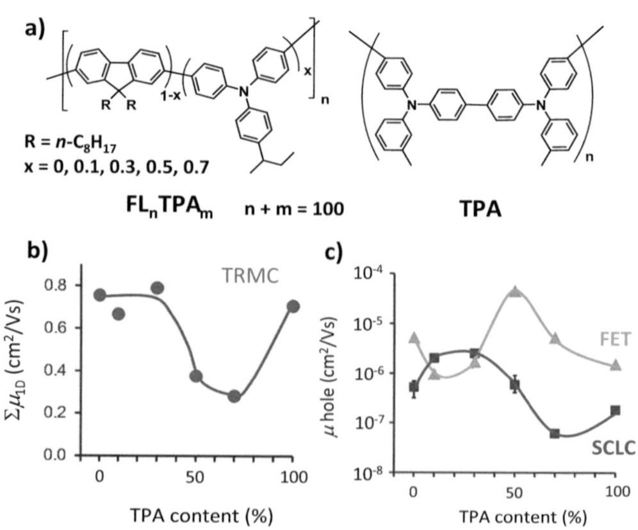

Fig. 36.5 (**a**) Chemical structures of fluorene–triphenylamine copolymers (FL$_n$TPA$_m$). (**b**) One-dimensional TRMC mobility ($\Sigma\mu_{1D}$) of FL$_n$TPA$_m$. (**c**) Hole mobility vs TPA contents measured by FET (*triangles*) and SCLC (*squares*) (Reproduced with permission from Ref. [56]. Copyright 2012 Japanese Society of Applied Physics)

plotted $\Sigma\mu_{1D}$ as a function of TPA content, where it clearly indicates an obvious drop at the TPA content of 50–70 %. The $\Sigma\mu_{1D}$ of the homopolymers (FL_{100} and TPA_{100}) is 0.7–0.8 cm^2/Vs, revealing the high intramolecular charge carrier mobility. The trend was further discussed by the conformation of molecular wire obtained from DFT calculations. Since the TPA has a large relative bond angle (ca. 120°) at the linkage, the conformation of the resultant $TPA_{50}FL_{50}$ polymer is in a zigzag form rather than straightforward like FL_{100}. This chain conformation leads to the decrease in $\Sigma\mu_{1D}$ probed by FP-TRMC. We next performed SCLC and FET measurements to reveal the long-range intermolecular charge carrier mobilities. Surprisingly, SCLC hole mobilities ($\mu_{h,\,SCLC}$) shown in Fig. 36.5c indicate almost the same trend with FP-TRMC mobilities, though the former are more than six orders smaller (10^{-8} to 10^{-6} cm^2/Vs) than the latter. In sharp contrast, FET hole mobilities ($\mu_{h,\,FET}$) exhibited the different dependence on the TPA contents. The $\mu_{h,\,FET}$ revealed the lowest value for $TPA_{10}FL_{90}$ (9.8×10^{-7} cm^2/Vs) and the highest value for $TPA_{50}FL_{50}$ (4.5×10^{-5} cm^2/Vs). After the maximum at TPA = 50 %, $\mu_{h,\,FET}$ decrease simply with the TPA content. Roughly speaking, $\mu_{h,\,SCLC}$ and $\mu_{h,\,FET}$ are in the mirror image on the TPA content. Although these mobilities are four to seven orders of magnitude smaller than those of FP-TRMC, it is of quite interest to find the same trend of FP-TRMC with SCLC and opposite one with FET. Our interest is focused on the reason why $\mu_{h,\,SCLC}$ and FP-TRMC mobility indicated the similar dependence on TPA content, although they are completely different measurements. It was suggested that intramolecular charge properties such as SOMO delocalization and polymer conformation are more predominant in the short-range charge transport for both ac and dc measurements. Another similarity between SCLC and FP-TRMC is that they probe bulk mobility, while the charge carriers in FET device are transported in a monomolecular layer at dielectric/semiconductor interface. Lucas et al. investigated SCLC, FET, and impedance spectroscopy of pentacene films and suggested that the observed difference in their mobilities arose from active layer morphology for different substrate and electric/electronic processes involved in the measurements [57]. The similarity of TRMC and SCLC mobilities has been also found in polycarbazole families, in spite of the orders of magnitude differences in their absolute values [58].

The last intuitive example of contrasting local and long-range mobilities is self-assembled organic nanostructure composed of π-electronic building blocks. Gemini-type amphiphilic hexa-*peri*-hexabenzocoronene (HBC) and a HBC derivative appended with an electron-accepting trinitrofluorenone (TNF) moiety have demonstrated a self-assembly to form a coaxial nanotubular structure in appropriate solvent mixture (Fig. 36.6a) [59]. Based on charge carrier generation efficiency independently determined by TAS, we simultaneously assessed the local and long-range charge carrier mobilities of the co-assembled nanotubes [60]. The former was found to be 3 cm^2V^{-1} s^{-1} for the co-assemblies, but dropped to 0.7 cm^2/Vs at HBC-TNF = 100 %, due to lowering of a structural integrity of π-stacked HBC arrays (Fig. 36.6b). In contrast, the latter evaluated by analyzing the FP-TRMC kinetic decays was less than 1.5×10^{-4} cm^2/Vs and monotonically decreased when the HBC-TNF content was increased (Fig. 36.6c). This large gap is rationalized by

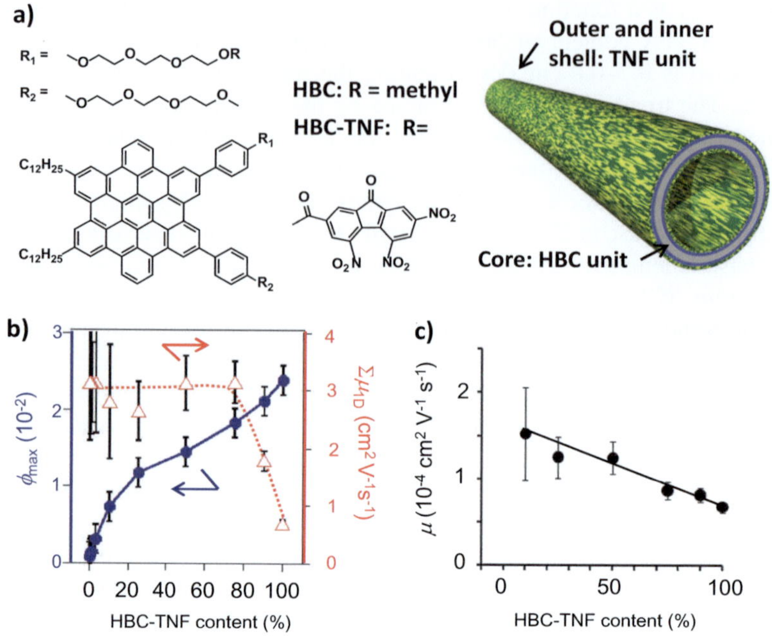

Fig. 36.6 (**a**) Chemical structures of HBC and HBC-TNF. The *right* image is co-assembled nanotube of HBC and HBC-TNF. (**b**) One-dimensional (intra-tubular) TRMC mobility ($\Sigma\mu_{1D}$, *right* axis) and charge carrier generation yield (ϕ_{max}, *left* axis). (**c**) Charge carrier mobilities obtained from the analysis of TRMC kinetic decays. This long-range mobility is assumed due to intertubular charge hopping (Reproduced with permission from Ref. [60]. Copyright 2011 American Chemical Society)

the incorporation of two charge-transport processes: intra-tubular and intertubular hopping. The former is likely dominated by free energy change and reorganization energy of the HBC molecules [27], while the latter is considerably affected by tubular outer-shell morphology. Another noteworthy finding is that the mobilities in these two processes are in a trade-off relationship; therefore, it is essential to improve the electrical connectivity among nanotubes without disrupting their pronounced local charge transport. Technological improvements in this regard might include post-modification to separate the formation processes of nanostructure and smooth film, macroscopic alignment, and dimensional control.

36.4 Conclusion

A brief overview of the recent evolution of TRMC technique is given toward facilitating the investigation of intrinsic electronic nature of organic electronics. Modulation of charge transport by frequency (FM), temperature, field induction

(FI), and high pressure (HP) allows us to perform systematic and detailed study on the semiconducting molecular wire. Exploiting light sources of not only a conventional nanosecond laser but also a white light pulse from a house-made Xe-flash lamp can accelerate the development of novel material and process for highly efficient OPV. There is still increasing concern about maximizing the mobility of organic electronics and revealing present drawbacks that limit the device performance. The two-dimensional graphene has already demonstrated more than 10^4 cm^2/Vs mobility [61], even beyond those of inorganic semiconductors. This might be indicative of a plenty of room for improving the electronic property of π-conjugated materials by controlling the dimensionality and spatial scale of π-electrons (π-space).

References

1. Klauk H (2010) Organic thin-film transistors. Chem Soc Rev 39:2643
2. Wang C, Dong H, Hu W, Liu Y, Zhu D (2012) Semiconducting π-conjugated systems in field-effect transistors: a material Odyssey of organic electronics. Chem Rev 112:2208
3. Shirota Y, Kageyama H (2007) Charge carrier transporting molecular materials and their applications in devices. Chem Rev 107:953
4. Xiao L, Chen Z, Qu B, Luo J, Kong S, Gong Q, Kido J (2011) Recent progresses on materials for electrophosphorescent organic light-emitting devices. Adv Mater 23:926
5. Hardin BE, Snaith HJ, McGehee MD (2012) The renaissance of dye-sensitized solar cells. Nat Photon 6:162
6. Credgington D, Durrant JR (2012) Insights from transient optoelectronic analyses on the open-circuit voltage of organic solar cells. J Phys Chem Lett 3:1465
7. Lipomi DJ, Tee BC-K, Vosgueritchian M, Bao Z (2011) Stretchable organic solar cells. Adv Mater 23:1771
8. Kaltenbrunner M, White MS, Głowacki ED, Sekitani T, Someya T, Sariciftci NS, Bauer S (2012) Ultrathin and lightweight organic solar cells with high flexibility. Nat Commun 3:770
9. Pai DM, Yanus JF, Stolka M (1984) Trap-controlled hopping transport. J Phys Chem 88:4714
10. Kudo K, Yamashina M, Morizumi T (1984) Field effect measurement of organic dye films. Jpn J Appl Phys 23:130
11. Rosen G (1966) Space-charge-limited currents in non-metallic solids. Phys Rev Lett 17:692
12. Takeya J, Kato J, Hara K, Yamagishi M, Hirahara R, Yamada K, Nakazawa Y, Ikehata S, Tsukagoshi K, Aoyagi Y, Takenobu T, Iwasa Y (2007) In-crystal and surface charge transport of electric-field-induced carriers in organic single-crystal semiconductors. Phys Rev Lett 98:196804
13. Juška G, Arlauskas K, Viliūnas M, Kočka J (2000) Extraction current transients: new method of study of charge transport in microcrystalline silicon. Phys Rev Lett 84:4946
14. Yang H, Shin TJ, Ling M-M, Cho K, Ryu CY, Bao Z (2005) Conducting AFM and 2D GIXD studies on pentacene thin films. J Am Chem Soc 127:11542
15. Lee S, Koo B, Shin J, Lee E, Park H, Kim H (2006) Effects of hydroxyl groups in polymeric dielectrics on organic transistor performance. Appl Phys Lett 88:162109
16. Zhang X-H, Tiwari SP, Kim S-J, Kippelen B (2009) Low-voltage pentacene organic field-effect transistors with high-k HfO_2 gate dielectrics and high stability under bias stress. Appl Phys Lett 95:223302
17. Jurchescu OD, Baas J, Palstra TTM (2004) Effect of impurities on the mobility of single crystal pentacene. Appl Phys Lett 84:3061

18. Yamagishi M, Takeya J, Tominari Y, Nakazawa Y, Kuroda T, Ikehata S, Uno M, Nishikawa T, Kawase T (2007) High-mobility double-gate organic single-crystal transistors with organic crystal gate insulators. Appl Phys Lett 90:182117
19. Minemawari H, Yamada T, Matsui H, Tsutsumi J, Haas S, Chiba R, Kumai R, Hasegawa T (2011) Inkjet printing of single-crystal films. Nature 475:364
20. Goldmann C, Krellner C, Pernstich KP, Haas S, Gundlach DJ, Batlogg B (2006) Determination of the interface trap density of Rubrene single-crystal field-effect transistors and comparison to the bulk trap density. J Appl Phys 99:034507
21. Umeda T, Kumaki D, Tokito S (2009) Surface-energy-dependent field-effect mobilities up to 1 cm ^{2}/V s for polymer thin-film transistor. J Appl Phys 105:024516
22. Guo X, Zhang M, Tan J, Zhang S, Huo L, Hu W, Li Y, Hou J (2012) Influence of D/A ratio on photovoltaic performance of a highly efficient polymer solar cell system. Adv Mater 24:6536
23. Lecover R, Williams N, Markovic N, Reich DH, Naiman DQ, Katz HE (2012) Next-generation polymer solar cell materials: designed control of interfacial variables. ACS Nano 6:2865
24. Garcia-Belmonte G, Munar A, Barea EM, Bisquert J, Ugarte I, Pacios R (2008) Charge carrier mobility and lifetime of organic bulk heterojunctions analyzed by impedance spectroscopy. Org Electron 9:847
25. Ai X, Beard MC, Knutsen KP, Shaheen SE, Rumbles G, Ellingso RJ (2006) Photoinduced charge carrier generation in a poly(3-hexylthiophene) and methanofullerene bulk heterojunction investigated by time-resolved terahertz spectroscopy. J Phys Chem B 110:25462
26. Saeki A, Koizumi Y, Aida T, Seki S (2012) Comprehensive approach to intrinsic charge carrier mobility in conjugated organic molecules, macromolecules, and supramolecular architectures. Acc Chem Res 45:1193
27. Seki S, Saeki A, Sakurai T, Sakamaki D (2014) Charge carrier mobility in organic molecular materials probed by electromagnetic waves. Phys Chem Chem Phys 16:11093
28. Grozema FC, Siebbeles LDA (2011) Charge mobilities in conjugated polymers measured by pulse radiolysis time-resolved microwave conductivity: from single chains to solids. J Phys Chem Lett 2:2951
29. Savenije TJ, Ferguson AJ, Kopidakis N, Rumbles G (2013) Revealing the dynamics of charge carriers in polymer: fullerene blends using photoinduced time-resolved microwave conductivity. J Phys Chem C 117:24085
30. Dunn HK, Peter LM, Bingham SJ, Maluta E, Walker AB (2012) In situ detection of free and trapped electrons in dye-sensitized solar cells by photo-induced microwave reflectance measurements. J Phys Chem C 116:22063
31. Warman JM, Gelinck GH, de Haas MP (2002) The mobility and relaxation kinetics of charge carriers in molecular materials studied by means of pulse-radiolysis time-resolved microwave conductivity: dialkoxy-substituted phenylene-vinylene polymers. J Phys Condens Matter 14:9935
32. Turner GM, Beard MC, Schmuttenmaer CA (2002) Carrier localization and cooling in dye-sensitized nanocrystalline titanium dioxide. J Phys Chem B 106:11716
33. Saeki A, Seki S, Sunagawa T, Ushida K, Tagawa S (2006) Charge-carrier dynamics in polythiophene films studied by in-situ measurement of flash-photolysis time-resolved microwave conductivity (FP-TRMC) and transient optical spectroscopy (TOS). Philos Mag 86:1261
34. Saeki A, Tagawa S (2009) Nanometer-scale dynamics of charges generated by radiations in condensed matter. Pure Appl Chem 81:45
35. Saeki A, Seki S, Takenobu T, Iwasa Y, Tagawa S (2008) Mobility and dynamics of charge carriers in Rubrene single crystals studied by flash-photolysis microwave conductivity and optical spectroscopy. Adv Mater 20:920
36. Okamoto T, Nakahara K, Saeki A, Seki S, Oh JH, Akkerman HB, Bao Z, Matsuo Y (2011) Aryl-perfluoroaryl substituted tetracene: induction of face-to-face π–π stacking and enhancement of charge carrier properties. Chem Mater 23:1646
37. Wakamiya A, Nishimura H, Fukushima T, Suzuki F, Saeki A, Seki S, Osaka I, Sasamori T, Murata M, Murata Y, Kaji H (2014) On-top π-stacking of quasiplanar molecules in hole-transporting materials: inducing anisotropic carrier mobility in amorphous films. Angew Chem Int Ed 53:5800

38. Yamamoto Y, Zhang G, Jin W, Fukushima T, Ishii N, Saeki A, Seki S, Tagawa S, Minari T, Tsukagoshi K, Aida T (2009) Ambipolar-transporting coaxial nanotubes with a tailored molecular graphene–fullerene heterojunction. Proc Natl Acad Sci 106:21051
39. Saeki A, Ohsaki S, Seki S, Tagawa S (2008) Electrodeless determination of charge carrier mobility in poly(3-hexylthiophene) films incorporating perylenediimide as photoconductivity sensitizer and spectroscopic probe. J Phys Chem C 112:16643
40. Saeki A, Yoshikawa S, Tsuji M, Koizumi Y, Ide M, Vijayakumar C, Seki S (2012) A versatile approach to organic photovoltaics evaluation using white light pulse and microwave conductivity. J Am Chem Soc 134:19035
41. Tsuji M, Saeki A, Koizumi Y, Matsuyama N, Vijayakumar C, Seki S (2014) Benzobisthiazole as weak donor for improved photovoltaic performance: microwave conductivity technique assisted molecular engineering. Adv Funct Mater 24:28
42. Ide M, Koizumi Y, Saeki A, Izumiya Y, Ohkita H, Ito S, Seki S (2013) Near-infrared absorbing thienoisoindigo-based copolymers for organic photovoltaics. J Phys Chem C 117:26859
43. Gopal A, Saeki A, Ide M, Seki S (2014) Fluorination of benzothiadiazole–benzobisthiazole copolymer leads to additive-free processing with meliorated solar cell performance. ACS Sustain Chem Eng 2:2613
44. Li H, Earmme T, Ren G, Saeki A, Yoshikawa S, Murari NM, Subramaniyan S, Crane MJ, Seki S, Jenekhe SA (2015) Beyond fullerenes: design of nonfullerene acceptors for efficient organic photovoltaics. J Am Chem Soc 136:14589
45. Honsho Y, Miyakai T, Sakurai T, Saeki A, Seki S (2013) Evaluation of intrinsic charge carrier transport at insulator-semiconductor interfaces probed by a non-contact microwave-based technique. Sci Rep 3:3182
46. Choi W, Miyakai T, Sakurai T, Saeki A, Yokoyama M, Seki S (2014) Non-contact, non-destructive, quantitative probing of interfacial trap sites for charge carrier transport at semiconductor-insulator boundary. Appl Phys Lett 105:033302
47. Murthy DHK, Gao M, Vermeulen MJW, Siebbeles LDA, Savenije TJ (2012) Mechanism of mobile charge carrier generation in blends of conjugated polymers and fullerenes: significance of charge delocalization and excess free energy. J Phys Chem C 116:9214
48. Oga H, Saeki A, Ogomi Y, Hayase S, Seki S (2014) Improved understanding of the electronic and energetic landscapes of perovskite solar cells: high local charge carrier mobility, reduced recombination, and extremely shallow traps. J Am Chem Soc 136:13818
49. Service RF (2014) Perovskite solar cells keep on surging. Science 344:458
50. Saeki A, Yasutani Y, Oga H, Seki S (2014) Frequency-modulated gigahertz complex conductivity of TiO_2 nanoparticles: interplay of free and shallowly trapped electrons. J Phys Chem C 118:22561
51. Beaujuge PM, Reynolds JR (2010) color control in π-conjugated organic polymers for use in electrochromic devices. Chem Rev 110:268–320
52. Mei J, Bao Z (2014) Side chain engineering in solution-processable conjugated polymers. Chem Mater 26:604–615
53. Noguchi Y, Saeki A, Fujiwara T, Yamanaka S, Kumano M, Sakurai T, Matsuyama N, Nakano M, Hirao N, Ohishi Y, Seki S (2015) Pressure modulation of backbone conformation and intermolecular distance of conjugated polymers toward understanding the dynamism of π-figuration of their conjugated system. J Phys Chem 119:7219
54. Grell M, Bradley DDC, Long X, Chamberlain T, Inbasekaran M, Woo EP, Soliman M (1998) Chain geometry, solution aggregation and enhanced dichroism in the liquidcrystalline conjugated polymer poly(9,9-dioctylfluorene). Acta Polym 49:439
55. Saeki A, Fukumatsu T, Seki S (2011) Intramolecular charge carrier mobility in fluorene-thiophene copolymer films studied by microwave conductivity. Macromolecules 44:3416
56. Fukumatsu T, Saeki A, Seki S (2012) Charge carrier mobilities in amorphous triphenylamine–fluorene copolymers: role of triphenylamine unit in intra- and intermolecular charge transport. Appl Phys Express 5:061701
57. Lucas B, El Amrani A, Moliton A, Skaiky A, El Hajj A, Aldissi M (2012) Charge transport properties in pentacene films: evaluation of carrier mobility by different techniques. Solid-State Electron 69:99

58. Yasutani Y, Honsho Y, Saeki A, Seki S (2012) Polycarbazoles: relationship between intra- and intermolecular charge carrier transports. Synth Met 162:1713
59. Yamamoto Y, Fukushima T, Saeki A, Seki S, Tagawa S, Ishii N, Aida T (2007) Molecular engineering of coaxial donor–acceptor heterojunction by coassembly of two different hexabenzocoronenes: graphitic nanotubes with enhanced photoconducting Properties. J Am Chem Soc 129:9276
60. Saeki A, Yamamoto Y, Koizumi Y, Fukushima T, Aida T, Seki S (2011) Photoconductivity of self-assembled hexabenzocoronene nanotube: insight into the charge carrier mobilities on local and long-range scales. J Phys Chem Lett 2:2549
61. Novoselov KS, Geim AK, Morozov SV, Jiang D, Zhang Y, Dubonos SV, Grigorieva IV, Firsov AA (2004) Electric field effect in atomically thin carbon films. Science 306:666

Chapter 37
STM Characterization of π-Electron Systems

Yutaka Majima

Abstract Molecular-resolution scanning tunneling microscopy (STM) and scanning tunneling spectroscopy (STS) are powerful tools for probing the electrical properties of individual π-electron systems, providing us with vital information on adsorption geometry, intermolecular patterns, and local conductivity of single molecules. These characteristics allow us to understand the function of single molecules and the mechanism of how they operate which in turn allow us to intelligently design more complicated multicomponent architectures. In this chapter, characterizations of π-electron systems by STM and STS are demonstrated. Experimental conditions of STM and STS for the measurements of π-electron systems are summarized. Molecular-resolution STM images of endohedral metallofullerenes and subporphyrins are shown. STS measurements reveal electrical properties on π-electron systems such as HOMO and LUMO. Molecular resonant tunneling behaviors of tribenzosubporphyrin are presented.

Keywords Scanning tunneling microscopy (STM) • Scanning tunneling spectroscopy (STS) • Resonant tunneling diode • Molecular device • HOMO-LUMO gap

37.1 Experimental Conditions of STM and STS

In this section, experimental conditions of STM and STS for the measurements of π-electron systems are demonstrated [1].

Alkanethiol self-assembled monolayer (SAM) on the Au(111) substrate is good platform to observe STM images of individual π-electron systems, since it behaves as tunneling barrier which prevents interferences between the molecular orbitals of π-electron systems and the Au(111) surface. Figure 37.1 shows a typical schematic of STM image observation for an individual π-electron system on alkanethiol SAM. The STM tip was an electropolished W probe and a mechanically cut PtIr probe. The Au(111) substrate was fabricated by thermal evaporation of gold onto

Y. Majima (✉)
Materials and Structures Laboratory, Tokyo Institute of Technology, Yokohama 226-8503, Japan
e-mail: majima.y.aa@m.titech.ac.jp

T. Akasaka et al. (eds.), *Chemical Science of π-Electron Systems*,
DOI 10.1007/978-4-431-55357-1_37

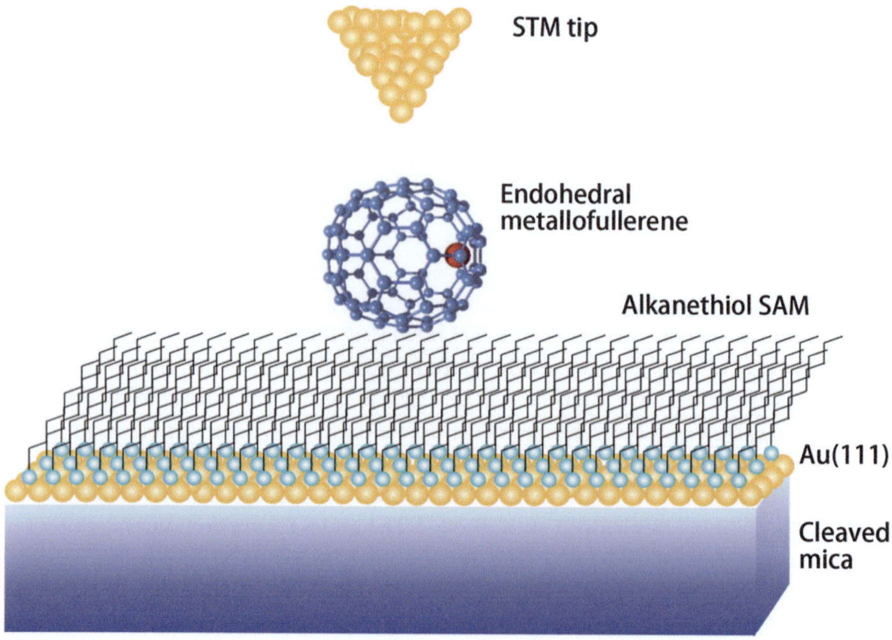

Fig. 37.1 Schematic diagram of STM image observation of π-electron system on alkanethiol SAM

freshly cleaved mica. Prior to gold evaporation, the mica substrates were held in a vacuum at 773 K for 4 h. The substrate temperature was maintained at 773 K during gold evaporation, and postannealing was performed at 748 K for 1 h. The substrate was briefly flame annealed and quenched in ethanol to form an atomically flat Au(111) surface. Alkanethiol SAM is prepared by immersing the Au (111) substrate in a 1 mM solution of alkanethiol in ethanol for 24 h, and the samples were rinsed with ethanol two times and dried in a pure N_2 flow. The STM images were measured at sample bias voltages (V) in the constant-current mode at 68 K by the ultrahigh vacuum (UHV) low-temperature STM (Unisoku, USM-501 modified) with a base pressure below 3.0×10^{-8} pA. The minimum set point current for the STM image observations is 2 pA that is enabled by using the low-noise preamplifier (DL instruments, 1211, battery mode). Figure 37.2a shows a typical STM image of hexanethiol (C6S) SAM on Au (111) surface. White dots correspond to hexanethiol molecule. Hexanethiols are chemically adsorbed at threefold hollows on the Au (111) surface forming the $\sqrt{3} \times \sqrt{3}R30°$ structure (Fig. 37.2b).

Molecular-resolution STM images of the Lu@C_{82} endohedral metallofullerene on C6S SAM at 65 K are shown in Fig. 37.3 [2]. Lu@C_{82} was sublimated on C6S SAM on Au(111)surface. Lu@C_{82} molecules tend to be trapped along the Au(111) monatomic step edge, at etch pits that arise from the rearrangement of surface gold atoms, and at the grain boundary of the octanethiol SAM. These

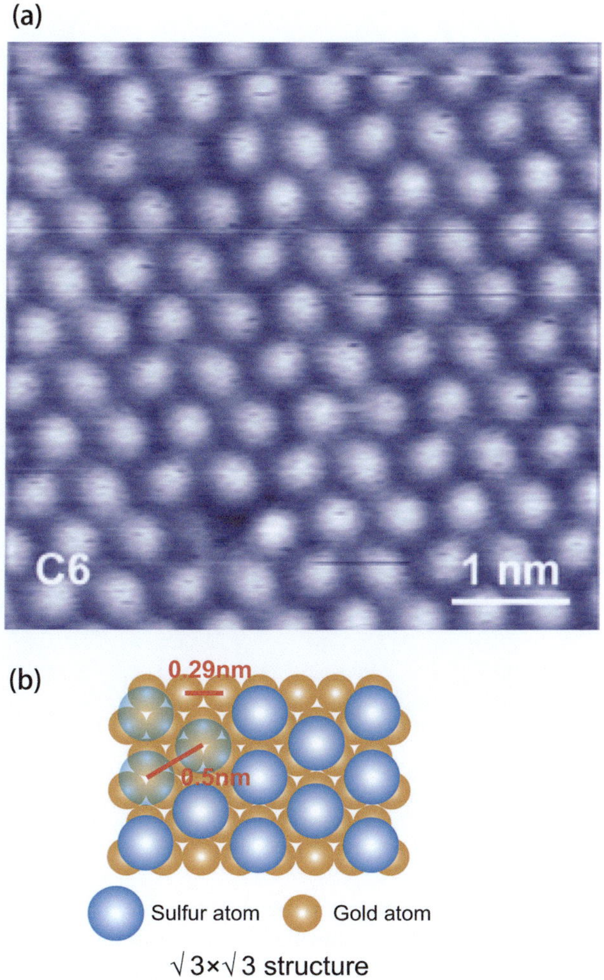

Fig. 37.2 (**a**) Typical STM image of hexanethiol (C6S) SAM on Au(111)/mica substrate. (**b**) Schematic of $\sqrt{3} \times \sqrt{3}R30°$ structure of alkanethiol on Au(111)

molecular diffusions show that a weak interaction exists between the molecule and the alkanethiol SAM due to van der Waals attraction. In Fig. 37.3, a well-ordered C6S SAM can be observed as the $\sqrt{3} \times \sqrt{3}R30°$ structure with a distance of 0.50 nm between octanethiol molecules. In Fig. 37.3, the bright spherical spots are Lu@C_{82} molecules. A striped structure of the Lu@C_{82} molecule is observed, suggesting that thermal rotations of Lu@C_{82} adsorbed on hexanethiol SAMs are prevented. The clear striped patterns of Lu@C_{82} which represent the internal structure of Lu@C_{82} were observed when images with a low sample bias voltage in the range of ±0.8 V. The striped structure corresponds to the molecular orbital of the fullerene shell in an individual Lu@C_{82} molecule. By comparing these molecular-resolution STM images with Kohn-Sham molecular orbitals of Lu@C_{82} calculated by density functional theory (DFT), we identified the molecular orientation of Lu@C_{82} [2].

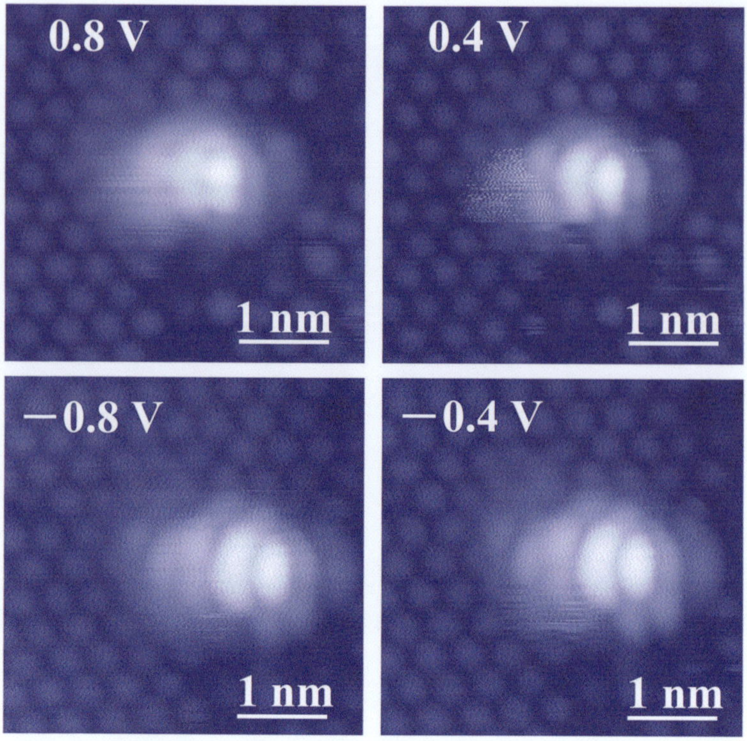

Fig. 37.3 Molecular-resolution STM images of single-metal endohedral metallofullerene Lu@C_{82} on C6S SAM/Au(111). Sample bias voltage dependence at 0.8, 0.4, −0.8, and −0.4 V are shown

We have also demonstrated the single molecular orientation switching of the single-metal endohedral metallofullerene by using low-temperature STM [3].

The tunneling current I-distance (d) characteristics of alkanethiol SAMs were obtained by using a scanning vibrating probe at the STM tip bias voltage of 2 V [3]. Figure 37.4a shows the experimental results of semilogarithmic $I-d$ characteristics. The kinks in the two tunneling current slopes are clearly observed in the $I - d$ characteristics of both hexanethiol (C6S) and octanethiol (C8S) SAMs, suggesting the contact between the top of the STM tip and the end of hexanethiol or octanethiol SAMs. Because the relative probe-substrate distance is set to zero at the kink, the top of the STM probe is located in a vacuum in the positive region of relative distance, and the STM probe penetrates the alkanethiol SAM in the negative region. In the positive region, the tunneling current decreases two digit per only 2 Å. As the diameter of metals is about 3 Å, the tunneling current flows only from the one metal atom at the tip-top. This is the reason why STM image has a molecular resolution.

The tunneling current at the kink can be estimated to be 7.0 nA and 680 pA for C6S and C8S SAMs, respectively. Figure 37.4b shows schematic of cross-sectional view of C6S and C8S SAMs. The vertical height difference between C8S and C6S

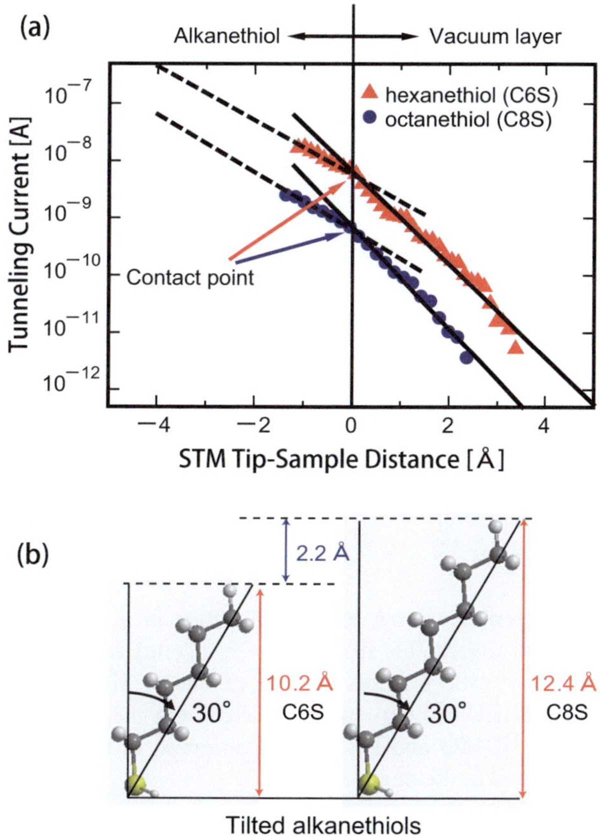

Fig. 37.4 (**a**) Semilogarithmic plot of $I-d$ characteristics above C6S and C8S SAMs on Au(111). The STM tip bias voltage is set to 2 V in both these measurements (Reprinted (adapted) with permission from Ref. [3]. Copyright 2005 American Chemical Society). (**b**) Vertical height difference between C6S and C8S, that tilt 30° from vertical axis

SAMs is 2.2 Å. As the tunneling current at the kink of C8S is one digit smaller than that of C6S, the change in the alkane chain of two methylene group results in one digit reduction of the tunneling current at the kink. Because the relative STM tip-sample distance is set to zero at the kink, choice of an alkanethiol chain length is important to observe good molecular-resolution STM images of π-electron systems.

During STS measurements, the tip has to be held at a fixed position above the sample surface. Considering this tunneling current dependence on the distance, the reproducibility of STS measurements has to be checked. Figure 37.5 shows the typical $I-V$ characteristics of heptanethiol (C7S) on Au(111) measured by STS. In this STS measurement, consecutive bidirectional sample bias voltage sweeps were made. As the $I-V$ characteristics of backward bias voltage sweep are the same as that of forward bias voltage sweep, the reproducibility can be confirmed within this STS measurement.

The tunneling current depends on the electronic local density of states (DOS) of the tip and the sample. The differential conductance (dI/dV) provides a direct measurement of the DOS of the sample surface. As shown in Fig. 37.5, the $I-V$

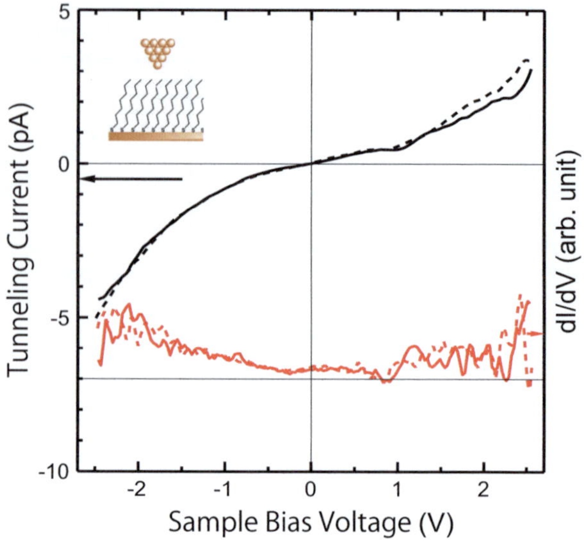

Fig. 37.5 Typical $I - V$ and $\mathrm{d}I/\mathrm{d}V - V$ characteristics of C7S SAMs with consecutive bidirectional sample bias voltage sweeps from −2.5 to 2.5 V (*solid line*) and from 2.5 to −2.5 V (*dashed line*)

characteristics show low conductance in a wide range between 2.5 and −2.5 V without discernible dI/dV peak in accord with no electroactive material of C7S. These are the features of STS results of alkanethiol that are derived from large HOMO-LUMO gap energy of 6 eV. On the contrary, dI/dV peaks corresponding to the HOMO and LUMO are observed in the case of the π-electron systems as described in the next section.

37.2 Molecular Resonant Tunneling Diode Demonstrated by STS

In this section, molecular electronic device operations of tribenzosubporphine in a single molecular level are demonstrated by combination of STM and STS. Negative differential resistance (NDR) phenomena have been clearly observed, which have been attributed to molecular resonant tunneling between a neutral tribenzosubporphine anchored to the metal surface and a tribenzosubporphine cation adsorbed on a W tip [4].

Prior to the construction of molecular electronic devices that contain multiple components, it is necessary to understand the electrical properties of individual molecules. While various mechanisms have been proposed for NDR, the most important mechanism is a molecular-origin resonant tunneling diode (RTD). In conventional RTD with compound semiconductors, there is a quantum well between two tunnel barriers with doped contacts on either side to form reservoirs of electrons. When a voltage is applied, a current can only flow when electrons can tunnel through the subband state in the quantum well. For NDR behavior, these subbands have to constitute discrete energy levels. Therefore, π-conjugated molecules with discrete energy levels originating from molecular orbitals can be used as quantum wells in molecular RTD.

Tribenzosubporphines are an interesting class of π-conjugated molecules in light of their 14π-aromatic circuits, triangular bowl-shaped structures, intense green fluorescence, and nonlinear optical properties [5–7]. In addition, due to the lack of substituents at the meso position, tribenzosubporphines possess streamlined contours which are ideal for single molecular observation by STM. Density functional theory (DFT) calculations reveal that tribenzosubporphines have molecular orbital diagrams with discrete energy levels of frontier orbitals such as HOMO-1, HOMO, and nearly generate LUMO, reflecting their compact aromatic system. These features are considered to be ideal for detection of RTD phenomena.

Figure 37.6b shows molecular-resolution STM image ($2.5 \times 2.5\,\text{nm}^2$) of a single molecule of B-(5-mercaptopentoxy)tribenzosubporphine inserted in C7S SAM at various sample voltage with a set point current of $-4\,\text{pA}$. A periodic structure behind B-(5-mercaptopentoxy)tribenzosubporphine was due to the head part of C7S SAM. These STM images show a triangle-like structure. A valley structure was observed at $-1.6\,\text{V}$ [4]. To demonstrate the Kohn-Sham molecular orbitals of B-(5-mercaptopentoxy)tribenzosubporphine, we have carried out ab initio pseudopotential DFT calculations [8, 9]. Geometry optimizations were carried out using the B3LYP functional and 6-31G (d) basis set with Gaussian 03 [10–12]. Graphical presentations of the computational results were generated by using GaussView 4.1 (Gaussian Inc.). Figure 37.6c shows calculated Kohn-Sham HOMO orbitals of isolated B-(5-mercaptopentoxy)tribenzosubporphine. It is important to note that the valley agrees with the Kohn-Sham molecular orbitals of HOMO of B-(5-mercaptopentoxy)tribenzosubporphine. It notes that the energy levels of calculated Kohn-Sham molecular orbitals indicate degenerate of LUMO and LUMO+1.

The $I - V$ and $\mathrm{d}I/\mathrm{d}V - V$ characteristics of B-(5-mercaptopentoxy)tribenzosubporphine are clearly different from those of C7S SAM (shown in Fig. 37.5). Figure 37.7 shows the representative $I - V$ and $\mathrm{d}I/\mathrm{d}V - V$ characteristics of B-(5-mercaptopentoxy)tribenzosubporphine and exhibits low conductance in a range between 2 and -1 V, but displays discrete $\mathrm{d}I/\mathrm{d}V$ peaks at $V = 2.4, -1.3$, and -2.1 V, apparently via the availability of the conductance between the tip and the molecular orbitals of B-(5-mercaptopentoxy)tribenzosubporphine. We have measured $I - V$ and $\mathrm{d}I/\mathrm{d}V - V$ characteristics in all the 8×8 divided parts of the STM image ($3 \times 3\,\text{nm}^2$) of an individual molecule of B-(5-mercaptopentoxy)tribenzosubporphine inserted in C7S SAM with consecutive bidirectional sample bias voltage sweeps from -2.5 to 2.5 V and from 2.5 to -2.5 V [4]. Spatial mapping STS is not easy and time consuming, but we employed this in order to reveal detailed molecular orbital features in a single molecular level. Importantly, however, we have confirmed that the measurement results of backward sweeps agree well with those of forward sweeps for all of 64 parts, which assures the reliability of the measurement results of $I - V$ and $\mathrm{d}I/\mathrm{d}V - V$ curves.

In order to understand the observed $\mathrm{d}I/\mathrm{d}V$ peaks quantitatively, a voltage distribution of W tip/vacuum/B-(5-mercaptopentoxy)tribenzosubporphine/insulating

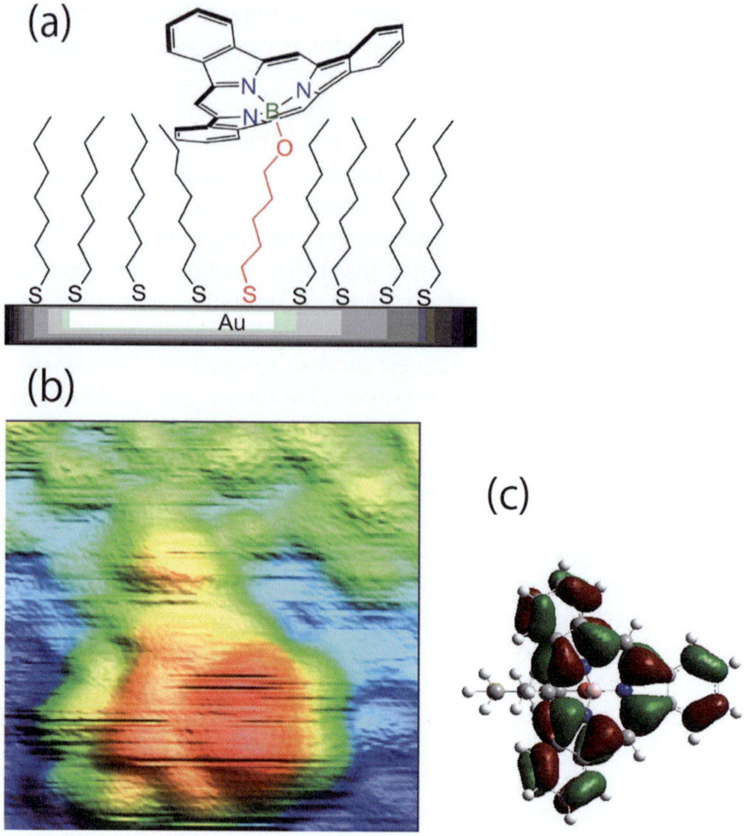

Fig. 37.6 (**a**) Schematic side view consisting of B-pentanethioloxy tribenzosubporphine: C7S mixed SAM/Au(111). (**b**) STM image of single B-pentanethioloxy tribenzosubporphine inserted in C7S SAM under the sample bias voltage of −1.6 V, at the set point current of −4 pA. STM image size is $2.5 \times 2.5\,\mathrm{nm}^2$. (**c**) Kohn-Sham molecular orbital of B-pentanethioloxy tribenzosubporphine with isovalue of 0.02 e/A^3 at HOMO (Majima et al. [4]. Copyright 2013 American Chemical Society)

layer/Au(111) surface was taken up, where insulating layer is C7S SAM. When the tunneling current flows through a molecular orbital, the positive and negative tunneling threshold voltages V_+ and V_- are expressed by Eqs. 37.1 and 37.2,

$$V_+ = \frac{C_1 + C_2}{eC_2} E_{\mathrm{dis}}^+ \tag{37.1}$$

and

$$V_- = -\frac{C_1 + C_2}{eC_2} E_{\mathrm{dis}}^-, \tag{37.2}$$

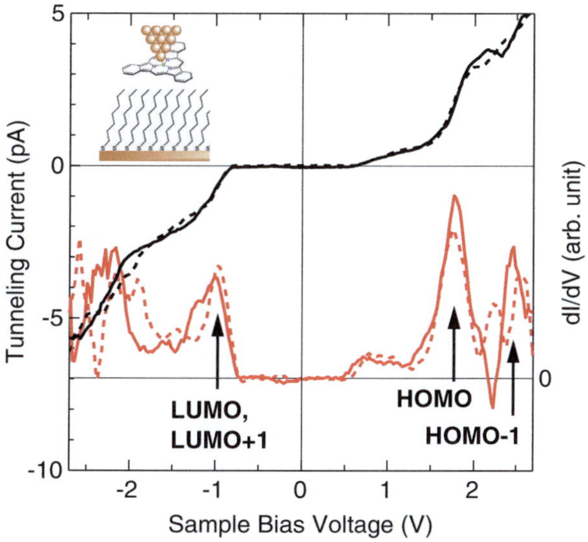

Fig. 37.7 Representative $I - V$ and $\mathrm{d}I/\mathrm{d}V - V$ characteristics measured on B-pentanethioloxy tribenzosubporphine with a bare W tip (Majima et al. [4]. Copyright 2013 American Chemical Society)

where C_1 is the capacitance of the vacuum layer between the W tip and B-(5-mercaptopentoxy)-tribenzosubporphine and C_2 is the capacitance between B-(5-mercaptopentoxy)tribenzosubporphine and Au(111) surface. Here, E^+_{dis} and E^-_{dis} express the energy differences between LUMO of B-(5-mercaptopentoxy)tribenzosubporphine and Fermi level (E_{F}) and between E_{F} and HOMO (or HOMO-1) of **1**, respectively [2, 13]. Using the electrical image method [14], C_1 and C_2 were estimated to be 0.11 and 0.23 aF, respectively, provided that the relative permittivity of alkanethiol SAM is 2.6 [15, 16], the distance between the W tip and B-(5-mercaptopentoxy)tribenzosubporphine is 0.35 nm [3], and the C7S SAM height above the Au(111) surface is 1.09 nm [17].

Considering the electron-donating properties of B-(5-mercaptopentoxy)tribenzosubporphine [5], the $\mathrm{d}I/\mathrm{d}V$ peak at $V = -1.3$ V in Fig. 37.7 is attributed to the HOMO level (E_{HOMO}) of **1**. Energy difference E^-_{dis} ($= E_{\text{F}} - E_{\text{HOMO}}$) is evaluated to be 0.88 eV by Eq. 37.2. As the second peak at $V = -2.1$ V in Fig. 37.7 is attributed to the HOMO-1 level ($E_{\text{HOMO}-1}$) of B-(5-mercaptopentoxy)-tribenzosubporphine, the energy difference E^-_{dis} ($= E_{\text{F}} - E_{\text{HOMO}-1}$) is evaluated to be 1.42 eV. As a result, $E_{\text{HOMO}} - E_{\text{HOMO}-1}$ is estimated as 0.54 eV.

On the other hand, the $\mathrm{d}I/\mathrm{d}V$ peak at $V = 2.4$ V in Fig. 37.7 is attributed to the LUMO level (E_{LUMO}) of B-(5-mercaptopentoxy)tribenzosubporphine. Energy difference E^+_{dis} ($= E_{\text{LUMO}} - E_{\text{F}}$) is evaluated to be 1.62 eV by Eq. 37.1. Consequently, the HOMO-LUMO gap is evaluated as 2.5 eV from the STS measurements. Therefore, the energy diagram of W tip/vacuum/B-(5-mercaptopentoxy)tribenzosubporphine:C7S SAM/Au(111) structure is described as shown in Fig. 37.8.

During the STM observations, we found that the tribenzosubporphine cation was often liberated from a 5-mercaptopentoxy anchor and adsorbed on to the W tip.

Fig. 37.8 Energy diagrams of W tip/vacuum/ B-pentanethioloxy tribenzosubporphine:C7S SAM/Au(111) structure

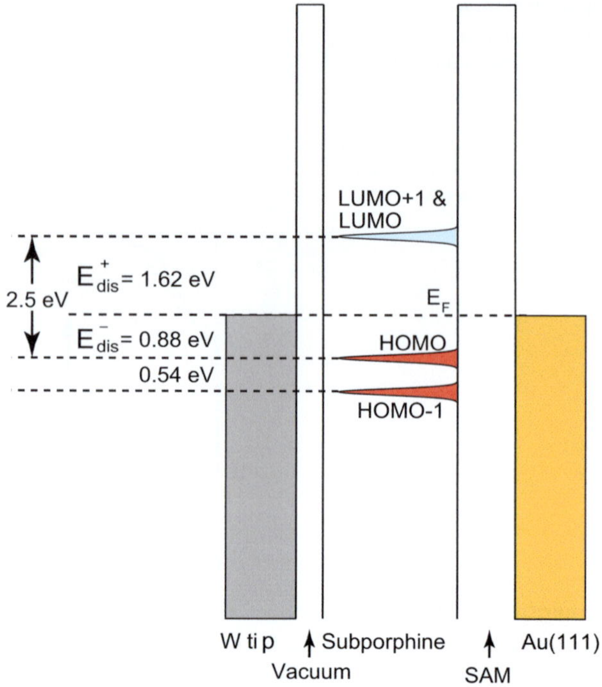

In this case, a liberated tribenzosubporphine cation was thought to be adsorbed on to the W tip on the basis of the following observations. We found that the cation-adsorbed W tip was very robust, and thus once the cation was adsorbed, it was rather difficult to detach the cation from the tip. The typical $I - V$ and $dI/dV - V$ characteristics of the electroactive subporphine cation-adsorbed W tip on C7S SAM are shown in Fig. 37.9. Discrete dI/dV main peaks are observed at $V = 2.4$, 1.8, −0.9, and −2.0 V [4]. When the electroactive subporphine cation is adsorbed at the tip, the capacitance C_1 of the vacuum layer between the W tip and tribenzosubporphine cation becomes much larger than that C_2 between tribenzosubporphine cation and Au(111) surface ($C_1 \gg C_2$). Consequently, dI/dV main peaks at $V = 2.4, 1.8, and - 0.9$ are attributed to HOMO-1, HOMO, and LUMO levels, respectively. It notes that the peak voltage differences between HOMO-1 and HOMO and between HOMO and LUMO are 0.6 and 2.7 V, respectively, which values are almost equal to the energy differences of $E_{HOMO} - E_{HOMO-1}$ of 0.54 eV and HOMO-LUMO gap of 2.5 eV, respectively. As a result, HOMO and LUMO levels of tribenzosubporphine cation are pinned to Fermi level of the W tip. Energy differences $E_F - E_{HOMO}$, $E_F - E_{HOMO-1}$, and $E_{LUMO} - E_F$ are evaluated to be 1.8, 2.4, and 0.9 eV, respectively. It notes that the HOMO-1, HOMO, and LUMO levels are approximately 0.9 eV shifted owing to the adsorbed electroactive subporphine cation. Consequently, energy levels of subporphyrin cation are pinned by the W tip due to the adsorption and are shifted 0.9 eV as compared with those of neutral

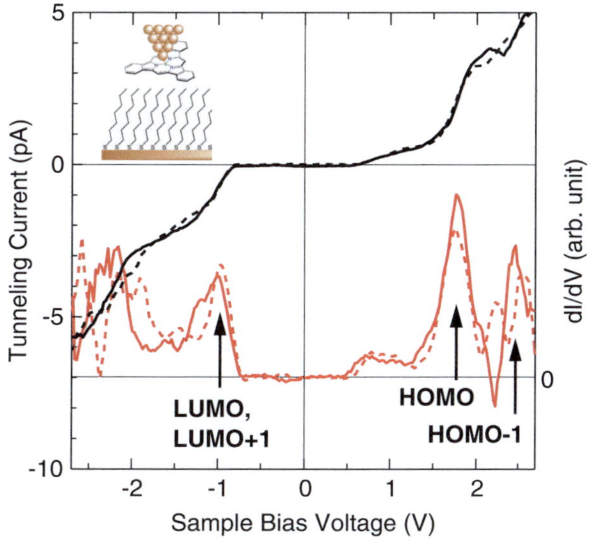

Fig. 37.9 Representative $I-V$ and $\mathrm{d}I/\mathrm{d}V-V$ characteristics measured on C7S SAM with tribenzosubporphine-cation attached W tip (Majima et al. [4]. Copyright 2013 American Chemical Society)

subporphyrin. By comparing STS results of Figs. 37.5 and 37.9, differentiation of tribenzosubporphine cation-absorbed W tip from the bare W tip is possible.

Figure 37.10 shows the typical $I-V$ and $\mathrm{d}I/\mathrm{d}V-V$ characteristics of tribenzosubporphine cation-absorbed W tip/vacuum/B-(5-mercaptopentoxy) tribenzosubporphine/Au(111) structure, which exhibits distinct NDR peak at $V=-1.9$ V with a peak-to-valley (PV) ratio of 2.6 in both forward and backward voltage sweeps. These NDR peaks were repeatedly observed not only at $V=-1.9$ V but also at $V=-2.8$ V with consecutive bidirectional sample voltage sweep from −3 to 3 V and from 3 to −3 [4].

Here, we discuss on the mechanism of the NDR phenomena based on molecular resonant tunneling. Energy difference between E_{LUMO} and E_{F} of the subporphyrin cations is 0.9 eV and that between E_{F} and E_{HOMO} is 1.8 eV. These results indicate that Fermi energy of the W tip absorbing subporphyrin cations is 0.9 eV close to LUMO level rather than LUMO level of B-(5-mercaptopentoxy) tribenzosubporphine. An energy diagram of W tip/tribenzosubporphirine cation/vacuum/B-(5-mercaptopentoxy)tribenzosubporphine/C7S SAM/Au(111) is shown in Fig. 37.9.

When negative sample voltage was applied to Au(111) substrate and reached at $V=-1.9$ V, an energy level alignment between the HOMO level of B-(5-mercaptopentoxy)tribenzosubporphine and the LUMO level of the tribenzosubporphirine cation resulted in molecular resonant tunneling, and electrons were easy to tunnel between W tip and Au(111) through the HOMO level of B-(5-mercaptopentoxy)tribenzosubporphine and the LUMO level of the tribenzosubporphirine cation. If larger negative sample bias voltage of −1.9 V was applied, the energy level overlap between the HOMO level of B-(5-mercaptopentoxy)tribenzosubporphine and the LUMO level of tribenzosubporphirine cation was reduced, which resulted in a reduction in current.

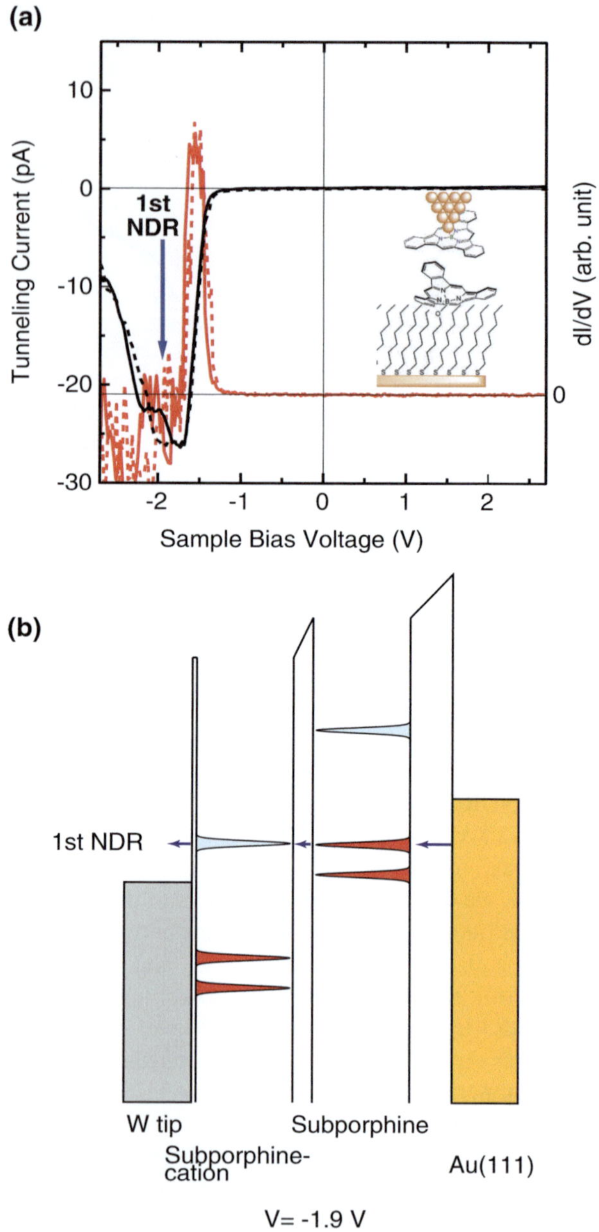

Fig. 37.10 (**a**) Representative $I - V$ and $dI/dV - V$ characteristics measured on B-(5-mercaptopentoxy) tribenzosubporphine SAM with tribenzosubporphine cation-attached W tip. (**b**) Energy diagrams of W tip/tribenzosubporphine-cation/vacuum/ B-pentanethioloxy tribenzosubporphine:C7S mixed SAM/Au(111) structure at −1.9 (Majima et al. [4]. Copyright 2013 American Chemical Society)

As a result, the first NDR with the peak and the valley was observed. More larger negative sample bias voltage was applied up to −2.8 V; the energy level overlap between the HOMO-1 level of B-(5-mercaptopentoxy)tribenzosubporphine and the LUMO level of the tribenzosubporphirine cation increased again, which allowed for a second molecular resonant tunneling event. As a result, the NDR was observed. If the HOMO and HOMO-1 energy levels were close, this NDR with the peak-to-valley ratio of 2.6 would not be observed since the continuous energy state prevented a conductance reduction. Therefore, it is concluded that the isolated energy levels of B-(5-mercaptopentoxy)tribenzosubporphine play crucial roles in the NDR phenomena.

In summary, the STS measurements of tribenzosubporphine **1** anchored to the Au(111) surface via a 5-mercaptopentoxy chain but embedded in the heptanethiol SAM (C7S SAM) were performed with bare W tip or a tribenzosubporphines cation-adsorbed W tip. The method allowed us to probe the inherent electronic properties of **1**, recording a triangular shape for the HOMO and a circular shape for the HOMO-1 by molecular-resolution STM images and evaluating the HOMO-LUMO gap to be 2.5 eV. In the spatial mapping STM, a tribenzosubporphine cation part of **1** was often found to be liberated from the 5-mercaptopentoxy anchor and adsorbed on to the W tip. When the STS of **1** in C7S SAM was measured by using the cation-adsorbed W tip, NDR phenomena were repeatedly observed with a peak-to-valley ratio of 2.6. The observed NDR phenomena have been attributed to the molecular resonant tunneling diode operation in which the electrons tunnel between the W tip and Au(111) surface through discrete frontier molecular orbitals of **1** and its cation. Through this study, the potential of tribenzosubporphines in molecular electronic devices has been underlined in terms of compact aromatic property with discrete frontier molecular orbitals, smoothly curved molecular shape, and availability of the axial group as an anchor that allows their fixation and electronic decoupling from the metal surface through the use of SAM-coated surface.

Acknowledgements The author thanks Prof. A. Osaka for synthesis of tribenzosubporphines and fruitful discussions. The work was partially supported by a Grant-in-Aid for Scientific Research on Innovative Areas (No. 20108011, π – Space) from the Ministry of Education, Culture, Sports, Science and Technology (MEXT), and MEXT Elements Strategy Initiative to Form Core Research Center, and the BK Plus program, Basic Science Research program (NRF-2014R1A1030419).

References

1. Kano S, Tada T, Majima Y (2015) Nanoparticle characterization based on STM and STS. Chem Soc Rev 44:970
2. Iwamoto M, Ogawa D, Yasutake Y, Azuma Y, Umemoto H, Ohashi K, Izumi N, Shinohara H, Majima Y (2010) Molecular orientation of individual Lu@C82 molecules demonstrated by scanning tunneling microscopy. J Phys Chem C 114:14704
3. Yasutake Y, Shi Z, Okazaki T, Shinohara H, Majima Y (2005) Single molecular orientation switching of an endohedral metallofullerene. Nano Lett 5:1057–1060
4. Majima Y, Ogawa D, Iwamoto M, Azuma Y, Tsurumaki E, Osuka A (2013) Negative differential resistance by molecular resonant tunneling between neutral tribenzosubporphine anchored to a Au(111) surface and tribenzosubporphine cation adsorbed on to a Tungsten tip. J Am Chem Soc 135:14159–14166
5. Inokuma Y, Kwon JH, Ahn TK, Yoon M-C, Kim D, Osuka A (2006) Tribenzosubporphines: synthesis and characterization. Angew Chem Int Ed 45:961

6. Inokuma Y, Osuka A (2008) Subporphyrins: emerging contracted porphyrins with aromatic 14 pi-electronic systems and bowl-shaped structures: rational and unexpected synthetic routes. Dalton Trans 19:2517
7. Osuka A, Tsurumaki E, Tanaka T (2011) Subporphyrins: a legitimate ring-contracted porphyrin with versatile electronic and optical properties. Bull Chem Soc Jpn 84:679
8. Kohn W, Sham L (1965) Quantum density oscillations in an inhomogeneous electron gas. Phys Rev 140:A1133
9. Parr RG, Yang W (1994) Density-functional theory of atoms and molecules. Oxford Science, Oxford
10. Becke AD (1993) Density – functional thermochemistry. III. The role of exact exchange. J Chem Phys 98:5648
11. Lee C, Yang W, Parr RG (1988) Development of the Colle-Salvetti correlation-energy formula into a functional of the electron density. Phys Rev B 37:785
12. Gaussian 09, Revision D. 01 Frisch MJ, Trucks GW, Schlegel HB, Scuseria GE, Robb MA, Cheeseman JR, Scalmani G, Barone V, Mennucci B, Petersson GA, Nakatsuji H, Caricato M, Li X, Hratchian HP, Izmaylov AF, Bloino J, Zheng G, Sonnenberg JL, Hada M, Ehara M, Toyota K, Fukuda R, Hasegawa J, Ishida M, Nakajima T, Honda Y, Kitao O, Nakai H, Vreven T, Montgomery JA, Peralta JE Jr, Ogliaro F, Bearpark M, Heyd JJ, Brothers E, Kudin KN, Staroverov VN, Kobayashi R, Normand J, Raghavachari K, Rendell A, Burant JC, Iyengar SS, Tomasi J, Cossi M, Rega N, Millam JM, Klene M, Knox JE, Cross JB, Bakken V, Adamo C, Jaramillo J, Gomperts R, Stratmann RE, Yazyev O, Austin AJ, Cammi R, Pomelli C, Ochterski JW, Martin RL, Morokuma K, Zakrzewski VG, Voth GA, Salvador P, Dannenberg JJ, Dapprich S, Daniels AD, Farkas O, Foresman JB, Ortiz JV, Cioslowski J, Fox DJ (2009). Gaussian, Inc., Wallingford CT
13. Zhao J, Zeng C, Cheng X, Wang K, Wang G, Yang J, Hou JG, Zhu Q (2005) Single C59N molecule as a molecular rectifier. Phys Rev Lett 95:045502
14. Oyama Y, Majima Y, Iwamoto M (1999) Tunneling resistance of double-barrier tunneling structures with an alkanethiol-protected Au nanoparticle. J Appl Phys 86:7087
15. Zhang H, Yasutake Y, Shichibu Y, Teranishi T, Majima Y (2005) Phys Rev B 72:205441
16. Porter MD, Bright TB, Allara DL, Chidsey CED (1987) Spontaneously organized molecular assemblies. 4. Structural characterization of n-alkyl thiol monolayers on gold by optical ellipsometry, infrared spectroscopy, and electrochemistry. J Am Chem Soc 109:3559
17. Bumm LA, Arnold JJ, Dunbar TD, Allara DL, Weiss PS (1999) Electron transfer through organic molecules. J Phys Chem B 103:8122

Chapter 38
Light Amplification in Low-Dimensional Crystals of Thiophene/Phenylene Co-oligomer Derivatives

Hisao Yanagi, Hitoshi Mizuno, Fumio Sasaki, and Shu Hotta

Abstract Optically pumped light amplification properties are investigated for low-dimensional organic crystals of thiophene/phenylene co-oligomer (TPCO) derivatives substituted with methoxy and cyano groups. Two-dimensional platelet single crystals of those TPCO derivatives are grown from solution and/or vapor phases. Their crystal structures are dependent on the molecular symmetry and substituent groups. The unsubstituted TPCOs belong to monoclinic crystals, while the methoxy-substituted ones crystallize in orthorhombic forms. The cyano-substituted TPCOs form triclinic or monoclinic structure which gives rise to coexistence of one-dimensional rodlike crystals along with platelet crystals in the vapor-phase growth. One-dimensional TPCO crystals are also prepared by hot-wall epitaxy on potassium chloride (KCl) surface. The unsubstituted TPCO molecules epitaxially orient along the <110> direction of KCl and crystallize in an orthogonally aligned needlelike morphology. Under optical pumping, their fluorescence is amplified by light confinement and self-waveguiding effects in the TPCO crystal cavities. Above a certain threshold of excitation fluence, amplified spontaneous emission (ASE) with spectrally narrowed bands appears. When a pair of parallel facets of the platelet crystal acts as a Fabry–Pérot (F-P) resonator, multimode laser oscillation is observed

H. Yanagi (✉)
Graduate School of Materials Science, Nara Institute of Science and Technology (NAIST), 8916-5 Takayama, Ikoma, Nara 630-0192, Japan
e-mail: yanagi@ms.naist.jp

H. Mizuno
Graduate School of Materials Science, Nara Institute of Science and Technology (NAIST), 8916-5 Takayama, Ikoma, Nara 630-0192, Japan

Interdisciplinary Graduate School of Science and Engineering, Shimane University, 1060 Nishi-Kawatsu-cho, Matsue, Shimane 690-8504, Japan

F. Sasaki
Electronics and Photonics Research Institute, National Institute of Advanced Industrial Science and Technology (AIST), 1-1-1 Umezono, Tsukuba, Ibaraki 305-8568, Japan

S. Hotta
Department of Macromolecular Science and Engineering, Graduate School of Science and Technology, Kyoto Institute of Technology, Matsugasaki, Sakyo-ku, Kyoto 606-8585, Japan

T. Akasaka et al. (eds.), *Chemical Science of π-Electron Systems*,
DOI 10.1007/978-4-431-55357-1_38

in the ASE bands. Similar F-P lasing is also obtained from one-dimensional rodlike and needlelike crystals. Moreover, time-resolved spectroscopy reveals unique pulse-shaped emission with time delay up to 300 ps for the two-dimensional platelet crystals at prethreshold excitation fluence. The origin of this time-delayed emission is discussed in terms of exited state coherence among uniaxially oriented TPCO molecules in the crystals.

Keywords Light amplification • Low-dimensional crystal • Thiophene/phenylene co-oligomer • Organic laser

38.1 Introduction

π-Electronic systems in small molecules and polymers govern optical functions of organic materials mostly in a visible wavelength region. In particular, electronic transition between the highest occupied molecular orbital (HOMO) and the lowest unoccupied one (LUMO) contributes to light absorption and emission. Moreover, the energy levels of HOMO and LUMO are crucial for charge carrier generation, injection, and transport in organic optoelectronic devices such as a light-emitting diode (LED), field-effect transistor (FET), and photovoltaic (PV) cell. In principle, a simple elongation of π-conjugated system rises the HOMO energy, while lowers the LUMO energy so that a narrowed HOMO/LUMO gap results in a redshift of electronic transition spectrum. Such an energy modification promotes electron injection to LUMO and hole injection to HOMO. Another control of π-electronic energy is carried out by the introduction of electron-donating or electron-withdrawing substituents into the molecular framework. For example, electron donation from a methoxy group rises both HOMO and LUMO energies, while electron withdrawal to a cyano group lowers them. Thus risen HOMO makes hole injection easier so that the molecule tends to be p-type semiconducting, while the lowered LUMO enhances electron injection making the molecule n-type. Thus, optoelectronic characteristics of π-conjugated molecules are flexibly designed by comprehensive modifications.

In addition to molecule character itself, molecular packing and orientation are crucial for their optoelectronic functions in the solid state. In particular, anisotropic natures such as linear and planar shapes typical for π-conjugated molecules make those factors very significant. Axial orientation of HOMO/LUMO transition dipoles in linear molecules determines polarization of light absorption and emission. Cofacial stacking of π-electronic planes enhances charge carrier transports along the stacking direction. Such anisotropic molecular interactions are also reflected to morphology of organic crystals. A bundle of linear molecules originates in two-dimensional nucleation of platelet crystals. By contrast, one-directional stacking of planar molecules produces rod- and needlelike crystals. Moreover, those dimensions of organic crystals affect their optical behaviors such as light confinement and waveguiding. A platelet crystal confines the light one dimensionally and waveguides it in a two-dimensional plane [21]. The light two dimensionally confined in a needlelike crystal propagates in one-dimensional needle-axis direction [25, 26].

The optical anisotropy based on molecular orientation and crystal dimensions becomes very essential when we focus on light amplification in organic fluorescent media. Under a spontaneous emission regime, excited molecules in a crystal independently emit photons in random phases and broad spectral energies. Once an inverted population is formed in singlet excited states, photon radiation moves to stimulated emission regime upon intense excitation. The emitted light is amplified in phase and spectrally gain-narrowed, resulting in so-called amplified spontaneous emission (ASE) [7]. This stimulated emission-based ASE effectively occurs when the emitted light is linearly polarized and confined in a low-dimensional cavity. Molecular crystals with low-dimensional morphology and one-directional oriented transition dipoles provide a favorable condition as light amplification media. Furthermore, crystal facets growing in parallel sometimes function as feedback mirrors to cause organic lasing [14].

From temporal viewpoint, a time constant of spontaneous emission is typically $\sim$1 ns for π-conjugated fluorescent molecules. Upon resonant transition, the HOMO electron is pumped up to vibrationally excited states ($\nu' > 1$) of LUMO and then instantaneously relaxed to the vibrational ground state much faster than 1 ps. Therefore, a time-decay profile of spontaneous fluorescence shows initial maximum intensity very close to the time origin of excitation pulse followed by subsequent exponential decay in ns scale. With increasing excitation fluence, the decay constant is shortened, and above the ASE threshold, the time profile is narrowed to be overlapping with that of excitation pulse. This pulse-shaped collapse to the time origin is a result of stimulated emission in lasing media. In contrast to such typical temporal behavior of ASE and lasing, however, time-delayed emission known as superfluorescence has been reported for inorganic crystals like O_2^--doped KCl [5]. Under a specific excitation density, individual emitters begin to correlate and build up a phase-locked macroscopic dipole. This macroscopic correlation of emitters in the crystal gives rise to pulse-shaped emission with time delay of ns order. We have recently observed resembling phenomena for π-conjugated molecular crystals although the time delay is shorter and the origin of time delay is still an open question [23, 27].

38.2 Crystal Characterization of Thiophene/Phenylene Co-oligomer Derivatives

38.2.1 Molecular Modification of Thiophene/Phenylene Co-oligomers

In order to realize organic lasing under control of molecular modification and crystal design as mentioned above, we use thiophene/phenylene co-oligomers

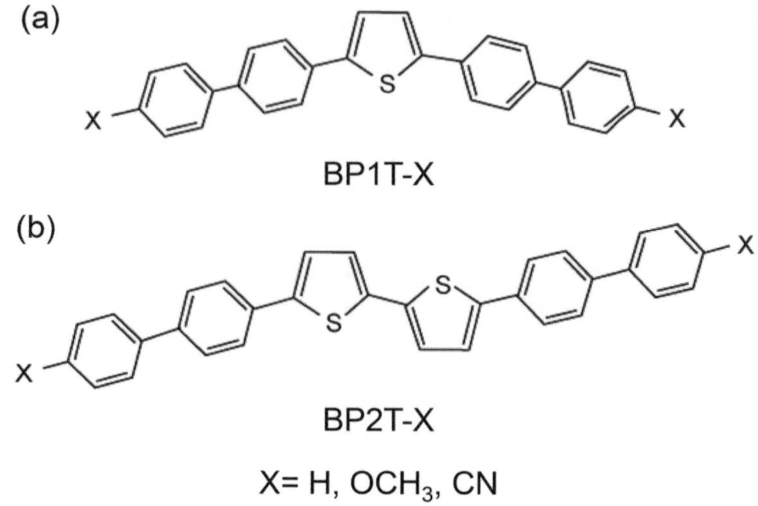

Fig. 38.1 Molecular structures of BP1T (**a**) and BP2T (**b**) derivatives

(TPCOs). TPCOs are built with arbitral number and sequence of thiophene and phenylene units in a chain form. Figure 38.1 shows chemical structures of representative TPCO derivatives with two types of molecular frameworks. The synthesis of these molecules is carried out according to literature procedures [9, 10]. 2,5-Bis(4-biphenylyl)thiophene (BP1T, Fig. 38.1a) is a biphenyl-capped thiophene with terminal X=H which has molecular symmetry of point group C_{2v}. 5,5′-Bis(4-biphenylyl)-2,2′-bithiophene (BP2T, Fig. 38.1b) is a biphenyl-capped bithiophene with terminal X=H which has molecular symmetry of point group C_{2h}. Substitution with methoxy (X=CH_3) and cyano (X=–CN) groups at their both terminal ends provides 2,5-bis(4′-methoxybiphenyl-4-yl)thiophene (BP1T-OMe), 2,5-bis(4′-cyanobiphenyl-4-yl)thiophene (BP1T-CN), 5,5′-bis(4′-methoxybiphenyl-4-yl)-2,2′-bithiophene (BP2T-OMe), and 5,5′-bis(4′-cyanobiphenyl-4-yl)-2,2′-bithiophene (BP2T-CN).

38.2.2 Methoxy-Substituted Thiophene/Phenylene Co-oligomers

Single crystals of those TPCOs are grown from solution and/or vapor phases [13, 19, 24]. For example, the procedure of solution growth for BP1T-OMe crystals is as follows. A crude powder of 10 mg BP1T-OMe is dissolved in 20 ml 1,2,4-trichlorobenzene by heating at 153 °C. The solution is slowly cooled down to 43 °C at a rate of ~11 °C/h. By filtrating the resulting precipitates and after dryness at atmosphere, thin platelet crystals are obtained on a filter paper. A well-

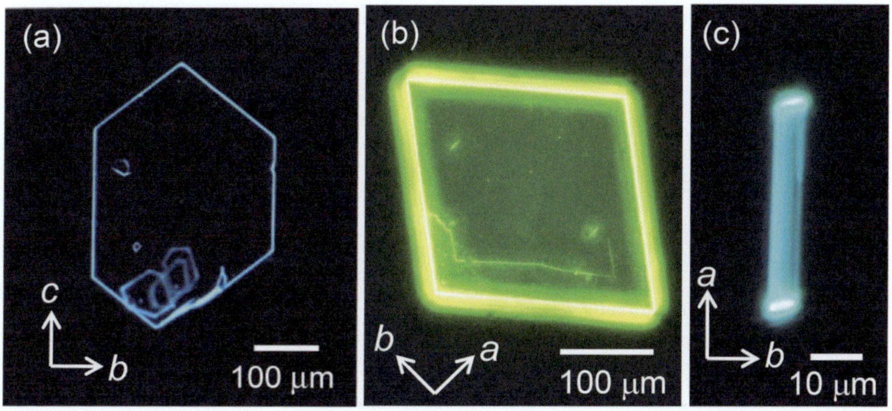

Fig. 38.2 Fluorescence micrographs of single crystals of BP1T-OMe (**a**), BP2T-OMe (**b**), and BP1T-CN (**c**). Reproduced from Yanagi et al. [27] with permission from the Japan Society of Applied Physics; Mizuno et al. [18] with permission from WILEY-VCH Verlag GmbH& Co. KGaA, Weinheim; Mizuno et al. [20] with permission from WILEY-VCH Verlag GmbH& Co. KgaA, Weinheim

Table 38.1 HOMO/LUMO energies of TPCO derivatives evaluated by DFT calculation

	HOMO (eV)	LUMO (eV)	ΔE (eV)
BP1T	−4.802	−2.417	2.385
BP1T-OMe	−4.580	−2.257	2.323
BP1T-CN	−5.305	−3.000	2.305
BP2T	−4.612	−2.568	2.044
BP2T-OMe	−4.456	−2.448	2.008
BP2T-CN	−5.018	−3.089	1.929

defined single crystal is selected and transferred onto a glass substrate by using a tungsten tip. Figure 38.2a shows a fluorescence micrograph of a representative BP1T-OMe crystal excited with a Hg lamp ($\lambda = 365$ nm) [27]. Typically, the crystal grows in a hexagonal platelet morphology with a few hundred μm size and a few μm thickness. The BP1T-OMe exhibits blue fluorescence according to the HOMO/LUMO transition energy of $\Delta E = 2.323$ eV estimated by density function theory (DFT) calculation using a numerical basis set with GGA/BP functional shown in Table 38.1. This ΔE value is close to that (2.385 eV) of the unsubstituted BP1T. However, due to the electron-donating effect, the HOMO/LUMO energies of BP1T-OMe are considerably higher than those of BP1T.

Note that in Fig. 38.2a the blue emission is observed only at the hexagonal contour of the crystal. It suggests that the emitted light is confined within the platelet slab and not leaked from the crystal surface. This effective confinement is critical for realizing organic lasing as mentioned in the following section and is attributed to molecular alignment in the BP1T-OMe crystal. The BP1T-OMe molecules are crystallized in an orthorhombic form with space group $Cmc2_1$ and lattice constants of $a = 4.964$, $b = 0.743$, and $c = 0.588$ nm [12]. As summarized in Table 38.2, most

Table 38.2 Crystallographic parameters of TPCO derivatives

	Crystal system	Space group	Lattice constants
BP1T	Monoclinic	$P2_1/n$	$a = 0.761$, $b = 0.582$, $c = 4.376$ nm
			$\beta = 93.5°$
BP1T-OMe	Orthorhombic	$Cmc2_1$	$a = 4.964$, $b = 0.743$, $c = 0.588$ nm
BP1T-CN	Triclinic	$P\text{-}1$	$a = 0.384$, $b = 1.615$, $c = 1.802$ nm
			$\alpha = 111.9$, $\beta = 94.3$, $\gamma = 90.4°$
BP2T	Monoclinic	$P2_1/c$	$a = 0.571$, $b = 0.760$, $c = 5.287$ nm
			$\beta = 97.1°$
BP2T-OMe	Orthorhombic	$Pbca$	$a = 0.738$, $b = 0.594$, $c = 5.632$ nm
BP2T-CN	Monoclinic	$P2_1$	$a = 1.834$, $b = 0.724$, $c = 1.8446$ nm
			$\beta = 100.5°$

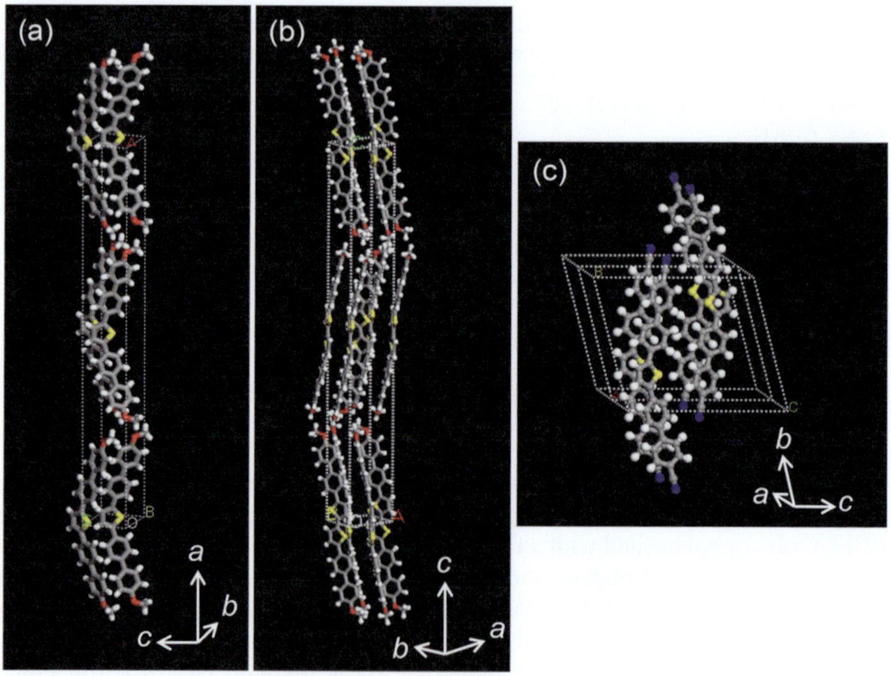

Fig. 38.3 Crystal structures of BP1T-OMe (**a**), BP2T-OMe (**b**), and BP1T-CN (**c**). Those structures are produced from crystallographic data reported in Cambridge Crystallographic Data Centre as registration codes of CCDC-630600, CCDC-883460, and CCDC-897042, respectively

of unsubstituted TPCOs such as BP1T and BP2T crystallize in a monoclinic form [8, 11]; therefore, the molecular chain axis is slightly tilted against the normal of the basal crystal plane. By contrast, the long axis of the bend-shaped BP1T-OMe molecule is upright against the basal *bc*-plane as shown in Fig. 38.3a. Since the HOMO/LUMO transition dipole moment is parallel to the line connecting the peripheral methoxy groups in BP1T-OMe, the emitted light propagates in parallel direction to the crystal plane by the transverse magnetic (TM) mode. Therefore, the platelet crystal of BP1T-OMe serves as a perfect slab-type self-waveguide for emitted light.

As in the case of BP1T-OMe, the substitution with electron-donating methoxy groups into BP2T rises the HOMO/LUMO energies. Moreover, the elongation of π-electronic conjugation in BP2T with an additional thiophene ring into BP1T results in a remarkable decrease of HOMO/LUMO transition energy of BP2T (Table 38.1). As shown in Fig. 38.2b, therefore, BP2T-OMe crystals exhibit green fluorescence. Platelet crystals of BP2T-OMe were grown from a trichlorobenzene solution according to the similar procedure for BP1T-OMe as mentioned above. X-ray diffraction (XRD) analysis determines that the BP2T-OMe molecules crystallize in an orthorhombic system the same as BP1T-OMe (Table 38.2). However, their molecular packing manners are different: the BP2T-OMe molecules are alternately inclined with respect to the *c*-axis. This inclined orientation is attributed to intermolecular interaction between the peripheral methoxy groups. Note that this interaction between methoxy groups plays the same role in the BP1T-OMe crystal; therefore, the peripheral biphenyl groups make the same angle between the adjacent molecules in BP1T-OMe and BP2T-OMe crystals. Since the HOMO/LUMO transition dipole moment of BP2T-OMe is parallel to its quasi-linear molecular axis, the emitted light is propagating along the crystal plane with a grazing angle. Therefore, the emitted light is not perfectly confined within the crystal slab, but some portion of fluorescence is leaked from the crystal face. Thus, bright emission at the crystal edges and weak emission on the basal face in Fig. 38.2b reasonably reflect the molecular packing in the BP2T-OMe crystal.

38.2.3 Cyano-Substituted Thiophene/Phenylene Co-oligomers

Introduction of electron-withdrawing cyano groups into BP1T and BP2T also changes their electronic characteristics as well as crystal structure. The HOMO/LUMO energies of BP1T-CN and BP2T-CN are drastically deepened by 0.40–0.58 eV, while their ΔE is slightly decreased as compared to those of unsubstituted BP1T and BP2T (Table 38.1). The slight decrease of ΔE is attributed to π-electronic delocalization in the molecular chain due to electron-withdrawing effect of the peripheral cyano groups. Figure 38.2c shows a fluorescence micrograph of a vapor-grown single crystal of BP1T-CN [20]. BP1T-CN is hardly crystallized from a solution phase due to its poor solubility. Vapor growth of BP1T-CN crystals was carried out by the following method. 5 mg of BP1T-CN powder was sealed in a glass tube with N_2 gas at a pressure of 170 mmHg. By heating in an air oven at 330 °C, crystal nucleation occurred on the upper part of the glass tube due to thermal convection. After heating for 30 h, BP1T-CN crystals were obtained sticking on the glass tube wall. Most of vapor-grown crystals of BP1T-CN formed rodlike morphology besides a few platelet crystals being also grown. The rodlike crystal in Fig. 38.2c shows blue emission at the rod ends as well as on the rod surface. This emission occurrence is related to a peculiar structure of the BP1T-CN crystal as schematically shown in Fig. 38.3c. XRD analysis determines that the BP1T-CN molecules crystallize in a triclinic form with lattice constants of $a = 0.384, b = 1.615, c = 1.802$ Å, $\alpha = 111.9, \beta = 94.3$, and $\gamma = 90.4°$. The BP1T-CN molecules are packed intertwining each other in the *bc*-plane and piled up along the a-axis to form rodlike morphology. The co-facial stacking of the BP1T-CN molecules along the *a*-axis results in one-dimensional rodlike morphology rather than two-dimensional platelet morphology. Since the top surface of the rodlike crystal is assigned to the (025) plane, it is evaluated that the HOMO/LUMO transition dipole moment of BP1T-CN is tilting by 62.4° against this basal plane and 70.4° against the *a*-axis of the crystal. This oblique molecular orientation causes light propagation along the *a*-axis and radiation from the rod ends, while a considerable portion of emitted light is leaked from the rod surface as seen in Fig. 38.2c.

38.2.4 Epitaxial Growth of Unsubstituted Thiophene/Phenylene Co-oligomers

As mentioned above, the molecular axes of TPCO derivatives are usually standing upright or obliquely tilting against the basal plane of the platelet and rodlike crystals. The morphology of those free-standing crystals grown from solution and vapor phases reflects crystal habits depending on the self-assembling nature of molecules. However, the crystal morphology drastically changes when the molecules are epitaxially grown on an appropriate substrate. Commensurate interaction of deposited molecules with the substrate lattice results in epitaxially oriented nucleation of molecular crystal. For TPCO molecules, a cleaved (001) surface of single-crystal potassium chloride (KCl) is served as a substrate since the lattice spacing along the <110> direction well matches with the repeating intervals of the thiophene/phenylene monomer units [25, 26]. Figure 38.4a shows a fluorescence micrograph of BP2T crystals epitaxially grown on the KCl substrate [17]. Using the hot-wall-epitaxy technique [1], BP2T was vapor-deposited onto the KCl (001) surface at a deposition rate of 1 nm/min under a vacuum of 2×10^{-5} mbar. During the deposition, the substrate temperature and the chamber wall temperature were kept individually at 140 °C. The deposited BP2T molecules crystallize in needlelike morphology which orthogonally orients along the [110] and [−110] directions on the KCl (001) surface.

In order to characterize the molecular orientation in the needlelike crystals, the KCl substrate was dissolved away and the crystal sample was provided for XRD and electron diffraction (ED) analyses. The XRD pattern shown in Fig. 38.4b indicates two diffraction peaks which are assigned to the (020) and (12–1) spacings. According to the crystal structure of BP2T as shown in Table 38.2, the BP2T

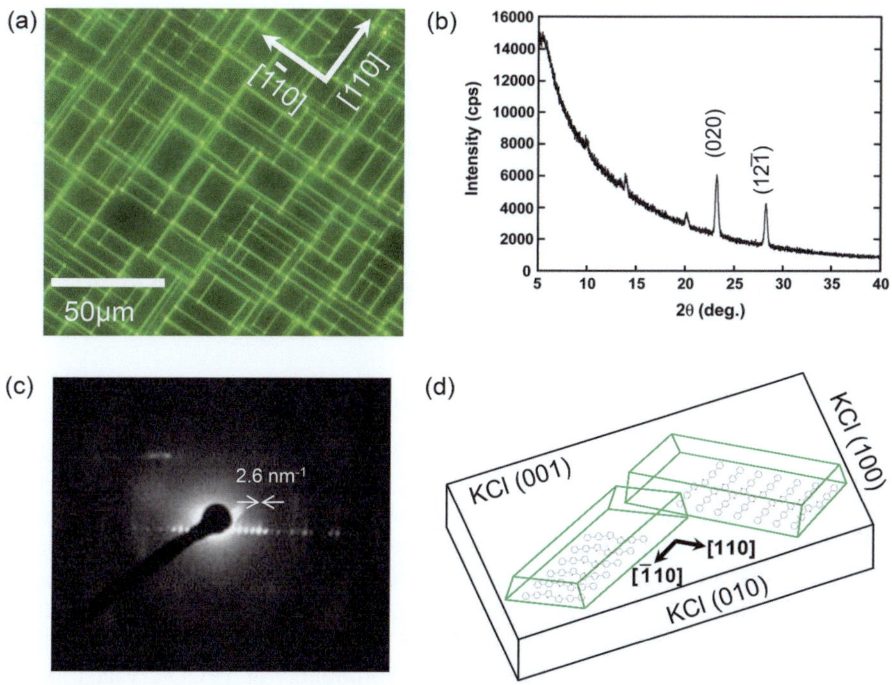

Fig. 38.4 (**a**) Fluorescence micrographs of BP2T needle crystals epitaxially grown on KCl (001) surface. XRD (**b**) and ED (**c**) patterns taken from the needle crystals. (**d**) Schematic diagram for epitaxial orientation of the needle crystals (Reproduced from Mizuno et al. [17] with permission from WILEY-VCH Verlag GmbH& Co. KgaA, Weinheim)

molecules are aligning almost parallel to these two planes suggesting that the molecular axis is lying in contact with the KCl (001) surface. The ED pattern in Fig. 38.4c was taken from a selected area of the needle crystals aligning in the [110] direction which was identical to the vertical direction in the figure. The ED pattern indicates a reflection series on the equator axis, the spacing (2.6 nm^{-1}) of which approximately equals to a half of the lattice constant *c*. Since the molecular axis of BP2T is almost parallel to the c-axis, the spacing of 2.6 nm corresponds to the molecular length of BP2T. It suggests that the molecular axis aligns in the horizontal direction of the ED pattern that is perpendicular to the long axis of the needle crystal. Based on these XRD and ED results, the epitaxial orientation of the needle crystal is schematically depicted in Fig. 38.4d. The ionic raw of K^+ ions along the [110] and [−110] interacts with the π-electronic chain of BP2T, and the interval of the K^+ ions (0.44 nm) along these directions commensurately matches with the repeating interval of thiophene/phenylene units of BP2T. Consequently, the BP1T molecules are flat-lying on the (001) surface of KCl and epitaxially orient along the [110] and [−110] directions of KCl. The molecules are piled up along the direction perpendicular to the molecular axis so that the needlelike crystals orthogonally grow in the [−110] and [110] directions. The cross-sectional dimension of the needlelike crystals is typically a few μm in width and a few hundred nm in thickness. Due to the lying molecular orientation, the polarized fluorescence is emitted from the top surface of the needlelike crystals. The emitted light is also propagated along the needle axis and scattered at the needle ends which are terminated by the orthogonally aligned needles.

38.3 Light Amplification in Low-Dimensional Crystals of TPCO Derivatives

38.3.1 Light Amplification and Lasing in Platelet Crystal of BP2T-OMe

As mentioned in the previous section, electronic characteristics of TPCO derivatives are controlled by molecular modification, and their light-emitting behaviors in crystal media are dependent on the molecular orientation and crystallographic morphology. In particular, light confinement in the crystal cavity is indispensable for achieving light amplification and lasing in the TPCO crystals. As seen in Fig. 38.2a and b, two-dimensional platelets are typical morphology of TPCO crystals. Bright emission at the crystal contour edges reveals that the fluorescence is effectively confined in the single-crystal self-cavity. The vertically standing orientation of molecules in the crystal enhances propagation of fluorescence light along the crystal plane. This anisotropic light propagation is favorable for light amplification in the crystal cavity under optical pumping since the polarized fluorescence in the TM mode is effectively amplified by the stimulated emission process. Furthermore, the parallel pair of side facets can act as a Fabry–Pérot (F-P) resonator [14, 15] to realize edge-emitting lasers in the platelet crystal.

Optical measurements were performed to observe light amplification in the platelet crystal of TPCO derivatives as follows. A stripe-shaped excitation beam of a frequency-doubled Ti:S pulsed laser ($\lambda = 397$ nm, 1 kHz, 200 fs duration) was focused on the crystal plane with an incident angle of 20°. The light emitted from the crystal edge was detected along the stripe direction which was set perpendicular to the parallel side facets of the platelet crystal. Figure 38.5a shows excitation density dependence of photoluminescence (PL) spectra taken from the platelet crystal of BP2T-OMe shown in Fig. 38.2b. At a weak excitation density of 3.6 μJ/cm^2, two broad fluorescence bands appear at $\lambda = 525$ and 565 nm which are assigned vibrational progressions of 0–1 and 0–2 transitions. Since the neighboring BP2T-OMe molecules in the crystal align in parallel configuration, the lowest 0–0 transition is forbidden due to the antiparallel exciton coupling [2]. With increasing excitation density, the PL intensity increases as shown in Fig. 38.5b where integrated intensities of the 0–1 and 0–2 bands are plotted as a function of the excitation density. At excitation density of 178 μJ/cm^2, the PL intensities start to nonlinearly

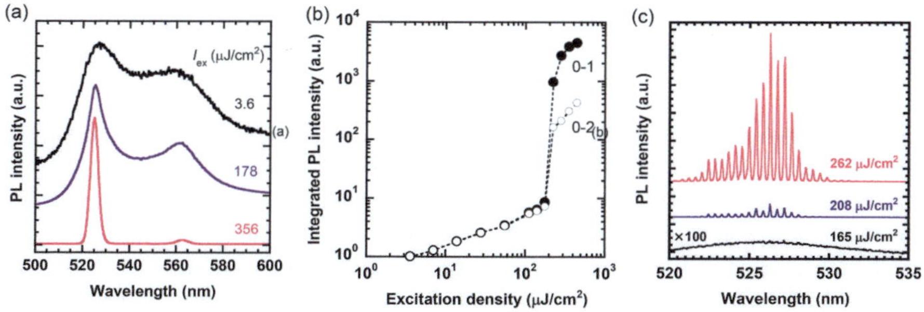

Fig. 38.5 (**a**) Excitation density dependence of PL spectra taken from single-crystal BP2T-OMe under optical excitation with fs-pulsed laser at $\lambda = 397$ nm. (**b**) Integrated intensities of the 0–1 and 0–2 bands as a function of excitation density. (**c**) Excitation density dependence of the 0–1 band spectra showing laser oscillations (Reproduced from Mizuno et al. [18] with permission from WILEY-VCH Verlag GmbH& Co. KGaA, Weinheim)

increase and the spectral bands are narrowed. This is a typical behavior of amplified spontaneous emission (ASE) showing a threshold excitation density of spectral gain narrowing. At 356 μJ/cm^2, the 0–1 band spectrum is remarkably amplified with a full width of half maximum (FWHM) of ~5 nm.

Detailed spectral changes in the 0–1 band region ($\lambda = 525$–535 nm) were observed as a function of excitation density using a high-resolution grating. As shown in Fig. 38.5c, the 0–1 band indicates no structure at excitation density of 165 μJ/cm^2 below the ASE threshold. With increasing excitation density at 208 μJ/cm^2 beyond the threshold, lasing oscillations are observed in the 0–1 band. At 262 μJ/cm^2, the multimode lasing clearly appears on the ASE spectral envelope. This lasing oscillation is based on F-P resonation inside the crystal cavity where the parallel facets of the platelet crystal act as a pair of feedback mirrors. The cavity length L of the used crystal is estimated to be 90 μm, and the lasing mode interval $\Delta\nu$ =16 cm^{-1} is obtained from Fig. 38.5c. The group refractive index n_g is calculated to be 3.5 according to the relation, $n_g = 1/2L\Delta\nu$. This n_g value is high enough to confine the emitted light in the crystal slab. The reflectance at the crystal facet is estimated to be 31 % from the equation, $R = (n_1 - n_2)^2/(n_1 + n_2)^2$, by substitution with $n_1 = n_g = 3.5$ and $n_2 = 1$ (air). The parallel facet of the crystal is defined as the {110} faces from the interior angles of platelet corners. Since the BP2T-OMe crystal has the orthorhombic form, the {110} faces are perpendicular to the basal *ab*-plane of the crystal. Due to those high reflectivity and vertical configuration of the parallel {110} facets, the platelet crystal of BP2T-OMe acts as a high-quality F-P resonator. The quality factor Q estimated from the mode peak width of FWHM = ~4 cm^{-1} at the ASE peak frequency is ~4500. This high Q value proves that the BP2T-OMe crystal has a potential self-cavity medium for organic lasing since this lasing is stably obtained without quenching in air even at room temperature.

38.3.2 Light Amplification and Lasing in Rodlike Crystal of BP1T-CN

Optically pumped ASE and lasing based on self-cavity effect have been also obtained from rodlike crystals of BP1T-CN shown in Fig. 38.2c. As mentioned in Section **2.3**, the BP1T-CN crystallizes in the unique triclinic form in which the molecular axis (i.e., the HOMO/LUMO transition dipole moment) is tilting by 62.4° against the basal (025) plane. Therefore, the light confinement is not perfect so that a considerable amount of fluorescence light is leaked from the rodlike crystal surface. Nevertheless, its fluorescence band is gain-narrowed to ASE when the emitted light under optical pumping is detected along the direction parallel to the long axis (*a*-axis) of the rodlike crystal. As shown in Fig. 38.6a, the integrated intensity of the 0–1 band appearing at 500 nm nonlinearly increases above a threshold excitation density of 138 μJ/cm^2. This light amplification along the rodlike crystal is attributed to the optical waveguiding effect along the *a*-axis in which π-conjugated BP1T-CN molecules are densely stacking in this direction. It is noted that the threshold excitation density of 138 μJ/cm^2 is slightly lower than that of the platelet crystal of BP2T-OMe (178 μJ/cm^2) mentioned in the previous section. This lower threshold for the BP1T-CN crystal is attributed to the inclined orientation of the molecular axis which increases absorption of excitation light since the inclining angle of the BP1T-CN molecule is larger than that of the BP2T-OMe molecules in the platelet crystal.

Figure 38.6b shows excitation density dependence of the 0–1 band spectra taken with a higher spectral resolution. With increasing excitation density, lasing oscillations are clearly observed along with the gain-narrowed ASE band above the threshold. These longitudinal multimode lasing is ascribed to the F-P resonation in the rodlike crystal. According to the relationship, $n_g = 1/2L\Delta\nu$, the group refractive

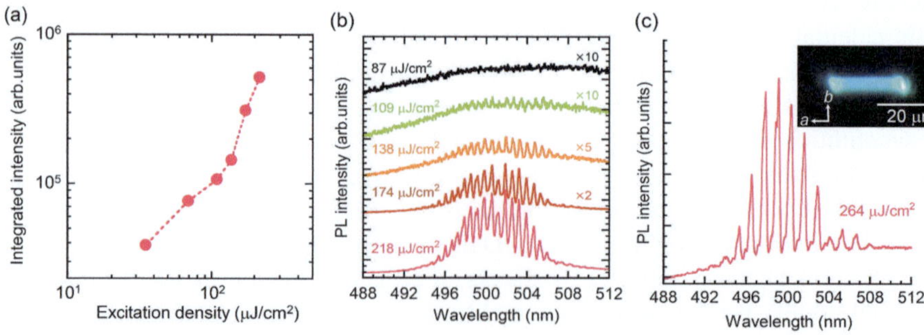

Fig. 38.6 (**a**) Excitation density dependence of integrated 0–1 band intensity in PL spectrum taken from the rodlike BP1T-CN crystal shown in Fig. 38.2c under optical excitation with fs-pulsed laser at $\lambda = 397$ nm. (**b**) Lasing spectrum of the 0–1 band taken from the rodlike BP1T-CN crystal shown in Fig. 38.2c. (**c**) Excitation density dependence of the 0–1 band lasing spectra taken from another rodlike crystal of BP1T-CN (Reproduced from Mizuno et al. [20] with permission from WILEY-VCH Verlag GmbH& Co. KGaA, Weinheim)

index n_g of the BP1T-CN is estimated to be 4.18 at $\lambda = 500$ nm using $L = 43$ μm and $\Delta\nu = 27.8$ cm^{-1}. This high n_g supports effective light confinement along the a-axis of the rodlike crystal. The Q factor of the crystal cavity is calculated to be ~1860 from the $\Delta\nu$ and the FWHM of the lasing peak. This Q value is lower than that (~4500) of the platelet crystal of BP2T-OMe. In the rodlike crystal of BP1T-CN, the F-P mirror facets at both ends of the rod are assigned to the (1-1-3) plane [20]. This plane is inclined against the basal (025) plane of the rodlike crystal, while the {110} mirror facets in the platelet crystal of BP2T-OMe are upright to its basal plane. This difference in their crystal cavity structures results in the lower Q factor of the BP1T crystal.

In the vapor growth of BP1T-CN crystals, there are different sizes of rodlike crystals sticking on the glass tube wall. We chose another rodlike crystal with a smaller size and measured the lasing spectrum as shown in Fig. 38.6c. At an excitation density of 264 μJ/cm^2 (above a lasing threshold of 132 μJ/cm^2), the lasing oscillation with the mode interval of $\Delta\nu = 50.2$ cm^{-1} was obtained. Since the cavity length is smaller (L=~20 μm), the mode interval of the laser oscillation is increased inversely proportional to L. Therefore, the mode number of lasing oscillation in the 0–1 band is decreased as compared to that in the lasing spectra shown in Fig. 38.6b.

38.3.3 Light Amplification and Lasing in Epitaxially Grown BP2T Crystals

As compared to light amplification in the platelet crystal and rodlike crystals of the TPCO derivatives, the fluorescence is more low dimensionally confined in the needlelike crystals of BP2T epitaxially grown on the KCl substrate (Fig. 38.4a). Since the molecular axis of BP2T is lying on the KCl (001) surface and stacking along the <110> direction of KCl in perpendicular to the needle crystal axis, the emitted light is propagated along the needle crystal with the transverse electric (TE) mode. Due to this lying molecular orientation, light amplification properties of the needle crystals are affected by the polarization angle of excitation. In the PL measurements, the stripe-shaped excitation beam ($\lambda = 397$ nm, incident angle: 20°) was set along the KCl [110] direction, i.e., parallel to one set of the oriented needle crystals (see Fig. 38.4a), and the emitted light was collected along the KCl [110] direction which was parallel to the KCl (001) surface. Figure 38.7a and b shows PL spectra taken from the needle crystals as a function of excitation density where the polarization of the excitation laser was set perpendicular (s) or parallel (p) to the incident plane, respectively. In both figures, the 0–1 and 0–2 emission bands appear around 520 nm, 570 nm, respectively, and the former band is gain-narrowed to ASE with increasing excitation density. From excitation density dependence of the integrated intensity as shown in the insets, the threshold excitation density is estimated to be 8.7 μJ/cm^2 for s-polarization and 35 μJ/cm^2 for p-polarization.

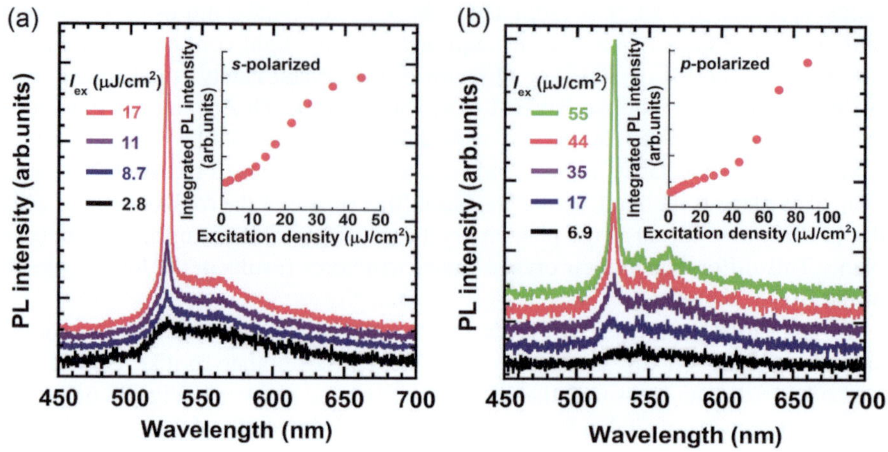

Fig. 38.7 Excitation density dependence of PL spectra taken from needle crystals of BP2T epitaxially grown on KCl (001) surface under *s*-polarized (**a**) and *p*-polarized (**b**) optical excitations at $\lambda = 397$ nm. The insets show integrated intensity of the 0–1 band as a function of excitation density (Reproduced from Mizuno et al. [17] with permission from WILEY-VCH Verlag GmbH& Co. KgaA, Weinheim)

Under *s*-polarized excitation, the polarization of the excitation laser is parallel to the HOMO/LUMO transition dipole of the BP2T molecules in the needlelike crystals orienting along the [110] direction. Therefore, the excitation laser is highly absorbed by thus oriented crystals, and the emitted light along the needle axis is effectively collected by the detector. Under *p*-polarized excitation, the excitation laser is not absorbed by the needle crystals orienting in the [110] direction. On the other hand, the *p*-polarized excitation laser is parallel to the molecular axis in the needlelike crystals orienting along the [−110] direction. The propagated light along those needle crystals is not collected directly, but the light scattered at the terminated ends is detected. Consequently, the threshold excitation density is lower for *s*-polarized excitation than *p*-polarized excitation. It is noted, however, that those threshold values are considerably low as compared to those for the platelet and rodlike crystals of the TPCO derivatives mentioned in the previous sections. Their low threshold value is again attributed to the efficient absorption of the excitation light by the lying molecules with polarization-matched HOMO/LUMO transition dipoles

In order to investigate lasing oscillation, high-resolution PL spectra were taken from the needle crystals of BP2T. However, we could not observe lasing from those crystals representatively shown in Fig. 38.4a. In the fluorescence micrograph in Fig. 38.4a, there are a variety of needle crystals with different needle lengths which are terminated by the orthogonally oriented needles. If the terminated ends of needle crystals act as the F-P mirror, a large number of different modes arising from those various cavity lengths are superposed in the 0–1 emission band region since the excitation area in our experiment is much larger than the size of individual needles. As shown in Fig. 38.8a, we found some needle crystals align one directionally

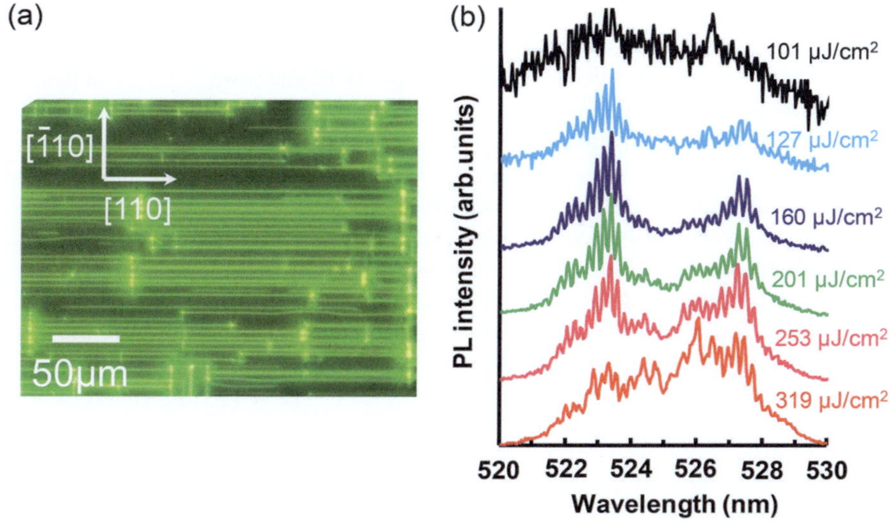

Fig. 38.8 (**a**) Fluorescence micrograph of needlelike BP2T crystals mainly oriented along the KCl [110] direction. (**b**) Excitation density dependence of high-resolution 0–1 band spectra taken from the epitaxially oriented needle crystals shown in (**a**) (Reproduced from Mizuno et al. [17] with permission from WILEY-VCH Verlag GmbH& Co. KgaA, Weinheim)

and their lengths are identical for a bundle of needles which are terminated by the same orthogonal needles. Figure 38.8b shows high-resolution PL spectra taken from the needle crystals orienting along the KCl [110] direction shown in Fig. 38.8a. With increasing excitation density, lasing oscillation appeared in the 0–1 emission band. Fourier-transformed patterns of those oscillating spectra give a frequency peak at $\Delta\nu$ of 8 cm^{-1}. Taking into account the group refractive indices of $n_g \sim 4$, the frequency $\Delta\nu = 8$ cm^{-1} results in a cavity length of $L = \sim 156$ μm. This L value almost corresponds to the lengths of the center bundle of needle crystals in Fig. 38.8a. It reveals that the epitaxially oriented needle crystals act as a one-dimensional F-P cavity and the lying orientation is suitable for optical amplification at low-threshold excitation. The molecular stacking along the needle axis enhances the light amplification in this direction supported by high stimulated emission rate of the TE mode.

38.3.4 Time-Delayed Emission in Single Crystal of BP1T-OMe

In the PL spectroscopy measurements with fs-pulsed laser excitation, we have simultaneously observed time-resolved spectra from the TPCO derivative crystals using a streak camera. In the spontaneous emission regime under weak excitation, the emission intensity of broad fluorescence bands decays exponentially with a time constant of ~1 ns. Once stimulated emission occurs under increased excitation above a threshold density reaching inversion population, the decay time is shortened and the decay profile changes to a pulsed shape. With increasing excitation density, the pulsed profile is getting close to the profile of the excitation pulse since the relaxation in the excited state and the stimulated process occur very fast within a picosecond. We have observed such a typical time profile for ASE and lasing of the TPCO derivative crystals when the excitation density is increased above their threshold. However, at a specific range just below the threshold density, called as prethreshold, we have found time responses with unusually long delay of a few hundred ps accompanied with peculiar band splitting [27].

Figure 38.9 shows excitation density dependences of the 0–2 band emission taken from a platelet crystal of BP1T-OMe shown in Fig. 38.2a. In Fig. 38.9a, the single crystal was excited with a stripe-shaped ns-pulsed laser ($\lambda = 420$ nm, 16 ns duration,

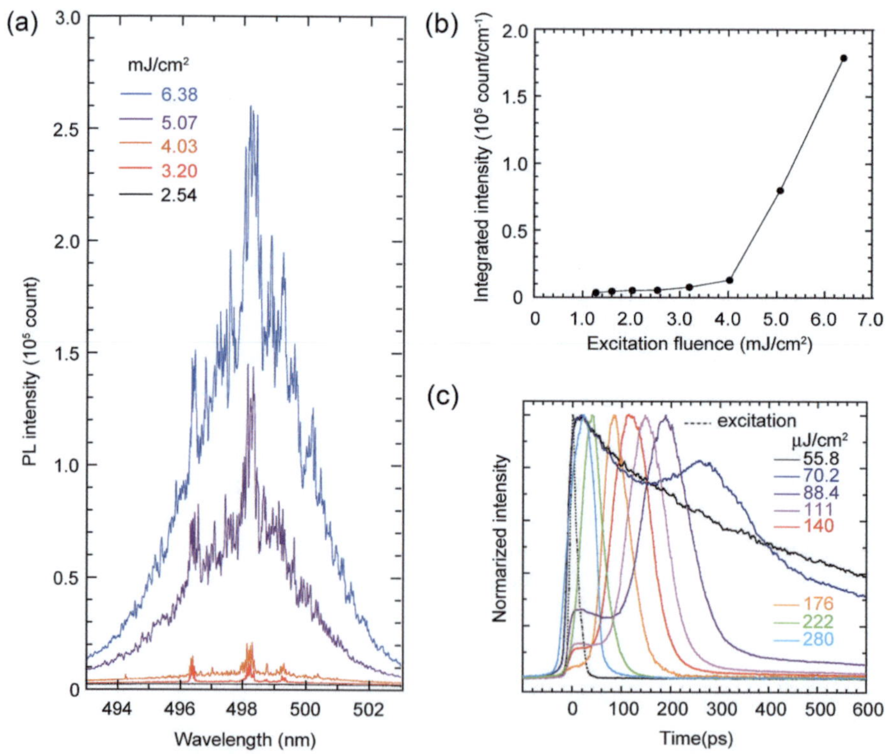

Fig. 38.9 (**a**) Excitation density dependence of PL spectra taken from single-crystal BP1T-OMe shown in Fig. 38.2a under optical excitation with ns-pulsed laser at $\lambda = 420$ nm. (**b**) Integrated emission intensities of the PL spectra shown in (**a**) as a function of excitation density. (**c**) Normalized time-delay profiles of the 0–2 emission band at $\lambda = 484$–514 nm as a function of excitation density (Reproduced from Yanagi et al. [27] with permission from The Japan Society of Applied Physics)

10 Hz repetition). With increasing excitation density, the integrated intensity of the 0–2 band nonlinearly increased above a threshold of around 4.0 mJ/cm^2 as is shown in Fig. 38.9b. Above this threshold, fine oscillations are observed on the ASE envelope of the 0–2 band. This oscillation is ascribed to multimode lasing due to the F-P facets of parallel crystal edges with a cavity length of $L = 280$ μm. It is noted, however, that below the threshold, unusual splitting peaks are found at $\lambda = 496.3$, 498.2, 499.2 nm in the ASE profile. In order to elucidate the origin of such band splitting in the prethreshold excitation, we observed time-decay profiles of the 0–2 band emissions using a fs-pulsed laser ($\lambda = 397$ nm, 150 fs duration, 1 kHz repetition) and a synchro-scan streak camera (5 ps resolution) at room temperature. Under weak excitation at 55.8 μJ/cm^2, the profile shows an exponential decay typical for spontaneous fluorescence. When the excitation density is increased to 70.2 μJ/cm^2, the profile suddenly changes to have a pulsed peak superimposed on the spontaneous decay. The time of this peak is delayed approximately by 300 ps with respect to the time origin of the excitation pulse. With increasing excitation density, the pulsed peak becomes prominent, its time delay is getting shorter, and the pulse width is narrower. The excitation density dependence of the 0–2 band spectra simultaneously observed reveals that these time-delayed pulse emissions appear in the prethreshold excitation density of 70.2–176 μJ/cm^2, and the band splitting is also observed in this prethreshold range as in the case of ns-laser excitation.

The origin of this time-delayed emission is still an open question; however, a similar phenomenon has been known as “superfluorescence” for a gas phase [6] and a solid state of O_2^- ions doped in KCl [[5], [16]] at low temperature. Superfluorescence is defined as cooperative emission from an ensemble of emitters, which are initially uncorrelated but subsequently generate a phase-matched macroscopic dipole. The buildup of the correlated state causes a time delay. Once it is built up, the macroscopic dipole radiates a pulse-shaped emission. Our results resemble this superfluorescence feature; however, a striking contrast is that our experiments are performed at room temperature. Another mechanism for time-delayed emission has been reported for exciton–polariton in semiconductor microcavities [4]. Strong coupling between an exciton and a photon typically generated in a semiconductor layer sandwiched with two distributed Bragg reflectors (DBRs) leads to new eigenstates called cavity polariton. In the relaxation process of exciton–polariton, the excited populations are preserved at a bottleneck of the lower branch of exciton–polariton. Then, stimulated cooling from the bottleneck to the lowest condensed state gives rises to a time delay before photoemissive transition to the ground state. Moreover, it is theoretically reported that triplet splitting, called Mollow triplet, appears in the emission spectrum due to strong coupling between electron–hole pairs and cavity photons [3]. Exciton–polaritons have been demonstrated even at room temperature for materials with a high exciton binding energy such as copper halides [22]. From those findings, the exciton–polariton mechanism is one of plausible origins to explain the time-delayed emission of the BP1T-OMe crystal observed at room temperature even though the strong coupling in such a macroscopic crystal cavity has to be further investigated in detail.

38.4 Conclusions

The presented TPCO derivatives are promising candidates for realizing organic lasers due to their versatility in molecular modification, color tunability, and ambient stability. In particular, the low-dimensional crystallization into platelet and rodlike and needlelike morphologies is significant for light confinement required for achieving ASE and lasing. The axially oriented molecules emit highly polarized luminescence which is suitable for efficient stimulated emission in the crystal cavity. The regular facets of the crystal edges can serve as a F-P resonator for multimode lasing. In order to further improve their laser performances, in particular toward electrically pumped lasing, it is necessary to optimize designing of those crystal dimensionality, morphology, and molecular orientation as well as electrical properties depending on their molecular modification and crystallographic structures.

The platelet crystals with the orthorhombic forms of BP1T-OMe and BP2T-OMe act as a two-dimensional slab cavity. The standing molecular orientation especially in the BP1T-OMe crystal is favorable for in-plane propagation of light with the TM mode. The vertical side facets of the orthorhombic crystal serve as the high-quality resonator for edge-emitting lasing. High carrier injection and transport in the in-plane direction due to π-electronic interaction are also favorable for edge-emitting electroluminescence devices. The modification with cyano-substitution resulted in the rodlike crystal of BP1T-CN with the unique triclinic structure. The oblique orientation of BP1T-CN molecules also enabled in-plane ASE and lasing from the rod edges. Their horizontal component of the HOMO/LUMO transition dipoles decreases the lasing threshold owing to increased absorption of excitation light. The π-electronic stacking along the rod axis is also favorable for stimulated amplification in the edge-emitting configuration. In the epitaxially oriented needlelike crystals of BP2T, the molecular axis is flat-lying on the substrate so that the lasing threshold is minimized under normal incidence of polarized excitation. The parallel stacking of π-electronic molecules along the needle axis enhances the light amplification in the one-dimensional direction, and the F-P lasing is achieved by the terminated needle ends. It is noted, however, that the oblique and lying orientation is also potential of surface-emitting configuration. In particular, the thin thickness of the rodlike and needlelike crystals is suitable for microcavity formation if their in-plane growth is improved.

The microcavity study with TPCO derivatives is also important to further elucidate the origin of the time-delayed pulse emission observed in the BP1T-OMe crystal. A plausible mechanism of exciton–polariton has been proved in the semiconductor layer of sub-μm thickness sandwiched between DBR mirrors. Therefore, it is required to fabricate wavelength-scale cavity with surface-emitting TPCO crystals and to study thickness dependence of exciton–photon coupling. Lasing from exciton–polaritons, called polariton lasing, occurs at lower-threshold excitation than conventional photon lasing. Actually, the time-delayed emission of the BP1T-OMe crystal was observed in the prethreshold excitation density at room temperature. Further experiments and theoretical interpretation of the TPCO crystals in microcavity structures would lead to exploration of a new class of molecular photonic devices such as an organic polariton laser.

References

1. Andreev A, Matt G, Brabec CJ, Sitter H, Badt D, Seyringer H, Sariciftci NS (2000) Highly anisotropically self-assembled structures of para-sexiphenyl grown by hot-wall epitaxy. Adv Mater 12:629–317
2. Bando K, Nakamura T, Fujiwara S, Matsumoto Y, Sasaki F, Kobayashi S, Shimoi Y, Hotta S (2008) Optical selection rule for the lower Davydov excitons in co-oligomer single crystals. Phys Rev B 77:045205–045206
3. Byrnes T, Horikiri T, Ishida N, Yamamoto Y (2010) BCS wave-function approach to the BEC-BCS crossover of exciton-polariton condensates. Phys Rev Lett 105:186402–186404
4. Deng H, Haug H, Yamamoto Y (2010) Exciton-polariton Bose-Einstein condensation. Rev Mod Phys 82:1489–1537
5. Florian R, Schwan LO, Schmid D (1984) Time-resolving experiments on Dicke superfluorescence of O_2^- centers in KCl. Phys Rev A 29:2709–2715
6. Gibbs HM, Vrehen QHF, Hikspoors HMJ (1977) Single-pulse superfluorescence in cesium. Phys Rev Lett 39:547–550
7. Gupta R, Stevenson M, Dogariu A, McGehee MD, Park JY, Srdanov A, Heeger AJ, Wang H (1998) Low-threshold amplified spontaneous emission in blend of conjugated polymers. Appl Phys Lett 73:3492–3494
8. Hotta S, Goto M (2002) Crystal structure analysis of 2,5-bis(4-biphenyl)thiophene. Adv Mater 14:498–501
9. Hotta S, Kimura H, Lee SA, Tamaki T (2000) Synthesis of thiophene/phenylene co-oligomers. II Block and alternating co oligomers. J Heterocycl Chem 37:281–286
10. Hotta S, Katagiri T (2003) Synthesis of thiophene/phenylene co-oligomers. IV. 6- to 8-ring molecules. J Heterocycl Chem 40:845–850
11. Hotta S, Goto M, Azumi R, Inoue M, Ichikawa M, Taniguchi Y (2004) Crystal structure of thiophene/phenylene co-oligomer with different molecular shapes. Chem Mater 16:237–241
12. Hotta S, Goto M, Azumi R (2007) Peculiar crystal structure of a thiophene/phenylene co-oligomer of 2,5-bis(4′-methoxybiphenyl-4-yl)thiophene. Chem Lett 36: 270–271
13. Ichikawa M, Hibino R, Inoue M, Haritani T, Hotta S, Koyama T, Taniguchi Y (2003) Improved crystal growth and emission gain-narrowing of thiophene/phenylene co-oligomers. Adv Mater 15: 213–217
14. Ichikawa M, Hibino R, Inoue M, Haritani T, Hotta S, Araki K, Koyama T, Taniguchi Y (2005) Laser oscillation in monolithic molecular single crystals. Adv Mater 17:2073–2077
15. Ichikawa M, Nakamura K, Inoue M, Mishima H, Haritani T, Hibino R, Koyama T, Taniguchi Y (2005) Photopumped laser oscillation and charge injected luminescence from organic semiconductor single crystals of a thiophene/phenylene co-oligomer. Appl Phys Lett 87:221113–3
16. Malcuit MS, Maki JJ, Simkin DJ, Boyd RW (1987) Transition from superfluorescence to amplified spontaneous emission. Phys Rev Lett 59:1189–1192
17. Mizuno H, Ohnishi I, Yanagi H, Sasaki F, Hotta S (2012) Lasing from epitaxially oriented needle crystals of a thiophene/phenylene co-oligomer. Adv Mater 24:2404–2408
18. Mizuno H, Haku U, Marutani Y, Ishizumi A, Yanagi H, Sasaki F, Hotta S (2012) Single crystals of 5,5′-bis (4′-methoxybiphenyl-4-yl)-2,2′-bithiophene for organic laser medium. Adv Mater 24:5744–5749
19. Mizuno H, Yanagi H, Ishizumi A, Sasaki F, Hotta S (2012) Optically pumped lasing from vapor-grown crystals of methoxy-substituted thiophene/phenylene co-oligomer. Phys Status Solidi A 209:2437–2440

20. Mizuno H, Maeda T, Yanagi H, Katsuki H, Aresti M, Quochi F, Saba M, Mura A, Bongiovanni G, Sasaki F, Hotta S (2014) Optically pumped lasing from single crystals of a cyano-substituted thiophene/phenylene co-oligomer. Adv Opt Mater 2:529–534
21. Nagawa M, Hibino R, Hotta S, Yanagi H, Ichikawa M, Koyama T, Taniguchi Y (2002) Emission gain-narrowing from single crystals of a thiophene/phenylene co-oligomer. Appl Phys Lett 80:544–546
22. Nakayama M, Kameda M, Kawase T, Kim D (2011) Cavity polaritons of heavy-hole and light-hole excitons in a CuI microcavity. Phys Rev B 83:235325–5
23. Sasaki F, Kobayashi S, Haraichi S, Yanagi H, Hotta S, Ichikawa M, Taniguchi Y (2006) Pulse-shaped emission with the time delay in single crystals of thiophene/phenylene co-oligomers. Jpn J Appl Phys 45:L1206–L1208
24. Yamao T, Miki T, Akagami H, Nishimoto Y, Ota S, Hotta S (2007) Direct formation of thin single crystal of organic semiconductor onto a substrate. Chem Mater 19:3748–3753
25. Yanagi H, Morikawa T, Hotta S, Yase K (2001) Epitaxial growth of thiophene/*p*-phenylene co-oligomers for highly polarized light-emitting crystals. Adv Mater 13:313–317
26. Yanagi H, Ohara T, Morikawa T (2001) Self-waveguided gain-narrowing of blue light-emission from epitaxially oriented *p*-sexiphenyl crystals. Adv Mater 13:1452–1455
27. Yanagi H, Marutani Y, Sasaki F, Makino Y, Yamao T, Hotta S (2011) Prethreshold lasing with time-delayed pulse emission from a single crystal of thiophene/phenylene co-oligomer. Appl Phys Express 4:062601–062603

Part VIII
π-Electron Systems in Biosystems and Biomimetics

Chapter 39
Electron Transfer Pathway Analysis in Bacterial Photosynthetic Reaction Center

Hirotaka Kitoh-Nishioka and Koji Ando

Abstract A new computational scheme to analyze electron transfer (ET) pathways in large biomolecules is presented with applications to ETs in bacterial photosynthetic reaction center. It consists of a linear combination of fragment molecular orbitals and an electron tunneling current analysis, which enables an efficient first-principle analysis of ET pathways in large biomolecules. The scheme has been applied to the ET from menaquinone to ubiquinone via nonheme iron complex in bacterial photosynthetic reaction center. It has revealed that not only the central Fe^{2+} ion but also particular histidine ligands are involved in the ET pathways in such a way to mitigate perturbations that can be caused by metal ion substitution and depletion, which elucidates the experimentally observed insensitivity of the ET rate to these perturbations.

Keywords Long-distance electron transfer • Photosynthetic reaction center • Fragment molecular orbitals • Electron transfer pathway analysis

39.1 Introduction

Long-distance electron transfers (ET) play essential roles in biological energy conversion [1–5]. The most fundamental are those in photosynthesis. In photosynthesis, the photon energy is captured in the form of electronic excitation energy by light-harvesting antenna systems composed of aggregates of pigments. The energy is then funneled to the "special pair", a pair of chromophores, in the reaction center embedded in the membrane protein. From the electronic excited state of the

H. Kitoh-Nishioka (✉)
Department of Chemistry, Graduate School of Science, Nagoya University, Furo-cho, Chikusa-ku, Nagoya 464-8601, Japan
e-mail: kito.hirotaka@b.mbox.nagoya-u.ac.jp

K. Ando
Department of Chemistry, Graduate School of Science, Kyoto University, Sakyo-ku, Kyoto 606-8502, Japan
e-mail: ando@kuchem.kyoto-u.ac.jp

T. Akasaka et al. (eds.), *Chemical Science of π-Electron Systems*,
DOI 10.1007/978-4-431-55357-1_39

special pair, a series ET processes occurs, which are followed by transmembrane proton pumps. The electrochemical energy thus generated by the gradient of proton concentration is utilized at the ATP synthase embedded in the same membrane. The trinity of long-distance ET, proton pump, and ATP synthesis is also functioning in cellular respiration for energy metabolism. Thus, the chemical transfers of electrons and protons are at the core of biological energy conversion. Both electrons and protons are highly quantum mechanical particles. Moreover, many degrees of freedom in the proteins are involved. Therefore, quantum statistical mechanical description is essential for microscopic understanding of their transfer processes. This posts a marked challenge to theoretical and computational chemical physics and has been an active area of research along with the rapid advances of computer technologies in recent years.

There exists a simple but fundamental question in biological ETs on the roles of protein environments: Are they playing only "passive" roles of simply holding the redox moieties at appropriate spatial and orientational arrangements, or do they play some "active" roles to mediate the ETs by involving their electronic wave functions as bridge states? Furthermore, if the protein environments play some active roles, their details are still unclear; for instance, whether the bridge states contribute mainly to the coherent superexchange mechanism or to the incoherent stepwise hopping ETs. This aspect of quantum coherence depends on the subtle microscopic mechanism involving competition between the ETs and the nuclear vibrational and conformational relaxations. Although it remains as an important open question, it will be out of the scope of this paper. We rather focus on the aspect of electronic coupling under the assumption of coherent superexchange mechanism.

In this chapter, we review our recent works on long-distance biological ETs with development of a new scheme for theoretical and computational analysis. In Sect. 39.2, we briefly overview the ETs in bacterial photosynthetic reaction center. After identifying the key quantities in the ET rate formula in Sect. 39.3.1, we outline in Sects. 39.3.2, 39.3.3 and 39.3.4 the computations of electronic structure of large molecular systems, ET matrix elements, and ET pathways. Applications to the ETs in bacterial photosynthetic center are discussed in Sect. 39.4. Section 39.5 concludes.

39.2 Electron Transfer in Photosynthetic Reaction Center

In biological energy conversion processes, π-electron systems play essential roles. For instance, in bacterial photosynthetic reaction center, one of the key molecules is bacteriochlorophyll (BChl) designated by B_L and B_M in Fig. 39.1. (In more details, the structure of BChl is similar to that designated by H_L in Fig. 39.2 but with a Mg^{2+} ion at the center of the tetrapyrrole ring.) The free-base form indicated by H_L in Figs. 39.1 and 39.2 is called bacteriopheophytin (BPhe). Other key molecules

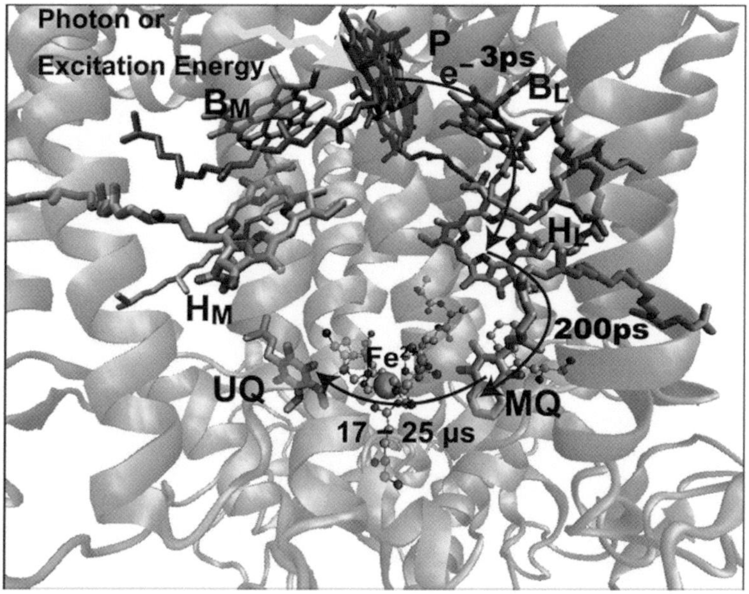

Fig. 39.1 Structure of bacterial photosynthetic reaction center P is the special pair of bacteriochlorophyll, B_L and B_M are bacteriochlorophylls, H_L and H_M are bacteriopheophytins, MQ is menaquinone, and UQ is ubiquinone. Between MQ and UQ is a nonheme iron ion complex

are menaquinone (MQ) and ubiquinone (UQ), in Fig. 39.2, and carotenoids. All these contain π-electrons which dominate the major chemical functions including the redox properties.

In the photosynthesis, the photons are first captured by antenna systems that consist of aggregates of BChl molecules, in which the energies are stored in a form of electronic excitation. The excitation energies are then funneled to the BChl dimer, called "special pair" (P in Fig. 39.1), in the reaction center. The electronically excited special pair then ejects an electron to one of the adjacent BChl molecules (B_L), which is followed, coherently or incoherently, by a sequence of ETs to BPhe (H_L) and to UQ via MQ. The molecules involved in these last two ET steps are displayed in Fig. 39.2. As presented in Fig. 39.1, these redox centers are embedded in the membrane proteins. This is thus an intriguing prototype system to examine the questions on the roles of protein environments in mediating biological ETs, as described in Sect. 39.1. To address these, it is essential to carry out quantum mechanical analysis at the electronic and atomic level, for which there exist two-stage basic tasks: the first is to determine accurate electronic structures of large molecules such as membrane proteins, at least with qualitative accuracy. The second is to establish analysis method to clarify the microscopic mechanism of ETs from the computed electronic wave functions.

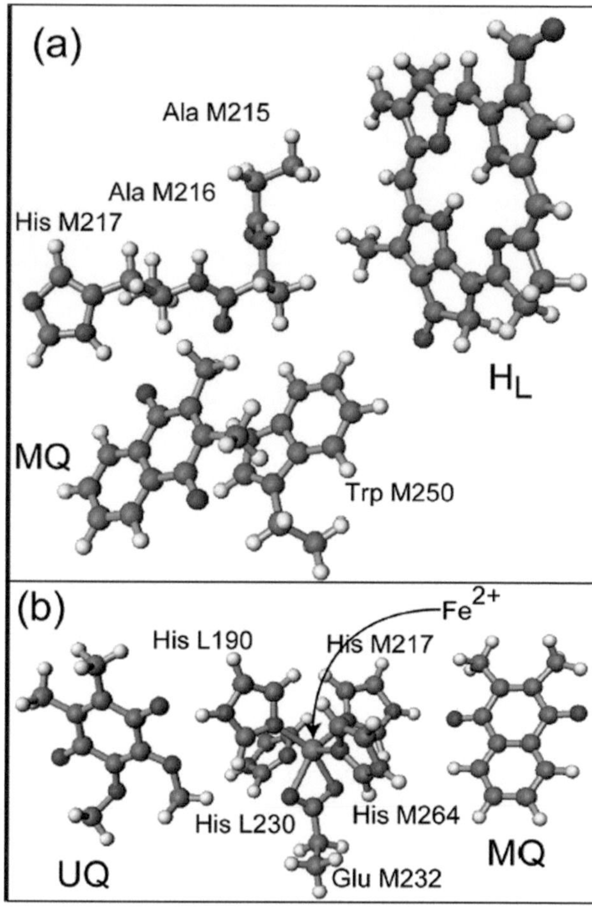

Fig. 39.2 Molecular structures (**a**) around the bacteriopheophytin (H_L) and menaquinone (MQ) and (**b**) the menaquinone and ubiquinone (UQ) via the nonheme iron complex with four histidine (His) ligands and one glutamate (Glu) ligand

39.3 Theory and Computational Methods

39.3.1 Rate of Electron Transfer

An essential key to elucidating ET mechanism is the measurement of rate constant. The current standard and practically useful theory of ET rate is the Marcus theory [6], which describes the ET rate k_{ET} with only a few parameters, the electronic transfer integral T_{DA}, the reaction free energy ΔG^0, and the nuclear reorganization energy λ,

$$k_{\mathrm{ET}} = \frac{2\pi}{\hbar} \frac{|T_{\mathrm{DA}}|^2}{\sqrt{4\pi\lambda k_{\mathrm{B}}T}} \exp\left[-\frac{(\Delta G^0 + \lambda)^2}{4\lambda k_{\mathrm{B}}T}\right], \tag{39.1}$$

in which $\hbar$ is the Planck constant divided by 2π, k_B is the Boltzmann constant, and T is the absolute temperature. This rate formula was derived as the semiclassical (or high temperature) limit of thermally averaged Fermi's golden rule which is based on the time-dependent perturbation theory with respect to the electronic coupling T_{DA}. Therefore, application of this formula to long-distance ETs assumes that the ET occurs as a quantum mechanically coherent process through the distance of ET. This aspect still deserves a careful scrutiny [7], but in this work we assume Eq. (39.1) is appropriate.

Among the three parameters in Eq. (39.1), the reaction free energy ΔG^0 can be measured by experiments most straightforwardly. The experimental evaluation of the reorganization energy λ is less trivial but still feasible in various ways. By contrast, the electronic transfer integral T_{DA} is the least trivial quantity, which has been estimated in practice only indirectly from other measurements. Therefore, theoretical evaluation of T_{DA} is of particular importance. Since Eq. (39.1) was derived with separation of electronic and nuclear degrees of freedom, that is, with the Condon approximation, the computation of transfer integral T_{DA} has been based solely on the electronic structure calculation under the Born-Oppenheimer approximation. This adiabaticity aspect again deserves careful examination but shall be assumed in this work. Therefore, the essential problem now is to compute the electronic wave functions of large molecular systems involved in the biological ETs.

39.3.2 Electronic Structure Calculation of Large Molecules

Although ET reactions can be described basically as a one-electron process, it is still essential to solve the many-electron problem in order to determine the major one-electron orbitals. This is because the most relevant orbitals are normally near the highest-occupied molecular orbitals (HOMO). The computation is highly demanding for large biomolecules but is becoming more and more feasible due to the recent advances of hardware power and numerical algorithms. Two representative methods of large-scale electronic structure calculations are the fragment molecular orbital (FMO) method [8, 9] and the divide-and-conquer method [10, 11]. Here we describe the former which was mainly employed in this work.

39.3.2.1 FMO Method

The FMO method [8, 9] first decomposes the total system into fragments. The MOs of each fragment are optimized self-consistently under the Coulomb field of other fragments. Then, dimer or trimer calculations are carried out under the Coulomb field of the optimized monomer fragments in order to take account of the exchange interactions. The total energy of the system is computed with

$$E_{\mathrm{total}}^{\mathrm{FMO2}} = \sum_{I>J} E_{IJ} - (N_{\mathrm{f}} - 2) \sum_{I} E_I \qquad (39.2)$$

when up to dimer exchange interaction was considered (FMO2), and with

$$E_{\text{total}}^{\text{FMO3}} = \sum_{I>J>K} E_{IJK} - (N_{\text{f}} - 3)\sum_{I>J} E_{IJ} - \frac{1}{2}(N_{\text{f}} - 2)(N_{\text{f}} - 3)\sum_{I} E_{I} \qquad (39.3)$$

when up to trimer exchange interaction was considered (FMO3). In Eqs. (39.2) and (39.3), N_{f} is the number of fragments, E_I, E_{IJ}, and E_{IJK} denote the energies of fragment monomer, dimer, and trimer, respectively. These formulas (39.2) and (39.3) are designed to remove the excess countings of energies with proper account of self-consistent electronic polarizations. That is, by carrying out the dimer and trimer calculations under the Coulomb field of optimized monomer fragments, the excess countings of polarization energies are also properly removed. The FMO method gives accurate energies and properties, such as the dipole moments, but not the MOs, of the total system. However, as will be discussed in Sect. 39.3.3, what we need in order to compute the transfer integral T_{DA} are the electronic Hamiltonian matrix and the MOs of the total system. This requirement is fulfilled by the method of linear combinations of FMOs, the FMO-LCMO method [12, 13], which shall be outlined next.

39.3.2.2 FMO-LCMO Method

In the FMO-LCMO method, the "intra-fragment" parts of the Hamiltonian matrix elements are computed from the results of FMO calculations in a form similar to Eqs. (39.2) and (39.3), whereas the "inter-fragment" parts are constructed with the matrices from the dimer or trimer calculations projected to the monomer FMO space. For instance, in the FMO2 version of the method, the intra-fragment Hamiltonian matrix elements are computed with a formula analogous to Eq. (39.2),

$$H_{Ip,Iq}^{(\text{total})} = \sum_{J \neq I} \langle \varphi_p^I | \boldsymbol{h}^{IJ} | \varphi_q^I \rangle - (N-2) \langle \varphi_p^I | \boldsymbol{h}^I | \varphi_q^I \rangle, \qquad (39.4)$$

and the inter-fragment matrix elements are defined as

$$H_{Ip,Jq}^{(\text{total})} = \langle \varphi_p^I | \boldsymbol{h}^{IJ} | \varphi_q^J \rangle \qquad (I \neq J), \qquad (39.5)$$

in which φ_p^I is the p-th orbital of fragment I. The Fock (or Kohn-Sham) matrices of fragment I and fragment dimer IJ are denoted by $\boldsymbol{h}^I$ and $\boldsymbol{h}^{IJ}$, respectively. Thus, the notation $\langle \varphi_p^I | \boldsymbol{h}^{IJ} | \varphi_q^J \rangle$ represents the dimer Fock (or Kohn-Sham) matrix projected to the monomer FMO space.

Because the FMOs of each monomer fragment are optimized independently from other fragments, the FMOs of different fragments are generally not orthogonal. This non-orthogonality should be taken into account at the diagonalization. This has been demonstrated to give accurate approximations to the canonical MOs and their energies of the total system [12–15].

By deploying the total Hamiltonian matrix and the MOs of large systems thus obtained from the FMO-LCMO method, we have developed a scheme to analyze the long-distance ET pathways with the bridge Green function method and the tunneling current method.

39.3.3 Electronic Coupling Matrix Elements

39.3.3.1 Two-State Picture in Nonadiabatic Regime

Because the electron transfer integral T_{DA} is approximately proportional to the overlap between the donor and acceptor orbitals, it decays rapidly along their distance. Thus, for long-distance biological ETs, the transfer integral T_{DA} is normally small such that the electronically nonadiabatic regime is appropriate. The ET rate of Eq. (39.1) assumes this nonadiabatic limit and is based on the time-dependent first-order perturbation theory in which the amplitude of the acceptor state $C_{\mathrm{A}}(t)$ is proportional to the perturbation T_{DA}. Therefore, the rate of ET, the raise of the population $|C_{\mathrm{A}}(t)|^2$, is proportional to $|T_{\mathrm{DA}}|^2$. Another assumption behind Eq. (39.1) is that the ET is considered as an effective two-state problem. This aspect will be considered in Sect. 39.3.3.3. After the reduction to the effective two-state problem, the static adiabatic energies, E_1 and E_2, are obtained by solving the 2×2 secular equation with the diagonal matrix elements H_{D} and H_{A} and the off-diagonal element T_{DA},

$$E_{2,1} = \frac{H_{\mathrm{D}} + H_{\mathrm{A}}}{2} \pm \frac{1}{2}\sqrt{(H_{\mathrm{D}} - H_{\mathrm{A}})^2 + 4T_{\mathrm{DA}}^2}. \tag{39.6}$$

Here we have omitted the dependence on the nuclear coordinates $\boldsymbol{R}$ for simplicity: the electronic Hamiltonian matrix elements, $H_{\mathrm{D}}(\boldsymbol{R})$, $H_{\mathrm{A}}(\boldsymbol{R})$, and $T_{\mathrm{DA}}(\boldsymbol{R})$, and hence the adiabatic energies, $E_{1,2}(\boldsymbol{R})$, all depend on R. Thus, the adiabatic energy splitting defined by

$$\Delta\varepsilon_{12}(\boldsymbol{R}) \equiv E_2(\boldsymbol{R}) - E_1(\boldsymbol{R}) = \sqrt{(H_{\mathrm{D}}(\boldsymbol{R}) - H_{\mathrm{A}}(\boldsymbol{R}))^2 + 4T_{\mathrm{DA}}(\boldsymbol{R})^2} \tag{39.7}$$

is twice the transfer integral T_{DA} at nuclear configurations $\boldsymbol{R}_c$ of the diabatic surface crossing that gives $H_{\mathrm{D}}(\boldsymbol{R}_c) = H_{\mathrm{A}}(\boldsymbol{R}_c)$,

$$T_{\mathrm{DA}} = \Delta\varepsilon_{12}(\boldsymbol{R}_c)/2. \tag{39.8}$$

Nevertheless, it is not a trivial task to find the diabatic surface crossing configurations $\boldsymbol{R}_c$, especially for large proteins that involve many degrees of freedom. This is the reason why we consider the generalized Mulliken-Hush (GMH) analysis and the bridge Green function method that are described next.

39.3.3.2 Generalized Mulliken-Hush Analysis

The GMH method [16] scales the energy splitting $\Delta\varepsilon_{12}$ at nuclear configurations *off* the surface crossing by a formula

$$T_{\mathrm{DA}} = \frac{|\mu_{12}|\ \Delta\varepsilon_{12}}{\sqrt{(\mu_1 - \mu_2)^2 + 4|\mu_{12}|^2}}, \tag{39.9}$$

in which μ_1 and μ_2 are the dipole moments of the adiabatic states with E_1 and E_2, and μ_{12} is the off-diagonal element. Thus, the quantities in the right-hand side are obtained straightforwardly from the standard electronic structure calculations at any, normally the equilibrium, nuclear configurations. The idea behind this formula is an assumption that the Hamiltonian matrix elements and dipole matrix elements scale similarly for states involved in ETs, such that the former elements in Eq. (39.7) are replaced by the latter elements to assume the scaling factor. Despite its simplicity, the GMH formula (39.9) has been successfully applied to a number of ET reactions. In applications to large systems where the computational cost for the electronic excited state shall be the bottleneck, the energy splitting $\Delta\varepsilon_{12}$ is replaced by that of the donor and acceptor MOs from the ground-state calculation. Accordingly, the dipole matrix elements of the donor-acceptor MOs are also applied. This provides a computationally feasible and reasonably accurate method for the transfer integrals of large ET systems.

39.3.3.3 Bridge Green Function Method

For long-distance ETs in biomolecules, the electronic Hamiltonian matrix can be very large involving the intervening molecular parts between the donor and acceptor sites. In the bridge Green function (BGF) method [14], the electronic Hamiltonian matrix is projected onto the space of a two-level system consisting of the donor and acceptor states. The reduced information in the remaining part of the system, the molecular parts that bridge and mediate the ET, is taken into account via the BGF matrix. This will be formulated below in a generalized form of the effective Hamiltonian method.

We first divide the total Hamiltonian matrix to submatrices of a target space (P-space), the remaining space (Q-space), and their off-diagonal space. The eigenvalue problem with non-orthogonal basis is thus described by

$$\begin{bmatrix} H_{PP} - ES_{PP} & H_{PQ} - ES_{PQ} \\ H_{QP} - ES_{QP} & H_{QQ} - ES_{QQ} \end{bmatrix} \begin{bmatrix} c_P \\ c_Q \end{bmatrix} = \begin{bmatrix} 0 \\ 0 \end{bmatrix}, \tag{39.10}$$

in which S denotes the overlap matrix. The non-orthogonal formulation is essential because the FMOs of different fragments are generally not orthogonal, as noted in Sect. 39.3.2.2. Inserting the formal solution for c_Q from the second line of Eq. (39.10) into the first line, the problem is reduced to the projected smaller P-space with the effective Hamiltonian

$$H_{\text{eff}} = H_{PP} + (H_{PQ} - ES_{PQ})G(E)(H_{QP} - ES_{QP}), \tag{39.11}$$

in which $G(E)$ is the Green function matrix representing the contribution from the Q-space (bridge space),

$$G(E) \equiv (ES_{QQ} - H_{QQ})^{-1}. \tag{39.12}$$

With the definition of the P-space to be the donor-acceptor states or orbitals, $(\phi_\text{D}, \phi_\text{A})$, the off-diagonal element of H_{eff} corresponds to the transfer integral T_{DA},

$$T_{\text{DA}} = \sum_I^N \sum_J^N \sum_{I_p}{}' \sum_{J_q}{}' (H_{\text{D},I_p} - E_{\text{tun}} S_{\text{D},I_p}) G^{\text{B}}(E_{\text{tun}})_{I_p,J_q} (H_{J_q,\text{A}} - E_{\text{tun}} S_{J_q,\text{A}}), \tag{39.13}$$

in which the sums over I_p and J_q both exclude ϕ_D and ϕ_A.

In the effective Hamiltonian of Eq. (39.11), the energy E is generally unknown. However, in the application to the two-state ET case of Eq. (39.13), the energy E_{tun} is the electron tunneling energy that is most naturally defined as the average of the orbital energies of donor and acceptor orbitals, $E_{\text{tun}} = (\varepsilon_\text{D} + \varepsilon_\text{A})/2$.

Because our method employs the FMOs as the basis functions, in contrast with the previous methods that employ atomic orbitals, we can directly extract pictures that reflect the chemical properties of the molecular fragments. This also applies to the tunneling current analysis to be described next.

39.3.4 Electron Transfer Pathway Analysis

There exist a number of ET pathway analysis methods. For reviews, see Refs. [17–19]. The most primitive (and thus useful in practice) would be the *Pathways* model [20] based on an empirical Green function method, whereas one of the most sophisticated at present would be that based on the ab initio multi-configuration self-consistent field (MCSCF) method with occupation-restricted multiple active space (ORMAS) model [21]. Here we employ the tunneling current method [19] originally developed with the semiempirical MOs and based on atomic orbitals (AOs). By contrast, our implementation is based on the ab initio FMO-LCMO calculations which provides direct picture based on the molecular fragments with reduction of the number of basis functions compared to the AO-based methods. In addition, it allows systematic improvements of approximations, for instance, with exploits of recently emerging developments of density functional theories (DFTs). This last aspect is, however, out of the scope of this article.

In the tunneling current analysis, the transfer integral $T_{\rm DA}$ is expressed in terms of contributions from electron current J_{I_p,J_q} between basis FMOs $\{\varphi_p^I\}$,

$$T_{\rm DA} = \hbar \sum_{I\in\Omega_{\rm D}} \sum_{J\notin\Omega_{\rm D}} J_{I,J}, \tag{39.14}$$

$$J_{I,J} = \sum_{I_p} \sum_{J_q} J_{I_p,J_q}, \tag{39.15}$$

in which the summation over I_p and J_q is over the FMOs within fragments I and J, and $\Omega_{\rm D}$ denotes the spatial region assigned to the donor of ET. The interorbital current J_{I_p,J_q} is computed from the electronic Hamiltonian and overlap matrices and the coefficients of FMO-LCMO, $\{C_{I_p}^{\rm i}\}$ and $\{C_{I_p}^{\rm f}\}$, which represent the mixing of bridge FMOs to the donor and acceptor FMOs, $\varphi_{\rm D}$ and $\varphi_{\rm A}$, in the initial (i) and final (f) states of the ET, $\psi^{\rm i}$ and $\psi^{\rm f}$. They are expressed as

$$|\psi^{\rm i}\rangle = C_{\rm D}^{\rm i}|\varphi_{\rm D}\rangle + \sum_{I}^{N} \sum_{I_p} C_{I_p}^{\rm i}|\varphi_p^I\rangle, \tag{39.16}$$

$$|\psi^{\rm f}\rangle = C_{\rm A}^{\rm f}|\varphi_{\rm A}\rangle + \sum_{I}^{N} \sum_{I_p} C_{I_p}^{\rm f}|\varphi_p^I\rangle, \tag{39.17}$$

$$J_{I_p,J_q} = \frac{1}{\hbar}\left(H_{I_p,J_q} - E_{\rm tun}S_{I_p,J_q}\right)\left(C_{I_p}^{\rm i}C_{J_q}^{\rm f} - C_{I_p}^{\rm f}C_{J_q}^{\rm i}\right). \tag{39.18}$$

All these are thus computed straightforwardly from the FMO-LCMO calculation.

The tunneling current method enables us to analyze long-distance ET pathways in real space. As noted above, an advantage of our implementation stems from the use of FMOs as the basis functions. Although it is possible to carry out similar analysis with the conventional AO-based methods by simply taking the sum within each fragments, the advantage of the FMOs is that the dimension of the basis functions is about one order of magnitude smaller, and thereby the computational cost is significantly reduced. Therefore, large ET systems that have been only tractable by semiempirical MO calculations are now accessible straightforwardly with the ab initio MO and DFT methods.

39.4 Applications to Bacterial Photosynthetic Reaction Center

We have applied the computational strategy described in the previous section to the ETs in bacterial photosynthetic reaction center. In this paper, we shall focus on the ET step between the two quinones, MQ and UQ, via the nonheme Fe^{2+} ion complex, displayed in the lower part of Fig. 39.1 and in Fig. 39.2b. As presented in Fig. 39.2b, the Fe^{2+} ion is surrounded by five ligands from amino-acid side chains, four histidines (His), and one glutamate (Glu). Because the imidazole rings of the His ligands contain spatially delocalized low-energy π-electrons, they are likely to form the bridge states to mediate the ET.

39.4.1 Experimental Facts

First we summarize the experimental findings. Because the bacterial photosynthetic reaction center is a prototype system for studying biological energy conversions, there exist a wealth of experimental works. The key findings to be treated in this work pertaining to the ET from MQ to UQ via the nonheme iron complex are:

- The ET time, defined by the inverse of ET rate constant $1/k_{ET}$, is in the range 25–36 μs for *Blastochloris viridis* [22, 23].
- According to the electron paramagnetic resonance (EPR) experiment, the Fe^{2+} ion of the nonheme iron complex is in the high-spin ($S = 2$) state [24].
- Substituting the Fe^{2+} ion by a Zn^{2+} did not much alter the ET rate: the ET time of 150 μs for the Fe^{2+} complex was reduced slightly to 140 μs for the Zn^{2+} complex in *Rhodobacter sphaeroides* R-26.1 [25].

  Note that this experiment was performed on a different bacteria so that the ET time for the Fe^{2+} complex is different from that in *Blastochloris viridis*.
- Depletion of the Fe^{2+} ion slowed down the ET only approximately twice, from 150 to 350 μs of the ET time, in *Rhodobacter sphaeroides* R-26.1 [25].

From these findings, it has been conjectured that the Fe^{2+} ion is not playing vital roles to mediate the ET from MQ to UQ.

39.4.2 Computational Results

Now we present the results of computational analysis. The focus of this paper will be on the key experimental findings listed in Sect. 39.4.1. Further details have been presented in Ref. [15]. The ET pathway analysis will be demonstrated particularly useful for elucidating the mechanism behind the insensitivity of ET rate on the metal ion substitution and depletion. Before proceeding to the pathway analysis, we shall assess the accuracy of the computed ET rate.

39.4.2.1 Evaluation of Electron Transfer Rate

First we evaluate the ET rate of Eq. (39.1). As noted in Sect. 39.3.1, the formula depends on three parameters, the reaction free energy ΔG, the nuclear reorganization energy λ, and the electronic transfer integral T_{DA}. While T_{DA} is a purely electronic quantity and λ originates mostly from the nuclear rearrangements, ΔG includes both the electronic redox energies and the nuclear relaxation energies. Although λ and ΔG can be evaluated in principle with use of molecular dynamics simulations, their accuracy depends crucially on the reliability of the force-field model for the entire protein system, which has not been well established. Therefore, we employ experimentally evaluated values of $\lambda = 1.0\,\text{eV}$ and $\Delta G = -0.07\,\text{eV}$ for a particular bacteria *Blastochloris viridis* [1, 26] and place our focus on T_{DA} which is the least straightforward for experimental evaluations.

The nuclear coordinates were taken from the X-ray crystallographic data in the Protein Data Bank (code 1PRC) [27]. Those of hydrogen atoms were optimized with the semiempirical PM3 quantum chemical calculations, in which the heavier atoms were fixed at the crystallographic coordinates. For the wild-type (WT) complex with the high-spin ($S = 2$) Fe^{2+} ion, the unrestricted Hartree-Fock (UHF) method was used. For other systems in the low-spin ($S = 0$) state, the restricted HF (RHF) method was used. In addition, for these low-spin cases, calculations with the FMO-LCMO method described in Sect. 39.3.2.2 were carried out in an aim to assess its accuracy. Moreover, we also assessed the FMO-LCVMO method [14] that limits the monomer FMO space to the minimal valence (VMO) space. The 6-31G(d) basis set was employed throughout.

The computed results of the transfer integral T_{DA} and the ET time $1/k_{ET}^{calc}$ are listed in Table 39.1. For the WT system with high-spin Fe^{2+}, the computed T_{DA} with the GMH and the BGF methods were 1.33 and $1.39\,\text{cm}^{-1}$, respectively. The agreement between the two methods supports their accuracy and consistency. It demonstrates in particular that the BGF method with the appropriate tunneling energy E_{tun} is capable of capturing the T_{DA} value corresponding to the diabatic surface crossing, similarly to the GMH method that was thus designed.

With use of these T_{DA} values, together with ΔG and λ from the experimental evaluation as described above, the ET time $1/k_{ET}^{calc}$ was calculated to be 8.7–9.5 μs at the room temperature. As noted in Sect. 39.4.1, the observed ET time was in the range 25–36 μs. Thus, the computed values a few times underestimate the experimental. However, the agreement of this order is sufficiently reasonable since the rate constant is exponentially sensitive to the thermal activation factor. Indeed, higher accuracy cannot be expected even with the state-of-the-art quantum chemical computation. In this regard, it is important not to rely too much on a single number but to carry out analysis from multiple viewpoints. To this end, we next discuss analysis on the spin-state alteration and metal ion substitution.

39.4.2.2 Role of Nonheme Iron Complex: Spin State and Substitution

While the EPR experiment indicated that the WT complex is in the high-spin ($S = 2$) state, it would be still intriguing to examine different spin states in an aim to explore their functional significance. This is straightforward for the computational analysis. In general, the metal-ligand distances in Fe^{2+} complexes can vary by a

Table 39.1 Computed electron transfer integral $|T_{DA}|$ (in cm^{-1}) and electron transfer time $1/k_{ET}^{calc}$ (in μs) for *Blastochloris viridis* with various metal ions and various computational methods. For the observed electron transfer time $1/k_{ET}^{obs}$, the values in the parentheses are for another bacteria *Rhodobacter sphaeroides* R-26.1. See the text for details

Metal ion	Method		$\|T_{DA}\|$	$1/k_{ET}^{calc}$	$1/k_{ET}^{obs}$
Fe^{2+}	UHF	GMH	1.33	9.48	25–36[a]
(high spin)		BGF	1.39	8.68	(150)[b]
Fe^{2+}	RHF	GMH	0.944	18.7	
(low spin)	FMO-LCMO	GMH	0.955	18.3	
		BGF	0.979	17.4	
	FMO-LCVMO	GMH	0.766	28.4	
		BGF	0.777	27.6	
Zn^{2+}	RHF	GMH	1.30	9.93	(140)[b]
	FMO-LCMO	GMH	1.31	9.65	
		BGF	1.34	9.21	
	FMO-LCVMO	GMH	1.20	11.5	
		BGF	1.23	11.1	
None	RHF	GMH	0.610	44.8	(350)[b]
	FMO-LCMO	GMH	0.649	39.4	
		BGF	0.655	38.8	
	FMO-LCVMO	GMH	0.731	31.2	
		BGF	0.737	30.7	

[a] For *Blastochloris viridis* [22, 23]
[b] For *Rhodobacter sphaeroides* R-26.1 [25]

few tenth of Å in different spin states. However, we used here the same molecular structure as that of the high-spin state in order to focus on the effect of spin state without introducing extra factors.

In the low-spin ($S = 0$) state, the transfer integral T_{DA} was calculated to be in a range 0.94–0.98 cm^{-1} with the full RHF and the FMO-LCMO methods. On the technical aspect, we note the accuracy of the FMO-LCMO methods in comparison with the full RHF reference. The smaller T_{DA} of the low-spin state than that of the high-spin state results in approximately twice slower ET rate with the ET time of 17–19 μs. It is unclear at present if the faster ET rate in the WT high-spin state has some physiological significance. The difference of the factor 2 is rather modest. Nevertheless, because the Fe^{2+} complex can change the spin state depending on the ligand structure, this small difference may have a functional significance to feature a possibility to control the ET rate. This aspect would be open for further investigations. Finally, we note that the FMO-LCVMO method, limited to the minimal valence MO space, gives reasonable values of $T_{DA} = 0.77$–$0.78\,cm^{-1}$.

Next we substituted the Fe^{2+} ion by a Zn^{2+} ion. For the same reason as that noted above for the spin-state variation, we used the same molecular structure as that of the high-spin WT state. The computed transfer integral T_{DA} was 1.30–1.34 cm^{-1} with the full RHF and the FMO-LCMO methods. These are very close to the values for the high-spin WT case. As a result, the computed ET times for the Zn^{2+} complex, 9.2–9.9 μs, were also very close to those for the high-spin WT case. This is in good accord with the experimental findings for *Rhodobacter sphaeroides* R-26.1: the ET time of 150 μs for the WT system compared to 140 μs for the Zn^{2+}-substituted system. Thus, the insensitivity of the ET rate against the metal ion substitution is well reproduced. (Because the experimental ET rate of Zn^{2+}-substituted system was unavailable for *Blastochloris viridis*, we compared the ratio of the ET times for *Rhodobacter sphaeroides* R-26.1. This also applies to the metal-depleted case discussed next.)

Finally, we carried out computation without the metal ion. Again, we used the same molecular structure as that of the high-spin WT state in an aim to extract the essential roles of the Fe^{2+} ion, even though the molecular structure must have reorganized in the actual metal-depleted system in the experiment. The computed transfer integral T_{DA} was 0.61–0.66 cm^{-1}, about twice smaller than that for the WT high-spin Fe^{2+} system. Consequently, the computed ET time was 39–45 μs, about four times slower than the WT system. This is again in qualitative accord with the experimental finding for *Rhodobacter sphaeroides* R-26.1: the ET time of 350 μs for the metal-depleted system compared to 150 μs for the WT system.

39.4.2.3 Electron Transfer Pathway Analysis

To gain further insights into the mechanism of ET, particularly with respect to the role of the nonheme iron complex, we carried out ET pathway analysis described in Sect. 39.3.4. Figure 39.3 displays the computed tunneling currents, in which the width of the arrows represents the relative contribution of the path measured by the normalized tunneling current $\hbar J_{I,J}/T_{DA}$. As noted in Sect. 39.4.1, the experimentally found insensitivity of the ET rate on the substitution and depletion of the Fe^{2+} ion has raised a conjecture that the Fe^{2+} ion does not contribute to the major ET pathway. However, the computational result in Fig. 39.3a indicates the contrary: the major ET pathway consists of MQ $\rightarrow$ Fe^{2+} $\rightarrow$ His L190 $\rightarrow$ UQ. The resultant patterns of pathways for the high-spin (a) and low-spin (b) cases are basically very similar, but the widths of the main arrows are overall wider in the low-spin (b) case. Nonetheless, because the contribution of the back flow from UQ to Fe^{2+} is also larger in (b) than in (a), the net ET current is smaller in the low-spin system.

In the Zn^{2+}-substituted system displayed in Fig. 39.3c, both the forward current from MQ to Zn^{2+} and the backward current from UQ to Zn^{2+} are reduced compared to the Fe^{2+} cases in Fig. 39.3a, b. This implies that the filled $(3d)^{10}$ AO configuration of Zn^{2+} is less effective to mediate the ET than the unfilled $(3d)^6$ of Fe^{2+}. However, the contribution of a pathway from MQ to His M217 becomes larger in the Zn^{2+} system, as if to compensate the reduced current through the metal ion. In addition, the direction of the current from the metal ion to His M217 is reversed such that the current is now directed from His M217 to Zn^{2+}. These elucidate the electronic mechanism behind the insensitivity of the transfer integral T_{DA} and the ET rate on the substitution of Fe^{2+} by Zn^{2+}.

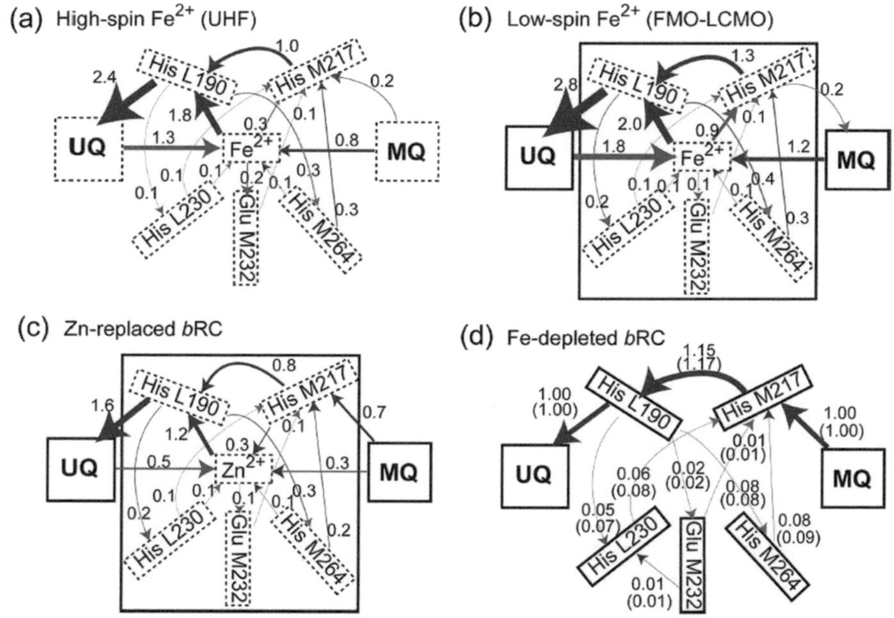

Fig. 39.3 Results of electron transfer pathway analysis for (**a**) wild-type high-spin ($S = 2$) Fe^{2+} complex, (**b**) low-spin ($S = 0$) Fe^{2+} complex, (**c**) the ion complex with the Fe^{2+} ion replaced by a Zn^{2+} ion, and (**d**) the complex without the metal ion. The numbers near the *arrows* denote the normalized ET currents $\hbar J_{I,J}/T_{DA}$ between the fragments I and J, which are also approximately represented by the thickness of the arrows

Figure 39.3d displays the corresponding pathway analysis for the metal ion-depleted system. Now it is clear that the pathway via His M217 and His L190 becomes dominant. Consequently, the reduction of the ET rate is only by a factor of 3–5, in accord with the experimental observation.

In this way, an intriguing picture emerges such that the His ligands play the role to secure the robustness of ET by providing an auxiliary ET pathway channel even in cases of disturbances such as the metal ion substitution or depletion.

39.5 Concluding Remarks

A new computational scheme to carry out ET pathway analysis in large biomolecules has been developed and applied to the ET from MQ to UQ via a nonheme iron ion complex in bacterial photosynthetic reaction center. The scheme consists of a combination of the FMO-LCMO method that enables ab initio electronic structure calculations of large biomolecules and the tunneling current analysis that provides pictorial understanding of ET mechanism. Nevertheless, since any computational studies on realistic molecular systems generally involve assortments of theoretical and computational approximations, it is essential to proceed with sufficient care checking the consistency with experimental findings. Then, it will become possible to discuss detailed microscopic mechanism at atomic and electronic levels that are inaccessible by experiments. With such intimate collaborations of experimental and theoretical studies, the progress of our understanding of biological energy conversions will be secured.

References

1. Moser CC, Keske JM, Warncke K, Farid RS, Dutton PL (1992) Nature of biological electron transfer. Nature 355:796
2. Dutton PL, Mosser CC (1994) Quantum biomechanics of long-range electron transfer in protein: hydrogen bonds and reorganization energies. Proc Natl Acad Sci USA 91:10247
3. Winkler JR, Di Bilio AJ, Farrow NA, Richards JH, Gray HB (1999) Electron tunneling in biological molecules. Pure Appl Chem 71:1753
4. Gray HB, Winkler JR (2005) Long-range electron transfer. Proc Natl Acad Sci USA 102:3534
5. Farver O, Pecht I (2011) Electron transfer in blue copper proteins. Coord Chem Rev 255:757
6. Marcus RA, Sutin N (1985) Electron transfers in chemistry and biology. Biochim Biophys Acta 811:265
7. Sumi H (1997) Electron transfer via a midway molecule as seen in primary processes in photosynthesis: superexcharge or sequential, or unified? J Electroanal Chem 438:11
8. Fedorov DG, Kitaura K (2007) Extending the power of quantum chemistry to large systems with the fragment molecular orbital method. J Phys Chem A 111:6904
9. Tanaka S, Mochizuki Y, Komeiji Y, Okiyama Y, Fukuzawa K (2014) Electron-correlated fragment-molecular-orbital calculations for biomolecular and nano systems. Phys Chem Chem Phys 16:10310
10. Yang W, Lee TS (1995) A density-matrix divide-and-conquer approach for electronic structure calculations of large molecules. J Chem Phys 103:5674
11. Akama T, Kobayashi M, Nakai H (2009) Electronic temperature in divide-and-conquer electronic structure calculation revisited: assessment and improvement of self-consistent field convergence. Int J Quantum Chem 109:2706
12. Tsuneyuki S, Kobori T, Akagi K, Sodeyama K, Terakura K, Fukuyama H (2009) Molecular orbital calculation of biomolecules with fragment molecular orbitals. Chem Phys Lett 476:104
13. Kobori T, Sodeyama K, Otsuka T, Tateyama Y, Tsuneyuki S (2013) Trimer effects in fragment molecular orbital-linear combination of molecular orbitals calculation of one-electron orbitals for biomolecules. J Chem Phys 139(9):094113
14. Nishioka H, Ando K (2011) Electronic coupling calculation and pathway analysis of electron transfer reaction using ab initio fragment-based method. I. FMO–LCMO approach. J Chem Phys 134:204109
15. Kitoh-Nishioka H, Ando K (2012) Fragment molecular orbital study on electron tunneling mechanisms in bacterial photosynthetic reaction center. J Phys Chem B 116(43):12933
16. Cave RJ, Newton MD (1997) Calculation of electronic coupling matrix elements for ground and excited state electron transfer reactions: comparison of the generalized Mulliken-Hush and block diagonalization methods. J Chem Phys 106:9213
17. Skourtis SS, Beratan DN (1999) Theories of structure–function relationships for bridge-mediated electron transfer reactions. Adv Chem Phys 106:377
18. Regan JJ, Onuchic JN (1999) Electron-transfer tubes. Adv Chem Phys 107:497

19. Stuchebrukhov AA (2003) Long-distance electron tunneling in proteins. Theor Chem Acc 110:291
20. Beratan DN, Betts JN, Onuchic JN (1991) Protein electron transfer rates set by the bridging secondary and tertiary structure. Science 252:1285
21. Nishioka H, Ando K (2011) Pathway analysis of super-exchange electronic couplings in electron transfer reactions using a multi-configuration self-consistent field method. Phys Chem Chem Phys 13:7043
22. Leibl W, Breton J (1991) Kinetic properties of the acceptor quinone complex in Rhodopseudomonas viridis. Biochemistry 30:9634
23. Mathis P, Sinning I, Michel H (1992) Kinetics of electron transfer from the primary to the secondary quinone in Rhodopseudomonas viridis. Biochim Biophys Acta 1098:151
24. Feher G, Okamura MY (1999) The primary and secondary acceptors in bacterial photosynthesis: II. The structure of the $Fe^{2+} - Q^{-}$ complex. Appl Magn Reson 16:63
25. Debus RJ, Feher G, Okamura MY (1986) Iron-depleted reaction centers from Rhodopseudomonas sphaeroides R-26.1: characterization and reconstitution with iron (2+), manganese (2+), cobalt (2+), nickel (2+), copper (2+), and zinc (2+). Biochemistry 25:2276
26. Moser CC, Page CC, Farid R, Dutton PL (1995) Biological electron transfer. J Bioenerg Biomembr 27:263
27. Deisenhofer J, Epp O, Sinning I, Michel H (1995) Crystallographic refinement at 2.3 Å resolution and refined model of the photosynthetic reaction centre from Rhodopseudomonas viridis. J Mol Biol 246:429

Chapter 40
Spectroscopic Analysis of the Redox Reactions of π-Conjugated Cofactors in Photosynthetic Reaction Center

Takumi Noguchi

Abstract Photosynthesis is a biological process in which solar energy is converted into a storable form of chemical energy. Light-driven electron transfer in a photosynthetic reaction center is performed by various types of redox cofactors with π-electron conjugation. In this chapter, spectroscopic investigations of the electron transfer reactions of π-conjugated cofactors in photosystem II (PSII), which has a function of water oxidation to obtain electrons necessary for CO_2 fixation, are introduced. Light-induced Fourier transform infrared difference spectroscopy was used to detect the structures and reactions of π-conjugated cofactors in PSII, i.e., tyrosine Y_Z, chlorophyll dimer P680, pheophytin electron acceptor $Pheo_{D1}$, and plastoquinone electron acceptors Q_A and Q_B. Detailed spectral analyses and estimations of redox potentials were further performed by quantum chemical calculations using the density functional theory and quantum mechanics/molecular mechanics methods. Studies of the regulation mechanisms of forward and backward electron transfer relevant to photoprotection of PSII are also described.

Keywords Fourier transform infrared spectroscopy • Quantum chemical calculation • Photosystem II • Photosynthesis • Electron transfer

40.1 Introduction

Photosynthesis is a biological process of energy conversion, in which solar energy is converted into a storable form of chemical energy [5]. Particularly, in oxygenic photosynthesis performed by plants and cyanobacteria, only water and CO_2 are used as chemical resources for the synthesis of sugars [52]. In this process, water functions as an ultimate electron donor for reduction of CO_2, and as the result of water oxidation, O_2 is produced as a byproduct and liberated into the air. This

T. Noguchi (✉)
Division of Material Science, Graduate School of Science, Nagoya University, Furo-cho, Chikusa-ku, Nagoya 464-8602, Japan
e-mail: tnoguchi@bio.phys.nagoya-u.ac.jp

T. Akasaka et al. (eds.), *Chemical Science of π-Electron Systems*,
DOI 10.1007/978-4-431-55357-1_40

photosynthetically produced O_2, which is the source of the oxygenic atmosphere (21 % O_2 content), is then used for respiration. The O_2 is also converted into ozone, which protects life from harmful UV radiation. Therefore, photosynthesis plays essential roles in sustaining the global environment and life on earth as a source of both energy and O_2.

Light energy conversion in oxygenic photosynthesis is preformed mainly by four large protein complexes, i.e., photosystems (PS) I and II, cytochrome (Cyt) b_6/f, and ATPase, embedded in thylakoid membranes [5]. Electrons abstracted from water in PSII are transferred to PSI via Cytb_6/f and then stored in NADPH. In each protein complex, several redox cofactors, most of which are π-electron conjugated compounds, are bound to form an electron transfer chain. Water oxidation and a series of electron transfer reactions are coupled with proton transfer, which produces a proton gradient across the thylakoid membrane. The electrochemical potential of the proton gradient is then used to produce ATP as a small unit of chemical energy in ATPase. These products of the primary energy-conversion process, NADPH as a reducing power and ATP as a chemical energy, are used to synthesize sugars in the subsequent carbon fixation process preformed in the stroma.

In this chapter, molecular mechanisms of the electron transfer reactions of π-conjugated cofactors in PSII are described. The PSII protein complex [77], which consists of more than 20 protein subunits and more than 50 pigments, including 35 chlorophyll (Chl) and 11 β-carotene molecules [17, 35, 73], has a function of light-driven water oxidation, which requires extremely high oxidation power. Most of the pigments serve as light-harvesting antenna, whereas several pigments bound to a reaction center (RC), which mainly consists of the D1 and D2 subunits, serve as redox cofactors in the electron transfer chain. The cofactors in the PSII RC (Fig. 40.1A) are arranged in a pseudo-C_2 symmetry except for the Mn_4Ca cluster, the catalytic center of water oxidation, which is located on the D1 side [17, 35, 73]. Despite this symmetric arrangement of cofactors, electron transfer takes an asymmetric pathway. The excited singlet state of Chl (a coupled excited state of monomeric Chl_{D1} and a Chl dimer P680) [16], which is formed by excitation transfer from antenna pigments, ejects an electron to the pheophytin electron acceptor on the D1 subunit, $Pheo_{D1}$, resulting in the formation of a $P680^+Pheo_{D1}^-$ charged pair [57]. An electron on $Pheo_{D1}^-$ is then transferred to the primary plastoquinone (PQ) electron acceptor, Q_A, and then to the secondary PQ acceptor, Q_B [53]. When the second electron transfer occurs, doubly reduced Q_B takes up two protons to become plastoquinol (PQH_2), which is then released into the thylakoid membranes. On the electron donor side, $P680^+$ abstracts an electron from the redox-active tyrosine Y_Z (D1-Tyr161) and then from the Mn_4Ca cluster. At the Mn_4Ca cluster, two water molecules are converted into one O_2 molecule and four protons, upon abstraction of four electrons, through the cycle of five intermediates called S_i state ($i = 0$–4) [22, 38].

Water oxidation in PSII corresponds to the reaction of an anode in electrolysis of water. The electrode is the Mn_4Ca cluster, instead of a Pt electrode in water electrolysis, while other redox cofactors, which are all π-conjugated compounds, function as a "wire" to transport electrons. Among them, however, P680 plays a

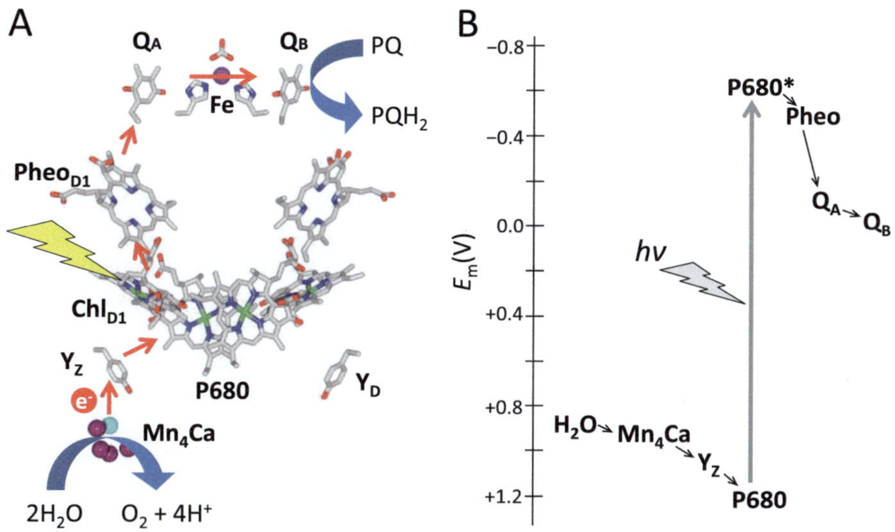

Fig. 40.1 (**A**) Spatial arrangement of the redox cofactors and electron transfer pathway in photosystem II. (**B**) Redox potentials of the cofactors in the electron transfer chain

role of a "battery" with a voltage of 1.8 V, which corresponds to a 680-nm photon that P680 can absorb. Efficient water oxidation at the Mn_4Ca cluster and smooth electron transfer to Q_B are realized by appropriate spatial arrangements of the redox cofactors along the electron transfer pathway (Fig. 40.1A) and the redox potentials of the cofactors to satisfy the energetic requirement for electron transfer at each step (Fig. 40.1B). The determination of the appropriate redox potentials is achieved by proper selection of chemical species as cofactors and an elaborate tuning of the potentials by interactions with surrounding proteins. Not only the forward electron transfer but also the backward electron transfer is often important for protection of the PSII protein from photodamage. PSII has mechanisms to regulate the redox potentials of $Pheo_{D1}$ and Q_A responding to the changes in the environment such as light intensity [33, 74].

Light-induced Fourier transform infrared (FTIR) difference spectroscopy is a powerful method to investigate the structure-function relationship of redox cofactors in photosynthetic proteins [2, 6–8, 11, 12, 36, 40–45]. Using this technique, detailed atomic-level information on the structures and reactions of the redox cofactors along with surrounding proteins and water molecules has been obtained. Accurate assignments of the FTIR bands and their structural implications as well as energetic information are available using quantum chemical calculations with the density functional theory (DFT) and quantum mechanics/molecular mechanics (QM/MM) methods. This chapter summarizes our recent studies on the mechanisms of the reaction control of the π-conjugated cofactors in PSII using FTIR spectroscopy assisted with quantum chemical calculations.

40.2 Tyrosine Y_Z: Proton-Coupled Electron Transfer Leading to Water Oxidation

The Mn_4Ca cluster, an "electrode" for water oxidation, is connected to a "π-electron wire" at the redox-active tyrosine Y_Z, which mediates electron transfer from the Mn_4Ca cluster to $P680^+$. Although another redox-active tyrosine Y_D (D2-Tyr160) is symmetrically located in PSII (Fig. 40.1A), the kinetics of its oxidation is much slower than that of Y_Z and hence serves only as a peripheral electron donor to $P680^+$. When Y_Z is oxidized, its phenolic proton is released, because of the extremely low pK_a (~ -2) of the radical cation of a phenolic group, forming a neutral $Y_Z^\bullet$ radical [15, 64]. The X-ray structure of PSII revealed that D1-His190 is located at a hydrogen bond distance from the phenolic oxygen of Y_Z [17, 35, 73], and hence this His has been suggested to be an immediate proton acceptor of Y_Z. The X-ray structure also showed that Y_Z interacts with the Mn_4Ca cluster via a water cluster, which consists of several water molecules, and is located in a hydrogen bond network connected from the Mn_4Ca cluster to the lumen [73]. It is thus possible that Y_Z functions not only as an electron mediator but also as a proton mediator during water oxidation.

A flash-induced $Y_Z^\bullet/Y_Z$ FTIR difference spectrum was obtained using PSII complexes in which the Mn_4Ca cluster was removed to achieve a longer lifetime of $Y_Z^\bullet$ (Fig. 40.2A) [4, 39]. The spectrum showed a number of signals in the 1800–1000 cm^{-1} region, e.g., the CO stretching/COH bending band of Y_Z (1256 cm^{-1}), the CO stretch of $Y_Z^\bullet$ (1514 cm^{-1}), the CN stretch of the coupled His (1109 cm^{-1}), the amide I (1700–1600 cm^{-1}) and II ($\sim$1550 cm^{-1}) bands of protein backbones, and the electrochromic shift of the keto C$=$O of P680 (1705/1697 cm^{-1}).

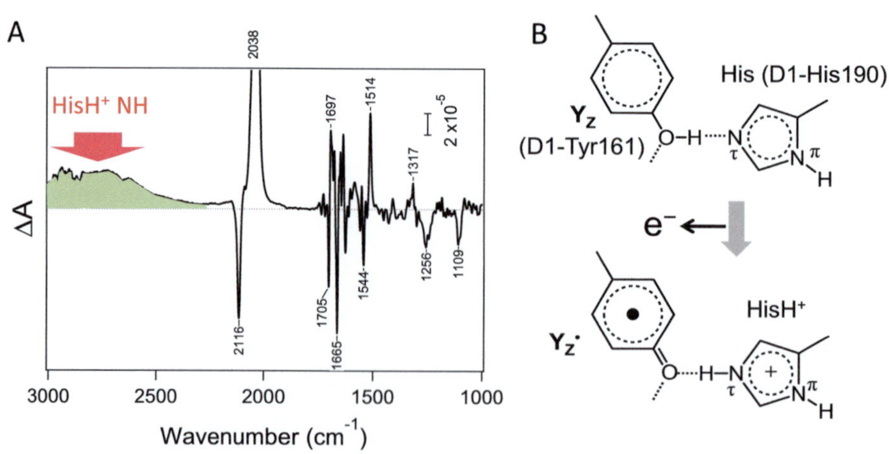

Fig. 40.2 (**A**) FTIR difference spectrum of Y_Z upon its photooxidation to the neutral radical $Y_Z^\bullet$. (**B**) Change in the protonation and hydrogen-bonded structures of Y_Z and the coupled D1-His190 upon photooxidation

A strong signal at 2116/2038 cm^{-1} due to the CN stretching vibrations of ferricyanide/ferrocyanide represents electron abstraction by ferricyanide on the electron acceptor side. In addition to these signals, a broad positive feature was observed around 2800 cm^{-1} (Fig. 40.2A) [39]. This band was assigned to the N_τ-H vibration of a protonated $HisH^+$ cation hydrogen-bonded with $Y_Z^\bullet$ from downshifts by H/D exchange and global ^{15}N substitution, and this assignment was confirmed by DFT and QM/MM calculations [39]. This observation indicates that the proton released from Y_Z is trapped by the neighboring D1-His190 and not liberated into the lumen (Fig. 40.2B). It is noted that this broad feature in the $Y_Z^\bullet/Y_Z$ spectrum was not observed in the $Y_D^\bullet/Y_D$ spectrum [39], which is consistent with the view that a proton from Y_D is not trapped by the neighboring His (D2-His189) but either transferred to a remote protonatable residue or released into the lumen [61].

The observed N-H frequency of ~2800 cm^{-1}, which is much lower than the usual NH frequencies of His and $HisH^+$ (3600–3000 cm^{-1}), reflects the presence of a strong hydrogen bond, the so-called charge-assisted hydrogen bond, between $Y_Z^\bullet$ and $HisH^+$ [39]. In addition, the broad feature of the N-H band indicates that this proton has a large polarizability [78], that is, it readily moves responding to a change in the electrostatic field, such as a change in the charge distribution of the Mn_4Ca cluster and the fluctuation of the protein and water environment.

Further QM/MM calculation of the Y_Z-His couple involving surrounding amino acid residues and water molecules in the QM region showed that the hydrogen bond network around Y_Z is rearranged upon its oxidation, and the locations and orientations of some of the water molecules are significantly altered (Fig. 40.3A) [39]. Specifically, upon forming a strong hydrogen bond at the Y_Z oxygen with $HisH^+$, the hydrogen bond with W4 is broken and then the W4 proton forms a new hydrogen bond with another water W_A. This hydrogen bond formation promotes the shift of W_A toward D1-His190. The distance between N_τ of His190 and the W_A oxygen is now shortened from 4.57 to 3.28 Å. From this rearrangement of the hydrogen bond network as well as the large polarizability of the proton between $Y_Z^\bullet$ and D1-His190, a novel proton transfer mechanism through $Y_Z^\bullet$-$HisH^+$ was proposed (Fig. 40.3B) [39]. In this mechanism, proton hopping from the polarizable proton of $HisH^+$ to W_A triggers immediate proton transfer from substrate water on the Mn_4Ca cluster to $Y_Z^\bullet$ by the "Grotthuss mechanism" [13], while the proton on W_A is transferred to the lumen through water chains near D1-Asn298. Thus, a proton is eventually transported from the substrate to the lumen. The rate-limiting step of this entire process is proton hopping from $HisH^+$ to W_A, and thus the Y_Z-His site functions as a "gate" in this proton transfer pathway.

This type of proton transfer should take place before electron transfer from the Mn_4Ca cluster to $Y_Z^\bullet$. Proton release prior to electron transfer has been proposed to occur at the S_2 and S_3 states, which have an excess positive charge on the Mn_4Ca cluster [20, 29, 49, 55]. Because the $S_3 \rightarrow S_0$ transition seems to use another putative pathway via D1-Asp61 and a nearby Cl ion [14, 54, 68], it is speculated that the proton pathway through $Y_Z^\bullet$-$HisH^+$ is functional in the $S_2 \rightarrow S_3$ transition [39]. In addition, this mechanism can also work in concerted proton-coupled electron transfer,

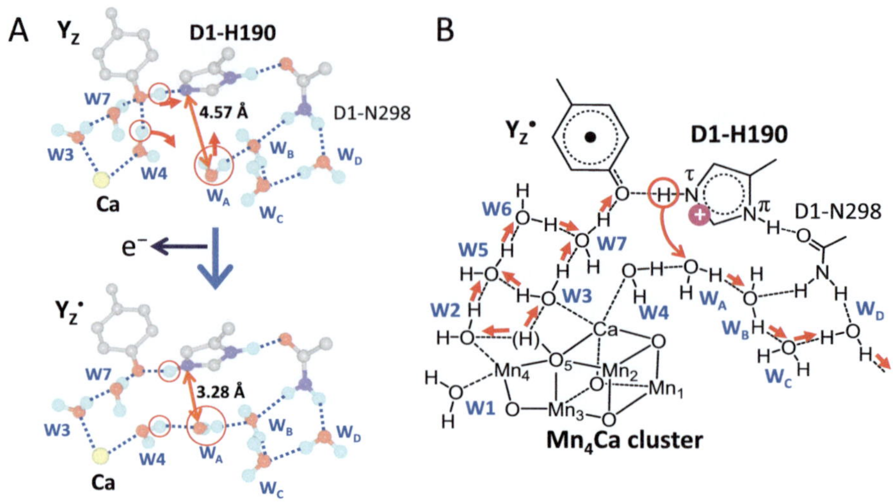

Fig. 40.3 (**A**) Rearrangement of the hydrogen bond network around Y_Z upon its oxidation estimated by QM/MM calculations. (**B**) Proposed mechanism of proton release from substrate water on the Mn_4Ca cluster via the $Y_Z^•$-HisH$^+$ moiety. Hopping of the high polarizability proton of HisH$^+$ to W_A triggers immediate proton transfer from substrate water to Y_Z and from W_A to the lumen through the water clusters by the Grotthuss mechanism

in which a proton and an electron are transferred simultaneously. If the $S_0 \rightarrow S_1$ transition has a concerted mechanism, as suggested by TRIR measurements [49], it is possible that this transition also uses the $Y_Z^•$-HisH$^+$ pathway for proton exit.

40.3 Chlorophyll Dimer P680: Source of Strong Oxidation Power

Oxidation of water to evolve O_2 requires a strong oxidation power ($E_m \sim 0.88$ V at pH 6). The source of this strong oxidation power in PSII is the Chl dimer P680, which has an extremely high redox potential (E_m) of ~1.2 V [28, 56]. Because the bacteriochlorophyll (Bchl) dimer P870 in a bacterial reaction center (bRC), which is thought to be the ancestor of PSII, has a much lower E_m value of ~0.5 V [34], how P680 gained this strong oxidation power is an essential question for understanding the evolutionary process of oxygenic photosynthesis, which occurred ~25 billion years ago [51]. Several factors have been proposed as causes of the high E_m of P680: (1) a change in the constituent pigment from Bchl to Chl, (2) a hydrophobic environment at the P680 site [18], (3) electrostatic interactions with proteins [24, 60], and (4) charge localization on one side of the Chl dimer [58, 60, 69].

In organic solvents, isolated Chl has E_m of ~0.8 V, which is higher by 0.1–0.2 V than that of Bchl [76], and hence this pigment change seems to contribute, to some extent, to the increase in the E_m of P680 from that of bacterial P870. Here, another question is why the E_m of P680 (~1.2 V) is even higher than that of isolated Chl (~0.8 V). Hasegawa and Noguchi [18] assessed the dependency of a dielectric constant (ε) on the E_m of Chl using DFT calculation and showed that E_m sharply increases as ε decreases in the relatively low ε region ($\varepsilon < 5$). The E_m value as high as ~1.4 V was estimated in an ideal hydrophobic environment ($\varepsilon \sim 2$), and hence it was proposed that a highly hydrophobic environment at the P680 site in PSII contributes to the high E_m. Ishikita and coworkers [24, 60], on the other hand, estimated the effect of the partial charges of amino acid residues and cofactors on the P680 E_m, using a linearized Poisson-Boltzmann equation, and well reproduced the high E_m value.

If the positive charge on $P680^+$ is delocalized over the two Chl molecules, its E_m should decrease. The effect of charge delocalization on the P680 E_m was estimated by DFT calculation, in which the E_m values of a monomeric Chl and a charge-delocalized Chl dimer were compared (Fig. 40.4A) [69]. It was found that charge delocalization decreases the E_m by 0.14 V, indicating in turn that localization of the charge on one side of the dimer can increase the E_m by the same extent (Fig. 40.4B).

The charge distribution of $P680^+$ was experimentally assessed by the measurement of $P680^+$/P680 FTIR difference spectra [47, 50, 65]. The spectra of PSII membranes of spinach showed a broad intervalence band with a maximum around 3000 cm^{-1} (Fig. 40.5Aa), indicative of hole transfer between the two Chl molecules of P680. The band feature was however significantly different in isolated PSII RC, which is a minimum photoactive unit consisting of the D1, D2, and Cyt*b*559 subunits (corresponding to bRC consisting of the L and M subunits)

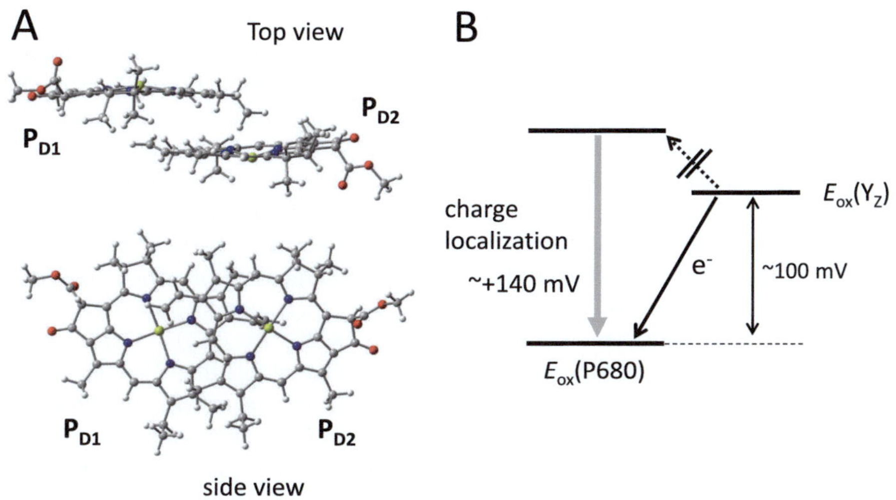

Fig. 40.4 (**A**) The model of the Chl dimer P680 used for DFT calculation. (**B**) The shift of the redox potential by charge localization on $P680^+$ estimated by DFT calculation and the relationship with the redox potential of Y_Z

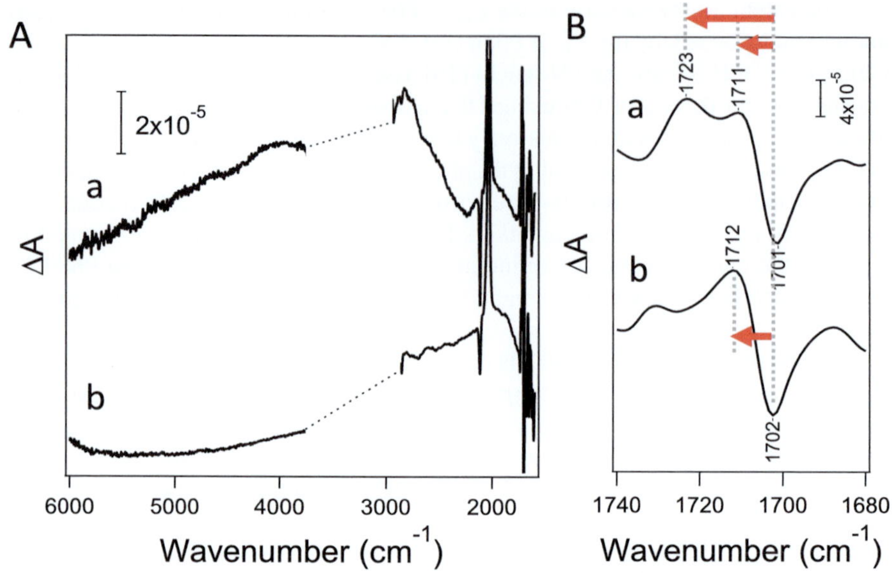

Fig. 40.5 Light-induced P680$^+$/P680 FTIR difference spectra of the membrane preparation (**a**) and the isolated reaction center complex (**b**) of PSII from spinach in the regions of the intervalence electronic bands (A) and the keto C = O stretching bands (B). The region connected by a *dotted line* in (A) is saturated by strong water absorption. The shifts of the keto C = O bands from neutral P680 to cationic P680$^+$ are expressed by *arrows* in (B)

without major core antenna proteins (PC43 and CP47) as well as numerous small subunits, showing a narrower band around 2000 cm^{-1} (Fig. 40.5Ab). This difference indicates that the charge distribution on the Chl dimer is different between the two preparations. In addition, the features of the keto C = O bands of P680$^+$ were also significantly different [50]. Two positive bands were observed at 1723 and 1711 cm^{-1} with comparative intensities in the PSII membranes, whereas only one strong band was detected at 1712 cm^{-1} in the RC complex (Fig. 40.5B). In contrast, neutral P680 showed a similar single peak at 1702–1701 cm^{-1} in both preparations. From these results, it was suggested that 70–80 % of the positive charge is localized on one Chl in P680$^+$ in more intact PSII, whereas the charge is evenly delocalized over the dimer in the smaller RC complex.

This result is consistent with the decrease in E_m by ~0.14 V by charge delocalization over a Chl dimer [69] and the experimental fact that Y_Z, which has a E_m value lower than that of intact P680 by ~0.1 V, cannot be oxidized by P680$^+$ in the isolated RC complex (Fig. 40.4B). These data in FTIR measurement and DFT calculation therefore indicate that localization of a positive charge on one side of P680 by forming a larger proton complex induces an E_m increase by 0.1–0.2 V, which is crucial for the water oxidation capability of PSII. Recent QM/MM and Poisson-Boltzmann calculations by Saito et al. [60] reproduced a significantly localized charge distribution with 80 % on P_{D1} (D1-side Chl of P680). They also showed that the difference in amino acid residues between the D1 and D2 subunits significantly contributes to the asymmetric charge distribution, which is contradict to the FTIR data of the RC complex showing a significant charge delocalization over the dimer [50]. This discrepancy may indicate that the interactions of the D1 and D2 subunits with the surrounding proteins such as CP43 and CP47 are important to retain the proper RC conformation, which is required for the asymmetric charge distribution of P680 and hence its high E_m value.

40.4 Pheophytin: Primary Electron Acceptor Regulating Forward and Backward Electron Transfer

$Pheo_{D1}$ functions as a primary electron acceptor upon initial charge separation (Fig. 40.1A). It also regulates forward and backward electron transfer by tuning its E_m [56, 74]. The conjugated keto C = O of Pheo forms a hydrogen bond with Gln130 of the D1 (PsbA in another nomenclature) polypeptide in the X-ray structure of *Thermosynechococcus elongatus* [17, 35] and *Thermosynechococcus vulcanus* [73]. In *T. elongatus*, which has three different *psbA* genes, *psbA1*, *psbA2*, and *psbA3*, encoding the PsbA (D1) protein, *psbA1* is usually expressed under a normal condition, while *psbA3* is mainly expressed under a high-light condition [30]. Gln130 in PsbA1 is replaced with Glu130 in PsbA3. A similar acclimation mechanism of exchanging PsbA polypeptides responding to the light condition has been reported also in other cyanobacterial species [9]. Site-directed mutagenesis studies have shown that the replacement of Gln130 to Glu increases the E_m($Pheo^-$/Pheo) by 30–40 mV [37, 66], which facilitates backward electron transfer from Q_A^- to Pheo, promoting energy dissipation by charge recombination. Thus, the change in the amino acid residue interacting with $Pheo_{D1}$ by replacing the D1 subunit is thought to be one of the protection mechanisms of PSII against photodamage induced by strong light [74].

The mechanism of the E_m shift by exchanging the amino acid hydrogen-bonded to the keto C = O of $Pheo_{D1}$ was investigated using FTIR spectroscopy in combination with DFT calculation [63]. Figure 40.6Aa shows FTIR difference spectra of $Pheo_{D1}$ upon its photoreduction measured using the A1-PSII (blue line) and A3-PSII (red line) core complexes of *T. elongatus*, which have PsbA1 and PsbA3, respectively, as a D1 subunit. It was shown that the keto C = O stretching band of neutral $Pheo_{D1}$ at 1682 cm^{-1} downshifted only slightly to 1681 cm^{-1} upon changing Gln130 to Glu, whereas that of the $Pheo_{D1}^-$ anion was largely shifted down from 1605 to 1588 cm^{-1} by 17 cm^{-1}. This larger downshift of the keto C = O frequency in anionic $Pheo_{D1}^-$ than neutral $Pheo_{D1}$ was well reproduced by DFT calculation (Fig. 40.6Ab) using model Pheo complexes hydrogen-bonded with acetamide as a model of Gln (A1 type) and with acetic acid as a model of Glu (A3 type) (Fig. 40.6B) [63]. Energy calculation also reproduced a higher E_m value in the A3-type model than the A1-type model, although the calculated E_m gap of 84 mV was much larger than the experimental value of 30–40 mV due probably

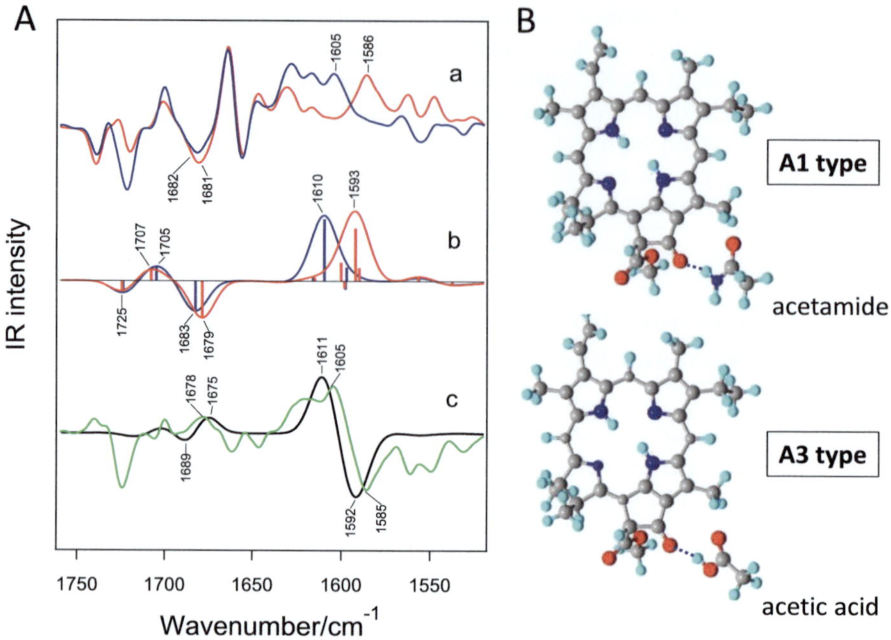

Fig. 40.6 (**A**) Light-induced $Pheo_{D1}^-/Pheo_{D1}$ FTIR difference spectra of A1-PSII (*blue line*) and A3-PSII (*red line*) from *T. elongatus* (*a*), calculated spectra of the model complexes of Pheo hydrogen-bonded with acetamide (A1 type: *blue line*) and acetic acid (A3 type: *red line*) (*b*), and A1-minus-A3 double difference spectra of the experimental (*green line*) and calculated (*black line*) spectra. (**B**) Optimized geometries of the A1- and A3-type Pheo models

to the calculation in vacuum. These data indicate that the change in the hydrogen bond partner at the conjugated keto C = O from an amide group to a carboxylic group induces a stronger hydrogen bonding interaction in the $Pheo^-$ anion than in the neutral state. It is therefore concluded that the mechanism of the E_m shift is that the stronger hydrogen bond with Glu stabilizes the anionic $Pheo_{D1}^-$ more than the neutral $Pheo_{D1}$, which causes an increase in the E_m in A3-PSII compared with that in A1-PSII.

40.5 Quinone-Iron Complex: Electron Acceptors Controlling PSII Reactions

Two PQ electron acceptors, Q_A and Q_B, are located on the electron acceptor side of PSII (Fig. 40.1A), connected with each other by the His-Fe-His π-electron bridge (Fig. 40.7). Although Q_A and Q_B both consist of PQ-9, the functions of Q_A and Q_B are significantly different. Q_A is only singly reduced and mediates an electron

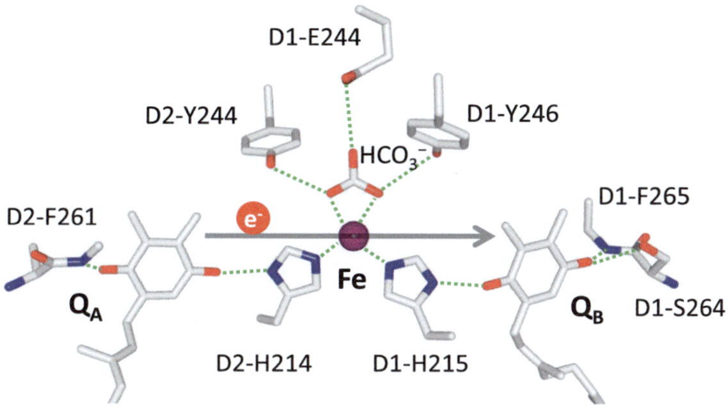

Fig. 40.7 Structure of the Q_A-Fe-Q_B site in PSII

from $Pheo^-$ to Q_B, while Q_B can be doubly reduced and then protonated to become PQH_2 [53]. Thus, Q_B functions as a substrate in PSII. This is consistent with the result of fragment molecular orbital (FMO) calculation that the binding energy of PQ-9 in each biding site of the PSII complex is much smaller for Q_B than for Q_A [19]. The E_m of one electron reduction of Q_A is lower by ~80 mV than that of Q_B, which facilitates electron transfer from Q_A to Q_B [10]. In addition, the $E_m(Q_A^-/Q_A)$ is known to increase by ~150 mV upon depletion of the Mn_4Ca cluster [26, 31, 62] and to shift within ±50 mV by binding of different types of herbicide at the Q_B site [32]. These E_m shifts change the kinetics of redox reactions and charge recombination pathways and thus control the electron flow in PSII to a forward direction for a higher efficiency and to a backward direction for photoprotection [33, 59].

40.5.1 Effect of Hydrogen Bonding Interaction on the Redox Potential and Molecular Vibrations of Plastoquinone

The E_m difference between Q_A and Q_B should originate from the interactions of individual PQ molecules with surrounding proteins. In the X-ray crystallographic structure of PSII [73], the conjugated C=O bonds of Q_A are within hydrogen bond distance of two nitrogen atoms (the imidazole of D2-H214 and the backbone amide of D2-F261), whereas those of Q_B are within hydrogen bond distance of two nitrogen atoms (the imidazole of D1-H215 and the backbone amide of D1-F265) and one oxygen atom (the hydroxide of D1-S264) (Fig. 40.7). Thus, the primary candidate for the factors to differentiate the E_m values of Q_A and Q_B is different hydrogen bonding interactions at their C=O bonds. To understand the effect of hydrogen bonding interaction on the redox potential of PQ, DFT calculations were

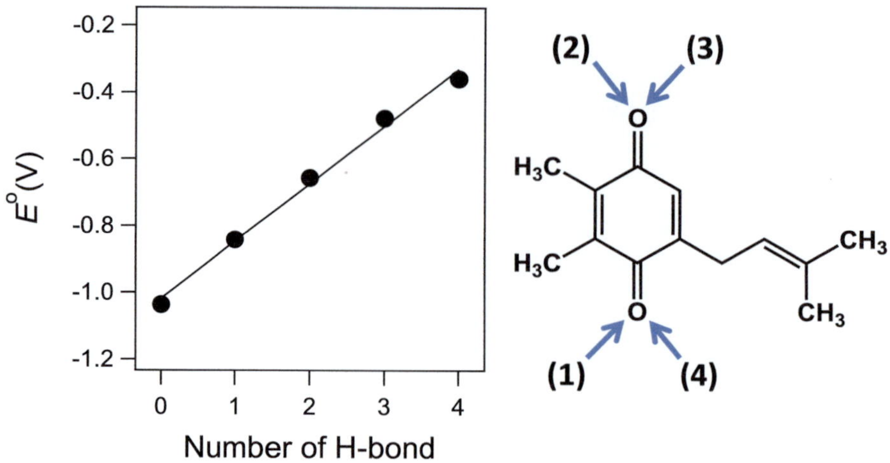

Fig. 40.8 Calculated redox potentials of PQ-1 as a function of the number of hydrogen bonds. The chemical structure of PQ-1 with the positions of hydrogen bond(s) with water molecule(s) is also shown

performed for the neutral and semiquinone anion forms of PQ-1 models, which have a different number (0–4) of hydrogen bonds with water molecules [1].

The E_m of PQ calculated at $\varepsilon = 4.9$ assuming the environment of protein showed an increase proportional to the number of hydrogen bonds (Fig. 40.8). The slope of the regression line was +172 mV per one hydrogen bond. Calculations using PQ-1 complexes hydrogen-bonded with model amino acids mimicking the hydrogen bond interactions of Q_A and Q_B showed a consistent result; A PQ model with three hydrogen bonds (Q_B model) provided a larger E_m value by 112 mV than the model with two hydrogen bonds (Q_A model). The experimental E_m gap between Q_A and Q_B, ~80 mV [10], is comparable to this E_m difference, suggesting that an additional hydrogen bond from the D1-Ser264 OH to Q_B is an important factor for the E_m gap.

Prediction of the hydrogen bond interactions from the X-ray structure [73], which did not resolve the positions of protons, can be confirmed by vibrational spectroscopy. Normal mode calculations of the PQ models using the DFT method showed that the C = O frequencies of neutral PQ in the 1700–1600 cm^{-1} region are lowered by 10–30 cm^{-1} upon hydrogen bond formation, and two separate frequencies are found when PQ has an asymmetric hydrogen bond interaction [1]. In spite of this sensitivity of the C = O vibrations to the hydrogen bond structure, severe overlap with strong amide I bands in the Q_A^-/Q_A and Q_B^-/Q_B FTIR difference spectra [3, 48, 67] hampered clear assignments of the C = O bands of neutral Q_A and Q_B.

In contrast, the CO stretching vibrations, coupled with the CC stretches, of the semiquinone anion show bands at ~1480 cm^{-1}, the frequency much lower than the C = O frequencies of neutral PQ, due to significantly lengthened C = O and C = C bonds upon anion formation [1, 72]. This frequency region is devoid of the overlap of main protein vibrations, and hence the strong CO/CC bands of Q_A^- and Q_B^- can be readily recognized in the FTIR difference spectra [3, 48, 67]. However, it was shown that the CO/CC modes of a PQ anion are not much sensitive to hydrogen bond interactions and are found only in the narrow range of 1479 ± 5 cm^{-1} for all the hydrogen-bonded model complexes examined [1, 72]. In addition, the direction of shift did not show a clear tendency in the relationship with hydrogen bonds. Nevertheless, a small shift of the prominent CO/CC band is a good marker of the change in the hydrogen bond interaction at the CO bonds of a PQ anion.

In the Q_A and Q_B models, the CO/CC bands of semiquinone anions are further coupled with amino acid vibrations such as the NH and OH bends of His and Ser side chains, respectively. These couplings provide a complex spectral feature with a larger number of vibrational modes in the CO/CC region compared with simpler PQ models hydrogen-bonded with water molecules [1]. The calculated band features of the CO/CC vibrations of the semiquinone anions of the Q_A and Q_B models fairly reproduced the experimental FTIR spectra of Q_A^- and Q_B^-. This result is consistent with the view that Q_A and Q_B have two and three hydrogen bonds, respectively, with amino acid residues in the PSII proteins.

40.5.2 E_m *Shift of* Q_A *by Herbicide Binding*

Binding of a herbicide to the Q_B site affects the E_m of Q_A. A urea-type herbicide, 3-(3,4-dichlorophenyl)-1,1-dimethylurea (DCMU), raises the E_m by ~50 mV, whereas a phenolic herbicide, bromoxynil, lowers the E_m by ~45 mV [32]. FTIR measurements of Q_A^-/Q_A difference spectra showed that PSII samples treated with phenolic herbicides such as bromoxynil and ioxynil induced a small but clear downshift of the CO/CC peak of Q_A^-, whereas samples treated with other types of herbicides, i.e., DCMU, atrazine, terbutryn, and bromacil, induced a small upshift (Fig. 40.9A) [72]. The former group of herbicide exhibited lower thermoluminescence temperatures than the latter group, indicative of lower E_m values as expected [72]. This small frequency shift in the CO/CC peak of Q_A^- was interpreted as reflecting the change in the hydrogen bond strength without changing the hydrogen bonding pattern from the results of DFT calculations of model PQ complexes [1, 72]. Frequencies of the CN (2215–2200 cm^{-1}) and CO (1516–1505 cm^{-1}) stretching bands of bromoxynil detected in Q_A^-/Q_A and/or Fe^{2+}/Fe^{3+} FTIR difference spectra in comparison with those of bromoxynil measured in ethanol and calculated using DFT strongly suggest that bromoxynil takes a deprotonated form in the Q_B binding pocket [71, 72]. Docking calculations confirmed the direct interaction of the phenolate CO of deprotonated bromoxynil with D1-His215, whereas DCMU was found to bind to the opposite side of the pocket without an interaction with D1-His215 (Fig. 40.9B) [71]. From these results along with the correlation between hydrogen bond strength and E_m (see Sect. 40.5.1), it is proposed that the effect of the formation of a strong hydrogen bond between D1-His215 and bromoxynil is transferred to the Q_A site via the His-Fe-His bridge (Fig. 40.7),

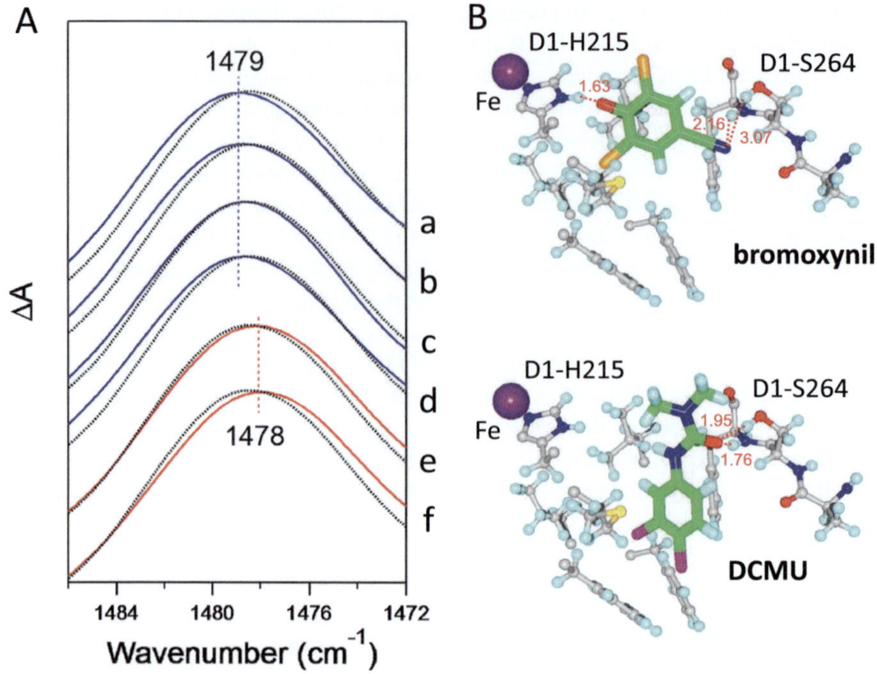

Fig. 40.9 (**A**) CO/CC stretching bands of Q_A^- in light-induced Q_A^-/Q_A FTIR difference spectra of PSII membranes in the presence of herbicides (*solid lines*) in comparison with the spectrum of the control sample without herbicide (*dotted lines*). (*a*) DCMU, (*b*) atrazine, (*c*) terbutryn, (*d*) bromacil, (*e*) bromoxynil, (*f*) ioxynil. (**B**) The binding sites of bromoxynil and DCMU in the Q_B pocket of PSII obtained by docking calculations

resulting in the weakened hydrogen bond of the Q_A CO with D2-His214 and the lowered E_m [71]. This view was confirmed by theoretical calculations using DFT and Poisson-Boltzmann approaches [25].

The change in the stability of the Q_A^- semiquinone anion by binding of different herbicides, bromoxynil and DCMU, at the Q_B site influences the rate of photodamage to PSII under strong light [23]. Degradation of Q_A, the Mn cluster, and P680 during illumination of strong light (2000 μE m^{-2} s^{-1}) was monitored by FTIR measurement, and it was found that photodamage to Q_A was suppressed and accelerated in the presence of DCMU and bromoxynil, respectively. Inactivation of the Mn cluster and P680 followed that of Q_A, supporting the "acceptor-side mechanism" of photodamage [75], in which the release of Q_A from the protein by its overreduction triggers charge recombination leading to the formation of harmful $^1O_2^*$ via the triplet state of Chl. The herbicide effect on the photodamage was interpreted that destabilization of Q_A^- by bromoxynil binding promotes the formation of Q_AH_2 by double reduction, whereas DCMU provides an opposite effect [23].

40.5.3 *Long-Range Interaction Between the Mn_4Ca Cluster and the Quinone-Fe Complex*

Upon depletion of the Mn_4Ca cluster, the $E_m(Q_A^-/Q_A)$ positively shifts by ~150 mV [26, 31 62]. The magnitude of this shift is even larger than the shift by herbicide binding at the Q_B site, despite the fact that the Mn_4Ca cluster is significantly distant from Q_A (~40 Å) (Fig. 40.1A). This long-range interaction between the Mn_4Ca cluster and Q_A plays an important role in the photoprotection mechanism of PSII. The upshift of $E_m(Q_A^-/Q_A)$ suppresses the electron transfer from Q_A^- to Q_B, preventing the accumulation of excessive oxidants on the electron donor side. It also makes the E_m gap between $Pheo_{D1}$ and Q_A larger, which suppresses charge recombination via $Pheo_{D1}$ forming harmful $^1O_2^*$ and promotes direct relaxation of Q_A^- [33, 59]. Thus, while the Mn_4Ca cluster is inactive due to, e.g., lowered pH in the lumen or the photoactivation process, PSII is protected from photodamage. Because of the long distance between Q_A and the Mn_4Ca cluster, however, it is not easy to rationalize the effect of Mn_4Ca depletion on $E_m(Q_A^-/Q_A)$. In addition, the Mn_4Ca cluster mostly interacts with the D1 subunit, whereas Q_A is surrounded by the D2 subunit. It is thus possible that the perturbation of the Mn_4Ca cluster is first transferred to the non-heme iron and Q_B sites through the transmembrane helices of the D1 subunit and then to the Q_A site via the His-Fe-His bridge.

To assess the mechanism of the long-range interaction between the donor and acceptor sides of PSII, the influence of Mn_4Ca depletion on the Fe-Q_B site was investigated using FTIR spectroelectrochemistry [27]. The non-heme iron, which is usually in a reduced Fe^{2+} form, is oxidized to Fe^{3+} at certain electrode potentials using a ferricyanide/ferrocyanide couple as a mediator, and single-flash illumination reduces it back to Fe^{2+}. Utilizing this redox reaction of the non-heme iron, Fe^{2+}/Fe^{3+}FTIR difference spectra have been obtained [21, 46, 70]. Prominent signals in the spectra arise from the bicarbonate ion and His side chains that are ligated to the iron, as well as from the Tyr and Glu residues in the hydrogen bond network (Fig. 40.7) [21, 27, 70]. The plot of the signal intensity as a function of the electrode potential provided a Nernst curve (Fig. 40.10A), from which $E_m(Fe^{2+}/Fe^{3+})$ was estimated for intact and Mn_4Ca-depleted PSII preparations [27]. Mn_4Ca depletion shifted the $E_m(Fe^{2+}/Fe^{3+})$ from +468 to +486 mV by +18 mV, which is ~8 times smaller than the shift of $E_m(Q_A^-/Q_A)$ by Mn_4Ca depletion. The changes in the spectral features in the regions of the His CN stretch (Fig. 40.10C) and the C = O stretch of COOH groups (Fig. 40.10B) in the Fe^{2+}/Fe^{3+} difference spectra suggested that Mn_4Ca depletion altered the pK_a values of a His side chain, most probably from D1-His215 interacting Q_B, and a carboxylate group, possibly from D1-Glu244, which are coupled with the non-heme iron [27]. It was concluded that the observed $E_m(Fe^{2+}/Fe^{3+})$ shift results from the additive effects of the changes in the protonation/deprotonation states of these amino acid residues surrounding the non-heme iron. The electrostatic interactions of these residues with Q_A may partly contribute to the $E_m(Q_A^-/Q_A)$ shift by Mn_4Ca depletion. However, a relatively small effect on $E_m(Fe^{2+}/Fe^{3+})$ (18 mV) compared

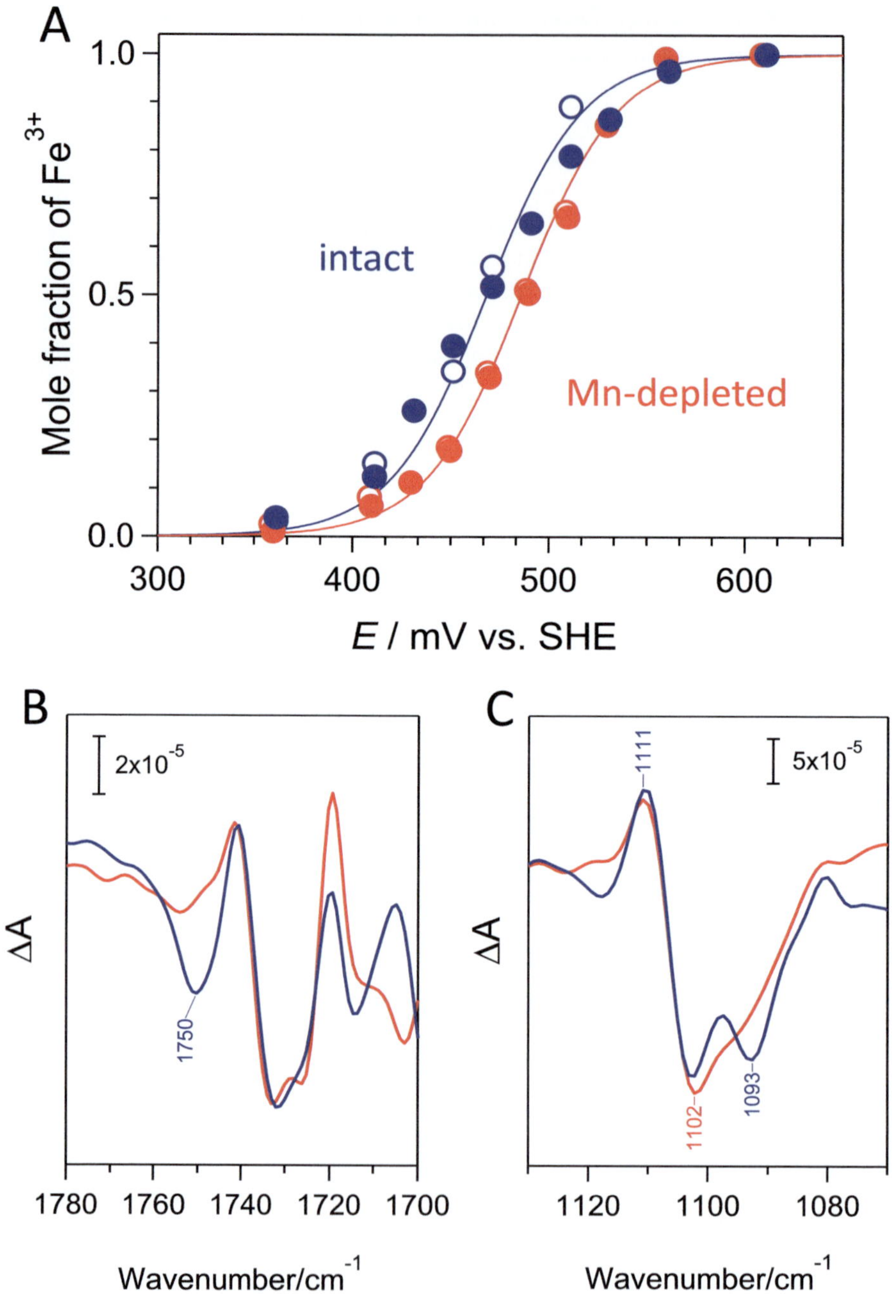

Fig. 40.10 (**A**) Nernst plots of the redox reactions of the non-heme iron in the intact (*blue symbols*) and Mn-depleted (*red symbols*) PSII membranes from spinach. The closed and open symbols are the data of oxidative and reductive potential steps, respectively. Theoretical Nernst curves assuming a one-electron redox process with $E_m = +468$ (intact) and +486 mV (Mn-depleted) are also presented. (**B**, **C**) Comparison of the Fe^{2+}/Fe^{3+} spectra of the intact (*blue line*) and Mn-depleted (*red line*) PSII membranes in the regions the C = O stretch of COOH (**B**) and the CN stretch of His (**C**)

with the effect on $E_m(Q_A^-/Q_A)$ (150 mV) implies that other direct pathways, which do not pass through the non-heme iron, may exist for the long-range interaction between the Mn_4Ca cluster and Q_A.

References

1. Ashizawa R, Noguchi T (2014) Effects of hydrogen bonding interactions on the redox potential and molecular vibrations of plastoquinone as studied by density functional theory calculations. Phys Chem Chem Phys 16:11864–11876
2. Berthomieu C, Hienerwadel R (2009) Fourier transform infrared (FTIR) spectroscopy. Photosynth Res 101:157–170
3. Berthomieu C, Nabedryk E, Mäntele W, Breton J (1990) Characterization by FTIR spectroscopy of the photoreduction of the primary quinone acceptor Q_A in photosystem II. FEBS Lett 269:363–367
4. Berthomieu C, Hienerwadel R, Boussac A, Breton J, Diner BA (1998) Hydrogen bonding of redox-active tyrosine Z of photosystem II probed by FTIR difference spectroscopy. Biochemistry 37:10547–10554
5. Blankenship RE (2014) Molecular mechanisms of photosynthesis, 2nd edn. Wiley-Blackwell, Hoboken
6. Breton J (2001) Fourier transform infrared spectroscopy of primary electron donors in type I photosynthetic reaction centers. Biochim Biophys Acta 1507:180–193
7. Chu H-A (2013) Fourier transform infrared difference spectroscopy for studying the molecular mechanism of photosynthetic water oxidation. Front Plant Sci 4:146
8. Chu H-A, Hillier W, Law NA, Babcock GT (2001) Vibrational spectroscopy of the oxygen-evolving complex and of manganese model compounds. Biochim Biophys Acta 1503:69–82
9. Clarke AK, Soitamo A, Gustafsson P, Oquist G (1993) Rapid interchange between two distinct forms of cyanobacterial photosystem II reaction-center protein D1 in response to photoinhibition. Proc Natl Acad Sci U S A 90:9973–9977
10. Crofts AR, Wraight CA (1983) The electrochemical domain of photosynthesis. Biochim Biophys Acta 726:149–185
11. Debus RJ (2008) Protein ligation of the photosynthetic oxygen-evolving center. Coord Chem Rev 252:244–258
12. Debus RJ (2015) FTIR Studies of metal ligands, networks of hydrogen bonds, and water molecules near the active site Mn_4CaO_5 cluster in photosystem II. Biochim Biophys Acta 1847:19–34
13. de Grotthuss CJT (1806) Sur la décomposition de l'eau et des corps qu'elle tient en dissolution à l'aide de l'électricité galvanique. Ann Chim 58:54–73
14. Dilbeck PL, Hwang HJ, Zaharieva I, Gerencser L, Dau H, Burnap RL (2012) The D1-D61N mutation in *Synechocystis* sp. PCC 6803 allows the observation of pH-sensitive intermediates in the formation and release of O_2 from photosystem II. Biochemistry 51:1079–1091
15. Diner B, Britt RD (2005) The redox-active tyrosine Y_Z and Y_D. In: Wydrzynski T, Satoh K (eds) Photosystem II: the light-driven water:plastoquinone oxidoreductase. Springer, Dordrecht, pp 207–233
16. Diner BA, Rappaport F (2002) Structure, dynamics, and energetics of the primary photochemistry of photosystem II of oxygenic photosynthesis. Annu Rev Plant Biol 53:551–580
17. Ferreira KN, Iverson TM, Maghlaoui K, Barber J, Iwata S (2004) Architecture of the photosynthetic oxygen-evolving center. Science 303:1831–1838
18. Hasegawa K, Noguchi T (2005) Density functional theory calculations on the dielectric-constant dependence of the oxidation potential of chlorophyll: implication for the high potential of P680 in photosystem II. Biochemistry 44:8865–8872

19. Hasegawa K, Noguchi T (2014) Molecular interactions of the quinone electron acceptors Q_A, Q_B and Q_C in photosystem II as studied by the fragment molecular orbital method. Photosynth Res 120:113–123
20. Haumann M, Liebisch P, Müller C, Barra M, Grabolle M, Dau H (2005) Photosynthetic O_2 formation tracked by time-resolved X-ray experiments. Science 310:1019–1021
21. Hienerwadel R, Berthomieu C (1995) Bicarbonate binding to the non-heme iron of photosystem II investigated by FTIR difference spectroscopy and ^{13}C-labeled bicarbonate. Biochemistry 34:16288–16297
22. Hillier W, Messinger J (2005) Mechanism of photosynthetic oxygen production. In: Wydrzynski T, Satoh K (eds) Photosystem II: the light-driven water:plastoquinone oxidoreductase. Springer, Dordrecht, pp 567–608
23. Idedan I, Tomo T, Noguchi T (2011) Herbicide effect on the photodamage process of photosystem II: Fourier transform infrared study. Biochim Biophys Acta 1807:1214–1220
24. Ishikita H, Saenger W, Biesiadka J, Loll B, Knapp E-W (2006) How photosynthetic reaction centers control oxidation power in chlorophyll pairs P680, P700, and P870. Proc Natl Acad Sci U S A 103:9855–9860
25. Ishikita H, Hasegawa K, Noguchi T (2011) How does the Q_B site influence propagate to the Q_A site in Photosystem II? Biochemistry 50:5436–5442
26. Johnson GN, Rutherford AW, Krieger A (1995) A change in the midpoint potential of the quinone Q_A in photosystem II associated with photoactivation of oxygen evolution. Biochim Biophys Acta 1229:202–207
27. Kato Y, Noguchi T (2014) Long-range interaction between the Mn_4CaO_5 cluster and the non-heme iron center in photosystem II as revealed by FTIR spectroelectrochemistry. Biochemistry 53:4914–4923
28. Kato Y, Sugiura M, Oda A, Watanabe T (2009) Spectroelectrochemical determination of the redox potential of pheophytin *a*, the primary electron acceptor in photosystem II. Proc Natl Acad Sci U S A 106:17365–17370
29. Klauss A, Haumann M, Dau H (2012) Alternating electron and proton transfer steps in photosynthetic water oxidation. Proc Natl Acad Sci U S A 109:16035–16040
30. Kos PB, Deak Z, Cheregi O, Vass I (2008) Differential regulation of *psbA* and *psbD* gene expression, and the role of the different D1 protein copies in the cyanobacterium *Thermosynechococcus elongatus* BP-1. Biochim Biophys Acta 1777:74–83
31. Krieger A, Rutherford AW, Johnson GN (1995) On the determination of redox midpoint potential of the primary quinone electron-acceptor, Q_A, in photosystem II. Biochim Biophys Acta 1229:193–201
32. Krieger-Liszkay A, Rutherford AW (1998) Influence of herbicide binding on the redox potential of the quinone acceptor in photosystem II. Relevance to photodamage and phytotoxicity. Biochemistry 37:17339–17344
33. Krieger-Liszkay A, Fufezan C, Trebst A (2008) Singlet oxygen production in photosystem II and related protection mechanism. Photosynth Res 98:551–564
34. Lin X, Murchison HA, Nagarajan V, Parson WW, Allen JP, Williams JC (1994) Specific alteration of the oxidation potential of the electron-donor in reaction centers from *Rhodobacter sphaeroides*. Proc Natl Acad Sci U S A 91:10265–10269
35. Loll B, Kern J, Saenger W, Zouni A, Biesiadka J (2005) Towards complete cofactor arrangement in the 3.0 Å resolution structure of photosystem II. Nature 438:1040–1044
36. Mäntele W (1993) Reaction-induced infrared difference spectroscopy for the study of protein function and reaction-mechanisms. Trends Biochem Sci 18:197–202
37. Merry SAP, Nixon PJ, Barter LMC, Schilstra M, Porter G, Barber J, Durrant JR, Klug DR (1998) Modulation of quantum yield of primary radical pair formation in photosystem II by site-directed mutagenesis affecting radical cations and anions. Biochemistry 37:17439–17447
38. Messinger J, Noguchi T, Yano J (2012) Photosynthetic O_2 evolution. In: Wydrzynski T, Hillier W (eds) Molecular solar fuels, Chapter 7. Royal Society of Chemistry, Cambridge, pp 163–207

39. Nakamura S, Nagao R, Takahashi R, Noguchi T (2014) Fourier transform infrared detection of a polarizable proton trapped between photooxidized tyrosine Y_Z and a coupled histidine in photosystem II: relevance to the proton transfer mechanism of water oxidation. Biochemistry 53:3131–3144
40. Noguchi T (2007) Light-induced FTIR difference spectroscopy as a powerful tool toward understanding the molecular mechanism of photosynthetic oxygen evolution. Photosynth Res 91:59–69
41. Noguchi T (2008) Fourier transform infrared analysis of the photosynthetic oxygen-evolving center. Coord Chem Rev 252:336–346
42. Noguchi T (2008) FTIR detection of water reactions in the oxygen-evolving center of photosystem II. Philos Trans R Soc B 363:1189–1195
43. Noguchi T (2013) Monitoring the reactions of photosynthetic water oxidation using infrared spectroscopy. Biomed Spectrosc Imaging 2:115–128
44. Noguchi T (2015) Fourier transform infrared difference and time-resolved infrared detection of the electron and proton transfer dynamics in photosynthetic water oxidation. Biochim Biophys Acta 1845:35–45
45. Noguchi T, Berthomieu C (2005) Molecular analysis by vibrational spectroscopy. In: Wydrzynski T, Satoh K (eds) Photosystem II: the light-driven water:plastoquinone oxidoreductase. Springer, Dordrecht, pp 367–387
46. Noguchi T, Inoue Y (1995) Identification of FTIR signals from the non-heme iron in photosystem II. J Biochem 118:9–12
47. Noguchi T, Tomo T, Inoue Y (1998) Fourier transform infrared study of the cation radical of P680 in the photosystem II reaction center: evidence for charge delocalization on the chlorophyll dimer. Biochemistry 37:13614–13625
48. Noguchi T, Inoue Y, Tang X-S (1999) Hydrogen bonding interaction between the primary quinone acceptor Q_A and a histidine side chain in photosystem II as revealed by Fourier transform infrared spectroscopy. Biochemistry 38:399–403
49. Noguchi T, Suzuki H, Tsuno M, Sugiura M, Kato C (2012) Time-resolved infrared detection of the proton and protein dynamics during photosynthetic oxygen evolution. Biochemistry 51:3205–3214
50. Okubo T, Tomo T, Sugiura M, Noguchi T (2007) Perturbation of the structure of P680 and the charge distribution on its radical cation in isolated reaction center complexes of photosystem II as revealed by Fourier transform infrared spectroscopy. Biochemistry 46:4390–4397
51. Olson JM, Blankenship RE (2004) Thinking about the evolution of photosynthesis. Photosynth Res 80:373–386
52. Ort DR, Yocum CF (eds) (1996) Oxygenic photosynthesis: the light reactions. Kluwer Academic Publishers, Dordrecht
53. Petrouleas V, Crofts AR (2005) The quinone iron acceptor complex. In: Wydrzynski T, Satoh K (eds) Photosystem II: the light-driven water:plastoquinone oxidoreductase. Springer, Dordrecht, pp 177–206
54. Pokhrel R, Service RJ, Debus RJ, Brudvig GW (2013) Mutation of lysine 317 in the D2 subunit of photosystem II alters chloride binding and proton transport. Biochemistry 52:4758–4773
55. Rappaport F, Blanchard-Desce M, Lavergne J (1994) Kinetics of electron transfer and electrochromic change during the redox transitions of the photosynthetic oxygen-evolving complex. Biochim Biophys Acta 1184:178–192
56. Rappaport F, Guergova-Kuras M, Nixon PJ, Diner BA, Lavergne J (2002) Kinetics and pathways of charge recombination in photosystem II. Biochemistry 41:8518–8527
57. Renger G, Holzwarth AR (2005) Primary electron transfer. In: Wydrzynski T, Satoh K (eds) Photosystem II: the light-driven water:plastoquinone oxidoreductase. Springer, Dordrecht, pp 139–175
58. Rutherford AW, Faller P (2003) Photosystem II: evolutionary perspectives. Philos Trans R Soc Lond B 358:245–253
59. Rutherford AW, Krieger-Liszkay A (2001) Herbicide-induced oxidative stress in photosystem II. Trends Biochem Sci 26:648–653

60. Saito K, Ishida T, Sugiura M, Kawakami K, Umena Y, Kamiya N, Shen JR, Ishikita H (2011) Distribution of the cationic state over the chlorophyll pair of the photosystem II reaction center. J Am Chem Soc 133:14379–14388
61. Saito K, Rutherford AW, Ishikita H (2013) Mechanism of tyrosine D oxidation in Photosystem II. Proc Natl Acad Sci U S A 110:7690–7695
62. Shibamoto T, Kato Y, Sugiura M, Watanabe T (2009) Redox potential of the primary plastoquinone electron acceptor Q_A in photosystem II from *Thermosynechococcus elongatus* determined by spectroelectrochemistry. Biochemistry 48:10682–10684
63. Shibuya Y, Takahashi R, Okubo T, Suzuki H, Sugiura M, Noguchi T (2010) Hydrogen bond interactions of the pheophytin electron acceptor and its radical anion in photosystem II as revealed by Fourier transform infrared difference spectroscopy. Biochemistry 49:493–501
64. Styring S, Sjöholm J, Mamedov F (2012) Two tyrosines that changed the world: interfacing the oxidizing power of photochemistry to water splitting in photosystem II. Biochim Biophys Acta 1817:76–87
65. Sugiura M, Boussac A, Noguchi T, Rappaport F (2008) Influence of histidine-198 of the D1 subunit on the properties of the primary electron donor, P680, of photosystem II in *Thermosynechococcus elongatus*. Biochim Biophys Acta 1777:331–342
66. Sugiura M, Azami C, Koyama K, Rutherford AW, Rappaport F, Boussac A (2014) Modification of the pheophytin redox potential in *Thermosynechococcus elongatus* Photosystem II with PsbA3 as D1. Biochim Biophys Acta 1837:139–148
67. Suzuki H, Nagasaka M, Sugiura M, Noguchi T (2005) Fourier transform infrared spectrum of the secondary quinone electron acceptor Q_B in photosystem II. Biochemistry 44:11323–11328
68. Suzuki H, Yu J, Kobayashi T, Nakanishi H, Nixon P, Noguchi T (2013) Functional roles of D2-Lys317 and the interacting chloride ion in the water oxidation reaction of photosystem II as revealed by Fourier transform infrared analysis. Biochemistry 52:4748–4757
69. Takahashi R, Hasegawa K, Noguchi T (2008) Effect of charge distribution over a chlorophyll dimer on the redox potential of P680 in photosystem II as studied by density functional theory calculations. Biochemistry 47:6289–6291
70. Takahashi R, Boussac A, Sugiura M, Noguchi T (2009) Structural coupling of a tyrosine side chain with the non-heme iron center in photosystem II as revealed by light-induced Fourier transform infrared difference spectroscopy. Biochemistry 48:8994–9001
71. Takahashi R, Hasegawa K, Takano A, Noguchi T (2010) The structures and binding sites of phenolic herbicides in the Q_B pocket of photosystem II. Biochemistry 49:5445–5454
72. Takano A, Takahashi R, Suzuki H, Noguchi T (2008) Herbicide effect on the hydrogen-bonding interaction of the primary quinone electron acceptor Q_A in photosystem II as studied by Fourier transform infrared spectroscopy. Photosynth Res 98:159–167
73. Umena Y, Kawakami K, Shen JR, Kamiya N (2011) Crystal structure of oxygen-evolving photosystem II at a resolution of 1.9 Å. Nature 473:55–60
74. Vass I, Cser K (2009) Janus-faced charge recombinations in photosystem II photoinhibition. Trends Plant Sci 14:200–205
75. Vass I, Styring S, Hundal T, Koivuniemi A, Aro E-M, Andersson B (1992) Reversible and irreversible intermediates during photoinhibition of photosystem II: stable reduced Q_A species promote chlorophyll triplet formation. Proc Natl Acad Sci U S A 89:1408–1412
76. Watanabe T, Kobayashi M (1991) Electrochemistry of chlorophylls. In: Scheer H (ed) Chlorophylls. CRC Press, Boca Raton, pp 287–315
77. Wydrzynski T, Satoh K (eds) (2005) Photosystem II: the light-driven water:plastoquinone oxidoreductase. Springer, Dordrecht
78. Zundel G (2000) Hydrogen bonds with large proton polarizability and proton transfer processes in electrochemistry and biology. Adv Chem Phys 111:1–217

Chapter 41
Protein-Controlled Isomerization in Rhodopsins

Hideki Kandori

Abstract Rhodopsins are photoactive proteins containing a retinal chromophore in animals and microbes. In animal and microbial rhodopsins, 11-*cis* and all-*trans* retinal, respectively, forms a Schiff base linkage with a Lys of the 7th helix, which is mostly protonated. Upon light absorption, ultrafast photoisomerization takes place from the 11-*cis* to the all-*trans* form in animal rhodopsins and from the all-*trans* to the 13-*cis* form in microbial rhodopsins. Retinal isomerization in the restricted protein environment causes light energy to be stored in the primary intermediate states, leading to each function in a late timescale. π-Electrons along the retinal chromophore play crucial roles in the functions of rhodopsins through their specific interactions with the protein moiety. First, a specific chromophore-protein interaction determines the energy gap of light absorption, i.e., the color of rhodopsins. Second, the isomerization reaction is fast and efficient in protein, even though the retinal-binding pocket is structurally restricted. Third, light energy is efficiently stored by the chromophore-protein interaction, to which steric effect and hydrogen-bonding alteration contribute. Finally, the relaxation from such a high-energy state leads to each function in much slower timescales. In this chapter, the ultrafast isomerization processes and their energy storages in animal and microbial rhodopsins are reviewed. The molecular mechanism of highly efficient photoisomerization will also be discussed.

Keywords Animal rhodopsin • Microbial rhodopsin • Retinal • Photoisomerization • Hydrogen bond

41.1 Retinal Chromophore in Animal and Microbial Rhodopsins

Photoreceptive proteins, which convert light into energy or a signal, have a chromophore molecule inside the protein. It is well known that specific chromophore-protein interactions result in unique photophysical and photochemical processes,

H. Kandori (✉)
Department of Frontier Materials, Nagoya Institute of Technology, Showa-ku, Nagoya 466-8555, Japan
e-mail: kandori@nitech.ac.jp

T. Akasaka et al. (eds.), *Chemical Science of π-Electron Systems*,
DOI 10.1007/978-4-431-55357-1_41

leading to functional optimization for light-energy and light-signal conversions. Rhodopsins are photoreceptive proteins that contain a retinal molecule [1]. The word "rhodopsin" originates from the Greek words "rhodo" and "opsis," which indicate rose and sight, respectively. Thus, the classical meaning of rhodopsin is the red-colored pigment in the retina of the eyes. After similar colored retinal-binding proteins were found in microbes, the definition of the word rhodopsin was expanded. The modern meaning of rhodopsin now encompasses photoactive proteins containing a retinal chromophore in animals and microbes [1].

Rhodopsins are now found in all domains of life and are classified into two groups. While lower organisms utilize the family of microbial rhodopsins for light-energy conversion and intracellular signaling, animals use the photosensory functions of a different family of rhodopsins (animal rhodopsin), a specialized subset of G protein-coupled receptors (GPCRs) [1–3]. Animal and microbial rhodopsins share a common architecture of seven-transmembrane α-helices with the N- and C-terminus facing outward from and inside of the cell, respectively, but have almost no sequence homology and differ largely in their functions [1]. Retinal, the aldehyde of vitamin A, is derived from β-carotene and is bound to the protein in the shape of 11-*cis* and all-*trans* forms in animal and microbial rhodopsins, respectively (Fig. 41.1). Retinal is attached by a Schiff base linkage to the ε-amino group of a Lysine side chain in the middle of the 7th helix, and this retinal Schiff base (RSB) is protonated in most cases (Fig. 41.1).

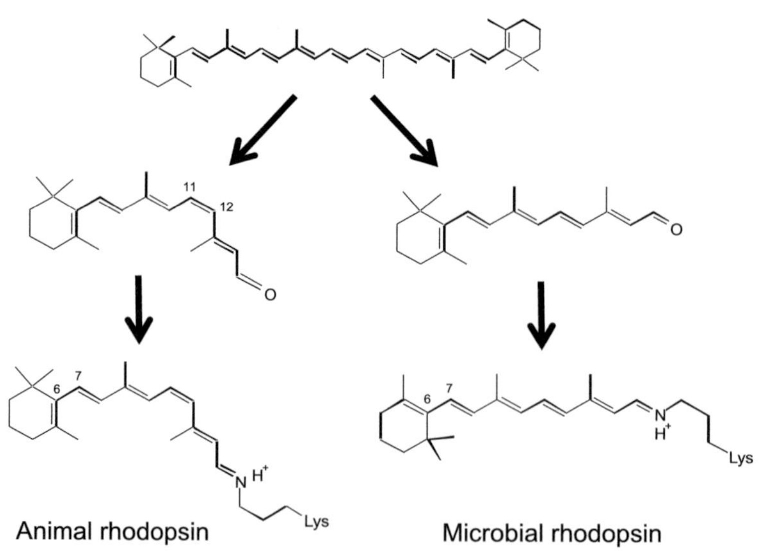

Fig. 41.1 Chromophore molecules in animal (*left*) and microbial (*right*) rhodopsins. β-carotene (*top*) is the source of the chromophore, and 11-*cis* and all-*trans* retinal is bound to protein through the Schiff base linkage to form animal and microbial rhodopsin, respectively

Color regulation is one of the most important effects of the π-electron system in the chromophore molecule. Color is owing to protein control of the energy gap between electronically ground and excited states of the retinal chromophore. For color regulation, protonation of the chromophore plays crucial roles. While absorption of unprotonated RSB takes place in the limited UV region (λ_{max} 350 nm), protonated RSB exhibits a large variation of absorption that covers the visible region (400–700 nm), which determines our "visible" region [1]. Wide color tuning in rhodopsins implies that the $\pi-\pi$ transition of the retinal chromophore is highly sensitive to the protein environment. In fact, such wide color tuning is not the case for protonated RSB in solution. The absorption spectra of protonated RSB in solution have been limited (λ_{max}: 430–480 nm) under various counterions and solvents tested, which are also similar between 11-*cis* and all-*trans* forms [4]. This indicates that wide color tuning is caused by the specific effect of protein. Among various interactions between the retinal chromophore and protein such as the electrostatic effect of charged groups, dipolar amino acids, aromatic amino acids, hydrogen-bonding interactions, and steric contact effect, the effect of a counterion near RSB is significant. To maintain the protonation of RSB, a negatively charged counterion is necessary inside the protein, which is generally hydrophobic. We now know that a negatively charged aspartate, glutamate, or chloride ion is located near the protonated RSB of animal and microbial rhodopsins and that the ion-pair is further stabilized by protein-bound water molecules (Fig. 41.2) [1]. Upon light absorption, the positive charge originally located at protonated RSB is more delocalized in the excited state (Fig. 41.3), and a higher or lower charge delocalization results in a smaller or larger energy gap, respectively. Accordingly, the presence of a counterion near the RSB that localizes the positive charge results in a large energy gap, i.e., a spectral blueshift.

In solution, the counterion is located at the most stable position relative to the protonated RSB, where the Schiff base N-H group presumably forms a direct hydrogen bond. Under such strong interactions, the λ_{max} of the retinal chromophore is located at 430–480 nm [5]. On the other hand, measurement of the gas-phase absorption spectrum of protonated RSB indicated a λ_{max} at 610 nm [6]. To understand the color tuning mechanism in protein, it is important to study the mechanism of the chromophore in other systems. Although artificial construction of wide color tuning of the rhodopsin chromophore using other materials had been unsuccessful for a long time, we found that the protonated RSB of 11-*cis* retinal exhibits various colors when adsorbed onto the clay, montmorillonite [7]. The absorption spectra cover the entire visible ranges in clays from three different mining locations, and silicate-based clay can mimic the carbon-based protein for color tuning of the chromophore in human vision. Another example is human cellular retinol-binding protein II, which was elegantly engineered to bind all-*trans*-retinal, in which the absorption maximum could be changed via mutations from 420 to 644 nm [8]. Interactions of retinal with aromatic amino acids play a role as well, as do hydrogen-bonding interactions and steric contact effects. Strong hydrogen bonds can lead to a transfer of charge, and steric contacts can lead to a twist of retinal.

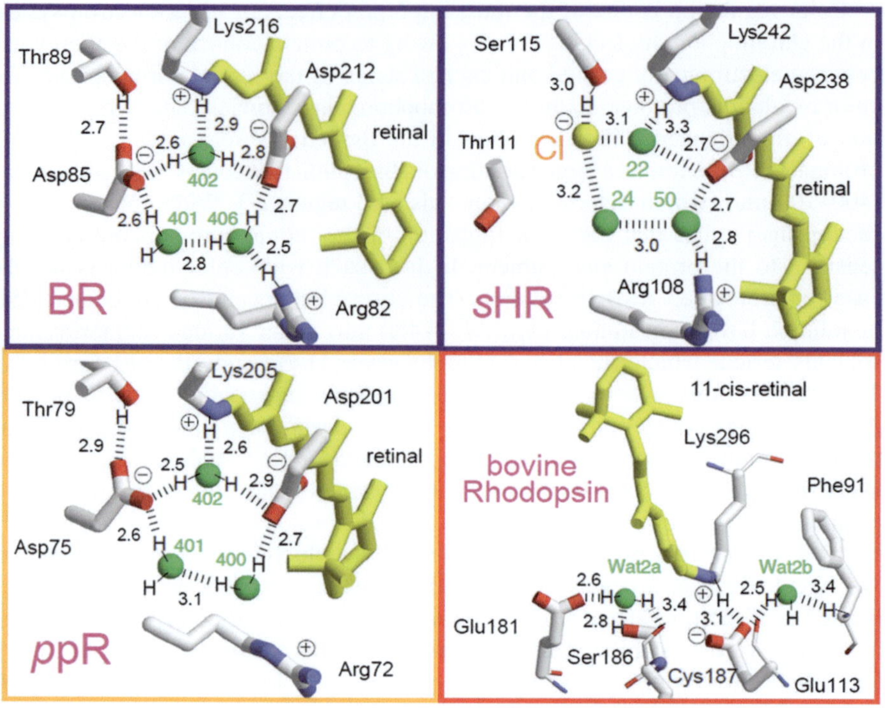

Fig. 41.2 X-ray crystallographic structures of the Schiff base region in BR, *pp*R, *s*HR, and bovine rhodopsin (PDB entries are 1C3W, 1JGJ, 1E12, and 1L9H, respectively). Each membrane normal is approximately in the vertical direction of the figures. *Upper* and *lower* regions correspond to the cytoplasmic and extracellular sides, respectively. *Green spheres* represent water molecules. Hydrogen atoms and hydrogen bonds (*dotted lines*) are supposed from the structures, while the numbers are the hydrogen-bonding distances in Å. This figure is reproduced from Furutani et al. [60]

This chapter provides recent knowledge on the photoreaction mechanism in animal and microbial rhodopsins. Ultrafast photoisomerization, which is particularly characteristic for these proteins, takes place in a protein environment. It was argued 40 years ago that product formation in a picosecond regime is too fast for isomerization, although now isomerization in a sub-picosecond regime has been established [1, 4]. This was achieved by means of ultrafast spectroscopy with femtosecond time resolution, which allowed the detection of photophysics and photochemistry of the retinal chromophore in real time. In addition, atomic structures have been determined for animal and microbial rhodopsins over the last decade, which enabled us to understand the atomic details underlying the mechanism of ultrafast photoisomerization [1]. The structures of the photoproduct states in animal and microbial rhodopsins are also reported. On the basis of these structures, a theoretical hypothesis is also challenged to explain the mechanism of ultrafast

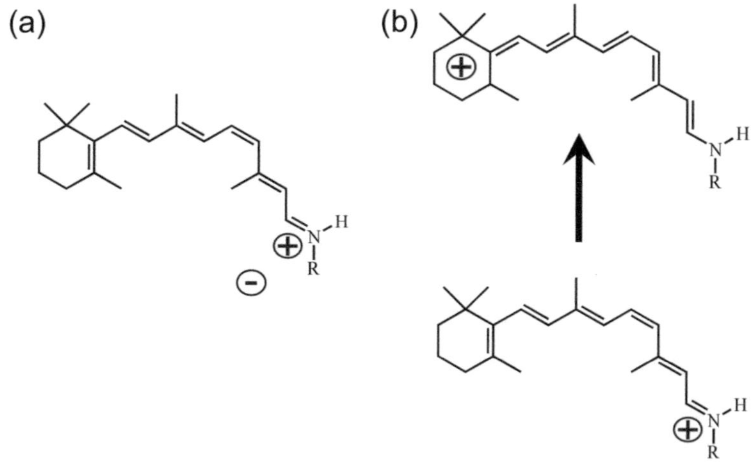

Fig. 41.3 Color tuning mechanism exemplified by animal rhodopsins. (**a**) The 11-*cis* retinal forms a Schiff base linkage that is protonated. To stabilize the positive charge, a negatively charged counterion is present near the Schiff base, causing the pKa of the Schiff base to be high. (**b**) Photoexcitation causes bond alteration, which leads, in the electronically excited state, to movement of the positive charge from the Schiff base to the β-ionone ring. If the counterion is located near the Schiff base or the β-ionone ring, electrostatic interactions lower the energy level of the ground state or excited state, yielding a spectral blueshift or redshift, respectively

photoisomerization. These studies emphasize the importance of the chromophore-protein interaction, leading to functional expression in a late timescale.

41.2 Unique Photochemistry of the Retinal Chromophore in Protein

Photochemistry of the rhodopsin chromophore is unique in protein. For instance, the quantum yield of the photoreaction of rhodopsins is high, and this serves as the molecular basis of high sensitivity underlying human vision [1]. Quantum yield is essentially independent of temperature and excitation wavelength. The fluorescence quantum yield of the rhodopsin chromophore is very low ($\phi \quad 10^{-5}$), and an ultrafast reaction was inferred through a barrierless excited-state potential surface. Product formation also takes place for rhodopsins at low temperatures, including at 77 K (liquid nitrogen) or 4 K (helium), where molecular motions are frozen. In early stages of investigation, these observations questioned *cis-trans* isomerization as the primary event in vision, because isomerization needs certain molecular motion of the chromophore. In fact, the first ultrafast spectroscopy of bovine rhodopsin led to a conclusion that favors a reaction mechanism other than isomerization as the primary reaction of rhodopsin, but ultrafast spectroscopy concluded that *cis-trans* photoisomerization is indeed a primary event [1, 4].

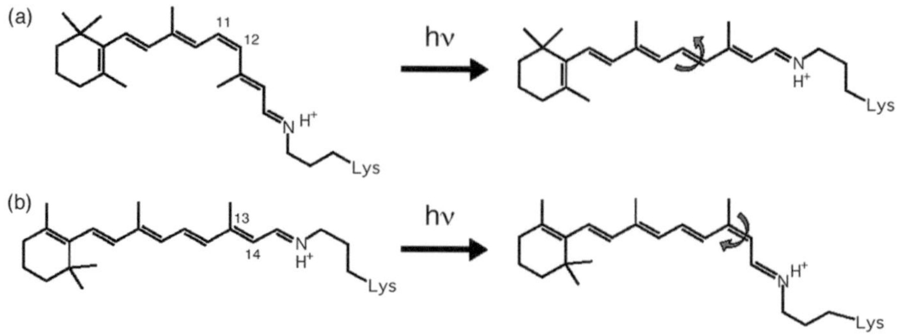

Fig. 41.4 Photochemical reactions in (**a**) visual and (**b**) archaeal rhodopsins. *Cis-trans* photoisomerization is a common reaction

Figure 41.4 shows photochemical reactions in animal (a) and microbial (b) rhodopsins. In animal rhodopsin, 11-*cis* retinal is isomerized into an all-*trans* form. Selectivity is 100 % and quantum yield is 0.67 for bovine rhodopsin [9]. In microbial rhodopsins, the all-*trans* retinal is isomerized into the 13-*cis* form. Selectivity is 100 % and quantum yield is 0.64 for bacteriorhodopsin (BR) [10]. Squid and octopus possess a photoisomerase called retinochrome, which supplies an 11-*cis* retinal for their rhodopsins through a specific photoreaction. Retinochrome possesses an all-*trans* retinal as the chromophore, and the all-*trans* retinal is isomerized into the 11-*cis* form with 100 % selectivity [11]. Thus, the photoproduct between microbial rhodopsins and retinochrome is different, the all-*trans* form being converted into the 13-*cis* and 11-*cis* forms, respectively. This fact implies that the protein environment determines the reaction pathways of photoisomerization in their excited states.

HPLC analysis revealed that the protonated Schiff base of 11-*cis* retinal in solution is almost predominantly isomerized into the all-*trans* form, indicating that the reaction pathway in rhodopsin is the nature of the chromophore itself [12]. On the other hand, the quantum yield was found to be 0.15 in methanol solution. Therefore, the isomerization reaction is 4–5 times more efficient in protein than in solution. The HPLC analysis also revealed that the protonated Schiff base of all-*trans* retinal in solution is predominantly isomerized into the 11-*cis* form (82 % 11-*cis*, 12 % 9-*cis*, and 6 % 13-*cis* in methanol) [12]. The 11-*cis* form as a photoproduct is the nature of retinochrome, but not that of microbial rhodopsins. This suggests that the protein environment of retinochrome serves as the intrinsic property of the photoisomerization of retinal chromophore. In contrast, it seems that the protein environment of microbial rhodopsins forces a change in the reaction pathway of isomerization into the 13-*cis* form. In this regard, it is interesting that the quantum yield of BR (0.64) is 4–5 times higher than that in solution (0.15) [1]. The efficiency of the altered excited-state reaction pathways in BR is never reduced. Rather, BR has an efficient reaction pathway from the all-*trans* to 13-*cis* form. Consequently, the isomerization reaction is achieved as efficiently as in animal rhodopsins.

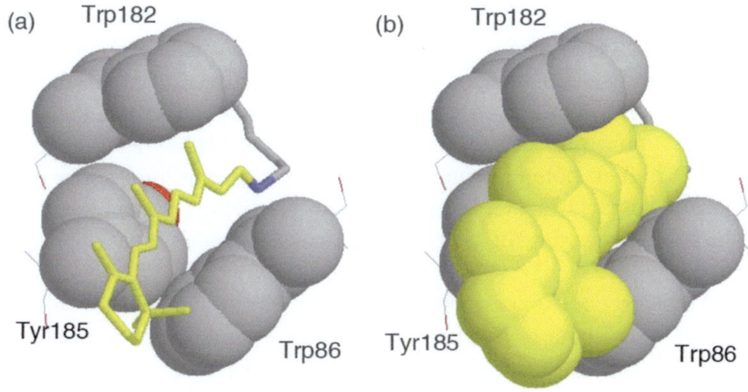

Fig. 41.5 Structure of bacteriorhodopsin (BR). Trp86, Trp182, and Tyr185 are shown in a space-filling model together with the retinal chromophore in a stick (**a**) or a space-filling (**b**) drawing

The water-containing hydrogen-bonding network in Fig. 41.2 must surely affect such a unique reaction pathway in BR. Figure 41.5 illustrates another structural feature of BR. The linear polyene chain of all-*trans* retinal is sandwiched vertically by two tryptophans, Trp86 and Trp182, while the phenol ring of Tyr185 is located parallel to the polyene chain of retinal chromophore. Thr89 is located on the other side (not shown). The presence of three bulky groups, Trp86, Trp182, and Tyr185, presumably determines the isomerization pathway from an all-*trans* to a 13-*cis* form by strongly affecting π-electrons in the retinal chromophore. As described above, the 11-*cis* form is the main photoproduct of the protonated RSB of all-*trans* retinal in solution. Therefore, the reaction pathway is altered in the protein environment of BR, whereas the quantum yield of isomerization is 4–5 times higher in BR (0.64) than that in solution (0.15).

In both animal and microbial rhodopsins, the first singlet excited state (S1) is the Bu-like state. After Franck-Condon excitation, stretching and tortional vibrations are coupled to relax in the S1 potential surface. The evolution along the tortional coordinate at C11=C12 in animal rhodopsins and C13=C14 in microbial rhodopsins leads to a conical intersection (CI) funnel, where the chromophore displays about a 90° twisted double bond, and fully efficient decay takes place to the ground state [1]. Monitoring excited-state dynamics is challenging both experimentally and theoretically.

41.3 Photoisomerization and Energy Storage in Animal Rhodopsins

The first study of the primary photoreaction processes in rhodopsins was performed by low-temperature spectroscopy, which can detect primary photointermediate states by reducing the thermal reaction rate at a low temperature. In 1963, a low-temperature spectroscopic study of bovine rhodopsin at 87 K detected a redshifted product [13]. Since this photoproduct (now called bathorhodopsin) is converted to lumirhodopsin upon warming and finally decomposed to all-*trans* retinal and opsin through several intermediates, it was proposed that bathorhodopsin has a "highly constrained and distorted" all-*trans* retinal as its chromophore and is on a higher potential energy level than bovine rhodopsin and subsequent intermediates [13]. According to their prediction, the process of animal rhodopsin to bathorhodopsin should be a *cis-trans* isomerization of the chromophore.

The first picosecond experiment of bovine rhodopsin in 1972 observed the formation of bathorhodopsin within 6 ps after excitation at room temperature and interpreted that its formation would be too fast to be attributed to such a conformational change as the *cis-trans* isomerization of the retinal chromophore [14]. This is reasonable because isomerization accompanies molecular motion that must take some time, but it is now known that the primary event is indeed *cis-trans* isomerization. The experimental evidence of the 11-*cis* to all-*trans* isomerization as the primary reaction in animal rhodopsin was provided by our time-resolved study of rhodopsin analogs possessing 11-*cis*-locked ring retinals. In the case of five-membered rhodopsin, only a long-lived excited state ($\tau = 85$ ps) formed without any ground-state photoproduct (Fig. 41.6d), providing direct evidence that *cis-trans* isomerization is the primary event in vision [15]. On the other hand, excitation of seven-membered rhodopsin yielded a ground-state photoproduct having a spectrum similar to photorhodopsin (Fig. 41.6c). These different results were interpreted in terms of the rotational flexibility along the C11-C12 double bond [15]. This hypothesis was further supported by the results with an 8-membered rhodopsin that possesses a more flexible ring. Upon excitation of eight-membered rhodopsin with a 21 ps pulse, two photoproducts, photorhodopsin-like and bathorhodopsin-like products, were observed (Fig. 41.6b) [16]. Photorhodopsin is a precursor of bathorhodopsin found by picosecond transient absorption spectroscopy [4]. Thus, picosecond absorption studies directly elucidated the correlation between the primary processes of rhodopsin and the flexibility of the C11-C12 double bond of the chromophore, allowing us to eventually conclude the respective potential surfaces, as shown in Fig. 41.6. The structure of the intermediate states in Rh7 and Rh8 has been reported by theoretical investigation [17].

Direct observation of the *cis-trans* isomerization process in real time was possible by using femtosecond pulses in 1991. Femtosecond transient absorption spectroscopy of bovine rhodopsin observed that product formation was completed within 200 fs, suggesting that isomerization is an event that takes place in femtoseconds [18]. In addition, oscillatory features with a period of 550 fs (60 cm^{-1}) were observed on the kinetics at probe wavelengths within the photoproduct absorption band of rhodopsin, whose phase and amplitude demonstrate that they are the result of nonstationary vibrational motion in the ground state of photorhodopsin [19]. The observation of coherent vibrational motion in photorhodopsin supports the idea that the primary step in vision is a vibrationally coherent process and that the high quantum yield of the *cis-trans* isomerization in animal rhodopsins is a consequence

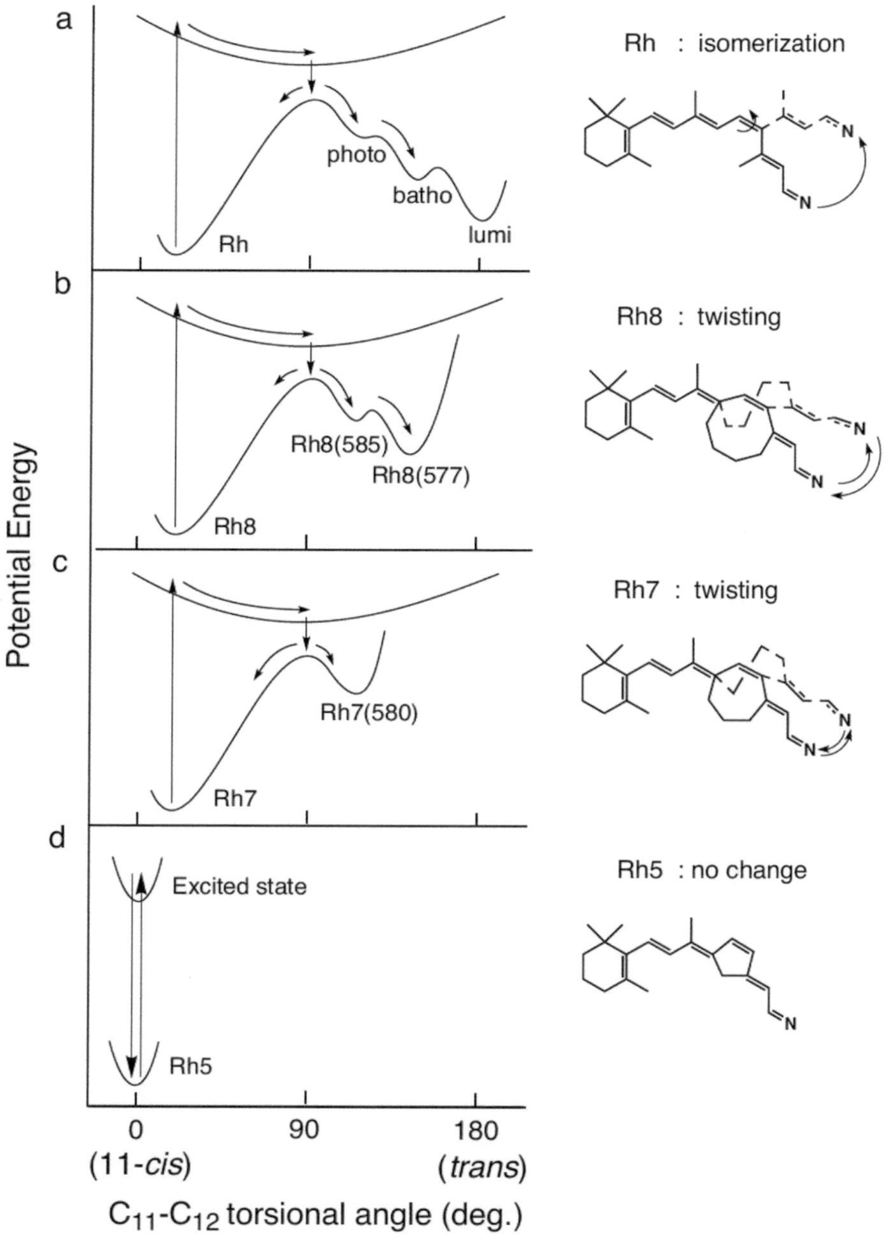

Fig. 41.6 Schematic drawing of ground- and excited-state potential surfaces along the 11-ene torsional coordinates of the chromophore of rhodopsin (**a**), 8-membered rhodopsin (**b**), 7-membered rhodopsin (**c**), and 5-membered rhodopsin (**d**). This figure is modified from Mizukami et al. [16]

of the extreme speed of the excited-state torsional motion [1]. Ultrafast photoisomerization in animal rhodopsins was also confirmed by other experimental techniques, such as femtosecond fluorescence spectroscopy and femtosecond stimulated Raman spectroscopy [1].

It is well known that rhodopsin is an excellent molecular switch to convert a light signal to the electric response of the photoreceptor cell. As already mentioned, the highly efficient photoisomerization of rhodopsin with a quantum yield of 0.67 is assured by the extremely fast *cis-trans* isomerization of the chromophore that is facilitated by the protein environment [1]. What then is the molecular mechanism of such photoisomerization? Figure 41.4a shows that the RSB side actually rotates, while no change takes place on the side of the β-ionone ring. This is reasonable because the β-ionone ring is bulky and hardly moves. However, the molecular motion of RSB in Fig. 41.4a is also significant, and an entire rotation is likely not the case for bathorhodopsin. Figure 41.7 shows the structure of bovine rhodopsin in which 11-*cis* retinal is embedded in the biding pocket of the protein. Therefore, upon photoisomerization, overall molecular shape must be unchanged before and after photoisomerization from the 11-*cis* to all-*trans* form.

Important experimental information was first obtained by resonance Raman spectroscopy, in which the C=N stretching vibration of RSB for animal rhodopsin and bathorhodopsin was observed. The C=N stretch in H_2O is considerably upshifted from the intrinsic stretching mode by coupling with the N-H in-plane bending

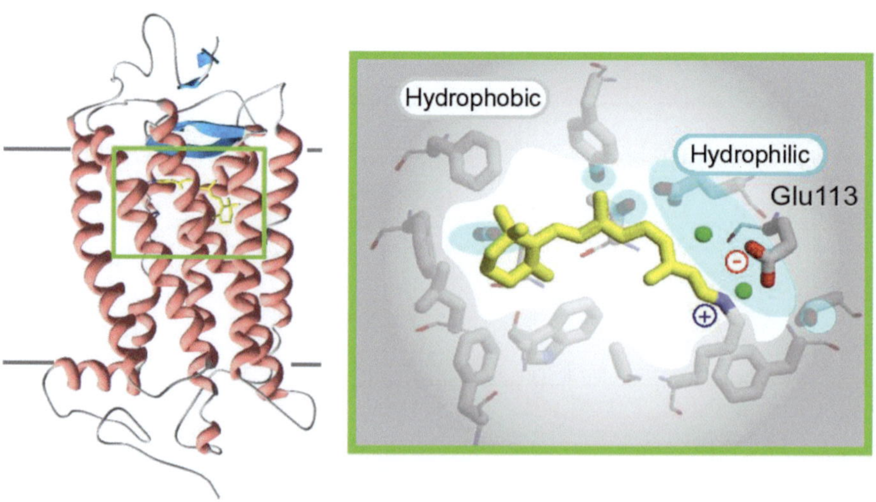

Fig. 41.7 (*Left*) Protein structure of bovine rhodopsin. The *upper* and *lower* regions correspond to the intradiscal and cytoplasmic sides, respectively. (*Right*) The highlighted structure of the chromophore (yellow stick drawing) and protein environment. While hydrophobic amino acid residues surround the β-ionone ring and the polyene chain of the 11-*cis* retinal chromophore, the retinal Schiff base region is highly hydrophilic. The Schiff base is protonated and negatively charged Glu113 (3.2 Å) stabilizes the protonation state of the chromophore

vibration. The intrinsic C=N stretch can be measured in D_2O, where the coupling of the N-H bending vibration is removed. Since the frequency of the N-H in-plane bending increases if the hydrogen bond of RSB is strengthened, the frequency difference of the C=N stretch in H_2O and D_2O has been regarded as the marker of hydrogen-bonding strength of RSB. Namely, this large difference corresponds to a strong hydrogen bond. The differences of the C=NH and C=ND frequencies are identical (32 cm^{-1}) in bovine rhodopsin and bathorhodopsin, implying that no change of the RSB hydrogen bond takes place upon retinal isomerization [20].

About 30 years later, X-ray crystallography was applied to bathorhodopsin by illuminating the bovine rhodopsin crystal at 105 K [21]. Difference electron density between rhodopsin and bathorhodopsin was small, as expected. Regarding the retinal chromophore, displacement was very small on both sides, both in the β-ionone ring and in RSB, the latter being consistent with Raman data [20], which showed no change in the hydrogen-bonding strength of RSB. In contrast, the dihedral angle around the C11=C12 bond changes from −40° to −155°, indicating that isomerization certainly takes place. Thus, a largely distorted polyene chain results in a minimally changed overall chromophore shape. This view is supported by femtosecond stimulated Raman spectroscopy [22] and quantum chemical calculations [23].

A low-temperature photocalorimetric study revealed that about 60 % of light energy (150 kJ/mol) is stored in the structure of bathorhodopsin [24]. Highly twisted retinal chromophore contributes to this energy storage, and such structural deformation of the polyene chain has been experimentally monitored by enhanced hydrogen-out-of-plane (HOOP) vibrational modes at 1000–800 cm^{-1} [25, 26]. It should be noted that electrostatic and hydrogen-bonding interactions are small, because the hydrogen-bonding strength of RSB is unchanged before and after isomerization [20]. This is an interesting contrast to the case in microbial rhodopsins in which relaxation of the distorted polyene chain leads to the formation of lumirhodopsin, which itself accompanies the motion of the β-ionone ring toward the 4th helix, which would be important for G-protein activation [1].

The protein environment thus beautifully controls efficient *cis-trans* photoisomerization in animal rhodopsins. The role of two methyl groups at the C9 and C13 positions is also noteworthy. Upon photoisomerization, the C9-methyl does not change its position, while the C13-methyl group actually moves (Fig. 41.4). The C13-methyl group has a steric conflict with the retinal C10-hydrogen in animal rhodopsins, which causes the twists of the C11=C12 and C12-C13 bonds even in the dark, and the important role of C13-methyl for efficient isomerization has been reported experimentally [27] and theoretically [28]. In contrast, the C9-methyl group appears to function as a scaffold for the primary isomerization [21], while its motion is important for activation of G-protein in late intermediates [29].

The chromophore molecule in solution (particularly methanol) is useful to evaluate the protein effect, termed the "opsin shift," in the color tuning mechanism. Similarly, excited-state isomerization dynamics of the chromophore of animal rhodopsins between protein and solution were compared by means of femtosecond fluorescence measurements, where five-membered 11-*cis*-locked retinal was also used. In methanol, the fluorescence lifetime of the five-membered chromophore was five times longer than that of the protonated Schiff base of 11-*cis* retinal [30]. This is in remarkable contrast to the chromophore molecule in protein, because the fluorescence lifetime of five-membered rhodopsin (85 ps) is two orders of magnitude longer than that of native rhodopsin [15]. These facts suggest that the protein moiety is enhancing the isomerization rates of the chromophore, where functional optimization has been achieved during evolution.

41.4 Photoisomerization and Energy Storage in Microbial Rhodopsins

Microbial rhodopsins exhibit a photocycle unlike animal rhodopsins that bleach upon illumination. This is highly advantageous in ultrafast spectroscopic studies. Therefore, various ultrafast spectroscopic techniques have been extensively applied in addition to low-temperature spectroscopy. In particular, BR has historically been regarded as the model system to test new spectroscopic methods. Like animal rhodopsins, light absorption of microbial rhodopsins causes the formation of redshifted primary intermediates [1]. The primary K intermediate can be stabilized at 77 K. Time-resolved visible spectroscopy of BR revealed the presence of a precursor, which was termed the J intermediate. The J intermediate is more redshifted (λ_{max} 625 nm) than the K intermediate (λ_{max} 590 nm). The excited state of BR possesses blueshifted absorption, which decays in about 200 and 500 fs [31]. Since the J intermediate is formed in <500 fs, it is reasonable to consider that photoisomerization takes place in femtoseconds.

To test when isomerization occurs, all-*trans*-locked five-membered retinal was incorporated into BR [32–34], as in animal rhodopsins, but the results were very different. In the experiments with a picosecond time resolution, an intermediate was found with properties similar to those of the J intermediate [32]. Together with the ultrafast pump-probe [33] and coherent anti-Stokes Raman [34] spectroscopic results, it was concluded that isomerization around C13=C14 is not a prerequisite for producing the J intermediate. More importantly, since the J intermediate is a ground-state species, isomerization does not take place in the excited state of BR according to the interpretation of these studies [32–34].

However, other experimental data favor a common mechanism between animal and microbial rhodopsins in which isomerization takes place in the excited state. Using a femtosecond visible-pump and infrared-probe spectroscopy, the characteristic 13-*cis* vibrational band was observed at 1190 cm^{-1}, appearing with a time constant of 0.5 ps, indicating that the all-*trans* to 13-*cis* isomerization takes place in femtoseconds [35]. This timescale is coincident with the formation of the J intermediate. Fourier transform of the transient absorption data with a less than 5-fs resolution also showed the appearance of the 13-*cis* form in less than 1 ps, supporting that the all-*trans* to 13-*cis* isomerization takes place in femtoseconds

[36]. Previous anti-Stokes resonance Raman spectroscopy proposed that the J intermediate is a vibrational hot state of the K intermediate [37]. Thus, many experimental results are consistent with the isomerization model in the excited state.

Photoisomerization takes place in the excited state in animal and microbial rhodopsins in a similar way, but relaxation from the Franck-Condon state may be different. In animal rhodopsins, the excited wave packet slides down the barrierless potential surface. In contrast, several experiments of BR favor the three-state model that postulates a small potential barrier along the isomerization coordinate [1]. Actually, real-time spectroscopy of BR with a less than 5-fs pulse showed that isomerization does not occur instantly, but instead involves transient formation of a tumbling state. This clearly supports the three-state model and discounts the initially suggested two-state model. It should be however be noted that the three-state model is even not sufficient, because both models adopt only one coordinate upon isomerization. Fluorescence spectroscopy of bovine rhodopsin [38] and transient absorption spectroscopy of halorhodopsin [39] clearly observed a prolonged excited state after photoisomerization is complete, which can be explained by neither the two-state nor three-state models. Another coordinate uncoupled to isomerization has to be taken into account, and BR would also possess such a state.

Unlike animal rhodopsins, ultrafast spectroscopy has also been applied to various BR mutants, revealing that only the replacements of the charged residues reduced the photoisomerization rate, leading to less efficient photoisomerization [40]. This observation explains an important role of the electrostatic interaction of the counterion complex in the primary photoisomerization mechanism. The prompt response of tryptophans (Fig. 41.5) after retinal photoexcitation was also reported by monitoring the absorption of tryptophans [41]. The excited state is more long-lived in other microbial rhodopsins [1] and hence less efficient for photoisomerization. These observations suggest that BR possesses an optimized mechanism for primary photoisomerization.

Ultrafast isomerization from the all-*trans* to 13-*cis* forms yields formation of the primary intermediate, which must store light energy in the structure. It should be noted that light-energy storage in microbial rhodopsins is lower than in bovine rhodopsin. Low-temperature photocalorimetric studies reported the energy stored in primary intermediates to be 150 kJ/mol and 67 kJ/mol for bovine rhodopsin [24] and BR [42], respectively. This indicates that only 30 % of light energy is stored in the structure of the primary K intermediate, which accounts for about half of the animal rhodopsins. In other words, about 70 % of light energy is dissipated in the formation of the K intermediate. Nevertheless, such energy loss may not be serious from a functional point of view, because BR pumps a single proton by using one photon, and free energy gain by pumping a proton is about 25 kJ/mol. Interestingly, quantum yields of photoisomerization are not so different between bovine rhodopsin (0.67) and BR (0.64). Therefore, such a difference in energy storage must be correlated with the structures of the primary intermediates.

X-ray crystallographic structures of the primary K intermediate have been reported by three groups [43–45]. As expected, protein structures changed little before and after isomerization. One group concluded that energy storage in the K

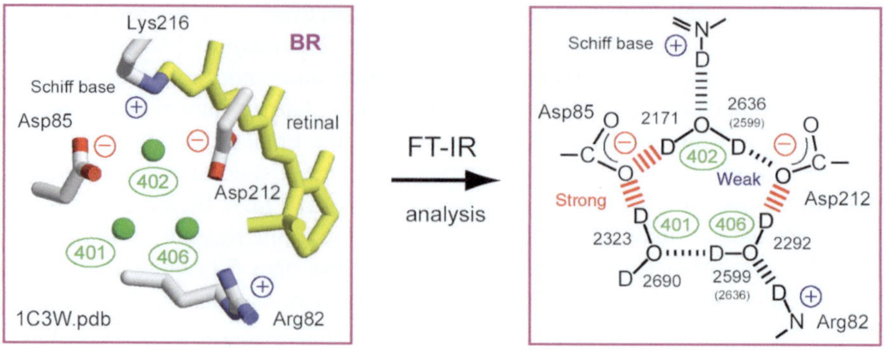

Fig. 41.8 Schematic drawing of the assignment of the water molecules in the Schiff base region of bacteriorhodopsin. This figure is reproduced from Kandori [61]

state of BR is almost completely explained by the distortion at position C13 [44]. However, it would be difficult to determine the bond angle accurately under the current resolution (>2.0 Å). In fact, the three reported structures of the K intermediate of BR are considerably different among these groups [43–45]. Vibrational spectroscopy reported that the enhanced HOOP vibrations in the K intermediate are mostly H-D exchangeable, implying that chromophore distortion is localized at the RSB moiety [1]. The stretching vibration of RSB is significantly upshifted (350 cm^{-1} for the N-D stretch and 500 cm^{-1} for the N-H stretch), indicating a weakened hydrogen bond [46]. Thus, in the case of BR, retinal isomerization accompanies rotational motions of the RSB side, so that the hydrogen bond is significantly weakened. This is in contrast with animal rhodopsins.

Figure 41.8 shows that the hydrogen-bonding acceptor of RSB is a water molecule, which bridges the ion-pair. The motion of RSB probably enforces rearrangement of the water-containing hydrogen-bonding network in the RSB region, and this was indeed probed by FTIR spectroscopy. We observed the presence of strongly hydrogen-bonded water molecules in BR, and the frequencies are upshifted upon formation of the K intermediate [47, 48]. This observation strongly suggests that the hydrogen-bonding interaction is highly destabilized in the K state, which possibly contributes to the high-energy state. In fact, normal mode calculation of quantum chemical/molecular mechanics successfully reproduced the prominent spectral features for BR and the K intermediate [49]. This further suggests that the contribution of weakened hydrogen bonds to energy storage is 46 kJ/mol, which is more than half of the total energy storage (67 kJ/mol). Energy storage through the distorted retinal chromophore has been well accepted and can be adopted for animal rhodopsins, but energy storage in hydrogen-bonding destabilization is a rather new concept.

The importance of hydrogen-bonding alteration in energy storage provided an unexpected finding, namely, a positive correlation between the strong hydrogen bond of water and proton-@ @pump activity (Fig. 41.9) [1]. As already mentioned,

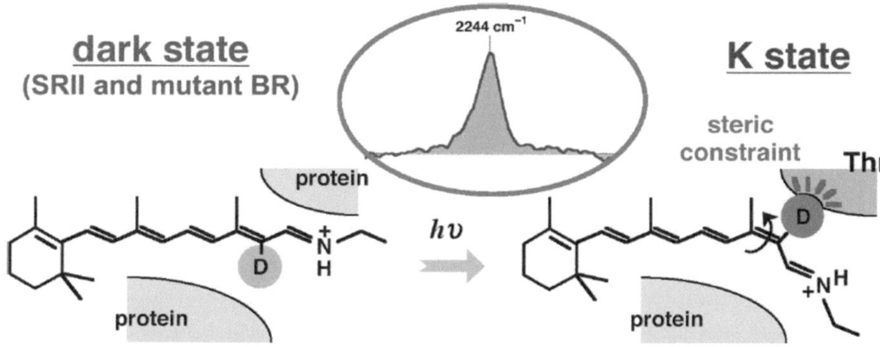

Fig. 41.9 Schematic drawing of the steric constraint formed in the K state of SRII and a mutant BR converted into a sensor. Among all seven monodeuterated all-*trans* retinal analogs (positions 7, 8, 10, 11, 12, 14, 15), only the C14-D stretching vibration at 2244 cm^{-1} is significantly enhanced upon formation of the K state of SRII [57], where the counterpart in the protein side is Thr204 [58]

we measured FTIR difference spectra of internal water molecules at 77 K, where measurement in D_2O is advantageous, because X-H and X-D stretching can be separated [4]. Water O-D stretching vibrations that appear at 2700–2100 cm^{-1} are dependent on their hydrogen-bonding strength, and we defined the strong hydrogen bond at <2400 cm^{-1}. Mutation studies of BR showed that D85N and D212N only exhibit no water bands under strongly hydrogen-bonding conditions [50, 51], suggesting that such water molecules may be a prerequisite for the proton-pump function. Since then, we tested various rhodopsins, including animal rhodopsins, with a view to assessing whether the protein contains strongly hydrogen-bonded water or not in the unphotolyzed state (O-D stretch at <2400 cm^{-1}). The results clearly showed a strong correlation between the presence of strongly hydrogen-bonded water molecule(s) and proton-pumping activity (Fig. 41.9). They were interpreted by a proposal in which the presence of a strong hydrogen bond of water is a prerequisite for proton pumping in rhodopsins [4]. It is likely that destabilization of the water-containing hydrogen-bonding network plays an important role for light-energy storage in this case.

This empirical rule explained well the observed asymmetric functional conversion between light-driven proton and chloride-ion pumps. BR and halorhodopsin (HR) are light-driven outward proton and inward chloride pumps, respectively. They have similar protein architecture, being composed of seven-transmembrane helices that bind an all-*trans* retinal. BR can be converted into a chloride pump by a single amino acid replacement at position 85 [52], suggesting that BR and HR share a common transport mechanism and the ionic specificity is determined by the amino acid at that position. However, HR cannot be converted into a proton pump by the corresponding reverse mutation. Although six and ten amino acids of HR were mutated into BR-like, such multiple HR mutants never pump protons [53]. Light-induced FTIR spectroscopy revealed that hydrogen bonds of RSB and water are both strong in BR and both weak in HR. Multiple HR mutants exhibit a strong hydrogen bond in RSB, but the hydrogen bond of water is still weak. It was concluded that the cause of the nonfunctional conversion of HR is the lack of strongly hydrogen-bonded water, the functional determinant of a proton pump [53].

As described previously, hydrogen-bonding alteration plays important roles in the energy storage of ion pumps. In the case of signaling microbial rhodopsins, a steric effect contributes more significantly. Sensory rhodopsin II (SRII) [54] and the complex with its transducer protein (HtrII) [55] both possess strongly hydrogen-bonded water molecules, but the latter does not pump protons. Presumably the specific interaction between SRII and HtrII ceases the proton-pump activity. Instead, the SRII/HtrII complex functions in light-signal transduction. Phylogenetic analysis suggests that rhodopsin photosensors like SRI and SRII evolved from light-driven proton pumps [56]. Therefore, sensory rhodopsins can be thought of as pumps that have gained a mechanism to activate transducer proteins. We previously observed enhancement of the C14-D stretching vibration of the retinal chromophore at 2244 cm^{-1} upon formation of the K intermediate and interpreted that a steric constraint occurs at the C14-D group in $SRII_K$ (Fig. 41.10) [57]. By using a mutant, the counterpart of the C14-D group was determined to be Thr204 [58]. Although the K state of wild-type BR does not possess the 2244-cm^{-1} band, the band newly appeared for the K state of a triple mutant of BR that functions as a light sensor (P200T/V210Y/A215T) [59]. We found a positive correlation between the vibrational amplitude of the C14 atom at 77 K and the physiological phototaxis response. These observations strongly suggest that the steric constraint between the C14 group of retinal and Thr204 of the protein (Thr215 for a triple mutant of BR) is a prerequisite for light-signal transduction by SRII [58]. Thus, the photoisomerization-induced steric constraint between the C14-H of the retinal chromophore and Thr204 in SRII is one such mechanism. It is likely that a water-containing hydrogen-bonding network is important for energy storage in proton pumps like BR, while a steric constraint is important for energy storage in sensor functions as seen for animal rhodopsins, SRII, and a BR mutant, as a light sensor.

41.5 Conclusion and Perspectives

This chapter compiles a current understanding of retinal photoisomerization in animal and microbial rhodopsins. Extensive studies by means of ultrafast spectroscopy of animal and microbial rhodopsins have shown that the primary process is isomerization, from an 11-*cis* to an all-*trans* form in animal rhodopsins and from an all-*trans* to a 13-*cis* form in microbial rhodopsins. Femtosecond spectroscopy of animal and microbial rhodopsins eventually captured their excited states, and, as a consequence, we now know that unique photochemistry takes place in human eyes and in the eyes of archaea. Such unique reactions are facilitated by the interaction between π-electrons of the chromophore and the protein environment, and recent structural determination further expanded our understanding on the

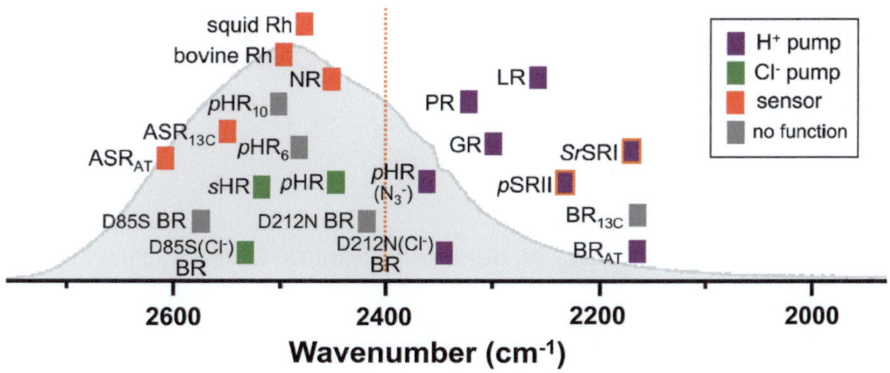

Fig. 41.10 In the light-induced difference FTIR spectra of various rhodopsins at 77 K, the lowest frequency among the observed water O-D stretching vibrations is plotted. *Purple squares* represent the results of the proton pump, while red and green squares are those of the sensor and chloride pump, respectively. The proteins of *dark gray squares* have no function. Abbreviations are as follows: BR_{AT} (BR containing all-*trans* retinal), BR_{13C} (BR containing 13-*cis* retinal), D85S BR, D85S(Cl^-) BR, D212N BR, D212N(Cl^-) BR, *Sr*SRI (*Salinibacter ruber* sensory rhodopsin I), *p*SRII (*Natronomonas pharaonis* sensory rhodopsin II), LR (*Leptosphaeria* rhodopsin), GR (*Gloeobacter* rhodopsin), PR (proteorhodopsin), ASR_{AT} (*Anabaena* sensory rhodopsin containing all-*trans* retinal), ASR_{13C} (*Anabaena* sensory rhodopsin containing 13-*cis* retinal), NR (*Neurospora* rhodopsin), bovine Rh (bovine rhodopsin), squid Rh (squid rhodopsin), *s*HR (*Halobacterium salinarum* halorhodopsin), *p*HR (*Natronomonas pharaonis* halorhodopsin), *p*HR(N_3^-) (*Natronomonas pharaonis* halorhodopsin that binds azide), *p*HR_6 (sextet mutant of *Natronomonas pharaonis* halorhodopsin in [53]) and *p*HR_{10} (tenfold mutant of *Natronomonas pharaonis* halorhodopsin in [53]). SRI and SRII are sensors, but it is known that they pump protons in the absence of transducer proteins (*purple squares* with *orange* frames). Light gray spectrum shows the water O-D stretching vibration, and only proton-pumping rhodopsins possess strongly hydrogen-bonded waters (O-D stretch at $<2400\ cm^{-1}$). This figure is modified from Muroda et al. [53]

basis of structure. In solution, photochemical properties are similar between the protonated RSB of 11-*cis* and all-*trans* retinal. Nevertheless, there are considerable difference in the photochemistry of 11-*cis* (animal) and all-*trans* (microbial) forms in protein (rhodopsin). In animal rhodopsins, conformational distortion takes place at the center of the chromophore, whereas no changes occur at RSB. Such changes may be leading to coherent product formation in femtoseconds. In contrast, structural changes take place only at the RSB region of microbial rhodopsins, which accompanies changes in the hydrogen-bonding network. Hydrogen-bonding alteration also plays important roles in the function of ion pumps. Since the atomic structures of animal and microbial rhodopsins are now at hand, theoretical investigations will become more important in the future. A combination of three methods, diffraction, spectroscopy, and theory, will lead to a real understanding of the mechanism of isomerization in rhodopsins.

Acknowledgments I thank the members of my laboratory and many collaborators. I also thank the project members of "Emergence of Highly Elaborated π-Space and its Function" for a fruitful discussion.

References

1. Ernst OP, Lodowski DT, Elstner M, Hegemann P, Brown LS, Kandori H (2014) Chem Rev 114:126–163
2. Spudich JL, Yang C-S, Jung K-H, Spudich EN (2000) Annu Rev Cell Dev Biol 16:365–392
3. Palczewski K (2006) Annu Rev Biochem 75:743
4. Kandori H (2011) Protein-controlled ultrafast photoisomerization in rhodopsin and bacteriorhodopsin. In: Ramamurthy V, Inoue Y (eds) Supramolecular photochemistry. Wiley, Hoboken, pp 571–595
5. Blatz PE, Mohler JH, Navangul HV (1972) Biochemistry 11:848
6. Andersen LH, Nielsen IB, Kristensen MB, El Ghazaly MOA, Haacke S, Nielsen MB, Petersen MA (2005) J Am Chem Soc 127:12347–12350
7. Furutani Y, Ido K, Sasaki M, Ogawa M, Kandori H (2007) Angew Chem Int Ed 46:8010–8012
8. Wang W, Nossoni Z, Berbasova T, Watson CT, Yapici I, Lee KS, Vasileiou C, Geiger JH, Borhan B (2012) Science 338:1340
9. Dartnall HJA (1967) Vis Res 8:339–358
10. Tittor J, Oesterhelt D (1990) FEBS Lett 263:269–273
11. Furutani Y, Terakita A, Shichida Y, Kandori H (2005) Biochemistry 44:7988–7997
12. Koyama Y, Kubo K, Komori M, Yasuda H, Mukai Y (1991) Photochem Photobiol 54:433–443
13. Yoshizawa T, Wald G (1963) Nature 197:1279–1286
14. Busch GE, Applebury ML, Lamola AA, Rentzepis PM (1972) Proc Natl Acad Sci U S A 69:2802–2806
15. Kandori H, Matuoka S, Shichida Y, Yoshizawa T, Ito M, Tsukida K, Balogh-Nair V, Nakanishi K (1989) Biochemistry 28:6460–6467
16. Mizukami T, Kandori H, Shichida Y, Chen A-H, Derguini F, Caldwell CG, Bigge C, Nakanishi K, Yoshizawa T (1993) Proc Natl Acad Sci U S A 90:4072–4076
17. De Vico L, Garavelli M, Bernardi F, Olivucci M (2005) J Am Chem Soc 127:2433–2442
18. Schoenlein RW, Peteanu LA, Mathies RA, Shank CV (1991) Science 254:412–415
19. Wang Q, Schoenlein RW, Peteanu LA, Mathies RA, Shank CV (1994) Science 266:422–424
20. Eyring G, Mathies RA (1979) Proc Natl Acad Sci U S A 76:33–37
21. Nakamichi H, Okada T (2006) Angew Chem Int Ed 45:1–5
22. Kukura P, McCamant DW, Yoon S, Wandschneider DB, Mathies RA (2005) Science 310:1006–1009
23. Frutos LM, Andruniow T, Santoro F, Ferre N, Olivucci M (2007) Proc Natl Acad Sci U S A 104:7764–7769
24. Cooper A (1979) Nature 282:531–533
25. Eyring G, Curry B, Broek A, Lugtenburg J, Mathies RA (1982) Biochemistry 21:384–393
26. Palings I, van den Berg EMM, Lugtenburg J, Mathies RA (1989) Biochemistry 28:1498–1507
27. Kochendoerfer GG, Verdegem PJE, van der Hoef I, Lugtenburg J, Mathies RA (1996) Biochemistry 35:16230–16240
28. Sugihara M, Hufen J, Buss V (2006) Biochemistry 45:801–810
29. Ganter UM, Schmid ED, Perez-Sala D, Rando RR, Siebert F (1989) Biochemistry 28:5954–5962
30. Kandori H, Katsuta Y, Ito M, Sasabe H (1995) J Am Chem Soc 117:2669–2670
31. Dobler J, Zinth W, Kaiser W (1988) Chem Phys Lett 144:215–220
32. Delaney JK, Brack TL, Atkinson GH, Ottolenghi M, Steinberg G, Sheves M (1995) Proc Natl Acad Sci U S A 92:2101–2105

33. Zhong Q, Ruhman S, Ottolenghi M, Sheves M, Friedman N (1996) J Am Chem Soc 118:12828–12829
34. Atkinson GH, Ujj L, Zhou Y (2000) J Phys Chem A 104:4130–4139
35. Herbst J, Heyne K, Diller R (2002) Science 297:822–825
36. Kobayashi T, Saito T, Ohtani H (2001) Nature 414:531–534
37. Doig SJ, Reid PJ, Mathies RA (1991) J Phys Chem 95:6372–6379
38. Kandori H, Furutani Y, Nishimura S, Shichida Y, Chosrowjan H, Shibata Y, Mataga N (2001) Chem Phys Lett 334:271
39. Kandori H, Yoshihara K, Tomioka H, Sasabe H (1992) J Phys Chem 96:6066–6071
40. Song L, El-Sayed MA, Lanyi JK (1993) Science 261:891–894
41. Schenkl S, van Mourik F, van der Zwan G, Haacke S, Chergui M (2005) Science 309:917–920
42. Birge RR, Cooper TM (1983) Biophys J 42:61–69
43. Edman K, Nollert P, Royant A, Belrhali H, Pebey-Peyroula E, Hajdu J, Neutze R, Landau EM (1999) Nature 401:822–826
44. Schobert B, Cupp-Vickery J, Hornak V, Smith SO, Lanyi JK (2002) J Mol Biol 321:715–726
45. Matsui Y, Sakai K, Murakami M, Shiro Y, Adachi S, Okumura H, Kouyama T (2002) J Mol Biol 324:469–481
46. Kandori H, Belenky M, Herzfeld J (2002) Biochemistry 41:6026–6031
47. Kandori H, Shichida Y (2000) J Am Chem Soc 122:11745
48. Shibata M, Tanimoto T, Kandori H (2003) J Am Chem Soc 125:13312
49. Hayashi S, Tajkhorshid E, Kandori H, Schulten K (2004) J Am Chem Soc 126:10516–10517
50. Shibata M, Ihara K, Kandori H (2006) Biochemistry 45:10633–10640
51. Shibata M, Yoshitsugu M, Mizuide N, Ihara K, Kandori H (2007) Biochemistry 46:7525–7535
52. Sasaki J, Brown LS, Chon Y-S, Kandori H, Maeda A, Needleman R, Lanyi JK (1995) Science 269:73–75
53. Muroda K, Nakashima K, Shibata M, Demura M, Kandori H (2012) Biochemistry 51:4677–4684
54. Kandori H, Furutani Y, Shimono K, Shichida Y, Kamo N (2001) Biochemistry 40:15693–15698
55. Furutani Y, Sudo Y, Kamo N, Kandori H (2003) Biochemistry 42:4837–4842
56. Sharma AK, Spudich JL, Doolittle WF (2006) Trends Microbiol 14:463–469
57. Sudo Y, Furutani Y, Wada A, Ito M, Kamo N, Kandori H (2005) J Am Chem Soc 127:16036–16037
58. Ito M, Sudo Y, Furutani Y, Okitsu T, Wada A, Homma M, Spudich JL, Kandori H (2008) Biochemistry 47:6208–6215
59. Sudo Y, Spudich JL (2006) Proc Natl Acad Sci U S A 103:16129–16134
60. Furutani Y, Shibata M, Kandori H (2005) Photochem Photobiol Sci 4:661–666
61. Kandori H (2010) Hydrogen bonds of protein-bound water molecules in rhodopsins. In: Han K-L, Zhao G-J (eds) Hydrogen bonding and transfer in the excited state. Wiley, Chichester, pp 377–391

Chapter 42
β-Diketiminates as Redox Non-innocent Supporting Ligands in Coordination Chemistry

Shinobu Itoh and Yuma Morimoto

Abstract Recent studies of redox non-innocence of β-diketiminate ligands are introduced. Reduction of Yb^{3+} and Al^{3+} as well as Li^{+} and Be^{2+} complexes using strong electron-transfer reductants induces the ligand-based one- and/or two-electron reduction, whereas electron-transfer oxidation of bis(β-diketiminato)-Ni^{2+} complexes gives Robin and Day class III mixed-valence complexes, having an unpaired electron completely delocalized between the two ligands. Furthermore, a new trianionic tetradentate ligand is developed by introducing two phenolate moieties on the nitrogen atoms of a β-diketiminate ligand to mimic the active site of galactose oxidase. In the copper(II) complex of this ligand, oxidation reaction also takes place at the ligand moiety to provide phenoxyl–copper(II) complexes.

Keywords β-diketiminate • Redox non-innocent ligand • Mixed valence • Inter-valence charge transfer

42.1 Introduction

β-Diketiminates are the important class of monoanionic didentate ligands, which have widely been adopted for the synthesis of a variety of metal complexes with various oxidation states throughout the periodic table [1–8]. Such complexes have attracted much recent attention as polymerization catalysts, novel organometallic compounds having an unusual coordination geometry and reactivity, and active site models of metalloenzymes especially for small molecular activation.

The ligand has generally a $[R^1\text{–N–C}(R^2)\text{–C}(R^3)\text{–C}(R^4)\text{–N–}R^5]^-$ (L^-) structural framework, generated by deprotonation of the NH group of pro-ligand LH (Scheme 42.1). The generated anionic charge is delocalized onto the conjugated π orbitals of the –N–C–C–C–N– moiety (Scheme 42.1). The ligands are basically sigma donor, but interaction between the d and/or f orbitals of the metal ions and the

S. Itoh (✉) • Y. Morimoto
Department of Material and Life Science, Division of Advanced Science and Biotechnology, Graduate School of Engineering, Osaka University, 2-1 Yamada-oka, Suita, Osaka 565-0871, Japan
e-mail: shinobu@mls.eng.osaka-u.ac.jp

T. Akasaka et al. (eds.), *Chemical Science of π-Electron Systems*,
DOI 10.1007/978-4-431-55357-1_42

Scheme 42.1 Formation of β-diketiminate ligand

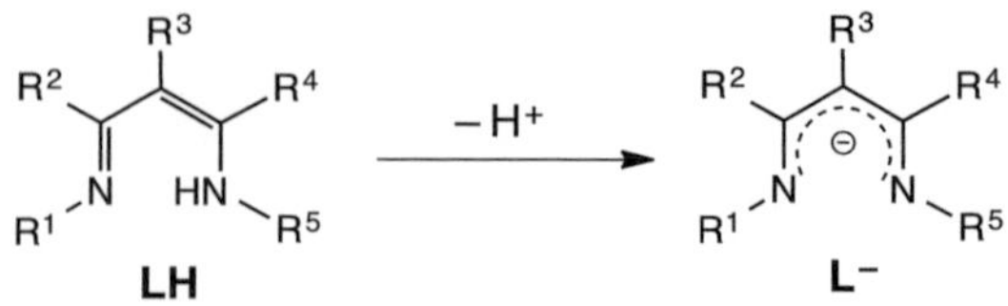

π orbitals of the conjugated system of the ligand backbone may dictate interesting physicochemical properties and chemical reactivity.

Structure, electronic and magnetic properties, and reactivity of the metal centers can be controlled by the substituents introduced not only on the nitrogen atoms (R^1 and R^5) but also on the carbon framework (R^2–R^4). In most cases, steric effects by the nitrogen substituents (R^1 and R^5) are considered to control the steric demand of the secondary coordination sphere. For example, coordinatively unsaturated organometallic and biomimetic complexes can be stabilized by steric protection using bulky substituents R^1 and R^5. The carbon substituents (R^2–R^4) are also important in tuning the electron donor ability of the supporting ligands. However, little attention has been focused on the redox activity (redox non-innocence) of the supporting ligand itself. *β*-Diketiminate could be reduced or oxidized in the redox reactions of the metal complexes, acting as non-innocent supporting ligand. In this chapter, we introduce recent advances on such chemistry of *β*-diketiminate ligand systems.

42.2 Ligand Reduction

In the early 2000s, Lappert and his coworkers reported that the reaction of $YbCl_3$ and two equivalent of KL (L = $([N(SiMe_3)C(Ph)]_2CH)^-$) in THF and subsequent reduction with Li yielded a four-coordinate bis(*β*-diketiminato)-Yb(II) complex, in which each supporting ligand is also connected to a Li^+ atom coordinated by a THF solvent molecule, $[Yb(L(\mu\text{-}Li(THF)))_2]$ (**1**) (Fig. 42.1) [9]. In this particular case, supporting ligand was reduced to the dianionic radical state $L^{2-\bullet}$ (Scheme 42.2). On the other hand, reduction of an Yb(II) *β*-diketiminate complex, YbL_2 supported by the same ligand using Yb-naphthalene complex, $Yb(C_{10}H_8)(THF)_3$, gave a mixed-valence $Yb(II)_2/Yb(III)$ trinuclear cluster, $[(YbL)_3(THF)]$ (**2**), in which two of the three ligands are reduced to the trianionic form L^{3-} (Scheme 42.2) [10]. The structural parameters of these complexes obtained by the X-ray crystallographic analysis were consistent with the assignment of $L^{2-\bullet}$ and L^{3-} forms of the ligands, which were supported by theoretical (DFT) calculations [9, 10]. The di- and trianionic states are probably stabilized by electron delocalization into the phenyl and $SiMe_3$ substituents as indicated in Scheme 42.3 [10, 11].

Afterward, they reported that the Li-complexes themselves of the similar *β*-diketiminate ligands can also be reduced to generate dinuclear and multinuclear Li complexes involving the paramagnetic dianionic ($L^{2-\bullet}$) or diamagnetic trianionic

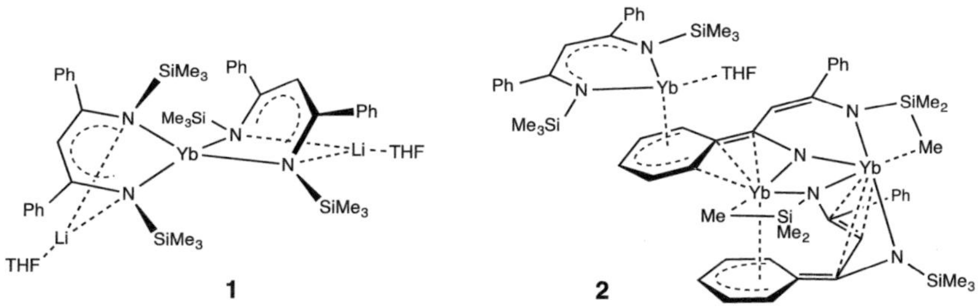

Fig. 42.1 Structure of (1) $[Yb(L(m\text{-}Li(THF)))_2]$ and (2) $[(YbL)_3(THF)]$

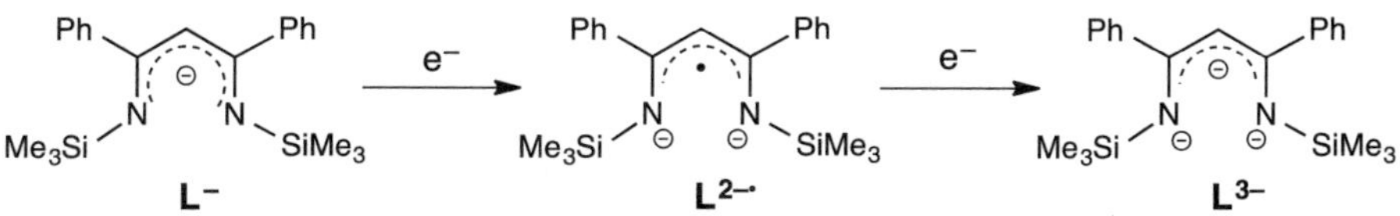

Scheme 42.2 Consecutive reduction of β-diketiminate ligand

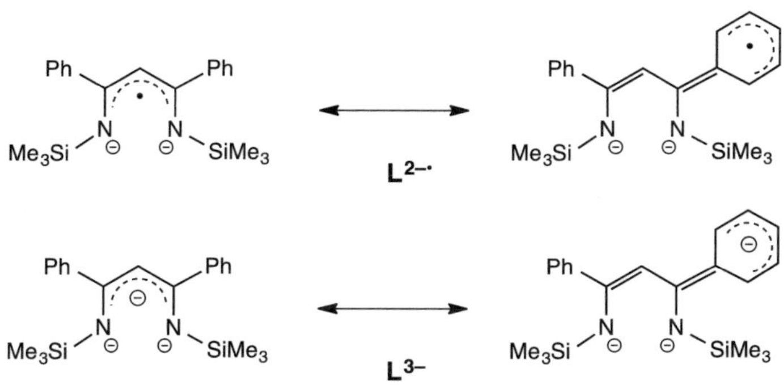

Scheme 42.3 Resonance structures of β-diketiminate dianion ($L^{2-\bullet}$) and trianion (L^{3-}) with phenyl substituents

(L^{3-}) ligands [11]. In order to investigate the redox properties of the ligands, an electrochemical study of the bis(β-diketiminato)complexes of type $M(L)_2$ ($M = Li^+$, K^+, Mg^{2+}, Ca^{2+}, and Yb^{2+}) was undertaken [12]. However, neither the spectroscopic nor the magnetic properties of these reduced ligands were examined in detail.

Radical complexes of the Group 13 elements have attracted attention due to their unusual structure, spectroscopic features, and reactivity [13]. Thus, the reactions of a β-diketiminatoaluminium(I) complex **3** with hydrogen atom and electron were examined theoretically (Scheme 42.4) [14]. It was suggested that the dianionic radical ligand ($L^{2-\bullet}$) is involved in both products **3H**$^{\bullet}$ and **3**$^{-\bullet}$. Tuononen and coworkers demonstrated that the reduction of bis(β-diketiminato)aluminum(III) complex **4**$^+$ with $CoCp^*_2$ (decamethylcobaltocene) gave a neutral paramagnetic complex **4**$^{\bullet}$, in which one of the ligands was reduced to be a dianionic radical

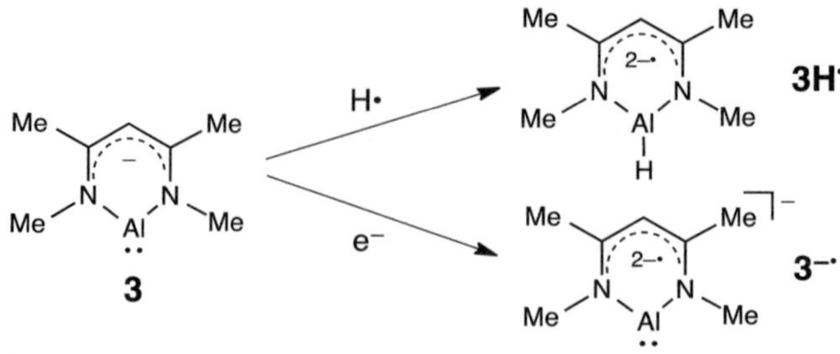

Scheme 42.4 Reduction of (β-diketiminato)aluminum(III) complex

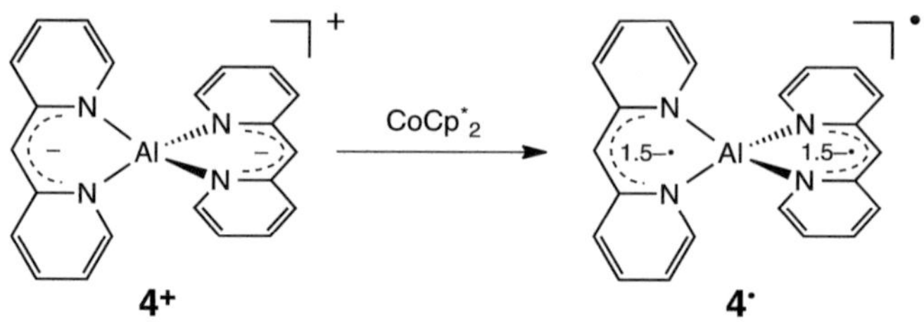

Scheme 42.5 Reduction of a bis(β-diketiminato)aluminum(III) complex

(Scheme 42.5) [15]. Although DFT calculation and EPR studies suggested that the generated spin is delocalized over the two ligands, further characterization of the radical complex was not performed.

Hill and coworkers demonstrated that the reduction of a mononuclear three coordinate beryllium complex **5** supported by a *β*-diketiminate ligand $(HC(C(Me)N(Dipp)))_2$ (Dipp=2,6-diisopropylphenyl) with potassium metal resulted in apparent hydrogen-atom transfer between two *β*-diketiminate backbones, yielding two dimeric, potassium chloride bridged diamidoberyllium complex **6** [16]. They proposed that this process occurs by the formation of a radical anion species $\mathbf{5}^{-\bullet}$ and intermolecular hydrogen-atom transfer by hydrogen abstraction mechanism (Scheme 42.6).

As a related system, Otten and coworkers examined the reduction of bis(formazanato)zinc(II) complexes, $[Zn^{II}(L^-)_2]$ (**7**), to find successive one-electron and two-election reduction to give $\mathbf{7}^{-\bullet}$ and $\mathbf{7}^{2(-\bullet)}$ (Scheme 42.7) [17]. The X-ray crystallographic analysis as well as the UV–vis and EPR spectroscopic analyses of the reduced complexes $\mathbf{7}^{-\bullet}$ and $\mathbf{7}^{2(-\bullet)}$ unambiguously indicated that they involved one and two reduced ligand $(L^{2-\bullet})$, respectively. In both cases, there is no electronic communication between the two ligands, allowing the spin density generated by the reduction localized on one ligand. This notion was supported by DFT calculation. The redox non-innocence of a formazan ligand was also suggested in the thermal decomposition of a bis(μ-oxido)dicopper(III) complex supported by a nitro-formazanate ligand [18].

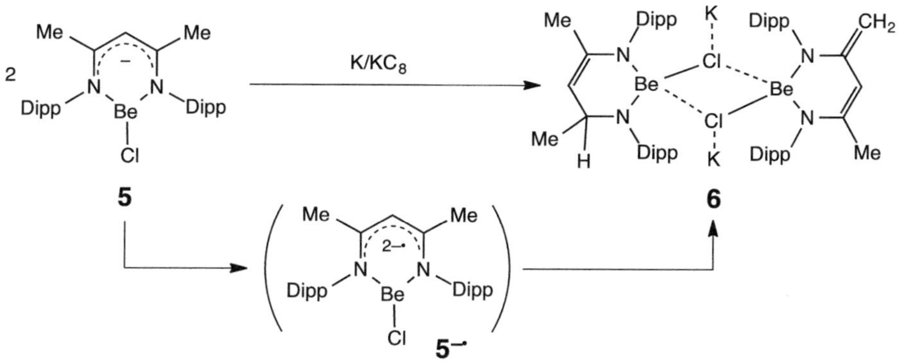

Scheme 42.6 Reduction of a (β-diketiminato)beryllium complex

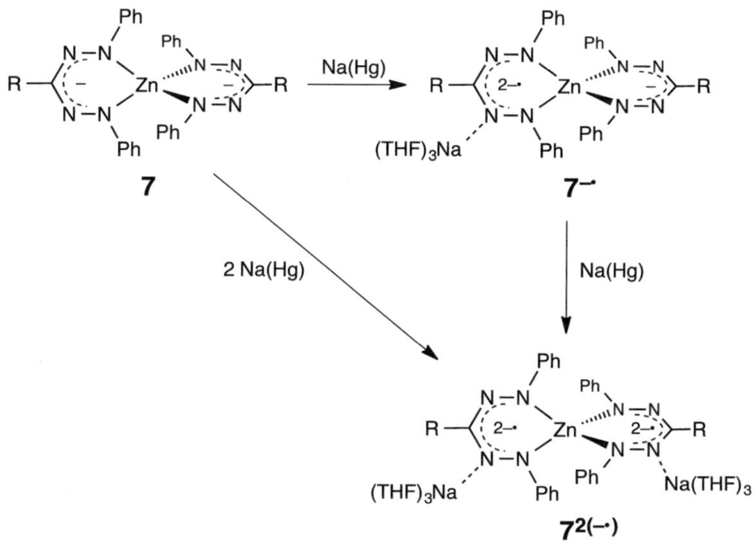

Scheme 42.7 Reduction of bis(formazanato)zinc(II) complex

42.3 Ligand Oxidation

Khusniyarov, Wieghardt, and coworkers examined oxidation reaction of bis(β-diketiminato)nickel(II) complex **8** to find a ligand-based oxidation, demonstrating that β-diketiminates can behave as redox non-innocent ligands in the oxidation reaction (Scheme 42.8) [19]. The electrochemically oxidized complex at –25 °C exhibited a strong MLCT band at 591 nm and a broad ligand-to-ligand inter-valence charge-transfer band in the near-IR (NIR) region (λ = 850 nm). The EPR spectrum of $\mathbf{8}^{+\bullet}$ measured in frozen MeCN solution shows a rhombic signal with $g_1 = 2.25$, $g_2 = 2.09$, and $g_3 = 2.03$. The large anisotropy of the g tensor

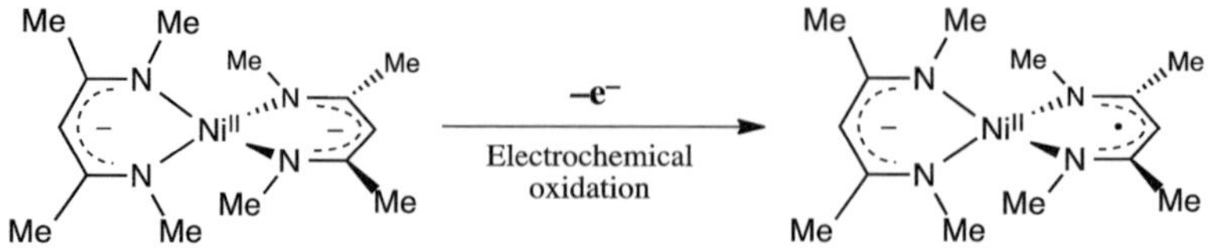

Scheme 42.8 Oxidation of a bis(β-diketiminato)nickel(II) complex reported by Khusniyarov, Wieghardt, and coworkers

Chart 42.1 β-diketiminate ligands with a series of substituents

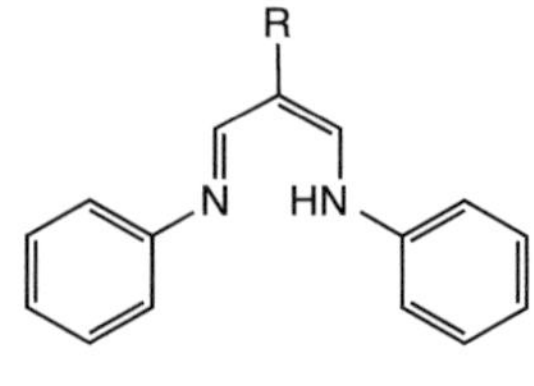

($\Delta g = 0.22$), position of the signals near $g_e = 2.00$, and the absence of a half-field signal can only be ascribed to a doublet ground state ($S = 1/2$) of $\mathbf{8}^{+\bullet}$ with spin density predominantly located at the metal center. On the basis of these spectroscopic characteristics together with the DFT studies, they concluded that an open-shell ligand L$^{\bullet}$ ($S_L = 1/2$) is strongly antiferromagnetically coupled to a high-spin Ni(II) ion ($S_{Ni} = 1$) to give an overall doublet ground state ($S_T = 1/2$). Due to the thermal instability of the oxidized complex $\mathbf{8}^{+\bullet}$, the structural features were discussed mainly on the basis of DFT calculations [19].

The authors and coworkers investigated the similar system in more detail using a series of β-diketiminate ligands ^{R}LH shown in Chart 42.1 [20]. The neutral bis(β-diketiminato)nickel(II) complexes [Ni^{II}(^{R}L^{-})$_2$] ($\mathbf{9^R}$) exhibited a reversible redox couple due to one-electron oxidation of the complex in CH_2Cl_2, which shifted negative direction as the electron-donating nature of R increases; $E^{ox}{}_{1/2} = 0.90$ V (NO_2) > 0.75 V (CN) > 0.27 V (Br) > 0.08 V (H) > –0.05 V (Me) vs. Fc/Fc$^+$. The nickel(II) complex **8** (see Scheme 42.8) was reported to exhibit $E^{ox}{}_{1/2}$ at –0.33 V vs. Fc/Fc$^+$, which is significantly more negative as compared to those of $\mathbf{9^R}$. This is due to electron donation by the four methyl substituents of the ligand in **8** (Scheme 42.8) [19]. Then, the one-electron oxidized complexes $\mathbf{9}^{\mathbf{R}+\bullet}$ were generated by chemical oxidation using $AgSbF_6$ or [Ru^{III}(bpy)$_3$](PF$_6$)$_3$ (bpy = 2,2′-bipyridine) as the oxidant.

Figure 42.2 shows UV–vis spectral changes for the redox titration of $\mathbf{9^H}$ with $AgSbF_6$ as a typical example [20]. The absorption band at 395 nm ($\varepsilon = 35{,}400$ M^{-1} cm^{-1}) due to the neutral starting complex decreases with concomitant increase in the absorption bands at 355 ($\varepsilon = 35{,}300$ M^{-1} cm^{-1}) and 620 nm ($\varepsilon = 13{,}300$ M^{-1} cm^{-1}) with clear isosbestic points at 373 and 450 nm. The spectral change completed, when an equimolar amount of $AgSbF_6$ was added,

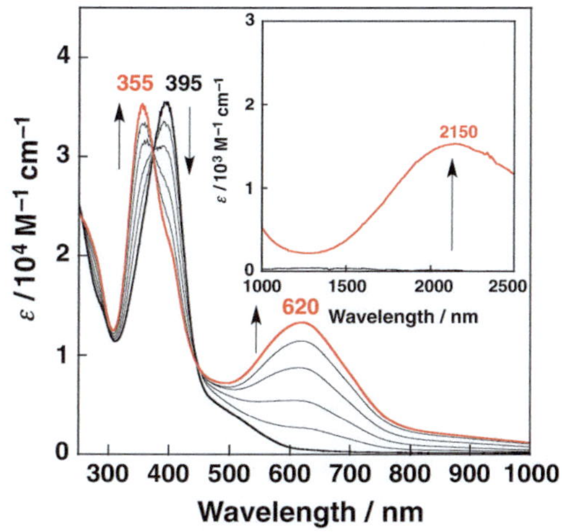

Fig. 42.2 UV–vis–NIR spectral change for the titration of $\mathbf{9^H}$ (5.9×10^{-5} M) with $AgSbF_6$ ($0–8.3 \times 10^{-5}$ M) in CH_2Cl_2 at 30 °C. Inset: expanded spectra in NIR region (Reprinted with permission from Inorg Chem 53:6159–6169. Copyright (2014) American Chemical Society)

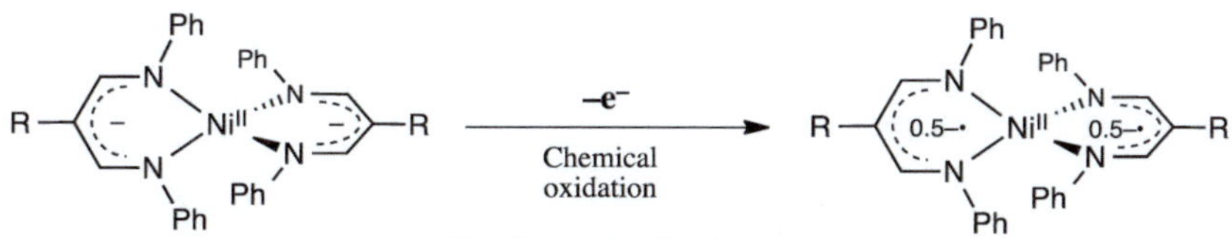

9R 9R+•

Scheme 42.9 Oxidation of a series of bis(β-diketiminato)nickel(II) complexes to form class III mixed-valence complexes

clearly demonstrating that the process was one-electron oxidation of $\mathbf{9^H}$, giving $\mathbf{9^{H+\bullet}}$. The oxidized complex also showed a broad and relatively intense absorption band in near-IR region (centered at $\lambda_{max} = 2{,}150$ nm, $\varepsilon = 1{,}530$ M^{-1} cm^{-1}) attributable to a ligand-to-ligand intervalence charge-transfer (LLIVCT) transition (inset of Fig. 42.2) [20]. Since such an LLIVCT band in the longer wavelength region (above 2000 nm) was not reported for $\mathbf{8^{+\bullet}}$ [19], this is the first example of the LLIVCT band directly observed in the β-diketiminate complexes. The result suggests that the one-electron oxidation occurred at the ligand moiety to give a nickel(II) organic radical complex. The one-electron oxidation of other complexes also gave a similar absorption band in the NIR region: $\mathbf{9^{Me+\bullet}}$, 2260 nm; $\mathbf{9^{Br+\bullet}}$, 2300 nm; $\mathbf{9^{CN+\bullet}}$, 2180 nm; and $\mathbf{9^{NO2+\bullet}}$, 2130 nm [20]. Detailed inspection of the NIR absorption band using Marcus–Hush theory [21, 22] together with DFT calculations suggested that the oxidized complexes $\mathbf{9^{R+\bullet}}$ are class III mixed-valence compounds according to the Robin and Day classification [23], where the generated unpaired electron is completely delocalized between the two ligands (Scheme 42.9) [20].

$\mathbf{9^{R+\bullet}}$ exhibited EPR spectra due to doublet species ($S = 1/2$) [20]. Figure 42.3 shows the EPR spectrum of $\mathbf{9^{H+\bullet}}$ as a typical example. Thus, the spin on the radical ligand ($S_L = 1/2$) strongly antiferromagnetically coupled to a high-spin Ni^{II} ion

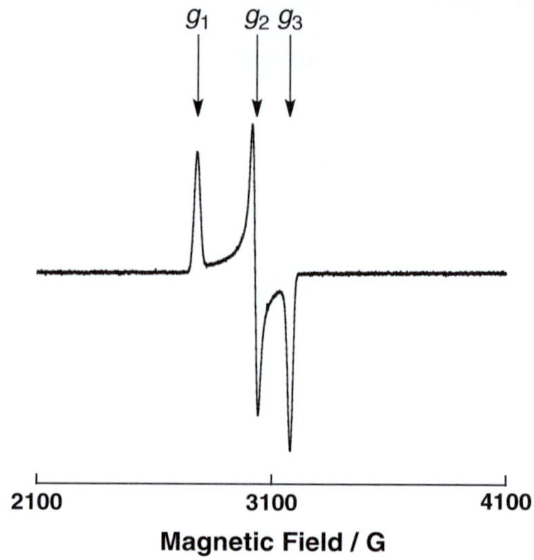

Fig. 42.3 EPR spectrum of $\mathbf{9^{H+\bullet}}$ in CH_2Cl_2/CH_3CN ($v:v=3:1$) at –196 °C. $g_1=2.301$, $g_2=2.134$, and $g_3=2.027$ [20] (Reprinted with permission from Inorg Chem 53:6159–6169. Copyright (2014) American Chemical Society)

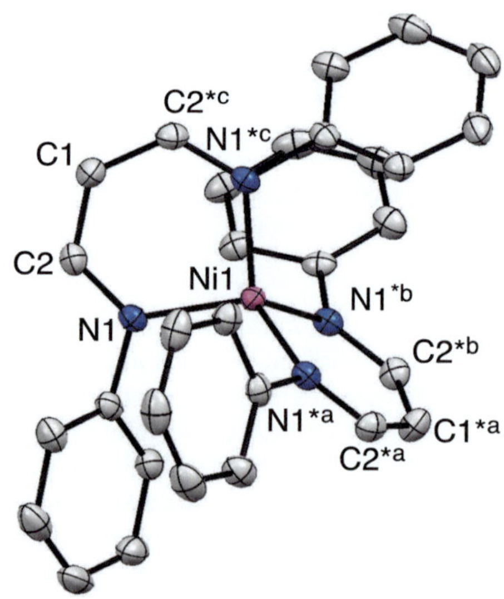

Fig. 42.4 ORTEP drawing of $\mathbf{9^{H+\bullet}}$ showing 50 % probability thermal ellipsoids. Hydrogen atoms are omitted for clarity (Reprinted with permission from Inorg Chem 53:6159–6169. Copyright (2014) American Chemical Society)

($S_{Ni}=1$) to give an overall doublet ground state ($S_T=1/2$) as in the case of $\mathbf{8^{+\bullet}}$ [19, 20].

A single crystal structure of one of the cationic complexes ($\mathbf{9^{H+\bullet}}$) was successfully determined as shown in Fig. 42.4 [20]. The crystal structure exhibited a centrosymmetry, indicating that all the Ni–N bonds and the four C–N–Ph moieties are structurally equivalent. This is consistent with the assignment as the class III mixed-valence complex as mentioned above.

The Ni–N distances (1.8914 Å) of $\mathbf{9^{H+\bullet}}$ become shorter as compared to those of the neutral starting complex $\mathbf{9^{H}}$ (averaged value, 1.947 Å), whereas the twist

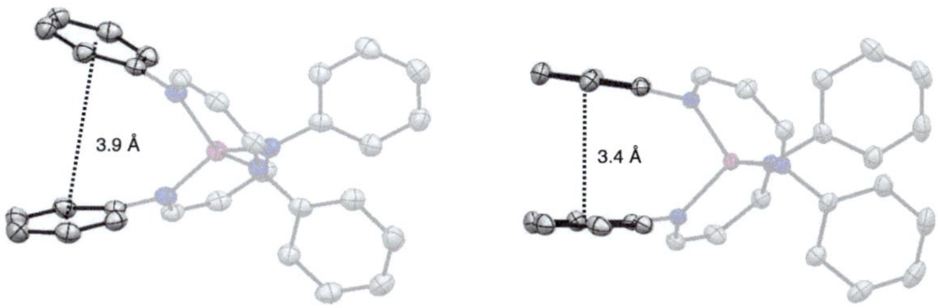

Fig. 42.5 Comparison of the steric relation between the two phenyl substituents in the crystal structures of $\mathbf{9^H}$ (*left*) and $\mathbf{9^{H+\bullet}}$ (*right*) (Reprinted with permission from Inorg Chem 53:6159–6169. Copyright (2014) American Chemical Society)

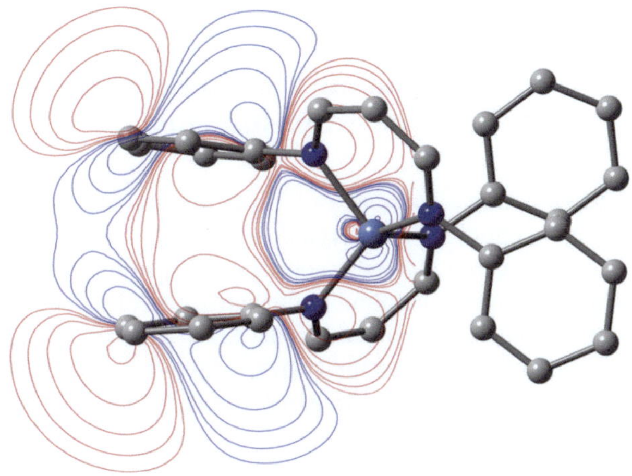

Fig. 42.6 Calculated contour map of electron density of SOMO of $\mathbf{9^{H+\bullet}}$ (Reprinted with permission from Inorg Chem 53:6159–6169. Copyright (2014) American Chemical Society)

angle θ (73.09°) of the former is larger than that of the latter (70.66°) [20]. These structural features around the metal center of the one-electron oxidized complex are consistent with those predicted by the DFT calculations [19, 20]. Interestingly, π–$\pi^{\bullet+}$ interaction was found between the two phenyl groups attached on the nitrogen atoms of different β-diketiminate ligands (Fig. 42.5) [20]. The distance between the two phenyl groups gets closer (3.4 Å) to have a typical value for such interaction, and their alignment is more paralleled as compared to that in the corresponding neutral complex (Fig. 42.5) [24–28]. The higher thermal stability of $\mathbf{9^{H+\bullet}}$ as compared to $\mathbf{8^{+\bullet}}$ can be attributed to such an intramolecular π–$\pi^{\bullet+}$ stacking interaction. The DFT calculations also suggested the presence of such an interaction in the one-electron oxidized complex (Fig. 42.6). The bond lengths between the nickel ion and the nitrogen atoms get slightly shorter (0.056 Å) despite the fact that the electron density of the β-diketiminate ligands decreases. This may be attributed to the formation of bonding interaction between the unpaired electrons on the nickel ion and the ligand radical.

Appearance of the π–π^* and the LLIVCT bands in the electronic absorption spectra as well as the EPR spectra due to $S = 1/2$ species clearly demonstrates that the one-electron oxidation of all other complexes (R = Me, Br, CN, and NO_2) occurs at the supporting ligand, affording similar type complexes in the class III mixed-valence category $\mathbf{9}^{\mathbf{R}+\bullet}$ [20]. Similar mixed-valence *copper(II)* complexes were generated using the same series of β-diketiminate ligands (unpublished results).

One-electron reduction of the neutral complexes with decamethylcobaltocene gave the anionic complexes, when the ligand has the electron-withdrawing substituent (R = CN, NO_2, Br) [20]. The generated anionic complexes exhibited EPR spectra due to a doublet species ($S = 1/2$), but showed no LLIVCT band in the near-IR region. Thus, the reduced complexes are best described as the d^9 nickel(I) complexes supported by two anionic β-diketiminate ligands, $[Ni^{I}(^{R}L^{-})_2]^{-}$ ($\mathbf{9}^{\mathbf{R}-}$). This conclusion was also supported by DFT calculations [20].

42.4 β-Diketiminate Ligand Carrying Additional Redox-Active Substituents

Transition-metal complexes with redox non-innocent ligands have attracted much recent attention as an inorganic–organic hybrid catalyst, where both the metal ion and the organic molecule cooperatively participate to the redox reactions [29]. A typical examples is found in the active site of galactose oxidase, where copper(II)–tyrosyl radical **A** (oxidized form) is involved as the active oxidant for the oxidation of primary alcohols to the corresponding aldehydes with concomitant reduction of molecular oxygen to hydrogen peroxide (Scheme 42.10, Eq. 42.1) [30, 31]:

$$RCH_2OH + O_2 \rightarrow RCHO + H_2O_2 \tag{42.1}$$

In this catalytic reaction, proton migration from the alcohol substrate to the Tyr495 at the axial position takes place to generate a copper(II)–alkoxide intermediate **B** (Scheme 42.10). Then, hydrogen-atom abstraction by the phenoxyl radical species of Tyr272, which is covalently bound to Cys228, occurs to give ketyl radical intermediate **C** [32–34]. In the final step, electron transfer from the ketyl radical intermediate to Cu^{II} proceeds to give aldehyde product and the reduced form of the enzyme **D**, which is converted to the oxidized form **A** by the reaction with O_2, generating H_2O_2. The order of the hydrogen-atom abstraction from the alkoxide by the phenoxyl radical and the electron transfer from the ketyl radical intermediate to Cu^{II} could be reverse. The covalent bonding between Tyr272 and Cys228 is thought to stabilize the phenoxyl radical intermediate [35]. One of the unique features of the active site of galactose oxidase is the existence of two phenol moieties (Tyr272 and Tyr495) bridged by the copper ion (Scheme 42.10). Thus, the oxidized form exhibits a strong absorption band in the near-IR region probably due to the ligand-to-ligand intervalence charge-transfer (LLIVCT) band [31].

In order to mimic such an interesting structural feature of galactose oxidase, we have examined redox behavior of a β-diketiminate ligand carrying two redox-active

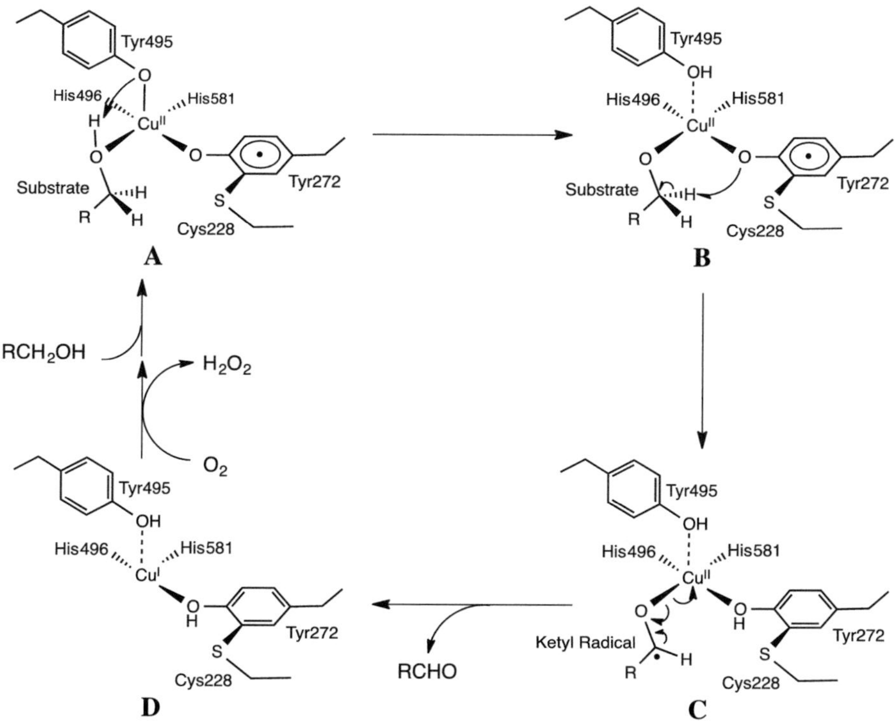

Scheme 42.10 Reaction mechanism of alcohol oxidation occurring on the active site of galactose oxidase

Chart 42.2 Schematic structure of LH3

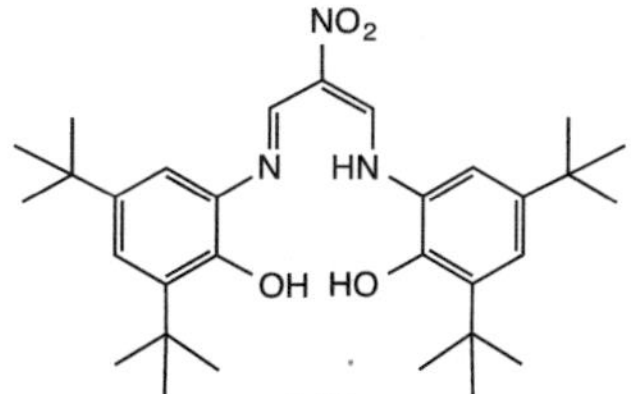

LH_3

phenolate substituents (full protonated ligand LH_3 is shown in Chart 42.2) [36]. Removal of all the proton of LH_3 gives the trianionic ligand L^{3-}, which may also stabilize the higher oxidation states of the supported metal complexes.

The anionic copper(II) complex **10**$^-$ was obtained by treating ligand **LH_3** with an equimolar amount of $Cu(OAc)_2$•H_2O in the presence of three equivalent of tetraethylammonium hydroxide (Et_4NOH) as a base. Figure 42.7a shows the crystal structure of **10**$^-$. The Cu–O (1.901(3) and 1.902(2) Å) and the Cu–N (1.933(2) and 1.931(3) Å) lengths are slightly longer that those of Cu^{II}–salen complex [$Cu^{II}(Sal^{2-})$] (Sal^{2-} = diphenolate form of $SalH_2$), *N,N′*-bis(3,5-di-*tert*-

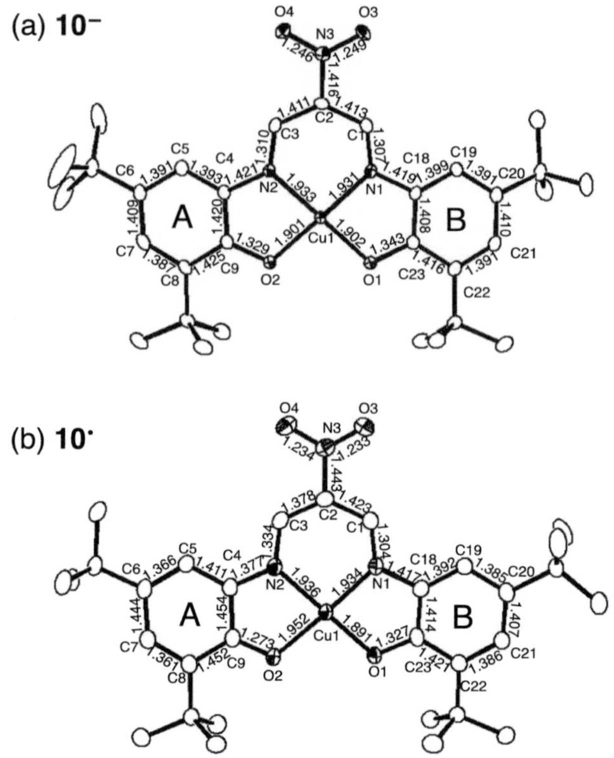

Fig. 42.7 ORTEP drawings of (**a**) $(Et_4N)[Cu^{II}(\mathbf{L}^{3-})]CH_2Cl_2$ (**10**$^-$) and (**b**) $Cu^{II}(\mathbf{L}^{\bullet 2-})$ (**10**$^\bullet$) showing 50 % probability thermal ellipsoids [36] (Reproduced by permission of The Royal Society of Chemistry)

butylsalicylidene)-1,2-cyclohexane-(1*R*,2*R*)-diamine (Cu–O, 1.886 and 1.887 Å; Cu–N, 1.904 and 1.915 Å) [37].

Complex **10**$^-$ exhibited two reversible redox couples at $E_{1/2} = -0.06$ and +0.46 V vs. Fc^+/Fc in CH_2Cl_2 containing 0.1 M $^n Bu_4NPF_6$. These redox couples are significantly negative as compared to those of $[Cu^{II}(Sal^{2-})]$, +0.45 and +0.65 V vs. Fc^+/Fc in CH_2Cl_2 [38]. Such a negative shift of the oxidation potentials of **10**$^-$ is apparently due to the electron-donating nature of the anionic β-diketiminato moiety.

The one-electron oxidation product **10**$^\bullet$ can be prepared by the chemical oxidation using an equimolar amount of $AgSbF_6$ in CH_2Cl_2, and its crystal structure is also shown in Fig. 42.7b. The crustal structure clearly indicated that one-electron oxidation of **10**$^-$ resulted in a distinctive structural change in the aminophenol moiety of ring A of the ligand (Fig. 42.7). Namely, structures of the aminophenol moiety of rings A and B are nearly equivalent in **10**$^-$, whereas contribution of a quinonoid canonical form is increased in the aminophenol moiety of ring A in the oxidized complex **10**$^\bullet$. In addition, the distance of Cu1–O2 increases significantly

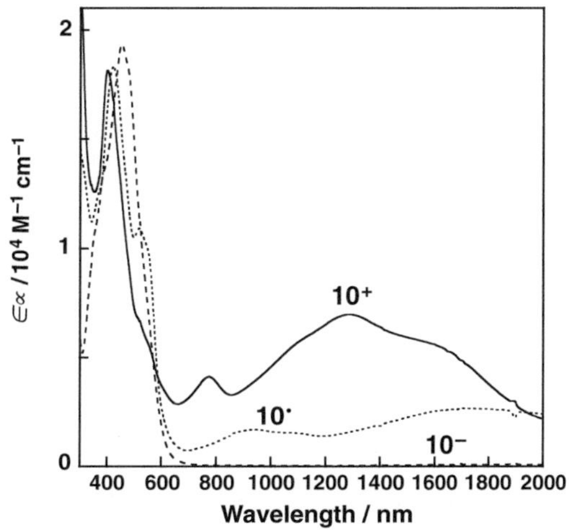

Fig. 42.8 Absorption spectra of $[Cu^{II}(\mathbf{L}^{3-})]^-$ ($\mathbf{10}^-$) (*dashed line*), $[Cu^{II}(\mathbf{L}^{\bullet 2-})]$ ($\mathbf{10}^\bullet$) (*dotted line*), and $[Cu^{II}(\mathbf{L}^-)]^+$ ($\mathbf{10}^+$) (*solid line*) in CH_2Cl_2 at 30 °C [36] (Reproduced by permission of The Royal Society of Chemistry)

as compared with that of Cu1–O1. Based on these results, the oxidation product can be best described as the Cu^{II}–phenoxyl radial complex $[Cu^{II}(\mathbf{L}^{\bullet 2-})]$, in which one-electron oxidation takes place at the ligand to give $\mathbf{L}^{\bullet 2-}$ rather than at the metal.

In Fig. 42.8, the absorption spectrum of the radical complex $\mathbf{10}^\bullet$ is shown (dotted line) together with that of the starting material $\mathbf{10}^-$ (dashed line). There is an intense absorption band at 423 nm ($\varepsilon = 18{,}400\ M^{-1}\ cm^{-1}$) together with a shoulder around 518 nm ($10{,}900\ M^{-1}\ cm^{-1}$) and very broad absorption bands at 939 nm ($1{,}630\ M^{-1}\ cm^{-1}$) and 1767 nm ($2{,}570\ M^{-1}\ cm^{-1}$) in the near-IR region.

The radical complex $\mathbf{10}^\bullet$ was silent in the X-band EPR, whereas it exhibited ^1H-NMR signals in the paramagnetic region. The magnetic susceptibility measurement unambiguously indicated that the copper(II)–radical complex has a triplet ground state ($S = 1$). DFT calculation study has suggested that the α spin on SOMO-155, which is mainly consisted of π-orbitals delocalized on the aminophenol rings and the β-diketiminate ligand framework, and another α spin on SOMO-153, which is mainly developed on the d_{x2-y2} orbital of copper (orthogonal to SOMO-155), are ferromagnetically coupled (Fig. 42.9). Time-dependent DFT (TD-DFT) calculations have suggested that the near-IR absorption bands are attributable to ligand-to-ligand charge-transfer (LLCT) transitions and the intense visible absorption bands are assignable to the charge-transfer transitions from the ligand to Cu^{II} (LMCT) as well as π–π* transitions of the ligand.

Further one-electron oxidation of $\mathbf{10}^\bullet$ by $AgSbF_6$ (1 equiv) in CH_2Cl_2 gave two-electron oxidized complex $[Cu^{II}(\mathbf{L}^-)]^+$ ($\mathbf{10}^+$) exhibiting an intense absorption band at 400 nm ($18{,}200\ M^{-1}\ cm^{-1}$) together with broad bands at 774 nm ($4{,}130\ M^{-1}\ cm^{-1}$) and 1288 nm ($6{,}970\ M^{-1}\ cm^{-1}$) (solid line spectrum in Fig. 42.8). This compound shows a typical EPR signals ($g_{//} = 2.222$, $g_\perp = 2.059$, $A_{//} = 188$ G) attributable to a copper(II) ion having tetragonal geometry, thus indicating that the second oxidation also took place at the ligand. TD-DFT calculations suggest that the NIR absorption bands also arise from the delocalized π-orbitals

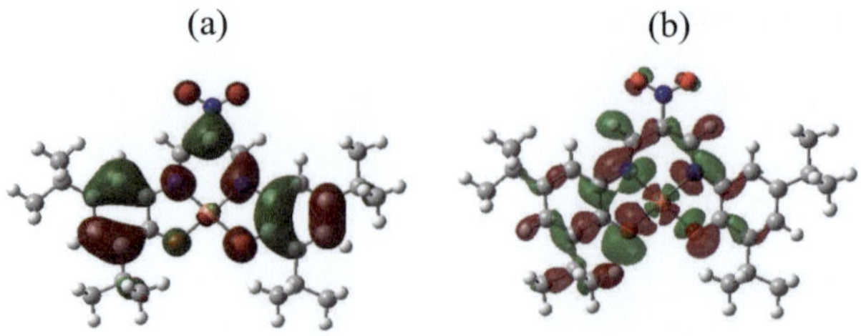

Fig. 42.9 SOMOs [(**a**)155a and (**b**) 153a] of **10**$^{\bullet}$ calculated by DFT using UCAM-B3LYP/6-311G(d) basis set with the triplet configurations [36] (Reproduced by permission of The Royal Society of Chemistry)

on the monoanionic ligand framework including the two aminophenol rings and β-diketiminate ligand framework and that the bands around 400 nm are attributable to metal-to-ligand charge-transfer (MLCT) transitions. The DFT calculations have also suggested that both the aminophenol rings take a quinonoid canonical form in the two-electron oxidized ligand $\mathbf{L}^{-}$ [39].

The reactivity of two-electron oxidized complex $\mathbf{10}^{+}$ in the C–H bond activation reaction of 1,4-cyclohexadiene is much higher than that of the one-electron oxidized complex **10**$^{\bullet}$. On the other hand, both complexes did not induce alcohol oxidation in contrast to the case of $[Cu^{II}(Sal^{2-})]$ exhibiting reactivity toward benzyl alcohol oxidation [38, 40].

42.5 Concluding Remarks

The monoanionic didentate ligands, β-diketiminates, have been adopted for the synthesis of the wide variety of metal complexes with various oxidation states throughout the periodic table. In most cases, much attention has been paid to the Lewis acidity and the redox behavior of the metal center, but little attention has been focused on the redox non-innocence of the β-diketiminate supporting ligands. However, recent studies introduced in this chapter clearly demonstrated that the redox function of the β-diketiminate ligands should also be considered in the redox functions of β-diketiminato complexes.

References

1. Bourget-Merle L, Lappert MF, Severn JR (2002) The chemistry of β-diketiminatometal complexes. Chem Rev 102:3031–3066
2. Mindiola DJ (2006) Oxidatively induced abstraction reactions. A synthetic approach to low-coordinate and reactive early transition metal complexes containing metal–ligand multiple bonds. Acc Chem Res 39:813–821

3. Wu J, Yu T-L, Chen C-T, Lin C-C (2006) Recent developments in main group metal complexes catalyzed/initiated polymerization of lactides and related cyclic esters. Coord Chem Rev 250:602–626
4. Cramer CJ, Tolman WB (2007) Mononuclear Cu–O2 Complexes: geometries, spectroscopic properties, electronic structures, and reactivity. Acc Chem Res 40:601–608
5. Holland PL (2008) Electronic structure and reactivity of three-coordinate iron complexes. Acc Chem Res 41:905–914
6. Tsai Y-C (2012) The chemistry of univalent metal β-diketiminates. Coord Chem Rev 256:722–758
7. Roesky HW, Singh S, Jancik V, Chandrasekhar V (2004) A paradigm change in assembling oh functionalities on metal centers. Acc Chem Res 37:969–981
8. Piers WE, Emslie DJH (2002) Non-cyclopentadienyl ancillaries in organogroup 3 metal chemistry: a fine balance in ligand design. Coord Chem Rev 233–234:131–155
9. Avent AG, Khvostov AV, Hitchcock PB, Lappert MF (2002) Unusual crystalline heterobimetallic trinuclear β-diketiminates [Yb(L(μ-Li(thf))2)] and [Yb(L′(μ-Li(thf))2)]•thf [L, L′ = {N(SiMe3)C(R)}2CH, R = Ph, C6H4Ph-4]. Chem Commun:1410–1411
10. Eisenstein O, Hitchcock PB, Khvostov AV, Lappert MF, Maron L, Perrin L, Protchenko AV (2003) Mono-, di-, and trianionic β-diketiminato ligands: a computational study and the synthesis and structure of [(YbL)3(THF)], L = [{N(SiMe3)C(Ph)}2CH]. J Am Chem Soc 125:10790–10791
11. Avent AG, Hitchcock PB, Khvostov AV, Lappert MF, Protchenko AV (2004) Reactions of Li- and Yb-coordinated N,N′-bis(trimethylsilyl)-β-diketiminates: one- and two-electron reductions, deprotonation, and C–N bond cleavage. Dalton Trans:2272–2280
12. Ibrahim SK, Khvostov AV, Lappert MF, Maron L, Perrin L, Pickett CJ, Protchenko AV (2006) An electrochemical and DFT study on selected β-diketiminato metal complexes. Dalton Trans: 2591–2596. doi:10.1039/B513378E:2591-2596
13. Power PP (2003) Persistent and stable radicals of the heavier main group elements and related species. Chem Rev 103:789–810
14. McKenzie I, Percival PW, Clyburne JAC (2005) A computational study of the reactions of a β-diketiminatoaluminium(I) complex with the hydrogen atom and the electron. Chem Commun 1134–1136. doi:10.1039/B414665D:1134-1136
15. Moilanen J, Borau-Garcia J, Roesler R, Tuononen HM (2012) Paramagnetic aluminium β-diketiminate. Chem Commun 48:8949–8951
16. Arrowsmith M, Hill MS, Kociok-Köhn G, MacDougall DJ, Mahon MF, Mallov I (2012) Three-coordinate beryllium β-diketiminates: synthesis and reduction chemistry. Inorg Chem 51:13408–13418
17. Chang M-C, Dann T, Day DP, Lutz M, Wildgoose GG, Otten E (2014) The formazanate ligand as an electron reservoir: Bis(Formazanate) zinc complexes isolated in three redox states. Angew Chem Int Ed 53:4118–4122
18. Hong SJ, Hill LMR, Gupta AK, Naab BD, Gilroy JB, Hicks RG, Cramer CJ, Tolman WB (2009) Effects of electron-deficient β-diketiminate and Formazan supporting ligands on copper(i)-mediated dioxygen activation. Inorg Chem 48:4514–4523
19. Khusniyarov MM, Bill E, Weyhermüller T, Bothe E, Wieghardt K (2011) Hidden noninnocence: theoretical and experimental evidence for redox activity of a β-diketiminate(1−) ligand. Angew Chem Int Ed 50:1652–1655
20. Takaichi J, Morimoto Y, Ohkubo K, Shimokawa C, Hojo T, Mori S, Asahara H, Sugimoto H, Fujieda N, Nishiwaki N, Fukuzumi S, Itoh S (2014) Redox chemistry of nickel(II) complexes supported by a series of noninnocent β-diketiminate ligands. Inorg Chem 53:6159–6169
21. Hush NS (1968) Electron transfer at electrodes and in solution: comparison of theory and experiment. Electrochim Acta 13:1005
22. Hush NS (2007) Progress in inorganic chemistry. Wiley, Hoboken NJ, pp 391–444
23. Robin MB, Day P (1968) In: Emeléus HJ, Sharpe AG (eds) Advances in inorganic chemistry and radiochemistry. Academic Press, Waltham, MA, pp 247–422

24. Takai A, Yasuda T, Ishizuka T, Kojima T, Takeuchi M (2013) A directly linked ferrocene–naphthalenediimide conjugate: precise control of stacking structures of π-systems by redox stimuli. Angew Chem Int Ed 52:9167–9171
25. Small D, Zaitsev V, Jung Y, Rosokha SV, Head-Gordon M, Kochi JK (2004) Intermolecular π-to-π bonding between stacked aromatic dyads. Experimental and theoretical binding energies and near-IR optical transitions for phenalenyl radical/radical versus radical/cation dimerizations. J Am Chem Soc 126:13850–13858
26. Sun DL, Rosokha SV, Lindeman SV, Kochi JK (2003) Intervalence (charge-resonance) transitions in organic mixed-valence systems. through-space versus through-bond electron transfer between bridged aromatic (redox) centers. J Am Chem Soc 125:15950–15963
27. Li M, Neal TJ, Wyllie GRA, Oliver AG, Schulz CE, Scheidt WR (2011) Metalloporphyrin mixed-valence π-cation radicals: [Fe(oxoOEC•/2)(Cl)]2SbCl6, structure, magnetic properties, and near-IR spectra. Inorg Chem 50:9114–9121
28. Herwig PT, Enkelmann V, Schmelz O, Müllen K (2000) Synthesis and structural characterization of hexa-tert-butyl- hexa-peri-hexabenzocoronene, its radical cation salt and its tricarbonylchromium complex. Chem Eur J 6:1834–1839
29. Chirik PJ, Wieghardt K (2010) Radical ligands confer nobility on base-metal catalysts. Science 327:794–795
30. Whittaker JW (2005) The radical chemistry of galactose oxidase. Arch Biochem Biophys 433:227–239
31. Whittaker JW (2003) Free radical catalysis by galactose oxidase. Chem Rev 103:2347–2364
32. Itoh S, Taki M, Fukuzumi S (2000) Active site models for galactose oxidase and related enzymes. Coord Chem Rev 198:3–20
33. Jazdzewski BA, Tolman WB (2000) Understanding the copper-phenoxyl radical array in galactose oxidase: contributions from synthetic modeling studies. Coord Chem Rev 200:633–685
34. Lyons CT, Stack TDP (2013) Recent advances in phenoxyl radical complexes of salen-type ligands as mixed-valence galactose oxidase models. Coord Chem Rev 257:528–540
35. Itoh S, Taki M, Takayama S, Nagatomo S, Kitagawa T, Sakurada N, Arakawa R, Fukuzumi S (1999) Oxidation of benzyl alcohol with CuII and ZnII complexes of the phenoxyl radical as a model of the reaction of galactose oxidase. Angew Chem Int Ed 38:2774–2776
36. Takaichi J, Ohkubo K, Sugimoto H, Nakano M, Usa D, Maekawa H, Fujieda N, Nishiwaki N, Seki S, Fukuzumi S, Itoh S (2013) Copper complexes of the non-innocent β-diketiminate ligand containing phenol groups. Dalton Trans 42:2438–2444
37. Bunce S, Cross RJ, Farrugia LJ, Kunchandy S, Meason LL, Muir KW, O'Donnell M, Peacock RD, Stirling D, Teat SJ (1998) Chiral Schiff base complexes of copper (II), vanadium (IV) and nickel (II) as oxidation catalysts. X-ray crystal structures of [Cu (R-salpn) (OH2)] and [Cu (±-busalcx)]. Polyhedron 17:4179–4187
38. Pratt RC, Stack TDP (2003) Intramolecular charge transfer and biomimetic reaction kinetics in galactose oxidase model complexes. J Am Chem Soc 125:8716–8717
39. The DFT calculations using UCAM-B3LYP/6-311G(d) basis set well reproduced the structures of anionic and neutral forms of copper(II) complexes
40. Pratt RC, Stack TDP (2005) Mechanistic insights from reactions between copper(II)—phenoxyl complexes ad substrates with activated C—H bonds. Inorg Chem 44:2367–2375

Chapter 43
Novel Functions of π-Electron Systems in a Heme-DNA Complex

Yasuhiko Yamamoto and Tomokazu Shibata

Abstract While DNA is largely double helical, guanine-rich sequences can exist in an alternative structural form known as the G-quadruplex. The G-quadruplex is stabilized by the formation of the G-quartet composed of four guanine bases that are cyclically associated through Hoogsteen hydrogen bonding. The size and planarity of the G-quartet are well suited for interaction with heme, the most ubiquitous cofactor found in nature, through π-π stacking, and hemes and various G-quadruplex DNAs have been reported to form stable heme-DNA complexes that exhibit spectroscopic and functional properties remarkably similar to those of hemoproteins. These findings paved the way to a new research field involving the exploitation of the heme reactivity in the scaffold of the DNA structure. The heme electronic structure can be indirectly tuned through DNA sequence alterations that change the interaction between the heme and the surrounding DNA moiety and be directly affected by chemical modification of heme peripheral side chains. Therefore, the combined use of these two approaches enables us to control the function of a heme-DNA complex, as the function of hemoproteins is controlled through the heme environment furnished by nearby amino acid residues and electronic tuning of the intrinsic heme reactivity.

Keywords Heme • G-quadruplex DNA • π-π stacking

43.1 Introduction

Metalloporphyrins are quite frequently found as the prosthetic groups of a variety of hemoproteins exhibiting diverse biological functions. In hemoproteins, the electronic nature of the metalloporphyrin is finely tuned through interaction with the surrounding protein moiety to allow a specific biological function. Despite the ubiquitous occurrence of various hemoproteins in nature, no functional role

Y. Yamamoto (✉) • T. Shibata
Department of Chemistry, University of Tsukuba, Tsukuba 305-8571, Japan
e-mail: yamamoto@chem.tsukuba.ac.jp

T. Akasaka et al. (eds.), *Chemical Science of π-Electron Systems*,
DOI 10.1007/978-4-431-55357-1_43

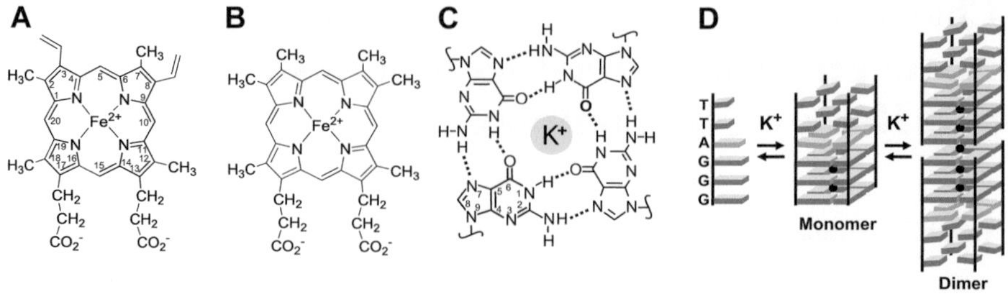

Fig. 43.1 Molecular structures of heme(Fe^{2+}) (**A**), iron(II) 3,8-dimethyl deuteroporphyrin complex (DMheme(Fe^{2+})) (**B**), and the G-quartet (**C**). The numbering systems for the heme and guanine bases are indicated with the structures. Schematic representation of the formation of parallel G-quadruplex DNA from DNA sequence d(TTAGGG) and the dimerization of the G-quadruplex DNA (**D**)

in vivo has yet been attributed to DNAs or RNAs possessing metalloporphyrins as prosthetic groups [1]. DNA aptamers capable of recognizing heme (Fe(II)- or Fe(III)-protoporphyrin IX complex (heme(Fe^{2+}) or heme(Fe^{3+}), respectively; Fig. 43.1A)) have been obtained by means of in vitro selection techniques [1], and the obtained aptamers exhibit remarkably high peroxidase activity [2–6]. Furthermore, Poon et al. [7] reported that the heme bound to the G-quartet of G-quadruplex DNAs and RNAs exhibits catalytic activity in oxygen transfer reactions. These findings implied that primordial ribozymes might have used prosthetic groups to increase their functional diversity, contemporary hemoproteins thereby having evolved [7]. Studies on the functional and structural properties of hemes in G-quadruplex DNAs (or RNAs) are expected to lead to a deeper understanding of the biological versatility of the heme π-electron system.

The G-quartet is formed from four guanine bases that are cyclically associated through Hoogsteen hydrogen bonding (Fig. 43.1C) [8, 9]. The size and planarity of the G-quartet are well suited for interaction with hemes through π-π stacking [10]. Characterization of the molecular recognition between hemes and the G-quartet is expected to provide knowledge for not only designing a stable complex between a heme and a DNA, but also for tuning of the heme reactivity in the scaffold of the DNA structure. The solution structure of a single repeat sequence of the human telomere, d(TTAGGG), has been shown to form a parallel G-quadruplex DNA, i.e., $(d(TTAGGG))_4$, in the presence of an appropriate K^+ concentration ($[K^+]$) [11, 12], which further assembles into a "dimer" through an intermolecular π-π stacking interaction between the 3′-terminal G-quartets [12] (Fig. 43.1D). We found that heme(Fe^{3+}) and $(d(TTAGGG))_4$ form a stable complex, i.e., the heme(Fe^{3+})-DNA complex [13–18]. Interestingly, the heme-DNA complex exhibited various spectroscopic and functional properties remarkably similar to those of hemoproteins such as myoglobin (Mb) [13–18]. Thus, the heme(Fe^{3+})-DNA complex appeared to possess the structural essentials of the "heme-deoxyribozymes" that mimic hemoproteins in function.

43.2 π-Electron Systems in G-Quadruplex DNAs

DNA is a linear polymer of four distinct nucleotides of which the phosphates bridge the 3′ and 5′ positions of successive sugar residues. As illustrated in Fig. 43.2a, the nucleotides consist of heteroatomic rings called bases, i.e., adenine (A), guanine (G), cytosine (C), and thymine (T); a sugar moiety, i.e., deoxyribose; and a phosphate group. Two polynucleotide strands wind around a common axis with a right-handed twist to form a double helix structure with a diameter of 2 nm. In the double helix structure, the two strands are antiparallel and wrap around each other [19] (Fig. 43.2b).

The bases are accommodated inside the double helix composed of sugar-phosphate chains. The planes of the bases are nearly perpendicular to the helix axis. Each base in the double helix is hydrogen bonded to a base on the opposite strand to form a base pair and only two types of base pairs known as Watson-Crick base pairs [19], i.e., A-T and G-C ones (Fig. 43.2c), can be formed in the DNA double helix, and hence the π-electron systems of the plane base pairs of DNA form extended stacking along the helix axis. The hydrophobic core inside the DNA helix, formed from the extended stacking of the π-electron systems of the plane base pairs, contributes significantly to the stability of the DNA double helix. Although the hydrogen bonds between the bases are required for the formation of Watson-Crick base pairs, and hence the DNA double helix, the hydrogen bonds contribute little to the stability of the DNA helical structure, as is true for proteins and various biopolymers. As described above, in addition to the usual double helix structure, DNA also forms a quadruplex one (Fig. 43.1D). G-quadruplex DNAs can adopt

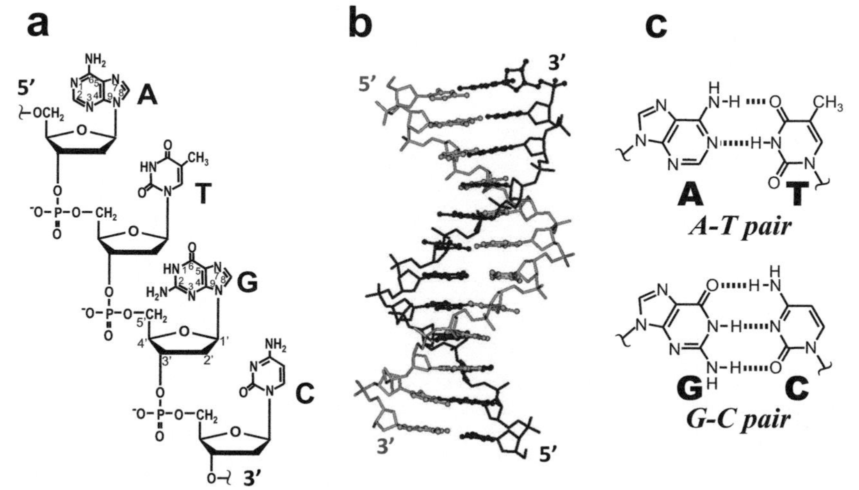

Fig. 43.2 Molecular structure of a portion of an oligonucleotide, i.e., d(ATGC) (**a**). The numbering systems for adenine and guanine bases and the sugar moiety are indicated with the structure. Schematic representation of the DNA duplex (**b**) and Watson-Crick base pairs (**c**)

several morphologies that include parallel or antiparallel configurations through either intramolecular or intermolecular organization [11, 12, 20–24].

43.3 Spectroscopic Properties of Heme-DNA Complexes

Optical studies indicated that heme(Fe^{3+}) and $(d(TTAGGG))_4$ interact to form a heme(Fe^{3+})-DNA complex with a binding constant of $(1.6 \pm 0.2) \times 10^7\ M^{-1}$ (Fig. 43.3). The high stability of the heme(Fe^{3+})-DNA complex was manifested in the large binding constant. The circular dichroism spectrum of the heme(Fe^{3+})-DNA complex exhibited a positive maximum at 260 nm [17], which is characteristic of a parallel-quadruplexed DNA [25], and induced negative intensity at 400 nm

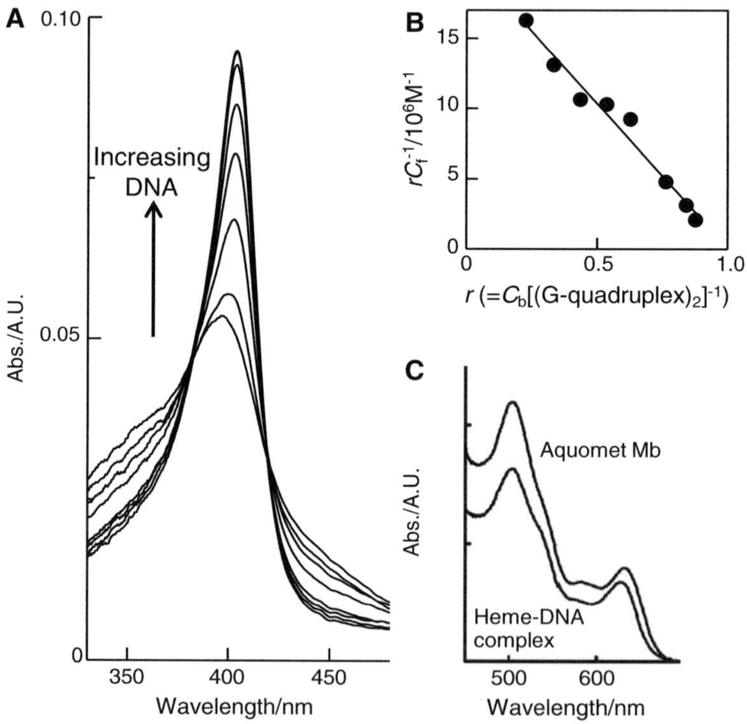

Fig. 43.3 Soret absorption of 1.0 μM heme(Fe^{3+}) in 300 mM KCl and 50 mM phosphate butter at pH 7.00, in the presence of various $(d(TTAGGG))_4$ concentrations, 0–5.0 μM, at room temperature (**A**). 0.08 w/v% TritonX-100 and 0.5 v/v% DMSO were added to the solution mixture to prevent heme(Fe^{3+}) aggregation. Scatchard plots for the absorbance at 404 nm, which yielded the binding constant of $(1.6 \pm 0.2) \times 10^7\ M^{-1}$ and the stoichiometric ratio of about 1:1 between heme(Fe^{3+}) and the quadruplexed DNA (**B**). Portions of the absorption spectra of the heme(Fe^{3+})-DNA complex (bottom) and aquomet Mb (top) in 300 mM KCl and 50 mM phosphate butter at pH 7.00 and room temperature (**C**)

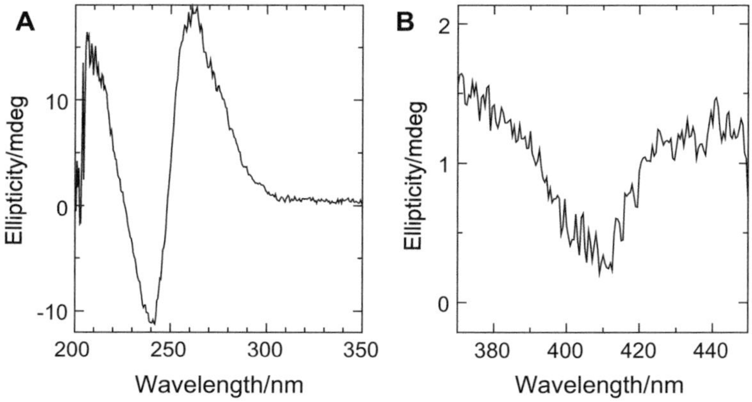

Fig. 43.4 Portions of CD spectra of 5.0 μM $(d(TTAGGG))_4$ in 300 mM KCl, 50 mM phosphate buffer at pH 7.0 and room temperature (**A**), and a mixture of 5.0 μM heme(Fe^{3+}) and 150 μM $(d(TTAGGG))_4$ complex in 300 mM KCl, 50 mM phosphate buffer, pH 7.0, 0.08 w/v% TritonX-100, and 0.5 v/v% DMSO at room temperature (**B**)

[17], which is indicative of a π-π stacking interaction between the porphyrin ring of heme(Fe^{3+}) and the base pairs of DNA [26] (Fig. 43.4). Additionally, as shown in Fig. 43.3C, the absorption spectrum of the heme(Fe^{3+})-DNA complex highly resembled that of aquomet Mb [27], in which heme(Fe^{3+}) is coordinated to a histidyl imidazole and a water molecule as axial ligands, suggesting that the heme(Fe^{3+}) environments in the heme(Fe^{3+})-DNA complex and aquomet Mb are highly alike.

The 600 MHz ^{1}H NMR spectrum of the heme(Fe^{3+})-DNA complex exhibited well-resolved paramagnetically shifted signals arising from heme(Fe^{3+}) side chain protons [14] (Fig. 43.5A). The specific complexation between heme(Fe^{3+}) and the DNA was manifested in the appearance of a single set of paramagnetically shifted signals, which reflected a unique heme(Fe^{3+}) electronic structure in the complex. The Curie plots, shifts vs. reciprocal of absolute temperature, for the resolved signals over the temperature range examined, 278–338 K, gave a straight line (results not shown), indicating that they obey the Curie law. This finding is consistent with the high stability of the heme(Fe^{3+})-DNA complex revealed by the optical studies. The four relatively large peaks at 60–80 ppm in Fig. 43.5A were reasonably assignable to heme(Fe^{3+}) side chain methyl protons, and the shift pattern of the heme(Fe^{3+}) methyl proton signals of the complex resembled that of aquomet Mb in Fig. 43.5A′, also demonstrating that structural features around heme(Fe^{3+}) in the two systems are alike. Furthermore, the low-temperature EPR spectrum of the complex exhibited signals at $g = 2$ and 6 (Fig. 43.5D), which are characteristic of a ferric high spin heme complex with the spin quantum number $S = 5/2$ [28], was similar to that of aquomet Mb at neutral pH. These findings also supported that heme(Fe^{3+}) in the heme(Fe^{3+})-DNA complex is coordinated to H_2O at neutral pH. In the case of a complex between heme(Fe^{3+}) and a DNA aptamer [5], a guanine residue and H_2O have been proposed to be coordinated to heme(Fe^{3+}) as axial ligands.

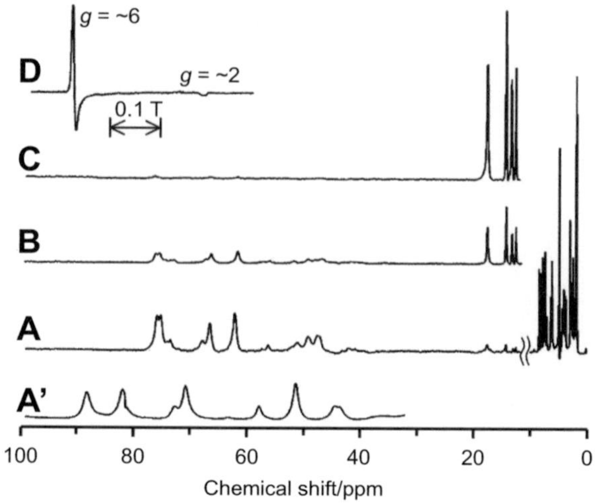

Fig. 43.5 NMR and EPR spectra of the heme(Fe^{3+})-(d(TTAGGG))$_4$ complex. The 600 MHz ^{1}H NMR spectra of the heme(Fe^{3+})-(d(TTAGGG))$_4$ complex in 300 mM KCl in 90 % H_2O/10 % D_2O at 298 K and pH 7.04 (**A**), pH 7.81 (**B**), pH 8.58 (**C**), aquomet Mb at 298 K and pH 7.00 (**A′**), and low-temperature (15 K) EPR spectrum of the heme(Fe^{3+})-(d(TTAGGG))$_4$ complex in 300 mM KCl, 50 mM phosphate buffer at pH 7.0 (**D**). In trace A, the y-gain for the spectral region of 0–10 ppm was reduced by 1/200 relative to that below 10 ppm (Image from Published Paper [14])

43.4 Interaction Between a Heme and the G-Quartet in Heme-DNA Complexes

43.4.1 Heme(Fe^{2+})-DNA Complex

Carbon monoxide (CO) has been shown to bind to a complex between heme(Fe^{2+}) and (d(TTAGGG))$_4$ (heme(Fe^{2+})-DNA complex) to form a CO adduct [16]. Heme(Fe^{2+}) coordinated to CO adopts a low spin d^6 system with $S = 0$, i.e., a diamagnetic form, and hence the CO adduct is better suited for NMR structural characterization than its paramagnetic counterpart, i.e., the heme(Fe^{3+})-DNA complex (see below). NMR structural characterization of the CO adduct revealed that heme(Fe^{2+}) stacks onto the G6 G-quartet of the DNA, and the heme plane is located at 0.40 nm from the G6 G-quartet one, the two orientations differing by 180° rotation about the 5-H-15-H axis with respect to the DNA, i.e., the obverse and reverse heme orientations (see inset in Fig. 43.6) [18].

The ^{1}H NMR spectra of (d(TTAGGG))$_4$ and the CO adduct are compared in Fig. 43.6. The formation of an all-parallel G-quadruplex DNA was manifested in the observation of three exchangeable proton signals, due to guanine imino NH protons, in the chemical shift region characteristic of G-quartets, i.e., below 11 ppm [11, 12] (Fig. 43.6A, A′). The spectrum in Fig. 43.6A is primarily due to the (d(TTAGGG))$_4$

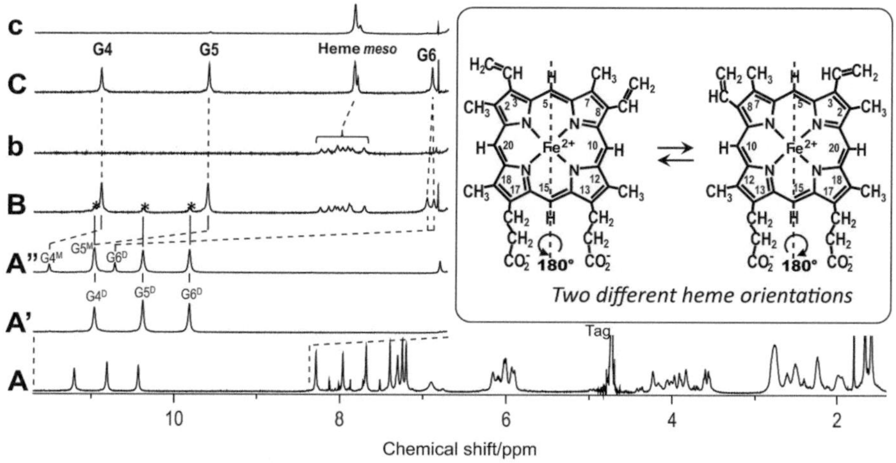

Fig. 43.6 600 MHz ^{1}H NMR spectrum of (d(TTAGGG)$_4$) (DNA) in 90 % H_2O/10 % ^2H_2O (**A**), and downfield portions, 8.33–11.70 ppm, of the spectra of DNA in 90 % H_2O/10 % ^2H_2O (**A′** and **A″**), CO adduct of heme(Fe^{2+})-DNA complex in 90 % H_2O/10 % ^2H_2O (**B**) and in 100 % ^2H_2O (**b**), and CO adduct of DMheme(Fe^{2+})-DNA complex in 90 % H_2O/10 % ^2H_2O (**C**) and in 100 % ^2H_2O (**c**). Sample A″ was prepared in 50 mM KCl, 50 mM potassium phosphate buffer, and pH 7.0, and the others in 300 mM KCl, 50 mM potassium phosphate buffer, pH 7.0. All the spectra were recorded at 25 °C. Gn^D and Gn^M, where $n = 4$, 5, or 6, represent the dimer and monomer of the DNA, respectively (see text). Peaks denoted by * in B are due to the DNA dimer remaining in the sample. The two heme orientations differing by 180° rotation about the 5-H-15-H axis, with respect to the DNA, are illustrated in the inset. These heme orientations interexchange with each other (Image from Published Paper [18])

dimer (Fig. 43.1D), and guanine imino NH proton signals arising from the monomer appeared at low [K^+] (Fig. 43.6A″) [12]. The guanine imino NH proton signals of the CO adduct could be readily identified through the analysis of the effects of the ^2H_2O content in the sample solutions on the spectra (Fig. 43.6 B, b, C, c).

Comparison of the shifts between the corresponding guanine imino NH proton signals of (d(TTAGGG))$_4$ and the CO adduct indicated that these signals, particularly the G6 one, exhibited upfield shifts upon the complexation with heme (Fig. 43.6B). In the absence of an unpaired electron in the heme(Fe^{2+}) of the complex, the upfield shifts observed for the guanine imino NH proton signals are attributed to the ring current effect of the porphyrin moiety of the heme. The guanine imino NH protons are ranked as G4 < G5 < G6, in order of increasing heme ring current-induced shift change ($\Delta\delta_{hisc}$), relative to the corresponding signals of the (d(TTAGGG))$_4$ monomer. These results indicated not only that the heme stacks onto the G6 G-quartet of the DNA in the complex, but also that the time scale for the heme binding to the DNA is $\ll 240\ s^{-1}$. Furthermore, analysis of the intensities of the NMR signals arising from the heme *meso* 5-, 10-, 15-, and 20-H and DNA A3 base H8 protons indicated a stoichiometric ratio of 1:1 for heme-DNA in the complex (Fig. 43.6B).

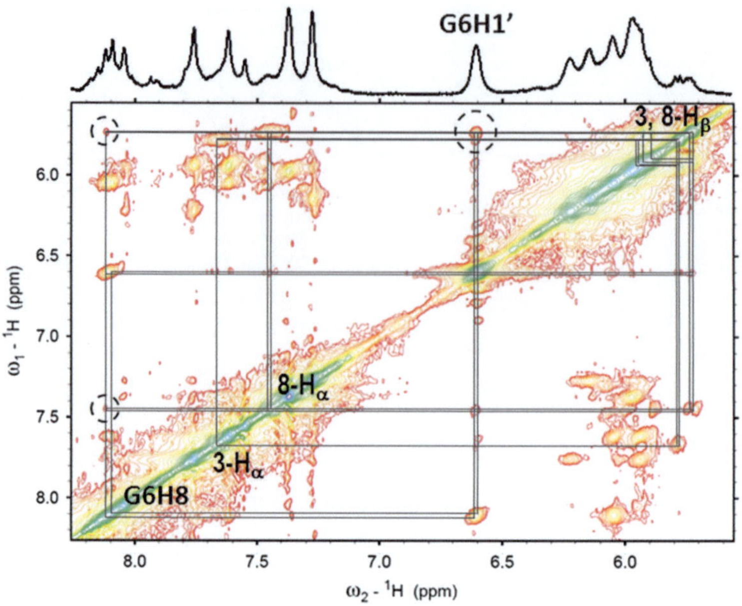

Fig. 43.7 A portion of the NOESY spectrum of CO adduct of heme(Fe^{2+})-(d(TTAGGG))$_4$ complex in ^2H_2O, 300 mM KCl, 50 mM potassium phosphate buffer, pH 7.0, at 25 °C. A mixing time of 150 ms was used to record the spectrum. Intermolecular NOE connectivities between heme vinyl, DNA G6 sugar H1′, and base H8 protons are indicated (Image from Published Paper [18])

In the NOESY spectrum of the CO adduct (Fig. 43.7), heme vinyl proton signals exhibited NOE connectivities with the G6 sugar H1′ and base H8 protons (G6H1′ and G6H8, respectively) of the DNA. These intermolecular NOE connectivities provided the direct evidence that the heme binds to the G6 G-quartet in the complex. This structure was fully supported through analysis of the $\Delta\delta_{\mathrm{hisc}}$ values for the guanine imino NH proton signals. Analysis of the $\Delta\delta_{\mathrm{hisc}}$ values, i.e., −0.37, −0.87, and −2.54 (or −2.49 (see below)) ppm for the G4, G5, and G6 signals, respectively, using the reported equation for the ring current effect of a porphyrin [29], together with a simple model for the DNA structure, suggested that the heme located at 0.40 nm from the G6 G-quartet reasonably satisfies the $\Delta\delta_{\mathrm{hisc}}$ values for the guanine imino NH proton signals. With such a heme orientation, relative to the DNA, the distances between heme vinyl and G6H8/G6H1′ protons could be small enough to allow intermolecular NOE connectivities, as illustrated in Fig. 43.7. In addition, the density functional theory calculation indicated the value of 0.40 nm as the optimal distance between the CO adduct of the porphinatoiron(II) complex and the G-quartet composed of four guanines (Fig. 43.8), also supporting the structure of the heme-DNA complex determined through the NMR structural characterization.

Interestingly, the G6 imino NH proton signal of the complex splits into two peaks with an intensity ratio of 1:1 (Fig. 43.6B). In addition, signal assignments using standard 2D NMR revealed that the G6H1′ 6H8 proton signals also appeared

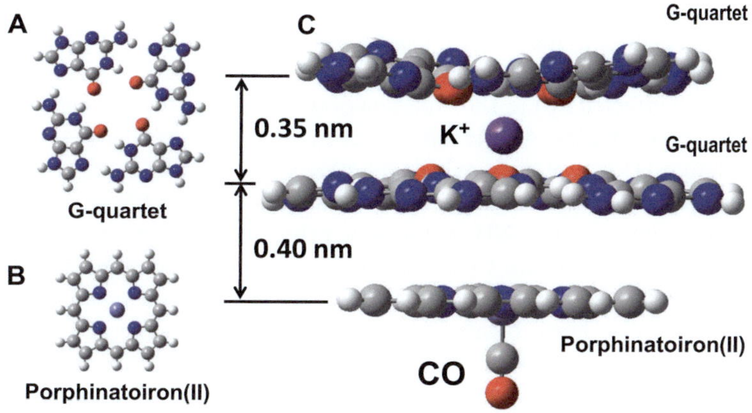

Fig. 43.8 Models used for the density functional theory calculation. The structures of porphinatoiron(II) complex (**A**), the G-quartet (**B**), and the molecular assembly composed of one porphinatoiron(II) complex possessing CO as an axial ligand, two G-quartets, and one potassium ion (K^+) (C) obtained on geometry optimization. As illustrated in (**C**), optimal values of 0.40 and 0.35 nm were obtained for the distances between the porphinatoiron(II) complex and the G-quartet and between G-quartets (Image from Published Paper [18])

as 1:1 doublet peaks (Fig. 43.7). Similarly, two sets of signals, in a ratio of 1:1, were also observed for most of the heme protons (see inset in Fig. 43.6). In order to determine the reason for the appearance of the two sets of signals for these protons, the spectrum of the CO adduct of the complex between C_2 symmetric DMheme (Fig. 43.1B) and $(d(TTAGGG))_4$ (DMheme-DNA complex) was analyzed (Fig. 43.6C, c). The analysis revealed that, in contrast to the case of the heme-DNA complex, only a single set of signals was observed for the DMheme-DNA complex, as reflected in the absence of the splitting of the G6 imino NH proton signal (Fig. 43.6C). These results demonstrated that the 1:1 splitting of some ^1H NMR signals of the CO adduct of the heme-DNA complex is due to the pseudo-C_2 symmetry of the heme molecular structure. Such heterogeneity of the heme orientation is well known as heme orientational disorder [30, 31] and has been shown to be a common structural feature of *b*-type hemoproteins [32–34].

Furthermore, the heme side chain proton signals of the CO adduct of the heme-DNA complex were slightly upfield-shifted relative to those of the corresponding signals of the zinc(II)-protoporphyrin IX complex [35], which has been used as an appropriate reference compound for the analysis of heme NMR shifts. The observed upfield shifts of the heme side chain proton signals of the complex would be due mainly to the ring current effects of guanines of the G6 G-quartet, as a result of the π-π stacking interaction between the heme and the G-quartet in the complex. Upon the formation of the heme-DNA complex, the pseudo-C_2 symmetric heme stacks onto the C_4 symmetric G-quartet. Hence, it is expected that four different sites with 4-fold rotational symmetry are available on the G6 G-quartet for heme binding in the obverse heme form as well as the reverse one. Furthermore, it is also expected that, in the heme-DNA complex, the orientation of the porphyrin moiety of the heme on the G6 G-quartet in the obverse heme form is slightly different from that in the reverse one, because the steric contacts between heme vinyl side chains and DNA G6 sugars are affected by the 180° rotation of the pseudo-C_2 symmetric heme about the *meso* 5-H-15-H axis. As a result, the heme-induced shift changes between the corresponding DNA proton signals of the complexes possessing the two different heme orientations ($\Delta\delta_{or}$) are expected to be slightly different from each other. In fact, not only the G6 imino NH proton signal (Fig. 43.6B), but also the G6H8 and G6H1' ones (Fig. 43.7) of the CO adduct of the heme-DNA complex were observed as 1:1 doublet peaks. These results indicated that the interconversion between the obverse and reverse heme forms of the complex is rather slow and also that the thermodynamic stabilities of the two forms are essentially identical. The $\Delta\delta_{or}$ value of 0.01 ppm for the G6H1' signal dictated that the timescale of the interconversion between the obverse and reverse heme forms of the complex is $\ll 6\ s^{-1}$. Remarkably slow 180° rotation of heme about the *meso* 5-H-15-H axis has been reported for various *b*-type hemoproteins [32–34]. Furthermore, considering the contacts between the heme and the G-quartet in the heme-DNA complex, the thermodynamic stabilities of the obverse and reverse heme forms should be almost identical. On the other hand, the DNA signals, other than the G6 imino NH, G6H1', and G6H8 ones, did not exhibit splitting, indicating that their $\Delta\delta_{or}$ values are quite small. By the way, the degeneracy among the four G6 imino NH signals is expected to be abolished upon the heme binding, because their shifts are greatly affected by the ring current of the heme porphyrin moiety, as reflected in the large $\Delta\delta_{hisc}$ value, and the equivalence among the four signals is readily perturbed by the orientation of the heme relative to the G6 G-quartet. Contrary to this expectation, the degeneracy among the G6 imino NH signals was retained in both the obverse and reverse heme forms (Fig. 43.6B). This finding indicated that the removal of the degeneracy among the four G6 imino NH signals is small and hence is averaged out by the interconversion among the 4-fold symmetric heme orientations within the obverse (or reverse) heme form. These conclusions were confirmed by the spectrum of the CO adduct of the DMheme-DNA complex, which exhibited only a single set of heme and DNA proton signals.

Thus, both experimental and theoretical studies indicated that the heme is located at 0.40 nm from the G6 G-quartet of the heme-DNA complex. Considering that the Fe-CO fragment is 0.43 nm in length, it is unlikely that the axial CO is coordinated to the Fe atom on the G-quartet side of the heme. Consequently, the axial CO is coordinated to the Fe atom, as illustrated in Fig. 43.9. Although another axial ligand is expected to be coordinated to heme(Fe^{2+}) in the CO adduct of the heme-DNA complex, the ligand remains to be identified. But, since there is only limited space available between the heme and the G6 G-quartet in the complex, the size of a ligand that can be accommodated in the space and then coordinated to the heme Fe atom should be quite small. There is a possibility of a direct interaction between the heme Fe and the G-quartet, presumably through an "electrostatic interaction" between the Fe and the four carbonyl oxygen atoms, although the calculation indicated that these Fe and O atoms are rather distant from each other, i.e., the interatomic distance was estimated to be 0.49 nm. Such electrostatic interaction might be the origin of the high peroxidase activities of heme-DNA complexes [2–6].

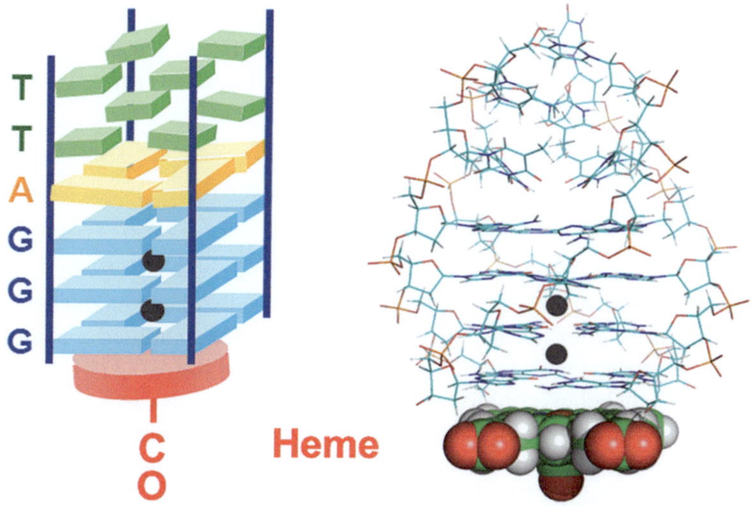

Fig. 43.9 Schematic representation of the structure of the CO adduct of the heme(Fe^{2+})-(d(TTAGGG))$_4$ complex (Image from Published Paper [18])

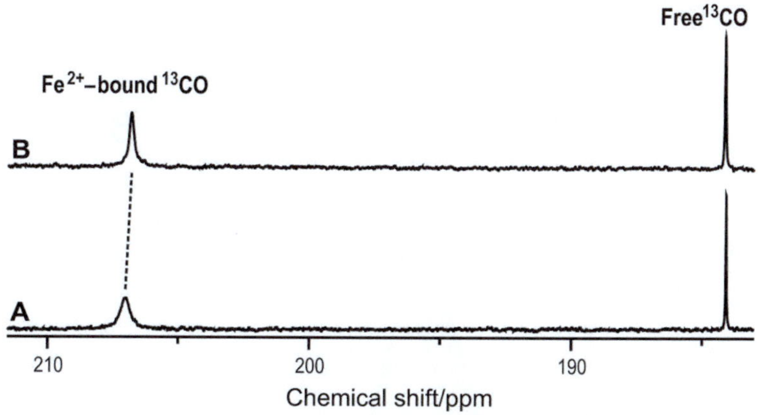

Fig. 43.10 125 MHz ^{13}C NMR spectra of ^{13}CO form of Mb (**A**) and ^{13}CO adduct of the heme(Fe^{2+})-(d(TTAGGG))$_4$ complex (**B**) in 90 % H_2O/10 % ^2H_2O, 300 mM KCl, 50 mM potassium phosphate buffer, pH 7.0, at 25 °C. The signal assignments are indicated with the spectra (Image from Published Paper [18])

Finally,the ^{13}C NMR spectrum of the ^{13}CO adduct of the heme(Fe^{2+})-DNA complex is compared with that of the ^{13}CO form of Mb in Fig. 43.10. The heme(Fe^{2+})-bound ^{13}CO signals of the heme(Fe^{2+})-DNA complex and Mb were observed at 206.81 and 207.03 ppm, and the similarity in the shifts between the two systems indicated that the nature of the Fe^{2+}-^{13}CO bond in the complex is similar to that in Mb [36]. Furthermore, the line widths of the heme(Fe^{2+})-bound ^{13}CO signals of the heme(Fe^{2+})-DNA complex and Mb were 35 and 70 Hz, respectively, and the larger line width for the signal of Mb than that of the heme(Fe^{2+})-DNA complex could be due to the lower molecular tumbling rate for the ^{13}CO bound to the protein [36, 37].

43.4.2 Heme(Fe^{3+})-DNA Complex

The structure of the heme(Fe^{3+})-DNA complex has also been characterized [17]. The Soret absorption of the heme(Fe^{3+})-DNA complex exhibited a characteristic pH-dependent change, with a midpoint at the pH value of 8.6 ± 0.3 [14] (Fig. 43.11). The pH-dependent behavior of the heme(Fe^{3+})-DNA complex was also manifested in its ^{1}H NMR spectrum [14] (Fig. 43.5B, C). With increasing pH, a new set of NMR signals appeared in the chemical shift range of 10–20 ppm at the expense of the signals below 40 ppm. Judging from the shifts of the signals, heme(Fe^{3+}) in the heme(Fe^{3+})-DNA complex adopts a high spin d^5 system with $S = 5/2$ at low pH and a low spin one with $S = 1/2$ at high pH. Since signals due to the two different spin states of the heme(Fe^{3+})-DNA complex were separately observed, the timescale of the pH-dependent interconversion between the two states is rather slow.

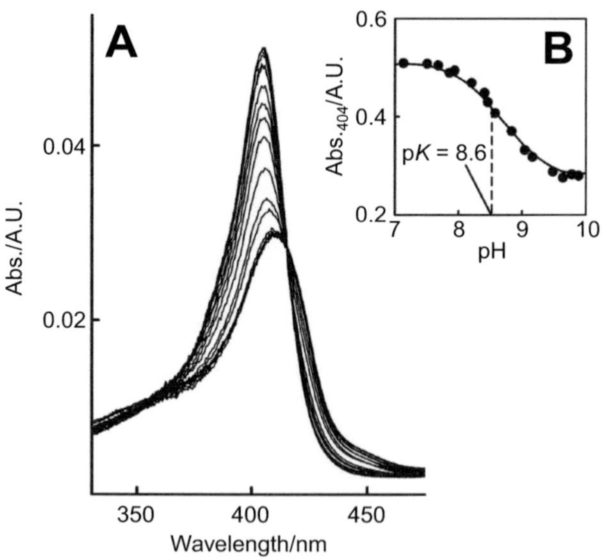

Fig. 43.11 pH dependence of Soret absorption of the heme(Fe^{3+})-(d(TTAGGG))$_4$ complex in 300 mM KCl at room temperature (**A**). The pH was varied from 6.83 to 10.04. A plot of the absorbance at 404 nm against pH (**B**). The midpoint was observed at pH 8.6 ± 0.2 (Image from Published Paper [14])

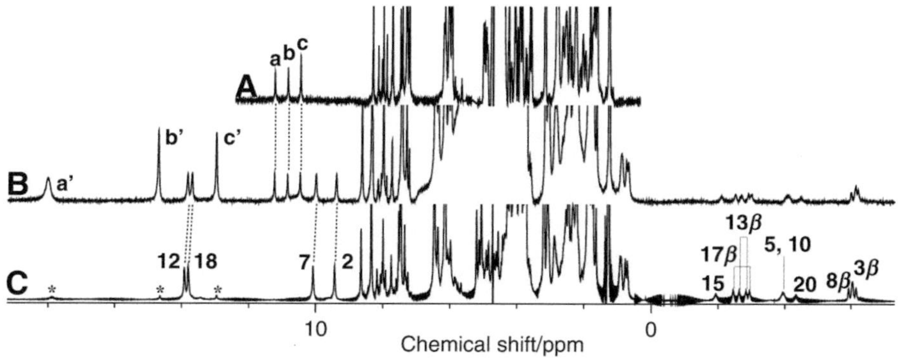

Fig. 43.12 The 600 MHz ^{1}H NMR spectra of G-quadruplex DNA, $(d(TTAGGG))_4$, in the absence (**A**) and presence of a 0.5 equivalent of heme(Fe^{3+}) (**B** and **C**) in 90 % ^1H_2O/10 % ^2H_2O (**A** and **B**) or 2 % ^1H_2O/98 % ^2H_2O (**C**), 300 mM KCl, 50 mM potassium phosphate buffer at pH 9.80 and 25 °C. Peaks *a–c* and *a′–c′* are due to guanine imino protons of the free and heme(Fe^{3+})-bound DNAs, respectively. The signal assignments of heme side chain protons are given in trace C. The small differences in shift between the corresponding signals in traces B and C are due to a solvent $^1H/^2H$ isotope effect (Image from Published Paper [17])

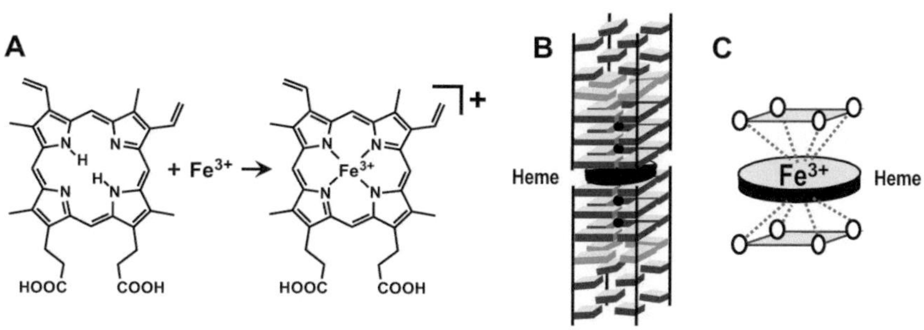

Fig. 43.13 Schematic representation of the incorporation of iron(III) into protoporphyrin IX (**A**), the 1:2 heme(Fe^{3+})-$(d(TTAGGG))_4$ complex (**B**), and the "electrostatic interaction" between heme(Fe^{3+}) and G6 G-quartets in the 1:2 complex (**C**). Heme(Fe^{3+}) possesses a net +1 charge (**A**). In the 1:2 complex, heme(Fe^{3+}) is sandwiched between the 3′-terminal G-quartets of $(d(TTAGGG))_4$ (**B**), and heme(Fe^{3+}) interacts with each of the eight carbonyl oxygen atoms of the G-quartets (**C**)

The NMR spectral parameters of the low spin system allowed detailed structural study of the complex [17] (Fig. 43.12). The study revealed that the heme(Fe^{3+}) is sandwiched between the 3′-terminal G-quartets of $(d(TTAGGG))_4$ to form a 1:2 complex between heme(Fe^{3+}) and $(d(TTAGGG))_4$ (Fig. 43.13). The interaction between the heme(Fe^{3+}) and 3′-terminal (G6) G-quartet in the complex was supported by the observation of intermolecular NOEs between the heme side chain (methyl and vinyl) and G6 (H8 and H1′–H4′) proton signals (Fig. 43.14).

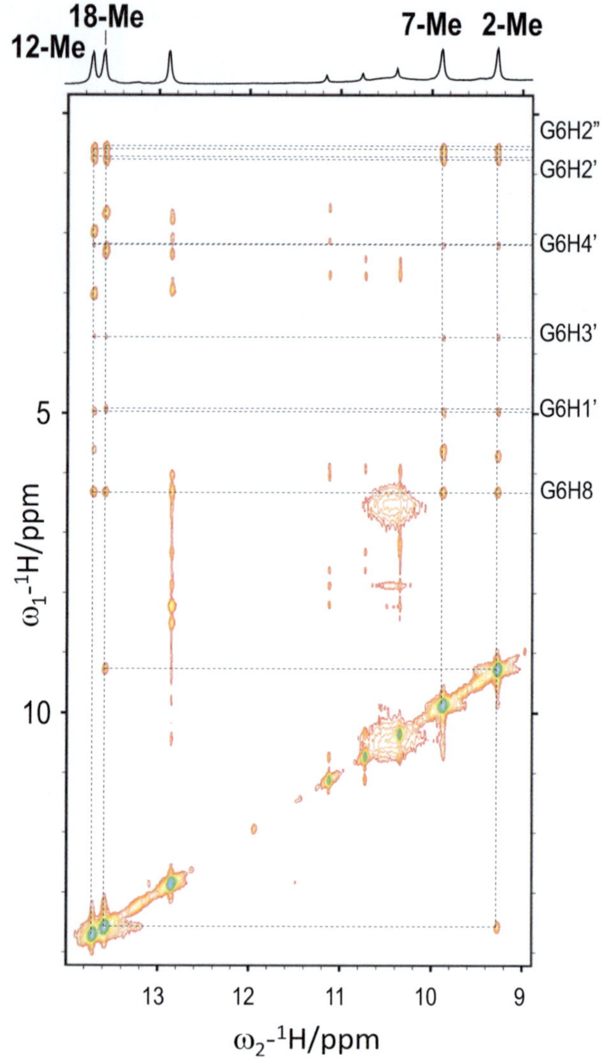

Fig. 43.14 A portion of the NOESY spectrum of $(d(TTAGGG))_4$ in the presence of a 0.5 equivalent of heme(Fe^{3+}) in 90% ^1H_2O/10% ^2H_2O, 300 mM KCl, 50 mM potassium phosphate buffer at pH 9.80 and 25 °C. A mixing time of 150 ms was used to record the spectrum. The signal assignments of four heme methyl and selected G6 proton signals are shown with the spectrum. The intermolecular NOE connectivities between the heme and G6 proton signals are indicated by broken lines (Image from Published Paper [17])

Furthermore, the shift difference between the G4 imino NH proton signals of the free and heme(Fe^{3+})-bound DNAs, i.e., 1.73 ppm, indicated that the timescale of the heme(Fe^{3+}) binding to the DNA is 10^2 s^{-1}.

The $(d(TTAGGG))_4$ dimer possesses a six G-quartet stack, with a linear array of at least five K^+ between the G-quartets [38, 39] (Fig. 43.1d). The K^+ between the G-quartets interacts electrostatically with the carbonyl oxygen atoms of the nearby eight guanine bases to stabilize the G-quartet structure through reduction of the possible electrostatic repulsion among the electrons of these oxygen atoms. Since heme(Fe^{3+}) possesses a net +1 charge (Fig. 43.13A), it could act as a cation like K^+ to stabilize the G-quartet structure. Consequently, although the K^+ between the

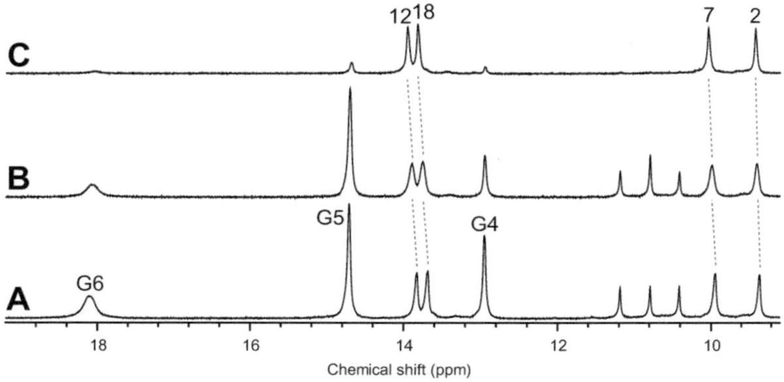

Fig. 43.15 Downfield-shifted portions of the 600 MHz ^{1}H NMR spectra of (d(TTAGGG))$_4$ in the presence of a 0.3 equivalent of heme(Fe^{3+}) in 90 % ^1H_2O/10 % ^2H_2O (**A**), 50 % ^1H_2O/50 % ^2H_2O (**B**), and 2 % ^1H_2O/98 % ^2H_2O (**C**), 300 mM KCl, 50 mM potassium phosphate buffer at pH 9.80 and 25 °C. The shifts and line widths of all four heme side chain methyl proton signals were affected by the ^2H_2O content (Image from Published Paper [17])

G6 G-quartets at the interface of the dimer is thought to be replaced by heme(Fe^{3+}) upon the formation of the 1:2 heme(Fe^{3+})-DNA complex, the electrostatic repulsion among the oxygen atoms of the G6 G-quartets could be in part decreased by the positive charge of heme(Fe^{3+}). Thus, the binding of the heme(Fe^{3+}) between the G6 G-quartets in the DNA dimer can be stabilized through the electrostatic interactions as well as a π-π stacking interaction between the porphyrin moiety of the heme and the G-quartet. Consequently, judging from the stability of the complex, the possibility of OH^- coordination to heme(Fe^{3+}) can also be precluded, because the neutralization of the net +1 charge of heme(Fe^{3+}) by the coordinated OH^- could result in a sizable loss of the stabilization energy of the complex.

In addition, an interaction between heme(Fe^{3+}) and the G6 G-quartets in the complex was clearly manifested in the solvent $^1H/^2H$ isotope effects observed on the shifts and line widths of the heme methyl proton signals of the complex (Fig. 43.15). The heme electronic structures of paramagnetic hemoproteins have been shown to be affected by the solvent $^1H/^2H$ isotope effect [40, 41]. Lecomte and La Mar [40] demonstrated that the in-plane asymmetry of the heme electronic structure in a ferric low spin cyanide complex of Mb (metMbCN$^-$) is larger in ^1H_2O than in ^2H_2O. In metMbCN$^-$, the distal His $N_\varepsilon H$ is hydrogen bonded to Fe^{3+}-bound CN^- [42], and hence electronic perturbation induced by the effect of the $^1H/^2H$ replacement on the hydrogen bond between Fe^{3+}-bound CN^- and the distal His $N_\varepsilon H$ hydrogen atom results in alteration of the coordination structure of CN^- and heme Fe, which in turn influences the in-plane asymmetry of the heme electronic structure in the protein. In addition, La Mar et al. [41] also detected the influence of the solvent $^1H/^2H$ isotope composition on the NMR parameters of heme methyl proton signals of high spin ferric hemoproteins possessing H_2O as an exogenous ligand, which is hydrogen bonded to distal His N_ε [43]. The substantial solvent $^1H/^2H$ isotope effect observed on both the shifts and line widths of heme methyl proton signals of the proteins has been attributed to the electronic perturbation induced by the effect of the $^1H/^2H$ replacement on the hydrogen bond between the Fe^{3+}-bound ligand and distal His. Thus, the heme electronic structures in the proteins are perturbed by the electronic nature of the Fe^{3+}-bound ligands, which is affected by the solvent $^1H/^2H$ isotope composition through the effect of the $^1H/^2H$ replacement on the hydrogen bond between the Fe^{3+}-bound ligand and distal His. Therefore, since a heme proton signal exhibiting a larger paramagnetic shift is in general more highly sensitive to the heme electronic structure, the magnitude of the solvent $^1H/^2H$ isotope effect on a heme proton signal is expected to depend upon its paramagnetic shift. In fact, the shifts of all four heme methyl proton signals of the heme(Fe^{3+})-DNA complex were affected by the $^1H_2O/^2H_2O$ composition, and the shift changes due to the solvent $^1H/^2H$ isotope effect correlated well with the paramagnetic shifts. These results indicated that the heme electronic structure in the complex is perturbed by the $^1H_2O/^2H_2O$ composition-dependent electronic nature of the Fe^{3+}-bound ligand. As described above, the G-quartet itself can be considered as a potential candidate for the Fe^{3+}-bound ligand in the heme(Fe^{3+})-DNA complex. With the structure of the complex in Fig. 43.13A, electrons of the carbonyl oxygen atoms of G6 G-quartets can interact with the positive charge of heme(Fe^{3+}). Since the carbonyl oxygen atoms of the G-quartet participate in the hydrogen bond networks of Hoogsteen-type base pairings (Fig. 43.1C), their electronic nature is altered by the effect of the $^1H/^2H$ replacement on the hydrogen bond networks.

The heme(Fe^{3+}) in the heme(Fe^{3+})-DNA complex can interact with each of the eight carbonyl oxygen atoms of the G6 G-quartets. Simultaneous interaction of the heme(Fe^{3+}) with all the oxygen atoms possibly provides a strong and axially symmetric ligand field surrounding the heme(Fe^{3+}), although the field due to each Fe^{3+}-O interaction may not be necessarily strong. Furthermore, the solvent $^1H/^2H$ isotope effect observed on the line widths of the heme methyl proton signals of the complex could be also attributed to the $^1H/^2H$ replacement on the hydrogen bond networks of G-quartet. Since the heme methyl proton shift of the complex is affected by the effect of the $^1H/^2H$ replacement on the hydrogen bond networks of the G6 G-quartets, the number of observed signals for a given heme methyl proton and the deviation in the shifts and intensities of the signals are determined by the probability distribution of the $^1H/^2H$ replacement of the Hoogsteen-type base pairings associated with each G6 G-quartet. This could explain the large line width of the heme methyl proton signals of the complex observed with ^2H_2O contents of 50 %. Thus, the broadening of the heme methyl proton signals of the complex with ^2H_2O contents of 50 % is not due to exchange broadening, but to overlapping of several signals exhibiting slightly different shifts to each other.

43.5 Functional Properties of the Heme-DNA Complex

As described above, the heme(Fe^{3+})-(d(TTAGGG))$_4$ complex exhibits various spectroscopic properties similar to those of Mb [13–17], manifested by the similarity in the heme structural environments between the two systems. Hence, considering the structure-function relationship of a hemoprotein, the heme(Fe^{3+})-(d(TTAGGG))$_4$ complex is expected to exhibit functional properties similar to the case of Mb. In fact, we have demonstrated that imidazole and carbon monoxide (CO) bind, as exogenous ligands, to the heme(Fe^{3+})-(d(TTAGGG))$_4$ and heme(Fe^{2+})-(d(TTAGGG))$_4$ complexes, respectively [15, 16, 18].

A variety of complexes between heme(Fe^{3+}) and G-quadruplex DNAs have been reported to exhibit peroxidase activity [2–6, 44]. A reaction mechanism has been proposed for the peroxidase activity of the complexes. In the absence of detailed structural properties of the complexes, however, the proposed mechanism relied heavily on the catalytic cycle of heme enzymes such as peroxidase, catalase, and cytochrome P450 [45]. The interaction between heme and the G-quartet in the complexes, described above, is expected to provide important clues for elucidating the role of the DNA in the catalytic activity of the complexes.

Based on the reaction mechanism proposed for the catalytic activity of heme-containing enzymes such as peroxidase and cytochrome P450, the coordination of oxygen, hydroperoxides, or other oxygen donors to heme(Fe^{3+}) initiates their catalytic cycles [44, 45]. In the 1:2 heme(Fe^{3+})-(d(TTAGGG))$_4$ complex, both the axial coordination sites of the heme(Fe^{3+}) are fully occupied, and hence no site is available for the coordination of exogenous ligands. Hence, from the standpoint of exploring the peroxidase activity of the heme(Fe^{3+})-(d(TTAGGG))$_4$ complex, the heme(Fe^{3+}) of the complex ought to possess a coordination site for H_2O_2. Such molecular design of the complex is possibly achieved through a change in the DNA sequence used to prepare G-quadruplex DNA because the addition of a DNA base other than guanine at the 3′-terminal of the DNA sequence is thought to result in inhibition of the dimerization reaction [12], which in turn is expected to eliminate the possibility of the formation of the 1:2 heme-DNA complex and hence to allow the formation of the 1:1 heme-DNA one possessing a coordination site available for exogenous ligands. Furthermore, it should be interesting to investigate the effects of the DNA sequence used to prepare G-quadruplex DNA on the reactivity of heme(Fe^{3+}) in the complex. The catalytic activity of the heme-DNA complex is also thought to be controlled through chemical modification of the heme peripheral side chains, as demonstrated for Mb [46–48]. Both the strategies, i.e., the DNA sequence alteration and heme chemical modification, could be combined to optimize the function of heme-DNA complexes.

References

1. Li Y, Geyer CR, Sen D (1996) Recognition of anionic porphyrins by DNA aptamers. Biochemistry 35(21):6911–6922
2. Li Y, Sen D (1997) Toward an efficient DNAzyme. Biochemistry 36(18):5589–5599
3. Travascio P, Li Y, Sen D (1998) DNA-enhanced peroxidase activity of a DNA-aptamer-hemin complex. Chem Biol 5(9):505–517
4. Travascio P, Bennet AJ, Wang DY, Sen D (1999) A ribozyme and a catalytic DNA with peroxidase activity: active sites versus cofactor-binding sites. Chem Biol 6(11):779–787

5. Travascio P, Witting PK, Mauk AG, Sen D (2001) The peroxidase activity of a hemin–DNA oligonucleotide complex: free radical damage to specific guanine bases of the DNA. J Am Chem Soc 123(7):1337–1348
6. Witting PK, Travascio P, Sen D, Mauk AG (2001) A DNA oligonucleotide-hemin complex cleaves t-butyl hydroperoxide through a homolytic mechanism. Inorg Chem 40(19):5017–5023
7. Poon LCH, Methot SP, Morabi-Pazooki W, Pio F, Bennet AJ, Sen D (2011) Guanine-rich RNAs and DNAs that bind heme robustly catalyze oxygen transfer reactions. J Am Chem Soc 133(6):1877–1884
8. Sen D, Gilbert W (1988) Formation of parallel four-stranded complexes by guanine-rich motifs in DNA and its implications for meiosis. Nature 334(6180):364–366
9. Laughlan G, Murchie AIH, Norman DG, Moore MH, Moody PCE, Lilley DMJ, Luisi B (1994) The high-resolution crystal structure of a parallel-stranded guanine tetraplex. Science 265(5171):520–524
10. Han H, Hurley LH (2000) G-quadruplex DNA: a potential target for anti-cancer drug design. Trends Pharmacol Sci 21(4):136–142
11. Wang Y, Patel DJ (1992) Guanine residues in d(T2AG3) and d(T2G4) form parallel-stranded potassium cation stabilized G-quadruplexes with anti glycosidic torsion angles in solution. Biochemistry 31(35):8112–8119
12. Kato Y, Ohyama T, Mita H, Yamamoto Y (2005) Dynamics and thermodynamics of dimerization of parallel G-quadruplexed DNA formed from d(TTAGn) (n = 3-5). J Am Chem Soc 127(28):9980–9981
13. Mikuma T, Terui N, Yamamoto Y, Hori H (2002) A novel heme-DNA coordination complex and its stability. Nucleic Acids Res Suppl 2(1):285–296
14. Mikuma T, Ohyama T, Terui N, Yamamoto Y, Hori H (2003) Coordination complex between haemin and parallel-quadruplexed d(TTAGGG). Chem Commun 14:1708–1709
15. Ohyama T, Kato Y, Mita H, Yamamoto Y (2006) Exogenous ligand binding property of a heme-DNA coordination complex. Chem Lett 35(1):126–127
16. Saito K, Nakano Y, Tai H, Nagatomo S, Hemmi H, Mita H, Yamamoto Y (2009) Characterization of heme coordination structure in heme-DNA complex possessing gaseous molecule as an exogenous ligand. Nucleic Acids Symp Ser 53:241–242
17. Saito K, Tai H, Hemmi H, Kobayashi N, Yamamoto Y (2012) Interaction between the heme and a G-quartet in a heme-DNA complex. Inorg Chem 51(15):8168–8176
18. Saito K, Tai H, Fukaya M, Shibata T, Nishimura R, Neya S, Yamamoto Y (2012) Structural characterization of a carbon monoxide adduct of a heme-DNA complex. J Biol Inorg Chem 17(3):437–445
19. Watson JD, Crick FH (1952) Molecular structure of nucleic acids; a structure for deoxyribose nucleic acid. Nature 171(4356):737–738
20. Sundquist WI, Klug A (1988) Telomeric DNA dimerizes by formation of guanine tetrads between hairpin loops. Nature 342(6251):825–829
21. Smith FW, Feigon J (1992) Quadruplex structure of Oxytricha telomeric DNA oligonucleotides. Nature 356(6365):164–168
22. Parkinson GN, Lee MPH, Neidle S (2002) Crystal structure of parallel quadruplexes from human telomeric DNA. Nature 417(6891):876–880
23. Sen D, Gilbert W (1992) Novel DNA superstructures formed by telomere-like oligomers. Biochemistry 31(1):65–70
24. Krishnan-Ghosh Y, Liu D, Balasubramanian S (2004) Formation of an interlocked quadruplex dimer by d(GGGT). J Am Chem Soc 126(35):11009–11016
25. Miyoshi D, Nakano A, Toda T, Sugimoto N (2001) Effect of divalent cations on antiparallel G-quartet structure of $d(G_4T_4G_4)$. FEBS Lett 496(2-3):128–133
26. Pastenack RF, Gibbs EJ, Villafranca JJ (1983) Interactions of porphyrins with nucleic acids. Biochemistry 22(23):2406–2414
27. Shikama K, Matsuoka A (1989) Spectral properties unique to the myoglobins lacking the usual distal histidine residue. J Mol Biol 209(3):489–491

28. Ikeda-Saito M, Hori H, Andersson LA, Prince RC, Pickering IJ, George GN, Sanders CR, Lutz RS, McKelvey EJ, Mattera R (1992) Coordination structure of the ferric heme iron in engineered distal histidine myoglobin mutants. J Biol Chem 267(32):22843–22852
29. Abraham RJ, Medforth CJ (1988) The NMR spectra of the porphyrins. 36-Ring currents in octaethylporphyrin, *meso*-tetraphenylporphyrin and phthalocyanine complexes. Magn Reson Chem 26(9):803–812
30. La Mar GN, Davis NL, Parish DW, Smith KM (1983) Heme orientational disorder in reconstituted and native sperm whale myoglobin. Proton nuclear magnetic resonance characterizations by heme methyl deuterium labeling in the Met-cyano protein. J Mol Biol 168(4):887–896
31. La Mar GN, Toi H, Krishnamoorthi R (1984) Proton NMR investigation of the rate and mechanism of heme rotation in sperm whale myoglobin: evidence for intramolecular reorientation about a heme two-fold axis. J Am Chem Soc 106(21):6395–6401
32. La Mar GN, Yamamoto Y, Jue T, Smith KM, Pandey RK (1985) ^{1}H NMR characterization of metastable and equilibrium heme orientational heterogeneity in reconstituted and native human hemoglobin. Biochemistry 24(15):3826–3831
33. McLachlan SJ, La Mar GN, Burns PD, Smith KM, Langry KC (1986) ^{1}H-NMR assignments and the dynamics of interconversion of the isomeric forms of cytochrome b_5 in solution. Biochim Biophys Acta 874(3):274–284
34. Du W, Syvitski R, Dewilde S, Moens L, La Mar GN (2003) Solution ^{1}H NMR characterization of equilibrium heme orientational disorder with functional consequences in mouse neuroglobin. J Am Chem Soc 125(27):8080–8081
35. Wüthrich K, Baumann R (1974) Hyperfine shifts of the ^{13}C NMR in protoporphyrin IX iron(III) dicyanide and deuteroporphyrin IX iron(III) dicyanide. Helv Chim Acta 57(2):336–350
36. Park KD, Guo K, Adebodun F, Chiu ML, Sligar SG, Oldfield E (1991) Distal and proximal ligand interactions in heme proteins: correlations between C-O and Fe-C vibrational frequencies, oxygen-17 and carbon-13 nuclear magnetic resonance chemical shifts, and oxygen-17 nuclear quadrupole coupling constants in $C^{17}O$- and ^{13}CO-labeled species. Biochemistry 30(9):2333–2347
37. Moon RB, Dill K, Richards JH (1977) Magnetic resonance studies of the binding of 13C-labeled carbon monoxide to myoglobins and hemoglobins containing modified hemes. Biochemistry 16(2):221–228
38. Haider S, Parkinson GN, Neidle S (2002) Crystal structure of the potassium form of an Oxytricha nova G-quadruplex. J Mol Biol 320(2):189–200
39. Hazel P, Parkinson GN, Neidle S (2006) Topology variation and loop structural homology in crystal and simulated structures of a bimolecular DNA quadruplex. J Am Chem Soc 128(16):5480–5487
40. Lecomte JTJ, La Mar GN (1987) Proton NMR probe for hydrogen bonding of distal residues to bound ligands in heme proteins: isotope effect on heme electronic structure of myoglobin. J Am Chem Soc 109(23):7219–7220
41. La Mar GN, Chatfield MJ, Peyton DH, de Ropp JS, Smith WS, Krishnamoorthi R, Satterlee JD, Erman JE (1988) Solvent isotope effects on NMR spectral parameters in high-spin ferric hemoproteins: an indirect probe for distal hydrogen bonding. Biochim Biophys Acta 956(3):267–276
42. Yamamoto Y, Iwafune K, Chujo R, Inoue Y, Imai K, Suzuki T (1992) Molecular mechanism for ligand stabilization in the mollusc myoglobin possessing the distal Val residue. J Mol Biol 228(2):343–346
43. Tai H, Nagatomo S, Mita H, Sambongi Y, Yamamoto Y (2005) Characterization of non-native heme coordination structures emerging upon guanidine hydrochloric acid-induced unfolding of *Pseudomonas aeruginosa* ferricytochrome c_{551}. Bull Chem Soc Jpn 78(11):2019–2025
44. Sen D, Poon LCH (2011) RNA and DNA complexes with hemin [Fe(III) heme] are efficient peroxidases and peroxygenases: How do they do it and what does it mean? Crit Rev Biochem Mol Biol 46(6):478–492

45. Stefan L, Denat F, Monchaud D (2012) Insights into how nucleotide supplements enhance the peroxidase-mimicking DNAzyme activity of the G-quadruplex/hemin system. Nucleic Acids Res 40(17):8759–8772
46. Shibata T, Nagao S, Fukaya M, Tai H, Nagatomo S, Morihashi K, Matsuo T, Hirota S, Suzuki A, Imai K, Yamamoto Y (2010) Effect of heme modification on oxygen affinity of myoglobin and equilibrium of the acid-alkaline transition in metmyoglobin. J Am Chem Soc 132(17):6091–6098
47. Nishimura R, Shibata T, Ishigami I, Ogura T, Tai H, Nagao S, Matsuo T, Hirota S, Shoji O, Watanabe Y, Imai K, Neya S, Suzuki A, Yamamoto Y (2014) Electronic control of discrimination between O_2 and CO in myoglobin lacking the distal histidine residue. Inorg Chem 53(2):1091–1099
48. Nishimura R, Matsumoto D, Shibata T, Yanagisawa S, Ogura T, Tai H, Matsuo T, Hirota S, Neya S, Suzuki A, Yamamoto Y (2014) Electronic control of ligand-binding preference of a myoglobin mutant. Inorg Chem 53(17):9156–9165

Chapter 44
Increasing the Hole Transfer Rate Through DNA by Chemical Modification

Kiyohiko Kawai and Tetsuro Majima

Abstract The self-assembly of hundreds of well-designed oligonucleotides, the so-called DNA origami method, recently enabled the creation of a variety of two- and three-dimensional nanostructures of defined size. Thus, if the DNA can efficiently conduct an electrical current, DNA can serve as an interesting bottom-up material for the construction of nano-electronic sensors and devices. However, experimental and theoretical efforts of the last two decades have demonstrated that DNA itself does not serve as a molecular wire. In this chapter, we showed that hole transfer rate through DNA can be increased by tuning its physicochemical properties by proper chemical modification. By using various natural and artificial nucleobases with different HOMO energy levels, we demonstrate that hole transfer rate strongly depends on HOMO energy gap between nucleobases (Δ_{HOMO}). The change in the Δ_{HOMO} value from 0.78 to 0 eV resulted in an increase in the hole transfer rate by more than three orders of magnitude. We also demonstrated that hole transfer rate through DNA is affected by the rigidity or the flexibility of DNA. 5-Me-2′-deoxyzebularine (**B**), which is supposed to increase the flexibility of DNA by weakening the base-pair association, increased hole transfer rate by more than 20-fold, thus demonstrating that hole transfer rate through DNA can be increased by making it more flexible. This was consistent with the fact that locked nucleic acid (LNA), which is known to reduce the flexibility of DNA making it more rigid, caused a decrease of more than two orders of magnitude in hole transfer rate. These results presented in this chapter may help in the future development of programmable DNA-based nano-electronics, which would require higher hole transport properties than that consists of natural nucleobases.

Keywords DNA • Hole transfer • HOMO energy level • DNA flexibility

---

K. Kawai (✉) • T. Majima
The Institute of Scientific and Industrial Research (SANKEN), Osaka University, Mihogaoka 8-1, Ibaraki, Osaka 567-0047, Japan
e-mail: kiyohiko@sanken.osaka-u.ac.jp

T. Akasaka et al. (eds.), *Chemical Science of π-Electron Systems*,
DOI 10.1007/978-4-431-55357-1_44

44.1 Introduction

DNA consists of four nucleobases, guanine (G), adenine (A), cytosine (C), and thymine (T), where G pairs with C and A pairs with T to make so-called base pairs. By programming the sequences of these four components, DNA can be used to construct various self-assembled nanostructures like cubes as well as two-dimensional molecular sheets [1]. Thus, if DNA serve as a molecular wire, it would become possible to construct nano-circuits by using DNA. At the end of 1990s, a hole (a positive charge) generated in DNA has been demonstrated to migrate along DNA over 60 base pairs, whose length is longer than 20 nm [2]. As a consequence, the hole transfer process in DNA has attracted wide attention in the past two decades [3, 4]. Our group and Lewis and Wasielewski's group investigated the hole transfer mechanism by determining the hole transfer rate through DNA [5, 6]. Hole transfer rate constants were determined by a combination of the synthesis of DNA site specifically modified with an adequate photosensitizer and transient absorption measurements. A hole transfer proceeded rapidly for homo-sequences consisting of either only A-T base pairs or G-C base pairs. Hole transfer rate through consecutive A-T base pairs and consecutive G-C base pairs were demonstrated to be larger than $1 \times 10^9\ s^{-1}$ [7, 8]. Among these two base pairs, G-C base pair has a higher HOMO energy than that of A-T base pair, and thus a hole initially generated on A-T base pair will be trapped by G-C base pair after hole transfer process between A-T base pairs. It was demonstrated that once a hole is trapped on G-C base pair, a hole transfers through DNA via series of short-range hole transfer between G-C base pairs. This hole transfer rate between G-C base pairs decreases significantly with increase in the intervening A-T base pairs between G-C base pairs. Hole transfer rate between G-C base pairs becomes slower than $1 \times 10^6\ s^{-1}$ with insertion of only two A-T base pairs between G-C base pairs [6]. This limits the sequences that can be used to construct nano-electronic circuits, and thus DNA itself cannot be used to construct programmable DNA-based nano-electronics such as based on DNA origami technique in which the use of various sequence patterns is indispensable.

DNA is one of the well-studied organic molecules, and there are plenty of reports on the synthesis of artificial nucleobase analogs with various properties. In this chapter, we describe our efforts to increase the hole transfer rate through DNA by utilizing various nucleobase analogs.

44.2 Method for the Measurement of Hole Transfer Rate Through DNA

The kinetics of the hole transfer process in DNA was examined by transient absorption measurements using a synthetic DNA having naphthalimide (NI) as a photosensitizer at one end and phenothiazine (PTZ) as a hole trap at the other end of DNA [9]. Photo-irradiation of NI triggers the hole injection into DNA, and the hole transfer rate through DNA was measured by monitoring the rate that a hole travelled to PTZ attached to the other end of DNA. Photo-irradiation of NI triggered the immediate formation of NI radical anion ($NI^{\bullet-}$) with a peak at around 400 nm showing that a hole is injected into DNA upon laser flash excitation. This absorption band remained almost unchanged during the time course of measurement (200 μs), thus charge-recombination process can be ignored under the present experimental time window. When a hole comes up to the PTZ moiety, it would be trapped by PTZ to form the PTZ radical cation ($PTZ^{\bullet+}$). Accordingly, the hole transfer process from one end of DNA to the other end can be monitored by the formation of the $PTZ^{\bullet+}$ which shows an absorption peak at around 520 nm during the laser flash photolysis measurement (Fig. 44.1). Kinetic modeling was employed to determine the hole transfer rate constants of each individual hole transfer step between G-C

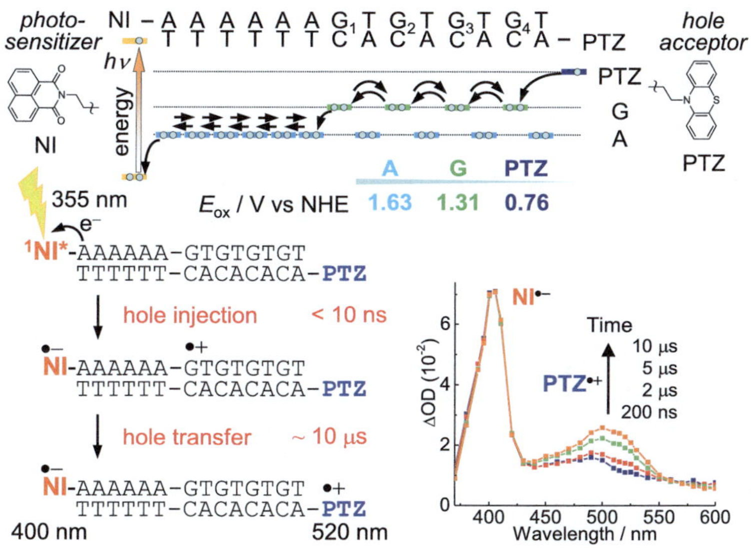

$$\frac{dG_1}{dt} = k_{ht}[G_2] - k_{ht}[G_1]$$

$$\frac{dG_2}{dt} = k_{ht}[G_1] - 2k_{ht}[G_2] + k_{ht}[G_3]$$

$$\frac{dG_3}{dt} = k_{ht}[G_2] - 2k_{ht}[G_3] + k_{ht}[G_4]$$

$$\frac{dG_4}{dt} = k_{ht}[G_3] - k_{ht}[G_4] - k_{PTZ}[G_4]$$

$$\frac{dPTZ}{dt} = k_{PTZ}[G_4]$$

Numerical analysis using MATLAB

Fig. 44.1 Strategy for the measurement of hole transfer rate through DNA

base pairs (k_{ht}). Numerical analysis was performed for the formation of $PTZ^{\bullet +}$ via the multi-step hole transfer mechanism between G-C base pairs using MATLAB software [6].

44.3 Modulating the HOMO Energy of Nucleobases to Increase the Hole Transfer Rate Through DNA

The kinetic measurement of the charge transfer in DNA showed that that a hole efficiently migrates through the base pairs of the same HOMO levels, i.e., G-C and A-T consecutive sequences. Inspired by these results, we hypothesized that the hole transfer rate through DNA can be increased by decreasing the HOMO energy gap (Δ_{HOMO}) between nucleobases. In this section, we synthesized DNA containing various natural and artificial nucleobases of different HOMO energy levels and discussed the hole transfer mechanism by determining the hole transfer rate through DNA.

44.3.1 Hole Transfer Rate Strongly Depends on the HOMO Energy Gap Between Nucleobases (Δ_{HOMO})

It was experimentally and theoretically demonstrated that hole transfer efficiency is sensitive to the HOMO level of nucleobases [4, 10, 11]. We synthesized DNAs having bases with different HOMO levels as a bridge between Gs: G, A, T, inosine (I), 2-aminopurine (^{a}P), and 8-bromoadenine (^{b}A). The HOMO level of the nucleobases was calculated at B3LYP/6-31G(d) by replacing a sugar moiety with a methyl group. The HOMO level of the nucleobases increased along the order of $T < I < {}^{b}A < A < {}^{a}P < G$, and the Δ_{HOMO} between G varied from 0 eV to 0.78 eV (Fig. 44.2). Interestingly, the formation rate of $PTZ^{\bullet +}$ correlated well with Δ_{HOMO} between G and the intervening base between Gs in the same strand of alternating sequence, and the hole transfer rate increased with decreasing Δ_{HOMO} (Fig. 44.2). In the case that the intervening base between two Gs is G, i.e., $\Delta_{HOMO} = 0$, the hole transfer proceeded the fastest and was accomplished faster than the time resolution of our setup (<50 ns), showing that hole transfer proceeded faster than $1 \times 10^{9}\ s^{-1}$. This was consistent with the hole transfer rate of $1.1 \times 10^{9}\ s^{-1}$ which was obtained by applying the random-walk hopping model to the hole transfer rate between adjacent guanines ($4.3 \times 10^{9}\ s^{-1}$) determined by Lewis and Wasielewski [7]. These results clearly demonstrated that the hole transfer rate can be increased by decreasing the Δ_{HOMO} and that it can also be finely tuned over three orders of magnitude by varying the Δ_{HOMO} [12].

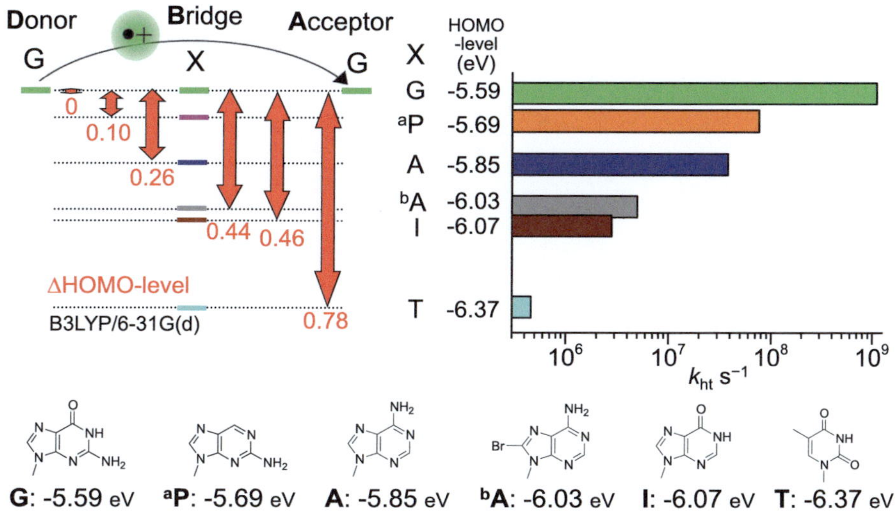

Fig. 44.2 Hole transfer rate through DNA can be controlled by modulating the HOMO energy level of nucleobases

44.3.2 Less Sequence-Dependent Rapid Hole Transfer Through DNA Achieved by Replacing A-T Base Pairs with Deazaadenosine (^{Z}A)-T Base Pairs

In the previous section, we described that the decrease in Δ_{HOMO} significantly increases the hole transfer rate through DNA, which is consistent with the fact that the hole transfer rate is inherently small except for consecutive G-C or A-T sequences. We hypothesized that it may be possible to increase the hole transfer rate in less sequence-dependent manner by decreasing Δ_{HOMO} between the two types of base pairs, i.e., G-C base pair (or its analog) and A-T base pair (or its analog). Nakatani and Saito reported that deazaadenosine (^{Z}A), having a higher HOMO level than A to be closer to that of G, serves as a stepping stone for increasing the hole transfer efficiency between G-C base pairs separated by intervening A-T base pairs [4]. We next replaced A-T base pairs with ^{z}A-T base pairs to increase the hole transfer rate through DNA in less sequence-dependent fashion.

We tested whether a hole can go over an A-T tract by replacing A-T base pair(s) with ^{z}A-T base pair(s). The formation of PTZ$^{\bullet+}$ was not observed in the experimental time window of 200 μs for DNA having five consecutive A-T base pairs. Of special interests, hole transfer rate from one end of the DNA to the other end ($k_{overall}$) dramatically increased with the increasing number of replaced ^{z}A-T base pair(s) in the A-T tract (Fig. 44.3). When all the A-T base pairs in the A-T tract were replaced with ^{z}A-T base pairs, the hole transfer across ca. 50 Å occurred faster than the time resolution of our experimental setup (<50 ns). Thus, $k_{overall}$ increased by more than three orders of magnitude by the replacement of A-T base pairs with

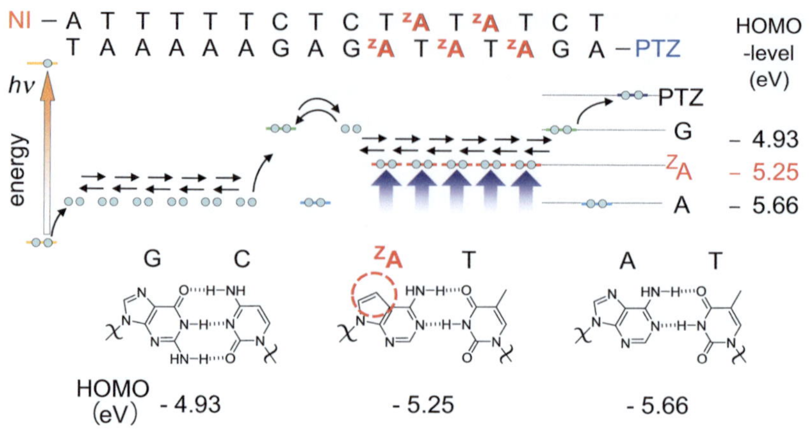

Fig. 44.3 Less sequence-dependent rapid hole transfer through DNA by replacing A-T base pairs with deazaadenosine (**ᶻA**)-T base pairs

ᶻA-T base pairs. These results clearly demonstrated that hole transfer rate through DNA can be increased in less sequence-dependent manner by decreasing the HOMO energy gap of two types of base pairs [13].

44.4 Modulating the Flexibility of DNA to Increase the Hole Transfer Rate Through DNA

DNA is a flexible, dynamic molecule, and thus changes in the distance and orientation between nucleobases due to thermal fluctuations can significantly alter orbital overlap between nucleobases and in turn alter the hole transfer rate through DNA. Barton and coworkers suggested that hole transfer through DNA is affected by the frequency of occurrence of structures that are particularly amenable to hole transfer [14]. Actually, the use of multiple conformations has now gained popularity to calculate the hole transfer rate through DNA in MD simulations [11, 15]. We next investigated whether the hole transfer rate through DNA can be increased by increasing the flexibility of DNA based on the chemical modification of DNA.

In order to investigate the influence of the local conformational flexibility of DNA on the hole transfer rate through DNA, we synthesized DNA modified with locked nucleic acids (LNA) and 5-Me-2′-deoxyzebularine (B) (Fig. 44.4). LNA is a bicyclic nucleic acid analog where a ribonucleoside is linked between the 2′-oxygen and the 4′-carbon atoms with a methylene unit [16]. This locks the sugar conformation into a C3′-endo conformation, which results in the decrease of the flexibility of nucleic acids. The lifetime of the base-pair opening of G-C base pair in the DNA/LNA hybrid was reported to be longer than those of the corresponding base pairs in the DNA/DNA duplex [17]. Thus, base-pair breathing rates in DNA

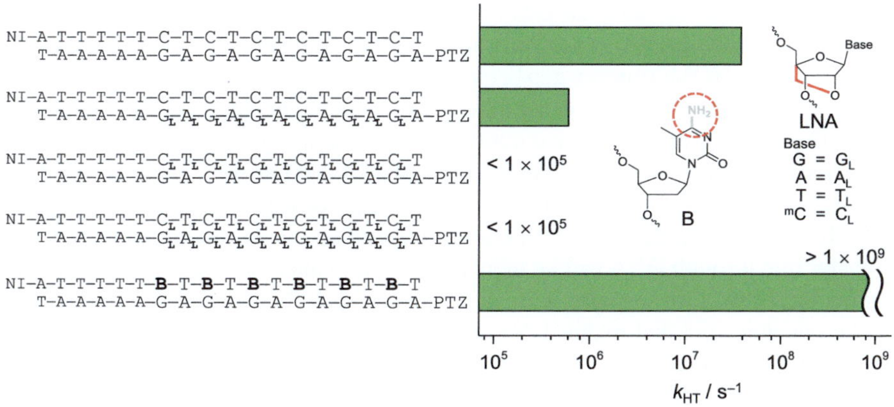

Fig. 44.4 Hole transfer rate through DNA can be controlled by changing the flexibility of DNA

decrease upon LNA modifications, leading to a loss of the flexibility of DNA or an increase in the rigidity of DNA. 5-Me-2′-deoxyzebularine is an analog of C lacking an amino group at C4 position and thus forms a reduced number of hydrogen bonds with G. Therefore, the replacement of C with 5-Me-2′-deoxyzebularine would lead to faster breathing of the base pair and thus increases the flexibility of DNA. The LNA modification of all the nucleotides in the G-A or G-T repeat region completely suppressed the hole transfer through DNA ($<10^5$ s^{-1}). Though the X-ray analysis showed that LNA confers a more efficient base stacking than natural B-DNA due to LNA's smaller helical twist [18], to our surprise, the loss of flexibility resulted in a significant decrease in the hole transfer rate through DNA. In sharp contrast, the replacement of C with 5-Me-2′-deoxyzebularine resulted in a significant increase in the hole transfer rate. When all Cs in the G-A repeat were replaced with 5-Me-2′-deoxyzebularines, the hole transfer occurred faster than the time resolution of our setup ($>10^9$ s^{-1}). Thus, in the case of G-A repeating sequences, LNA modification resulted in a k_{HT} decrease of more than two orders of magnitude, whereas 5-Me-2′-deoxyzebularine modification caused more than a 20-fold increase in hole transfer rate through DNA. These results clearly demonstrated that hole transfer rate through DNA is strongly dependent on the flexibility of DNA and that it can be increased by increasing the flexibility of DNA which can be controlled by the chemical modification of DNA [19].

44.5 Conclusions

In this chapter, we described our efforts to increase the hole transfer rate through DNA which is based on the chemical modification of DNA. By using various natural and artificial nucleobases that have different HOMO energy level, we showed that the hole transfer rate through DNA strongly depends on the HOMO energy gap between nucleobases. It was demonstrated that hole transfer rate can be increased by decreasing the HOMO energy gap between nucleobases and can be finely tuned over three orders of magnitude by using natural and artificial nucleobases having various HOMO energy level. We pointed out that low efficient hole transfer in G-C and A-T mixed random sequences originates from the HOMO energy gap between these two types of base pairs. By using the analog of A, deazaadenosine that has lower HOMO energy level than that of A closer to that of G, it was demonstrated that hole transfer rate through DNA can be increased by more than three orders of magnitude by replacing A-T base pairs with deazaadenosine-T base pairs. By introducing locked nucleic acid that makes DNA more rigid and 5-Me-2′-deoxyzebularine that is supposed to increase the flexibility of DNA, we demonstrated that hole transfer rate through DNA is strongly affected by the rigidity or the flexibility of its duplex structure. It was demonstrated that hole transfer rate through DNA can be increased by increasing the flexibility of DNA. While DNA itself cannot be used to construct nano-circuits, our results suggest that the artificial DNA which properties were finely tuned by chemical modification may serve as a potential bottom-up molecule in nano-electronic sensors and devices.

Acknowledgments We are deeply indebted to Mr. Mitsuo Hayashi for his contributions to this study during his master course studies. This work has been partly supported by a Grants-in-Aid for Scientific Research (Project 17105005, 18750148, 21750170, 22245022, and others) from MEXT of Japanese government.

References

1. Endo M, Sugiyama H (2014) Single-molecule imaging of dynamic motions of biomolecules in DNA origami nanostructures using high-speed atomic force microscopy. Acc Chem Res 47:1645. doi:10.1021/ar400299m; Kuzuya A, Ohya Y (2014) Nanomechanical molecular devices made of DNA Origami. Acc Chem Res 47:1742. doi:10.1021/ar400328v; Rangnekar A, LaBean TH (2014) Building DNA nanostructures for molecular computation, templated assembly, and biological applications. Acc Chem Res 47:1778. doi:10.1021/ar500023b; Simmel SS, Nickels PC, Liedl T (2014) Wireframe and tensegrity DNA nanostructures. Acc Chem Res 47:1691. doi:10.1021/ar400319n; Tan LH, Xing H, Lu Y (2014) DNA as a powerful tool for morphology control, spatial positioning, and dynamic assembly of nanoparticles. Acc Chem Res 47:1881. doi:10.1021/ar500081k; Wang Z-G, Ding B (2014) Engineering DNA self-assemblies as templates for functional nanostructures. Acc Chem Res 47:1654. doi:10.1021/ar400305g; Wu TC, Rahman M, Norton ML (2014) From nonfinite to finite 1D arrays of origami tiles. Acc Chem Res 47:1750. doi:10.1021/ar400330y; Rothemund PWK, Andersen ES (2012) Nanotechnology: the importance of being modular. Nature 485:584. doi:10.1038/485584a
2. Meggers E, Michel-Beyerle ME, Giese B (1998) Sequence dependent long range hole transport in DNA. J Am Chem Soc 120:12950. doi:10.1021/ja983092p; Henderson PT, Jones D, Hampikian G, Kan Y, Schuster GB (1999) Long-distance charge transport in duplex DNA: the phonon-assisted polaron-like hopping. Proc Natl Acad Sci USA 96:8353. doi:10.1073/pnas.96.15.8353; Nakatani K, Dohno C, Saito I (1999)Chemistry of

sequence-dependent remote guanine oxidation: photoreaction of duplex DNA containing cyanobenzophenone-substituted uridine. J Am Chem Soc 121:10854. doi:10.1021/ja9920213
3. Giese B, Biland A (2002) Recent developments of charge injection and charge transfer in DNA. Chem Commun 667. doi:10.1039/b111044f; Genereux JC, Barton JK (2010) Mechanisms for DNA charge transport. Chem Rev 110:1642. doi:10.1021/cr900228f; Genereux JC, Boal AK, Barton JK (2010) DNA-mediated charge transport in redox sensing and signaling. J Am Chem Soc 132:891. doi:10.1021/ja907669c; Kanvah S, Joseph J, Schuster GB, Barnett RN, Cleveland CL, Landman U (2010) Oxidation of DNA: damage to nucleobases. Acc Chem Res 43:280. doi:10.1021/ar900175a; Schuster GB (2000) Long-range charge transfer in DNA: transient structural distortions control the distance dependence. Acc Chem Res 33:253. doi:10.1021/ar980059z; Senthilkumar K, Grozema FC, Guerra CF, Bickelhaupt FM, Lewis FD, Berlin YA, Ratner MA, Siebbeles LDA (2005) Absolute rates of hole transfer in DNA. J Am Chem Soc 127:14894. doi:10.1021/ja054257e; Bixon M, Giese B, Wessely S, Langenbacher T, Michel-Beyerle ME, Jortner J (1999) Long-range charge hopping in DNA. Proc Natl Acad Sci USA 96:11713. doi:10.1073/pnas.96.21.11713; Jortner J, Bixon M, Langenbacher T, Michel-Beyerle ME (1998) Charge transfer and transport in DNA. Proc Natl Acad Sci USA 95:12759. doi:10.1073/pnas.95.22.12759; Giese B, Amaudrut J, Kohler AK, Spormann M, Wessely S (2001) Direct observation of hole transfer through DNA by hopping between adenine bases and by tunneling. Nature 412:318
4. Nakatani K, Dohno C, Saito I (2000) Modulation of DNA-mediated hole-transport efficiency by changing superexchange electronic interaction. J Am Chem Soc 122:5893. doi:10.1021/ja994113y
5. Kawai K, Majima T (2013) Hole transfer kinetics of DNA. Acc Chem Res 46:2616. doi:10.1021/ar400079s; Lewis FD, Letsinger RL, Wasielewski MR (2001) Dynamics of photoinduced charge transfer and hole transport in synthetic DNA hairpins. Acc Chem Res 34:159. doi:10.1021/ar0000197; Osakada Y, Kawai K, Fujitsuka M, Majima T (2008) Kinetics of charge transfer in DNA containing a mismatch. Nucleic Acids Res 36:5562. doi:10.1093/nar/gkn505; Lewis FD, Liu J, Zuo X, Hayes RT, Wasielewski MR (2003) Dynamics and energetics of single-step hole transport in DNA hairpins. J Am Chem Soc 125:4850. doi:10.1021/ja029390a; Lewis FD, Liu J, Weigel W, Rettig W, Kurnikov IV, Beratan DN (2002) Donor-bridge-acceptor energetics determine the distance dependence of electron tunneling in DNA. Proc Natl Acad Sci USA 99:12536. doi:10.1073/pnas.192432899; Lewis FD, Liu X, Liu J, Miller SE, Hayes RT, Wasielewski MR (2000) Direct measurement of hole transport dynamics in DNA. Nature 406:51. doi:10.1038/35017524
6. Takada T, Kawai K, Fujitsuka M, Majima T (2005) Contributions of the distance-dependent reorganization energy and proton-transfer to the hole-transfer process in DNA. Chem Eur J 11:3835. doi:10.1002/chem.200500052
7. Conron SMM, Thazhathveetil AK, Wasielewski MR, Burin AL, Lewis FD (2010) Direct measurement of the dynamics of hole hopping in extended DNA G-tracts. An unbiased random walk. J Am Chem Soc 132:14388–14390
8. Takada T, Kawai K, Cai X, Sugimoto A, Fujitsuka M, Majima T (2004) Charge separation in DNA via consecutive adenine hopping. J Am Chem Soc 126:1125. doi:10.1021/ja035730w
9. Takada T, Kawai K, Fujitsuka M, Majima T (2004) Direct observation of hole transfer through double-helical DNA over 100 Å. Proc Natl Acad Sci USA 101:14002. doi:10.1073/pnas.0402756101; Osakada Y, Kawai K, Fujitsuka M, Majima T (2006) Charge transfer through DNA nanoscaled assembly programmable with DNA building blocks. Proc Natl Acad Sci USA 103:18072. doi:10.1073/pnas.0607148103
10. Voityuk AA, Roesch N (2002) Quantum chemical modeling of electron hole transfer through π stacks of normal and modified pairs of nucleobases. J Phys Chem B 106:3013. doi:10.1021/jp013417f; Bixon M, Jortner J (2001) Charge transport in DNA via thermally induced hopping. J Am Chem Soc 123:12556. doi:10.1021/ja010018p; Venkatramani R, Davis KL, Wierzbinski E, Bezer S, Balaeff A, Keinan S, Paul A, Kocsis L, Beratan DN, Achim C, Waldeck DH (2011) Evidence for a near-resonant charge transfer mechanism for double-stranded peptide nucleic acid. J Am Chem Soc 133:62. doi:10.1021/ja107622m;

Paul A, Bezer S, Venkatramani R, Kocsis L, Wierzbinski E, Balaeff A, Keinan S, Beratan DN, Achim C, Waldeck DH (2009) Role of nucleobase energetics and nucleobase interactions in single-stranded peptide nucleic acid charge transfer. J Am Chem Soc 131:6498. doi:10.1021/ja9000163
11. Venkatramani R, Keinan S, Balaeff A, Beratan DN (2011) Nucleic acid charge transfer: black, white and gray. Coord Chem Rev 255:635. doi:10.1016/j.ccr.2010.12.010
12. Kawai K, Hayashi M, Majima T (2012) HOMO energy gap dependence of hole-transfer kinetics in DNA. J Am Chem Soc 134:4806. doi:10.1021/ja2109213
13. Kawai K, Kodera H, Majima T (2010) Long-range charge transfer through DNA by replacing adenine with diaminopurine. J Am Chem Soc 132:627. doi:10.1021/ja907409z; Kawai K, Kodera H, Osakada Y, Majima T (2009) Sequence-independent and rapid long-range charge transfer through DNA. Nat Chem 1:156. doi:10.1038/nchem.171
14. O'Neill MA, Barton JK (2004) DNA charge transport: conformationally gated hopping through stacked domains J Am Chem Soc 126:11471. doi:10.1021/ja048956n; O'Neill MA, Barton JK (2004) DNA-mediated charge transport requires conformational motion of the DNA bases: elimination of charge transport in rigid glasses at 77 K. J Am Chem Soc 126:13234. doi:10.1021/ja0455897; O'Neill MA, Becker H-C, Wan C, Barton JK, Zewail AH (2003) Ultrafast dynamics in DNA-mediated electron transfer: base gating and the role of temperature. Angew Chem Int Ed 42:5896. doi:10.1002/anie.200352831
15. Valis L, Wang Q, Raytchev M, Buchvarov I, Wagenknecht H-A, Fiebig T (2006) Base pair motions control the rates and distance dependencies of reductive and oxidative DNA charge transfer. Proc Natl Acad Sci USA 103:10192. doi:10.1073/pnas.0600957103; Kubar T, Elstner M (2008) What governs the charge transfer in DNA? The role of DNA conformation and environment. J Phys Chem B 112:8788. doi:10.1021/jp803661f; Hatcher E, Balaeff A, Keinan S, Venkatramani R, Beratan DN (2008) PNA versus DNA: effects of structural fluctuations on electronic structure and hole-transport mechanisms. J Am Chem Soc 130:11752. doi:10.1021/ja802541e
16. Hariharan M, McCullagh M, Schatz GC, Lewis FD (2010) Conformational control of thymine photodimerization in single-strand and duplex DNA containing locked nucleic acid TT steps. J Am Chem Soc 132:12856. doi:10.1021/ja106203q
17. Bruylants G, Boccongelli M, Snoussi K, Bartik K (2009) Comparison of the thermodynamics and base-pair dynamics of a full LNA: DNA duplex and of the isosequential DNA: DNA duplex. Biochemistry 48:8473. doi:10.1021/bi900615z
18. Eichert A, Behling K, Betzel C, Erdmann VA, Fuerste JP, Foerster C (2010) The crystal structure of an 'All Locked' nucleic acid duplex. Nucleic Acids Res 38:6729. doi:10.1093/nar/gkq505
19. Kawai K, Hayashi M, Majima T (2012) Hole transfer in LNA and 5-Me-2′-deoxyzebularine-modified DNA. J Am Chem Soc 134:9406. doi:10.1021/ja302641e

Chapter 45
Theoretical Calculations of Excitation Energy Transfer

Kazuhiro J. Fujimoto

Abstract A transition–density–fragment interaction combined with a transfer integral (TDFI-TI) method is explained. The basis of the TDFI-TI method is to describe the elements of the Hamiltonian matrix, and it gives quantitative descriptions of the elcctronic coupling energies and the absorption spectra. The TDFI-TI method allows us to perform the decomposition analysis, which provides the components of the electronic coupling energy. Moreover, the TDFI-TI method can reduce the computational costs, compared with the full quantum-mechanical (QM) method. The TDFI-TI approach is useful for analyzing and understanding the mechanism of excitation energy transfer and crystallochromy.

Keywords Excitation energy transfer • Electronic coupling • Transition density • Crystallochromy

45.1 Introduction

Excitation energy transfer (EET) is a well-known phenomenon observed in pair or aggregates of molecules [1, 2], and its feature is widely used in chemical and biological systems. One example of EET is light-harvesting complex (LHC) in green-plant photosynthesis [3, 4]. A large number of photosynthetic pigments are included in LHC, and sunlight energy captured by these pigments is transferred to reaction center using EET [5–7]. The fundamental understanding of EET in photosynthetic system is expected to lead to more efficient artificial solar cells [8].

The mechanism of EET has been investigated with Förster [9] and Dexter [10] theories, in which electronic coupling required for the rate constant of EET is calculated with the Coulomb and exchange interactions between two interacting molecules, respectively. In common, the EET process is explained by the Förster-type mechanism in the case of the large separation of two molecules and by the Dexter-type mechanism for the short separation [11, 12]. Both theories are based

K.J. Fujimoto (✉)
Department of Computational Science, Graduate School of System Informatics, Kobe University, Kobe, Japan
e-mail: fujimoto@ruby.kobe-u.ac.jp

T. Akasaka et al. (eds.), *Chemical Science of π-Electron Systems*,
DOI 10.1007/978-4-431-55357-1_45

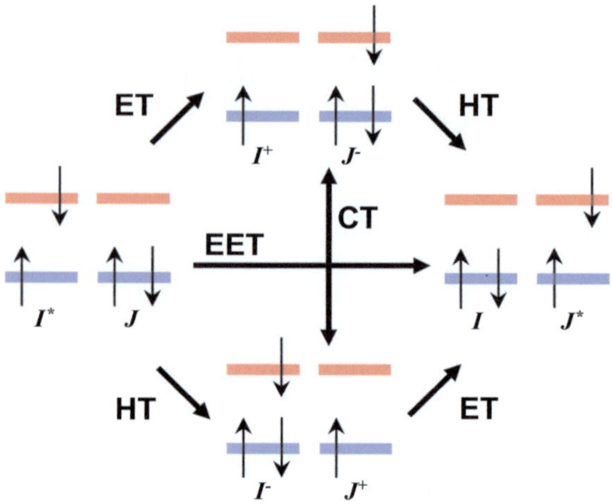

Fig. 45.1 Schematic illustration of the EET via intermediate states. *Blue* and *red lines* show the HOMO and LUMO of the monomer (*I* or *J*), respectively

on a two-state model with the initial (excitation of donor molecule) and final (de-excitation of donor molecule and excitation of acceptor molecule) states [1]. In contrast, Harcourt et al. presented an alternative view of the mechanism of EET [13]. They focused on a two-electron transfer process as well as the direct coupling between the initial and final states and showed a significant influence of charge-resonance (CR) states on the EET mechanism. Figure 45.1 shows a schematic illustration for this process. In their theory, electronic coupling was described with two kinds of configurations. One is the local excitations (LE) within a molecule, and the other is the CR states between two molecules. Since the Förster and Dexter theories were based only on the LE states, the admixture of the CR states provided a general framework for electronic coupling.

The magnitude of electronic coupling strongly influences the EET rate, so that the accurate description of electronic coupling plays a significant role in the EET study. In Förster theory, electronic coupling reduced to the Coulomb interaction was approximated to a dipole–dipole (dd) interaction using transition dipole moments for interacting molecules. The dd method is easily applicable to the electronic coupling calculations, because of its simplicity. In addition, it provides an intuitive interpretation of electronic coupling. Owing to these features, the dd method has been applied to various EET studies. However, the dd method must be used with care because the utility of this method is limited to the case where the intermolecular separation between two interacting molecules is larger than their molecular sizes [1, 2]. To circumvent this problem, many theoretical studies have been performed [14–30]. A transition–density–fragment interaction (TDFI) method was developed also for this purpose [31, 32]. The basic idea of TDFI is similar to a transition density cube (TDC) method [17] proposed by Krueger et al. The difference between the two methods is in the treatment of transition densities. The TDFI method employs the self-consistent transition densities of the interacting molecules obtained with a density–fragment interaction (DFI) scheme [33], whereas the TDC method uses the transition densities that are separately calculated in gas phase. In the previous studies, the TDFI method was applied to the EET in xanthorhodopsin [31] and to an exciton-coupled circular dichroism spectrum observed in a dimerized retinal chromophore [32]. As a result, TDFI could give a much improved description of the electronic coupling, compared with the dd method. Subsequently, the TDFI method was extended to the exchange interaction, and furthermore, it was combined with a transfer integral method (TDFI-TI) [34]. The TDFI-TI method provides electronic coupling on the basis of the LE and CR configurations, so that it is also applicable to EET via CR states proposed by Harcourt et al. Moreover, the decomposition analysis of the electronic coupling energy can be performed with the TDFI-TI method, which is useful for understanding the mechanism of EET.

In this chapter, I explain the TDFI-TI approaches for the electronic coupling energies [34] and absorption spectra [35]. For checking the accuracy of the TDFI-TI method, several test calculations are performed to an ethylene dimer. In these calculations, the components of the electronic coupling energy are also analyzed. Next, the TDFI-TI method is applied to two kinds of solid-state tetracenes, red and yellow solids of tetracenes [36]. Based on the successful reproduction of the experimental absorption energies, the spectral tuning mechanism on the two solids is also explored.

45.2 TDFI-TI Method

First, let us consider the total Hamiltonian for IJ molecular complex

$$\widehat{H} = \widehat{H}_I + \widehat{H}_J + \widehat{V}_{IJ}, \tag{45.1}$$

and the basis functions

$$\begin{aligned} |\Phi_1\rangle &= \left|\Psi_I^e \cdot \Psi_J^g\right\rangle, \\ |\Phi_2\rangle &= \left|\Psi_I^g \cdot \Psi_J^e\right\rangle, \\ |\Phi_3\rangle &= \left|\Psi_I^+ \cdot \Psi_J^-\right\rangle, \\ |\Phi_4\rangle &= \left|\Psi_I^- \cdot \Psi_J^+\right\rangle. \end{aligned} \tag{45.2}$$

In Eq. (45.1), $\hat{H}_X$ denotes the local Hamiltonian for the molecule X ($X = I$ or J), and $\widehat{V}_{IJ}$ is the Coulomb interaction between the molecules I and J. In Eq. (45.2), Ψ_X^g and Ψ_X^e represent the ground (g) and excited (e) states for the molecule X, respectively, and Ψ_X^+ and Ψ_X^- are the cationic (+) and anionic (−) states for the molecule X, respectively. The basis functions Φ_1 and Φ_2 represent the initial (I) and final (F) states of the EET process, respectively, and they correspond to the LE states within the molecule X. On the other hand, Φ_3 and Φ_4 express the CR states between the molecules I and J. The basic structure of the Hamiltonian matrix is represented by

$$\mathbf{H} = \begin{pmatrix} \mathbf{H}_{\mathrm{LE}} & \mathbf{H}_{\mathrm{LE-CR}} \\ \mathbf{H}_{\mathrm{CR-LE}} & \mathbf{H}_{\mathrm{CR}} \end{pmatrix}, \tag{45.3}$$

where $\mathbf{H}_{\mathrm{LE}}$ and $\mathbf{H}_{\mathrm{CR}}$ are the block matrices for the LE and CR states, respectively, and $\mathbf{H}_{\mathrm{LE-CR}}$ and $\mathbf{H}_{\mathrm{CR-LE}}$ are the mixing blocks between the LE and CR states. It is noted that the Förster and Dexter theories focus only on the block matrix $\mathbf{H}_{\mathrm{LE}}$. Now, a symmetric homodimer is considered as a total system for simplicity. In this case, the Hamiltonian matrix given by Eq. (45.3) is rewritten as the 4×4 matrix

$$\mathbf{H} = \begin{pmatrix} E_{\mathrm{LE}} & V_{\mathrm{EET}} & V_{\mathrm{ET}} & V_{\mathrm{HT}} \\ V_{\mathrm{EET}} & E_{\mathrm{LE}} & V_{\mathrm{HT}} & V_{\mathrm{ET}} \\ V_{\mathrm{ET}} & V_{\mathrm{HT}} & E_{\mathrm{CR}} & V_{\mathrm{CT}} \\ V_{\mathrm{HT}} & V_{\mathrm{ET}} & V_{\mathrm{CT}} & E_{\mathrm{CR}} \end{pmatrix}. \tag{45.4}$$

Electronic coupling between the molecules I and J is calculated with the elements of the Hamiltonian matrix. Using the perturbation technique, electronic coupling based on Eq. (45.4) is represented by [37–39]

$$\begin{aligned} T_{\mathrm{IF}} &= H_{12} - \frac{H_{13}H_{32}}{H_{33} - H_{11}} - \frac{H_{14}H_{42}}{H_{44} - H_{11}} \\ &\quad + \frac{H_{13}H_{34}H_{42}}{(H_{33} - H_{11})(H_{44} - H_{11})} + \frac{H_{14}H_{43}H_{32}}{(H_{44} - H_{11})(H_{33} - H_{11})}. \\ &= V_{\mathrm{EET}} - 2\frac{V_{\mathrm{ET}}V_{\mathrm{HT}}}{E_{\mathrm{CR}} - E_{\mathrm{LE}}} + \frac{V_{\mathrm{ET}}^2 V_{\mathrm{CT}} + V_{\mathrm{HT}}^2 V_{\mathrm{CT}}}{(E_{\mathrm{CR}} - E_{\mathrm{LE}})^2} \\ &\equiv T_{\mathrm{Direct}} + T_{\mathrm{Indirect}}^{(1)} + T_{\mathrm{Indirect}}^{(2)} \end{aligned} \tag{45.5}$$

Here, the first term T_{Direct} is called the direct coupling, and the second and third terms are the indirect coupling. Hence, the indirect coupling is separately described with the second and third order interactions ($T_{\mathrm{Indirect}}^{(2)}$ and $T_{\mathrm{Indirect}}^{(3)}$). The direct coupling is related only to the initial and final states (Φ_1 and Φ_2), and it is considered in the Förster and Dexter theories. On the other hand, the indirect coupling represents the sequential EET process via the intermediate states (Φ_3 and Φ_4). The second-order indirect coupling $T_{\mathrm{Indirect}}^{(2)}$ is regarded as the two-step pathway through electron transfer (ET) and hole transfer (HT), and the third-order indirect coupling $T_{\mathrm{Indirect}}^{(3)}$ corresponds to the three-step pathway related to charge transfer (CT) as well as ET and HT. At a first glance, the expression of Eq. (45.5) looks like the so-called superexchange (SX) coupling [1, 37]. However, the SX coupling is associated with donor–bridge–acceptor molecules, not with the intermediate states. Since the basis functions given by Eq. (45.2) are composed of the electronic states of the donor and acceptor molecules, Eq. (45.5) is different from the SX coupling. As found from Eq. (45.5), the electronic coupling calculation requires four types of the off-diagonal elements, i.e., V_{EET}, V_{ET}, V_{HT}, and V_{CT}. In this study, the element V_{EET} is obtained with the TDFI method [31, 32], and the estimations of V_{ET}, V_{HT}, and V_{CT} are performed with the TI method [34]. Each calculation scheme is explained in the following paragraphs.

Now, we focus on the calculation of V_{EET}, which corresponds to the off-diagonal element with Φ_1 and Φ_2. This matrix element is described as

$$\begin{aligned} V_{\mathrm{EET}} &= \langle \Phi_1 | \widehat{V}_{IJ} | \Phi_2 \rangle \\ &= \int \mathrm{d}\mathbf{r}_1 \int \mathrm{d}\mathbf{r}'_1 \frac{\rho_I^{t*}(\mathbf{r}_1, \mathbf{r}_1)\, \rho_J^t(\mathbf{r}'_1, \mathbf{r}'_1)}{|\mathbf{r}_1 - \mathbf{r}'_1|} - \frac{1}{2} \int \mathrm{d}\mathbf{r}_1 \int \mathrm{d}\mathbf{r}'_1 \frac{\rho_I^{t*}(\mathbf{r}_1, \mathbf{r}'_1)\, \rho_J^t(\mathbf{r}'_1, \mathbf{r}_1)}{|\mathbf{r}_1 - \mathbf{r}'_1|}. \\ &\equiv V_{\mathrm{Coul}} + V_{\mathrm{Exch}} \end{aligned} \tag{45.6}$$

Here, a one-electron transition density for the molecule X is introduced, which is defined by

$$\rho_X^t(\mathbf{r}_1, \mathbf{r}'_1) \equiv N_{\mathrm{e}}^X \int \mathrm{d}s_1 \int \mathrm{d}\mathbf{x}_2 \ldots \int \mathrm{d}\mathbf{x}_{N_{\mathrm{e}}^X} \Psi_X^e(\mathbf{x}_1, \mathbf{x}_2, \ldots, \mathbf{x}_{N_{\mathrm{e}}^X})\, \Psi_X^{g*}(\mathbf{x}_1, \mathbf{x}_2, \ldots, \mathbf{x}_{N_{\mathrm{e}}^X}), \tag{45.7}$$

where N_{e}^X denotes the number of electrons in the molecule X, and $\mathbf{x}_i$ expresses space and spin coordinates of the i-th electron, i.e., $\mathbf{x}_i \equiv (\mathbf{r}_i, s_i)$. The first and second terms in Eq. (45.6) represent the Coulomb and exchange interactions, respectively. As mentioned in the Introduction, the Förster and Dexter types of EET are explained by the Coulomb and exchange interactions, respectively. To convert Eq. (45.6) to a matrix form, the transition density given by Eq. (45.7) is expanded in terms of atomic orbitals (AOs), $\chi_\nu(\mathbf{r}_1)$, as

$$\begin{aligned} \rho_X^t(\mathbf{r}_1, \mathbf{r}'_1) &= \sum_{\nu,\mu \in X} |\chi_\nu(\mathbf{r}_1)\rangle \langle \chi_\nu(\mathbf{r}_1)|\, \rho_X^t(\mathbf{r}_1, \mathbf{r}'_1)\, |\chi_\mu(\mathbf{r}'_1)\rangle \langle \chi_\mu(\mathbf{r}'_1)| \\ &= \sum_{\nu,\mu \in X} |\chi_\nu(\mathbf{r}_1)\rangle\, P_{\nu\mu}^{tX} \langle \chi_\mu(\mathbf{r}'_1)|. \end{aligned} \tag{45.8}$$

Using this equation, the Coulomb and exchange interactions are rewritten in the matrix form

$$V_{\mathrm{EET}} = V_{\mathrm{Coul}} + V_{\mathrm{Exch}} = \sum_{\nu,\mu \in I} \sum_{\lambda,\sigma \in J} P_{\nu\mu}^{tI} P_{\lambda\sigma}^{tJ} \left[(\mu\nu|\sigma\lambda) - \frac{1}{2} (\mu\lambda|\sigma\nu) \right], \tag{45.9}$$

where $\left(\mu\nu\middle|\sigma\lambda\right)$ expresses a two-electron integral in the AO representation, which is defined by

$$\left(\mu\nu\middle|\sigma\lambda\right) \equiv \int d\mathbf{r}_1 \int d\mathbf{r}_2 \frac{\chi_\mu^*(\mathbf{r}_1)\,\chi_\nu(\mathbf{r}_1)\,\chi_\sigma^*(\mathbf{r}_2)\,\chi_\lambda(\mathbf{r}_2)}{|\mathbf{r}_1 - \mathbf{r}_2|}. \tag{45.10}$$

For a negligible overlap between the basis functions Φ_1 and Φ_2, Eq. (45.9) becomes the TDFI expression of the Coulomb and exchange interactions. If the orthogonality between Φ_1 and Φ_2 is not satisfied, Eq. (45.9) is modified such as [37]

$$\begin{aligned} V_{\mathrm{EET}}^{\mathrm{TDFI}} &= \frac{V_{\mathrm{EET}} - E_{\mathrm{LE}} S_{12}}{1 - S_{12}^2} \\ &= \frac{V_{\mathrm{Coul}}}{1 - S_{12}^2} + \frac{V_{\mathrm{Exch}}}{1 - S_{12}^2} - \frac{E_{\mathrm{LE}} S_{12}}{1 - S_{12}^2} \equiv V_{\mathrm{Coul}}^{\mathrm{TDFI}} + V_{\mathrm{Exch}}^{\mathrm{TDFI}} + V_{\mathrm{Ovlp}}^{\mathrm{TDFI}}, \end{aligned} \tag{45.11}$$

where E_{LE} is the energy of the LE state, and it corresponds to the diagonal element of Eq. (45.4). S_{12} denotes the overlap between Φ_1 and Φ_2, which is written by

$$S_{12} = \langle \Phi_1 | \Phi_2 \rangle = -\frac{1}{N_{\mathrm{e}}^{IJ}} \sum_{\nu,\mu \in I} \sum_{\lambda,\sigma \in J} P_{\nu\mu}^{tI} S_{\mu\lambda}^{\mathrm{AO}} P_{\lambda\sigma}^{tJ} S_{\sigma\nu}^{\mathrm{AO}}, \tag{45.12}$$

with the AO overlap matrix

$$S_{\mu\lambda}^{\mathrm{AO}} \equiv \int d\mathbf{r}_1 \chi_\mu^*(\mathbf{r}_1)\,\chi_\lambda(\mathbf{r}_1). \tag{45.13}$$

Here, N_{e}^{IJ} is the number of electrons in the IJ complex. The energy E_{LE} in Eq. (45.11) is obtained with the DFI method [33], which enables to estimate a polarization effect coming from the counterpart molecule. In the DFI method, the energy of the LE state is given by

$$\begin{aligned} E_{\mathrm{LE}} &= \langle \Phi_1 | \widehat{H} - E_0 | \Phi_1 \rangle \\ &= E_{\mathrm{ex}}^I + \sum_{\nu,\mu \in I} \left(P_{\nu\mu}^{eI} - P_{\nu\mu}^{gI}\right) \left\{ v_{\mu\nu}^J + \sum_{\lambda,\sigma \in J} P_{\lambda\sigma}^{gJ} \left[\left(\mu\nu\middle|\sigma\lambda\right) - \frac{1}{2} \left(\mu\lambda\middle|\sigma\nu\right) \right] \right\}, \end{aligned} \tag{45.14}$$

where E_0 and E_{ex}^I denote the ground-state energy of the IJ complex and the excitation energy of the non-interacting molecule I, respectively. $P_{\nu\mu}^{gX}$ and $P_{\nu\mu}^{eX}$ are one-electron density matrices of the molecule X in the ground and excited states, respectively. These density matrices are obtained in the similar way to Eq. (45.8)

$$P_{\nu\mu}^{sX} = \int d\mathbf{r}_1 \int d\mathbf{r}'_1 \chi_\nu^* (\mathbf{r}_1) \rho_X^s (\mathbf{r}_1, \mathbf{r}'_1) \chi_\mu (\mathbf{r}'_1), \tag{45.15}$$

using the one-electron density for the state s ($s = g$ or e)

$$\rho_X^s (\mathbf{r}_1, \mathbf{r}'_1) \equiv N_e^X \int ds_1 \int d\mathbf{x}_2 \ldots \int d\mathbf{x}_{N_e^X} \Psi_X^s (\mathbf{x}_1, \mathbf{x}_2, \ldots, \mathbf{x}_{N_e^X}) \Psi_X^{s*} (\mathbf{x}_1, \mathbf{x}_2, \ldots, \mathbf{x}_{N_e^X}). \tag{45.16}$$

Here, Ψ_X^s denotes the electronic state s for the molecule X. In Eq. (45.14), $v_{\mu\nu}^J$ expresses a nuclear potential generated by the molecule J, which is defined by

$$v_{\mu\nu}^J \equiv \int d\mathbf{r}_1 \chi_\mu^* (\mathbf{r}_1) \left[\sum_{A \in J} \frac{-Z_A}{|\mathbf{r}_1 - \mathbf{R}_A|} \right] \chi_\nu (\mathbf{r}_1), \tag{45.17}$$

where Z_A and $\mathbf{R}_A$ denote a charge and a space coordinate of the nucleus A, respectively. In this study, the symmetric homodimer is taken into account, and thus both the diagonal elements H_{11} and H_{22} are given by Eq. (45.14), i.e., $H_{11} = H_{22} = E_{\mathrm{LE}}$. In this way, the direct coupling ($T_{\mathrm{Direct}} = V_{\mathrm{EET}}$) can be calculated with Eq. (45.11).

Next, let us consider the calculations of V_{ET}, V_{HT}, and V_{CT}. These calculations are performed with the TI method [37]. The basis of the TI method is to describe an intermolecular orbital interaction such as

$$\langle \Phi_{i\to a} | \hat{H} | \Phi_{j \to b} \rangle = \delta_{ij} \langle \varphi_a | \hat{F} | \varphi_b \rangle - \delta_{ab} \langle \varphi_i | \hat{F} | \varphi_j \rangle + 2 (\varphi_a \varphi_i | \varphi_j \varphi_b) - (\varphi_a \varphi_b | \varphi_j \varphi_i). \tag{45.18}$$

Here, φ_i and φ_j are occupied molecular orbitals (MOs), and φ_a and φ_b are virtual MOs. δ_{xy} represents Kronecker delta, and $\langle \varphi_x | \hat{F} | \varphi_y \rangle$ is a Fock matrix element in the MO representation. For simplicity, we assume that highest occupied MOs (HOMOs) and lowest unoccupied MOs (LUMOs) are mainly involved in the EET process. Then, the Hamiltonian matrix elements V_{ET}, V_{HT}, and V_{CT} are described as

$$\begin{aligned} V_{\mathrm{ET}} &= t_{\mathrm{H}\to\mathrm{L}}^I \left\{ \langle \varphi_{\mathrm{L}}^I | \hat{F} | \varphi_{\mathrm{L}}^J \rangle + 2 (\varphi_{\mathrm{L}}^I \varphi_{\mathrm{H}}^I | \varphi_{\mathrm{H}}^I \varphi_{\mathrm{L}}^J) - (\varphi_{\mathrm{L}}^I \varphi_{\mathrm{L}}^J | \varphi_{\mathrm{H}}^I \varphi_{\mathrm{H}}^I) \right\} \\ &= t_{\mathrm{H}\to\mathrm{L}}^I \left\{ \sum_{\alpha \in I} \sum_{\beta \in J} c_{\mathrm{L}\alpha}^I F_{\alpha\beta} c_{\beta\mathrm{L}}^J \right. \\ &\left. + \sum_{\mu \in I} \sum_{\nu \in I} c_{\nu\mathrm{H}}^I c_{\mathrm{L}\mu}^I \sum_{\sigma \in J} \sum_{\lambda \in I} c_{\lambda\mathrm{L}}^J c_{\mathrm{H}\sigma}^I \left[2 (\mu\nu | \sigma\lambda) - (\mu\lambda | \sigma\nu) \right] \right\}, \end{aligned} \tag{45.19}$$

$$
\begin{aligned}
V_{\mathrm{HT}} &= t^{I}_{\mathrm{H}\to\mathrm{L}}\left\{-\left\langle\varphi^{I}_{\mathrm{H}}\right|\widehat{F}\left|\varphi^{J}_{\mathrm{H}}\right\rangle + 2\left(\varphi^{I}_{\mathrm{H}}\varphi^{I}_{\mathrm{L}}\middle|\,\varphi^{I}_{\mathrm{L}}\varphi^{J}_{\mathrm{H}}\right) - \left(\varphi^{I}_{\mathrm{H}}\varphi^{J}_{\mathrm{H}}\middle|\,\varphi^{I}_{\mathrm{L}}\varphi^{I}_{\mathrm{L}}\right)\right\} \\
&= t^{I}_{\mathrm{H}\to\mathrm{L}}\left\{-\sum_{\alpha\in I}\sum_{\beta\in J} c^{I}_{\mathrm{H}\alpha}F_{\alpha\beta}c^{J}_{\beta\mathrm{H}}\right. \\
&\quad\left.+\sum_{\mu\in I}\sum_{\nu\in I} c^{I}_{\nu\mathrm{L}}c^{I}_{\mathrm{H}\mu}\sum_{\sigma\in J}\sum_{\lambda\in I} c^{J}_{\lambda\mathrm{H}}c^{I}_{\mathrm{L}\sigma}\left[2\left(\mu\nu\middle|\sigma\lambda\right) - \left(\mu\lambda\middle|\sigma\nu\right)\right]\right\},
\end{aligned}
\tag{45.20}
$$

$$
\begin{aligned}
V_{\mathrm{CT}} &= 2\left(\varphi^{I}_{\mathrm{H}}\varphi^{J}_{\mathrm{L}}\middle|\,\varphi^{I}_{\mathrm{L}}\varphi^{J}_{\mathrm{H}}\right) - \left(\varphi^{I}_{\mathrm{H}}\varphi^{J}_{\mathrm{H}}\middle|\,\varphi^{I}_{\mathrm{L}}\varphi^{J}_{\mathrm{L}}\right) \\
&= \sum_{\mu\in I}\sum_{\nu\in J} c^{J}_{\nu\mathrm{L}}c^{I}_{\mathrm{H}\mu}\sum_{\sigma\in I}\sum_{\lambda\in J} c^{J}_{\lambda\mathrm{H}}c^{I}_{\mathrm{L}\sigma}\left[2\left(\mu\nu\middle|\sigma\lambda\right) - \left(\mu\lambda\middle|\sigma\nu\right)\right].
\end{aligned}
\tag{45.21}
$$

In the individual second lines of Eqs. (45.19, 45.20, and 45.21), the matrix elements are written in the AO representation. $c^{X}_{\nu i}$ denotes a linear combination of atomic orbital–molecular orbital (LCAO-MO) coefficient for the molecule X, i.e., $\varphi^{X}_{i} = \sum_{\nu\in X} c^{X}_{\nu i}\chi_{\nu}$. $F_{\mu\nu}$ is the Fock matrix given by

$$
F_{\mu\nu} = h^{\mathrm{core}}_{\mu\nu} + \sum_{\sigma,\lambda}\sum_{k} 2c_{\lambda k}c_{k\sigma}\left[\left(\mu\nu\middle|\sigma\lambda\right) - \frac{1}{2}\left(\mu\lambda\middle|\sigma\nu\right)\right],
\tag{45.22}
$$

with the one-electron core Hamiltonian matrix

$$
h^{\mathrm{core}}_{\mu\nu} = \int d\mathbf{r}_1\chi^{*}_{\mu}(\mathbf{r}_1)\left[-\frac{1}{2}\nabla^{2}_{1}\right]\chi_{\nu}(\mathbf{r}_1) + \int d\mathbf{r}_1\chi^{*}_{\mu}(\mathbf{r}_1)\left[\sum_{A}\frac{-Z_A}{|\mathbf{r}_1 - \mathbf{R}_A|}\right]\chi_{\nu}(\mathbf{r}_1).
\tag{45.23}
$$

It is noted that Eqs. (45.19) and (45.20) include not only the intermolecular orbital interaction given by Eq. (45.18) but also a configuration-interaction (CI) coefficient $t^{I}_{\mathrm{H}\to\mathrm{L}}$. Owing to this treatment, the sequential EET process from the initial state Φ_1 is taken into account for the electronic coupling calculation. Since the element V_{CT} is not related to the initial state, the CI coefficient is not included in Eq. (45.21). It is also noted that Eq. (45.21) has no contribution from the one-electron core Hamiltonian matrix $h^{\mathrm{core}}_{\mu\nu}$ because of the Slater–Condon rules [40, 41]. As mentioned above, Eqs. (45.19, 45.20, and 45.21) express the orbital interaction with respect to HOMOs and LUMOs. In the cases of the other orbital interactions, the appropriate expression is obtained by replacing the LCAO-MO and CI coefficients in Eqs. (45.19, 45.20, and 45.21) by the corresponding ones. The overlap correction is also applied to the off-diagonal elements V_{ET}, V_{HT}, and V_{CT}. Similar to the matrix element V_{EET}, the element V_{ET} is modified as

$$
V^{\mathrm{TI}}_{\mathrm{ET}} = \frac{V_{\mathrm{ET}} - E_{\mathrm{LE}}S_{13}}{1 - S^{2}_{13}}.
\tag{45.24}
$$

In the cases of V_{HT} and V_{CT}, the overlap corrections are performed with S_{14} and S_{34}, respectively, instead of S_{13}.

In the above schemes, all of the off-diagonal elements required for Eq. (45.5) can be estimated. The diagonal element E_{LE} is also obtained with Eq. (45.14). The unknown term in Eq. (45.5) is the diagonal element E_{CR}, which corresponds to the energy of the CR state. The calculation of E_{CR} is performed with ionization potential (IP), electron affinity (EA), and electron–hole attraction. According to Koopmans' theorem [42], IP and EA are given by

$$E_{\mathrm{IP},i}^{X} = {}^{N_{\mathrm{e}}^{X}-1}E_{i}^{X} - {}^{N_{\mathrm{e}}^{X}}E_{0}^{X} = -\varepsilon_{i}^{X}, \tag{45.25}$$

$$E_{\mathrm{EA},a}^{X} = {}^{N_{\mathrm{e}}^{X}}E_{0}^{X} - {}^{N_{\mathrm{e}}^{X}+1}E_{a}^{X} = -\varepsilon_{a}^{X}, \tag{45.26}$$

where ε_i^X and ε_a^X denote orbital energies of the occupied MO φ_i^X and the virtual MO φ_a^X for the molecule X, respectively. The electron–hole attraction is described as the Coulomb interaction between a hole in the occupied MO for the molecule I and an electron in the virtual MO for the molecule J

$$E_{\mathrm{Coul},ia}^{+-} = -\left(\varphi_i^I\varphi_i^I \middle| \varphi_a^J\varphi_a^J\right) = -\sum_{\mu,\nu\in I}\sum_{\lambda,\sigma\in J} c_{\nu i}^{I} c_{i\mu}^{I} c_{\lambda a}^{J} c_{a\sigma}^{J} \left(\mu\nu\middle|\sigma\lambda\right). \tag{45.27}$$

Using Eqs. (45.25, 45.26, and 45.27), the energy of the CR state is represented by

$$\begin{aligned} E_{\mathrm{CR}} &= \langle\Phi_3|\,\widehat{H} - E_0\,|\Phi_3\rangle \\ &= E_{\mathrm{IP,H}}^{I} - E_{\mathrm{EA,L}}^{J} + E_{\mathrm{Coul,HL}}^{+-} \\ &= -\varepsilon_{\mathrm{H}}^{I} + \varepsilon_{\mathrm{L}}^{J} - \sum_{\mu,\nu\in I}\sum_{\lambda,\sigma\in J} c_{\nu\mathrm{H}}^{I} c_{\mathrm{H}\mu}^{I} c_{\lambda\mathrm{L}}^{J} c_{\mathrm{L}\sigma}^{J} \left(\mu\nu\middle|\sigma\lambda\right). \end{aligned} \tag{45.28}$$

This calculation is performed also with the DFI method for including the polarization effect caused by the interfragment interaction. In this way, the TDFI-TI method allows us to perform the electronic coupling calculation of Eq. (45.5).

45.3 Accuracy of the TDFI-TI Method

For checking the accuracy of the TDFI-TI method, several test calculations were performed to an ethylene dimer. In the first step, the optimized structure of an ethylene dimer was selected as the computational model. To estimate the error associated with the TDFI-TI method, the following reference value was also calculated:

$$T_{\mathrm{IF}}^{\mathrm{FullQM}} = \frac{1}{2}\left(E_a' - E_b'\right), \tag{45.29}$$

Table 45.1 Electronic coupling energies of the ethylene dimer (cm^{-1})

	CIS	TD-rCAM-B3LYP
Full-QM (ESD)	1976	1454
TDFI-TI	1997	1417
TDFI[a]	1649	1097

[a]The Coulomb interaction is estimated

where E_i' denotes the i-th excited state energy for the dimer. This equation is known as an energy splitting in dimer (ESD) method [43], and it is applicable only to symmetric dimer with identical monomer energies [19].

Table 45.1 summarizes the results of the electronic coupling calculations. In these calculations, the CI-singles (CIS) method [44] and time-dependent density functional theory [45] with rCAM-B3LYP functional [46] (TD-rCAM-B3LYP) were employed to examine the dependence on the electronic structure method. The TDFI-TI calculations gave the electronic coupling energies of 1997 cm^{-1} for CIS and of 1417 cm^{-1} for TD-rCAM-B3LYP, which indicated that the absolute values of the electronic coupling energies depend on the electronic structure methods (580 cm^{-1}). However, the TDFI-TI method successfully reproduced the individual reference values (1976 cm^{-1} for CIS and 1454 cm^{-1} for TD-rCAM-B3LYP) with the deviations of 21 cm^{-1} and 37 cm^{-1} for CIS and TD-rCAM-B3LYP, respectively. These results showed that the TDFI-TI method could give a quantitative value of the electronic coupling energy without respect to the electronic structure method. Table 45.1 also lists the electronic coupling energy obtained with the TDFI method, in which only the Coulomb interaction (V_{Coul}) was estimated. In this case, the calculated electronic coupling energies were to be 1649 cm^{-1} for CIS and 1097 cm^{-1} for TD-rCAM-B3LYP. Therefore, these calculations led to the smaller values by 348 cm^{-1} for CIS and by 320 cm^{-1} for TD-rCAM-B3LYP than the corresponding TDFI-TI values. These results clearly showed that the electronic coupling energy of the ethylene dimer cannot be described only with the Coulomb interaction.

To gain further insight into the accuracy of the TDFI-TI method, the dependence on the intermolecular separation was investigated. Figure 45.2 shows the evolutions of the electronic coupling energies as a function of the intermolecular distance of the ethylene dimer. The definition of the x-axis coordinate is also shown in Fig. 45.2. The optimized structure corresponds to the intermolecular distance of 4.17 Å. As a result of the calculations, the TDFI-TI method gave almost the same values as the full-QM (ESD) method in all the regions calculated. The deviations from the reference values were to be 78 cm^{-1}, which were only 0.8 % of the amount of the electronic coupling energies. From these results, it was confirmed that the TDFI-TI method could quantitatively describe the reference values without respect to the intermolecular separation.

Based on the successful reproduction of the reference values, the electronic coupling energy of the ethylene dimer was analyzed in terms of five contributions

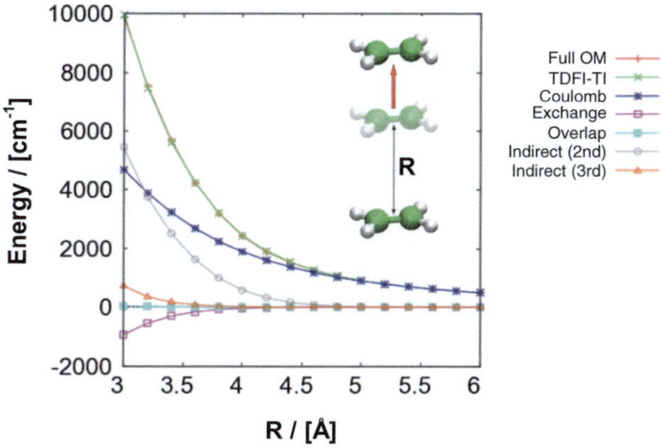

Fig. 45.2 Electronic coupling energies between two ethylene molecules as a function of intermolecular distance. The TDFI-TI calculations are performed with CIS/6-31G*

(i.e., V_{Coul}, V_{Exch}, V_{Ovlp}, $T^{(2)}_{\mathrm{Indirect}}$, and $T^{(2)}_{\mathrm{Indirect}}$). Each contribution is also shown in Fig. 45.2. As found here, the Coulomb interaction predominantly contributed to the amount of the electronic coupling energy in most of the regions. In particular, the large contribution from the Coulomb interaction was confirmed at the large separation. At more than 4.5 Å, the Coulomb interaction showed almost the same values as the amount of the electronic coupling energy. With the decrease in the intermolecular separation, the contribution from the Coulomb interaction gradually became small. In contrast, the second-order indirect coupling gave a large contribution at the small separation. At 3.0 Å, the value of the second-order indirect coupling was to be 5446 cm^{-1}, which was larger than the Coulomb interaction (4689 cm^{-1}). These results revealed that the Coulomb interaction mainly contributed to the electronic coupling energy in most of the cases, and the second-order indirect coupling gave a significant contribution at the small separation. Similar to the second-order indirect coupling, the values of the third-order indirect coupling gradually became large with the decrease in the intermolecular separation. However, the magnitudes of the third-order indirect coupling were smaller than those of the second-order indirect coupling. These results indicated that the two-step pathways through ET and HT are preferred than the three-step pathways through ET, CT, and HT. The exchange interaction also contributed to the electronic coupling energy at the small separation. However, these values (-943 cm^{-1}) provided a negative contribution to the amount of the electronic coupling energy. In common, the exchange interaction is considered as a main factor for EET at small separation. However, the present calculations showed a larger contribution from the second-order indirect coupling than that from the exchange interaction at the small separation. These results supported the previous study by Scholes et al [39]. In all the regions, the overlap effect gave negligible contributions (40 cm^{-1}). In this way, the decomposition analysis based on the TDFI-TI method clarified the main contribution to the amount of the electronic coupling energy.

45.4 Application to Crystallochromy

In the above section, the TDFI-TI method was applied to the electronic coupling calculations of the ethylene dimer. These calculations were performed with Eq. (45.5), in which the elements of the Hamiltonian matrix given by Eq. (45.4) are employed. Using Eq. (45.4), we can calculate the excited states of molecular aggregates. Based on this idea, the TDFI-TI method was applied to solid-state tetracenes. It is noted that the electronic coupling energy given by Eq. (45.5) is not used in this section.

Similar to the above section, we assume the molecular complex IJ as a total system, and thus the Hamiltonian matrix is given by Eq. (45.4). The K-th excited state for the total system is represented by a linear combination of Eq. (45.2)

$$|\Psi_K\rangle = \sum_l C_{lK} |\Phi_l\rangle, \tag{45.30}$$

for $l = 1$–4. To determine the expansion coefficients C_{lK}, the Hamiltonian matrix by Eq. (45.4) must be diagonalized. However, the standard diagonalization procedure is not applicable to this Hamiltonian matrix because the basis functions given by Eq. (45.2) are not orthogonal. To circumvent this problem, the Löwdin symmetric orthogonalization method [47] is employed. First, the Hamiltonian matrix is transformed to the orthogonal representation such as

$$\mathbf{S}^{-1/2\dagger}\mathbf{H}\mathbf{S}^{-1/2} = \mathbf{H}_{\mathrm{s}}, \tag{45.31}$$

where $\mathbf{H}_{\mathrm{s}}$ denotes the Hamiltonian matrix in the symmetrically orthogonal representation. The transformation matrix is represented by $\mathbf{S}^{-1/2}$, in which $\mathbf{S}$ is the overlap of the basis functions, i.e., $S_{kl} \equiv \langle \Phi_k | \Phi_l \rangle$. This transformation allows us to perform the diagonalization of $\mathbf{H}_{\mathrm{s}}$

$$\mathbf{U}^{\dagger}\mathbf{H}_{\mathrm{s}}\mathbf{U} = \mathbf{H}_{\mathrm{d}}. \tag{45.32}$$

Following this procedure, the state energies E_K and the expansion coefficients C_{lK} are obtained from the diagonal elements of $\mathbf{H_d}$ and from the elements of the transformation matrix $\mathbf{C}$ $\left(\mathbf{C} = \mathbf{S}^{-1/2}\mathbf{U}\right)$, respectively. It is noted that E_K directly corresponds to the excitation energies for the total system because the ground-state energy is taken to be zero in this scheme. Using the matrix $\mathbf{C}$, the transition dipole moments for the total system are obtained as

$$\mathbf{D}_K^t = \sum_l C_{lK}\mathbf{d}_l^t, \tag{45.33}$$

Table 45.2 TDFI-TI absorption energies of the BUR and BUY N-mers (eV)[a]

	Monomer	Dimer	Trimer	Tetramer	Pentamer	Hexamer
BUR	2.56	2.37	2.29	2.25	2.23	2.23 (2.23[b])
BUY	2.68	2.65	2.62	2.61	2.61	2.61 (2.62[b])

[a]SAC-CI/6-31G* data are employed
[b]Absorption energy calculated in crystal

where $\mathbf{d}_l^t$ denotes the noninteracting electric transition dipole moments. The oscillator strength for the total system is calculated with $\mathbf{D}_K^t$

$$f_K = \frac{2}{3} E_K \left| \mathbf{D}_K^t \right|^2. \tag{45.34}$$

This oscillator strength is used for the spectral calculations.

The above scheme was applied to two kinds of the solid-state tetracenes, yellow (BUY) and red (BUR) solids [36]. Although these solids have the same molecular composition (butyl derivative), they show different colors depending on the molecular packing arrangement in solid state. This phenomenon is called crystallochromy [48]. In this section, the symmetry-adapted cluster-CI (SAC-CI) method [49] was employed for constructing the Hamiltonian matrix, and then the absorption energies of BUR and BUY were calculated.

Table 45.2 summarizes the TDFI-TI absorption energies of the BUR and BUY. These calculations were performed from the individual monomers to hexamers. In the BUR calculations, the TDFI-TI absorption energies were to be 2.56~2.23 eV, which indicated that the absorption energies were largely redshifted (0.33 eV) by the multimerization effect. The energy convergence less than 0.01 eV was satisfied by the hexamer calculation. In this case, the calculated absorption energy of BUR (2.23 eV) was well in agreement with the experimental value (2.18 eV). On the other hand, the calculated absorption energies of BUY were to be 2.68~2.61 eV. Similar to BUR, the TDFI-TI absorption energy of BUY (2.61 eV) successfully reproduced the experimental value (2.56 eV). The multimerization effect for BUY was estimated to be 0.07 eV, and therefore the difference in the multimerization effects between BUR and BUY was to be 0.26 eV. From the comparison of the monomer absorption energies between BUR (2.56 eV) and BUY (2.68 eV), the monomer conformational effect was estimated to be 0.12 eV. These results revealed that the multimerization effect (0.26 eV) largely contributed to the spectral shift between BUR and BUY (0.38 eV), compared with the monomer conformational effect (0.12 eV).

To evaluate the electrostatic solid-state effect, the TDFI-TI calculations in solid state were also performed to the hexamer conformations of BUR and BUY. These results are also listed in Table 45.2. The solid-state absorption energies for BUR and BUY were calculated to be 2.23 eV and 2.62 eV, respectively. Compared with the gas-phase absorption energies (2.23 eV for BUR and 2.61 eV for BUY), the electrostatic solid-state effects were estimated to be 0.00 eV and 0.01 eV for BUR

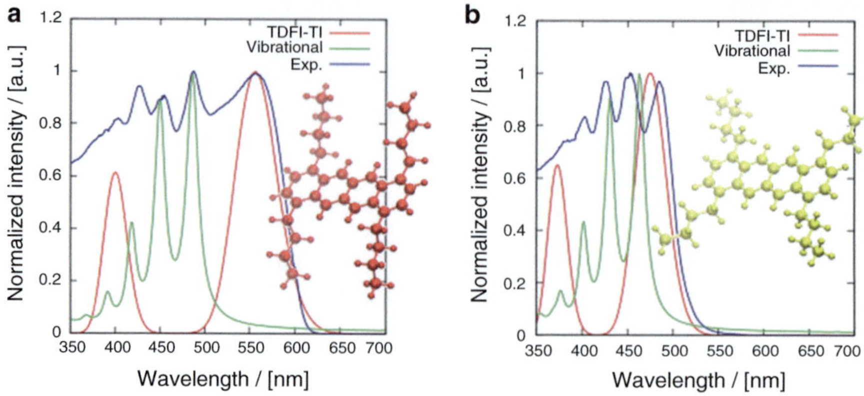

Fig. 45.3 Absorption spectra of (**a**) BUR and (**b**) BUY

and BUY, respectively. Therefore, the contribution from the electrostatic solid-state effect to the spectral shift was estimated to be 0.01 eV by subtracting the spectral shift in gas phase (0.38 eV) from that in solid state (0.39 eV). From these results, the electrostatic solid-state effect was found to give a small contribution to the spectral shift between BUR and BUY.

Following these procedures, the TDFI-TI calculations successfully reproduced the experimental absorption energies (2.18 eV for BUR and 2.56 eV for BUY). The deviations from the experimental values were to be 0.05 eV for BUR and 0.06 eV for BUY. Hence, these calculations (0.39 eV) quantitatively reproduced the experimental spectral shift (0.38 eV) with the deviation of 0.01 eV. As mentioned above, this spectral shift resulted mainly from the difference in the multimerization effects between BUR and BUY. Figure 45.3 shows the calculated absorption spectra for BUR and BUY. Here, two kinds of spectral shapes, Gaussian and Lorentzian functions, were employed. The Gaussian function was used for describing the absorption spectra for the hexamer results, which are shown by red lines in Fig. 45.3. As found here, the calculated spectra successfully reproduced the first peaks in the experimental spectra of BUR and BUY. In particular, this calculation clearly represented the redshifted peak in the BUR spectrum. We next focused on the rest of the experimental peaks. In the experimental study, these peaks were considered to be due to the vibrational progressions [36]. Following this reason, the vibrational spectra were also calculated. These calculations were performed with the monomer absorption energies of BUR and BUY, and the spectral shapes were described with the Lorentzian function. As shown by green lines in Fig. 45.3, the calculated vibrational spectra were well in agreement with the positions of the second, third, and fourth peaks in the experimental spectra of BUR and BUY. Therefore, the influences of the vibrational progressions were confirmed also from the calculated spectra. These results clearly showed that the present approach is useful for investigating the experimental spectra of BUR and BUY.

45.5 Conclusions

In this chapter, I explained the TDFI-TI calculations for the electronic coupling energies and the absorption spectra. The basis of the TDFI-TI method is to describe the elements of the Hamiltonian matrix. The TDFI-TI scheme can be combined with other electronic structure methods, even though only the CIS, TD-rCAM-B3LYP, and SAC-CI methods were employed in the present study. The decomposition analysis with the TDFI-TI method provides the components of the electronic coupling energy, whereas such an analysis cannot be performed with the full-QM method. Moreover, the TDFI-TI method for the absorption spectra can reduce the computational costs compared with the full-QM method, so that it is applicable to large systems. These advantages of the TDFI-TI method are expected to be useful for analyzing and understanding the mechanisms of EET and crystallochromy.

Acknowledgements This study was supported by MEXT KAKENHI Grant No. 23108709 and JSPS KAKENHI Grant No. 25810004

References

1. May V, Kühn O (2011) Charge and energy transfer dynamics in molecular systems, 3rd edn. Wiley-VCH, Weinheim
2. Scholes GD (2003) Long-range resonance energy transfer in molecular systems. Annu Rev Phys Chem 54:57
3. Beljonne D, Curutchet C, Scholes GD, Silbey RJ (2009) Beyond Förster resonance energy transfer in biological and nanoscale systems. J Phys Chem B 113:6583
4. König C, Neugebauer J (2012) Quantum chemical description of absorption properties and excited-state processes in photosynthetic systems. ChemPhysChem 13:386
5. Sumi H (2001) Bacterial photosynthesis begins with quantum-mechanical coherence. Chem Rec 1:480
6. Ritz T, Damjanović A, Schulten K (2002) The quantum physics of photosynthesis. ChemPhysChem 3:243
7. Renger T (2009) Theory of excitation energy transfer: from structure to function. Photosynth Res 102:471
8. Scholes GD, Fleming GR, Olaya-Castro A, van Grondelle R (2011) Lessons from nature about solar light harvesting. Nat Chem 3:763
9. Förster T (1948) Zwischenmolekulare energiewanderung und fluoreszenz. Ann Phys 437:55
10. Dexter DL (1953) A theory of sensitized luminescence in solids. J Chem Phys 21:836
11. Berezin MY, Achilefu S (2010) Fluorescence lifetime measurements and biological imaging. Chem Rev 110:2641
12. Speiser S (1996) Photophysics and mechanisms of intramolecular electronic energy transfer in bichromophoric molecular systems: solution and supersonic jet studies. Chem Rev 96:1953
13. Harcourt RD, Scholes GD, Ghiggino KP (1994) Rate expressions for excitation transfer. II. Electronic considerations of direct and through–configuration exciton resonance interactions. J Chem Phys 101:10521
14. Warshel A, Parson WW (1987) Spectroscopic properties of photosynthetic reaction centers. 1. Theory. J Am Chem Soc 109:6143

15. Nagae H, Kakitani T, Katoh T, Mimuro M (1993) Calculation of the excitation transfer matrix elements between the S2 or S1 state of carotenoid and the S2 or S1 state of bacteriochlorophyll. J Chem Phys 98:8012
16. Scholes GD, Ghiggino KP, Oliver AM, Paddon-Row MN (1993) Intramolecular electronic energy transfer between rigidly linked naphthalene and anthracene chromophores. J Phys Chem 97:11871
17. Krueger BP, Scholes GD, Fleming GR (1998) Calculation of couplings and energy-transfer pathways between the pigments of LH2 by the ab initio transition density cube method. J Phys Chem B 102:5378
18. Damjanović A, Ritz T, Schulten K (1999) Energy transfer between carotenoids and bacteriochlorophylls in light-harvesting complex II of purple bacteria. Phys Rev E 59:3293
19. Tretiak S, Middleton C, Chernyak V, Mukamel S (2000) Exciton Hamiltonian for the bacteriochlorophyll system in the LH2 antenna complex of purple bacteria. J Phys Chem B 104:4519
20. Hsu C-P, Fleming GR, Head-Gordon M, Head-Gordon T (2001) Excitation energy transfer in condensed media. J Chem Phys 114:3065
21. Iozzi MF, Mennucci B, Tomasi J, Cammi R (2004) Excitation energy transfer (EET) between molecules in condensed matter: a novel application of the polarizable continuum model (PCM). J Chem Phys 120:7029
22. Wong KF, Bagchi B, Rossky PJ (2004) Distance and orientation dependence of excitation transfer rates in conjugated systems: beyond the Förster theory. J Phys Chem A 108:5752
23. Madjet ME, Abdurahman A, Renger T (2006) Intermolecular Coulomb couplings from ab initio electrostatic potentials: application to optical transitions of strongly coupled pigments in photosynthetic antennae and reaction centers. J Phys Chem B 110:17268
24. Neugebauer J (2007) Couplings between electronic transitions in a subsystem formulation of time-dependent density functional theory. J Chem Phys 126:134116
25. Fückel B, Köhn A, Harding ME, Diezemann G, Hinze G, Basché T, Gauss J (2008) Theoretical investigation of electronic excitation energy transfer in bichromophoric assemblies. J Chem Phys 128:074505
26. Fink RF, Pfister J, Zhao HM, Engels B (2008) Assessment of quantum chemical methods and basis sets for excitation energy transfer. Chem Phys 346:275
27. Chen H-C, You Z-Q, Hsu C-P (2008) The mediated excitation energy transfer: effects of bridge polarizability. J Chem Phys 129:084708
28. Madjet ME, Müh F, Renger T (2009) Deciphering the influence of short-range electronic couplings on optical properties of molecular dimers: application to "special pairs" in photosynthesis. J Phys Chem B 113:12603
29. Yeganeh S, Van Voorhis T (2010) Triplet excitation energy transfer with constrained density functional theory. J Phys Chem C 114:20756
30. Kawatsu T, Matsuda K, Hasegawa J (2011) Bridge-mediated excitation energy transfer pathways through protein media: a Slater determinant-based electronic coupling calculation combined with localized molecular orbitals. J Phys Chem A 115:10814
31. Fujimoto KJ, Hayashi S (2009) Electronic Coulombic coupling of excitation-energy transfer in xanthorhodopsin. J Am Chem Soc 131:14152
32. Fujimoto KJ (2010) Transition-density-fragment interaction approach for exciton-coupled circular dichroism spectra. J Chem Phys 133:124101
33. Fujimoto K, Yang W-T (2008) Density-fragment interaction approach for quantum-mechanical/molecular-mechanical calculations with application to the excited states of a Mg2+-sensitive dye. J Chem Phys 129:054102
34. Fujimoto KJ (2012) Transition-density-fragment interaction combined with transfer integral approach for excitation-energy transfer via charge-transfer states. J Chem Phys 137:034101
35. Fujimoto KJ, Kitamura C (2013) A theoretical study of crystallochromy: spectral tuning of solid-state tetracenes. J Chem Phys 139:084511
36. Kitamura C, Abe Y, Ohara T, Yoneda A, Kawase T, Kobayashi T, Naito H, Komatsu T (2010) Synthesis and crystallochromy of 1,4,7,10-tetraalkyltetracenes: tuning of solid-state optical properties of tetracenes by alkyl side-chain length. Chem Eur J 16:890

37. Newton MD (1991) Quantum chemical probes of electron-transfer kinetics: the nature of donor-acceptor interactions. Chem Rev 91:767
38. Ratner MA (1990) Bridge-assisted electron transfer: effective electronic coupling. J Phys Chem 94:4877
39. Scholes GD, Harcourt RD, Ghiggino KP (1995) Rate expressions for excitation transfer. III. An ab initio study of electronic factors in excitation transfer and exciton resonance interactions. J Chem Phys 102:9574
40. Slater JC (1929) The theory of complex spectra. Phys Rev 34:1293
41. Condon EU (1930) The theory of complex spectra. Phys Rev 36:1121
42. Koopmans T (1934) Über die Zuordnung von Wellenfunktionen und Eigenwerten zu den Einzelnen Elektronen Eines Atoms. Physica 1:104
43. Coropceanu V, Cornil J, da Silva Filho DA, Olivier Y, Silbey R, Brédas J-L (2007) Charge transport in organic semiconductors. Chem Rev 107:926
44. Foresman JB, Head-Gordon M, Pople JA, Frisch MJ (1992) Toward a systematic molecular orbital theory for excited states. J Chem Phys 96:135
45. Runge E, Gross EKU (1984) Density-functional theory for time-dependent systems. Phys Rev Lett 52:997
46. Cohen AJ, Mori-Sánchez P, Yang W-T (2007) Development of exchange-correlation functionals with minimal many-electron self-interaction error. J Chem Phys 126:191109
47. Löwdin P-O (1950) On the non-orthogonality problem connected with the use of atomic wave functions in the theory of molecules and crystals. J Chem Phys 18:365
48. Klebe G, Graser F, Hädicke E, Berndt J (1989) Crystallochromy as a solid-state effect: Correlation of molecular conformation, crystal packing and colour in perylene-3,4:9,10-bis(dicarboximide) pigments. Acta Crystallogr Sect B 45:69
49. Nakatsuji H (1978) Cluster expansion of the wavefunction. Excited states. Chem Phys Lett 59:362